Lecture Notes in Computer Science 13135

More information about this subseries at https://link.springer.com/bookseries/7407

Ding-Zhu Du · Donglei Du · Chenchen Wu ·
Dachuan Xu (Eds.)

Combinatorial Optimization and Applications

15th International Conference, COCOA 2021
Tianjin, China, December 17–19, 2021
Proceedings

Editors
Ding-Zhu Du
University of Texas at Dallas
Richardson, TX, USA

Donglei Du
University of New Brunswick
Fredericton, NB, Canada

Chenchen Wu
Tianjin University of Technology
Tianjin, China

Dachuan Xu
Beijing University of Technology
Beijing, China

ISSN 0302-9743 ISSN 1611-3349 (electronic)
Lecture Notes in Computer Science
ISBN 978-3-030-92680-9 ISBN 978-3-030-92681-6 (eBook)
https://doi.org/10.1007/978-3-030-92681-6

LNCS Sublibrary: SL1 – Theoretical Computer Science and General Issues

This Springer imprint is published by the registered company Springer Nature Switzerland AG
The registered company address is: Gewerbestrasse 11, 6330 Cham, Switzerland

Preface

The 15th Annual International Conference on Combinatorial Optimization and Applications (COCOA 2021) took place in Tianjin, China, during December 17–19, 2021. COCOA 2021 provided an excellent venue for researchers in the area of combinatorial optimization and its applications, including algorithm design, theoretical and experimental analysis, and applied research of general algorithmic interest.

The Program Committee received a total of 122 submissions, among which 55 were accepted for presentation at the conference. Each contributed paper was subject to a rigorous peer review process, with reviewers drawn from a large group of members of the Program Committee.

We would like to express our sincere appreciation to everyone who made COCOA 2021 a success by volunteering their time and effort: the authors, the Program Committee members, and the reviewers. We thank Springer for accepting the proceedings of COCOA 2021 for publication in the Lecture Notes in Computer Science (LNCS) series. Our special thanks also extend to the other chairs and the conference Organizing Committee members for their excellent work.

October 2021

Ding-Zhu Du
Donglei Du
Chenchen Wu
Dachuan Xu

Organization

General Chair

Ding-Zhu Du — University of Texas at Dallas, USA

Program Committee Co-chairs

Donglei Du — University of New Brunswick, Canada
Chenchen Wu — Tianjin University of Technology, China
Dachuan Xu — Beijing University of Technology, China

Local Organizing Chairs

Yongtang Shi — Nankai University, China
Xujian Huang — Tianjin University of Technology, China

Finance Chairs

Yicheng Xu — Shenzhen Institute of Advanced Technology, Chinese Academy of Sciences, China
Jun Yue — Shandong Normal University, China

Publication Chairs

Lu Han — Beijing University of Posts and Telecommunications, China
Hui Lei — Nankai University, China

Web Chairs

Xinxin Zhong — Tianjin University of Technology, China
Rui Li — Tianjin University of Technology, China

Program Committee

Zhipeng Cai — Georgia State University, USA
Vincent Chau — Southeast University, China
Xujin Chen — Academy of Mathematics and Systems Science, Chinese Academy of Sciences, China

Yukun Cheng	Suzhou University of Science and Technology, China
Bhaskar Dasgupta	University of Illinois at Chicago, USA
Neng Fan	University of Arizona, USA
Qilong Feng	Central South University, China
Longkun Guo	Fuzhou University, China
Michael Khachay	Russian Academy of Sciences, Russia
Joong-Lyul Lee	University of North Carolina at Pembroke, USA
Jianping Li	Yunnan University, China
Minming Li	City University of Hong Kong, Hong Kong
Guohui Lin	University of Alberta, Canada
Bin Liu	Ocean University of China, China
Xiwen Lu	East China University of Science and Technology, China
Kameng Nip	Xiamen University, China
Weitian Tong	Eastern Michigan University, USA
Boting Yang	University of Regina, Canada
Yitong Yin	Nanjing University, China
Jinjiang Yuan	Zhengzhou University, China
An Zhang	Hangzhou Dianzi University, China
Peng Zhang	Shandong University, China
Xiaoyan Zhang	Nanjing Normal University, China
Yong Zhang	Shenzhen Institute of Advanced Technology, Chinese Academy of Sciences, China
Zhao Zhang	Zhejiang Normal University, China
Martin Ziegler	KAIST, South Korea
Vassilis Zissimopoulos	National and Kapodistrian University of Athens, Greece

Additional Reviewers

Sijia Dai
Weiming Feng
Guichen Gao
Jiaming Hu
Lingxiao Huang
Yuping Ke
Daniel Khachay
Ivan Adrian Koswara
Ioannis Lamprou
Yongxin Lan
Shi Li
Bingkai Lin
Mingmou Liu
Katherine Neznakhina
Yuri Ogorodnikov
Chunying Ren
John Sigalas
Xin Sun
Xiaoyun Tian
Ioannis Vaxevanakis
Changjun Wang
Chenhao Wang
Kai Wang
Weiwei Wu
Xiaowei Wu
Jie Xue
Yongjie Yang
Fan Yuan
Chihao Zhang
Hongxiang Zhang
Xinbo Zhang
Yan Zhao
Yingchao Zhao
Chaodong Zheng

Contents

Routing Among Convex Polygonal Obstacles in the Plane

R. Inkulu(✉) and Pawan Kumar

Department of Computer Science and Engineering, IIT Guwahati, Guwahati, India
{rinkulu,p.kumar16}@iitg.ac.in

Abstract. Given a set $\mathcal{P}$ of h pairwise disjoint convex polygonal obstacles in the plane, defined with n vertices, we preprocess $\mathcal{P}$ and compute one routing table at each vertex in a subset of vertices of $\mathcal{P}$. For routing a packet from any vertex $s \in \mathcal{P}$ to any vertex $t \in \mathcal{P}$, our scheme computes a routing path with a multiplicative stretch $1+\epsilon$ and an additive stretch $2k\ell$, by consulting routing tables at only a subset of vertices along that path. Here, k is the number of obstacles of $\mathcal{P}$ the routing path intersects, and ℓ depends on the geometry of obstacles in $\mathcal{P}$. During the preprocessing phase, we construct routing tables of size $O(n + \frac{h^3}{\epsilon^2}\text{polylog}(\frac{h}{\epsilon}))$ in $O(n + \frac{h^3}{\epsilon^2}\text{polylog}(\frac{h}{\epsilon}))$ time, where $\epsilon < 1$ is an input parameter.

Keywords: Computational geometry · Shortest paths · Approximation algorithms

1 Introduction

The *routing problem* is popular in both graph algorithms and computational geometry. This problem seeks to find a path from the source of a packet to its destination such that (i) along the path, next hop or the subpath is decided from the local information stored with the current hop, (ii) distance along the routing path is upper bounded by a multiplicative factor times the shortest distance between the source and destination plus possibly with an additive factor, and (iii) the space occupied by the preprocessed data structures (routing tables) is small. In the case of graphs, hops are vertices of the graph. And, in the geometric version of this problem, hops are vertices defining the scene.

Using compact data structures, answering approximate distance queries in a graph is introduced by Thorup and Zwick in [45,46]. Later, the compact routing schemes for graphs has been extensively studied [1,4,13,18,20,38–40]. The routing schemes for special graphs, such as trees [21,41], planar graphs [44], unit disk graphs [28,48], networks of low doubling dimension [33], and for graphs embedded in geometric spaces [7,8,11] are also considered. In [38], Peleg and

R. Inkulu—This research is supported in part by SERB MATRICS grant MTR/2017/000474.

D.-Z. Du et al. (Eds.): COCOA 2021, LNCS 13135, pp. 1–11, 2021.
https://doi.org/10.1007/978-3-030-92681-6_1

Upfal had shown that any routing scheme with constant stretch factor needs to store $\Omega(n^c)$ bits per node, for some constant $c > 0$.

The polygonal domain comprises pairwise-disjoint simple polygons (known as obstacles) in the plane. For convenience, we assume obstacles in the polygonal domain are placed in a large bounding box. For any polygonal domain $\mathcal{P}$, the *free space* $\mathcal{F}(\mathcal{P})$ is the closure of the bounding box without the union of the interior of all the obstacles in $\mathcal{P}$. Any two points $p, q \in \mathcal{F}(\mathcal{P})$ are *visible* to each other if the open line segment joining p and q lies entirely in $\mathcal{F}(\mathcal{P})$. A vertex $v \in \mathcal{P}$ is said to be a *visible vertex* to a point $q \in \mathcal{F}(\mathcal{P})$ whenever v is visible to q. For any obstacle O of a polygonal domain, the boundary of O is denoted by $bd(O)$. We denote the number of vertices of $\mathcal{P}$ with n and the number of obstacles of $\mathcal{P}$ by h.

The shortest distance between two nodes s, t of a graph G is denoted by $d_G(s,t)$. The Euclidean distance between any two points p and q is denoted by $\|pq\|$. The obstacle-avoiding geodesic shortest distance between any two points $p, q \in \mathcal{F}(\mathcal{Q})$ amid a set $\mathcal{Q}$ of polygonal obstacles is denoted by $dist_{\mathcal{Q}}(p,q)$.

Computing a shortest path between two given points in a polygonal domain is a fundamental problem in computational geometry. This problem is primarily studied using two approaches. In one approach, by constructing a graph in $\mathcal{F}(\mathcal{P})$, called visibility graph, whose nodes are the vertices of $\mathcal{P}$ and edges are the line segments between mutually visible vertices (refer to Ghosh [22]). Then, a shortest path of interest is determined in the visibility graph [31,32,47]. In the other approach [24,27,29,30,34], a Dijkstra wavefront is expanded in $\mathcal{F}(\mathcal{P})$, starting from the source until it strikes the destination. Significantly, using Dijkstra wavefront expansion approach, for computing a shortest path between two points in $\mathcal{F}(\mathcal{P})$, Hershberger and Suri devised an $O(n \lg n)$ time algorithm in [24]. Further, by extending the algorithm by Kapoor [29] (which also expands Dijkstra wavefront in $\mathcal{F}(\mathcal{P})$, Inkulu, Kapoor, and Maheshwari [27] devised an $O(n + h((\lg h)^{\delta} + (\lg n)(\lg h)))$ time algorithm. Here, δ is a small positive constant (resulting from the time for triangulating $\mathcal{F}(\mathcal{P})$ with the algorithm in [6]). The shortest path problem in polygonal domains is extensively studied [3,16,23,25,26,37,42,43]. A survey of shortest path algorithms in geometric domains can be found in [35].

A closely related problem is computing a spanner. Given a graph G, a subgraph H of G is a t-spanner of G for $t \geq 1$ whenever for all pairs of vertices u and w in G, $d_G(u,w) \leq d_H(u,w) \leq t \cdot d_G(u,w)$. The geometric spanner is a spanner of a graph embedded in a geometric domain. For a comprehensive survey of results on geometric spanners, refer to monograph [36] by Narasimhan and Smid and the recent article [12] by Bose and Smid. The spanners of points located in the free space of a polygonal domain are studied in [2,14,17,26].

Consider the set of rays: for $0 \leq i \leq \kappa$, the ray r_i passes through the origin and makes an angle $i\theta$ with the positive x-axis, with $(\kappa + 1)\theta = 2\pi$. Each pair of successive rays defines a cone whose apex is at the origin. This collection of κ cones is denoted by $\mathcal{C}$. It is clear that the cones of $\mathcal{C}$ partition the plane. The Yao-graphs [49] and Θ-graphs [17] compute a spanner by including an edge between each vertex v and a nearest vertex to v in every cone C_v, where C_v is a cone in $\mathcal{C}$ translated so that its apex is at v. These graphs differ with respect to how the nearest neighbor in each cone C_v with the apex at v is chosen: in case of Θ-graphs, the nearest neighbor to v is the one that is closest to v in C_v (with ties broken arbitrarily); In the case of Yao-graphs, the nearest neighbor to v is the one whose projection onto the line that bisects C_v is closest to v (with ties broken arbitrarily). In these graphs, the stretch factor depends on the number of cones in $\mathcal{C}$. Significantly, both the Yao and Θ-graphs are used in designing routing schemes [8–10].

The routing scheme for a polygonal domain is considered by Banyassady et al. [5]. When the polygonal domain comprises convex polygonal obstacles, the result here improves [5] with respect to the size of routing tables as well as the preprocessing time. When every vertex stores only the edges that incident to it and stores no routing table, Bose et al. [8] shown that no geometric routing scheme can achieve a stretch factor $o(\sqrt{n})$. This lower bound applies regardless of the amount of information that may be stored in the message.

In the rest of the paper, the polygonal domain comprises convex polygonal obstacles in a large-sized bounded box.

Our Contributions

By preprocessing the input polygonal domain $\mathcal{P}$ in $O(n+\frac{h^3}{\epsilon^2}\text{polylog}(\frac{h}{\epsilon}))$ time, we compute one routing table at each of the vertices in a subset of vertices of $\mathcal{P}$, and all the routing tables' together are of size $O(n+\frac{h^3}{\epsilon^2}\text{polylog}(\frac{h}{\epsilon}))$. Following [26], we compute a sketch Ω of the input polygonal domain $\mathcal{P}$: each convex polygon P in $\mathcal{P}$ is approximated with another convex polygon Q such that $Q \subseteq P$ and the number of vertices of Q depends only on the input parameter ϵ. This is accomplished by partitioning the boundary of each obstacle of $\mathcal{P}$ into patches. Let $\mathcal{C}$ be a set of cones that partition the plane, wherein the apex of each cone in $\mathcal{C}$ is at the origin and the cone angles of all the cones in $\mathcal{C}$ are the same. For every convex polygonal obstacle Q in Ω, using the ideas from [17], for every vertex v of Q and for every cone $C \in \mathcal{C}$, we compute a closest vertex of v in Ω that is visible to v in cone C_v, where C_v is the cone resulting from translating C so that the apex of the translated cone is at v. Further, by using a property from [5], for every vertex $v \in \Omega$, we partition the boundaries of obstacles in Ω into a set of $O(\frac{1}{\epsilon}+h)$ pieces. For each such piece of v and the cone C_v that it corresponds to, we introduce an entry into the routing table at v, which considers $\mathcal{F}(\mathcal{P}) \subseteq \mathcal{F}(\otimes)$ and every routing path must belong to $\mathcal{F}(\mathcal{P})$. For any two successive vertices v', v'' along the boundary of an obstacle $Q \in \Omega$, for Q corresponds to an obstacle $P \in \mathcal{P}$, when the section of $bd(P)$ from v' and v'' occurs while traversing $bd(P)$ in clockwise direction, we save the labels of vertices of P that occur between v' and v'' in the routing table at v'. This helps in routing the packet to a destination node along a patch of $bd(P)$ whenever the destination node belongs to that patch.

In the packet routing phase, a packet is routed from any vertex s of $\mathcal{P}$ to any vertex t of $\mathcal{P}$. Our routing scheme computes a routing path from s to t

incrementally, that is, by computing successive subpaths of a routing path. Let s belongs to an obstacle P of $\mathcal{P}$. If $s \notin \Omega$, we first route the packet from s to an endpoint v of patch to which s belongs to, where v is a closest vertex in Ω to s along the boundary of P. In the other case, v is the same as s. By consulting the routing table at v, we forward the packet along a path located in $\mathcal{F}(\mathcal{P})$ to next hop $v' \in \Omega$. Upon packet reaching v', our routing scheme checks whether v' is equal to t. If it is, the algorithm terminates. Otherwise, our algorithm consults the routing table at v' and forwards the packet to the next hop along a geodesic subpath in $\mathcal{F}(\mathcal{P})$. That is, depending on whether the packet belongs to patch S, where $v' \in S$. If $t \in S$, then the packet is forwarded along a section of S. If t does not belong to S, we forward the packet from v' using an analogous algorithm to forward the packet from v to v'. We prove that this scheme computes a routing path with an multiplicative stretch $1 + \epsilon$ and an additive stretch $2k\ell$. That is, $d(s,t) \leq r(s,t) \leq (1+\epsilon)d(s,t) + 2k\ell$, where $r(s,t)$ is the distance along the routing path between s and t, and $d(s,t)$ is the geodesic shortest distance among obstacles in $\mathcal{P}$ between s and t. Here, k is the number of obstacles intersected by the routing path, and ℓ is a parameter that depends on the geometry of polygonal obstacles in $\mathcal{P}$. Further, our algorithm does $O(\lg(n + \frac{h}{\epsilon}))$ amount of work at $O(1)$ vertices of each of the k obstacles that the routing path intersects.

The algorithm in [5] computes a routing path with stretch $1+\epsilon$ with preprocessing time $O(n^2 \lg n + \frac{n}{\epsilon})$ and it computes routing tables of size $O(n(\frac{1}{\epsilon}+h)\lg n)$. Our algorithm substantially improves the space of the routing tables' and the time to compute these tables, when h is small compared to n (which is typically true). However, in this paper, only the polygonal domain with convex obstacles is considered; and, unlike the result in [5], the routing path obtained here has an additive stretch as well.

2 A Few Structures

In the following subsections, we prove a few structures needed to describe our algorithm. These include a sketch of $\mathcal{P}$, routing tables stored at a select set of vertices of $\mathcal{P}$, geodesic cones introduced at these vertices, and pieces defined with respect to vertices and cones of Ω.

2.1 Sketch of $\mathcal{P}$

For any obstacle $P_i \in \mathcal{P}$ and any two points p' and p'' on the boundary of P_i, the section of boundary of P_i that occurs while traversing from p' to p'' in counterclockwise order is termed a *patch* of P_i. In specific, we partition the boundary of each $P_i \in \mathcal{P}$ into a collection of patches Γ_i such that for any two points p', p'' belonging to any patch $\alpha \in \Gamma_i$, the angle between the outward (w.r.t. the centre of P_i) normals to respective edges at p' and p'' is upper bounded by $\frac{\epsilon}{2}$. The maximum angle between the outward normals to any two edges that belong to a patch α constructed in our algorithm is the *angle subtended by* α. To facilitate in computing patches of any obstacle P_i, we partition the unit circle

$\mathbb{S}^2$ centered at the origin into a minimum number of segments such that each circular segment is of length at most $\frac{\epsilon}{2}$. For every such segment s of $\mathbb{S}^2$, a patch (corresponding to s) comprises a maximal set of the contiguous sequence of edges of P_i whose outward normals intersect s when each of these normals is translated to the origin. In particular, for each patch $\alpha \in \Gamma_i$, the first and last vertices of α that occur while traversing the boundary of P_i are chosen to be in the *coreset* S_i *of* P_i. The *coreset* $\mathcal{S}$ *of* $\mathcal{P}$ is then simply $\bigcup_i S_i$. For $1 \leq i \leq h$, the *core-polygon* Q_i of P_i is $CH(S_i)$. Let Ω be the set comprising of core-polygons corresponding to each of the polygons in $\mathcal{P}$. The set Ω is called a *sketch of* $\mathcal{P}$.

Proposition 1 (Lemma 1, [26]). *Let Γ_i be a partition of the boundary of a convex polygon P_i into a collection of $O(\frac{1}{\epsilon})$ patches as described above. The geodesic distance between any two points p, q belonging to any patch $\alpha \in \Gamma_i$, the geodesic distance between p and q along α is upper bounded by $(1+\epsilon)\|pq\|$, for $\epsilon < 1$.*

Proof: For any two points p and q, respectively located on edges e_p and e_q of a patch, the angle between e_p and e_q is upper bounded by $\pi - \frac{\epsilon}{2}$. Hence, the geodesic length of patch between p and q is upper bounded by $\frac{\|pq\|}{\sin(\frac{\pi}{2} - \frac{\epsilon}{4})} \leq (1+\epsilon)\|pq\|$, when $\epsilon < 1$. □

The h convex polygons in $\mathcal{P}$ (resp. Ω) are denoted by $P_1, \ldots, P_h$ (resp. $Q_1, \ldots, Q_h$). The convex polygon in Ω that is a corepolygon of convex polygon $P_i \in \mathcal{P}$ is denoted by Q_i. Analogously, the convex polygon in $\mathcal{P}$ from which the corepolygon $Q_i \in \Omega$ is computed is denoted by P_i. For any patch α, the endpoint that occurs last while traversing α in the counterclockwise direction is called the *owner* of α.

2.2 Routing Path and Its Stretch

Every vertex v of $\mathcal{P}$ is associated with a unique label $\ell(v)$, a binary number. And, every obstacle P of $\mathcal{P}$ is associated with a unique label $\ell(P)$, which is also a binary number. We assume the packet needs to be transferred from any vertex, called a *source vertex*, of $\mathcal{P}$ to any other vertex, called a *destination vertex*, of $\mathcal{P}$. We denote the source and destination vertices by s and t, respectively. The packet stores the label of the destination, $\ell(t)$, with it. In the preprocessing phase, with each vertex $v \in \Omega$, we store a *routing table* $\rho(v)$ comprising unique labels of vertices. The routing tables' together help in guiding the packet to reach its destination efficiently. Specifically, at a subset of vertices v along the path the packet travels, using only $\rho(v)$ and t, the algorithm determines the geodesic path to reach a specific vertex in Ω.

Suppose t belongs to a patch α and the owner of α is t'. Then, our routing scheme first routes packet to t', and then the packet gets routed from t' to t along a geodesic shortest path on α between t' to t. In every iteration of the algorithm, the packet is routed from one vertex of Ω to another vertex of Ω along a geodesic path located in $\mathcal{F}(\mathcal{P})$, until the packet reaches t'. This is accomplished by consulting routing tables at vertices of Ω that occur along that path. The piecewise linear s-t path computed by our routing scheme is guaranteed to belong to $\mathcal{F}(\mathcal{P})$, and we call this path the *routing path*. The distance along the routing path is the *routing distance*.

For every two vertices $v', v'' \in \mathcal{P}$, suppose $d(v', v'') \leq r(v', v'') \leq \delta \cdot d(v', v'') + \rho$, for $\delta, \rho > 1$. Here, $r(v', v'')$ is the routing distance between v' and v'' determined by an algorithm, and $d(v', v'')$ is the geodesic shortest distance between v' and v'' in $\mathcal{P}$. Then, the routing path computed by that algorithm is said to be an (δ, ρ)-approximation to the geodesic shortest path. Specifically, δ is termed the multiplicative stretch and ρ is called the additive stretch of the routing path.

2.3 Geodesic Cones

Let r' and r'' be two rays with origin at p. Let $\vec{v_1}$ and $\vec{v_2}$ be the unit vectors along the rays r' and r'' respectively. A *cone* $C_p(r', r'')$ is the set of points defined by rays r' and r'' such that a point $q \in C_p(r', r'')$ if and only if q can be expressed as a convex combination of vectors $\vec{v_1}$ and $\vec{v_2}$ with positive coefficients. When the rays r' and r'' are evident from the context, we denote the cone with C_p. The counterclockwise angle from the positive x-axis to the line that bisects the cone angle of C_p is termed as the *orientation of the cone* C_p. The angle between rays r' and r'' is the *cone angle* of C_p.

We denote the set of cones, each with cone angle ϵ and each with the apex at the origin, which together partitions the plane by $\mathcal{C}$. We note the number of cones in $\mathcal{C}$ is $O(\frac{1}{\epsilon})$. Any cone $C \in \mathcal{C}$ translated so that its apex is at a point $p \in \mathbb{R}^2$ is denoted by C_p.

Let $C \in \mathcal{C}$ be a cone with orientation θ and let $C' \in \mathcal{C}$ be the cone with orientation $-\theta$. For each cone $C \in \mathcal{C}$ and a set K of points, the set of cones resultant from introducing a cone C_p for every point $p \in K$ is the *conic Voronoi diagram* with respect to C and K. Using the algorithm from [17], for every cone $C \in \mathcal{C}$, a conic Voronoi diagram (CVD) is computed using a plane sweep (refer to [19]). And, the planar point location data structure is used to locate the region in that CVD to which a given query point belongs. In specific, we compute a closest vertex of v in C_v for every vertex $v \in \Omega$ using the $CVD(C, V_\Omega)$, where V_Ω is the set of vertices that define Ω. Among the points in C_v, if more than one point is close to v, then we arbitrarily pick one of those points.

Further, [17] computes a geodesic spanner by joining each vertex v to a closest point in each cone C_v for every $C \in \mathcal{C}$. And, it proves that this construction indeed yields a spanner with a multiplicative stretch $(1 + \epsilon)$. Our algorithm implicitly relies on this construction.

2.4 Piece

A *piece of* C_v is a section of the boundary of an obstacle such that the first edge of the shortest path from v to any vertex of that section lies in C_v. Naturally, a piece is always with respect to a vertex v and a cone C_v. For any $C \in \mathcal{C}$, if β is a piece of C_v, then β is called a *piece of* v. Among all the vertices of Ω in C_v, let

$r \in \Omega$ be the vertex closest to v. Any packet at v that is destined for any vertex belonging to a piece of C_v, our routing scheme forwards it to r along a geodesic shortest path in $\mathcal{F}(\mathcal{P})$.

The following property from [5] (Lemma 4.2) proves that every piece is contiguous along the boundary of an obstacle.

Proposition 2. *Let $e = (v, s)$ be an edge in shortest path tree T rooted at v. Also, let S be the set of all vertices belonging to any obstacle $Q \in \Omega$ whose first edge in the shortest path from v is e. Then, all the vertices in S occur contiguously along the boundary of Q. Furthermore, let $f = (v, s')$ another edge in T, such that e and f are consecutive edge in T around v. Let S' be the set of all vertices belonging to any obstacle $Q \in \Omega$ whose first edge in the shortest path from v is either e or f. Then, again all the vertices in S' occur contiguously along the boundary of Q.*

For any vertex $v \in \Omega$, from the non-crossing property of shortest paths, there is at most one section of the boundary of any obstacle that is part of a piece of more than one cone with apex at v. From this, the following is an immediate upper bound on the number of pieces.

Lemma 1 (Lemma 5.2, [5]). *For any vertex $v \in \Omega$, the number of pieces of v is $O(\frac{1}{\epsilon} + h)$.*

3 Algorithm

In the preprocessing phase, we compute Ω and build a routing table at each vertex of Ω. In the packet routing phase, a packet located at any given source vertex $s \in \mathcal{P}$ is routed to any given destination node $t \in \mathcal{P}$ along a path located in $\mathcal{F}(\mathcal{P})$.

As described above, we compute a sketch Ω of $\mathcal{P}$ that has h convex polygonal obstacles, defined with $O(\frac{h}{\epsilon})$ vertices. To find pieces of vertices of Ω, for every vertex $v \in \Omega$, using the algorithm from [24], we compute the shortest path tree in $\mathcal{F}(\Omega)$ that contains a shortest path from v to every other vertex of Ω. As described in Subsect. 2.3, using CVDs from [17], we compute a nearest visible vertex of v in C_v for every vertex $v \in \Omega$ and every $C \in \mathcal{C}$. Further, we build data structures with the processing algorithm for answering ray-shooting queries from [15].

By exploiting Proposition 2, for every piece β of C_v, we could store in $\rho(v)$ the label of obstacle P to which β belongs, the endpoints of β, together with a nearest visible vertex $r \in \Omega$ in C_v among obstacles in Ω. However, since the routing path must belong to $\mathcal{F}(\mathcal{P})$ and since $\mathcal{F}(\mathcal{P}) \subseteq \mathcal{F}(\Omega)$, instead of saving r in $\rho(v)$, we ray-shoot with ray vr among obstacles of $\mathcal{P}$. Suppose this query returns a point p located on an obstacle $P' \in \mathcal{P}$. Then the line segment pr intersects P' with the other endpoint, say $p' \in bd(P')$. We note that both p and p' belong to the same patch α'. Among the two endpoints of α', let v' be the endpoint that has the shortest geodesic distance along $bd(P')$ to p'. We save v'

and p in $\rho(v)$, noting the geodesic path in $\mathcal{F}(\mathcal{P})$ from v to v' passes through p as part of reaching to r. Essentially, packet is transferred from v to p along line segment vp and then it is transferred from p to v' along the geodesic shortest path on α'.

In other words, during the routing phase, if a packet reaches vertex v, our scheme checks whether (i) v is the destination of the packet (i.e., $t = v$), (ii) packet needs be routed to t that is located on patch α to which v belongs, or (iii) needs to be transferred from v by consulting $\rho(v)$. For below description, let v be located on a patch α and let v' be located on a patch α'. There are three subcases to (iii): In Subcase (a), the labels of p, v' and r are stored with the entry of interest in $\rho(v)$. In this subcase, the packet is routed to p along a line segment vp first, and then it is transferred from p to v' along the geodesic shortest path on patch α'. In Subcase (b), the labels of p, p', and r are present in the entry. Then, we transfer it from v to p' along α, from p' to p along the line segment $p'p$, and from p to v' along the geodesic shortest path on α'. In Subcase (c), only r is present; hence, we directly transfer the packet from v to r along line segment vr. The Subcase (c) implies r is visible to v, and r is not visible from v in subcases (a) and (b).

Theorem 1. *Given a polygonal domain $\mathcal{P}$ comprising convex polygonal obstacles, our algorithm preprocesses $\mathcal{P}$ in $O(n + \frac{h^3}{\epsilon^2}\text{polylog}(\frac{h}{\epsilon}))$ time and construct routing tables of size $O(n+\frac{h^3}{\epsilon^2}\text{polylog}(\frac{h}{\epsilon}))$ so that given any two vertices $s, t \in \mathcal{P}$ algorithm outputs a routing path between s and t located in $\mathcal{F}(\mathcal{P})$ with a multiplicative stretch $1+\epsilon$ and an additive stretch $2k\ell$, while the routing scheme makes the routing decisions by searching the routing tables located at $O(k)$ nodes along the routing path. Here, k is the number of obstacles the routing path intersects, ℓ is the maximum length of any patch in $\mathcal{P}$, and $\epsilon < 1$ is an input parameter.*

Proof: Computing Ω from $\mathcal{P}$ takes $O(\frac{h}{\epsilon})$ time. Using the algorithm from [24], to compute a SPT_v in Ω from any vertex $v \in \Omega$ takes $O(\frac{h}{\epsilon}\lg\frac{h}{\epsilon})$ time. The time involved in computing shortest path trees rooted at all the vertices of Ω takes $O(\frac{h^2}{\epsilon^2}\lg\frac{h}{\epsilon})$. For any cone $C \in \mathcal{C}$, for every vertex v of Ω, determining a vertex in Ω that is closest in C_v to v takes $O(\frac{h}{\epsilon}\lg\frac{h}{\epsilon})$ time using a plane sweep, that is, by building a conic Voronoi diagram. Since there are $O(\frac{1}{\epsilon})$ cones in $\mathcal{C}$, the total time to compute closest neighbor of every vertex in every cone together takes $O(\frac{h}{\epsilon^2}\lg\frac{h}{\epsilon})$ time. The preprocessing time for ray-shooting query algorithm from [15] is $O(n+h^2\text{polylog}(h))$ and the space of data structures that it constructs is $O(n+h^2)$. Since we invoke $O(h)$ ray-shoot queries for each vertex $v \in \Omega$ and cone $C \in \mathcal{C}$ combination, since Ω has $O(\frac{h}{\epsilon})$ vertices, and since $\mathcal{C}$ has $O(\frac{1}{\epsilon})$ cones, it takes $O(\frac{h^2}{\epsilon^2}\lg n)$ time to compute p, p', v', r. From [5], the size of routing tables is $O((\frac{h}{\epsilon}(\frac{1}{\epsilon}+h)\lg\frac{h}{\epsilon})$, i.e., $O((\frac{h}{\epsilon^2}+\frac{h^2}{\epsilon})\lg\frac{h}{\epsilon})$. And, since each such closest point may intersect h obstacles, the number of entries in all the routing tables together is $O(\frac{h^3}{\epsilon^2}\text{polylog}(\frac{h}{\epsilon}))$ for $\epsilon < 1$. Due to Subcases (a) and (b) of Case (iii), if the path intersects k obstacles and the maximum length of any patch it intersects is ℓ, then there is an additive factor of $2k\ell$. □

References

1. Abraham, I., Gavoille, C.: On approximate distance labels and routing schemes with affine stretch. In: Peleg, D. (ed.) DISC 2011. LNCS, vol. 6950, pp. 404–415. Springer, Heidelberg (2011). https://doi.org/10.1007/978-3-642-24100-0_39
2. Arikati, S., Chen, D.Z., Chew, L.P., Das, G., Smid, M., Zaroliagis, C.D.: Planar spanners and approximate shortest path queries among obstacles in the plane. In: Diaz, J., Serna, M. (eds.) ESA 1996. LNCS, vol. 1136, pp. 514–528. Springer, Heidelberg (1996). https://doi.org/10.1007/3-540-61680-2_79
3. Asano, T., Asano, T., Guibas, L.J., Hershberger, J., Imai, H.: Visibility of disjoint polygons. Algorithmica **1**(1), 49–63 (1986). https://doi.org/10.1007/BF01840436
4. Awerbuch, B., Bar-Noy, A., Linial, N., Peleg, D.: Improved routing strategies with succinct tables. J. Algorithms **11**(3), 307–341 (1990)
5. Banyassady, B., et al.: Routing in polygonal domains. Comput. Geom. **87**, 101593 (2020)
6. Bar-Yehuda, R., Chazelle, B.: Triangulating disjoint Jordan chains. Int. J. Comput. Geom. Appl. **4**(4), 475–481 (1994)
7. Bose, P., Fagerberg, R., van Renssen, A., Verdonschot, S.: Optimal local routing on Delaunay triangulations defined by empty equilateral triangles. SIAM J. Comput. **44**, 1626–1649 (2015)
8. Bose, P., Fagerberg, R., van Renssen, A., Verdonschot, S.: Competitive local routing with constraints. J. Comput. Geom. **8**(1), 125–152 (2017)
9. Bose, P., Korman, M., van Renssen, A., Verdonschot, S.: Routing on the visibility graph. In: Proceedings of International Symposium on Algorithms and Computation, pp. 18:1–18:2 (2017)
10. Bose, P., Korman, M., van Renssen, A., Verdonschot, S.: Constrained routing between non-visible vertices. Theor. Comput. Sci. **861**, 144–154 (2021)
11. Bose, P., Morin, P.: Competitive online routing in geometric graphs. Theor. Comput. Sci. **324**(2), 273–288 (2004)
12. Bose, P., Smid, M.: On plane geometric spanners: a survey and open problems. Comput. Geom. **46**(7), 818–830 (2013)
13. Chechik, S.: Compact routing schemes with improved stretch. In: Proceedings of Symposium on Principles of Distributed Computing, pp. 33–41 (2013)
14. Chen, D.Z.: On the all-pairs Euclidean short path problem. In: Proceedings of Symposium on Discrete Algorithms, pp. 292–301 (1995)
15. Chen, D.Z., Wang, H.: Visibility and ray shooting queries in polygonal domains. Comput. Geom. **48**(2), 31–41 (2015)
16. Chiang, Y.-J., Mitchell, J.S.B.: Two-point Euclidean shortest path queries in the plane. In: Proceedings of Symposium on Discrete Algorithms, pp. 215–224 (1999)
17. Clarkson, K.L., Kapoor, S., Vaidya, P.M.: Rectilinear shortest paths through polygonal obstacles in $O(n(\lg n)^2)$ time. In: Proceedings of Symposium on Computational Geometry, pp. 251–257 (1987)
18. Cowen, L.: Compact routing with minimum stretch. J. Algorithms **38**, 170–183 (1999)
19. de Berg, M., Cheong, O., van Kreveld, M., Overmars, M.: Computational Geometry: Algorithms and Applications, 3rd edn. Springer, Heidelberg (2008). https://doi.org/10.1007/978-3-540-77974-2
20. Eilam, T., Gavoille, C., Peleg, D.: Compact routing schemes with low stretch factor. J. Algorithms **46**(2), 97–114 (2003)

21. Fraigniaud, P., Gavoille, C.: Routing in trees. In: Orejas, F., Spirakis, P.G., van Leeuwen, J. (eds.) ICALP 2001. LNCS, vol. 2076, pp. 757–772. Springer, Heidelberg (2001). https://doi.org/10.1007/3-540-48224-5_62
22. Ghosh, S.K.: Visibility Algorithms in the Plane. Cambridge University Press, New York (2007)
23. Guibas, L.J., Hershberger, J., Leven, D., Sharir, M., Tarjan, R.E.: Linear-time algorithms for visibility and shortest path problems inside triangulated simple polygons. Algorithmica **2**, 209–233 (1987). https://doi.org/10.1007/BF01840360
24. Hershberger, J., Suri, S.: An optimal algorithm for Euclidean shortest paths in the plane. SIAM J. Comput. **28**(6), 2215–2256 (1999)
25. Inkulu, R., Kapoor, S.: Planar rectilinear shortest path computation using corridors. Comput. Geom. **42**(9), 873–884 (2009)
26. Inkulu, R., Kapoor, S.: Approximate Euclidean shortest paths amid polygonal obstacles. In: Proceedings of Symposium on Algorithms and Computation (2019)
27. Inkulu, R., Kapoor, S., Maheshwari, S.N.: A near optimal algorithm for finding Euclidean shortest path in polygonal domain. CoRR 1011.6481 (2010)
28. Kaplan, H., Mulzer, W., Roditty, L., Seiferth, P.: Routing in unit disk graphs. Algorithmica **80**(3), 830–848 (2018). https://doi.org/10.1007/s00453-017-0308-2
29. Kapoor, S.: Efficient computation of geodesic shortest paths. In: Proceedings of Symposium on Theory of Computing, pp. 770–779 (1999)
30. Kapoor, S., Maheshwari, S.N.: Efficient algorithms for Euclidean shortest path and visibility problems with polygonal obstacles. In: Proceedings of Symposium on Computational Geometry, pp. 172–182 (1988)
31. Kapoor, S., Maheshwari, S.N.: Efficiently constructing the visibility graph of a simple polygon with obstacles. SIAM J. Comput. **30**(3), 847–871 (2000)
32. Kapoor, S., Maheshwari, S.N., Mitchell, J.S.B.: An efficient algorithm for Euclidean shortest paths among polygonal obstacles in the plane. Discrete Comput. Geom. **18**(4), 377–383 (1997). https://doi.org/10.1007/PL00009323
33. Konjevod, G., Richa, A.W., Xia, D.: Scale-free compact routing schemes in networks of low doubling dimension. ACM Trans. Algorithms **12**(3), 1–29 (2016)
34. Mitchell, J.S.B.: Shortest paths among obstacles in the plane. Int. J. Comput. Geom. Appl. **6**(3), 309–332 (1996)
35. Mitchell, J.S.B.: Shortest paths and networks. In: Handbook of Discrete and Computational Geometry, pp. 811–848. CRC Press (2017)
36. Narasimhan, G., Smid, M.H.M.: Geometric Spanner Networks. Cambridge University Press, Cambridge (2007)
37. Overmars, M.H., Welzl, E.: New methods for computing visibility graphs. In: Proceedings of the Fourth Annual Symposium on Computational Geometry, pp. 164–171 (1988)
38. Peleg, D., Upfal, E.: A trade-off between space and efficiency for routing tables. J. ACM **36**(3), 510–530 (1989)
39. Roditty, L., Tov, R.: New routing techniques and their applications. In: Proceedings of ACM Symposium on Principles of Distributed Computing, pp. 23–32 (2015)
40. Roditty, L., Tov, R.: Close to linear space routing schemes. Distrib. Comput. **29**(1), 65–74 (2015). https://doi.org/10.1007/s00446-015-0256-5
41. Santoro, N., Khatib, R.: Labelling and implicit routing in networks. Comput. J. **28**(1), 5–8 (1985)
42. Sharir, M., Schorr, A.: On shortest paths in polyhedral spaces. SIAM J. Comput. **15**(1), 193–215 (1986)
43. Storer, J.A., Reif, J.H.: Shortest paths in the plane with polygonal obstacles. J. ACM **41**(5), 982–1012 (1994)

44. Thorup, M.: Compact oracles for reachability and approximate distances in planar digraphs. J. ACM **51**(6), 993–1024 (2004)
45. Thorup, M., Zwick, U.: Compact routing schemes. In: Proceedings of Symposium on Parallel Algorithms and Architectures, pp. 1–10 (2001)
46. Thorup, M., Zwick, U.: Approximate distance oracles. J. ACM **52**(1), 1–24 (2005)
47. Welzl, E.: Constructing the visibility graph for n-line segments in $O(n^2)$ time. Inf. Process. Lett. **20**(4), 167–171 (1985)
48. Yan, C., Xiang, Y., Dragan, F.F.: Compact and low delay routing labeling scheme for unit disk graphs. Comput. Geom. **45**(7), 305–325 (2012)
49. Yao, A.C.: On constructing minimum spanning trees in k-dimensional spaces and related problems. SIAM J. Comput. **11**(4), 721–736 (1982)

Target Coverage with Minimum Number of Camera Sensors

Pei Yao[1], Longkun Guo[1(✉)], Shuangjuan Li[2], and Huihong Peng[1]

[1] Fuzhou University, Fuzhou, China
lkguo@fzu.edu.cn
[2] South China Agricultural University, Guangzhou, China

Abstract. With the development of the smart city, camera sensors have attracted more and more research interests from both academic researchers and industrial engineers. Given a set of points of interests (POI) and a set of cameras, practical applications require to deploy these cameras with the minimum cost so that these POIs can be fully covered by these cameras. In this paper, we study a problem called Min-Num LTC-CS, which is, given a set of POIs located on a line segment and a set of cameras distributed on the plane, to choose a minimum number of cameras so that these POIs can be fully covered by the sensing ranges of these cameras. We first propose a grouping algorithm by grouping the POIs according to whether they can be covered by the same camera with certain rotation angle and then construct a graph using these POI groups. We show that there exists a feasible constrained *st*-flow if and only if there exists a subset of cameras that can completely cover these POIs. Then we propose an LP formulation for the constrained flow problem and prove that any basic solution of the LP formulation is integral, which consequently leads to an optimal solution to Min-Num LTC-CS by solving this LP formulation. Lastly, extensive numerical experiments are conducted to demonstrate the practical performance of our algorithms.

Keywords: Camera sensor network · Constrained flow · Linear programming · Integer optimal solution

1 Introduction

Wireless sensor network has been attracting lots of research interest from many researchers, since it has many practical applications, including monitoring the gas pipeline such that the pipeline leakage can be timely found by these sensors, and monitoring the border of the country or a building to prevent illegal entrance.

Compared with traditional sensors, camera sensors can obtain more rich digital information such as the pictures and videos. Different from traditional sensors, camera sensors have some unique coverage characteristics [16]. Based on computer vision technology, camera sensors can be widely used in lots of applications, such as coal mine monitoring, urban underground engineering, and online virtual roaming. However, in some cases the pictures or videos captured by

D.-Z. Du et al. (Eds.): COCOA 2021, LNCS 13135, pp. 12–24, 2021.
https://doi.org/10.1007/978-3-030-92681-6_2

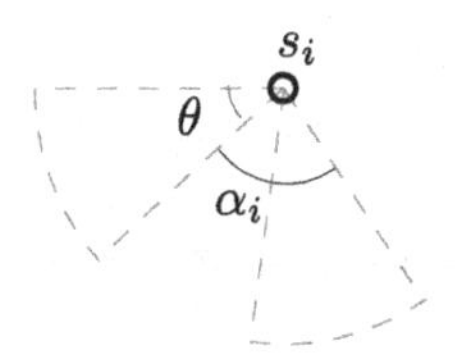

Fig. 1. Coverage angle θ and rotation angle α for s_i.

cameras are useless if these cameras are deployed with wrong rotation angles. Therefore, for different coverage requirements, how to choose the rotation angles of cameras for intrusion detection is a challenging problem. Different from traditional sensors, once the camera sensor is deployed, its position is fixed and its covering region can be changed only by rotation. Similar to stationary sensor and mobile sensor, there are three types of sensor coverage [4]: region coverage [10], barrier coverage [9], and point coverage [19]. In this paper, we study a special target problem, called on-a-Line Target Coverage of Camera Sensors (LTC-CS), which is formally defined as follows:

Definition 1 *(On-a-Line Target Coverage of Camera Sensors problems, LTC-CS). Let Π be a set of points of interests (POIs) in which each POI $p_j \in \Pi$ has a position with $(p_j, 0)$. Let Γ be a set of camera sensors each of which $s_i \in \Pi$ has a position (x_i, y_i), a sensing radius r, and a coverage angle θ. The Min-Num LTC-CS problem aims to choose a minimum number of camera sensors from Γ such that each POI in Π is covered by the sensing range of at least one camera sensor.*

Figure 1 illustrates an example of coverage angle θ and rotation angle α_i. As a generalization of Min-Num LTC-CS, the Min-Sum LTC-CS problem aims to minimize the sum of the rotation angles of all sensors used for covering all the POIs. Formally, let γ be the sum of the rotation angles, i.e. $\gamma = \sum_{s_i \in \Gamma'} \alpha_i$ where $\Gamma' \subseteq \Gamma$ where each Γ' is the set of used cameras. Then Min-Sum LTC-CS is to minimize γ.

1.1 Related Work

Coverage using directional wireless network which consists of directional sensors was firstly discussed by *Ma* and *Liu* [13]. When all POIs are distributed on a plane, the sensor coverage of these POIs can be regarded as a set cover problem, and *Fowler et al.* [5] proved that the planar geometric covering problem is an NP-complete problem by reduction to 3-SAT problem. *Cai et al.* [2,3] studied the multiple directional coverages sets problem (MDCS), which is to find K cover set $D_1, \cdots, D_K \subseteq D$, where D is a collection of a subset of A and D_i has a nonnegative weight t_i, such that $\sum_{i=1}^{K} t_i$ is maximized and $\sum_{i=1}^{K} |s \cap D_i| \cdot t_i \leq L$ for each $s \in S$ and a given positive number L, where S is a collection of subsets of D. The authors proved that the MDCS problem is NP-complete and proposed some centralized algorithms and distributed algorithms for MDCS.

Ai and *Abouzeid* [1] proposed a maximum coverage with minimum sensors problem and proved this problem is NP-complete by reduction to the maximum cover problem. They proposed an integer linear programming formulation and also presented an approximation centralized greedy algorithm. Other problems about directional sensors were also studied, such as the k-coverage problem [6,8], service delay minimization problem [21], rotatable and directional sensor deployment problem [23].

Camera sensor network, a special directional sensor networks, has also been extensively studied by lots of researchers. *Liu et al.* [11] studied the directional k-coverage problem in a camera sensor network and the aim of the problem is to cover each point in the given region by at least k different cameras. *Wang* and *Cao* [22] studied full-view coverage in a camera sensor network, where full-view coverage requires that an object is always covered by at least one camera for any direction from 0 to 2π and the facing direction of the object is sufficiently close to the viewing direction of the sensor. Later, *Ma et al.* [14] studied the minimum camera barrier coverage problem in a camera sensor network based on the definition of full-view coverage in [22] and proposed an optimal algorithm to solve this problem. *Jia et al.* [7] designed a $(1 - \frac{1}{e})$-approximation algorithm and an efficient heuristic algorithm for the maximum full-view target coverage problem in a camera sensor network.

Recently, *Liu* and *Ouyang* [12] derived a k-coverage probabilistic expression to estimate the minimum number of camera sensors when the k-coverage can be achieved while all camera sensors are randomly deployed outside the field of interest. *Si et al.* [18] proposed a realistic resolution criterion to capture the intruder's face for a three-Dimensional (3D) sensing model of a camera sensor and they are the first to study the barrier coverage of camera sensor network in 3D setting. Later, *Wang et al.* [20] studied the fundamental problem of the placement of unmanned aerial vehicles for achieving 3D directional coverage and proposed a greedy algorithm to solve the problem with a $(1 - 1/e)$ approximation ratio. *Saeed et al.* [17] proposed an autonomous system called Argus which aims to minimize the number of drones required to cover a set of targets and proved the problem is NP-hard by reduced it to a polygon illumination problem. *Mao et al.* [15] proposed an alternating optimization algorithm with guaranteed convergence based on block coordinate descent and successive convex approximation to minimize the maximum computation delay among internet of things devices.

1.2 Our Results

In this paper, we devise an approach for optimally solving Min-Num LTC-CS in polynomial time. The contribution of the paper can be summarized as follows:

- Propose a grouping algorithm that clusters the given POIs according to whether a set of POIs can be covered by a camera sensor.
- Propose an LP formulation for the constrained flow problem for modeling the placement problem.

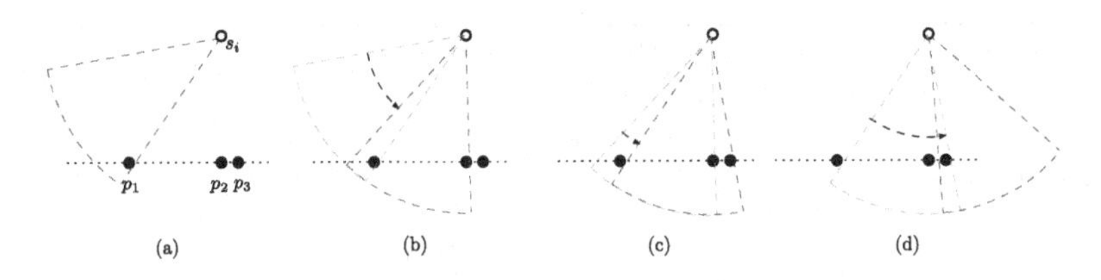

Fig. 2. An example of a POI groups. In (a), there exist three POIs $\Pi = \{p_1, p_2, p_3\}$. As a consequence, there can be four POI groups $\mathcal{G} = \{G_1, G_2, G_3, G_4\}$, where $G_1 = \{p_1\}$, $G_2 = \{p_1, p_2\}$, $G_3 = \{p_2, p_3\}$ and $G_4 = \{p_3\}$.

- Prove that any basic solution of the LP formulation can be rounded to an integral solution.

1.3 Organization

The remainder of the paper is organized as follows: Sect. 2 presents a grouping algorithm and the integer linear program formulations of two problems: Min-Num LBC-CS; Sect. 3 proposes an LP-rounding algorithms to solve the Min-Num LBC-CS; Sect. 4 demonstrates experimental results; Sect. 5 lastly concludes this paper.

2 Integer Linear Programs

In this section, we propose an Integer Linear Programming (ILP) formulation for Min-Num on-a-Line Target Coverage of Camera Sensor (LTC-CS). First of all, we propose a grouping algorithm for clustering the set of points of interests (POIs) into some groups such that each group can be covered by a camera sensor under a rotation angle. Then, we formulate an ILP based on these groups to solve Min-Num LTC-CS.

2.1 Camera Sensor Group

For a camera sensor s_i, its coverage region can be modeled as a sector such that s_i can cover multiple POIs at the same time. For two POIs p_i and p_j and camera sensor s_i, we denote $\angle p_i s_i p_j$ as a angle. Then, we say a set of POIs Π_1 can be ***completely covered*** by sensor s_i if and only if any pair of POIs $p_i, p_j \in \Pi_1$ satisfies $\angle p_i s_i p_j \leq \theta$ and $|p_i s_i| \leq r$, $|p_j s_i| \leq r$, where θ is the coverage angle of camera sensor s_i and r is the radius of s_i. We say such set of POIs Π_1 is a POI group, whose formal definition is as follows:

Definition 2. *(POIs group) Let $\Pi = \{p_1, \cdots, p_m\}$ be a set of POIs, where each $p_i \in \Pi$ is with the position $(x_i, 0)$ on a line. Let Π_1 be a subset of Π. If Π_1 is exactly a set of target* ***completely covered*** *by one camera sensor, then we say Π_1 is a POI group.*

Algorithm 1. A grouping algorithm for POIs groups.

Input: A set of camera sensors Γ and each camera sensor $s_i \in \Gamma$ is with the position (x_i, y_i); the coverage angle θ; the sensing radius r; a set of POIs Π, in which each $p_j \in \Pi$ has a position $(p_j, 0)$.
Output: a set of POIs groups $\mathcal{G}$.

Phase I: For each camera sensor s_i, find all POIs that it can cover;
1: Set $\mathcal{G} := \emptyset$, $\Lambda := \{\Lambda_1, \cdots, \Lambda_n\}$ where $\Lambda_i = \emptyset$;
2: **For** $i = 1$ **to** n **do**
3: Set $v_i = x_i - \sqrt{r^2 - y_i^2}$ and $w_i = x_i + \sqrt{r^2 - y_i^2}$;
/* The maximum range at which camera sensor s_i intersects the line during rotation. */
4: Set $p_m = \min\{p_j | v_i \leq p_j \leq w_i\}$, $p_n = \max\{p_j | v_i \leq p_j \leq w_i\}$;
5: Set $\Lambda_i := \{p_m, p_{m+1}, \cdots, p_n\}$;
6: **Endfor**

Phase II: For each camera sensor s_i, group all the POIs that s_i can cover
7: **For** $i = 1$ to n **do**
8: Select the first element of Λ_i as the first POIs group, i.e. $G := \{\Lambda_i[1]\}$;
9: Set $\mathcal{G} := \mathcal{G} \cup \{G\}$, and let $j_w := 1$, $j_u = 2$;
10: **While** $j_u \leq |\Lambda_i|$ **do**
11: Find $\theta_{min} := \min\{\rho(\theta, j_w), \rho'(\theta, j_u)\}$;
/* $\rho(\theta, j_w)$ is the angle at which element $\Lambda_i[j_w]$ is exactly deleted from G and $\rho'(\theta, j_u)$ is the angle at which element $\Lambda_i[j_u]$ is exactly added to G. */
12: **If** $\theta_{min} = \rho(\theta, j_w) \neq \rho'(\theta, j_u)$ **then**
13: Set $j_w := j_w + 1$, $G := G \backslash \{\Lambda_i[j_w]\}$, $\mathcal{G} := \mathcal{G} \cup \{G\}$;
14: **If** $\theta_{min} = \rho'(\theta, j_u) \neq \rho(\theta, j_w)$ **then**
15: Set $j_u := j_u + 1$, $G := G \cup \{\Lambda_i[j_u]\}$, $\mathcal{G} := \mathcal{G} \cup \{G\}$;
16: **If** $\rho(\theta, j_w) = \rho'(\theta, j_u)$ **then**
17: Set $j_w := j_w + 1$, $j_u := j_u + 1$, $G := G \backslash \{\Lambda_i[j_w]\} \cup \{\Lambda_i[j_u]\}$, $\mathcal{G} := \mathcal{G} \cup \{G\}$;
18: **Endwhile**
19: **Endfor**
20: **return** $\mathcal{G}$.

For briefness, we use $\mathcal{G} = \{G_1, \cdots, G_k\}$ to represent all POI groups of Π, where each $G_i \in \mathcal{G}$ is a POI group of Π and contains at least one POI. Obviously, $G_i \in \mathcal{G}$ can be covered by one camera sensor. Figure 2 shows an example of grouping POIs. In the figure, there exist three POIs $\Pi = \{p_1, p_2, p_3\}$, which can produce a family of four POI groups $\mathcal{G} = \{G_1, G_2, G_3, G_4\}$, where $G_1 = \{p_1\}$ (Fig. 2(a)), $G_2 = \{p_1, p_2\}$ is obtained by adding p_2 to G_1 (Fig. 2(b)), $G_3 = \{p_2, p_3\}$ is obtained by adding p_3 to G_2 and deleting p_1 at the same time (Fig. 2(c)), and $G_4 = \{p_3\}$ is acquired by deleting p_2 from G_3 (Fig. 2(d)).

Based on Definition 2, we can propose a grouping algorithm of grouping the given set of POIs Π into several groups. Two POI groups G_i and G_j are different if and only if there exists at least a POI p_i that satisfies $p_i \in G_i$, $p_i \notin G_j$ or $p_i \notin G_i$, $p_i \in G_j$. The grouping algorithm runs in two steps as follows:

1. Find a subset of Π denoted as Π_i which can be covered by camera sensor s_i when s_i rotates 360°, i.e. the distance $d(s_i, p_j) \leq r$ for $\forall p_j \in \Pi_i$;
2. Find two POIs p_i and p_j, where $p_i \in \Pi_i$ is the leftmost POI that is covered by sensor s_i and $p_j \in \Pi_i$ is the rightmost POI that can be covered on the line at the same time by s_i, i.e. the angle $\angle p_i s_i p_j \leq \theta$.

For a POI group G_i, let $\rho(\theta, j_w)$ and $\rho'(\theta, j_w)$ be the minimum angle at which element $j_w \in G_i$ is deleted from G_i or added to G_i, respectively. Then the detailed algorithm is shown in Algorithm 1.

Lemma 1. *Algorithm 1 takes $O(nm)$ time to construct the POI groups $\mathcal{G}$.*

Proof. From Step 2 to Step 6, Algorithm 1 needs $O(nm)$ time to compute a set of POIs Λ_i which is a set of POIs that camera sensor s_i may cover. Steps 7–18 take $O(nm)$ times to construct the POI group $\mathcal{G}$.

2.2 Integer Linear Programming Formulation

Let Π be a set of POIs and $\mathcal{G} = \{G_1, \cdots, G_k\}$ be a set of POI groups, where $\mathcal{G}$ is formed by grouping all POIs of Π. Let x_{ij} represent whether camera sensor s_i cover POI group G_j, then $x_{ij} \in \{0, 1\}$. Let y_j indicate whether POI group G_j is covered, i.e. $y_j \in \{0, 1\}$. Then the ILP formulation of Min-Num can be described as follows:

$$\begin{aligned}
\min \quad & \sum_i x_{ij} & & \\
s.t. \quad & \sum_j x_{ij} \leq 1 & & \forall s_i \in \Gamma \\
& y_j \leq \sum_i x_{ij} & & \forall g_j \in \mathcal{G} \\
& \sum_{j:\, t \in G_j} y_j \geq 1 & & \forall G_j \in \mathcal{G} \\
& x_{ij} \in \{0, 1\} & & \forall s_i \in \Gamma, G_j \in \mathcal{G} \\
& y_j \in \{0, 1\} & & \forall G_j \in \mathcal{G}
\end{aligned} \tag{1}$$

where the first constraint means each camera sensor s_i can be used at most once, the second guarantees that the portion of the POI group G_j is covered does not exceed the total portion of all camera sensors covering G_j, and the third inequality ensures that each POI must be covered at least one in total.

Moreover, let α_{ij} be the rotation angle of camera sensor s_i when s_i is used to cover the POI group G_j. Then we can generalize the above ILP to Min-Sum LTC-CS by setting the objective function as $\min \sum_i x_{ij}\alpha_{ij}$, while retaining the constraints the same as those of the Min-Num version.

3 LP-Rounding Algorithm via Transformation to the Shortest Matching-Path Problem

In this section, we first transform the Min-Num for the on-a-Line Target Coverage of Camera Sensor (LTC-CS) problem into a Shortest Matching-Path (SMP) problem over a weighted graph. Secondly, the Linear Program (LP) of SMP was given. Thirdly, we prove there exists an integer solution for SMP iff SMP is feasible.

3.1 The Construction

To solve Min-Num LTC-CS, we will construct a graph based on Algorithm 1. In the construction, it is important to judge whether two points of interest (POI) groups are adjacent, where the definition of the adjacent POI groups is as follows:

Definition 3. *(The adjacent POI groups). Let $\mathcal{G} = \{G_1, \cdots, G_k\}$ be a set of POI groups, where $G_i \in \mathcal{G}$ contains at least one POI. Let l_i and g_i be the leftmost and rightmost POIs of $G_i \in \mathcal{G}$, respectively. Then for two POI group $G_i, G_j \in \mathcal{G}$ with $l_i \leq g_j$, we say they are adjacent if and only if there exists no targets between g_i and l_j or $l_j \leq g_i$ holds.*

The directed auxiliary graph can be constructed in the following main steps:

1. **Vertices:** For each POI group, we add a corresponding vertex, and collectively obtain V as the set of vertices $V = \{G_1, \cdots, G_m\}$, where m is the number of POI groups.
2. **Edges:** For two adjacent points G_i and G_j, we add an arc $\langle G_i, G_j \rangle$ to the graph, in which G_i is the tail of the arc and G_j is the head of the arc.
   (a) If $p_1 \in G_i$ holds, there exists an arc $\langle s, G_i \rangle$, and if $p_m \in G_j$, there exists an arc $\langle G_j, t \rangle$, where $i, j \in [n]^+$ and m is the number of POI groups;
   (b) For any two POI groups G_i and G_j, if G_i is adjacent to G_j then there exists an arc $\langle G_i, G_j \rangle$;
   (c) For each POI group G_i, if a camera sensor s_j can ***completely cover*** G_i then there exists a flow from s_j to G_i.

Figure 3 shows an example executing the above construction in details, in which $\Gamma = \{s_1, s_2, s_3, s_4, s_5\}$ and $\Pi = \{p_1, \cdots, p_{10}\}$ (Fig. 3(a)). By Algorithm 1, seventeen POI groups are computed as in Fig. 3(b). Then we can construct the directed graph, as part of the auxiliary graph, for all POI groups as in Fig. 3(c). Finally, we add the points corresponding to the sensors in the graph as in Fig. 3(d), which corresponds for the counterpart of matching.

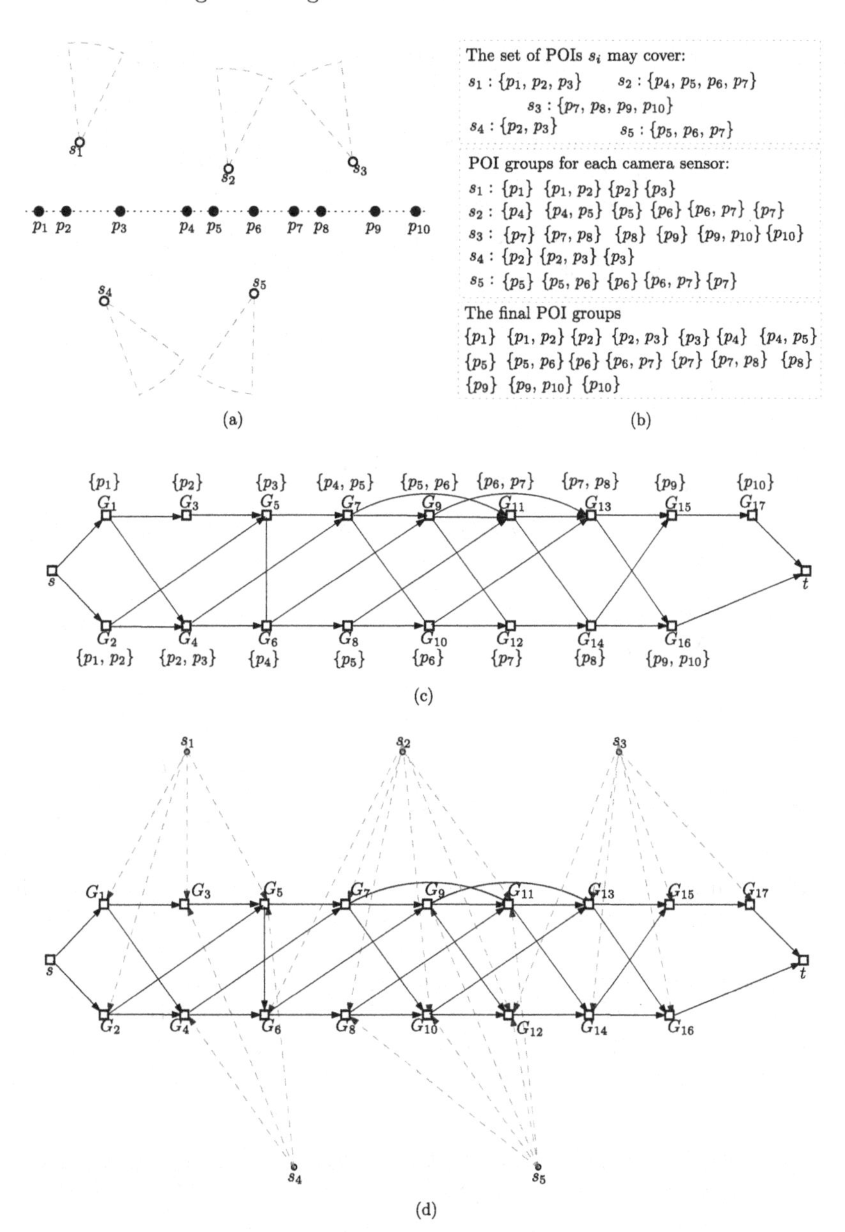

The set of POIs s_i may cover:
$s_1 : \{p_1, p_2, p_3\}$ $s_2 : \{p_4, p_5, p_6, p_7\}$
$s_3 : \{p_7, p_8, p_9, p_{10}\}$
$s_4 : \{p_2, p_3\}$ $s_5 : \{p_5, p_6, p_7\}$
POI groups for each camera sensor:
$s_1 : \{p_1\}\ \{p_1, p_2\}\ \{p_2\}\ \{p_3\}$
$s_2 : \{p_4\}\ \{p_4, p_5\}\ \{p_5\}\ \{p_6\}\ \{p_6, p_7\}\ \{p_7\}$
$s_3 : \{p_7\}\ \{p_7, p_8\}\ \{p_8\}\ \{p_9\}\ \{p_9, p_{10}\}\ \{p_{10}\}$
$s_4 : \{p_2\}\ \{p_2, p_3\}\ \{p_3\}$
$s_5 : \{p_5\}\ \{p_5, p_6\}\ \{p_6\}\ \{p_6, p_7\}\ \{p_7\}$
The final POI groups
$\{p_1\}\ \{p_1, p_2\}\ \{p_2\}\ \{p_2, p_3\}\ \{p_3\}\ \{p_4\}\ \{p_4, p_5\}$
$\{p_5\}\ \{p_5, p_6\}\ \{p_6\}\ \{p_6, p_7\}\ \{p_7\}\ \{p_7, p_8\}\ \{p_8\}$
$\{p_9\}\ \{p_9, p_{10}\}\ \{p_{10}\}$
(a)
(b)
(c)
(d)

Fig. 3. An example of transformation. (a) The instance contains ten POIs and five camera sensors; (b) Seventeen groups are generated; (c) The corresponding auxiliary graph with 17 vertices for the groups; (d) is the graph in (c) plus another five points corresponding to the sensors.

3.2 Linear Programming for Shortest Matching-Path Problem

Let E_1 be a set of edges in which each edge is between two POI groups. Let E_2 be a set of edges in which each edge $(i, j) \in E_2$ represents a flow from a camera sensor s_i to a POI group G_j. For Min-Num LTC-CS, the aim is to use a minimum number of camera sensors to cover all POIs, so the aim of SMP is to find the path from s to t and minimize the sum of flows. The Integer LP (ILP) of SMP which is transformed from Min-Num LTC-CS is shown as follows (ILP(2)):

$$
\begin{aligned}
&\min \sum_{(i,j)\in E_2} x_{ij} \\
&s.t \sum_{e\in\delta^+\cap E_1(j)} y_e - \sum_{e\in\delta^-\cap E_1(j)} y_e = \begin{cases} 1 & j = s \\ -1 & j = t \\ 0 & j \neq s, t \end{cases} \\
&\sum_{j:(i,j)\in E_2} x_{ij} \leq 1 && \forall i \\
&\sum_{i:(i,j)\in E_2} x_{ij} \geq \sum_{e\in\delta^+(j)\cap E_1} y_e && \forall j \qquad (2)\\
&x_{ij} \in \{0, 1\} && (i, j) \in E_2 \\
&y_e \in \{0, 1\} && e \in E_1
\end{aligned}
$$

where the first constraint is to ensure a feasible flow over the edges of E_1, the second guarantees that the total outflows from s_i is no more than 1 over the edges of E_2 since each camera sensor s_i can be used for at most once. The third ensures that the total inflows to j over edges of E_2 must be larger than the total outflows of j in E_1. By relaxing $0 \leq x_{ij} \leq 1$ and $y_e \geq 0$, we can get LP(1) a linear programming formulation relaxing ILP (2). As one of the main results of the paper, we conclude that there is a connection between the above ILP formulation ILP(2) and its LP relaxation LP(1), as stated in the following theorem whose proof is omitted due to the length limitation.

Theorem 1. *The value of an optimal fractional solution of LP(1) equals to that of an optimal solution of ILP(2). Moreover, an optimal fractional solution of LP(1) can be rounded to an integral solution of Min-Num LTC-CS.*

Note that when the objective function of the LP is $\min \sum_i x_{ij}\alpha_{ij}$, the above formulations can be extended to Min-Sum LTC-CS. In addition, the above theorem can also be extended to Min-Sum LTC-CS.

4 Numerical Experiments

In this section, we will evaluate the practical performance and runtime of the linear program of the shortest matching-path problem (denoted by ILP-SMP) by

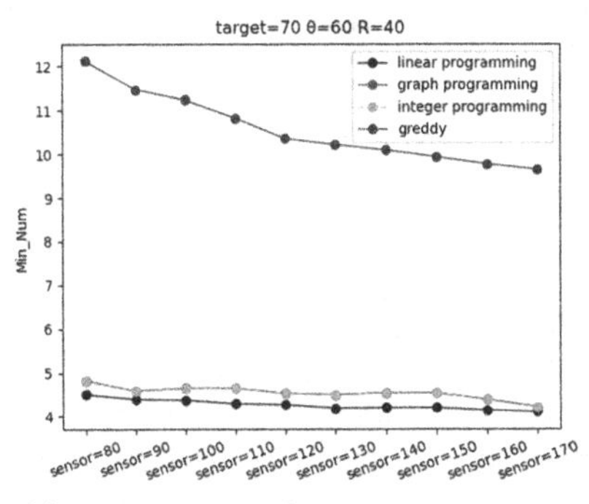

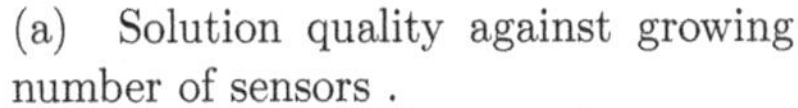

(a) Solution quality against growing number of sensors .

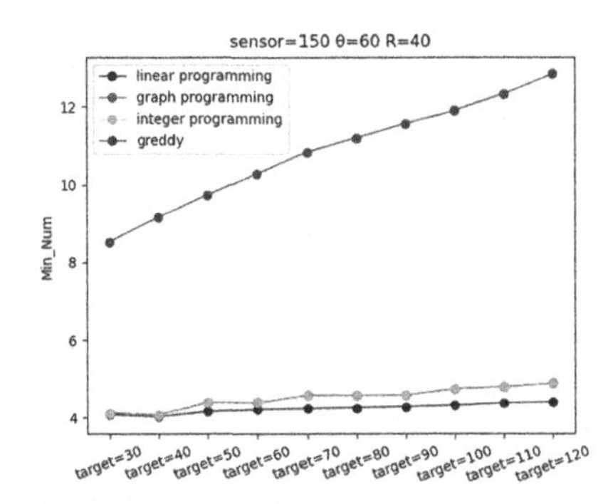

(b) Solution quality against increasing number of targets.

Fig. 4. Solution quality of ILP-LTC-CS, LP-LTC-CS, ILP-SMP and GA in comparison.

comparing with other baselines: the integer linear program of LTC-CS (denoted by ILP-LTC-CS), the linear program of LTC-CS (denoted by LP-LTC-CS), and the greedy algorithm (denoted by GA). All experiments are carried out on a Win 10 platform with Intel Core i5-6200U CPU, 8.0G RAM. All algorithms were implemented with Java. In our experiments, the radius of the camera sensor is set as 40 and the angle is set as 60.

4.1 Solution Quality in Comparison

Figure 4 is the experimental results of Min-Num LTC-CS. In Fig. 4a, the number of POIs is fixed at 70 and the number of camera sensors increases from 80 to 170 and are divided into ten groups. It can be seen that the results of ILP-SMP and ILP-LTC-CS are always the same with the increasing number of sensors. Besides, the results of GA are larger than ILP-LTC-CS which is at least 2 times and the results of LP-LTC-CS are less than ILP-LTC-CS. Figure 4b is the comparison of the results of ILP-LTC-CS, LP-LTC-CS, ILP-SMP, and GA with the increasing number of target points for Min-Num LTC-CS. In these experiments, the number of camera sensors is 150 and the number of POIs increases from 30 to 120 with the step 10. Regardless of the number of POIs, the results of ILP-SMP and ILP-LTC-CS are the same. It can be seen that the results of GA are larger than that of ILP-LTC-CS (at least 2 times), and the results of LP-LTC-CS are smaller than that of ILP-LTC-CS in all the experimental results.

4.2 Runtime Comparison

Figure 5 shows the running time of ILP-LTC-CS, LP-LTC-CS, ILP-SMP, and GA of Min-Num LTC-CS. In the experiments of Fig. 5a, there are 70 POIs that need to cover. The figure shows the results of ten groups of experiments, in which the number of camera sensors increases from 80 to 170. First, the runtime of ILP-LTC-CS is the largest among that of LP-LTC-CS, ILP-SMP, and GA in each

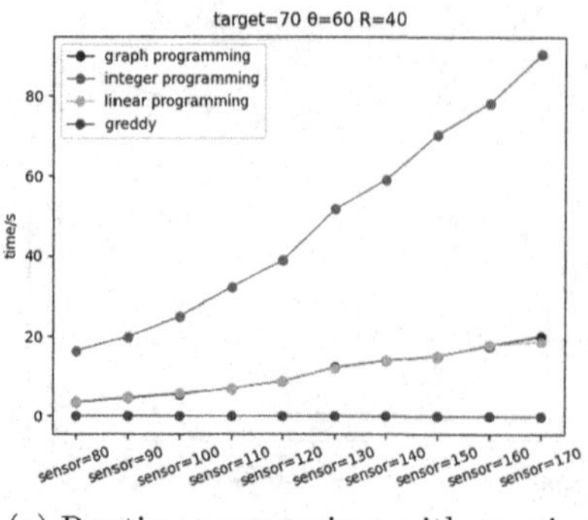

(a) Runtimes comparison with growing number of sensors.

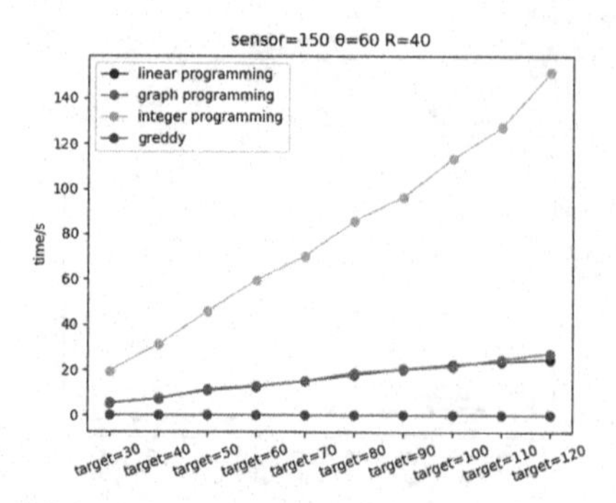

(b) Runtimes comparison with growing number of targets.

Fig. 5. Runtimes comparison of ILP-LTC-CS, LP-LTC-CS, ILP-SMP and GA.

group of experiments. Second, the runtimes of ILP-LTC-CS, LP-LTC-CS, and ILP-SMP increase with the increasing number of sensors, and the runtime of GA T is constant as the number of sensors increases. Figure 5b compares the running time of ILP-LTC-CS, LP-LTC-CS, ILP-SMP, and GA. In all experiments, the number of camera sensors is 180 and the number of POIs increases from 30 to 120 with a step 10. As for the running time of ILP-LTC-CS, LP-LTC-CS, ILP-SMP, and GA, ILP-LTC-CS is the highest, and GA is the lowest. When more POIs need to cover, the runtime of ILP-LTC-CS, LP-LTC-CS, and ILP-SMP will increase, but the runtime of GA will always be fixed,

5 Conclusion

In this paper, we proposed the linear programmings for the on-a-Line Target Coverage with Minimized Number of Camera Sensors (Min-Num LTC-CS) and the on-a-Line Target Coverage with Minimizing the Sum of rotation angles of Camera Sensors (Min-Sum LTC-CS). It has been proved that the solutions of the two linear programmings are integer solutions via transformed to the shortest matching-path problem. We first grouped the given set of Points Of Interest (POIs) and formulated the linear programmings of Min-Num LTC-CS and Min-Sum LTC-CS. Then, we constructed a graph based on the grouping of the above POIs. Finally, we proposed a rounding method to obtain integer solutions of Min-Num LTC-CS and Min-Sum LTC-CS. Numerical experiments demonstrate that the experiment results are consistent with the theoretical analysis.

Acknowledgements. The authors are supported by Natural Science Foundation of China (No. 61772005), Outstanding Youth Innovation Team Project for Universities of Shandong Province (No. 2020KJN008), Natural Science Foundation of Fujian Province (No. 2020J01845) and Educational Research Project for Young and Middle-aged Teachers of Fujian Provincial Department of Education (No. JAT190613).

References

1. Ai, J., Abouzeid, A.A.: Coverage by directional sensors in randomly deployed wireless sensor networks. J. Comb. Optim. **11**(1), 21–41 (2006). https://doi.org/10.1007/s10878-006-5975-x
2. Cai, Y., Lou, W., Li, M., Li, X.Y.: Target-oriented scheduling in directional sensor networks. In: Infocom IEEE International Conference on Computer Communications. IEEE (2007)
3. Cai, Y., Lou, W., Li, M., Li, X.Y.: Energy efficient target-oriented scheduling in directional sensor networks. IEEE Trans. Comput. **58**(9), 1259–1274 (2009)
4. Cardei, M., Jie, W.: Coverage in wireless sensor networks. In: Handbook of Sensor Networks, vol. 21, pp. 201–202 (2004)
5. Fowler, R.J., Paterson, M.S., Tanimoto, S.L.: Optimal packing and covering in the plane are np-complete. Inf. Process. Lett. **12**(3), 133–137 (1981)
6. Fusco, G., Gupta, H.: Selection and orientation of directional sensors for coverage maximization. In: 2009 6th Annual IEEE Communications Society Conference on Sensor, Mesh and Ad Hoc Communications and Networks, pp. 1–9. IEEE (2009)
7. Jia, J., Dong, C., Hong, Y., Guo, L., Ying, Yu.: Maximizing full-view target coverage in camera sensor networks. Ad Hoc Netw. **94**, 101973 (2019)
8. Kasbekar, G.S., Bejerano, Y., Sarkar, S.: Lifetime and coverage guarantees through distributed coordinate-free sensor activation. IEEE/ACM Trans. Netw. **19**(2), 470–483 (2010)
9. Kumar, S., Lai, T.H., Arora, A.: Barrier coverage with wireless sensors. In: Proceedings of the 11th Annual International Conference on Mobile Computing and Networking, pp. 284–298 (2005)
10. Li, X., Frey, H., Santoro, N., Stojmenovic, I.: Localized sensor self-deployment with coverage guarantee. ACM SIGMOBILE Mob. Comput. Commun. Rev. **12**(2), 50–52 (2008)
11. Liu, L., Ma, H., Zhang, X.: On directional k-coverage analysis of randomly deployed camera sensor networks. In: 2008 IEEE International Conference on Communications, pp. 2707–2711. IEEE (2008)
12. Liu, Z., Ouyang, Z.: *k*-coverage estimation problem in heterogeneous camera sensor networks with boundary deployment. IEEE Access **6**, 2825–2833 (2017)
13. Ma, H., Liu, Y.: On coverage problems of directional sensor networks. In: Jia, X., Wu, J., He, Y. (eds.) MSN 2005. LNCS, vol. 3794, pp. 721–731. Springer, Heidelberg (2005). https://doi.org/10.1007/11599463_70
14. Ma, H., Yang, M., Li, D., Hong, Y., Chen, W.: Minimum camera barrier coverage in wireless camera sensor networks. In: 2012 Proceedings IEEE INFOCOM, pp. 217–225. IEEE (2012)
15. Mao, S., He, S., Wu, J.: Joint UAV position optimization and resource scheduling in space-air-ground integrated networks with mixed cloud-edge computing. IEEE Syst. J. **15**(3), 3992–4002 (2021)
16. Puvvadi, U.L.N., Di Benedetto, K., Patil, A., Kang, K.-D., Park, Y.: Cost-effective security support in real-time video surveillance. IEEE Trans. Ind. Inform. **11**(6), 1457–1465 (2015)
17. Saeed, A., Abdelkader, A., Khan, M., Neishaboori, A., Harras, K.A., Mohamed, A.: On realistic target coverage by autonomous drones. ACM Trans. Sens. Networks **15**(3), 32:1–32:33 (2019)
18. Si, P., Chengdong, W., Zhang, Y., Jia, Z., Ji, P., Chu, H.: Barrier coverage for 3D camera sensor networks. Sensors **17**(8), 1771 (2017)

19. Wang, J., Zhong, N.: Efficient point coverage in wireless sensor networks. J. Comb. Optim. **11**(3), 291–304 (2006). https://doi.org/10.1007/s10878-006-7909-z
20. Wang, W., et al.: PANDA: placement of unmanned aerial vehicles achieving 3D directional coverage. In: 2019 IEEE Conference on Computer Communications, INFOCOM 2019, Paris, France, 29 April–2 May 2019, pp. 1198–1206. IEEE (2019)
21. Wang, Y., Cao, G.: Minimizing service delay in directional sensor networks. In: 2011 Proceedings of the IEEE INFOCOM, pp. 1790–1798. IEEE (2011)
22. Wang, Y., Cao, G.: On full-view coverage in camera sensor networks. In: 2011 Proceedings of the IEEE INFOCOM, pp. 1781–1789. IEEE (2011)
23. Wang, Y.-C., Chen, Y.-F., Tseng, Y.-C.: Using rotatable and directional (R&D) sensors to achieve temporal coverage of objects and its surveillance application. IEEE Trans. Mob. Comput. **11**(8), 1358–1371 (2011)

Two-Stage Submodular Maximization Under Curvature

Yanzhi Li[1], Zhicheng Liu[2], Chuchu Xu[3], Ping Li[4], Hong Chang[3], and Xiaoyan Zhang[3](✉)

[1] School of Mathematical Sciences, University of Science and Technology of China, Hefei 230026, Anhui, China
davidlee@mail.ustc.edu.cn

[2] College of Taizhou, Nanjing Normal University, Taizhou 225300, China

[3] School of Mathematical Science and Institute of Mathematics, Nanjing Normal University, Jiangsu 210023, China
{changh,zhangxiaoyan}@njnu.edu.cn

[4] Huawei Technologies Co. Ltd., Theory Lab, Central Research Institute, 2012 Labs, Hongkong 9990777, China
liping129@huawei.com

Abstract. Submodular function optimization has been widely studied in machine learning and economics, which is a relatively new research field in the context of big data and has attracted more attention. In this paper, we consider a two-stage submodular maximization problem subject to cardinality and p-matroid constraints, and propose an approximation algorithm with constant approximation ratio depends on the maximum curvature of the submodular functions involved, which generalizes the previous bound.

Keywords: Two-stage submodular maximization · Submodular term · Approximation algorithm · Curvature

1 Introduction

We consider a two-stage submodular maximization problem. Given a ground set $V = \{1, \ldots, n\}$, let $F = (f_1, f_2, \ldots, f_m)$ be a set of functions such that each $f_j : 2^V \to \mathbb{R}_+$ $(j = 1, \ldots, m)$ is a nonnegative monotone submodular function. For a given nonnegative integer k, the two-stage problem is as follows

$$\max_{|S|\leq k} F(S) = \max_{|S|\leq k} \sum_{i=1}^{m} \max_{T\in\mathcal{I}(S)} (f_j(T)). \tag{1}$$

where $\mathcal{I}(S)$ is the family of the common independent sets of p-matroid $\mathcal{M} = (S, \mathcal{I}(S))$ over the same ground set $S \subseteq V$. Any matriod satisfies three properties: (i) $\emptyset \in \mathcal{I}$; (ii) If $J' \subseteq J \in \mathcal{I}(S)$, then $J' \in \mathcal{I}(S)$; and (iii) $\forall A, B \in \mathcal{I}(S)$, if $|A| < |B|$, then there exists an element $u \in B \setminus A$ such that $A + u \in I$.

D.-Z. Du et al. (Eds.): COCOA 2021, LNCS 13135, pp. 25–34, 2021.
https://doi.org/10.1007/978-3-030-92681-6_3

A set function $f : 2^V \to \mathbb{R}_+$ is normalized if $f(\emptyset) = 0$. It is non-decreasing if $f(S) \le f(T), \forall S \subseteq T \subseteq V$. It is submodular if $f(S) + f(T) \ge f(S \cap T) + f(S \cup T), \forall S, T \subseteq V$.

In terms of relevant work, two-stage submodular maximization problems have been investigated in the literature. The two-stage submodular maximization problem subject to uniform and general matroid constraints was first proposed in [2] with applications in machine learning, in particular, dictionary learning [6,11], topic modelling [5], and (convolutional) auto encoders [9], among others. They proposed a continuous optimization method and get an approximation ratio which asymptotically approaches $1 - 1/e$. For the case where the asymptote does not work, they design a local search algorithm whose approximation ratio is close to 1/2. [8] extended the two-stage problem to more general monotone submodular functions and more general matroid constraints to give a $\frac{1}{2}(1-e^{-2})$-approximation algorithm.

Recently, Yang et al. [10] presented a $\frac{1}{p+1}(1 - e^{-(p+1)})$-approximation algorithm for the two-stage problem (1.1). We show that the bound can be generalized if we make further assumptions on the total curvature for the submodular function [1], which is defined as follows

$$k_f = 1 - \min_{v \in V} \frac{f(V) - f(V \backslash \{v\})}{f(v)}. \tag{2}$$

Curvature is attractive since it is linear time computable with only oracle function access [3]. Our contribution is to design a $\left(\frac{1-k_f}{p}(1 - e^{-p}) + \frac{k_f}{p+1}(1 - e^{-(p+1)})\right)$-approximation algorithm for cardinality and p-matroid constrained two-stage submodular maximization problem. Note that $\frac{1-k_f}{p}(1-e^{-p})+\frac{k_f}{p+1}(1-e^{-(p+1)}) > \frac{1}{p+1}(1 - e^{-(p+1)})$ when $k_f \neq 1$.

The remainder of our paper is organized as follows. Section 2 present the algorithms along with its analysis for this problems and Sect. 3 gives some concluding remarks.

2 Two-Stage Submodular Maximization

We present a replacement greedy algorithm in Sect. 2.1 and analyze its approximation ratio in Sect. 2.2.

2.1 Algorithm for Two-Stage Submodular Maximization

We first consider the problem of maximizing the sum of submodular and modular functions. Construct the following functions:

$$\ell_i(T) = \sum_{t \in T} f_i(t|V \setminus \{t\}),$$
$$g_i(T) = f_i(T) - \ell_i(T).$$

Note that $g_i(T)$ is a monotone nonnegative submodular function, and $\ell_i(T)$ is a modular function.

Given a ground set $V = \{1, \dots, n\}$, the problem is to select a set $S \subseteq V$ of cardinality no more than a given parameter k and $T \in \mathcal{I}(S)$ to maximize the following objective function:

$$\sum_{i=1}^{m} \max_{T \in \mathcal{I}(S)} (g_i(T) + \ell_i(T)).$$

The main idea of Algorithm 1 is as follows. Our replacement greedy algorithm works in k rounds, and in each round it tries to increase the solution in a specific greedy way. It starts with an empty set $S^0 = \emptyset$ and checks (in each round) whether new elements can be added to the set without violating the matroid constraints or it can be replaced with the elements in the current solution while increasing the value of the objective function.

Algorithm 1. Replacement Greedy

1: $S^0 \leftarrow \emptyset, T_i^0 \leftarrow \emptyset (\forall 1 \leq i \leq m)$
2: **for** $1 \leq j \leq k$ **do**
3: $\quad t^* \leftarrow \arg\max_{t \in V} \sum_{i=1}^{m} \nabla_i(t, T_i^{j-1})$
4: $\quad$ **for** $\forall 1 \leq i \leq m$ **do**
5: $\quad\quad$ **if** $\nabla_i(t^*, T_i^{j-1}) > 0$ **then**
6: $\quad\quad\quad T_i^j \leftarrow T_i^{j-1} \cup \{t^*\} \setminus \mathsf{Rep}_i(t^*, T_i^{j-1})$
7: $\quad\quad$ **else**
8: $\quad\quad\quad T_i^j \leftarrow T_i^{j-1}$
9: $\quad\quad$ **end if**
10: $\quad$ **end for**
11: **end for**
12: Return sets S^k and $T_1^k, T_2^k, \cdots, T_m^k$

For convenience, we define the following notations. Let OPT be the optimal solution. Let $\Delta_i^g(x, T_i^j) = g_i(\{x\} \cup T_i^j) - g_i(T_i^j)$ denote the marginal contribution of an element x to the set T_i^j when we consider function g_i, where these marginal contributions are nonnegative due to the monotonicity of g_i. Similarly, we use $\nabla_i^g(x, y, T_i^j) = g_i(\{x\} \cup T_i^j \setminus \{y\}) - g_i(T_i^j)$ to define the gain of removing an element y and replacing it with x. Here, $\nabla_i^g(x, y, T_i^j)$ may not be positive. In addition, we also denote

$$\Delta_i(x, T_i^j) = \left(1 - \frac{p+1}{k}\right)^{k-j} \Delta_i^g(x, T_i^j) + \left(1 - \frac{p}{k}\right)^{k-j} \ell_i(x),$$

$$\nabla_i(x, y, T_i^j) = \left(1 - \frac{p+1}{k}\right)^{k-j} \nabla_i^g(x, y, T_i^j) + \left(1 - \frac{p}{k}\right)^{k-j} (\ell_i(x) - \ell_i(y)).$$

Consider the set T_i^j. When we replace the element of T_i^j with x, we will not violate the p-matroid constraint, i.e., $\mathcal{I}(x, T_i^j) = \{y \in T_i^j : T_i^j \cup \{x\} \setminus \{y\} \in \mathcal{I}(S)\}$.

So, we define the replacement gain of x w.r.t. a set T_i^j as follows:

$$\nabla_i(x, T_i^j) = \begin{cases} \Delta_i(x, T_i^j) & \text{if } T_i^j \cup \{x\} \in \mathcal{I}(S), \\ \max\{0, \max_{y \in \mathcal{I}(x, T_i^j)} \nabla_i(x, y, T_i^j)\} & \text{otherwise.} \end{cases}$$

Finally, we use $\mathsf{Rep}_i(x, T_i^j)$ to denote the element that should be replaced by x as follows:

$$\mathsf{Rep}_i(x, T_i^j) = \begin{cases} \emptyset & \text{if } T_i^j \cup \{x\} \in \mathcal{I}(S), \\ \arg\max_{y \in \mathcal{I}(x, T_i^j)} \nabla_i(x, y, T_i^j) & \text{otherwise.} \end{cases}$$

2.2 Analysis of the Algorithm

In this section, we analyze the approximation ratio of Algorithm 1. Our analysis relies on the following distorted objective function Φ. Let k denote the cardinality constraint. For any $j = 1, \cdots, k$ and any set T_i $(i = 1, \ldots, m)$, we define

$$\Phi_j(S^j) = \sum_{i=1}^{m} \left(\left(1 - \frac{p+1}{k}\right)^{k-j} g_i(T_i^j) + \left(1 - \frac{p}{k}\right)^{k-j} \ell_i(T_i^j) \right).$$

Lemma 1. *In each iteration of Algorithm 1,*

$$\begin{aligned} & \Phi_j(S^j) - \Phi_{j-1}(S^{j-1}) \\ = & \sum_{i=1}^{m} \left(\nabla_i(t^j, T_i^{j-1}) + \frac{p+1}{k}\left(1 - \frac{p+1}{k}\right)^{k-j} g_i(T_i^{j-1}) + \frac{p}{k}\left(1 - \frac{p}{k}\right)^{k-j} \ell_i(T_i^{j-1}) \right). \end{aligned}$$

Proof.

$$\begin{aligned} & \Phi_j(S^j) - \Phi_{j-1}(S^{j-1}) \\ = & \sum_{i=1}^{m} \left(\left(1 - \frac{p+1}{k}\right)^{k-j} g_i(T_i^j) + \left(1 - \frac{p}{k}\right)^{k-j} \ell_i(T_i^j) \right) \\ & - \sum_{i=1}^{m} \left(\left(1 - \frac{p+1}{k}\right)^{k-(j-1)} g_i(T_i^{j-1}) + \left(1 - \frac{p}{k}\right)^{k-(j-1)} \ell_i(T_i^{j-1}) \right) \\ = & \sum_{i=1}^{m} \left(1 - \frac{p+1}{k}\right)^{k-j} \left(g_i(T_i^j) - (1 - \frac{p+1}{k}) g_i(T_i^{j-1}) \right) \\ & + \sum_{i=1}^{m} \left(1 - \frac{p}{k}\right)^{k-j} \left(\ell_i(T_i^j) - (1 - \frac{p}{k}) \ell_i(T_i^{j-1}) \right) \\ = & \sum_{i=1}^{m} \left((1 - \frac{p+1}{k})^{k-j} (g_i(T_i^j) - g_i(T_i^{j-1})) + (1 - \frac{p}{k})^{k-j} (\ell_i(T_i^j) - \ell_i(T_i^{j-1})) \right) \\ & + \sum_{i=1}^{m} \left(\frac{p+1}{k}\left(1 - \frac{p+1}{k}\right)^{k-j} g_i(T_i^{j-1}) \right) + \sum_{i=1}^{m} \left(\frac{p}{k}\left(1 - \frac{p}{k}\right)^{k-j} \ell_i(T_i^{j-1}) \right) \\ = & \sum_{i=1}^{m} \left(\nabla_i(t^j, T_i^{j-1}) + \frac{p+1}{k}\left(1 - \frac{p+1}{k}\right)^{k-j} g_i(T_i^{j-1}) + \frac{p}{k}\left(1 - \frac{p}{k}\right)^{k-j} \ell_i(T_i^{j-1}) \right). \end{aligned}$$

Lemma 2. *If we add $t^j \in V$ to S^{j-1}, then*

$$\sum_{i=1}^{m} \nabla_i \left(t^j, T_i^{j-1}\right) \geq \frac{1}{k} \sum_{i=1}^{m} \sum_{t \in T_i^* \setminus T_i^{j-1}} \nabla_i(t, T_i^{j-1}),$$

where $S^* = \arg\max_{S \subseteq V, |S| \leq k} \sum_{i=1}^{m} \max_{T \in \mathcal{I}(S)} f_i(T)$, $T_i^* = \arg\max_{A \in \mathcal{I}(S^*)} f_i(A)$.

Proof.

$$\begin{aligned}
\sum_{i=1}^{m} \nabla_i(t^j, T_i^{j-1}) &\geq \frac{1}{|S^* \setminus S^{j-1}|} \sum_{t \in S^* \setminus S^{j-1}} \sum_{i=1}^{m} \nabla_i(t, T_i^{j-1}) \\
&\geq \frac{1}{|S^* \setminus S^{j-1}|} \sum_{i=1}^{m} \sum_{t \in T_i^* \setminus S^{j-1}} \nabla_i(t, T_i^{j-1}) \\
&= \frac{1}{|S^* \setminus S^{j-1}|} \sum_{i=1}^{m} \sum_{t \in T_i^* \setminus T_i^{j-1}} \nabla_i(t, T_i^{j-1}) \\
&\geq \frac{1}{k} \sum_{i=1}^{m} \sum_{t \in T_i^* \setminus T_i^{j-1}} \nabla_i(t, T_i^{j-1}),
\end{aligned}$$

where the first inequality holds because the righthand side is the average increment of values of optimal elements at step j if we add them instead of t^j, which is the maximum one. The second inequality holds because $T_i^* \setminus S^{j-1} \subseteq T^* \setminus T^{j-1}$. The last inequality follows from $|S^* \setminus S^{j-1}| \leq k$.

Property 1. *([4])* Let $\mathcal{M}_j = (V, \mathcal{I}_j)$ be a matroid for every $j \in \{1, \ldots, k\}$. For any two independent sets $A, B \in \mathcal{I}_j$, there exists a mapping $\pi_j : B \setminus A \to A \setminus B \cup \{\emptyset\}$ such that

- $(A \setminus \pi_j(b)) \cup b \in \mathcal{I}_j$ for all $b \in B \setminus A$;
- $|\pi_j^{-1}(a)| \leq 1$ for all $a \in A \setminus B$;
- let $A_b = \{\pi_1(b), \ldots, \pi_k(b)\}$, $(A \setminus A_b) \cup b \in \cap_{j=1}^{k} \mathcal{I}_j$ for all $b \in B \setminus A$.

Lemma 3. *For $j = 1, 2, \ldots, k$, we have*

$$\begin{aligned}
&\sum_{i=1}^{m} \nabla_i(t^j, T_i^{j-1}) \\
&\geq \frac{1}{k} \left(1 - \frac{p+1}{k}\right)^{k-j} \sum_{i=1}^{m} \left(g_i(T_i^*) - (p+1) g_i(T_i^{j-1})\right) \\
&+ \frac{1}{k} \left(1 - \frac{p}{k}\right)^{k-j} \sum_{i=1}^{m} \left(\ell(T_i^*) - p \ell_i(T_i^{j-1})\right).
\end{aligned}$$

Proof. Based on Property 1, there exist mappings $\pi_t : T_i^*\backslash T_i^{j-1} \to T_i^{j-1}\backslash T_i^* \cup \{\emptyset\}$ such that $(T_j^i\backslash A_e) \cup \{e\} \in \cap_{t=1}^p \mathcal{I}_t$ $(t \in \{1,2,...,k\})$, where $e \in T_i^*\backslash T_i^{j-1}$ and $A_e = \{\pi_1(e), ..., \pi_k(e)\}$.

$$\begin{aligned}
&\sum_{i=1}^m \sum_{t\in T_i^*\backslash T_i^{j-1}} \nabla_i(t, T_i^{j-1}) \\
\geq &\sum_{i=1}^m \sum_{t\in T_i^*\backslash T_i^{j-1}} \left(1-\frac{p+1}{k}\right)^{k-j} \left(g_i(\{t\}\cup T_i^{j-1}\backslash\{\pi(t)\}) - g_i(T_i^{j-1})\right) \\
&+\sum_{i=1}^m \sum_{t\in T_i^*\backslash T_i^{j-1}} \left(1-\frac{p}{k}\right)^{k-j} \left(\ell_i(\{t\}\cup T_i^{j-1}\backslash\{\pi(t)\}) - \ell_i(T_i^{j-1})\right) \\
= &\sum_{i=1}^m \sum_{t\in T_i^*\backslash T_i^{j-1}} \left(1-\frac{p+1}{k}\right)^{k-j} \varphi_{(i,t)} + \sum_{i=1}^m \sum_{t\in T_i^*\backslash T_i^{j-1}} \left(1-\frac{p}{k}\right)^{k-j} \psi_{(i,t)} \\
= &\sum_{i=1}^m \sum_{t\in T_i^*\backslash T_i^{j-1}} \left(1-\frac{p+1}{k}\right)^{k-j} \left(\Delta_i^g(t, T_i^{j-1}) - \Delta_i^g(\pi(t), \{t\}\cup T_i^{j-1}\backslash\{\pi(t)\})\right) \\
&+\sum_{i=1}^m \sum_{t\in T_i^*\backslash T_i^{j-1}} \left(1-\frac{p}{k}\right)^{k-j} \left(\ell_i(t) - \ell_i(\pi(t))\right),
\end{aligned}$$

where

$$\varphi_{(i,t)} = \left(g_i(\{t\}\cup T_i^{j-1}) - g_i(T_i^{j-1}) + g_i(\{t\}\cup T_i^{j-1}\backslash\{\pi(t)\}) - g_i(\{t\}\cup T_i^{j-1})\right),$$

$$\psi_{(i,t)} = \left(\ell_i(\{t\}\cup T_i^{j-1}) - \ell_i(T_i^{j-1}) + \ell_i(\{t\}\cup T_i^{j-1}\backslash\{\pi(t)\}) - \ell_i(\{t\}\cup T_i^{j-1})\right).$$

Together with Lemma 2 implies that

$$\begin{aligned}
&\sum_{i=1}^m \nabla_i(t^j, T_i^{j-1}) \\
\geq &\frac{1}{k}\sum_{i=1}^m \sum_{t\in T_i^*\backslash T_i^{j-1}} \left(1-\frac{p+1}{k}\right)^{k-j} \left(\Delta_i^g(t, T_i^{j-1}) - \Delta_i^g(\pi(t), \{t\}\cup T_i^{j-1}\backslash\{\pi(t)\})\right) \\
&+\frac{1}{k}\sum_{i=1}^m \sum_{t\in T_i^*\backslash T_i^{j-1}} \left(1-\frac{p}{k}\right)^{k-j} \left(\ell_i(t) - \ell_i(\pi(t))\right).
\end{aligned}$$

We know that

$$\begin{aligned}
&\frac{1}{k}\sum_{i=1}^m \sum_{t\in T_i^*\backslash T_i^{j-1}} \left(1-\frac{p+1}{k}\right)^{k-j} \Delta_i^g(t, T_i^{j-1}) \\
\geq &\frac{1}{k}\left(1-\frac{p+1}{k}\right)^{k-j} \sum_{i=1}^m \left(g_i(T_i^*\cup T_i^{j-1}) - g_i(T_i^{j-1})\right)
\end{aligned}$$

$$\geq \frac{1}{k}\left(1-\frac{p+1}{k}\right)^{k-j}\sum_{i=1}^{m}\left(g_i(T_i^*)-pg_i(T_i^{j-1})\right),$$

where the first inequality holds because of the submodularity of g_i and the second inequality follows from the monotonicity of g_i.

We also have

$$\begin{aligned}
&\frac{1}{k}\sum_{i=1}^{m}\sum_{t\in T_i^*\backslash T_i^{j-1}}\left(1-\frac{p+1}{k}\right)^{k-j}\Delta_i^g(\pi(t),\{t\}\cup T_i^{j-1}\backslash\{\pi(t)\})\\
&\leq \frac{1}{k}\sum_{i=1}^{m}\sum_{y\in T_i^{j-1}}\left(1-\frac{p+1}{k}\right)^{k-j}\Delta_i^g(y,\{y^{-1}\}\cup T_i^{j-1}\backslash\{y\})\\
&\leq \frac{1}{k}\left(1-\frac{p+1}{k}\right)^{k-j}\sum_{i=1}^{m}g_i(T_i^{j-1}),
\end{aligned}$$

where the first inequality holds since the range of mapping π is a subset of T_i^{j-1} and no two elements in $T_i^*\backslash T_i^{j-1}$ are mapped to the same $y\in T_i^{j-1}$, while the second inequality follows from the submodularity of g_i.

So, we have,

$$\begin{aligned}
&\frac{1}{k}\sum_{i=1}^{m}\sum_{t\in T_i^*\backslash T_i^{j-1}}\left(1-\frac{p+1}{k}\right)^{k-j}\left(\Delta_i^g(t,T_i^{j-1})-\Delta_i^g(\pi(t),\{t\}\cup T_i^{j-1}\backslash\{\pi(t)\})\right)\\
&\geq \frac{1}{k}\left(1-\frac{p+1}{k}\right)^{k-j}\sum_{i=1}^{m}\left(g_i(T_i^*)-(p+1)g_i(T_i^{j-1})\right).
\end{aligned}$$

We also have

$$\begin{aligned}
&\frac{1}{k}\sum_{i=1}^{m}\sum_{t\in T_i^*\backslash T_i^{j-1}}\left(1-\frac{p}{k}\right)^{k-j}(\ell_i(t)-\ell_i(\pi(t)))\\
&=\frac{1}{k}\left(1-\frac{p}{k}\right)^{k-j}\sum_{i=1}^{m}\left(\ell(T_i^*)-\ell_i(T_i^{j-1})\right).
\end{aligned}$$

Hence,

$$\begin{aligned}
&\sum_{i=1}^{m}\nabla_i(t^j,T_i^{j-1})\\
&\geq \frac{1}{k}\left(1-\frac{p+1}{k}\right)^{k-j}\sum_{i=1}^{m}\left(g_i(T_i^*)-(p+1)g_i(T_i^{j-1})\right)\\
&\quad+\frac{1}{k}\left(1-\frac{p}{k}\right)^{k-j}\sum_{i=1}^{m}\left(\ell(T_i^*)-p\ell_i(T_i^{j-1})\right).
\end{aligned}$$

Finally, the main result is summarized as follows.

Theorem 1. *Algorithm 1 returns a set S^k of size k such that*

$$\sum_{i=1}^{m}(g_i(T_i^k)+\ell_i(T_i^k)) \geq \left(\frac{1}{p+1}(1-e^{-(p+1)})\right)\sum_{i=1}^{m} g_i(T_i^*) + \left(\frac{1}{p}(1-e^{-p})\right)\sum_{i=1}^{m}\ell_i(T_i^*).$$

Proof. According to the definition of Φ, we have

$$\begin{aligned}
\Phi_0(S^0) &= 0,\\
\Phi_k(S^k) &= \sum_{i=1}^{m}\left(\left(1-\frac{p+1}{k}\right)^{k-k} g_i(T_i^k) + \left(1-\frac{p}{k}\right)^{k-k}\ell_i(T_i^k)\right)\\
&= \sum_{i=1}^{m}\left(g_i(T_i^k)+\ell_i(T_i^k)\right).
\end{aligned}$$

Applying Lemma 1 and Lemma 3, we have

$$\begin{aligned}
&\Phi_j(S^j) - \Phi_{j-1}(S^{j-1})\\
&= \sum_{i=1}^{m}\left(\nabla_i(t^j, T_i^{j-1}) + \frac{p+1}{k}\left(1-\frac{p+1}{k}\right)^{k-j} g_i(T_i^{j-1}) + \frac{p}{k}\left(1-\frac{p}{k}\right)^{k-j}\ell_i(T_i^{j-1})\right)\\
&\geq \frac{1}{k}\left(1-\frac{p+1}{k}\right)^{k-j}\sum_{i=1}^{m} g_i(T_i^*) + \frac{1}{k}\left(1-\frac{p}{k}\right)^{k-j}\sum_{i=1}^{m}\ell(T_i^*).
\end{aligned}$$

Finally,

$$\begin{aligned}
&\sum_{i=1}^{m}(g_i(T_i^k)+\ell_i(T_i^k))\\
&= \sum_{j=1}^{k}(\Phi_j(S^j) - \Phi_{j-1}(S^{j-1}))\\
&\geq \sum_{j=1}^{k}\left(\frac{1}{k}\left(1-\frac{p+1}{k}\right)^{k-j}\sum_{i=1}^{m} g_i(T_i^*) + \frac{1}{k}\left(1-\frac{p}{k}\right)^{k-j}\sum_{i=1}^{m}\ell(T_i^*)\right)\\
&\geq \left(\frac{1}{p+1}(1-e^{-(p+1)})\right)\sum_{i=1}^{m} g_i(T_i^*) + \left(\frac{1}{p}(1-e^{-p})\right)\sum_{i=1}^{m}\ell_i(T_i^*).
\end{aligned}$$

Theorem 2. *There exists an algorithm returning a set S^k of size k such that*

$$F(S^k) \geq \left(\frac{1}{p}(1-e^{-p}) - k_f\left(\frac{1}{p}(1-e^{-p}) - \frac{1}{p+1}(1-e^{-(p+1)})\right)\right) OPT.$$

Proof. According to the submodularity of g_i, we have,

$$\ell_i(T) = \sum_{t \in T} f_i(t|V \setminus \{t\}) \geq (1 - k_{f_i}) f_i(T),$$

So,

$$\begin{aligned}
&\sum_{i=1}^{m} f_i(T_i^k) = \sum_{i=1}^{m} \left(g_i(T_i^k) + \ell_i(T_i^k)\right) \\
&\geq \left(\frac{1}{p+1}(1 - e^{-(p+1)})\right) \sum_{i=1}^{m} g_i(T_i^*) + \left(\frac{1}{p}(1 - e^{-p})\right) \sum_{i=1}^{m} \ell_i(T_i^*) \\
&= \left(\frac{1}{p+1}(1 - e^{-(p+1)})\right) \sum_{i=1}^{m} (f_i(T_i^*) - \ell_i(T_i^*)) + \left(\frac{1}{p}(1 - e^{-p})\right) \sum_{i=1}^{m} \ell_i(T_i^*) \\
&= \left(\frac{1}{p+1}(1 - e^{-(p+1)})\right) \sum_{i=1}^{m} f_i(T_i^*) + \left(\frac{1}{p}(1 - e^{-p}) - \frac{1}{p+1}(1 - e^{-(p+1)})\right) \sum_{i=1}^{m} \ell_i(T_i^*) \\
&\geq \left(\frac{1}{p+1}(1 - e^{-(p+1)})\right) \sum_{i=1}^{m} f_i(T_i^*) + \left(\frac{1}{p}(1 - e^{-p}) - \frac{1}{p+1}(1 - e^{-(p+1)})\right) (1 - k_f) \sum_{i=1}^{m} f_i(T_i^*) \\
&= \left(\frac{1}{p}(1 - e^{-p}) - k_f \left(\frac{1}{p}(1 - e^{-p}) - \frac{1}{p+1}(1 - e^{-(p+1)})\right)\right) \sum_{i=1}^{m} f_i(T_i^*).
\end{aligned}$$

Hence, we can get,

$$\begin{aligned}
F(S^k) &\geq \left(\frac{1}{p}(1 - e^{-p}) - k_f \left(\frac{1}{p}(1 - e^{-p}) - \frac{1}{p+1}(1 - e^{-(p+1)})\right)\right) OPT \\
&= \left(\frac{1 - k_f}{p}(1 - e^{-p}) + \frac{k_f}{p+1}(1 - e^{-(p+1)}\right) OPT,
\end{aligned}$$

where $F(S^k) = \sum_{i=1}^{m} \max_{T \in \mathcal{I}(S^k)} (f_i(T_i^k)$.

3 Conclusion

Many researchers made substantial contribution in submodular maximization problem, but there are not much research done on the two-stage submodular maximization problem. In the present paper, we consider the two-stage submodular maximization problem subject to cardinality and p-matroid constraints, and we propose a $\left(\frac{1-k_f}{p}(1 - e^{-p}) + \frac{k_f}{p+1}(1 - e^{-(p+1)})\right)$-approximation algorithm for it by replacement greedy method under curvature. In the future, we believe that there will be more substantial progress in approximation algorithms for the two-stage submodular maximization problem and the variant of it, and it will be interesting to further improve the approximation ratios of these problems.

Acknowledgements. The research is supported by NSFC (Nos. 11871280,11971349, 12101314), Qinglan Project, Natural Science Foundation of Jiangsu Province (No. BK20200723), and Natural Science Foundation for institutions of Higher Learning of Jiangsu Province (No.20KJB110022).

References

1. Bai, W., Bilmes J.A.: Greed is still good: maximizing monotone submodular+Supermodular (BP) functions. In: ICML, pp. 304–313 (2018)
2. Balkanski, E., Krause, A., Mirzasoleiman, B., Singer, Y.: Learning sparse combinatorial representations via two-stage submodular maximization. In: ICML, pp. 2207–2216 (2016)
3. Conforti, M., Cornuejols, G.: Submodular set functions, matroids and the greedy algorithm: tight worst-case bounds and some generalizations of the Rado-Edmonds theorem. Discrete Appl. Math. **7**(3), 251–274 (1984)
4. Lee, J., Mirrokni, V.S., Nagarajan, V., Sviridenko, M.: Maximizing nonmonotone submodular functions under matroid or knapsack constraints. SIAM J. Discrete Math. **23**(4), 2053–2078 (2010)
5. Maas, A.L., Daly, R.E., Pham, P.T., Huang, D., Ng, A.Y., Potts, C.: Learning word vectors for sentiment analysis. In: ACL-HLT,vol. 1, pp. 142–150 (2011)
6. Mairal, J., Bach, F., Ponce, J., Sapiro, G.: Online dictionary learning for sparse coding. In: ICML, pp. 689–696 (2009)
7. Schrijver, A.: Combinatorial optimization-polyhedra and efficiency. Algor. Combin. **24**, 1–1881 (2003)
8. Stan, S., Zadimoghaddam, M., Krause, A., Karbasi, A.: Probabilistic submodular maximization in sub-linear time. In: ICML, pp. 3241–3250 (2017)
9. Vincent, P., Larochelle, H., Lajoie, I., Bengio, Y., Manzagol, P.: Stacked denoising autoencoders: learning useful representations in a deep network with a local denoising criterion. J. Mach. Learn. Res. **11**, 3371–3408 (2010)
10. Yang, R., Gu, S., Gao, C., Wu, W., Wang, H., Xu, D.: A constrained two-stage submodular maximization. Theor. Comput. Sci. **853**, 57–64 (2021)
11. Zhou, M., Chen, H., Ren, L., Sapiro, G., Carin, L., Paisley, J.W.: Non-parametric Bayesian dictionary learning for sparse image representations. In: NIPS, pp. 2295–2303 (2009)

An Improved Approximation Algorithm for Capacitated Correlation Clustering Problem

Sai Ji[1], Yukun Cheng[2(✉)], Jingjing Tan[3], and Zhongrui Zhao[4]

[1] Academy of Mathematics and Systems Science, Chinese Academy of Sciences, Beijing 100190, People's Republic of China

[2] Suzhou University of Science and Technology, Suzhou 215009, People's Republic of China
ykcheng@amss.ac.cn

[3] School of Mathematics and Information Science, Weifang University, Weifang 261061, People's Republic of China

[4] Department of Operations Research and Information Engineering, Beijing University of Technology, Beijing 100124, People's Republic of China

Abstract. Correlation clustering problem is a classical clustering problem and has many applications in protein interaction networks, cross-lingual link detection, communication networks, etc. In this paper, we discuss the capacitated correlation clustering problem on labeled complete graphs, in which each edge is labeled $+$ or $-$ to indicate two endpoints are "similar" or "dissimilar", respectively. Our objective is to partition the vertex set into several clusters, subject to an upper bound on cluster size, so as to minimize the number of disagreements. Here the number of disagreements is defined as the total number of the edges with positive labels between clusters and the edges with negative labels within clusters. The main contribution of this work is providing a 5.37-approximation algorithm for the capacitated correlation clustering problem, improving the current best approximation ratio of 6 [21]. In addition, we have conducted a series of numerical experiments, which effectively demonstrate the effectiveness of our algorithm.

Keywords: Correlation clustering · Capacitated correlation clustering · Approximation algorithm · LP-rounding

1 Introduction

Clustering problems arise in many applications such as machine learning, computer vision, data mining and data compression. These problems have been widely studied in the literatures [5,9,11,13,19]. Compared with clustering problems, which need to specify the number of clusters in advance, the correlation clustering problem does not place this constraint on the clustering task. The correlation clustering problem was first introduced by Bansal et al. [6], which has applications in protein interaction networks, cross-lingual link detection, and communication networks, etc. Generally, a correlation clustering problem is modeled on a labeled complete graph $G = (V, E)$, in which each edge (u, v) is labeled $+$ or $-$ to indicate the two vertices u and v are "similar" or "dissimilar", respectively. The goal is to partition the vertices into several disjoint subsets,

D.-Z. Du et al. (Eds.): COCOA 2021, LNCS 13135, pp. 35–45, 2021.
https://doi.org/10.1007/978-3-030-92681-6_4

each being called a cluster, so that the edges within clusters are mostly positive and the edges between clusters are mostly negative. However the perfect clustering may not exist because of the similarity assessments between vertices. Thus plenty of work turned to seek an optimal clustering by considering two kinds of objectives of correlation clustering. One is to maximize the number of agreements, which is defined as the total number of positive edges within clusters and negative edges between clusters. The other is to minimize the number of disagreements, that is the number of negative edges within clusters plus the number of positive edges across clusters.

The correlation clustering problem is proved to be NP-hard [6], meaning that one cannot obtain an optimal solution in polynomial time under the assumption that $P \neq NP$. Different approximation algorithms have been proposed for the correlation clustering problem [1,7,18,22,23]. To be specific, for the minimization version, Bansal et al. [6] provided a first constant-factor approximation algorithm. Charikar et al. [12] proved that minimizing the number of disagreements is APX-hard. They gave a nature integer programming and proved that the integrality gap of the LP formulation is 2. Then, they presented a 4-approximation algorithm based on LP-rounding technique. They also studied the correlation clustering problem on general graphs and in addition provided an $O(\log n)$-approximation algorithm. Later, Ailon et al. [3] introduced a simple randomized 3-approximation algorithm for the optimal problem to minimize the number of disagreements. Until now, the best deterministic approximation algorithm for the minimizing disagreements is an LP-rounding algorithm with 2.06-approximation ratio, which was provided by Chawla et al. [14].

Besides the study on the correlation clustering problems, there are several interesting variants of the correlation clustering problem, including overlapping correlation clustering [8], correlation clustering in data streams [2], high-order correlation clustering [17], correlation clustering with noisy input [20], correlation clustering problem with constrained cluster sizes [21], min-max correlation clustering [4], and so on. In this paper, we would discuss the correlation clustering problem with constrained cluster sizes.

The correlation clustering problem with constrained cluster size was first introduced by Puleo and Milenkovic [21]. In this problem, we are given a complete graph $G = (V, E)$ and an integer U. The goal of this problem is to partition the vertices into several clusters that contain no more than U vertices so as to minimize the number of disagreements. Referring to other clustering problems [10,15,16], there are two kinds of constraints on cluster size: lower bound constraint and capacity constraint. Thus the problem introduced by Puleo and Milenkovic [21] is called the capacitated correlation clustering problem more appropriately. Puleo and Milenkovic [21] introduced a penalty parameter μ_v for each vertex v, if vertex v is clustered into a cluster C with more than U vertices, then it occurs a penalty cost $\mu_v(|C| - U)$. They provided a 6-approximation algorithm for the weighted capacitated correlation clustering problem by setting $\mu_v = 1, \forall v \in V$ and splitting each cluster which contains more than U vertices into several clusters that satisfy the size constraint. In this paper, we propose a 5.37-approximation algorithm for the capacitated correlation clustering problem, which improves the current best approximation ratio. The improvement comes from following two perspectives of innovations.

(1) **Innovation in Algorithm Design:** Different from the previous work [21] by introducing a penalty cost for each vertex, we first model the capacitated correlation clustering problem as an integer programming, and then design an algorithm based on LP-rounding technique to directly output the feasible clusters, each subject to the upper bound.
(2) **Innovation in Theoretically Analysis:** Based on our proposed algorithm, we compute the number of disagreements corresponding to a cluster, by distinguishing three cases. Especially, when analyzing the number disagreements caused by positive edges across clusters for one of cases, we innovatively expand the scope of positive edges and skillfully use their corresponding values in the fractional optimal solution of LP relaxation. Such an operation thus contributes the constant approximation ratio.

The rest of this paper is organized as follows. In Sect. 2, we provide the definition of the capacitated correlation clustering problem, formulate its integer programming as well as the corresponding LP relaxation. The approximation algorithm and the theoretical analysis are proposed in Sect. 3. Moreover, the numerical experiments are conducted in Sect. 4 and the conclusions are given in the last section.

2 Preliminaries

In this section, we formulate the capacitated correlation clustering problem as an integer programming as well as its corresponding relaxation.

Definition 1 (Capacitated Correlation Clustering Problem). *Given a labeled complete graph $G = (V, E)$ as well as an integer U. A capacitated correlation clustering problem is to partition the vertex set V into several clusters, subject to an upper bound U on each cluster's size, such that the number of disagreements is minimized.*

For each edge $(u, v) \in E$, we introduce a binary decision variable x_{uv} to indicate whether two vertices u and v are in a same cluster. To be specific, if u and v lie in a same cluster, then $x_{uv} = 0$; otherwise $x_{uv} = 1$. Then the capacitated correlation clustering problem can be formulated as follows:

$$\begin{aligned} \min \quad & \sum_{(u,v)\in E^+} x_{uv} + \sum_{(u,v)\in E^-} (1 - x_{uv}) \\ \text{s. t.} \quad & x_{uv} + x_{vw} \geq x_{uw}, && \forall u, v, w \in V, \\ & \sum_{v\in V} (1 - x_{uv}) \leq U, && \forall u \in V, \\ & x_{uu} = 0, && \forall u \in V, \\ & x_{uv} \in \{0, 1\}, && \forall u, v \in V. \end{aligned} \tag{1}$$

The value of the objective function is just equal to the number of disagreements, which contains two parts. The first part is the number of disagreements caused by the positive edges between clusters and the second part is the number of disagreements coming from the negative edges whose endpoints lie in the same cluster. There are three types of constraints in Programming (1). The first type of constraint ensures that the solution output by Programming (1) is a feasible clustering of the *correlation clustering problem*. Since for any three vertices $u, v, w \in V$, if one cluster contains vertices u, v and vertices v, w lie in a same cluster, then all of three vertices u, v, w must be in this cluster, showing $x_{uw} = 0$. The second type of constraint indicates the solution satisfies the capacitated constraint and the third type is a natural one. By relaxing the variables, we obtain the following LP relaxation of (1):

$$\begin{aligned} \min \quad & \sum_{(u,v)\in E^+} x_{uv} + \sum_{(u,v)\in E^-} (1 - x_{uv}) \\ \text{s. t.} \quad & x_{uv} + x_{vw} \geq x_{uw}, && \forall u, v, w \in V, \\ & \sum_{v\in V} (1 - x_{uv}) \leq U, && \forall u \in V, \\ & x_{uu} = 0, && \forall u \in V, \\ & x_{uv} \in [0, 1], && \forall u, v \in V. \end{aligned} \tag{2}$$

3 Algorithm and Analysis

This section is one of crucial parts of this paper, in which we first provide our approximation Algorithm 1 in Subsect. 3.1 and then present the theoretical analysis for constant approximation ratio in Subsect. 3.2.

3.1 Iterative Clustering Algorithm

Algorithm 1. Iterative clustering algorithm

Input: A labeled complete graph $G = (V, E)$, positive integer U, parameter $\alpha \in (0, 1/2)$
Output: A partition of vertices

1: Solve (2) to obtain the optimal fractional solution x^*
2: Let $S := V$
3: **while** $S \neq \emptyset$ **do**
4: Select a vertex v_i from S randomly. Let $T_{v_i} := \{u \in S : x^*_{uv_i} \leq \alpha\}$, and obtain $T^*_{v_i} = \{u_1, \cdots, u_{\min\{U, |T_{v_i}|\}}\} \subseteq T_{v_i}$, satisfying $x^*_{v_i u_j} \leq x^*_{v_i u_{j+1}}, j = \{1, \cdots, |T_{v_i}| - 1\}$
5: **if** $\frac{\sum_{u\in T^*_{v_i}} x^*_{uv_i}}{|T^*_{v_i}|} \geq \frac{\alpha}{2}$, **then**
6: Let $V_i = \{v_i\}$
7: **else**
8: Let $V_i = T^*_{v_i}$
9: **end if**
10: Update $S := S - V_i$
11: **end while**
12: **return** the partition of V

Before providing Algorithm 1, let us introduce a high level description for it. There are two main phases in this algorithm. In the first phase, we solve Programming (2) to obtain the optimal fractional solution x^* (Line 1 in Algorithm 1). The value x^*_{uv} can be viewed as the distance between vertex u and vertex v. The second phase is an iterative process, which is the core of our algorithm (Lines 3–11 in Algorithm 1). In each iteration, a vertex is selected randomly from the un-clustered vertices as a center, and at most $(U-1)$ other vertices are also chosen from un-clustered vertices, based on the distances between them and the current center, to form a new cluster. Such a step is repeated until all the vertices are clustered, and therefore a feasible clustering for the capacitated correlation problem is obtained ultimately.

3.2 Theoretical Analysis

Assume that there are k iterations in Algorithm 1, and thus vertex set V is partitioned into k clusters, denoted by $V_1, \cdots, V_k$. At the end of the i-th iteration, the cluster V_i must be one of three types:

- Type 1: $V_i := \{v_i\}$;
- Type 2: $V_i := T^*_{v_i}$ with $T^*_{v_i} \subset T_{v_i}$;
- Type 3: $V_i := T^*_{v_i}$ with $T^*_{v_i} = T_{v_i}$.

Some useful properties can be explored based on the construction of V_i.

Property 1. Let V_i be the i-th cluster output from Algorithm 1. Then we have

(1) $x^*_{v_iu} \geq \alpha \geq x^*_{v_iv}$, for any $v \in T^*_{v_i}$ and any $u \in V_i \backslash T^*_{v_i}$;
(2) $1 - x^*_{qv} \geq \frac{1-2\alpha}{2\alpha} x^*_{qv}$, for any $q, v \in V_i$.

The first property can be derived from the construction of $T^*_{v_i}$ directly, and the second one is correct because $x^*_{qv} \leq x^*_{qv_i} + x^*_{vv_i} \leq 2\alpha$ for any $q, v \in V_i$.

Let us denote $[i] = \{1, 2, \cdots, i\}$, for any positive integer i. Based on the partition, the number of the disagreements from negative edges is

$$\sum_{i \in [k]} |(w,p) \in E^-, w, p \in V_i|, \tag{3}$$

and the disagreements caused by positive edges is

$$\sum_{i \in [k]} \left|(q,v) \in E^+, q \in V_i, v \in \cup_{t \in [k] \backslash [i]} V_t\right|. \tag{4}$$

Subsequently, we would compute the upper bounds on the number of disagreements by distinguishing three types of clusters, respectively.

Type 1 of Cluster. Because cluster V_i belongs to Type 1 (as shown in Fig. 1), the disagreements generated by V_i must be the positive edges between v_i and vertices in other clusters. Therefore, the number of new disagreements contributed by cluster V_i of Type 1 is $|(v_i, v) \in E^+, v \in \cup_{t \in [k] \backslash [i]} V_t|$ and the upper bound on number of disagreements is shown in Lemma 1.

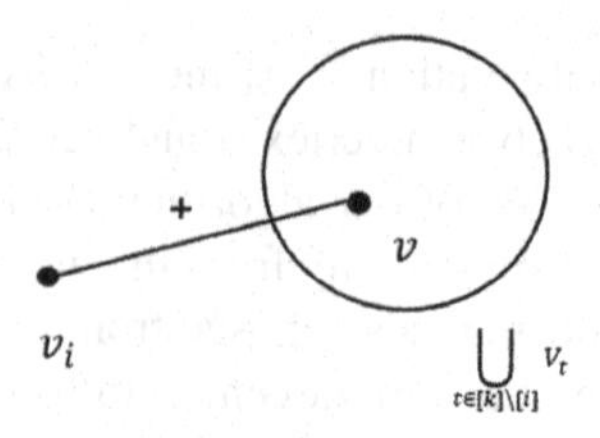

Fig. 1. Type 1 of cluster.

Lemma 1. *If cluster V_i returned by Algorithm 1 is of Type 1, then the number of disagreements from edges $(v_i, v) \in E^+, v \in \cup_{t\in[k]\setminus[i]} V_t$ has an upper bound of*

$$\frac{2}{\alpha}\left[\sum_{(v_i,v)\in E^+, v\in\cup_{t\in[k]\setminus[i]} V_t} x^*_{v_i v} + \sum_{(v_i,v)\in E^-, v\in T^*_{v_i}} (1 - x^*_{v_i v})\right].$$

Type 2 of Cluster. As V_i is of Type 2 (as shown in Fig. 2), we have $V_i = T^*_{v_1} \subset T_{v_i}$, indicating $|V_i| = U$ and $\emptyset \neq T_{v_i}\setminus T^*_{v_i} \subseteq \cup_{t\in[k]\setminus[i]} V_t$. Different with Type 1, it is possible that $|V_i| \geq 2$, and thus there are two kinds of disagreements. One is from the positive edges $(q, v) \in E^+$ with $q \in V_i$ and $v \in V\setminus V_i$, and the other is from the negative edges $(w, p) \in E^-$ with $w, p \in V_i$. For each $q \in V_i$, we use all positive edges $(q, v) \in E^+, v \in V\setminus V_i$ to upper bound the number of the first kind of disagreements caused by vertex q, and the corresponding upper bound is shown in Lemma 2.

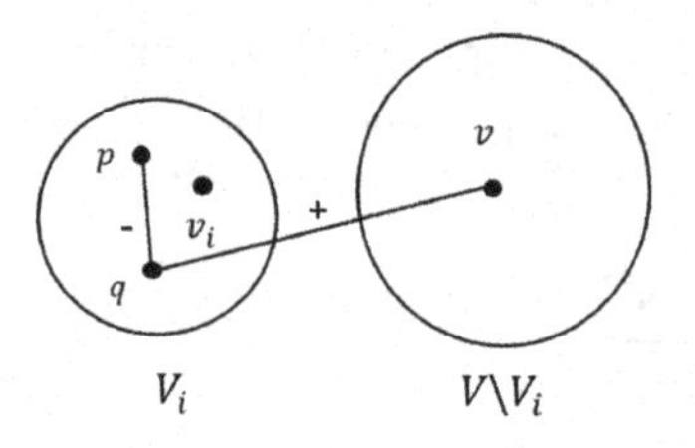

Fig. 2. Type 2 of cluster.

Lemma 2. *Suppose that V_i is of Type 2. For each vertex $q \in V_i$, the number of disagreements caused by positive edges $(q, v) \in E^+, v \in \cup_{t\in[k]\setminus[i-1]} V_t$ is upper bounded by*

$$\sum_{(q,v)\in E^+, v\in\cup_{t\in[k]\setminus[i-1]} V_t} x^*_{qv} + \frac{2\alpha}{1-2\alpha} \sum_{(q,v)\in E^-, v\in V_i} (1 - x^*_{qv}).$$

To explore the upper bound of disagreements from negative edges, we can observe that for any two vertices $w, p \in V_i$, $x^*_{w,p} \leq \alpha$ and thus $1 - x^*_{wp} \geq 1 - 2\alpha$, which indicates $|(w, p)| = 1 \leq (1 - x^*_{wp})/(1 - 2\alpha)$. Therefore, following lemma can be derived.

Lemma 3. *If V_i is of Type 2, then for each negative edge $(w,p) \in E^-$ with $w,p \in V_i$, the number of disagreement caused by (w,p) can be bounded by $(1-x^*_{wp})/(1-2\alpha)$.*

Type 3 of Cluster. As V_i is of Type 3 (as shown in Fig. 3), we have $V_i = T_{v_i}$ and $|V_i| = |T_{v_i}| \le U$. Similar to Type 2, there are also two kinds of disagreements, caused by the negative edges within V_i and the positive edges between V_i and another cluster. The disagreement of a negative edge $(w,p) \in E^-, w,p \in V_i$ is upper bounded by $\frac{1}{1-2\alpha}x^*_{wp}$ by Lemma 3. Next, we would analyze the upper bound on the number of disagreements caused by positive edges $(q,v) \in E^+$ for any $q \in V_i$ and $v \in \cup_{t\in[k]\setminus[i]} V_t$, under two conditions: $x^*_{v_i v} \ge 3\alpha/2$ and $\alpha \le x^*_{v_i v} < 3\alpha/2$, respectively. By applying the similar proof in [12], we obtain the following lemma.

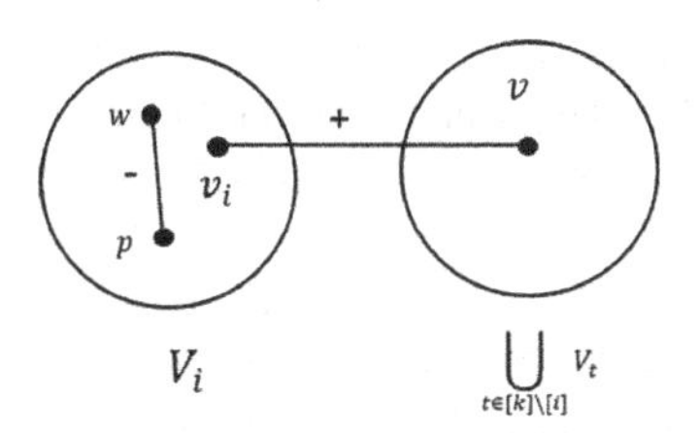

Fig. 3. Type 3 of cluster.

Lemma 4. *If V_i is of Type 3, then the upper bound on the number of disagreements caused by the positive edges satisfies:*

(1) The number of disagreements from positive edges $(q,v) \in E^+$, satisfying $q \in V_i$, $v \in \cup_{t\in[k]\setminus[i]} V_t$ and $x^*_{v_i v} \ge 3\alpha/2$, is upper bounded by $2x^*_{qv}/\alpha$;
(2) The number of disagreements from positive edges $(q,v) \in E^+$, satisfying $q \in V_i$, $v \in \cup_{t\in[k]\setminus[i]} V_t$ and $\alpha \le x^*_{v_i v} < 3\alpha/2$, is upper bounded by

$$\frac{2}{\alpha}\left[\sum_{(q,v)\in E^+, q\in V_i} x^*_{qv} + \sum_{(q,v)\in E^-, q\in V_i} (1-x^*_{qv})\right].$$

Combining Lemma 1 - Lemma 4, we obtain Theorem 1.

Theorem 1. *The number of disagreements returned by Algorithm 1 is bounded by*

$$\max\left\{\frac{2}{\alpha}, \frac{1+2\alpha}{1-2\alpha}\right\}\left[\sum_{(u,v)\in E^+} x^*_{uv} + \sum_{(u,v)\in E^-} (1-x^*_{uv})\right],$$

where x^* is the optimal fractional solution of (2). *By setting $\alpha = \frac{\sqrt{33}-5}{2}$, Algorithm 1 has an approximation ratio of* 5.37.

4 Numerical Experiments

In this section, we would explore the practicality of our proposed algorithm for three kinds of data. Moreover, we verify the effectiveness of Algorithm 1 through different values of the total number N of vertices and the upper bound U of cluster size.

4.1 Datasets

We test the performance of the Algorithm 1 by using three data sets. For the first data set, we use the "Iris" data set, which contains 150 data samples divided into 3 categories with 50 data in each category, each containing 4 attributes. Where the samples for the same species are denoted as "+" and between different species as "–" to initialize the graph.

Secondly, we run Algorithm 1 on the Census1990 data set. It consists of $2,458,285$ data points with 68 attributes. The graph is initialized by using the attribute "age" for clustering, marking people in the same age group as "+" and those in different age groups as "–".

Finally, we execute Algorithm 1 with "Heart Disease" data set. This data set is integer valued from 0 (no presence) to 4. Experiments with the Cleveland database have concentrated on simply attempting to distinguish presence (values $1,2,3,4$) from absence (value 0). Thus samples of patients with the same type of heart disease (including non-presence) are marked as "+" and different as "–".

4.2 Experimental Setup and Results

Implementation Details. The code is really implemented on Pycharm 2017.3.2 using Python 3.6. The entire experiment is implemented on a single process with an Intel(R) Xeon(R) E5-2620 v4 CPU at 2.10 GHz and 256 GB RAM.

Setting the Experiments Parameters. The way we set the hyperparameters is the same regardless of which dataset is mentioned above. As mentioned before, we construct data of different sizes to test the performance of Algorithm 1. Let the total number of vertices of the undirected graph, N, increase gradually from 10 to 100 in steps of 10. Let the capacity U of each cluster be 15%, 20%, 25%, 30% and 35% of the total number of vertices respectively. Particularly, if U is a decimal, we shall round U to make sure that U is an integer. Finally, we set the parameter $\alpha = (-5+\sqrt{33})/2$ in all data experiments.

Results. The results of the numerical experiment are shown in Table 1, Table 2 and Table 3. In these three tables, $appro_{U=15\%}$ represents the ratio of output from Algorithm 1 to the optimal solution of (2) when U is taken to be 15%. From Tables Table 1, Table 2 and Table 3, we can observe that the approximation ratio of the instances after running Algorithm 1 is much better than the one from the theoretical analysis. In the three data sets experiments, the majority of approximation ratio, overwhelmingly, is stabilized between 1 and 3.

Table 1. "Iris" data. If U is a decimal, we round U to make sure that U is an integer.

N	$appro_{U:15\%}$	$appro_{U:20\%}$	$appro_{U:25\%}$	$appro_{U:30\%}$	$appro_{U:35\%}$
10	1.0000	1.1644	1.1088	1.7429	1.5849
20	1.2121	1.2914	1.3329	1.5855	2.0648
30	1.2373	1.3405	1.4910	1.7638	2.0028
40	1.3490	1.3211	1.6923	2.1276	1.7020
50	1.4204	1.6462	2.0924	2.0537	2.0969
60	1.5667	1.9695	2.4923	2.0704	1.8284
70	1.5898	2.0750	2.6694	1.7248	1.8674
80	1.7341	2.2943	2.9256	2.0912	1.7977
90	1.6923	2.3839	3.1530	1.7315	1.9228
100	1.7569	2.3796	2.8714	1.9916	1.9182

Table 2. "USCensus1990"data. If U is a decimal, we round U to make sure that U is an integer.

N	$appro_{U:15\%}$	$appro_{U:20\%}$	$appro_{U:25\%}$	$appro_{U:30\%}$	$appro_{U:35\%}$
10	1.0000	1.7111	1.3434	2.0000	1.5333
20	1.2988	1.7415	1.5400	1.8857	1.3944
30	1.4966	1.5413	1.4406	1.2400	1.2667
40	1.3817	1.8711	1.6540	1.7143	1.8667
50	1.3740	1.6954	2.0748	1.7270	1.8889
60	1.6924	1.7694	1.4809	1.7188	1.0000
70	1.8057	2.0932	2.4369	1.1818	1.0000
80	1.8541	2.3153	2.6482	3.4470	1.8065
90	1.7774	2.4853	2.0879	1.8000	1.8235
100	2.0046	2.8666	2.3304	1.8750	1.9444

Table 3. "Heart Disease"data. If U is a decimal, we round U to make sure that U is an integer.

N	$appro_{U:15\%}$	$appro_{U:20\%}$	$appro_{U:25\%}$	$appro_{U:30\%}$	$appro_{U:35\%}$
10	1.0000	1.6000	1.3612	1.2400	1.4000
20	1.3561	1.2580	1.1429	1.4857	1.2227
30	1.3628	1.6538	1.4185	1.4400	1.1636
40	1.4218	1.6566	1.7910	1.6505	1.1733
50	1.4339	1.8565	1.9156	1.0000	1.0000
60	1.6539	2.1737	1.7174	1.1600	1.1500
70	1.7318	2.2765	2.3931	1.4990	1.1840
80	1.9869	2.4635	2.0114	1.1692	1.0000
90	1.8665	2.7867	2.3616	1.2660	1.0000
100	2.3083	2.2376	2.2748	1.1636	1.1889

5 Conclusions

In this article, we study the capacitated correlation clustering problem and give a 5.37-approximation algorithm for this problem. There are two directions for the future research on correlation clustering problem. From the results of numerical experiments, we realize that there is plenty of room for the improvement of the approximation ratio. So one direction is to continue the study this problem and to obtain a better approximation ratio by innovating the algorithm. The other is to study other variants of the correlation clustering problem, such as capacitated min-max correlation clustering problem and capacitated correlation clustering problem in data streams.

Acknowledgements. The first author is supported by National Natural Science Foundation of China (No. 12101594) and the Project funded by China Postdoctoral Science Foundation (No. 2021M693337). The second author is supported by National Nature Science Foundation of China (No. 11871366), Qing Lan Project for Young Academic Leaders and Qing Lan Project for Key Teacher. The third author is supported by Natural Science Foundation of Shandong Province (No. ZR2017LA002). The fourth author is supported by National Natural Science Foundation of China (No. 12131003) and Beijing Natural Science Foundation Project No. Z200002.

References

1. Ailon, N., Avigdor-Elgrabli, N., Liberty, E., Zuylen, A.V.: Improved approximation algorithms for bipartite correlation clustering. SIAM J. Comput. **41**(5), 1110–1121 (2012)
2. Ahn, K.J., Cormode, G., Guha, S., Mcgregor, A., Wirth, A.: Correlation clustering in data streams. In: Proceedings of the 32nd International Conference on Machine Learning, pp. 2237–2246 (2015)
3. Ailon, N., Charikar, M., Newman, A.: Aggregating inconsistent information: ranking and clustering. J. ACM **55**(5), 1–27 (2008)
4. Ahmadi, S., Khuller, S., Saha, B.: Min-max correlation clustering via multicut. In: Proceedings of the 20th International Conference on Integer Programming and Combinatorial Optimization, pp. 13–26 (2019)
5. Ahmadian, S., Norouzi-Fard, A., Svensson, O., Ward, J. : Better guarantees for k-means and Euclidean k-median by primal-dual algorithms. In: Proceedings of the 58th Annual Symposium on Foundations of Computer Science, pp. 61–72 (2017)
6. Bansal, N., Blum, A., Chawla, S.: Correlation clustering. Mach. Learn. **56**(1–3), 89–113 (2004)
7. Bressan, M., Cesa-Bianchi, N., Paudice, A., Vitale, F.: Correlation clustering with adaptive similarity queries. In: Proceedings of the 32nd Annual Conference on Neural Information Processing Systems, pp. 12510–12519 (2019)
8. Bonchi, F., Gionis, A., Ukkonen, A.: Overlapping correlation clustering. Knowl. Inf. Syst. **35**(1), 1–32 (2013)
9. Backurs, A., Indyk, P., Onak, K., Schieber, B., Vakilian, A., Wagner, T.: Scalable fair clustering. In: Proceedings of the 37th International Conference on Machine Learning, pp. 405–413 (2019)
10. Cohen-Addad, V.: Approximation schemes for capacitated clustering in doubling metrics. In: Proceedings of the 30th Annual ACM-SIAM Symposium on Discrete Algorithms, pp. 2241–2259 (2020)

11. Choo, D., Grunau, C., Portmann, J., Rozhon, V.: k-means++: few more steps yield constant approximation. In: Proceedings of the 37th International Conference on Machine Learning, pp. 1909–1917 (2020)
12. Charikar, M., Guruswami, V., Wirth, A.: Clustering with qualitative information. J. Comput. Syst. Sci. **3**(71), 360–383 (2005)
13. Cohen-Addad, V., Klein, P.N., Mathieu, C.: Local search yields approximation schemes for k-means and k-median in Euclidean and minor-free metrics. SIAM J. Comput. **48**(2), 644–667 (2019)
14. Chawla, S., Makarychev, K., Schramm, T., Yaroslavtsev, G.: Near optimal LP rounding algorithm for correlation clustering on complete and complete k-partite graphs. In: Proceedings of the 47th ACM Symposium on Theory of Computing, pp. 219–228 (2015)
15. Castro, J., Nasini, S., Saldanha-Da-Gama, F.: A cutting-plane approach for large-scale capacitated multi-period facility location using a specialized interior-point method. Math. Program. **163**(1–2), 411–444 (2021)
16. Filippi, C., Guastaroba, G., Speranza, M.G.: On single-source capacitated facility location with cost and fairness objectives. Eur. J. Oper. Res. **289**(3), 959–974 (2021)
17. Kim, S., Yoo, C.D., Nowozin, S., Kohli, P.: Image segmentation using higher-order correlation clustering. IEEE Trans. Patt. Anal. Mach. Intell. **36**(9), 1761–1774 (2014)
18. Lange, J.H., Karrenbauer, A., Andres, B.: Partial optimality and fast lower bounds for weighted correlation clustering. In: Proceedings of the 35th International Conference on International Conference on Machine Learning, pp. 2892–2901 (2018)
19. Li, S., Svensson, O.: Approximating k-median via pseudo-approximation. SIAM J. Comput. **45**(2), 530–547 (2016)
20. Mathieu, C., Schudy, W.: Correlation clustering with noisy input. In: Proceedings of the 21th Annual ACM-SIAM Symposium on Discrete Algorithms, pp. 712–728 (2010)
21. Puleo, G.J., Milenkovic, O.: Correlation clustering with constrained cluster sizes and extended weights bounds. SIAM J. Optim. **25**(3), 1857–1872 (2015)
22. Thiel, E., Chehreghani, M.H., Dubhashi, D.: A non-convex optimization approach to correlation clustering. In: Proceedings of the 33rd AAAI Conference on Artificial Intelligence, pp. 5159–5166 (2019)
23. Veldt, N., Wirth, A., Gleich, D.F.: Parameterized correlation clustering in hypergraphs and bipartite graphs. Proceedings of the 26th ACM SIGKDD International Conference on Knowledge Discovery and Data Mining, pp. 1868–1876 (2020)

The Selection of COVID-19 Epidemic Prevention and Control Programs Based on Group Decision Making

Chunsheng Cui, Baiqiu Li(✉), and Liu Wang

Henan University of Economics and Law, Zhengzhou 450046, Henan, China

Abstract. COVID-19 has been sweeping the world for nearly two years. As the virus continues to mutate, epidemic prevention and control has become a long and experienced war. In the face of the sudden spread of virus strains, how to quickly and effectively formulate prevention and control plans is the most important to ensure the safety and social stability of cities. This paper is based on the characteristics of the persistence of the epidemic and the rapid transmission of the mutant strain, as well as the database of epidemic prevention and control plans formed by the existing prevention and control. Then, epidemic prevention experts select effective alternatives from the program database and rank their preferences through the preliminary analysis of the local epidemic situation. The process of the integration scheme is to minimize the differences to maximize the needs of the local epidemic, and then obtain the consensus ranking of the scheme and determine the final prevention and control scheme. The proposed method of this paper, on the one hand, can optimize the opinions of the epidemic prevention expert group and form a consensus decision. On the other hand, it can save time and carry out the work effectively, which is of certain practical significance to the prevention and control work of local outbreaks.

Keywords: Epidemic prevention and control · Group decision making · Alternative ranking · Consensus reaching

1 Introduction

At the end of 2019, COVID-19 erupted in Wuhan, China, spreading to many countries and regions around the world [13]. At present, novel coronavirus produces many mutated strains that have posed a serious threat to world health safety, such as Alpha, Beta, Gamma, Delta and Lambda, and so on. The transmission rate and carrying capacity of the mutated virus far exceed the original virus, which increases the difficulty of the epidemic prevention and control work. Judging from the current international epidemic situation, human beings will coexist with the novel coronavirus for a long time, and the work of epidemic prevention and control is becoming the norm. In May 2021, Guangzhou became the first city in China to fight Delta variant strains and handled the outbreak

D.-Z. Du et al. (Eds.): COCOA 2021, LNCS 13135, pp. 46–60, 2021.
https://doi.org/10.1007/978-3-030-92681-6_5

perfectly within a month. In August of the same year, the epidemic broke out and spread in Nanjing, Yunnan, Yangzhou, Zhengzhou and other cities. The speed of transmission was as fast as possible, which once again triggered the new thinking of epidemic prevention and control experts and even people in all countries. Based on this, this paper considers how to quickly and effectively formulate prevention and control solutions to ensure urban safety and social stability in view of the continuous variation of novel coronaviruses at this stage.

In the face of major issues, group decision making is a crucial step in determining the final programme. It is to integrate the preferences of multiple decision makers into group preferences [8,9]. What's more, it is considered an efficient and accurate means to make optimal decisions quickly, so many experts in different fields devote time and energy to study group decision making in depth [18,21]. The first group of experts on group decision making focused on voting [14]. And the simple majority principle is the most typical and widely used in group decision making [1,5]. Group decision making is designed to achieve the consensus of group opinions, so the use of a consensus mechanism in group decision making is conducive to the smooth implementation of group programs, but also conducive to building a more harmonious interpersonal relationship within the organization [16]. Then, experts studying group decision making try to maximize consensus to rank alternatives, and many scholars measure the preferences of experts based on fuzzy quantitative [6,15]. Due to the influence of the objective factors such as the uncertainty of things themselves and the subjective factors such as the knowledge structure and judgment level of decision-makers, the views of decision-makers tend to be greatly different [7]. Cook et al. [4] proposed the Borda-Kendall method to measure consensus for ranking alternatives, but distance-based approaches sometimes fail to properly reflect consensus in group decision making. Huo et al. [12] put forward a concept based on premetric to express different opinions of experts, identify the differences among experts when ranking the alternatives, negotiate and adjust the preferences of experts with the largest differences, and finally obtain the ranking of the alternatives with the smallest differences, which makes up for the deficiency of distance-based identification method. Next, Hou's subsequent paper followed a post-consensus analysis of the methodology to facilitate new insights into the alternatives [17]. Group decision making has been widely used in failure mode and impacts analysis, supply chain management, water resources management and other fields. At present, group decision-making has also been studied in emergency decision-making and disaster management [19,20]. However, there is little research on how to make decisions on many prevention and control solutions in the face of the changing COVID-19 situation.

With the rapid development of 2019 novel coronavirus mutation strains, rapid prevention and control of the new outbreak point is the focus of the current epidemic prevention and control. Because of the rapid spread of variant strains, once it is found that the source of infection in the outbreak is variant strains, the government departments should quickly analyze the situation and organize experts to put forward prevention and control solutions for the local epidemic.

In this context, this paper studies that the expert group selects the alternatives suitable for this epidemic according to the existing epidemic prevention schemes. Then, according to the method proposed by Huo et al. [10–12], select the solution that the expert group considers most to meet the needs of the local epidemic. Experts express their preferences for a program faster than scoring it, which can save time, improve efficiency and reduce the spread of population in decision making on epidemic programs.

The rest of the paper is arranged as follows: The second part introduces the group decision making methods used in this paper, including the concepts of preference map, consensus gap, and consensus evaluation sequence and so on. The third part introduces the general process of group decision making of epidemic control programs from the perspective of epidemic persistence and rapid transmission of variant strains. The fourth part applies the decision making method proposed in the third part to the selection of control schemes for the spread of mutant strains, taking nucleic acid testing (NAT) as an example. At the same time, compared with the existing group decision making methods, the method used in this paper is more suitable for the decision making of epidemic prevention and control programs. Finally, the fifth part is the conclusions and prospects of this paper.

2 Theoretical Basis

This paper studies how to rank the epidemic preventive measures based on expert preferences. This part briefly describes the basic theories and concepts used [10–12].

Let $E = \{e_1, e_2, \ldots, e_m\}$ be the set of the expert group and $A = \{a_1, a_2, \ldots, a_n\}$ be the alternatives to be ranked, where $1 < m < +\infty$ and $1 < n < +\infty$. Assuming that expert preferences are considered to allow a parallel sequencing, and the alternatives in a tie are arranged in the same positions, which are continuous positive integers.

Definition 1 [10]. *A sequence $(S_i)_{n\times 1}$ is called the preference map (PM) of the alternation set A with respect to the order relation $\preceq$, if and only if the following is true: $S_i = \{|P_i| + 1, |P_i| + 2, \ldots, |P_i| + |Q_i|\}$, where $P_i = \{a_j | a_j \in A, a_j \succ a_i\}$ and $Q_i = \{a_k | a_k \in A, a_i \sim a_k\}$.*

Definition 1 is based on the following two definitions:

A sequence $(P_i)_{n\times 1}$ is called the predominance sequence of the alternation set A with respect to the order relation $\preceq$, if and only if the following is true: $P_i = \{a_j | a_j \in A, a_j \succ a_i\}$.

A sequence $(Q_i)_{n\times 1}$ is called the indifference sequence of the alternation set A with respect to the order relation $\preceq$, if and only if the following is true: $Q_i = \{a_k | a_k \in A, a_i \sim a_k\}$.

Definition 2 [10]. *Assume that* $V^{(1)} = (V_i^{(1)})_{n\times1}$, $V^{(2)} = (V_i^{(2)})_{n\times1}$ *are two PMs of the experts, then the consensus gap between them is defined as follows:*

$$\begin{aligned}\Delta(V^{(1)}, V^{(2)}) &= \sum_{i=0}^{n} \delta(V^{(1)}, V^{(2)}) \\ &= \sum_{i=1}^{n} \max\{0, \min V_i^{(1)} - \max V_i^{(2)}, \min V_i^{(2)} - \max V_i^{(1)}\}.\end{aligned}$$

The consensus gap index is a premetric, which satisfies only the properties non-negativity and symmetry, so as to represent the disagreement between the two preference maps.

Moreover, a dispute matrix that is associated with the expert's disagreements on alternatives is defined by $DispM = (S_{ik})_{n\times n}$, where $DispM = (S_{ik})_{n\times n}$ represents the total gap of the experts if a_i is to be ranked at position k.

Definition 3 [10,11]. *Assume that* $V^{(1)} = (V_i^{(1)})_{n\times1}, V^{(2)} = (V_i^{(2)})_{n\times1}, \dots, V^{(m)} = (V_i^{(m)})_{n\times1}$ *are the PMs of the experts. The experts are in consensus if, and only if* $\forall i(V_i^{(1)} \cap V_i^{(2)} \cap \cdots \cap V_i^{(m)} \neq \varnothing)$. *The consensus ranking is* (W_i), *where*

$$\begin{pmatrix} W_1 \\ W_2 \\ \vdots \\ W_n \end{pmatrix} = \begin{pmatrix} V_1^{(1)} \\ V_2^{(1)} \\ \vdots \\ V_n^{(1)} \end{pmatrix} \cap \begin{pmatrix} V_1^{(2)} \\ V_2^{(2)} \\ \vdots \\ V_n^{(2)} \end{pmatrix} \cap \cdots \cap \begin{pmatrix} V_1^{(m)} \\ V_2^{(m)} \\ \vdots \\ V_n^{(m)} \end{pmatrix} = \begin{pmatrix} \bigcap_{k=1}^{m} V_1^{(k)} \\ \bigcap_{k=1}^{m} V_2^{(k)} \\ \vdots \\ \bigcap_{k=1}^{m} V_n^{(k)} \end{pmatrix}.$$

The consensus gap between each pair of PMs represents differences between the two experts, and the disagreement matrix represents the disagreement among all experts.

The disagreement matrix is defined as:

$$D = (\Delta_{jk})_{(m\times m)}, \text{ where } \Delta_{jk} = \Delta(V^{(j)}, V^{(k)}).$$

Definition 4 [12]. *The Consensus Evaluation Sequence (CES) is defined as follows:*

$$CES = [GCI; MDP, PDisaI; MDA, MDispI].$$

The consensus evaluation sequence (CES) represents the degree of expert consensus on the ranking of alternatives. It contains the group consensus index (GCI), the maximum disagreement pairs (MDP), the pairwise disagreement index (PDisaI), the maximum dispute alternatives (MDA) and the maximum dispute index (MDisaI).

(1) GCI indicates the proportion of the number of expert pairs that reach consensus among all possible expert pairs. It is defined as follows:

$$GCI = \frac{2\sum_{i=1}^{m-1}\sum_{j=j+1}^{m}\rho_{ij}}{m(m-1)}, \text{where } \rho_{ij} = \begin{cases} 1, & \Delta_{ij} = 0, \\ 0, & \text{others.} \end{cases} \tag{1}$$

The range of values of the GCI is $[0, 1]$. $GCI = 1$ expresses that the experts reach a complete consensus, and the bigger GCI expresses that the higher the consensus level of the experts.

(2) PDisaI indicates the biggest disagreement among the experts. It is defined as follows:

$$PDisaI = \max_{j}\{\max_{k}\{\Delta_{jk} | j < k\}\}. \tag{2}$$

$PDisaI = 0$ indicates that the experts have a complete consensus on each choice; otherwise, PDisaI represents the largest inconsistency value of the expert group.

(3) MDP implies the expert pairs with the biggest differences. It is defined as follows:

$$MDP = \{(j, k) | \Delta_{jk} = PDisaI, j < k, \Delta_{jk} < 0\}. \tag{3}$$

MDP represents the subscript-pair set of expert pairs with that index value, if PDisaI is not 0.

(4) MDispI indicates the most controversial index of experts on alternatives. It is defined as follows:

$$MDispI = \max_{i} \min_{k}\{S_{ik}\}. \tag{4}$$

$MDispI = 0$ shows that the experts have no controversy about the ranking of each alternative; otherwise, MDispI represents the maximum controversial value of the experts for the alternatives.

(5) MDA implies the alternative with the biggest disagreement. It is defined as follows:

$$MDA = \{i | S_{ik} = MDispI, S_{ik} > 0\}. \tag{5}$$

MDA represents the subscript set of alternatives with that index value, if MDisaI is not 0.

Decision-makers can identify whether the group of experts fully reaches consensus from the CES. Also, decision-makers can get expert pairs who have the maximum disagreement and the alternatives which have the maximum controversial value, if the group of experts is not in consensus.

3 General Process of Group Decision Making in Epidemic Prevention and Control Programs

At present, the epidemic is usually caused by the spread of new coronavirus, showing new characteristics, such as long duration, wide range, strong transmission ability, high viral load of infected patients, the rapid development of patients and so on. The measures of epidemic prevention and control need to be formulated and implemented according to the changing situation of the epidemic. Therefore, the epidemic prevention and control programs require to be developed rapidly, effective and enforceable. Since the novel coronavirus mutates rapidly, the control of the epidemic is related to the immediate safety of the people and social stability. When formulating epidemic prevention and control policies, the government ought to screen out the existing epidemic prevention and control schemes according to the local epidemic situation, rather than simply copying the solutions adopted at the last outbreak. Then ask the experts to make group decisions and choose the best alternative to meet local needs.

The goal of group decision making is to comprehensively consider the opinions of various experts, integrate individual decision making into group decision making, and ultimately reach expert consensus to the maximum extent to determine the most feasible alternative. Group decision making can adopt the opinions of experts from various aspects, so as to break the limitations of individual knowledge and thinking, and reduce the error rate in decision making of epidemic prevention and control.

Based on the selection of epidemic prevention and control plans, this paper puts forward the following steps:

Step1: Propose feasible solutions.

Epidemic prevention and control policies involve all aspects, each of which is an independent decision. The final promulgated prevention and control policies are the sum of all aspects of decision making. Among them, the control policies, the nucleic acid testing policies and the traffic control programs are the three aspects that should be first decided in the early outbreak of the epidemic. After nearly two years of experience, a database of epidemic prevention and control plans has been formed. Select a set of options suitable for the local epidemic in the program library, denoted $X_i(i = 1, 2, .., n)$.

Step 2: Select experts and rank the alternatives in preference.

First, identify the experts involved in the decision making of the scheme according to their professional direction (if necessary, experts with similar or less divergent preferences are grouped according to their previous ranking preferences). The expert group is denoted as $E_i(i = 1, 2, ..., m)$. Then, based on the feasible alternatives proposed in step 1, each expert makes a comprehensive ranking according to the feasibility, implementation effectiveness, control strength and other aspects of the alternatives.

Step 3: Form expert preference maps (PMs) for the ranking of alternatives.

Based on expert ranking, expert preferences are transformed into PMs according to Definition 1. Define Consensus Evaluation Sequence (CES) and confirm acceptable thresholds for the Group Consensus Index (GCI).

Step 4: Build up a dispute matrix, calculate the pairwise disagreement index (PDisaI) and determine whether the experts reach a consensus.

The dispute matrix is constructed according to the PMs. And then obtain the PDisaI. If $PDisaI = 0$ or GCI reaches an acceptable threshold, it indicates that experts reach a consensus on the ranking of epidemic prevention and control schemes, which is solved according to Definition 3; otherwise, turn to step 5.

Step 5: Iterate over the preference ranking of experts for alternatives.

Since experts do not reach a consensus on the order of the epidemic prevention and control plan, choose a $\Delta_{ij}(i < j \text{ and } \Delta_{ij} > 0)$ in descending order of the dispute index. Through negotiating with experts related to the selected

Δ_{ij}, modify their preferences for the order of alternatives. If the two experts do not agree to modify their rankings, then move to another pair of experts with $\Delta_{ij} > 0$ and $i < j$. If none of these experts are willing to change their rankings, go to Step 6; otherwise, after obtaining the corrected preference order obtained by the experts, go to step 3.

Step 6: Build an assignment model to minimize the divergence.

The assignment model is established as follows:

$$\begin{aligned} \min \quad & f = \sum_i \sum_k x_{ik} S_{ik} \\ \text{s.t.} \quad & \sum_{i=1}^{n} x_{ik} = 1, \quad k = 1, 2, \ldots, n, \\ & \sum_{i=1}^{k} x_{ik} = 1, \quad i = 1, 2, \ldots, n, \\ & x_{ik} = 0, 1, \quad i, k = 1, 2, \ldots, n. \end{aligned}$$

S_{ik} is obtained by Definition 2, indicating the total consensus gap for experts if X_i is to be ranked at position k. Then solve the assignment model.

Step 7: Analyze the results.

According to the results of the allocation model, the final program ranking is obtained. Then, analyze the consensus results and determine the selected epidemic prevention and control solution.

4 The Realization of Group Decision Making for NAT Solutions in Epidemic Prevention and Control

In view of the new situation of the novel coronavirus epidemic, the mutant strains suddenly spread in a city, and the epidemic prevention and control center is expected to make decisions quickly.

Next, the selection of nucleic acid testing scheme is taken as an example to introduce the application of group decision making method in the formulation of novel coronavirus prevention and control policies.

Step 1: Propose feasible solutions.

In view of the current situation of COVID-19, epidemic prevention and control experts have screened six alternative plans based on the existing database of epidemic prevention and control plans. The above six alternatives are denoted as $\{X_1, X_2, X_3, X_4, X_5, X_6\}$.

X_1: Organize citizens to carry out six nucleic acid testing on the 1st, 4th, 7th, 10th, 13th, and 16th days after the closure of the area. In nucleic acid sampling, nucleic acid testing in medium and high-risk areas and the containment area is performed by one-to-one testing method, and priority is given. Testing of the key population is adopted 5 individual test samples mixed into one reagent, as well as the testing of the common population is adopted 10 individual test samples mixed into one reagent. All work need to be done at off-peak.

X_2: Organize a round of nucleic acid testing every three days until no new cases and asymptomatic infections. Nucleic acid testing needs to be done at least 6 times per person. Single sample testing is adopted in key control areas. The testing of the common population is adopted 10 individual test samples mixed into one reagent.

X_3: Four nucleic acid testings need to be conducted for citizens in high-risk areas on the 1st, 4th, 7th and 14th days after the closure of the area. Three nucleic acid testings (each time is separated at least 24 h apart) need to be conducted for citizens in medium risk areas on the 1st, 4th, 7th and 14th days after the closure of the area. Other people in the city needed to be tested once. A single check is used for everyone.

X_4: Four times (on the 1st, 4th, 7th, 14th days of isolation) of one-to-one nucleic acid testing is organized. The detection objects are confirmed cases, suspected cases, asymptomatic cases, close contacts and resident population, migrant workers and foreigners within their scope of activities.

X_5: Organize citizens in the closed area to carry out NAT five times, on the 1st, 4th, 7th, 10th, 14th days of quarantine. Organize citizens in the control area to carry out NAT four times, on the 1st, 4th, 7th and 14th days of quarantine. Other people in the city do not go out unless necessary and stay at home to quarantine.

X_6: Organize confirmed cases and close contacts to carry out NAT five times on the 1st, 4th, 7th, 10th and 14th days of quarantine. At the same time, organize the second close contacts to carry out NAT three times, on the 1st, 4th and 7th days of quarantine. Single check is used for everyone.

Step 2: Select experts and rank the alternatives in preference.

Among the 30 experts in the Center for Epidemic Prevention and Control, select 6 experts in the direction of nucleic acid detection to make decisions on the nucleic acid testing scheme in the outbreak area. Six experts are denoted as $\{X_1, X_2, X_3, X_4, X_5, X_6\}$.

The group of experts sorts the six alternatives according to their preferences, and the ranking is allowed parallel. The experts' preferences are as follows:

$$\begin{aligned}
&E_1 : X_1 \sim X_2 \succ X_5 \succ X_3 \sim X_4 \sim X_6,\\
&E_2 : X_1 \succ X_2 \succ X_5 \succ X_3 \succ X_4 \sim X_6,\\
&E_3 : X_1 \sim X_3 \sim X_4 \succ X_2 \succ X_5 \sim X_6,\\
&E_4 : X_1 \sim X_2 \succ X_3 \sim X_4 \succ X_5 \succ X_6,\\
&E_5 : X_1 \sim X_2 \succ X_3 \sim X_4 \sim X_5 \succ X_6,\\
&E_6 : X_1 \succ X_2 \succ X_3 \sim X_5 \succ X_4 \sim X_6.
\end{aligned}$$

Step 3: Form expert preference maps (PMs) for the ranking of alternatives.

According to Definition 1, the PMs of the experts based on the above experts' preferences are obtained as follows:

$$\begin{matrix} \\ X_1 \\ X_2 \\ X_3 \\ X_4 \\ X_5 \\ X_6 \end{matrix} \begin{matrix} V^{(1)} \\ \begin{pmatrix} \{1,2\} \\ \{1,2\} \\ \{4,5,6\} \\ \{4,5,6\} \\ \{3\} \\ \{4,5,6\} \end{pmatrix} \end{matrix}, \begin{matrix} V^{(2)} \\ \begin{pmatrix} \{1\} \\ \{2\} \\ \{4\} \\ \{5,6\} \\ \{3\} \\ \{6\} \end{pmatrix} \end{matrix}, \begin{matrix} V^{(3)} \\ \begin{pmatrix} \{1,2,3\} \\ \{4\} \\ \{1,2,3\} \\ \{1,2,3\} \\ \{5,6\} \\ \{6,6\} \end{pmatrix} \end{matrix}, \begin{matrix} V^{(4)} \\ \begin{pmatrix} \{1,2\} \\ \{1,2\} \\ \{3,4\} \\ \{3,4\} \\ \{5\} \\ \{6\} \end{pmatrix} \end{matrix}, \begin{matrix} V^{(5)} \\ \begin{pmatrix} \{1,2\} \\ \{1,2\} \\ \{3,4,5\} \\ \{3,4,5\} \\ \{3,4,5\} \\ \{6\} \end{pmatrix} \end{matrix}, \begin{matrix} V^{(6)} \\ \begin{pmatrix} \{1\} \\ \{2\} \\ \{3,4\} \\ \{5,6\} \\ \{3,4\} \\ \{5,6\} \end{pmatrix} \end{matrix}.$$

Identify 1 as the acceptable GCI threshold.

Step 4: Build up a dispute matrix, calculate the pairwise disagreement index (PDisaI) and determine whether the experts reach a consensus.

According to Definition 2, the disagreement matrix D_0 is obtained as follows:

$$D_0 = (\Delta_{ij})_{6\times 6} = \begin{pmatrix} 0 & 0 & 6 & 2 & 0 & 0 \\ 0 & 0 & 7 & 3 & 0 & 0 \\ 6 & 7 & 0 & 2 & 2 & 5 \\ 2 & 3 & 2 & 0 & 0 & 2 \\ 0 & 0 & 2 & 0 & 0 & 0 \\ 0 & 0 & 5 & 2 & 0 & 0 \end{pmatrix}.$$

The pairwise disagreement index (PDisaI) is calculated by formula (2):

$$PDisaI = 7 \neq 0$$

The group consensus index (GCI) is calculated by using formula (1):

$$GCI = \frac{7}{15}$$

Obviously, the group of experts is not in consensus. The consensus evaluation sequence (CES) at this time is:

$$CES = [GCI = \frac{7}{15}; MDP = \{(2,3)\}, PDisaI = 7;$$
$$MDA = \{5\}, MDispI = 4]$$

It shows that a pair of experts with the maximum controversy is (E_2, E_3), and the experts have the maximum disagreement on alternative X_5.

Step 5: Iterate over the preference ranking of experts for alternatives.

Iteration 1:

Because expert 2 and expert 3 have the maximum disagreement, we ask experts 2 and 3 to modify their preferences by negotiating. The changed preferences by them are as follows:

$$E_2 : X_1 \succ X_2 \succ X_3 \sim X_4 \succ X_5 \sim X_6,$$
$$E_3 : X_1 \sim X_2 \succ X_3 \sim X_4 \succ X_5 \sim X_6.$$

Based on Definition 1, the PMs of the experts are as follows:

$$\begin{array}{c} \\ X_1 \\ X_2 \\ X_3 \\ X_4 \\ X_5 \\ X_6 \end{array}
\begin{array}{c} V^{(1)} \\ \begin{pmatrix} \{1,2\} \\ \{1,2\} \\ \{4,5,6\} \\ \{4,5,6\} \\ \{3\} \\ \{4,5,6\} \end{pmatrix} \end{array},
\begin{array}{c} V^{(2)} \\ \begin{pmatrix} \{1\} \\ \{2\} \\ \{3,4\} \\ \{5,6\} \\ \{3,4\} \\ \{5,6\} \end{pmatrix} \end{array},
\begin{array}{c} V^{(3)} \\ \begin{pmatrix} \{1,2\} \\ \{1,2\} \\ \{3,4\} \\ \{3,4\} \\ \{5,6\} \\ \{6,6\} \end{pmatrix} \end{array},
\begin{array}{c} V^{(4)} \\ \begin{pmatrix} \{1,2\} \\ \{1,2\} \\ \{3,4\} \\ \{3,4\} \\ \{5\} \\ \{6\} \end{pmatrix} \end{array},
\begin{array}{c} V^{(5)} \\ \begin{pmatrix} \{1,2\} \\ \{1,2\} \\ \{3,4,5\} \\ \{3,4,5\} \\ \{3,4,5\} \\ \{6\} \end{pmatrix} \end{array},
\begin{array}{c} V^{(6)} \\ \begin{pmatrix} \{1\} \\ \{2\} \\ \{3,4\} \\ \{5,6\} \\ \{3,4\} \\ \{5,6\} \end{pmatrix} \end{array}.$$

Based on Definition 2, the disagreement matrix D_1 is built as follows:

$$D_1 = (\Delta_{ij})_{6\times 6} = \begin{pmatrix} 0 & 0 & 2 & 2 & 0 & 0 \\ 0 & 0 & 2 & 0 & 0 & 0 \\ 2 & 2 & 0 & 0 & 0 & 2 \\ 2 & 0 & 0 & 0 & 0 & 2 \\ 0 & 0 & 0 & 0 & 0 & 0 \\ 0 & 0 & 2 & 2 & 0 & 0 \end{pmatrix}.$$

PDisaI is calculated by using formula (2):

$$PDisaI = 2 \neq 0$$

CGI is calculated by using formula (1):

$$CGI = \frac{10}{15} = \frac{1}{3}$$

Obviously, the group of experts is still not in consensus. The consensus evaluation sequence (CES) at this time is:

$$CES = [GCI = \frac{1}{3}; MDP = \{(1,3),(2,3),(1,4),(3,6),(4,6)\}, PDisaI = 2;$$
$$MDA = \{5\}, MDispI = 4].$$

We can obtain that there are five pairs of experts that have the maximum disagreement: E_1 and E_3; E_2 and E_3; E_1 and E_4; E_3 and E_6; E_4 and E_6. Meanwhile, the experts have the maximum disagreement on alternative X_5.

Iteration 2:

According to the comprehensive consideration of the parameter values in the above CES, we randomly selected a pair of experts in the six most controversial experts in this iteration, and finally we selected experts 1 and 3 for consultation.

The preferences modified after consultation between experts 1 and 3 are as follows:

$$E_1 : X_1 \sim X_2 \succ X_5 \succ X_3 \sim X_4 \sim X_6,$$
$$E_3 : X_1 \sim X_2 \succ X_3 \sim X_4 \sim X_5 \sim X_6.$$

According to Definition 1, at this time, the PMs of the experts are as follows:

$$\begin{array}{c} \\ X_1 \\ X_2 \\ X_3 \\ X_4 \\ X_5 \\ X_6 \end{array}\begin{array}{c} V^{(1)} \\ \begin{pmatrix} \{1,2\} \\ \{1,2\} \\ \{4,5,6\} \\ \{4,5,6\} \\ \{3\} \\ \{4,5,6\} \end{pmatrix} \end{array}, \begin{array}{c} V^{(2)} \\ \begin{pmatrix} \{1\} \\ \{2\} \\ \{3,4\} \\ \{5,6\} \\ \{3,4\} \\ \{5,6\} \end{pmatrix} \end{array}, \begin{array}{c} V^{(3)} \\ \begin{pmatrix} \{1,2\} \\ \{1,2\} \\ \{3,4,5,6\} \\ \{3,4,5,6\} \\ \{3,4,5,6\} \\ \{3,4,5,6\} \end{pmatrix} \end{array}, \begin{array}{c} V^{(4)} \\ \begin{pmatrix} \{1,2\} \\ \{1,2\} \\ \{3,4\} \\ \{3,4\} \\ \{5\} \\ \{6\} \end{pmatrix} \end{array}, \begin{array}{c} V^{(5)} \\ \begin{pmatrix} \{1,2\} \\ \{1,2\} \\ \{3,4,5\} \\ \{3,4,5\} \\ \{3,4,5\} \\ \{6\} \end{pmatrix} \end{array}, \begin{array}{c} V^{(6)} \\ \begin{pmatrix} \{1\} \\ \{2\} \\ \{3,4\} \\ \{5,6\} \\ \{3,4\} \\ \{5,6\} \end{pmatrix} \end{array}.$$

According to Definition 2, the disagreement matrix D_2 is built as follows:

$$D_2 = (\Delta_{ij})_{6\times 6} = \begin{pmatrix} 0 & 0 & 0 & 2 & 0 & 0 \\ 0 & 0 & 0 & 0 & 0 & 0 \\ 0 & 0 & 0 & 0 & 0 & 0 \\ 2 & 0 & 0 & 0 & 0 & 2 \\ 0 & 0 & 0 & 0 & 0 & 0 \\ 0 & 0 & 0 & 2 & 0 & 0 \end{pmatrix}.$$

Then, PDisaI is calculated by using formula (2) and CGI is calculated by using formula (1):

$$PDisaI = 2 \neq 0, GCI = \frac{13}{15}$$

Now, we can obtain that CES is as follows:

$$CES = [GCI = \frac{1}{3}; MDP = \{(1,3),(2,3),(1,4),(3,6),(4,6)\}, PDisaI = 2;$$
$$MDA = \{5\}; MDispI = 4].$$

Based on the Iteration 2, there are two pairs of experts having the maximum disagreement. They are experts 1 and 4 and experts 4 and 6. And the experts still have the maximum disagreement on alternative X_5.

Iteration 3: We ask experts 4 and 6 to modify their preferences by negotiating, but neither of them wants to compromise.

Iteration 4: We ask experts 4 and 6 to modify their preferences by negotiating, but neither of them wants to modify.

Step 6: Build an assignment model to minimize the divergence.

The expert group has not yet reached a consensus on the sequencing of alternatives following the iterative consultations. We minimize differences by constructing an assignment model:

$$\begin{aligned}
\min \quad & f=\sum_{i=1}^{6}\sum_{k=1}^{6} x_{ik}S_{ik} \\
\text{s.t.} \quad & \sum_{i=1}^{6} x_{ik}=1, \quad k=1,2,3,4,5,6, \\
& \sum_{i=1}^{6} x_{ik}=1, \quad i=1,2,3,4,5,6, \\
& x_{ik}=0,1, \quad i,k=1,2,3,4,5,6.
\end{aligned}$$

Solve the model and the final ranking with minimal disagreement is as follows:

$$X_1 \succ X_2 \succ X_4 \succ X_5 \succ X_3 \succ X_6.$$

Step 7: Analyze the results.

It shows that the NAT scheme X_1 is most suitable for the local epidemic in the face of sudden transmission of mutant strains in the city.

Next, the Cook-Seiford method [2,3] is used to solve the decision making problem of the above epidemic prevention and control, and the results are compared with the results obtained in this paper.

Firstly, the sequential ranking is indicated by the Cook-Seiford Vector (CSV) based on the initial expert ranking of the alternatives.

CSV is represented by medians for parallel rankings, and ultimately assigns an ordinary single number to all experts' ordering for each alternative. For example, $\{1,2\}$ is represented as $\{1.5\}$, and $\{1,2,3\}$ is represented as $\{2\}$. Thus, the original expert preferences for alternatives in the above examples can be expressed as:

$$\begin{array}{c} \\ X_1\\X_2\\X_3\\X_4\\X_5\\X_6\end{array}
\begin{array}{c} CSV^{(1)} \\ \begin{pmatrix}1.5\\1.5\\5\\5\\3\\5\end{pmatrix}\end{array},
\begin{array}{c} CSV^{(2)} \\ \begin{pmatrix}1\\2\\4\\5.5\\3\\5.5\end{pmatrix}\end{array},
\begin{array}{c} CSV^{(3)} \\ \begin{pmatrix}2\\4\\2\\2\\5.5\\5.5\end{pmatrix}\end{array},
\begin{array}{c} CSV^{(4)} \\ \begin{pmatrix}1.5\\1.5\\3.5\\3.5\\5\\6\end{pmatrix}\end{array},
\begin{array}{c} CSV^{(5)} \\ \begin{pmatrix}1.5\\1.5\\4\\4\\4\\6\end{pmatrix}\end{array},
\begin{array}{c} CSV^{(6)} \\ \begin{pmatrix}1\\2\\3.5\\5.5\\3.5\\5.5\end{pmatrix}\end{array}.$$

Secondly, calculate the Cook-Seiford distance between every two CSVs and represent the problem as assigned problem:

$$d_{ik}=\sum_{i=1}^{6}|a_i^l-k|$$

The distance matrix is obtained as follows:

$$(d_{ik})_{6\times 6} = \begin{pmatrix} 2.5 & 3.5 & 9.5 & 15.5 & 21.5 & 27.5 \\ 6.5 & 3.5 & 7.5 & 11.5 & 17.5 & 23.5 \\ 16 & 10 & 6 & 4 & 8 & 14 \\ 19.5 & 13.5 & 9.5 & 6.5 & 6.5 & 10.5 \\ 18 & 12 & 6 & 5 & 7 & 12 \\ 27.5 & 21.5 & 15.5 & 9.5 & 3.5 & 2.5 \end{pmatrix}.$$

Thirdly, minimize the distance among experts:

$$\begin{aligned} \min \quad & f = \sum_{i=1}^{6}\sum_{k=1}^{6} x_{ik} d_{ik} \\ \text{s.t.} \quad & \sum_{i=1}^{6} x_{ik} = 1, \quad k = 1,2,3,4,5,6, \\ & \sum_{i=1}^{6} x_{ik} = 1, \quad i = 1,2,3,4,5,6, \\ & x_{ik} = 0,1, \quad i,k = 1,2,3,4,5,6. \end{aligned}$$

Finally, the final consensus ranking is as follows:

$$X_1 \succ X_2 \succ X_5 \succ X_3 \succ X_4 \succ X_6.$$

Comparison with the results obtained in this paper:

$$X_1 \succ X_2 \succ X_4 \succ X_5 \succ X_3 \succ X_6.$$

We can obtain two conclusions:

First, it finds that the two methods are controversial about the ranking position of alternative X_4. The method presented in this paper moves the location of X_4 forward after consultation by experts. In the case of an outbreak, the order in which NAT should be conducted among the local population should be confirmed cases, suspected cases, asymptomatic cases, close contacts, the persons living and active within the scope of the activities of the confirmed cases, the persons living and active within the scope of the activities of the close contacts, all personnel in the region around the affected area. X_4 means that organize to carry out nucleic acid screening for confirmed patients, close contacts and personnel within their activities at first. X_5 and X_3 are the nucleic acid screening plans for all people around the outbreak of the epidemic. Considering the limited human and material resources in the early outbreak, it is necessary and reasonable to give priority to key populations. This just proves that the method proposed in this paper is helpful for the prevention and control decision.

Second, obviously, Cook-Seiford is a single iterative approach that terminates at the first decision. This approach is equivalent to having only one vote, without iterations of expert preferences. It is unable to identify disputes and then negotiate with controversial experts to minimize disputes, so it cannot guide the expert group to adjust their preferences. Due to the rapid changes in the situation of the novel coronavirus and the uncertainty of the spread of the mutant virus at present, the error of using the decision method in this paper is smaller than that of using the existing decision method in the face of sudden epidemics. Therefore, the method proposed in this paper is feasible and more conducive to dealing with uncertain outbreaks.

In summary, the method presented in this paper is more suitable for the decision making of COVID-19 prevention and control schemes.

5 Conclusions

Considering the current international situation of COVID-19, this paper uses the method of preference ranking in group decision making to formulate prevention and control programs, minimizing the differences of the expert group to quickly and effectively formulate prevention and control schemes to ensure citizen safety and social stability. On the one hand, the application of this method can optimize the opinions of the epidemic prevention expert group and form a consensus scheme. On the other hand, it can save time in epidemic prevention and control and carry out epidemic prevention and control quickly and effectively. However, there are still some problems in the research of this paper. Firstly, the consensus evaluation sequence index used in this paper still has room for improvement. Secondly, the decision making method of epidemic prevention and control proposed in this paper is an empirical decision making based on the existing epidemic prevention and control scheme database and experts' preference. How to use intelligent decision making to determine the prevention and control solution in the epidemic outbreak area is the direction that needs to be further studied and improved.

Acknowledgement. This study was supported by 2020 Henan University Philosophy and Social Sciences Applied Research Major Project Plan (NO. 2020-YYZD-02), Humanities and Social Science Research General Project of Henan Provincial Department of Education in 2021 (NO. 2021-ZZJH-020), 2020 Henan Province Philosophy and Social Science Planning Project (NO. 2020BJJ041), 2021 Key Scientific Research Projects of Colleges and Universities in Henan Province (NO. 21A520021).

References

1. Busetto, F., Codognato, G., Tonin, S.: Simple majority rule and integer programming. Math. Soc. Sci. **113**, 160–163 (2021)
2. Cook, W.D.: Distance-based and ad hoc consensus models in ordinal preference ranking. Eur. J. Oper. Res. **172**(2), 369–385 (2006)
3. Cook, W.D., Seiford, L.M.: Priority ranking and consensus formation. Manag. Sci. **24**(16), 1721–1732 (1978)
4. Cook, W.D., Seiford, L.M.: On the Borda-Kendall consensus method for priority ranking problems. Manag. Sci. **28**(6), 621–637 (1982)

5. Eraslan, H., Merlo, A.: Majority rule in a stochastic model of bargaining. J. Econ. Theory **103**(1), 31–48 (2002)
6. Feng, X., Shang, X., Xu, Y., Wang, J.: A method to multi-attribute decision-making based on interval-valued q-rung dual hesitant linguistic Maclaurin symmetric mean operators. Complex Intell. Syst. **6**(3), 447–468 (2020)
7. Fu, C., Yang, S.L.: The group consensus based evidential reasoning approach for multiple attributive group decision analysis. Eur. J. Oper. Res. **206**(3), 601–608 (2010)
8. Gehrlein, W.V.: Social choice and individual values. RAIRO - Oper. Res. **15**(3), 287–296 (1981)
9. Gou, X., Xu, Z.: Managing noncooperative behaviors in large-scale group decision-making with linguistic preference orderings: the application in Internet Venture Capital. Inf. Fusion **69**, 142–155 (2021)
10. Hou, F.: A consensus gap indicator and its application to group decision making. Group Decis. Negot. **24**(3), 415–428 (2015). https://doi.org/10.1007/s10726-014-9396-4
11. Hou, F.: The prametric-based GDM selection procedure under linguistic assessments. In: 2015 IEEE International Conference on Fuzzy Systems (FUZZ-IEEE), pp. 1–8. IEEE (2015)
12. Hou, F., Triantaphyllou, E.: An iterative approach for achieving consensus when ranking a finite set of alternatives by a group of experts. Eur. J. Oper. Res. **275**(2), 570–579 (2019)
13. Klavinskis, L.S., Liu, M.A., Lu, S.: A timely update of global COVID-19 vaccine development. Emerg. Microbes Infect. **9**(1), 2379–2380 (2020)
14. Rae, D.W.: Decision-rules and individual values in constitutional choice. Am. Polit. Sci. Rev. **63**(1), 40–56 (1969)
15. Rani, D., Garg, H.: Complex intuitionistic fuzzy preference relations and their applications in individual and group decision-making problems. Int. J. Intell. Syst. **36**(4), 1800–1830 (2021)
16. Susskind, L.E., McKearnen, S., Thomas-Lamar, J.: The Consensus Building Handbook: A Comprehensive Guide to Reaching Agreement. Sage Publications, London (1999)
17. Triantaphyllou, E., Hou, F., Yanase, J.: Analysis of the final ranking decisions made by experts after a consensus has been reached in group decision making. Group Decis. Negot. **29**(2), 271–291 (2020). https://doi.org/10.1007/s10726-020-09655-5
18. Wallenius, J., Dyer, J.S., Fishburn, P.C., Steuer, R.E., Zionts, S., Deb, K.: Multiple criteria decision making, multiattribute utility theory: recent accomplishments and what lies ahead. Manag. Sci. **54**(7), 1336–1349 (2008)
19. Wan, Q., Xu, X., Chen, X., Zhuang, J.: A two-stage optimization model for large-scale group decision-making in disaster management: minimizing group conflict and maximizing individual satisfaction. Group Decis. Negot. **29**(5), 901–921 (2020). https://doi.org/10.1007/s10726-020-09684-0
20. Xu, X.H., Du, Z.J., Chen, X.H.: Consensus model for multi-criteria large-group emergency decision making considering non-cooperative behaviors and minority opinions. Decis. Support Syst. **79**, 150–160 (2015)
21. Zhan, Q., Fu, C., Xue, M.: Distance-based large-scale group decision-making method with group influence. Int. J. Fuzzy Syst. **23**(2), 535–554 (2021). https://doi.org/10.1007/s40815-020-00993-9

Which Option Is a Better Way to Improve Transfer Learning Performance?

Honghui Xu, Zhipeng Cai(✉), and Wei Li

Department of Computer Science, Georgia State University,
Atlanta, GA 30302-3965, USA
hxu16@student.gsu.edu, {zcai,wli28}@gsu.edu

Abstract. Transfer learning has been widely applied in Artificial Intelligence of Things (AIoT) to support intelligent services. Typically, collection and collaboration are two mainstreaming methods to improve transfer learning performance, whose efficiency has been evaluated by real-data experimental results but lacks validation of theoretical analysis. In order to provide guidance of implementing transfer learning in real applications, a theoretical analysis is in desired need to help us fully understand how to efficiently improve transfer learning performance. To this end, in this paper, we conduct comprehensive analysis on the methods of enhancing transfer learning performance. More specifically, we prove the answers to three critical questions for transfer learning: i) by comparing collecting instances and collecting attributes, which collection approach is more efficient? ii) is collaborative transfer learning efficient? and iii) by comparing collection with collaboration, which one is more efficient? Our answers and findings can work as fundamental guidance for developing transfer learning.

Keywords: AIoT · Transfer learning · Collaborative transfer learning

1 Introduction

A compound annual growth rate of 47% in network traffic indicates the proliferation of connected Internet of Things (IoT) devices in recent years [9]. Meanwhile, known as the next generation IoT, Artificial Intelligence of Things (AIoT) [19] relies on machine learning (ML) techniques to extract information from data collected by a number of IoT devices [24] and has been leveraged by real-world services, such as Facebook, Google, Amazon, and Microsoft, to provide individuals with more efficient services. Transfer learning, which takes an advantage of learned knowledge from known categories to novel/unknown categories with limited training instances collected by IoT devices for prediction, is one of the widely-used machine learning models in AIoT [3]. Therefore, improving transfer learning performance plays a critical role to promote the quality of AIoT-based services, which has attracted lots of research attention.

D.-Z. Du et al. (Eds.): COCOA 2021, LNCS 13135, pp. 61–74, 2021.
https://doi.org/10.1007/978-3-030-92681-6_6

Currently, researchers have realized the performance enhancement of transfer learning through collection and collaboration. On the one hand, the transfer learning performance can be improved by collecting more training instances [12,18,34] and more attributes since transfer learning's prediction is based on shared attributes [11,17,32]. Even though these two collection methods have been validated by conducting real-data experiments, no study has been carried out to theoretically investigate which collection option is more efficient. On the other hand, inspired by the idea of collaborative learning [10,14], collaborative transfer learning models have been proposed to improve transfer learning performance by sharing data among institutions [4,5,13,20,21,30,35]. Although these works have demonstrated the efficiency of collaborative transfer learning through the real-data experimental results [4,20], there lacks theoretically analysis whether collaboration can indeed bring the performance enhancement for transfer learning. Therefore, in literature, the challenging question, "how to efficiently improve transfer learning performance", has not been answered yet. However, with the wide applications of transfer learning models, a thorough analysis is urgent to be done before implementing these models so as to provide individuals, institutions, and organizations with theoretical guidance illustrating how to choose collection or collaboration for enhancing transfer learning performance efficiently.

To fill this gap, in this paper, a series of theoretical analysis is well implemented: i) first of all, we study how to select a more efficient collection option from two collection ways (*i.e.* collecting training instances and collecting more attributes) for performance enhancement in transfer learning; ii) next, we investigate whether collaborative transfer learning in a two-party collaboration scenario is worth doing to enhance performance; and iii) finally, we compare collection and collaboration to analyze which one is more efficient for promoting transfer learning. In our analysis, the conclusions are obtained through rigorous proof and can be used as fundamental guidance for applying transfer learning in real applications. Our multi-fold contributions are addressed as below.

- To the best of our knowledge, this is the first work to perform theoretical analysis on the methods of improving transfer learning performance.
- We provide a theoretical basis for entities to select a more efficient collection option from two collection ways to improve transfer learning performance.
- We prove whether the collaboration in a two-party scenario is efficient to promote transfer learning performance.
- We offer guidance to choose collection or collaboration for performance improvement in transfer learning.

The remainder of this paper is organized as follows. Related works are briefly summarized in Sect. 2. After formulating problems in Sect. 3, we detail our analysis process in Sect. 4, Sect. 5, and Sect. 6. Finally, Sect. 7 concludes this paper and discusses our future work.

2 Related Works

Transfer learning [3,31], which refers to domain adaptation or sharing of knowledge and representations, is a kind of learning model to achieve the goal of knowledge transfer by transferring information of learned models to changing or unknown data distributions. Transfer learning approaches can be broadly classified into four categories: instance-based, feature-based, parameter-based, and relational-based approaches. i) Instance-based transfer learning approaches [12,18,34] are mainly based on the instance weighting strategy. ii) Feature-based approaches [11,17,32] map the original features into a new feature space, which can be further divided into two subcategories, including asymmetric and symmetric feature-based transfer learning. The asymmetric approaches [2,17] transform the features in the source domain to match the features in the target domain, while symmetric approaches [28,32] attempt to find a common feature space and then map the source and target features into the common feature space. iii) Parameter-based transfer learning approaches [7,16,29] keep an eye on transferring the shared knowledge at the parameter level. iv) Relational-based approaches [6,22] focus on transferring the logical relationship learned in the source domain to the target domain. These transfer learning approaches reduce the effort to collect new labeled training samples and have been widely applied in different real applications, such as disease prediction [23], sign language recognition [8], and target online display advertising [26]. However, no existing work theoretically investigates how to enhance transfer learning performance efficiently.

Collaborative transfer learning was proposed to promote transfer learning performance by sharing data among institutions [4,5,20,21,30], which is motivated by the idea of collaborative learning [10,14]. For instance, in the process of healthcare and financial marketing analysis, sharing data among institutions can help enhance the accuracy of medical diagnosis [4] and financial marketing forecasting [20]. However, these works only validate the efficiency of collaboration through the real-data experimental results but lack theoretical analysis.

In a nutshell, the question of how to enhance transfer learning performance efficiently has not been answered with theoretical analysis. To fill this blank, in this paper, we will conduct comprehensively and deeply analysis on the methods of enhancing transfer learning performance, aiming to offer theoretical guidance for the improvement of transfer learning.

3 Problem Formulation

In this section, we mathematically formulate the problem of transfer learning based on PAC learning framework [15].

Transfer learning essentially trains a classifier by using the training instances associated with known labels to predict the testing instances associated with novel/unknown labels [33]. Inspired by the idea of [25], transfer learning can be defined to learn a classifier as follows.

Definition 1. *A classifier consists of two mapping functions, where the first one is the map from a training instance space $\mathcal{X}$ to an attribute space $\mathcal{T}$ (i.e. $\mathcal{X} \rightarrow \mathcal{T}$), and the second one is the map from $\mathcal{T}$ to a label space $\mathcal{Y}$ (i.e. $\mathcal{T} \rightarrow \mathcal{Y}$).*

In the learning process of $\mathcal{X} \rightarrow \mathcal{T}$, we aim to train T binary classifiers using $\mathcal{X}$, where $T \in \mathbb{N}^+$ is the number of attributes in $\mathcal{T}$; and in the learning process of $\mathcal{T} \rightarrow \mathcal{Y}$, we use these T binary classifiers to predict class labels in $\mathcal{Y}$. Especially, every instance's label is represented by a 0–1 binary attribute vector, where a vector element is set as 1 if and only if the element's corresponding attribute exists in this instance. Let $L_{\mathcal{X}\rightarrow\mathcal{Y}}$ denote the training loss (*e.g.*, prediction error) of the entire process $\mathcal{X} \rightarrow \mathcal{Y}$. Suppose the upper bound of $L_{\mathcal{X}\rightarrow\mathcal{Y}}$ is $\tau \in (0,1)$. The probability that the training loss does not exceed τ can be represented as:

$$\mathbb{P}(L_{\mathcal{X}\rightarrow\mathcal{Y}} \leq \tau) = 1 - \gamma, \tag{1}$$

where $\gamma \in (0,1)$ is the error probability. Different from traditional learning processes, transfer learning technically focuses on how to use the training instances to learn T binary classifiers instead of a classifier for prediction. Thus, considering the requirement of $L_{\mathcal{X}\rightarrow\mathcal{Y}} \leq \tau$, the training loss of each binary classifier in the process $\mathcal{X} \rightarrow \mathcal{T}$, denoted by $L_{\mathcal{X}\rightarrow\mathcal{T}}$, should not be larger than $\frac{\tau}{T}$. The probability of holding $L_{\mathcal{X}\rightarrow\mathcal{T}} \leq \frac{\tau}{T}$ for any a classifier in $\mathcal{T}$ is represented by:

$$\mathbb{P}(L_{\mathcal{X}\rightarrow\mathcal{T}} \leq \frac{\tau}{T}) = 1 - \delta, \tag{2}$$

where $\delta \in (0,1)$ is the error probability. Accordingly, the probability to ensure $L_{\mathcal{X}\rightarrow\mathcal{T}} \leq \frac{\tau}{T}$ for T classifiers is $(1-\delta)^T$, and the probability to ensure $L_{\mathcal{X}\rightarrow\mathcal{Y}} \leq \tau$ with T classifiers is $\text{BinoCDF}(\tau; T, \frac{\tau}{T})$ that is the binomial cumulative distribution probability. On the other hand, according to PAC learning framework [15], there should be at least N training instances in $\mathcal{X}$ such that Eq. (1) and Eq. (2) can be satisfied, in which N can be computed as follows:

$$N = \frac{T^2}{\tau^2} \lceil \ln(\frac{2}{\delta}) \rceil. \tag{3}$$

From Eq. (3), we have $\delta = \frac{2}{e^{\frac{N\tau^2}{T^2}}}$.

To sum up, with N training instances and T classifiers, the performance of the transfer learning process $\mathcal{X} \rightarrow \mathcal{Y}$ can be defined as the total probability $P(N,T)$ that the training loss does not exceed the upper bound τ, *i.e.*,

$$P(N,T) = (1 - \frac{2}{e^{\frac{N\tau^2}{T^2}}})^T \cdot \text{BinoCDF}(\tau; T, \frac{\tau}{T}) \cdot (1 - \gamma). \tag{4}$$

Generally speaking, there are two major methods to improve the performance of transfer learning. i) **Collection:** collecting more training instances for learning better binary classifiers and/or collecting more attributes for predicting novel labels more accurately. ii) **Collaboration:** collaborative transfer learning aims to enhance the learning performance by sharing collected training instances and attributes among participants.

To thoroughly understand how to improve transfer learning performance efficiently, the intent of this paper is to deeply investigate the following three problems:

- By comparing collecting instances and collecting attributes, which collection approach is more efficient?
- Is collaborative transfer learning efficient?
- By comparing collection with collaboration, which one is more efficient?

In the following, we elaborate on our theoretical analysis on these three problems in Sect. 4, Sect. 5, and Sect. 6 in order.

4 Collecting Instances vs. Collecting Attributes

First of all, for presentation simplicity, we let

$$f(N,T) = (1 - \frac{2}{e^{\frac{N\tau^2}{T^2}}})^T, \tag{5}$$

and

$$g(T) = \text{BinoCDF}(\tau; T, \frac{\tau}{T}). \tag{6}$$

Then, we can rewrite $P(N,T)$ as below,

$$P(N,T) = f(N,T)g(T)(1-\gamma). \tag{7}$$

The benefit of collecting one more instance r_n and the benefit of collecting one more attribute r_t can be computed by Eq. (8) and Eq. (9), respectively.

$$r_n = \Delta P_N(N,T) = P(N,T) - P(N-1,T). \tag{8}$$

$$r_t = \Delta P_T(N,T) = P(N,T) - P(N,T-1). \tag{9}$$

The benefit difference $k(N,T)$ is calculated as:

$$\begin{aligned} k(N,T) = r_t - r_n =& P(N,T) - P(N,T-1) - [P(N,T) - P(N-1,T)] \\ =& P(N-1,T) - P(N,T-1) \\ =& \left[f(N-1,T)g(T) - f(N,T-1)g(T-1)\right](1-\gamma). \end{aligned} \tag{10}$$

We further use Taylor Theorem [27] and Newton's forward interpolation formula [1] to get a bivar linear function $\tilde{k}(N,T)$ to approximate $k(N,T)$. Meanwhile, to guarantee the existence of gradients in the following calculation process, we approximate $k(N,T)$ at the point $(3,3)$, *i.e.*,

$$\tilde{k}(N,T) = k(3,3) + (N-3)\Delta k_N(3,3) + (T-3)\Delta k_T(3,3). \tag{11}$$

From Eq. (10), we have

$$k(3,3) = \left[f(2,3)g(3) - f(3,2)g(2)\right](1-\gamma). \tag{12}$$

According to Eq. (10) and Newton's forward interpolation formula [1], we can obtain the calculation of gradients in the following.

$$\begin{aligned}\Delta k_N(3,3) &= \frac{k(3,3)-k(2,3)}{3-2} = k(3,3)-k(2,3)\\ &= [f(2,3)g(3)-f(3,2)g(2)-f(1,3)g(3)+f(2,2)g(2)](1-\gamma).\end{aligned} \tag{13}$$

$$\begin{aligned}\Delta k_T(3,3) &= \frac{k(3,3)-k(3,2)}{3-2} = k(3,3)-k(3,2)\\ &= [f(2,3)g(3)-f(3,2)g(2)-f(2,2)g(2)+f(3,1)g(1)](1-\gamma).\end{aligned} \tag{14}$$

By substituting Eq. (12), Eq. (13), and Eq. (14), we can rewrite $\tilde{k}(N,T)$ as:

$$\begin{aligned}\tilde{k}(N,T) =& [f(2,3)g(3)-f(3,2)g(2)]\,(1-\gamma)\\ &+(N-3)[f(2,3)g(3)-f(3,2)g(2)-f(1,3)g(3)+f(2,2)g(2)](1-\gamma)\\ &+(T-3)[f(2,3)g(3)-f(3,2)g(2)-f(2,2)g(2)+f(3,1)g(1)](1-\gamma).\end{aligned} \tag{15}$$

If $\tilde{k}(N,T) \leq 0$, collecting one more training instance is more efficient for performance enhancement; otherwise, collecting one more attribute is more efficient. Thus, we can compare $\tilde{k}(N,T)$ with 0 to decide which collection option is more efficient for one party to improve transfer learning performance.

Theorem 1. *Suppose that one party has N training instances and T attributes. Collecting one more training instance is more efficient than collecting one more attribute to improve transfer learning performance, when any one of the following two conditions holds:*

(i) Condition 1: N, T, and τ satisfy Eq. (16) and Eq. (17);
(ii) Condition 2: N, T, and τ satisfy Eq. (18) and Eq. (19).

$$\begin{aligned}N \leq &-(T-3)\frac{f(2,3)g(3)-f(3,2)g(2)-f(2,2)g(2)+f(3,1)g(1)}{f(2,3)g(3)-f(3,2)g(2)-f(1,3)g(3)+f(2,2)g(2)}\\ &-\frac{f(2,3)g(3)-f(3,2)g(2)}{f(2,3)g(3)-f(3,2)g(2)-f(1,3)g(3)+f(2,2)g(2)}+3.\end{aligned} \tag{16}$$

$$f(2,3)g(3)-f(3,2)g(2)-f(1,3)g(3)+f(2,2)g(2)>0. \tag{17}$$

$$\begin{aligned}N \geq &-(T-3)\frac{f(2,3)g(3)-f(3,2)g(2)-f(2,2)g(2)+f(3,1)g(1)}{f(2,3)g(3)-f(3,2)g(2)-f(1,3)g(3)+f(2,2)g(2)}\\ &-\frac{f(2,3)g(3)-f(3,2)g(2)}{f(2,3)g(3)-f(3,2)g(2)-f(1,3)g(3)+f(2,2)g(2)}+3.\end{aligned} \tag{18}$$

$$f(2,3)g(3)-f(3,2)g(2)-f(1,3)g(3)+f(2,2)g(2)<0. \tag{19}$$

Note that $f(2,3) = (1-\frac{2}{e^{\frac{2\tau^2}{9}}})^3$, $f(3,2) = (1-\frac{2}{e^{\frac{3\tau^2}{4}}})^2$, $f(2,2) = (1-\frac{2}{e^{\frac{\tau^2}{2}}})^2$, $f(3,1) = (1-\frac{2}{e^{3\tau^2}})$, $f(1,3) = (1-\frac{2}{e^{\frac{\tau^2}{9}}})^3$ *according to Eq. (5), and* $g(1) = \mathrm{BinoCDF}(\tau; 1, \frac{\tau}{T})$, $g(2) = \mathrm{BinoCDF}(\tau; 2, \frac{\tau}{T})$, $g(3) = \mathrm{BinoCDF}(\tau; 3, \frac{\tau}{T})$ *according to Eq. (6).*

Proof. The requirement of collecting one more instance to be more efficient is $\tilde{k}(N,T) \leq 0$. According to Eq. (15), we have

$$\begin{aligned}&[f(2,3)g(3) - f(3,2)g(2)]\,(1-\gamma)\\&+ (N-3)[f(2,3)g(3) - f(3,2)g(2) - f(1,3)g(3) + f(2,2)g(2)](1-\gamma)\\&+ (T-3)[f(2,3)g(3) - f(3,2)g(2) - f(2,2)g(2) + f(3,1)g(1)](1-\gamma) \leq 0.\end{aligned} \tag{20}$$

When Eq. (20) holds, there are three cases for consideration:
(i) if $f(2,3)g(3) - f(3,2)g(2) - f(1,3)g(3) + f(2,2)g(2) > 0$ (*i.e.*, Eq. (17) is satisfied), then we can get Eq. (16);
(ii) if $f(2,3)g(3) - f(3,2)g(2) - f(1,3)g(3) + f(2,2)g(2) < 0$ (*i.e.*, Eq. (19) is satisfied), then Eq. (18) is obtained;
and (iii) if $f(2,3)g(3) - f(3,2)g(2) - f(1,3)g(3) + f(2,2)g(2) = 0$, this is meaningless for our investigated problem.

Thus, Theorem 1 is proved.

Theorem 2. *Suppose that one party has N training instances and T attributes. Collecting one more attribute is more efficient than collecting one more training instance to enhance transfer learning performance, when any one of the following two conditions holds:*

(i) Condition 1: N, T, and τ satisfy Eq. (21) and Eq. (22);
(ii) Condition 2: N, T, and τ satisfy Eq. (23) and Eq. (24).

$$\begin{aligned}N > &-(T-3)\frac{f(2,3)g(3) - f(3,2)g(2) - f(2,2)g(2) + f(3,1)g(1)}{f(2,3)g(3) - f(3,2)g(2) - f(1,3)g(3) + f(2,2)g(2)}\\&-\frac{f(2,3)g(3) - f(3,2)g(2)}{f(2,3)g(3) - f(3,2)g(2) - f(1,3)g(3) + f(2,2)g(2)} + 3.\end{aligned} \tag{21}$$

$$f(2,3)g(3) - f(3,2)g(2) - f(1,3)g(3) + f(2,2)g(2) > 0. \tag{22}$$

$$\begin{aligned}N < &-(T-3)\frac{f(2,3)g(3) - f(3,2)g(2) - f(2,2)g(2) + f(3,1)g(1)}{f(2,3)g(3) - f(3,2)g(2) - f(1,3)g(3) + f(2,2)g(2)}\\&-\frac{f(2,3)g(3) - f(3,2)g(2)}{f(2,3)g(3) - f(3,2)g(2) - f(1,3)g(3) + f(2,2)g(2)} + 3.\end{aligned} \tag{23}$$

$$f(2,3)g(3) - f(3,2)g(2) - f(1,3)g(3) + f(2,2)g(2) < 0. \tag{24}$$

Note that $f(2,3) = (1 - \frac{2}{e^{\frac{2\tau^2}{9}}})^3$, $f(3,2) = (1 - \frac{2}{e^{\frac{3\tau^2}{4}}})^2$, $f(2,2) = (1 - \frac{2}{e^{\frac{\tau^2}{2}}})^2$, $f(3,1) = (1 - \frac{2}{e^{3\tau^2}})$, $f(1,3) = (1 - \frac{2}{e^{\frac{\tau^2}{9}}})^3$ *according to Eq. (5), and* $g(1) = \mathrm{BinoCDF}(\tau; 1, \frac{\tau}{T})$, $g(2) = \mathrm{BinoCDF}(\tau; 2, \frac{\tau}{T})$, $g(3) = \mathrm{BinoCDF}(\tau; 3, \frac{\tau}{T})$ *according to Eq. (6).*

Proof. Collecting one more attribute is a more efficient option, which is equivalent to $\tilde{k}(N,T) > 0$. From Eq. (15), we have

$$[f(2,3)g(3) - f(3,2)g(2)](1-\gamma)$$
$$+ (N-3)[f(2,3)g(3) - f(3,2)g(2) - f(1,3)g(3) + f(2,2)g(2)](1-\gamma)$$
$$+ (T-3)[f(2,3)g(3) - f(3,2)g(2) - f(2,2)g(2) + f(3,1)g(1)](1-\gamma) > 0. \quad (25)$$

There are three cases for solving Eq. (25):
(i) $f(2,3)g(3) - f(3,2)g(2) - f(1,3)g(3) + f(2,2)g(2) > 0$ (*i.e.*, Eq. (22) is met), then Eq. (21) is obtained;
(ii) $f(2,3)g(3) - f(3,2)g(2) - f(1,3)g(3) + f(2,2)g(2) < 0$ (*i.e.*, Eq. (24) is satisfied), then we can gain Eq. (23);
and (iii) $f(2,3)g(3) - f(3,2)g(2) - f(1,3)g(3) + f(2,2)g(2) = 0$, this is a meaningless case for our investigated problem.

Therefore, Theorem 2 is proved.

5 Whether to Collaboration

Similar to the analysis in Sect. 4, we use Taylor Theorem [27] and Newton's forward interpolation formula [1] to obtain a bivar linear function $\tilde{P}(N,T)$ to approximate $P(N,T)$. Additionally, to ensure the existence of gradients for $\tilde{P}(N,T)$, we approximate $P(N,T)$ at $(3,3)$ as follows,

$$\tilde{P}(N,T) = P(3,3) + (N-3)\Delta P_N(3,3) + (T-3)\Delta P_T(3,3). \quad (26)$$

Based on Eq. (7) and Newton's forward interpolation formula [1], there exist

$$\begin{aligned}\Delta P_N(3,3) &= \frac{P(3,3)-P(2,3)}{3-2} = P(3,3) - P(2,3) \\ &= f(3,3)g(3)(1-\gamma) - f(2,3)g(3)(1-\gamma),\end{aligned} \quad (27)$$

and

$$\begin{aligned}\Delta P_T(3,3) &= \frac{P(3,3)-P(3,2)}{3-2} = P(3,3) - P(3,2) \\ &= f(3,3)g(3)(1-\gamma) - f(3,2)g(2)(1-\gamma).\end{aligned} \quad (28)$$

With the substitution of Eq. (27) and Eq. (28), $\tilde{P}(N,T)$ can be rewritten as:

$$\begin{aligned}\tilde{P}(N,T) = &P(3,3) + (N-3)[f(3,3)g(3)(1-\gamma) - f(2,3)g(3)(1-\gamma)] \\ &+ (T-3)[f(3,3)g(3)(1-\gamma) - f(3,2)g(2)(1-\gamma)].\end{aligned} \quad (29)$$

In this section, our goal is to understand whether collaborative transfer learning is efficient for one party in a two-party (including party **A** and party **B**) collaboration scenario. Without loss of generality, party **A** has N_1 training instances and T_1 attributes, and the party **B** has N_2 training instances and T_2 attributes. If the party **A** does not collaborate with party **B**, according to Eq. (29), party **A** can obtain transfer learning performance $\tilde{P}(N_1,T_1)$ as below,

$$\begin{aligned}\tilde{P}(N_1,T_1) = &P(3,3) + (N_1-3)[f(3,3)g(3)(1-\gamma) - f(2,3)g(3)(1-\gamma)] \\ &+ (T_1-3)[f(3,3)g(3)(1-\gamma) - f(3,2)g(2)(1-\gamma)].\end{aligned} \quad (30)$$

If party **A** collaborates with party **B**, according to Eq. (29), party **A**'s transfer learning performance $\tilde{P}(N_1 + N_2, T_1 \cup T_2)$ becomes

$$\begin{aligned}\tilde{P}(N_1 + N_2, T_1 \cup T_2) = P(3,3) \\ &+ (N_1 + N_2 - 3)[f(3,3)g(3)(1-\gamma) - f(2,3)g(3)(1-\gamma)] \\ &+ (T_1 \cup T_2 - 3)[f(3,3)g(3)(1-\gamma) - f(3,2)g(2)(1-\gamma)].\end{aligned} \tag{31}$$

If $\tilde{P}(N_1 + N_2, T_1 \cup T_2) \geq \tilde{P}(N_1, T_1)$, collaboration is efficient for party **A** to improve transfer learning performance; otherwise, collaboration is not efficient for party **A**. In order words, we should compare $\tilde{P}(N_1, T_1)$ with $\tilde{P}(N_1 + N_2, T_1 \cup T_2)$ in order to judge whether collaboration is efficient for party **A** to enhance transfer learning performance.

Theorem 3. *In a two-party (party* **A** *and party* **B***) collaboration scenario, party* **A** *has* N_1 *training instances and* T_1 *attributes, and party* **B** *has* N_2 *training instances and* T_2 *attributes. Collaboration is more efficient for party* **A** *to improve transfer learning performance, when* T_1*,* N_2*,* T_2*, and* τ *satisfy Eq.* (32).

$$N_2 \geq -(T_1 \cup T_2 - T_1)\frac{f(3,3)g(3) - f(3,2)g(2)}{f(3,3)g(3) - f(2,3)g(3)}, \tag{32}$$

where $f(3,3) = (1 - \frac{2}{e^{\frac{\tau^2}{3}}})^3$, $f(3,2) = (1 - \frac{2}{e^{\frac{3\tau^2}{4}}})^2$, $f(2,3) = (1 - \frac{2}{e^{\frac{2\tau^2}{9}}})^3$ *according to Eq.* (5)*, and* $g(2) = \text{BinoCDF}(\tau; 2, \frac{\tau}{T_1})$, $g(3) = \text{BinoCDF}(\tau; 3, \frac{\tau}{T_1})$ *according to Eq.* (6).

Proof. Collaboration is more efficient for party **A**, which means $\tilde{P}(N_1 + N_2, T_1 \cup T_2) \geq \tilde{P}(N_1, T_1)$. By substituting Eq. (30) and Eq. (31), we gain Eq. (33).

$$\begin{aligned}&P(3,3) + (N_1 + N_2 - 3)[f(3,3)g(3)(1-\gamma) - f(2,3)g(3)(1-\gamma)] \\ &+ (T_1 \cup T_2 - 3)[f(3,3)g(3)(1-\gamma) - f(3,2)g(2)(1-\gamma)] \geq \\ &P(3,3) + (N_1 - 3)[f(3,3)g(3)(1-\gamma) - f(2,3)g(3)(1-\gamma)] \\ &+ (T_1 - 3)[f(3,3)g(3)(1-\gamma) - f(3,2)g(2)(1-\gamma)].\end{aligned} \tag{33}$$

Since $f(3,3) > f(2,3)$ and $g(3) > 0$, we have $[f(3,3)g(3) - f(2,3)g(3)] > 0$. Then, by solving Eq. (33), Eq. (32) is obtained. Thus, Theorem 3 is proved.

Theorem 4. *In a two-party (party* **A** *and party* **B***) collaboration scenario, party* **A** *has* N_1 *training instances and* T_1 *attributes, and party* **B** *has* N_2 *training instances and* T_2 *attributes. Collaboration cannot enhance transfer learning performance for party* **A** *if* T_1*,* N_2*,* T_2*, and* τ *satisfy Eq.* (34).

$$N_2 < -(T_1 \cup T_2 - T_1)\frac{f(3,3)g(3) - f(3,2)g(2)}{f(3,3)g(3) - f(2,3)g(3)}, \tag{34}$$

where $f(3,3) = (1 - \frac{2}{e^{\frac{\tau^2}{3}}})^3$, $f(3,2) = (1 - \frac{2}{e^{\frac{3\tau^2}{4}}})^2$, $f(2,3) = (1 - \frac{2}{e^{\frac{2\tau^2}{9}}})^3$ *according to Eq.* (5)*, and* $g(2) = \text{BinoCDF}(\tau; 2, \frac{\tau}{T_1})$, $g(3) = \text{BinoCDF}(\tau; 3, \frac{\tau}{T_1})$ *according to Eq.* (6).

Proof. The failure of collaboration to improve performance indicates $\tilde{P}(N_1 + N_2, T_1 \cup T_2) < \tilde{P}(N_1, T_1)$. Accordingly, Eq. (35) can be gained using Eq. (30) and Eq. (31).

$$\begin{aligned} &P(3,3) + (N_1 + N_2 - 3)[f(3,3)g(3)(1-\gamma) - f(2,3)g(3)(1-\gamma)] \\ &+ (T_1 \cup T_2 - 3)[f(3,3)g(3)(1-\gamma) - f(3,2)g(2)(1-\gamma)] < \\ &P(3,3) + (N_1 - 3)[f(3,3)g(3)(1-\gamma) - f(2,3)g(3)(1-\gamma)] \\ &+ (T_1 - 3)[f(3,3)g(3)(1-\gamma) - f(3,2)g(2)(1-\gamma)]. \end{aligned} \tag{35}$$

In Eq. (35), $[f(3,3)g(3) - f(2,3)g(3)] > 0$ because $f(3,3) > f(2,3)$ and $g(3) > 0$. Then, the solution to Eq. (35) is Eq. (34). Thus, Theorem 4 is proved.

From Theorem 3 and Theorem 4, any party in a two-party collaboration scenario can make a judgement whether collaboration is an efficient choice.

6 Collection vs. Collaboration

In this section, we further compare the efficiency of collection and collaboration. Considering a two-party (party **A** and party **B**) collaboration scenario, party **A** has N_1 training instances and T_1 attributes, and party **B** has N_2 training instances and T_2 attributes. Via collaboration, party **A** can increase transfer learning performance to $\tilde{P}(N_1 + N_2, T_1 \cup T_2)$. On the other hand, we assume that party **A** can increase N_1 to N_{NT} and increase T_1 to T_{NT} through collection. Correspondingly, party **A** can enhance transfer learning performance to $\tilde{P}(N_{NT}, T_{NT})$ as shown in Eq. (36).

$$\begin{aligned} \tilde{P}(N_{NT}, T_{NT}) = &P(3,3) + (N_{NT} - 3)[f(3,3)g(3)(1-\gamma) - f(2,3)g(3)(1-\gamma)] \\ &+ (T_{NT} - 3)[f(3,3)g(3)(1-\gamma) - f(3,2)g(2)(1-\gamma)]. \end{aligned} \tag{36}$$

If $\tilde{P}(N_1 + N_2, T_1 \cup T_2) \geq \tilde{P}(N_{NT}, T_{NT})$, collaboration is more efficient for party **A** to improve transfer learning performance; otherwise, collection is more efficient for party **A**. Thus, we need to compare $\tilde{P}(N_{NT}, T_{NT})$ and $\tilde{P}(N_1 + N_2, T_1 \cup T_2)$ so as to help party **A** compare the efficiency of collection and collaboration for performance improvement.

Theorem 5. *In a two-party (party **A** and party **B**) collaboration scenario, party **A** has N_1 training instances and T_1 attributes, and party **B** has N_2 training instances and T_2 attributes. Assume that party **A** can increase N_1 to N_{NT} and increase T_1 to T_{NT} via collection. Collaboration is more efficient than collection for party **A** to enhance transfer learning performance, when N_1, N_2, N_{NT}, T_1, T_2, T_{NT}, and τ satisfy Eq. (37).*

$$N_1 + N_2 - N_{NT} \geq -(T_1 \cup T_2 - T_{NT}) \frac{f(3,3)g(3) - f(3,2)g(2)}{f(3,3)g(3) - f(2,3)g(3)}, \tag{37}$$

where $f(3,3) = (1 - \frac{2}{e^{\frac{\tau^2}{3}}})^3$, $f(3,2) = (1 - \frac{2}{e^{\frac{3\tau^2}{4}}})^2$, *and* $f(2,3) = (1 - \frac{2}{e^{\frac{2\tau^2}{9}}})^3$ *according to Eq.* (5), *and* $g(2) = \text{BinoCDF}(\tau; 2, \frac{\tau}{T_1})$, $g(3) = \text{BinoCDF}(\tau; 3, \frac{\tau}{T_1})$ *according to Eq.* (6).

Proof. If collaboration is more efficient, we have

$$\tilde{P}(N_1 + N_2, T_1 \cup T_2) \geq \tilde{P}(N_{NT}, T_{NT}),$$

which is can be equivalently expressed by Eq. (38) via substituting Eq. (31) and Eq. (36).

$$\begin{aligned} &P(3,3) + (N_1 + N_2 - 3)[f(3,3)g(3)(1-\gamma) - f(2,3)g(3)(1-\gamma)] \\ &+ (T_1 \cup T_2 - 3)[f(3,3)g(3)(1-\gamma) - f(3,2)g(2)(1-\gamma)] \geq \\ &P(3,3) + (N_{NT} - 3)[f(3,3)g(3)(1-\gamma) - f(2,3)g(3)(1-\gamma)] \\ &+ (T_{NT} - 3)[f(3,3)g(3)(1-\gamma) - f(3,2)g(2)(1-\gamma)]. \end{aligned} \tag{38}$$

Since $f(3,3) > f(2,3)$ and $g(3) > 0$, $[f(3,3)g(3) - f(2,3)g(3)] > 0$. Then, by solving Eq. (38), Eq. (37) is obtained. Therefore, Theorem 5 is proved.

Theorem 6. *In a two-party (party* **A** *and party* **B***) collaboration scenario, party* **A** *has* N_1 *training instances and* T_1 *attributes, and party* **B** *has* N_2 *training instances and* T_2 *attributes. Assume that party* **A** *can increase* N_1 *to* N_{NT} *and increase* T_1 *to* T_{NT} *via collection. Collection is more efficient than collaboration for party* **A** *to enhance transfer learning performance, when* N_1, N_2, N_{NT}, T_1, T_2, T_{NT}, *and* τ *satisfy Eq.* (39).

$$N_1 + N_2 - N_{NT} < -(T_1 \cup T_2 - T_{NT})\frac{f(3,3)g(3) - f(3,2)g(2)}{f(3,3)g(3) - f(2,3)g(3)}, \tag{39}$$

where $f(3,3) = (1 - \frac{2}{e^{\frac{\tau^2}{3}}})^3$, $f(3,2) = (1 - \frac{2}{e^{\frac{3\tau^2}{4}}})^2$, $f(2,3) = (1 - \frac{2}{e^{\frac{2\tau^2}{9}}})^3$ *according to Eq.* (5), *and* $g(2) = \text{BinoCDF}(\tau; 2, \frac{\tau}{T_1})$, $g(3) = \text{BinoCDF}(\tau; 3, \frac{\tau}{T_1})$ *according to Eq.* (6).

Proof. Similar to the analysis in Theorem 5, $\tilde{P}(N_1+N_2, T_1 \cup T_2) < \tilde{P}(N_{NT}, T_{NT})$ is the condition of collection to be more efficient. By substituting Eq. (31) and Eq. (36), we obtain Eq. (40) as follows.

$$\begin{aligned} &P(3,3) + (N_1 + N_2 - 3)[f(3,3)g(3)(1-\gamma) - f(2,3)g(3)(1-\gamma)] \\ &+ (T_1 \cup T_2 - 3)[f(3,3)g(3)(1-\gamma) - f(3,2)g(2)(1-\gamma)] < \\ &P(3,3) + (N_{NT} - 3)[f(3,3)g(3)(1-\gamma) - f(2,3)g(3)(1-\gamma)] \\ &+ (T_{NT} - 3)[f(3,3)g(3)(1-\gamma) - f(3,2)g(2)(1-\gamma)]. \end{aligned} \tag{40}$$

As $f(3,3) > f(2,3)$ and $g(3) > 0$, $[f(3,3)g(3) - f(2,3)g(3)] > 0$. Thus, for Eq. (40), the solution is Eq. (39); that is, Theorem 6 is proved.

Theorem 5 and Theorem 6 demonstrate the conditions of selecting collaboration or collection as an efficient method for a party to improve transfer learning performance in a two-party scenario.

7 Conclusion and Future Work

In this paper, we comprehensively investigate how to improve transfer learning performance more efficiently. To the best of our knowledge, this is the first work to theoretically analyze the methods of improving transfer learning performance. Specifically, Theorem 1 and Theorem 2 are proposed in Sect. 4 to help select a more efficient collection option from two collection ways for promoting transfer learning performance. Besides, Theorem 3 and Theorem 4 are presented in Sect. 5 to help judge whether collaboration is more efficient to enhancing transfer learning performance without considering collection. Moreover, Theorem 5 and Theorem 6 are put forward in Sect. 6 to help judge whether collaboration is still efficient while considering both collection and collaboration. To sum up, our proposed theorems and conclusions provide the thoroughly theoretical decision-making guidance for improving transfer learning performance.

In our future work, we will advance our theoretical analysis for transfer learning performance by taking into account the cost of collection and collaboration.

Acknowledgment. This work was partly supported by the National Science Foundation of U.S. (2118083, 1912753, 1704287, 1741277, 1829674).

References

1. AL-Sammarraie, O.A., Bashir, M.A.: Generalization of Newton's forward interpolation formula. Int. J. Sci. Res. Publ. (2015)
2. Argyriou, A., Evgeniou, T., Pontil, M.: Convex multi-task feature learning. Mach. Learn. **73**, 243–272 (2008). https://doi.org/10.1007/s10994-007-5040-8
3. Blanke, U., Schiele, B.: Remember and transfer what you have learned-recognizing composite activities based on activity spotting. In: International Symposium on Wearable Computers, pp. 1–8. IEEE (2010)
4. Chen, Y., Qin, X., Wang, J., Yu, C., Gao, W.: FedHealth: a federated transfer learning framework for wearable healthcare. IEEE Intell. Syst. **35**(4), 83–93 (2020)
5. Daga, H., Nicholson, P.K., Gavrilovska, A., Lugones, D.: Cartel: a system for collaborative transfer learning at the edge. In: Proceedings of the ACM Symposium on Cloud Computing, pp. 25–37. ACM (2019)
6. Davis, J., Domingos, P.: Deep transfer via second-order Markov logic. In: Proceedings of the 26th Annual International Conference on Machine Learning, pp. 217–224. ACM (2009)
7. Evgeniou, T., Pontil, M.: Regularized multi-task learning. In: Proceedings of the Tenth ACM SIGKDD International Conference on Knowledge Discovery and Data Mining, pp. 109–117. ACM (2004)
8. Farhadi, A., Forsyth, D., White, R.: Transfer learning in sign language. In: 2007 IEEE Conference on Computer Vision and Pattern Recognition, pp. 1–8. IEEE (2007)
9. Global Forecast: Cisco visual networking index: global mobile data traffic forecast. Update 2017–2022 (2019)
10. Hsieh, K., et al.: Gaia: geo-distributed machine learning approaching LAN speeds. In: 14th USENIX Symposium on Networked Systems Design and Implementation, pp. 629–647. USENIX (2017)

11. Jebara, T.: Multi-task feature and kernel selection for SVMs. In: Proceedings of the 21st International Conference on Machine Learning, p. 55. ACM (2004)
12. Jiang, J., Zhai, C.: Instance weighting for domain adaptation in NLP. In: Proceedings of the 45th Annual Meeting of the Association of Computational Linguistics, pp. 264–271. ACL (2007)
13. Ju, C., Gao, D., Mane, R., Tan, B., Liu, Y., Guan, C.: Federated transfer learning for EEG signal classification. In: 42nd Annual International Conference of the IEEE Engineering in Medicine & Biology Society, EMBC 2020, Montreal, QC, Canada, 20–24 July 2020, pp. 3040–3045. IEEE (2020). https://doi.org/10.1109/EMBC44109.2020.9175344
14. Kang, Y., et al.: Neurosurgeon: collaborative intelligence between the cloud and mobile edge. ACM SIGARCH Comput. Archit. News **45**, 615–629 (2017)
15. Kearns, M.J., Vazirani, U.V., Vazirani, U.: An Introduction to Computational Learning Theory. MIT Press, Cambridge (1994)
16. Lawrence, N.D., Platt, J.C.: Learning to learn with the informative vector machine. In: Proceedings of the 21st International Conference on Machine Learning, p. 65. ACM (2004)
17. Lee, S.I., Chatalbashev, V., Vickrey, D., Koller, D.: Learning a meta-level prior for feature relevance from multiple related tasks. In: Proceedings of the 24th International Conference on Machine Learning, pp. 489–496. ACM (2007)
18. Liao, X., Xue, Y., Carin, L.: Logistic regression with an auxiliary data source. In: Proceedings of the 22nd International Conference on Machine Learning, pp. 505–512. ACM (2005)
19. Luo, C., et al.: AIoT bench: towards comprehensive benchmarking mobile and embedded device intelligence. In: Zheng, C., Zhan, J. (eds.) Bench 2018. LNCS, vol. 11459, pp. 31–35. Springer, Cham (2019). https://doi.org/10.1007/978-3-030-32813-9_4
20. Ma, Z., et al.: PMKT: privacy-preserving multi-party knowledge transfer for financial market forecasting. Future Gener. Comput. Syst. **106**, 545–558 (2020)
21. Ma, Z., et al.: PMKT: privacy-preserving multi-party knowledge transfer for financial market forecasting. Future Gener. Comput. Syst. **106**, 545–558 (2020). https://doi.org/10.1016/j.future.2020.01.007
22. Mihalkova, L., Huynh, T., Mooney, R.J.: Mapping and revising Markov logic networks for transfer learning. In: AAAI, vol. 7, pp. 608–614. AAAI (2007)
23. Ogoe, H.A., Visweswaran, S., Lu, X., Gopalakrishnan, V.: Knowledge transfer via classification rules using functional mapping for integrative modeling of gene expression data. BMC Bioinform. **16**, 1–15 (2015). https://doi.org/10.1186/s12859-015-0643-8
24. Olmedilla, D.: Applying machine learning to ads integrity at Facebook. In: Proceedings of the 8th ACM Conference on Web Science, p. 4. ACM (2016)
25. Palatucci, M., Pomerleau, D., Hinton, G.E., Mitchell, T.M.: Zero-shot learning with semantic output codes. In: Advances in Neural Information Processing Systems, pp. 1410–1418. MIT Press (2009)
26. Perlich, C., Dalessandro, B., Raeder, T., Stitelman, O., Provost, F.: Machine learning for targeted display advertising: transfer learning in action. Mach. Learn. **95**, 103–127 (2014). https://doi.org/10.1007/s10994-013-5375-2
27. Rababah, A.: Taylor theorem for planar curves. Proc. Am. Math. Soc. **119**, 803–810 (1993)
28. Raina, R., Battle, A., Lee, H., Packer, B., Ng, A.Y.: Self-taught learning: transfer learning from unlabeled data. In: Proceedings of the 24th International Conference on Machine Learning, pp. 759–766. ACM (2007)

29. Schwaighofer, A., Tresp, V., Yu, K.: Learning gaussian process kernels via hierarchical Bayes. In: Advances in Neural Information Processing Systems, pp. 1209–1216. MIT Press (2005)
30. Sharma, S., Xing, C., Liu, Y., Kang, Y.: Secure and efficient federated transfer learning. In: 2019 IEEE International Conference on Big Data, pp. 2569–2576. IEEE (2019)
31. Torralba, A., Murphy, K.P., Freeman, W.T.: Sharing visual features for multiclass and multiview object detection. IEEE Trans. Pattern Anal. Mach. Intell. **29**, 854–869 (2007)
32. Wang, C., Mahadevan, S.: Manifold alignment using procrustes analysis. In: Proceedings of the 25th International Conference on Machine Learning, pp. 1120–1127. ACM (2008)
33. Wang, W., Zheng, V.W., Yu, H., Miao, C.: A survey of zero-shot learning: settings, methods, and applications. ACM Trans. Intell. Syst. Technol. **10**, 1–37 (2019)
34. Wu, P., Dietterich, T.G.: Improving SVM accuracy by training on auxiliary data sources. In: Proceedings of the 21st International Conference on Machine Learning, p. 110. ACM (2004)
35. Yang, H., He, H., Zhang, W., Cao, X.: FedSteg: a federated transfer learning framework for secure image steganalysis. IEEE Trans. Netw. Sci. Eng. **14**, 78–88 (2018)

On Maximizing the Difference Between an Approximately Submodular Function and a Linear Function Subject to a Matroid Constraint

Yijing Wang[1], Yicheng Xu[2,3(✉)], and Xiaoguang Yang[1]

[1] Academy of Mathematics and Systems Science, Chinese Academy of Sciences, Beijing 100190, People's Republic of China
[2] Shenzhen Institute of Advanced Technology, Chinese Academy of Sciences, Shenzhen 518055, People's Republic of China
yc.xu@siat.ac.cn
[3] Guangxi Key Laboratory of Cryptography and Information Security, Guilin 541004, People's Republic of China

Abstract. In this paper, we investigate the problem of maximizing the difference between an approximately submodular function and a non-negative linear function subject to a matroid constraint. This model has widespread applications in real life, such as the team formation problem in labor market and the assortment optimization in sales market. We provide a bicriteria approximation algorithm with bifactor ratio $(\frac{\gamma}{1+\gamma}, 1)$, where $\gamma \in (0, 1]$ is a parameter to characterize the approximate submodularity. Our result extends Ene's recent work on maximizing the difference between a monotone submodular function and a linear function. Also, a generalized version of the proposed algorithm is capable to deal with huge volume data set.

Keywords: Approximately submodular · Matroid constraint · Bicriteria algorithm · Massive data

1 Introduction

The problem of maximizing the difference of a normalized monotone approximately submodular function and a non-negative linear function plays an important role in team formation problem [12,18] and assortment optimization [1,2,5,16]. In the sales market, the sellers wish to attract customers by displaying goods. There is a counter charge for the exhibition, but also, the profit can be made if the goods are sold. Thus the sellers are aiming to maximize the profits through the products exhibition. Since the display area is limited, it is of significance to select which goods to display. This problem can be characterized as $\max_{S \subseteq G} f(S) - \ell(S)$, where G is the ground set of goods, and $f, \ell : 2^G \rightarrow \mathbb{R}_{\geq 0}$ represent the benefit and cost functions respectively.

D.-Z. Du et al. (Eds.): COCOA 2021, LNCS 13135, pp. 75–85, 2021.
https://doi.org/10.1007/978-3-030-92681-6_7

When function f is strictly submodular, it is easy to find the objective function $d(S) := f(S) - \ell(S)$ may be negative and non-monotone but submodular. Maximizing a submodular set function problem is usually NP-hard, and designing approximation algorithms is a common way to solve it. However, the model that maximizes a potentially negative as well as non-monotone submodular function may do not get a traditional multiplicative factor approximation, even it is inapproximable [9,20]. For this case, it is effective and appropriate to design a bicriteria approximation algorithm [6] provided in Definition 2 to measure the quality of a solution set.

Sviridenko et al. [22] propose a $(1-1/e, 1)$-bicriteria approximation algorithm for the non-negative monotone submodular function minus non-negative linear function model subject to a matroid constraint. In their work, they first modified the model $\max_{S\in\mathcal{F}} f(S) - \ell(S)$ to $\max_{S\subseteq G}\{f(S) : c(S) \leq B, S \in \mathcal{F}\}$, where $\mathcal{F}$ is a family of independent sets, and B is a knapsack constraint which is constructed utilizing the value of $\ell(O^*)$. Based on the continuous greedy technique, they performed a variant greedy algorithm by guessing $\ell(O^*)$'s value approximately. Unfortunately, guessing the value of $\ell(O^*)$ attracted a amount of time complexity. Feldman [10] improves the time complexity by introducing a time weight vector to modify the objective function in the work of Sviridenko et al. further, and keeps the same approximation ratio as Sviridenko et al.

Different from the methods of Sviridenko et al. and Feldman, Ene [8] proposes a simple but standard greedy algorithm to study the model with a matroid constraint and obtains the bicriteria approximation ratio $(1/2, 1)$. Besides the matroid constraints, Nikolakaki et al. [19] also investigate the model with cardinality constraints under streaming setting, and acquire a $((3-\sqrt{5})/2, 1)$-bicriteria approximation algorithm. Without any constraints, they design an online approximation algorithm with bicriteria ratio $(1/2, 1)$.

As approximately submodular functions arise in numerous applications, such as influence maximization in social networks and interpretation of deep neural networks as well as high dimensional sparse regression in machine learning, more and more scholars focus on the research of approximately submodularity [3,11,13,15]. For the model a non-negative monotone approximately submodular function minus a non-negative modular function under cardinality constraints, Harshaw et al. [14] provide a $(1-1/e^{\gamma}-\epsilon, 1)$-bicriteria approximation algorithm, where $\gamma \in (0,1]$ is a parameter to characterize the approximately submodularity and $\epsilon > 0$. Instead of the greedy techniques in the above results, Qian [21] takes a multi-objective evolutionary method to design algorithm, and gets a bicriteria ratio $(1-1/e^{\gamma}, 1)$ for the problem of maximizing the difference of a monotone γ-weakly submodular and a non-negative modular function with a cardinality constraint.

Based on the above models and results, we extend the submodular functions to approximately submodular one, and study the model with matroid constraints. Firstly we provide a simple bicriteria algorithm on a given ground set with bifactor approximation ratio $(\gamma/(1+\gamma), 1)$, where $\gamma \in (0,1]$. When function f is submodular, our result coincides with Ene's recent work [8]. Faced with massive dataset, we improve the bicriteria algorithm to a general one which is applicable to any form datasets whether they are given finitely or presented dynamically. The bicriteria ratio is $(\frac{\beta-\alpha}{\beta}, \frac{\beta-\alpha}{\alpha\gamma})$, where $0 < \gamma \leq 1$ and parameters $0 < \alpha \leq \beta$ are used to balance the gain and cost.

The remainder of this work is structured as follows. In Sect. 2, we introduce some preliminaries including the definitions of approximately submodular function and bicriteria approximation algorithm. We describe a bicriteria algorithm as well as its corresponding analysis in Sect. 3. In Sect. 4, we give a general and novel approximation algorithm to deal with online and streaming data. The conclusion of this work is given in Sect. 5.

2 Preliminaries

The studied problem can be formulated as $\max_{S\subseteq G}\{f(S)-\ell(S) : S \in \mathcal{F}\}$, where function f is a non-negative normalized monotone approximately submodular function and ℓ is a non-negative linear function.

For a ground set $G = \{e_1, e_2, \ldots, e_n\}$ and a set function defined on G as $g : 2^G \rightarrow \mathbb{R}$, we call function g non-negative if it satisfies $g(S) \geq 0$ for any $S \subseteq G$; g is normalized if it satisfies $g(\emptyset) = 0$; g is monotone if it satisfies $g(S) \leq g(T)$ for any subsets $S \subseteq T \subseteq G$; g is linear if it satisfies $g(S) = \sum_{e\in S} g(e)$; and g is submodular if it satisfies $g(e|S) \leq g(e|T)$, for any $T \subseteq S \subset G$ and $e \in G \setminus S$, where $g(e|S) := g(S \cup \{e\}) - g(S)$ characterizes the marginal gain when adding e into S. Combining with the propositions of Lehmann et al. [17] and the definition in [15], the concept of approximately submodular is given as Definition 1.

Definition 1. *The approximately submodularity ratio of a non-negative normalized monotone set function $g : 2^G \rightarrow \mathbb{R}$ is the largest value $\gamma \in (0, 1]$ such that for any subsets $T \subseteq S \subset G$, element $e \in G\setminus S$, there are $\gamma g(e|S) \leq g(e|T)$. Function $g : 2^G \rightarrow \mathbb{R}$ is called γ-approximately submodular function or approximately submodular function. When $\gamma = 1$, function g is strictly submodular.*

Given the properties of functions f and ℓ, it is difficult to get a traditional multiplicative factor approximation for the studied model. The bicriteria approximation ratio is a natural and feasible way to estimate the performance of a given algorithm, which is defined as the follows.

Definition 2. *An algorithm is called a (σ, ρ)-approximation algorithm if it outputs a solution S such that $f(S)-\ell(S) \geq \sigma\cdot f(O^*)-\rho\cdot\ell(O^*)$ for $\max\{f(S)-\ell(S) : S \subseteq G\}$, where $0 < \sigma \leq 1$, $\rho \geq 0$ and O^* is an optimal solution to the original problem.*

In the model, our goal is to get a subset S subject to the matroid constraint $\mathcal{M} = (G, \mathcal{F})$ to maximize the objective value $d(S) := f(S) - \ell(S)$. By means of the introduction of Edmonds [7], we explain the definition of matroid $\mathcal{M} = (G, \mathcal{F})$ as follows.

Definition 3. *The pair* $(G, \mathcal{F})$ *is a matroid if the ground set* G *and its subset family* $\mathcal{F}$ *satisfies the following requirements simultaneously:*

- *nonempty:* $\emptyset \in \mathcal{F}$*;*
- *hereditary property: for any subsets* $T \in \mathcal{F}$*,* $S \subseteq T$*, there is* $S \in \mathcal{F}$*;*
- *augmentation property: for any subsets* $S, T \in \mathcal{F}$*, if* $|S| \leq |T|$*, then there is an element* $e \in T \setminus S$*, such that* $S \cup \{e\} \in \mathcal{F}$*.*

If $\mathcal{M} = (G, \mathcal{F})$ is a matroid, we call $\mathcal{F}$ an independent set family, and subset S is independent if $S \in \mathcal{F}$. In this paper, we assume there are two oracles which can be considered as two black boxes, where one is used to calculate the function value of a given subset S, and the other is to judge whether S is independent. We intend to minimize the number of oracle calls in our algorithm.

3 A Bicriteria Algorithm

In this section, we provide a bicriteria approximation algorithm accompany with its analysis for the model $\max_{S \subseteq G}\{f(S) - \ell(S) : S \in \mathcal{F}\}$, where f is γ-approximately submodular and ℓ is non-negative linear. Easy to observe that the objective $d(S) := f(S) - \ell(S)$ is also approximately submodular but not necessarily be nonnegative and monotone. Towards this end, we first construct a intermediary function $\hat{d}(S) := f(S) - (1+\gamma)\ell(S)$, where $\gamma \in (0, 1]$ is the submodularity ratio of function f. Then we iteratively select an element e whose marginal gain is maximal w.r.t. the intermediary function $\hat{d}(S)$. If element e brings a non-negative gain to the original target function $d(S)$ on the current solution S, we choose it, otherwise we give up selecting element e. The detailed presentation is shown in Algorithm 1.

Algorithm 1. A bicriteria algorithm for $f - \ell$

1: **Input:**
   Give a ground set $G = \{e_1, e_2, \ldots, e_n\}$, a γ-approximately submodular function f, a non-negative linear function ℓ, matroid $\mathcal{M} = (G, \mathcal{F})$.
2: **Output:**
   A subset $S \in \mathcal{F}$.
3: **Process:**
4: Initially set $S := \emptyset$, $R := G$
5: For $i = 1, 2, 3, \ldots$
6:   select $e_i = \arg\max_{e \in R: S \cup \{e\} \in \mathcal{F}} f(e|S) - (1+\gamma)\ell(e)$
7:   if $f(e_i|S) - \ell(e_i) < 0$
8:     update $R := R \setminus \{e_i\}$
9:   else
10:     update $S := S \cup \{e_i\}$
11:   remove every element e from R such that $S \cup \{e\} \notin \mathcal{F}$
12: Return S

Before analyzing the performance guarantee of Algorithm 1, we introduce a lemma with respect to the property of two independent subsets in matroid

$\mathcal{M} = (G, \mathcal{F})$, which is a natural extension of the result provided by Buchbinder et al. [4].

Lemma 4. *Let U and V be two independent subsets in matroid $\mathcal{M} = (G, \mathcal{F})$ such that $|U| = |V|$. There is a bijection $b : U \backslash V \to V \backslash U$ such that $V \backslash b(e) \cup \{e\} \in \mathcal{F}$ for every element $e \in U \setminus V$ and every $e \in V \setminus U$.*

Note that Lemma 4 is applicable to two bases of matroid $\mathcal{M} = (G, \mathcal{F})$, of which the proof is provided in [4]. Based on Lemma 4, we prove that Algorithm 1 outputs a solution set S such that $f(S) - \ell(S) \geq \frac{\gamma}{1+\gamma} f(O^*) - \ell(O^*)$, when the solution set S and optimal set O^* have the same cardinality. The detail is shown in the proof to Lemma 5.

Lemma 5. *If S satisfies $|S| = |O^*|$, then Algorithm 1 is a $(\frac{\gamma}{1+\gamma}, 1)$-approximation, where O^* is an optimal solution of the original model.*

Proof. Assume $|S| = |O^*| = m$. Let $S_i = \{e_1, e_2, \ldots, e_i\}$ represent the solution after i-th iteration. Denote $\{o_1, o_2, \ldots, o_{|\bar{O}|}\}$ as the difference set $\bar{O} = O^* \setminus S$, where the elements are arbitrarily ordered. Denote $\bar{O}_i = \{o_1, o_2, \ldots, o_i\}$ with $i \leq |O^* \setminus S|$.

Observe from Lemma 4 that there is a bijection $b : O^* \to S$ such that $S \setminus b(o) \cup \{o\}$ is independent for each elements $o \in O^*$. The detailed mapping between S and O^* is matched in the following way: $b(o) = e$ for each element $o \in O^* \backslash S$ and $b(o) = o$ for each element $o \in O^* \cap S$. Since $S = \{e_1, e_2, \ldots, e_m\}$, where the elements are ordered as they are added into S by the algorithm. Let $o_i = b(e_i)$ for all $i \in [m]$, and the optimal solution set O^* is ordered in $o_1, o_2, \ldots, o_m$. By the selection rule in Algorithm 1, we know the following inequality holds: $\hat{d}(e_i|S_{i-1}) \geq \hat{d}(o_i|S_{i-1})$, which is explained as follows.

In the bijection, if $o_i = e_i$, the above inequality is nature. If $o_i \neq e_i$, we know $S \setminus \{e_i\} \cup \{o_i\} \in \mathcal{F}$ by the mapping in Lemma 4. Since $S_{i-1} \subseteq S \setminus \{e_i\}$, there is $S_{i-1} \cup \{o_i\} \in \mathcal{F}$ by the hereditary property. By the selection rule in Algorithm 1, we choose the element e_i to add into subset S_{i-1}, whose marginal gain is maximal in iteration i-th. Then we get element o_i is a candidate for e_i, and the inequality $\hat{d}(e_i|S_{i-1}) \geq \hat{d}(o_i|S_{i-1})$ holds.

Thus, for any $i \in [m]$, it satisfies that

$$\begin{aligned} & f(S_{i-1} \cup \{e_i\}) - f(S_{i-1}) - (1+\gamma)\ell(e_i) \\ \geq & f(S_{i-1} \cup \{o_i\}) - f(S_{i-1}) - (1+\gamma)\ell(o_i). \end{aligned} \tag{1}$$

Since function f is approximately submodular, then $\gamma f(o_i|S \cup O_{i-1}) \leq f(o_i|S_{i-1})$, i.e.,

$$f(S_{i-1} \cup \{o_i\}) - f(S_{i-1}) \geq \gamma f(S \cup O_{i-1} \cup \{o_i\}) - \gamma f(S \cup O_{i-1}). \tag{2}$$

Summing up all indexes $i \in [m]$ and combining with Inequality (2), the left-hand side of Inequality (1) can be restated as:

$$
\begin{aligned}
LHS &= \sum_{i\in m}(f(S_i) - f(S_{i-1})) - (1+\gamma)\sum_{i\in m}\ell(e_i) \\
&= f(S) - (1+\gamma)\ell(S).
\end{aligned} \tag{3}
$$

Similarly, based on the monotonicity of function f, the right-hand side of Inequality (1) can be reformulated as:

$$
\begin{aligned}
RHS &= \sum_{i\in m}(f(S_{i-1}\cup\{o_i\}) - f(S_{i-1}) - (1+\gamma)\ell(o_i)) \\
&\geq \gamma\sum_{i\in m}(f(S\cup O_i) - f(S\cup O_{i-1})) - (1+\gamma)\sum_{i\in m}\ell(o_i) \\
&= \gamma(f(S\cup O^*) - f(S)) - (1+\gamma)\ell(O^*) \\
&\geq \gamma f(O^*) - \gamma f(S) - (1+\gamma)\ell(O^*).
\end{aligned} \tag{4}
$$

Rearranging Inequalities (1)–(4), we obtain the final solution set S satisfies

$$
f(S) - (1+\gamma)\ell(S) \geq \gamma f(O^*) - \gamma f(S) - (1+\gamma)\ell(O^*).
$$

Therefore, when solution set S returned by Algorithm 1 has the same cardinality as an optimal solution, it can do that $f(S) - \ell(S) \geq \frac{\gamma}{1+\gamma}f(O^*) - \ell(O^*)$, completing the proof. □

Therefore, we obtain the following conclusion.

Theorem 6. *Algorithm 1 outputs a solution S with $f(S) - \ell(S) \geq \frac{\gamma}{1+\gamma}f(O^*) - \ell(O^*)$, where $\gamma \in (0,1]$.*

Proof. Assume set $O^* = \{o_1, o_2, \dots, o_p\}$ is an optimal solution to the problem, where $p = |O^*|$, and $S = \{e_1, e_2, \dots, e_q\}$ is the solution returned by Algorithm 1, where $q = |S|$. Let $\bar{O} = O^* \setminus S$ and the set $\bar{O} = \{o_1, o_2, \dots, o_{|\bar{O}|}\}$ is ordered arbitrarily. Denote $S_i = \{e_1, e_2, \dots, e_i\}$ is the subset returned after iteration i-th and $\bar{O}_i = \{o_1, o_2, \dots, o_i\}, i \leq |O^* \setminus S|$ correspondingly. We expand the analysis of Theorem 6 from the perspective of the cardinalities of S and O^*. There are three relationships for p and q.

- Case 1: $p = q$.
  Rely on Lemma 5, we can directly find the quality of solution S such that $f(S) - \ell(S) \geq \frac{\gamma}{1+\gamma}f(O^*) - \ell(O^*)$ in this case, where $\gamma \in (0,1]$.
- Case 2: $p > q$.
  Utilizing the augmentation property in matroid $\mathcal{M} = (G, \mathcal{F})$, we know there is an element $o \in \bar{O} = O^* \setminus S$ satisfying $S \cup \{o\} \in \mathcal{F}$. As element o is not chosen by the algorithm, it must guarantee $f(o|S) - \ell(o) = d(o|S) < 0$, that is $d(S \cup \{o\}) < d(S)$. Repeat this augmentation until we get a set S' such that $|S'| = |O^*|$. By the argument process, we know $d(S) > d(S')$.

Combing the result of Case 1, it is easy to find the augmentation set S' satisfies $f(S') - \ell(S') \geq \frac{\gamma}{1+\gamma} f(O^*) - \ell(O^*)$, i.e., $d(S') = f(S') - \ell(S') \geq \frac{\gamma}{1+\gamma} f(O^*) - \ell(O^*)$. Since $d(S) > d(S')$, there is

$$\begin{aligned} d(S) = f(S) - \ell(S) > d(S') &= f(S') - \ell(S') \\ &\geq \frac{\gamma}{1+\gamma} f(O^*) - \ell(O^*). \end{aligned}$$

Thus the output set S guarantees $f(S) - \ell(S) \geq \frac{\gamma}{1+\gamma} f(O^*) - \ell(O^*)$ when $p = |O^*| > |S| = q$.

- Case 3: $p < q$.
Denote set $S' = \{e_1, e_2, \ldots, e_p\} \subset S$, which is consists of the first p elements added into S. Because the elements $e_{p+1}, e_{p+2}, \ldots, e_q$ are added into set S' sequentially, there are $d(S) \geq d(S')$ by the selection rule in Algorithm 1.
Combing the result of Case 1, we obtain $f(S') - \ell(S') \geq \frac{\gamma}{1+\gamma} f(O^*) - \ell(O^*)$, i.e., $d(S') = f(S') - \ell(S') \geq \frac{\gamma}{1+\gamma} f(O^*) - \ell(O^*)$. Since $d(S) \geq d(S')$, there is

$$\begin{aligned} d(S) = f(S) - \ell(S) \geq d(S') &= f(S') - \ell(S') \\ &\geq \frac{\gamma}{1+\gamma} f(O^*) - \ell(O^*). \end{aligned}$$

Thus the output set S guarantees $f(S) - \ell(S) \geq \frac{\gamma}{1+\gamma} f(O^*) - \ell(O^*)$ when $p = |O^*| < |S| = q$.

Combining the above three cases, we get the solution set S output by Algorithm 1 satisfying $f(S) - \ell(S) \geq \frac{\gamma}{1+\gamma} f(O^*) - \ell(O^*)$. □

4 An Extended Bicriteria Algorithm for Hiigh Volume Data

As shown in Sect. 3, we need to know the complete information of data when running Algorithm 1. So it does not work for huge volume dataset. Accompanied by the information age, a huge volume data is being generated every second. It is necessary to design an effective algorithm that is suitable for massive data. For massive data sets, we need online or streaming algorithms. Thus we extend Algorithm 1 to a more general one which is applicable to any volume dataset.

Based on a similar idea, we first construct a intermediary function $\hat{d}(S) := \alpha \cdot f(S) - \beta \cdot \ell(S)$, where parameters $0 < \alpha \leq \beta$ are used to balance the gain and cost. With the advent of dataset, we iteratively select the single element e that takes positive marginal gain to the intermediary function. The detailed presentation is shown in Algorithm 2.

Theorem 7. *Algorithm 2 outputs a solution S with $f(S) - \ell(S) \geq \frac{\beta-\alpha}{\beta} f(O^*) - \frac{\beta-\alpha}{\alpha\gamma} \ell(O^*)$, where $\gamma \in (0, 1]$ and $0 < \alpha \leq \beta$.*

Algorithm 2. A general bicriteria algorithm for $f - \ell$

1: **Input:**
Give a ground set $G = \{e_1, e_2, \dots, e_n\}$, a γ-approximately submodular function f, a non-negative linear function ℓ, matroid $\mathcal{M} = (G, \mathcal{F})$, parameters $0 < \alpha \leq \beta$.
2: **Output:**
A subset $S \in \mathcal{F}$.
3: **Process:**
4: Initially set $S := \emptyset$
5: For each arriving element e
6: if $S \cup \{e\} \in \mathcal{F}$ and $\alpha \cdot f(e|S) - \beta \cdot \ell(e) > 0$
7: set $S := S \cup \{e\}$
8: Return S

Proof. Let $p = |O^*|$ and $q = |S|$. Denote $S_i = \{e_1, e_2, \dots, e_i\}$ as the output set at i-th iteration. Assume that it is ordered as the sequence that the elements are selected by Algorithm 2. We start with the proof of the upper bound. For each element $e_i \in S$, it holds the following inequality

$$\alpha \cdot f(e_i|S_{i-1}) - \beta \cdot \ell(e_i) > 0.$$

Summing up all elements in S, we obtain

$$\sum_{e_i \in S} \alpha \cdot f(e_i|S_{i-1}) - \sum_{e_i \in S} \beta \cdot \ell(e_i) > 0.$$

Rearranging the inequality, then we have $\ell(S) \leq \frac{\alpha}{\beta} f(S)$.

For the lower bound, according to the selection rule in Algorithm 2, for each element $o_i \in O^* \setminus S$, since o_i is not added into set S_{i-1}, it must perform that

$$\alpha \cdot f(o_i|S_{i-1}) - \beta \cdot \ell(o_i) \leq 0.$$

Rewriting the inequality, we obtain

$$\alpha \cdot f(o_i|S_{i-1}) \leq \beta \cdot \ell(o_i).$$

Since f is approximately submodular, then

$$f(o_i|S_{i-1}) \geq \gamma \cdot f(o_i|S \cup O_{i-1}).$$

Therefore,

$$\alpha \cdot \gamma \cdot f(o_i|S \cup O_{i-1}) \leq \alpha \cdot f(o_i|S_{i-1}) \leq \beta \cdot \ell(o_i).$$

Summing up all elements $o_i \in O^* \setminus S$, we have

$$\alpha \cdot \gamma \cdot \sum_{o_i \in O^* \setminus S} f(o_i|S \cup O_{i-1}) \leq \beta \cdot \sum_{o_i \in O^* \setminus S} \ell(o_i).$$

Rearranging the above inequality yields

$$\alpha \cdot \gamma \cdot (f(S \cup O^*) - f(S)) \leq \beta \cdot \ell(O^* \setminus S). \tag{5}$$

Since f is monotone, the left-hand side of Inequality (5) can be lower bounded by $LHS \geq \alpha \cdot \gamma \cdot (f(O^*) - f(S))$. As function ℓ is non-negative linear, the right-hand side of Inequality (5) can be upper bounded by $RHS \leq \beta \cdot \ell(O^*)$. Then Inequality (5) can be reduced into

$$f(S) \geq f(O^*) - \frac{\beta}{\alpha \cdot \gamma} \ell(O^*).$$

Based on the lower bound of $f(S)$ and the upper bound of $\ell(S)$, we get

$$\begin{aligned} f(S) - \ell(S) &\geq f(S) - \frac{\alpha}{\beta} f(S) = (1 - \frac{\alpha}{\beta}) f(S) \\ &\geq \frac{\beta - \alpha}{\beta} \left(f(O^*) - \frac{\beta}{\alpha \cdot \gamma} \ell(O^*) \right) \\ &= \frac{\beta - \alpha}{\beta} f(O^*) - \frac{\beta - \alpha}{\alpha \cdot \gamma} \ell(O^*), \end{aligned}$$

completing the proof. □

5 Conclusion

In this work, we consider the maximization of the difference between a normalized non-negative monotone approximately submodular function and a non-negative linear function subject to a matroid constraint. This model captures several typical models, for example, the maximization of a submodular function minus a linear function. As our main contribution, we provide a $(\frac{\gamma}{1+\gamma}, 1)$-approximation algorithm, where $\gamma \in (0, 1]$ is the submodularity parameter of function f. This result also extends several previous work, for example the Ene's recent work [8]. We then modify Algorithm 1 to deal with the huge volume dataset and obtain a $(\frac{\beta-\alpha}{\beta}, \frac{\beta-\alpha}{\alpha\gamma})$-approximation algorithm, where $0 < \alpha \leq \beta$ are parameters to balance the gain and cost.

Acknowledgements. This work was supported by the National Key Research and Development Program of China under Grants 2018AAA0101000. Yicheng Xu was supported by Guangxi Key Laboratory of Cryptography and Information Security (No. GCIS202116).

References

1. Anagnostopoulos, A., Becchetti, L., Castillo, C., Gionis, A., Leonardi, S.: Online team formation in social networks. In: Proceedings of the 21st International Conference on World Wide Web, pp. 839–848 (2012)
2. Anagnostopoulos, A., Castillo, C., Fazzone, A., Leonardi, S., Terzi, E.: Algorithms for hiring and outsourcing in the online labor market. In: Proceedings of the 24th ACM SIGKDD International Conference on Knowledge Discovery and Data Mining, pp. 1109–1118 (2018)

3. Bian, A.A., Buhmann, J.M., Krause, A., Tschiatschek, S.: Guarantees for greedy maximization of non-submodular functions with applications. In: Proceedings of the 34th International Conference on Machine Learning, pp. 498–507 (2017)
4. Buchbinder, N., Feldman, M., Garg, M.: Deterministic $(1/2 + \epsilon)$-approximation for submodular maximization over a matroid. In: Proceedings of the 30th Annual ACM-SIAM Symposium on Discrete Algorithms, pp. 241–254 (2019)
5. Désir, A., Goyal, V., Segev, D., Ye, C.: Constrained assortment optimization under the Markov chained-based choice model. Manag. Sci. **66**(2), 698–721 (2020)
6. Du, D., Li, Y., Xiu, N., Xu, D.: Simultaneous approximation of multi-criteria submodular functions maximization. J. Oper. Res. Soc. China **2**(3), 271–290 (2014)
7. Edmonds, J.: Submodular functions, matroids, and certain Polyhedra. In: Jünger, M., Reinelt, G., Rinaldi, G. (eds.) Combinatorial Optimization—Eureka, You Shrink! LNCS, vol. 2570, pp. 11–26. Springer, Heidelberg (2003). https://doi.org/10.1007/3-540-36478-1_2
8. Ene, A.: A note on maximizing the difference between a monotone submodular function and a linear function. arXiv preprint arXiv: 2002.07782 (2020)
9. Feige, U.: A threshold of $\ln n$ for approximating set cover. J. ACM **45**, 634–652 (1998)
10. Feldman, M.: Guess free maximization of submodular and linear sums. In: Proceedings of the 16th International Conference Workshop on Algorithms and Data Structures, pp. 380–394 (2019)
11. Friedrich, T., Göbel, A., Neumann, F., Quinzan, F., Rothenberger, R.: Greedy maximization of functions with bounded curvature under partition matroid constraints. In: Proceedings of the 33rd AAAI Conference on Artificial Intelligence, pp. 2272–2279 (2019)
12. Golshan, B., Lappas, T., Terzi, E.: Profit-maximizing cluster hires. In: Proceedings of the 20th ACM SIGKDD International Conference on Knowledge Discovery and Data Mining, pp. 1196–1205 (2014)
13. Gong, S., Nong, Q., Liu, W., Fang, Q.: Parametric monotone function maximization with matroid constraints. J. Glob. Optim. **75**(3), 833–849 (2019)
14. Harshaw, C., Feldman, M., Ward, J., Karbasi, A.: Submodular maximization beyond non-negativity: guarantees, fast algorithms, and applications. In: Proceedings of the 36th International Conference on Machine Learning, pp. 2634–2643 (2019)
15. Kuhnle, A., Smith, J.D., Crawford, V.G., Thai, M.T.: Fast maximization of non-submodular, monotonic functions on the integer lattice. In: Proceedings of the 35th International Conference on Machine Learning, pp. 2791–2800 (2018)
16. Lappas, T., Liu, K., Terzi, E.: Finding a team of experts in social networks. In: Proceedings of the 15th ACM SIGKDD International Conference on Knowledge Discovery and Data Mining, pp. 467–476 (2009)
17. Lehmann, B., Lehmann, D., Nisan, N.: Combinatorial auctions with decreasing marginal utilities. Games Econ. Behav. **55**, 270–296 (2006)
18. Liu, S., Poon, C.K.: A simple greedy algorithm for the profit-aware social team formation problem. In: Proceedings of the 11th International Conference on Combinatorial Optimization and Applications, pp. 379–393 (2017)
19. Nikolakaki, S.M., Ene, A., Terzi, E.: An efficient framework for balancing submodularity and cost. In: Proceedings of the 27th ACM SIGKDD Conference on Knowledge Discovery and Data Mining, pp. 1256–1266 (2021)
20. Papadimitriou, C., Yannakakis, M.: Optimization, approximation, and complexity classes. J. Comput. Syst. Sci. **43**, 425–440 (1991)

21. Qian, C.: Multi-objective evolutionary algorithms are still good: maximizing monotone approximately submodular minus modular functions. arXiv preprint arXiv: 1910.05492 (2019)
22. Sviridenko, M., Vondrák, J., Ward, J.: Optimal approximation for submodular and supermodular optimization with bounded curvature. Math. Oper. Res. **42**(4), 1197–1218 (2017)

On Various Open-End Bin Packing Game

Ling Gai[1], Weiwei Zhang[2], Wenchang Luo[3], and Yukun Cheng[4](✉)

[1] Glorious Sun School of Business & Management, Donghua University, Shanghai 200051, China
lgai@dhu.edu.cn

[2] School of Management, Shanghai University, Shanghai 201444, China
zthomas@shu.edu.cn

[3] School of Mathematics and Statistics, Ningbo University, Ningbo 315211, China
luowenchang@nbu.edu.cn

[4] Business School, Suzhou University of Science and Technology, Suzhou 215009, China
ykcheng@amss.ac.cn

Abstract. In this paper, we introduce various open-end bin packing problems in a game theoretic setting. The items (as agents) are selfish and intelligent to minimize the cost they have to pay, by selecting a proper bin to fit in. For both general open-end bin packing game and minimum open-end bin packing game, we prove the existence of the pure Nash Equilibrium and study the Price of Anarchy. We prove the upper bound to be approximately 2 and show a corresponding tight lower bound for both models. Furthermore, we study the open-end bin packing game with conflict and also give the proof for the existence of Nash Equilibrium. Under multipartite and simple conflict graph, we study the upper bound of Price of Anarchy separately.

Keywords: Open-end bin packing · Game · Price of anarchy · Conflict

1 Introduction

Open-end bin packing problem was first introduced by Leung et al. [8,9], then its parameterized online version was studied by Zhang [13]. In this problem, a bin is allowed to be filled to a level exceeding one, as long as the bin is not full (the content is strictly less than one) before the last item is packed. A nice example were presented in [8,9,13] for the problem.

In Hong Kong, passengers use magnetic card to pay for their subway trips. Generally there is a standard denomination in the card and the fare is deducted from the card every time the passenger passes the toll gate in the subway station. The passenger can pass the gate as long as the remaining balance is still positive, even if it is less than the fare needed. So the passenger can "gain" by taking a last long trip with the card of small positive balance. We can see that here the cards correspond to the bins and the fares of each trip correspond to the items. How to minimize the number of cards for a passenger who makes several trips corresponds to the open-end bin packing problem.

D.-Z. Du et al. (Eds.): COCOA 2021, LNCS 13135, pp. 86–95, 2021.
https://doi.org/10.1007/978-3-030-92681-6_8

The open-end bin packing problem is shown to be strongly NP-hard and any on-line algorithm must have an asymptotic worst-case performance ratio at least two [8,9]. The parameterized version was studied by [13] where the items are with size less than (or equal to) $1/m$ and a best possible algorithm was presented. In 2003 and 2008, the ordered open-end bin packing was studied by [1,12] where the items are indexed, and the packing of each bin has to follow the index such that the largest-indexed item is the last item in the bin. Both worst case and average case were analyzed, approximation algorithms and optimal algorithms were presented for this problem. After that, several variants of open-end bin packing problem with different names appeared in literatures [2–5]. They are the minimum open-end bin packing, the maximum open-end bin packing and the maximum indexed open-end bin packing. In the minimum version, removing the smallest item would bring the total size to a value below 1; in the maximum version, the largest item's removing could bring the total size back to below 1. Since the size of the largest item is at least that of the last packed item, we can see that the maximum open-end bin packing is the general open-end bin packing model introduced by [8,9,13]; As for the maximum indexed open-end bin packing, it is another name of the ordered open-end bin packing.

Recent years there is a trend to combine the well-defined combinatorial optimization problem with game issue, like bin packing game [6,11], bin covering game [10] etc. Taking the bin packing game as an example, different from the classical case with a central decision maker, the items are rational and eager to get higher utility by selecting bins. This make the problem more complicated while more realistic. To our knowledge, there is not game issue consideration on the various open-end bin packing problems.

In this paper, we aim to study various open-end bin packing problems under game circumstance. We are curious about the existence of Nash Equilibrium for these problems, and the performance of the Equilibria. The methodology of the Price of Anarchy is applied, which compares the output of the worst Nash Equilibrium with the optimal solution.

2 Definitions and Terminologies

Suppose there is a set of items to be packed into some unit capacity bins. The items are with size $s_i \in (0, 1]$, $i = 1, \ldots, n$. Each of the items is owned by a rational agent. The agent has to pay for his packed item, which is proportional to the item size. For an item i packed in bin B_h, let $S(B_h)$ denote the load of bin B_h. Then the cost of item i is $c_i = \frac{s_i}{\sum_{j \in B_h} s_j} = \frac{s_i}{S(B_h)}$. The agent could move the item unilaterally if that could reduce his cost. A Nash Equilibrium is reached if no agent would like to move unilaterally. The Price of Anarchy [7], often abbreviated as PoA, is a concept in economics and game theory that measures how the efficiency of a system degrades due to selfish behavior of its agents. Let G denote a game, NE denote a solution of a Nash Equilibrium, OPT denote the solution of the corresponding optimization problem. Then $PoA(G) = sup\frac{NE(G)}{OPT}$.

Corresponding to the various versions of open-end bin packing problem, the models studied in this paper are named as general open-end bin packing game, minimum open-end bin packing game. We also study another model which is named as open-end bin packing game with conflict. In this problem, items are conflict to some of the other, which means they could not be packed in a same bin. We use a conflict graph to denote these relations. The node represents the items, and if there is an edge between two nodes, they are conflict.

3 General Open-End Bin Packing Game

In the general open-end bin packing game, except at most one bin, the load of each bin minus the size of the last item is strictly less than 1. First we check the existence of a Nash Equilibrium in this problem.

3.1 Existence of Nash Equilibrium in the General Open-End Bin Packing Game

Define the potential function as $P = \sum_{h \in t} S(B_h)^2 \leq 4t$, where t is the number of bins used. We can see that this potential function is upper-bounded and its value strictly increases with the item's cost reduction unilateral deviation. Specifically, suppose item i deviates to bin B_k from bin B_h to decrease its cost, then

$$\begin{aligned} \Delta &= P' - P \\ &= S(B_k + s_i)^2 + S(B_h - s_i)^2 - S(B_k)^2 - S(B_h)^2 \\ &= 2s_i(S(B_k) - S(B_h)) + 2s_i^2 \\ &= 2s_i(S(B_k) - S(B_h) + s_i) > 0 \end{aligned}$$

So, begining with any feasible packing, the maximum of potential function could be reached after limited steps. Hence the existence of Nash Equilibrium is proved.

3.2 The Upper Bound of Price of Anarchy

Lemma 1. *There exists at most one bin with load less than 1.*

Proof. Suppose there are two bins B_h and B_k, $|B_h| = |B_k| = 1$, $S(B_h) < 1$ and $S(B_k) < 1$ in a Nash Equilibrium packing. For $s_i \in B_h$ and $S(B_k) < 1$, s_i can be packed into bin B_k. This is a contradiction.

Theorem 1. *The upper bound of the Price of Anarchy for general open-end bin packing game is 2.*

Proof. The load of each bin in an optimal packing is less than 2, hence $2OPT \geq \sum s_i$. According to lemma 1, $\sum s_i \geq NE - 1$. Thus, $2OPT \geq NE - 1$.

3.3 The Lower Bound of Price of Anarchy

Given an instance with kn items of size $1-\varepsilon$, kn items of size ε, $k^2n - kn$ items of size 2ε, where $\varepsilon = \frac{1}{2k}$, we can show that the Price of Anarchy of the Nash Equilibrium is greater than or equal to $\frac{2k-1}{k}$.

As shown in Figs. 1 and 2, the items are separated to n groups. Under proportional cost sharing mechanism, there is a Nash Equlibrium packing with $2k-1$ bins used for each group, while an optimal packing only uses k bins. So when k is large enough, the lower bound of Price of Anarchy is 2.

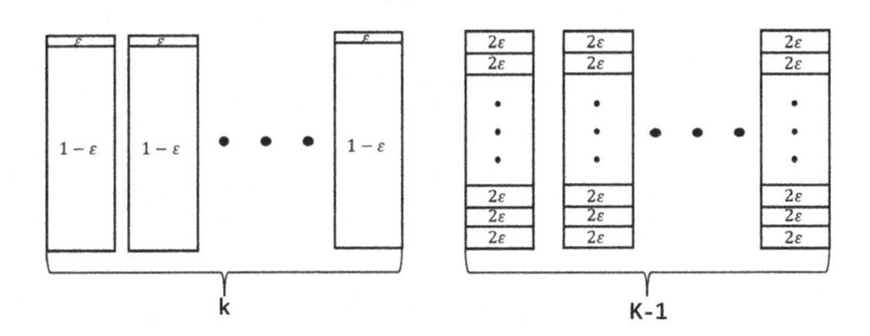

Fig. 1. A Nash equlibrium packing for the general open-end bin packing game

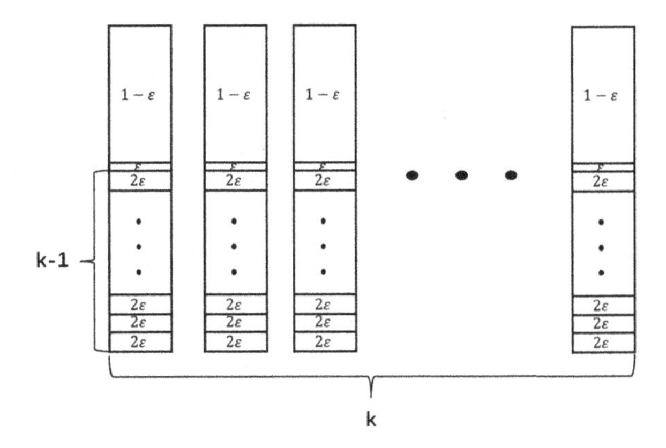

Fig. 2. An optimal packing of the general open-end bin packing problem

4 Minimum Open-End Bin Packing Game

In the definition of minimum open-end bin packing problem, a packing is feasible only if the load of bins is less than 1 or less than 1 after removing the smallest item. When we consider the game issue of the problem, we have to notice the limitations of this definition. If we follow this definition, a trivial case would be incurred as shown in Figs. 3 and 4. As stated before, the cost that each item has to share is proportional to its size, $c_i = \frac{s_i}{\sum_{s_j \in B_h} s_j}$.

If the movement of item i from bin B_h to bin B_k is prohibited, where $\exists s_j \in B_k$ $s_i \geq s_j$ and $s_i + S(B_k) > 1 + s_j$, there are some special NE packings where the load of any bin is below 1. As shown in the following Figs. 5, there is a family of such NE packings for $2^k \times \epsilon = 1$.

So we modified the definition of minimum open-end bin packing game as follows. We force the smallest item in the bin, which doesn't satisfy the assumption of minimum open-end bin packing, to move into other bin, or open a new bin. To achieve this effect, define $c(s_{min}) = 1 + \varepsilon$ if $S(B_h) - s_{min} \geq 1$, where s_{min} is the smallest item in bin B_h, $\varepsilon > 0$. Moreover, if the smallest item $s_{min} \in B_h$ satisfies that $S(B_h) - s_{min} \geq 1$, s_{min} has the priority to move.

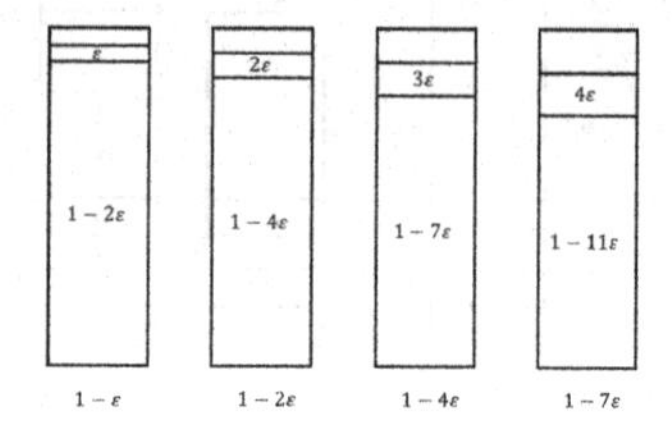

Fig. 3. An example of Nash Equlibrium packing under minimum open-end packing definition

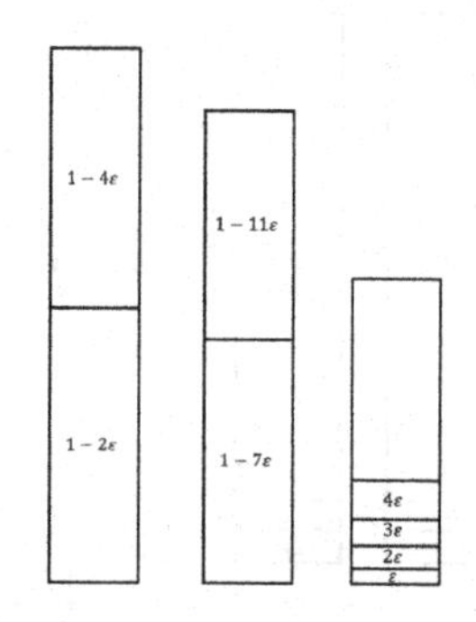

Fig. 4. An optimal packing

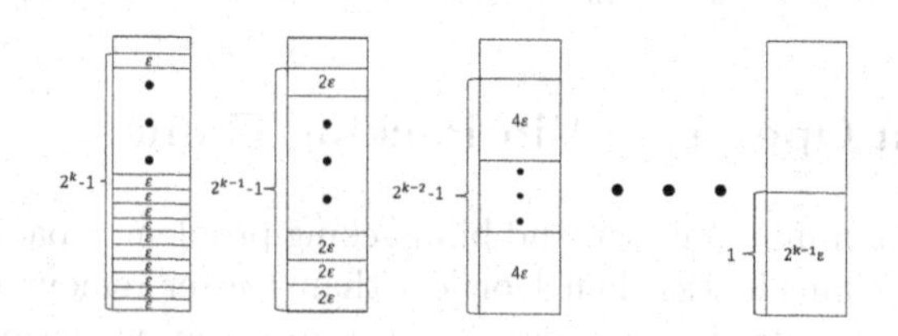

Fig. 5. A Nash Equlibrium packing with $2^k \times \epsilon = 1$

4.1 Existence of Nash Equilibrium in Minimum Open-End Bin Packing Game

Define the potential function as $P = \sum_{h \in t} S(B_h)^2 < 4t$, where t is the set of bins. Some movement will force the smallest item in the bin move into other bin, and therefore there are two different movements– regular movement and forced movement. During the regular movement, no item will leave the bin. Thus, the regular movement consists of the movement of item i deviating to bin B_k from bin B_h.

$$\begin{aligned}
\Delta &= P' - P \\
&= S(B_k + s_i)^2 + S(B_h - s_i)^2 - S(B_k)^2 - S(B_h)^2 \\
&= 2s_i(S(B_k) - S(B_h)) + 2s_i^2 \\
&= 2s_i(S(B_k) - S(B_h) + s_i) > 0
\end{aligned}$$

During the forced movement, there will be more than one item , which have to leave the bin. Let a set $S_{min} = s_1, s_2, s_3, ..., s_m$ denote the smallest items leaving the bin one by one. Hence, the forced movement consists of the movement of item j deviating to bin B_k from bin B_h and the movement of item $s_i \in S_min$. To simplify the proof, the exchange of potential function in former movement is called δ_0 and the exchange in latter movement is called δ_i. Obviously, $\Delta_0 = 2s_j(S(B_k) - S(B_h)) + 2s_j \times s_j$. And the item i will move into bin B_i. If $S(B_i) \geq S(B_k)$, then $\Delta_i = 2s_i(S(B_k) - S(B_h)) + 2s_i \times s_i > 0$. $\forall s_i \in S_{min}$,

$$\begin{aligned}
\Delta_i = &\left(S(B_k) + s_j - \sum_{m \leq i} s_m \right)^2 + (S(B_i) + s_i)^2 - \left(S(B_k) + s_j - \sum_{m < i} s_m \right)^2 - S(B_i)^2 \\
&= -s_i \left(2S(B_k) + 2s_j - 2\sum_{m \leq i} s_m - s_i \right) + s_i \left(2S(B_i) + s_i \right) \\
&= 2s_i \left(S(B_i) - S(B_k) - s_j + \sum_{m \leq i} s_m \right) \\
&> 2s_i \left(S(B_h) - S(B_k) - s_j \right)
\end{aligned}$$

$$\begin{aligned}
\Delta &= \Delta_0 + \sum_{S_{min}} \Delta_i \\
&= (S(B_k) - S(B_h)) \left(s_j - \sum_{s_i \in S_{min} s_i} \right) + 2s_i \left(s_j - \sum_{s_i \in S_{min} s_i} \right) \\
&> 0
\end{aligned}$$

Therefore, for any type of movement, there is always $\delta > 0$. So begining with any feasible packing, the maximum of potential function could be reached after limited steps. Hence the existence of Nash Equilibrium is proved.

4.2 The Upper Bound of Price of Anarchy

Lemma 2. *There is at most one bin with load less than 1.*

Proof. Suppose there are two bins with load less than 1 in a Nash Equilibrium packing, then $S(B_h) \geq S(B_k)$. $\forall s_i \in B_k$, $S(B_h) + s_i > S(B_k)$. A contradiction.

Theorem 2. *The upper bound of the Price of Anarchy for minimum open-end bin packing game is 2.*

Proof. The load of bins in an optimal packing is less than 2, hence $2Opt \geq \sum s_i$. According to lemma 1, $\sum s_i \geq NE - 1$. Thus, $2Opt \geq NE - 1$.

4.3 The Lower Bound of Price of Anarchy

In the following we show that under the proportional cost sharing mechanism, the Price of Anarchy of the Nash Equilibrium is greater than or equal to 2.

Proof. Here we are given an instance with $kn - n$ items of size $1 - \epsilon$, kn items of size ε, $kn - n$ items of size 2ϵ and $kn - n$ items of size $1 - 2\varepsilon$, where $\varepsilon = \frac{1}{3k}$. The items are separated into n groups. We can see that for each group items, there is a Nash Equlibrium packing with $2k - 2$ bins used, while an optimal packing only uses k bins, as showed in Fig. 6 and 7. So the lower bound of PoA is at least $\frac{2k-2}{k}$. When k is large enough, the lower bound is 2.

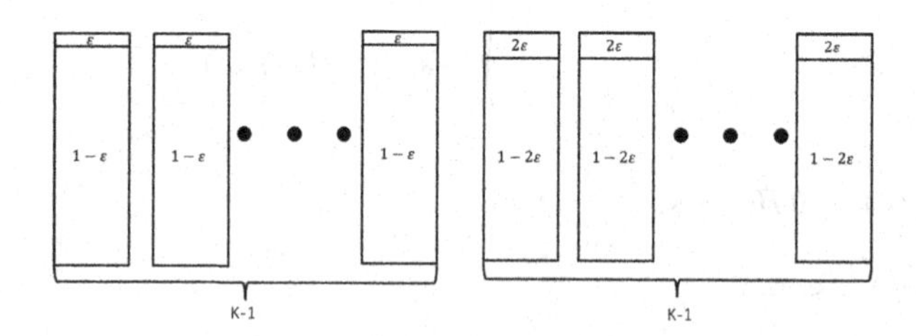

Fig. 6. A Nash Equlibrium packing for minimum open-end bin packing game

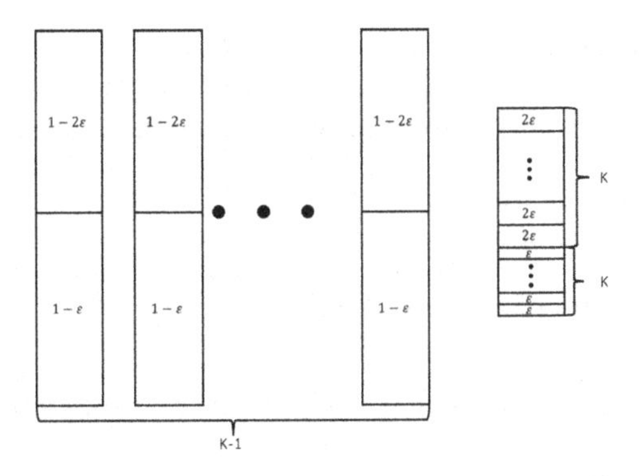

Fig. 7. An optimal packing for minimum open-end bin packing

5 Open-End Bin Packing Game with Conflict

In the open-end bin packing game with conflict, a graph $G = (V, E)$ is given together to present the conflict between items. If for two items s_i and s_j, there is an edge $(s_i, s_j) \in E$, then s_i and s_j could not be packed into a same bin. Define the cost of item $s_i \in B_h$ equals to $1 + \epsilon$, if there is an item $s_j \in B_h$, $(s_i, s_j) \in E$.

5.1 Existence of Nash Equilibrium

A bin is called conflict-free, if for any $s_i, s_j \in G$, $(s_i, s_j) \notin E$. Otherwise, define the *deconflict move* as following: For $s_i, s_j \in B_h$ and $(s_i, s_j) \in E$, let s_j move into some bin B_k, such that $\forall s_l \in B_k$, $(s_j, s_l) \notin E$.

From the definition we know that, each deconflict move can reduce at least one pair of conflict items. Suppose that there are K pairs of conflict items in the packing, then after at most K times deconflict move, each bin in the packing is conflict-free. Then define the potential function as $P = \sum_{h \in t} S(B_h)^2$. Similar to the proof in general open-end bin packing game, the existence of Nash Equilibrium can be proved.

5.2 Open-End Bin Packing Game with a Complete Multipartite Conflict Graph

We consider the case that the conflict graph is a complete k-partite graph, which implies that for any two items in two different independent sets respectively, they cannot be packed into a same bin.

5.3 The Upper Bound of the Price of Anarchy

Lemma 3. *There are at most k bins with load less than 1.*

Proof. For each independent set, it can be treated as an open-end bin packing. According to lemma 1, there is at most one bin with load less than 1 for each independent set. Hence, there are at most k bins with load less than 1.

Theorem 3. *The upper bound of the Price of Anarchy for open-end bin packing game with a complete multipartite conflict graph is 2.*

Proof. Similar to the proof of Thoerem 2, $2OPT \geq \sum s_i \geq (NE - K)$.

5.4 Open-End Bin Packing with a Simple Conflict Graph

Observation. For any bins B_h, B_k in a Nash Equilibrium packing, if $S(B_h) \leq S(B_k) \leq 1$, then for any $s_i \in B_h$, there exists an item $s_j \in B_k$, $(s_i, s_j) \in G$.

Lemma 4. *There are at most $\delta + 1$ bins with load less than 1 in a Nash Equilibrium packing, where δ is the maximum degree of the vertices in graph G.*

Proof. According to observation 1, for an item i in bin B_h, there must be at least one item connecting with item i in some other bins of load less than 1. Since δ is the maximum degree of vertices in graph G, there are at most $\delta + 1$ bins.

Theorem 4. $2OPT \geq NE - \delta$.

The proof is abbreviated due to similarity.

Remark 1. Note that if the maximum degree of vertices equals the number of vertices minus 1, then δ seems to be invalidate for the estimation of the upper bound. In fact, according to observation 1, the degree of each item in the bins with load less than 1 is at least the number of bins with load less than 1. Therefore, δ is the solution of following model:

$$\max y$$

$$\sum_{i \in V} f(i) \geq y$$

$$f(i) = \begin{cases} 1, \text{ if } \delta(i) \geq y; \\ 0, \text{ otherwise.} \end{cases} \tag{1}$$

where $\delta(i)$ denotes the degree of vertex i.

Remark 2. The Lower Bound of Price of Anarchy for above open-end bin packing games with conflict can be similarly reached as that of the general open-end bin packing game.

6 Conclusion

In this paper we consider several versions of open-end bin packing game, the general open-end bin packing game, the minimum open-end bin packing game and the open-end bin packing game with conflict. For each of the problem, we prove the existence of Nash Equilibrium. We also prove the tight bound of Price of Anarchy. As for the future work, we are considering a dual version of the open-end bin packing problem. Given the predetermined trips, how to minimize total fare if the passengers can buy cards with different amount of denomination. We are also interested in the performance under game situation.

References

1. Ceselli, A., Righini, G.: An optimization algorithm for the ordered open-end bin-packing problem. Oper. Res. **56**(2), 425–436 (2008)
2. Epstein, L., Levin, A.: Asymptotic fully polynomial approximation schemes for variants of open-end bin packing. Inf. Process. Lett. **109**, 32–37 (2008)
3. Epstein, L.: Open-end bin packing: new and old analysis approaches. https://arxiv.org/abs/2105.05923v1 [cs.DS] (2021)
4. Epstein, L., Levin, A.: A note on a variant of the online open end bin packing problem. Oper. Res. Lett. **48**, 844–849 (2020)
5. Balogh, J., Epstein, L., Levin, A.: More on ordered open end bin packing. https://arxiv.org/abs/2010.07119v1 [cs.DS] (2020)
6. Gai, L., Wu, C., Xu, C., Zhang. W.: Selfish bin packing under Harmonic mean cost sharing mechanism. Optim. Lett. 1–12 (2021)
7. Koutsoupias, E., Papadimitriou, C.: Worst-case equilibria. Comput. Sci. Rev. **3**(2), 65–69 (2009)
8. Joseph, Y.-T., Leung, M.D., Young, G.H.: A note on an open-end bin packing problem. J. Sched. Manuscriot, 1996
9. Leung, J.Y.T., Dror, M., Young, G.H.: A note on an open-end bin packing problem. J. Sched. **4**(4), 201–207 (2001)
10. Li, W., Fang, Q., Liu, W.: An incentive mechanism for selfish bin covering. In: Chan, T.-H.H., Li, M., Wang, L. (eds.) COCOA 2016. LNCS, vol. 10043, pp. 641–654. Springer, Cham (2016). https://doi.org/10.1007/978-3-319-48749-6_46
11. Wang, Z., Han, X., Dosa, G., Tuza, Z.: A general bin packing game: interest taken into account. Algorithmica **80**, 1534–1555 (2018)
12. Yang, J., Leung, T.Y.T.: The ordered open-end bin-packing problem. Oper. Res. **51**(5), 759–770 (2003)
13. Zhang, G.: Parameterized on-line open-end bin packing. Computing **60**, 267–273 (1998)

A Linear-Time Streaming Algorithm for Cardinality-Constrained Maximizing Monotone Non-submodular Set Functions

Min Cui[1], Donglei Du[2], Ling Gai[3](✉), and Ruiqi Yang[4]

[1] Department of Operations Research and Information Engineering, Beijing University of Technology, 100 Pingleyuan, Chaoyang District, Beijing 100124, People's Republic of China
B201840005@emails.bjut.edu.cn

[2] Faculty of Management, University of New Brunswick, Fredericton, NB E3B 5A3, Canada
ddu@unb.ca

[3] Glorious Sun School of Business and Management, Donghua University, Shanghai 200051, People's Republic of China
lgai@dhu.edu.cn

[4] School of Mathematical Sciences, University of Chinese Academy Sciences, Beijing 100049, People's Republic of China
yangruiqi@ucas.ac.cn

Abstract. Nowadays, massive amounts of data are growing at a rapid rate every moment. If data can be processed and analyzed promptly as they arrive, they can bring huge added values to the society. In this paper, we consider the problem of maximizing a monotone non-submodular function subject to a cardinality constraint under the streaming setting and present a linear-time single-pass deterministic algorithm for this problem. We analyze the algorithm using the parameter of the generic submodularity ratio γ to achieve an approximation ratio of $\left[\frac{\gamma^4}{c(1+\gamma+\gamma^2+\gamma^3)} - \varepsilon\right]$ for any $\varepsilon \geq 0$ with the query complexity $\lceil n/c \rceil + c$, and the memory complexity is $O(ck\log(k)\log(1/\varepsilon))$, where c is a positive integer. When $\gamma = 1$, the algorithm achieves the same ratio for the submodular version of the problem with the matching query complexity and memory complexity.

Keywords: Non-submodular · Streaming · Linear-time · Cardinality-constrained

1 Introduction

The problem of submodular optimization can be regarded as a subset selection problem, with a growing number of applications, especially in artificial intelligence [31], data mining [17], document summarization [21], boosting information spread [22], genomics [34], social network influence [27], recommender

D.-Z. Du et al. (Eds.): COCOA 2021, LNCS 13135, pp. 96–110, 2021.
https://doi.org/10.1007/978-3-030-92681-6_9

system [30], and securities market analysis [18], to name just a few. A set function $f : 2^N \to R^+$ is submodular if for any $S \subseteq T \subseteq N$ and $e \in N \setminus T$, $f(S \cup \{e\}) - f(S) \geq f(T \cup \{e\}) - f(T)$ holds. The cardinality-constrained submodular maximization problem is to select a subset S of N with cardinality at most k such that the objective value $f(S)$ is maximized:

$$\max_{S \subseteq N} \{f(S) : \ |S| \leq k\} \tag{1}$$

In the current era of digitized information, human activities are carried out through information technology, generating a humongous stack of data [4]. The unprecedented growth of text, image and video data requires technologies that can effectively process them at high speed. Submodular optimization under the massive data environment [19,26] is a relatively new area to which many stakeholders have begun to pay close attention. The streaming model [1,24,25,28] is a popular model for processing massive amounts of data.

When we have to process massive data, computing systems may quickly become overloaded if these data are stored at once. Streaming models overcome this difficulty by visiting the elements in a streaming fashion. According to [1], the performance guarantee of a streaming algorithm is defined by four parameters: pass time, memory complexity, update time and approximation ratio. Therefore any streaming algorithm aims to select appropriate elements from a large number of inputs to make the final output as good as possible. The main goal is to provide a good trade-off between the space to process the input stream and the accuracy of the solution. Other relevant parameters include the update time needed to make the estimate, and the number of passes ideally is equal to 1.

For the maximization of cardinality-constraint monotone submodular functions under the streaming setting, Chakrabarti and Kale et al. [5] provided the first single-pass streaming 1/4-approximation algorithm with $2n$ queries, $O(k)$ memory and $\Omega(n \log k)$ time complexity. Badanidiyuru et al. [1] introduced a threshold-based procedure, improving the single-pass algorithm to $1/2 - \varepsilon$ in $O(k \log(k)/\varepsilon)$ with $O(n \log(k)/\varepsilon)$ queries and time. Kazemi et al. [13] provided SieveStream++ algorithm which is a single pass algorithm that keeps $1/2 - \varepsilon$ approximation ration with memory complexity of $O(k/\varepsilon)$. Lately, Feldman et al. [11] showed that any single-pass algorithm with approximation guarantee of $1/2 + \varepsilon$ must essentially store all elements of the stream. Kuhnle [15] proposed the first deterministic streaming algorithms with linear time complexity.

However, submodularity is a demanding attribute in many applications [14,22] and hence may not be satisfied in practice. Das and Kempe [8] introduced the parameter-submodularity ratio $\gamma_{N,k}$, as a key quantity to capture how close a general set function is to the submodular function. Sviridenko et al. [32] used the notion of curvature c (a concept introduced for any non-decreasing submodular function by Conforti and Cornuéjols [6]) to obtain an $(1 - c/e)$-approximation algorithm for the monotone non-submodular functions maximization problem subject to a matroid constraint. Bian et al. [2] combined the curvature c and the submodularity ratio $\gamma_{N,k}$ to derive the first tight constant-factor approximation ratio of $\frac{1}{c}(1 - e^{-\gamma_{N,k}c})$ with $O(nk)$ oracle queries for the non-submodular cardinality-constraint maximization problem. Kuhnle et al. [16] adapted the threshold greedy framework to non-submodular functions, achieving an approximation ratio $1 - e^{-\nu\gamma_{N,k}} - \varepsilon$ for ν no less than $\gamma_{N,k}$ with query complexity $O(n \log k \log_{\kappa}(\varepsilon^2/k)$, where $\kappa, \varepsilon \in (0,1)$ are parameters of the algorithm. Gong et al. [12] provided a more practical measurement γ which is called generic submodularity ratio and gave the first $(1 - e^{-\gamma} - o(1))$ approximation continuous greedy algorithm for multilinear optimization problem under matroid constraints. Later, Nong et al. [29] used the generic submodularity ratio γ in the greedy algorithms for maximizing cardinality-constraint strictly-monotone set function, and provided a $(1 - e^{-\gamma})$-approximation algorithm. Cui et al. [7] and Liu et al. [23] both adapted the threshold technique from Fahrbach et al. [10] and the generic submodularity ratio γ for maximizing cardinality-constraint monotone non-submodular set function. Cui et al. [7] gave an adaptive distributed algorithm that achieved an approximation ratio $1 - e^{-\gamma^2} - \varepsilon$ with $O(\log(n/\eta)/\varepsilon^2)$ adaptive rounds and $O(n \log\log(k)/\varepsilon^3)$ oracle queries in expectation. Liu and Hu [23] developed a $1 - e^{-\gamma} - \varepsilon$-approximation continuous algorithm with $O(n/\varepsilon \log(n/\varepsilon))$ queries.

The research on non-submodular cardinality-constraint maximization problems in streaming setting started only recently. Elenberg et al. [9] adopted the high-level idea of Badanidiyuru et al. [1] to give a greedy deterministic streaming algorithm with an approximation ratio $(1 - \varepsilon)\frac{\gamma}{2}(3 - e^{-\gamma'/2} - 2\sqrt{2 - \gamma'/2})$ and $O(\varepsilon^{-1} k \log k)$ memory for maximizing γ'-weakly submodular functions. Wang et al. [33] used the weak submodularity ratio from [9] to present four sieve-streaming algorithms for maximizing cardinality-constraint monotone non-submodular set function. The best approximation ratio for single-pass algorithm is $\min\{\frac{\alpha\gamma}{2^{\gamma}}, 1 - \frac{1}{2^{\gamma}}\}$ with k memory where α is a parameter $\in [0,1]$. Li et al. [20] proposed two algorithms named Sieve-Streaming++ algorithm and Batch-Sieve-Streaming++, and the better approximation ratio is $\min\{\frac{(1-\varepsilon)\gamma}{2^{\gamma}}, 1 - \frac{1}{2^{\gamma}}\}$ with $O(k/\varepsilon)$ memory.

Contributions. Our main contributions are summarized as follows:

- We propose the first linear-time single-pass streaming algorithm for maximizing a monotone nonsubmodular function with a cardinality constraint and show that the approximation ratio is $\frac{\gamma^4}{c(1+\gamma+\gamma^2+\gamma^3)} - \varepsilon$. When $\gamma = 1$, the algorithm matches the same ratio for the submodular version of the problem.
- The number of queries of the algorithm is at most $\lceil n/c \rceil + c$ and the memory complexity of the algorithm is $O(ck \log(k) \log(1/\varepsilon))$. These ensure that a feasible solution can be obtained quickly without taking up too much space.

In this paper, we adopt the high-level idea in Kuhnle [15] with the generic submodularity ratio γ [12]. One of our main technologies is set-swapping, which is similar to but different from the previous swapping technique [3,5]. We only consider whether to add this newly arrived elements set when the number of newly arrived elements reaches c; we would delete some elements that have arrived in order when the number of stored elements reaches the upper bound. These methods indirectly reduce number of queries. The parameter c is the number of elements processed by the algorithm at one time when the stream arrives. While the approximation ratio and the query complexity of the algorithm are negatively correlated with the value of c, the memory complexity is positively correlated with the value of c. In practical applications, we can appropriately adjust the size of c according to the needs.

Organization. This paper is structured as follows: The first section is an introduction to this work, explaining its motivation, related works and main contributions. Section 2 introduces the basic definitions and notations used throughout this article, and gives some properties of the definitions. We provide the implementation details for the deterministic algorithm, and give the approximate ratio analysis in Sect. 3. Finally, Sect. 4 offers direction of future work.

2 Preliminaries

In this section, we give a detailed description on notations related to this paper.

We use N to denote the ground set and $|N| = n$. The function $f : 2^N \to R^+$ is a non-negative monotone non-submodular set function with $f(\emptyset) = 0$.

Definition 1. Monotone: *The set function f is monotone if for any two subset $S \subseteq T \subseteq N$, we have*

$$f(S) \leq f(T).$$

For any pair of $S, T \subseteq N$, $f(S \cup T) - f(S)$ denotes the marginal gain of adding T to S. Specially, for any $S \subseteq N$ and any $e \in N$, $f(S \cup \{e\}) - f(S)$ denotes the marginal gain of adding element e to the set S.

Next, we introduce a tool to connect the general function with the submodular function. The generic submodularity ratio γ defined below is the parameter to measure the multiple relationship between the marginal gains of adding a single element to the set and adding it to the proper subset.

Definition 2. Generic submodularity ratio [12]: *Given a ground set N and an increasing set function $f : 2^N \to R^+$, the generic submodularity ratio of f is the largest scalar γ such that for any $S \subseteq T \subseteq N$ and any $e \in N \setminus T$,*

$$f(S \cup \{e\}) - f(S) \geq \gamma \cdot [f(T \cup \{e\}) - f(T)].$$

Proposition 1 *(Property of generic submodularity ratio [12]). For an increasing set function $f : 2^N \to R^+$ with generic submodularity ratio γ, it holds that*

1. *$\gamma \in (0, 1]$;*
2. *The function f is submodular if and only if $\gamma = 1$.*

Lemma 1. *For a non-negative strictly-monotone set function $f : 2^N \to R^+$ with generic submodularity ratio γ, it holds that: for any $S, T \subseteq N$,*

1. $f(S) + \gamma f(T) \geq f(S \cap T) + \gamma f(S \cup T)$
2. Set $T \setminus S = \{T_1, T_2, \cdots, T_h\}$, $T_i \cap T_j = \emptyset (i, j \in [1, h], i \neq j)$

$$f(S \cup T) - f(S) \leq \frac{1}{\gamma} \sum_{j=1}^{h} [f(S \cup T_j) - f(S)]$$

Proof. 1. Let $X \subseteq Y, C = \{j_1, ..., j_l\}$, $C \subseteq N \backslash Y$, and $C \cap Y = \emptyset$. By the generic submodularity ratio,

$$\begin{aligned} f(X \cup \{j_1\}) - f(X) &\geq \gamma [f(Y \cup \{j_1\}) - f(Y)] \\ f(X \cup \{j_1, j_2\}) - f(X \cup \{j_1\}) &\geq \gamma [f(Y \cup \{j_1, j_2\}) - f(Y \cup \{j_1\})] \\ &\dots \\ f(X \cup \{j_1, ..., j_l\}) - f(X \cup \{j_1 ..., j_{l-1}\}) &\geq \gamma [f(Y \cup \{j_1, ..., j_l\}) \\ &\quad - f(Y \cup \{j_1, ..., j_{l-1}\})] \end{aligned} \tag{2}$$

Summing up those inequalities, we get

$$\begin{aligned} f(X \cup C) - f(X) &= f(X \cup \{j_1, ..., j_l\}) - f(X) \\ &\geq \gamma [f(Y \cup \{j_1, ..., j_l\}) - f(Y)] \\ &\geq \gamma [f(Y \cup C) - f(Y)] \end{aligned} \tag{3}$$

For any $S, T \subseteq N$, we set $X = S \cap T, C = S \setminus T, Y = T$. So we have $X \cup C = S, Y \cup C = S \cup T$.

$$f(S) - f(S \cap T) \geq \gamma [f(S \cup T) - f(T)]$$

2. From the setting of S, T, $T \setminus S = \cup_{j=1}^{h} T_j$.

$$\begin{aligned} f(S \cup T) - f(S) &= f(S \cup (\cup_{j=1}^{h} T_j)) - f(S) \\ &= f(S \cup (\cup_{j=1}^{h} T_j)) - f(S \cup (\cup_{j=1}^{h-1} T_j)) \\ &\qquad + f(S \cup (\cup_{j=1}^{h-1} T_j)) - f(S \cup (\cup_{j=1}^{h-2} T_j)) \\ &\qquad + \cdots \\ &\qquad + f(S \cup T_1) - f(S) \\ &\leq \frac{1}{\gamma} [f(S \cup T_h) - f(S)] + \frac{1}{\gamma} [f(S \cup T_{h-1}) - f(S)] \\ &\qquad + \cdots + [f(S \cup T_1) - f(S)] \\ &= \frac{1}{\gamma} \sum_{j=1}^{h} [f(S \cup T_j) - f(S)] \end{aligned} \tag{4}$$

where the inequality holds from Inequality (3).

In the rest of this paper, the algorithm outputs S as the final solution; OPT (i.e. $f(O) = OPT$) and O denote the values of the optimal sets and one of the optimal sets, respectively.

3 The Single-Pass Deterministic Algorithm

In this section, we propose a single-pass, deterministic streaming algorithm for maximizing a monotone non-submodular function subject to a cardinality constraint. The parameter c is the number of elements processed by the algorithm at one time as the streaming arrives, the approximation ratio, query complexity and memory complexity of the algorithm are all related to this parameter. Our algorithm is a set swapping algorithm, and uses the order in which elements are added to A to compare elements, instead of directly comparing with the marginal benefits of other elements in A; in addition, the algorithm does not query the function value of each element arriving, but only when c elements arrived. These indirect, reduced number of comparison methods leads to the algorithm's linear time complexity.

For the given c, as the streaming data arrive, the algorithm maintains a dynamic storage set A, whose length is at most $2cr(\lceil k\gamma \rceil + 1)\lceil \log_2(k) \rceil$. At the beginning of the algorithm, A is an empty set. H denotes the collection block that temporarily stores elements, and the size of H is no more than c. When the size of H is reached c or the streaming ends, the algorithm performs one query on $f(A \cup H)$. If the marginal benefit of adding set H to A is no less than $1/(\gamma k)$ multiplied by the function value of A, adds H into A; otherwise, discard these elements, and H restores the newly arriving elements. If the number of elements in A is no less than $2cr(\lceil k\gamma \rceil + 1)\lceil \log_2(k) \rceil$, removing $cr(\lceil k\gamma \rceil + 1)\lceil \log_2(k) \rceil$ elements from A. When the stream ends, the algorithm selects the last arrived ck elements of A, takes each k elements as a group according to the last arrival order, and adds them into the c sets. Finally, the algorithm compares the function values of these c sets and chooses the maximum value as the output.

Theorem 1. *Let $c \geq 1$, $\varepsilon \geq 0$, and $k \geq 2$, for any monotone, non negative non-submodular function f, the Single-pass Deterministic algorithm is a $[\frac{\gamma^4}{c(1+\gamma+\gamma^2+\gamma^3)} - \varepsilon]$-approximation algorithm, the query complexity is at most $\lceil n/c \rceil + c$, and the memory complexity is $O(ck \log(k) \log(1/\varepsilon))$.*

In order to prove the approximate ratio of the algorithm, we give some additional notations. According to the construction of A, initially $A_0 = \emptyset$. Let A_i $(0 \leq i < n)$ denote the state of A after the i-th iteration of the for loop. Let A_n denote the state of A in the end of the for loop. First, we need the following lemma [15]:

Lemma 2. *For $z \geq 1$, if $x \geq (1+k)\log_2 z$, then $(1+1/k)^x \geq z$.*

Proof. It follows from the inequality: for $y > 0$, $\log y \geq 1 - 1/y$.

$$\begin{aligned}(1+1/k)^x &= 2^{x \log_2(1+1/k)} \\ &\geq 2^{x(1-\frac{1}{(1+1/k)})} \\ &\geq 2^{(1+k)\log_2 z \cdot \frac{1}{1+k}} = z\end{aligned} \tag{5}$$

Algorithm 1. Single-pass Deterministic

Input: evaluation oracle $f : 2^N \to R^+$, constraint k, generic submodularity ratio γ, and error $0 < \varepsilon < \gamma k^2$

Output: the solution S

1: Set $r \leftarrow \lceil \log_2(1/(\varepsilon)) \rceil + 3$
2: Set $l \leftarrow cr(\lceil k\gamma \rceil + 1)\lceil \log_2(k) \rceil$
3: Initialize $A \leftarrow \emptyset$, $S \leftarrow \emptyset$, $H \leftarrow \emptyset$
4: **for** every element e arriving **do**
5: Set $H \leftarrow H \cup \{e\}$
6: **if** $| H |= c$ or all elements arrived **then**
7: **if** $f(A \cup H) - f(A) \geq f(A)/(\gamma k)$ **then**
8: Set $A \leftarrow A \cup H$
9: **if** $| A | \geq 2l$ **then**
10: Set $A \leftarrow \{$ the l elements most recently added to $A\}$
11: Set $H \leftarrow \emptyset$
12: Initialize $j \leftarrow 1$
13: **while** $j \leq c$ **do**
14: Set $S_j \leftarrow \{$ the k elements recently added to $A\}$
15: Set $A \leftarrow A \setminus S_j$
16: Update $S \leftarrow \arg\max\{f(S), f(S_j)\}$
17: Update $j \leftarrow j + 1$
18: **return** S

Next, we show the change of the value of $f(A)$ between iterations of the for loop.

Lemma 3. *For any i-th iteration* $(1 \leq i \leq n)$*, Lines 7–10 of the Single-pass Deterministic algorithm has one of the following properties:*

1. *If there is not delete operation in i-th iteration,* $f(A_{i-1}) \leq f(A_i)$*;*
2. *If there is a delete operation in i-th iteration,* $f(A_{i-1}) \leq \frac{1}{\gamma} f(A_i)$*;*
   2.1 *If i-th iteration is not the first round with deletion, set the last round recorded with deletion of i-th iteration is the* i'*-th iteration,* $f(A_{i'}) \leq f(A_i)$*.*

Proof. 1. If the elements are not deleted in i-th iteration, the number of the elements in A_{i-1} is less than $2cr(\lceil k\gamma \rceil + 1)\lceil \log_2(k) \rceil - c$. We only consider whether to add c elements to A_{i-1}. From the monotonicity of the function, $f(A_{i-1}) \leq f(A_i)$.

2. If the elements are deleted in i-th iteration, then at the beginning of i-th iteration, the number of the elements in A is at least $2cr(\lceil k\gamma \rceil + 1)\lceil \log_2(k) \rceil - c$; in this round, the algorithm performed two operations in A which are adding at most c elements recorded as H_i and deleting at least $cr(\lceil k\gamma \rceil + 1)\lceil \log_2(k) \rceil$ elements recorded as D_i, $A_i = (A_{i-1} \setminus D_i) \cup H_i$. Consequently, for every D_i $(1 \leq i \leq n)$, we can find $0 \leq l < i$ of the first for loop that $D_i = A_l$.
   For the relevance of $f(A_i)$ and $f(A_{i-1})$:

From the end of l-th iteration to the beginning of i-th iteration, there have been $cr(\lceil k\gamma\rceil + 1)\lceil \log_2(k)\rceil - c$ elements added in the A, which add precisely the elements in $A_{i-1} \setminus A_l$. It holds that

$$\begin{aligned} f(A_{i-1} \setminus D_i) \geq & \gamma[f(A_{i-1}) - f(D_i)] \\ \geq & \gamma \cdot (1 + \frac{1}{\gamma k})^{r(\lceil k\gamma\rceil+1)\lceil \log_2(k)\rceil - 1} . f(A_l) - \gamma f(A_l) \\ \geq & \gamma \cdot (1 + \frac{1}{\gamma k})^{(r-1)(\lceil k\gamma\rceil+1)\lceil \log_2(k)\rceil} . f(A_l) - \gamma f(A_l) \\ \geq & \gamma(k^{r-1} - 1) \cdot f(A_l) \end{aligned} \quad (6)$$

where the first inequality holds from the Lemma 1; the second inequality is obtained by the fact that from the l-th iteration to the $(i-1)$-th iteration, the added value of $f(A)$ in each round of added elements to A is at least $\frac{1}{\gamma k} f(A)$; the third inequality follows from the fact $r(\lceil k\gamma\rceil + 1)\lceil \log_2(k)\rceil - 1 \geq (r-1)(\lceil k\gamma\rceil + 1)\lceil \log_2(k)\rceil$ for any $k \geq 2$; and the last inequality are received by the Lemma 2. Therefore

$$\begin{aligned} f(A_{i-1}) \leq & f(D_i) + \frac{f((A_{i-1} \setminus D_i))}{\gamma} \\ = & f(A_l) + \frac{f((A_{i-1} \setminus D_i))}{\gamma} \\ \leq & (1 + \frac{1}{(k^{r-1} - 1)}) \frac{f((A_{i-1} \setminus D_i))}{\gamma} \end{aligned} \quad (7)$$

From Inequality (3), we get

$$\begin{aligned} & f((A_{i-1} \setminus D_i) \cup H_i) - f(A_{i-1} \setminus D_i) \\ \geq & \gamma[f(A_{i-1} \cup H_i) - f(A_{i-1})] \\ \geq & \frac{f(A_{i-1})}{k} \geq \frac{f(A_{i-1} \setminus D_i)}{k}, \end{aligned} \quad (8)$$

where the second inequality is obtained by the condition of the addition in the round. Finally, applying Inequalities (7) and (8), we obtain

$$\begin{aligned} f(A_i) = & f((A_{i-1} \setminus D_i) \cup H_i) \\ \geq & (1 + \frac{1}{k}) f(A_{i-1} \setminus D_i) \\ \geq & \frac{1 + \frac{1}{k}}{1 + \frac{1}{(k^{r-1}-1)}} \gamma f(A_{i-1}) \geq \gamma f(A_{i-1}), \end{aligned} \quad (9)$$

where the last inequality holds because of $r \geq 3$ and $k \geq 2$.
If i-th iteration is not the first round with deletion, the last round recorded with deletion of i-th iteration is the i'-th iteration. We discuss the change of function value that the set from $A_{i'}$ to A_i:

From the end of the i'-th iteration to the end of the i-th iteration, there have been $cr(\lceil \gamma k \rceil + 1)log_2(k)$ elements added in the A, which add precisely the elements in $A_i \setminus A_i'$. It is easy to see that $D_i = A_i'$.

$$\begin{aligned} f(A_i) - f(A_i') =& f(A_{i-1} \cup H_i \setminus D_i) - f(A_i') \\ \geq& \gamma[f(A_{i-1} \cup H_i) - f(D_i)] - f(A_i') \\ \geq& \gamma \cdot (1 + \frac{1}{\gamma k})^{r(\lceil k\gamma \rceil + 1)\lceil \log_2(k) \rceil} . f(A_i') - (\gamma + 1) f(A_i') \\ \geq& [\gamma(k^r - 1) - 1] \cdot f(A_i') \end{aligned} \tag{10}$$

where the first and the second inequalities are established on the same basis as the first two inequalities of inequality (7); the third inequality is obtained by Lemma 2 because f is non-negative function, $\varepsilon \leq \gamma k^2$, $\gamma(k^r - 1) - 1 \geq 0$, the inequality at least is 0.

Next, we limit the total value of $f(A)$ lost from deleted elements during the entire algorithm run. Let $A^* = \cup_{0 \leq i \leq n} A_i$ denote all stored elements of A and e_i denote the element received in i-th iteration.

Lemma 4. $f(A^*) \leq \frac{1}{\gamma^2}(1 + \frac{1}{\gamma k^{r-1} - 1}) f(A_n)$

Proof. Suppose there were m sets deleted from A, recorded as $A^* \setminus A_n = \{D_j, 1 \leq j \leq m\}$ where each D_j is deleted on Line 10 of the algorithm, and they are sorted with the reverse order in which they were deleted such that $j_1 < j_2$ implies D_{j_2} was deleted before D_{j_1}. Set $D_0 = A_n$. For any $j \in [0, m-1]$, from $A = D_{j+1}$ to $A = D_j$, there are at least $cr(\lceil k\gamma \rceil + 1)\lceil \log_2(k) \rceil$ elements added to A for every number of c elements arrived and deleted exactly once. Moreover, each addition increases the value of $f(A)$ by at least $\frac{f(A)}{\gamma k}$ except in the deletion case. Hence, by Lemma 3,

$$\begin{aligned} f(D_{i-1}) \geq& \cdot (1 + \frac{1}{\gamma k})^{r(\lceil k\gamma \rceil + 1)\lceil \log_2(k) \rceil - 1} . \gamma \cdot f(D_i) \\ \geq& \gamma k^{r-1} \cdot f(D_i) \end{aligned} \tag{11}$$

Therefore, for any $0 \leq j \leq m$, $f(A_n) = f(D_0) \geq (\gamma k^{r-1})^j \cdot f(D_j)$. Then,

$$\begin{aligned} f(A^*) \leq& f(A_n) + \frac{1}{\gamma} f(A^* \setminus A_n) \\ \leq& \frac{1}{\gamma^2} \sum_{j=0}^{m} f(D_j) \\ \leq& \frac{1}{\gamma^2} \sum_{j=0}^{m} \frac{1}{(\gamma k^{r-1})^j} f(A_n) \\ \leq& \frac{1}{\gamma^2} \cdot \frac{1}{1 - \frac{1}{\gamma k^{r-1}}} \cdot f(A_n) = \frac{1}{\gamma^2}(1 + \frac{1}{\gamma k^{r-1} - 1}) \cdot f(A_n), \end{aligned} \tag{12}$$

where the first and second inequalities are received by Lemma 2 and $\gamma \in (0, 1]$, while the forth inequality is the summation of the geometric series.

Then, we bound the value of OPT with the $f(A_n)$.

Lemma 5. $OPT \leq \frac{1}{\gamma^2}(1+\gamma^2+\frac{1}{\gamma k^{r-1}-1})f(A_n)$

Proof. O is one of the optimal solutions for the problem. Suppose for each element $o \in O$, $i(o)$ is the iteration in the for loop that processes the element o, and we let $\overline{i(o)}$ be the first iteration with delete operation after the $i(o)$-th iteration. Therefore, we have

$$f(A_{i(o)-1}) \leq f(A_{\overline{i(o)}-1}) \leq \frac{1}{\gamma} f(A_{\overline{i(o)}}) \leq \frac{1}{\gamma} f(A_n)$$

where all inequalities hold from the Lemma 3. Then,

$$\begin{aligned} f(O) - f(A^*) \leq & f(O \cup A^*) - f(A^*) \\ \leq & \sum_{o \in O \setminus A^*} \frac{1}{\gamma}[f(A^* \cup \{o\}) - f(A^*)] \\ \leq & \sum_{o \in O \setminus A^*} \frac{1}{\gamma^2}[f(A_{i(o)-1} \cup \{o\}) - f(A_{i(o)-1})] \\ \leq & \sum_{o \in O \setminus A^*} \frac{1}{\gamma^2}[f(A_{i(o)-1} \cup H_i(o)) - f(A_{i(o)-1})] \\ < & \sum_{o \in O \setminus A^*} \frac{f(A_{i(o)-1})}{\gamma^3 k} \\ \leq & \sum_{o \in O \setminus A^*} \frac{f(A_n)}{\gamma^4 k} \\ \leq & \frac{f(A_n)}{\gamma^4}, \end{aligned} \tag{13}$$

where the first and the forth inequalities are received by the monotonicity of f; the second inequality is obtained by the Lemma 2; the third inequality holds from the definition of generic submodularity ratio; the condition of added element in A in the algorithm makes the fifth inequality true; and the last inequality holds because of the size of O. Finally, from Lemma 4 and inequality (13), we have

$$OPT \leq \frac{1}{\gamma^2}(1+\gamma^2+\frac{1}{\gamma k^r - 1})f(A_n)$$

Recall that the final solution of the Single-pass Deterministic algorithm has the largest function value of each k elements from the ck elements recently added to A_n. We record the ck elements which are recently added to A_n as $\widehat{A}$. Then, we discuss the relationship of the value of $\widehat{A}$ and OPT.

Lemma 6.

$$OPT \leq \left(\frac{1+\gamma+\gamma^2+\gamma^3}{\gamma^3} + \frac{1+\gamma}{\gamma^3(\gamma k^{r-1}-1)})f(\widehat{A})\right)$$

Proof. First, we discuss the relationship of the values of $f(\widehat{A})$ and $f(A_n)$.

If $\mid A \mid \leq ck$, $f(\widehat{A}) = f(A_n)$, the lemma holds.

Suppose $\mid A \mid > ck$. Let $\widehat{A} = \{D_1^{'}, D_2^{'}, \cdots, D_k^{'}\}$, $\mid D_j^{'} \mid = c(j \in [1,k])$, in the order these sets were added to A. Then,

$$\begin{aligned} f(\widehat{A}) \geq & \gamma[f(A_n) - f(A_n \setminus \widehat{A})] \\ = & \gamma \sum_{i=1}^{k} [f((A_n \setminus \widehat{A}) \cup \{D_1^{'}, D_2^{'}, \cdots, D_i) - f((A_n \setminus \widehat{A}) \cup \{D_1^{'}, D_2^{'}, \cdots, D_{i-1})] \\ \geq & \gamma \sum_{i=1}^{k} \frac{f((A_n \setminus \widehat{A}) \cup \{D_1^{'}, D_2^{'}, \cdots, D_{i-1})}{\gamma k} \\ \geq & \sum_{i=1}^{k} \frac{f(A_n \setminus \widehat{A})}{k} = f(A_n \setminus \widehat{A}), \end{aligned} \tag{14}$$

where the first inequality is obtained by the Lemma 1; the condition of added element in A in the algorithm makes the second inequality true; and the third inequality is received by the monotonicity of f. Thus

$$f(A_n) \leq f(A_n \setminus \widehat{A}) + \frac{f(\widehat{A})}{\gamma} \leq \frac{1+\gamma}{\gamma} f(\widehat{A})$$

Finally, from Lemma 5 and the above inequality, we get

$$OPT \leq \left(\frac{1+\gamma+\gamma^2+\gamma^3}{\gamma^3} + \frac{1+\gamma}{\gamma^3(\gamma k^{r-1}-1)}\right) f(\widehat{A})$$

Then, we can give the proof of Theorem 1.

Proof. First, we show the approximation ratio: find the relationship between $f(S)$ and $f(\widehat{A})$, and bring in Lemma 6 to obtain the approximate ratio.

When $c = 1$, $S = \widehat{A}$,

$$\begin{aligned} f(S) = f(\widehat{A}) \geq & \frac{1}{\frac{1+\gamma+\gamma^2+\gamma^3}{\gamma^3} + \frac{1+\gamma}{\gamma^3(\gamma k^{r-1}-1)}} OPT \\ \geq & \left[\frac{\gamma^3}{1+\gamma+\gamma^2+\gamma^3} - \frac{\gamma^2}{(1+\gamma)(1+\gamma^2)(k^{r-1}+\gamma^2 k^{r-1}-\gamma)}\right] OPT \\ \geq & \left[\frac{\gamma^3}{1+\gamma+\gamma^2+\gamma^3} - \frac{1}{k^{r-1}-1}\right] OPT. \end{aligned} \tag{15}$$

Suppose $c > 1$, and $\{S_1, \cdots, S_c\}$ is the partition of $\widehat{A}$. Then

$$\begin{aligned} f(\widehat{A}) \leq & \frac{1}{\gamma} \sum_{j=1}^{c} f(S_j) \\ \leq & \frac{c}{\gamma} \max_{1 \leq j \leq c} f(S_j) = \frac{c}{\gamma} f(S) \end{aligned} \tag{16}$$

Consequently,

$$f(S) \geq \frac{\gamma}{c} f(\widehat{A}) \geq [\frac{\gamma^4}{c(1+\gamma+\gamma^2+\gamma^3)} - \frac{1}{k^{r-1}-1}]OPT.$$

By the choice of r, $f(S) \geq [\frac{\gamma^4}{c(1+\gamma+\gamma^2+\gamma^3)} - \varepsilon]OPT$.
Then, we consider the queries of the algorithm by two parts. The first part is to obtain the value when every c elements arrive. So the number of queries in this part is $\lceil n/c \rceil$; the second part is the number of queries made while comparing the values of at most c candidate solutions; so, the number of queries is at most $\lceil n/c \rceil + c$.
Finally, we show the memory complexity of the algorithm. It depends on the size of A, S and H. Obviously, the lengths of S and H are at most k and c, respectively. The length of A could reach $2cr(\lceil k\gamma \rceil + 1)\lceil \log_2(k) \rceil$ depending on the choice of r. Recall that $r = \lceil \log_2(1/(\varepsilon)) \rceil + 3$. Therefore, the memory complexity of algorithm is $O(ck \log(k) \log(1/\varepsilon))$.

From [15], we know the approximation ratio of the latest deterministic, single-pass streaming algorithm with linear time complexity for the submodular maximization subject to a cardinality constraint problem is $1/(4c) - \varepsilon$. When $\gamma = 1$, our algorithm has an approximation ratio consistent with the above results, also has the same query complexity and almost consistent storage memory complexity.

4 Discussion

In this paper, we propose the first linear-time single-pass deterministic streaming algorithm for the maximization of a monotone non-submodular function with a cardinality constraint. The key idea to reducing the time complexity of this algorithm is to replace the querying function value of each arriving element with querying the function value of c elements, and to compare the marginal gain of adding these c elements with the function value of the entire storage set rather than comparing the function value of each element in the storage set. Our algorithm, while effective in query and run in a reasonable time on large instance, may output a less ideal solution. One future work is to improve the approximation ratio of the algorithm without increasing the time complexity and memory complexity.

Acknowledgements. The first author is supported by Beijing Natural Science Foundation Project No. Z200002 and National Natural Science Foundation of China (No. 12131003). The second author is supported by the Natural Sciences and Engineering Research Council of Canada (NSERC) grant 06446, and Natural Science Foundation of China (Nos. 11771386, 11728104). The third author is supported by National Natural Science Foundation of China (No. 11201333). The fourth author is supported by the Fundamental Research Funds for the Central Universities (No. E1E40108X2) and National Natural Science Foundation of China (No. 12101587).

References

1. Badanidiyuru, A., Mirzasoleiman, B., Karbasi, A., Krause, A.: Streaming submodular maximization: massive data summarization on the fly. In: 20th International Proceedings on SIGKDD, pp. 671–680. ACM, New York, USA (2014)
2. Bian, A.A., Buhmann, J.M., Krause, A., Tschiatschek, S.: Guarantees for greedy maximization of non-submodular functions with applications. In: 34th International Proceedings on ICML, pp. 498–507. PMLR, Sydney, NSW, Australia (2017)
3. Buchbinder, N., Feldman, M., Schwartz, R.: Online submodular maximization with preemption. ACM Trans. Algorithms **15**(3), 1–31 (2019)
4. Caldarola, E.G., Rinaldi, A.M.: Big data: a survey. In: 4th International Proceedings on DATA, pp. 362–370. SciTePress, Colmar, Alsace, France (2015)
5. Chakrabarti, A., Kale, S.: Submodular maximization meets streaming: matchings, matroids, and more. Math. Program. **154**, 225–247 (2015)
6. Conforti, M., Cornuéjols, G.: Submodular set-functions, matroids and the greedy algorithm - tight worst-case bounds and some generalizations of the rado-edmonds theorem. Discrete Appl. Math. **7**(3), 251–274 (1984)
7. Cui, M., Xu, D., Guo, L., Wu, D.: Approximation guarantees for parallelized maximization of monotone non-submodular function with a cardinality constraint. In: Zhang, Z., Li, W., Du, D.-Z. (eds.) AAIM 2020. LNCS, vol. 12290, pp. 195–203. Springer, Cham (2020). https://doi.org/10.1007/978-3-030-57602-8_18
8. Das, A., Kempe, D.: Submodular meets spectral: greedy algorithms for subset selection, sparse approximation and dictionary selection. In: 28th International Proceedings of ICML, pp. 1057–1064. Omnipress, Bellevue, Washington, USA (2011)
9. Elenberg, E. R., Dimakis, A. G., Feldman, M., Karbasi, A.: Streaming weak submodularity: interpreting neural networks on the fly. In: 31st International Proceedings on NIPS, pp. 4044–4054. Long Beach, CA, USA (2017)
10. Fahrbach, M., Mirrokn, V., Zadimoghaddam, M.: Submodular maximization with nearly optimal approximation, adaptivity and query complexity. In: 30th International Proceedings on SODA, pp. 255–273. SIAM, San Diego, CA, USA (2019)
11. Feldman, M., Norouzi-Fard, A., Svensson, O., Zenklusen, R.: The one-way communication complexity of submodular maximization with applications to streaming and robustness. In: 52nd International Proceedings on STOC, pp. 1363–1374. ACM, Chicago, IL, USA (2020)
12. Gong, S., Nong, Q., Liu, W., Fang, Q.: Parametric monotone function maximization with matroid constraints. J. Glob. Optim. **75**(3), 833–849 (2019). https://doi.org/10.1007/s10898-019-00800-2
13. Kazemi, E., Mitrovic, M., Zadimoghaddam, M., Lattanzi, S., Karbasi, A.: Submodular streaming in all its glory: tight approximation, minimum memory and low adaptive complexity. In: 36th International Proceedings of ICML, pp. 3311–3320. PMLR, Long Beach, California, USA (2019)
14. Kempe, D., Kleinberg, J., Tardos, E.: Maximizing the spread of influence through a social network. Theor. Comput. **11**(1), 105–147 (2015)
15. Kuhnle, A.: Quick streaming algorithms for maximization of monotone submodular functions in linear time. In: 24th International Proceedings on AISTATS, pp. 13–15. PMLR, Virtual Event (2021)
16. Kuhnle, A., Smith, J., Crawford, V.G., Thai, M.T.: Fast maximization of non-submodular, monotonic functions on the integer lattice. In: 35th International Proceedings of ICML, pp. 2786–2795. PMLR, Stockholm, Sweden (2018)

17. Lakkaraju, H., Bach, S.H., Leskovec, J.: Interpretable decision sets: a joint framework for description and prediction. In: 22nd International Proceedings of SIGKDD, pp. 1675–1684. ACM, San Francisco, CA, USA (2016)
18. Larcker, D.F., Watts, E.M.: Where's the greenium? J. Account. Econ. **69**(2), 101312 (2020)
19. Levin, R., Wajc, D.: Streaming submodular matching meets the primal-dual method. In: 32nd International Proceedings on SODA, pp. 1914–1933. SIAM, Virtual Conference (2021)
20. Li, M., Zhou, X., Tan, J., Wang, W.: Non-submodular streaming maximization with minimum memory and low adaptive complexity. In: Zhang, Z., Li, W., Du, D.-Z. (eds.) AAIM 2020. LNCS, vol. 12290, pp. 214–224. Springer, Cham (2020). https://doi.org/10.1007/978-3-030-57602-8_20
21. Lin, H., Bilmes, J. A.: A class of submodular functions for document summarization. In: 49th International Proceedings on ACL, pp. 510–520. The Association for Computer Linguistics, Portland, Oregon, USA (2011)
22. Lin, Y., Chen, W., Lui, J.C.: Boosting information spread: an algorithmic approach. In: 33rd International Proceedings on ICDE, pp. 883–894. IEEE Computer Society, San Diego, CA, USA (2017)
23. Liu, B., Hu, M.: Fast algorithms for maximizing monotone nonsubmodular functions. In: Zhang, Z., Li, W., Du, D.-Z. (eds.) AAIM 2020. LNCS, vol. 12290, pp. 204–213. Springer, Cham (2020). https://doi.org/10.1007/978-3-030-57602-8_19
24. McGregor, A.: Graph stream algorithms: a survey. In: 31st International Proceedings of ICML, pp. 9–20. JMLR.org, Beijing, China (2014)
25. Mirzasoleiman, B., Karbasi, A., Badanidiyuru, A., Krause, A.: Distributed submodular cover: succinctly summarizing massive data. In: 29th International Proceedings on NIPS, pp. 2881–2889. Montreal, Quebec, Canada (2015)
26. Mirzasoleiman, B., Karbasi, A., Sarkar, R., Krause, A.: Distributed submodular maximization: identifying representative elements in massive data. In: 27th International Proceedings on NIPS, pp. 2049–2057. Lake Tahoe, Nevada, USA (2013)
27. Mossel, E., Roch, S.: On the submodularity of influence in social networks. In: 39th International Proceedings on STOC, pp. 128–134. ACM, San Diego, California, USA (2007)
28. Muthukrishnan, S.: Data streams: algorithms and applications. Theoret. Comput. Sci. **1**(2), 117–236 (2005)
29. Nong, Q., Sun, T., Gong, S., Fang, Q., Du, D., Shao, X.: Maximize a monotone function with a generic submodularity ratio. In: Du, D.-Z., Li, L., Sun, X., Zhang, J. (eds.) AAIM 2019. LNCS, vol. 11640, pp. 249–260. Springer, Cham (2019). https://doi.org/10.1007/978-3-030-27195-4_23
30. Norouzi-Fard, A., Tarnawski, J., Mitrovic, S., Zandieh, A., Mousavifar, A., Svensson, O.: Beyond 1/2-Approximation for submodular maximization on massive data streams. In: 35th International Proceedings of ICML, pp. 3826–3835. PMLR, Stockholm, Sweden (2018)
31. Ribeiro, M. T., Singh, S., Guestrin, C.: "Why should I trust you?": explaining the predictions of any classifier. In: 22nd International Proceedings of SIGKDD, pp. 1135–1144. ACM, San Francisco, CA, USA (2016)
32. Sviridenko, M., Vondrák, J., Ward, J.: Optimal approximation for submodular and supermodular optimization with bounded curvature. In: 26th International Proceedings on SODA, pp. 1134–1148. SIAM, San Diego, CA, USA (2015)

33. Wang, Y., Xu, D., Wang, Y., Zhang, D.: Non-submodular maximization on massive data streams. J. Glob. Optim. **76**(4), 729–743 (2019). https://doi.org/10.1007/s10898-019-00840-8
34. Wei, K., Libbrecht, M.W., Bilmes, J.A., Noble, W.S.: Choosing panels of genomics assays using submodular optimization. Genom. Biol. **17**(1), 229 (2016)

Approximation Algorithms for Two Parallel Dedicated Machine Scheduling with Conflict Constraints

An Zhang[1(✉)], Liang Zhang[1], Yong Chen[1], Guangting Chen[2(✉)], and Xing Wang[1]

[1] Department of Mathematics, Hangzhou Dianzi University, Hangzhou 310018, China
{anzhang,liangzhang,chenyong,wx198491}@hdu.edu.cn
[2] Taizhou University, Linhai 317000, China
gtchen@hdu.edu.cn

Abstract. We investigate two parallel dedicated machine scheduling with conflict constraints. The problem of minimizing the makespan has been shown to be NP-hard in the strong sense under the assumption that the processing sequence of jobs on one machine is given and fixed a priori. The problem without any fixed sequence was previously recognized as weakly NP-hard. In this paper, we first present a $\frac{9}{5}$-approximation algorithm for the problem with a fixed sequence. Then we show that the tight approximation ratios of the algorithm are $\frac{7}{4}$ and $\frac{5}{3}$ for two subproblems which remain strongly NP-hard. We also send an improved algorithm with approximation ratio $3 - \sqrt{2} \approx 1.586$ for one subproblem. Finally, we prove that the problem without any fixed sequence is actually strongly NP-hard, and design a $\frac{5}{3}$-approximation algorithm to solve it.

Keywords: Parallel dedicated machines · Conflict graph · Approximation algorithm · NP-hard

1 Introduction

In the problem of scheduling with conflict constraints (SWC in short), a set N of jobs is supposed to be processed by a set of machines subject to a conflict graph $G = (N, E)$ in which the conflict constraints are specified among jobs (vertices). Two jobs connected by an edge in the graph are called in conflict in the sense that they cannot be processed concurrently. The objective is to find a conflict-free schedule with the minimum makespan. The SWC problem arises in a more general scheduling model where a set of renewable resources is provided for processing the jobs along with the set of machines [1]. Specifically, there are λ resources each with a total availability capacity of σ units. Each job consumes at most ρ ($\leq \sigma$) units of each resource at any time of its processing, and the availability of that amount will be released at the completion of the job. The general resource constraints can be represented by a conflict hypergraph while a conflict graph suffices for the special case where resources are non-shareable,

D.-Z. Du et al. (Eds.): COCOA 2021, LNCS 13135, pp. 111–124, 2021.
https://doi.org/10.1007/978-3-030-92681-6_10

that is, the case of $\sigma = \rho = 1$. Some applications of this model are mentioned by Baker and Coffman [2] in balancing the load in a parallel computation, and by Halldórsson et al. [3] in traffic intersection control, frequency assignment in cellular networks, and session management in local area networks.

Scheduling jobs of arbitrary processing time on a constant number of parallel machines is weakly NP-hard even if the conflict graph is empty. Even et al. [4] proposed a polynomial time algorithm for the two machine case where jobs have processing time either 1 or 2. The problem with the same settings except that jobs might have different release time in $\{0, r\}$ is shown to be NP-hard in the strong sense even in complements of bipartite graphs [11]. In the same paper [11], Bendraouche and Boudhar also proved that the two machine SWC problem is strongly NP-hard in complements of bipartite graphs when all jobs have processing time in $\{1, 2, 3\}$. Even et al. [4] proved that the two machine SWC problem with processing time in $\{1, 2, 3, 4\}$ is APX-hard. Bendraouche et al. [10] extended a result of [4] by sending a strong NP-hardness proof for all cases with processing time in $\{a, 2a + b\}$ $(b \neq 0)$ in complements of bipartite graphs. Mohabeddine and Boudhar [12] showed that the two machine SWC problem with arbitrary processing times is polynomial solvable in complements of caterpillars or cycles, and weakly NP-hard in complements of trees.

Scheduling jobs of unit processing time on m parallel machines are considered by Baker and Coffman [2] which they called the mutual exclusion scheduling (MES) problem. As Even et al. observed [4], the MES problem is equivalent to finding a minimum coloring of the conflict graph in which every color class contains at most m vertices, known as m bounded coloring. In particular, an optimal schedule for $m = 2$ can be obtained from a maximum matching on the complement of the conflict graph. It has been shown to be strongly NP-hard in general conflict graphs for $m \geq 3$ [2] and in various special conflict graphs, such as chordal graphs [6], complements of line graphs [7] and complements of comparability graphs [8] for $m \geq 3$, bipartite graphs, cographs and interval graphs for $m \geq 4$ [5], and permutation graphs for $m \geq 6$ [8]. Even et al. [4] further pointed out that the hardness proof given by Petrank [9] on a related problem can be applied to the MES problem so that the latter turns out to be APX-hard even for $m = 3$.

In [1], Garey and Graham analyzed a type of greedy list scheduling algorithm for the more general scheduling problem in conflict hypergraphs, which implies an approximation ratio of $\frac{m+1}{2}$. The tightness is confirmed by Even et al. [4] in conflict graphs. The same paper [4] also proposed an improved algorithm with approximation ratio $\frac{4}{3}$ for the two machine SWC problem with processing time in $\{1, 2, 3\}$. The SWC problem with unit processing time can be formulated as the m-set cover problem for which Fürer and Yu [13] presented a packing-based algorithm with approximation ratio $\mathcal{H}_m - 0.642 + \Theta(\frac{1}{m})$, where $\mathcal{H}_m = \sum_{i=1}^{m} \frac{1}{i}$ is the harmonic number. The best known approximation results on this problem are claimed by Yu in his dissertation thesis [14]. For mixed integer programming formulations and exact solutions of the SWC problem, we refer to a recent paper given by Hà et al. [15].

Hong and Lin [16] studied a SWC problem where jobs are allocated to machines in advance so that each machine is dedicated for processing a fixed set of jobs. They [16] observed that the two machine SWC problem $PD2|G = (N, E)|C_{max}$ can be regarded as two machine scheduling under multiple resource constraints, that is, $PD2|res\lambda\sigma\rho|C_{max}$, and hence is weakly NP-hard [17]. In [16], Hong and Lin proved that the problem becomes strongly NP-hard if the sequence of jobs on one machine is given and fixed a priori, and the problem with fixed sequences on both machines admits a polynomial time optimal algorithm. As special cases of the SWC problem on parallel dedicated machines, the resource constrained scheduling problems $PD2|res1\sigma\rho|C_{max}$ and $PD2|res211|C_{max}$ are polynomially solvable [17], and $PD2|res222|C_{max}$, $PD2|res311|C_{max}$ and $PD3|res111|C_{max}$ are weakly NP-hard [17,18]. Kellerer and Strusevich [18] proved that the problem $PDm|res111|C_{max}$ is strongly NP-hard if the number m of machines is an input. They presented a group technique algorithm with approximation ratio $\frac{3}{2} - \frac{1}{2m}$ for odd m and $\frac{3}{2} - \frac{1}{2(m-1)}$ for even m, and improved the algorithm to approximation ratio $\frac{5}{4}$ for $m = 3, 4$, and finally a PTAS for any constant m. In [17], Kellerer and Strusevich designed a 2-approximation algorithm and a PTAS for a special case of the problem $PDm|res\lambda11|C_{max}$ where each job consumes at most one of the λ resources.

In this paper, we revisit the SWC problem on two parallel and dedicated machines [16], that is, $PD2|G = (N, E)|C_{max}$. Our first contribution is a $\frac{9}{5}$-approximation algorithm for the strong NP-hard case where the processing sequence on one machine is given and fixed a priori. We also analyze the tight approximation ratios of this algorithm for two subproblems of the case which remain strongly NP-hard. An improved algorithm with approximation ratio $(3 - \sqrt{2})$ is further designed for one subproblem. Our next contribution is a strong NP-hardness proof for the problem without any fixed sequence, which is previously known weakly NP-hard [16]. This also gives the complexity of the resource constrained scheduling problem $PDm|res\lambda11|C_{max}$ for any constant $m \geq 2$ and an input number of non-sharable resources each with unit availability capacity [17]. Finally, we provide a $\frac{5}{3}$-approximation algorithm to solve the SWC problem without any fixed sequence. The next section gives a formal definition of the problem. Section 3 and Sect. 4 considers the problem with a fixed sequence and without any fixed sequences respectively. Some concluding remarks are made in Sect. 5.

2 Problem Statement

Given two parallel dedicated machines and a set N of jobs which is partitioned in advance into two disjoint subsets, N_1 and N_2, so that the i-th machine is dedicated for processing jobs in $N_i, i = 1, 2$. Let $N_1 = \{J_{1,1}, J_{1,2}, \cdots, J_{1,n_1}\}$ and $N_2 = \{J_{2,1}, J_{2,2}, \cdots, J_{2,n_2}\}$, a job $J_{1,s} \in N_1$ and a job $J_{2,t} \in N_2$ are called in conflict if they are not allowed to be processed concurrently. The conflict constraints are specified by an undirected bipartite graph $G = (N, E)$, where the vertex set $N = N_1 \cup N_2$ is composed of all the jobs in N_1 and N_2, and an edge connecting two jobs $J_{1,s}$ and $J_{2,t}$ indicates that they are conflicting. Any job must be processed non-preemptively on its machine till completion and any machine can process at most one job at a time. Let $p_{i,j}$ denote the processing time of job $J_{i,j}$, and $C_{i,j}$ be its completion time. Then the makespan of a schedule is defined by the maximum completion time of all jobs, that is, $C_{max} = \max_{i=1,2;j=1,2,\cdots,n_i}\{C_{i,j}\}$. The problem $PD2|G = (N, E)|C_{max}$ asks for a conflict-free schedule with the minimum makespan.

The problem has been recognized weakly NP-hard [16] since its special cases, $PD2|res222|C_{max}$ and $PD2|res311|C_{max}$, are already weakly NP-hard [17]. By polynomially reducing the 3-partition problem (3PP in short) to the problem where the processing sequence on one machine is given and fixed a priori, Hong and Lin [16] proved that the problem in this case becomes NP-hard in the strong sense. One sees from the reduction that the problem remains strongly NP-hard even if each job in N_1 has no larger processing time than each job in N_2, or vice versa, each job in N_1 has strictly larger processing time than each job in N_2. In reminder of the paper, we denote the problem with a fixed sequence by $PD2|G = (N, E), seq1|C_{max}$, and refer to the two cases defined above as Subproblem 1 and Subproblem 2 of it respectively.

Let $T_i = \sum_{j=1}^{n_i} p_{i,j}, i = 1, 2$ and let C^*_{max} denote the optimal makespan, then the following lower bounds for C^*_{max} trivially hold from the dedicated machine settings:

$$C^*_{max} \geq \max\{T_1, T_2\}. \tag{1}$$

3 The SWC Problem with a Fixed Sequence

This section focuses on the SWC (sub)problem where the processing sequence of jobs on one machine is given and fixed a priori. Without loss of generality, let the sequence be fixed to $J_{1,1}, J_{1,2}, \cdots, J_{1,n_1}$ on the first machine.

3.1 A $\frac{9}{5}$-Approximation Algorithm

The schedule output by the algorithm Approx1 for the problem $PD2|G = (N, E), seq1|C_{max}$ is composed of a series of blocks which can be roughly divided into two types: type-A blocks and type-B blocks. In a type-A block denoted as $[A_t; J_{2,t}]$, there are a set A_t of consecutive jobs from N_1 and a single job $J_{2,t}$ from N_2. Jobs in A_t are processed one after another without any idle time in their fixed sequence, starting from the same time when the job $J_{2,t}$ starts to be processed (see Fig. 1 for an illustration), which is called the start time of the block. When Approx1 generates a type-A block, it also ensures that $\sum_{J_{1,j}\in A_t} p_{1,j} \leq p_{2,t} \leq 2\sum_{J_{1,j}\in A_t} p_{1,j}$. In a type-$B$ block denoted as $[J_{1,s}; B_s]$, there are a set B_s of jobs from N_2 and a single job $J_{1,s}$ from N_1. Jobs in B_s are processed one after another without any idle time, starting from the same time when the job $J_{1,s}$ starts to be processed (see Fig. 1 too). When Approx1 generates a type-B block, it also ensures that $\sum_{J_{2,j}\in B_s} p_{2,j} \leq p_{1,s}$.

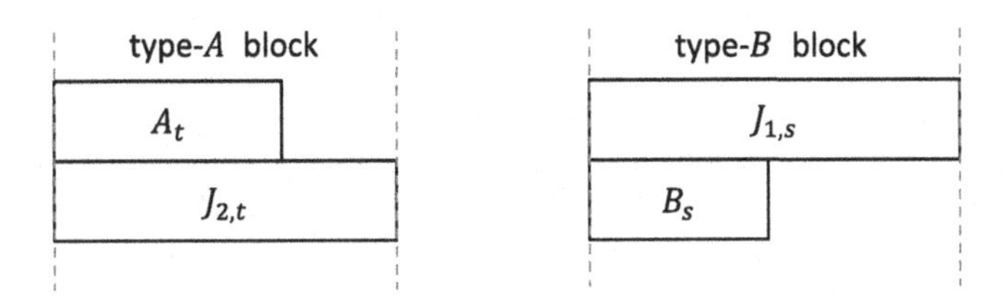

Fig. 1. Two types of blocks.

Each round the algorithm strives to find a type-A block, and replaces it with a type-B block if it fails. The start time of the newly found block is made no earlier than the completion time of the previous block, that is, the time when all the jobs in the previous block have been finished. Figure 2 shows an instance from which one can see how the algorithm APPROX1 outputs a conflict-free schedule with blocks. A high-level description of APPROX1 is depicted in Fig. 3.

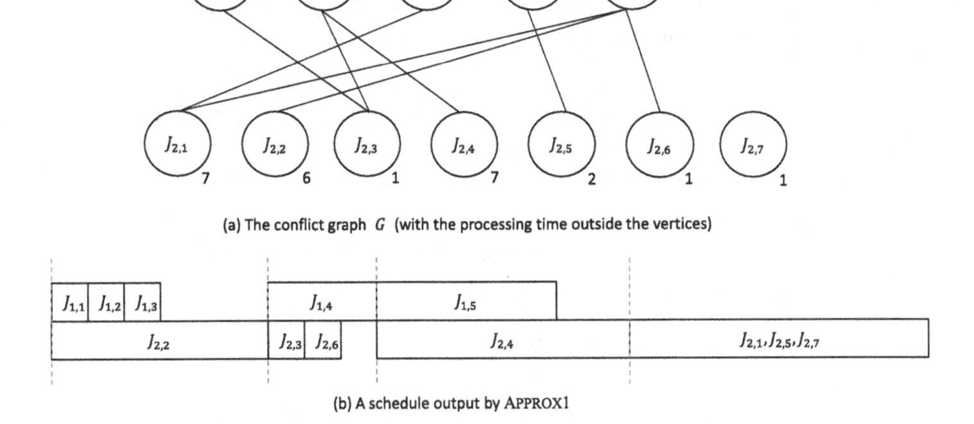

(a) The conflict graph G (with the processing time outside the vertices)
(b) A schedule output by APPROX1

Fig. 2. An instance solved by APPROX1.

Let $\mathcal{A}_i$ ($\mathcal{B}_i$), $i = 1, 2, 3$ be disjoint subsets of N_1 (N_2) defined as follows. For each type-A block $[A_t; J_{2,t}]$, $A_t \subseteq \mathcal{A}_1$ and $J_{2,t} \in \mathcal{B}_1$. For each type-$B$ block $[J_{1,s}; B_s]$ with $p_{1,s}/2 \leq \sum_{J_{2,j} \in B_s} p_{2,j} \leq p_{1,s}$, $J_{1,s} \in \mathcal{A}_2$ and $B_s \subseteq \mathcal{B}_2$. And for each type-$B$ block $[J_{1,s}; B_s]$ with $\sum_{J_{2,j} \in B_s} p_{2,j} < p_{1,s}/2$, $J_{1,s} \in \mathcal{A}_3$ and $B_s \subseteq \mathcal{B}_3$. Note $\cup_{i=1}^{3} \mathcal{A}_i = N_1$. Let $\mathcal{B}_4 = N_2 \setminus \cup_{i=1}^{3} \mathcal{B}_i$. Note $\mathcal{B}_4$ consists of the remaining jobs of B in Step 4 of APPROX1. For simplicity, denote by a_i (b_i) the total processing time of jobs in $\mathcal{A}_i$ ($\mathcal{B}_i$). By Step 2 and Step 3 of APPROX1, and the above definitions, we have

$$a_1 \leq b_1 \leq 2a_1, \ \frac{a_2}{2} \leq b_2 \leq a_2, \ 0 \leq \ b_3 < \frac{a_3}{2},$$

Algorithm APPROX1:

1. Initially let $A = N_1$, $B = N_2$ and $s = 1$.
2. If there exists a job $J_{2,t} \in B$ which does not conflict with a set of consecutive jobs beginning with $J_{1,s}$, say $J_{1,s}, J_{1,s+1}, \cdots, J_{1,s'}$, and satisfies that $\sum_{j=s}^{s'} p_{1,j} \leq p_{2,t} \leq 2\sum_{j=s}^{s'} p_{1,j}$ (choose one arbitrarily if more than one candidate is found), then let $A_t = \{J_{1,s}, J_{1,s+1}, \cdots, J_{1,s'}\}$ and generate a type-A block $[A_t; J_{2,t}]$. Let $A = A \setminus A_t, B = B \setminus \{J_{2,t}\}$ and $s = s' + 1$. Go to Step 4.
3. Find all jobs in B with processing time smaller than $p_{1,s}$ and not conflicting with $J_{1,s}$. Denote them by a set B'_s (Note that B'_s might be empty).
   3.1 if $\sum_{J_{2,j} \in B'_s} p_{2,j} \leq p_{1,s}$, then let $B_s = B'_s$.
   3.2 if $\sum_{J_{2,j} \in B'_s} p_{2,j} > p_{1,s}$, then let B_s be a subset of B'_s such that $p_{1,s}/2 \leq \sum_{J_{2,j} \in B_s} p_{2,j} \leq p_{1,s}$. [a]
   3.3 generate a type-B block $[J_{1,s}; B_s]$. Let $A = A \backslash \{J_{1,s}\}, B = B \backslash B_s$ and $s = s + 1$.
4. If $s < n_1 + 1$, return Step 2. Otherwise, process the blocks as early as possible in the fixed sequence given by the jobs from N_1 in the blocks, provided that each block can start after the previous block is completed. Process all remaining jobs in B (if any) one after another without any idle time from the completion of the last block.

[a] We remark that B_s exists and can be found in polynomial time in this case. For example, we can reorder jobs in B'_s in the nondecreasing order of their processing time and select jobs sequentially until the total processing time reaches or exceeds $p_{1,s}/2$ for the first time.

Fig. 3. A high-level description of the algorithm APPROX1.

Which, together with (1), yields that

$$C^*_{max} \geq T_1 = a_1 + a_2 + a_3 \geq \frac{b_1}{2} + a_2 + a_3, \tag{2}$$

and

$$C^*_{max} \geq T_2 = b_1 + b_2 + b_3 + b_4 \geq b_1 + \frac{a_2}{2} + b_4. \tag{3}$$

Lemma 1. $C_{max} = b_1 + a_2 + a_3 + b_4$.

Proof. Since $\sum_{J_{1,j} \in A_t} p_{1,j} \leq p_{2,t}$ in any type-A block $[A_t; J_{2,t}]$, its completion time is determined by $J_{2,t} \in \mathcal{B}_1$. And similarly since $\sum_{J_{2,j} \in B_s} p_{2,j} \leq p_{1,s}$ in any type-B block $[J_{1,s}; B_s]$, its completion time is determined by $J_{1,s} \in \mathcal{A}_2 \cup \mathcal{A}_3$. Therefore, we can conclude

$$C_{max} = \sum_{J_{2,t} \in \mathcal{B}_1} p_{2,t} + \sum_{J_{1,s} \in \mathcal{A}_2 \cup \mathcal{A}_3} p_{1,s} + \sum_{J_{2,t} \in \mathcal{B}_4} p_{2,t} = b_1 + a_2 + a_3 + b_4.$$

Lemma 2. *For each type-B block* $[J_{1,s}; B_s]$ *with* $\sum_{J_{2,j}\in B_s} p_{2,j} < p_{1,s}/2$, *any job in* $\mathcal{B}_4$ *that does not conflict with* $J_{1,s}$ *(if any) must have processing time no smaller than* $p_{1,s}$.

Proof. Suppose not, let $J_{2,k} \in \mathcal{B}_4$ be the job with $p_{2,k} < p_{1,s}$ and not conflicting with $J_{1,s}$. Clearly this job has not been assigned yet when APPROX1 starts to generate this block. If $p_{2,k} + \sum_{J_{2,j}\in B_s} p_{2,j} \le p_{1,s}$, then Step 3.1 of the algorithm must assign $J_{2,k}$ into this block. If $p_{2,k} + \sum_{J_{2,j}\in B_s} p_{2,j} > p_{1,s}$, then Step 3.2 of the algorithm must update the block so that $p_{1,s}/2 \le \sum_{J_{2,j}\in B_s} p_{2,j} \le p_{1,s}$. Both implies a contradcition. This proves the lemma.

Lemma 3. *In an optimal schedule, the total processing overlap time of a job* $J_{2,t} \in \mathcal{B}_4$ *and the jobs in* $\mathcal{A}_3$ *cannot exceed* $\frac{p_{2,t}}{2}$.

Proof. Since the order for processing jobs on the first machine is fixed to $(J_{1,1}, J_{1,2}, \cdots, J_{1,n_1})$, all jobs that overlap with the processing of $J_{2,t}$ must follow that sequence. In an optimal schedule, let $J_{1,s}$ and $J_{1,s'}$ $(s' \ge s)$ be the first and the last job from $\mathcal{A}_3$ that overlap with $J_{2,t}$. If no such job exists or $\sum_{j=s}^{s'} p_{1,j} < \frac{p_{2,t}}{2}$, then we are already done. Due to $J_{1,s}, J_{1,s'} \in \mathcal{A}_3$ and $s' \ge s$, the algorithm APPROX1 must first generate the type-B block $[J_{1,s}; B_s]$ with $\sum_{J_{2,j}\in B_s} p_{2,j} < p_{1,s}/2$, and then the type-$B$ block $[J_{1,s'}; B_{s'}]$ with $\sum_{J_{2,j}\in B_{s'}} p_{2,j} < p_{1,s'}/2$, but not necessarily immediately. Since $J_{1,s}$ and $J_{1,s'}$ overlap with the processing of $J_{2,t}$ in the optimal schedule, none of the jobs between $J_{1,s}$ and $J_{1,s'}$ conflict with $J_{2,t}$. Hence, if $\sum_{j=s}^{s'} p_{1,j} \le p_{2,t} \le 2\sum_{j=s}^{s'} p_{1,j}$, then APPROX1 must find the job $J_{2,t}$ in Step 2 when $J_{1,s}$ is involved by the algorithm, and thus generate a type-A block so that $J_{2,t}$ is included in $\mathcal{B}_1$. It is a contradiciton since $J_{2,t}$ is actually included in $\mathcal{B}_4$. Therefore, we suppose in what follows that $\sum_{j=s}^{s'} p_{1,j} > p_{2,t}$. If $p_{1,s} \le p_{2,t} \le 2p_{1,s}$, then the similar argument can lead to a contradiction. By Lemma 2, we are only left with the case when $p_{2,t} > 2p_{1,s}$.

Let k be the smallest index such that $\sum_{j=s}^{k} p_{1,j} \ge \frac{p_{2,t}}{2}$. Then k is well-defined and $s < k \le s'$, since we have $p_{1,s} < \frac{p_{2,t}}{2}$ and $\sum_{j=s}^{s'} p_{1,j} > p_{2,t}$. If $\sum_{j=s}^{k} p_{1,j} \le p_{2,t}$, then we must obtain a contradition by the similar argument as above. Otherwise, $\sum_{j=s}^{k} p_{1,j} > p_{2,t}$. From the definition of k, we obtain $\sum_{j=s}^{k-1} p_{1,j} < \frac{p_{2,t}}{2}$, which follows that

$$p_{1,k} = \sum_{j=s}^{k} p_{1,j} - \sum_{j=s}^{k-1} p_{1,j} > \frac{p_{2,t}}{2}. \tag{4}$$

If $J_{1,k} \in \mathcal{A}_3$, then by Lemma 2, $p_{2,t} \ge p_{1,k}$. Together with (4), APPROX1 must find the job $J_{2,t}$ in Step 2 when $J_{1,k}$ is involved by the algorithm, and thus generate a type-A block so that $J_{2,t}$ is included in $\mathcal{B}_1$ too, contradicting with the fact that $J_{2,t} \in \mathcal{B}_4$. Therefore, $J_{1,k} \notin \mathcal{A}_3$.

Since $J_{1,s'} \in \mathcal{A}_3$, $s < k < s'$. Note in the optimal schedule, $J_{1,s}$ and $J_{1,s'}$ are the first and the last job from $\mathcal{A}_3$ that overlap with $J_{2,t}$. This means that the job $J_{1,k}$ must be entirely processed within the processing interval of $J_{2,t}$. By $J_{1,k} \notin \mathcal{A}_3$ and (4), the total processing overlap time of $J_{2,t}$ and the jobs in $\mathcal{A}_3$ is no more than $p_{2,t} - p_{1,k} < \frac{p_{2,t}}{2}$. This proves the lemma.

Lemma 4.

$$C^*_{max} \geq a_3 + \frac{b_4}{2}. \tag{5}$$

Proof. By Lemma 3, the total overlap time of jobs in $\mathcal{B}_4$ and $\mathcal{A}_3$ does not exceed $\frac{b_4}{2}$. Then the total processing time of jobs in $\mathcal{A}_3$ that do not overlap with any job in $\mathcal{B}_4$ must be at least $\max\{a_3 - \frac{b_4}{2}, 0\}$. Thus $C^*_{max} \geq \max\{a_3 - \frac{b_4}{2}, 0\} + b_4 \geq a_3 + \frac{b_4}{2}$.

Theorem 1. APPROX1 *is a* $\frac{9}{5}$*-approximation algorithm for* $PD2|G = (N, E), seq1|C_{max}$.

Proof. Let us multiply both sides of inequalities (2), (3), and (5) by $\frac{3}{5}$, $\frac{4}{5}$ and $\frac{2}{5}$, respectively. Then, one can easily verify that the summation of both sides of the resulting inequalities is

$$\begin{aligned}\frac{9}{5}C^*_{max} &\geq \frac{3}{5} \times (\frac{b_1}{2} + a_2 + a_3) + \frac{4}{5} \times (b_1 + \frac{a_2}{2} + b_4) + \frac{2}{5} \times (a_3 + \frac{b_4}{2})\\ &= \frac{11}{10}b_1 + a_2 + a_3 + b_4 \geq b_1 + a_2 + a_3 + b_4 = C_{max},\end{aligned}$$

where the last inequality is due to Lemma 1. This proves the theorem.

3.2 Tight Analysis of APPROX1 for Two Subproblems

In the Subproblem 1, each job in N_1 has no larger processing time than each job in N_2. Therefore, APPROX1 must always find $B_s = B'_s = \emptyset$ when generating a type-B block $[J_{1,s}; B_s]$ in Step 3. Accordingly, we obtain $\mathcal{A}_2 = \mathcal{B}_2 = \mathcal{B}_3 = \emptyset$, implying that $a_2 = b_2 = b_3 = 0$. Thus we rewrite the inequalities (2) and (3), and Lemma 4 as follows,

$$C^*_{max} \geq \max\{\frac{b_1}{2} + a_3, b_1 + b_4, a_3 + \frac{b_4}{2}\}. \tag{6}$$

Theorem 2. APPROX1 *is a* $\frac{7}{4}$*-approximation algorithm for Subproblem 1, and the ratio is tight.*

Proof. Similar to Theorem 1, we multiply the three terms in (6) by $\frac{1}{2}$, $\frac{3}{4}$ and $\frac{1}{2}$ respectively, and compute the summation of the resulting inequalities, obtaining

$$\frac{7}{4}C^*_{max} \geq \frac{1}{2} \times (\frac{b_1}{2} + a_3) + \frac{3}{4} \times (b_1 + +b_4) + \frac{1}{2} \times (a_3 + \frac{b_4}{2}) = b_1 + a_3 + b_4 \geq C_{max},$$

where the last inequality is due to Lemma 1 and the fact $a_2 = 0$.

The tightness instance is shown in Fig. 4, where there are three jobs in N_1 and two jobs in N_2. Figure 4(a) gives the conflict graph with the processing time outside the vertices. The algorithm APPROX1 may generate a type-A block $[\{J_{1,1}\}; J_{2,1}]$ in Step 2, and two type-B blocks, $[J_{1,2}; \emptyset]$ and $[J_{1,3}; \emptyset]$, in Step 3. This means that $\mathcal{A}_1 = \{J_{1,1}\}$, $\mathcal{B}_1 = \{J_{2,1}\}$, $\mathcal{A}_3 = \{J_{1,2}, J_{1,3}\}$ and $\mathcal{B}_4 = \{J_{2,2}\}$. Thus by Step 4, one gets a feasible schedule with $C_{max} = b_1 + a_3 + b_4 = p_{2,1} + p_{1,2} + p_{1,3} + p_{2,2} = 7 + \epsilon$ (see Fig. 4(b) for an illustration). In an optimal schedule, the second machine processes $J_{2,2}$ followed by $J_{2,1}$ so that both machines can keep busy till completion (see Fig. 4(c) for an illustration). Therefore $C^*_{max} = T_1 = T_2 = 4 + \epsilon$. Then we have $\frac{C_{max}}{C^*_{max}} \to \frac{7}{4}$ ($\epsilon \to 0$). This proves the theorem.

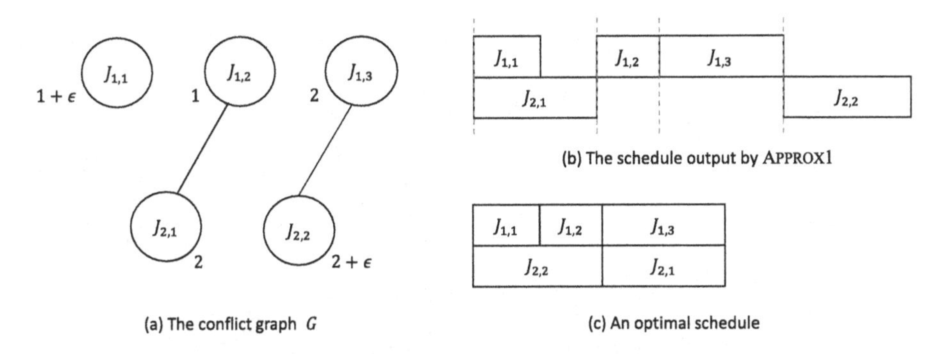

Fig. 4. A tight instance of APPROX1 for Subproblem 1.

Subproblem 2 assumes that each job in N_1 has processing time larger than each job in N_2. With this assumption, APPROX1 never goes into Step 2. Hence only type-B blocks are generated by the algorithm. This means that $\mathcal{A}_1 = \mathcal{B}_1 = \emptyset$, or equivalently, $a_1 = b_1 = 0$. Rewrite (2) and (3) accordingly, one can get $C^*_{max} \geq \max\{a_2 + a_3, \frac{a_2}{2} + b_4\}$. Besides, applying the assumption to Lemma 2, one also obtains that for each type-B block $[J_{1,s}; B_s]$ with $\sum_{J_{2,j} \in B_s} p_{2,j} < p_{1,s}/2$, any job in $\mathcal{B}_4$ must conflict with $J_{1,s}$. Therefore the lower bound given by Lemma 4 can be updated, that is, $C^*_{max} \geq a_3 + b_4$. In summary, we come up with the following lower bound.

$$C^*_{max} \geq \max\{a_2 + a_3, \frac{a_2}{2} + b_4, a_3 + b_4\}. \tag{7}$$

Theorem 3. APPROX1 *is a* $\frac{5}{3}$*-approximation algorithm for Subproblem 2, and the ratio is tight.*

Proof. By Lemma 1, $C_{max} \leq a_2 + a_3 + b_4$. Similarly, by multiplying the three terms in (7) by $\frac{2}{3}$, $\frac{2}{3}$ and $\frac{1}{3}$ respectively, and computing the summation of the resulting inequalities, we obtain

$$\frac{5}{3} C^*_{max} \geq \frac{2}{3} \times (a_2 + a_3) + \frac{2}{3} \times (\frac{a_2}{2} + +b_4) + \frac{1}{3} \times (a_3 + b_4) = a_2 + a_3 + b_4 \geq C_{max}.$$

The tightness instance is shown in Fig. 5, where there are two jobs in N_1 and three jobs in N_2. Figure 5(a) gives the conflict graph with the processing time outside the vertices. The algorithm APPROX1 may generate two type-B blocks, $[J_{1,1}; \{J_{2,1}\}]$ and $[J_{1,2}; \emptyset]$, in Step 3, resulting that $\mathcal{A}_2 = \{J_{1,1}\}$, $\mathcal{B}_2 = \{J_{2,1}\}$, $\mathcal{A}_3 = \{J_{1,2}\}$ and $\mathcal{B}_4 = \{J_{2,2}, J_{2,3}\}$. Thus Step 4 gives a feasible schedule with $C_{max} = a_2 + a_3 + b_4 = p_{1,1} + p_{1,2} + p_{2,2} + p_{2,3} = 5 + \epsilon$ (see Fig. 5(b) for an illustration). In an optimal schedule, the second machine processes $J_{2,2}$ followed by $J_{2,1}$ so that both machines can keep busy till completion (see Fig. 5(c) for an illustration). Therefore $C^*_{max} = T_1 = 3 + \epsilon$. Then we have $\frac{C_{max}}{C^*_{max}} \to \frac{5}{3}$ $(\epsilon \to 0)$. This proves the theorem.

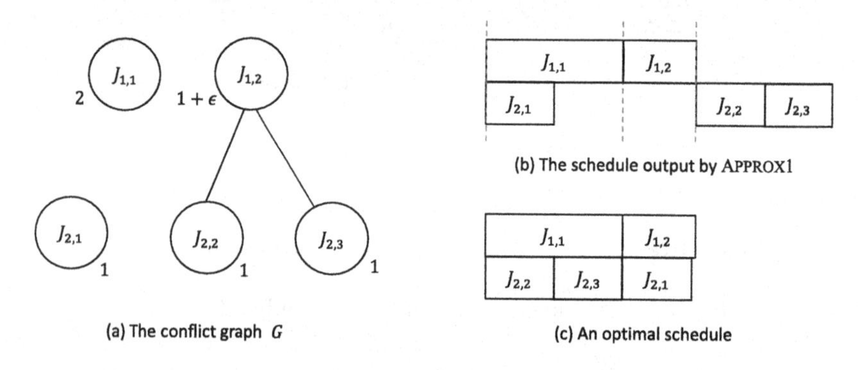

Fig. 5. A tight instance of APPROX1 for Subproblem 2.

3.3 An Improved Algorithm for Subproblem 2

By simply modifying APPROX1, we obtain an improved algorithm APPROX2 for Subproblem 2. Compared with APPROX1, the new algorithm allows that $\sum_{J_{2,j} \in B_s} p_{2,j} > p_{1,s}$ in a type-B block $[J_{1,s}; B_s]$. In fact, it first sequences the jobs in N_2 in the non-increasing order of their processing time. Then a set B_s of jobs from N_2 are selected along that sequence when generating a type-B block $[J_{1,s}; B_s]$, provided that all jobs in B_s do not conflict with $J_{1,s}$, and whenever possible, it ensures that the total processing time of them can reach the threshold $\frac{\sqrt{2}}{2} p_{1,s}$. A high-level description of APPROX2 is depicted in Fig. 6.

When APPROX2 stops, it generates a total of n_1 type-B blocks. Let $\mathcal{A}_i$ $(\mathcal{B}_i)$, $i = 1, 2, 3$ be disjoint subsets of N_1 (N_2) defined as follows. For a block with $\sum_{J_{2,j} \in B_s} p_{2,j} > p_{1,s}$, $J_{1,s} \in \mathcal{A}_1$ and $B_s \subseteq \mathcal{B}_1$; for a block with $\frac{\sqrt{2}}{2} p_{1,s} \le \sum_{J_{2,j} \in B_s} p_{2,j} \le p_{1,s}$, $J_{1,s} \in \mathcal{A}_2$ and $B_s \subseteq \mathcal{B}_2$; and for a block with $\sum_{J_{2,j} \in B_s} p_{2,j} < \frac{\sqrt{2}}{2} p_{1,s}$, $J_{1,s} \in \mathcal{A}_3$ and $B_s \subseteq \mathcal{B}_3$. Let $\mathcal{B}_4 = N_2 \setminus \cup_{i=1}^{3} \mathcal{B}_i$. Let a_i (b_i) denote the total processing time of jobs in $\mathcal{A}_i$ $(\mathcal{B}_i)$. Thus we have $T_1 = a_1 + a_2 + a_3$ and $T_2 = b_1 + b_2 + b_3 + b_4$. Besides, the makespan of APPROX2 can be calculated by $C_{max} = b_1 + a_2 + a_3 + b_4$.

Algorithm APPROX2:

1. Initially let $B = N_2$ and $s = 1$.
2. Reorder the jobs in N_2 so that $p_{2,j-1} \leq p_{2,j}$ for any $j = 2, 3, \cdots n_2$.
3. Find all jobs in B that do not conflict with $J_{1,s}$, denote them by a set B'_s (It allows that $B'_s = \emptyset$ too).
   3.1 if $\sum_{J_{2,j} \in B'_s} p_{2,j} < \frac{\sqrt{2}}{2} p_{1,s}$, then let $B_s = B'_s$.
   3.2 if $\sum_{J_{2,j} \in B'_s} p_{2,j} \geq \frac{\sqrt{2}}{2} p_{1,s}$, then find the smallest index k such that $\sum_{J_{2,j} \in B'_s, j \leq k} p_{2,j} \geq \frac{\sqrt{2}}{2} p_{1,s}$. Let $B_s = \{J_{2,j} \in B'_s | j \leq k\}$.
   3.3 generate a type-B block $[J_{1,s}; B_s]$ and start to process it at the earliest time when the $(s-1)$-th block is completed. Let $B = B \setminus B_s$ and $s = s + 1$.
4. If $s < n_1 + 1$, return Step 3. Otherwise, process all remaining jobs in B (if any) one after another without any idle time from the completion of the n_1-th block.

Fig. 6. A high-level description of the algorithm APPROX2.

Lemma 5. $a_1 < b_1 < \sqrt{2} a_1$, $\frac{\sqrt{2}}{2} a_2 \leq b_2 \leq a_2$ *and* $0 \leq b_3 < \frac{\sqrt{2}}{2} a_3$.

Proof. The inequalities $a_1 < b_1$, $\frac{\sqrt{2}}{2} a_2 \leq b_2 \leq a_2$ and $0 \leq b_3 < \frac{\sqrt{2}}{2} a_3$ trivially hold from the definition of $\mathcal{A}_i$ and $\mathcal{B}_i, i = 1, 2, 3$. For each $B_s \subseteq \mathcal{B}_1$, it's clear that $\sum_{J_{2,j} \in B'_s} \geq \sum_{J_{2,j} \in B_s} p_{2,j} > p_{1,s}$. From Step 3.2 of APPROX2, $B_s = \{J_{2,j} \in B'_s | j \leq k\}$, where k is the smallest index such that $\sum_{J_{2,j} \in B'_s, j \leq k} p_{2,j} \geq \frac{\sqrt{2}}{2} p_{1,s}$. This means that $J_{2,k} \in B_s$ and

$$\sum_{J_{2,j} \in B_s, j \leq k} p_{2,j} > p_{1,s} > \frac{\sqrt{2}}{2} p_{1,s} > \sum_{J_{2,j} \in B_s, j < k} p_{2,j}. \tag{8}$$

Note the assumption of Subproblem 2 and Step 2 of APPROX2 together gives that $p_{1,s} > p_{2,j} \geq p_{2,k}$ for any $J_{2,j} \in B_s$. Then $B_s \setminus \{J_{2,k}\} \neq \emptyset$, and hence we derive $p_{2,k} \leq \sum_{J_{2,j} \in B_s, j < k} p_{2,j} < \frac{\sqrt{2}}{2} p_{1,s}$. Together with (8),

$$\sum_{J_{2,j} \in B_s, j \leq k} p_{2,j} = p_{2,k} + \sum_{J_{2,j} \in B_s, j < k} p_{2,j} < \sqrt{2} p_{1,s}.$$

Accordingly, we can conclude that $b_1 < \sqrt{2} a_1$. This proves the lemma.

Lemma 6. $C^*_{max} \geq a_3 + b_4$.

Proof. For each $B_s \subseteq \mathcal{B}_3$, $\sum_{J_{2,j} \in B_s} p_{2,t} < \frac{\sqrt{2}}{2} p_{1,s}$. Thus the algorithm APPROX2 must go to Step 3.1, which implies that $B'_s = B_s$. Note B'_s contains all jobs in the current $B = N_2 \setminus \cup_{j=1}^{s-1} B_j$ that do not conflict with $J_{1,s}$. In other words, all jobs in $N_2 \setminus \cup_{j=1}^{s} B_j$ must conflict with $J_{1,s}$. Recall $\mathcal{B}_4 = N_2 \setminus \cup_{i=1}^{3} \mathcal{B}_i = N_2 \setminus \cup_{j=1}^{n_1} B_j$. Thus each job in $\mathcal{B}_4$ must conflict with $J_{1,s}$. Then we can conclude that each job in $\mathcal{B}_3$ must conflict with each job in $\mathcal{A}_3$. Therefore, $C^*_{max} \geq a_3 + b_4$.

One sees that Lemma 5 and 6 together update the lower bound of the optimal makespan,

$$C^*_{max} \geq \max\{T_1, T_2, a_3+b_4\} \geq \max\{\frac{\sqrt{2}}{2}b_1+a_2+a_3, b_1+\frac{\sqrt{2}}{2}a_2+b_4, a_3+b_4\}. \quad (9)$$

Theorem 4. APPROX2 *is a* $(3-\sqrt{2})$*-approximation algorithm for Subproblem 2, and the ratio is tight.*

Proof. Note APPROX2 outputs of a schedule with makespan $C_{max} = b_1 + a_2 + a_3 + b_4$. We then multiply the rightmost terms in (9) by $2-\sqrt{2}$, $2-\sqrt{2}$ and $\sqrt{2}-1$ respectively, and compute the summation of the resulting inequalities, obtaining

$$(3-\sqrt{2})C^*_{max} \geq (2-\sqrt{2}) \times (\frac{\sqrt{2}}{2}b_1 + a_2 + a_3) + (2-\sqrt{2}) \times (b_1 + \frac{\sqrt{2}}{2}a_2 + b_4)$$
$$+(\sqrt{2}-1) \times (a_3 + b_4) = b_1 + a_2 + a_3 + b_4 = C_{max}.$$

The tightness instance is shown in Fig. 7, where there are two jobs in N_1 and four jobs in N_2. Figure 7(a) gives the conflict graph with the processing time outside the vertices. The algorithm APPROX2 generates two type-B blocks, $[J_{1,1}; \{J_{2,1}, J_{2,2}\}]$ and $[J_{1,2}; \emptyset]$, in Step 3, resulting that $\mathcal{A}_1 = \{J_{1,1}\}$, $\mathcal{B}_1 = \{J_{2,1}, J_{2,2}\}$, $\mathcal{A}_2 = \mathcal{B}_2 = \mathcal{B}_3 = \emptyset$, $\mathcal{A}_3 = \{J_{1,2}\}$ and $\mathcal{B}_4 = \{J_{2,3}, J_{2,4}\}$. Accordingly, the makespan of the output schedule equals $C_{max} = b_1 + a_2 + a_3 + b_4 = p_{2,1} + p_{2,2} + p_{1,2} + p_{2,3} + p_{2,4} = 2\sqrt{2}+1$ (see Fig. 7(b) for an illustration). On the other hand, the optimal algorithm schedules $J_{2,3}, J_{2,4}$ before $J_{2,1}, J_{2,2}$ on the second machine so that both machines can keep busy till completion (see Fig. 7(c) for an illustration). Therefore $C^*_{max} = T_1 = \sqrt{2}+1+\epsilon$. Then we have $\frac{C_{max}}{C^*_{max}} \rightarrow \frac{2\sqrt{2}+1}{\sqrt{2}+1} = 3-\sqrt{2}$ $(\epsilon \rightarrow 0)$. This proves the theorem.

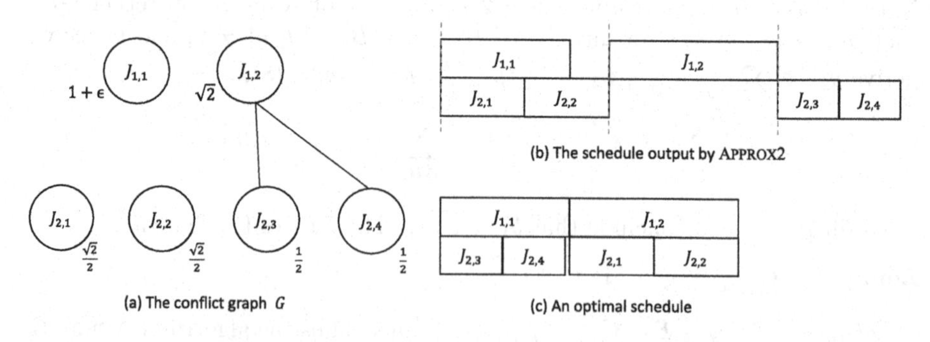

Fig. 7. A tight instance of APPROX2 for Subproblem 2.

4 The SWC Problem Without Any Fixed Sequence

This section considers the two machine SWC problem where no processing sequence is fixed in advance. This problem was previously recognized weakly NP-hard [16]. We give a strong NP-hardness proof and present an approximation algorithm to solve it. The main results are as follows.

Theorem 5. *The problem $PD2|G = (N,E)|C_{max}$ is NP-hard in the strong sense, and there exists a $\frac{5}{3}$-approximation algorithm.*

5 Conclusions

We studied the problem $PD2|G = (N,E)|C_{max}$ of scheduling with conflict constraints on two parallel dedicated machines. The problem was previously known weakly NP-hard, and shown to be NP-hard in the strong sense under the assumption that jobs on one machine must follow a given and fixed processing sequence [16]. We proposed several approximation results for this problem, including a $\frac{9}{5}$-approximation algorithm for the SWC problem with a fixed sequence as well as its tight analysis on two strong NP-hard subproblems, a $(3-\sqrt{2})$-approximation algorithm for one subproblem, and a $\frac{5}{3}$-approximation algorithm for the SWC problem without any fixed sequence. Our approximation results are mostly based on the idea of sequentially generating and processing blocks each with a set of conflict-free jobs and meeting certain processing time requirements. In addition, we proved the problem $PD2|res\lambda 11|C_{max}$ of scheduling jobs on two parallel dedicated machines with non-sharable resources is NP-hard in the strong sense. This answers that the two machine SWC problem without any fixed sequence is strongly NP-hard too.

A natural question is designing better approximation algorithms or even PTASes for all strong NP-hard variants of the problem $PD2|G = (N,E)|C_{max}$. In particular, the problem of scheduling with multiple non-shareable resources can be viewed as a special case. The previous approximation results on this problem is given under the assumption that each job consumes at most one resource [17]. Thus the next improvement may start by removing this assumption. Another interesting question is the SWC problem on m parallel dedicated machines. Note that in this case, the conflict constraints will imply a multipartite conflict graph instead.

Acknowledgements. This research is supported by the Zhejiang Provincial NSF Grant LY21A010014 and the NSFC Grants 11771114, 11971139.

References

1. Garey, M.R., Graham, R.L.: Bounds for multiprocessor scheduling with resource constraints. SIAM J. Comput. **4**(2), 187–200 (1975)
2. Baker, B.S., Jr., Coffman, E.G.: Mutual exclusion scheduling. Theor. Comput. Sci. **162**(2), 225–243 (1996)

3. Halldórsson, M.M., Kortsarz, G., Proskurowski, A., Salman, R., Shachnai, H., Telle, J.A.: Multicoloring trees. Inf. Comput. **180**(2), 113–129 (2003)
4. Even, G., Halldórsson, M.M., Kaplan, L., Ron, D.: Scheduling with conflicts: online and offline algorithms. J. Sched. **12**(2), 199–224 (2009)
5. Bodlaender, H.L., Jansen, K.: Restrictions of graph partition problems part I. Theor. Comput. Sci. **148**, 93–109 (1995)
6. Corneil, D.G.: The complexity of generalized clique packing. Discrete Appl. Math. **12**(3), 233–239 (1985)
7. Cohen, E., Tarsi, M.: NP-completeness of graph decomposition problem. J. Complex. **7**(2), 200–212 (1991)
8. Jansen, K.: The mutual exclusion scheduling problem for permutation and comparability graphs. Inf. Comput. **180**(2), 71–81 (2003)
9. Petrank, E.: The hardness of approximation: gap location. Comput. Complex. **4**, 133–157 (1994)
10. Bendraouche, M., Boudhar, M., Oulamara, A.: Scheduling: agreement graph vs resource constraints. Eur. J. Oper. Res. **240**, 355–360 (2015)
11. Bendraouche, M., Boudhar, M.: Scheduling jobs on identical machines with agreement graph. Comput. Oper. Res. **39**(2), 382–390 (2012)
12. Mohabeddine, A., Boudhar, M.: New results in two identical machines scheduling with agreement graphs. Theor. Comput. Sci. **779**, 37–46 (2019)
13. Fürer, M., Yu, H.: Packing-Based approximation algorithm for the k-Set cover problem. In: Asano, T., Nakano, S., Okamoto, Y., Watanabe, O. (eds.) ISAAC 2011. LNCS, vol. 7074, pp. 484–493. Springer, Heidelberg (2011). https://doi.org/10.1007/978-3-642-25591-5_50
14. Yu, H.W.: Combinatorial and algebraic algorithms in set covering and partitioning problems. The Pennsylvania State University. ProQuest Dissertations Publishing, 3647540 (2014)
15. Hà, M.H., Ta, D.Q., Nguyen, T.T.: Exact algorithms for scheduling problems on parallel identical machines with conflict jobs. arXiv: 2102.06043 (2021)
16. Hong, H.C., Lin, B.M.T.: Parallel dedicated machine scheduling with conflict graphs. Comput. Ind. Eng. **124**, 316–321 (2018)
17. Kellerer, H., Strusevich, V.A.: Scheduling problems for parallel dedicated machines under multiple resource constraints. Discrete Appl. Math. **133**(1), 45–68 (2004)
18. Kellerer, H., Strusevich, V.A.: Scheduling parallel dedicated machines under a single non-shared resource. Eur. J. Oper. Res. **147**, 345–364 (2003)
19. Garey, M.R., Johnson, D.S.: Computers and Intractability: A Guide to the Theory of NP-Completeness. W. H. Freeman & Co., New York (1979)

Computing the One-Visibility Cop-Win Strategies for Trees

Boting Yang(✉)

Department of Computer Science, University of Regina, Regina, SK, Canada
Boting.Yang@uregina.ca

Abstract. We investigate the one-visibility cops and robber game on trees. For a tree, we use copnumbers of its subtrees to characterize a key structure, called road. We give an $O(n \log n)$ time algorithm to compute an optimal cop-win strategy for a tree with n vertices.

1 Introduction

Nowakowski and Winkler [5] and Quilliot [6] introduced the cops and robber game which is a well-known graph searching model played on graphs. They characterized the graphs in which one cop can capture the robber. Megiddo et al. [4] introduced the edge searching model. They gave an $O(n)$ time algorithm to compute the edge search number of a tree and an $O(n \log n)$ time algorithm to find an optimal search strategy, where n is the number of vertices in the tree. The zero-visibility cops and robber game was introduced by Tošić [7], which is a hybrid of the cops and robber game [5,6] and the edge searching model [4]. Recent works on this model include results on a variety of graph classes [8]. Clarke et al. [3] considered the ℓ-visibility cops and robber game. Yang and Akter [10] gave a linear-time algorithm for computing the one-visibility copnumber of trees.

In the one-visibility cops and robber game, we have a graph, a set of cops, and a single robber. The robber has full information about the cops. But the cops have the information about the robber's location only at the moment when there is a cop who is located on a neighbouring vertex of the robber or located on the same vertex as the robber (the robber is captured in the latter case). The game is played over a sequence of rounds. At round 0, after the cops choose a set of vertices to occupy, the robber chooses a vertex to occupy. At each of the following rounds, the cops move first and the robber moves next; in a cops' turn, each cop either moves to a neighbouring vertex or stays still, then the robber does the same in his turn. Only when the distance between the cops and the robber is at most one, we say that the cops *see* the robber, that is, the cops have the information about the location of the robber. The cops *capture* the robber if one of them occupies the same vertex as the robber. If a cop eventually occupies

Research supported in part by an NSERC Discovery Research Grant, Application No.: RGPIN-2018-06800.

D.-Z. Du et al. (Eds.): COCOA 2021, LNCS 13135, pp. 125–139, 2021.
https://doi.org/10.1007/978-3-030-92681-6_11

the same vertex as the robber at some moment in the game, then the *cops win*. The *one-visibility copnumber* of a graph G, denoted as $\mathsf{c}_1(G)$, is the minimum number of cops required to capture the robber on G.

A *cop-win strategy* is a sequence of cops' actions to capture the robber. A cop-win strategy for G is *optimal* if it uses $\mathsf{c}_1(G)$ cops to capture the robber. If a cop alternates two actions between two adjacent vertices u and v, i.e., "moving from u to v" and "moving from v to u", for at least three consecutive rounds, then we say that this cop *vibrates* between u and v in these rounds; we also call the edge uv a *vibrating edge*. A subgraph known to not contain the robber is called *cleared*.

Let G be a graph with vertex set $V(G)$ and edge set $E(G)$. For $u, v \in V(G)$, we use uv to denote the edge between them and also use $u \sim v$ to denote a path between them. The *distance* between u and v, denoted by $\text{dist}_G(u, v)$, is the length of the shortest path between u and v in G, where the length of a path is the number of edges on the path. The *k-th closed neighbourhood* of v is defined as $N_G^k[v] = \{u \in V(G) \mid \text{dist}_G(u, v) \leq k\}$. For $U \subseteq V(G)$, we use $G - U$ to denote the subgraph obtained from G by deleting the vertices of U from G.

In [10], Yang and Akter explored one-visibility cops and robber game on trees. They proposed a linear-time algorithm for computing the one-visibility copnumber of trees. In this paper, we focus on computing optimal cop-win strategies efficiently. For this goal, we introduce a key structure for trees, called *road*, to provide essential structural information for optimal cop-win strategies. Using the roadmaps, we present an approach for computing an optimal cop-win strategy in $O(n \log n)$ time, where n is the number of vertices on the tree.

2 Structures of Trees by Copnumbers

The following three kinds of vertices are introduced for trees in [10].

Definition 1. Let T be a tree with $\mathsf{c}_1(T) = k$. If there is a vertex v in T such that each component X in the forest $T - N_T^3[v]$ has $\mathsf{c}_1(X) < k$, then v is a *hub* of T. If there is a vertex v in T such that $T - N_T^3[v]$ has two components with copnumber k satisfying that the path in T connecting them contains v, then v is an *avenue vertex* of T. If T does not have a hub or an avenue vertex and there is a vertex v in T such that $T - N_T^2[v]$ has two components with copnumber k satisfying that the path in T connecting them contains v, then v is a *street vertex* of T.

Note that hubs, avenue vertices or street vertices can help to make cop-win search plans. But sometimes it takes extra time to recognize these vertices if each of them corresponds to a distinct strategy. In this paper, we introduce the road to unify these structures; this will bypass the cases where it is not easy to distinguish whether a subtree contains hubs, avenue vertices, or street vertices.

As observed in [3,10], the one-visibility copnumber of a tree is always greater than or equal to that of any of its subtrees. The following structural lemma is crucial to Algorithm 1.

Lemma 1. *Let T be a tree that contains a vertex $v \in V(T)$ such that each component in the forest $T - N_T^3[v]$ has copnumber at most k. If in $T - N_T^3[v]$ there are at least three components with copnumber exactly k such that the path in T connecting any pair of these three components contains v, then $\mathsf{c}_1(T) = k+1$.*

Yang and Akter [10] showed that any tree that has neither a hub nor an avenue vertex must contain exactly two street vertices, which are adjacent to each other. The edge whose endpoints are the street vertices of T is called the *street* of T.

For avenue vertices, we have the following properties.

Lemma 2. *Let T be a tree with $\mathsf{c}_1(T) = k$.* (i) *If T contains only one avenue vertex u, then u has exactly two neighbours p_1 and p_2, each of which satisfies that $T - N_T^2[p_i]$, $i \in \{1,2\}$, has two components with copnumber k such that the path in T connecting them contains p_i.* (ii) *If T contains m avenue vertices, where $m \geq 2$, then the subgraph induced by all the avenue vertices of T is a path, denoted by $u_1 \cdots u_m$, and moreover, u_1 (resp. u_m) has a unique neighbour $p_1 \neq u_2$ (resp. $p_2 \neq u_{m-1}$) such that $T - N_T^2[p_1]$ (resp. $T - N_T^2[p_2]$) has two components with copnumber k satisfying that the path in T connecting them contains p_1 (resp. p_2).*

The path $u_1 \cdots u_m$ in Lemma 2(ii) is called the *avenue* of T; the unique avenue vertex u in Lemma 2(i) is also defined as the *avenue* of T. The vertices p_1 and p_2 in Lemma 2(i)(ii) are called *pre-avenue vertices* of T.

We now introduce the road which unifies the above structures.

Definition 2. Let T be a tree with $\mathsf{c}_1(T) = k$. A path P in T is called a *road* of T, denoted by $\mathsf{R}(T)$, if each component X in the forest $T - N_T^3[V(P)]$ has $\mathsf{c}_1(X) < k$.

The relations between the road and the structures of hub, street and avenue are shown in the following theorem.

Theorem 1. *Let T be a tree.* (i) *If T contains a hub, then any path containing a hub is a road.* (ii) *If T contains a street vertex, then any path containing the street is a road.* (iii) *If T contains an avenue vertex, then any path containing the two pre-avenue vertices is a road.*

Since every tree must contain a hub, or an avenue vertex, or a street vertex [10], the next result is an immediate consequence of Theorem 1.

Corollary 1. *Every tree contains at least one road.*

3 Computing Copnumbers and Roadmaps

A *rooted tree* T with root r is denoted by $T^{[r]}$. Each vertex $v \neq r$ of $T^{[r]}$ is connected with r by a unique path, where the *parent* of v is the neighbour of v on this path; v is also called a *descendant* of each vertex on this path except v.

For any vertex $v \in V(T^{[r]})$, we use $T^{[v]}$ to denote the subtree of $T^{[r]}$ induced by v and all its descendants. Thus v is the root of the subtree $T^{[v]}$. Note that in the forest $T^{[v]} - v$, each component is a rooted subtree whose root is a child of v.

We first describe the general idea of an optimal cop-win strategy, called *road-tactic*. Let T be a tree with $\mathsf{c}_1(T) = k$, and let $P = p_1 \cdots p_m$ be a road of T. We first place all k cops on p_1. Let Y_i, $1 \leq i \leq m$, be the component containing p_i in the forest $T - E(P)$. The subtree Y_i is called a *branch* of the road P. For each $i = 1, \ldots, m$, we use the following strategy, called *branch-tactic*, to clear Y_i; after Y_i is cleared, all cops move back to p_i, then move forward to p_{i+1} along the edge $p_i p_{i+1}$ until Y_m is cleared. In the *branch-tactic* for clearing Y_i, let $v_1, \ldots, v_d$ be all children of p_i. For each v_j, $1 \leq j \leq d$, let v_{jh}, $1 \leq h \leq m_j$, be all children of v_j. Notice that all k cops are located on p_i at the start of clearing Y_i. For each v_j $(1 \leq j \leq d)$ and for each v_{jh} $(1 \leq h \leq m_j)$, let one cop vibrate between v_j and v_{jh}, in the meantime, let other cops move into each subtree rooted at a grandchild of v_{jh} until all the subtrees rooted at the grandchildren of v_{jh} are cleared. Finally, let all cops move back to p_i after all vertices of Y_i are cleared.

Since the road-tactic described above will be an optimal cop-win strategy, we have to compute $\mathsf{c}_1(T)$ and the road P, which implies that we must compute $\mathsf{c}_1(Y_i)$ $(1 \leq i \leq m)$ and the road in each Y_i before we can find $\mathsf{c}_1(T)$ and P. We will use a bottom-up dynamic programming idea to compute the copnumbers of the subtrees and their roads. The bottom case is to find the copnumber of each leaf which is trivial. Then we consider each subtree in which all children are leaves. We can continue this process until the copnumber and a road of the whole tree is computed from the copnumbers and roads of the children of the root. In this section, we design an algorithm to compute $\mathsf{c}_1(T)$ and the roadmaps. In the next section, we will give an algorithm to implement the road-tactic strategy using the copnumbers and roadmaps of the subtrees obtained from this section.

The definitions and notations of the following terms can be found in Definitions 5.1–5.3 in [10]: *k-pre-branching vertex, k-weakly-branching vertex, k-branching vertex, k-pre-branching indicator* $I^k_{\mathrm{pb}}(v)$, *k-initial-counter* $J^k(v)$, *k-weakly-branching indicator* $I^k_{\mathrm{wb}}(v)$, *and k-weakly-counter* $J^k_{\mathrm{w}}(v)$.

Let u be a descendant of v in $T^{[v]}$. If u is a k-pre-branching (resp. k-weakly-branching, k-branching) vertex in $T^{[u]}$, then we say that u is a *k-pre-branching* (resp. *k-weakly-branching*, *k-branching*) *descendant* of v.

In the following definition, we generalize the label in [10] so that the new label contains roads, which will be used to construct cop-win strategies.

Definition 3. Let $T^{[v]}$ be a tree with root v. The *label* of v in $T^{[v]}$, denoted by $\mathtt{L}(v, T^{[v]})$, is a sequence with the form

$$\mathtt{L}(v, T^{[v]}) = (t_m, x_m, P_m; \ldots; t_1, x_1, P_1; I^{t_1}_{\mathrm{wb}}(v), J^{t_1}_{\mathrm{w}}(v); I^{t_1}_{\mathrm{pb}}(v), J^{t_1}(v)) \quad (1)$$

where t_i, x_i and P_i are defined as follows:

1. If $T^{[v]}$ contains only one vertex, then $t_1 = 1$, $x_1 = \perp$, $P_1 = v$, and $\mathtt{L}(v, T^{[v]}) = (1, \perp, v; 0, 0; 0, 0)$; otherwise, set $i \leftarrow 1$ and $T^{[v]}_a \leftarrow T^{[v]}$.

2. Set $s_i \leftarrow \mathtt{c}_1(T_a^{[v]})$, and go to exactly one of the following subcases:
   (a) If v has an s_i-branching descendant in $T_a^{[v]}$, let v_i be this descendant. Set $R_i \leftarrow \mathtt{R}(T_a^{[v_i]})$. Update $T_a^{[v]} \leftarrow T_a^{[v]} - V(T_a^{[v_i]})$, $i \leftarrow i+1$, and go back to Case 2.
   (b) If v does not have an s_i-branching descendant in $T_a^{[v]}$, set $m \leftarrow i$ and $R_m \leftarrow \mathtt{R}(T_a^{[v]})$. If v is an s_m-branching vertex in $T_a^{[v]}$, then $v_m \leftarrow v$; otherwise, $v_m \leftarrow \perp$. Determine $I_{\text{wb}}^{s_m}(v)$, $J_{\text{w}}^{s_m}(v)$, $I_{\text{pb}}^{s_m}(v)$ and $J^{s_m}(v)$ of $T_a^{[v]}$ by their definitions.
3. For each $i \in \{1, \ldots, m\}$, let $t_i = s_{m-i+1}$, $x_i = v_{m-i+1}$, and $P_i = R_{m-i+1}$.

The set of roads $\{P_1, \ldots, P_m\}$ in $\mathtt{L}(v, T^{[v]})$ is called the *roadmap* of $T^{[v]}$.

Lemma 3. *The label in Eq. (1) has the following properties.* (i) *P_m is a road of $T^{[v]}$, and $\mathtt{c}_1(T^{[v]}) = t_m$.* (ii) *For each $i \in \{2, \ldots, m\}$, $x_i \neq \perp$.* (iii) *If $x_1 \neq \perp$, then $x_1 = v$; otherwise, in the rooted subtree $T^{[v]} - \bigcup_{i=2}^{m} V(T^{[x_i]})$, neither v nor its descendants is a t_1-branching vertex in this subtree.*

The last seven components in the label Eq. (1) can be determined as follows.

Lemma 4. *Let $T^{[v]}$ be a tree with root v whose label has the form of Eq. (1), where $m \geq 2$. Let $T_1^{[v]} = T^{[v]} - \bigcup_{i=2}^{m} V(T^{[x_i]})$. Then $t_1 = \mathtt{c}_1(T_1^{[v]})$, P_1 is a road of $T_1^{[v]}$, and moreover,* (i) *if v is a t_1-branching vertex in $T_1^{[v]}$, then $x_1 = v$, and $I_{\text{wb}}^{t_1}(v) = J_{\text{w}}^{t_1}(v) = I_{\text{pb}}^{t_1}(v) = J^{t_1}(v) = 0$;* (ii) *if v is a t_1-weakly-branching vertex in $T_1^{[v]}$, then $x_1 = \perp$, $I_{\text{wb}}^{t_1}(v) = 1$, $I_{\text{pb}}^{t_1}(v) = J^{t_1}(v) = 0$ and $0 \leq J_{\text{w}}^{t_1}(v) \leq 2$;* (iii) *if v is a t_1-pre-branching vertex in $T_1^{[v]}$, then $x_1 = \perp$, $I_{\text{wb}}^{t_1}(v) = J_{\text{w}}^{t_1}(v) = J^{t_1}(v) = 0$ and $I_{\text{pb}}^{t_1}(v) = 1$;* (iv) *if v is not any of the above three kinds of vertices in $T_1^{[v]}$, then $x_1 = \perp$, $I_{\text{wb}}^{t_1}(v) = J_{\text{w}}^{t_1}(v) = I_{\text{pb}}^{t_1}(v) = 0$ and $0 \leq J^{t_1}(v) \leq 2$.*

As distinct from the label in [10], $\mathtt{L}(v, T^{[v]})$ in Eq. (1) contains the roadmap of $T^{[v]}$. It also contains the structure information shown in the following lemma.

Lemma 5. *Let $T^{[v]}$ be a tree such that the label of its root is of the form in Eq. (1). If $m \geq 2$, then there must exist an avenue of $T^{[x_m]}$, and for each $2 \leq i < m$, there is also an avenue of the subtree $T^{[x_i]} - \bigcup_{j=i+1}^{m} V(T^{[x_j]})$.*

The terminals of roads in Eq. (1) are related to the utmost-k-pre-branching vertices that will be defined in Definition 4. Before we give the definition, we need the following property of k-pre-branching vertices.

Lemma 6. *Let $T^{[v]}$ be a tree with $\mathtt{c}_1(T^{[v]}) = k$. Suppose the root v is a k-pre-branching vertex in $T^{[v]}$. Let $U = \{u \in V(T^{[v]}) \mid u$ is a k-pre-branching vertex in the rooted subtree $T^{[u]}\}$. Then the graph H induced by all vertices of U is a path and v is a terminal of this path.*

Definition 4. Let $T^{[r]}$ be a rooted tree and $\mathtt{c}_1(T^{[r]}) = k$. Suppose that there is a vertex $v \in V(T^{[r]})$ that is a k-pre-branching vertex in the rooted subtree $T^{[v]}$. By Lemma 6, the graph H induced by all k-pre-branching vertices in $T^{[v]}$ is a path. Let $H = v \sim p$ be this path, where $p = v$ if v is the only vertex in H. Then we say that p is an *utmost-k-pre-branching vertex* of v, which is also called an *utmost-k-pre-branching vertex* in $T^{[r]}$.

The next result follows from Lemma 6 and Definition 4.

Lemma 7. *Let $T^{[v]}$ be a rooted tree with $\mathtt{c}_1(T^{[v]}) = k$.* (i) *If v is a k-pre-branching vertex in $T^{[v]}$, then there is a unique utmost-k-pre-branching vertex in $T^{[v]}$.* (ii) *If v is a k-weakly-branching vertex or k-branching vertex in $T^{[v]}$, then there are exactly two utmost-k-pre-branching vertices in $T^{[v]}$.* (iii) *If v is not any of the above three kinds of vertices in $T^{[v]}$, then there is no utmost-k-pre-branching vertex in $T^{[v]}$.*

Definition 5. In Eq. (1), we call the first three components (t_m, x_m, P_m) an *item* associated with $T^{[v]}$, and call each triple (t_i, x_i, P_i), $1 \leq i < m$, an *item* associated with subtree $T^{[v]} - \bigcup_{j=i+1}^{m} V(T^{[x_j]})$. For each item (t_i, x_i, P_i), $1 \leq i \leq m$, t_i is called the *key* of the item and x_i and P_i are its *attributes*.

For the label in the form of Eq. (1), if $x_1 \neq \perp$, it follows from Definition 3 that $x_1 = v_m = v$.

Definition 6. Let $T^{[u]}$ be a tree with root u and let Y be the set of all children of u. Suppose that for each child $y \in Y$, its label $\mathtt{L}(y, T^{[y]})$ is of the form

$$(t^y_{m(y)}, x^y_{m(y)}, P^y_{m(y)}; \ldots; t^y_1, x^y_1, P^y_1; I^{t^y_1}_{\text{wb}}(y), J^{t^y_1}_{\text{w}}(y); I^{t^y_1}_{\text{pb}}(y), J^{t^y_1}(y)) \tag{2}$$

where $m(y)$ is a positive integer corresponding to m in Eq. (1). Let $I_\perp = \{y \in Y \mid x^y_1 = \perp \text{ and } m(y) \geq 2\}$ and let $I_b = \{y \in Y \mid x^y_1 = y\}$. The *kernel subtree* of $T^{[u]}$, denoted by $T^{[u]}_{\text{ker}}$, is defined as $T^{[u]}_{\text{ker}} = T^{[u]} - \bigcup_{y \in I_b} V(T^{[y]}) - \bigcup_{y \in I_\perp} V(T^{[x^y_2]})$.

Yang and Akter [10] gave an algorithm for computing the one-visibility cop-number of trees in linear time. In this section, we modify that algorithm so that the new algorithm can compute roadmaps, which will play an important role in constructing cop-win strategies. The input of Algorithm 1 can be any tree T. We first pick a vertex of T as its root. This root induces the parent-child relation, by which we sort all vertices in a topological order. For each vertex u whose children have been labelled, we cut off those subtrees rooted at the branching descendants of u to construct the kernel subtree $T^{[u]}_{\text{ker}}$. We then call Algorithm 2 to compute the label of u in $T^{[u]}_{\text{ker}}$ which contains a road of $T^{[u]}_{\text{ker}}$. We finally merge the label of u in $T^{[u]}_{\text{ker}}$ with the labels of the branching descendants of u to obtain the label of u in $T^{[u]}$.

In order to describe Algorithm 2, we need more notations. For a graph G, let $\mathtt{c}^*_1(G) = \max\{\mathtt{c}_1(G') \mid G' \text{ is a component in } G\}$. Let $T^{[u]}$ be a tree with root

Algorithm 1. Computing vertex labels with roadmaps

Input: A tree T.

Output: The labels of all vertices with roadmaps.

1: Arbitrarily pick a vertex v of T as its root; compute the parent-child relation in the rooted tree $T^{[v]}$. If $T^{[v]}$ contains only one vertex, then output $\mathtt{L}(v, T^{[v]}) = (1, \perp, v; 0, 0; 0, 0)$; if $T^{[v]}$ contains only two vertices v and v', then output $\mathtt{L}(v', T^{[v']}) = (1, \perp, v'; 0, 0; 0, 0)$ and $\mathtt{L}(v, T^{[v]}) = (1, \perp, vv'; 0, 0; 0, 1)$; otherwise, topologically sort all vertices of $T^{[v]}$ such that v is the last vertex in the list.

2: For each vertex u in the sorted list, repeat Steps 3 – 7 until the label of the root v is computed; then output the labels of all vertices.

3: If u has no child, set its label as $(1, \perp, u; 0, 0; 0, 0)$, and go to Step 2. Let u_j, $1 \leq j \leq t$, be all children of u with labels $\mathtt{L}(u_j, T^{[u_j]})$ in the form of Eq. (2). Construct the kernel subtree $T^{[u]}_{\text{ker}}$. If u is the only vertex in $T^{[u]}_{\text{ker}}$, then $\mathtt{L}(u, T^{[u]}_{\text{ker}}) = (1, \perp, u; 0, 0; 0, 0)$; otherwise, for each child w of u in $T^{[u]}_{\text{ker}}$, assign the last seven components of $\mathtt{L}(w, T^{[w]})$ to $\mathtt{L}(w, T^{[w]}_{\text{ker}})$, and call Algorithm 2 to compute $\mathtt{L}(u, T^{[u]}_{\text{ker}})$. Let $\kappa = \mathtt{c}_1(T^{[u]}_{\text{ker}})$.

4: For each u_j, $1 \leq j \leq t$, if its label in $T^{[u_j]}$ contains $\perp$, let L_j be a list obtained from $\mathtt{L}(u_j, T^{[u_j]})$ by deleting its last seven components and items whose key is less than κ (L_j can be an empty list); otherwise, let L_j be a list obtained from $\mathtt{L}(u_j, T^{[u_j]})$ by deleting its last four components and items whose key is less than κ. Let L_{t+1} be a list containing only the first item of $\mathtt{L}(u, T^{[u]}_{\text{ker}})$.

5: If no key in $L_1, \ldots, L_t, L_{t+1}$ is repeated, then $\mathtt{L}(u, T^{[u]}) \leftarrow \mathtt{L}(u, T^{[u]}_{\text{ker}})$, update the road by Theorem 3, and insert the items of $L_1, \ldots, L_t$ into $\mathtt{L}(u, T^{[u]})$. Go to Step 2.

6: Find the largest repeated key k^* in the lists $L_1, \ldots, L_t, L_{t+1}$. Let $K = (k_1, \ldots, k_\ell)$ be a list containing the distinct keys from $L_1, \ldots, L_{t+1}$ satisfying that the keys in K are decreasing and are greater than or equal to k^*.

7: Find the smallest index h in K such that $k_h = k_{h+1} + 1 = \cdots = k_\ell + (\ell - h)$. Update $K \leftarrow (k_1, \ldots, k_{h-1}, k'_h)$ where $k'_h = k_h + 1$. Create a list $X = (Q_1, \ldots, Q_{h-1}, Q_h)$, where each Q_i, $1 \leq i \leq h-1$, is an item with key k_i and attributes x_i and P_i (i.e., x_i is a k_i-branching vertex in some subtree and P_i is the road in that subtree) and Q_h is an item with key k'_h and attributes $\perp$ and the road u (i.e., this subtree contains no k'_h-branching vertex and u is the road, referring to Theorem 4). Insert $(0, 0; 0, 0)$ at the end of X. Set $\mathtt{L}(u, T^{[u]}) \leftarrow X$. Go to Step 2.

u whose children are $v_1, \ldots, v_d$. Suppose $c_1^*(T^{[u]} - u) = k \geq 1$. We define the following counters:

$$n^k_{\text{c}}(T^{[u]} - u) = \left|\left\{j \mid \mathtt{c}_1(T^{[v_j]}) = k \text{ for } j \in \{1, \ldots, d\}\right\}\right|,$$

$$n^k_{\text{pb}}(T^{[u]} - u) = \sum_{j=1}^{d} I^k_{\text{pb}}(v_j), \qquad n^k_{\text{wb}}(T^{[u]} - u) = \sum_{j=1}^{d} I^k_{\text{wb}}(v_j),$$

$$h^k(T^{[u]} - u) = \max_{1 \leq j \leq d} \{J^k(v_j)\}, \qquad h^k_{\text{w}}(T^{[u]} - u) = \max_{1 \leq j \leq d} \{J^k_{\text{w}}(v_j)\}.$$

Similarly to Theorem 5.7 in [10], we have the following result.

Theorem 2. *Let $T^{[u]}$ be a rooted tree and let $T^{[u]}_{\text{ker}}$ be the kernel subtree with $c^*_1(T^{[u]}_{\text{ker}} - u) = k$. Let $v_1, \ldots, v_d$ be all children of u in $T^{[u]}_{\text{ker}}$. Suppose each label $\mathtt{L}(v_j, T^{[v_j]}_{\text{ker}})$, $1 \le j \le d$, is of the form*

$$\mathtt{L}(v_j, T^{[v_j]}_{\text{ker}}) = (t^{v_j}, \perp, P^{v_j}; I^{t^{v_j}}_{\text{wb}}(v_j), J^{t^{v_j}}_{\text{w}}(v_j); I^{t^{v_j}}_{\text{pb}}(v_j), J^{t^{v_j}}(v_j))). \quad (3)$$

Then the label $\mathtt{L}(u, T^{[u]}_{\text{ker}})$ can be computed by Algorithm 2.

Remark 1. From Lemmas 4.5 and 4.7 in [10], we know that if a tree contains a street vertex or an avenue vertex, then this tree contains a unique street or avenue. Unlike street or avenue, roads are flexible. The motivation of this flexibility is to avoid the cost of distinguishing the different structures (i.e., hub, street, avenue), which is a time-consuming process in a recursive algorithm. For example, consider Lines 22–23 in Algorithm 2 where $n^k_{\text{wb}}(T^{[u]}_{\text{ker}} - u) = 0$ and $n^k_{\text{pb}}(T^{[u]}_{\text{ker}} - u) = 1$. If $T^{[u]}_{\text{ker}}$ contains an avenue, then the road $z^{v_i} \sim u$ contains this avenue of $T^{[u]}_{\text{ker}}$; if $T^{[u]}_{\text{ker}}$ contains a street, then the road $z^{v_i} \sim u$ contains this street; or, if $T^{[u]}_{\text{ker}}$ contains a hub, then the road $z^{v_i} \sim u$ also contains a hub. So in Algorithm 2, we bypass these complicated subcases by using the concept of road.

Lemma 8. *Suppose that $T^{[u]}$ is a tree and the label of its root in the output of Algorithm 1 is*

$$\mathtt{L}(u, T^{[u]}) = (t_m, x_m, P_m; \ldots; t_1, x_1, P_1; I^{t_1}_{\text{wb}}(u), J^{t_1}_{\text{w}}(u); I^{t_1}_{\text{pb}}(u), J^{t_1}(u)),$$

where $m \ge 2$. Then (i) *the terminals of P_m are two utmost-t_m-pre-branching vertices in $T^{[x_m]}$ and $\mathtt{L}(x_m, T^{[x_m]}) = (t_m, x_m, P_m; 0, 0; 0, 0)$;* (ii) *the terminals of each P_i, $2 \le i \le m-1$, are two utmost-t_i-pre-branching vertices in the subtree $T^{[x_i]}_i = T^{[x_i]} - \bigcup_{j=i+1}^m V(T^{[x_j]})$ and $\mathtt{L}(x_i, T^{[x_i]}_i) = (t_i, x_i, P_i; 0, 0; 0, 0)$; and* (iii) *each P_i, $2 \le i \le m$, contains at least three vertices.*

Theorem 3. *Let $T^{[u]}$ be a tree with root u whose children are $u_1, \ldots, u_h$. Suppose that for $1 \le j \le h$, $\mathtt{L}(u_j, T^{[u_j]}) =$*

$$(t^{u_j}_{m_j}, x^{u_j}_{m_j}, P^{u_j}_{m_j}; \ldots; t^{u_j}_1, x^{u_j}_1, P^{u_j}_1; I^{t^{u_j}_1}_{\text{wb}}(u_j), J^{t^{u_j}_1}_{\text{w}}(u_j); I^{t^{u_j}_1}_{\text{pb}}(u_j), J^{t^{u_j}_1}(u_j)) \quad (4)$$

and these labels satisfy

$$\begin{cases} m_j = 1 \text{ and } x^{u_j}_1 = \perp, & \text{if } 1 \le j \le h_1, \\ m_j \ge 2 \text{ and } x^{u_j}_1 = \perp, & \text{if } h_1 < j \le h_2, \\ x^{u_j}_1 = u_j, & \text{if } h_2 < j \le h. \end{cases} \quad (5)$$

Let $T^{[u]}_{\text{ker}}$ be the kernel subtree of $T^{[u]}$ with $c^*_1(T^{[u]}_{\text{ker}} - u) = k$ and let P be the road in the label $\mathtt{L}(u, T^{[u]}_{\text{ker}})$ output by Algorithm 2.

Algorithm 2. Computing vertex labels of a kernel subtree

Input: A kernel subtree $T_{\text{ker}}^{[u]}$ with $c_1^*(T_{\text{ker}}^{[u]} - u) = k$. Let v_j, $1 \leq j \leq d$, be all children of u in $T_{\text{ker}}^{[u]}$; each v_j has a label in the form of Eq. (3).
Output: The label $\mathtt{L}(u, T_{\text{ker}}^{[u]})$.

1: **if** $n_{\text{wb}}^k(T_{\text{ker}}^{[u]} - u) \geq 2$ **then return** $(k+1, \perp, u; 0, 0; 0, 0)$.
2: **if** $n_{\text{wb}}^k(T_{\text{ker}}^{[u]} - u) = 1$ and $n_{\text{pb}}^k(T_{\text{ker}}^{[u]} - u) \geq 1$ **then**
3: **return** $(k+1, \perp, u; 0, 0; 0, 0)$.
4: **if** $n_{\text{wb}}^k(T_{\text{ker}}^{[u]} - u) = 1$, where v_i is the k-weakly-branching child of u, $n_{\text{pb}}^k(T_{\text{ker}}^{[u]} - u) = 0$ and $n_{\text{c}}^k(T_{\text{ker}}^{[u]} - u) \geq 2$ **then**
5: **if** $h_{\text{w}}^k(T_{\text{ker}}^{[u]} - u) = 2$ **then return** $(k+1, \perp, u; 0, 0; 0, 0)$.
6: **else if** $h_{\text{w}}^k(T_{\text{ker}}^{[u]} - u) = 1$ and $h^k(T_{\text{ker}}^{[u]} - u) \geq 1$ **then**
7: **return** $(k+1, \perp, u;\ 0, 0; 0, 0)$
8: **else if** $h_{\text{w}}^k(T_{\text{ker}}^{[u]} - u) = 1$ and $h^k(T_{\text{ker}}^{[u]} - u) = 0$ **then**
9: **return** $(k, \perp, P^{v_i};\ 1, 2; 0, 0)$
10: **else if** $h_{\text{w}}^k(T_{\text{ker}}^{[u]} - u) = 0$ and $h^k(T_{\text{ker}}^{[u]} - u) = 2$ **then**
11: **return** $(k+1, \perp, u;\ 0, 0; 0, 0)$
12: **else if** $h_{\text{w}}^k(T_{\text{ker}}^{[u]} - u) = 0$ and $h^k(T_{\text{ker}}^{[u]} - u) \leq 1$ **then**
13: **return** $(k, \perp, P^{v_i};\ 1, 1; 0, 0)$
14: **if** $n_{\text{wb}}^k(T_{\text{ker}}^{[u]} - u) = 1$, where v_i is the k-weakly-branching child of u, and $n_{\text{c}}^k(T_{\text{ker}}^{[u]} - u) = 1$ **then**
15: **if** $h_{\text{w}}^k(T_{\text{ker}}^{[u]} - u) = 2$ **then return** $(k, u, P^{v_i};\ 0, 0; 0, 0)$
16: **else if** $h_{\text{w}}^k(T_{\text{ker}}^{[u]} - u) = 1$ **then return** $(k, \perp, P^{v_i};\ 1, 2; 0, 0)$
17: **else if** $h_{\text{w}}^k(T_{\text{ker}}^{[u]} - u) = 0$ **then return** $(k, \perp, P^{v_i};\ 1, 1; 0, 0)$
18: **if** $n_{\text{wb}}^k(T_{\text{ker}}^{[u]} - u) = 0$ and $n_{\text{pb}}^k(T_{\text{ker}}^{[u]} - u) \geq 3$ **then**
19: **return** $(k+1, \perp, u;\ 0, 0; 0, 0)$
20: **if** $n_{\text{wb}}^k(T_{\text{ker}}^{[u]} - u) = 0$ and $n_{\text{pb}}^k(T_{\text{ker}}^{[u]} - u) = 2$ **then**
21: **return** $(k, \perp, z^{v_i} \sim z^{v_j}; 1, 0; 0, 0)$, where v_i and v_j are k-pre-branching children of u, and z^{v_i} and z^{v_j} are the utmost-k-pre-branching vertices in $T_{\text{ker}}^{[v_i]}$ and $T_{\text{ker}}^{[v_j]}$ respectively.
22: **if** $n_{\text{wb}}^k(T_{\text{ker}}^{[u]} - u) = 0$ and $n_{\text{pb}}^k(T_{\text{ker}}^{[u]} - u) = 1$ **then**
23: **return** $(k, \perp, z^{v_i} \sim u;\ 0, 0; 1, 0)$, where v_i is a k-pre-branching child of u and z^{v_i} is the utmost-k-pre-branching vertex in $T_{\text{ker}}^{[v_i]}$.
24: **if** $n_{\text{wb}}^k(T_{\text{ker}}^{[u]} - u) = 0$ and $n_{\text{pb}}^k(T_{\text{ker}}^{[u]} - u) = 0$ **then**
25: **if** $h^k(T_{\text{ker}}^{[u]} - u) = 2$ **then return** $(k, \perp, u;\ 0, 0; 1, 0)$
26: **else if** $h^k(T_{\text{ker}}^{[u]} - u) = 1$ **then return** $(k, \perp, u;\ 0, 0; 0, 2)$
27: **else if** $h^k(T_{\text{ker}}^{[u]} - u) = 0$ **then return** $(k, \perp, u;\ 0, 0; 0, 1)$

(i) *If* $\mathsf{c}_1(T_{\text{ker}}^{[u]}) = k+1$ *and* $c_1^*(T^{[u]} - u) \leq k$*, then* $\mathsf{c}_1(T^{[u]}) = k+1$ *and* u *is a road of* $T^{[u]}$.
(ii) *If* $\mathsf{c}_1(T_{\text{ker}}^{[u]}) = k$ *and* $t_{m_j}^{u_j} < k$ *for each* $j \in \{h_1+1, \ldots, h\}$*, then* $\mathsf{c}_1(T^{[u]}) = k$ *and* P *is a road of* $T^{[u]}$.

Theorem 4. *Let $T^{[u]}$ be a tree with root u whose children are $u_1, \ldots, u_h$. Suppose that for $1 \le j \le h$, $\mathtt{L}(u_j, T^{[u_j]})$ has the form of Eq. (4) and satisfies condition (5). Let $T^{[u]}_{\text{ker}}$ be the kernel subtree of $T^{[u]}$.*

(i) *Suppose $\mathtt{c}_1(T^{[u]}_{\text{ker}}) = k$ and $\mathtt{c}_1^*(T^{[u]} - u) \le k$. If there is a $j \in \{h_1+1, \ldots, h\}$ such that $t^{u_j}_{m_j} = k$, then $\mathtt{c}_1(T^{[u]}) = k+1$ and u is a road of $T^{[u]}$.*

(ii) *Suppose $\mathtt{c}_1(T^{[u]}_{\text{ker}}) < k$ and $\mathtt{c}_1^*(T^{[u]} - u) = k$. If there are two indices $i, j \in \{h_1+1, \ldots, h\}$ such that $t^{u_i}_{m_i} = t^{u_j}_{m_j} = k$, then $\mathtt{c}_1(T^{[u]}) = k+1$ and u is a road of $T^{[u]}$.*

From Theorems 2, 3 and 4, we can prove the correctness of Algorithm 1.

Theorem 5. *For a tree T, Algorithm 1 computes the labels of all vertices with roadmaps.*

Lemma 9. *For a tree T with n vertices, Algorithm 1 can be implemented in a way such that the length of each vertex label is $O(\log n)$.*

Theorem 6. *For a tree T with n vertices, Algorithm 1 computes the labels of all vertices with roadmaps in $O(n \log n)$ time.*

Proof. In Step 1 of Algorithm 1, after the root of T is selected arbitrarily, it takes $O(n)$ time to compute the the parent-child relation induced by the root. It also takes $O(n)$ time to topologically sort all vertices of T such that the root is the last vertex in the list.

In each iteration of the loop from Step 2 to 7, we compute the label of each vertex in the sorted list. Suppose that u is the current vertex in the sorted list for which we want to compute $\mathtt{L}(u, T^{[u]})$. Let $u_1, \ldots, u_t$ be all the children of u. Since these children are listed before u in the list, their labels have been computed in the previous iterations. It follows from Lemma 9 that the length of any vertex label is $O(\log n)$. So in Step 3, constructing the kernel subtree $T^{[u]}_{\text{ker}}$ needs $O(t \log n)$ time. For each child w of u in $T^{[u]}_{\text{ker}}$, we can obtain $\mathtt{L}(w, T^{[w]}_{\text{ker}})$ by taking the last seven components of $\mathtt{L}(w, T^{[w]})$. Note that when we construct $T^{[u]}_{\text{ker}}$, if a child w of u in $T^{[u]}$ is a branching vertex in $T^{[w]}$, that is, $x^w_1 = w$ in $\mathtt{L}(w, T^{[w]})$ with the form of Eq. (2), then w is not a child of u in $T^{[u]}_{\text{ker}}$. So in Algorithm 2, we have $d \le t$.

In Algorithm 2, it is easy to see that all lines except Lines 20–23 takes $O(d)$ time because each label $\mathtt{L}(w, T^{[w]}_{\text{ker}})$, where w is a child of u in $T^{[u]}_{\text{ker}}$, contains only seven components. In Lines 22–23, in order to find the utmost-k-pre-branching vertex in $T^{[u]}_{\text{ker}}$, we check the pre-branching indicators in the labels of the d children of u and find the child v_i with $I^k_{\text{pb}}(v_i) = 1$. Then we check the label $\mathtt{L}(v_i, T^{[v_i]}_{\text{ker}})$ with the form of Eq. (3). If P^{v_i} contains only v_i, then v_i itself is the utmost-k-pre-branching vertex in $T^{[v_i]}_{\text{ker}}$. Thus uv_i is a road in $T^{[u]}_{\text{ker}}$. If P^{v_i} contains two terminals z^{v_i} and v_i, i.e., $P^{v_i} = z^{v_i} \sim v_i$, then z^{v_i} is the utmost-k-pre-branching vertex in $T^{[v_i]}_{\text{ker}}$, and furthermore, z^{v_i} is also the utmost-k-pre-branching vertex in $T^{[u]}_{\text{ker}}$. Hence $z^{v_i} \sim u$ is a road in $T^{[u]}_{\text{ker}}$. So to find a road for

$T^{[u]}_{\text{ker}}$ in Lines 22–23, we only need to check the labels of the d children of u, which can be done in $O(d)$ time because these labels contain only seven components. In Lines 20–21, similarly, we first check the labels of the d children of u to find the children v_i and v_j satisfying that $I^k_{\text{pb}}(v_i) = I^k_{\text{pb}}(v_j) = 1$; we then check the labels $\mathtt{L}(v_i, T^{[v_i]}_{\text{ker}})$ and $\mathtt{L}(v_j, T^{[v_j]}_{\text{ker}})$ to find the utmost-k-pre-branching vertices z^{v_i} and z^{v_j}. In this way, we can find the road $z^{v_i} \sim z^{v_j}$ of $T^{[u]}_{\text{ker}}$ in $O(d)$ time. Thus Algorithm 2 can be implemented in $O(d)$ time. Therefore, the total running time of Step 3 is bounded by $O(t \log n)$.

In Step 4, it takes $O(t \log n)$ time to create the lists $L_1, \ldots, L_{t+1}$. From Lemma 5.12 in [10], Steps 5 and 6 need $O(\max\{t^1_{m_1}, \ldots, t^{t+1}_{m_{t+1}}\} + t)$ time to check if all keys in $L_1, \ldots, L_{t+1}$ are different or find the largest repeated key k^* in the lists, where $t^i_{m_i}$, $1 \le i \le t+1$, is the first component in L_i. Thus the complexity of Steps 5 and 6 is $O(t + \log n)$. In Step 7, it follows from Theorem 4 that the road in item $Q_h = (k'_h, \perp; u)$ contains only one vertex which is the root of $T^{[u]}$. So this road can be found in $O(1)$ time. From Lemma 9, the length of X is $O(\log n)$. Thus Step 7 needs $O(\log n)$ time.

Since every vertex of $T^{[v]}$ has only one parent except the root v, from the above analysis we know that the complexity of the loop from Step 2 to 7 is $O(n \log n)$. Therefore, the total runtime of Algorithm 1 is $O(n \log n)$. □

Remark 2. Note that in [10], only the root's label is required because it contains the copnumber of the tree. It is not necessary to store the labels of other vertices. This is one of the reasons that the running time of Algorithm 1 in [10] is $O(n)$. However, in Algorithm 1 of this paper, we need to compute the roadmaps of all rooted subtrees of $T^{[v]}$, where v is the root of T, because we will need these roadmaps to construct an optimal cop-win strategy for T in Sect. 4. From Lemma 9, we know that the total size of the output of Algorithm 1 can be $\theta(n \log n)$ in the worst case. So it is impossible to design an algorithm for computing the roadmaps of a tree with n vertices in $o(n \log n)$ time.

4 Computing Optimal Cop-Win Strategies

For graph searching models, computing an optimal search strategy usually takes more time than computing a search number. For the cops and robber game, it is even harder to compute an optimal cop-win strategy because a cop-win strategy can change dynamically depending on the robber's action in each round. In this section, we consider the problem of finding an optimal cop-win strategy for a tree in the one-visibility cops and robber game. We first give an $O(n \log n)$ time algorithm to compute an ordering of vertices and edges which is used as a timeline to protect vertices and edges. Note that the input size of this algorithm is $O(n \log n)$, where n is the number of vertices on the tree. We then present an $O(n)$ time algorithm to construct an optimal cop-win strategy that is a sequence of instructions to guide the cops in their search for the robber.

Note that in a tree, once the robber has been seen by a cop, this cop can chase the robber and eventually force the robber to a leaf in a finite number of rounds; then this cop can capture the robber on the leaf. Note that the number of rounds from seeing to capturing is bounded by the diameter of the tree. So in this section, the cops' goal is to find the robber, that is, if the robber is seen by a cop, then the cops win and the game is over.

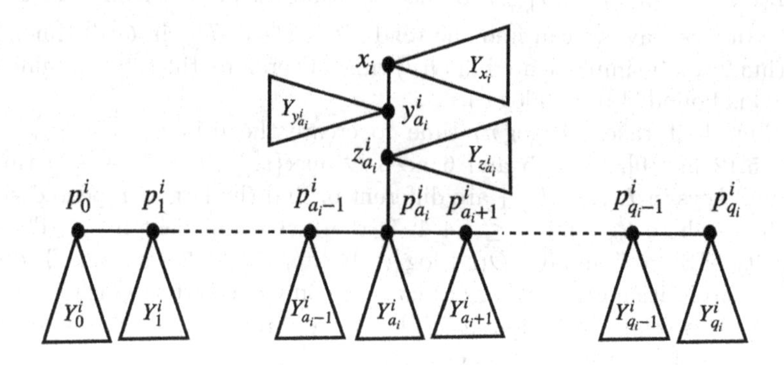

Fig. 1. A tree $T_i^{[x_i]}$, $2 \le i \le m$, with road $P_i = p_0^i \dots p_{q_i}^i$.

Definition 7. Let $T^{[u]}$ be a rooted tree. Suppose that the label $\mathtt{L}(u, T^{[u]})$ is computed by Algorithm 1 and has the form

$$(t_m, x_m, P_m; \dots; t_1, x_1, P_1; I_{\text{wb}}^{t_1}(u), J_{\text{w}}^{t_1}(u); I_{\text{pb}}^{t_1}(u), J^{t_1}(u)). \tag{6}$$

For $2 \le i < m$, let $T_i^{[x_i]} = T^{[x_i]} - \bigcup_{j=i+1}^{m} V(T^{[x_j]})$, and let $T_m^{[x_m]} = T^{[x_m]}$. For each $2 \le i \le m$, let $P_i = p_0^i p_1^i \dots p_{q_i}^i$ be a road in $T_i^{[x_i]}$. It follows from Lemma 8(iii) that each P_i, $2 \le i \le m$, contains at least three vertices. For each $i \in \{2, \dots, m\}$, let $p_{a_i}^i$, $1 \le a_i \le q_i - 1$, be the great-grandchild of x_i in $T_i^{[x_i]}$ (see Fig. 1); let $x_i y_{a_i}^i z_{a_i}^i p_{a_i}^i$ be the path connecting x_i with $p_{a_i}^i$; and let Y_{x_i} (resp. $Y_{y_{a_i}^i}$, $Y_{z_{a_i}^i}$ and $Y_{a_i}^i$) be the component containing x_i (resp. $y_{a_i}^i$, $z_{a_i}^i$ and $p_{a_i}^i$) in the forest $T_i^{[x_i]} - \{x_i y_{a_i}^i\}$ (resp. $T_i^{[x_i]} - \{x_i y_{a_i}^i, y_{a_i}^i z_{a_i}^i\}$, $T_i^{[x_i]} - \{y_{a_i}^i z_{a_i}^i, z_{a_i}^i p_{a_i}^i\}$ and $T_i^{[x_i]} - \{z_{a_i}^i p_{a_i}^i, p_{a_i-1}^i p_{a_i}^i, p_{a_i}^i p_{a_i+1}^i\}$). For each $i \in \{2, \dots, m\}$, let Y_0^i be the component in the forest $T_i^{[x_i]} - \{p_0^i p_1^i\}$ containing p_0^i; let $Y_{q_i}^i$ be the component in $T_i^{[x_i]} - \{p_{q_i-1}^i p_{q_i}^i\}$ containing $p_{q_i}^i$; and let Y_j^i, $j \in \{1, \dots, q_i - 1\} \setminus \{a_i\}$, be the component containing p_j^i in the forest $T_i^{[x_i]} - \{p_{j-1}^i p_j^i, p_j^i p_{j+1}^i\}$. Let $T_1^{[u]} = T^{[u]} - \bigcup_{j=2}^{m} V(T^{[x_j]})$ and $P_1 = p_0^1 \dots p_{q_1}^1$. When $q_1 \ge 2$, it follows from Algorithm 1 that $p_1^1 \dots p_{q_1-1}^1$ is an avenue of $T_1^{[u]}$, and p_0^1 and $p_{q_1}^1$ are two pre-avenue vertices; then let Y_0^1 be the component in the forest $T_1^{[u]} - \{p_0^1 p_1^1\}$ containing p_0^1, let $Y_{q_1}^1$ be the component in $T_1^{[u]} - \{p_{q_1-1}^1 p_{q_1}^1\}$ containing $p_{q_1}^1$, and let Y_j^1, $j \in \{1, \dots, q_1 - 1\}$, be the component containing p_j^1 in the forest $T_1^{[u]} - \{p_{j-1}^1 p_j^1, p_j^1 p_{j+1}^1\}$. When $q_1 = 1$, from Algorithm 1, $P_1 = p_0^1 p_1^1$ is a street of $T_1^{[u]}$; then let Y_0^1 (resp. Y_1^1) be the component in the forest $T_1^{[u]} - \{p_0^1 p_1^1\}$ containing p_0^1 (resp. p_1^1). When $q_1 = 0$, i.e., $P_1 = p_0^1$ is a hub of $T_1^{[u]}$, then let $Y_0^1 = T_1^{[u]}$.

Note that in Definition 7, each subtree Y_j^i, $1 \leq i \leq m$, $0 \leq j \leq q_i$, is a rooted subtree where p_j^i is the root. Similarly, Y_{x_i} (resp. $Y_{y_{a_i}^i}$, and $Y_{z_{a_i}^i}$) is a rooted subtree where x_i (resp. $y_{a_i}^i$, and $z_{a_i}^i$) is the root.

Let $T^{[u]}$ be a rooted tree. To construct an optimal cop-win strategy, it is essential to give an ordering of the vertices by which the cops will clear these vertices. In Algorithm 3, we use the vertex labels computed by Algorithm 1 to traverse $T^{[u]}$, during which, for each vertex v, we stamp it with a *start-time* $s(v)$ that is the time when we start to protect $N[v]$ in Algorithm 4, and stamp it with a *finish-time* $f(v)$ that is the time when we finish protecting $N[v]$. In Algorithm 3, we also timestamp some edges, which will be the potential vibrating edges in our optimal cop-win strategy presented in Algorithm 4. For each of those edges vv' that will be a potential vibrating edge, we stamp it with a *start-time* $s(vv')$ that is the time when we start to protect $N[v] \cup N[v']$ in Algorithm 4, and stamp it with a *finish-time* $f(vv')$ that is the time when we finish vibrating between v and v'. In particular, `stamp` is an operation to assign the current *time* to an object; *time* is a global variable that is 0 initially and is incremented by 1 after each assignment. StampSubtree($T^{[x]}, x$) is a function to timestamp vertices and potential vibrating edges in the rooted subtree $T^{[x]}$.

Let $T^{[u]}$ be a rooted tree with $\mathsf{c}_1(T^{[u]}) = k$. After we obtain the timestamps of vertices and some edges of $T^{[u]}$ from Algorithm 3, Algorithm 4 describes how these timestamps are used to guide the k cops to clear $T^{[u]}$. In this algorithm, we say that a cop on a vertex v is *free* if he finishes protecting $N[v]$; otherwise, we say that the cop on the vertex v is *busy*. We say that a cop vibrating between vertices v and v' is *free* if he finishes protecting $N[v] \cup N[v']$; otherwise, we say that the cop vibrating between v and v' is *busy*. We call a vertex *internal* if this vertex has at least one child; we call an edge vv' *internal* if v is a parent of v' and v' has at least one grandchild.

Theorem 7. *For a tree T, the strategy output by Algorithm 4 is an optimal cop-win strategy for T.*

Theorem 8. *For a tree T with n vertices, an optimal cop-win strategy for T can be computed in $O(n \log n)$ time.*

For a tree $T^{[u]}$, it follows from Algorithm 4 that each vibrating edge vv' is an internal edge which is stamped a start-time $s(vv')$ and a finish-time $f(vv')$ in Algorithm 3. Note that when a cop starts vibrating between v and v' at the time $s(vv')$ and finishes vibrating at the time $f(vv')$, this cop alternates two sliding actions between v and v' for at least three consecutive rounds. So $f(vv') - s(vv') \geq 2$, and thus the open interval $(s(vv'), f(vv'))$ contains at least one timestamp.

Algorithm 3. TIMESTAMP($T^{[u]}$)

Input: $T^{[u]}$ with the vertex labels from Algorithm 1, where the notations of labels and subtrees are given in Eq. (6) and Definition 7.
Output: Timestamps of all vertices and potential vibrating edges in $T^{[u]}$.

```
From L(u, T^[u]), construct T_1^[u] and P_1(= p_0^1 ... p_{q1}^1) defined in Definition 7.
if m = 1 (defined in Definition 7) then
    for j = 0 to q_1 do
        stamp s(p_j^1); StampSubtree(Y_j^1, p_j^1); stamp f(p_j^1).
    return
for i = m down-to 2 do
    for j = 0 to a_i - 1 do
        stamp s(p_j^i); StampSubtree(Y_j^i, p_j^i); stamp f(p_j^i).
    stamp s(p_{a_i}^i), s(z_{a_i}^i), s(y_{a_i}^i), s(y_{a_i}^i z_{a_i}^i) and s(x_i).
for j = 0 to q_1 do
    stamp s(p_j^1); StampSubtree(Y_j^1, p_j^1); stamp f(p_j^1).
for i = 2 to m do
    StampSubtree(Y_{x_i}, x_i); stamp f(x_i); stamp f(y_{a_i}^i z_{a_i}^i);
    StampSubtree(Y_{y_{a_i}^i}, y_{a_i}^i); stamp f(y_{a_i}^i);
    StampSubtree(Y_{z_{a_i}^i}, z_{a_i}^i); stamp f(z_{a_i}^i);
    StampSubtree(Y_{a_i}^i, p_{a_i}^i); stamp f(p_{a_i}^i).
    for j = a_i + 1 to q_i do
        stamp s(p_j^i); StampSubtree(Y_j^i, p_j^i); stamp f(p_j^i).
function StampSubtree(T^[x], x)
  if T^[x] contains only the vertex x then return
  for each child v of x do
     stamp s(v);
     if v is not a leaf then
        for each child v' of v do
           stamp s(v') and s(vv');
           if v' is not a leaf then
              for each child v'' of v' do
                 if v'' is a leaf then stamp s(v'') and f(v'');
                 else TimeStamp(T^[v'']).
           stamp f(v') and f(vv').
     stamp f(v).
```

Definition 8. For a tree $T^{[u]}$ with the timestamps from TIMESTAMP($T^{[u]}$), let $E_{\text{vib}} = \{vv' \in E(T^{[u]}) \mid \text{there is a cop vibrating on } vv' \text{ in Algorithm 4}\}$. Let $\nu(t, T^{[u]}) = |\{(s(vv'), f(vv')) \mid t \in (s(vv'), f(vv')), vv' \in E_{\text{vib}}\}|$, and $\nu(T^{[u]}) = \max\{\nu(t, T^{[u]}) \mid t \text{ is a timestamp computed by Algorithm 3}\}$.

We have the following relation between $\nu(T^{[u]})$ and $\mathtt{c}_1(T^{[u]})$.

Theorem 9. *For a tree* $T^{[u]}$, $\mathtt{c}_1(T^{[u]}) = \nu(T^{[u]}) + 1$.

Algorithm 4. Computing an optimal cop-win strategy

Input: $T^{[u]}$ with $k = \mathsf{c}_1(T^{[u]})$ and the timestamps from Algorithm 3.
Output: An optimal cop-win strategy for $T^{[u]}$.

1: Create a linked list $\mathcal{S}$, which initially contains the instruction "If a cop can see the robber at any timestamp, then the game is over".
2: Insert "Place k cops on the vertex whose start-time is 0" into $\mathcal{S}$ and set the cops free.
3: **for** each timestamp in increasing order **do** insert one of the following instructions into $\mathcal{S}$:
4:     **if** the timestamp is the start-time of an internal vertex v that is not adjacent to a vertex occupied by a cop **then**
5:         move a free cop to v and set this cop busy.
6:     **if** the timestamp is the start-time of an internal edge vv' **then**
7:         let a cop start vibrating between v and v' and set this cop busy.
8:     **if** the timestamp is the finish-time of an internal vertex v **then**
9:         set all cops on v free.
10:     **if** the timestamp is the finish-time of an internal edge vv' **then**
11:         set the cop vibrating between v and v' free.
12:     **if** the timestamp is for a non-internal vertex or edge **then**
13:         no new actions for the cops.
14: **return** $\mathcal{S}$.

References

1. Bonato, A., Nowakowski, R.J.: The Game of Cops and Robbers on Graphs. American Mathematical Society, Providence (2011)
2. Bonato, A., Yang, B.: Graph searching and related problems. In: Pardalos, P.M., Du, D.-Z., Graham, R.L. (eds.) Handbook of Combinatorial Optimization, pp. 1511–1558. Springer, New York (2013). https://doi.org/10.1007/978-1-4419-7997-1_76
3. Clarke, N.E., Cox, D., Duffy, C., Dyer, D., Fitzpatrick, S., Messinger, M.-E.: Limited visibility Cops and Robbers, 2017. Discrete Appl. Math. **282**, 53–64 (2020)
4. Megiddo, N., Hakimi, S.L., Garey, M., Johnson, D., Papadimitriou, C.H.: The complexity of searching a graph. J. ACM **35**, 18–44 (1988)
5. Nowakowski, R.J., Winkler, P.: Vertex-to-vertex pursuit in a graph. Discrete Math. **43**, 235–239 (1983)
6. Quilliot, A.: Jeux et pointes fixes sur les graphes, Thèse de 3ème cycle, Université de Paris VI, 131–145 (1978)
7. Tošić, R.: Vertex-to-vertex search in a graph. In: Proceedings of the Sixth Yugoslav Seminar on Graph Theory, pp. 233–237, University of Novi Sad (1985)
8. Xue, Y., Yang, B., Zilles, S.: A simple method for proving lower bounds in the zero-visibility cops and robber game. J. Combin. Optim. (2021). http://link.springer.com/article/10.1007/s10878-021-00710-8
9. Yang, B., Zhang, R., Cao, Y., Zhong, F.: Search numbers in networks with special topologies. J. Interconnect. Netw. **19**, 1–34 (2019)
10. Yang, B., Akter, T.: One-visibility cops and robber on trees. Theor. Comput. Sci. **886**, 139–156 (2021)

Complexity and Approximation Results on the Shared Transportation Problem

Tom Davot(✉), Rodolphe Giroudeau, and Jean-Claude König

LIRMM, Univ Montpellier, CNRS, Montpellier, France
{davot,rgirou,konig}@lirmm.fr

Abstract. In our modern societies, a certain number of people do not own a car, by choice or by obligation. For some trips, there is no or few alternatives to the car. One way to make these trips possible for these people is to be transported by others who have already planned their trips. We propose to model this problem using as path-finding problem in a list edge-colored graph. This problem is a generalization of the $s-t$-path problem, studied by Böhmová et al. We study two optimization functions: minimizing the number of color changes and minimizing the number of colors. We study the complexity and the approximation of this problem. We show the existence of polynomial cases. We show that this problem is NP-complete and hard to approximate, even in restricted cases. Finally, we provide an approximation algorithm.

Keywords: Complexity · Approximation · List edge coloring

1 Introduction

Shared mobility received a lot of attention in the last decades, both from industry and academics. The motivation behind this is ecological awareness, savings and social benefits. The rise of this kind of transportation is traduced by the apparition of mobility platforms and the emergence of scientific studies focusing on the different various relative questions. In particular, researchers in the field of operational research have been interested in studying various optimization problems resulting from shared mobility systems. In these systems, we seek to match people having similar itineraries on the same dates. A survey on ride-sharing systems can be found in [10]. The authors present a classification of different existing ride-sharing systems and identify some challenges. In [19], the authors present dynamic ride-sharing systems. The authors show the need of optimization technologies for the success of this type of ride-sharing systems.

Different types of mobilities sharing systems exist. Carpooling is proposed by large companies to encourage their employees to share itineraries to and from work, in order to reduce the use of private cars. In dial-a-ride problems (DARP) [6,7,12], schedules and vehicles routes are designed based on user requests. In Vanpool problem [12], passengers drive to a park-and-ride location then they

D.-Z. Du et al. (Eds.): COCOA 2021, LNCS 13135, pp. 140–151, 2021.
https://doi.org/10.1007/978-3-030-92681-6_12

share their trips with a van to the target location. An exhaustive survey on optimization for shared mobility can be found in [18].

In this paper, we consider a problem where one person aims to travel from a place to another and can not make the trip by their own means. This can be due to several reasons: disability, absence of driving license, personal choice... In order to make the trip possible, we can use the help of drivers that have already planned their travel and offer to transport another person. We aim to match one or more driver that can share its/their trip. We model this problem as an $s-t$-path problem on list edge-colored digraph. We consider two objective functions. The first objective function is a color minimization which is a common objective for optimization problems on colored graphs. The second function, rather less classical, is to minimize the number of color changes along the path.

Results and Related Works. Similar problems have been studied in the literature. In [2], the authors showed that the Minimum Label Path/Cycle Problem in undirected graph is NP-hard. The authors also provide some exact exponential-time and approximation algorithms to solve the problem. Another approximation algorithm and approximation hardness results have been presented in [11]. Some parameterized intractability results for minimum label path and other different minimum labeling problems have been presented in [8]. Other optimization problems on edge-labeled graphs have been considered in the literature. The minimum labeling spanning tree is widely studied [4,5,15]. The objective is to find a spanning tree such that the number of labels is the smallest possible. Another variant of the problem is considered in [21]. The problem is called Label-Constrained Minimum Spanning Tree Problem and the objective is to find the minimum weight spanning tree using at most k labels.

The number of problems in the literature using an edge-coloring or a list edge-coloring is large. We can cite the classic proper edge-coloring as example [3]. A close related problem has been proposed by Broersma et al. [2]: the aim is to find a path/cycle in a colored graph with a minimum number of colors. This problem is NP-hard even in bipartite planar. The authors also propose several exact and approximation algorithms. Finally, the complexity of the exact algorithms and the performance ratio of the approximation algorithms are also analyzed.

Our problem is equivalent to finding a path from s to t, using at most k colors, in an oriented graph G with a list of allowed colors for each arc. Each color represents a driver. It is a generalization of the st-Path problem studied in [1]. Whereas in the version studied by Böhmová et al. each subgraph induced by a color must be a path (representing a subway line), we have chosen not to put any restriction on these subgraphs in the problem formulation. This enables us to model the route options that drivers could propose. However, in most of our results this restriction still holds.

In [1], some complexity results related to a subway network are presented: they propose an efficient algorithm for finding an st-route according to the number of line changes plus one. A non-approximation result, for the minimization of the used lines is proposed. Lastly, a polynomial-time algorithm is developed for the problem of enumerating all st-paths with a bounded length.

In the following we extend the complexity results of Böhmová et al. by considering severals topologies or restricted cases.

Organization of the Article. The next section is devoted to notation and to the presentation of the two problems studied in this article. In Sect. 3, we present some restricted polynomial cases. Section 4 is devoted to the computational complexity. We develop a polynomial-time approximation algorithm in Sect. 5. Finally, in Sect. 6, we show the non-existence of a subexponential-time algorithm.

2 Problems Description

2.1 Notation

In this article, we consider a specific oriented graph called *list arc-colored graph.* A list arc-colored graph $G = (V, A, \chi)$ is a graph with a set of vertices V, a set of arcs A and a function $\chi : A \mapsto \mathcal{P}(\mathbb{N}) \setminus \{\emptyset\}$ that associates a (sub-)set of colors to each arc. We denote by n and m the numbers of vertices and arcs of G, respectively. For each color i, we denote $G_i = (V_i, A_i)$ the subgraph induced by the arcs colored with color i. Formally, $A_i = \{e \mid i \in \chi(e)\}$ and $V_i = \{v \mid \exists uv \in A_i\}$. Let $\tilde{G}_i$ be the transitive closure of G_i (*i.e.* the arc uv belongs to G_i if and only if there is an oriented path from u to v in G_i) and we denote $\tilde{G} = \cup_i \tilde{G}_i$.

In the problems studied in this paper, given a list arc-colored graph G, the aim is to construct a *colored path* $\pi = (P, \lambda)$ where P is an oriented path of G and $\lambda : V(P) \mapsto \mathbb{N}$ is an arc-coloring function. For each arc e of P, we attribute a unique color among $\chi(e)$, that is, $\lambda(e) \in \chi(e)$. The number of colors of π is denoted $\lambda_{\#}(\pi) = |\{\lambda(e) \mid e \in P\}|$. The number of *color changes*, denoted $\lambda_c(\pi)$ is the number of pairs of consecutive arcs of π that have different colors. Formally, let $P = (e_1, \ldots, e_k)$ be a path, we have $\lambda_c(\pi) = |\{(e_i, e_{i+1}) \mid \lambda(e_i) \neq \lambda(e_{i+1})\}|$. Finally, we denote by $\pi[i]$ the subgraph of π induced by the arcs with color i. For simplicity, we sometimes denote $\pi[i]$ as the subgraph induced by the color i.

2.2 Objective Functions

In this article, we consider two objective functions consisting of minimizing the number of colors or the number of color changes of a colored path. Hence, we define the two following problems.

k-COLORED PATH (k-CP)
Input: A list arc-colored oriented graph $G = (V, A, \chi)$, two given vertices s and t and a positive integer k.
Question: Is there a colored path π between s and t in G such that $\lambda_{\#}(\pi) \leq k$?

Table 1. Overview of complexity results

Problem	Topology	Complexity	
k-COLORED PATH	G_i is a bounded-length path	NP-C	Theorem 3
	Planar	NP-C	Theorem 5
	Bipartite	NP-C	Theorem 4
	Planar and bipartite	NP-C	Theorem 6
	General	Non-approximable	See [1]
	Path	Non-approximable	Corollary 3
	G_i is strongly connected	P	Theorem 2
	G_i is a path of length two	P	Lemma 2
k-COLOR CHANGE PATH	General	P	Theorem 1

k-COLOR CHANGE PATH (k-CCP)
Input: A list arc-colored oriented graph $G = (V, A, \chi)$, two given vertices s and t and a positive integer k.
Question: Is there a colored path π between s and t such that $\lambda_c(\pi) \leq k$?

We study the complexity of these problems according to some graph topologies. An overview of the result is available in Table 1.

3 Polynomial Cases

We present in this section some polynomial time algorithms for some specific cases related to the connectivity of colored subgraphs $G_i, \forall i$. For this, we show that the research for a shortest path in the modified graph $\tilde{G}$ guarantees obtaining a optimal solution.

Theorem 1. k-COLOR CHANGE PATH *admits a polynomial-time algorithm in* $O(n^3)$ *time.*

Proof. Let G be a list arc-colored oriented graph. First, note that $\tilde{G}$ can be constructed in $O(n^3)$. We show that G contains a colored path π between s and t with $\lambda_c(\pi) = k$ if and only if there is an oriented path p of length at most $k+1$ between s and t in $\tilde{G}$.

- Let π be a colored path between s and t in G. For each monochromatic subpath $(u, \ldots, v)$ of π, introduce the arc uv in p (uv exists in $\tilde{G}$ by definition). Since $\lambda_c(\pi) = k$, it exists $(k+1)$ monochromatic subpaths in π. Thus, we construct a oriented path of length $k+1$ in $\tilde{G}$.
- Let p be an oriented path of length $k+1$ between s and t in $\tilde{G}$. For each arc uv of p, it exists a monochromatic path p' between u and v in G, by construction of $\tilde{G}$. We add p' in π. Therefore, we obtain a colored path π between s and t in G that is constituted by at most $k+1$ maximal monochromatic paths. Hence, we obtain a path π with $\lambda_c(\pi) = k$.

Thus, we can construct an optimal colored path in G by computing a shortest path in $\tilde{G}$ and then apply the transformation described above. Since, a shortest path can be computed in $O(n^2)$ using Dijkstra's algorithm [9], the overall complexity is $O(n^3)$.

We propose to extend the previous result to k-COLORED PATH in some restricted case. So, we introduce the following lemma.

Lemma 1. *Let G be a list arc-colored oriented graph. k-COLORED PATH can be solved in time $O(n^3)$ if it exists an optimal colored path π such that for each color i, $\pi[i]$ is connected.*

Proof. Let G be a list arc-colored oriented graph. Using the same argument as in the proof of Theorem 1, we can show that G contains a colored path π respecting lemma's property with $\lambda_{\#}(\pi) = k$ between s and t if and only if there is a path p of length at most k between s and t in $\tilde{G}$. Hence, we can derive an optimal colored path in G by computing a shortest path in $\tilde{G}$ and by applying the transformation described in the proof of Theorem 1.

Theorem 2. *k-COLORED PATH admits an $O(n^3)$ time algorithm if for each color i of χ, G_i is strongly connected.*

Proof. Let G be a list arc-colored oriented graph such that each subgraph G_i is strongly connected and let π be a colored path between s and t. If there is a color i such that $\pi[i]$ contains two non-connected subpaths $(u, \ldots, v)$ and $(u', \ldots, v')$, then since G_i is strongly connected, we can replace the subpath $(u, \ldots, v')$ in π by a path in G_i from u to v' without increasing $\lambda_{\#}(\pi)$. By doing that, we ensure that for each color i, $\pi[i]$ is connected. Thus, by Lemma 1, we conclude that k-COLORED PATH can be solved in $O(n^3)$-time in G.

Corollary 1. *k-COLOR CHANGE PATH in a non-oriented graph G admits a polynomial-time algorithm if $\forall i, G_i$ is connected.*

Hereafter, we propose a polynomial-time algorithm for the case of each subgraph G_i induced by a color i is a path of length at most two.

Lemma 2. *k-COLORED PATH can be solved in $O(n^3)$-time in graphs for which each color induces a path of length at most two.*

Proof. Let π be an optimal colored path. Suppose that there is a color i such that $\pi[i]$ is not connected. Let (v_1, v_2, v_3) be the path constituting G_i. Since both arcs v_1v_2 and v_2v_3 appears in π, then π is not an elementary path, contradicting its optimality. Hence, for each color i, $\pi[i]$ is connected and by Lemma 1, k-COLORED PATH can be solved in $O(n^3)$ in G.

4 Computational Hardness

In this section, we consider k-COLOR CHANGE PATH problem in which each graph induced by a color is a path. We show that in that case, k-COLOR CHANGE PATH is NP-complete. We then show that it remains NP-complete even if the graph is bipartite or planar.

4.1 Each Color Induces a Path of Bounded Length

We now show that k-COLORED PATH is NP-complete. We use a similar idea as the proof proposed for the NP-completeness of the problem of minimizing the number of used colored in an edge-colored graph [2]. We reduce from the following classical NP-complete problem.

3-SAT
Input: A Boolean formula φ where each clause contains exactly three literals
Question: Is φ satisfiable?

Construction 1. *Let φ be an instance of* 3-SAT *with m' clauses $C_0, \ldots, C_{m'-1}$ and n' variables $x_0, \ldots, x_{n'-1}$. For each variable x_i, let ψ_i (resp. $\overline{\psi}_i$) be the list of clauses where x_i appears positively (resp. negatively). We construct a list arc-colored graph $G = (V, A, \chi)$ as follows:*

- *create a vertex Q_m,*
- *for each variable x_i, create a vertex v_i,*
- *for each clause C_j create a vertex Q_j and create a vertex q_j^i, for each literal ℓ_i of C_j,*
- *for each variable x_i and for each clause C_j in which x_i appears, introduce an oriented path $(Q_j, q_j^i, v_i, v_{i+1})$ (or $(Q_j, q_j^i, v_{n-1}, Q_0)$ if $i = n-1$) with color c_j^i,*
- *for each variable x_i and for each pair of clauses (C_j, C_k) in ψ_i (resp. $\overline{\psi}_i$) such that $j < k$, introduce an oriented path $(Q_k, q_k^i, q_j^i, C_{j+1})$ with color $t_{j,k}^i$, and*
- *finally, for each variable x_i, let C_j be the last clause of ψ_i (resp. $\overline{\psi}_i$), introduce the arc (q_j^i, Q_{j+1}) with color z_i (resp. $\bar{z}_i$).*

Notice that the graph induced by each color is a path of length exactly one or three.

Theorem 3. k-COLOR CHANGE PATH *remains* NP*-complete even if each color induces a path of length at most three.*

Proof. Let φ be 3-SAT formula and G the graph obtained by Construction 1. Clearly, k-COLOR CHANGE PATH is in NP. We show that φ is satisfiable if and only if it exists a colored path between v_0 and Q_m with $\lambda_{\#}(\pi) = n + m$.

- Suppose that φ is satisfiable and consider ϕ a satisfying assignment for φ. Let $f_\phi : \{C_0, \ldots, C_{m-1}\} \mapsto \{x_0, \ldots, x_{n-1}\}$ be a function that assigns to each clause C_j a unique variable x_i such that the assignment of x_i satisfies C_j. We suppose that $f_\phi^{-1}(x_i)$ is an ascending ordered list according to the indices. We construct π as follows. For each variable x_i and for each clause $C_j \in f_\phi^{-1}(x_i)$:
  - If C_j is the first clause of $f_\phi^{-1}(x_i)$, add in π the outgoing arc of x_i with color c_j^i and the arc (C_j, q_j^i) with color c_j^i. Otherwise, let C_k be the clause that precedes C_j in $f_\phi^{-1}(x_i)$, add in π the two arcs (q_k^i, C_{k+1}) and (C_j, q_j^i) with color t_j^k.

  - If C_j is the last clause of $f_\phi^{-1}(x_i)$, add in π the arc (q_j^i, C_{j+1}) with color z_j.

  Since each clause C_j belongs to a list f_ϕ^{-1}, we construct a path between x_0 and C_m using $n+m$ colors.

- Let π be a path between x_0 and C_m using $n+m$ colors. Consider the function $g_\pi : \{x_0, ..., x_{n-1}\} \mapsto \mathcal{P}(\{C_0, \dots, C_{m-1}\})$ defined by $g_\pi(x_i) = \{C_j \mid q_j^i \in \pi\}$. First, by construction for each vertex $x_i \in \pi$, we have $x_{i+1} \in \pi$ (or C_0 if $i = n-1$). Thus, π contains the subpath $P_{literals} = (x_0, \dots, x_{n-1}, C_0)$ and uses clearly n colors in it. Moreover, always by construction, for each vertex $C_j \in \pi$, the only way to reach a vertex C_k, with $k > j$ from C_j is to take a path (C_j, q_j^i, C_{j+1}). Thus, by extension, π contains a subpath $P_{clauses} = (C_0, q_0^i, C_1, q_1^{i'}, \dots, C_m)$ and so, each clause is contained in a set $g_\pi(x_i)$. Clearly, π uses a new color for each arc (q_j^i, C_{j+1}). Hence, since π can not use more than m colors in $P_{clauses}$, each arc $a_j = (C_j, q_j^i)$ is colored with a color already used in the subpath $(x_0, \dots, C_j)$. Hence, if $\lambda(a_j) = t_j^k$, then $(q_k^i, C_{k+1}) \in \pi$ and therefore, by construction, we have $\{C_k, C_j\} \subseteq \psi_i$ or $\{C_k, C_j\} \subseteq \overline{\psi}_i$. If $\lambda(a_j) = c_j^i$, then the arc (x_i, x_{i+1}) is colored with c_j^i in π. Since only one outgoing arc of x_i appears in π, by induction, we have $g_\pi(x_i) \subseteq \psi_i$ or $g_\pi(x_i) \subseteq \overline{\psi}_i$. If $g_\pi(x_i) \subseteq \psi_i$, we assign $x_i = \texttt{true}$ in ϕ and $x_i = \texttt{false}$ otherwise. The assignment of x_i satisfies every clause in $g_\pi(x_i)$ and since for each clause C_j, $g_\pi^{-1}(C_j)$ is defined, then ϕ is a satisfying assignment of φ.

From Construction 1, it is easy to extend this result for the case where each path induced by a color has length exactly three: if a path has length one, we can extend it by adding two new vertices. Samewise, we can show that this problem remains NP-complete if every graph G_i is a disjoint union of arcs, by simply removing the backward arcs in the construction.

Corollary 2. k-COLOR CHANGE PATH *remains* NP-*complete even if:*

- *each color induces a path of length exactly three, or*
- *each color induces a collection of disjoint arcs.*

4.2 In Bipartite and Planar Graphs

In the following, we reuse Construction 1 to extend the previous hardness result to planar and bipartite graphs.

Let G be a graph resulting of Construction 1. We can make it bipartite by applying the following transformation. Let P_1 and P_2 be any partition of the vertices of G. For each arc $a = (v_1, v_2)$ such that $v_1 \in P_1$ and $v_2 \in P_2$, we can introduce a new vertex u and replace (a) by the arcs (v_1, u) and (u, v_2).

Notice that since this transformation can only be applied to some backward arcs of G, the length of each path induced by a color is bounded by four. Hence, we obtain the following result.

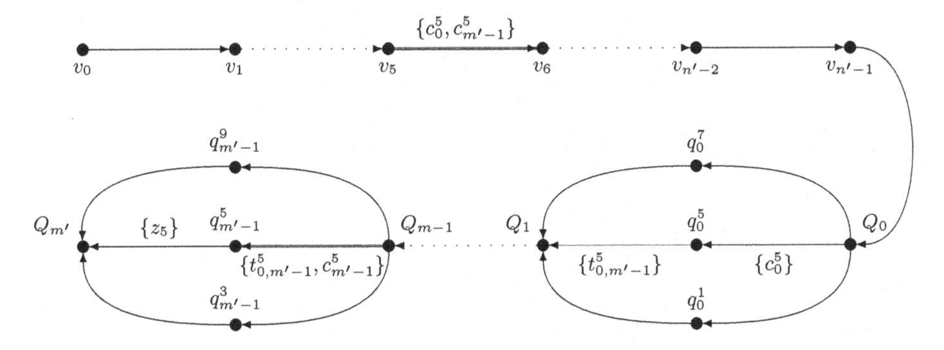

Fig. 1. Illustration of Construction 1. In this example, the literal x_5 appears in clauses $C_0 = (x_1, x_5, x_7)$ and $C_{m'-1} = (\bar{x}_3, x_5, \bar{x}_9)$. For simplicity, backward arcs $(q_0^5, v_5), (q_{m'-1}^5, v_5)$ and $(q_{m'-1}^5, q_0^5)$ are not drawn.

Theorem 4. *k-COLOR CHANGE PATH is NP-complete even in bipartite graphs where each color induces a path of length at most four.*

Further, if we draw G as in Fig. 1, only backward arcs can cross. In order to make such graph planar, we apply the classical technique consisting of adding a vertex for each arc intersection. Formally, let (u, u') and (v, v') be two intersecting backward arcs with color i and j respectively. We introduce a new vertex v, remove (u, u') and (v, v') and we construct two paths (u, x, u') and (v, x, v') with color i and j, respectively. Therefore, we add at most $O(m'^2)$ vertices to the construction. Notice that the lengths of the paths induced by a color are no longer bounded.

Theorem 5. *k-COLOR CHANGE PATH is NP-complete in planar graphs.*

Finally, by combining the two previous techniques, we obtain the following result.

Theorem 6. *k-COLOR CHANGE PATH remains NP-complete for planar bipartite graphs.*

4.3 In Paths

We now show that k-COLOR CHANGE PATH in paths is equivalent to the classical problem SET COVER, defined as follows.

SET COVER (SC)
Input: A univers $U = (e_1, \ldots, e_{n'})$ of n' elements, a collection $C = \{S_1, \ldots, S_{m'}\}$ of m' subsets of U and a positive integer k.
Question: Is there a collection $C' \subseteq C$ such that $\bigcup_{S_i \in C'} S_i = U$ and $|C'| \leq k$?

Construction 2. *Let (U, C) be an instance of* SET COVER, *we construct a list-arc colored graph G as follows:*

- *construct an oriented path* $(v_1, v_{n'+1})$, *and*
- *for each subset* $S_j \in C$ *and each element* $e_i \in S_j$, *color the arc* (v_i, v_{i+1}) *with color* j.

An example of graph produced by Construction 2 is depicted in Fig. 2.

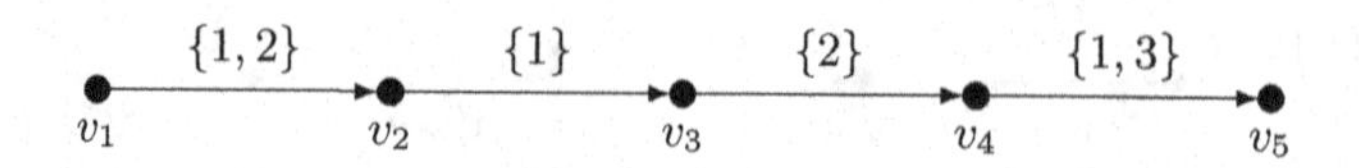

Fig. 2. Example of graph produced by Construction 2 on the univers containing the sets $S_1 = \{1, 2, 4\}, S_2 = \{1, 3\}$ and $S_3 = \{4\}$.

Theorem 7. *The optimization version of* k-COLORED PATH *is LOG-APX-hard even in paths.*

Proof. Let (U, C) be an instance of SET COVER and let G be its list arc-colored graph resulting from Construction 2. We show that (U, C) admits a set cover of size k if and only if G contains a colored path π between v_1 and $v_{n'+1}$ with $\lambda_{\#}(\pi) = k$.

- Let C' be a minimal set cover of (U, C) of size k. We construct π as follows. For each element e_i, let $S_j \in C'$ containing x_i. Add the arc $v_i v_{i+1}$ with color j in π. Clearly π cannot use more colors than $|C'|$, thus we obtain a colored path π between v_1 and $v_{n'+1}$ with $\lambda_{\#}(\pi) \leq |C'|$.
- Let π be a colored path between v_1 and $v_{n'+1}$ such that $\lambda_{\#}(\pi) = k$. We construct the following set cover $C' = \{S_j \mid \lambda(v_i v_{i+1}) = j\}$. Since G is an oriented path, for each vertex v_i, the arc $v_i v_{i+1}$ belongs to π. Thus, by construction, for each element $e_i \in U$, e_i is contained in the subset S_j, where $\lambda(x_i, x_{i+1}) = j$. Hence, C' is a set cover of (C, U).

Suppose it exists a polynomial-time algorithm A that can approximate k-COLORED PATH with a factor $R(G)$. Then, we can obtain a polynomial-time algorithm with the same approximation factor for SET COVER by applying successively Construction 2, A and the transformation to obtain a set cover from a colored path described above. Lund and Yannakakis show that SET COVER can not be approximated with a factor better than a logarithmic function [17] if $P \neq$ NP. Hence, it implies that k-COLORED PATH can not be approximated in polynomial-time with a factor better than a logarithmic function if $P \neq$ NP.

We now reduce k-COLORED PATH in paths to SET COVER.

Construction 3. *Let* G *be a list-arc colored path on the vertices* $(v_1, \ldots, v_n)$. *We construct an instance of* SET COVER (U, C) *as follows:*

- *construct the univers* $U = \{e_1, \ldots, e_{n-1}\}$,
- *for each color* j, *introduce a set* S_j, *and*

– *for each arc* $a = (v_i, v_{i+1})$ *and each color* $j \in \chi(a)$*, emplace the element* e_i *in* S_j*.*

Notice, that the previous construction is the inverse function of Construction 2. Thus, we can reuse the same argument as in Theorem 7 to show the following.

Lemma 3. *Let* G *be a list-arc colored path and* (U, C) *be its instance of* SET COVER *resulting of Construction 3. It exists a colored path* π *between* v_1 *and* v_n *with* $\lambda_{\#}(\pi) = k$ *if and only if it exists a set cover of* (U, C) *of size* k*.*

Corollary 3. SET COVER $\equiv$ k-COLORED PATH *in paths.*

5 Approximation Results

In the following, we consider the problem in which each subgraph G_i is an oriented path of length at most $\ell_i \leq \ell$.

We develop a polynomial-time approximation algorithm based on the computation of a shortest path in $\tilde{G}$. As for the algorithms of Sect. 3, the overall time complexity of this approximation algorithm is $O(n^3)$.

Lemma 4. *Let* G *be a list arc-colored graph such that each subgraph* G_i *is an oriented path of length at most* ℓ *and let* π *be a colored path between* s *and* t*. For each color* i*,* $\pi[i]$ *contains at most* $\lceil \frac{\ell}{2} \rceil$ *connected components.*

Proof. Let $v_i v_j$ be an arc of π colored with color c. Let $v_j v_k$ be the outgoing arc of v_j in G_i. Since π is elementary, either π contains the subpath (v_i, v_j, v_k), or π does not contain $v_j v_k$. Hence, two consecutive arcs of G_i cannot appear in different connected components of $\pi[i]$ and then $\pi[i]$ contains at most $\lceil \frac{\ell}{2} \rceil$ connected components.

Theorem 8. *Let* G *be a list arc-colored graph such that each subgraph* G_i *is an oriented path of length at most* ℓ*. An optimal solution of* k-COLOR CHANGE PATH *in* G *is a* $\lceil \frac{l}{2} \rceil$*-approximation of* k-COLOR CHANGE PATH*.*

Proof. Let π be a colored path between s and t. By Lemma 4, we have

$$\lambda_c(\pi) \leq \lceil \frac{\ell}{2} \rceil \cdot \lambda_{\#}(\pi). \tag{1}$$

Let π_{opt} be an optimal solution of k-COLOR CHANGE PATH and π_{app} be an optimal solution of k-COLOR CHANGE PATH. We have

$$\lambda_{\#}(\pi_{app}) \leq \lambda_c(\pi_{app}) \tag{2}$$

and

$$\lambda_c(\pi_{app}) \leq \lambda_c(\pi_{opt}). \tag{3}$$

Thus,

$$\lambda_{\#}(\pi_{app}) \overset{(2)}{\leq} \lambda_c(\pi_{app}) \overset{(3)}{\leq} \lambda_c(\pi_{opt}) \overset{(1)}{\leq} \lceil \frac{\ell}{2} \rceil \cdot \lambda_{\#}(\pi_{opt}). \tag{4}$$

Therefore, we obtain

$$\frac{\lambda_{\#}(\pi_{app})}{\lambda_{\#}(\pi_{opt})} \leq \lceil \frac{\ell}{2} \rceil.$$

6 Lower Bounds for Exact Algorithms

We propose some negative results for k-COLOR CHANGE PATH about the existence of subexponential-time algorithms under ETH [13,14].

Corollary 4. *There is no $2^{o(n)}$ (resp. $2^{o(\sqrt{n+m})}$)-time algorithm for the optimization version of k-COLORED PATH even in graphs where each color induces a path of length at most three or in bipartite graphs where each color induces a path of length at most four (resp. in bipartite planar graphs).*

Proof. Let φ be a 3-SAT formula with n' variables and m' clauses and G be its list arc-colored graph resulting from Construction 1. By construction, the number of arcs and vertices of G is $O(n' + m')$, even if we make the graph bipartite. Thus, since 3-SAT does not admit a $2^{o(n'+m')}$-time algorithm, k-COLORED PATH does not admit a $2^{o(n+m)}$-time algorithm [13,16,20]. Making the graph planar as described for Theorem 5 adds $O(m'^2)$ vertices in G. Thus, we can conclude that k-COLORED PATH does not admit a $2^{o(\sqrt{|V|+|E|})}$-time algorithm in bipartite planar graphs.

7 Conclusion

In this paper, we tackle the trip sharing problem in complexity and approximation viewpoints. We show that in the case of each input colored graph, for a fixed color, has a length at most two, the problem is polynomial whereas the problem becomes NP-complete for each colored path has length three. The complexity results are supplemented by hardness results according to topology (planar, bipartite and bipartite planar). On positive side, we develop a polynomial-time approximation algorithm a ratio at most $\lceil \frac{l}{2} \rceil$ with l the length of each input colored path. Next step could be to develop exact algorithms using for example a tree decomposition.

References

1. Böhmová, K., Häfliger, L., Mihalák, M., Pröger, T., Sacomoto, G., Sagot, M.-F.: Computing and listing ST-paths in public transportation networks. Theory Comput. Syst. **62**(3), 600–621 (2018)
2. Broersma, H., Li, X., Woeginger, G.J., Zhang, S.: Paths and cycles in colored graphs. Electron. J. Comb. **31**, 299–312 (2005)
3. Cao, Y., Chen, G., Jing, G., Stiebitz, M., Toft, B.: Graph edge coloring: a survey. Graphs Combin. **35**(1), 33–66 (2019)
4. Captivo, M.E., Clímaco, J.C.N., Pascoal, M.M.B.: A mixed integer linear formulation for the minimum label spanning tree problem. Comput. Oper. Res. **36**(11), 3082–3085 (2009)
5. Chwatal, A.M., Raidl, G.R.: Solving the minimum label spanning tree problem by mathematical programming techniques. Adv. Oper. Res. **2011** (2011)

6. Cordeau, J.-F., Laporte, G.: The dial-a-ride problem (DARP): variants, modeling issues and algorithms. Q. J. Belg. Fr. Ital. Oper. Res. Soc. **1**(2), 89–101 (2003)
7. Cordeau, J.-F., Laporte, G.: The dial-a-ride problem: models and algorithms. Ann. Oper. Res. **153**(1), 29–46 (2007)
8. Fellows, M., Guo, J., Kanj, I.: The parameterized complexity of some minimum label problems. J. Comput. Syst. Sci. **76**(8), 727–740 (2010)
9. Fredman, M.L., Tarjan, R.E.: Fibonacci heaps and their uses in improved network optimization algorithms. In: 25th Annual Symposium on Foundations of Computer Science, pp. 338–346 (1984)
10. Furuhata, M., Dessouky, M., Ordonez, F., Brunet, M.-E., Wang, X., Koenig, S.: Ridesharing: the state-of-the-art and future directions. Transport. Res. B Meth. **57**, 28–46 (2013)
11. Hassin, R., Monnot, J., Segev, D.: Approximation algorithms and hardness results for labeled connectivity problems. J. Comb. Optim. **14**(4), 437–453 (2007)
12. Ho, S.C., Szeto, W.Y., Kuo, Y.-H., Leung, J.M.Y., Petering, M., Tou, T.W.H.: A survey of dial-a-ride problems: literature review and recent developments. Transport. Res. B Meth. **111**, 395–421 (2018)
13. Impagliazzo, R., Paturi, R.: On the complexity of k-SAT. J. Comput. Syst. Sci. **62**(2), 367–375 (2001)
14. Impagliazzo, R., Paturi, R., Zane, F.: Which problems have strongly exponential complexity? J. Comput. Syst. Sci. **63**(4), 512–530 (2001)
15. Krumke, S.O., Wirth, H.-C.: On the minimum label spanning tree problem. Inf. Process. Lett. **66**(2), 81–85 (1998)
16. Lokshtanov, D., Marx, D., Saurabh, S.: Lower bounds based on the Exponential Time Hypothesis. Bull. EATCS **105**, 41–72 (2011)
17. Lund, C., Yannakakis, M.: On the hardness of approximating minimization problems. J. ACM **41**(5), 960–981 (1994)
18. Mourad, A., Puchinger, J., Chu, C.: A survey of models and algorithms for optimizing shared mobility. Transport. Res. B Meth. **123**, 323–346 (2019)
19. Niels, A., Agatz, H., Erera, A.L., Savelsbergh, M.W.P., Wang, X.: Optimization for dynamic ride-sharing: a review. Eur. J. Oper. Res. **223**(2), 295–303 (2012)
20. Woeginger, G.J.: Exact algorithms for NP-hard problems: a survey. In: Jünger, M., Reinelt, G., Rinaldi, G. (eds.) Combinatorial Optimization — Eureka, You Shrink! LNCS, vol. 2570, pp. 185–207. Springer, Heidelberg (2003). https://doi.org/10.1007/3-540-36478-1_17
21. Xiongm, Y., Golden, B., Wasil, E., Chen, S.: The label-constrained minimum spanning tree problem. In: Raghavan, S., Golden, B., Wasil, E. (eds.) Telecommunications Modeling, Policy, and Technology, vol. 44, pp. 39–58. Springer, Boston (2008). https://doi.org/10.1007/978-0-387-77780-1_3

The Complexity of Finding Optimal Subgraphs to Represent Spatial Correlation

Jessica Enright[1], Duncan Lee[2], Kitty Meeks[1], William Pettersson[1(✉)], and John Sylvester[1]

[1] School of Computing Science, University of Glasgow, Glasgow, UK
{jessica.enright,kitty.meeks,william.pettersson, john.sylvester}@glasgow.ac.uk
[2] School of Mathematics and Statistics, University of Glasgow, Glasgow, UK
duncan.lee@glasgow.ac.uk

Abstract. Understanding spatial correlation is vital in many fields including epidemiology and social science. Lee, Meeks and Pettersson (Stat. Comput. 2021) recently demonstrated that improved inference for areal unit count data can be achieved by carrying out modifications to a graph representing spatial correlations; specifically, they delete edges of the planar graph derived from border-sharing between geographic regions in order to maximise a specific objective function. In this paper we address the computational complexity of the associated graph optimisation problem. We demonstrate that this problem cannot be solved in polynomial time unless P = NP; we further show intractability for two simpler variants of the problem. We follow these results with two parameterised algorithms that exactly solve the problem in polynomial time in restricted settings. The first of these utilises dynamic programming on a tree decomposition, and runs in polynomial time if both the treewidth and maximum degree are bounded. The second algorithm is restricted to problem instances with maximum degree three, as may arise from triangulations of planar surfaces, but is an FPT algorithm when the maximum number of edges that can be removed is taken as the parameter.

Keywords: Parameterised complexity · Treewidth · Colour coding · Spatial statistics

1 Introduction

Spatio-temporal count data relating to a set of n non-overlapping areal units for T consecutive time periods are prevalent in many fields, including epidemiology [11] and social science [1]. As geographical proximity can often indicate correlation, such data can be modelled as a graph, with vertices representing areas and edges between areas that share a geographic boundary and so are assumed to be correlated. The count data is then represented as a weight assigned to each vertex. However, such models are often not ideal representations as geographical

D.-Z. Du et al. (Eds.): COCOA 2021, LNCS 13135, pp. 152–166, 2021.
https://doi.org/10.1007/978-3-030-92681-6_13

proximity does not always imply correlation [9]. Instead, Lee, Meeks and Pettersson [7] recently proposed a new method for addressing this issue by deriving a specific objective function (given in full in Sect. 2.2), and then searching for a spanning subgraph with no isolated vertices which maximises this function. Maximising this objective function corresponds to maximising the natural log of the product of full conditional distributions over all vertices (corresponding to spatial units) in a conditional autoregressive model. Such models are typically written as a series of univariate full conditional distributions rather than a joint distribution. This objective function is highly non-linear, and rewards removing as few edges as possible, while applying a penalty that (non-linearly) increases as the difference between the weight of each vertex and the average weight over its neighbours increases. Due to the size of the data, exhaustive searches for optimal subgraphs are intractable and so efficient algorithms are required for this problem. Lee, Meeks and Pettersson [7] gave a heuristic for this problem, but point out that many standard techniques are not applicable to this problem, suggesting that this problem is hard to solve efficiently in general.

1.1 Our Contribution

We show that the problem is indeed NP-hard, even on planar graphs, and provide examples that illustrate two of the major challenges inherent in the problem: we cannot optimise independently in disjoint connected components and we cannot iterate towards a solution. We also show that the decision variant of minimising the penalty portion of the objective function is NP-complete even when restricted to planar graphs with maximum degree at most five. We then investigate a simplification in which the goal is to find a subgraph with a penalty term of zero. We show that this is solvable linear time and space in the number of edges of the graph, and we completely characterise all such subgraphs. However, we also show that finding a subgraph with a penalty term of zero on all vertices of degree two or more is NP-complete.

In the positive direction, we give two exact algorithms that are tractable in their respective restricted settings. These both require that the input graph have bounded maximum degree: we note that graphs arising from areal studies will often have small maximum degree. The first algorithm runs in polynomial time if both the maximum degree and treewidth of the underlying graph are bounded. The second algorithm is only guaranteed to be correct if the underlying graph has maximum degree three, but is fixed-parameter tractable when parameterised by the maximum number of edges that can be removed.

1.2 Paper Outline

Section 2 gives notation and definitions, the formal problem definition, and examples that illustrate two of the major challenges inherent in the problem. We then prove in Sect. 3 that, unless P = NP, there is no polynomial-time algorithm to solve the main optimisation problem, even when restricted to planar graphs. Section 4 then examines three simplifications of the problem. In Sect. 5 we introduce two algorithms to exactly solve the problem in certain special cases, and we finish with concluding thoughts and open problems in Sect. 6. Note that some details and proofs are omitted due to space constraints.

2 Background

In this section we give the notation we need for this paper, define the problem, and then demonstrate why some common techniques from graph theory are not applicable to this problem.

2.1 Notation and Definitions

A graph is a pair $G = (V, E)$, where the *vertex set* V is a finite set, and the *edge set* $E \subseteq V^{(2)}$ is a set of unordered pairs of elements of V. Two vertices u and v are said to be *adjacent* if $e = uv \in E$; u and v are said to be the *endpoints* of e. The *neighbourhood* of v in G is the set $N_G(v) := \{u \in V : uv \in E\}$, and the *degree* of v in G is $d_G(v) := |N_G(v)|$. An *isolated vertex* is a vertex of degree zero, and a *leaf* is a vertex of degree one. The *maximum degree* of a graph G is $\Delta(G) := \max_{v \in V} d_G(v)$. A graph $H = (V_H, E_H)$ is a *subgraph* of G if $V_H \subseteq V$ and $E_H \subseteq E$; H is a *spanning subgraph* of G if $V_H = V$ so that H is obtained from G by deleting a (possibly empty) subset of edges. Given an edge e in $E(G)$ (respectively a set $E' \subseteq E(G)$) we write $G \setminus e$ (respectively $G \setminus E'$) for the subgraph of G obtained by deleting e (respectively deleting every element of E'). A graph G is *planar* if it can be drawn in the plane (i.e. vertices can be mapped to points in the plane, and edges to curves in the plane whose extreme points are the images of its endpoints) in such a way that no two edges cross. Given any partition of a subset of the plane into regions, we can define a planar graph whose vertices are in bijection with the set of regions, in which two regions are adjacent if and only if they share a border of positive length. In particular, if each region has three sides (i.e., the partition is a triangulation of a subset of the plane) then the resulting graph will have maximum degree three.

2.2 The Optimisation Problem

Following Lee, Meeks and Pettersson [7], we are concerned with the following optimisation problem.

Correlation Subgraph Optimisation
Input: A graph $G = (V, E)$ where $|V| = n$, and function $f : V \to \mathbb{Q}$.
Question: What is the maximum value of

$$\text{score}(H, f) := \sum_{v \in V} \ln d_H(v) - n \ln \left[\sum_{v \in V} d_H(v) \left(f(v) - \frac{\sum_{u \in N_H(v)} f(u)}{d_H(v)} \right)^2 \right],$$

taken over all spanning subgraphs H of G such that $d_H(v) \geq 1$ for all $v \in V$?

We will say that a subgraph H of G is *valid* if H is a spanning subgraph of G and $d_H(v) \geq 1$ for all $v \in V$. Given a vertex v in the input graph G, we will sometimes refer to $f(v)$ as the *weight* of v. We also define the neighbourhood discrepancy of a vertex f in a graph H with weight function f (written $\mathrm{ND}_H(v, f)$) as

$$\mathrm{ND}_H(v, f) := \left(f(v) - \frac{\sum_{u \in N_H(v)} f(u)}{d_H(v)} \right)^2 .$$

2.3 Why Common Graph Algorithm Techniques Fail

This problem is particularly resistant to many approaches common in algorithmic graph theory. We will describe two of these now. Firstly, on a disconnected graph G, combining optimal solutions on each connected component is not guaranteed to find an optimal solution on G. This is true even if there are only two disconnected components, one of which is an isolated edge and the other being a path, as illustrated in the following example.

Example 1. Consider the graph G consisting of a path on four vertices (v_1, v_2, v_3, v_4) along with an isolated edge between vertices v_a and v_b, as shown in Fig. 1, and let $H = G \setminus \{v_2v_3\}$. Note that H is the only proper subgraph of G which has no isolated vertices. Let f be defined as follows: $f(v_1) = 0$, $f(v_2) = 1$, $f(v_3) = 10$, $f(v_4) = 11$, $f(v_a) = 0$, and $f(v_b) = x$ for some real x. If $x = 1$ then $\mathrm{score}(G, f) < \mathrm{score}(H, f)$ but if $x = 1000$ then $\mathrm{score}(G, f) > \mathrm{score}(H, f)$.

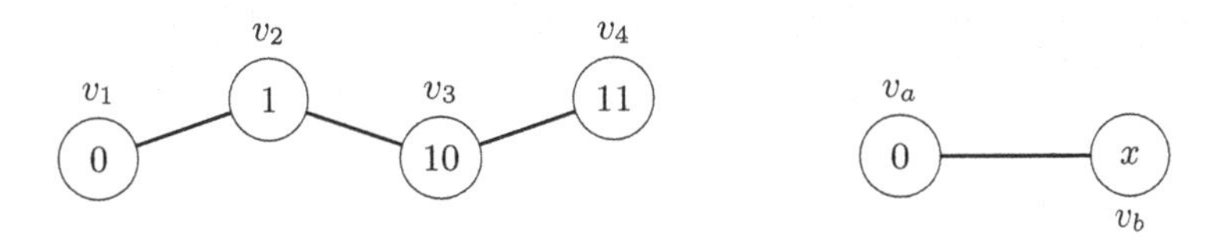

Fig. 1. Graph for Example 1. The value of the function at each vertex is shown inside the respective vertex.

To understand why disconnected components can affect each other in such a manner, note that the negative term in the score function contains a logarithm of a sum of neighbourhood discrepancies. This means that the relative importance of the neighbourhood discrepancy of any set of vertices depends on the total sum of the neighbourhood discrepancies across the whole graph. In other words, the presence of a large neighbourhood discrepancy elsewhere (even in a separate component) in the graph can reduce the impact of the neighbourhood discrepancy at a given vertex or set of vertices. However, the positive term in the score function is a sum of logarithms, so the contribution to the positive term from the degree of one vertex does not depend on any other part of the graph.

A reader might also be tempted to tackle this problem by identifying a "best" edge to remove and proceeding iteratively. The following example highlights that any algorithm using such a greedy approach may, in some cases, not find an optimal solution.

Example 2. Consider the graph G being a path on six vertices labelled v_1, v_2, v_3, v_4, v_5, and v_6 with $f(v_1) = 1000$, $f(v_2) = 2000$, $f(v_3) = 1999$, $f(v_4) = 1001$, $f(v_5) = 2019$, and $f(v_6) = 981$ as shown in Fig. 2. Let $H = G \setminus \{v_2v_3, v_4v_5\}$, and let $H' = G \setminus \{v_3v_4\}$. The maximum score that can be achieved with the removal of only one edge is achieved by removing edge v_3v_4 and creating H'. However, the optimal solution to CORRELATION SUBGRAPH OPTIMISATION on G is H, and involves removing edges v_2v_3 and v_4v_5.

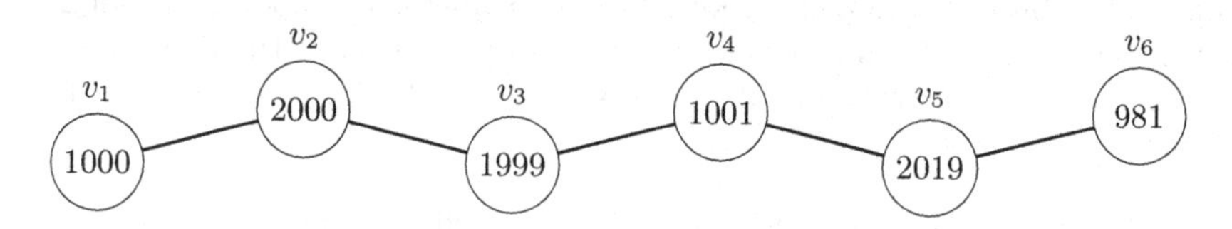

Fig. 2. Graph for Example 2. The value of the function at each vertex is shown inside the respective vertex.

3 Hardness on Planar Graphs

In this section we prove NP-hardness of CORRELATION SUBGRAPH OPTIMISATION on planar graphs.

Theorem 1. *There is no polynomial-time algorithm to solve* CORRELATION SUBGRAPH OPTIMISATION *on planar graphs unless P=NP.*

We prove this result by means of a reduction from the following problem, shown to be NP-complete in [10]; the *incidence graph* G_Φ of a CNF formula Φ is a bipartite graph whose vertex sets correspond to the variables and clauses of Φ respectively, and in which a variable x and clause C are connected by an edge if and only if x appears in C.

CUBIC PLANAR MONOTONE 1-IN-3 SAT
Input: A 3-CNF formula Φ in which every variable appears in exactly three clauses, variables only appear positively, and the incidence graph G_Φ is planar.
Question: Is there a truth assignment to the variables of Φ so that exactly one variable in every clause evaluates to TRUE?

We begin by describing the construction of a graph G and function $f : V(G) \to \mathbb{N}$ corresponding to the formula Φ in an instance of CUBIC PLANAR MONOTONE 1-IN-3 SAT; the construction will be defined in terms of an integer parameter $t \geq 1$ whose value we will determine later. Note that G is not the incidence graph G_Φ of Φ.

Suppose that Φ has variables $x_1, \ldots, x_n$ and clauses $C_1, \ldots, C_m$. Since every variable appears in exactly three clauses and each clause contains exactly three variables, we must have $m = n$. For each variable x_i, G contains a variable gadget on $3t^2 + 6t + 8$ vertices. The non-leaf vertices of the gadget are:

- u_i, with $f(u_i) = 7t$,
- v_i, with $f(v_i) = 4t$,
- z_i with $f(z_i) = t$,
- z_i' with $f(z_i') = 4t$, and
- $w_{i,j}$ for each $j \in \{1, 2, 3\}$, with $f(w_{i,j}) = 3t$.

The vertex v_i is adjacent to u_i, z_i and each $w_{i,j}$ with $i \in \{1, 2, 3\}$; z_i is adjacent to z_i'. We add leaves to this gadget as follows:

- u_i has $3t$ pendant leaves, each assigned value $7t + 1$ by f;
- z_i has $3t$ pendant leaves, each assigned value $t - 1$ by f;
- z_i' has $3t^2$ pendant leaves, each assigned value $4t$ by f;
- each vertex $w_{i,j}$ has exactly one pendant leaf, assigned value $3t$ by f.

For each clause C_j, G contains a clause gadget on $t^2 + 2$ vertices: a_j and a_j', which are adjacent, and t^2 pendant leaves adjacent to a_j'. We set $f(a_j) = 2t$, and f takes value t on a_j' and all of its leaf neighbours. We complete the definition of G by specifying the edges with one endpoint in a variable gadget and the other in a clause gadget: if the variable x_i appears in clauses C_{r_1}, C_{r_2} and C_{r_3}, with $r_1 < r_2 < r_3$, then we have edges $w_{i,1}a_{r_1}$, $w_{i,2}a_{r_2}$ and $w_{i,3}a_{r,3}$. The construction of the variable and clause gadgets is illustrated in Fig. 3.

Recall that a subgraph H of G is *valid* if H is a spanning subgraph of G and $d_H(v) \geq 1$ for all $v \in V$. Recall that the *neighbourhood discrepancy* of a vertex v with respect to f in a valid subgraph H, written $\mathrm{ND}_H(v, f)$, is

$$\mathrm{ND}_H(v, f) := \left(f(v) - \frac{\sum_{u \in N_H(v)} f(v)}{d_H(v)} \right)^2 .$$

The goal of CORRELATION SUBGRAPH OPTIMISATION is therefore to maximise

$$\mathrm{score}(H, f) := \sum_{v \in V} \ln d_H(v) - n \ln \left[\sum_{v \in V} d_H(v) \, \mathrm{ND}_H(f, v) \right],$$

over all valid subgraphs H of G. We now give several results that are necessary; the proofs of these are omitted due to space constraints but can be found in [3]. This first set of results give several properties of valid subgraphs of G.

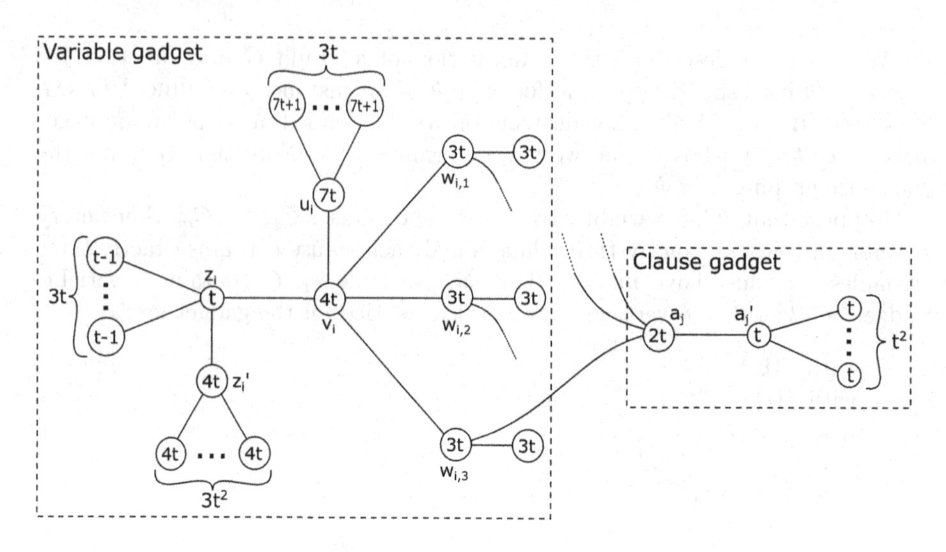

Fig. 3. Construction of the variable and clause gadgets.

Lemma 1. *For any valid subgraph H,*

$$\sum_{u \text{ a leaf in } G} \mathrm{ND}_H(u, f) = 6nt.$$

Lemma 2. *For any valid subgraph H,*

$$0 \leq \mathrm{ND}_H(z_i', f), \mathrm{ND}_H(a_i', f) < 1/t^2.$$

Lemma 3. *For any valid subgraph H,*

$$\sum_{v \in V} \ln d_H(v) \geq 6n \ln t + 2n.$$

Lemma 4. *Let H be any subgraph of G (not necessarily valid). Then*

$$\sum_{v \in V} \ln d_H(v) \leq 6n \ln t + 20n.$$

We now give two lemmas that relate the existence of truth assignments of a 3-CNF formulae to bounds on the neighbourhood discrepancies of some vertices within a valid subgraph H.

Lemma 5. *If Φ is satisfiable, there is a valid subgraph H such that for all $v \in V \setminus \{z_i', a_i' : 1 \leq i \leq n\}$ with $d_G(v) > 1$ we have $\mathrm{ND}(v, H) = 0$.*

Lemma 6. *If Φ is not satisfiable, then for any valid subgraph H, there exists a vertex $v \in V \setminus \{z_i', a_i' : 1 \leq i \leq n\}$ with $d_G(v) > 1$ such that*

$$\mathrm{ND}_H(v, f) \geq t^2/9.$$

We now give bounds on the possible values for $\text{score}(H, f)$ depending on whether or not Φ is satisfiable.

Lemma 7. *If Φ is satisfiable, there is a valid subgraph H with*

$$\text{score}(H, f) \geq 6n \ln t - n \ln(12nt).$$

Lemma 8. *If Φ is not satisfiable, then for every valid subgraph H we have*

$$\text{score}(H, f) \leq 6n \ln t + 20n - n \ln(t^2/9).$$

We are now ready to prove Theorem 1, which we restate here for convenience.

Theorem 1. *There is no polynomial-time algorithm to solve* CORRELATION SUBGRAPH OPTIMISATION *on planar graphs unless P = NP.*

Proof. We suppose for a contradiction that there is a polynomial-time algorithm $\mathcal{A}$ to solve CORRELATION SUBGRAPH OPTIMISATION on planar graphs, and show that this would allow us to solve CUBIC PLANAR MONOTONE 1-IN-3 SAT in polynomial time.

Given an instance Φ of CUBIC PLANAR MONOTONE 1-IN-3 SAT, where we will assume without loss of generality that Φ has $n > e^{47}$ variables, we proceed as follows. First construct (G, f) as defined above, taking $t = n^2$; it is clear that this can be done in polynomial time in $|\Phi|$. Note that G is planar: to see this, observe that repeatedly deleting vertices of degree one gives a subdivision of the incidence graph which is planar by assumption. We then run $\mathcal{A}$ on (G, f) and return YES if the output is at least $\frac{17}{2} n \ln n$, and NO otherwise.

It remains to demonstrate that this procedure gives the correct answer. Suppose first that Φ is satisfiable. In this case, by Lemma 7, we know that there exists a subgraph H of G with

$$\begin{aligned} \text{score}(H, f) &\geq 6n \ln t - n \ln(12nt) \\ &= 6n \ln n^2 - n \ln(12n^3) \\ &= 12n \ln n - 3n \ln n - n \ln 12 \\ &\geq 9n \ln n - 3n \\ &> \frac{17}{2} n \ln n, \end{aligned}$$

since $3 < \ln n/2$, so our procedure returns YES.

Conversely, suppose that Φ is not satisfiable. In this case, by Lemma 8 we know that, for every valid subgraph H we have

$$\begin{aligned} \text{score}(H, f) &\leq 6n \ln t + 20n - n \ln(t^2/9) \\ &= 6n \ln n^2 + 20n - n \ln(n^4/9) \\ &= 12n \ln n + 20n - 4n \ln n + n \ln 9 \\ &\leq 8n \ln n + 23n \\ &< \frac{17}{2} n \ln n, \end{aligned}$$

since $23 < \ln n/2$, so our procedure returns NO. □

4 Simplifications of the Problem

One may wonder if the hardness of CORRELATION SUBGRAPH OPTIMISATION is due to the interplay between the two parts of the objective function. We show in Sect. 4.1 that just determining if there is a valid subgraph with total neighbourhood discrepancy below some given constant is NP-complete, even if the input graph is planar and has maximum degree at most five. In Sect. 4.2 we that show that subgraphs that have zero neighbourhood discrepancy everywhere (if they exist) can be found in time linear in the number of edges, however determining if there exists a subgraph that has zero neighbourhood discrepancy everywhere excluding leaves is NP-complete.

4.1 Minimising Neighbourhood Discrepancy

Consider the following problem, which questions the existence of a subgraph whose total neighbourhood discrepancy is below a given constant.

AVERAGE VALUE NEIGHBOURHOOD OPTIMISATION
Input: A graph $G = (V, E)$, a function $f : V \rightarrow \mathbb{Q}$, and $k \in \mathbb{Q}$.
Question: Is there a spanning subgraph H of G such that $d_H(v) \geq 1$ for all $v \in V$ and

$$\sum_{v \in V} \left(f(v) - \frac{\sum_{u \in N_H(v)} f(u)}{d_H(v)} \right)^2 \leq k \text{ ?}$$

First observe that the AVERAGE VALUE NEIGHBOURHOOD OPTIMISATION is clearly in NP. The NP-hardness of AVERAGE VALUE NEIGHBOURHOOD OPTIMISATION can be shown by giving a reduction from CUBIC PLANAR MONOTONE 1-IN-3 SAT, which we used earlier in Sect. 3. The full proof is omitted due to space constraints but can be found in [3].

Theorem 2. AVERAGE VALUE NEIGHBOURHOOD OPTIMISATION *is NP-complete, even when restricted to input graphs G that are planar and have maximum degrees at most five.*

4.2 Ideal and Near-Ideal Subgraphs

An obvious upper-bound to score(H, f) is given by $\sum_{v \in V(H)} \ln d_H(v)$ (i.e. assume every vertex has zero neighbourhood discrepancy), so a natural question to ask is whether, for a given graph G and function f, a valid subgraph H of G can be found that achieves this bound. In such a graph, it must hold that $\text{ND}_H(v, f) = 0$ for every $v \in V(H)$. We say such a graph H is f*-ideal* (or simply ideal, if f is clear from the context). We now show that this definition is equivalent to saying that a graph H is f-ideal if and only the restriction of f to any connected component of H is a constant-valued function.

Theorem 3. *A graph H is f-ideal if and only if for each connected component C_i in H there exists some constant c_i such that $f(v) = c_i$ for all $v \in V(C_i)$.*

Proof. Let P denote a path of maximal length in an f-ideal graph such that the weights of the vertices of P strictly increase as one follows the path. In an ideal graph, any edge between vertices of different weights means that P must contain at least two distinct vertices, however the first and last vertices in such a path cannot have zero neighbourhood discrepancy. Thus, no such path on one or more edges can exist in an ideal graph, so a graph G is ideal if and only if for each connected component C_i in G there exists some constant c_i such that $f(v) = c_i$ for all $v \in V(C_i)$. □

Thus, ideal subgraphs can be found by removing any edge uv if $f(u) \neq f(v)$ (in $O(|E|) = O(n^2)$ time), and if necessary we can test if such a graph has no isolated vertices (and thus is valid) quickly. The proof of Theorem 3 highlights that maximal paths with increasing weights must start and end on vertices that do not have zero neighbourhood discrepancy, so one might be tempted to relax the ideal definition to only apply on vertices that are not leaves. We therefore say a graph H is *f-near-ideal* if $\mathrm{ND}_H(v, f) = 0$ for every $v \in V(H)$ with $d_H(v) \geq 2$. In other words, we now allow non-zero neighbourhood discrepancy, but only at leaves, motivating the following problem.

> **NEAR IDEAL SUBGRAPH**
> *Input:* A graph $G = (V, E)$ where $|V| = n$, and a function $f : V \mapsto \mathbb{Q}$.
> *Question:* Is there a valid subgraph H of G such that H is f-near-ideal?

While an ideal subgraph (if one exists) can be found quickly, it turns out that solving NEAR IDEAL SUBGRAPH is NP-complete, even on trees. We reduce from subset-sum, which is NP-complete [6], and which we define as follows.

> **SUBSET SUM**
> *Input:* An integer k, and a set of integers $S = \{s_1, s_2, \ldots, s_n\}$.
> *Question:* Is there a subset $U \subseteq \{1, 2, \ldots, n\}$ such that $\sum_{u \in U} s_u = k$?

Given an instance (S, k) of SUBSET SUM, we will construct a graph G with weight function f such that (G, f) has a near-ideal subgraph if and only if there is a solution to our instance of SUBSET SUM.

The graph G contains $3n + 3$ vertices labelled as follows:

- v_t for the target value, v_s for a partial sum, and v_z for a pendant, and
- v_p^j for $p \in \{1, \ldots, n\}$ and $j \in \{1, 2, 3\}$.

Vertex v_s is adjacent to vertices v_t, v_z, and v_p^1 for $p \in \{1, \ldots, n\}$. For each $p \in \{1, \ldots, n\}$, v_p^1 is adjacent to v_p^2, and v_p^2 is adjacent to v_p^3. This graph can be seen in Fig. 4. We then define f as follows:

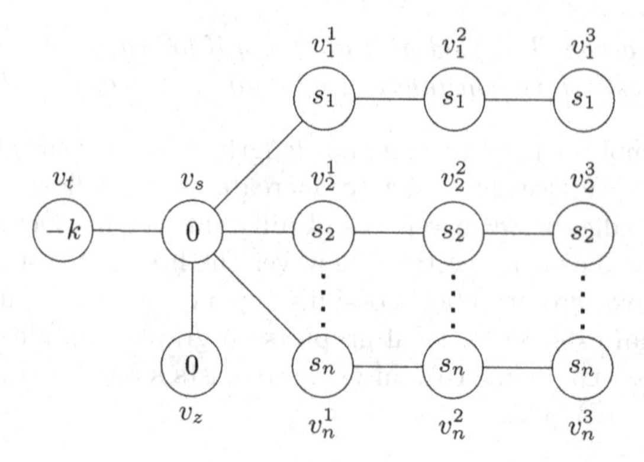

Fig. 4. Diagram of graph for reduction from SUBSET SUM. The values inside the vertices are their associated weights.

- $f(v_t) = -k$,
- $f(v_s) = f(v_z) = 0$, and
- $f(v_p^j) = s_p$ for $p \in \{1, \ldots, n\}$, and for $j \in \{1, 2, 3\}$.

Note that for the condition $d_H(v) \geq 1$ to hold for our subgraph H, the only edges in G that might not be in H are of the form $v_s v_p^1$ or $v_p^1 v_p^2$ for some $p \in \{1, \ldots, n\}$. Additionally, for any $p \in \{1, \ldots, n\}$, at most of one of $v_s v_p^1$ or $v_p^1 v_p^2$ can be removed. We can then show that G has a near-ideal subgraph if and only if it is constructed from a yes-instance of SUBSET SUM. The complete proof is omitted due to space constraints but can be found in [3].

Theorem 4. NEAR IDEAL SUBGRAPH *is NP-complete, even if the input graph* G *is a tree.*

5 Parameterised Results

In this section we describe two parameterised algorithms for CORRELATION SUBGRAPH OPTIMISATION. We make use of two parameterised complexity problem classes to describe these. A problem is in the *fixed parameter tractable* (or FPT) class with respect to some parameter k if the problem can be solved on inputs of size n in time $f(k) \cdot n^{O(1)}$ for some computable function f. Note in particular that the exponent of n is constant. Another class of parameterised problems is XP: a problem is in XP with respect to some parameter k if the problem can be solved on inputs of size n in time $O(n^{f(k)})$. In XP problems, the exponent of n may change for different values of k, but if an upper bound on k is given then this also upper bounds the exponent of n. For further background on parameterised complexity, see [2].

In Sect. 5.1 we show that CORRELATION SUBGRAPH OPTIMISATION is in XP parameterised by the maximum degree when treewidth is bounded, and is in FPT parameterised by treewidth when the maximum degree is bounded. Then in Sect. 5.2 we consider the more restricted case where G has maximum degree three, and show that with this restriction CORRELATION SUBGRAPH OPTIMISATION is in FPT parameterised by the number of edges that are removed. We highlight that this restriction on the maximum degree occurs naturally in triangulations of surfaces, such as can occur when discretising geographic maps.

5.1 An Exact XP Algorithm Parameterised by Treewidth and Maximum Degree

We now briefly describe an exact XP algorithm for solving CORRELATION SUBGRAPH OPTIMISATION on arbitrary graphs that leads to the following result.

Theorem 5. CORRELATION SUBGRAPH OPTIMISATION *can be solved in time*

$$O(2^{2\Delta(G)(tw(G)+1)} \cdot n^{2\Delta(G)+1}).$$

The algorithm follows fairly standard dynamic programming techniques on a nice tree decomposition T of G with treewidth $tw(G)$ that is rooted at some arbitrary leaf bag. A *nice tree decomposition* is a tree decomposition with one leaf bag selected as a root bag so that the children of a bag are adjacent bags that are further from the root, and the additional property that each leaf bag is empty, and each non-leaf bag is either a introduce bag, forget bag, or join bag, which are defined as follows. An introduce bag ν has exactly one child below it, say μ, such that ν contains every element in μ as well as precisely one more element. A forget bag ν has exactly one child below it, say μ, such that ν contains every element in μ except one. A join bag λ has exactly two children below it, say μ and ν, such that λ, μ, and ν, all have precisely the same elements. See [2], in particular Chap. 7, for an introduction to tree decompositions, and a formal definition of nice tree decompositions.

We will outline the core ideas here; full details are omitted due to space constraints but can be found in [3]. We first define some specific terminology that will be useful when describing the algorithm. Let T be a tree decomposition (not necessarily nice) with an arbitrary bag labelled as the root. For each bag $\nu \in T$, denote by G_ν the induced subgraph of G consisting precisely of vertices that appear in bags below ν but do not appear in ν, where we take below to mean further away from the root bag. The set of edges between a vertex in ν and a vertex in G_ν will be important to our algorithm, so we will write $E_\nu = \{uv \in E(G) \mid u \in \nu \wedge v \in G_\nu\}$ to be the set of edges with one endpoint in G_ν and the other in ν. An example of a graph, a tree decomposition, G_ν, and E_ν are shown in Fig. 5.

Our algorithm will process each bag, from the leaves towards the root, determining a set of states for each bag such that we can guarantee that the optimal solution will correspond to a state in the root bag. Given a bag ν of a tree decomposition and a set of edges $I \subseteq E_\nu$, define $\mathcal{G}'_{\nu,I}$ to be the set of graphs G' with $V(G') = V(G_\nu) \cup \nu$, $E(G') \subseteq E(G)$, and for any edge $uv \in E_\nu$, $uv \in E(G')$ if and only if $uv \in I$.

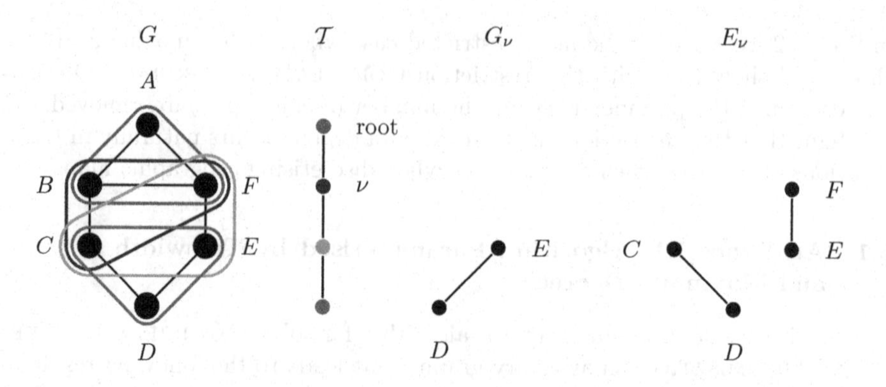

Fig. 5. From left to right, we have a graph G with a tree decomposition displayed by circling vertices, the tree indexing a tree decomposition of G drawn as a graph with the root and the bag ν labelled, the graph G_ν consisting of the induced subgraph on vertices D and E, and the the set of edges $E_\nu = \{CD, EF\}$ (i.e., the edges of G that are between a vertex in G_ν and a vertex in ν).

Definition 1. *For a bag ν, the set of all valid states at ν is*

$$S_\nu = \{(I, D) \,|\, I \subseteq E_\nu,\ \mathcal{G}'_{\nu,I} \neq \emptyset, \text{ and there exists a graph } H \in \mathcal{G}'_{\nu,I} \\ \text{with } D = \sum_{v \in G_\nu} \ln d_H(v), \text{ and } d_H(v) \geq 1\ \forall v \in V(G_\nu)\}.$$

Each state corresponds to at least one graph (H in the definition) but there may be multiple graphs that all lead to the same state. For each state we will also store the best possible (i.e., lowest) value of $\sum_{v \in H} d_H(v)\,\mathrm{ND}_H(v, f)$ (i.e., total neighbourhood discrepancy summed over vertices that only appear below the current bag) over all of the graphs H that correspond to a given state. This allows us to compute the contribution to the penalty portion of the objective function from the subtree under consideration.

5.2 Parameterisation by k in Low Degree Graphs

We also study the problem when G has maximum degree three and we want to bound the maximum number of edges that can be removed. In this setting we define k-Correlation Subgraph Optimisation and show that it is in FPT when parameterised by k, the maximum number of edges that can be removed from G to create H. This setting is of interest as the dual graph of any triangulation has maximum degree three and triangulations are often used to represent discretised surfaces [5,8].

k-CORRELATION SUBGRAPH OPTIMISATION
Input: A graph $G = (V, E)$ where $|V| = n$, an integer k, and a function $f : V \to \mathbb{Q}$.
Question: What is the maximum value of

$$\mathrm{score}(H, f) := \sum_{v \in V} \ln d_H(v) - n \ln \left[\sum_{v \in V} d_H(v) \left(f(v) - \frac{\sum_{u \in N_H(v)} f(u)}{d_H(v)} \right)^2 \right],$$

taken over all spanning subgraphs H of G such that $|E(G \setminus H)| \leq k$ and $d_H(v) \geq 1$ for all $v \in V$?

Theorem 6. *For an integer $k \geq 1$, k-CORRELATION SUBGRAPH OPTIMISATION can be solved on graphs with maximum degree three in time $2^{k(2\log k + O(1))} n \log n$.*

This can be proven using the following guide; full details are omitted due to space constraints but can be found in [3]. Consider in turn each possibility R for the graph consisting of deleted edges, and for each such graph we consider in turn the possibilities of the degree sequence of the remaining graph. The number of distinct graphs R that must be considered is independent of n, and for each R the number of degree sequences of $G \setminus R$ is linear in n. As R has maximum degree two and therefore consists only of paths and cycles, it has treewidth at most two. We can therefore adapt well-known colour-coding methods (see [4, Section 13.3] for more details) for finding subgraphs with bounded treewidth in FPT time so that we can identify a subgraph R in G whose removal gives the biggest improvement to the neighbourhood discrepancy term while still maintaining the correct degree sequence of $G \setminus R$.

6 Discussion and Conclusions

CORRELATION SUBGRAPH OPTIMISATION is a graph optimisation problem arising from spatial statistics with direct applications to epidemiology and social science that we show is intractable unless P = NP. We also show that it is resistant to common techniques in graph algorithms, but can be solved in polynomial time if both the treewidth and maximum degree of G are bounded, or if G has maximum degree three and we bound the maximum number of edges that can be removed. However the question still remains as to whether CORRELATION SUBGRAPH OPTIMISATION itself is hard when the maximum degree of the input graph is bounded. We also note as an interesting open problem whether CORRELATION SUBGRAPH OPTIMISATION admits efficient parameterised algorithms with respect to (combinations of) parameters other than the maximum degree. Additionally, the original paper that introduced CORRELATION SUBGRAPH OPTIMISATION gives one heuristic for solving the problem, but leaves open any guarantee on the performance of this heuristic. Thus the investigation of the performance of this heuristic, or indeed of any new approximation algorithms, form two other significant open problems for correlation subgraph optimisation.

Acknowledgements. All authors gratefully acknowledge funding from the Engineering and Physical Sciences Research Council (ESPRC) grant number EP/T004878/1 for this work, while Meeks was also supported by a Royal Society of Edinburgh Personal Research Fellowship (funded by the Scottish Government).

References

1. Bradley, J.R., Wikle, C.K., Holan, S.H.: Bayesian spatial change of support for count-valued survey data with application to the American community survey. J. Am. Statist. Assoc. **111**(514), 472–487 (2016)
2. Cygan, M., et al.: Parameterized Algorithms, vol. 5. Springer, Cham (2015). https://doi.org/10.1007/978-3-319-21275-3
3. Enright, J., Lee, D., Meeks, K., Pettersson, W., Sylvester, J.: The complexity of finding optimal subgraphs to represent spatial correlation (2021). arXiv:2010.10314
4. Flum, J., Grohe, M.: The parameterized complexity of counting problems. SIAM J. Comput. **33**(4), 892–922 (2004)
5. He, Q., Zeng, C., Xie, P., Liu, Y., Zhang, M.: An assessment of forest biomass carbon storage and ecological compensation based on surface area: a case study of Hubei Province. China Ecol. Indicat. **90**, 392–400 (2018)
6. bibitemch13Karp72 Karp, R.M.: Reducibility among combinatorial problems. In: Miller, R.E., Thatcher, J.W. (eds.) Proceedings of a Symposium on the Complexity of Computer Computations, The IBM Research Symposia Series, pp. 85–103. Plenum Press, New York (1972)
7. Lee, D., Meeks, K., Pettersson, W.: Improved inference for areal unit count data using graph-based optimisation. Stat. Comput. **31**(4), 1–17 (2021). https://doi.org/10.1007/s11222-021-10025-7
8. Mindell, J.S., et al.: Using triangulation to assess a suite of tools to measure community severance. J. Transp. Geogr. **60**, 119–129 (2017)
9. Mitchell, R., Lee, D.: Is there really a "wrong side of the tracks" in urban areas and does it matter for spatial analysis? Ann. Assoc. Am. Geogr. **104**(3), 432–443 (2014)
10. Cristopher Moore and John Michael Robson: Hard tiling problems with simple tiles. Discrete Comput. Geom. **26**(4), 573–590 (2001)
11. Stoner, O., Economou, T., Marques da Silva, G.D.: A hierarchical framework for correcting under-reporting in count data. J. Am. Stat. Assoc. **114**(528), 1481–1492 (2019)

New Approximation Algorithms for the Rooted Budgeted Cycle Cover Problem

Jiangkun Li and Peng Zhang(✉)

School of Software, Shandong University, Jinan 250101, Shandong, China
lijk_lee@mail.sdu.edu.cn, algzhang@sdu.edu.cn

Abstract. The rooted Budgeted Cycle Cover (BCC) problem is a fundamental optimization problem arising in wireless sensor networks and vehicle routing. Given a metric space (V, w) with vertex set V consisting of two parts D (containing depots) and $V \setminus D$ (containing nodes), and a budget $B \geq 0$, the rooted BCC problem asks to find a minimum number of cycles such that each cycle has length at most B and must contain a depot in D, and that these cycles collectively cover all the nodes in $V \setminus D$. In this paper, we give new approximation algorithms for the rooted BCC problem. For the rooted BCC problem with single depot, we give an $O(\log \frac{B}{\mu})$-approximation algorithm, where μ is the minimum distance. For the rooted BCC problem with multiple depots, we give an $O(\log n)$-approximation algorithm, where n is the number of vertices. Experiments show that our algorithms have good performance in practice.

Keywords: Budgeted cycle cover · Wireless sensor network · Graph algorithm · Approximation algorithm · Combinatorial optimization

1 Introduction

Given an undirected graph, the Cycle Cover problem [7,12–14,18,23,24,27,28] uses cycles to cover all the vertices of the graph. Cycle Cover is a fundamental combinatorial optimization problem in operations research and approximation algorithms. This problem arises from many application fields, including particularly wireless sensor networks.

From the viewpoint of applications, Cycle Cover is a problem belonging to the class of *vehicle routing* problems (see, e.g., [5,9,10,20]). In this scenario, cycles represent the trajectories of vehicles, and vertices represent the places that need to be visited. We came across the Cycle Cover problem in the study of wireless sensor networks, which we will discuss in details below.

In a wireless sensor network, sensors usually work in a large open region, and they are supported by rechargeable batteries with limited energy. So, sensors should be recharged periodically to avoid their energy expiration. To this aim, multiple mobile chargers are employed to traverse within the network and charge the sensors [6,11,15,19,21,22]. The mobile chargers have a fixed capacity

D.-Z. Du et al. (Eds.): COCOA 2021, LNCS 13135, pp. 167–179, 2021.
https://doi.org/10.1007/978-3-030-92681-6_14

(i.e., amount of carried energy). As the mobile chargers themselves need to be recharged at depots, the trajectory of a mobile charger is a cycle which must include a depot. The task is, how to locate a minimum number of mobile chargers at several depots, and how to plan the traveling trajectory for each charger, so that they can charge all the sensors before returning to the depots?

Moreover, it is a basic task in wireless sensor networks to transmit data from sensors to sinks (usually a few). Since sensors locate in a large open region and energy consumption is very crucial to sensors, data transmission is performed by employing mobile sinks to travel around the sensors to gather their data [16,17,23,26], so as to reduce the energy consumption of sensors as much as possible. To finish this task, we also need to locate as few as possible mobile sinks at depots and to plan their trajectories.

We can use a metric space (V, w) to model the depots (to keep mobile sinks or mobile chargers) and the sensors in a wireless sensor network: The depots and sensors constitute the vertex set V. The distance function w on $V \times V$ satisfies the triangle inequality. The above applications in wireless sensor networks exactly suggest the following rooted Budgeted Cycle Cover (BCC) problems.

Definition 1. The Single Depot Budgeted Cycle Cover **Problem.**

(Instance) *We are given a metric space (V, w), where V is a set of n vertices and $w \colon V \times V \to \mathbb{R}^+$ is a metric function. We are also given a depot (a.k.a. root) $r \in V$ and a budget $B \geq 0$.*

(Goal) *The problem is to find a minimum number of cycles such that (i) each vertex in V is contained in some cycle, (ii) each cycle has length at most B, and (iii) each cycle must contain the depot r.*

In Definition 1, by metric space (V, w) we mean that (i) $\forall v \in V$, $w(v, v) = 0$, (ii) $\forall u, v \in V$, $w(u, v) = w(v, u)$, and (iii) $\forall t, u, v \in V$, $w(t, v) \leq w(t, u) + w(u, v)$, where the third property is often called the triangle inequality. A metric space (V, w) can also be represented by a complete graph $G = (V, E)$ with metric weight w defined on edges.

A cycle that contains the depot r is called a *r-rooted cycle.* A cycle of length at most B is called a B-budgeted cycle. So, a solution to the single depot BCC problem is constituted of r-rooted B-budgeted cycles.

Definition 2. The Multi-depot Budgeted Cycle Cover **Problem.**

(Instance) *We are given a metric space (V, w), where V is a set of n vertices and $w \colon V \times V \to \mathbb{R}^+$ is a metric function. We are also given a depot subset $D \subset V$ and a budget $B \geq 0$.*

(Goal) *The problem is to find a minimum number of rooted cycles such that (i) each vertex in $V \setminus D$ is contained in some cycle, (ii) each cycle has length at most B, and (iii) each cycle must contain at least one depot in D.*

A cycle that contains a depot in D is called a *D-rooted cycle.* A solution to the multi-depot BCC problem is constituted of D-rooted B-budgeted cycles.

For clarity, we also call the vertices in $V \setminus D$ *nodes,* while the vertices in D are obviously called depots. Nodes and depots are all vertices. For simplicity, let $N = V \setminus D$.

1.1 Related Work

The rooted BCC problem is NP-hard, since it contains the decision version of the famous TSP problem as a special case. Therefore, approximation is an essential way to deal with the problem. Nagarajan et al. [18] gave a bicriteria approximation algorithm for the single-depot BCC problem. In polynomial time, their algorithm finds r-rooted cycles covering all nodes in $V \setminus \{r\}$, such that each cycle has length at most $(1+\epsilon)B$, and that the number of cycles is at most $O(\log \frac{1}{\epsilon})OPT$, where OPT is the number of cycles in the optimal solution. That is, $O(\log \frac{1}{\epsilon})$ is the approximation ratio, and $1+\epsilon$ is the violation of the budget.

Khuller et al. [14] studied the Single Gas Station problem, which is also referred to as the single depot *min-sum* BCC problem. In this problem, we are given a metric space (V, w), a depot $r \in V$, and a bugdet B. The task is to find r-rooted B-budgeted cycles to cover all nodes in $V \setminus \{r\}$, such that the total weight of these cycles is minimized. For the single depot min-sum BCC problem, as well as the single depot BCC problem, Khuller et al. [14] gave an $O(\log n)$-approximation algorithm.

Given a metric space (V, w) and a budget B, the (un-rooted) BCC problem asks to find a set of B-budgeted cycles to cover all the vertices, such that the number of cycles is minimized [13,27]. The current best known approximation ratio for BCC is 32/7, given by Yu et al. [28].

1.2 Our Results

In this paper, we give new approximation algorithms for the single depot BCC problem and the multi-depot BCC problem.

For the the single depot BCC problem, we design an $O(\log \frac{B}{\mu})$ approximation algorithm, where μ is the minimum distance in the problem instance. Our technique to achieve this result is by classifying the vertices to many consecutive layers. For each layer, we use several path segments to cover the vertices in this layer. Each path segment is converted into a cycle by connecting its two endpoints to the depot. Our layering technique is inspired by Nagarajan et al. [18], where the technique is used to give a bicriteria approximation algorithm for the single depot BCC problem. We observed that by carefully adjusting the layering method, we can get a true approximation algorithm for the single depot BCC problem. The previous known approximation results for the single depot BCC problem are the $O(\log \frac{1}{\epsilon}, 1+\epsilon)$-approximation algorithm [18] and the $O(\log n)$-approximation algorithm [14], as mentioned in Sect. 1.1.

For the multi-depot BCC problem, we design an $O(\log n)$-approximation algorithm. This is a new result for the multi-depot BCC problem. The technique is a greedy approach, which repeatedly covers some nodes using a length bounded cycle, until all nodes are covered. Our technique is inspired by Khuller et al. [14], where the authors use the greedy cover technique to give an approximation algorithm for the single depot min-sum BCC problem. We observe that this method can be adapted to give an approximation algorithm for the multi-depot BCC problem. To the best of our knowledge, our $O(\log n)$-approximation algorithm is the first nontrivial approximation algorithm for the multi-depot BCC problem.

Let Algorithm $\mathcal{S}$ be the $O(\log \frac{B}{\mu})$-approximation algorithm for the single depot BCC problem, and Algorithm $\mathcal{M}$ be the $O(\log n)$-approximation algorithm for the multi-depot BCC problem. We test Algorithm $\mathcal{S}$ and Algorithm $\mathcal{M}$ on randomly generated instances.

The experimental results show that both algorithms have good practice performance. In particular, Algorithm $\mathcal{M}$ behaves rather well in the experiments. The approximation ratios of Algorithm $\mathcal{S}$ on the test instances are about 2 to 4, while their analytical counterparts (i.e., $O(\log \frac{B}{\mu})$) are about 65 to 79. The approximation ratios of Algorithm $\mathcal{M}$ on the test instances are about only 1.1 to 1.5, while their analytical counterparts are about 13 to 18. These results show that the approximation ratios on test instances are much less than the analytical approximation ratios. We remark that since we do not know the optimal values of the test instances, we use a lower bound of the optimal value to compute the approximation ratios on the test instances. In other words, the real approximation ratios of the two algorithms on the test instances may be even smaller.

Our experiments show that the algorithms proposed in the paper can be used as subroutines with theoretical performance guarantees for solving the related problems in practice.

1.3 More Related Work

Actually, there are several types of the Cycle Cover problem. Besides the Budgeted Cycle Cover problem, two common types of the problem are the Minimum Cycle Cover (MCC) problem and the Min-Max Cycle Cover (MMCC) problem.

Given a metric space (V, w) and an integer k, the MCC problem asks to find at most k cycles to cover all the vertices in V, such that the total weight of the cycles is minimized. Recently, Khachay et al. [12] gave a 2-approximation algorithm for this problem.

Then let us focus on the (unrooted) MMCC problem [7,13,23,27]. Given a metric space (V, w) and an integer k, the MMCC problem asks to find k cycles to cover all the vertices in V such that the maximum weight of the cycles is minimized. The best known approximation ratio for this problem is 5, given by Yu et al. [27].

The single-depot MMCC problem can be approximated within 5/2 [8]. The multi-depot MMCC problem has been studied in [23,24,27]. Recently, Yu et al. [27] gave a 6-approximation algorithm for the multi-depot MMCC problem by reducing it to the unrooted MMCC problem. This is the currently best approximation ratio for multi-depot MMCC.

In the capacitated multi-depot MMCC problem [23,24], each depot has a *capacity* indicating the maximum number of vehicles (i.e., cycles) it can hold. Different depots may have different capacities. The current best ratio for this problem is $7 + \epsilon$ for any $\epsilon > 0$, given by Xu et al. [23].

From the viewpoint of algorithmic graph theory, Cycle Cover is a cover-like combinatorial optimization problem. There are actually many cover problems in algorithmic graph theory, such as Vertex Cover (using vertices to cover edges), Edge Cover (using edges to cover vertices), Tree Cover (using trees to cover vertices) [7,13], Path Cover (using paths to cover vertices) [1,25], and Star Cover (using stars to cover vertices) [1,29], etc.

2 Single Depot Budgeted Cycle Cover

First note that if there is any vertex v whose distance to the root r is $B/2$, we have to use a single cycle (r, v, r) to cover such vertex. (Note that the problem is defined in the metric space (V, w), so there is an edge between any two different vertices.) So, as a pre-processing step, we can find all vertices that are of distance $B/2$ from the root, and use separate cycles to cover these vertices. Then, they can be removed from the input safely. Therefore, in the following we may assume that there is no vertex with distance $B/2$ to r.

2.1 Layering of Vertices and the Algorithm

Let μ_0 be the minimum distance between two vertices. Suppose that there are q different distances from vertices in $V \setminus \{r\}$ to r, and they are denoted as w_1, w_2, ..., w_q. Let $S = \{w_1, w_2, \ldots, w_q\} \cup \{0\} \cup \{B/2\}$. Let μ_1 be the minimum difference of any two different weights in S, that is, $\mu_1 = \min\{|w - w'| \mid \forall w, w' \in S, \text{s.t. } w \neq w'\}$. Then μ is defined as the minimum of μ_0 and μ_1:

$$\mu = \min\{\mu_0, \mu_1\}. \tag{1}$$

The definition of μ will be clear when the layering of vertices is presented.

The parameter ϵ is defined as

$$\epsilon = \frac{\mu}{B}, \tag{2}$$

and the parameter t is defined as

$$t = \left\lceil \log \frac{1}{\epsilon} \right\rceil. \tag{3}$$

We partition all the vertices in V into $t + 1$ layers $\{V_0, V_1, \ldots, V_t\}$ according to their distances to the root r. See Fig. 1 for an illustration. The layer V_j ($j = 0, 1, \ldots, t$) is defined as follows.

$$V_j = \left\{ v \colon (1 - 2^j \epsilon)\frac{B}{2} < w(r, v) \leq (1 - 2^{j-1}\epsilon)\frac{B}{2} \right\}. \tag{4}$$

The furthest layer to r is layer V_0, which contains vertices whose distances to r are in $((1 - \epsilon)\frac{B}{2}, (1 - \frac{1}{2}\epsilon)\frac{B}{2}]$. Note that $(1 - \frac{1}{2}\epsilon)\frac{B}{2} = \frac{B}{2} - \frac{\mu}{4}$. By the definition (1) of μ, there is no vertex with distance in $(\frac{B}{2} - \frac{\mu}{4}, \frac{B}{2})$ to the root r.

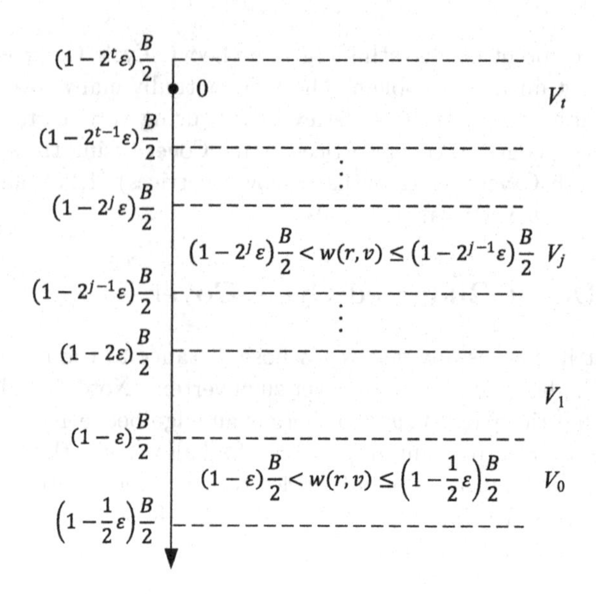

Fig. 1. Layering of vertices. The depot r is at the origin.

The nearest layer to r is layer V_t, containing vertices whose distances to r are in $((1-2^t\epsilon)\frac{B}{2}, (1-2^{t-1}\epsilon)\frac{B}{2}]$. Note that since $t = \lceil \log \frac{1}{\epsilon} \rceil$, we have $(1-2^t\epsilon)\frac{B}{2} \le 0$. All vertices (if any) in V_t are very close to r. If $(1-2^t\epsilon)\frac{B}{2} < 0$, the root r will be contained in V_t.

The approximation algorithm for single depot BCC is given in Algorithm 1. The algorithm partitions vertices into t layers according to (4). For each layer, the algorithm finds some $2^{j-1}\epsilon B$-budgeted paths to cover all the vertices in the layer. The found paths are converted into cycles finally. To find the $2^{j-1}\epsilon B$-budgeted paths, the algorithm call an approximation algorithm for the (unrooted) Budgeted Path Cover (BPC) problem as a subroutine. Given a metric space and a budget λ, the BPC problem asks to find a minimum number of λ-budgeted paths to cover all the vertices in the space. This problem can be approximated within a factor of 3 [1].

Algorithm 1 Algorithm $\mathcal{S}$ for the single depot BCC problem.
Input: An instance (V, w, r, B) of single depot BCC.
Output: A set of r-rooted B-budgeted cycles covering all nodes in $V \setminus \{r\}$.

1 If there is any vertex v whose distance to r is $\frac{B}{2}$, then use a single cycle (r, v, r) to cover v.
2 Partition vertices in V into layers $\{V_0, V_1, \ldots, V_t\}$ according to (4).
3 For each layer $0 \le j \le t$, run the approximation algorithm [1] for the BPC problem on metric space (V_j, w) with budget $2^{j-1}\epsilon B$, obtaining a set of paths Π_j.

4 Convert each found path $\pi \in \Pi_j$ into a cycle by adding respectively two edges from the root r to its two endpoints.
5 **return** all the obtained cycles.

2.2 Analysis

Lemma 1. *Each cycle found by Algorithm $\mathcal{S}$ has length at most B.*

Proof. Fix a layer V_j $(1 \leq j \leq t)$ and assume that V_j is not empty. Let $\pi \in \Pi_j$ be a path found in step 3 of Algorithm $\mathcal{S}$. Let u and v be the two endpoints of π. Note that u and v may be the same vertex (when π contains only one vertex). By the definition of V_j, we have $w(r,u) \leq (1-2^{j-1}\epsilon)\frac{B}{2}$ and $w(r,v) \leq (1-2^{j-1}\epsilon)\frac{B}{2}$. So, the length of the cycle (r, π, r) is at most $(1-2^{j-1}\epsilon)\frac{B}{2} \cdot 2 + 2^{j-1}\epsilon B = B$.

Let k^* be the optimum of the single depot BCC problem, that is, $k^* = OPT$ is the minimum number of B-budgeted r-rooted cycles that cover all nodes in $V \setminus \{r\}$.

Lemma 2. *For each $j = 0, 1, \ldots, t$, Algorithm $\mathcal{S}$ uses at most $6k^*$ paths to cover all vertices in layer V_j.*

Proof. Let us fix a layer number j in $\{0, 1, \ldots, t\}$.

Let Γ^* be an optimal solution to the single depot BCC problem. We know that Γ^* contains k^* cycles which are B-budgeted r-rooted and cover all vertices in $V \setminus \{r\}$. Especially, Γ^* covers all vertices in V_j. Here we assume that V_j is not empty, since otherwise there is nothing to do.

Let $\sigma^* \in \Gamma^*$ be a cycle covering some vertices in V_j. Note that σ^* may cross V_j several times. Let us imagine a walk along σ^* from the root r. Let $f \in V_j$ be the vertex that σ^* enters V_j at the first time, and $l \in V_j$ be the vertex that σ^* leaves V_j at the last time. The part of σ^* from f to l is a path, denoted by σ^*_{fl}. Note that σ^*_{fl} may visit vertices outside V_j. We convert σ^*_{fl} to a path by short-cutting vertices of σ^*_{fl} not in V_j. Let π_{fl} be the resulting path.

If $0 \leq j \leq t-1$, by the definition of V_j, the length of σ^*_{fl} is at most

$$B - (1 - 2^j\epsilon)\frac{B}{2} \cdot 2 = 2^j\epsilon B.$$

This is also an upper bound of the length of π_{fl}. If $j = t$, the length of σ^*_{fl} is $< B$. (Note that the root $r \in V_t$.) The length of π_{fl} is obviously $< B$, too. Since $t = \lceil \log \frac{1}{\epsilon} \rceil$, we have $B \leq 2^j\epsilon B$ in this case. So, in both cases we can use

$$2^j\epsilon B$$

as an upper bound of the length of π_{fl}, whatever the layer number j is.

Now we claim that we can decompose π_{fl} into at most two sub-paths each of which has length at most $2^{j-1}\epsilon B$. The decomposition method will be given below. If so, then we will have constructed a feasible solution to the BPC problem on vertex set V_j, which contains $2k^*$ paths. Since the BPC problem admit 3-approximation [1], we know that $\forall j$, $|\Pi_j| \leq 6k^*$, proving the lemma.

Now, we show how to decompose π_{fl} into at most two short sub-paths. If π_{fl} only contains one vertex (namely, $f = l$), we need not to separate π_{fl}. It is already a path of length zero. If π_{fl} contains exactly two vertices, namely, f and l, we just separate π_{fl} into two zero-length sub-paths (f) and (l).

Then consider the case that π_{fl} contains ≥ 3 vertices. Begining from f, we walk along π_{fl} to find an edge (u, v) on the path such that the sub-path from f to u (denoted by π_{fu}) has length $\leq 2^{j-1}\epsilon B$ and the sub-path from f to v has length $> 2^{j-1}\epsilon B$. If such an edge (u, v) can be found, then removing (u, v) from π_{fl} gets two sub-paths each of which has length $\leq 2^{j-1}\epsilon B$, since the length of π_{fl} is $\leq 2^j\epsilon B$. If such an edge cannot be found, then we break π_{fl} into two sub-paths π_{fu} and (l), where u is the previous vertex of l on path π_{fl}. Obviously, the length of π_{fu} is $\leq 2^{j-1}\epsilon B$ in this case.

Theorem 1. *The single depot* Budgeted Cycle Cover *problem can be approximated within $O(\log \frac{B}{\mu})$ in polynomial time, where B is the length bound of each cycle, and μ is the minimum distance (as defined in (1)).*

Proof. By Lemma 1, the solution found by Algorithm $\mathcal{S}$ is a feasible solution. By Lemma 2, the solution contains at most $6(t+1)k^* = O(\log \frac{B}{\mu})k^*$ cycles, where the equality is due to (2) and (3). The algorithm obviously runs in polynomial time.

When all distances are integers, we have $\mu \geq 1$. In this case, the approximation ratio of Algorithm $\mathcal{S}$ is $O(\log B)$.

3 Multi-depot Budgeted Cycle Cover

The multi-depot BCC problem asks to find a minimum number of D-rooted B-budgeted cycles to cover all nodes in $V \setminus D$. Recall that a D-rooted cycle means a cycle contains *at least one depot* in D. For clarity, let $N = V \setminus D$. The vertices in N are all nodes.

3.1 The Algorithm

For the multi-depot BCC problem, we use a greedy approach to deal with it. Our technique is inspired by Khuller et al. [14]. The high-level idea of the approach is very simple. We repeatedly cover some nodes using a length bounded cycle, until all nodes are covered. To find the length bounded cycles, we call an approximation algorithm for the s-t Orienteering problem as a subroutine. Given a metric space (V, w), two vertices s and t, and a budget B, where each vertex in V is with a prize $p(v) \in \mathbb{R}^+$, the s-t Orienteering problem asks to find a B-budgeted s-t path such that the total prize of the path is maximized. The s-t Orienteering problem has been studied in [2,3] and [4]. The problem is NP-hard. Its current best approximation ratio is $2 + \epsilon$ due to [4].

The approximation algorithm for the multi-depot BCC problem is given as Algorithm 2.

Algorithm 2 Algorithm $\mathcal{M}$ for the multi-depot BCC problem.
Input: An instance (V, w, D, B) of multi-depot BCC.
Output: A set of D-rooted B-budgeted cycles covering all vertices in N.
1 Let U be the set of not covered nodes. Initially $U \leftarrow N$. Let $\mathcal{C}$ be the set of found cycles. Initially $\mathcal{C} \leftarrow \emptyset$.
2 **while** $U \neq \emptyset$ **do**
3 Let the prize of every not covered node (i.e., node in U) be one, and the prize of every covered node (i.e., node in $N \setminus U$) be zero.
4 For every depot $d \in D$, build an instance for the s-t Orienteering problem with $s = d$, $t = d$, and budget being B. Call the algorithm in [4] for s-t Orienteering to find a path starting from d and ending at d (i.e., a cycle containing d).
5 Let C be the cycle with the maximum total prize (i.e., number of covered nodes) among all cycles found in Step 4. Remove all nodes in C from U. Add C to $\mathcal{C}$.
6 **endwhile**
7 **return** $\mathcal{C}$.

3.2 Analysis

Let k^* be the value of the optimal solution to the multi-depot BCC problem, that is, $k^* = OPT$ is the number of cycles used in the optimal solution to cover all the nodes in N. By sligthly abusing the notation, we also use OPT to denote the optimal solution. Let s_i be the total number of nodes covered from the first round to the i-th round of Algorithm $\mathcal{M}$. Let α be the (best) approximation ratio of the s-t Orienteering problem. By [4], $\alpha = 2 + \epsilon$. Actually, for our result for the multi-depot BCC problem, we only need α to be a constant. Then we have the following Lemma .

Lemma 3. *For the node number s_i, we have*

$$s_1 \geq \frac{1}{\alpha} \cdot \frac{|N|}{k^*}, \tag{5}$$

$$s_i \geq s_{i-1} + \frac{1}{\alpha} \cdot \frac{|N| - s_{i-1}}{k^*}, \qquad i \geq 2. \tag{6}$$

Proof. Let $O_{\max}$ be the cycle with the maximum prize in the optimal solution. Then its prize is at least the average prize of cycles in the optimal solution, which is $\frac{|N|}{k^*}$. Suppose d is a depot contained in $O_{\max}$. So, the optimal solution to the s-t Orienteering problem corresponding to d has prize at least $\frac{|N|}{k^*}$. Since the s-t Orienteering problem admits $1/\alpha$-approximation, for depot d, Step 4 will find a cycle whose prize is $\geq \frac{1}{3} \cdot \frac{|N|}{k^*}$. Since step 5 of Algorithm $\mathcal{M}$ chooses the cycle with the maximum prize ever found in this round, we know that (5) holds.

Then consider the i-th round. In this round, the number of nodes to be covered is $|N| - s_{i-1}$. Moreover, the cycles in the optimal solution k^* still cover all these nodes. By similar reasoning, the number of nodes covered in the i-round of the algorithm is $\geq \frac{1}{\alpha} \cdot \frac{|N|-s_{i-1}}{k^*}$. So, we know that (6) holds.

Lemma 4. *The number of iterations of Algorithm $\mathcal{M}$ is $O(\log n)k^*$.*

Proof. Solving the recurrence inequality (6), we get

$$
\begin{aligned}
s_i &\geq s_{i-1} + \frac{1}{\alpha} \cdot \frac{|N| - s_{i-1}}{k^*} = \frac{|N|}{\alpha k^*} + \left(1 - \frac{1}{\alpha k^*}\right) s_{i-1} \\
&\geq \frac{|N|}{\alpha k^*} + \left(1 - \frac{1}{\alpha k^*}\right)\left(\frac{|N|}{\alpha k^*} + \left(1 - \frac{1}{\alpha k^*}\right) s_{i-2}\right) \\
&= \left(1 + \left(1 - \frac{1}{\alpha k^*}\right)\right)\frac{|N|}{\alpha k^*} + \left(1 - \frac{1}{\alpha k^*}\right)^2 s_{i-2} \\
&\geq \cdots \\
&\geq \left(\left(1 - \frac{1}{\alpha k^*}\right)^0 + \cdots + \left(1 - \frac{1}{\alpha k^*}\right)^{i-2}\right)\frac{|N|}{\alpha k^*} + \left(1 - \frac{1}{\alpha k^*}\right)^{i-1} s_1 \\
&\geq \left(\left(1 - \frac{1}{\alpha k^*}\right)^0 + \cdots + \left(1 - \frac{1}{\alpha k^*}\right)^{i-1}\right)\frac{|N|}{\alpha k^*} \\
&= \left(1 - \left(1 - \frac{1}{\alpha k^*}\right)^i\right)|N|.
\end{aligned}
$$

So, the number of cycles found by Algorithm $\mathcal{M}$ is no more than the least i such that

$$\left(1 - \left(1 - \frac{1}{\alpha k^*}\right)^i\right)|N| > |N| - 1, \tag{7}$$

since the left hand side of (7) is the number of nodes covered from the first round to the i-th round. Solving (7), we get that i can be the least value satisfying

$$i > (\alpha \ln |N|)k^*. \tag{8}$$

That is to say, when $i = O(\log |N|)k^*$, the algorithm will cover all the nodes. Therefore, the number of iterations of Algorithm $\mathcal{M}$ is $O(\log n)k^*$, by noticing that $n = |N \cup D|$ is the number of vertices in the problem.

Theorem 2. *The multi-depot* BCC *problem can be approximated within $O(\log n)$ in polynomial time.*

Proof. At each iteration of Algorithm $\mathcal{M}$, the algorithm puts uses a new cycle to cover some nodes. By Lemma 4, the iteration number is $O(\log n)k^*$. So, the approximation ratio of Algorithm $\mathcal{M}$ is $O(\log n)$. Finally, it is obvious that Algorithm $\mathcal{M}$ runs in polynomial time. The theorem follows.

4 Experiments

For Algorithm $\mathcal{S}$ and Algorithm $\mathcal{M}$, we performed several experiments to test their performance and to investigate the impact of some important parameters on the algorithmic performance, including the budget B and the network size n.

In our experiments, the test instances are randomly generated. For each test instance, the generation steps are as follows. First, we generate a complete graph $G = (V, E)$ with n vertices. Initially, each edge in E has a weight uniformly distributed in $[100, 1000]$. Then, to make the edge weights metric, we use Floyd's Algorithm to compute the minimum distance between any two vertices in graph G and use this value as the final weight of the edge between this two vertices. For the single depot BCC problem, we choose a vertex as the depot uniformly at random. For the multi-depot BCC problem, we randomly choose 5% of the vertices to constitute the depot set D.

For each algorithm, we do two groups of experiments. In the first group, vertex number n is fixed to be 500, and budget B varies from 2000 to 10000. This is aimed to see the impact of budget B on the number of cycles. In the second group, budget B is fixed to be 5000, and vertex number n varies from 100 to 500. This is aimed to see the impact of network size n on the number of cycles. For each experiment, in which n and B are both fixed, we generate 100 random instances and calculate the average of the outputs (i.e., number of cycles) as the result of the experiment. Therefore, we do four groups of experiments in total.

From the experimental results we can see that both Algorithm $\mathcal{S}$ and Algorithm $\mathcal{M}$ have good practical performance. Meanwhile, these two algorithms are steady and scalable. Their approximation ratios on the test instances are much better than the corresponding analytical ratios. Among them Algorithm $\mathcal{M}$ is rather better. Note that all the test instances have large size. This means that we cannot calculate their optimal values effectively. Instead, we infer lower bounds of the optimal values for the single depot BCC problem and the multi-depot BCC problem. The approximation ratios of our algorithms reported in the experiments are calculated using these lower bounds. This also means that the true ratios of the algorithms on the test instances are even better. The detailed experimental results and analysis will be given in the full version of the paper. They are omitted in this preliminary version due to space limitation.

5 Conclusions

The rooted Budgeted Cycle Cover problem is a fundamental problem arising in wireless sensor networks and vehicle routing. This problem is NP-hard, thus people usually pursue approximation algorithms and heuristics for this problem. In this paper, we design an $O(\log \frac{B}{\mu})$-approximation algorithm for the single depot BCC problem, and an $O(\log n)$-approximation algorithm for the multi-depot BCC problem. These ratios are the currently best for the respective problems. The approximation algorithms are purely combinatorial and easy to implement. We test these two algorithms on randomly generated instances. The experimental results show that both of them have good practice performance. In particular, the $O(\log n)$-approximation algorithm behaves rather well in the experiments. The algorithms can be used as subroutines with theoretical performance guarantees for solving the related problems in practice.

Acknowledgements. This work is supported by the National Natural Science Foundation of China (61972228 and 61672323), and the Natural Science Foundation of Shandong Province (ZR2019MF072 and ZR2016AM28).

References

1. Arkin, E., Hassin, R., Levin, A.: Approximations for minimum and min-max vehicle routing problems. J. Algorithms **59**, 1–18 (2006)
2. Bansal, N., Blum, A., Chawla, S., Meyerson, A.: Approximation algorithms for deadline-TSP and vehicle routing with time-windows. In: Proceedings of the 36th Annual ACM Symposium on Theory of Computing (STOC), pp. 166–174 (2004)
3. Blum, A., Chawla, S., Karger, D.R., Lane, T., Meyerson, A., Minkoff, M.: Approximation algorithms for orienteering and discounted-reward TSP. SIAM J. Comput. **37**(2), 653–670 (2007)
4. Chekuri, C., Korula, N., Pál, M.: Improved algorithms for orienteering and related problems. ACM Trans. Algorithms **8**(3), 23:1–23:27 (2012)
5. Cordeau, J.F., Laporte, G., Savelsbergh, M.W., Vigo, D.: Vehicle routing. In: Barnhart, C., Laporte, G. (eds.) Handbook in OR & MS, vol. 14, pp. 367–428 (2007)
6. Erol-Kantarci, M., Mouftah, H.T.: Suresense: sustainable wireless rechargeable sensor networks for the smart grid. IEEE Wireless Commun. **19**(3), 30–36 (2012)
7. Even, G., Garg, N., Könemann, J., Ravi, R., Sinha, A.: Min-max tree covers of graphs. Oper. Res. Lett. **32**(4), 309–315 (2004)
8. Frederickson, G., Hecht, M., Kim, C.: Approximation algorithms for some routing problems. SIAM J. Comput. **7**(2), 178–193 (1978)
9. Golden, B., Raghavan, S., Wasil, E. (eds.): The Vehicle Routing Problem: Latest Advances and New Challenges. Springer, Boston (2008). https://doi.org/10.1007/978-0-387-77778-8
10. Golden, B.L., Assad, A.A. (eds.): Veh. Rout Methods Stud. North-Holland, Amsterdam (1988)
11. Guo, S., Wang, C., Yang, Y.: Mobile data gathering with wireless energy replenishment in rechargeable sensor networks. In: Proceedings of IEEE INFOCOM, pp. 1932–1940 (2013)
12. Khachay, M., Neznakhina, K.: Approximability of the minimum-weight k-size cycle cover problem. J. Glob. Optim. **66**(1), 65–82 (2016)
13. Khani, M.R., Salavatipour, M.R.: Improved approximation algorithms for the min-max tree cover and bounded tree cover problems. In: Goldberg, L.A., Jansen, K., Ravi, R., Rolim, J.D.P. (eds.) APPROX/RANDOM -2011. LNCS, vol. 6845, pp. 302–314. Springer, Heidelberg (2011). https://doi.org/10.1007/978-3-642-22935-0_26
14. Khuller, S., Malekian, A., Mestre, J.: To fill or not to fill: the gas station problem. ACM Trans. Algorithms **7**(3), 36:1–36:16 (2011)
15. Li, Z., Peng, Y., Zhang, W., Qiao, D.: J-RoC: a joint routing and charging scheme to prolong sensor network lifetime. In: Proceedings of the 19th IEEE International Conference on Network Protocols (ICNP), pp. 373–382 (2011)

16. Liang, W., Luo, J., Xu, X.: Prolonging network lifetime via a controlled mobile sink in wireless sensor networks. In: Proceedings of the Global Communications Conference (GLOBECOM), pp. 1–6 (2010)
17. Liang, W., Schweitzer, P., Xu, Z.: Approximation algorithms for capacitated minimum forest problems in wireless sensor networks with a mobile sink. IEEE Trans. Comput. **62**(10), 1932–1944 (2013)
18. Nagarajan, V., Ravi, R.: Approximation algorithms for distance constrained vehicle routing problems. Networks **59**(2), 209–214 (2012)
19. Shi, Y., Xie, L., Hou, Y.T., Sherali, H.: On renewable sensor networks with wireless energy transfer. In: Proceedings of IEEE INFOCOM, pp. 1350–1358 (2011)
20. Toth, P., Vigo, D. (eds.): The Vehicle Routing Problem. SIAM Monographs on Discrete Mathematics and Applications. SIAM, Philadelphia (2002)
21. Wang, C., Li, J., Ye, F., Yang, Y.: Multi-vehicle coordination for wireless energy replenishment in sensor networks. In: Proceedings of the 27th IEEE International Symposium on Parallel and Distributed Processing (IPDPS), pp. 1101–1111 (2013)
22. Xie, L., Shi, Y., Hou, Y.T., Sherali, H.D.: Making sensor networks immortal: an energy-renewal approach with wireless power transfer. IEEE/ACM Trans. Netw. **20**(6), 1748–1761 (2012)
23. Xu, W., Liang, W., Lin, X.: Approximation algorithms for min-max cycle cover problems. IEEE Trans. Comput. **64**(3), 600–613 (2015)
24. Xu, Z., Xu, D., Zhu, W.: Approximation results for a min-max location-routing problem. Discrete Appl. Math. **160**, 306–320 (2012)
25. Xu, Z., Xu, L., Li, C.L.: Approximation results for min-max path cover problems in vehicle routing. Naval Res. Logist. **57**, 728–748 (2010)
26. Xu, Z., Liang, W., Xu, Y.: Network lifetime maximization in delay-tolerant sensor networks with a mobile sink. In: Proceedings of IEEE 8th International Conference on Distributed Computing in Sensor Systems (DCOSS), pp. 9–16 (2012)
27. Yu, W., Liu, Z.: Improved approximation algorithms for some min-max and minimum cycle cover problems. Theor. Comput. Sci. **654**, 45–58 (2016)
28. Yu, W., Liu, Z., Bao, X.: New approximation algorithms for the minimum cycle cover problem. Theoretical Computer Science **793**, 44–58 (2019)
29. Zhao, W., Zhang, P.: Approximation to the minimum rooted star cover problem. In: Proceedings of the 4th International Conference of Theory and Applications of Models of Computation (TAMC), pp. 670–679 (2007)

Evolutionary Equilibrium Analysis for Decision on Block Size in Blockchain Systems

Jinmian Chen, Yukun Cheng(✉), Zhiqi Xu, and Yan Cao

School of Business, Suzhou University of Science and Technology, Suzhou 215009, China
{jinmian_chen,joisexu}@post.usts.edu.cn, ykcheng@amss.ac.cn, cy@usts.edu.cn

Abstract. In a PoW-based blockchain network, mining pools (the solo miner could be regarded as a mining pool containing one miner) compete to successfully mine blocks to pursue rewards. Generally, the rewards include the fixed block subsidies and time-varying transaction fees. The transaction fees are offered by the senders whose transactions are packaged into blocks and is positively correlated with the block size. However, the larger size of a block brings the longer latency, resulting in a smaller probability of successfully mining. Therefore, finding the optimal block size to trade off these two factors is a complex and crucial problem for the mining pools. In this paper, we model a repeated mining competition dynamics in blockchain system as an evolutionary game to study the interactions among mining pools. In this game, each pool has two strategies: to follow the default size $\bar{B}$, i.e., the upper bound of a block size, or not follow. Because of the bounded rationality, each mining pool pursues its evolutionary stable block size (ESS) according to the mining pools' computing power and other factors by continuous learning and adjustments during the whole mining process. A study framework is built for the general evolutionary game, based on which we then theoretically explore the existence and stability of the ESSs for a case of two mining pools. Numerical experiments with real Bitcoin data are conducted to show the evolutionary decisions of mining pools and to demonstrate the theoretical findings in this paper.

Keywords: Blockchain · Block size · Transaction fee · Mining competition · Evolutionary game

1 Introduction

Bitcoin is a decentralized payment system [11], based on a public transaction ledger, which is called the blockchain. Generally, a block is composed of a block header and a block body, which contains a certain amount of transactions. Each transaction is composed of the digital signature of the sender, the transaction

D.-Z. Du et al. (Eds.): COCOA 2021, LNCS 13135, pp. 180–194, 2021.
https://doi.org/10.1007/978-3-030-92681-6_15

data, such as the value of digital tokens, the addresses of the sender and the receiver, as well as the corresponding transaction fee. With the bitcoin system developing, the number of transactions in the whole network increases quickly, while the block size currently is limited to 1 MB. Such a bounded block size results in the congestion of the blockchain network. To alleviate this situation, *Segregated Witness* (SegWit) [10] is brought up and applied to segregate the witness (digital signatures) from the transactions. Then the witness is put into the "extended block", which has no impact on the original block size. By SegWit, a block is able to contain more transactions, enhancing the transaction processing efficiency. This effect is equivalent to expanding the block size to 2 MB. Thus the block size of a newly mining block may be more than 1 MB.

Proof-of-work (PoW) is the most popular consensus applied in bitcoin blockchain system, which reaches a consensus based on miners computing power. Under the PoW-based consensus protocol, the process of successfully mining a block includes two steps, i.e., solving the PoW puzzle and propagating the block to be verified. During the propagation, the block is likely to be discarded because of long latency which depends on the size of the block. The larger size of block brings the longer latency, leading to a higher chance that the block suffers *orphaned* [6]. So, besides raising income from transaction fees by packaging more transactions in a block, miners need to consider a suitable total size of the block that would not deeply increase the probability to be orphaned. With the incentive of transaction fees and the long latency resulting from large block size, how to select transactions and decide the total block size for maximum payoff is critical for every miner.

Game theory has been widely applied in mining management, such as computational power allocation, fork chain selection, block size setting and pool selection. In terms of block size setting, [14] analyzed the quantity setting of block space with the effect of a transaction fee market, in which a block space supply curve and a mempool demand curve were introduced to find the optimal block space for the maximum payoff of miners. By proposing a Bitcoin-unlimited scheme, [16] modeled a non-cooperative game to examine the interaction among the miners, each of whom chooses its own upper bound of the block size while it invalidates and discards the excessive block that is larger than its upper bound. And the game was proved to exist an unique Nash equilibrium where all miners choose the same upper bound. Given the limitation on the number of transactions included in the block, the interaction on choosing transactions between the miners and the users was modeled as a non-cooperative game by [2]. The unique Nash equilibrium of this game can be obtained when satisfying certain conditions, which is related to the number of miners, the hash rate, bitcoin value, the transaction fees, the block subsidy and the cost of the mining. As for dynamic evolutionary behaviors in blockchain, [9] respectively modeled the dynamics of block mining selection among pools and pool selection among individual miners as an evolutionary game in a proof-of-work (PoW) blockchain network. It identified the hash rate and the block propagation delay as two major factors resulting in the mining competition outcome. Also, in [13], evolutionary game was applied to examine the process dynamics of selecting super nodes for transaction verification in the Delegated proof of stake (DPoS) blockchain. The authors found that the strategy of candidates has to do with how much reward they can obtain from the blockchain platform. Inspired by [7], we adopt its novel expected payoff function different with that of [9], to study the block size determination of mining pools in a dynamic process.

In this work, we assume that the mining pools are bounded rational and can adapt their strategy on different block sizes according to the received rewards. Note that the total computing power of each mining pool and the long latency also affect the choices of pools. Accordingly, we model the repeated mining competition as an evolutionary game, where each pool controlling a certain amount of computing power has to decide whether following the default size—2 MB or not, for maximum payoffs. Evolutionary stable strategies are considered to be the solutions of this game. Then we perform theoretical analysis on the existences and corresponding conditions of the ESS for a special case of two mining pools. Finally, simulations are performed to verify the proposed schemes. In addition, we discuss the impact of the hash rate of pools, the unit transaction fee, the unit propagation delay,'as well as the default block size on the strategy decision of mining pools.

The rest of this paper is organized as follows. Section 2 introduces the system model and the reward function of pools. In Sect. 3, we formulate an evolutionary game model to study the block size selection problem, and particularly analyze a case of two mining pools. Section 4 presents the numerical results and some additional analysis on different factors, and concludes our study.

2 System Model and and Mining Pool's Expected Reward

In this paper, we consider the PoW-based blockchain system where there are n mining pools, denoted by $N = \{1, 2, \cdots, n\}$, and each contains several miners. All mining pools compete to mine blocks by costing an amount of computing power, and thus to pursue the corresponding rewards. Similar to [4] and [5], we assume the whole system is in a quasi-static state, meaning no miners join in or leave the system. Under this assumption, each miner keeps its state unchanged, including which mining pool it is in, and how much computing power it has. This leads to the mining pools' constant scale and their total computing power.

By the consensus protocol of PoW, a mining pool, who obtains the reward, must satisfy the following two conditions: it is the first one to solve a proof-of-work puzzle by consuming an amount of computing power and it is also the first one to make its mined block reach the consensus. The expected average block arriving interval is about of $T = 600\,\mathrm{s}$, by adjusting the difficulty of the proof-of-work puzzle. The whole mining process in a blockchain system consists of a series of one-shot competitions, in each of which one block is mined. A one-shot competition can be viewed as a non-cooperative game, in which all mining pools are the players and they shall make decisions on the mined blocks' size to maximize their own rewards.

Similar to the model in [7], we compute the expected reward of each mining pool i by regarding its *block finding time* in one-shot competition as a random variable, denoted by X_i, which follows the exponential distribution. To be specific, let B_i be the block size decided by pool i. Denote h_i $(0 \leq h_i \leq 1)$ to be the relative computing power of pool i, that is the ratio of pool i's computing power to the total computing power in blockchain system. The propagation time of pool i's block is linear with its size B_i, that is $q_i = \rho B_i$. It is obvious that pool i's block cannot reach the consensus if the block finding time is less than the propagation time q_i. The mining rate of pool i is denoted by $\lambda_i = \frac{h_i}{T}$, where $T = 600\,\mathrm{s}$ is the average block arriving interval. By the definition of the exponential distribution, the probability density function (PDF) of X_i is

$$f_{X_i}(t; B_i, \lambda_i) = \begin{cases} 0, t < q_i; \\ \lambda_i e^{-\lambda_i(t-q_i)}, t \geq q_i. \end{cases} \tag{1}$$

and the cumulative distribution function (CDF) of X_i is

$$F_{X_i}(t; B_i, \lambda_i) = Pr(X_i \leq t) = \begin{cases} 0, \ t < q_i; \\ 1 - e^{-\lambda_i(t-q_i)}, \ t \geq q_i. \end{cases}$$

So, the probability that block finding time of pool i is larger than t is

$$Pr(X_i > t) = 1 - F_{X_i}(t; B_i, \lambda_i) = \begin{cases} 1, \ t < q_i; \\ e^{-\lambda_i(t-q_i)}, \ t \geq q_i. \end{cases}$$

Define X to be the block finding time among all mining pools. Then $X = \min_{i \in N}\{X_i\}$, i.e. the first time to find a block, and hence

$$\begin{aligned} Pr(X > t) &= \Pi_{i=1,\cdots,n} Pr(X_i > t) = \Pi_{i \in Active(t)} Pr(X_i > t) \\ &= e^{\sum_{i \in Active(t)}[-\lambda_i(t-q_i)]}, \end{aligned} \tag{2}$$

where $Active(t) = \{i | q_i \leq t\}$ is the pool set, each pool i in which has the propagation time less than time t. For convenience, we call each pool i in $Active(t)$ an active pool at time t.

From the probability function of (2), it is not hard to derive the CDF and PDF of random variable X as the follows

$$F_X(t; \mathbf{B}, \lambda) = 1 - Pr(X > t) = 1 - e^{\sum_{i \in Active(t)}[-\lambda_i(t-q_i)]}; \tag{3}$$

$$f_X(t; \mathbf{B}, \lambda) = (\sum_{i \in Active(t)} \lambda_i) e^{\sum_{i \in Active(t)}[-\lambda_i(t-q_i)]}, \tag{4}$$

where $\mathbf{B} = (B_1, \cdots, B_n)$ and $\lambda = (\lambda_1, \cdots, \lambda_\mathbf{n})$.

As stated before, the reward of a mined block comes from two aspects: the fixed subsidies R (e.g., 6.25 BTC for one block currently), and a variable amount of transaction fees. Particularly, the transaction fees are more dependent on the size of a block, since a block with a larger size contains more transactions. For the sake of simplicity, we assume the total transaction fee is linearly dependent on the block size, i.e., αB_i. This is similar to the suggested pricing standard of transaction fee for users in some token wallets, such as 0.0005 BTC per KB [12]. So the total reward for a block mined by pool i is $R + \alpha B_i$. In addition, the probability that a pool i solves the proof-of-work puzzle at time t is the ratio of its computing power to all other active mining pools at time t. Then the reward of mining pool i in expectation at time t is

$$E(reward_i|X = t) = \begin{cases} 0, \ t < q_i; \\ \frac{h_i}{\sum_{j \in Active(t)} h_j}(R + \alpha B_i), \ t \geq q_i. \end{cases} \tag{5}$$

and then its reward in expectation is expressed as

$$\begin{aligned} U_i &= E[E(reward_i|X = t)] \\ &= \int_{-\infty}^{+\infty} E(reward_i|X = t) \cdot f_X(t; \mathbf{B}, \lambda)\mathbf{dt} \\ &= \lambda_i(R + \alpha B_i) \sum_{l=i}^{n} \frac{e^{\sum_j \lambda_j (q_j - q_l)} - e^{\sum_j \lambda_j (q_j - q_{l+1})}}{\sum_{j \in Active(q_l)} \lambda_j}, \end{aligned} \tag{6}$$

where $q_{n+1} = +\infty$.

3 Evolutionary Game Model for Decision on Block Size

In a PoW-based blockchain system, we suppose that there are n independent mining pools, each pool i owning an amount of relative computing power h_i. The whole mining process is a series of one-shot competitions, and all mining pools compete to mine a block to win the reward in each one-shot. In this paper, we model the mining competition dynamics as an evolutionary game to study the dynamic interactions among mining pools. In our evolutionary game model, each pool has two kinds of strategies: to follow the default size $\bar{B}$, i.e., the upper bound of a block size, or to choose the block size less than $\bar{B}$. For simplicity, these two strategies is named as "following" strategy and "not following" strategy, respectively. Because of the bounded rationality, each mining pool pursues its evolutionary stable block size (ESS) through continuous learning and adjustments.

In this section, we first propose the analysis scheme for general case, and then theoretically analyze the existence and stability of the ESS for a case of two mining pools.

3.1 Analysis Scheme

In our evolutionary game model, there is a crucial problem for each mining pool i that is how to decide the optimal block size to maximize its payoff in expectation. Note that each pool i has two kinds of strategies: one is to fix the block to default size, e.g., $\bar{B} = 2$ MB, and the other is to choose a block size $B_i < \bar{B}$. So in the k-th shot, let us define two subsets,

$$N^1(k) = \{i \in N | B_i(k) < \bar{B}\} \text{ and } N^2(k) = N - N^1(k) = \{i \in N | B_i(k) = \bar{B}\},$$

and call $(N^1(k), N^2(k))$ a *subset profile.* Clearly, subset profile $(N^1(k), N^2(k))$ is determined after all mining pools making decisions on their block sizes. There are 2^n subset profiles totally in each one-shot, and hence we denote the collection of subset profiles in the k-th slot by $\mathcal{N}(k) = \{(N^1(k), N^2(k))\}$.

Suppose that a subset profile $(N^1(k), N^2(k))$ in the k-th shot is given. Each pool $i \in N^1(k)$ selects the "not following" strategy. In addition, it continues to decide the optimal block size $B_i^* < \bar{B}$ by maximizing its expected payoff under a given subset profile $(N^1(k), N^2(k))$.

$$\begin{aligned} B_i^* &= \arg \pi^i_{(N^1(k),N^2(k))} = \arg \max_{B_i < \bar{B}} U_i \\ &= \arg \max_{B_i < \bar{B}} \left\{ \lambda_i (R + \alpha B_i) \sum_{l=i}^{n} \frac{e^{\sum_j \lambda_j (q_j - q_l)} - e^{\sum_j \lambda_j (q_j - q_{l+1})}}{\sum_{j \in Active(q_l)} \lambda_j} \right\}. \end{aligned} \tag{7}$$

Each mining pool $i \in N^2(t)$ sets its block size as $\bar{B}$ and has its payoff

$$\pi^i_{(N^1(k),N^2(k))} = \lambda_i (R + \alpha \bar{B}) \sum_{l=i}^{n} \frac{e^{\sum_j \lambda_j (q_j - q_l)} - e^{\sum_j \lambda_j (q_j - q_{l+1})}}{\sum_{j \in Active(q_l)} \lambda_j}. \tag{8}$$

During the evolutionary game, the mining pools keep learning to adjust their low-income strategies to a higher-income one dynamically. Until n mining pools reach a stable strategy profile, at which no one would like to change its strategy, an equilibrium state of block size $(B_1^*, B_2^*, \cdots, B_n^*)$ is obtained. Though all the mining competitions are carried out during a series of discrete slots, we can view each block generating slot as a very small interval with respect to the whole mining process, and hence deal with it as a continuous version. It allows us to apply the standard technique to study the evolutionary process for the decisions on block size.

Let $x_i(k)$, $0 \leq x_i(k) \leq 1$, represent the probability of mining pool $i \in N$ to choose the "not following" strategy at the k-th slot. Correspondingly, the probability of pool i to choose the default size is $1 - x_i(k)$. If the choice of pool i is not to follow the default size, then its conditional expected payoff is

$$E_i^1(k) = \sum_{\substack{(N^1(k),N^2(k)) \in \mathcal{N}(k), \\ i \in N^1(k)}} \left(\prod_{l \in N^1(k), l \neq i} x_l(k) \prod_{l \in N^2(k)} (1 - x_l(k)) \cdot \pi^i_{(N^1(k),N^2(k))} \right). \tag{9}$$

If mining pool i selects the "following default size" strategy, then its conditional expected payoff is

$$E_i^2(k) = \sum_{\substack{(N^1(k),N^2(k)) \in \mathcal{N}(k), \\ i \in N^2(k)}} \left(\prod_{l \in N^1(k)} x_l(k) \prod_{l \in N^2(k), l \neq i} (1 - x_l(k)) \cdot \pi^i_{(N^1(k),N^2(k))} \right). \tag{10}$$

Combining (9) and (10), the average payoff of mining pool i is

$$\bar{E}_i(k) = x_i(k) E_i^1(k) + (1 - x_i(k)) E_i^2(k). \tag{11}$$

By [3], the growth rate of a strategy selected by a participant is just equal to the difference between the payoff of this strategy and its average payoff. Then the replicator dynamic equations for all mining pools are as follows:

$$f_i(\mathbf{x}) = \dot{x}_i(k) = x_i(k)(E_i^1(k) - \bar{E}_i(k)), \quad \forall i \in N. \tag{12}$$

According to the replicator dynamics (12), a mining pool would like to choose a smaller block size, when its conditional payoff $E_i^1(k)$ is larger than the average payoff $\overline{E}_i(k)$. Otherwise, it will set its block size as $\bar{B}$. A state is stable if no mining pool would like to change its strategy over time in the replicator dynamics, and such a stable state is considered to be the evolutionary equilibrium [8]. The strategies in this state are evolutionary stable, called ESS. Specifically speaking, when the payoff of "not following" strategy is equal to the average payoff for each pool, all mining pools reaches the ESS and no one has incentive to change its current strategy. Therefore, the ESS can be obtained by solving $\dot{x}_i(k) = 0$ for all $i \in N$, whose solution is called the *fixed equilibrium point* of replicator dynamics.

3.2 A Case Study of Two Mining Pools

Based on the analysis scheme for general case in previous subsection, we continue to study the case of two mining pools ($n = 2$) to exemplify the equilibrium analysis for the decision on block size. We normalize the whole computing power in system, thus mining pool i's relative computing power is $h_i \in \mathbf{h} = \{h_1, h_2\}$ and $\sum_{i=1}^{2} h_i = 1$. The whole mining process contains a series of one-shot competitions, and pool 1 and pool 2 need to decide their block sizes B_1 and B_2 to pursue the optimal payoffs in each one-shot competition. Without loss of generality, we concentrate on the case of $0 \leq h_1 \leq \frac{1}{2} \leq h_2 \leq 1$. The analysis for the case of $h_2 \leq h_1 \leq$ is symmetric, and thus we omit the discussion. As stated in [7], "a miner with less mining power prefers a smaller block size in order to optimize his payoff". Thus the case of $0 \leq h_1 \leq h_2 \leq 1$ leads to $B_1 \leq B_2$ in one-shot competition and then the propagation time $q_1 = \rho B_1 \leq q_2 = \rho B_2$. In a one-shot mining competition, if $B_1 \leq B_2 \leq \bar{B}$, then

$$\begin{aligned} U_1 &= \lambda_1(R + \alpha B_1) \sum_{l=1}^{2} \frac{e^{\sum \lambda_j (q_j - q_l)} - e^{\sum \lambda_j (q_j - q_{l+1})}}{\sum_{j \in Active(q_l)} \lambda_j} \\ &= (R + \alpha B_1)[1 - h_2 e^{\lambda_1 \rho (B_1 - B_2)}]; \end{aligned} \tag{13}$$

$$\begin{aligned} U_2 &= \lambda_2(R + \alpha B_2) \sum_{l=2}^{2} \frac{e^{\sum \lambda_j (q_j - q_l)} - e^{\sum \lambda_j (q_j - q_{l+1})}}{\sum_{j \in Active(q_l)} \lambda_j} \\ &= (R + \alpha B_2) h_2 e^{\lambda_1 \rho (B_1 - B_2)}. \end{aligned} \tag{14}$$

Since each mining pool has two kinds of strategies, i.e., to follow the default size $\bar{B}$, and not to follow, in a one-shot competition, there are four strategy profiles: (B_1, B_2), $(B_1, \bar{B})$, $(\bar{B}, B_2)$, and $(\bar{B}, \bar{B})$. Note that $(\bar{B}, B_2)$ either does not exist, or equals to $(\bar{B}, \bar{B})$ under the condition of $B_1 \leq B_2 \leq \bar{B}$. So we do not discuss this strategy profile any more.

Clearly, each pool would receive different payoffs, under different strategy profiles. For the strategy profile (B_1, B_2), meaning that both pools choose the "not following" strategy, we define the payoffs of two pools π_{11}^1 and π_{11}^2 are their optimal payoffs subject to the conditions of $B_1 < \bar{B}$ and $B_2 < \bar{B}$. The corresponding optimal block sizes (B_1^*, B_2^*) can be obtained by solving

$$\frac{\partial U_1(B_1, B_2)}{\partial B_1} = 0, \text{ and } \frac{\partial U_2(B_1, B_2)}{\partial B_2} = 0,$$

simultaneously. For the strategy profile $(B_1, \bar{B})$, showing that pool 1 would not follow the default size and pool 2's block size is $\bar{B}$, π_{12}^1 and π_{12}^2 are denoted to be the payoffs of pool 1 and 2. To be specific, π_{12}^1 is defined to be the optimal payoff of pool 1 under the condition of $B_1 < \bar{B}$ and the corresponding optimal block size B_1^* can be determined by solving $\frac{dU_1(B_1,\bar{B})}{dB_1} = 0$. For the strategy profile of $(\bar{B}, \bar{B})$, both of two pools set their block sizes as $\bar{B}$, then their payoffs are denoted by π_{22}^1 and π_{22}^2. We illustrate the payoffs of two pools under different strategy profiles in the following payoff matrix (Table 1).

Table 1. Payoff matrix of the case of two mining pools.

	Mining pool 2	
Mining pool 1	$B_2(x_2)$	$\bar{B}(1-x_2)$
$B_1(x_1)$	(π_{11}^1, π_{11}^2)	(π_{12}^1, π_{12}^2)
$\bar{B}(1-x_1)$	$(\backslash, \backslash)$	(π_{22}^1, π_{22}^2)

Lemma 1. *In a one-shot mining competition, if $B_1 \leq B_2 \leq \bar{B}$, then*

1. *For strategy profile (B_1, B_2) with $0 \leq B_1 < B_2 < \bar{B}$, the optimal block size of pool 2 is $B_2^* = \frac{1}{\lambda_1 \rho} - \frac{R}{\alpha}$, if $0 < \frac{1}{\lambda_1 \rho} - \frac{R}{\alpha} < \bar{B}$. Let $\widehat{B}_1^*$ be the solution satisfying $\frac{dU_1(B_1,B_2^*)}{dB_1} = 0$. If $0 \leq \widehat{B}_1^* < B_2^*$, then the optimal block size of pool 1 is $\widehat{B}_1^*$. Then the payoffs are*

$$\pi_{11}^1 = (R + \alpha \widehat{B}_1^*)[1 - h_2 e^{\lambda_1 \rho(\widehat{B}_1^* - B_2^*)}], \ \pi_{11}^2 = [R + \alpha B_2^*] h_2 e^{\lambda_1 \rho(\widehat{B}_1^* - B_2^*)}. \quad (15)$$

If $\widehat{B}_1^* < 0$, then the best choice of pool 1 is to set its block size as zero and the payoffs are

$$\pi_{11}^1 = R[1 - h_2 e^{-\lambda_1 \rho B_2^*}], \ \pi_{11}^2 = (R + \alpha B_2^*) h_2 e^{-\lambda_1 \rho B_2^*}. \quad (16)$$

2. *For strategy profile $(B_1, \bar{B})$ with $0 < B_1 < B_2 = \bar{B}$, the block size of pool 2 is $\bar{B}$. Let $\widetilde{B}_1^*$ be the solution satisfying $\frac{dU_1(B_1,\bar{B})}{dB_1} = 0$. If $0 \leq \widetilde{B}_1^* < \bar{B}$, then the optimal block size of pool 1 is $\widetilde{B}_1^*$. Then the payoffs are*

$$\pi_{12}^1 = (R + \alpha \widetilde{B}_1^*)[1 - h_2 e^{\lambda_1 \rho(\widetilde{B}_1^* - \bar{B})}], \ \pi_{12}^2 = (R + \alpha \bar{B}) h_2 e^{\lambda_1 \rho(\widetilde{B}_1^* - \bar{B})}. \quad (17)$$

If $\widetilde{B}_1^* < 0$, then the best choice of pool 1 is to set its block size as zero and the payoffs are

$$\pi_{12}^1 = R[1 - h_2 e^{-\lambda_1 \rho \bar{B}}],\ \pi_{12}^2 = (R + \alpha\bar{B})h_2 e^{-\lambda_1 \rho \bar{B}}. \tag{18}$$

3. For strategy profile $(\bar{B}, \bar{B})$, the block sizes of two pools are both equal to $\bar{B}$ and the corresponding payoffs are

$$\pi_{22}^1 = h_1(R + \alpha\bar{B}) \ \ and \ \ \pi_{22}^2 = h_2(R + \alpha\bar{B}). \tag{19}$$

Remark 1: By Lemma 1-(1) and (2), the optimal block size of pool 1 depends on the block size of pool 2 in strategy profiles $(\widehat{B}_1^*, B_2^*)$ and $(\widetilde{B}_1^*, \bar{B})$, and can be obtained from equations $\frac{dU_1(B_1, B_2^*)}{dB_1} = 0$ and $\frac{dU_1(B_1, \bar{B})}{dB_1} = 0$, if $0 \le \widehat{B}_1^* < B_2^*$ and $0 \le \widetilde{B}_1^* < \bar{B}$, respectively. For convenience, we denote $g(B_2)$ to be the implicit function, satisfying

$$\frac{dU_1(B_1, B_2)}{dB_1} = \alpha - [\alpha + \lambda_1\rho(R + \alpha g(B_2)]h_2 e^{\lambda_1\rho(g(B_2) - B_2)} = 0,$$

Therefore, $\widehat{B}_1^* = g(B_2^*) = g(\frac{1}{\lambda_1\rho} - \frac{R}{\alpha})$ and $\widetilde{B}_1^* = g(\bar{B})$.

Lemma 2. *Let $g(B_2)$ be the implicit function satisfying $\frac{dU_1(g(B_2), B_2)}{dB_1} = 0$. Then $g(B_2)$ is monotone increasing with B_2 and $g(B_2) < B_2$ for all $B_2 \ge 0$.*

We leave the proof of Lemma 2 in online full version [1]. Based on the monotonicity of $g(B_2)$, we have $\widehat{B}_1^* < \widetilde{B}_1^*$, if $\frac{1}{\lambda_1\rho} - \frac{R}{\alpha} < \bar{B}$. Moreover, the property of $g(B_2) \le B_2$ ensures $\widehat{B}_1^* < \frac{1}{\lambda_1\rho} - \frac{R}{\alpha}$ and $\widetilde{B}_1^* < \bar{B}$.

In the following, we would analyze the strategy selections of two mining pools by distinguishing two conditions: (1) $\frac{1}{\lambda_1\rho} - \frac{R}{\alpha} \ge \bar{B}$; (2) $0 \le \frac{1}{\lambda_1\rho} - \frac{R}{\alpha} < \bar{B}$; and (3) $\frac{1}{\lambda_1\rho} - \frac{R}{\alpha} < 0$, and then explore the equilibrium solutions in the evolutionary game in the following.

Theorem 1. *In a one-shot mining competition, if $0 \le B_1 \le B_2 \le \bar{B}$ and $\frac{1}{\lambda_1\rho} - \frac{R}{\alpha} \ge \bar{B}$, then*

1. $(\widetilde{B}_1^*, \bar{B})$ is a strict Nash equilibrium, if $0 \le \widetilde{B}_1^* < \bar{B}$; or
2. $(0, \bar{B})$ is a strict Nash equilibrium, if $\widetilde{B}_1^* < 0$; or
3. $(\bar{B}, \bar{B})$ is a strict Nash equilibrium, if $\widetilde{B}_1^* \ge \bar{B}$.

Theorem 1 illustrates that the dominant strategy of pool 2 is to follow the default size $\bar{B}$ when $\frac{1}{\lambda_1\rho} - \frac{R}{\alpha} \ge \bar{B}$, while the optimal strategy of pool 1 depends on the value of $\widetilde{B}_1^*$. The detailed proof is provided in online full version [1].

Next, we concentrate on the condition of $0 \le \frac{1}{\lambda_1\rho} - \frac{R}{\alpha} < \bar{B}$, under which pool 2 may set its block size as $B_2^* = \frac{1}{\lambda_1\rho} - \frac{R}{\alpha}$ or $\bar{B}$.

Recall that $x_i(k)$, $i = 1, 2$, is the probability of mining pool i to adopt the "following" strategy in the k-th slot competition, and thus $1 - x_i(k)$ is the probability of mining pool i to follow the default size. The expected payoffs of mining pool 1 in the k-th slot competition, when it chooses "not following" or "following" strategy, are

$$E_1^1(k) = x_2(k)\pi_{11}^1 + (1 - x_2(k))\pi_{12}^1; \quad E_1^2(k) = (1 - x_2(k))\pi_{22}^1.$$

The average payoff of mining pool 1 is

$$\overline{E}_1(k) = x_1(k)E_1^1(k) + (1 - x_1(k))E_1^2(k).$$

Similarly, we can derive the expected payoffs of mining pool 2 as follows,

$$E_2^1(k) = x_1(k)\pi_{11}^2; \quad E_2^2(k) = x_1(k)\pi_{12}^2 + (1 - x_1(k))\pi_{22}^2.$$

The average payoff of mining pool 2 is

$$\overline{E}_2(k) = x_2(k)E_2^1(k) + (1 - x_2(k))E_2^2(k).$$

Based on the analysis scheme (12) for the general case, the replicator dynamic system of pool 1 and 2 for the case of two mining pools are:

$$\begin{cases} f_1(\mathbf{x}) = \dot{x}_1(k) = x_1(1-x_1)(E_1^1 - \overline{E}_1) = x_1(1-x_1)\left[(\pi_{11}^1 - \pi_{12}^1 + \pi_{22}^1)x_2 + (\pi_{12}^1 - \pi_{22}^1)\right]; \\ f_2(\mathbf{x}) = \dot{x}_2(k) = x_2(1-x_2)(E_2^1 - \overline{E}_2) = x_2(1-x_2)\left[(\pi_{11}^2 - \pi_{12}^2 + \pi_{22}^2)x_1 - \pi_{22}^2\right]. \end{cases} \tag{20}$$

Note that all the solutions satisfying $f_1(\mathbf{x}) = \dot{x}_1(k) = 0$ and $f_2(\mathbf{x}) = \dot{x}_2(k) = 0$ are the fixed equilibrium points of the replicator dynamic system. It is not hard to see that there exist four fixed equilibrium points of this system under the condition of $B_1 \leq B_2 \leq \bar{B}$: $(0,0)$, $(1,0)$, $(1,1)$ and (x_1^*, x_2^*), where

$$x_1^* = \frac{\pi_{22}^2}{\pi_{11}^2 - \pi_{12}^2 + \pi_{22}^2}, \quad x_2^* = \frac{\pi_{22}^1 - \pi_{12}^1}{\pi_{11}^1 - \pi_{12}^1 + \pi_{22}^1}. \tag{21}$$

To fulfill the condition for probability vector $\mathbf{x}$, x_1^* and x_2^* must be in $[0, 1]$.

Theorem 2. *For the evolutionary game between two mining pools, if* $0 \leq B_1 \leq B_2 \leq \bar{B}$ *and* $0 \leq \frac{1}{\lambda_1 \rho} - \frac{R}{\alpha} < \bar{B}$, *then*

- $(1,0)$ *is an ESS, when (1)* $0 \leq \widehat{B}_1^* < \widetilde{B}_1^* < \bar{B}$ *or (2)* $\widehat{B}_1^* < 0 < \widetilde{B}_1^* < \bar{B}$ *and* $\lambda_1\rho(\frac{R}{\alpha} + \bar{B})e^{\lambda_1\rho(\widetilde{B}_1^* - \bar{B} - \frac{R}{\alpha})+1} > 1$;
- $(1,1)$ *is an ESS, when (1)* $\widehat{B}_1^* < \widetilde{B}_1^* < 0$ *or (2)* $\widehat{B}_1^* < 0 < \widetilde{B}_1^* < \bar{B}$ *and* $\lambda_1\rho(\frac{R}{\alpha} + \bar{B})e^{\lambda_1\rho(\widetilde{B}_1^* - \bar{B} - \frac{R}{\alpha})+1} < 1$;
- $(0,0)$ *and* (x_1^*, x_2^*) *cannot be ESSs.*

Proof. To obtain the ESS of the evolutionary game for block size selection, we first compute the Jacobian matrix of the replicator dynamic system (20),

$$J = \begin{bmatrix} \frac{\partial f_1(\mathbf{x})}{\partial x_1} & \frac{\partial f_1(\mathbf{x})}{\partial x_2} \\ \frac{\partial f_2(\mathbf{x})}{\partial x_1} & \frac{\partial f_2(\mathbf{x})}{\partial x_2} \end{bmatrix},$$

where

$$\frac{\partial f_1(\mathbf{x})}{\partial x_1} = (1-2x_1)\{[\pi_{11}^1 - (\pi_{12}^1 - \pi_{22}^1)]x_2 + (\pi_{12}^1 - \pi_{22}^1)\};$$
$$\frac{\partial f_1(\mathbf{x})}{\partial x_2} = x_1(1-x_1)[\pi_{11}^1 - (\pi_{12}^1 - \pi_{22}^1)];$$
$$\frac{\partial f_2(\mathbf{x})}{\partial x_1} = x_2(1-x_2)[(\pi_{11}^2 - \pi_{12}^2) + \pi_{22}^2];$$
$$\frac{\partial f_2(\mathbf{x})}{\partial x_2} = (1-2x_2)\{[(\pi_{11}^2 - \pi_{12}^2) + \pi_{22}^2)]x_1 - \pi_{22}^2\}.$$

Table 2. The determinants and traces of Jacobian matrix J at fixed equilibrium points.

	$Det(J)$	$Tr(J)$
$(0,0)$	$K \cdot (-M)$	$K - M$
$(1,0)$	$(-K) \cdot N$	$N - K$
$(1,1)$	$(-L) \cdot (-N)$	$-L - N$
(x_1^*, x_2^*)	$\frac{KLMN}{(L-K)(M+N)}$	0

By the results in [3], if a fixed equilibrium point (x_1, x_2) is an ESS, then the Jacobian matrix of the replicator dynamic system is negative definite at (x_1, x_2), equivalent to determinant $Det(J(x_1, x_2)) > 0$ and trace $Tr(J(x_1, x_2)) < 0$.

To simplify the discussion, let us denote $K = \pi_{12}^1 - \pi_{22}^1, L = \pi_{11}^1, M = \pi_{22}^2, N = \pi_{11}^2 - \pi_{12}^2$. Table 2 shows the determinants and the traces of Jacobian matrix at different fixed equilibrium points. Next we propose the fact, based on which it is easy for us obtain this theorem.

Fact 1. *Based on the expressions of* π_{11}^i, π_{12}^i *and* π_{22}^i, $i = 1, 2$, *in (15)–(19),*

$$K = \pi_{12}^1 - \pi_{22}^1 > 0,\ L = \pi_{11}^1 > 0,\ M = \pi_{22}^2 > 0,$$
$$N = \pi_{11}^2 - \pi_{12}^2$$
$$= \begin{cases} \left[R + \alpha(\frac{1}{\lambda_1\rho} - \frac{R}{\alpha})\right] h_2 e^{\lambda_1\rho(\widehat{B}_1^* - \frac{1}{\lambda_1\rho} + \frac{R}{\alpha})} - (R+\alpha\bar{B})h_2 e^{\lambda_1\rho(\widetilde{B}_1^* - \bar{B})} < 0, \\ \textit{if } 0 \le \widehat{B}_1^* < \widetilde{B}_1^* < \bar{B}; \\ \left[R + \alpha(\frac{1}{\lambda_1\rho} - \frac{R}{\alpha})\right] h_2 e^{\lambda_1\rho(-\frac{1}{\lambda_1\rho} + \frac{R}{\alpha})} - (R+\alpha\bar{B})h_2 e^{\lambda_1\rho(\widetilde{B}_1^* - \bar{B})} < 0 (> 0), \\ \textit{if } \widehat{B}_1^* < 0 < \widetilde{B}_1^* < \bar{B} \textit{ and } \lambda_1\rho(\frac{R}{\alpha} + \bar{B})e^{\lambda_1\rho(\widetilde{B}_1^* - \bar{B} - \frac{R}{\alpha})+1} > 1 (< 1); \\ \left[R + \alpha(\frac{1}{\lambda_1\rho} - \frac{R}{\alpha})\right] h_2 e^{\lambda_1\rho(-\frac{1}{\lambda_1\rho} + \frac{R}{\alpha})} - (R+\alpha\bar{B})h_2 e^{-\lambda_1\rho\bar{B}} > 0, \\ \textit{if } \widetilde{B}_1^* < \widetilde{B}_1^* \le 0. \end{cases}$$

Because $Tr(J(x_1^*, x_2^*)) = 0$ and $Det(J(0,0)) = -K \cdot M < 0$, (x_1^*, x_2^*) and $(0,0)$ cannot be ESS. Moreover, according to the results of Fact 1, we have $Det(J(1,0)) > 0$ and $Tr(J(1,0)) < 0$ when $0 \le \widetilde{B}_1^* < \bar{B}$, and then the fixed equilibrium points $(1,0)$ is an ESS. When $\widetilde{B}_1^* < 0$, $Det(J(1,1)) > 0$ and $Tr(J(1,1)) < 0$. So the fixed equilibrium points $(1,1)$ is an ESS. □

Corollary 1. *For the evolutionary game between two mining pools, if* $0 \le B_1 \le B_2 \le \bar{B}$ *and* $0 \le \frac{1}{\lambda_1 \rho} - \frac{R}{\alpha} < \bar{B}$, *then*

- $(\widetilde{B}_1^*, \bar{B})$ *is an evolutionary stable strategy profile, when* $0 \le \widehat{B}_1^* < \widetilde{B}_1^* < \bar{B}$, *or subject to* $\lambda_1 \rho(\frac{R}{\alpha} + \bar{B}) e^{\lambda_1 \rho(\widetilde{B}_1^* - \bar{B} - \frac{R}{\alpha})+1} > 1$ *when* $0 \le \widehat{B}_1^* < 0 < \widetilde{B}_1^* < \bar{B}$;
- $(0, \frac{1}{\lambda_1 \rho} - \frac{R}{\alpha})$ *is an evolutionary stable strategy profile, when* $\widehat{B}_1^* < \widetilde{B}_1^* < 0$, *or subject to* $\lambda_1 \rho(\frac{R}{\alpha} + \bar{B}) e^{\lambda_1 \rho(\widetilde{B}_1^* - \bar{B} - \frac{R}{\alpha})+1} < 1$ *when* $0 \le \widehat{B}_1^* < 0 < \widetilde{B}_1^* < \bar{B}$.

At last, let us discuss the case that $\frac{1}{\lambda_1 \rho} - \frac{R}{\alpha} < 0$.

Theorem 3. *If* $0 \le B_1 \le B_2 < \bar{B}$ *and* $\frac{1}{\lambda_1 \rho} - \frac{R}{\alpha} < 0$, *then*

- $(0, 0)$ *is a strict Nash equilibrium when* $\widetilde{B}_1^* \le 0$;
- $(\bar{B}, \bar{B})$ *is a strict Nash equilibrium when* $\widetilde{B}_1^* \ge \bar{B}$.
- *when* $0 < \widetilde{B}_1^* < \bar{B}$, *then*
  - $(0, 0)$ *is an evolutionary stable strategy profile if* $R > \frac{e^{\lambda_1 \rho(\widetilde{B}_1^* - \bar{B})}}{1 - e^{\lambda_1 \rho(\widetilde{B}_1^* - \bar{B})}} \alpha \bar{B}$;
  - $(\widetilde{B}_1^*, \bar{B})$ *is an evolutionary stable strategy profile if* $R < \frac{e^{\lambda_1 \rho(\widetilde{B}_1^* - \bar{B})}}{1 - e^{\lambda_1 \rho(\widetilde{B}_1^* - \bar{B})}} \alpha \bar{B}$.

From the condition of $\frac{1}{\lambda_1 \rho} - \frac{R}{\alpha} < 0$, we can see that the fixed subsidy R is quite high ($> \frac{\alpha}{\lambda_1 \rho}$). Under this condition, Theorem 3 states an possibility that neither of pools would like to choose transactions into their blocks. It means that the mining pools may give up the available transaction fees in hopes of enhancing their chances to win the high fixed subsidy. The detailed proof can be found in online full version [1].

4 Numerical Experiments and Conclusions

4.1 Numerical Experiments

In this section, we would analyze the case of two mining pools for the decision on block size. Based on the statistic data about Bitcoin blockchain on [15], we firstly consider the following setting to discuss the influence of default size on the strategy selection of mining pools: the unit transaction fee $\alpha = 1 \times 10^{-6}$ BTC, the propagation speed $\rho = 2.9 \times 10^{-4}$ s/Byte, the block subsidy $R = 6.25$ BTC and the average mining time $T = 600$ s, and three upper bounds of block size, i.e., $\bar{B} = 1$ MB, 2 MB and 3 MB, as well as the relative computing power of each pool is $h_i \in (0, 1)$ and $h_1 + h_2 = 1$. Let us set $h_1 = 0.3$ and $h_2 = 0.7$ and take 0.2, 0.5 and 0.8 respectively as the initial values of $x_i, (i = 1, 2)$. Under the setting, we get $\widehat{B}_1^* = -8.70 \times 10^5$, $\frac{T}{h_1 \rho} - \frac{R}{\alpha} = 6.47 \times 10^5$, and $\widetilde{B}_1^* = -6.43 \times 10^5, 6.95 \times 10^3, 6.69 \times 10^5$ corresponding to $\bar{B} = 1 \times 10^6, 2 \times 10^6, 3 \times 10^6$, respectively. These results satisfy the conditions of $\widehat{B}_1^* < \bar{B}, \widetilde{B}_1^* < \bar{B}$ and $\frac{T}{h_1 \rho} - \frac{R}{\alpha} < \bar{B}$. Figure 1 illustrates the evolution processes of behaviors of two pools with different default block sizes, verifying the result that $(1, 0)$ is an ESS in Theorem 2. It is clear that the speed of convergence to fixed equilibrium point $(1, 0)$, i.e., the strategy profile $(\widetilde{B}_1^*, \bar{B})$, becomes faster with the increasing of the maximum capacity of a block. Hence larger upper bound $\bar{B}$ brings pool 2 more payoff, stimulating the speed of convergence to the strategy of $\bar{B}$. At the same time, $g'(B_2) > 0$ (shown in Lemma 2) shows the optimal size of B_1 increases with B_2, and thus the rate of convergence to $\widehat{B}_1^*$ is accelerated by the increasing of $\bar{B}$.

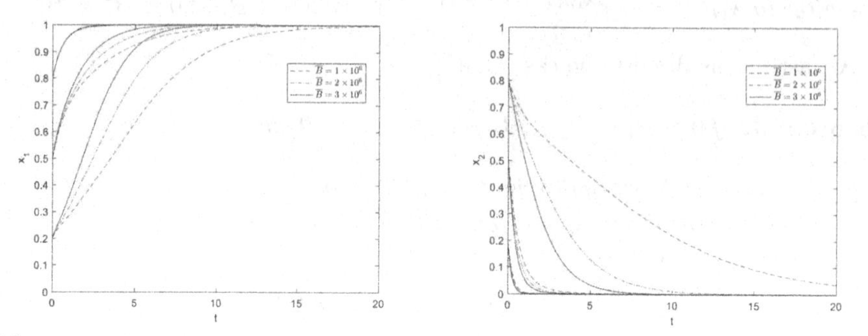

(a) The evolutionary behavior of pool 1. (b) The evolutionary behavior of pool 2.

Fig. 1. The impact of the upper bound of block size $\bar{B}$ on pool's behaviors.

Next, we change the value of α and ρ, i.e., $\{\alpha, \rho\} = \{0.8\times10^{-6}, 3\times10^{-4}\}$, and take $h_1 = 0.24$. Under this setting, $0 < B_2^* = \frac{1}{\lambda_1\rho} - \frac{R}{\alpha} < \bar{B}$ and $\widehat{B}_1^* < \widetilde{B}_1^* < 0$. Figure 2 illustrates the evolutionary behaviors of the two pools are shown, in which the ESS is $(1,1)$, i.e., $(0, \frac{1}{\lambda_1\rho} - \frac{R}{\alpha})$ is the evolutionary stable strategy. From Fig. 2(a), x_1 converges to 1 in a relatively short time, while x_2 takes much longer time to converge to 1 shown in Fig. 2(b). In addition, when $x_2(0) \leq 0.5$, pool 2 has a strong tendency to take $B_2 = \bar{B}$ initially, considering the small mining rate of pool 1 and its own large winning probability. However, as time

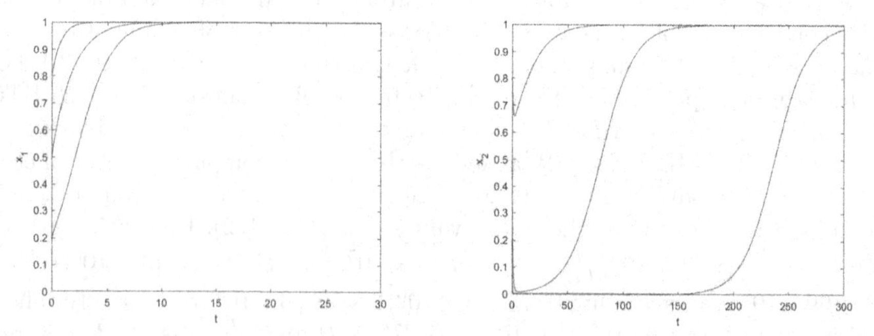

(a) The evolutionary behavior of pool 1. (b) The evolutionary behavior of pool 2.

Fig. 2. The evolution of two pools when (1,1) is ESS.

goes by, pool 2 tends to realize the best response is just $B_2^* = \frac{1}{\lambda_1 \rho} - \frac{R}{\alpha}$. This explains the transition from $x_2 = 0 \dashrightarrow x_2 = 1$ when $x_2(0) \leq 0.5$. Hence, if $\widehat{B}_1^* < \widetilde{B}_1^* < 0$, the best response of pool 1 is $B_1 = 0$, meaning a block without any transactions just for faster propagation process.

4.2 Conclusion

In this paper, the issue of selecting appropriate block sizes by mining pools in a blockchain system is discussed. We model this block size determination problem as an evolutionary game, in which each pool may follow the upper bound of block size, i.e., the default size $\bar{B}$, or not. In addition, if a mining pool chooses not to follow $\bar{B}$, then it shall continue to decide its optimal block size under different strategy profile. In our evolutionary game model, all mining pools are supposed to be bounded rational and each pool switches the low-payoff strategy to a higher one on and on by learning others' better strategies, until the whole network reaches an evolutionary stable state (ESS). The theoretical analysis has been done, particularly for a case of two mining pools, we prove the existence of different ESS under different conditions. In addition to verify the results in our work, several numerical experiments by using real Bitcoin data are conducted to show the evolutionary decisions of mining pools.

Acknowledgments. This work is partially supported by the National Nature Science Foundation of China (No. 11871366), Qing Lan Project of Jiangsu Province, the Research Innovation Program for College Graduate Students of Jiangsu Province (No. KYCX21-2998 and KYCX20-2790), and the project of the philosophy & social sciences of higher education in Jiangsu province. (No. 2018SJA1347).

References

1. Chen, J., Cheng, Y., Xu, Z., Cao, Y.: Evolutionary equilibrium analysis for decision on block size in blockchain systems (full version) (2021). https://arxiv.org/abs/2110.09765
2. Easley, D., OHara, M., Basu, S.: From mining to markets: the evolution of bitcoin transaction fees. J. Financ. Econ. **134**, 91–109 (2019)
3. Friedman, D.: On economic applications of evolutionary game theory. J. Evol. Econ. **8**(1), 15–43 (1998)
4. Garay, J., Kiayias, A., Leonardos, N.: The bitcoin backbone protocol: analysis and applications. In: Oswald, E., Fischlin, M. (eds.) EUROCRYPT 2015. LNCS, vol. 9057, pp. 281–310. Springer, Heidelberg (2015). https://doi.org/10.1007/978-3-662-46803-6_10
5. Gervais, A., Karame, G.O., Wüst, K., Glykantzis, V., Ritzdorf, H., Capkun, S.: On the security and performance of proof of work blockchains. In: Proceedings of the 2016 ACM SIGSAC Conference on Computer and Communications Security, CCS 2016, p. 3C16. Association for Computing Machinery (2016)
6. Houy, N.: The bitcoin mining game. Lcloud/fogr J. **1**(13), 53–68 (2016)
7. Jiang, S., Wu, J.: Bitcoin mining with transaction fees: a game on the block size. In: 2019 IEEE International Conference on Blockchain (Blockchain), pp. 107–115 (2019)

8. Li, J., Kendall, G., John, R.: Computing nash equilibria and evolutionarily stable states of evolutionary games. IEEE Trans. Evol. Comput. **20**(3), 460C469 (2016)
9. Liu, X., Wang, W., Niyato, D., Zhao, N., Wang, P.: Evolutionary game for mining pool selection in blockchain networks. IEEE Wirel. Commun. Lett. **7**(5), 760–763 (2018)
10. Lombrozo, E., Lau, J., Wuille, P.: Segregated witness (consensus layer) (2015). https://github.com/bitcoin/bips/wiki/Comments:BIP-0141
11. Nakamoto, S.: Bitcoin: a peer-to-peer electronic cash system (2008). http://bitcoin.org
12. NervosFans: Bitcoin transaction fee rules. https://zhuanlan.zhihu.com/p/38479785
13. Pan, D., Zhao, J.L., Fan, S., Zhang, Z.: Dividend or no dividend in delegated blockchain governance: a game theoretic analysis. J. Syst. Sci. Syst. Eng. **30**, 1861–9576 (2021)
14. Rizun, P.R.: A transaction fee market exists without a block size limit, block Size Limit Debate Working Paper (2015)
15. TOKENVIEW: Bitcoin browser. https://btc.tokenview.com/
16. Zhang, R., Preneel, B.: On the necessity of a prescribed block validity consensus: analyzing bitcoin unlimited mining protocol. In: Proceedings of the 13th International Conference on Emerging Networking Experiments and Technologies, p. 108C119. Association for Computing Machinery, New York (2017)

Efficient Algorithms for Scheduling Parallel Jobs with Interval Constraints in Clouds

Xuanming Xu[1] and Longkun Guo[2(✉)]

[1] College of Mathematics and Statistics, Fuzhou University, Fuzhou, China
[2] College of Computer and Data Science, Fuzhou University, Fuzhou, China

Abstract. Given a set of time slots and jobs, each of which has a degree of parallelism, an interval constraint (typically an arrival time and a deadline), and a processing time, we consider the problem of minimizing the number of machines for scheduling all the jobs while satisfying the constraints. This paper starts with a Linear Programming (LP) formulation and argues that its decision version is with an integral polyhedron. Meanwhile, we determine the problem's feasibility with the LP, whether the given jobs can be accommodated by a given number of machines over allocated time slots, based on the property of the integrality. Then, we show that the feasibility leads to an LP-based algorithm, in which the optimal solution can be obtained in polynomial time. Moreover, our algorithm can also optimally compute a schedule to the problem rather than merely determining the minimum number of required machines, which compares favorably to the Earliest Deadline First (EDF) algorithm and the Least Laxity First (LLF) algorithm in solution quality.

Keywords: Machine minimization · Integral polyhedron · Optimality · Parallel processing

1 Introduction

Applications in the cloud imposed the problem of minimizing the number of machines for executing a given batch of parallelable jobs. It was known NP-hard to determine whether a given set of jobs without parallelism can be feasibly scheduled in two machines [9,15]. In general, the scheduling problem can be divided into two categories depending on whether the jobs are preemptive or not [10]. With preemption, any process can be preempted at any time and resumed later on the same or a different machine. The non-preemptive scheduling problem has attracted many research interests in relevant fields. In contrast, there exist fewer results in the literature for the preemptive version due to the overhead of interrupting and resuming jobs. With the development of task management technology, the overhead of preemption drops dramatically. Hence, many applications in clouds require scheduling preemptive jobs. This brings our problem formally defined as follows:

D.-Z. Du et al. (Eds.): COCOA 2021, LNCS 13135, pp. 195–202, 2021.
https://doi.org/10.1007/978-3-030-92681-6_16

Definition 1. *(Machine minimization for Parallelable Interval Constrained Jobs (MPI-CJ)) Let T and J are a set of time slots and a set of jobs, respectively. Each job's summary (a_j, d_j, W_j, p_j) is given in advance, where a_j is the arrival time, d_j is the deadline, W_j is the workload and p_j is the parallelism degree. Each job must be executed preemptively in one of the given machines within time interval $T_j = (a_j, d_j]$, and each machine can only process one job at a time slot. The problem is to minimize the total number of machines used and process every job successfully in its time interval.*

To the best of our knowledge, Phillips *et al.* [17] were the first to significantly address the online job scheduling problem with machine minimization in preemptive setting. To determine a feasible preemptive schedule on a minimum number of machines, they presented the Least Laxity First (LLF) algorithm with competitive ratio $\mathcal{O}\left(\log \frac{t_{\max}}{t_{\min}}\right)$, where $t_{\max}$ and $t_{\min}$ are the maximum and minimum processing times of jobs, respectively. However, it remained a considerable gap to the best known lower bound on the competitive ratio of $\frac{5}{4}$. Since then, no significant improvement had been proposed until nearly two decades later when Chen *et al.* [4] effectively improved the competitive ratio to $\mathcal{O}(\log m)$, where m is the minimum number of machines in the offline setting. The result depends only on the optimum value m, instead of other input parameters. Based on the algorithm and analysis [4], the result was improved to $\mathcal{O}\left(\frac{\log m}{\log \log m}\right)$ by Azar *et al.* [1] and to $\mathcal{O}(\log \log m)$ by Im *et al.* [12]. In contrast, for the offline setting, the optimal solution can be obtained in polynomial time without parallelism degree by rounding the solution of an LP through distributing fractionally assigned workload in a round-robin fashion to all machines within a time slot [11].

The non-preemptive machine minimization problem is considerably harder than the preemptive version. In the offline setting, Raghavan and Tompson [18] used a standard LP formulation and randomized rounding to obtain an approximation factor of $\mathcal{O}\left(\frac{\log n}{\log \log n}\right)$, where n is the number of jobs. Using a more sophisticated LP formulation and rounding procedure, the result was improved by Chuzhoy *et al.* [5] to $\mathcal{O}\left(\sqrt{\frac{\log n}{\log \log n}}\right)$. Combining with three parameters: the number of machines m, the looseness λ, and the slackness σ, Van *et al.* [2] considered interval constrained scheduling on a minimum number of machines, and they refined known results of Cieliebak *et al.* [6] by using tools of parameterized complexity analysis. In the online setting, Saha [19] proposed an algorithm with the competitive ratio of $\mathcal{O}\left(\log \frac{t_{\max}}{t_{\min}}\right)$, which has been the best result so far. Since then, more researches have been focused on special cases of the problem. With uniform deadlines (but arbitrary job lengths), Devanur *et al.* [8] gave a constant-competitive online algorithm for the case with an upper bound of 16. With unit processing lengths, Kao *et al.* [13] proposed an algorithm that produces an upper bound of 5.2 and a lower bound of 2.09. To investigate how the lookahead ability improves the performance of online algorithms, Chen *et al.* [3] proposed the competitive online algorithm for the case with uniform processing times and a common deadline in two lookahead models.

2 Polyhedron and LP-Based Algorithm for MPI-CJ

In this section, we propose an LP relaxation for the MPI-CJ problem as well as for its decision version of determining whether there exists feasible scheduling regarding a given number of machines. Then we show the linear program of decision MPI-CJ is with an integral polyhedron, which means there exist integer basic optimum solutions for the linear program. Moreover, we devise an LP-based algorithm to get the minimum number of machines through incorporating binary search with the LP based on its integrality. Besides, the output schedule of our algorithm outperforms the Earlier Deadline First (EDF) algorithm and the Least Laxity First (LLF) algorithm via giving examples.

2.1 Polyhedron of MPI-CJ

Let $x_j(t)$ indicate the number of processing units of the tth time slot assigned to job j. Then the LP relaxation for the MPI-CJ problem is as in LP (1):

$$\begin{aligned} \min \quad & \gamma \\ s.t. \quad & \sum_{t\in T_j} x_j(t) \geq W_j \quad \forall j \in J & (1)\\ & \sum_{j:t\in T_j} x_j(t) - \gamma \leq 0 \quad \forall t \in \bigcup_{j\in J} T_j & (2)\\ & 0 \leq x_j(t) \leq p_j \quad \forall j \in J, t \in T_j & (3)\\ & \gamma \geq 0, \end{aligned}$$

where Constraint (1) guarantees that the processing units assigned to job j are no fewer than the requirement of its workload, Constraint (2) ensures that the number of machines γ is sufficient for the requirement in time slot t, and Constraint (3) requires that each job satisfies the parallelism degree bound.

One straightforward idea is to identify whether LP (1) is integral, i.e., whether a basic solution of LP (1) is a vector of integers. Unfortunately, there exist fractions in the basic solution of LP (1) as depicted in the following: Given 4 time slots and 5 jobs, each of which is with arrival time $a_j = 0$, deadline $d_j = 4$, workload $W_j = 2$ and parallelism degree $p_j = 1$. Through solving LP (1), we get $\gamma = 2.5$ as the minimum number of machines. Thus, the variables of a basic optimal solution of LP (1) are with fractional values, so LP (1) is not integral. Thus, we have the negative result as below:

Proposition 1. *LP (1) is not integral as it admits a fractional basic solution.*

So we instead investigate the following polyhedron for the decision version LP (2) of MPI-CJ:

$$-\sum_{t\in T_j} x_j(t) \le -W_j \quad \forall j \in J \tag{4}$$

$$\sum_{j:t\in T_j} x_j(t) \le C \quad \forall t \in \bigcup_{j\in J} T_j \tag{5}$$

$$x_j(t) \le p_j \quad \forall j \in J, t \in T_j \tag{6}$$

$$x_j(t) \ge 0 \quad \forall j \in J, t \in T_j, \tag{7}$$

where C is a fixed number, and the polyhedron is the set of feasible solutions concerning C. Different from LP (1), we can show the vertex solutions (or namely basic solutions) of the LP polyhedron are integral as in the theorem below:

Theorem 1. *The polyhedron of LP (2) is integral. That is, each $x_j(t)$ in any basic solution of LP (2) is an integer.*

The proof of the above theorem starts from the following famous property:

Lemma 1 [16]**.** *Let $A \in \mathbb{R}^{m\times n}$ be an integer matrix. Any basic optimum solution to this linear inequality equations LP (3)*

$$\begin{cases} Ax \le b \\ x \ge 0 \end{cases} \tag{8}$$

is an integer vector if A is a Totally Unimodular Matrix (TUM) and b is an integer vector, where the integer matrix A is a TUM if the determinant of every square submatrix of A is 0 or 1 or -1.

Let matrix B be of the form $\begin{pmatrix} A \\ I \end{pmatrix}$, where $A = (a_{ij})$ from the Constraint (4) and (5), and I comes from the upper bound Constraint (6). Then by Lemma 1, we only need to show the correctness of the following lemma to finish the proof:

Lemma 2. *The matrix B, the coefficient matrix of LP (2), is a totally unimodular matrix.*

The proof of the above lemma is omitted due to the length limitation.

2.2 LP-Based Algorithm for MPI-CJ

Following Theorem 1, for any given instance of MPI-CJ and any fixed integer C, we can determine its feasibility by verifying whether the polyhedron of LP (2) is an empty set. Therefore, we can immediately obtain an algorithm for MPI-CJ by employing the binary search against a range containing the minimum machine number and incorporating LP (2) to verify the feasibility. The range is initially $[L, R]$, where $L = 1$ and $R = |J| \cdot \max_{j\in J}\{p_j\} = n \cdot \max_{j\in J}\{p_j\}$ are the lower and upper bound of the minimum machine number, respectively. The binary search

Algorithm 1. LP-based Binary Search Algorithm for MPI-CJ

Input: A set of time slots T, a set of jobs J ($|J| = n$), in which each job $j \in J$ is with summary (a_j, d_j, W_j, p_j);
Output: The minimum number of required parallel machines $C \in \mathbb{Z}^+$.
1: Set $R = n \cdot \max_{j \in J} \{p_j\}$ and $L = 1$;
2: **While** $R - L > 1$ **do**
3: Set $C = \lfloor \frac{R+L}{2} \rfloor$;
4: **If** LP (2) is feasible wrt C **then**
5: Set $R = C$;
6: **Else** set $L = C$;
7: **Endwhile**
8: Return $C = R$.

reiterates the process of decreasing the size of the range until $R - L \leq 1$ holds. The layout is formally shown in Algorithm 1.

For the correctness of Algorithm 1, following Theorem 1, we can determine whether the instance of decision MPI-CJ is feasible with respect to C by solving LP (2). On the other hand, binary search guarantees to find the minimum C under which decision MPI-CJ is feasible. Then from the fact that LP (2) is polynomially solvable [14], we have:

Corollary 1. *MPI-CJ is optimally solvable in polynomial time.*

2.3 Performance Comparison of Our Algorithm with EDF and LLF

Earlier Deadline First (EDF) and Least Laxity First (LLF) [7] are two well-studied scheduling algorithms in the context of the machine minimization problems. EDF algorithm executes at any time the job whose deadline is the closest, that is, the released job with the earliest absolute deadline has the highest priority. LLF algorithm executes the job with the smallest laxity at any time, i.e., the released job with the smallest laxity is assigned the highest priority. Our Algorithm 1 is to find the minimal number of machines required to finish all jobs in their time intervals T_j and provide a solution for scheduling. Then, we show that our algorithm can schedule jobs on the minimal number of machines over allocated time slots, but EDF and LLF can not in two examples as below:

Example 1. EDF, using 5 machines, can not schedule the input instance which can be scheduled on 5 machines in the given 6 time slots. Consider the following input instance I_1 consisting of 6 jobs all with the arrival times 0: The first 5 jobs $\{j_1, j_2, \cdots, j_5\}$ have workloads 4, parallelism degrees 4 and deadlines 5; The 6th job j_6 has workload 5, parallelism degree 1, and deadline 6. EDF, the fractional and integral schedule for I_1 are shown in Fig. 1.

As illustrated in Fig. 1a, the instance I_1 of MPI-CJ can not be successfully scheduled in 6 time slots over 5 machines by the EDF algorithm. According to

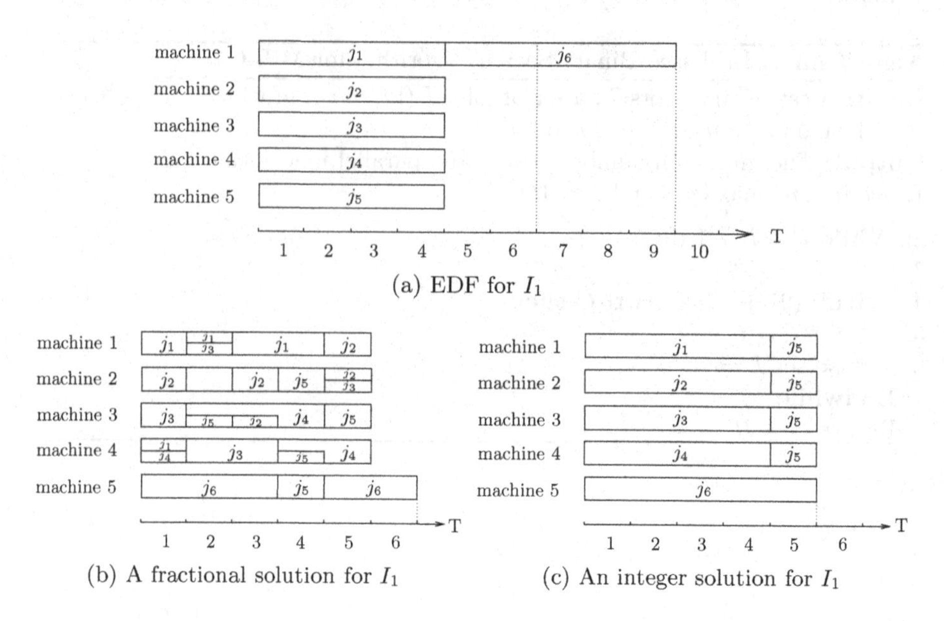

(a) EDF for I_1

(b) A fractional solution for I_1

(c) An integer solution for I_1

Fig. 1. The instance I_1 of MPI-CJ in $T = \{1, 2, \cdots, 6\}$, $C = 5$

the criterion that the job with the earliest deadline has the highest priority, the first five jobs $j_1, j_2, \cdots, j_5$ must be processed by the 4th time slot, and job j_6 can only start processing at the 5th time slot, which means it can not be processed within the given time slot due to its parallelism degree is 1 and workload is 5. As illustrated in Fig. 1b, all jobs are successfully completed before their deadlines in the given time slot and machines. In particular, there exist fractions in the solution. Moreover, Fig. 1c shows an integer solution where all jobs are finished by the 5th time slot, and it is the optimal solution among these three solutions. As shown in Fig. 1b and 1c, our algorithm can finish the instance I_1 of MPI-CJ. Therefore, our algorithm is better than EDF in this instance.

Example 2. LLF, using 4 machines, can not schedule the input instance which can be scheduled on 4 machines in the given 10 time slots. Consider the following input instance I_2: There are 2 workload-8 jobs $\{j_1, j_2\}$ with arrival times 0, deadlines 12, parallelism degree 1, and thus initial laxities of 4. Meanwhile, for $0 \leq i \leq 2$, there are 4 workloads-1 jobs with arrival times $2i$, deadlines $2(i+1)$, parallelism degrees 1, and thus laxities of 1. Altogether, 12 workloads-1 jobs $\{j_3, j_4, \cdots, j_{14}\}$ are arrived from timestamp 0 to timestamp 4. LLF, the fractional and integral schedule for I_2 are depicted in Fig. 2.

As illustrated in Fig. 2a, the LLF algorithm can not successfully finish the instance I_2 of MPI-CJ in 4 machines over 10 time slots. In LLF, the job with the smallest laxity is assigned the highest priority. Jobs j_1 and j_2 are processed completely in the 11th time slot due to the parallelism are 1, and the workload is 8. As illustrated in Fig. 2b, the fractional solution led to a successful schedule for the instance I_2 of MPI-CJ and finished all the jobs at the 10th time slot.

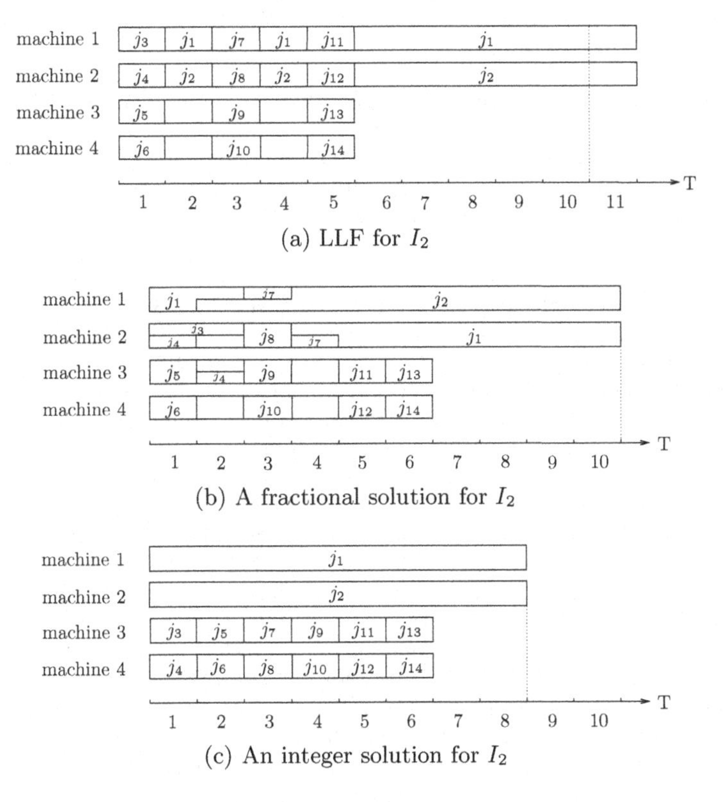

(a) LLF for I_2

(b) A fractional solution for I_2

(c) An integer solution for I_2

Fig. 2. The instance I_2 of MPI-CJ in $T = \{1, 2, \cdots, 10\}$, $C = 4$

In contrast, Fig. 2c shows the integer solution that completes all jobs by the 8th time slot. Apparently, the last solution outperforms the other two solutions, demonstrating that our algorithm is more efficient than LLF in the instance.

3 Conclusion

In this paper, we proved that the problem of Machine minimization against Parallel Interval Constrained Jobs (MPI-CJ) is polynomially solvable. The proof is based on the fact that the relaxing LP is with an integral polyhedron, which also leads to an LP-based polynomial algorithm. Then, we showed that our algorithm produces optimal solutions while the Earliest Deadline First (EDF) algorithm and the Least Laxity First (LLF) algorithm only produce approximation solutions.

Acknowledgements. The authors are supported by Natural Science Foundation of China (No. 61772005), Outstanding Youth Innovation Team Project for Universities of Shandong Province (No. 2020KJN008), Natural Science Foundation of Fujian Province (No. 2020J01845) and Educational Research Project for Young and Middle-aged Teachers of Fujian Provincial Department of Education (No. JAT190613).

References

1. Azar, Y., Cohen, S.: An improved algorithm for online machine minimization. Oper. Res. Lett. **46**(1), 128–133 (2018)
2. van Bevern, R., Niedermeier, R., Suchý, O.: A parameterized complexity view on non-preemptively scheduling interval-constrained jobs: few machines, small looseness, and small slack. J. Scheduling **20**(3), 255–265 (2016). https://doi.org/10.1007/s10951-016-0478-9
3. Chen, C., Zhang, H., Xu, Y.: Online machine minimization with lookahead. J. Comb. Optim. (2020). https://doi.org/10.1007/s10878-020-00633-w
4. Chen, L., Megow, N., Schewior, K.: An $\mathcal{O}$(log m)-competitive algorithm for online machine minimization. SIAM J. Comput. **47**(6), 2057–2077 (2018)
5. Chuzhoy, J., Guha, S., Khanna, S., Naor, J.S.: Machine minimization for scheduling jobs with interval constraints. In: 45th Annual IEEE Symposium on Foundations of Computer Science, pp. 81–90. IEEE (2004)
6. Cieliebak, M., Erlebach, T., Hennecke, F., Weber, B., Widmayer, P.: Scheduling with release times and deadlines on a minimum number of machines. In: Levy, J.-J., Mayr, E.W., Mitchell, J.C. (eds.) TCS 2004. IIFIP, vol. 155, pp. 209–222. Springer, Boston, (2004). https://doi.org/10.1007/1-4020-8141-3_18
7. Dertouzos, M.L., Mok, A.K.: Multiprocessor online scheduling of hard-real-time tasks. IEEE Trans. Softw. Eng. **15**(12), 1497–1506 (1989)
8. Devanur, N., Makarychev, K., Panigrahi, D., Yaroslavtsev, G.: Online algorithms for machine minimization. CoRR abs/1403.0486 (2014)
9. Garey, M.R., Johnson, D.S.: Two-processor scheduling with start-times and deadlines. SIAM J. Comput. **6**(3), 416–426 (1977)
10. George, L., Rivierre, N., Spuri, M.: Preemptive and non-preemptive real-time uniprocessor scheduling. In: Ph.D. thesis, Inria (1996)
11. Horn, W.: Some simple scheduling algorithms. Naval Res. Logistics Q. **21**(1), 177–185 (1974)
12. Im, S., Moseley, B., Pruhs, K., Stein, C.: An $\mathcal{O}$(log log m)-competitive algorithm for online machine minimization. In: 2017 IEEE Real-Time Systems Symposium (RTSS), pp. 343–350. IEEE (2017)
13. Kao, M.-J., Chen, J.-J., Rutter, I., Wagner, D.: Competitive design and analysis for machine-minimizing job scheduling problem. In: Chao, K.-M., Hsu, T., Lee, D.-T. (eds.) ISAAC 2012. LNCS, vol. 7676, pp. 75–84. Springer, Heidelberg (2012). https://doi.org/10.1007/978-3-642-35261-4_11
14. Korte, B., Vygen, J., Korte, B., Vygen, J.: Combinatorial optimization, vol. 2. Springer, Berlin (2012)
15. Lewis, H.R.: Computers and intractability: a guide to the theory of np-completeness Siam Rev. **24**(1), 90 (1983)
16. Papadimitriou, C.H., Steiglitz, K.: Combinatorial Optimization: Algorithms and Complexity. Courier Corporation, New York (1998)
17. Phillips, C.A., Stein, C., Torng, E., Wein, J.: Optimal time-critical scheduling via resource augmentation. In: Proceedings of the Twenty-ninth Annual ACM Symposium on Theory of Computing, pp. 140–149 (1997)
18. Raghavan, P., Tompson, C.D.: Randomized rounding: a technique for provably good algorithms and algorithmic proofs. Combinatorica **7**(4), 365–374 (1987)
19. Saha, B.: Renting a cloud. In: IARCS Annual Conference on Foundations of Software Technology and Theoretical Computer Science (FSTTCS 2013), vol. 24, pp. 437–448. Schloss Dagstuhl-Leibniz-Zentrum fuer Informatik (2013)

Two-Stage Stochastic Max-Weight Independent Set Problems

Min Li, Qian Liu, and Yang Zhou(✉)

School of Mathematics and Statistics, Shandong Normal University, Jinan 250014, People's Republic of China
{liminEmily,zhouyang}@sdnu.edu.cn

Abstract. The two-stage stochastic maximum-weight independent set problem extends the classical independent set problem. Given an independent system associated with one deterministic weight function and a random weight function both defined over the same ground set, the problem is to select two nonoverlapping independent subsets, one in the first stage and the other in the second stage, whose union has the maximum total expected weight. In this paper, we show that this problem can be formulated as a submodular function maximization subject to a matroid constraint if the independent system is a matroid. Furthermore, we also show that the two-stage stochastic maximum-weight knapsack independent set problem is neither submodular nor supermodular maximization problem by designing a counterexample.

Keywords: Submodular · Matroid · Knapsack · Two-stage stochastic programming

1 Introduction

The maximization problem for independence systems is one of the most influential problems in combinatorial optimization: given an independence system $\mathcal{M} = (V, \mathcal{I})$, and a weight function $h : V \to \mathbb{R}_+$, the objective is to find an independent set $S \in \mathcal{I}$ maximizing

$$h(S) := \sum_{i \in S} h_i.$$

There is a so-called Best-In-Greedy algorithm to solve this problem approximately [13]. Especially, an independence system is a matroid if and only if the Best-In-Greedy finds an optimum solution for this maximization problem [7].

The main focus of our work is to extend this single-stage problem to a two-stage problem, where the first stage weight h is known, but the second stage weight $u(\omega)$ is unknown, whose value will be realized at stage 2. Assume that $|V| = n$, then $u(\omega) = (u_1(\omega), u_2(\omega), \ldots, u_n(\omega))$ is an n-dimensional random vector defined on a

This paper is supported by National Science Foundation of China (No. 12001335) and Natural Science Foundation of Shandong Province (Nos. ZR2019PA004, ZR2020MA029) of China.

D.-Z. Du et al. (Eds.): COCOA 2021, LNCS 13135, pp. 203–213, 2021.
https://doi.org/10.1007/978-3-030-92681-6_17

probability space $(\Omega, \mathcal{F}, P)$. We need to select a subset $S \subseteq V$ at stage 1 and another nonoverlapping subset $T \subseteq V \backslash S$ at stage 2 such that $S \cup T \in \mathcal{I}$ maximizes

$$f(S) = h(S) + \mathbb{E}\left[Q_{u(\omega)}(S)\right], \tag{1}$$

where $Q_{u(\omega)}(S) = \max_{T \subseteq V \backslash S, T \cup S \in \mathcal{I}} \sum_{i \in T} u_i(\omega)$. We will mainly study the case that the independence system is a matroid or a knapsack constraint from the view of submodularity. In fact, He et al. have shown that this problem is NP-hard even if the independence system $\mathcal{M}$ is a uniform matroid, which is also known as sell or hold problem [11]. The motivation of our work comes from the hot study and extensive application of two-stage stochastic combinatorial optimization problems and submodular maximization problems.

Stochasticity is opposed to determinacy and means that some data are random, also the aim of stochastic programming is precisely to find an optimal decision in problems where some of the data are unknown or uncertain [6]. When random parameters are introduced, there are many two-stage stochastic combinatorial optimization problems presented, which are usually NP-hard even if the original deterministic decision problems are easy. For example, the maximum-weight matching problem is polynomially solvable, but the two-stage stochastic maximum-weight matching problem is shown to be NP-hard and a factor $\frac{1}{2}$-approximation algorithm is designed [12]. The study of 2-stage minimum spanning tree is also introduced in [8]. The Steiner tree (network) problem as well as the vertex cover problem has been studied and the algorithms over two (or multiple) stages of these problems have been presented [9,10]. Moreover, the two-stage knapsack problem with random weights is studied under the aspect of approximability [14].

Submodularity is a very important property in many fields, including combinatorial optimization, machine learning, economics, etc. Submodular maximization problems are normally hard from a computational point of view. Therefore there are extant literatures focusing on designing approximation algorithms. There is a $(1-\frac{1}{e})$-approximation algorithm for monotone submodular maximization with a cardinality constraint [16]. Furthermore, by applying adaptive-sampling algorithm, Balkanski et al. present a novel approach that yields an approximation algorithm with $(1 - \frac{1}{e} - \epsilon)$-approximation and $O(\frac{\log n}{\epsilon^2})$-adaptivity [1]. For monotone submodular maximization problem with a matroid constraint, the best deterministic approximation ratio is 0.5008 [3] and the randomized performance guarantee is $(1 - \frac{1}{e})$ [4]. And by using adaptive-sampling algorithm, the performance guarantee can be obtained as $(1 - \frac{1}{e} - \epsilon)$ with $O(\frac{\log^2 n}{\epsilon^3})$-adaptivity [5]. When the objective function is non-monotone, the best deterministic approximation ratio is $(\frac{1}{4} - \epsilon)$ [15] and the randomized approximation factor is 0.385 [2]. For monotone submodular maximization problem with a knapsack constraint, the best approximation ratio is $(1 - \frac{1}{e})$ [17]. If the objective function is non-monotone, both the best deterministic and randomized approximation ratios are $\frac{1}{2}$ [15].

In this paper, we continue revealing the good properties of the two-stage stochastic maximum-weight independent problems in view of submodularity, and give the following contributions:

- The instance of two-stage stochastic maximum-weight matroid problem can be formulated as a submodular maximization problem with a matroid constraint, where the commutative theory of bases and the good properties of circuits in matroids are mainly used.
- A $\frac{1}{2}$-approximation algorithm is designed for discrete two-stage stochastic maximum-weight matroid problem.
- The instance of two-stage stochastic maximum-weight knapsack problem cannot be formulated as a submodular (or supermodular) maximization problem by showing a counterexample.

The rest of this paper is organized as follows. In Sect. 2, we introduce some basic notations and results. In Sect. 3, we formalize the two-stage maximum-weight matroid independent set problem and present our main result about the two-stage maximum-weight matroid independent set problem and its analysis. And we present a counterexample to show that the two-stage maximum-weight problem with a knapsack constraint is neither a submodular function nor a supermodular function. In this part, we also design a $\frac{1}{2}$-approximation algorithm for the discrete two-stage maximum-weight matroid independent set problem.

2 Preliminaries

Assume that $V = \{1, 2, \ldots, n\}$ is a finite set and let $\mathcal{I}$ be a family of subsets of V, we will firstly present some definitions and results about matroids as well as submodular functions. For more information on matroid theory, one can refer to [13].

Definition 1. *$\mathcal{M} = (V, \mathcal{I})$ is an independence system if*

1. $\emptyset \in \mathcal{I}$,

2. for any subset $A \in \mathcal{I}$, if $B \subseteq A$, we have $B \in \mathcal{I}$.

If $\mathcal{M} = (V, \mathcal{I})$ is an independence system, we call each element of $\mathcal{I}$ as an independent set, and the subsets of V but not included by $\mathcal{I}$ as dependent sets. The minimal dependent sets are called circuits and the maximal independent sets are called bases. For $S \subseteq V$, the maximal independent subsets of S are called bases of S. Moreover, for a basis B of S, B adding any element $x \in S$ (if there are more elements) becomes a circuit, which will be denoted by $C(B, x)$.

Definition 2. *Given an independence system $\mathcal{M} = (V, \mathcal{I})$, the following function $R : 2^V \to \mathbb{R}$ is called the rank function of $\mathcal{M}$*

$$R(A) = \max\{|I| : I \subseteq A, I \in \mathcal{I}\}, \forall A \subseteq V.$$

Definition 3. *An independence system $(V, \mathcal{I})$ is a matroid if for all $A, B \in \mathcal{I}$ with $|B| = |A| + 1$, there exists one element $e \in B \setminus A$ such that $A \cup \{e\} \in \mathcal{I}$.*

Let $(V, \mathcal{I})$ be a matroid and $p : V \to \mathbb{R}_+$, we have known that the Best-In-Greedy algorithm can return one optimal solution of the following problem:

$$\max\Big\{\sum_{i \in S} p_i : S \in \mathcal{I}\Big\}.$$

Moreover, there is the following result presented in [13].

Lemma 1. *Let $(V,\mathcal{I})$ be a matroid and $p : V \to \mathbb{R}_+$, $k \in \mathbb{N}$ and $X \in \mathcal{I}$ with $|X| = k$. Then $p(X) = \max\{\sum_{i \in Y} p_i : Y \in \mathcal{I}, |Y| = k\}$ if and only if the following two conditions hold:*

1. For all $y \in V \setminus X$ with $X \cup \{y\} \notin \mathcal{I}$ and all $x \in C(X,y)$ we have $p(x) \geq p(y)$;
2. For all $y \in V \setminus X$ with $X \cup \{y\} \in \mathcal{I}$ and all $x \in X$ we have $p(x) \geq p(y)$.

For a given matroid, the cardinalities of all its bases are the same. There are also axiom system defining matroids by circuits, bases and rank functions as follows [13].

Lemma 2. *Assume that $(V,\mathcal{I})$ is an independence system and $\mathfrak{C}$ is the set of its circuits, then the following statements are equivalent:*

1. $(V,\mathcal{I})$ is a matroid;
2. For any $X \in \mathcal{I}$ and $e \in V$, $X \cup \{e\}$ contains at most one circuit;
3. For any $C_1, C_2 \in \mathfrak{C}$ with $C_1 \neq C_2$ and $e \in C_1 \cap C_2$, there exists a $C_3 \in \mathfrak{C}$ with $C_3 \subseteq (C_1 \cup C_2) \setminus \{e\}$;
4. For any $C_1, C_2 \in \mathfrak{C}$ with $C_1 \neq C_2$ and $e \in C_1 \cap C_2$ and $f \in C_1 \setminus C_2$, there exists a $C_3 \in \mathfrak{C}$ with $f \in C_3 \subseteq (C_1 \cup C_2) \setminus \{e\}$.

For any $A \in \mathcal{I}$ and $r \notin A$, if $A \cup \{r\} \notin \mathcal{I}$, there is a unique circuit in $A \cup \{r\}$, which will be denoted by $C(A,r)$.

Lemma 3. *Let V be a finite set and $\mathfrak{B} \subseteq 2^V$. $\mathfrak{B}$ is the set of bases of some matroid $\mathcal{M} = (V,\mathcal{I})$ if and only if the following holds:*

1. $\mathfrak{B} \neq \emptyset$;
2. For any $B_1, B_2 \in \mathfrak{B}$ and $x \notin B_1 \setminus B_2$, there exists a $y \in B_2 \setminus B_1$ with $(B_1 \setminus \{x\}) \cup \{y\} \in \mathfrak{B}$.

Lemma 4. *Given a matroid $\mathcal{M} = (V,\mathcal{I})$ and a set function $R : 2^V \to \mathbb{R}$, then R is a rank function of $\mathcal{M}$ if only if the following conditions hold.*

1. $0 \leq R(A) \leq |A|$, for any $A \subseteq V$;

2. $R(A) \leq R(B)$, for any $A \subseteq B \subseteq V$;

3. $R(A) + R(B) \geq R(A \cup B) + R(A \cap B)$, $\forall A, B \subseteq V$.

In fact, the rank function of a matroid is a special kind of submodular functions defined as follows.

Definition 4. *Given a finite set V, the set function $f : 2^V \to \mathbb{R}$ is submodular if*

$$f(A) + f(B) \geq f(A \cap B) + f(A \cup B), \forall A, B \subseteq V.$$

If $-f(\cdot)$ is a submodular function, then $f(\cdot)$ is called supermodular. There are many equivalent definitions of submodular functions, and we just present two of them in the following lemma.

Lemma 5. *Given a finite set V, the set function $f : 2^V \to \mathbb{R}$ is submodular if and only if one of the following result holds:*

1. $f(A \cup \{x\}) - f(A) \geq f(B \cup \{x\}) - f(B)$, for any $A \subseteq V$ and $x \in V \setminus B$.
2. $f(A \cup \{x\}) - f(A) \geq f(A \cup \{x,y\}) - f(A \cup \{y\})$, for any $A \subseteq V$ and $x, y \in V \setminus A$.

3 The Two-Stage Maximum-Weight Independent Set Problem

In this part, we mainly introduce the results of two special independent systems: one is matroid, and the other is knapsack. For the first case, we will show that the two-stage maximum-weight matroid problem is a submodular maximization problem with a matroid constraint. For the two-stage maximum-weight with a knapsack constraint, we will show it is neither a submodular problem nor a supermodular problem by presenting a counter example.

3.1 Main Result and Analysis for Independent Set Problems

In this part, we assume that $\mathcal{M} = (V, \mathcal{I})$ is a matroid, and for any $S \subseteq V$, we denote F_S as the set of feasible solutions in the second stage with respect to S,

$$F_S = \{T \subseteq V : S \cup T \in \mathcal{I}, S \cap T = \emptyset\}.$$

Since $\mathcal{M} = (V, \mathcal{I})$ is a matroid, F_S as well as $Q_{u(\omega)}(S)$ is well-defined if $S \in \mathcal{I}$. Without loss of generality, we denote $u(\omega) = u$ in the following discussion. When $S \notin \mathcal{I}$ we define $Q_u(S)$ as follows. If S is a circuit,

$$Q_u(S) := \begin{cases} Q_u(\emptyset), & \text{if } |S| = 1, \\ \min\limits_{r,q \in S} \{Q_u(S \setminus \{r\}) + Q_u(S \setminus \{q\}) - Q_u(S \setminus \{r, q\})\}, & \text{otherwise.} \end{cases}$$

Then define

$$Q_u(S) := \min_{r,q \in S} \{Q_u(S \setminus \{r\}) + Q_u(S \setminus \{q\}) - Q_u(S \setminus \{r, q\})\}$$

sequentially for S when S is not a circuit but $|S| = R(S) + 1, R(S) + 2, \ldots, |V|$.

Our main result is to show that the two-stage problem is a submodular maximization problem subject to a matroid constraint, by presenting the submodularity of the second-stage optimal objective function $Q_u(S)$.

Theorem 1. *Assume that $(V, \mathcal{I})$ is a matroid, then the function $f(S)$ defined in (1) is submodular.*

Proof. If we can prove that the function $Q_{u(\omega)}(S)$ is submodular, then

$$\mathbb{E}\left[Q_{u(\omega)}(S)\right] = \int_{\Omega} Q_{u(\omega)}(S) dP(\omega),$$

is submodular, since integration preserves submodularity. Therefore, $f(S)$ is submodular since it is the sum of a modular function and a submodular function. In the following part, we will prove $Q_u(S)(= Q_{u(\omega)}(S))$ is submodular by showing that for any $S \subseteq V$ and $r, q \in V \setminus S$,

$$Q_u(S \cup \{r\}) - Q_u(S) \geq Q_u(S \cup \{r, q\}) - Q_u(S \cup \{q\}). \tag{2}$$

It is easy to conclude (2) in the case that S, $S \cup \{r\}$, $S \cup \{q\}$ and $S \cup \{r, q\}$ are not all independent sets.

In the following part, we just need to prove the correctness of (2) when $S \cup \{r, q\} \in \mathcal{I}$. For any $S \in \mathcal{I}$, we define a new matroid $\mathcal{M}_S = (V \setminus S, \mathcal{I}_S)$, where $\mathcal{I}_S = \{T \subseteq V \setminus S : S \cup T \in \mathcal{I}\}$. We denote T_S as the optimal solution of the second stage obtained by the Best-In-Greedy algorithm with respect to $\mathcal{M}_S$. That is, assume $|S| = l$, we sort the rest items in $V \setminus S$ according to their prices at the second stage in non-increasing order: $u_1 \geq u_2 \geq \cdots \geq u_{n-l}$. Then T_S can be obtained from an empty set by individually checking i from 1 to $n - l$: If $S \cup T_S \cup \{i\}$ is an independent set, then add i to T_S. By this algorithm, we can easily get the following results by Lemma 1.

(i) $S \cup T_S$ is a basis of the matroid $(V, \mathcal{I})$, denoted as B_S and assume that $k = |B_S|$.

(ii) If for some j, which is not added to T_S, by Lemma 2 there must be a unique circuit in $S \cup (T_S \cap \{1, 2, \ldots, j-1\}) \cup \{j\}$, denoted by $C(S \cup (T_S \cap \{1, 2, \ldots, j-1\}), j)$. Furthermore, if c_j denotes the item with largest subscript except j in this circuit, i.e., with second-to-last subscript, then $C(S \cup (T_S \cap \{1, 2, \ldots, j-1\}), j) \setminus \{c_j\} \in \mathcal{I}$.

We first present the following claim.

Claim 1. *For any $S \cup \{r\} \in \mathcal{I}$, assume $B_S \neq B_{S \cup \{r\}}$, then $B_{S \cup \{r\}} = B_S \setminus \{c_r\} \cup \{r\}$, where c_r is the element with second-to-last subscript in the unique circuit $C(S \cup (T_S \cap \{1, 2, \ldots, r-1\}), r)$. Please see Fig. 1 and Fig. 2.*

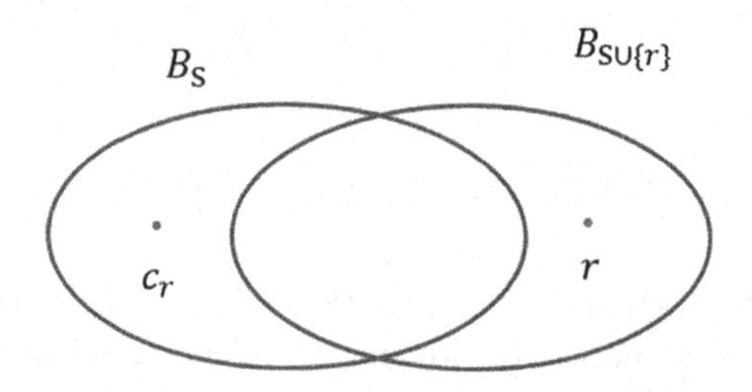

Fig. 1. The relationship between B_S and $B_{S \cup \{r\}}$.

In the following parts, we will prove (2) holds in four cases according to the relationship among the bases produced by S, $S \cup \{r\}$, $S \cup \{q\}$ and $S \cup \{r, q\}$.

Case 1. $B_S = B_{S \cup \{r\}}$ and $B_{S \cup \{q\}} = B_{S \cup \{r,q\}}$.
In this case, it is trivial to see that both sides of (2) is $-u_r$.

Case 2. $B_S \neq B_{S \cup \{r\}}$ and $B_{S \cup \{q\}} = B_{S \cup \{r,q\}}$.
We will show that this case is impossible to happen. If not, we first know that $B_S \neq B_{S \cup \{q\}}$. By Claim 1, we have $B_{S \cup \{q\}} \setminus B_S = \{q\}$, which contradicts that $r \in B_{S \cup \{q\}} \setminus B_S$.

Case 3. $B_S = B_{S \cup \{r\}}$ and $B_{S \cup \{q\}} \neq B_{S \cup \{r,q\}}$.
In this case, the left hand of (2) is equal to $-u_r$. Replacing S by $S \cup \{q\}$ in Claim 1, we know there is a circuit $C(S \cup \{q\} \cup (T_{S \cup \{q\}} \cap \{l+1, \ldots, r-1\}, \{r\})$, and assume the

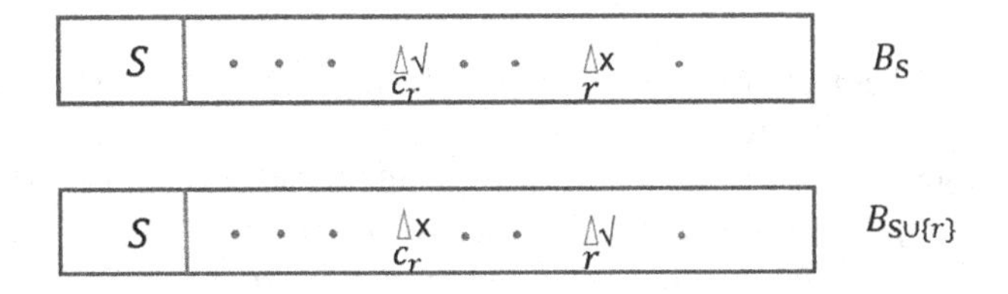

Fig. 2. The relationship between B_S and $B_{S\cup\{r\}}$.

second-to-last element is still c_r such that $u_{c_r} \geq u_r$, we know that $B_{S\cup\{r,q\}} \setminus B_{S\cup\{q\}} = \{r\}$ and $B_{S\cup\{q\}} \setminus B_{S\cup\{r,q\}} = \{c_r\}$. Therefore, the right hand of (2) is $-u_{c_r}$. Then (2) holds.

Case 4. $B_S \neq B_{S\cup\{r\}}$ and $B_{S\cup\{q\}} \neq B_{S\cup\{r,q\}}$.
First, we denote c_r as the second-to-last element of the circuit $C_1 := C(S \cup (T_S \cap \{1, \ldots, r-1\}), \{r\})$ (see Fig. 3), and c_q as the second-to-last element of the circuit $C_2 := C(S \cup (T_S \cap \{1, \ldots, q-1\}), \{q\})$ in the case that $B_S \neq B_{S\cup\{q\}}$ (see Fig. 4).

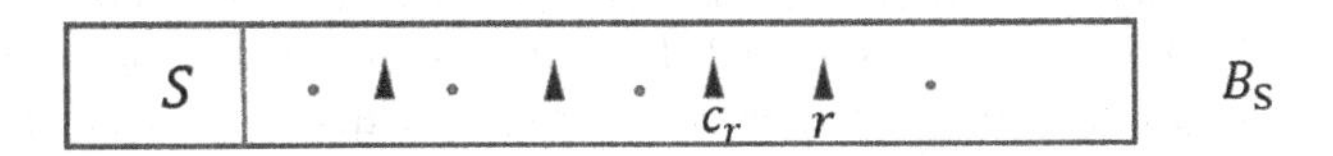

Fig. 3. The circuit C_1 produced by adding r to $S \cup (T_S \cap \{1, 2, \ldots, r-1\})$.

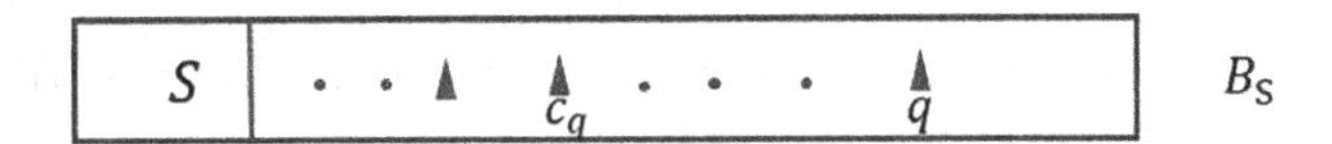

Fig. 4. The circuit C_2 produced by adding q to $S \cup (T_S \cap \{1, 2, \ldots, q-1\})$.

In fact, if $B_S = B_{S\cup\{q\}}$, it is trial to find that both hands of (2) are equal to $-u_{c_r}$. Therefore, we assume that $B_S \neq B_{S\cup\{q\}}$ in the following part. According to Claim 1, we have known that $B_{S\cup\{r\}} = B_S \setminus \{c_r\} \cup \{r\}$, $B_{S\cup\{q\}} = B_S \setminus \{c_q\} \cup \{q\}$, which implies $c_r \neq q$ and $c_q \neq r$ since $\{r, q\} \in \mathcal{I}$. In the following part, we will prove the result is correct depending on the relationship between c_r and c_q.

Case 4.1. $c_r = c_q$.
In this subcase, we denote c_r and c_q by $c_{r,q}$. Combining $r \notin C_2$ and $q \notin C_1$, by Lemma 2 we know that there exists a circuit $C \subseteq (C_1 \cup C_2) \setminus \{c_{r,q}\}$ contains r. Since C_1 is the unique circuit containing r before $c_{r,q}$ in $B_S \cap \{1, 2, \ldots, r-1\}$, we have q must be contained in C. Moreover, we know that C must also be the unique circuit produced by adding r to $B_{S\cup\{q\}}$ in the first stage. Assume c'_r is the element with largest subscript in circuit C except r, q, then $u_{c'_r} \geq u_{c_{r,q}}$ by the rules of choosing $c_{r,q}$. Therefore, the left hand of (2) is $-u_{c_{r,q}}$ and the right hand of (2) is $-u_{c'_r}$, which implies that (2) holds.

Case 4.2. $c_r \neq c_q$.
At first, we remember that $B_{S\cup\{r\}} = (B_S \cup \{r\}) \setminus \{c_r\}$, $B_{S\cup\{q\}} = (B_S \cup \{q\}) \setminus \{c_q\}$ as well as $c_r, c_q \in B_S$. Therefore, $B_{S\cup\{r\}} \setminus B_{S\cup\{q\}} = \{r, c_q\}$ and $B_{S\cup\{q\}} \setminus B_{S\cup\{r\}} = \{q, c_r\}$. By Lemma 3, for $c_r \in B_{S\cup\{q\}}$, there should be an element $y \in B_{S\cup\{r\}}$ such that $B_{S\cup\{q\}} \setminus \{c_r\} \cup \{y\}$ is a basis. According to the above discussion, we know that $y \in \{r, c_q\}$. If $y = c_q$, then $B_{S\cup\{q\}} \setminus \{c_r\} \cup \{y\} = B_S \setminus \{c_r\} \cup \{q\}$ is a basis, which contradicts to the fact that $S \cup (T_S \cap \{1, 2, \ldots, q-1\}) \cup \{q\}$ contains a circuit. Thus $y = r$. That is, $B_S \setminus \{c_r, c_q\} \cup \{r, q\}$ is a basis. Therefore, $B_S \setminus \{c_r, c_q\} \cup \{r, q\}$ is a feasible solution to the second stage based on $S \cup \{r, q\}$. Now, we pay attention to the relationship between bases $B_{S\cup\{q\}}$ and $B_{S\cup\{r,q\}}$ by taking $S \cup \{q\}$ instead of S in Claim 1. If we denote c''_r as the last item except r and q in the circuit $C_3 := C(S \cup \{q\} \cup (T_{S\cup\{q\}} \cap \{1, \ldots, r-1\}), \{r\})$, which is $C(S \cup \{q\} \cup (T_S \cap \{1, \ldots, r-1\}) \setminus \{c_q\}, \{r\})$, then $B_{S\cup\{r,q\}} = B_S \setminus \{c''_r, c_q\} \cup \{r, q\}$. So we will finish our proof if we can show that $B_{S\cup\{r,q\}} = B_S \setminus \{c_r, c_q\} \cup \{r, q\}$ by proving that $u_{c_r} = u_{c''_r}$. First, since $B_S \setminus \{c_r, c_q\} \cup \{r, q\}$ is a feasible solution and $B_S \setminus \{c''_r, c_q\} \cup \{r, q\}$ is an optimal solution to the second stage based on $S \cup \{r, q\}$, we obtain that $u_{c_r} \geq u_{c''_r}$. Next, we will prove $u_{c_r} \leq u_{c''_r}$ depending on whether $c_q \in C_1$ or not. If $c_q \notin C_1$, we know that C_1 is the unique circuit in $(B_S \setminus \{c_q\}) \cup \{r\}$. Moreover, $(B_S \setminus \{c_q\}) \cup \{q\}$ is an independent set and $(B_S \setminus \{c_q\}) \cup \{r, q\}$ is a dependent set. That is, there is a unique circuit in $(B_S \setminus \{c_q\}) \cup \{r, q\}$ by Lemma 2. Thus, this unique circuit should be the same as C_1, as well as C_3. Therefore, $u_{c_r} = u_{c''_r}$. So we assume that $c_q \in C_1$, then $u_{c_q} \geq u_{c_r}$ (or $c_q < c_r$) because of the choice of c_r, which implies that $r \notin C_2$ since $c_r < r$ and the choice of c_q. Moreover, with $c_q \in C_1 \cap C_2$ and $r \in C_1 \setminus C_2$, there is a circuit $C' \subseteq (C_1 \cup C_2) \setminus \{c_q\}$ by Lemma 2, which contains r. Without loss of generality, we can assume the elements in the two circuits C_1 and C_2 with non-increasing weights as follows: $C_1 = \{i_1, \ldots, i_l, c_q, i_{l+2}, \ldots, c_r, r\}$ and $C_2 = \{j_1, j_2, \ldots, c_q, q\}$. It is easy to know that both $C_1 \setminus \{c_q\}$ and $(C_1 \cup C_2) \setminus \{c_q, q\}$ are independent sets. Then $q \in C'$. Since there is only one circuit C'' in $S \cup (T_{S\cup\{r,q\}} \cap \{1, 2, \ldots, \max\{r, q\}\}) \cup \{r, q\}$, thus $C'' = C'$. That is, $c''_r \in C'$. Then by the rule of chosen c_r, we know that $u_{c_r} \leq u_{c''_r}$. ■

Since uniform matroid is a special matroid, the two-stage stochastic maximum-weight problem with cardinality constraints is submodular. Therefore, we can get the following corollary immediately.

Corollary 1 [11]. *Assume that V is a finite ground set, the function*

$$\max_{S\subseteq V, |S|\leq k} \sum_{i\in S} r_i + \mathbb{E}\left[\max_{T\subseteq V\setminus S, |T\cup S|\leq k} \sum_{i\in T} u_i(\omega)\right],$$

is submodular.

Table 1. The input of Example 1 with $L = 100$.

V	1	2	3	4	5	6
size	38	13	25	35	45	48
u	55	14	15	26	84	25

In this part, we can show that if the independent set is not a matroid, such as knapsack constraint, the submodularity may be destroyed by showing a counterexample in the following section. Given $V = \{1, 2, \ldots, n\}$, nonnegative size $s_i (1 \le i \le n)$ and L, as well as the independence system $(V, \mathcal{I})$ defined by $\mathcal{I} := \{S \subseteq V | \sum_{i \in S} s_i \le L\}$, then we can similarly use the function defined in (1) to describe the two-stage stochastic maximum-weight knapsack independent set problem. In the following part, we will show that $Q_{u(\omega)}(S)$ is neither submodular nor supermodular by a counterexample. Therefore, the same conclusion about $f(S)$ can be obtained.

Example 1. The information of this example is listed in Table 1. We first take $S_1 = \{1\}$, $r_1 = \{2\}, q_1 = \{3\}$, then

$$T_{S_1} = \{2, 5\},\ T_{S_1 \cup \{r_1\}} = \{5\},$$
$$T_{S_1 \cup \{q_1\}} = \{4\},\ T_{S_1 \cup \{r_1, q_1\}} = \emptyset.$$

Thus,

$$Q_u(S_1 \cup \{r_1\}) - Q_u(S_1) = u_5 - (u_2 + u_5) = -14,$$
$$Q_u(S_1 \cup \{r_1, q_1\}) - Q_u(S_1 \cup \{q_1\}) = 0 - (u_4) = -26.$$

Therefore, $Q_u(S_1 \cup \{r_1\}) - Q_u(S_1) > Q_u(S_1 \cup \{r_1, q_1\}) - Q_u(S_1 \cup \{q_1\})$.

Then take $S_2 = \{1\}, r_2 = \{3\}, q_2 = \{4\}$, then

$$T_{S_2} = \{2, 5\},\ T_{S_2 \cup \{r_2\}} = \{4\},$$
$$T_{S_2 \cup \{q_2\}} = \{3\},\ T_{S_2 \cup \{r_2, q_2\}} = \emptyset.$$

Thus,

$$Q_u(S_2 \cup \{r_2\}) - Q_u(S_2) = u_4 - (u_2 + u_5) = -72,$$
$$Q_u(S_2 \cup \{r_2, q_2\}) - Q_u(S_2 \cup \{q_2\}) = 0 - (u_3) = -15.$$

Therefore, $Q_u(S_2 \cup \{r_2\}) - Q_u(S_2) < Q_u(S_2 \cup \{r_2, q_2\}) - Q_u(S_2 \cup \{q_2\})$.

3.2 A $\frac{1}{2}$-Approximation Algorithm for the Discrete Two-Stage Stochastic Maximum-Weight Matroid Independent Set Problem

In this subsection, we further introduce the discrete two-stage matroid independent set problem, which mainly follows [11]. Compared to two-stage matroid independent set problem, the second stage weight $u(\omega)$ here only has values chosen from a finite set $\{u_1, u_2, \ldots, u_m\}$ with n-dimensional vectors and the probability of taking u_i is $p_i (1 \le i \le m)$. We have known that the best approximation ratio for non-monotone submodular maximization problem with a matroid constraint is $\frac{1}{4} - \epsilon$. In this part, we can present a $\frac{1}{2}$-approximation algorithm for discrete two-stage stochastic maximum-weight matroid independent set problem. When $u(\omega)$ is a general random variable, the solution of two-stage stochastic maximum-weight matroid independent set problem could be approximated by the solution of a sequence of discrete two-stage stochastic maximum-weight matroid independent set problems by the sample average approximation methods [11]. Let $\bar{u}_i = \mathbb{E}[u_i(\omega)]$ denote the expected weight of element i in the second stage.

Algorithm 1. Greedy algorithm for the discrete two-stage stochastic maximum-weight matroid independent set problem.

1: Apply Best-In-Greedy algorithm to the problem according to the weights h of the first stage and obtain the revenue h_1;
2: Apply Best-In-Greedy algorithm to the problem according to the weights $\bar{u}$ of the second stage and obtain the revenue h_2;
3: **if** $h_1 \geq h_2$ **then**
4:     Choose the elements obtained in the first stage;
5: **else**
6:     Choose the elements obtained in the second stage;
7: **end if**

Theorem 2. *Algorithm 1 is a $\frac{1}{2}$-approximation algorithm for discrete two-stage stochastic maximum-weight matroid independent set problem.*

References

1. Balkanski, E., Rubinstein, A., Singer, Y.: An exponential speedup in parallel running time for submodular maximization without loss in approximation. In: Proceedings of SODA, pp. 283–302 (2019)
2. Buchbinder N., Feldman M.: Deterministic algorithms for submodular maximization problems. ACM Transactions on Algorithms **14**(3), 1–20 (2018)
3. Buchbinder N., Feldman M., Garg M.: Deterministic (1/2+ϵ)-approximation for submodular maximization over a matroid. In: Proceedings of SODA, pp. 241–254 (2019)
4. Calinescu, G., Chekuri, C., Pál, M., Vondrák, J.: Maximizing a monotone submodular function subject to a matroid constraint. SIAM J. Comput. **40**, 1740–1766 (2011)
5. Chekuri C., Quanrud K.: Parallelizing greedy for submodular set function maximization in matroids and beyond. In: Proceedings of STOC, pp. 78–89 (2018)
6. Dyer, M., Stougie, L.: Computational complexity of stochastic programming problems. Math. Program. **106**, 423–432 (2006)
7. Edmonds, J.: Matroids and the greedy algorithm. Math. Program. **1**, 127–136 (1971)
8. Flaxman A.D., Frieze A., Krivelevich M.: On the random 2-stage minimum spanning tree. In: Proceedings of SODA, pp. 919–926 (2005)
9. Gupta, A., Kumar, A., Pál, M., Roughgarden, T.: Approximations via cost-sharing: Simpler and better approximation algorithms for network design. J. ACM **54**, 1–38 (2007)
10. Gupta, A., Pál, M., Ravi, R., Sinha, A.: Sampling and cost-sharing: approximation algorithms for stochastic optimization problems. SIAM J. Comput. **40**, 1361–1401 (2011)
11. He, Q., Ahmed, S., Nemhauser, G.L.: Sell or hold: a simple two-stage stochastic combinatorial optimization problem. Oper. Res. Lett. **40**, 69–73 (2012)
12. Kong, N., Schaefer, A.J.: A factor $\frac{1}{2}$ approximation algorithm for two-stage stochastic matching problems. Eur. J. Oper. Res. **172**, 740–746 (2006)
13. Korte B., Vygen J.: Combinatorial Optimization: Theory and Algorithms, Fifth Edition. Springer (2011)
14. Kosuch, S.: Approximability of the two-stage stochastic knapsack problem with discretely distributed weights. Disc. Appl. Math. **165**, 195–204 (2014)
15. Lee, J., Mirrokni, V.S., Nagarajan, V., Sviridenko, M.: Maximizing nonmonotone submodular functions under matroid or knapsack constraints. SIAM J. Disc. Math. **23**, 2053–2078 (2010)

16. Nemhauser, G.L., Wolsey, L.A., Fisher, M.L.: An analysis of approximations for maximizing submodular set functions-I. Math. Program. **14**, 265–294 (1978)
17. Sviridenko, M.: A note on maximizing a submodular set function subject to a knapsack constraint. Oper. Res. Lett. **32**, 41–43 (2004)

Routing and Scheduling Problems with Two Agents on a Line-Shaped Network

Hao Yan and Xiwen Lu(✉)

Department of Mathematics, East China University of Science and Technology, Shanghai 200237, China
xwlu@ecust.edu.cn

Abstract. We consider routing and scheduling problems with two agents on a line-shaped network in this paper. There are two agents and each agent has some jobs which are located in the network. Let $L = (V, E)$ be a line-shaped network, where $V = \{v_0\} \bigcup V^A \bigcup V^B$ is the set of $n + 1$ vertices and E is a set of edges. A job v is located at some vertex v, which is also denoted as v. The travel time $d(u, v)$ is associated with each edge $\{u, v\} \in E$. The vehicle starts from an initial vertex $v_0 \in V$ and visits all jobs for their processing. The objective is to find a route of the vehicle that minimizes the completion time of agent A under the constraint condition that the completion time of agent B is no more than the threshold value Q. We express this problem as $line - 1|C^B_{max} \leq Q|C^A_{max}$. For the problem without release time, an $O(n)$ time algorithm is provided. For the problem with release time, we show that this problem is NP-hard even though there is only one job in agent B and the jobs in agent A have no release time. Finally we give a (3, 3)-approximation algorithm for the before general problem.

Keywords: Network scheduling · Agent scheduling · Polynomial time algorithm · Approximation algorithm

1 Introduction

In recent years vehicle scheduling problems(VSP) have become an important research area, due to their various applications in some applied disciplines. For the single-vehicle scheduling problems, a single-vehicle starts from the depot and visits the jobs at different vertices in order to process them. Each job has its own release time, processing time. The objective is to find a routing schedule of the vehicle that minimizes the completion time.

For the single agent problem on a line (L-VSP), it has been shown to be NP-hard by Tsitsiklis [1]. Psaraftis et al. [2] showed that there was a 2- approximation algorithm for this problem, and if all the processing times are zero (L-VRP), both the tour version and the path version can be solved in polynomial time. If

This research was supported by the National Natural Science Foundation of China under Grant 11871213.

D.-Z. Du et al. (Eds.): COCOA 2021, LNCS 13135, pp. 214–223, 2021.
https://doi.org/10.1007/978-3-030-92681-6_18

the jobs have additional deadline, L-VRP is shown to be strongly NP-hard by Tsitsiklis [1]. But if we don't consider the release time, L-VRP can be solved in polynomial time. A $\frac{3}{2}$-approximation algorithm was provided by Karuno et al. [3] for tour version of L-VSP in which the depot is at one of the two extreme points and Gaur et al. [4] presented a $\frac{5}{3}$-approximation algorithm for the problem with the same setting except that the depot is located arbitrarily on the line. For the path-version of L-VSP, a $\frac{3}{2}$-approximation algorithm was given by Yu and Liu [5].

Agent scheduling problems have been studied 18 years since they were introduced by Baker and Smith [6] and Agnetis et al. [7]. During this period, various problems have been studied by a large number of researchers, for example Agnetis et al. [8] extended some problems which have different functions for single agent to multi-agent version, and Cheng et al. [9] provided complexity results for some special cases of multi-agent scheduling problems. But there is little research on multi-agent vehicle scheduling problems that is widely used in real life such as an AGV on a shop floor which always belongs to different customers and in the aviation network, aircraft also belongs to many operators. The research of the multi-agent vehicle scheduling problems has strong theoretical and practical significance.

In this paper, we study the two-agent vehicle scheduling problems on a line-shaped network. There are two competing agents in a given network, called A and B. Let $L = (V, E)$ be a line-shaped network, where $V = \{v_0\} \bigcup V^A \bigcup V^B$ is a set of $n+1$ vertices and E is a set of edges. A job v is located at some vertex v, which is also denoted as v. There is a single vehicle, which is initially situated at the depot v_0 at time 0. Agent A and agent B has n_A and n_B jobs to be processed by the single vehicle respectively. We also denote the job set of agent A and agent B by $V^A = \{v_1^A, v_2^A, ..., v_{n_A}^A\}$ and $V^B = \{v_1^B, v_2^B, ..., v_{n_B}^B\}$. Let $n = n_A + n_B$. The vehicle starting from the depot v_0 visits all the jobs on the line-shaped network to process them. The travel time $d(u, v)$ is associated with each edge $\{u, v\} \in E$. Each job v has its release time $r(v)$ and processing time $h(v)$. For any arbitrary schedule π, let $C_{max}^A(\pi)$ and $C_{max}^B(\pi)$ be the makespan for agent A and agent B, respectively. The objective of the problem is to find an optimal schedule that minimizes the makespan of agent A under the constraint condition that the completion time of agent B is no more than the threshold value Q. We can express this problem as $line-1|C_{max}^B \leq Q|C_{max}^A$.

The rest of the paper is organized as follows. In Sect. 2, we provide a polynomial algorithm for the problem without processing time. In Sect. 3, we show the problem is NP-hard even though there is only one job in agent B and the jobs in agent A have no release time. Furthermore, we give a $(3, 3)$-approximation algorithm for general problem.

2 Problem Without Release Time

In this section, we will give a polynomial algorithm for the problem $line-1|r(v_j) = 0, C_{max}^B \leq Q|C_{max}^A$. According to the different value ranges of Q, we will give different properties of the optimal schedule. Then we enumerate all the schedules that satisfy those properties, and select the best feasible schedule which is the optimal schedule under the constraint.

For simplicity, we only consider the case that the rightmost endpoint of the line-shaped network belongs to agent B. For the case that the rightmost endpoint belongs to agent A, we can give the similar lemmas and a homologous algorithm to solve it. The detailed algorithm will be given in the Appendix at the end of the paper.

Next, we give some notations used in following.

π^*: the optimal schedule of the problem.

L^X: the length from depot v_0 to vertex $v_{n_X}^X$ for $X \in \{A, B\}$.

$h(v)$: the processing time of the job v.

H: total processing time of jobs in V.

H^X: total processing time of jobs in V^X for $X \in \{A, B\}$.

$W = \{v_1^B, ..., v_i^B\}$: vertex set of agent B located before $v_{n_A}^A$ on the line-shaped network L.

$C(v)$: the completion time of the job v.

Before giving the following lemmas, we provide a trivial lower bound of this problem. That is $C_{max}^A(\pi^*) \geq L^A + H^A$.

Lemma 2.1. *If $L^B + H \leq Q$, there is $C_{max}^A(\pi^*) < C_{max}^B(\pi^*)$ in the optimal schedule π^*.*

Proof. Assume that there is $C_{max}^B(\pi^*) \leq C_{max}^A(\pi^*)$ in the optimal schedule π^*. Let v_x^A to be the last job completed in agent A. We know $C_{max}^B(\pi^*) \leq C(v_x^A)$. Thus $C_{max}^A(\pi^*) = C(v_x^A) \geq L^B + H^B + d(v_{n_B}^B, v_x^A) + H^A$.

We construct a new schedule π such that the vehicle processes all jobs from left to right one by one on the line-shaped network. Because the rightmost endpoint belongs to agent B, we gain that $C_{max}^A(\pi) \leq L^A + H^A + H^B$. Hence we get that $C_{max}^A(\pi) < C_{max}^A(\pi^*)$. This is a contradiction with the optimal schedule π^*.

Lemma 2.2. *If $L^B + H^B \leq Q < L^B + H$, there is $C_{max}^B(\pi^*) < C_{max}^A(\pi^*)$ in the optimal schedule π^*.*

Proof. We will prove this lemma by counter evidence method. Assume that there is $C_{max}^A(\pi^*) \leq C_{max}^B(\pi^*)$ in the optimal schedule π^*. We know that the jobs in agent A are completed before the jobs in agent B. Thus we have $C_{max}^B(\pi^*) \geq L^B + H > Q$. This is a contradiction with the constraint.

Lemma 2.3. *If $C_{max}^A(\pi^*) < C_{max}^B(\pi^*)$ and $W \neq \emptyset$, there are results as follows.*

(1) If $v_l^B (l \leq i)$ is completed before $C_{max}^A(\pi^*)$, then $v_k^B (1 \leq k \leq l-1)$ must be completed before $C_{max}^A(\pi^*)$.

(2) All jobs in W completed before $C_{max}^A(\pi^*)$ must be processed continuously from left to right according to the index on the line-shaped network.

(3) For agent B, all the jobs in $\{V^B \setminus W\}$ are processed after $C_{max}^A(\pi^*)$. In other words, only the jobs in W can be processed before $C_{max}^A(\pi^*)$.

Proof. (1) Assume that v_k^B is the first job in $\{v_1^B, ..., v_{l-1}^B\}$ processed after $C_{max}^A(\pi^*)$ in π^*. Then we have the following inequality $C_{max}^A(\pi^*) \geq L^A + H^A + \sum_{j=1}^{k-1} h(v_j^B) + h(v_l^B)$.

Next, we construct a new schedule π. In the first phase, the vehicle starts from v_0 to $v_{n_A}^A$ processing the jobs in $V^A \bigcup \{v_1^B, ..., v_{k-1}^B\}$. In the second phase, the vehicle processes the remaining jobs along the shortest path.

Because v_k^B is processed after $C_{max}^A(\pi)$ and the jobs in $\{v_k^B, ..., v_{n_B}^B\}$ are processed along the shortest path in π, we know that $C_{max}^B(\pi) \leq C_{max}^B(\pi^*) \leq Q$. Thus π is a feasible schedule. Besides we have $C_{max}^A(\pi) = L^A + H^A + \sum_{j=1}^{k-1} h(v_j^B) < C_{max}^A(\pi^*)$. Hence we know π is a better feasible schedule, which is contrary to the optimal schedule π^*.

Thus v_k^B is processed before $C_{max}^A(\pi^*)$, and we can prove that all the jobs in $\{v_1^B, ..., v_{l-1}^B\}$ are processed before $C_{max}^A(\pi^*)$ in π^* by the similar method.

(2) Assume that $\{v_1^B, ..., v_k^B, v_{k+1}^B, ..., v_l^B\}$ are completed before $C_{max}^A(\pi^*)$ according to (1) and $C(v_{k+1}^B) < C(v_k^B)$. Let H' be the total processing time of the remaining jobs in agent B completed before $C_{max}^A(\pi^*)$. Then we get that $C_{max}^A(\pi^*) \geq L^A + H^A + h(v_{k+1}^B) + h(v_k^B) + 2d(v_k^B, v_{k+1}) + H'$.

Construct a new schedule π such that the vehicle processes the jobs in $V^A \bigcup \{v_1^B, ..., v_k^B, v_{k+1}^B, ..., v_l^B\}$ one by one from v_0 to $v_{n_A}^A$ and processes the remaining jobs along the shortest path.

Because the jobs in $\{v_1^B, ..., v_l^B\}$ are processed before $C_{max}^A(\pi)$ and the jobs in $\{v_{l+1}^B, ..., v_{n_B}^B\}$ are processed along the shortest path in π, we get that $C_{max}^B \leq Q$. Hence π is a feasible schedule.

For the completion time of agent A in π, we have $C_{max}^A(\pi) = L^A + H^A + h(v_{k+1}^B) + h(v_k^B) + H' < C_{max}^A(\pi^*)$. So π is a better feasible schedule than π^*. This is a contradiction.

(3) We can use a method similar to that in (1) (2) to prove the conclusion in (3). For simplicity, the proof is omitted.

Lemma 2.4. *If $C_{max}^B(\pi^*) < C_{max}^A(\pi^*)$ in the optimal schedule π^*, all jobs in V^A completed before $C_{max}^B(\pi^*)$ must be processed continuously from left to right on the line-shaped network L.*

Proof. The proof method is similar to Lemma 2.3 (2).

We will use above Lemmas to give the following polynomial algorithm A for the problem $line - 1|C_{max}^B \leq Q|C_{max}^A$. We enumerate all the schedules that satisfy those Lemmas, and select the best feasible schedule as the optimal schedule.

In the following, for the convenience of description, we denote the set $\{a_1, a_2, ..., a_h\}$ as $\emptyset$ when $h = 0$.

Algorithm A

Step1. According to the following candidate schedules to construct the schedule π^B.

For $s = 0, 1, ..., i - 1$,

π_s^{1B}: first the vehicle processes the jobs in $V^A \bigcup \{v_1^B, ..., v_s^B\}$ from the depot v_0 to $v_{n_A}^A$. Next, the vehicle goes to $v_{n_B}^B$ without processing any job, and finally the vehicle returns to vertex v_{s+1}^B to process the remaining jobs in J^B one by one.

π_s^{2B}: first the vehicle processes the jobs in $V^A \bigcup \{v_1^B, ..., v_s^B\}$ from the depot v_0 to $v_{n_A}^A$. Next, the vehicle goes to v_{s+1}^B without processing any job, and finally the vehicle processes the remaining jobs in V^B from v_{s+1}^B to $v_{n_B}^B$ in turn.

For $s = i$,

π_i^B: the vehicle processes all the jobs one by one v_0 to $v_{n_B}^B$.

Step2. Choose the best feasible schedule from $\{\pi_s^{1B}, s = 0, 1, ..., i-1\}$, $\{\pi_s^{2B}, s = 0, 1, ..., i-1\}$ and π_i^B as π^B with the makespan no more than Q for agent B.

Step3. By the following candidate schedules to construct the schedule π^A.

For $y = 0, 1, ..., n_A - 1$,

π_y^A: first the vehicle processes the jobs in $V^B \bigcup \{v_1^A, ..., v_y^A\}$ from the depot v_0 to $v_{n_B}^B$. Next, the vehicle returns to vertex v_{y+1}^A to process the remaining jobs in V^A in turn.

Step4. Choose the best feasible schedule from $\{\pi_y^A, y = 0, 1, ..., n_A - 1\}$ as π^A under the threshold value constraint.

Step5. If $C_{max}^A(\pi^B) \leq C_{max}^A(\pi^A)$, select π^B as the optimal schedule π. Otherwise, choose π^A as the optimal schedule π.

Theorem 2.1. *The two-agent problem line* $- 1|C_{max}^B \leq Q|C_{max}^A$ *can be solved in* $O(n)$ *time by Algorithm A.*

Proof. According to different value ranges of Q, we will prove this theorem by the following four cases respectively.

Case1. $L^A + \min\{d(v_{n_A}^A, v_1^B) + d(v_1^B, v_{n_B}^B), d(v_{n_A}^A, v_{n_B}^B) + d(v_{n_B}^B, v_1^B)\} + H \leq Q$.

If $C_{max}^A(\pi_0^{1B}) < C_{max}^A(\pi_0^{2B})$, let $\pi = \pi_0^{1B}$. Otherwise let $\pi = \pi_0^{2B}$. Thus there is

$$C_{max}^A(\pi) = L^A + H^A \leq C_{max}^A(\pi^*)$$
$$C_{max}^B(\pi) = L^A + \min\{d(v_{n_A}^A, v_1^B) + d(v_1^B, v_{n_B}^B), d(v_{n_A}^A, v_{n_B}^B) + d(v_{n_B}^B, v_1^B)\} + H \leq Q \tag{1}$$

So π is an optimal schedule in this case.

Case2. $L^B + H \leq Q < L^A + \min\{d(v_{n_A}^A, v_1^B) + d(v_1^B, v_{n_B}^B), d(v_{n_A}^A, v_{n_B}^B) + d(v_{n_B}^B, v_1^B)\} + H$.

According to Lemma 2.1, we know $C_{max}^A(\pi^*) < C_{max}^B(\pi^*)$ in the optimal schedule π^* in this case. Thus we can construct the following candidate schedules $\{\pi_s^{1B}, s = 0, 1, ..., i-1\}$, $\{\pi_s^{2B}, s = 0, 1, ..., i-1\}$ and π_i^B by Lemma 2.3, and calculate the completion time of agent B for each candidate.

$$\begin{cases} C_{max}^B(\pi_s^{1B}) = L^A + d(v_{n_A}^A, v_{s+1}^B) + d(v_{s+1}^B, v_{n_B}^B) + H \\ C_{max}^B(\pi_s^{2B}) = L^A + d(v_{n_A}^A, v_{n_B}^B) + d(v_{n_B}^B, v_{s+1}^B) + H \\ C_{max}^B(\pi_i^B) = L^B + H \end{cases} \tag{2}$$

Thus there is at least one schedule π_i^B satisfying the constraint. Therefore, there must be an optimal schedule π from the candidate schedules which satisfy the constraint. Because $\{\pi_s^{1B}, s = 0, 1, ..., i-1\} \bigcup \{\pi_s^{2B}, s = 0, 1, ..., i-1\} \bigcup \pi_i^B$ are all the candidate schedules meeting Lemma 2.3 and π is optimal of those schedules, then π is an optimal schedule in this case.

Case3. $L^B + H^B \leq Q < L^B + H$.

By Lemma 2.2, we gain that $C_{max}^B(\pi^*) < C_{max}^A(\pi^*)$ in the optimal schedule π^* in this case. Similarly we can construct the following candidate schedules $\{\pi_y^A, y = 0, 1, ..., n_A - 1\}$ by Lemma 2.4.

When $1 \leq y \leq n_A - 1$, there is $C_{max}^B(\pi_y^A) = L^B + H^B + \sum_{j=1}^{y} h(v_j^A)$. When $y = 0$, we have $C_{max}^B(\pi_0^A) = L^B + H^B$. So there is at least one feasible schedule π_0^A satisfying the constraint. Thus there must be a feasible optimal schedule π from the candidate schedules. By the same method in Case2, we know that π is an optimal schedule in this case.

Case4. When $Q < L^B + H^B$.

There is no feasible schedule in this case.

Through the construction process of the schedule π in algorithm A, the computation in Step1 and Step3 is at most $O(n)$. So this problem can be solved in $O(n)$ time by Algorithm A.

3 Problem with Release Time

3.1 Special Case

We will investigate the complexity of the problem. We only study a special case of the problem. For the special problem, there are n_A jobs in V^A without release time and there is only one job v_1^B in V^B which has the release time $r(v_1^B)$. This problem can be described as $line - 1|r(v_j^A) = 0, r(v_1^B), C_{max}^B \leq Q|C_{max}^A$. Next, we will prove even this special problem is NP-hard.

Theorem 3.1. *The problem* $line - 1|r(v_j^A) = 0, r(v_1^B), C_{max}^B \leq Q|C_{max}^A$ *is NP-hard.*

Proof. We can prove this theorem by reduction from Partition-Problem.

Partition-Problem: given a set $A = \{a_1, a_2, ..., a_m\}$ with m positive integers and its index set is $M = \{1, 2, ..., m\}$. Let $\sum_{i=1}^{m} a_i = F$. Is there a subset $S \subseteq M$ such that $\sum_{i \in S} a_i = \sum_{i \in M \setminus S} a_i = \frac{1}{2}F$?

For any instance of Partition-Problem, we can construct an instance of $line - 1|r(v_j^A) = 0, r(v_1^B), C_{max}^B \leq Q|C_{max}^A$ as follows. There is a line-shaped network with length L. For the job set, V^A contains m jobs located at the leftmost vertex of the line. For any job $v_i^A (i = 1, 2, ..., m)$, the release time is $r(v_i^A) = 0$ and the processing time is $h(v_i^A) = a_i$. V^B has only one job v_1^B located at the rightmost vertex of the line, and the release time is $r(v_1^B) = \frac{1}{2}F + L$ and the processing time is $h(v_1^B) = b$. Let $Q = L + \frac{1}{2}F + b$ and $\tau = 2L + F + b$. Is there a feasible schedule π such that $C_{max}^A(\pi) \leq \tau$ holds under the condition of $C_{max}^B(\pi) \leq Q$?

If there is a subset $S \subseteq M$ such that $\sum_{i \in S} a_i = \sum_{i \in M \setminus S} a_i = \frac{1}{2}F$, we can construct a schedule π such that in the first phase, the vehicle processes the jobs in $\{v_i^A, i \in S\} \bigcup \{v_1^B\}$ from left to right and in the second phase, the vehicle processes the remaining jobs in V^A from v_1^B to the depot. Because $\sum_{i \in S} h(v_i^A) = \frac{1}{2}F$ and the length of the line is L, we know that the arriving time of the vehicle at v_1^B is $\frac{1}{2}F + L$. Thus there is no waiting before processing v_1^B in π. So we can calculate that

$$\begin{cases} C_{max}^B(\pi) = L + \dfrac{1}{2}F + b \leq Q \\ C_{max}^A(\pi) = 2L + F + b \leq \tau \end{cases} \tag{3}$$

Hence we know that π is a feasible schedule and $C_{max}^A(\pi) \leq \tau$ for decision version of the routing and scheduling problem.

If there is a feasible schedule π, we have $C_{max}^B(\pi) \leq Q = L + \frac{1}{2}F + b$. Because the release time of v_1^B is $r(v_1^B) = \frac{1}{2}F + L$ and the processing time of v_1^B is $h(v_1^B) = b$, we know that the arriving time of the vehicle at v_1^B is earlier than $L + \frac{1}{2}F$. Assume that the jobs in $\{v_i^A, i \in S_1\}$ are completed before $C_{max}^B(\pi)$ in the schedule π, we have $\sum_{i \in S_1} h(v_i^A) \leq \frac{1}{2}F$. Because $C_{max}^A(\pi) \leq \tau = 2L + F + b$, we can obtain $\sum_{i \in M \setminus S_1} h(v_i^A) \leq \frac{1}{2}F$. Thus we can get that $\sum_{i \in S_1} h(v_i^A) = \sum_{i \in M \setminus S_1} h(v_i^A) = \frac{1}{2}F$.

So we have the conclusion that the problem $line-1|r(v_j^A) = 0, r(v_1^B), C_{max}^B \leq Q|C_{max}^A$ is NP-hard.

According to Theorem 3.1, we can obtain the following corollary.

Corollary 3.1. *The problem* $line - 1|r(v_j), C_{max}^B \leq Q|C_{max}^A$ *is NP-hard.*

3.2 General Case

In this subsection, we will give an approximation algorithm for the general problem. Let π^* be the optimal schedule of the problem and $L = \max\{L^A, L^B\}$. Assume $r_{max}^X = \max\{r(v), v \in V^X\}$ be the maximal release time in V^X and $r_{max} = \max\{r(v), v \in V\}$ be the maximal release time in V.

By the definition of $(\beta_1, \beta_2, ..., \beta_g)$-approximation algorithm about agent scheduling problems [10]. We can give the following definition of (β_A, β_B)-approximation algorithm for the problem $line - 1|r_j, C_{max}^B \leq Q|C_{max}^A$.

Definition 3.1. Let $C_{max}^A(\pi^*)$ be the optimal makespan and Q be the threshold value for an instance of the problem $line - 1|r_j, C_{max}^B \leq Q|C_{max}^A$. An algorithm is called a (β_A, β_B)- approximation algorithm if it provides a schedule π such that $C_{max}^A(\pi) \leq \beta_A C_{max}^A(\pi^*)$ and $C_{max}^B(\pi) \leq \beta_B Q$. Moreover, if $\beta_A = \beta_B = \beta$, we say that the algorithm is a β-approximation algorithm.

Next, we provide several lower bounds used in this subsection.

Lemma 3.1. *If* $C_{max}^A(\pi^*) < C_{max}^B(\pi^*)$ *in the optimal schedule* π^*, *then*

$$\begin{aligned} &(1)\, C_{max}^B(\pi^*) \geq L^B + H^A + H^B, \; for\; L^A < L^B \\ &(2)\, C_{max}^B(\pi^*) \geq L^A + d(v_{n_A}^A, v_{n_B}^B) + H^A + H^B, \; for\; L^A > L^B \end{aligned} \tag{4}$$

Proof. (1) This is a trivial case. Then the proof is omitted.

(2) Since the last completed job belongs to agent B and $L^A > L^B$, then the total traveling time is no less than $L^A + d(v_{n_A}^A, v_{n_B}^B)$. Thus $C_{max}^B(\pi^*) \geq L^A + d(v_{n_A}^A, v_{n_B}^B) + H^A + H^B$.

Lemma 3.2. *If $C_{max}^B(\pi^*) < C_{max}^A(\pi^*)$ in the optimal schedule π^*, then*

$$\begin{aligned} &(1)\, C_{max}^A(\pi^*) \geq L^A + H^A + H^B,\ for\ L^A > L^B \\ &(2)\, C_{max}^A(\pi^*) \geq L^B + d(v_{n_B}^B, v_{n_A}^A) + H^A + H^B,\ for\ L^A < L^B \end{aligned} \tag{5}$$

Proof. The proof method is similar to Lemma 3.1.

An approximation algorithm B for $line{-}1|r(v_j), C_{max}^B \leq Q|C_{max}^A$ is described as follows.

Algorithm B

Step1. Schedule the jobs in J^A one by one from v_0 to $v_{n_A}^A$, then schedule the jobs in J^B along the shortest path. Denote the schedule by π_1.

Step2. Schedule the jobs in J^B one by one from v_0 to $v_{n_B}^B$, then schedule the jobs in J^A along the shortest path. Denote the schedule by π_2.

Step3. Choose the better feasible schedule from π_1 and π_2 as π with the makespan no more than Q for agent B.

Theorem 3.2. *For the problem $line - 1|r(v_j), C_{max}^B \leq Q|C_{max}^A$, the Algorithm B is a 3-approximation algorithm.*

Proof. We will prove that the conclusion is true by following two cases.

When $C_{max}^A(\pi^*) < C_{max}^B(\pi^*)$, let $\pi = \pi_1$.

By Step 1 of Algorithm B, we have $C_{max}^A(\pi_1) \leq r_{max}^A + L^A + H^A \leq 2C_{max}^A(\pi^*)$.

If there is no waiting time for processing the jobs in agent B in π_1, then we divide the discussion into two cases.

(1) $L^A < L^B$. We have $C_{max}^B(\pi_1) \leq r_{max}^A + L^A + H^A + \min\{d(v_{n_A}^A, v_1^B), d(v_{n_A}^A, v_{n_B}^B)\} + L^B + H^B$. Because $L^A < L^B$, then $L^A + \min\{d(v_{n_A}^A, v_1^A), d(v_{n_A}^A, v_{n_B}^B)\} \leq L^B$. Thus we have $C_{max}^B(\pi_1) \leq r_{max}^A + L^B + L^B + H^A + H^B$. According to Lemma 3.1(1), we gain $C_{max}^B(\pi_1) \leq 3C_{max}^B(\pi^*) \leq 3Q$.

(2) $L^A > L^B$. We know $C_{max}^B(\pi_1) \leq r_{max}^A + L^A + H^A + d(v_{n_A}^A, v_{n_B}^B) + L^B + H^B$. By Lemma 3.1(2), we have $C_{max}^B(\pi_1) \leq r_{max}^A + L^B + C_{max}^B(\pi^*) \leq 3C_{max}^B(\pi^*) \leq 3Q$.

If there is some waiting time for processing the jobs in agent B in π_1, then we can assume v_k^B is the last waiting job. Thus we know that $C_{max}^B(\pi_1) \leq r(v_k^B) + h(v_k^B) + H^B + L^B \leq 2C_{max}^B(\pi^*) \leq 3Q$.

Therefore we know $C_{max}^A(\pi_1) \leq 2C_{max}^A(\pi^*)$ and $C_{max}^B(\pi_1) \leq 3Q$ when $C_{max}^A(\pi^*) < C_{max}^B(\pi^*)$.

When $C_{max}^A(\pi^*) > C_{max}^B(\pi^*)$, let $\pi = \pi_2$. By a similar way, we gain $C_{max}^A(\pi_2) \leq 3C_{max}^A(\pi^*)$ and $C_{max}^B(\pi_2) \leq 2C_{max}^B(\pi^*) \leq 2Q$.

So we have

$$\begin{cases} C^A_{max}(\pi) \leq 3C^A_{max}(\pi^*) \\ C^B_{max}(\pi) \leq 3Q \end{cases} \tag{6}$$

Thus by the Definition 3.1 we know that the Algorithm B is a 3-approximation algorithm.

4 Conclusions

In this paper we consider single vehicle scheduling problem with two-agent on a line-shaped network. We prove the problem $line - 1|r(v_j), C^B_{max} \leq Q|C^A_{max}$ is NP-hard. Moreover we have presented a polynomial Algorithm A for $line - 1|r(v_j) = 0, C^B_{max} \leq Q|C^A_{max}$ and a 3-approximation algorithm for $line - 1|r(v_j), C^B_{max} \leq Q|C^A_{max}$.

However, it is still an open problem that whether the approximation bounds for the approximation algorithm is tight. (β_A, β_B)-approximation algorithms would be very interesting in future for the multi-agent vehicle scheduling problem on a given network. We will try to provide a better approximation algorithm for the problem. In this paper, we only consider the two-agent scheduling problem on a line-shaped network due to the complexity of the problem. Approximation algorithms for multi-agent scheduling problems may be considered as well in the future.

Appendix

For the problem $line - 1|r(v_j) = 0, C^B_{max} \leq Q|C^A_{max}$, when the rightmost endpoint belongs to agent A, we will give the following Algorithm C. This algorithm is actually similar to Algorithm A. Let $M = \{v^A_1, ..., v^A_j\}$ be the vertex set of agent A located before $v^B_{n_B}$ on the line-shaped network L.

Algorithm C

Step1. According to the following candidate schedules to construct the schedule σ^A.

For $h = 0, 1, ..., j - 1$,

σ^{1A}_h: first the vehicle processes the jobs in $V^B \bigcup \{v^A_1, ..., v^A_h\}$ from the depot v_0 to $v^B_{n_B}$. Next, the vehicle goes to $v^A_{n_A}$ without processing any job, and finally the vehicle returns to vertex v^A_{h+1} to process the remaining jobs in J^A one by one.

σ^{2A}_h: first the vehicle processes the jobs in $V^B \bigcup \{v^A_1, ..., v^A_h\}$ from the depot v_0 to $v^B_{n_B}$. Next, the vehicle goes to v^A_{h+1} without processing any job, and finally the vehicle processes the remaining jobs in V^A from v^A_{h+1} to $v^A_{n_A}$ in turn.

For $h = j$,

σ^A_j: the vehicle processes all the jobs one by one v_0 to $v^A_{n_A}$.

Step2. Choose the best feasible schedule from $\{\sigma^{1A}_h, h = 0, 1, ..., j - 1\}$, $\{\sigma^{2A}_h, h = 0, 1, ..., j - 1\}$ and σ^A_j as σ^A with the makespan no more than Q for agent B.

Step3. By the following candidate schedules to construct the schedule σ^B. For $l = 0, 1, ..., n_B - 1$,

σ_l^B: first the vehicle processes the jobs in $V^A \bigcup \{v_1^B, ..., v_l^B\}$ from the depot v_0 to $v_{n_A}^A$. Next, the vehicle returns to vertex v_{l+1}^B to process the remaining jobs in V^B in turn.

Step4. Choose the best feasible schedule from $\{\sigma_l^B, l = 0, 1, ..., n_B - 1\}$ as σ^B under the threshold value constraint.

Step5. If $C_{max}^A(\sigma^B) \le C_{max}^A(\sigma^A)$, select σ^B as the optimal schedule σ. Otherwise, choose σ^A as the optimal schedule σ.

According to different value ranges of Q, we can give following four cases. (1) $2L^A - d(v_0, v_1^B) + H \le Q$; (2)$L^A + d(v_{n_A}^A, v_{n_B}^B) + H \le Q < 2L^A - d(v_0, v_1^B) + H$; (3) $L^B + H^B \le Q < L^A + d(v_{n_A}^A, v_{n_B}^B) + H$; (4) $Q < L^B + H^B$. By the similar method to Theorem 2.1, we can prove that the problem $line - 1|r(v_j) = 0, C_{max}^B \le Q|C_{max}^A$ is solvable in $O(n)$ time, when the rightmost endpoint belongs to agent A.

References

1. Tsitsiklis, J.N.: Special cases of traveling salesman and repairman problems with time windows. Networks **22**(3), 263–282 (1992)
2. Psaraftis, H., Solomon, M.M., Magnanti, T.L., Kim, T.U.: Routing and scheduling on a shoreline with release times. Manag. Sci. **36**, 212–223 (1990)
3. Karuno, Y., Nagamochi, H., Ibaraki, T.: A 1.5-approximation for single vehicle scheduling problem on a line with release and handling times. In: Proceedings ISCIE/ASME 1998 Japan-USA Symp on Flexible Automation, Ohtsu, Japan, pp. 1363–1368 (1998)
4. Gaur, D.R., Gupta, A., Krishnamurti, R.: A $\frac{5}{3}$-approximation algorithm for scheduling vehicles on a path with release and handling times. Inf. Process. Lett. **86**(2), 87–91 (2003)
5. Yu, W., Liu, Z.: Single-vehicle scheduling problems with release and service times on a line. Networks **57**(2), 128–134 (2010)
6. Baker, K.R., Smith, J.C.: A multiple-criterion model for machine scheduling. J. Sched. **6**(1), 7–16 (2003)
7. Agnetis, A., Mirchandani, P.B., Pacciarelli, D., Pacifici, A.: Scheduling problem with two competing agents. Oper. Res. **52**(2), 229–242 (2004)
8. Agnetis, A., Pacciarelli, D., Pacifici, A.: Multi-agent single machine scheduling. Ann. Oper. Res. **150**(1), 3–15 (2007)
9. Cheng, T.C.E., Ng, C.T., Yuan, J.J.: Multi-agent scheduling on a single machine with max-form criteria. Eur. J. Oper. Res. **188**(2), 603–609 (2008)
10. Lee, K., Choi, B.C., Leung, J.Y.T., Pinedo, M.L.: Approximation algorithms for multi-agent scheduling to minimize total weighted completion time. Inf. Process. Lett. **109**(16), 913–917 (2009)

The Price of Anarchy of Generic Valid Utility Systems

Yin Yang[1], Qingqin Nong[1(✉)], Suning Gong[1], Jingwen Du[1], and Yumei Liang[2]

[1] School of Mathematical Science, Ocean University of China, Qingdao 266100, Shandong, People's Republic of China
qqnong@ouc.edu.cn

[2] Shanghai Lixin University of Accounting and Finance, Shanghai 201620, People's Republic of China

Abstract. In this paper we introduce an (a, b)-generic $(a, b \in \mathbb{R}^+)$ valid utility system, a class of non-cooperative games with n players. The social utility of an outcome is measured by a submodular function. The private utility of a player is at most a times the change in social utility that would occur if the player declines to participate in the game. For any outcome, the total amount of the utility of all players is at most b times the social utility. We show that the price of anarchy of the system is at least $\frac{a}{a+b}$ if there exist pure strategy Nash equilibria. For the case that there does not exist a pure strategy Nash equilibrium, we design a mechanism to output an outcome that gives a social utility within $\frac{a}{2bn+a-b}$ times of the optimal.

Keywords: Valid utility system · Price of anarchy · Nash equilibrium · Best response

1 Introduction

Non-cooperative game is a widely studied game in which there is no cooperation, prior information exchange, transmission, or mandatory agreement among players. Nash equilibrium is a special outcome that no one can get more utility by unilaterally changing his strategy. To evaluate the efficiency of Nash equilibria, Roughgarden and Tardos [14] propose the concept of *price of anarchy*, which is the ratio of the social utility of a worst Nash equilibria to that of the optimal outcome in the game.

In 1999, Koutsoupias and Papadimitriou [8] firstly use the price of anarchy as a measurement of a network routing problem. Later, a lot of studies extensively apply this measure to more complex games, such as congestion games [2,9], auction games [5,7,10], influence maximization games [1,6], valid utility systems [11,15], and so on.

This research was supported in part by the National Natural Science Foundation of China under grant number 12171444 and 11871442, and was also supported in part by the Natural Science Foundation of Shandong Province under grant number ZR2019MA052.

D.-Z. Du et al. (Eds.): COCOA 2021, LNCS 13135, pp. 224–233, 2021.
https://doi.org/10.1007/978-3-030-92681-6_19

In 2002, Vetta [15] firstly proposes the concept of valid utility system. It is a class of games with submodular social utility functions. Each player's reward is not less than the difference of social utility caused by his absence. In the meantime, the reward of all players does not exceed the social utility. Vetta shows that the price of anarchy of that system is $\frac{1}{2}$. Valid utility system can be applied to a wide range of games, including facility location games [3,15], traffic routing games [15] and market sharing games [12]. But most of the existing results concentrate on the applications and ignore the improvement of the system structure.

In 2020, Grimsman et al. [4] extend valid utility system to incomplete information, in which a subset of players either blind (cannot observe any other players' choices) or isolated (blind, and cannot be observed by others). Their main results show that when k ($1 \leq k \leq n$) players are compromised (in any combination of blind or isolated), the price of anarchy is $\frac{1}{2+k}$. They show that if players use marginal utility functions, and at least one compromised player is blind (not isolated), the price of anarchy is $\frac{1}{1+k}$.

In 2020, Ma et al. [11] extend the valid utility system to α-scalable valid utility system from another point of view. Under this situation, the reward of each player is not less than $\frac{1}{\alpha}$ times the difference of social utility caused by his absence, where α is a positive real number and other constrains are the same as those in the valid utility system. In the context of cache network, they prove that the selfish cache game belongs to a class of α-scalable valid utility system, and conclude that the price of anarchy of this system is $\frac{1}{1+\alpha}$.

Main Contribution. We have following contributions.

1. We introduce an (a, b)-generic ($a, b \in \mathbb{R}^+$) valid utility system, a class of non-cooperative games with n players. The social utility of an outcome is measured by a submodular function. The private utility of a player is at most a times the change in social utility that would occur if the player declines to participate in the game. For any outcome, the total amount of the utility of all players is at most b times the social utility (Sect. 3.1).
2. We prove the existence of pure strategy Nash equilibria in an (a, b)-generic basic utility system, and show that the lower bound of the price of anarchy of this system and (a, b)-generic valid utility system is $\frac{a}{a+b}$ (Sect. 3.2).
3. We design a mechanism for the case that there is no pure strategy Nash equilibria in an (a, b)-generic valid utility system. It outputs an outcome whose social utility is at least $\frac{a}{2bn+a-b}$ times of the optimal (Sect. 3.3).

2 Preliminaries

2.1 Game Theory

We consider a non-cooperative game as follows. Let $[n] = \{1, 2, \ldots, n\}$ be the set of players. Let V be a ground set and $V_1, V_2, \ldots, V_n$ be a partition of V. Let $\mathcal{S}_i = \{s_i \mid s_i \subseteq V_i\}$ ($i = 1, \ldots, n$) be the pure strategy set of player i. Each player aims to maximize his private utility function $\alpha_i : 2^V \to \mathbb{R}^+$, which in general depends on the strategies of all the players. The social utility function is $\gamma : 2^V \to \mathbb{R}^+$. We denote the game by a tuple $G = ([n], \{\mathcal{S}_i\}_{i\in[n]}, \{\alpha_i\}_{i\in[n]}, \gamma)$. Note that the social utility function $\gamma : 2^V \to \mathbb{R}^+$ and private utility functions $\alpha_i : 2^V \to \mathbb{R}^+$ $(i \in [n])$ can be regarded as set functions, owing to the disjoint setting of $V_1, V_2, \ldots, V_n$.

Denote strategy space as $\mathcal{S} = \mathcal{S}_1 \times \mathcal{S}_2 \times \cdots \times \mathcal{S}_n$. Given an outcome $\boldsymbol{s} = (s_1, s_2, \ldots, s_n) \in \mathcal{S}$, let $\boldsymbol{s} \otimes \tilde{s}_i = (s_1, \ldots, s_{i-1}, \tilde{s}_i, s_{i+1}, \ldots, s_n)$ be a new outcome that player i unilaterally changes his strategy from s_i to $\tilde{s}_i$. We use the standard notation $(s_i, \boldsymbol{s}_{-i})$ to denote the outcome where player i chooses strategy s_i and the other players select strategies $\boldsymbol{s}_{-i} = (s_1, \ldots, s_{i-1}, s_{i+1}, \ldots, s_n)$. We use $\emptyset_i$ to denote the null strategy of player i, and let $\emptyset = (\emptyset_1, \emptyset_2, \ldots, \emptyset_n)$ be an outcome that each player has a null strategy. For the rest of paper, we assume that $\gamma(\emptyset) = 0$.

The outcome $\boldsymbol{s} = (s_1, s_2, \ldots, s_n) \in \mathcal{S}$ is a *pure strategy Nash equilibrium*, if

$$\alpha_i(\tilde{s}_i, \boldsymbol{s}_{-i}) \le \alpha_i(\boldsymbol{s}), \ \forall i \in [n], \ \forall \tilde{s}_i \in \mathcal{S}_i. \tag{1}$$

In 1951, Nash [13] presents the following theorem. But it holds for mixed strategy Nash equilibria rather than pure strategy Nash equilibria (abbreviated as PNE).

Theorem 1. *A finite non-cooperative game always has at least one Nash equilibrium.*

The efficiency of the Nash equilibria reached by a game is measured by the *price of anarchy* [8].

Definition 1. *In a non-cooperative game $\mathcal{G}$ with PNE, the price of anarchy*

$$PoA = \min_{\boldsymbol{s}\in\mathcal{N}} \frac{\gamma(\boldsymbol{s})}{\gamma(\boldsymbol{s}^*)}, \tag{2}$$

where $\mathcal{N} \subseteq \mathcal{S}$ is the set of PNE in the instance I, and $\boldsymbol{s}^*$ is the optimal outcome in the instance I.

2.2 Submodular Function

Given a ground set V, consider a set function $f : 2^V \to \mathbb{R}$. f is *monotone*, if $f(A) \le f(B)$, $\forall A \subseteq B \subseteq V$. f is *submodular*, if $f(A) + f(B) \ge f(A \cup B) + f(A \cap B)$, $\forall A, B \subseteq V$. Essentially, submodularity reflects the property of diminishing returns and possesses the following equivalent definition.

A set function f is *submodular* if and only if

$$f(A \cup \{e\}) - f(A) \ge f(B \cup \{e\}) - f(B), \forall A \subseteq B, \forall e \in V \setminus B. \tag{3}$$

3 Generic Valid Utility Systems

In Sect. 3.1, we present our system and prove its existence. In Sect. 3.2, we show that the existence of a PNE for a special generic basic utility system and give a conclusion that its price of anarchy is at least $\frac{a}{a+b}$. Since generic valid utility system may not have PNE, in Sect. 3.3 we design a best response mechanism and give a $\frac{a}{2bn+a-b}$-approximate outcome.

3.1 Definitions and Properties

We first present the definition of (a, b)-generic valid utility system.

Definition 2. *Given a social utility function* $\gamma : 2^V \to \mathbb{R}^+$ *and* n *private utility functions* $\alpha_i : 2^V \to \mathbb{R}^+$ $(i \in [n])$*, the game* $([n], \{\mathcal{S}_i\}_{i\in[n]}, \{\alpha_i\}_{i\in[n]}, \gamma)$ *is said to be an* (a, b)*-generic valid utility system, if for any* $\boldsymbol{s} \in \mathcal{S}$ *and any* $a, b \in \mathbb{R}^+$ *such that* $a \le b$*, it satisfies that:*

1. γ *and* α_i $(i \in [n])$ *are measured in the same criteria;*
2. γ *is monotone and submodular;*
3. $\alpha_i(\boldsymbol{s}) \ge a \cdot (\gamma(\boldsymbol{s}) - \gamma(\boldsymbol{s} \otimes \emptyset_i))$*;*
4. $\sum_i \alpha_i(\boldsymbol{s}) \le b \cdot \gamma(\boldsymbol{s})$*.*

Remark 1. In particular, the *Valid Utility System* proposed in Vetta [15] is a special case of the (a, b)-generic valid utility system with $a = b = 1$.

Specially, we say that an (a, b)-Generic Valid Utility System is basic if the third constraint holds in equation, i.e., $\alpha_i(\boldsymbol{s}) = a \cdot (\gamma(\boldsymbol{s}) - \gamma(\boldsymbol{s} \otimes \emptyset_i))$ for any $\boldsymbol{s} \in \mathcal{S}$.

Next, we show that there indeed exist (a, b)-generic valid utility systems. Due to the non-negative of $\gamma(\cdot)$ and $\alpha_i(\cdot)$ $(i \in [n])$, we only consider the meaningful case where $0 < a \le b$.

Theorem 2. *For any parameters* a *and* b $(a, b \in \mathbb{R}^+)$ *and any monotone submodular function* γ*, if* $a \le b$*, then there exist set functions* $\alpha_i (i \in [n])$*, such the game* $([n], \{\mathcal{S}_i\}_{i\in[n]}, \{\alpha_i\}_{i\in[n]}, \gamma)$ *is an* (a, b)*-generic valid utility system.*

Proof. It is sufficient to prove that there exists an (a, b)-generic basic utility system, a special case of (a, b)-generic valid utility systems. Given a pair of reals $a, b \in \mathbb{R}^+$ and a submodular function $\gamma : 2^V \to \mathbb{R}$, we define $\alpha_i(\boldsymbol{s}) = a \cdot (\gamma(\boldsymbol{s}) - \gamma(\boldsymbol{s} \otimes \emptyset_i))$ for any $\boldsymbol{s} \in \mathcal{S}$. If $a \le b$, we have

$$\begin{aligned} \textstyle\sum_{i=1}^n \alpha_i(\boldsymbol{s}) &= \textstyle\sum_{i=1}^n a \cdot [\gamma(\boldsymbol{s}) - \gamma(\boldsymbol{s} \otimes \emptyset_i)] \le \textstyle\sum_{i=1}^n a \cdot [\gamma(\boldsymbol{s}^i) - \gamma(\boldsymbol{s}^{i-1})] \\ &\le \textstyle\sum_{i=1}^n b \cdot [\gamma(\boldsymbol{s}^i) - \gamma(\boldsymbol{s}^{i-1})] \le b \cdot \gamma(\boldsymbol{s}). \end{aligned} \tag{4}$$

Therefore, the game $([n], \{\mathcal{S}_i\}_{i\in[n]}, \{\alpha_i\}_{i\in[n]}, \gamma)$ is an (a, b)-generic basic utility system. □

3.2 PoA of a Generic Basic Utility System

For an (a, b)-generic basic utility system $([n], \{\mathcal{S}_i\}_{i\in[n]}, \{\alpha_i\}_{i\in[n]}, \gamma)$, we can obtain the following results.

Theorem 3. *An* (a, b)*-generic basic utility system* $([n], \{\mathcal{S}_i\}_{i\in[n]}, \{\alpha_i\}_{i\in[n]}, \gamma)$ *always has at least one PNE.*

Proof. Consider an (a,b)-generic basic utility system $([n], \{\mathcal{S}_i\}_{i\in[n]}, \{\alpha_i\}_{i\in[n]}, \gamma)$. We construct a corresponding directed graph $G = (V_e, E)$ as follows. For any outcome $\boldsymbol{s} \in \mathcal{S}$, there is a vertex $v_{\boldsymbol{s}}$ in V_e. For any player $i \in [n]$, if

$$\alpha_i(s_1,\ldots,s_i,\ldots,s_n) < \alpha_i(s_1,\ldots,\tilde{s}_i,\ldots,s_n), \tag{5}$$

then there is a directed edge from vertex $(s_1,\ldots,s_i,\ldots,s_n)$ to vertex $(s_1,\ldots,\tilde{s}_i,\ldots,s_n)$.

Note that vertex $a \in V_e$ corresponds to a PNE if the out-degree of a equals to zero. Thus, it is enough to prove that directed graph G is acyclic. Assuming that directed graph G is cyclic. Let D be a directed cycle in G, and D contains vertices $a^0, a^1, \ldots, a^{t-1}, a^t$ sequentially. Here $a^i = (s_1^i, s_2^i, \ldots, s_n^i)$, for $i = 0, 1, \ldots, t$ and $a^0 = a^t$. On the one hand, from the construction of G, a^r and a^{r+1} have only one distinction, which is one player's strategy, denoted as i_r. In other words, for any $i \neq i_r$, $s_i^r = s_i^{r+1}$ holds. And for $r = 0, 1, \ldots, t-1$, there is

$$\alpha_{i_r}(s_1^r,\ldots,s_{i_r}^r,\ldots,s_n^r) < \alpha_{i_r}(s_1^{r+1} = s_1^r,\ldots,s_{i_r}^{r+1},\ldots,s_n^{r+1} = s_n^r). \tag{6}$$

Adding up these t terms, we have

$$\sum_{r=0}^{t-1}[\alpha_{i_r}(a^{r+1}) - \alpha_{i_r}(a^r)] > 0. \tag{7}$$

On the other hand, by the definition of (a,b)-generic basic utility system, we have $\alpha_{i_r}(a^r) = a \cdot [\gamma(a^r) - \gamma(a^r \otimes \emptyset_{i_r})]$ and $\alpha_{i_r}(a^{r+1}) = a \cdot [\gamma(a^{r+1}) - \gamma(a^{r+1} \otimes \emptyset_{i_r})]$. Then

$$\begin{aligned}\alpha_{i_r}(a^{r+1}) - \alpha_{i_r}(a^r) &= a \cdot [\gamma(a^{r+1}) - \gamma(a^r)] + a \cdot [\gamma(a^r \otimes \emptyset_{i_r}) - \gamma(a^{r+1} \otimes \emptyset_{i_r})] \\ &= a \cdot [\gamma(a^{r+1}) - \gamma(a^r)],\end{aligned} \tag{8}$$

where the last equality follows from that $s_i^r = s_i^{r+1}$, for any $i \neq i_r$. Adding up these t terms, we have

$$\begin{aligned}&\sum_{r=0}^{t-1}[\alpha_{i_r}(a^{r+1}) - \alpha_{i_r}(a^r)] \\ &= \sum_{r=0}^{t-1} a \cdot [\gamma(a^{r+1}) - \gamma(a^r)] \\ &= a \cdot [\gamma(a^t) - \gamma(a^0)] \\ &= 0,\end{aligned} \tag{9}$$

which is contradicted with inequality (7). Thus, we complete the proof. □

Then we discuss the efficiency of any Nash equilibrium in an (a,b)-generic basic utility system.

Theorem 4. *For any (a,b)-generic basic utility system $([n], \{\mathcal{S}_i\}_{i\in[n]}, \{\alpha_i\}_{i\in[n]}, \gamma)$, there holds $PoA \geq \frac{a}{a+b}$.*

Proof. Let $\Omega = (\sigma_1, \sigma_2, \dots, \sigma_n)$ be an optimal outcome, and $\boldsymbol{s} = (s_1, s_2, \dots, s_n)$ be any Nash equilibrium. Then, we have

$$\begin{aligned}
\gamma(\Omega) &\le \gamma(\Omega \cup \boldsymbol{s}) \\
&\le \gamma(\boldsymbol{s}) + \sum_{i:\sigma_i \in \Omega \setminus \boldsymbol{s}} [\gamma(\boldsymbol{s} \cup \sigma_i) - \gamma(\boldsymbol{s})] \\
&\le \gamma(\boldsymbol{s}) + \sum_{i:\sigma_i \in \Omega \setminus \boldsymbol{s}} [\gamma(\boldsymbol{s} \otimes \sigma_i) - \gamma(\boldsymbol{s} \otimes \emptyset_i)] \\
&= \gamma(\boldsymbol{s}) + \tfrac{1}{a} \cdot \sum_{i:\sigma_i \in \Omega \setminus \boldsymbol{s}} \alpha_i(\boldsymbol{s} \otimes \sigma_i) \\
&\le \gamma(\boldsymbol{s}) + \tfrac{1}{a} \cdot \sum_{i:\sigma_i \in \Omega \setminus \boldsymbol{s}} \alpha_i(\boldsymbol{s}) \\
&\le \gamma(\boldsymbol{s}) + \tfrac{1}{a} \cdot \sum_{i=1}^{n} \alpha_i(\boldsymbol{s}) \\
&\le \gamma(\boldsymbol{s}) + \tfrac{b}{a} \cdot \gamma(\boldsymbol{s}) \\
&= \tfrac{a+b}{a} \cdot \gamma(\boldsymbol{s})
\end{aligned} \tag{10}$$

where the first inequality follows from the monotonicity of γ, the next two inequalities follow from the submodularity of γ, the first equality follows from the definition of (a, b)-generic basic utility system, the fourth inequality holds since $\boldsymbol{s}$ is a PNE, the fifth inequality holds by the non-negativity of α_i ($i \in [n]$), and the last inequality follows from the definition of (a, b)-generic basic utility system. □

And we have a theorem as follows.

Theorem 5. *For any (a, b)-generic valid utility system $([n], \{\mathcal{S}_i\}_{i\in[n]}, \{\alpha_i\}_{i\in[n]}, \gamma)$ with at least one PNE, there holds PoA $\ge \frac{a}{a+b}$.*

Its proof is similar to that of Theorem 4. The difference is that the first equality sign in formula (10) becomes less-than-equal.

3.3 Mechanism of Generic Valid Utility System

Consider any (a, b)-generic valid utility system $([n], \{\mathcal{S}_i\}_{i\in[n]}, \{\alpha_i\}_{i\in[n]}, \gamma)$ with positive a, b. We cannot make sure that there exists a PNE or not in the system. This requires an alternative approach to evaluate these types of games. In this section, we design a mechanism to output an outcome which is not necessarily a PNE, but whose efficiency is guaranteed.

Sort the players in an arbitrary order. Without loss of generality, we denote them as $1, 2, \dots, n$. In this order and starting from any initially outcome $\boldsymbol{s} = (s_1, \dots, s_n)$, players change their strategies sequentially. The decision rule is that each player makes the best response to the current outcome. Let $\tilde{\boldsymbol{s}}^{(i)} = (\tilde{s}_1, \dots, \tilde{s}_i, s_{i+1}, \dots, s_n)$ represent the intermediate outcome after the change of the i-th player.

Let $\boldsymbol{t}$ be the output of Algorithm 1 and Ω be the optimal solution. In the following we estimate the efficiency of $\boldsymbol{t}$, the ratio between $\gamma(\boldsymbol{t})$ and $\gamma(\Omega)$.

Algorithm 1. Best Response Mechanism

Input: An (a, b)-generic valid utility system $([n], \{\mathcal{S}_i\}_{i\in[n]}, \{\alpha_i\}_{i\in[n]}, \gamma)$.
Output: An outcome $\boldsymbol{t}$.

1: Initialize: Sort the players in an arbitrary order and denote them by $1, 2, \ldots, n$; let $\boldsymbol{s} = (s_1, s_2, \ldots, s_n)$ be an arbitrary initial outcome; set $\boldsymbol{t} = \tilde{\boldsymbol{s}}^{(0)} = \boldsymbol{s}$.
2: **for** $i = 1$ **to** n **do**
3: $\tilde{s}_i \leftarrow \arg\max_{s\in\mathcal{S}_i}\{\alpha_i(\tilde{\boldsymbol{s}}^{(i-1)} \otimes s)\}$, $\tilde{\boldsymbol{s}}^{(i)} = \tilde{\boldsymbol{s}}^{(i-1)} \otimes \tilde{s}_i$.
4: **if** $\gamma(\tilde{\boldsymbol{s}}^{(i)}) > \gamma(\boldsymbol{t})$.
$\boldsymbol{t} \leftarrow \tilde{\boldsymbol{s}}^{(i)}$.
5: **end if**
6: **end for**
7: Return $\boldsymbol{t}$.

Theorem 6. *Consider an (a, b)-generic valid utility system $([n], \{\mathcal{S}_i\}_{i\in[n]}, \{\alpha_i\}_{i\in[n]}, \gamma)$. Let $\boldsymbol{t}$ be the outcome returned by Algorithm 1 and Ω be the optimal outcome. Then $\gamma(\boldsymbol{t}) \geq \frac{a}{2bn+a-b} \cdot \gamma(\Omega)$.*

Proof. Denote the optimal outcome by $\Omega = (\sigma_1, \ldots, \sigma_n)$. For the initial outcome $\boldsymbol{s}$, if $\gamma(\boldsymbol{s}) \geq \frac{a}{2bn+a-b} \cdot \gamma(\Omega)$, we have $\gamma(\boldsymbol{t}) \geq \gamma(\boldsymbol{s}) \geq \frac{a}{2bn+a-b} \cdot \gamma(\Omega)$ holds trivially. Thus, we only need to concentrate on the case that $\gamma(\boldsymbol{s}) < \frac{a}{2bn+a-b} \cdot \gamma(\Omega)$. We claim that there is some $l \in [n]$ such that $\tilde{\boldsymbol{s}}^{(l)}$ is the first outcome satisfying $\alpha_l(\tilde{\boldsymbol{s}}^{(l)}) \geq \frac{ab}{2bn+a-b} \cdot \gamma(\Omega)$. If the claim holds, we obtain

$$\gamma(\boldsymbol{t}) \geq \gamma(\tilde{\boldsymbol{s}}^{(l)}) \geq \frac{1}{b} \cdot \sum_{i=1}^{n} \alpha_i(\tilde{\boldsymbol{s}}^{(l)}) \geq \frac{1}{b} \cdot \alpha_l(\tilde{\boldsymbol{s}}^{(l)}) \geq \frac{a}{2bn+a-b} \cdot \gamma(\Omega), \tag{11}$$

where the first inequality follows from our mechanism, the second inequality follows from the definition of (a, b)-generic valid utility system, the third inequality follows from the nonnegativity of $\alpha_i(\cdot)$ $(i \in [n])$, and the last inequality follows from the claim. Then we are done if we can show that the claim is true.

We now prove the correctness of the claim. Firstly, let $m \in [n-1]$ be a positive integer such that $\alpha_i(\tilde{\boldsymbol{s}}^{(i)}) < \frac{ab}{2bn+a-b} \cdot \gamma(\Omega)$ for any $i \in [m]$. Then

$$\begin{aligned}
&\gamma(\boldsymbol{s} \cup \tilde{\boldsymbol{s}}^{(m)}) - \gamma(\boldsymbol{s}) \\
&= \sum_{i=1}^{m} [\gamma(\boldsymbol{s} \cup \tilde{\boldsymbol{s}}^{(i)}) - \gamma(\boldsymbol{s} \cup \tilde{\boldsymbol{s}}^{(i-1)})] \\
&\leq \sum_{i=1}^{m} [\gamma(\tilde{\boldsymbol{s}}^{(i)}) - \gamma(\tilde{\boldsymbol{s}}^{(i)} \otimes \emptyset_i)] \\
&\leq \sum_{i=1}^{m} \frac{1}{a} \cdot \alpha_i(\tilde{\boldsymbol{s}}^{(i)}) \\
&< \frac{bm}{2bn+a-b} \cdot \gamma(\Omega),
\end{aligned} \tag{12}$$

where the first inequality follows from the submodularity of γ, the second inequality follows from the definition of (a, b)-generic valid utility system. We have

$$\begin{aligned}
&\gamma(\boldsymbol{s}\cup\tilde{\boldsymbol{s}}^{(m)}\cup\Omega)-\gamma(\boldsymbol{s}\cup\tilde{\boldsymbol{s}}^{(m)})\\
&=[\gamma(\boldsymbol{s}\cup\tilde{\boldsymbol{s}}^{(m)}\cup\Omega)-\gamma(\boldsymbol{s})]-[\gamma(\boldsymbol{s}\cup\tilde{\boldsymbol{s}}^{(m)})-\gamma(\boldsymbol{s}))]\\
&>\gamma(\Omega)-\gamma(\boldsymbol{s})-\tfrac{bm}{2bn+a-b}\cdot\gamma(\Omega)\\
&\geq\tfrac{2bn-b-bm}{2bn+a-b}\cdot\gamma(\Omega),
\end{aligned}\tag{13}$$

where the first inequality follows from the monotonicity of γ and (12), and the last inequality follows from our assumption that $\gamma(\boldsymbol{s}) < \frac{a}{2bn+a-b}\cdot\gamma(\Omega)$. Moreover, there is

$$\begin{aligned}
&\gamma(\boldsymbol{s}\cup\tilde{\boldsymbol{s}}^{(m)}\cup\{\sigma_1,\ldots,\sigma_m\})-\gamma(\boldsymbol{s}\cup\tilde{\boldsymbol{s}}^{(m)})\\
&\leq\gamma(\boldsymbol{s}\cup\tilde{\boldsymbol{s}}^{(m)}\cup\{\sigma_1,\ldots,\sigma_m\})-\gamma(\boldsymbol{s})\\
&=\sum_{i=1}^{m}[\gamma(\boldsymbol{s}\cup\tilde{\boldsymbol{s}}^{(i)}\cup\{\sigma_1,\ldots,\sigma_i\})-\gamma(\boldsymbol{s}\cup\tilde{\boldsymbol{s}}^{(i-1)}\cup\{\sigma_1,\ldots,\sigma_{i-1}\})]\\
&\leq\sum_{i=1}^{m}[\gamma(\boldsymbol{s}\cup\tilde{\boldsymbol{s}}^{(i)}\cup\{\sigma_i\})-\gamma(\boldsymbol{s}\cup\tilde{\boldsymbol{s}}^{(i-1)})]\\
&\leq\sum_{i=1}^{m}[\gamma(\tilde{\boldsymbol{s}}^{(i)}\cup\{\sigma_i\})-\gamma(\tilde{\boldsymbol{s}}^{(i-1)})]\\
&=\sum_{i=1}^{m}[\gamma(\tilde{\boldsymbol{s}}^{(i)}\otimes(\sigma_i\cup\tilde{s}_i))-\gamma(\tilde{\boldsymbol{s}}^{(i-1)})]\\
&\leq\sum_{i=1}^{m}[\gamma(\tilde{\boldsymbol{s}}^{(i)}\otimes(\sigma_i\cup\tilde{s}_i))-\gamma(\tilde{\boldsymbol{s}}^{(i)}\otimes\emptyset_i)]\\
&\leq\sum_{i=1}^{m}\tfrac{1}{a}\cdot\alpha_i(\tilde{\boldsymbol{s}}^{(i)}\otimes(\sigma_i\cup\tilde{s}_i))\\
&\leq\sum_{i=1}^{m}\tfrac{1}{a}\cdot\alpha_i(\tilde{\boldsymbol{s}}^{(i)})\\
&<\tfrac{bm}{2bn+a-b}\cdot\gamma(\Omega),
\end{aligned}\tag{14}$$

where the first inequality follows from the monotonicity of γ, the second and third inequalities follow from the submodularity of γ, the fourth inequality follows from the monotonicity of γ, the fifth inequality follows from the definition of (a, b)-generic valid utility system, the sixth inequality holds by the fact that $\tilde{\boldsymbol{s}}^{(i)}$ is the best response for i, and the last inequality follows from our assumption. Combining inequalities (13) and (14), we obtain

$$\begin{aligned}
&\textstyle\sum_{i=m+1}^{n}[\gamma(\boldsymbol{s}\cup\tilde{\boldsymbol{s}}^{(m)}\cup\{\sigma_i\})-\gamma(\boldsymbol{s}\cup\tilde{\boldsymbol{s}}^{(m)})]\\
&\geq\gamma(\boldsymbol{s}\cup\tilde{\boldsymbol{s}}^{(m)}\cup\{\sigma_{m+1},\ldots,\sigma_n\})-\gamma(\boldsymbol{s}\cup\tilde{\boldsymbol{s}}^{(m)})\\
&\geq\gamma(\boldsymbol{s}\cup\tilde{\boldsymbol{s}}^{(m)}\cup\Omega)-\gamma(\boldsymbol{s}\cup\tilde{\boldsymbol{s}}^{(m)}\cup\{\sigma_1,\ldots,\sigma_m\})\\
&=\gamma(\boldsymbol{s}\cup\tilde{\boldsymbol{s}}^{(m)}\cup\Omega)-\gamma(\boldsymbol{s}\cup\tilde{\boldsymbol{s}}^{(m)})-[\gamma(\boldsymbol{s}\cup\tilde{\boldsymbol{s}}^{(m)}\cup\{\sigma_1,\ldots,\sigma_m\})-\gamma(\boldsymbol{s}\cup\tilde{\boldsymbol{s}}^{(m)})]\\
&\geq\tfrac{2bn-b-2bm}{(2bn+a-b)}\cdot\gamma(\Omega),
\end{aligned}\tag{15}$$

where the first two inequalities follow from the submodularity of γ. Then there must exist one player l with $m+1 \leq l \leq n$ satisfying

$$\begin{aligned}\gamma(\boldsymbol{s} \cup \tilde{\boldsymbol{s}}^{(m)} \cup \sigma_l) - \gamma(\boldsymbol{s} \cup \tilde{\boldsymbol{s}}^{(m)}) &\geq \tfrac{2bn-b-2bm}{(2bn+a-b)(n-m)} \cdot \gamma(\Omega) \\ &= (\tfrac{2b}{2bn+a-b} - \tfrac{b}{(2bn+a-b)(n-m)}) \cdot \gamma(\Omega) \\ &\geq \tfrac{b}{2bn+a-b} \cdot \gamma(\Omega).\end{aligned} \tag{16}$$

where the first inequality follows from (15), and the last inequality follows from $n \geq 2$. If there is some $l' \in [l-1]$ with $\alpha_{l'}(\tilde{\boldsymbol{s}}^{(l')}) \geq \frac{ab}{2bn+a-b} \cdot \gamma(\Omega)$, the claim holds naturally. Otherwise, set $m = l-1$. Then for each $i \in [m]$, $\alpha_i(\tilde{\boldsymbol{s}}^{(i)}) < \frac{ab}{2bn+a-b} \cdot \gamma(\Omega)$ and the inequalities (12)–(16) hold. We can obtain

$$\begin{aligned}\alpha_l(\tilde{\boldsymbol{s}}^{(l)}) &= \alpha_l(\tilde{\boldsymbol{s}}^{(l-1)} \otimes \tilde{s}_l) \\ &\geq \alpha_l(\tilde{\boldsymbol{s}}^{(l-1)} \otimes \sigma_l) \\ &\geq a \cdot [\gamma(\tilde{\boldsymbol{s}}^{(l-1)} \otimes \sigma_l) - \gamma(\tilde{\boldsymbol{s}}^{(l-1)} \otimes \emptyset_l)] \\ &\geq a \cdot [\gamma((\boldsymbol{s} \cup \tilde{\boldsymbol{s}}^{(m)}) \cup \sigma_l) - \gamma(\boldsymbol{s} \cup \tilde{\boldsymbol{s}}^{(m)})] \\ &\geq \tfrac{ab}{2bn+a-b} \cdot \gamma(\Omega).\end{aligned} \tag{17}$$

where the first inequality holds since $\tilde{s}_l$ is the best response for l to the outcome $\tilde{\boldsymbol{s}}^{(l-1)}$, the second inequality follows from the definition of (a, b)-generic valid utility system, the third inequality follows from the submodularity of γ, and the last inequality follows from (16). Thus, the claim holds and this completes the proof of theorem. □

References

1. Bharathi, S., Kempe, D., Salek, M.: Competitive influence maximization in social networks. In: Deng, X., Graham, F.C. (eds.) WINE 2007. LNCS, vol. 4858, pp. 306–311. Springer, Heidelberg (2007). https://doi.org/10.1007/978-3-540-77105-0_31
2. Christodoulou, G., Koutsoupias, E.: The price of anarchy of finite congestion games. In: Proceedings of the Thirty-Seventh Annual ACM Symposium on Theory of Computing, pp. 67–73. Association for Computing Machinery, New York (2005)
3. Goemans, M., Mirrokni, V., Vetta, A.: Sink equilibria and convergence. In: 46th Annual IEEE Symposium on Foundations of Computer Science (FOCS 2005), pp. 142–151 (2005)
4. Grimsman, D., Seaton, H.-J., Marden, R.-J., Brown, N.-P.: The cost of denied observation in multiagent submodular optimization. In: 2020 59th IEEE Conference on Decision and Control (CDC), pp. 1666–1671 (2020)
5. Hartline, J., Hoy, D., Taggart, S.: Price of anarchy for auction revenue. In: Proceedings of the Fifteenth ACM Conference on Economics and Computation, pp. 693–710 (2014)
6. He, X., Kempe, D.: Price of anarchy for the N-player competitive cascade game with submodular activation functions. In: Chen, Y., Immorlica, N. (eds.) WINE 2013. LNCS, vol. 8289, pp. 232–248. Springer, Heidelberg (2013). https://doi.org/10.1007/978-3-642-45046-4_20
7. Johari, R., Tsitsiklis, N.-J.: Efficiency loss in a network resource allocation game. Math. Oper. Res. **29**(3), 407–435 (2004)
8. Koutsoupias, E., Papadimitriou, C.: Worst-case equilibria. In: Meinel, C., Tison, S. (eds.) STACS 1999. LNCS, vol. 1563, pp. 404–413. Springer, Heidelberg (1999). https://doi.org/10.1007/3-540-49116-3_38

problem, they presented an online algorithm with the best-possible competitive ratio $\frac{\sqrt{5}+3}{2} \approx 2.618$. For the off-line problem, they presented a $(2-\frac{1}{m})$– approximation algorithm and a polynomial-time approximation scheme (PTAS). Since then, scheduling with rejection has receive more and more attention in recent two decades. Shabtay et al. [24] pointed out, an important application of scheduling with rejection arises in make-to-order production systems with limited production capacity and tight delivery requirements. Another important application of scheduling with rejection occurs in scheduling with an outsourcing option. Moreover, there are many close connections between scheduling with rejection and other scheduling models such as scheduling with controllable processing times, scheduling with costs, and scheduling with due date assignment, etc.

For more problems and results on this topic, we refer the reader to a survey on off-line scheduling with rejection by Shabtay et al. [24]. More papers dealing scheduling with rejection are [1,3,8,12,14,15,18–20,23,27,29,30].

1.2 Scheduling with Generalized Due Dates

Scheduling with generalized due dates (GDD) are first introduced by Hall [10]. In the GDD model, there are n jobs $J_1, \cdots, J_n$ and n given due dates $d_{[1]}, \cdots, d_{[n]}$, where $[i]$ means the i-th position in a schedule. That is, if J_j is i-th scheduled job, then job J_j is assigned the due date $d_{[i]}$. Thus, in the GDD model, each due date is not for a specific job but for a position. Hall [10] described a number of GDD applications in different industries including public utility planning problems, survey design and manufacturing. In most cases, when changing the setting from standard (job-dependent) due dates to GDD, the problem becomes easier. For example, Hall [10] showed that problem $1|\text{GDD}|\sum T_j$ can be solved in polynomial time; however, the corresponding problem $1||\sum T_j$ is NP-hard (see Du and Leung [4]). Sriskandarajah [25] showed that problem $1|\text{GDD}|\sum w_jT_j$ is NP-hard. Furthermore, Gao and Yuan [5,6] showed that problems $1|\text{GDD}|\sum(E_j+T_j)$ and $1|\text{GDD}|\sum w_jT_j$ are strongly NP-hard, respectively. More papers on scheduling with generalized due dates can be found in [8,11,21,22,26,28].

1.3 Scheduling with Rejection and Generalized Due Dates

To our knowledge, only a few of papers studied several scheduling problems with rejection and generalized due dates. Gerstl and Mosheiov [8] studied the single machine scheduling problems with rejection and generalized due dates. Two objective functions are considered: maximum tardiness plus the total rejection cost, and total tardiness plus the total rejection cost. They showed that these two problems are NP-hard and proposed two pseudo-polynomial dynamic programmes and efficient heuristics for them. Mosheiov et al. [22] considered single machine scheduling with generalized due dates to minimize the total late work. For this problem, they provided a polynomial-time algorithm. Furthermore, the problem is extended to allow job rejection. The objective is to minimize the sum of the total late work and the total rejection cost. They proved that the latter problem is NP-hard and introduced pseudo-polynomial dynamic programming algorithms for this problem. Mor et al. [20] considered two scheduling problems with rejection and generalized due dates in a proportionate flow shop. The goal in the first problem is to minimize the sum of the total tardiness and the total rejection cost, while the goal in the second problem is to minimize the total rejection cost, given a bound on the total tardiness. They showed that both problems are NP-hard and designed some exact algorithms and approximation schemes for them.

The remaining parts are organised as follows: In Sect. 2, we provide the problem formulation on our problems. In Sect. 3, we show that the first problem with generalized release dates is binary NP-hard. Furthermore, we provide a pseudo-polynomial time algorithm, a 2-approximation algorithm and a full polynomial-time approximation scheme (FPTAS) for this problem. In Sects. 4 and 5, we show that the latter two problems with generalized processing times or generalized rejection costs can be solved in polynomial time, respectively.

2 Problem Formulation

The single machine scheduling problems with rejection and generalized parameters can be described as follows. There are n jobs $J_1, \ldots, J_n$ and a single machine. Each job J_j has a processing time p_j, a release date r_j and a rejection cost e_j. We assume that all p_j, r_j and e_j are integers. J_j is either rejected, in which case the rejection cost e_j has to be paid, or accepted and processed on the machine. Let A and R be the sets of accepted jobs and rejected job, respectively. For each job J_j with $J_j \in A$, let C_j be its completion time in a schedule. Furthermore, we define $C_{\max} = \max\{C_j : J_j \in A\}$ and $\sum_{J_j \in R} e_j$ by the makespan and the total rejection cost, respectively. The objective is to minimize the sum of the makespan and the total rejection cost.

Inspired by generalized due dates, we introduce three generalized parameters into our problems as follows: (1) generalized release dates (GRD), (2) generalized processing times (GPT), and (3) generalized rejection costs (GRC). In the GRD model, there are n given release dates $r_{[1]}, \cdots, r_{[n]}$. If J_j is i-th scheduled job in a schedule, then job J_j is assigned the release date $r_{[i]}$, i.e., $r_j = r_{[i]}$. That is, if J_j is i-th scheduled job, then it must be processed at or after time $r_{[i]}$. The GRD model can be applied in many service industries. For example, a patient usually make an appointment for a doctor in advance. According to his/her position among all patients in the appointment system, a conservatively estimated starting time is usually recommended to the patient such that the patient must be served by the doctor at or after this time. Thus, for each patient, this estimated starting time can be viewed as his/her generalized release date. Under the GRD modol, following the general notation introduced by Graham et al. [9], the corresponding problem can be denoted by

$$1|GRD, reject|C_{\max} + \sum_{J_j \in R} e_j. \qquad (1)$$

In the GPT model, if J_j is i-th scheduled job, then job J_j is assigned the processing time $p_{[i]}$. The GPT model might occur in many scheduling problems with position-dependent processing times (see Agnetis and Mosheiov [1]). Due to the learning or deterioration effects, the processing time of a job usually depend on its normal processing time and its position in a schedule. Furthermore, when all jobs have the same normal processing time, the processing time of each job will only depend on its position in a schedule. Thus, in this situation, the position-dependent processing time can be viewed as its generalized processing time. Under the GPT model, the corresponding problem can be denoted by

$$1|r_j, GPT, reject|C_{\max} + \sum_{J_j \in R} e_j. \tag{2}$$

Finally, in the GRC model, if J_j is i-th rejected job, then job J_j is assigned the rejection cost $e_{[i]}$. The GRC model might occur in some scheduling problems with outsourcing under different discount rates (see Lu et al. [16,17]). For example, we first assume that all jobs have the same original outsourcing cost e. Furthermore, if J_j is i-th outsourced job, then J_j has a corresponding discount rate DR_i. Thus, the actual outsourcing cost of J_j is $e_j = e \cdot DR_i$. Thus, the actual outsourcing cost $e_j = e \cdot DR_i$ can be viewed as the generalized rejection cost of J_j. Under the GRC modol, the corresponding problem can be denoted by

$$1|r_j, GRC, reject|C_{\max} + \sum_{J_j \in R} e_j. \tag{3}$$

3 Scheduling with Generalized Release Dates

3.1 NP-Hardness Proof

In this subsection, we show that problem $1|GRD, reject|C_{\max} + \sum_{J_j \in R} e_j$ is NP-hard. Here we modify the NP-hardness proof for problem $1|r_j, reject|C_{\max} + \sum_{J_j \in R} e_j$ in Zhang et al. [29].

Theorem 1. *Problem* $1|GRD, reject|C_{\max} + \sum_{J_j \in R} e_j$ *is NP-hard.*

Proof. The decision version of the problem is clearly in NP, we use the NP-complete Partition problem (see Garey and Johnson [7]) for the reduction.

Partition problem: Given $t+1$ non-negative integers $a_1, a_2, \cdots, a_t, B$ such that $\sum_{i=1}^{t} a_i = 2B$, are there two disjointed sets S_1 and S_2 such that $\sum_{a_i \in S_1} a_i = \sum_{a_i \in S_2} a_i = B$?

For a given instance of Partition problem, we construct an instance of the decision version of the problem $1|GRD, reject|C_{\max} + \sum_{J_j \in R} e_j$ as follows:

- $n = 2t + 1$ jobs.
- For $j = 1, \cdots, t$, we have $p_j = 2a_j$ and $e_j = a_j$.
- For $j = t + 1, \cdots, 2t + 1$, we have $p_j = 0$ and $e_j = 3B + 1$.

- The release dates are defined by $r_{[1]} = \cdots = r_{[t]} = 0$ and $r_{[t+1]} = \cdots = r_{[2t+1]} = 2B$.
- The threshold value is defined by $Y = 3B$.
- The decision version asks whether there is a schedule π such that $C_{\max} + \sum_{J_j \in R} e_j \leq Y$.

It can be observed that the above construction can be done in polynomial time. We show in the following that Partition problem has a solution if and only if there is a schedule π of the scheduling problem such that $C_{\max} + \sum_{J_j \in R} e_j \leq Y$.

($\Longrightarrow$) We first assume that Partition problem has a solution (S_1, S_2). Set $A = \{J_j : a_j \in S_1\} \cup \{J_j : j = t+1, \cdots, 2t+1\}$ and $R = \{J_j : a_j \in S_2\}$. We accept all jobs in A and reject all jobs in R. Furthermore, the jobs in $\{J_j : a_j \in S_1\}$ are processed before the jobs in $\{J_j : j = t+1, \cdots, 2t+1\}$. It is easy to verify that

$$C_{\max} + \sum\nolimits_{J_j \in R} e_j = \sum\nolimits_{a_j \in S_1} p_j + \sum\nolimits_{a_j \in S_2} e_j = 2\sum\nolimits_{a_j \in S_1} a_j + \sum\nolimits_{a_j \in S_2} a_j = 3B.$$

Thus, if Partition problem has a solution (S_1, S_2), we can construct a feasible schedule such that $C_{\max} + \sum_{J_j \in R} e_j = 3B$.

($\Longleftarrow$) Next, we suppose that there is a schedule π such that $C_{\max} + \sum_{J_j \in R} e_j \leq Y$. We need to show that Partition problem has a solution. Denote by A and R the sets of the accepted jobs and the rejected jobs in π, respectively. We have the following claims.

Claim 1. $\{J_j : j = t+1, \cdots, 2t+1\} \subseteq A$ and $C_{\max} \geq 2B$.

If there is a job J_j with $t+1 \leq j \leq 2t+1$ such that $J_j \in R$. Note that $e_j = 3B+1$ for each $j = t+1, \cdots, 2t+1$. Thus, we have $C_{\max} + \sum_{J_j \in R} e_j \geq 3B+1 > Y$, a contradiction. Thus, we have $\{J_j : j = t+1, \cdots, 2t+1\} \subseteq A$. Note further that $|A| \geq t+1$. Thus, there must be a job $J_j \in A$ which is processed at or after time $r_{[t+1]} = 2B$. It follows that $C_{\max} \geq 2B$. Claim 1 follows.

Claim 2. $\sum_{J_j \in R} a_j = B$.

Since $C_{\max} \geq 2B$ and $C_{\max} + \sum_{J_j \in R} e_j \leq Y = 3B$, we have $\sum_{J_j \in R} a_j = \sum_{J_j \in R} e_j \leq B$. If $\sum_{J_j \in R} a_j < B$, then we have

$$\begin{aligned} C_{\max} + \sum\nolimits_{J_j \in R} e_j &\geq \sum\nolimits_{J_j \in A} p_j + \sum\nolimits_{J_j \in R} e_j \\ &= 2\sum\nolimits_{J_j \in A} a_j + \sum\nolimits_{J_j \in R} a_j \\ &= 2(\sum\nolimits_{J_j \in A} a_j + \sum\nolimits_{J_j \in R} a_j) - \sum\nolimits_{J_j \in R} a_j \\ &= 4B - \sum\nolimits_{J_j \in R} a_j \\ &> 3B, \end{aligned}$$

a contradiction. Thus, we have $\sum_{J_j \in R} a_j = B$. Claim 2 follows.

Let $S_1 = \{a_j : J_j \in R\}$ and $S_2 = \{a_j : J_j \in A \cap \{J_j : j = 1, \cdots, t\}\}$. By Claim 2, we have $\sum_{J_j \in S_1} a_j = \sum_{J_j \in S_2} a_j = B$. Thus, (S_1, S_2) is a solution of Partition problem. Theorem 1 follows.

3.2 Dynamic Programming Algorithm

For problem $1|r_j, reject|C_{\max} + \sum_{J_j \in R} e_j$, Zhang et al. [29] shows that there exists an optimal schedule such that the accepted jobs are processed in the ERD-rule. Based on the ERD-rule, they provided two dynamic programming algorithms for the above problem. However, in our problem $1|GRD, reject|C_{\max} + \sum_{J_j \in R} e_j$, all release dates are position-dependent. Thus, the ERD-rule is invalid for our problem. Next, we will show that the LPT-rule is valid for problem $1|GRD, reject|C_{\max} + \sum_{J_j \in R} e_j$. This means that we can replace the ERD-rule by the LPT-rule for proposing our dynamic programming algorithm for problem $1|GRD, reject|C_{\max} + \sum_{J_j \in R} e_j$. We have the following lemma.

Lemma 1. *For problem $1|GRD, reject|C_{\max} + \sum_{J_j \in R} e_j$, there exists an optimal schedule such that all accepted jobs are processed in the LPT-rule.*

Proof. Lemma 1 can be proved by a pairwise exchange argument. Thus, we omit the detailed proof.

From Lemma 1, we first re-label the jobs in the LPT-rule such that $p_1 \geq p_2 \geq \cdots \geq p_n$. Write $E = \sum_{j=1}^{n} e_j$. Let $f_j(V, i)$ be the optimal value of the objective function when the jobs in consideration are $J_1, \cdots, J_j$, the total rejection cost is exactly V and the number of the current accepted jobs is i. Now, we consider any optimal schedule for the jobs $J_1, \cdots, J_j$ in which the total rejection cost is exactly V and the number of the current accepted jobs is i. In any such schedule, there are two possible cases: either J_j is rejected or J_j is accepted and processed on the machine.

Case 1. Job J_j is rejected. In this case, when only the jobs $J_1, \cdots, J_{j-1}$ is considered, the number of the current accepted jobs is still i and the total rejection cost is $V - e_j$. Thus, we have $f_j(V, i) = f_{j-1}(V - e_j, i) + e_j$.

Case 2. Job J_j is accepted. In this case, when only the jobs $J_1, \cdots, J_{j-1}$ is considered, the number of the current accepted jobs is $i-1$ and the total rejection cost is still V. Note that J_j is the $i-$th scheduled job and its release date is exactly $r_{[i]}$. Furthermore, the makespan for the accepted jobs among $J_1, \cdots, J_{j-1}$ is $f_{j-1}(V, i-1) - V$. Thus, we have $f_j(V, i) = \max\{f_{j-1}(V, i-1) - V, r_{[i]}\} + p_j + V$.

Combining the above two cases, we have the following dynamic programming algorithm DP1.

Dynamic Programming Algorithm DP1

The Boundary Conditions:

$$f_1(V, i) = \begin{cases} r_{[1]} + p_1, & \text{if } V = 0 \text{ and } i = 1; \\ e_1, & \text{if } V = e_1 \text{ and } i = 0; \\ +\infty, & \text{otherwise.} \end{cases}$$

The Recursive Function:

$$f_j(V, i) = \min\{f_{j-1}(V - e_j, i) + e_j, \max\{f_{j-1}(V, i-1) - V, r_{[i]}\} + p_j + V\}.$$

The optimal value is given by $\min\{f_n(V, i) : 0 \leq V \leq E, 0 \leq i \leq n\}$.

Theorem 2. *Algorithm DP1 solves problem* $1|GRD, reject|C_{\max} + \sum_{J_j \in R} e_j$ *in* $O(n^2 E)$ *time.*

Proof. The correctness of algorithm DP1 is guaranteed by the above discussion. The recursive function has at most $O(n^2E)$ states and each iteration costs a constant time. Thus, the running time of algorithm DP1 is bounded by $O(n^2E)$.

3.3 Approximation Algorithms

Assume that S is a set of some jobs. We define $P(S) = \sum_{J_j \in S} p_j$ and $E(S) = \sum_{J_j \in S} e_j$ by the total processing time and the total rejection cost of S, respectively. Now, we propose a simple 2-approximation algorithm.

Approximation Algorithm A

Step 1: We first re-label the jobs $J_1, \cdots, J_n$ such that $e_1 - p_1 \geq e_2 - p_2 \geq \cdots \geq e_n - p_n$.

Step 2: Set $S_0 = \emptyset$ and $S_i = \{J_1, \cdots, J_i\}$ for each $i = 1, 2, \cdots, n$. Furthermore, we also set $\overline{S_i} = \{J_{i+1}, \cdots, J_n\}$ for $i = 0, \cdots, n-1$ and $\overline{S_n} = \emptyset$.

Step 3: Accept all jobs in set S_i and reject the jobs in set $\overline{S_i}$. Assign the accepted jobs to be processed in the time interval $[r_{[i]}, r_{[i]} + P(S_i)]$ on the machine, where we assume that $r_{[0]} = 0$. The resulting schedule is denoted by $\pi(i)$.

Step 4: Let $Z(i)$ be the value of the objective function for each $\pi(i)$. Among all the schedules obtained above, select the one with the minimum $Z(i)$ value.

Let π be the schedule obtained from the above approximation algorithm A. Let Z and Z^* be the objective values of the schedule π and an optimal schedule π^*, respectively.

Theorem 3. $Z \leq 2Z^*$ *and the bound is tight.*

Proof. Let A^* and R^* be the sets of the accepted jobs and the rejected jobs in π^*, respectively. Let $|A^*| = k^*$ be the number of the jobs in A^*. By the definition of k^*, we have $Z^* \geq r_{[k^*]}$. Furthermore, we also have

$$\begin{aligned}
Z^* &\geq P(A^*) + E(R^*) \\
&= \textstyle\sum_{J_j \in A^*} p_j + \sum_{J_j \in R^*} e_j \\
&= \textstyle\sum_{J_j \in A^*} p_j + \sum_{j=1}^{n} e_j - \sum_{J_j \in A^*} e_j \\
&= \textstyle\sum_{j=1}^{n} e_j - \sum_{J_j \in A^*} (e_j - p_j) \\
&\geq \textstyle\sum_{j=1}^{n} e_j - \sum_{J_j \in S_{k^*}} (e_j - p_j) \\
&= \textstyle\sum_{J_j \in S_{k^*}} p_j + \sum_{J_j \in \overline{S_{k^*}}} e_j,
\end{aligned}$$

where the last inequation holds since S_{k^*} contains the jobs with the k^* maximum $e_j - p_j$ values. Thus, we have

$$Z \le Z(k^*) = r_{[k^*]} + P(S_{k^*}) + E(\overline{S_{k^*}}) = r_{[k^*]} + \sum\nolimits_{J_j \in S_{k^*}} p_j + \sum\nolimits_{J_j \in \overline{S_{k^*}}} e_j \le 2Z^*.$$

To show that the bound is tight, we consider the following instance with three jobs J_1, J_2, J_3 with $(p_1, e_1) = (0, 2)$, $(p_2, e_2) = (1, 2)$ and $(p_3, e_3) = (1, 0)$. Note that $e_1 - p_1 > e_2 - p_2 > e_3 - p_3$. The generalized release dates are defined by $r_{[1]} = 0$ and $r_{[2]} = r_{[3]} = 1$. If $i = 0$, then all jobs are rejected in schedule $\pi(0)$. Thus, we have $Z(0) = \sum_{j=1}^{3} e_j = 4$. If $i = 1$, then only J_1 is accepted and J_2, J_3 are rejected in $\pi(1)$. Thus, we have $Z(1) = r_{[1]} + p_1 + e_2 + e_3 = 2$. If $i = 2$, then jobs J_1, J_2 are accepted and J_3 is rejected in $\pi(2)$. Thus, we have $Z(2) = r_{[2]} + p_1 + p_2 + e_3 = 2$. If $i = 3$, then all jobs are accepted in $\pi(2)$. Thus, we have $Z(3) = r_{[3]} + p_1 + p_2 + p_3 = 3$. As a result, we have $Z = \min\{Z(i) : i = 0, 1, 2, 3\} = 2$. However, the optimal schedule is to accept J_2, J_1 (J_2 is scheduled before J_1) and reject J_3. That is, $Z^* = 1$. Thus, we have $Z = 2Z^*$.

By using the rounding technology to simplify the input data, Zhang et al. [29] provided a fully polynomial-time approximation scheme (FPTAS) for problem $1|r_j, reject|C_{\max} + \sum_{J_j \in R} e_j$. By borrowing their FPTAS, we can obtain an FPTAS for problem $1|GRD, reject|C_{\max} + \sum_{J_j \in R} e_j$. In fact, two FPTASs and their theoretical analysis are identical, only the running time is different. For the completeness, we repeat the detailed procedure of the FPTAS in the following.

Let Z and Z^* be the objective values of the schedule obtained by algorithm A and an optimal schedule π^*, respectively. By Theorem 3, we have $Z^* \le Z \le 2Z^*$. For any job J_j with $e_j > Z$, it is easy to see that $J_j \in A^*$, where A^* is the set of the accepted jobs in an optimal schedule π^*. Otherwise, we have $Z^* \ge e_j > Z \ge Z^*$, a contradiction. If we modify the rejection cost of such a job J_j such that $e_j = Z$, this does not change the optimal objective value. Thus without loss of generation, we can assume that $e_j \le Z$ for $j = 1, 2, \cdots, n$. Now, we propose an FPTAS A_ϵ for problem $1|GRD, reject|C_{\max} + \sum_{J_j \in R} e_j$.

FPTAS A_ϵ

Step 1: For any $\epsilon > 0$, set $M = \frac{\epsilon Z}{2n}$. Given an instance I, we define a new instance I' by rounding the rejection cost of the job in I such that $e_j{}' = \lfloor \frac{e_j}{M} \rfloor M$, for $j = 1, \cdots, n$.

Step 2: Apply the dynamic programming algorithm DP1 to instance I' to obtain an optimal solution $\pi^*(I')$ for instance I'.

Step 3: Replace the modified rejection cost $e_j{}'$ by the original rejection cost e_j in $\pi^*(I')$ for each $j = 1, \cdots, n$ to obtain a feasible solution π for instance I.

Let Z_ϵ be the objective value of the schedule π obtained from A_ϵ. We have the following theorem.

Theorem 4. *$Z_\epsilon \le (1 + \epsilon)Z^*$ and the running time of A_ϵ is $O(\frac{n^4}{\epsilon})$.*

Proof. Let $Z^*(I')$ be the optimal objective value of the schedule $\pi^*(I')$. From Step 1 of A_ϵ, we have $e_j' \le e_j < e_j' + M$. Thus, we have $Z^*(I') \le Z^*$. Replace e_j' by e_j for each $j = 1, \cdots, n$, we have

$$Z_\epsilon \le Z^*(I') + \sum_{j=1}^{n}(e_j - e_j') \le Z^* + nM \le Z^* + \frac{\epsilon Z}{2} \le (1+\epsilon)Z^*.$$

Since $e_j \le Z$ for $j = 1, \cdots, n$, we have $\sum_{j=1}^{n} \lfloor \frac{e_j}{M} \rfloor \le \frac{2n}{\epsilon} \sum_{j=1}^{n} \frac{e_j}{Z} \le \frac{2n^2}{\epsilon}$. Note that $e_j' = \lfloor \frac{e_j}{M} \rfloor M$, for $j = 1, \cdots, n$. Then, in the dynamic programming algorithm $DP1$, we have $V \in \{kM : 0 \le k \le \sum_{j=1}^{n} \lfloor \frac{e_j}{M} \rfloor\}$. That is, there are at most$\sum_{j=1}^{n} \lfloor \frac{e_j}{M} \rfloor = O(\frac{n^2}{\epsilon})$ choices for each V in DP1. Thus, the running time of A_ϵ is $O(\frac{n^4}{\epsilon})$. The theorem follows.

4 Scheduling with Generalized Processing Times

For problem $1|r_j|C_{\max}$, Lawer [13] show that it can be solved by ERD-rule in $O(n \log n)$ time. Note that the ERD-rule depends only on the release dates of the jobs, not on their processing times. Thus, the ERD-rule is still valid for our problem $1|r_j, GPT, reject|C_{\max} + \sum_{J_j \in R} e_j$. Thus, we have the following lemma.

Lemma 2. *For problem* $1|r_j, GPT, reject|C_{\max} + \sum_{J_j \in R} e_j$, *there exists an optimal schedule such that all accepted jobs are processed in the ERD-rule.*

Based on Lemma 2, we first re-label the jobs such that $r_1 \le r_2 \le \cdots \le r_n$. Let $f_j(C, i)$ be the optimal value of the objective function when the jobs in consideration are $J_1, \cdots, J_j$, the current makespan of the accepted jobs is exactly C and the number of accepted jobs is i. If $i = 0$, then we have $C = 0$. If $i > 0$, then the processing times of i accepted jobs are $p_{[1]}, \cdots, p_{[i]}$, respectively. Let t be minimum time such that all accepted jobs are processed consecutively in time $[t, C]$. Thus, we have $t \in \{r_k : 1 \le k \le j\}$. Furthermore, there must exist some i' with $i' \le i$ such that $C - t = p_{[i']} + \cdots + p_{[i]}$. From the above discussion, we can conclude that

$$C \in \{0\} \cup \{r_k + p_{[i']} + \cdots + p_{[i]} : 1 \le k \le j, 1 \le i' \le i \le j\}.$$

Now, we consider any optimal schedule for the jobs $J_1, \cdots, J_j$ in which the current makespan of the accepted jobs is exactly C and the number of accepted jobs is i. In any such schedule, there are two possible cases: either J_j is rejected or J_j is accepted and processed on the machine.

Case 1. Job J_j is rejected. In this case, when the jobs $J_1, \cdots, J_{j-1}$ are considered, the current makespan of the accepted jobs is still C and the number of accepted jobs is still i. Thus, we have $f_j(C, i) = f_{j-1}(C, i) + e_j$. For convenience, we set $V_R = f_{j-1}(C, i) + e_j$.

Case 2. Job J_j is accepted. In this case, when the jobs $J_1, \cdots, J_{j-1}$ are considered, the number of accepted jobs is $i-1$. Note that J_j is the i-th processed job. Thus, its processing time is $p_{[i]}$. It follows that $C \geq r_j + p_{[i]}$. Assume that the previous makespan is C' before J_j is scheduled. Thus, the previous total rejection cost is $f_{j-1}(C', i-1) - C'$. Note further that the starting time of J_j is exactly $\max\{C', r_j\}$. Thus, we have $C = \max\{C', r_j\} + p_{[i]}$. If $C > r_j + p_{[i]}$, then we have $C' = C - p_{[i]}$. If $C = r_j + p_{[i]}$, then we have $C' \leq r_j$. From the above discussion, we have

$$f_j(C,i) = \begin{cases} f_{j-1}(C - p_{[i]}, i-1) + p_j, & \text{if } C > r_j + p_{[i]}; \\ \min\{f_{j-1}(C', i-1) - C' + C : 0 \leq C' \leq r_j\}, & \text{if } C = r_j + p_{[i]}. \end{cases}$$

For convenience, we also set $V_A^1 = f_{j-1}(C - p_{[i]}, i-1) + p_j$ and $V_A^2 = \min\{f_{j-1}(C', i-1) - C' + C : 0 \leq C' \leq r_j\}$.

Combining the above two cases, we have the following dynamic programming algorithm DP2.

Dynamic Programming Algorithm DP2

The Boundary Conditions:

$$f_1(C,i) = \begin{cases} r_1 + p_{[1]}, & \text{if } C = r_1 + p_{[1]} \text{ and } i = 1; \\ e_1, & \text{if } C = 0 \text{ and } i = 0; \\ +\infty, & \text{otherwise.} \end{cases}$$

The Recursive Function

$$f_j(C,i) = \begin{cases} V_R, & \text{if } C < r_j + p_{[i]}; \\ \min\{V_R, V_A^1\}, & \text{if } C > r_j + p_{[i]}; \\ \min\{V_R, V_A^2\}, & \text{if } C = r_j + p_{[i]}. \end{cases}$$

The optimal value is given by

$$\min\{f_n(C,i) : C \in \{0\} \cup \{r_k + p_{[i']} + \cdots + p_{[i]} : 1 \leq k \leq n, 1 \leq i' \leq i \leq n\}, 0 \leq i \leq n\}.$$

Theorem 5. *Algorithm DP2 solves* $1|r_j, GPT, reject|C_{\max} + \sum_{J_j \in R} e_j$ *in* $O(n^5)$ *time.*

Proof. The correctness of the algorithm is guaranteed by the above discussion. Note that $C, C' \in \{0\} \cup \{r_k + p_{[i']} + \cdots + p_{[i]} : 1 \leq k \leq n, 1 \leq i' \leq i \leq n\}$ in each iteration. When $C \neq r_j + p_{[i]}$, the recursive function has at most $O(n^5)$ states and each iteration costs a constant time. When $C = r_j + p_{[i]}$, the recursive function has at most $O(n^2)$ states and each iteration costs an $O(n^3)$ time due to the choices of C'. Thus, the running time is bounded by $O(n^5)$.

5 Scheduling with Generalized Rejection Cost

Note that Lemma 2 is still valid for problem $1|r_j, GRC, reject|C_{\max} + \sum_{J_j \in R} e_j$. We first re-label the jobs such that $r_1 \leq r_2 \leq \cdots \leq r_n$. Let $f_j(i)$ be the optimal value of the objective function when the jobs in consideration are $J_1, \cdots, J_j$ and the current number of rejected jobs is i. Now, we consider any optimal schedule for the jobs $J_1, \cdots, J_j$ in which the current rejected number of jobs is i. In any such schedule, there are two possible cases: either J_j is rejected or J_j is accepted and processed on the machine.

Case 1. Job J_j is rejected. The number of rejected jobs is $i - 1$ before J_j is rejected. Note that, job J_j is the i–th rejected job, its rejection cost is $e_{[i]}$. Thus, we have $f_j(i) = f_{j-1}(i-1) + e_{[i]}$.

Case 2. Job J_j is accepted. In this case, before J_j is accepted, the number of rejected jobs is still i. Let V be the current total rejection cost. Clearly, we have $V = 0$ if $i = 0$, and $V = e_{[1]} + \cdots + e_{[i]}$ if $i > 0$. Thus, the previous makespan before J_j is scheduled is $f_{j-1}(i) - V$. Thus, we have $f_j(i) = \max\{f_{j-1}(i) - V, r_j\} + p_j + V$.

Combining the above two cases, we have the following dynamic programming algorithm DP3.

Dynamic Programming Algorithm DP3

The Boundary Conditions:

$$f_1(i) = \begin{cases} r_1 + p_1, & \text{if } i = 0; \\ e_{[1]}, & \text{if } i = 1; \\ +\infty, & \text{otherwise.} \end{cases}$$

The Recursive Function

$$f_j(i) = \min\{f_{j-1}(i-1) + e_{[i]}, \max\{f_{j-1}(i) - V, r_j\} + p_j + V\},$$

where $V = 0$ if $i = 0$, and $V = e_{[1]} + \cdots + e_{[i]}$ if $i > 0$.

The optimal value is given by $\min\{f_n(i) : 0 \leq i \leq n\}$.

Theorem 6. *Algorithm DP3 solves problem* $1|r_j, GRC, reject|C_{\max} + \sum_{J_j \in R} e_j$ *in* $O(n^2)$ *time.*

Proof. The correctness of the algorithm is guaranteed by the above discussion. The recursive function has at most $O(n^2)$ states and each iteration cost a constant time. Thus, the running time is bounded by $O(n^2)$.

6 Conclusions and Future Research

In this paper, we consider three single machine scheduling problems with rejection and generalized parameters. Inspired by generalized due dates, we introduce three generalized parameters: (1) generalized release dates, (2) generalized processing times, and (3) generalized rejection costs. We show that the first scheduling problem with generalized release dates is binary NP-hard. Furthermore, we provide a pseudo-polynomial time algorithm, a 2-approximation algorithm and a full polynomial-time approximation scheme (FPTAS) for this problem. For the latter two problems with generalized processing times or generalized rejection costs, we provide a polynomial-time optimal algorithm, respectively.

In future research, an interesting direction is to consider other generalized parameters such as: generalized weights or generalized delivery times. Moreover, it is also interesting to consider the online versions of these problems. Finally, we also plan to extend this problem into parallel machine scheduling or batch processing scheduling in the future.

Acknowledgments. The authors thank the anonymous reviewers for their constructive comments. This research was supported by NSFCs (11901168, 11971443 and 11771406).

References

1. Agnetis, A., Mosheiov, G.: Scheduling with job rejection and position-dependent processing times on proportionate flowshops. Optim. Lett. **11**, 885–892 (2017)
2. Bartal, Y., Leonardi, S., Spaccamela, A.M., Stougie, J.: Multi-processor scheduling with rejection. SIAM J. Discret. Math. **13**, 64–78 (2000)
3. Chen, R.-X., Li, S.-S.: Minimizing maximum delivery completion time for order scheduling with rejection. J. Comb. Optim. **40**(4), 1044–1064 (2020). https://doi.org/10.1007/s10878-020-00649-2
4. Du, J.Z., Leung, J.Y.T.: Minimizing total tardiness on one machine is NP-hard. Math. Oper. Res. **15**, 483–495 (1990)
5. Gao, Y., Yuan, J.J.: Unary NP-hardness of minimizing the total deviation with generalized or assignable due dates. Discret. Appl. Math. **189**, 49–52 (2015)
6. Gao, Y., Yuan, J.J.: Unary NP-hardness of minimizing total weighted tardiness with generalized due dates. Oper. Res. Lett. **44**, 92–95 (2016)
7. Garey, M.R., Johnson, D.S.: Computers and Intractablity: A Guide to the Theory of NP-Completeness. Freeman, San Francisco (1979)
8. Gerstl, E., Mosheiov, G.: Single machine scheduling problems with generalized due-dates and job-rejection. Int. J. Prod. Res. **55**, 3164–3172 (2017)
9. Graham, R.L., Lawler, E.L., Lenstra, J.K., Rinnooy Kan, A.H.G.: Optimization and approximation in deterministic sequencing and scheduling: a survey. Ann. Discrete Math. **5**, 287–326 (1979)
10. Hall, N.G.: Scheduling problems with generalized due dates. IIE Trans. **18**, 220–222 (1986)
11. Hall, N.G., Sethi, S.P., Srikandarajah, S.: On the complexity of generalized due date scheduling problems. Eur. J. Oper. Res. **51**, 100–109 (1991)

12. Hermelin, D., Pinedo, M., Shabtay, D., Talmon, N.: On the parameterized tractability of a single machine scheduling with rejection. Eur. J. Oper. Res. **273**, 67–73 (2019)
13. Lawer, E.L.: Optimal sequencing a single machine subject to precedence constraints. Manage. Sci. **19**, 544–546 (1973)
14. Liu, P., Lu, X.: New approximation algorithms for machine scheduling with rejection on single and parallel machine. J. Comb. Optim. **40**(4), 929–952 (2020). https://doi.org/10.1007/s10878-020-00642-9
15. Liu, Z.X.: Scheduling with partial rejection. Oper. Res. Lett. **48**, 524–529 (2020)
16. Lu, L.F., Zhang, L.Q., Zhang, J., Zuo, L.L.: Single machine scheduling with outsourcing under different fill rates or quantity discount rates. Asia-Pacific J. Oper. Res. **37**, 1950033 (2020)
17. Lu, L.F., Zhang, L.Q., Ou, J.W.: In-house production and outsourcing under different discount schemes on the total outsourcing cost. Ann. Oper. Res. **298**, 361–374 (2021)
18. Ma, R., Guo, S.N.: Applying "Peeling Onion" approach for competitive analysis in online scheduling with rejection. Eur. J. Oper. Res. **290**, 57–67 (2021)
19. Mor, B., Mosheiov, G., Shapira, D.: Flowshop scheduling with learning effect and job rejection. J. Sched. **23**(6), 631–641 (2019). https://doi.org/10.1007/s10951-019-00612-y
20. Mor B., Mosheiov G., Shabtay D.: Minimizing the total tardiness and job rejection cost in a proportionate flow shop with generalized due dates. J. Sched. (2021). https://doi.org/10.1007/s10951-021-00697-4
21. Mosheiov, G., Oron, D.: A note on the SPT heuristic for solving scheduling problems with generalized due dates. Comput. Oper. Res. **31**, 645–655 (2004)
22. Mosheiov, G., Oron, D., Shabtay, D.: Minimizing total late work on a single machine with generalized due-dates. Eur. J. Oper. Res. **293**, 837C846 (2021)
23. Oron, D.: Two-agent scheduling problems under rejection budget constraints. Omega **102**, 102313 (2021)
24. Shabtay, D., Gaspar, N., Kaspi, M.: A survey on off-line scheduling with rejection. J. Sched. **16**, 3–28 (2013)
25. Srikandarajah, S.: A note on the generalized due dates scheduling problem. Nav. Res. Logist. **37**, 587–597 (1990)
26. Tanaka, K., Vlach, M.: Minimizing maximum absolute lateness and range of lateness under generalized due dates on a single machine. Ann. Oper. Res. **86**, 507–526 (1999)
27. Wang, D.J., Yin, Y.Q., Jin, Y.: Parallel-machine rescheduling with job unavailability and rejection. Omega **81**, 246–260 (2018)
28. Yin, Y.Q., Cheng, S.R., Cheng, T.C.E., Wu, C.C., Wu, W.H.: Two-agent single-machine scheduling with assignable due dates. Appl. Math. Comput. **219**, 1674–1685 (2012)
29. Zhang, L.Q., Lu, L.F., Yuan, J.J.: Single machine scheduling with release dates and rejection. Eur. J. Oper. Res. **198**, 975–978 (2009)
30. Zou, J., Yuan, J.J.: Single-machine scheduling with maintenance activities and rejection. Discret. Optim. **38**, 100609 (2020)

Approximation Algorithm and Hardness Results for Defensive Domination in Graphs

Michael A. Henning[1], Arti Pandey[2(✉)], and Vikash Tripathi[2]

[1] Department of Pure and Applied Mathematics, University of Johannesburg, Auckland Park 2006, South Africa
mahenning@uj.ac.za

[2] Department of Mathematics, Indian Institute of Technology Ropar, Nangal Road, Rupnagar 140001, Punjab, India
{arti,2017maz0005}@iitrpr.ac.in

Abstract. In a graph $G = (V, E)$, a non-empty set A of k distinct vertices, is called a *k-attack* on G. The vertices in the set A is considered to be *under attack*. A set $D \subseteq V$ can defend or counter the attack A on G if there exists a one to one function $f : A \longmapsto D$, such that either $f(u) = u$ or there is an edge between u and it's image $f(u)$, in G. A set D is called a *k-defensive dominating set*, if it defends against any k-attack on G. Given a graph $G = (V, E)$, the minimum k-defensive domination problem requires us to compute a minimum cardinality k-defensive dominating set of G. When k is not fixed, it is co-NP-hard to decide if $D \subseteq V$ is a k-defensive dominating set. However, when k is fixed, the decision version of the problem is NP-complete for general graphs. On the positive side, the problem can be solved in linear time when restricted to paths, cycles, co-chain graphs and threshold graphs for any k. In this paper, we mainly focus on the problem when $k > 0$ is fixed. We prove that the decision version of the problem remains NP-complete for bipartite graphs, this answers a question asked by Ekim et al. (Discrete Math. 343 (2) (2020)). We give lower and upper bound on the approximation ratio for the problem. Further, we show that the minimum k-defensive domination problem is APX-complete for bounded degree graphs. On the positive side, we show that the problem is efficiently solvable for complete bipartite graphs for any $k > 0$.

Keywords: Domination · Defensive domination · NP-completeness · Graph algorithms · Approximation algorithms · APX-completeness

1 Introduction

A *dominating set* D of a graph $G = (V, E)$ is a set $D \subseteq V$, such that every vertex not in D has a neighbour in D. The classical MINIMUM DOMINATION problem requires to compute a dominating set of minimum size. The cardinality of a minimum dominating set is known as domination number, denoted by

D.-Z. Du et al. (Eds.): COCOA 2021, LNCS 13135, pp. 247–261, 2021.
https://doi.org/10.1007/978-3-030-92681-6_21

$\gamma(G)$. Due to wide applications, the concept of domination is extensively studied in literature. A detailed study on domination can be found in the books [10–13]. Depending on requirements in different applications, many variations of the classical domination problems were introduced and studied by several researchers. Such variations of domination are obtained by imposing additional conditions on the dominating set.

In this paper, we study a version of domination, which focuses on security in networks, known as *k-defensive domination.* This concept was introduced in [8] and studied further in [4,6,7]. A similar concept, *alliances in graphs,* which is well studied in literature, also focuses on security in networks. A recent survey of alliances in graphs can be found in [9], and a survey on algorithms and complexity of alliances in graphs can be found in [14]. The k-defensive domination problem can be defined as follows:

A *k-attack* on a graph $G = (V, E)$ is a set of k-distinct vertices, $A = \{v_1, v_2, \ldots, v_k\} \subseteq V$. The vertices in the set A are said to be *under attack.* The *k-attack* A, is said to be *defended* or *countered* by a set $D \subseteq V$, if there exists an injective map $f\colon A \longmapsto D$, such that if $f(v) = w$ where $v \in A$ and $w \in D$, then $w \in N_G[v]$. In this case, v is said to be defended by w. A set $D \subseteq V$, is called a k-defensive dominating set of G, if any k-attack A on G can be defended by set D. The minimum cardinality of a k-defensive dominating set is called the *k-defensive domination number,* denoted by $\gamma_k(G)$. For $k = n$, $\gamma_k(G) = n$. We observe that in the special case when $k = 1$, the k-defensive domination problem is equivalent to the classical domination problem. The problem is defined in the case when k is fixed and in the case when k is a part of input. Formally, the problems and decision version can be stated as follows:

MINIMUM FIXED k-DEFENSIVE DOMINATION problem

Instance: A graph $G = (V, E)$.
Solution: A k-defensive dominating set D of G.
Measure: Cardinality of the set D.

FIXED k-DEFENSIVE DOMINATION DECISION problem

Instance: A graph $G = (V, E)$ and a positive integer $d \leq |V|$.
Question: Does there exist a k-defensive dominating set D of G such that $|D| \leq d$?

MINIMUM k-DEFENSIVE DOMINATION problem

Instance: A graph $G = (V, E)$ and a positive integer $k \geq 1$.
Solution: A k-defensive dominating set D of G.
Measure: Cardinality of the set D.

k-DEFENSIVE DOMINATION DECISION problem

Instance: A graph $G = (V, E)$ and a positive integer $k, d \leq |V|$.
Question: Does there exist a k-defensive dominating set D of G such that $|D| \leq d$?

Note that for $k = 1$, the FIXED k-DEFENSIVE DOMINATION problem reduces to classical MINIMUM DOMINATION problem. Hence, we are more interested in the case when $k > 1$. We remark that even for $k = 2$, $\gamma_k(G)$ may be incomparable with $\gamma(G)$. For example, if we take a star graph $G = K_{1,n}$ on $n+1$ vertices, then $\gamma(G) = 1$ but $\gamma_k(G) = n$ for any $k > 1$. We also observe that if D is a k-defensive dominating set of a graph $G = (V, E)$, then $|D| \geq k$.

In [8], the authors studied the MINIMUM k-DEFENSIVE DOMINATION problem in complete graphs, cycles and paths. They describe the structure of optimal solution in complete graphs and cycles and give lower bound for paths. They also proposed an algorithm to compute a minimum cardinality k-defensive dominating set in trees. Ekim et al. [6] proved that the FIXED k-DEFENSIVE DOMINATION DECISION problem is NP-complete for general graphs and split graphs. They also proved that, when k is not fixed, it is co-NP-hard to decide, if a given set of vertices is a k-defensive dominating set or not. Further, they proved that the MINIMUM k-DEFENSIVE DOMINATION is efficiently solvable for paths, cycles, threshold graphs, and co-chain graphs. In [7], Ekim et al. proposed a greedy algorithm, which computes a minimum k-defensive dominating set in a proper interval graph $G = (V, E)$ in $O(|V| \cdot k)$-time. The same authors in [6,7] asked to determine the complexity of the problem in bipartite graphs.

In this paper, we continue the algorithmic study of the problem. The contribution of the paper is as follows. In Sect. 2, we discuss some basic notations. In Sect. 3, we prove that the FIXED k-DEFENSIVE DOMINATION DECISION problem is NP-complete for bipartite graphs, thereby providing an answer to a question asked by Ekim et al. in [6,7]. In Sect. 4 we prove that, for any fixed $k > 1$, the k-defensive domination problem cannot be approximated within $(1 - \frac{\epsilon}{2}) \ln |V|$ for any $\epsilon > 0$ unless P = NP. In Sect. 5, we propose an approximation algorithm with approximation ratio $1 + k \ln(n)$. In Sect. 6, we prove that for any fixed $k > 1$, the FIXED k-DEFENSIVE DOMINATION problem is APX-complete in $(k+2)$-bounded degree graphs. In Sect. 7, we study the MINIMUM FIXED k-DEFENSIVE DOMINATION problem in complete bipartite graphs. Finally, Sect. 8, concludes the paper with some open question and remarks.

2 Preliminaries

In this paper, we consider only simple and connected graphs with at least three vertices. Let $G = (V, E)$ be a simple and connected graph. Two vertices $u, v \in V$ are *adjacent* if $uv \in E$. For a vertex $u \in V$, the sets $N_G(u) = \{v \in V \mid uv \in E(G)\}$ and $N_G[u] = N_G(u) \cup \{u\}$ are called the *open* and *closed neighbourhoods of* u, respectively. For a set $A \subseteq V$, we define $N(A) = \cup_{v \in A} N_G(u)$ and $N[A] = \cup_{v \in A} N_G[u]$. For a vertex $u \in V$, $|N_G(u)|$ is called the *degree* of u in G, denoted by $d_G(u)$. A *path* P is a sequence of vertices $v_1 v_2, \dots v_k$ such that $v_i v_{i+1} \in E(G)$ for each $i \in [k-1]$. If $P = v_1 v_2, \dots v_k$ is a path, then we say, P is connecting v_1 and v_k, and length of P is $k - 1$. The distance between two distinct vertices $u, v \in V(G)$, denoted by $d_G(u, v)$, is the length of a shortest path connecting u and v.

A graph G is said to be *complete* if any pair of distinct vertices in G are adjacent. For a graph G, a *clique* of G is a complete subgraph of G. A set $I \subseteq V$ is *independent* if no vertices in I are adjacent. A graph $G = (V, E)$ is a *bipartite graph*, if the vertex set V can be partitioned into two non-empty sets X and Y such that each edge of G has one end in X and other end in Y. The sets X and Y are called the *partite sets* of G. The pair (X, Y) is called a *bipartition* of G, and we represent the resulting graph by $G = (X, Y, E)$. A bipartite graph $G = (X, Y, E)$ is a *complete bipartite graph*, if all possible pairs $(x, y) \in X \times Y$ are adjacent in G.

Let $G = (V, E)$ be a graph. The kth power of G, denoted by G^k, is the graph with same vertex set V and where two vertices u and v are adjacent in G^k if and only if $d_G(u, v) \leq k$. For a set $S \subset V$, the graph induced by S, denoted by $G[S]$, is a subgraph of G with vertex set S. Two vertices $x, y \in S$ are adjacent in $G[S]$ only if $xy \in E$. We use the standard notation $[n]$ to denote the set of integers $\{1, 2, \ldots, n\}$. For a k-defensive dominating set of G, we have the following observation.

Observation 1. *Let $G = (V, E)$ be a graph and D a k-defensive dominating set of G. If $u, v \in V(G)$ with $N_G[u] \subseteq N_G[v]$ such that $u \in D$ and $v \notin D$, then the set $(D \setminus \{u\}) \cup \{v\}$ is also a k-defensive dominating set of G.*

3 NP-Completeness Result for Bipartite Graphs

In this section, we show that the Fixed k-Defensive Domination Decision problem is NP-complete for bipartite graphs. This will be done in two steps. First, given a bipartite graph $G = (X, Y, E)$, we construct another bipartite graph $G' = (X', Y', E')$ such that any dominating set D' of G' has a non empty intersection with X' as well as Y'. That is, if D' is any dominating set of G', then $D' \cap X' \neq \emptyset$ and $D' \cap Y' \neq \emptyset$. We further claim that G has a dominating set of size at most d if and only if G' has a dominating set of size at most $d + 2$. This implies that, the hardness of the domination problem remains the same in the graph G' as well. In the second step, from G' we will construct another bipartite graph $H = (X'', Y'', E'')$ such that G' has a dominating set of size at most t if and only if H has a k-defensive dominating set of size at most $t + p$, where p will be defined during the proof.

Theorem 1. *For any k,* Fixed-k-DefDom *is NP-complete for bipartite graphs.*

Proof. We know that, for any fixed k, Fixed-k-Defensive Domination is in NP for general graphs. Therefore for any fixed k, Fixed-k-Defensive Domination problem is in NP for bipartite graphs as well. To prove the hardness, we will give a polynomial reduction from the Minimum Domination Problem which is already known to be NP-hard for bipartite graphs [2]. First, given a bipartite graph $G = (X, Y, E)$, we construct a bipartite graph $G' = (X', Y', E')$ as follows:

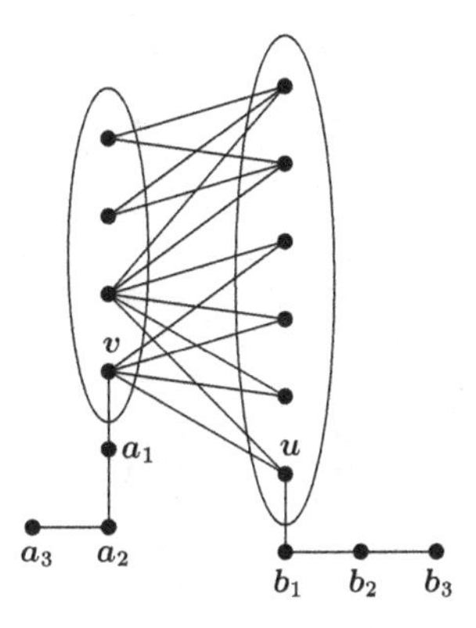

Fig. 1. An illustration to the construction of G' from G in the proof of Theorem 1.

Consider two paths $P = a_1a_2a_3$ and $P' = b_1b_2b_3$. Pick any vertex $v \in X$ and make a_1 adjacent to v. Similarly pick any vertex $u \in Y$ and make b_1 adjacent to u. Note that the new constructed graph is a bipartite graph with bipartitions $X' = X \cup \{a_2, b_1, b_3\}$ and $Y' = Y \cup \{a_1, a_3, b_2\}$ and edge set $E' = E \cup \{va_1, a_1a_2, a_2a_3, ub_1, b_1b_2, b_2b_3\}$. An illustration of the construction of G' from G is given in Fig. 1. We prove next the following claim:

Claim. The graph G has a dominating set of size at most d if and only if the graph G' has a dominating set of size at most $d + 2$.

Proof. If G has a dominating set D of size at most d, then the set $D' = D \cup \{a_2, b_2\}$ is a dominating set of size at most $d + 2$.

Conversely, suppose G' has a dominating set D' of size at most $d + 2$. If $a_3 \in D'$, then we can replace a_3 in D' with a_2. Hence, in order to dominate a_3, we can choose D' so that $a_2 \in D'$. If $a_1 \in D'$, then we can replace a_1 in D' with v. Hence, we may further choose D' so that $D' \cap \{a_1, a_2, a_3\} = \{a_2\}$. Similarly, we may choose D' so that $D' \cap \{b_1, b_2, b_3\} = \{b_2\}$. The set $D = D' \setminus \{a_2, b_2\}$ is a dominating set of G of size at most d. □

Hence, given a bipartite graph G, we can construct another bipartite graph G' such that G has a dominating set of size at most d if and only if G' has a dominating set of size at most $d + 2$. More importantly, any dominating set D' of $G' = (X', Y', E')$ contains at least one vertex from X' as well as at least one vertex from Y'.

From G', we construct next a bipartite graph $H = (X'', Y'', E'')$ as follows: for $i \in [2]$, let $X_i = \{x_1^i, x_2^i, \dots x_{k-1}^i\}$ and $Y_i = \{y_1^i, y_2^i, \dots y_{k-1}^i\}$. Let $X'' = X' \cup X_1 \cup Y_1$ and $Y'' = Y' \cup X_2 \cup Y_2$. We join each vertex of X_1 with each vertex of Y_2 so that the subgraph of G induced by $X_1 \cup Y_2$, that is $G[X_1 \cup Y_2]$, is a complete bipartite graph. Similarly, we join each vertex of X_2 with each vertex of Y_1 so that the induced graph $G[X_2 \cup Y_1]$ is a complete bipartite graph. Finally, we make vertex x_i^1 and x_i^2 adjacent to all vertices of Y' and X', respectively, for

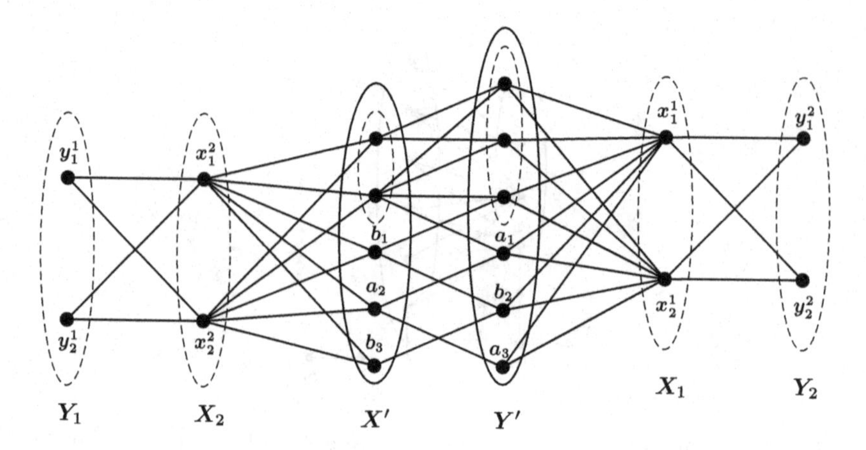

Fig. 2. An illustration of the construction of H from G' for $k = 3$ in the proof of Theorem 1.

each $i \in [k-1]$. Formally, $E'' = E' \cup \{(x_i^1, y) | y \in Y', i \in [k-1]\} \cup \{(x_i^2, x) | x \in X', i \in [k-1]\} \cup \{(x_i^1, y_j^2) \mid i, j \in [k-1]\} \cup \{(x_i^2, y_j^1) \mid i, j \in [k-1]\}$. An illustration of the construction of H from G' is given in Fig. 2. Now in order to show the hardness result, we will prove the following claim:

Claim. The graph $G' = (X', Y', E')$ has a dominating set of size at most t if and only if the graph $H = (X'', Y'', E'')$ has a k-defensive dominating set of size at most $t + 2(k-1)$.

Proof. The proof is omitted due to space constraints. □

The theorem now follows from above Claims. □

Since $k > 1$ is a constant, we note that the reduction is also a parameterized reduction. Hence, we have the following result.

Corollary 1. Fixed k-Defensive Domination *problem is W[2]-hard for bipartite graphs when parameterized by the solution size.*

4 Lower Bounds on Approximation Ratio

For the concepts related to approximation algorithms and approximation hardness, we refer the reader to [15]. In this section we will give a lower bound on the approximation ratio of the k-defensive domination problem in graphs. For this purpose we will give a polynomial time approximation preserving reduction from the Minimum Domination problem to k-defensive domination problem. We will use the same construction used by Ekim et al. [6], to show the NP-completeness of the Fixed k-Defensive Domination Decision problem. For sake of completeness, we will illustrate the construction here also.

Chlebík and Chlebíková in [2], proved that the MINIMUM DOMINATION problem can not be approximated within a factor of $(1-\epsilon)\ln(|V|)$ in polynomial time, for any $\epsilon > 0$, unless NP $\subseteq$ DTIME($n^{O(\log\log n)}$). They proved this result by designing a reduction from well known SET COVER problem. Later, in [5], Dinur and Steurer proved that the MINIMUM SET COVER problem can not be approximated within a factor of $(1-\epsilon)\ln(|V|)$ in polynomial time, for any $\epsilon > 0$, unless P=NP. Hence, the following lower bound on the approximation ratio of the MINIMUM DOMINATION problem is already known.

Theorem 2. *For a graph $G = (V, E)$, the* MINIMUM DOMINATION *problem cannot be approximated within $(1-\epsilon)\ln(|V|)$ for any $\epsilon > 0$ unless P= NP.*

Now we will prove the following result for k-defensive domination problem.

Theorem 3. *For a graph $G = (V, E)$,* FIXED k-DEFENSIVE DOMINATION DECISION *problem cannot be approximated within $(\frac{1}{2}-\epsilon)\ln(|V|)$ for any $\epsilon > 0$ unless P= NP.*

Proof. Note that for $k = 1$, the problem is equivalent to the MINIMUM DOMINATION problem and hence the results follows. Hence, we may assume that $k > 1$ otherwise, the result is immediate. Given an arbitrary instance $G = (V, E)$ of the MINIMUM DOMINATION problem, we will construct an instance $H = (V', E')$ of the k-defensive domination problem as follows. Consider two sets $U = \{u_1, \dots u_{k-1}\}$ and $W = \{w_1, \dots w_{k-1}\}$. Let H' be obtained from G by adding the vertices in the sets U and W, forming a clique on the vertices in U, adding a perfect matching between U and W, and joining every vertex in U to all vertices in V. Thus, $V' = V \cup U \cup W$ and $E' = E \cup \{u_i w_i \mid i \in [k-1]\} \cup \{u_i u_j \mid i, j \in [k-1] \text{ and } i \leq j\} \cup \{uv \mid u \in U \text{ and } v \in V\}$. We note that w_i has degree 1 in H with u_i as its unique neighbour for each $i \in [k-1]$. Further, W is an independent set in H.

We show that if S^* is a minimum dominating set of G, then the set $S = S^* \cup U$ is a k-defensive dominating set of H. Consider any k-attack A in H such that A contains r vertices from the set W with the remaining $k-r$ vertices from the set $U \cup V$ where $0 \leq r \leq k-1$. The r vertices of W can be defended by their r neighbours in U, and the $k-r$ vertices in A that belong to $V \cup U$ can be defended by the remaining $k-r-1$ vertices of U and the set S^*. Thus, $S = S^* \cup U$ is a k-defensive dominating set of H. In particular, if S_k^* is a minimum cardinality k-defensive dominating set, then $|S_k^*| \leq |S^*| + (k-1)$.

Suppose there exists an Algorithm A that approximates a k-defensive dominating set S' of the graph H in polynomial time within a factor $\alpha = (\frac{1}{2} - \epsilon)\ln(|V'|) \geq 1$, for some $\epsilon > 0$. Using ALGORITHM A we give an algorithm ALGORITHM B which approximates a dominating S set of G.

Algorithm 1: Algorithm B

Input: A graph $G = (V, E)$.
Output: A Dominating set S of G.
begin
    if *there exists a minimum dominating set S of G of cardinality less than k* **then**
        return S;
    else
        Construct graph H from G.
        Compute a k-defensive dominating set S' of H using algorithm ALGORITHM A.
        Let $S = S' \cap V$.
        for *each $v \in V$* **do**
            if ($v \in V$ *is not dominated by* S) **then**
                include v in S
        return S;

Clearly, the set S returned by ALGORITHM B is a dominating set of G. Next, we show that $|S| \leq |S'| - (k-1)$. We note that S' is a k-defensive dominating set of H computed by ALGORITHM A. Since w_i is a vertex of degree 1 with u_i as its unique neighbor, to defend an attack on w_i, either w_i or u_i must be in S'. Moreover if $w_i \in S'$ and $u_i \notin S'$, then since $N[w_i] \subseteq N[u_i]$, by Observation 1 the set $(S' \setminus \{w_i\}) \cup \{u_i\}$ is also a k-defensive dominating set of H. Therefore, we may choose S' so that $U \subseteq S'$. Now suppose $|(S' \setminus U) \cap W| = r$. Without loss of generality we assume that $S' \cap W = \{w_1, w_2, \ldots w_r\}$.

We show that at most r vertices of V are not dominated by the set $S_G = S' \cap V$. Let T be the set of vertices in G that are not dominated by set S_G. Suppose, to the contrary, that $|T| > r$. If $|T| \geq k$, then any subset $A \subseteq T$ with $|A| = k$ yields a k-attack which cannot be defended by S', a contradiction. Therefore, $|T| \leq k-1$. Let $W' = \{w_{|T|}, w_{|T|+1}, \ldots w_{k-1}\}$, and so $|W'| = k - |T|$. We now consider a k-attack $A = T \cup W$. Since S' is a k-defensive dominating set of H, the attack A must be defended by S'. Since every vertex in W' must be defended by its unique neighbor that belongs to U, we have only $|U| - |W'| = (k-1) - (k - |T|) = |T| - 1$ defenders left to defend T, which is not possible. This contradicts the fact that S' is a k-defensive dominating set of H. Therefore, $|T| \leq r$. We now let $S = S_G \cup T$. The set S is a dominating set of G. Moreover, $|S| = |S_G| + |T| = (|S'| - |U| - |S' \cap W|) + |T| \leq (|S'| - (k-1) - r) + r = |S'| - (k-1)$. Moreover, if S_k^* is a minimum cardinality k-defensive dominating set of H and S^* is a minimum cardinality dominating set of G, then we have $|S^*| \leq S_k^* - (k-1)$. Hence, we may conclude that $\gamma_k(H) = \gamma(G) + (k-1)$.

Hence, we conclude that ALGORITHM B returns a minimum dominating set S of G if $|S| < k$; otherwise, it returns a dominating set of G of cardinality at most $|S'| - (k-1) \leq \alpha|S_k^*| - (k-1)$, where S_k^* denotes a minimum k-defensive dominating set of H. Let S^* is a minimum dominating set of G. If $|S^*| < k$ then ALGORITHM B returns a minimum cardinality dominating set G. Hence, we assume that $|S^*| \geq k$. Now we have, $|S| \leq |S'| - (k-1) \leq \alpha|S_k^*| - (k-1) = \alpha(|S^*| + (k-1)) - (k-1) = \alpha|S^*| + (\alpha - 1)(k-1)$. Now, since $|S^*| \geq k$, $|S| \leq \alpha|S^*| + (\alpha-1)|S^*| \leq 2\alpha|S^*| = (2(\frac{1}{2} - \epsilon)\ln|V'|)|S^*| = ((1 - 2\cdot\epsilon)\ln|V'|)|S^*| = ((1-\epsilon')\ln|V'|)|S^*|$ where $\epsilon' = 2\cdot\epsilon$. Also for sufficiently large value of $|V|$, $\ln|V'| = \ln(|V| + 2(k-1)) \approx \ln|V|$. Therefore, if we take $\epsilon = \frac{\epsilon'}{2}$, the minimum dominating set of G can be approximated within a factor of $\alpha = (1-\epsilon')\ln|V|$, a contradiction to Theorem 2.

This completes the proof of Theorem 3. □

5 Approximation Algorithm

In this section we will give an approximation algorithm for FIXED k-DEFENSIVE DOMINATION in graphs. Before designing the algorithm, we first recall the following results:

Theorem 4 [7]. *A set $D \subseteq V$ in a graph $G = (V, E)$ is a k-defensive dominating set of G if and only if it defends against every k-attack A such that $G^2[A]$ is connected.*

Theorem 5 [7]. *A set $D \subseteq V$ in a graph $G = (V, E)$ is a k-defensive dominating set of G if and only if $|N[A] \cap D| \geq |A|$ for every set $A \subseteq V$ with $|A| \leq k$ and $G^2[A]$ is connected.*

Given a graph $G = (V, E)$, and a fixed positive integer $k > 1$, we approximate a k-defensive dominating set by reducing it to another problem known as *total vector domination*, which we define next.

Let $G = (V, E)$ be a graph and $\mathcal{K} = (k_v \mid v \in V)$ be a vector associated with G such that for all $v \in V$, we have $k_v \in \{0, 1, \ldots d_G(v)\}$. A *total vector dominating set* of G is a set $S \subseteq V$ such that $|S \cap N_G(u)| \geq k_u$ for all $u \in V$. The *total vector domination* problem for G is to find a minimum cardinality total vector dominating set of G. The minimum size of a total vector dominating set of G is the *total vector domination number*, denoted by $\gamma^t(G, \mathcal{K})$. An instance of a total vector domination problem is a pair $(G, \mathcal{K})$, where G is a graph, and $\mathcal{K}$ is a corresponding required vector. The following approximation result is already known for the total vector domination problem.

Theorem 6. ([3]) *The total vector domination problem in a graph G can be approximated in polynomial time within a factor of $\ln(\Delta(G)) + 1$, where $\Delta(G)$ denotes the maximum degree of a vertex in G.*

Before proving the main theorem of this section, we provide the following construction for a given graph G and a fixed positive integer k.

Construction $\mathcal{A}$: Let $G = (V, E)$ be a graph such that $|V| = n$ and $k > 1$ be a fixed integer. We construct another graph $G' = (V', E')$ as follows: let $\mathcal{S}$ be the set of all sets A such that $|A| \le k$ and $G^2[A]$ is connected. Let $|\mathcal{S}| = r$. Note that, $|\mathcal{S}| = r = O(n^k)$. More precisely, let $\mathcal{S} = \{S_1, \dots S_r\}$ be such that $S_i \subset V$, $|S_i| \le k$, and $G^2[S_i]$ is connected for each $i \in [r]$. Let $U = \{s_1, \dots, s_r\}$ be a set of new vertices, where the vertex s_i corresponding to the set S_i for each $i \in [r]$. Now we construct the graphs $G' = (V', E')$ where $V' = U \cup V$ and $E' = \{uv \colon u, v \in V \text{ and } u \neq v\} \cup \{s_i u \colon u \in N_G[S_i], i \in [r]\}$. We may note that $G'[V]$ is a clique in G' and U is an independent set of G'. Now construct a vector $\mathcal{K} = (k_v \mid v \in V')$ such that $k_v = 1$ for all $v \in V$ and $k_{s_i} = |S_i|$ for all $s_i \in U$. We note that the transformation of the graph G to the graph G' is polynomial time, if $k > 1$ is fixed.

Let $k > 1$ be a fixed integer. Given an instance $G = (V, E)$ of the k-Defensive Domination problem, in polynomial time we can construct an instance $(G' = (V', E'), \mathcal{K})$ of the total vector domination problem, using Construction $\mathcal{A}$. We now prove that a k-defensive dominating set D of G can be obtained from a total vector dominating set D' of G' such that $|D| = |D'|$, and vice versa.

Lemma 1. *Fix an integer $k > 1$. Let $G = (V, E)$ be a graph and $G' = (V', E')$ be the graph obtained from G using Construction $\mathcal{A}$ with vector $\mathcal{K}$. A k-defensive dominating set D of G can be obtained from a total vector dominating set D' of G' and vice versa.*

Proof. Let D' be a total vector dominating set of $G' = (V', E')$. We note that $V' = U \cup V$ where U contains the vertices s_i corresponding to sets $S_i \subseteq V$ where $|S_i| \le k$ and $G^2[S_i]$ is connected. Further, we know that $G'[V]$ is a clique and U is an independent set in G'. In addition, we have $\mathcal{K} = (k_v \mid v \in V')$ such that $k_v = 1$ for all $v \in V$ and $k_{s_i} = |S_i|$ for all $s_i \in U$. This implies that $|N_{G'}(s_i) \cap D'| \ge |S_i|$ for all $i \in [r]$. We now let $D = D' \cap V$. Since U is an independent set and $|N_{G'}(s_i) \cap D'| \ge |S_i|$, we have $|N_G[S_i] \cap D| \ge |S_i|$ for all $i \in [r]$. Hence, using Theorem 5, the set D is a k-defensive dominating set of G with $|D| \le |D'|$.

Now, suppose D is a k-defensive dominating set of G. Using Theorem 5, we have $|N_G[S_i] \cap D| \ge |S_i|$ for all $i \in [r]$. We now let $D' = D$. Using the fact that $|N_G[S_i] \cap D| \ge |S_i|$, we have that $|N_{G'}(s_i) \cap D'| \ge |S_i|$ for all $i \in [r]$. Also since $k > 1$, we have $|N_{G'}(v) \cap D'| \ge 1$. Hence, the set D' is a total vector dominating set of G' with $|D'| \le |D|$. □

In Lemma 1, a set D is a feasible solution of an instance $G = (V, E)$ of Fixed k-Defensive Domination problem if and only if D can be obtained from a feasible solution D' of the corresponding instance $(G' = (V', E'), \mathcal{K})$ of the total vector domination problem. Using this we prove the following result.

Theorem 7. *In any graph $G = (V, E)$ with $|V| = n$,* Min k-Defensive Dominating Set *problem can be approximated within an approximation of $1 + k \ln(n)$.*

Proof. Given a graph $G = (V, E)$ with $|V| = n$, we can construct a graph $G' = (V', E')$ and corresponding vector $\mathcal{K}$, using Construction $\mathcal{A}$. We note that the size of G' is polynomial in the size of G. Indeed, $|V'| = |V| + O(|V|^k) = O(|V|^k)$ and since G' is a simple graph, we have $|E'| = O(|V|^{2k})$. Now, using Theorem 6, we can approximate a total vector dominating set D' of G' with an approximation ratio of $1 + \ln(\Delta(G'))$. We note that $\Delta(G') \leq n^k$. Therefore, we have $1 + \ln(\Delta(G')) = 1 + \ln(n^k) = 1 + k\ln(n)$.

Using Lemma 1, we note that from D' we can obtain a k-defensive dominating set D of G. Consequently, we have a k-defensive dominating set D of G with an approximation ratio of $1 + k \cdot \ln(n)$, yielding the desired result. □

6 APX-Completeness

In this section, we will show that Fixed k-Defensive Domination problem is APX-complete for bounded degree graphs. For this purpose we recall the following definitions. An optimization problem Π is a tuple $(\mathcal{I}, \mathcal{S}, c, \text{opt})$, where $\mathcal{I}$ is the set of instances of Π, $\mathcal{S}(I)$ is the set of feasible solutions of an instance of $I \in \mathcal{I}$, $c : \mathcal{I} \times \mathcal{S} \to \mathbb{N}$ is called objective function, and opt is either minimize or maximize (in short, min or max respectively). We now define L-reduction.

Definition 1. ***L-reduction***
Let $\Pi_1 = (\mathcal{I}_1, \mathcal{S}_1, c_1, min)$ and $\Pi_2 = (\mathcal{I}_2, \mathcal{S}_2, c_2, min)$ be two minimization problems. Let $f : \mathcal{I}_1 \to \mathcal{I}_2$ be a polynomial time function which transforms each instance of Π_1 to an instance of Π_2. We say the function f is an L-reduction if there exists positive constants α and β such that for any instance $I \in \mathcal{I}_1$ the following holds:

1. $\min f(I) \leq \alpha \cdot \min(I)$.
2. *For every feasible solution $y \in \mathcal{S}_2(f(I))$, in polynomial time, we can find a solution $x \in \mathcal{S}_1(I)$ such that* $|\min(I) - c_1(I, x)| \leq \beta \cdot |\min(f(I)) - c_2(f(I), y)|$.

To show that an optimization problem $\Pi \in$ APX is APX-complete, we need to show the existence of an L-reduction from some known APX-complete to the problem Π.

Since every k-defensive dominating set is a dominating set, for any graph G, we have $\gamma_k(G) \geq \gamma(G)$. It is well know that if $G = (V, E)$ is a graph with maximum degree Δ, then $\gamma(G) \geq \frac{|V|}{\Delta+1}$. Therefore, $\gamma_k(G) \geq \frac{|V|}{\Delta+1}$. Thus, taking $D = V$ as a k-defensive dominating set, we have $|D| \leq \gamma_k(G)(\Delta + 1)$. Consequently, Fixed k-Defensive Domination problem is in APX.

To show the APX-completeness of the Minimum Fixed k-Defensive Domination problem, we give an L-reduction from the minimum domination problem. The following result is already known for the minimum domination problem.

Theorem 8 [1]. *The* Minimum Domination *problem is APX-complete for graphs with maximum degree* 3.

Now, we prove the APX-completeness of the Minimum Fixed k-Defensive Domination problem for graphs with maximum degree $k+2$.

Theorem 9. *The* Minimum Fixed k-Defensive Domination *is APX-complete for graphs with maximum degree* $k+2$.

Proof. Since by Theorem 8, the Minimum Domination problem is APX-complete for the graphs with maximum degree 3, to complete the proof of the theorem, it is enough to establish an L-reduction from the instances of the Minimum Domination problem for graphs with maximum degree 3 to the instances of the Minimum Fixed k-Defensive Domination problem for graphs with maximum degree $k+2$. Given a graph $G=(V,E)$, where $V=\{v_1, v_2, \dots, v_n\}$, we construct a graph $G'=(V',E')$ in the following way:

- Define a graph $H_i=(V_i,E_i)$, where the vertex set V_i and the edge set E_i are defined in the following way:
  $V_i = X_i \cup Y_i \cup \{z_i\}$, where $X_i=\{x_i^1, x_i^2, \dots, x_i^{k-1}\}$ and $Y_i=\{y_i^1, y_i^2, \dots, y_i^{k-1}\}$.
  $E_i=\{x_i^j y_i^j \mid j\in[k-1]\} \cup \{x_i^j x_i^p \mid 1\le j<p\le k-1\} \cup \{z_i x_i^j \mid j\in[k-1]\}$.
  We call H_i, a gadget.
- For each $v_i\in V$, add a gadget H_i and make v_i adjacent to all the vertices of X_i.

We note that the graph G' can be constructed from G in the polynomial time. Note that, for each $i\in[n]$, the degree of every vertex in Y_i is 1, degree of every vertex in X_i is $k+1$, and degree of z_i is $k-1$ in G'. Also for any vertex $v_i\in V$, degree of v_i in G' is at most $k+2$. Hence, $\Delta(G')\le k+2$. An example of the construction of G' from G is illustrated in Fig. 3.

Next, we prove the following claim:

Claim. If D^* is a minimum dominating set of G, and D_k^* is a minimum k-defensive dominating set of G', then $|D^*|=|D_k^*|-kn$.

Proof. The proof is omitted due to space constraints. □

We now return to the proof of our theorem. By above claim, if D^* is a minimum dominating set of G, and D_k^* is a minimum k-defensive dominating set of G, then $|D_k^*|=|D^*|+kn$. Since G is a graph of order n with maximum degree 3, we have $n\le 4\gamma(G)=4|D^*|$. Hence, we have $|D_k^*|\le|D^*|+4k|D^*|=(4k+1)|D^*|$. Hence, $|D_k^*|\le\alpha\cdot|D^*|$, where $\alpha=4k+1$.

Now consider a k-defensive dominating set, say D_k of G'. Without loss of generality, we may assume that $X_i\cup z_i\subseteq D_k$ and $Y_i\cap D_k=\emptyset$ for all $i\in[n]$. Similar to the proof of claim, we may argue that the set D is a dominating set of G of cardinality at most $|D_k|-kn$. Hence $|D|-|D^*|\le(|D_k|-kn)-(|D_k^*|-kn)=|D_k|-|D_k^*|$.

This proves that $|\,|D^*|-|D|\,|\le\beta\cdot|\,|D_k^*|-|D_k|\,|$, where $\beta=1$. This proves that f an L-reduction with $\alpha=4k+1$ and $\beta=1$.

This completes the proof of our theorem. □

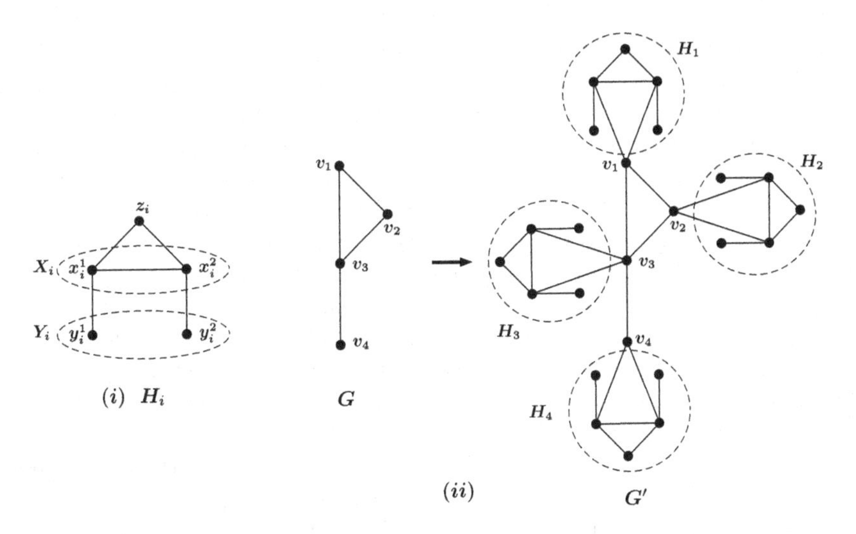

Fig. 3. (*i*) a gadget H_i for $k = 3$, (*ii*) an example of the construction of G' from G for $k = 3$ in the proof of Theorem 9.

7 Defensive Domination Complete Bipartite Graphs

Let $G = (V, E)$ be a graph and D be a k-defensive dominating set of G. Let A be any k-attack on G. We call a vertex u a *non-attacked defender* if $u \notin A$ and $u \in D$. Further, we say that a vertex u is *non-defender attacked* if $u \in A$ and $u \notin D$.

In this section, we study k-Defensive Domination problem in complete bipartite graphs. Let $G = (X, Y, E)$ be a complete bipartite graph of order n such that $|X| = n_1$ and $|Y| = n_2$, and so $n = n_1 + n_2$. If $n \leq k$, then $\gamma_k(G) = n$. If $k = 1$, then the k-Defensive Domination problem is equivalent to the classical domination problem, and therefore, if G is a complete bipartite graph then $\gamma_k(G) = \gamma(G) \leq 2$. Hence, it is only of interest to consider the case when $n > k \geq 2$. Without loss of generality we may assume that $n_1 \geq n_2$. Here, we have the following cases for k: (*i*) $k > n_1$, (*ii*) $n_2 \leq k \leq n_1$, (*iii*) $\frac{n_2}{2} \leq k < n_2$, and (*iv*) $k \leq \frac{n_2}{2}$.

Theorem 10. *If $G = (X, Y, E)$ is a complete bipartite graph, where $|X| = n_1$, $|Y| = n_2$ and $n_1 \geq n_2 \geq 1$, then the following holds:*

(a) If $n_1 < k$, then $\gamma_k(G) = k$.
(b) If $n_2 = k$, then $\gamma_k(G) = k$.
(c) If $n_2 < k \leq n_1$, then $\gamma_k(G) = n_1$.
(d) If $k < n_2 < 2k$, then $\gamma_k(G) = n_2$.
(e) If $n_2 \geq 2k$, then $\gamma_k(G) = 2k$.

Proof. Let $G = (X, Y, E)$ be a complete bipartite graph where $|X| = n_1$ and $|Y| = n_2$. Also assume that $n_1 + n_2 = n > k \geq 2$. Let $X = \{x_1, \ldots x_{n_1}\}$ and $X = \{y_1, \ldots, y_{n_2}\}$.

(a) Suppose that $n_1 < k$. For any graph G, we know that $\gamma_k(G) \geq k$. To show that $\gamma_k(G) = k$, it suffices for us to show that there exists a k-defensive dominating set D of G of cardinality k. Let $D = Y \cup \{x_1, \ldots, x_{k-n_2}\}$, and so $|D \cap X| = k - n_2$. We claim that the set D is a k-defensive dominating set of G. Since $n_1 < k$, we have $|X \setminus D| = |X| - |X \cap D| = n_1 - (k - n_2) = n_1 + n_2 - k < n_2$. Thus, there are at most $n_2 - 1$ vertices in X that do not belong to the set D.

Now consider any k-attack A consisting of l_1 and l_2 vertices from X and Y, respectively. That is, $|A \cap X| = l_1$ and $|A \cap Y| = l_2$, where $l_1 + l_2 = k$. Let $|A \cap (X \setminus D)| = r_1$ and $|A \cap (X \cap D)| = r_2$. Clearly, $r_1 + r_2 = l_1$.

Suppose that $r_1 \leq n_2 - l_2$. Thus, the number of non-attacked defenders in Y is at least r_1. In this case, these r_1 attacked vertices in $X \setminus D$ can be defended by the non-attacked defenders of Y. The remaining attacked vertices all belong to the set D, and can be defended by themselves.

Suppose that $r_1 > n_2 - l_2$. We note that the number of non-attacked defenders in X is given by $|(X \cap D) \setminus A| = (k - n_2) - r_2$. If $(k - n_2) - r_2 < r_1 - (n_2 - \ell_2)$, then $\ell_1 + \ell_2 = r_1 + r_2 + \ell_2 > k$, a contradiction. Hence, $(k - n_2) - r_2 \geq r_1 - (n_2 - \ell_2)$. Let Y_2 be a subset of vertices in $A \cap Y$ of cardinality $r_1 - (n_2 - \ell_2)$. The r_1 vertices in $X \setminus D$ can be defended by the set of $n_2 - l_2$ non-attacked defenders in Y together with the set of $r_1 - (n_2 - \ell_2)$ vertices in Y_2. The vertices in Y_2 can be defended by the set of non-attacked defenders in X, since by our earlier observations there are at least $|Y_2|$ non-attacked defenders in X. The remaining attacked vertices all belong to the set D, and can be defended by themselves.

In both cases, the set D is a k-defensive dominating set of G of cardinality k, implying that $\gamma_k(G) \leq |D| = k$. As observed earlier, $\gamma_k(G) \geq k$. Consequently, $\gamma_k(G) = k$.

The proof of part (b), (c), (d) and (e) of the theorem is omitted due to space constraints. □

8 Conclusion

In this paper, we studied the hardness and approximation results of the MINIMUM FIXED k-DEFENSIVE DOMINATION problem. We proved that the decision version of the problem remain NP-complete even for chordal bipartite graphs, thereby answering a question posed by Ekim et at. in [6]. We studied the approximation hardness of the problem. We proved that, for a fixed $k > 1$, a k-defensive dominating set of a graph $G = (V, E)$ cannot be approximated within a ratio of $(1 - \frac{\epsilon}{2}) \ln(|V|)$ unless P=NP. Using the same result, we claim that for fixed $k > 1$, the problem is W[2]-hard for bipartite graphs when parameterized by solution size. We also proposed a polynomial algorithm which approximates a k-defensive dominating set of a graph of order n with approximation ratio $1 + k\ln(n)$ for a fixed $k > 1$. We prove that the FIXED k-DEFENSIVE DOMINATION problem is APX-complete for $(k+2)$-bounded degree graphs, where $k > 1$. Further, investigating the hardness of the problem and designing polynomial time approximation algorithm for the problem in planar graphs is a good research direction.

In the last section, we show that the MINIMUM k-DEFENSIVE DOMINATION problem is polynomial time solvable for complete bipartite graphs. The problem is already polynomial time solvable for paths, cycles, co-chain graphs, threshold graphs and proper interval graphs. One can try to design polynomial time algorithms for the problem in interval graphs and block graphs.

References

1. Alimonti, P., Kann, V.: Some APX-completeness results for cubic graphs. Theoret. Comput. Sci. **237**(1–2), 123–134 (2000)
2. Chlebík, M., Chlebíková, J.: Approximation hardness of dominating set problems in bounded degree graphs. Inform. Comput. **206**(11), 1264–1275 (2008)
3. Cicalese, F., Milanič, M., Vaccaro, U.: On the approximability and exact algorithms for vector domination and related problems in graphs. Discrete Appl. Math. **161**(6), 750–767 (2013)
4. Dereniowski, D., Gavenčiak, T., Kratochvíl, J.: Cops, a fast robber and defensive domination on interval graphs. Theoret. Comput. Sci. **794**, 47–58 (2019)
5. Dinur, I., Steurer, D.: Analytical approach to parallel repetition. In: STOC 2014–Proceedings of the 2014 ACM Symposium on Theory of Computing, pp. 624–633. ACM, New York (2014)
6. Ekim, T., Farley, A.M., Proskurowski, A.: The complexity of the defensive domination problem in special graph classes. Discrete Math. **343**(2), 111665, 13 (2020)
7. Ekim, T., Farley, A.M., Proskurowski, A., Shalom, M.: Defensive domination in proper interval graphs. CoRR abs/2010.03865 (2020)
8. Farley, A.M., Proskurowski, A.: Defensive domination. In: Proceedings of the Thirty-Fifth Southeastern International Conference on Combinatorics, Graph Theory and Computing, vol. 168, pp. 97–107 (2004)
9. Haynes, T.W., Hedetniemi, S.T.: Alliances and related domination parameters. In: Haynes, T.W., Hedetniemi, S.T., Henning, M.A. (eds.) Structures of Domination in Graphs. DM, vol. 66, pp. 47–77. Springer, Cham (2021). https://doi.org/10.1007/978-3-030-58892-2_3
10. Haynes, T.W., Hedetniemi, S.T., Henning, M.A. (eds.): Topics in Domination in Graphs, Developments in Mathematics, vol. 64. Springer, Cham (2020). https://doi.org/10.1007/978-3-030-51117-3
11. Haynes, T.W., Hedetniemi, S.T., Henning, M.A. (eds.): Structures of Domination in Graphs, vol. 66. Springer, Cham (2021). https://doi.org/10.1007/978-3-030-58892-2
12. Haynes, T.W., Hedetniemi, S.T., Slater, P.J. (eds.): Domination in Graphs: Advanced Topics, Monographs and Textbooks in Pure and Applied Mathematics, vol. 209. Marcel Dekker, Inc., New York (1998)
13. Haynes, T.W., Hedetniemi, S.T., Slater, P.J.: Fundamentals of Domination in Graphs, Monographs and Textbooks in Pure and Applied Mathematics, vol. 208. Marcel Dekker Inc, New York (1998)
14. Hedetniemi, S.T.: Algorithms and complexity of alliances in graphs. In: Haynes, T.W., Hedetniemi, S.T., Henning, M.A. (eds.) Structures of Domination in Graphs. DM, vol. 66, pp. 521–536. Springer, Cham (2021). https://doi.org/10.1007/978-3-030-58892-2_17
15. Vazirani, V.V.: Approximation Algorithms. Springer, Heidelberg (2001). https://doi.org/10.1007/978-3-662-04565-7

An Improved Physical ZKP for Nonogram

Suthee Ruangwises(✉)

Department of Mathematical and Computing Science,
Tokyo Institute of Technology, Tokyo, Japan

Abstract. Nonogram is a logic puzzle consisting of a rectangular grid with an objective to color every cell black or white such that the lengths of blocks of consecutive black cells in each row and column are equal to the given numbers. In 2010, Chien and Hon developed the first physical zero-knowledge proof for Nonogram, which allows a prover to physically show that he/she knows a solution of the puzzle without revealing it. However, their protocol requires special tools such as scratch-off cards and a machine to seal the cards, which are difficult to find in everyday life, making the protocol impractical. Their protocol also has a nonzero soundness error. In this paper, we propose a more practical physical zero-knowledge proof for Nonogram that uses only a deck of regular paper cards and also has perfect soundness.

Keywords: Zero-knowledge proof · Card-based cryptography · Nonogram · Puzzle

1 Introduction

Nonogram, also known as *Picross* or *Pic-a-Pix*, is one of the most popular logic puzzles alongside Sudoku, Kakuro, and other puzzles. A large number of Nonogram mobile apps have been developed [7].

A Nonogram puzzle consists of a rectangular grid of size $m \times n$. The objective of this puzzle is to color every cell black or white according to the following constraints. In each row and each column, there is a sequence of numbers, say $(x_1, x_2, ..., x_k)$, assigned to it; this means the corresponding row (resp. column) must contain exactly k blocks of consecutive black cells with lengths $x_1, x_2, ..., x_k$ in this order from left to right (resp. from top to bottom), with at least one white cell separating consecutive blocks. For instance, in Fig. 1, the sequence $(3, 2)$ on the topmost row means that row must contain a block of three consecutive black cells, and another block of two consecutive black cells to the right of it, separated by at least one white cell.

Suppose that Percy created a difficult Nonogram puzzle and challenged his friend Violetta to solve it. After a while, Violetta could not solve his puzzle and began to doubt whether the puzzle has a solution. In order to convince her that his puzzle actually has a solution without revealing it (which would make the challenge pointless), Percy needs a *zero-knowledge proof (ZKP)*.

D.-Z. Du et al. (Eds.): COCOA 2021, LNCS 13135, pp. 262–272, 2021.
https://doi.org/10.1007/978-3-030-92681-6_22

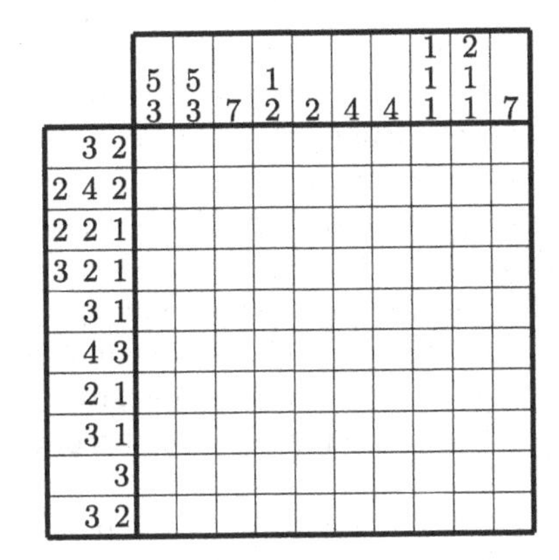

Fig. 1. An example of a Nonogram puzzle (left) and its solution (right)

1.1 Zero-Knowledge Proof

A prover P and a verifier V are given a computational problem x, but only P knows a solution w of x. A ZKP is an interactive proof introduced by Goldwasser et al. [6], which allows P to convince V that he/she knows w without revealing any information about it. A ZKP must satisfy the following properties.

1. **Completeness:** If P knows w, then P can convince V with high probability. (Here we consider the *perfect completeness* property where the probability is one.)
2. **Soundness:** If P does not know w, then P cannot convince V, except with a small probability called *soundness error*. (Here we consider the *perfect soundness* property where the soundness error is zero.)
3. **Zero-knowledge:** V learns nothing about w. Formally, there exists a probabilistic polynomial time algorithm S (called a *simulator*) that does not know w such that the outputs of S and the outputs of the real protocol follow the same probability distribution.

Goldreich et al. [5] showed that a computational ZKP exists for every NP problem. As Nonogram is NP-complete [25], one can construct a computational ZKP for it via a reduction to another problem. Such construction, however, is not intuitive and looks unconvincing. Hence, many recent results instead focused on constructing physical ZKPs using portable objects such as a deck of cards and envelopes. These protocols have benefits that they do not require computers and allow external observers to check that the prover truthfully executes the protocol (which is often a challenging task for digital protocols). These intuitive protocols can also be used to teach the concept of ZKP to non-experts.

1.2 Related Work

Development of physical ZKPs for logic puzzles began in 2009 when Gradwohl et al. [8] proposed six ZKP protocols for Sudoku. Each of these protocols either requires special tools such as scratch-off cards, or has a nonzero soundness error. Later, Sasaki et al. [22] improved the ZKP for Sudoku to achieve perfect soundness without using special tools. Very recently, Ruangwises [18] developed another ZKP for Sudoku that uses a deck of all different cards.

In 2010, Chien and Hon [3] proposed the first physical ZKP for Nonogram. Their protocol, however, requires scratch-off cards and a machine to seal the cards. These special tools cannot be easily found in everyday life, making the protocol very impractical. Moreover, their protocol has a nonzero soundness error. In fact, the soundness error is as high as 6/7, which means practically, the protocol has to be repeated for many times until the soundness error becomes reasonably low.

After Sudoku and Nonogram, physical ZKPs for several other logic puzzles have been developed, including Akari [1], Takuzu [1,12], Kakuro [1,13], KenKen [1], Makaro [2], Norinori [4], Slitherlink [11], Juosan [12], Numberlink [19], Suguru [17], Ripple Effect [20], Nurikabe [16], Hitori [16], and Bridges [21]. All of these subsequent protocols uses only regular paper cards without requiring special tools and have perfect soundness (except the ones in [1]).

1.3 Our Contribution

Despite Nonogram being the second logic puzzle after Sudoku to have a physical ZKP, it still lacks a practical protocol that do not require special tools, or the one with perfect soundness. The problem of developing such protocol has remained open for more than ten years.

In this paper, we propose a practical physical ZKP for Nonogram using only a deck of regular paper cards, which are easy to find in everyday life and can be reused. Our protocol also has perfect completeness and perfect soundness. In an $m \times n$ Nonogram puzzle with a total of w white cells, our protocol uses $2mn + 2\max(m, n) + 6$ cards and $mn + 2m + 2n + 2w$ shuffles.

2 Preliminaries

2.1 Cards

We use four types of cards in our protocol: $\boxed{\clubsuit}$, $\boxed{\heartsuit}$, $\boxed{\spadesuit}$, and $\boxed{\diamondsuit}$. The front sides of cards in the same type are identical. The back sides of all cards are identical and are denoted by $\boxed{?}$.

2.2 Random Cut

Given a sequence of k cards, a *random cut* rearranges the cards by a random cyclic shift, i.e. shifts the cards cyclically to the right by r cards for a uniformly random $r \in \{0, 1, ..., k-1\}$ unknown to all parties.

The random cut can be performed in real world by taking turns to apply several *Hindu cuts* (taking several cards from the bottom and putting them on the top) to the sequence of cards [24].

2.3 Pile-Shifting Shuffle

Given an $\ell \times k$ matrix of cards, a *pile-shifting shuffle* rearranges the columns of the matrix by a random cyclic shift, i.e. shifts the columns cyclically to the right by r columns for a uniformly random $r \in \{0, 1, ..., k-1\}$ unknown to all parties (see Fig. 2).

The pile-shifting shuffle was developed by Shinagawa et al. [23]. It can be performed in real world by putting the cards in each column into an envelope and applying the random cut to the sequence of envelopes.

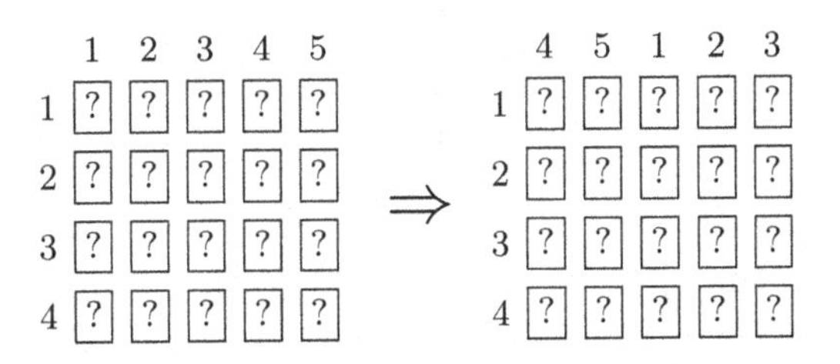

Fig. 2. An example of a pile-shifting shuffle on a 4×5 matrix with $r = 2$

2.4 Copy Protocol

Given a sequence of two face-down cards of either ♣♡ or ♡♣, a *copy protocol* creates an additional copy of the original sequence without revealing it. This protocol was developed by Mizuki and Sone [15].

Fig. 3. A 3×2 matrix constructed in Step 1 of the copy protocol

1. Construct a 3×2 matrix of cards by placing an original sequence in the first row, and a sequence ♣♡ in each of the second and third rows (see Fig. 3).
2. Turn over all face-up cards. Apply the pile-shifting shuffle to the matrix.
3. Turn over the two cards in the first row. If the revealed sequence is ♣♡, do nothing; if the sequence is ♡♣, swap the two columns of the matrix.
4. The sequences in the second and third rows are the two copies of the original sequence as desired.

Note that this protocol also verifies that the original sequence is either ♣♡ or ♡♣ (and not ♣♣ or ♡♡).

2.5 Chosen Cut Protocol

Given a sequence of k face-down cards $A = (a_1, a_2, ..., a_k)$, a *chosen cut protocol* allows the prover P to select a card a_i he/she wants without revealing i to the verifier V. This protocol was developed by Koch and Walzer [9].

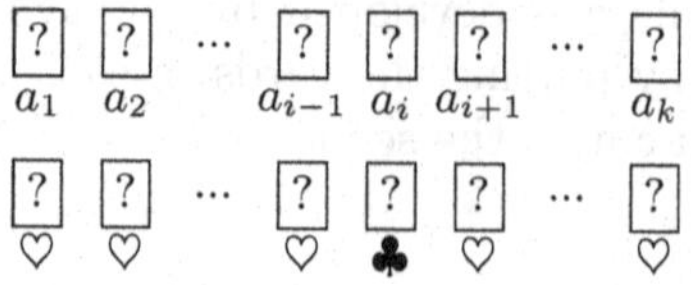

? ? ··· ? ? ? ··· ?
a_1 a_2 a_{i-1} a_i a_{i+1} a_k

? ? ··· ? ? ? ··· ?
♡ ♡ ♡ ♣ ♡ ♡

Fig. 4. A $2 \times k$ matrix constructed in Step 1 of the chosen cut protocol

1. Construct the following $2 \times k$ matrix (see Fig. 4).
   (a) In the first row, P publicly places the sequence A.
   (b) In the second row, P secretly places a face-down ♡ at every position except at the i-th column where P places a face-down ♣ instead.
2. Apply the pile-shifting shuffle to the matrix.
3. Turn over all cards in the second row. The card above the only ♣ will be the card a_i as desired.

3 Main Protocol

First, P secretly places a face-down sequence ♣♡ on each black cell, and a face-down sequence ♡♣ on each white cell in the Nonogram grid according to his/her solution. Then, P publicly applies the copy protocol to create an additional copy of the sequence on every cell. Each of the two copies will be used to verify the row and the column that cell is located on. Note that the copy protocol also verifies that each sequence is in a correct format (either ♣♡ or ♡♣).

From now on, we will show the verification of a row R with n cells and with a sequence $(x_1, x_2, ..., x_k)$ assigned to it. (The verification of a column works analogously in the direction from top to bottom.)

P takes only the left card from the sequence on each cell in R (which is a ♣ for a black cell and is a ♡ for a white cell) to form a sequence of n cards $S = (a_1, a_2, ..., a_n)$, where each card corresponds to each cell in R in this order from left to right. As R may start and end with a white or black cell, P publicly appends two face-down ♡s a_0 and a_{n+1} at the beginning and the end of S, respectively, to ensure that the sequence starts and ends with a ♡ (S now has length $n + 2$). Finally, P publicly appends a face-down marker card ♢, called a_{n+2}, at the end of S to mark the end of the sequence (S now has length $n+3$). See Fig. 5 for an example. The purpose of appending a ♢ is to allow all parties to locate the beginning and the end of S after S has been shifted cyclically.

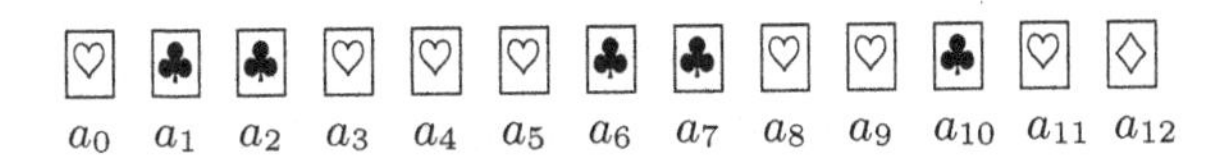

Fig. 5. A sequence S representing the third row of the solution in Fig. 2

The verification is divided into the following three phases.

3.1 Phase 1: Counting Black Cells

Currently, there are k blocks of consecutive $\clubsuit$s in S. In this phase, P will reveal the length of each block as well as replacing every $\clubsuit$ with a $\spadesuit$.

P performs the following steps for k rounds. In the i-th round:

1. Apply the chosen cut protocol to S to select a card corresponding to the leftmost cell of the i-th leftmost block of black cells in R (the block with length x_i). Let a_j denote the selected card.
2. Turn over cards $a_j, a_{j+1}, a_{j+2}, ..., a_{j+x_i-1}$ to reveal that they are all $\clubsuit$s (otherwise V rejects), where the indices are taken modulo $n + 3$.
3. Turn over cards a_{j-1} and a_{j+x_i} to reveal that they are both $\heartsuit$s (otherwise V rejects), where the indices are taken modulo $n + 3$.
4. Replace every face-up $\clubsuit$ with a $\spadesuit$. Turn over all face-up cards. The purpose of this step is to mark that this block of black cells has been selected.

After k rounds, V is now convinced that there are k different blocks of black cells with lengths $x_1, x_2, ..., x_k$, but still does not know about the order of these blocks, or whether there are any additional blocks of black cells besides these k blocks. Note that all $\clubsuit$s in S is now replaced by $\spadesuit$s. See Fig. 6 for an example.

Fig. 6. The sequence S from Fig. 5 at the end of Phase 1 (in a cyclic rotation)

3.2 Phase 2: Removing White Cells

Currently, there are $k+1$ blocks of consecutive $\heartsuit$s in S (including a block at the beginning containing a_0 and a block at the end containing a_{n+1}). In this phase, P will remove some $\heartsuit$s from S such that there will be exactly one remaining $\heartsuit$ in each block.

Let $X = x_1 + x_2 + ... + x_k$. There are $n - X$ white cells in R, so there are currently $n - X + 2$ $\heartsuit$s in S. P performs the following steps for $(n - X + 2) - (k + 1) = n - X - k + 1$ rounds.

1. Apply the chosen cut protocol to S to select any ♡ such that there are currently at least two remaining ♡s in a block it belongs to.
2. Turn over the selected card to reveal that it is a ♡ (otherwise V rejects).
3. Remove that card from S.

Now, each pair of consecutive blocks of ♠s are separated by exactly one ♡, and there is also a ♡ before the first block and after the last block (S now has length $X + k + 2$). See Fig. 7 for an example.

Fig. 7. The sequence S from Fig. 5 at the end of Phase 2 (in a cyclic rotation)

3.3 Phase 3: Verifying Order of Blocks

P applies the random cut to S, turns over all cards, and shifts the sequence cyclically such that the rightmost card is a ◇. V verifies that the remaining cards in S are: a ♡, a block of x_1 ♠s, a ♡, a block of x_2 ♠s, ..., a ♡, a block of x_k ♠s, a ♡, and a ◇ in this order from left to right. Otherwise, V rejects.

P performs the verification for every row and column of the grid. If all rows and columns pass the verification, then V accepts.

3.4 Optimization

In fact, at the beginning we do not need to copy a sequence on each cell since we only use one card per cell during the verification. Instead, we perform the copy protocol without putting cards in the third row of the matrix in Step 1. This modified protocol verifies that the original sequence is either ♣♡ or ♡♣, and then returns the original sequence in the second row. This modified protocol was developed by Mizuki and Shizuya [14].

After verifying that each sequence is in a correct format, we use the left card to verify a row, and the right card to verify a column that cell is located on. When verifying a column, the corresponding card is a ♡ for a black cell and a ♣ for a white cell, so we now treat ♣ and ♡ exactly the opposite way in the main protocol.

After the optimization, our protocol uses $mn + 1$ ♣s, $mn + \max(m, n) + 4$ ♡s, $\max(m, n)$ ♠s, and one ◇, resulting in a total of $2mn + 2\max(m, n) + 6$ cards. It uses $mn + 2m + 2n + 2w$ shuffles, where w is the total number of white cells in the grid.

4 Proof of Security

We will prove the perfect completeness, perfect soundness, and zero-knowledge properties of our protocol.

Lemma 1 (Perfect Completeness). *If P knows a solution of the Nonogram puzzle, then V always accepts.*

Proof. Suppose P knows a solution of the puzzle. Consider the verification of any row R.

In Phase 1, in each i-th round P selects a card a_j corresponding to the leftmost cell of the i-th leftmost block of black cells in R. Since that block has length x_i and has never been selected before, all of the cards $a_j, a_{j+1}, a_{j+2}, ..., a_{j+x_i-1}$ must be [♣]s. Also, since there is at least one white cell between two consecutive blocks of black cells (and at least one [♡] to the left of the leftmost block of [♣]s and to the right of the rightmost block of [♣]s), both of the cards a_{j-1} and a_{j+x_i} must be [♡]s. Hence, the verification will pass Phase 1.

At the start of Phase 2, there are a total of $n - X + 2$ [♡]s in $k + 1$ blocks of [♡]s, so P can remove a [♡] in each round for $(n-X+2)-(k+1) = n-X-k+1$ rounds such that there will be exactly one remaining [♡] in each block. Hence, the verification will pass Phase 2.

In Phase 3, there is exactly one [♡] between two consecutive blocks of [♠]s (and a [♡] at the beginning and the end of S). Moreover, the blocks of [♠]s are arranged in exactly the same order as the blocks of black cells, so the lengths of these blocks are $x_1, x_2, ..., x_k$ in this order from left to right. Hence, the verification will pass Phase 3.

Since this is true for every row (and analogously for every column), V will always accept. □

Lemma 2 (Perfect Soundness). *If P does not know a solution of the Nonogram puzzle, then V always rejects.*

Proof. We will prove the contrapositive of this statement. Suppose V accepts, meaning that the verification must pass for every row and column. We will prove that P must know a solution.

Consider the verification of any row R. In Phase 1, the steps in each i-th round ensure that there exists a block of exactly x_i consecutive black cells in R. Also, since all [♣]s selected in previous rounds have been replaced with [♠]s, this block must be different from the blocks revealed in previous rounds. Therefore, there must be at least k different blocks of black cells with lengths $x_1, x_2, ..., x_k$ (in some order) in R.

Moreover, in Phase 2 only [♡]s are removed from S, and when all cards in S are turned face-up in Phase 3 there is no [♣] remaining. This implies R has no other black cells besides the ones in these k blocks.

During Phase 3, the lengths of the blocks of [♠]s are $x_1, x_2, ..., x_k$ in this order from left to right. As the blocks of [♠]s are arranged in exactly the same order as the blocks of black cells, the lengths of the blocks of black cells must also be $x_1, x_2, ..., x_k$ in this order from left to right.

Since this is true for every row (and analogously for every column), P must know a solution of the puzzle. □

Lemma 3 (Zero-Knowledge). *During the verification, V learns nothing about P's solution.*

Proof. To prove the zero-knowledge property, it is sufficient to show that all distributions of cards that are turned face-up can be simulated by a simulator S that does not know P's solution.

- In Step 3 of the copy protocol in Sect. 2.4, the revealed sequence has an equal probability to be ♣♡ or ♡♣, so this step can be simulated by S.
- In Step 3 of the chosen cut protocol in Sect. 2.5, the ♣ has an equal probability to be at any of the k positions, so this step can be simulated by S.
- In the main protocol, when verifying each row (resp. column), there is only one deterministic pattern of the cards that are turned face-up in every phase. This pattern solely depends on the sequence $(x_1, x_2, ..., x_k)$ assigned to that row (resp. column), which is a public information, so the whole protocol can be simulated by S. □

5 Future Work

We developed a physical ZKP for Nonogram using $2mn + 2\max(m, n) + 6$ cards and $mn + 2m + 2n + 2w$ shuffles. A challenging future work is to develop a ZKP for this puzzle that can be performed using a standard deck (a deck containing all different cards with no duplicates) like the one for Sudoku in [18], an open problem posed in [10]. Other possible future work includes developing ZKPs for other popular logic puzzles or improving the already existing ones to become more practical, as well as exploring methods to physically verify other numerical functions.

References

1. Bultel, X., Dreier, J., Dumas, J.-G., Lafourcade, P.: Physical zero-knowledge proofs for Akari, Takuzu, Kakuro and KenKen. In: Proceedings of the 8th International Conference on Fun with Algorithms (FUN), pp. 8:1–8:20 (2016)
2. Bultel, X., et al.: Physical zero-knowledge proof for Makaro. In: Izumi, T., Kuznetsov, P. (eds.) SSS 2018. LNCS, vol. 11201, pp. 111–125. Springer, Cham (2018). https://doi.org/10.1007/978-3-030-03232-6_8
3. Chien, Y.-F., Hon, W.-K.: Cryptographic and physical zero-knowledge proof: from Sudoku to Nonogram. In: Boldi, P., Gargano, L. (eds.) FUN 2010. LNCS, vol. 6099, pp. 102–112. Springer, Heidelberg (2010). https://doi.org/10.1007/978-3-642-13122-6_12

4. Dumas, J.-G., Lafourcade, P., Miyahara, D., Mizuki, T., Sasaki, T., Sone, H.: Interactive physical zero-knowledge proof for Norinori. In: Du, D.-Z., Duan, Z., Tian, C. (eds.) COCOON 2019. LNCS, vol. 11653, pp. 166–177. Springer, Cham (2019). https://doi.org/10.1007/978-3-030-26176-4_14
5. Goldreich, O., Micali, S., Wigderson, A.: Proofs that yield nothing but their validity and a methodology of cryptographic protocol design. J. ACM **38**(3), 691–729 (1991)
6. Goldwasser, S., Micali, S., Rackoff, C.: The knowledge complexity of interactive proof systems. SIAM J. Comput. **18**(1), 186–208 (1989)
7. Google Play: Nonogram. https://play.google.com/store/search?q=Nonogram
8. Gradwohl, R., Naor, M., Pinkas, B., Rothblum, G.N.: Cryptographic and physical zero-knowledge proof systems for solutions of Sudoku puzzles. Theor. Comput. Syst. **44**(2), 245–268 (2009)
9. Koch, A., Walzer, S.: Foundations for actively secure card-based cryptography. In: Proceedings of the 10th International Conference on Fun with Algorithms (FUN), pp. 17:1–17:23 (2020)
10. Koyama, H., Miyahara, D., Mizuki, T., Sone, H.: A secure three-input AND protocol with a standard deck of minimal cards. In: Santhanam, R., Musatov, D. (eds.) CSR 2021. LNCS, vol. 12730, pp. 242–256. Springer, Cham (2021). https://doi.org/10.1007/978-3-030-79416-3_14
11. Lafourcade, P., Miyahara, D., Mizuki, T., Robert, L., Sasaki, T., Sone, H.: How to construct physical zero-knowledge proofs for puzzles with a "single loop" condition. Theor. Comput. Sci. **888**, 41–55 (2021)
12. Miyahara, D.: Card-Based ZKP protocols for Takuzu and Juosan. In: Proceedings of the 10th International Conference on Fun with Algorithms (FUN), pp. 20:1–20:21 (2020)
13. Miyahara, D., Sasaki, T., Mizuki, T., Sone, H.: Card-Based physical zero-knowledge proof for Kakuro. IEICE Trans. Fundam. Electron. Commun. Comput. Sci. **E102.A**(9), 1072–1078 (2019)
14. Mizuki, T., Shizuya, H.: Practical card-based cryptography. In: Ferro, A., Luccio, F., Widmayer, P. (eds.) Fun with Algorithms. FUN 2014. LNCS, vol 8496. Springer, Cham. https://doi.org/10.1007/978-3-319-07890-8_27
15. Mizuki, T., Sone, H.: Six-Card secure AND and Four-Card secure XOR. In: Deng, X., Hopcroft, J.E., Xue, J. (eds.) FAW 2009. LNCS, vol. 5598, pp. 358–369. Springer, Heidelberg (2009). https://doi.org/10.1007/978-3-642-02270-8_36
16. Robert, L., Miyahara, D., Lafourcade, P., Mizuki, T.: Interactive physical ZKP for connectivity: applications to Nurikabe and Hitori. In: Proceedings of the 17th Conference on Computability in Europe (CiE), pp. 373–384 (2021)
17. Robert, L., Miyahara, D., Lafourcade, P., Mizuki, T.: Physical zero-knowledge proof for Suguru Puzzle. In: Devismes, S., Mittal, N. (eds.) SSS 2020. LNCS, vol. 12514, pp. 235–247. Springer, Cham (2020). https://doi.org/10.1007/978-3-030-64348-5_19
18. Ruangwises, S.: Two standard decks of playing cards are sufficient for a ZKP for Sudoku. In: Chen, C.-Y., Hon, W.-K., Hung, L.-J., Lee, C.-W. (eds.) COCOON 2021. LNCS, vol. 13025, pp. 631–642. Springer, Cham (2021). https://doi.org/10.1007/978-3-030-89543-3_52
19. Ruangwises, S., Itoh, T.: Physical zero-knowledge proof for numberlink puzzle and k vertex-disjoint paths problem. New Gener. Comput. **39**(1), 3–17 (2021)
20. Ruangwises, S., Itoh, T.: Physical zero-knowledge proof for ripple effect. In: Uehara, R., Hong, S.-H., Nandy, S.C. (eds.) WALCOM 2021. LNCS, vol. 12635, pp. 296–307. Springer, Cham (2021). https://doi.org/10.1007/978-3-030-68211-8_24

21. Ruangwises, S., Itoh, T.: Physical ZKP for connected spanning subgraph: applications to bridges puzzle and other problems. In: Kostitsyna, I., Orponen, P. (eds.) UCNC 2021. LNCS, vol. 12984, pp. 149–163. Springer, Cham (2021). https://doi.org/10.1007/978-3-030-87993-8_10
22. Sasaki, T., Miyahara, D., Mizuki, T., Sone, H.: Efficient card-based zero-knowledge proof for Sudoku. Theor. Comput. Sci. **839**, 135–142 (2020)
23. Shinagawa, K., et al.: Card-Based protocols using regular polygon cards. IEICE Trans. Fundam. Electron. Commun. Comput. Sci. **E100.A**(9), 1900–1909 (2017)
24. Ueda, I., Miyahara, D., Nishimura, A., Hayashi, Y., Mizuki, T., Sone, H.: Secure implementations of a random bisection cut. Int. J. Inf. Secur. **19**(4), 445–452 (2019). https://doi.org/10.1007/s10207-019-00463-w
25. Ueda, N., Nagao, T.: NP-completeness Results for NONOGRAM via Parsimonious Reductions. Technical Report TR96-0008, Department of Computer Science, Tokyo Institute of Technology (1996)

Finding All Leftmost Separators of Size $\leq k$

Mahdi Belbasi(✉) and Martin Fürer(✉)

Department of Computer Science and Engineering, Pennsylvania State University, University Park, PA 16802, USA
{belbasi,fhs}@psu.edu

Abstract. We define a notion called leftmost separator of size at most k. A leftmost separator of size k is a minimal separator S that separates two given sets of vertices X and Y such that we "cannot move S more towards X" such that $|S|$ remains smaller than the threshold. One of the incentives is that by using leftmost separators we can improve the time complexity of treewidth approximation. Treewidth approximation is a problem which is known to have a linear time FPT algorithm in terms of input size, and only single exponential in terms of the parameter, treewidth. It is not known whether this result can be improved theoretically. However, the coefficient of the parameter k (the treewidth) in the exponent is large. Hence, our goal is to decrease the coefficient of k in the exponent, in order to achieve a more practical algorithm. Hereby, we trade a linear-time algorithm for an $\mathcal{O}(n \log n)$-time algorithm. The previous known $\mathcal{O}(f(k) n \log n)$-time algorithms have dependences of $2^{24k}k!$, $2^{8.766k}k^2$ (a better analysis shows that it is $2^{7.671k}k^2$), and higher. In this paper, we present an algorithm for treewidth approximation which runs in time $\mathcal{O}(2^{6.755k}\ n \log n)$,

Furthermore, we count the number of leftmost separators and give a tight upper bound for them. We show that the number of leftmost separators of size $\leq k$ is at most C_{k-1} (Catalan number). Then, we present an algorithm which outputs all leftmost separators in time $\mathcal{O}(\frac{4^k}{\sqrt{k}}n)$.

1 Introduction

Finding vertex separators that partition a graph in a "balanced" way is a crucial problem in computer science, both in theory and applications. For instance, in a divide and conquer algorithm, most of the time it is vital to have balanced subproblems. If we want to separate two subsets of vertices in a graph, we prefer the separator to be closer to the bigger side. In this work, we place the bigger set on the left side and the smaller one on the right. Before going into depth, we review and introduce some notations[1].

[1] Please note that the full version of this result with all the proofs and the figures can be found on the ArXiv version [2]. We have omitted some details here due to space constraints.

D.-Z. Du et al. (Eds.): COCOA 2021, LNCS 13135, pp. 273–287, 2021.
https://doi.org/10.1007/978-3-030-92681-6_23

1.1 Notation

W.l.o.g., assume that G is a connected graph. $S \subseteq V$ is a separator that separates two subsets of vertices $X, Y \subseteq V$ in G, if there is no $X - Y$ path in $G[V \setminus S]$, where $G[V \setminus S]$ is the induced graph on $V \setminus S$. In the following, we use $G - S$ instead of $G[V \setminus S]$ for the sake of simplicity. We call S an $(X, Y)^G$-separator. Later on, we drop the superscripts if it is obvious from the context.

Definition 1. *$S^G_{X,Y}$ is the set of all $(X, Y)^G$-separators.*

Definition 2. *The separator $S \in S^G_{X,Y}$ partitions $G - S$ into three sets $V_{X,S}$, $V_{S,Y}$, and V_Z, where $V_{X,S}$ is the set of vertices with a path from $X \setminus S$, $V_{S,Y}$ is the set of vertices with a path from $Y \setminus S$, and V_Z is the set of all vertices reachable from neither $X \setminus S$ nor $Y \setminus S$ in $G - S$.*

Having a non-empty V_Z set is only to our advantage. We think of X being on the left side and Y on the right side of S. Any of the three sets (X, Y, and S) might intersect.

Definition 3 ***Partial Ordering.*** *We say separator $S \in S^G_{X,Y}$ is at least as much to the left as separator $S' \in S^G_{X,Y}$ if $V_{X,S} \subseteq V_{X,S'}$. In this case, we use the notation $S \preceq S'$.*

Definition 4. *Separator $S \in S^G_{X,Y}$ is called an $(X, Y, \leq k)^G$-*separator *if $|S| \leq k$.*

Definition 5. *Separator $S \in S^G_{X,Y}$ is called a* leftmost $(X, Y, \leq k)^G$-separator *if it is minimal and there exists no other minimal $(X, Y, \leq k)^G$-separator S' such that $S' \preceq S$.*

Notice that the minimality is important here, otherwise according to the partial ordering definition, one can keep adding extra vertices to the left of S' (towards X) and artificially make it more to the left. In order to avoid this, we require all separators we work with to be minimal unless specified otherwise.

The notion of leftmost separator is closely related to the notion of *important separator.* Important separator has been defined in [10], and then used in [5] and [9].

The difference between a leftmost separator and an important separator comes from their corresponding partial orders. The partial order defined for important separators is as follows:

Definition 6 ***Partial Ordering used for Important Separators.*** *Separator $S \in S^G_{X,Y}$ dominates (or "is more important than") separator $S' \in S^G_{X,Y}$ if $|S| \leq |S'|$ and $V_{S,Y} \subset V_{S',Y}$.*

Definition 7. *Separator $S \in S^G_{X,Y}$ is an* important $(X, Y, \leq k)^G$-separator *if there exists no other minimal $(X, Y, \leq k)^G$-separator S' dominating S.*

As you see, when ordering important separators we also look at the relation between the sizes but in a leftmost separator, its size just has to be $\leq k$.

Lemma 1. *Every leftmost* $(X,\ Y,\ \leq\ k)^G$*-separator is an important* $(X,\ Y,$ $\leq k)^G$*-separator, but the converse does not hold.*

The proof is omitted due to space constraints.

Notice that not all the important separators are leftmost. Our purpose is to find a separator more towards the bigger side in order to have more balanced separators. For that reason, not all the important separators are good. For instance we do not need to consider S_1 because that is the most unbalanced separator one can find. This is the main reason that we defined the new notion of leftmost separators. As argued, leftmost separators are better candidates for our application. However, as the reader can see, there is a strong similarity between these two notions. We give tight upper bounds for the number of leftmost separators and a tight upper bound for the number of important separators.

Lemma 2. *Let* $\mathcal{A}^G_{X,Y,\leq k}$ *and* $\mathcal{B}^G_{X,Y,\leq k}$ *be the set of all leftmost* $(X, Y, \leq k)^G$*-separators and the set of all important* $(X, Y, \leq k)^G$*-separators, respectively. Then,*

$$\mathcal{B}^G_{X,Y,\leq k} = \bigcup_{i=1}^{k} \mathcal{A}^G_{X,Y,\leq i}$$

The proof has been omitted due to space constraints.

In this paper, we show that the number of leftmost $(X,\ Y,\ \leq k)$-separators is $\leq C_{k-1}$, where C_n is the n-th Catalan number. Furthermore, we close the gap and show that this upper bound is tight. Then, we give an $\mathcal{O}(4^k kn)$-time algorithm finding all minimal leftmost $(X,\ Y,\ \leq\ k)$-separators. Notice that $C_{k-1} \sim \frac{4^{k-1}}{\sqrt{\pi}(k-1)^{\frac{3}{2}}}$.

Based on Lemma 2, this implies that the number of important $(X,\ Y,\ \leq k)$-separators is $\leq \sum_{i=1}^{k-1} C_i$ and the bound is tight.

One of the important applications of the algorithm that finds all the leftmost separators is treewidth approximation. Treewidth approximation is a crucial problem in computer science. Courcelle's methatheorem [6] states that every problem which can be described in monadic second order logic has an FPT algorithm with the treewidth k as its parameter. An FPT algorithm is an algorithm that runs in time $\mathcal{O}(f(k)\,poly(n))$, where n is the input size, k is the parameter (here, treewidth), and $f(\cdot)$ is a computable function.

So, based on Courcelle's methatheorem, many NP-complete graph problems obtain polynomial algorithms (in terms of the input size), and hence they can be solved fast if the treewidth is small. These algorithms require access to tree decompositions of small width. However, finding the exact treewidth itself is another NP-complete problem [1]. Here, we look for an approximation algorithm to solve the treewidth problem.

Problem 0. Given a graph $G = (V, E)$, and an integer $k \in \mathbb{N}$, is the treewidth of G at most k? If yes, output a tree decomposition with width $\leq \alpha k$, where $\alpha \geq 1$ is a constant. Otherwise, output a subgraph which is the bottleneck.

There are various algorithms solving this problem for different α's (the approximation ratio). As mentioned above, we are interested in constant-factor approximation FPT algorithms.

Algorithms [4] and [8] both run in $2^{\mathcal{O}(k)}n$ time, which is linear in n. However, the coefficients of k in the exponent are large. The former one does not mention the exact coefficient and seems to have a very large coefficient. The latter one, which is a very recent paper, mentions that the coefficient of k in the exponent is some number between 10 and 11. Our goal is to make treewidth approximation more applicable by decreasing the coefficient of k in the exponent. We can afford an extra $\log n$ factor in the running time in order to reduce the huge dependence on k. We sacrifice the linear dependence on n, and give an algorithm which runs way faster in various cases. So, let us look at $n \log n$-time algorithms. Reed [12] gave the first $n \log n$-time algorithm. He did not mention the dependence on k precisely but a detailed analysis in [3] shows that it is $\mathcal{O}(2^{24k}k!)$. Here, even though $2^{24k} = o(k!)$, actually $k!$ is reasonable for small k's while 2^{24k} is not. Later on, the authors of this paper introduced an $\mathcal{O}(2^{8.766k}n \log n)$-algorithm [3]. The algorithm presented in this paper is based on [12] and [3]. In these papers, when it is known that a good separator S exists between X and Y, an efficient algorithm finds an arbitrary separator between X and Y. The ability to find leftmost separators allows for an improvement. If S is a good separator between X and Y, and $V_{X,S}$ is estimated to be at least as big as $V_{S,Y}$, then the best leftmost separator between X and Y has a definite advantage.

Instead of a balanced separator with minimum size, we consider all leftmost separators (closest possible to the bigger side). This helps us to obtain an $\mathcal{O}(2^{6.755k}n \log n)$-time algorithm with the same approximation ratio of 5 as in [12] and [3].

Before moving onto the next section, we have to mention that the algorithm to find all leftmost separators works for both directed and undirected graphs.

Below, we summarize our contributions.

1.2 Our Contributions

First, we give a tight upper bound on the number of the leftmost separators.

Theorem 1. *Let $G = (V, E)$ be a graph, $X, Y \subseteq V$, and $k \in \mathbb{N}$. The number of leftmost $(X, Y, \leq k)^G$-separators*[2] *is at most* C_{k-1}[3]$= \frac{1}{k}\binom{2(k-1)}{k-1} \sim \frac{4^{k-1}}{\sqrt{\pi}(k-1)^{3/2}}$.

[2] Notice that all leftmost separators are minimal per definition.
[3] C_n is the nth Catalan number.

Furthermore, the number of important $(X,\ Y,\ \leq k)^G$-separators is at most $\sum_{i=0}^{k-1} C_i$. *Both bounds are tight.*

Then, we give an algorithm finding all leftmost separators.

Theorem 2. *Let $G = (V, E)$ be a graph, $X, Y \subseteq V$, and $k \in \mathbb{N}$. There is an $\mathcal{O}(2^{2k}\sqrt{k}n)$-time algorithm which outputs all the leftmost $(X,\ Y,\ \leq k)^G$-separators.*

Now, we use the algorithm finding all the leftmost separators to solve treewidth approximation much faster.

Theorem 3. *Let $G = (V, E)$ be a graph, and $k \in \mathbb{N}$. There is an algorithm that either outputs a tree decomposition of G with width $\leq 5(k-1)$, or determines that $tw(G) > k-1$ in time $\mathcal{O}\left(2^{6.755k} n \log n\right)$.*

2 Finding the Leftmost Minimum Size $(X,\ Y,\ \leq k)^G$-Separator

This section is omitted due to space constraint but a variation of it can be found in [7].

Algorithm 1: *Left_Minimum_Sep*$(G, Y, \mathcal{P})$: FIND THE LEFTMOST MINIMUM SEPARATOR

Input: Graph $G = (V, E)$, a subset of vertices Y, and a set $\mathcal{P}$ of pairwise disjoint paths from X to Y

Output: The leftmost minimum $(X,\ Y,\ \leq k)^G$-separator, and an updated set of pairwise disjoint paths $\mathcal{P}$

1 **while** $\exists$ *a $\mathcal{P}$-augmenting walk Q* **do**

2   Update $\mathcal{P}$.

    `// by sending a unit flow through the edges of` $\mathcal{P}$ `and` Q`. Also,` $|\mathcal{P}|$ `is increased by 1`

3 Construct $R(\mathcal{P}) = \{v \in V \mid \exists$ a $\mathcal{P}$-augmenting walk from X to $v\}$ using DFS.

4 Initialize $C(\mathcal{P}) \leftarrow \emptyset$

5 **for** $P \in \mathcal{P}$ **do**

6     $c_P \leftarrow$ the first vertex $\in P$ and $\notin R(\mathcal{P})$.

7     $C(\mathcal{P}) \leftarrow C(\mathcal{P}) \cup \{c_P\}$.

8 **return** $(C(\mathcal{P}), \mathcal{P})$.

Algorithm 2: *Init*(G, Y): INITIALIZATION

```
Input: Graph G
1 {S, P} ← Left_Minimum_Sep(G, Y, ∅)
      // S is the minimum size leftmost (X, Y, ≤ k)-separator and P is
      a set of pairwise disjoint paths
2 push all the vertices in S onto empty stack R.
3 Branch(G[V_{X,S} ∪ S], S, ∅, P, R, True)
```

Algorithm 3: *Branch*$(G, S, Y, I, \mathcal{P}, R, leftmost)$: THE MAIN PROCEDURE FOR FINDING ALL LEFTMOST $(X, Y, \leq k)$-SEPARATORS

```
Input: Graph G, a separator S ∈ S_{X,Y}, a subset of vertices Y, I: the included
       vertices, P: a set of pairwise disjoint paths between X and Y, R: the
       stack to hold the order of handling vertices, and leftmost: a boolean
       indicating whether we have a leftmost separator.
1  if I == S then
2  │ A ← A ∪ {S}
3  else
4  │ pop v from R
5  │ Y' ← (S \ {v}) ∪ (N(v) ∩ V_{X,S})
6  │ {S', P'} ← Left_Minimum_Sep(G, Y', P)
7  │ if |S'| ≤ k ∧ I ⊆ S' then
8  │ │ leftmost ← False
9  │ │ if |S'| < k then
10 │ │ │ let R' be a copy of R. Push all vertices of (N(v) ∩ V_{X,S}) onto R'
11 │ │ │ Branch(G[V_{X,S'}], S', Y', I, P', R', True)
12 │ if (|S \ I| ≥ 2 ∨ leftmost) ∧ (|S| ≤ k) then
13 │ │ Branch(G[V_{X,S}], S, Y, I ∪ {v}, P, R, leftmost)
```

3 Finding All Minimal Leftmost $(X, Y, \leq k)$-Separators

In this section, we present our main algorithm. In the introduction, we mentioned why it is important to find the leftmost[4] separators. Also, in Sect. 2 we reviewed an algorithm (Algorithm 1) to find the leftmost minimum separator. We use this algorithm in ours.

In our problem, we have two subsets of vertices X and Y such that $|X| \geq |Y|$. W.l.o.g., assume X is the set on the left and Y is the set on the right.

> **Problem 2.** Given a graph $G = (V, E)$, sets $X, Y \subseteq V$, and $k \in \mathbb{N}$, what are the minimal leftmost $(X, Y, \leq k)^G$-separators?

[4] We drop the term "minimal" because it is the default for the separators throughout this paper unless mentioned otherwise.

Theorem 4. *Given a graph $G = (V, E)$, sets $X, Y \subseteq V$, and $k \in \mathbb{N}$, there exists an algorithm which solves Problem 2 in time $\mathcal{O}(2^{2k}\sqrt{k}n)$*

Proof. We present a recursive branching algorithm (Algorithm 3). Initially, it calls Algorithm 1 to find the leftmost minimum size $(X,\ Y,\ \leq k)$-separator (using a simple flow algorithm), namely S by feeding X as the left set, Y as the right set, and an empty set of pairwise disjoint paths (namely $\mathcal{P}$) from X to Y (inherited from the parent branch). Notice that the leftmost minimum size separator is unique. This is the root (namely r) of the computation tree (namely $\mathcal{T}$). Let us refer to the computation subtree rooted at node x, as $\mathcal{T}(x)$.

In each node x of $\mathcal{T}$ with the corresponding graph G_x, let Y_x, I_x, and $\mathcal{P}_x$ be the right set, the set of vertices that we require to be in the separator, and the set of disjoint paths inherited from the parent's node, respectively. Notice that we do not pass X_x (the left set) as an argument since it does not change throughout the algorithm and hence we have defined it as a global variable $(\forall x, X_x = X)$.

Claim. Let S be a minimal leftmost $(X,\ Y,\ \leq k)^G$-separator, and S' be the separator generated by Algorithm 1. Then, $S \subseteq V_{X,S'} \cup S'$.

Proof. For the sake of contradiction, assume that $\exists v \in S$ such that $v \notin V_{X,S'} \cup S'$. This means that $v \in V_{S',Y} \cup V_Z$.

If $v \in V_Z$, then $S \setminus \{v\}$ is still a leftmost $(X,\ Y,\ \leq k)^G$-separator which contradicts the minimality of S.

The other possibility is that $v \in V_{S',Y}$. Let S_{out} be the set of all such vertices; i.e., $S_{out} = \{v \in S \mid v \in V_{S',Y}\}$. Any $X - S_{out}$-path[5] goes through S', otherwise S' would not be a separator. Let S'_{in} be the set of all vertices of S' that are on a path from vertex of X to a vertex of S_{out}. Now, let $S'' = (S \setminus S_{out}) \cup S'_{in}$. Hence, S'' is an $(X, Y, \leq k)^G$-separator such that $S'' \preceq S$ with $|S''| \leq S \leq k$, which is a contradicts that S is a leftmost $(X,\ Y,\ \leq k)$-separator. □

Claim 3 allows us to ignore the subgraph $G[V_{S,Y} \cup V_Z]$ and focus only on the graph to the left of the current separator and keep moving towards left until it is impossible.

Let S_x be the separator found by Algorithm 1 while processing node x from the computation tree. If S_x is a leftmost $(X,\ Y,\ \leq k)$-separator, this branch terminates and we add S_x to the set of all the leftmost $(X,\ Y,\ \leq k)^G$-separators, namely $\mathcal{A}$ ($\mathcal{A}$ is a global variable). Otherwise, we keep pushing the separator to the left by branching 2-fold. Let us call the children of x by c_1 and c_2. If S_x was not a leftmost $(X,\ Y,\ \leq k)^{G_x}$-separator[6], this means that there exists at least one leftmost $(X,\ Y,\ \leq k)^{G_x}$-separator, namely S' such that $S' \neq S$ and $S' \preceq S$. Also, as a result of Claim 3, S' is a leftmost $(X,\ Y,\ \leq k)^G$-separator, too.

In each node x, we call Algorithm 1 to find the minimum separator S_x of size $\leq k$ between X and Y_x. Now, we push all the vertices of S_x onto stack R.

[5] any path from a vertex in X to a vertex in S_{out}.

[6] note that by Claim 3 $G_x = G[V_{X,S} \cup S]$.

Then, we pop vertex v which is on top of the stack R and consider the following two scenarios (corresponding to c_1 and c_2).

1. If we want v to belong to the leftmost separators. In this case, we add v to set I, which is the set of vertices that we require to be in all the leftmost separators in $\mathcal{T}_{c_1}$.
2. If v does not belong to the leftmost separators. Here, we pop v and push the left neighbors of v (i.e., $N(v) \cap V_{X,S_X}$) onto R (we just move to the left due to Claim 3).

Notice that the order of handling vertices is not important but we use stack because it simplifies the proof later on.

Every produced separator is a leftmost $(X, Y, \leq k)^G$-separator because the only time that one branch terminates is when it finds a leftmost separator. Now, we show that all the leftmost $(X, Y, \leq k)^G$-separators are generated by the algorithm given.

Let S_0 be an arbitrary leftmost $(X, Y, \leq k)^G$-separator, and as before let S be the separator generated by Algorithm 1. At this point R is filled with the vertices of S. Pop v from R.

- If $v \in S_0$, then put v in I, and recurs.
- If $v \notin S_0$, push $N(v) \cap V_{X,S}$ into R, recurs.

This determines an exact computation branch. All branches halt with a minimal leftmost separator since each time we go at least one more to the left. So, this branch terminates with minimal leftmost separator S_0' as well. All the vertices of S_0 are pushed into R at some point because otherwise this branch terminates with a separator which is not leftmost and we can push it more to the left. At the end of this branch, $I \subseteq S_0'$. On the other hand, $I = S_0 \subseteq S_0'$, which implies that $S_0 = S_0'$ because both of them should be minimal. □

In [11], the authors mention that the number of important $(X,Y,\leq k)^G$-separators is $\leq 4^k$, and they even mention that this upper bound can be tight by a polynomial factor. Here, we give a precise upper bound for the number of leftmost $(X,Y,\leq k)^G$-separators and the number of important $(X,Y,\leq k)^G$-separators and we show that both bounds are tight.

Before that, let us review the definition of the Catalan numbers.

Definition 8. Catalan numbers *is a sequence of numbers where the n-th Catalan number is:*

$$C_n = \binom{2n}{n} - \binom{2n}{n+1} = \frac{1}{n+1}\binom{2n}{n} \sim \frac{4^n}{n^{3/2}\sqrt{\pi}}.$$

Theorem 5. *Let $G = (V, E)$ be a graph and let $X, Y \subseteq V$ and let $k \in \mathbb{N}$. The number of leftmost $(X,Y,\leq k)^G$-separators is at most C_{k-1} and the number of important $(X,Y,\leq k)^G$-separators is at most $\sum_{i=0}^{k-1} C_i$. Furthermore, both upper bounds are tight.*

Before giving a proof, we mention some definitions.

Definition 9. *A* full k-paranthesization *is a string over the alphabet* $\{[,]\}$ *consisting of* k *"[" and* k *"]" having more "[" than "]" in every non-trivial prefix. This forms a language slightly different from the Dyck language (a well-known language). We call this language* restricted Dyck language *and any string in this language is called a* restricted Dyck word.

Notice that $[x]$ is a restricted Dyck word iff x is a Dyck word[7].

Full paranthesizations (the restricted Dyck language over the alphabet $\{[,]\}$) are generated by the following context-free grammar $G = (\{S, A\}, \{[,]\}, R, S)$, where the set of the rules R is:

$$S \rightarrow [A] \mid \epsilon$$
$$A \rightarrow AA \mid [A] \mid \epsilon$$

Definition 10. *The* k-parentheses tree *is the binary tree* (V, E) *with the following properties.*

- V *is the set of prefixes of full* k'*-parenthesizations for* $0 < k' \leq k$.
- *If "*x*" is in* V*, then "*$x[$*" is the left child of* x*, and if "*$x]$*" is in* V*, then "*$x]$*" is the right child of* x.
- *If "*$x[$*" or "*$x]$*" are not in* V*, then* x *has no left or right child respectively.*

Definition 11. *The* compact $\leq k$-parentheses tree *is obtained from the* $\leq k$*-parentheses tree by removing all nodes containing* k *"[" and ending in "]".*

Note that the nodes with k "[" ending in "]" have been removed, because no branching happens at their parents, as there are no left children.

The compact $\leq k$-parentheses tree is a full binary tree, every node has 0 or 2 children. The number of leaves in this tree is equal to the number of full $\leq k$-parenthesizations, which is equal to $\sum_{k'=1}^{k-1} C_{k'} = \mathcal{O}(C_k)$ where C_k is the k-th Catalan number. An immediate consequence is that the number of nodes in the compact $\leq k$-parentheses tree is $2\sum_{k'=1}^{k-1} C_{k'} - 1 = \mathcal{O}(C_k)$.

For the algorithm to find all leftmost $(X, Y, \leq k)$-separators, the worst case computation tree is the compact $\leq k$-parentheses tree with the nodes representing full k'-parenthesizations for $1 \leq k' \leq k - 1$ removed.

An arbitrary computation tree for the algorithm to find all leftmost $\leq k$-separators is obtained form the worst case computation tree by the following two operations.

(a) For any node x, the subtree rooted at the left child $c_1(x)$ may be just removed or replaced by the subtree rooted at $c_1(c_1(x))$. This splicing operation can be repeated to jump down arbitrarily far.

[7] Notice the difference between the restricted Dyck Language and the Dyck language itself. In every non-trivial prefix of a Dyck word, the number of "[" is $\geq$ the number of "]", whereas for the restricted version it should be strictly greater.

(b) For any node x with less than or equal to k "["s, let x_i be x concatenated with i "]"s. If x_{i^*} is a full k'-parenthesization for some k' with $0 < k' < k$, and the nodes $x_{i'}$ for $0 \leq i' < i^*$ have no left children, then the leaf x_{i^*} is added. (In the added node x_{i^*}, a minimal separator S of size $< k$ is picked, because there is no larger minimal $\leq k$-separator to the left of S.)

Note that the number of leaves in an arbitrary computation tree for the algorithm to find all $(X, Y, \leq k)$-separators is $\leq C_{k-1}$, because for every leaf "x]" added in (b), at least one leaf, namely "$x[\cdots[]\cdots]]$" (full parenthesization of length k) has been removed in (a).

For the recursive procedure, we introduce a boolean parameter leftmost. The parameter is originally true in the root and in every node that is a left child. If a node has a left child, i.e., excluding v, a new larger separator S' with $|S'| \leq k$ has been found, then leftmost is set to false. The current truth value of leftmost is passed to the right child (include v call). If x contains $k' < k$ "[" and $k' - 1$ "]", then a right call (including v) is only made if leftmost is true, i.e., there is no larger $(X, Y, \leq k)$-separator to the left of the current small (size $< k$) separator.

In every node, S is the minimum leftmost separator between and including X and the current Y.

- In the root, Y is the original Y. $I = \emptyset$, and *leftmost* $=$ *true*. The node is represented by a sequence of "["s of length $|S|$.
- A node x with $|S| = k' < k$, is represented by a prefix x of a full parenthesization with k' "[" and $|I|$ "]".

[Alternatively, one could include a root in the computation tree corresponding to the empty 0-parenthesization. In such a root, no S is defined. The minimum leftmost separator S' is computed. A left call is made if $|S'| \leq k$. No right call is made. In this view, the root node has only one child. Not including such a root node means that this node is not handled by the recursive procedure, but by a different calling procedure.]

Now, here is the proof of Theorem 5.

Proof. First, we prove the result corresponding to the leftmost separators. As explained above, we show excluding a vertex by a "[", which corresponds to a left branch. Analogously, "]" denotes including a vertex in I (requiring the leftmost separator to have v in it), which corresponds to the right branch.

Now, we see why we used a stack to handle branching on the vertices. Even though the order does not matter but we need a stack to have a nice correspondence between the algorithm behavior and the restricted Dyck words. It is well known that the number of the Dyck words of length $2k$ is C_k. So, the upper bound on the number of leftmost $(X,\ Y,\ \leq\ k)$-separators is $C_{k-1} = \frac{\binom{2(k-1)}{k-1}}{k} \sim \frac{4^{k-1}}{(k-1)^{3/2}\sqrt{\pi}}$. We also close the gap and show that this bound is tight.

Let $\hat{G}$ be a complete binary tree with depth at least $k+1$. Let X be the set of all the leaves of $\hat{G}$, and Y be the root. (see Fig. 1). The number of full subtrees with exactly k leaves is C_{k-1}. On the other hand in $\hat{G}$, no $(X, Y, \leq k-1)^{\hat{G}}$-separator is a leftmost $(X, Y, \leq k)^{\hat{G}}$-separator since we can replace a node with its both children and move more to the left. So, the number of minimal leftmost $(X, Y, \leq k)^{\hat{G}}$-separators is exactly C_{k-1}.

For the upper bound on the number of important separators, we use Lemma 2 and this immediately implies that the number of important $(X, Y, \leq k)^{\hat{G}}$-separators $\leq \sum_{i=1}^{i=k-1} C_i$.

For the tightness part, assume the same $\hat{G}$ with the same X and Y. The number of important $(X, Y, \leq k)^{\hat{G}}$-separators is $\sum_{i=1}^{k-1} C_i$ since based on Lemma 2, the number of important $(X, Y, \leq k)^{\hat{G}}$-separators is $\leq \sum_{i=1}^{k-1}$(the number of leftmost $(X, Y, \leq k)^{\hat{G}}$-separators). On the other hand, as we mentioned earlier, in $\hat{G}$, no leftmost $(X, Y, \leq k-1)^{\hat{G}}$-separator is a leftmost $(X, Y, \leq k)^{\hat{G}}$-separator. Hence, the number of important $(X, Y, \leq k)^{\hat{G}}$-separators is exactly $\sum_{i=1}^{k-1}$(the number of leftmost $(X, Y, \leq k)^{\hat{G}}$-separators), which is equal to $\sum_{i=1}^{k-1} C_i$. □

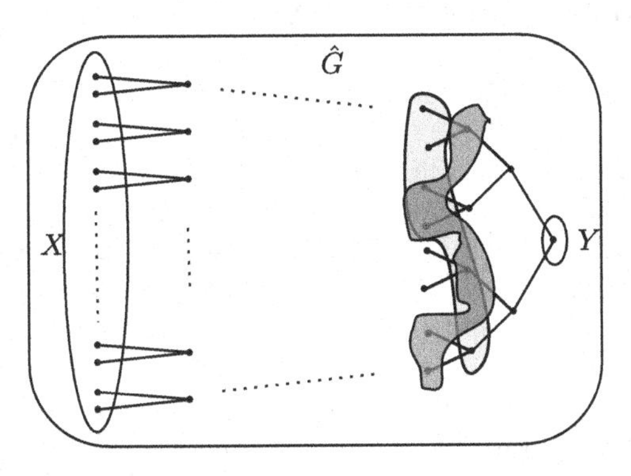

Fig. 1. Here, $k = 6$. We illustrate two out of 42 minimal leftmost $(X, Y, \leq 6)^{\hat{G}}$-separators. Also, there are $1+2+5+14+42 = 64$ important $(X, Y, \leq 6)^{\hat{G}}$-separators.

Notice that $\sum_{i=1}^{k-1} C_i = \Theta(\frac{4^{k-1}}{(k-1)^{3/2}})$.

Remark 1. Let $\mathcal{A}^G_{X,Y,\leq}$ and $\mathcal{B}^G_{X,Y,\leq}$ be the set of all leftmost $(X, Y, \leq k)^G$-separators and the set of Based on Lemma 2, the set of all important $(X, Y, \leq k)^G$-separators that are not leftmost $(X, Y, \leq k)^G$-separator is the union of the set of all leftmost $(X, Y, \leq i)^G$-separator, for $i = 1, \cdots, k-1$.

Remark 2. Let $\hat{G}$ be a complete binary tree with depth at least $k+1$, $\hat{X}$ be the set of all leaves of $\hat{G}$, and $\hat{Y}$ be the root. For a fixed $k \in \mathbb{N}$,

1. $\hat{G}$ has the highest number of leftmost $(X, Y, \leq k)^G$-separators among all graphs like G and for all $X, Y \subseteq V(G)$ and it happens by setting $X = \hat{X}$ and $Y = \hat{Y}$.
2. $\hat{G}$ has the highest number of important $(X, Y, \leq k)^G$-separators among all graphs like G and for all $X, Y \subseteq V(G)$ and it happens by setting $X = \hat{X}$ and $Y = \hat{Y}$.
3. $\hat{G}$ has the highest number of important but not leftmost $(X, Y, \leq k)^G$-separators among all graphs like G and for all $X, Y \subseteq V(G)$ and it happens by setting $X = \hat{X}$ and $Y = \hat{Y}$.

Theorem 6. *Let $G = (V, E)$ be a graph, $X, Y \subseteq V$, and $k \in \mathbb{N}$. There is an algorithm which finds all minimal leftmost $(X, Y, \leq k)^G$-separators in time $\mathcal{O}(\frac{4^k}{\sqrt{k}} n)$.*

Proof. Our algorithm searches all the possibilities and enumerates all the possible leftmost separators. We showed that the number of leaves, which is the number of leftmost separators is $\mathcal{O}(C_{k-1})$. So, we have $\mathcal{O}(C_{k-1})$ nodes in our computation tree and work done in every node is $\mathcal{O}(kn)$ (the running time of a simple flow algorithm). Hence, the total running time is $\mathcal{O}(\frac{2^{2k}}{\sqrt{k}} n)$. □

4 Application to Treewidth Approximation

As mentioned in the introduction, Reed's 5-approximation algorithm [12] for treewidth runs in time $\mathcal{O}(2^{24k}(k+1)!\, n \log n)$. This was improved to a 5-approximation algorithm running in $\mathcal{O}(k^2 2^{8.766k} n \log n)$ [3]. Here, we further improve the treewidth approximation algorithm. In order to do so, we briefly describe the algorithms given in [12], and [3]. Hence, we review some notations and for the others we give references.

For a graph $G = (V, E)$ and a subset W of the vertices, $G[W]$ is the subgraph induced by W. For the sake of simplicity throughout this paper, let $G - W$ be $G[V \setminus W]$ and $G - v$ be $G - \{v\}$ for any $W \subseteq V(G)$ and any $v \in V(G)$.

Also, in a weighted graph, a non-negative integer weight $w(v)$ is defined for each vertex v. For a subset W of the vertices, the weight $w(W)$ is simply the sum of the weights of all vertices in W. Furthermore, the total weight or the weight of G is the weight of V. Some of the common definitions have been omitted due to space constraints.

Lemma 3. [7, Lemma 11.16] *Let $G = (V, E)$ be a graph of treewidth at most $k-1$ and $W \subseteq V$. Then there exists a balanced W-separator of G of size at most k.*

Definition 12. *Let $G = (V, E)$ be a graph and $W \subseteq V$. A* weakly balanced separation *of W is a triple (X, S, Y), where $X, Y \subseteq W$, $S \subseteq V$ are pairwise disjoint sets such that:*

- $W = X \cup (S \cap W) \cup Y$.
- S *separates* X *from* Y.
- $0 < |X|, |Y| \leq \frac{2}{3}|W|$.

Lemma 4. [7, Lemma 11.19] *For* $k \geq 3$, *let* $G = (V, E)$ *be a graph of treewidth at most* $k-1$ *and* $W \subseteq V$ *with* $|W| \geq 2k+1$. *Then there exists a weakly balanced separation of* W *of size at most* k.

Theorem 7. [7, Corollary 11.22] *For a graph of treewidth at most* $k - 1$ *with a given set* $W \subseteq V$ *of size* $|W| = 3k - 2$, *a weakly balanced separation of* W *can be found in time* $O(2^{3k}k^2n)$.

4.1 Our Improvement

In this subsection, our goal is to use the algorithm for finding the leftmost separators to further improve the coefficient of k in the exponent of the tree decomposition algorithm to make it more applicable.

- For the analysis, we consider a centroid by volume, namely C. It has size $k' \leq k$.
- Each connected component of $G - C$ has volume at most $\frac{1}{2}(n - k')$.
- These connected components can be grouped into 3 parts, each with volume at most $\frac{1}{2}(n - k')$. (Just place the components by decreasing volume into the part with currently smallest volume.)
- Let the proper volume be the part of the volume that has its corresponding weight in the same part. In other words, the proper volume is the number of vertices whose representative is in the same part.
- Let t be the threshold for the size of the small trees. At most $k'(t - 2) = k'((\frac{1}{2} - \epsilon)n/k - 2) \leq (\frac{1}{2} - \epsilon)n - 2k'$ vertices can be in a different part than their representative. Therefore, the total proper volume is at least $n - k' - (\frac{1}{2} - \epsilon)n + 2k' \geq (\frac{1}{2} + \epsilon)n + k'$
- Of the proper volume, at least $(\frac{1}{2} + \epsilon)n + k' - \frac{1}{2}(n - k') > \epsilon n$ is not in the part with largest proper volume.
- Therefore, there are at least 2 parts with proper volume at least $\frac{\epsilon n}{2}$
- Of these 2 parts, we put the part with larger weight on the left side, the other one on the right side.
- We also put the third part on the left side.
- The left part has weight at least half the total weight, which is $\frac{1}{2}(n - k')$.
- The right part has weight at most $\frac{1}{2}(n - k')$ and (proper) volume at least $\frac{\epsilon n}{2}$

The algorithm tries all possible 2-partitions of the representatives. This includes the left-right partition that we are currently investigating. While searching for a leftmost separator, the centroid is a competitor. Thus the algorithm finds a separator that is equal to the centroid or is located strictly to the left of it. From now on, left (call it X) and right (namely Y) are defined by the leftmost $(X, Y, \leq k)^G$-separator found by Algorithm 2. This separator has size k'' with $k' \leq k'' \leq k$. It produces the same weight partition as the centroid, but part of the volume might shift to the right.

- Thus the left part has still weight at least $\frac{1}{2}(n-k')$ and therefore volume at least $\frac{1}{2}(n-k') - (\frac{1}{2}-\epsilon)n + 2k \geq \epsilon n + \frac{3}{2}k$.
- The right part has still volume at least $\frac{\epsilon n}{2}$
- The recursive calls are done with the subgraphs induced by the union of the vertices of a connected component with the vertices of the separator. Their number of vertices is upper bounded by n minus the volume of the smaller side. It is less than $(1-\frac{\epsilon}{2})n$.
- After $\frac{2\ln 2}{\epsilon}(\log n - \log b) = O(\log n)$ rounds for $b \geq k$, the largest volume of a recursive call is at most b.

In the worst case, the algorithm alternates between a split by volume and $\log k$ splits of W steps. Let the time spent between two splits by volume be at most $f(k)n$. Note that $f(k) \leq g(k) + 3^{3k} k \log k = O(g(k))$, where $g(k)$ is the time of one split by volume step. Then we get the following recurrence for an upper bound on the running time of the whole algorithm.

$$T(n) = \begin{cases} O(k) & \text{if } n \leq 3k \\ \max_{p,n_1,\ldots,n_p} \left\{\sum_{i\in[p]} T(n_i)\right\} + f(k)n & \text{otherwise,} \end{cases}$$

where the maximum is taken over $p \geq 2$ and $n_1,\ldots,n_p \in [n-1]$ such that $\sum_{i\in[p]}(n_i - k) = n - k$. Note that $\sum_{i\in[p]}(n_i - k'') = n - k''$ reflects that every recursive call includes a connected component of $G - S$ together with the separator S of size k''. We can round up k'' to k, because $T(n)$ is an increasing function. Because, the sum of the n_i's is more than n, it is beneficial to consider the following modified function $T'(n') = T(n+k)$. Then we get the simpler recursion

$$T'(n') = \begin{cases} O(k) & \text{if} n' \leq 2k \\ \max_{p,n'_1,\ldots,n'_p} \left\{\sum_{i\in[p]} T'(n'_i)\right\} + f(k)(n'+k) & \text{otherwise,} \end{cases}$$

where the maximum is taken over $p \geq 2$ and $n'_1,\ldots,n'_p \in [n'-1]$ such that $\sum_{i\in[p]}(n'_i) = n'$.

Now we prove

$$T'(n') \leq \frac{c}{\epsilon} f(k) n' \log n'$$

by induction, where c is minimal such that $c \geq 3$ and the base case ($n' \leq 2k$) is satisfied. Assume that the ith component of size n_i is on the side of the separator with smaller volume if and only if $1 \leq i \leq p'$. Let $n_S = \sum_{i\in[p']} n'_i$, and let $n_L = n' - n_S$. Furthermore, let

$$h_S = \sum_{i\in[p']} n'_i \log n'_i \leq \sum_{i\in[p']} n'_i \log \frac{n'}{2} = n_S(\log n' - 1),$$

and

$$h_L = \sum_{i\in\{p'+1,\ldots,p\}} n'_i \log n'_i \leq \sum_{i\in\{p'+1,\ldots,p\}} n'_i \log n' = n_L \log n'.$$

Recall that $\epsilon n'/2 \leq n_S \leq n'/2$. By the inductive hypothesis, for $n' > 2k$ we have

$$T'(n') \leq \frac{c}{\epsilon} f(k)(h_S + h_L) + f(k)(n' + k)$$

This implies $T'(n') \leq \frac{c}{\epsilon} f(k)(n' \log n' - n_S) + f(k)(n' + k)$

Thus $T'(n') \leq \frac{c}{\epsilon} f(k) n' \log n'$ if $\frac{c}{\epsilon} n_S \geq n' + k$. As $n_S \geq \epsilon n'/2$ and $n' > 2k$, this is the case when $c \geq 3$.

Each split by volume can be done by finding at most $C_{k-1} = \Theta(4^k/k^{3/2})$ separators in time $\mathcal{O}(C_{k-1}kn) = \mathcal{O}(4^k/\sqrt{k})$ for each placement of at most $n/t = k/(1/2 - \epsilon)$ representatives to the left or right side and the placement of at most k vertices into the centroid. These are at most $(2 + 8\epsilon)k$ representatives for $\epsilon \leq 1/4$. Choosing $\epsilon = \Theta(1/k)$ this results in a running time of $f(k)n = \mathcal{O}(\binom{(3+8\epsilon)k}{k} 2^{4k} k^{-1/2} n) = \mathcal{O}(\frac{3^{3k}}{2^{2k}k} 2^{4k} n$ for one split by volume in a graph of size n. Together with the solution of the previous recurrence, we obtain.

Theorem 8. *If a graph has treewidth at most k, then a tree decomposition of width at most $5(k-1)$ can be found in time $\mathcal{O}(2^{6.755k} n \log n)$.*

References

1. Arnborg, S., Corneil, D.G., Proskurowski, A.: Complexity of finding embeddings in a k-tree. SIAM J. Algebraic Discrete Methods **8**(2), 277–284 (1987)
2. Belbasi, M., Fürer, M.: Finding all leftmost separators of size $\leq k$. arXiv preprint arXiv:2111.02614 (2021)
3. Belbasi, M., Fürer, M.: An improvement of Reed's treewidth approximation. In: Uehara, R., Hong, S.-H., Nandy, S.C. (eds.) WALCOM 2021. LNCS, vol. 12635, pp. 166–181. Springer, Cham (2021). https://doi.org/10.1007/978-3-030-68211-8_14
4. Bodlaender, H.L., Drange, P.G., Dregi, M.S., Fomin, F.V., Lokshtanov, D., Pilipczuk, M.: A c^{kn} 5-approximation algorithm for treewidth. SIAM J. Comput. **45**(2), 317–378 (2016)
5. Chitnis, R., Hajiaghayi, M.T., Marx, D.: Fixed-parameter tractability of directed multiway cut parameterized by the size of the cutset. SIAM J. Comput. **42**(4), 1674–1696 (2013)
6. Courcelle, B.: The monadic second-order logic of graphs. I. Recognizable sets of finite graphs. Inf. Comput. **85**(1), 12–75 (1990)
7. Flum, J., Grohe, M.: Parameterized Complexity Theory (Texts in Theoretical Computer Science. An EATCS Series). TTCS, Springer, Heidelberg (2006). https://doi.org/10.1007/3-540-29953-X
8. Korhonen, T.: A single-exponential time 2-approximation algorithm for treewidth. arXiv e-prints [to appear in FOCS] arXiv:2104.07463 (2021)
9. Lokshtanov, D., Marx, D.: Clustering with local restrictions. Inf. Comput. **222**, 278–292 (2013)
10. Marx, D.: Parameterized graph separation problems. Theor. Comput. Sci. **351**(3), 394–406 (2006)
11. Marx, D., Razgon, I.: Fixed-parameter tractability of multicut parameterized by the size of the cutset. SIAM J. Comput. **43**(2), 355–388 (2014)
12. Reed, B.A.: Finding approximate separators and computing tree width quickly. In: Proceedings of the Twenty-Fourth Annual ACM Symposium on Theory of Computing (STOC), pp. 221–228 (1992)

Maximize the Probability of Union-Influenced in Social Networks

Guoyao Rao[1], Yongcai Wang[1], Wenping Chen[1], Deying Li[1](✉), and Weili Wu[2]

[1] School of Information, Renmin University of China, Beijing 100872, China
{gyr,ycw,wenpingchen,deyingli}@ruc.edu.cn
[2] Department of Computer Science, University of Texas at Dallas, Richardson, TX 75080, USA
weiliwu@utdallas.edu

Abstract. Nowadays, the social network plays an important role in advertisements and propaganda, and it creates the research of social influence. The prior works in social influence mainly consider the influence of individual or just the number of them. However, the union related is usual seen that is always together and each one is indispensable such as the team recruitment, in which a company or some business projections wish to recruit several candidates from different positions all to compose a team through the social networks. In this paper, different from targeted influence model, we consider such scenarios as an union-influence and propose the union-influence probability maximization problem (UIPM) to choose seeds to maximize the probability of the all nodes in an union are influenced. Unlike the most problems in previous social influence, the object function of UIPM is not submodularity or supermodularity. Then we design a data-driven $\beta(1-\frac{1}{\epsilon})$-approximation algorithm. At last we evaluate the performance on effectiveness and efficiency of the algorithms we proposed by the experiments in real-world social network datasets.

1 Introduction

Nowadays, the development of internet is profoundly influencing our lives. Internet also has changed the traditional business model and marketing strategy. Especially in the outbreak of COVID-19, many offline businesses and markets change to run online. At the same time, social platforms such as Facebook, Twitter, Weibo and Wechat is replacing traditional media and gradually become the main ways for information dissemination and communication. So the online social network also play an very important role in the online business such as famous viral marketing using the effect of word-of-mouth. For example, in viral marketing, many businesses would like to promote their products through social network platforms by choosing few costumers to experience firstly, and then

This work is supported by the National Natural Science Foundation of China (Grant NO. 12071478, 61972404), and partially by NSF 1907472.

D.-Z. Du et al. (Eds.): COCOA 2021, LNCS 13135, pp. 288–301, 2021.
https://doi.org/10.1007/978-3-030-92681-6_24

letting them spread the related positive information about their products to attract more latent costumers in social networks. It creates the research of social influence. Since Kemp et al. [9] al firstly formulated the influence maximization (IM) problem, many variants, extension and applications researches have been well studied. Most of the previous works are major in the individual influence, e.g., maximize the number of influenced individual in whole or target part or maximizing the probability of certain individual to be influenced.

However, in many scenarios, the entirety is more important than individual. For example, a successful online group-buying depends on whether all members have disputes in the shopping intentions, i.e., whether they can been simultaneously influenced in the marketing viral of a product. We further consider another scenario as following. A person or a company want to recruit a team with several intention candidates, and to build such special team successfully, and then it hopes all candidates will trust and have interest. However, this person or company usually can't contact them directly as they may not trust the strangers, so he want to some people assist him by using their relationship networks i.e., let some agents (e.g., their friends or their friends' friends, etc.) influence these target candidates through the influence spread in social networks. We note such team or group above which requires members' consistency (i.e., be influenced synchronicity) as an **union** which is also usual seen in scenarios such as group cooperation, government lobby. And we consider to use the method of social influence to achieve this goal.

In previous works, some researches like the target influence [11,18] is closed to solve our problem, in which they set the group members as targets and further maximize the number of them to be influenced, but such idea may be immature with unbalanced influence and we show it as the following example in Fig. 1. Let the direct edge in the figure represents the direct influence from the starting node to the ending node and the edge weight is the influence probability. When we chose a seed from $\{s_1, s_2\}$, if adopting the target maximization strategy to chose s_2 as seed, we have the successful probability to influence the union is 0.05 though there are expected 3.05 members influenced in the union. However the seed s_1 can make the probability larger to 0.125 though the expected number of influenced members is 2.6 less than that with s_1. It seems that the uniformity of seeds' influence is more important to achieve the goal of union to be influenced.

The big difference between our idea from the target influence is that we don't want to optimize the number and we consider to optimize the chance of target entirety influenced. So in this paper, we consider such entirety influenced problem that is to make all the members in other word an union be influenced as far as possible instead of making more members influenced. The main contributions are as followings:

- We propose the union-influence probability maximization (UIPM) problem and prove it's NP-hard and the probability computation is #P-hard.
- We prove the UIPM is not either submodularity or supmodularity. To address this problem, we design a $\beta(1-\alpha)$-approximation algorithm.

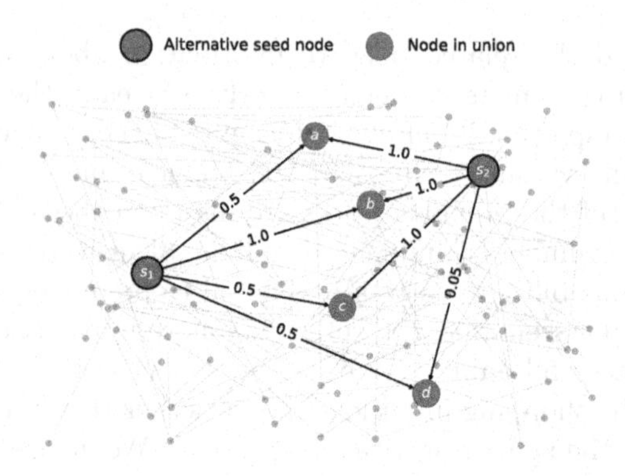

Fig. 1. Illustration of distinguishing the union influence and target influence.

- We conduct and designed experiments based on four real-world databases which validate our proposed algorithm performance well to solve the problem.

2 Related Work

Kempe et al. [9] firstly formulate the Influence Maximization problem (IM) based on two influence diffusion models they proposed, i.e., Independent Cascade (IC) and Linear Threshold (LT). The IM problem aims to select k nodes as seeds to maximize the expected number of influenced node through a stochastic diffusion process in social networks. This work has attracted and inspired many researches into social influence and brought many variants and extended works. Many researches major in the spread problem based on different and specific scenes, such as time-constrained [5,12], topic-aware [1,4], competition [2,13], rumor-control [8], multi-round [19] and so on.

Specially, the researches like [11] consider the number influence in a group of nodes as a target set, and as we said in the introduction, it's an immature strategies in our problem. In addition, the research [7] of personalize influence maximization and the works like [24] of the acceptance probability maximization (APM) problem all are based on the idea to maximize the probability of certain given node to be influenced alone. Then we still don't see any other researches closed to our problem which aims to influence several nodes synchronously.

There were two important properties used to solve related researches, i.e., submodularity and supmodularity [15] which are defined as following: Let S be a finite set and function $f : 2^S \rightarrow \mathbb{R}$, and if for any $A \subseteq B \subseteq S$ and any $x \notin S \backslash B$, we say that f is submodularity as long as $f(A \cup \{x\}) - f(A) \geq f(B \cup \{x\}) - f(B)$, and f is supermodularity if the inequality is reversed.

The basic IM problem is proved to be NP-hard and the influence computation is #P-hard. With the good property of submodularity for the target function, the basic greedy method can provide a $(1 - 1/e - \epsilon)$-approximation solution [15], where ϵ is the loss caused by influence estimation since it's hard to get the accurate influence.

Gneral greedy-based methods cost too much time using the heavy Monte Carlo simulations to estimate the marginal gain of node's influence, and it's hard to apply to the large scale network, although there are many improvements [6,10,17]. Then Tang et al. [22] and Borgs et al. [3] proposed the reverse influence set(RIS) sampling method to estimate the influence. The idea of RIS is using a revere propagate from a node v to get a random set of nodes that can influence v, hence through the number of set covered by the seeds set to estimate the influence which is more efficient than repeatable simulations. Then after transforming the IM problem to the classical set cover problem, there are many RIS-based extensions and improvements such as the IMM [21], SSA and D-SSA [16], OPIM [20].

3 Problems Formulation

Review of the Classical IC Model: IC model [1] is a classical diffusion model in social influence and in this paper, all of our work is based on this diffusion model. Let $G(V, E, p)$ be an influence graph which is a directed and weighted network where the weight of each direct edge $p_{uv} \in [0, 1]$ is the probability that a node u influences another node v, and then the influence spreads from a set of seeds in rounds. Initially, only all seeds are active while other nodes are inactive. In each diffusion round, each node becoming active in the previous round has one chance to influence each one of its inactive out-neighbours following the influence probability. The process terminates as long as no more inactive nodes can be influenced to become active.

There is also an equivalent generating formulation [1] for the diffusion proposed above as following: (1) Firstly, get a G's edge-induced subgraph noted as g by flipping[1] each edge e_{uv} randomly following the probability p_{uv}. Specially we write $g \sim G$ to mean that g is randomly realized from G by such way. (2) Secondly, we mark g_S to be the set of nodes which can be reached in the graph g by any seed node in seeds set S, and we naturally have g_S are the influenced nodes.

Problems Descriptions: Based on the diffusion formulation of IC model, here we give our UIPM problem definition as following.

Definition 1 ***Union-influenced Probability Maximization Problem.*** *Given the influence graph $G(V, E, p)$, an union set $\mathbb{U} \subseteq V$, a set of alternative nodes $\mathbb{S} \subseteq V$, and a budget k, the union-influenced probability maximization problem is to chose a set S^* of at most k nodes from $\mathbb{S}$ as seeds maximizing the probability that every node in $\mathbb{U}$ is influenced, i.e., $S^* := argmax_{S \subseteq \mathbb{S}, |S| \leq k} Pr_{g \sim G}\{\mathbb{U} \subseteq g_S\}$, and further we rewrite the problem as $S^* := argmax_{S \subseteq \mathbb{S}, |S| \leq k} \sigma(S)$, where $\sigma(S) = \alpha \cdot Pr_{g \sim G}\{\mathbb{U} \subseteq g_S\} + \beta$ called the linear enlarge measurement and*

[1] We said flipping an edge e_{uv} is that remove it with the probability $1 - p_{uv}$ from the graph and mark it to be "on" if the edge isn't removed and otherwise be "off".

$\alpha, \beta > 0$ is the parameters to enlarge the value of probabilities to a wider value range $[\beta, \alpha + \beta]$.

Note that we add the scaling linear parameters to expand the union-influenced probabilities as the value of probabilities is crowded in a value range of zero lower bound $[0, 1]$, but through linear expanding without the lost of total order(i.e., given two sets S_1 and S_2, $Pr\{\mathbb{U} \subseteq A_{S_1}\} > Pr\{\mathbb{U} \subseteq A_{S_2}\}$ i.f.f. $\sigma(S_1) > \sigma(S_2)$), $Pr\{\mathbb{U} \subseteq A_{S_1}\} < Pr\{\mathbb{U} \subseteq A_{S_2}\}$ i.f.f. $\sigma(S_1) < \sigma(S_2)$), and $Pr\{\mathbb{U} \subseteq A_{S_1}\} = Pr\{\mathbb{U} \subseteq A_{S_2}\}$ i.f.f. $\sigma(S_1) = \sigma(S_2)$)), we can more precisely to estimate the target function, hence more easily to design the algorithm to solve this problem shown later. We firstly introduce two obvious cases for the UIPM problem as following:

1. When $|\mathbb{U}| = 1$, the UIPM problem is the same as personal influence maximization problem [7].
2. When $k \geq |\mathbb{U}|$ and $\mathbb{U} \subseteq \mathbb{S}$, the UIPM problem is easily to be solved by selecting all the nodes in U as the seeds.

In this paper, we consider more about the general situation and nextly we analyse the hardness of the UIPM.

Theorem 1. *The UIPM problem is NP-hard.*

Proof. *Consider any instance of the NP-complete Set Cover problem with a set collection $\mathbf{C} = \{c_1, c_2, \ldots, c_m\}$, a set of nodes $T = \{t_1, t_2, \ldots, t_n\}$. We wish to know whether there exist k sets in $\mathbf{C}$ covering all nodes in T. We construct a special $UIPM$ problem with $G(V, E, p), \mathbb{U}, \mathbb{S}, k$ as following:*

(1) Create $\mathbb{U} = \{u_1, u_2, ..., u_n\}$, each $u_i (1 \leq i \leq n)$ corresponding t_i in T.
(2) Create $\mathbb{S} = \{s_1, s_2, ..., s_m\}$, each $s_i (1 \leq j \leq m)$ corresponding c_i in $\mathbf{C}$.
(3) Let $V = \mathbb{U} \cup \mathbb{S}$, and create an E's edge $< s_j, u_i >$ with influence probability 1 as long as $t_i \in c_j$.

It's easy to get that there exist k sets in $\mathbf{C}$ covering all nodes in T if and only if there exists at most k seeds in $\mathbb{S}$ with the union-influenced probability 1 and hence $\sigma = \alpha + \beta$. So we proved it.

Theorem 2. *The computation problem for σ is #P-hard.*

Proof. *Consider any instance of the #P-complete s-t connectedness counting problem with $G'(V', E')$ and two vertix s and t in V'. We wish to count the number of G''s subgraphs where s is connected to t, and we denote these subgraphs as a set $\mathcal{G}$. We show this problem is equivalent to the following computation problem of union-influenced probability with $G(V, E, p), \mathbb{U}$ where $V = V' \cup \mathbb{U}$, $E = E' \cup \{< t, u > |u \in \mathbb{U}\}$. Given a seed set $S = \{s\}$, we can easily have that $Pr\{\mathbb{U} \subseteq A_S\} = Pr\{t \in A_{\{s\}}\} \cdot p^{|\mathbb{U}|} = p^{|\mathbb{U}|} \sum_{g \in \mathcal{G}} Pr(g) = p^{|\mathbb{U}|} |\mathcal{G}| p^{|E|} = |\mathcal{G}| p^{|E|+|\mathbb{U}|}$. Thus we can get the size of $\mathcal{G}$ by the computation of $\frac{Pr\{\mathbb{U} \subseteq A_S\}}{p^{|E|+|\mathbb{U}|}}$ and further by $\frac{\sigma - \beta}{\alpha p^{|E|+|\mathbb{U}|}}$. We proved it.*

Many problems in social influence have good properties such as submodularity or supmodularity which is very useful in designing approximation algorithms. However, in our problems, these properties are lost.

Theorem 3. *The object function σ is either submodularity nor supmodularity.*

Proof. *We show it by a special case in Fig. 2 with a given union $\mathbb{U} = \{a, b, c\}$ and alternative seeds $\mathbb{S} = \{s_1, s_2, s_3, s_4\}$. let $\delta_{s'} Pr\{\mathbb{U} \subseteq A_S\}$ and $\delta_{s'}\sigma(S)$ correspond to the gain of union-influenced probability and σ respectively after add a seed s' into S. Then we have $\delta_{s_3} Pr\{\mathbb{U} \subseteq A_{\{s_1\}}\} = 0$ and $\delta_{s_3} Pr\{\mathbb{U} \subseteq A_{\{s_1,s_2\}}\} = 0.5$ and hence $\delta_{s_3}\sigma(s_1) \leq \delta_{s_3}\sigma(s_1, s_2)$, so it's not submodularity. We also have $\delta_{s_4} Pr\{\mathbb{U} \subseteq A_{\{s_1\}}\} = 0.5$ and $\delta_{s_4} Pr\{\mathbb{U} \subseteq A_{\{s_1,s_2\}}\} = 0.5$, and so $\delta_{s_4}\sigma(s_1) \geq \delta_{s_4}\sigma(s_1, s_2)$, and it's still not supmodularity.*

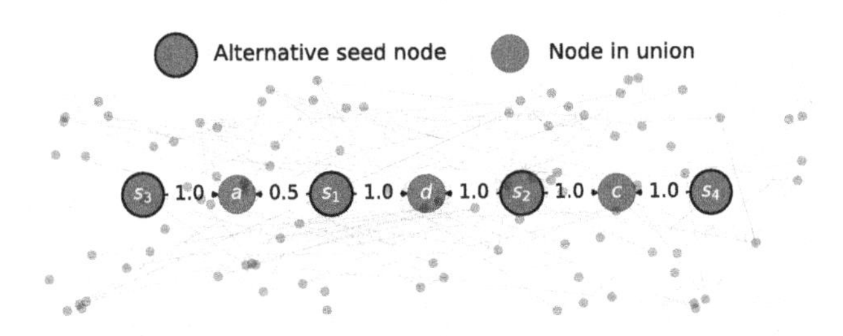

Fig. 2. A special case to illustrate of the properties of σ.

4 The UIS-Based Algorithms Designs

If we use the natural idea of greedy-climbing to solve the UIPM, unlike most of the social influence problem with submodularity, it has no approximation-guarantee. Besides since the computation of probability is #P-hard, the probability estimation method regularly using the simulation of Mont Carlo is very heavy. For the later algorithm designs, inspired by the RIS method of using a set cover to estimate the influence, we first introduce the union-reverse influence set-sequences (UIS) which can be used to estimate the union-influenced probability more efficiently. We first give our definition of union-reverse influence set-sequences as following:

Definition 2. ***Union-reverse Influence Set-sequences.*** *Given an influence Graph G and an Union $\mathbb{U}$, let g be an edge-induced subgraph of G by flipping edges randomnly with the probability , the union-reverse influence set-sequences is a set sequence $(r_{u_1}, r_{u_1}, \cdots, r_{u_{|\mathbb{U}|}})$ where set r_{u_i} is a subset of alternative seeds $\mathbb{S}$ in which every node can be reversely reached by $\mathbb{U}$'s node u_i over g.*

By the definition, as shown in Algorithm 1, a natural idea of sampling an UIS set sequence is firstly creating a G's edge-induced subgraph g by flipping all edges with the edge probability, and then doing multiple independent reverse BFSs (breadth-first search) over g sourced from each different union node u_i in $\mathbb{U}$ to get the corresponding nodes set r_{u_i} reached by u_i. Nextly, we will analyse how

Algorithm 1: SG-BFS($G(V, E, p), U, S$)

```
1 Get a G's random subgraph of g ;
2 for each u_i ∈ U do
3   Do a bfs search in g sourced from u_i and get the searched set r_{u_i}
4 return (r_{u_1}, r_{u_2}, ..., r_{u_|U|})
```

the UIS can be used to estimate our object function. Given a set sequence $\mathcal{R}$ and a set S, let $\mathcal{I}_{\mathcal{R}}$ be the indicator function where $\mathcal{I}_{\mathcal{R}}(S) = 0$ if $\exists r_j \in \mathcal{R}$, $r_j \cap S = \emptyset$ and $\mathcal{I}_{\mathcal{R}}(S) = 1$ if $\forall r_j \in \mathcal{R}$, $r_j \cap S \neq \emptyset$. Then we have the theorem as following:

Theorem 4. *Given an union U, let $\mathcal{R}$ be the related random UIS, the union-influenced probability of the seed set S equals to the expectation of $\mathcal{I}_{\mathcal{R}}(S)$, i.e., $Pr\{U \subseteq A_S\} = E(\mathcal{I}_{\mathcal{R}}(S))$.*

Proof. *By the generating model of IC and the definition of UIS, we have*

$$\begin{aligned}
Pr_{g \sim G}\{\mathbb{U} \subseteq g_S\} &= Pr_{g \sim G}\{\bigwedge_{u_j \in \mathbb{U}} (u_j \in g_S)\} \\
&= Pr_{g \sim G}\{\bigwedge_{u_j \in \mathbb{U}} (\bar{g}_{u_j} \cap S \neq \emptyset)\} \\
&= Pr_{\mathcal{R}}\{\bigwedge_{r_j \in \mathcal{R}} (r_j \cap S \neq \emptyset)\} \\
&= E(\mathcal{I}_{\mathcal{R}}(S)).
\end{aligned}$$

Then we can use the statistic method sampling UIS to estimate the expectation of the random variable $\mathcal{I}_{\mathcal{R}}(S)$ instead of Monte Carlo simulations in which it needs re-simulation once the seeds set is changed. Sampling θ UIS sequences independently and getting $\Re_\theta = \{\mathcal{R}_1, \mathcal{R}_2, \ldots, \mathcal{R}_\theta\}$, we have that $\bar{\sigma}(S, \Re_\theta) := \frac{\alpha}{\theta} \sum_{i \in [\theta]} \mathcal{I}_{\mathcal{R}_i}(S) + \beta$ is the unbiased estimation for $\sigma(S, U)$.

To analyse the gap of estimation error which is related to the sampling number θ, we can get first Lemma 1 by the Chernoff Bounds [14] and hence get the following gap Theorem 5.

Lemma 1. *If $\theta \geq \frac{-\ln(1-\delta)\alpha^2}{2\epsilon^2\beta^2}$, we have $\bar{\sigma} - \sigma \leq \epsilon\sigma$ with probability at least δ and if $\theta \geq \frac{-\ln(1-\delta)\alpha^2}{\epsilon^2\beta^2}$, we have $\bar{\sigma} - \sigma \geq -\epsilon\sigma$ with probability at least δ, For any $\epsilon > 0$ and $0 \leq \delta < 1$.*

Proof. *We first introduce the Chernoff Bounds [14]:*

Let $X_1, X_2, \cdots, X_\theta$ *be random variables such that* $a \leq X_i \leq b$ *for all* i. *Let* $X = \sum_{i=1}^{\theta} X_i$ *and* $\mu = E(X)$. *Then for any* $\epsilon \geq 0$,

$$Pr\{X \geq (1+\epsilon)\mu\} \leq exp(-\frac{2\epsilon^2\mu^2}{\theta(b-a)^2}), \tag{1}$$

$$Pr\{X \leq (1-\epsilon)\mu\} \leq exp(-\frac{\epsilon^2\mu^2}{\theta(b-a)^2}). \tag{2}$$

Let $X_i = \alpha\mathcal{I}_{R_i}(S) + \beta$, *for* $i = 1.2, \ldots, \theta$ *and then* $\beta \leq X_i \leq \alpha + \beta$. *We have* $X = \sum_{i=1}^{n} X_i = \theta\bar{\sigma}$, *and* $\mu = E(X) = E(\theta \cdot (\alpha(\mathcal{I}_{\mathcal{R}}(S) + \beta))) = \theta(\alpha E(\mathcal{I}_{\mathcal{R}}(S) + \beta) = \theta \cdot \sigma$. *By the Eq. 1, let* $\theta \geq \frac{-\ln(1-\delta)\alpha^2}{2\epsilon^2\beta^2}$ *and we have*

$$\begin{aligned} Pr\{\theta\bar{\sigma} - \theta\sigma \geq \epsilon \cdot \theta\sigma\} &= Pr\{\bar{\sigma} - \sigma \geq \epsilon\sigma\} \\ &\leq exp(-\frac{2\epsilon^2\theta\sigma^2}{\alpha^2}) \leq exp(-\frac{2\epsilon^2\theta\beta^2}{\alpha^2}) \leq 1-\delta. \end{aligned}$$

Then get $Pr\{\bar{\sigma} - \sigma \leq \epsilon\sigma\} \geq \delta$. *By the Eq. 2, let* $\theta \geq \frac{-\ln(1-\delta)\alpha^2}{\epsilon^2\beta^2}$ *and we have*

$$\begin{aligned} Pr\{\theta\bar{\sigma} - \theta\sigma \leq -\epsilon \cdot \theta\sigma\} &= Pr\{\bar{\sigma} - \sigma \leq -\epsilon\sigma\} \\ &\leq exp(-\frac{\epsilon^2\theta\sigma^2}{\alpha^2}) \leq exp(-\frac{\epsilon^2\theta\beta^2}{\alpha^2}) \leq 1-\delta. \end{aligned}$$

Then get $Pr\{\bar{\sigma} - \sigma \geq -\epsilon\sigma\} \geq \delta$.

Theorem 5. *If* $\theta \geq \frac{-\alpha^2}{\epsilon^2\beta^2} \cdot min\{\frac{\ln(1-\delta_1)}{2}, \ln(1-\delta_2)\}$, *for any* $\epsilon > 0$, *we have* $Pr\{|\bar{\sigma} - \sigma| \leq \epsilon\sigma\} \geq \delta$, *where* $\delta_1, \delta_2 \geq 0$ *and* $\delta_1 + \delta_2 = \delta$.

By above analysis such like Theorem 5, ignoring the loss of estimation after sampling sufficient number of UIS sequences, we can get a solution of UIPM by solving the maximization problem of $\bar{\sigma}$ instead of the origin target σ, i.e., find the nodes set $S^\star := argmax_{S \subseteq \mathcal{S}, |S| \leq k} \bar{\sigma}(S, \Re_\theta)$. Further back to our original problem, we have the following theorem.

Theorem 6. *If* $\theta \geq \frac{-\ln(1-\delta)\alpha^2}{\epsilon^2\beta^2}$, *we have* $\sigma(S^\star) \geq \frac{1-\epsilon}{1+\epsilon}\sigma(S^*)$. *with probability at least* $2\delta - 1$.

Proof. *We have* $\bar{\sigma}(S^\star) \geq \bar{\sigma}(S^*)$. *By the Lemma 1, if* $\theta \geq \frac{-\ln(1-\delta)\alpha^2}{\epsilon^2\beta^2}$, *we have* $\bar{\sigma}(S^*) \geq (1-\epsilon)\sigma(S^*)$ *and* $\sigma(S^\star) \geq \frac{1}{1+\epsilon}\bar{\sigma}(S^\star)$ *and the union probability of them is at least* $2\delta - 1$. *We proved it.*

The Theorem 6 tells that we can get a high accuracy and confidence solution for the UIPM by solving $S^\star$. So nextly, we consider how to solve this problem. We show that $\bar{\sigma}$ is also not submodularity w.r.t the seed set S by a special case as following: Given $\mathcal{R}_1 = (\{v_4\}, \{v_1\})$, $\mathcal{R}_2 = (\{v_1, v_3\}, \{v_2\})$, $\mathcal{R}_3 = (\{v_1\}, \{v_2, v_3\})$ and $T = \{v_1, v_2, v_3, v_4\}$, we have $\bar{\sigma}(\{v_1, v_2\}, \Re_\theta) - \bar{\sigma}(\{v_2\}, \Re_\theta) = 2 \geq \bar{\sigma}(\{v_1\}, \Re_\theta) - \bar{\sigma}(\emptyset, \Re_\theta) = 0$.

SA Algorithm: Considering the similar idea of Sandwich, and for our target function $\bar{\sigma}$ without submodularity, we find a lower bund $\bar{\sigma}_l$ and upper bound $\bar{\sigma}^u$, i.e., $\bar{\sigma}_l(S) \leq \bar{\sigma}(S) \leq \bar{\sigma}^u(S)$ for all $S \in \mathcal{S}$. Supposing that we get a α-approximation solution S^l and S^u β-approximation respectively corresponding to each maximization problem of the two bunds. Hence we can have the following Sandwich approximation lemma:

Lemma 2. *Let $S^+ = argmax_{S\in\{S^l,S^u\}}\bar{\sigma}(S)$, then we have $\bar{\sigma}(S^+) \geq max\{\alpha \cdot \frac{\bar{\sigma}(S^u)}{\bar{\sigma}^u(S^u)}, \beta \cdot \frac{\bar{\sigma}_l(S^l)}{\bar{\sigma}(S^\star)}\} \cdot \bar{\sigma}(S^\star)$.*

Proof. *We have*

$$\bar{\sigma}(S^u) = \frac{\bar{\sigma}(S^u)}{\bar{\sigma}^u(S^u)} \cdot \bar{\sigma}^u(S^u) \geq \frac{\bar{\sigma}(S^u)}{\bar{\sigma}^u(S^u)} \cdot \alpha \cdot \bar{\sigma}^u(S^\star) \geq \frac{\bar{\sigma}(S^u)}{\bar{\sigma}^u(S^u)} \cdot \alpha \cdot \bar{\sigma}(S^\star),$$

and

$$\bar{\sigma}(S^l) = \bar{\sigma}_l(S^l) \geq \beta \cdot \bar{\sigma}_l(S^\star) \geq \beta \cdot \frac{\bar{\sigma}_l(S^l)}{\bar{\sigma}(S^\star)}\bar{\sigma}(S^\star),$$

Then we proved it.

Nextly we will construct such upper and lower bounds for $\bar{\sigma}$. Firstly we get two collections of sets $R^l = \{r_1^l, r_2^l, \ldots, r_\theta^l\}$ and $R^u = \{r_1^u, r_2^u, \ldots, r_\theta^u\}$ where $r_i^l = \bigcap_{r_j \in R_i} r_j$ and $r_i^u = \bigcup_{R_j \in R} R_j$ corresponding to each $\mathcal{R}_i \in \Re_\theta$. Let $\mathcal{I}_r$ be the indicator function where $\mathcal{I}_r(S) = 0$ if $r \cap S = \emptyset$ and $\mathcal{I}_r(S) = 1$ if $r \cap S \neq \emptyset$

Further we can define two following functions $\eta_1(S, R^u) := \frac{\alpha}{\theta}\sum_{r_i^l \in R^u} \mathcal{I}_{r_i^u}(S) + \beta$, $\eta_2(S, R^l) := \frac{\alpha}{\theta}\sum_{r_i^u \in R^l} \mathcal{I}_{r_i^l}(S) + \beta$. Then we prove that η_1 is an upper bound of $\bar{\sigma}$ and η_2 is a lower bound of $\bar{\sigma}$ as shown in theorem 8 which can be naturally infered by following lemma.

Lemma 3. *For any $S \subseteq \mathbb{S}$, we have $\mathcal{I}_{r_i^l}(S) \leq \mathcal{I}_{R_i}(S) \leq \mathcal{I}_{r_i^u}(S)$.*

Proof. *For each R_i, we have the corresponding $r_i^l = \bigcap_{r_j \in R_i} r_j$, $r_i^u = \bigcup_{r_j \in R_i} r_j$. If $\mathcal{I}_{r_i^l}(S) = 1$, i.e., $r_i^l \cap S \neq \emptyset$, we have $r_j \cap S \neq \emptyset$ for each $r_j \in R_i$, and hence $\mathcal{I}_{R_i}(S) = 1$. So we have $\mathcal{I}_{r_i^l}(S) \leq \mathcal{I}_{R_i}(S)$. If $\mathcal{I}_{R_i}(S) = 1$, i.e., $\exists r_j \in R^i$ s.t. $r_j \cap S \neq \emptyset$, we have each $r_i^u \cap S \neq \emptyset$ as $r_j \subset r_i^u$, and hence $\mathcal{I}_{r_i^u}(S) = 1$. So we have $\mathcal{I}_{R_i}(S) \leq \mathcal{I}_{r_i^u}(S)$.*

Theorem 7. *For all $S \subseteq S$, we have $\eta_2 \leq \bar{\sigma} \leq \eta_1$.*

We naturally have the Theorem 7 from Lemma 3. Actually the sum of $I_r(S)$ is the number of set covered by S and it's submodular, hence we have η_1 and η_2 are also submodular. So standard greedy algorithm for maximum coverage [23] can provide a $(1 - 1/e)$-approximation solution for η_1 and η_2 respectively. Further we can get the SA-algorithm in Algorithm 2 to obtain an approximation solution S^+. and we have the approximation for the original problem as following theorem.

Theorem 8. *The solution S^+ given by algorithm SA can guarantee that $\sigma(S^+) \geq \beta(1-\frac{1}{e})\sigma(S^*)$ with probability at least $2\delta - 1$, where $\beta = \frac{1-\epsilon}{1+\epsilon} \cdot max\{\frac{\bar{\sigma}(S^u)}{\bar{\sigma}^u(S^u)}, \frac{\bar{\sigma}_l(S^l)}{\bar{\sigma}(S^\star)}\}$.*

Algorithm 2: UIS-SA($G_{U,S}, k, \epsilon, \delta$)

---

1. Sample $\theta = \lceil \frac{-\ln(1-\delta)\alpha^2}{\epsilon^2\beta^2} \rceil$ UIS sequences by Algorithm 1;
2. Using coverage [23] to get a solution S^l for η_2;
3. Using coverage [23] to get a solution S^u for η_1;
4. Let $S^+ = argmax\{\bar{\sigma}(S^l), \bar{\sigma}(S^u)\}$;
5. **return** S^+

Proof. *According Lemma 2, we have*

$$\bar{\sigma}(S^+) \geq (1-1/e) \cdot max\{\frac{\bar{\sigma}(S^u)}{\bar{\sigma}^u(S^u)}, \frac{\bar{\sigma}_l(S^l)}{\bar{\sigma}(S^\star)}\} \cdot \bar{\sigma}(S^\star)$$
$$\geq (1-1/e) \cdot max\{\frac{\bar{\sigma}(S^u)}{\bar{\sigma}^u(S^u)}, \frac{\bar{\sigma}_l(S^l)}{\bar{\sigma}(S^\star)}\} \cdot \bar{\sigma}(S^*),$$

and according Lemma 1, we have $\bar{\sigma}(S^*) \geq (1-\epsilon)\sigma(S^*)$ and $\sigma(S^+) \geq \frac{1}{1+\epsilon}\bar{\sigma}(S^+)$ with union probability at least $2\delta - 1$. Then we proved it.

Note that we can't get β exactly because of $\frac{\bar{\sigma}_l(S^l)}{\bar{\sigma}(S^\star)}$ since the optimum solution of $\bar{\sigma}(S^\star)$ is NP-hard to get, but we can get the value of $\frac{\bar{\sigma}(S^u)}{\bar{\sigma}^u(S^u)}$ after the computation, which is a lower bound for the β.

Actually for the lower bound and upper bound, we can find physical explanations. The node u in most of intersection set of UIS sequences means that it has high probability to influence all union nodes. The node v in most of union sets of UIS sequences means that it has high probability to influence at least one union node. So in the network, if there are more the nodes like u, the best solution may be closed to the one provided by lower bound, and if there are more nodes like v not u, the best solution may be closed to the one provided by upper bound.

5 Experiments

We have conduct an experimental study to evaluate the performance of our proposed methods over 4 real-world datasets[2] (BlogcCatalog, Flickr, DBLP, Twitter). The number of vertices corresponds to 10K, 80k, 203k, 580k and the number of edges corresponds to 333k, 5.9M, 382k, 717k. All codes of the experiments are written in C++ and all experiments run in a linux server with a 12 cores, 24 threads, 3.6 Hz, CPU and 64 G memory.

---

[2] http://networkrepository.com.

Experiment Setup: As the general setting in IC model, we set the influence probability i,e. the weight of each direct edge $<u, v>$, $p_{uv} := \frac{1}{d_{in}(v)}$ where $d_{in}(v)$ is the in-degree of node v. For the node choice of the union, we first exclude the nodes with zero in-degree as they can't be influenced. Since the node's influence probability is related to the in-degree, so to avoid low union influence probability of the union, we specially choice nodes for each union from the remaining nodes according to the in-degree such that is the node with lower in-degree has larger probability to be chosen. We use 10000 times of Menton Carlo simulations and count the proportion of union being influenced to get the evaluation of union-influenced probability for given seeds.

For alternative seeds setting, We exclude the nodes with zero out-degree as they can't influence anyone, we also exclude the union nodes and their one hop in-going neighbours as it's easy to make choice by selecting nodes from them. So we set the alternative seeds $\mathbb{S} = V \setminus (\mathbb{U} \bigcup (\cup_{u \in \mathbb{U}} N_u^{in}) \cup \{v | d_{ou}(v) = 0, v \in V\})$, where $d_{in}(v)$ is the out-degree of node v and N_u^{in} are u's one hop in-going neighbours. By default, we set $\alpha = 100$, $\beta = 10$, $\epsilon = 0.1$, $\delta = 0.99$, which can guarantee a high accuracy and confidence for the algorithm we designed.

Comparison Algorithms. We compare our algorithms with some baseline algorithms where **Target-IM** is the algorithm proposed [18] to solve the targeted influence in $\mathbb{U}$. **UIS-greedy** is the general greedy algorithm to solve the seeds selection problem to maximize $\bar{\sigma}$. and **Random** is the basic baseline algorithms by choosing seeds randomly.

Union Influence Probability Comparison: We conduct these experiments on two datasets of DBLP and Twitter. Firstly, for each dataset, we choose 10 nodes to compose an union to evaluate the performance of the different algorithms by running each algorithm 10 times. We vary the budgets k by 1,5,10,20,50,100 by setting half of the budgets are less than the union size and half are more than the union size. At last, we report the best one of the 10 runs for each algorithm. As shown in Fig. 3, the algorithms we proposed are significantly better than others and the improved algorithms UIS-SA have great improvement compared with the general UIS-greedy (nearly +30% in DBLP, +100% in Twitter).

Lower Bound of Budget Cost Comparison: Nextly we evaluate the lower bound of the budget cost (i.e., the necessary number of seeds) for certain union influence probability in all four datasets. We set 6 unions with different sizes of 1, 5, 10, 15, 20 and vary the budgets k from 1 to 500 to compare the lower bound of the budget with the union influence at least 0.1 under different algorithms by running each 10 times too. Special for the situation that all of the budgets k we provide can't achieve such influence, we record its lower bound as ∞. As shown in Fig. 4, to achieve the targeted union influence probability, the Random costs more than 500 budget of the seeds size. For the result in DBLP and Twitter,

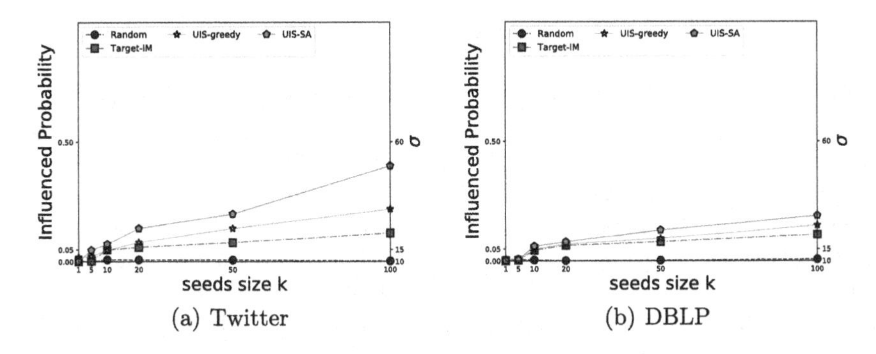

(a) Twitter (b) DBLP

Fig. 3. The performance comparisons achieved by different algorithms in different datasets with an union of size 10.

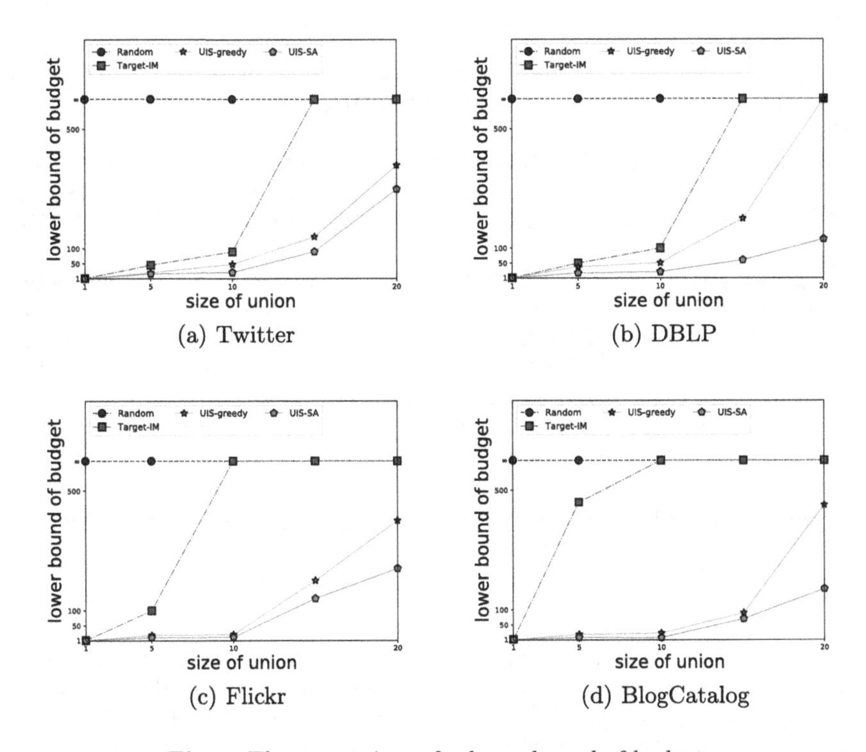

(a) Twitter (b) DBLP

(c) Flickr (d) BlogCatalog

Fig. 4. The comparisons for lower bound of budget

we have the budget cost comparison: Target-IM>UIS-greedy>UIS-SA when the union size under 10 and Target-IM≫UIS-greedy>UIS-SA when the union size above 10. Note that when the union size is one, the lower bound of them is same to be 1 as the problem is equal to the special case of personal influence maximization [7].

6 Conclusions

In this paper, we study the entirety influence problem, and propose the union-influence probability maximization (UIPM) problem in which the goal is to find a seed set with small size such that the probability of all nodes in a union being influenced is maximized. We show the UIPM is NP-hard and the computation of target function is #P-hard. Without the property of submodularity, we propose a data-driven $\beta(1-\frac{1}{e})$-approximation algorithm based on the union reverse Influence set-sequences. To analyse and evaluate proposed methods, a lot of experiments have been conducted on real-world datasets. The results show that the method we proposed solve the problems well.

References

1. Barbieri, N., Bonchi, F., Manco, G.: Topic-aware social influence propagation models. Knowl. Inf. Syst. **37**(3), 555–584 (2013). https://doi.org/10.1007/s10115-013-0646-6
2. Bharathi, S., Kempe, D., Salek, M.: Competitive influence maximization in social networks. In: Deng, X., Graham, F.C. (eds.) WINE 2007. LNCS, vol. 4858, pp. 306–311. Springer, Heidelberg (2007). https://doi.org/10.1007/978-3-540-77105-0_31
3. Borgs, C., Brautbar, M., Chayes, J., Lucier, B.: Maximizing social influence in nearly optimal time. In: Proceedings of the Twenty-Fifth Annual ACM-SIAM Symposium on Discrete Algorithms, pp. 946–957. SIAM (2014)
4. Chen, S., Fan, J., Li, G., Feng, J., Tan, K.I., Tang, J.: Online topic-aware influence maximization. Proc. VLDB Endow. **8**(6), 666–677 (2015)
5. Cohen, E., Delling, D., Pajor, T., Werneck, R.F.: Timed influence: computation and maximization. arXiv preprint arXiv:1410.6976 (2014)
6. Goyal, A., Lu, W., Lakshmanan, L.V.: CELF++: optimizing the greedy algorithm for influence maximization in social networks. In: Proceedings of the 20th International Conference Companion on World Wide Web, pp. 47–48. ACM (2011)
7. Guo, J., Zhang, P., Zhou, C., Cao, Y., Guo, L.: Personalized influence maximization on social networks. In: Proceedings of the 22nd ACM International Conference on Information & Knowledge Management, pp. 199–208 (2013)
8. He, X., Song, G., Chen, W., Jiang, Q.: Influence blocking maximization in social networks under the competitive linear threshold model. In: Proceedings of the 2012 SIAM International Conference on Data Mining, pp. 463–474. SIAM (2012)
9. Kempe, D., Kleinberg, J., Tardos, É.: Maximizing the spread of influence through a social network. In: Proceedings of the Ninth ACM SIGKDD International Conference on Knowledge Discovery and Data Mining, pp. 137–146. ACM (2003)
10. Leskovec, J., Krause, A., Guestrin, C., Faloutsos, C., VanBriesen, J., Glance, N.: Cost-effective outbreak detection in networks. In: Proceedings of the 13th ACM SIGKDD International Conference on Knowledge Discovery and Data Mining, pp. 420–429. ACM (2007)
11. Li, G., Chen, S., Feng, J., Tan, K.l., Li, W.S.: Efficient location-aware influence maximization. In: Proceedings of the 2014 ACM SIGMOD International Conference on Management of Data, pp. 87–98. ACM (2014)

12. Liu, B., Cong, G., Xu, D., Zeng, Y.: Time constrained influence maximization in social networks. In: 2012 IEEE 12th International Conference on Data Mining, pp. 439–448. IEEE (2012)
13. Lu, W., Chen, W., Lakshmanan, L.V.: From competition to complementarity: comparative influence diffusion and maximization. Proc. VLDB Endow. **9**(2), 60–71 (2015)
14. Motwani, R., Raghavan, P.: Randomized Algorithms. Cambridge University Press, New York (1995)
15. Nemhauser, G.L., Wolsey, L.A., Fisher, M.L.: An analysis of approximations for maximizing submodular set functions-I. Math. Program. **14**(1), 265–294 (1978)
16. Nguyen, H.T., Thai, M.T., Dinh, T.N.: Stop-and-stare: optimal sampling algorithms for viral marketing in billion-scale networks. In: Proceedings of the 2016 International Conference on Management of Data, pp. 695–710. ACM (2016)
17. Ohsaka, N., Akiba, T., Yoshida, Y., Kawarabayashi, K.I.: Fast and accurate influence maximization on large networks with pruned Monte-Carlo simulations. In: Twenty-Eighth AAAI Conference on Artificial Intelligence (2014)
18. Song, C., Hsu, W., Lee, M.L.: Targeted influence maximization in social networks. In: Proceedings of the 25th ACM International on Conference on Information and Knowledge Management, pp. 1683–1692. ACM (2016)
19. Sun, L., Huang, W., Yu, P.S., Chen, W.: Multi-round influence maximization. In: Proceedings of the 24th ACM SIGKDD International Conference on Knowledge Discovery & Data Mining, pp. 2249–2258. ACM (2018)
20. Tang, J., Tang, X., Xiao, X., Yuan, J.: Online processing algorithms for influence maximization. In: Proceedings of the 2018 International Conference on Management of Data, pp. 991–1005 (2018)
21. Tang, Y., Shi, Y., Xiao, X.: Influence maximization in near-linear time: a martingale approach. In: Proceedings of the 2015 ACM SIGMOD International Conference on Management of Data, pp. 1539–1554. ACM (2015)
22. Tang, Y., Xiao, X., Shi, Y.: Influence maximization: near-optimal time complexity meets practical efficiency. In: Proceedings of the 2014 ACM SIGMOD International Conference on Management of Data, pp. 75–86. ACM (2014)
23. Vazirani, V.V.: Approximation Algorithms. Springer, Heidelberg (2013). https://doi.org/10.1007/978-3-662-04565-7
24. Yang, D.N., Hung, H.J., Lee, W.C., Chen, W.: Maximizing acceptance probability for active friending in online social networks. In: Proceedings of the 19th ACM SIGKDD International Conference on Knowledge Discovery and Data Mining, pp. 713–721 (2013)

A Novel Algorithm for Max Sat Calling MOCE to Order

Daniel Berend[1,2], Shahar Golan[3], and Yochai Twitto[2(✉)]

[1] Department of Mathematics, Ben-Gurion University, 84105 Beer Sheva, Israel
berend@cs.bgu.ac.il

[2] Department of Computer Science, Ben-Gurion University, 84105 Beer Sheva, Israel
twittoy@cs.bgu.ac.il

[3] Department of Computer Science, Jerusalem College of Technology, 9116001 Jerusalem, Israel
sgolan@jct.ac.il

Abstract. In this paper, we present and study a new algorithm for the Maximum Satisfiability (Max Sat) problem. The algorithm, GO-MOCE, is based on the Method of Conditional Expectations (MOCE, also known as Johnson's Algorithm), and applies a greedy variable ordering to it. We conduct an extensive empirical evaluation on two collections of instances – instances from a past Max Sat competition and random instances. We show that GO-MOCE reduces the number of unsatisfied clauses by tens of percents as compared to MOCE. We prove that, using tailored data structures we designed, GO-MOCE retains the linear time complexity. Moreover, its runtime overhead in our experiments is at most 10%. We combine GO-MOCE with CCLS, a state-of-the-art solver, and show that the combined solver improves CCLS on the above mentioned collections.

1 Introduction

In the Maximum Satisfiability (Max Sat) problem [10], we are given a sequence of clauses over some Boolean variables. Each clause is a disjunction of literals over distinct variables. A literal is either a variable or its negation. We seek a truth (`true`/`false`) assignment for the variables, maximizing the number of satisfied (made `true`) clauses. In the Max r-Sat problem, each clause is restricted to consist of at most r literals. Here we restrict our attention to instances with clauses consisting of exactly r literals each.

Let n be the number of variables. Denote the variables by $x_1, x_2, \ldots, x_n$. The number of clauses is denoted by m, and the clauses by $C_1, C_2, \ldots, C_m$. We

We thank David Gamarnik, MohammadTaghi Hajiaghayi, Dmitry Panchenko, and Gregory Sorkin for helpful information and correspondence regarding bounds on the optimum of Max r-Sat. This research was partially supported by the Milken Families Foundation Chair in Mathematics, by the Lynne and William Frankel Foundation for Computer Science, and by the Israeli Council for Higher Education (CHE) via the Data Science Research Center, Ben-Gurion University of the Negev.

D.-Z. Du et al. (Eds.): COCOA 2021, LNCS 13135, pp. 302–317, 2021.
https://doi.org/10.1007/978-3-030-92681-6_25

denote the *density*, i.e., the clause-to-variable ratio, by $\alpha = m/n$. We use the terms "positive variable" and "negative variable" to refer to a variable and to its negation, respectively. Whenever we find it convenient, we consider the truth values `true` and `false` as binary 1 and 0, respectively. In pseudocodes we use T and F instead of `true` and `false`, respectively.

As Max r-Sat (for $r \geq 2$) is NP-hard [5], large instances cannot be exactly solved efficiently in the worst case unless $P = NP$, so one must resort to approximation algorithms and heuristics. Numerous methods have been suggested for solving Max Sat, e.g. [3,11,22,25,33,34], and an annual competition of solvers has been held since 2006 [4]. Overall, satisfiability related questions attracted a lot of attention from the scientific community. For a comprehensive overview of the whole domain of satisfiability, we refer to [9].

Various complete solvers for Max Sat have been developed during recent years, some of which were presented in the annual competition of Max Sat solvers [4]. Among these practical solvers, one can find Branch and Bound solvers (e.g., MaxSatz [21], Clone [27]), Satisfiability based solvers (e.g., SAT4J [20], QMaxSat [19]), Unsatisfiablity based solver (e.g., WPM1 [1,2]), etc. Complete solvers which participated in the last evaluations include MaxHS [15], Pacose [26], EvalMaxSAT [6], and more. Practical incomplete solvers for Max Sat are actively researched as well. Some of them competed in the incomplete track of the last evaluations. E.g., Loandra [8], TT-open-WBO-Inc [24], sls-mcs [17], and more.

Consider the naive randomized approximation algorithm, which assigns to each variable a truth value uniformly at random, independently of all other variables. It satisfies a proportion of $1 - 1/2^r$ of all clauses on the average. Furthermore, it can also be easily derandomized using the Method of Conditional Expectations (MOCE, also known as Johnson's Algorithm) [12,16,29], yielding an assignment guaranteed to satisfy at least this proportion of all clauses.

MOCE iteratively constructs an assignment by going over the variables in an arbitrary, usually random, order. At each step, it sets the seemingly better truth value to the currently considered variable. This is done by comparing the expected number of satisfied clauses under each of the two possible truth values.

For a given truth value, the expected number of satisfied clauses is the sum of three quantities. The first is the number of clauses already satisfied by the assignment to the previously assigned variables. The second is the additional number of clauses satisfied by the assignment of the given truth value to the current variable. The third is the expected number of clauses that will be satisfied by a uniformly random assignment to all other currently unassigned variables. The truth value for which this sum is the larger of the two is the one selected for the current variable. Ties are broken arbitrarily (usually randomly). The process is repeated until all variables are assigned.

In a sense, this method is optimal for Max 3-Sat, as no polynomial time algorithm for Max 3-Sat can guarantee a performance ratio exceeding $7/8$ unless $P = NP$ [18]. Typically, though, this method yields assignments that are much better than this worst-case bound. Theoretical and empirical works related to MOCE, and algorithms of the same spirit, include [13,28,30,31].

In [7] it was shown that state-of-the-art local search algorithms can be improved by supplying them a good starting point. In particular, combining MOCE with CCLS [22] results in a decrease in the number of unsatisfied clauses of up to 75%. This motivates us to develop improved linear-time algorithms that can supply even better starting points.

In Sect. 2 we introduce our algorithm GO-MOCE for Max Sat. A full time complexity analysis is provided in Sect. 3. Section 4 is dedicated to presenting an empirical study designed to evaluate the performance of GO-MOCE. In Sect. 5 we show how to use our algorithm to improve CCLS and demonstrate the improvement on some families of random instances and on public competition benchmarks. A conclusion is provided in Sect. 6.

2 The Greedy Order MOCE (GO-MOCE)

In this section we present a new algorithm, GO-MOCE, for Max Sat. The algorithm is based on the Method of Conditional Expectations, and in addition applies a greedy variable ordering to it. Subsection 2.1 is dedicated to the concept of gain and the basic idea behind GO-MOCE. In Subsect. 2.2 and 2.3, we describe how we efficiently maintain and update instances and gains during the execution of GO-MOCE. A pseudocode of GO-MOCE is presented in Subsect. 2.4.

2.1 The Concept of Gain and Its Usage in GO-MOCE

A gain conveys information on the profitability of assigning a given value to a given variable, namely the expected increase in the number of satisfied clauses.

Consider a variable x and a clause C of length $l = |C|$ in which x appears. The probability that C will be satisfied by a random assignment of the l variables appearing in it is $1 - 1/2^l$. If we assign the variable x a truth value, for which C is satisfied, the contribution of C to the expected number of satisfied clauses, increases by $1/2^l$ from $1 - 1/2^l$ to 1. On the other hand, if we assign the variable x a truth value, for which C is not satisfied, this literal is removed from it. Hence the contribution of C to the expected number of satisfied clauses decreases by $1/2^l$, from $1 - 1/2^l$ to $1 - 1/2^{l-1}$.

For a clause C, a variable x appearing in C, and a truth value b, put

$$\text{sign}(C, x) = \begin{cases} -1, & x \text{ appears negatively in } C, \\ 0, & x \text{ does not appear in } C, \\ 1, & x \text{ appears positively in } C, \end{cases}$$

and

$$\text{sat}(C, x, b) = \begin{cases} \text{sign}(C, x), & b = \texttt{true}, \\ -\text{sign}(C, x), & b = \texttt{false}. \end{cases}$$

Each clause C contributes $\mathrm{sat}(C, x, b) \cdot 1/2^{|C|}$ to the gain of assigning b to x. The total gain of assigning the truth value b to x is

$$\mathrm{gain}(x, b) = \sum_{C \in I} \mathrm{sat}(C, x, b) \cdot 1/2^{|C|}.$$

On each iteration, GO-MOCE uses the gains to assign the variable with the largest gain to the value that provides this gain, breaking ties randomly. The main challenge for the design of GO-MOCE is to efficiently maintain the instance and the gains throughout the execution.

2.2 Efficient Instance Representation and Residualization

In our algorithm, an instance is represented by means of two core data structures:

`clauses`– a mapping of clause indices to clauses. A clause is a mapping of variable indices to Boolean values, where `true` means the variable appears as is in the clause and `false` means it appears negated.

`variables`– a mapping of variable indices to variables. A variable is a mapping of clause indices to Boolean values, where `true` means the variable appears as is in the clause and `false` means it appears there negated.

Residualization is the operation in which information, regarding the assignment of a given truth value to a given variable, is used to simplify an instance. The simplification is done by removing assigned variables, satisfied clauses, and empty clauses (which are considered unsatisfied). The input to the residualization operation is a variable and its assigned truth value. The result of the operation is a simplified instance, which we refer to as the *residual instance*.

First, if a clause is satisfied by the variable assignment, we remove it from `clauses` and from its associated variables in `variables`. If no clauses are then associated with some variable, the variable itself is removed from `variables`. Then, we remove the literal associated with the assigned variable from the clauses it appears in (i.e., the clauses it does not satisfy). In case a clause becomes empty, we remove it from `clauses` and mark it as unsatisfied.

2.3 Efficient Maintenance and Update of Gains

As the number of variables is large, we should maintain the gains of all the variables in an efficient manner throughout the execution. Since $\mathrm{gain}(x, \texttt{true}) = -\mathrm{gain}(x, \texttt{false})$, we maintain only the gain of one truth value for each variable (the gain of `true`). We need to accommodate the following operations: find the gain of a given variable; update the gain of a given variable; find the largest gain over all variables; remove the largest gain from the collection of gains.

To allow performing these operations efficiently, our algorithm maintains a compound data structure `gains`, consisting of two elementary data structures: `gbv` (**g**ains **b**y **v**ariables), which maps each variable to its gain, and `gbm` (**g**ains **b**y **m**agnitudes), which allows fast access to the largest gain at any given moment.

The operations on gains are done in such a way that gbv and gbm are kept synchronized and consistent (partially sorted).

The following describes the gains maintenance. Initially, we go over all the variables, and for each one perform a full direct calculation of its gain as described in Subsect. 2.1. Then, we insert the calculated values into gains. At each step of the algorithm, once we have selected an assignment of a truth value to the current variable, we should update the gains of all affected variables. These are all neighbors of the currently assigned variable, namely all variables appearing with it in at least one clause. Recall that we only maintain the gain of assigning true. This process is done by iterating over all the clauses in which the currently assigned variable appears. For each such clause, we iterate over all its other variables. The magnitude of the change in the gain is always $1/2^l$, where l is the length of the clause we currently consider. Table 1 summarizes the changes in the gain of true of a variable neighboring the currently assigned variable.

Table 1. The change in the gain of true of a neighboring variable, for each clause that neighbor shares with the currently assigned variable.

Assigned\Neighbor	Appears positively in clause	Appears negatively in clause
Satisfies clause	$-1/2^l$	$1/2^l$
Unsatisfies clause	$1/2^l$	$-1/2^l$

2.4 Pseudocode

In this subsection we provide a pseudocode for GO-MOCE – see Algorithm 1. The pseudocode details the main procedure GO-MOCE and its sub-procedures: Calculate Gains – initializing the gains, Find Best Assignment – finding the best variable to be assigned and its "correct" value, Update Gains – updating the gains, and Residualize Instance – updating the instance. Each of the last three procedures is performed at each iteration of the main loop.

3 Time Complexity

We deal with instances of Max r-Sat over n variables, with density α. Thus, the number of clauses is $m = \alpha n$, and the overall size of the instance is $rm = r\alpha n$.

The key data structure for an efficient implementation is gbm, which supports three operations: add element, remove element and extract maximal element. We update the gain of a variable in terms of these operations by removing it from gbm and reinserting it with the updated gain. In Subsect. 3.1, we assume gbm is implemented by a balanced binary search tree. Thus, the complexity of each of these three operations is $O(\log n)$. In Subsect. 3.2, we present a tailored data

Algorithm 1. The Greedy Order MOCE

Input: An instance I over n variables.
Output: An assignment of truth values to the variables.

```
 1: procedure GO-MOCE(I)
 2:     gains ← CALCULATE GAINS(I)
 3:     while I has clauses do                                ▷ instance is not empty
 4:         variable, value ← FIND BEST ASSIGNMENT(gains)
 5:         variable ← value                                  ▷ assign variable
 6:         UPDATE GAINS(gains, I, variable, value)
 7:         RESIDUALIZE INSTANCE(I, variable, value, gains)
 8:     end while
 9: end procedure
```

```
10: procedure CALCULATE GAINS(I)
11:     for each variable x of I do
12:         g_x^T = 0                                ▷ expected gain of assigning x = T
13:         for each clause C of I in which x appears do
14:             g_x^T = g_x^T + sign(C, x) · 2^{-|C|}
15:         end for
16:         update gains (i.e., gbv and gbm) with g_x^T
17:     end for
18: end procedure
```

```
19: procedure FIND BEST ASSIGNMENT(gains)
20:     variable ← a maximum gain variable extracted from gbm
21:     if gbv.g_variable^T > 0 then
22:         value ← T
23:     else if gbv.g_variable^T < 0 then
24:         value ← F
25:     else
26:         value ← either T or F, selected uniformly at random
27:     end if
28: end procedure
```

```
29: procedure UPDATE GAINS(gains, I, variable, value)
30:     for each clause C of I, in which variable appears, do
31:         for each variable neighbor in C, except for variable, do
32:             gbv.g_neighbor^T = gbv.g_neighbor^T − sat(C, variable, value) ∗ sign(C, neighbor) ·
                2^{-|C|}
33:             gbm.g_neighbor^T = gbv.g_neighbor^T
34:         end for
35:     end for
36: end procedure
```

```
37: procedure RESIDUALIZE INSTANCE(I, variable, value, gains)
38:     for each clause C of I, in which variable appears, do
39:         if sat(C, variable, value) then
40:             remove C from clauses
41:             for each variable neighbor in C, except for variable, do
42:                 remove C from neighbor in variables
43:                 if neighbor has no clauses then
44:                     remove neighbor from variables and gains
45:                     assign neighbor to either T or F, uniformly at random
46:                 end if
47:             end for
48:         else
49:             remove variable from C in clauses
50:             if C has no variables then
51:                 remove C from clauses
52:             end if
53:         end if
54:     end for
55:     remove variable from variables
56: end procedure
```

structure for `gbm`, for which these operations take an amortized constant time. In our experiments, to be presented in Sects. 4 and 5, we used the latter version.

All the other data structures – `clauses`, `variables`, and `gbv` – can be implemented as simple arrays. A removal of an element from these arrays is done by marking it as deleted. In addition, for `clauses`, we hold a counter for the actual number of elements, so that we can terminate when it is depleted. With this implementation, all operations on these data structures take a constant time.

3.1 A Linearithmic Time Complexity Implementation

We analyze the asymptotic time complexity of GO-MOCE by inspecting each of its procedures line-by-line. As explained above, here we assume that `gbm` is implemented by a balanced binary search tree.

Lemma 1. *The time complexity of* CALCULATE GAINS *is* $O(n \log n + r\alpha n)$.

Proof. In CALCULATE GAINS, line 12 takes constant time and is executed $O(n)$ times. Line 14 takes constant time and is executed once for every appearance of every literal. In other words, it is executed r times for each clause, and rm times in total. Line 16 is executed n times and takes $O(\log n)$ time.

Lemma 2. *The overall time complexity of the calls to* FIND BEST ASSIGNMENT *is* $O(n \log n)$.

Proof. The procedure is called once in every iteration of the main loop. All the operations in this procedure take constant time, except for the extract maximum operation (line 20) that takes $O(\log n)$ time.

Lemma 3. *The overall time complexity of the calls to* UPDATE GAINS *is* $O(r^2 \alpha n \log n)$.

Proof. Each execution of line 32 takes $O(1)$ time, and of line 33 – $O(\log n)$ time. For each clause, we execute these lines $O(r)$ times. Hence, the processing of each clause takes $O(r \log n)$ time. The procedure is executed for each variable at most once. For each such variable, the procedure inspects all the clauses it appears in. Since all variables appear altogether rm times in the instance, and each such appearance adds $O(r \log n)$ time, the total time for all calls is $O(r^2 \alpha n \log n)$.

Lemma 4. *Throughout the execution of GO-MOCE, the overall time complexity of the calls to* RESIDUALIZE INSTANCE *is* $O(n \log n + r\alpha n)$.

Proof. Lines 40 and 51 are executed at most once for each clause, and lines 42 and 49 are executed at most once for each literal. Since each of these four lines takes $O(1)$ time, and there are m clauses and rm literals, the total time for these lines is $O(rm)$. Since a variable is removed from the instance at most once, lines 44 and 55 are executed at most once for each variable. This means that there are $O(n)$ executions of these two lines altogether, each taking $O(\log n)$ time.

Employing the lemmas above, we can easily prove the following result, asserted in the opening of the section.

Theorem 1. *The time complexity of GO-MOCE, with* `gbm` *implemented as a balanced binary search tree, is* $O(r^2 \alpha n \log n)$. *In particular, if* $r = O(1)$ *and* $\alpha = O(1)$, *it is* $O(n \log n)$.

Proof. As a direct consequence of the previous lemmas, the time complexity of UPDATE GAINS dominates all the other parts of the algorithm, and hence determines the time complexity of the algorithm.

3.2 A Linear Time Complexity Implementation

We now suggest a tailored implementation for `gbm` which yields constant amortized time for each of the required operations. Afterward, we show that this tailored data structure reduces the (worst-case) time complexity of GO-MOCE to $O(r^2 \alpha n)$, which is linear assuming r is bounded.

Our implementation for `gbm` comprises of 2 elementary data structures:

1. An array `levels`, such that `levels`[i] contains all variables whose current gain is $i2^{-r}$. Each level is implemented by a dynamic array. In addition, we maintain the maximal gain, $g_{\max}$.
2. An array `index`, such that `index`[i] is the location of variable x_i within `levels`[2^r`gbv.`i].

Note that a dynamic array supports addition and removal of elements at the end of the array in constant amortized time. An element may be deleted from the middle of the array in constant time by swapping it with the last element, updating `index` of the swapped element and deleting the currently last element.

Given the above data structures, the three required operations of `gbm` are implemented as follows:

1. **Add x_i to gbm**: Add x_i at the end of `levels[`2^r`gbv.`i`]`. If gbv.i is higher than $g_{\max}$, update $g_{\max}$ accordingly.
2. **Remove x_i from gbm**: Remove x_i from `levels[`2^r`gbv.`i`][index[`i`]]`. If the level of x_i is maximal and contains no other variables, scan the levels below it until finding a non-empty level, and update $g_{\max}$ accordingly. Note that an element may be deleted from the middle of the array in constant time by swapping it with the last element, updating `index` accordingly and deleting the currently last element.
3. **Extract a top-gain variable from gbm**: The top level variables are in `levels[`$2^r g_{\max}$`]`. Remove a random element from this level. As above, update $g_{\max}$ if necessary.

In the rest of this subsection we analyze the time complexity of GO-MOCE for the new implementation of gbm.

Lemma 5. *During the execution of GO-MOCE:*

1. *The maximal gain $g_{\max}$ does not exceed $2^{-r}m$.*
2. *The sum of all increases in the maximal gain between two consecutive steps does not exceed $2^{-r}m$.*

Remark 1. With high probability, for random instances of Max r-Sat, the number of gain levels is $O(\log n/\log\log n)$ [32], which is practically constant. For example, our empirical experiments show that, even for $r = \alpha = 10$ and $n = 1000000$, there are approximately 50 levels only. Another important property of GO-MOCE, both intuitively and experimentally is that the difference between maximal gains in two consecutive steps is $2^{-r}O(1)$ with high probability. In other words, this number is a bounded multiple of the "basic gain" 2^{-r}.

Proof (Lemma 5). A gain is defined as the increase in the expected number of satisfied clauses. Since this number is initially $(1-2^{-r})m$ and there are at most m satisfied clauses, no variable can have a gain of more than $2^{-r}m$. Since every time the maximal gain increases, the next variable assignment will increase the expectation at least by the increase size, the sum of all increases is at most $2^{-r}m$.

Theorem 2. *The time complexity of GO-MOCE, using our tailored data structure, is $O(r^2\alpha n)$. In particular, if $r = O(1)$ and $\alpha = O(1)$, it is $O(n)$.*

Proof. The two gbm operations that do not have constant time complexity are its initialization and finding of the maximal gain level. We will show that the amortized time required for these operations, though, is constant. Indeed, since the maximal possible gain is $2^{-r}m$ and each additional level represents a "step" of 2^{-r}, the initialization will require at most $2^{-r}m/2^{-r} = m$ space and time.

The time of finding the maximal gain level at any step is the number of levels between the current and the next level. The sum of all gain decreases is bounded by the sum of the initial maximal gain and all gain increases. As both the initial maximal gain and the sum of all increases are bounded by $2^{-r}m$, over the whole execution the total time complexity of finding the next maximal level is $O(m)$.

The instance size is $r\alpha n$. As any reasonable algorithm reads the whole instance, the only "non-linear" term in the expression for the complexity is r^2. Note that this is relevant only to the atypical case, when the clause length is unbounded. Let us now explain why the *average-case* time complexity is $O(r\alpha n)$.

The reason for the additional r factor is the number of clause shortenings, which is bounded by mr. Any clause may be shortened r times, and each time we need to deal with all $O(r)$ literals still in the clause. Next, we explain why the number of shortenings is usually $O(m)$, leading to total runtime of $O(r\alpha n)$.

If we set the variables randomly instead of by GO-MOCE (but perform all other operations as does GO-MOCE), then each clause has a probability of 1/2 of being satisfied the first time we set a variable appearing in it, a probability of 1/4 being satisfied the second time, and so forth. The number of times the clause is shortened is distributed geometrically with parameter 1/2, except that it is truncated at r. The average number of times a clause is shortened is $\sum_{i=1}^{r} i/2^i < 2$, so that the total number of clause shortenings is $O(m)$ on the average.

If we use MOCE, and more so if we use GO-MOCE, the situation is usually better. In fact, to each variable we assign the value satisfying a set of clauses weighing more than the set of clauses we shorten due to that value. Hence the number of times we shorten a clause should be stochastically smaller than the corresponding number for the algorithm setting the variables uniformly at random. This fact has been also verified in our simulations, and the observed number of clause shortenings was always less than m.

4 Performance Analysis

In this section, we perform a comparative evaluation of GO-MOCE versus its baseline algorithm MOCE. We compare the performance of the algorithms by comparing the number of unsatisfied clauses and the runtime. We conclude this section by comparing these two algorithms on public competition benchmarks.

We conducted experiments on several families of random instances. A family is defined by three parameters: r, α, and n. The families have been selected in a systematic way, so as to reveal trends in the performance. Instances are constructed as follows. The clauses are selected independently of each other. Each clause is generated by selecting r distinct variables uniformly at random, then negating each of them independently with probability 1/2.

We focus on random Max 3-Sat, with the number of variables ranging from 1000 to 1000000, and the density from 3 to 10. This range of densities allows us to study the performance of the algorithms at hand both below and above the satisfiability threshold density (which is approximately 4.27 for Max 3-Sat [14,23]). We executed both MOCE and GO-MOCE on 1000 instances of each of the families, and recorded the number of clauses unsatisfied by them.

Table 2 presents the average number of clauses unsatisfied by MOCE and GO-MOCE. The rows of the table (but the last row) record the number of variables, while the columns record the densities. For the sake of readability, all the numbers in those rows are rounded to the nearest integer. It turns out that, for any fixed density, the average number of clauses unsatisfied by each of the algorithms scales linearly with the number of clauses, and thus can be described as a proportion of the number of all clauses. This proportion is provided in the last row of the table. The data in the table indicates that the same holds for GO-MOCE. Figure 1b provides graphs of the average proportion of clauses unsatisfied by each of the two algorithms for $r = 3$ and α ranging from 3 to 10 in steps of size 0.25. For each density, we provide the average proportion of unsatisfied clauses over 1000 instances.

Table 2. The average number of clauses unsatisfied by MOCE and GO-MOCE.

$n \backslash \alpha$	4		5		7		10	
	MOCE	GO-MOCE	MOCE	GO-MOCE	MOCE	GO-MOCE	MOCE	GO-MOCE
1000	90	25	149	63	288	172	526	377
10000	899	250	1492	633	2886	1719	5262	3765
100000	8995	2508	14943	6331	28856	17199	52629	37646
1000000	89915	25056	149438	63316	288588	171955	526267	376454
% Unsat	2.25%	0.63%	2.99%	1.27%	4.12%	2.46%	5.26%	3.76%

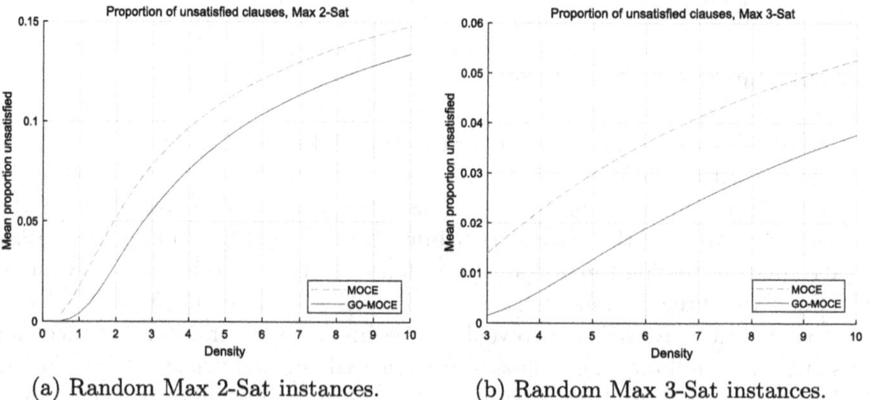

(a) Random Max 2-Sat instances.

(b) Random Max 3-Sat instances.

Fig. 1. The average proportion of clauses unsatisfied by MOCE and GO-MOCE.

Given the advantage of GO-MOCE over MOCE, one may want to consider the runtime overhead GO-MOCE introduces. Besides the time complexity provided in Subsect. 3, we measured the actual runtimes of both algorithms. In our experiments, we used the tailored data structure we suggested in Subsect. 3 for `gbm`. Table 3 lists these runtimes. For each family, we provide the mean runtime over the 1000 instances. Examining the runtimes, one can see that the additional time overhead introduced by GO-MOCE is negligible.

Table 3. The average runtime (CPU seconds) of MOCE and GO-MOCE.

$n\backslash\alpha$	4		5		7		10	
	MOCE	GO-MOCE	MOCE	GO-MOCE	MOCE	GO-MOCE	MOCE	GO-MOCE
1000	0.092	0.096	0.099	0.105	0.114	0.123	0.131	0.141
10000	0.355	0.363	0.407	0.430	0.509	0.535	0.651	0.686
100000	2.738	2.990	3.209	3.413	4.136	4.459	5.377	5.826
1000000	24.694	27.553	29.676	32.504	42.202	45.772	80.720	85.969

We have also examined instances of Max 2-Sat – see Fig. 1a. The pattern of performance, whereby the proportion of unsatisfied clauses is independent of n, holds here and for other values of r we have tested as well, indicating that this is a general property of MOCE and GO-MOCE. We also studied the asymptotic performance of GO-MOCE as the density grows larger. We showed that, for Max 3-Sat, GO-MOCE has an advantage of 21%–41% over MOCE.

We conclude this section by presenting a comparative evaluation of GO-MOCE versus MOCE on public competition benchmarks. An international evaluation of Max Sat solvers has been held annually since 2006 [4]. Here, we focus on the random and crafted categories of the 2016 evaluation. The random category has two benchmarks and the crafted category has three benchmarks.

We compare the solvers using the Instance Won measure, used in the 2016 evaluation. A solver *wins* an instance if no other solver has found a superior solution. Note that there may be several winners in an instance. The Instance Won measure of a solver is the number of instances it has won.

The results are presented in Table 4. In this table, one can see, for example, that the crafted category has a benchmark called Max Cut with 292 instances. GO-MOCE won in 284 instances, while MOCE won only in 18 instances (with a draw on 10 instances). The result of the competition between MOCE and GO-MOCE is a clear-cut victory to GO-MOCE over all benchmarks.

Table 4. The Instance Won for MOCE versus GO-MOCE over the instances of the Max Sat 2016 Evaluation.

Category	Benchmark	Size	MOCE	GO-MOCE
Random	Abrame-Habet	372	0	372
	High Girth	82	0	82
Crafted	Bipartite	100	0	100
	Max Cut	292	18	284
	Set Cover	10	0	10

5 Improving a State-of-the-Art Solver Using GO-MOCE

In this section, we suggest a combination of GO-MOCE with Configuration Checking Local Search (CCLS), a state-of-the-art solver, based on local search. We refer to this combined solver as GO-MOCE-CCLS. Thus, we first run GO-MOCE, and then run CCLS, starting from the assignment obtained by GO-MOCE instead of a random assignment. We compare the performance of this combined solver to two other solvers. The first is CCLS itself, which we denote here by RAND-CCLS. The second is MOCE-CCLS, which works analogously, starting the CCLS search from an assignment produced by MOCE. We evaluate GO-MOCE-CCLS on random instances as well as on public benchmarks.

Random Instances: We conduct experiments on diverse families of random instances. In all experiments we used $n = 100000$ variables and let CCLS perform $10n = 1000000$ flips. Table 5 presents the results. As one can see, GO-MOCE-CCLS improves on RAND-CCLS consistently by up to 75%. Note that when we are well above the satisfiability threshold, the optimum gets further from 0, and we cannot possibly improve by so much. Thus, the improvement we see is in fact only a lower bound on the real improvement. We also analyze the performance of GO-MOCE-CCLS as a function of the number of flips we allow CCLS to make. For some families the performance of GO-MOCE-CCLS after n flips is better than that of RAND-CCLS even after $10n$ flips.

Table 5. The average number and proportion of clauses unsatisfied by RAND-CCLS, MOCE-CCLS, and GO-MOCE-CCLS, over 100 random instances.

r	α	RAND-CCLS		MOCE-CCLS		GO-MOCE-CCLS	
2	2	4875	2.43%	4869	2.43%	4866	2.43%
2	3	16312	5.43%	16294	5.43%	16280	5.42%
3	4	352	0.08%	222	0.05%	87	0.02%
3	5	4127	0.82%	3893	0.77%	3606	0.72%
3	7	14896	2.13%	14497	2.07%	13996	1.99%
4	9	1151	0.12%	717	0.08%	245	0.02%
4	11	5108	0.46%	4412	0.40%	3581	0.32%
4	13	10124	0.77%	9231	0.71%	8130	0.62%

Max Sat 2016 Evaluation Instances: Similarly to the evaluation described in Sect. 4, we use the instances of the 2016 Max Sat international competition to evaluate GO-MOCE-CCLS. We compare RAND-CCLS and GO-MOCE-CCLS using the Instance Won measure described earlier. For each solver we give CCLS $10n$ flips. The results are presented in Table 6.

As one can see, GO-MOCE-CCLS achieved better results in all the benchmarks. For example, on the Abrame-Habet benchmark, the Instance Won measure of GO-MOCE-CCLS was 315, whereas the one of RAND-CCLS is only 250.

This implies that, on this benchmark, GO-MOCE-CCLS was the sole winner on 122 instances, whereas RAND-CCLS was the sole winner on only 57 instances. On 193 instances there was a draw.

Table 6. The Instance Won of RAND-CCLS and GO-MOCE-CCLS, over the instances of Max Sat 2016 Evaluation.

Category	Benchmark	Size	RAND-CCLS	GO-MOCE-CCLS
Random	Abrame-Habet	372	250	315
	High Girth	82	38	58
Crafted	Bipartite	100	49	69
	Max Cut	292	156	209
	Set Cover	10	3	9

6 Conclusion

We have studied a new algorithm, GO-MOCE, for Max Sat. We have shown that it performs much better than MOCE, yet keeps the runtime almost the same.

GO-MOCE may be seen as a derandomization of MOCE. It derandomizes the order in which the variables are assigned, but not completely. When selecting the variable to be assigned at a given step, it does not determine which of the variables with the same maximal gain will be assigned. As there are usually numerous variables with this gain, it opens a possibility for further improvement, by choosing the variable to be set in a smart way rather than randomly.

References

1. Ansótegui, C., Bonet, M.L., Gabàs, J., Levy, J.: Improving SAT-based weighted MaxSAT solvers. In: Milano, M. (ed.) CP 2012. LNCS, pp. 86–101. Springer, Heidelberg (2012). https://doi.org/10.1007/978-3-642-33558-7_9
2. Ansótegui, C., Bonet, M.L., Levy, J.: Solving (weighted) partial MaxSAT through satisfiability testing. In: Kullmann, O. (ed.) SAT 2009. LNCS, vol. 5584, pp. 427–440. Springer, Heidelberg (2009). https://doi.org/10.1007/978-3-642-02777-2_39
3. Ansótegui, C., Bonet, M.L., Levy, J.: SAT-based MaxSAT algorithms. Artif. Intell. **196**, 77–105 (2013)
4. Argelich, J., Li, C.M., Manyá, F., Planes, J.: MaxSAT Evaluations. http://www.maxsat.udl.cat/
5. Ausiello, G., et al.: Complexity and Approximation: Combinatorial Optimization Problems and Their Approximability Properties, 2nd edn. Springer-Verlag, Heidelberg (2003). https://doi.org/10.1007/978-3-642-58412-1
6. Avellaneda, F.: A short description of the solver EvalMaxSAT. MaxSAT Eval. **2020**, 8 (2020)
7. Berend, D., Twitto, Y.: Effect of initial assignment on local search performance for Max Sat. In: Faro, S., Cantone, D. (eds.) 18th International Symposium on Experimental Algorithms (SEA 2020). Leibniz International Proceedings in Informatics (LIPIcs), vol. 160, pp. 8:1–8:14 (2020)

8. Berg, J., Korhonen, T., Järvisalo, M.: Loandra: PMRES extended with preprocessing entering MaxSAT evaluation 2017. MaxSAT Evaluation 2017, p. 13 (2017)
9. Biere, A., Heule, M., van Maaren, H.: Handbook of Satisfiability, vol. 185. IOS Press, Amsterdam (2009)
10. Biere, A., Heule, M., van Maaren, H., Walsh, T. (eds.): MaxSAT, Hard and Soft Constraints, vol. 185, pp. 613–631. IOS Press, Amsterdam (2009)
11. de Boer, P.T., Kroese, D.P., Mannor, S., Rubinstein, R.Y.: A tutorial on the cross-entropy method. Ann. Oper. Res. **134**(1), 19–67 (2005). https://doi.org/10.1007/s10479-005-5724-z
12. Chen, J., Friesen, D.K., Zheng, H.: Tight bound on Johnson's algorithm for maximum satisfiability. J. Comput. Syst. Sci. **58**(3), 622–640 (1999)
13. Coppersmith, D., Gamarnik, D., Hajiaghayi, M., Sorkin, G.B.: Random MAX SAT, random MAX CUT, and their phase transitions. Random Struct. Algorithms **24**(4), 502–545 (2004)
14. Crawford, J.M., Auton, L.D.: Experimental results on the crossover point in random 3-SAT. Artif. Intell. **81**(1–2), 31–57 (1996)
15. Davies, J.: Solving MaxSAT by decoupling optimization and satisfaction. Ph.D. thesis, University of Toronto (2013)
16. Erdős, P., Selfridge, J.L.: On a combinatorial game. J. Comb. Theory Ser. A **14**(3), 298–301 (1973)
17. Guerreiro, A.P., Terra-Neves, M., Lynce, I., Figueira, J.R., Manquinho, V.: Constraint-based techniques in stochastic local search MaxSAT solving. In: Schiex, T., de Givry, S. (eds.) CP 2019. LNCS, vol. 11802, pp. 232–250. Springer, Cham (2019). https://doi.org/10.1007/978-3-030-30048-7_14
18. Håstad, J.: Some optimal inapproximability results. J. ACM (JACM) **48**(4), 798–859 (2001)
19. Koshimura, M., Zhang, T., Fujita, H., Hasegawa, R.: QMaxSAT: a partial max-sat solver. J. Satisf. Boolean Model. Comput. **8**(1–2), 95–100 (2012)
20. Le Berre, D., Parrain, A.: The SAT4J library, release 22. J. Satisf. Boolean Model. Comput. **7**(2–3), 59–64 (2010)
21. Li, C.M., Manya, F., Planes, J.: New inference rules for max-sat. J. Artif. Intell. Res. **30**, 321–359 (2007)
22. Luo, C., Cai, S., Wu, W., Jie, Z., Su, K.W.: CCLS: an efficient local search algorithm for weighted Maximum Satisfiability. IEEE Trans. Comput. **64**(7), 1830–1843 (2014)
23. Mertens, S., Mézard, M., Zecchina, R.: Threshold values of random k-SAT from the cavity method. Random Struct. Algorithms **28**(3), 340–373 (2006)
24. Nadel, A.: TT-open-WBO-Inc: tuning polarity and variable selection for anytime SAT-based optimization. In: Proceedings of the MaxSAT Evaluations (2019)
25. Narodytska, N., Bacchus, F.: Maximum satisfiability using core-guided MaxSAT resolution. In: Proceedings of the Twenty-Eighth AAAI Conference on Artificial Intelligence, pp. 2717–2723. AAAI Press (2014)
26. Paxian, T., Reimer, S., Becker, B.: Pacose: an iterative sat-based MaxSAT solver. MaxSAT Eval. **2018**, 20 (2018)
27. Pipatsrisawat, K., Darwiche, A.: Clone: solving weighted Max-SAT in a reduced search space. In: Orgun, M.A., Thornton, J. (eds.) AI 2007. LNCS (LNAI), vol. 4830, pp. 223–233. Springer, Heidelberg (2007). https://doi.org/10.1007/978-3-540-76928-6_24
28. Poloczek, M.: Bounds on greedy algorithms for MAX SAT. In: Demetrescu, C., Halldórsson, M.M. (eds.) ESA 2011. LNCS, vol. 6942, pp. 37–48. Springer, Heidelberg (2011). https://doi.org/10.1007/978-3-642-23719-5_4

29. Poloczek, M., Schnitger, G.: Randomized variants of Johnson's algorithm for MAX SAT. In: Proceedings of the Twenty-Second Annual ACM-SIAM Symposium on Discrete Algorithms, pp. 656–663. SIAM (2011)
30. Poloczek, M., Schnitger, G., Williamson, D.P., Van Zuylen, A.: Greedy algorithms for the maximum satisfiability problem: simple algorithms and inapproximability bounds. SIAM J. Comput. **46**(3), 1029–1061 (2017)
31. Poloczek, M., Williamson, D.P.: An experimental evaluation of fast approximation algorithms for the maximum satisfiability problem. In: Goldberg, A.V., Kulikov, A.S. (eds.) SEA 2016. LNCS, vol. 9685, pp. 246–261. Springer, Cham (2016). https://doi.org/10.1007/978-3-319-38851-9_17
32. Raab, M., Steger, A.: "Balls into Bins" — a simple and tight analysis. In: Luby, M., Rolim, J.D.P., Serna, M. (eds.) RANDOM 1998. LNCS, vol. 1518, pp. 159–170. Springer, Heidelberg (1998). https://doi.org/10.1007/3-540-49543-6_13
33. Selman, B., Kautz, H.A., Cohen, B.: Local search strategies for satisfiability testing. In: DIMACS Series in Discrete Mathematics and Theoretical Computer Science, pp. 521–532 (1996)
34. Selman, B., Levesque, H., Mitchell, D.: A new method for solving hard satisfiability problems. In: Proceedings of the Tenth National Conference on Artificial Intelligence, pp. 440–446. AAAI Press (1992)

The Smallest Number of Vertices in a 2-Arc-Strong Digraph Without Pair of Arc-Disjoint In- and Out-Branchings

Ran Gu[1], Gregory Gutin[2], Shasha Li[3], Yongtang Shi[4], and Zhenyu Taoqiu[4](✉)

[1] College of Science, Hohai University, Nanjing 210098, Jiangsu, People's Republic of China
rangu@hhu.edu.cn

[2] Department of Computer Science, Royal Holloway, University of London, Egham, Surrey TW20 0EX, UK
g.gutin@rhul.ac.uk

[3] Department of Mathematics, Ningbo University, Ningbo 315211, Zhejiang, China

[4] Center for Combinatorics and LPMC, Nankai University, Tianjin 300071, China
shi@nankai.edu.cn, tochy@mail.nankai.edu.cn

Abstract. Branchings play an important role in digraph theory and algorithms. In particular, a chapter in the monograph of Bang-Jensen and Gutin, Digraphs: Theory, Algorithms and Application, Ed. 2, 2009 is wholly devoted to branchings. The well-known Edmonds Branching Theorem provides a characterization for the existence of k arc-disjoint out-branchings rooted at the same vertex. A short proof of the theorem by Lovász (1976) leads to a polynomial-time algorithm for finding such out-branchings. A natural related problem is to characterize digraphs having an out-branching and an in-branching which are arc-disjoint. Such a pair of branchings is called a good pair.

Bang-Jensen, Bessy, Havet and Yeo (2020) pointed out that it is NP-complete to decide if a given digraph has a good pair. They also showed that every digraph of independence number at most 2 and arc-connectivity at least 2 has a good pair, which settled a conjecture of Thomassen for digraphs of independence number 2. Then they asked for the smallest number n_{ngp} of vertices in a 2-arc-strong digraph which has no good pair. They proved that $7 \leq n_{ngp} \leq 10$. In this paper, we prove that $n_{ngp} = 10$, which solves the open problem.

Gu was supported by Natural Science Foundation of Jiangsu Province (No. BK20170860), National Natural Science Foundation of China (No. 11701143), and Fundamental Research Funds for the Central Universities. Li was supported by National Natural Science Foundation of China (No. 11301480), Zhejiang Provincial Natural Science Foundation of China (No. LY18A010002), and the Natural Science Foundation of Ningbo, China. Shi and Taoqiu are supported by the National Natural Science Foundation of China (No. 11922112), the Natural Science Foundation of Tianjin (Nos. 20JCJQJC00090 and 20JCZDJC00840) and the Fundamental Research Funds for the Central Universities, Nankai University.

D.-Z. Du et al. (Eds.): COCOA 2021, LNCS 13135, pp. 318–332, 2021.
https://doi.org/10.1007/978-3-030-92681-6_26

Keywords: Arc-disjoint branchings · Out-branching · In-branching · Arc-connectivity

1 Introduction

Let $D = (V, A)$ be a digraph. For a non-empty subset $X \subset V$, the *in-degree* (resp. *out-degree*) of the set X, denoted by $d_D^-(X)$ (resp. $d_D^+(X)$), is the number of arcs with head (resp. tail) in X and tail (resp. head) in $V \setminus X$. The *arc-connectivity* of D, denoted by $\lambda(D)$, is the minimum out-degree of a proper subset of vertices. A digraph is *k-arc-strongly connected* (or, just *k-arc-strong*) if $\lambda(D) \geq k$. In particular, a digraph is *strongly connected* (or, just *strong*) if $\lambda(D) \geq 1$.

An *out-branching* (*in-branching*) of a digraph $D = (V, A)$ is a spanning oriented tree which has a vertex r called its root such that there is a directed path from (to) r to (from) every other vertex. Branchings play an important role in digraph theory and algorithms. In particular, Chap. 9 in the monograph [5] is wholly devoted to branchings. The well-known Edmonds Branching Theorem (see e.g. [5]) provides a characterization for the existence of k arc-disjoint out-branchings rooted at the same vertex. A short proof of the theorem by Lovász [11] leads to a polynomial-time algorithm for finding such out-branchings. A natural related problem is to characterize digraphs having an out-branching and an in-branching which are arc-disjoint. Such a pair of branchings is called a *good pair.*

Thomassen [12] conjectured the following:

Conjecture 1. There is a constant c, such that every digraph with arc-connectivity at least c has a good pair.

He also proved that it is NP-complete to decide whether a given digraph D has an out-branching and an in-branching both rooted at the same vertex such that these are arc-disjoint. This implies that it is NP-complete to decide if a given digraph has a good pair [2]. Conjecture 1 has been verified for semicomplete digraphs [1] and their genearlizations: locally semicomplete digraphs [7] and semicomplete compositions [6] (it follows from the main result in [6]).

An out-branching and an in-branching of D are *k-distinct* if each of them has at least k arcs, which are absent in the other. Bang-Jensen et al. [8] proved that the problem of deciding whether a strongly connected digraph D has k-distinct out-branching and in-branching is fixed-parameter tractable when parameterized by k. Settling an open problem in [8], Gutin et al. [10] extended this result to arbitrary digraphs.

In [2], Bang-Jensen et al. showed that every digraph of independence number at most 2 and arc-connectivity at least 2 has a good pair, which settles the conjecture for digraphs of independence number 2.

Theorem 1. *If D is a digraph with $\alpha(D) \leq 2 \leq \lambda(D)$, then D has a good pair.*

Moreover, they also proved that every digraph on at most 6 vertices and arc-connectivity at least 2 has a good pair and gave an example of a 2-arc-strong digraph D on 10 vertices with independence number 4 that has no good pair. They posed the following open problem.

Problem 1 ([2]). What is the smallest number n of vertices in a 2-arc-strong digraph which has no good pair?

In this paper, we prove that every digraph on at most 9 vertices and arc-connectivity at least 2 has a good pair, which answers this problem. The main results of the paper are as follows.

Theorem 2. *Every 2-arc-strong digraph on 7 vertices has a good pair.*

Theorem 3. *Every 2-arc-strong digraph on 8 vertices has a good pair.*

Theorem 4. *Every 2-arc-strong digraph on 9 vertices has a good pair.*

This paper is organised as follows. In the rest of this section, we provide further terminology and notation on digraphs. Undefined terms can be found in [4,5]. In Sect. 2, we outline the proofs of Theorems 2, 3 and 4 and state some auxiliary lemmas which we use in their proofs. Section 3 contains a number of technical lemmas which will be used in proofs of our main results. Then we respectively devote one section for proofs of each theorem and its relevant auxiliary lemmas. The proofs not given in this paper due to the space limit can be found in [9].

Additional Terminology and Notation. For a positive integer n, $[n]$ denotes the set $\{1, 2, \ldots, n\}$. Throughout this paper, we will only consider digraphs without loops and multiple arcs. Let $D = (V, A)$ be a digraph. We denote by uv the arc whose *tail* is u and whose *head* is v. Two vertices u, v are *adjacent* if at least one of uv and vu belongs to A. If u and v are adjacent, then we also say that u is a *neighbour* of v and vice versa. If $uv \in A$, then v is called an *out-neighbour* of u and u is called an *in-neighbour* of v. Moreover, we say uv is an *out-arc* of u and an *in-arc* of v and that u *dominates* v. The *order* $|D|$ of D is $|V|$.

In this paper, we will extensively use *digraph duality*, which is as follows. Let D be a digraph and let $D^{\rm rev}$ be the *reverse* of D, i.e., the digraph obtained from D by reversing every arc xy to yx. Clearly, D contains a subdigraph H if and only if $D^{\rm rev}$ contains $H^{\rm rev}$. In particular, D contains a good pair if and only if $D^{\rm rev}$ contains a good pair.

Let $N_D^-(X) = \{y : yx \in A, x \in X\}$ and $N_D^+(X) = \{y : xy \in A, x \in X\}$. Note that X may be just a vertex. For two non-empty disjoint subsets $X, Y \subset V$, we use $N_Y^-(X)$ to denote $N_D^-(X) \cap Y$ and $d_Y^-(X) = |N_Y^-(X)|$. Analogously, we can define $N_Y^+(X)$ and $d_Y^+(X)$. For two non-empty subsets $X_1, X_2 \subset V$, define $(X_1, X_2)_D = \{v_1v_2 \in A \colon v_1 \in X_1 \text{ and } v_2 \in X_2\}$ and $[X_1, X_2]_D = (X_1, X_2)_D \cup (X_2, X_1)_D$. We will drop the subscript when the digraph is clear from the context.

We write $D[X]$ to denote the subdigraph of D induced by X. A *clique* in D is an induced subdigraph $D[X]$ such that any two vertices of X are adjacent. We say that D contains K_p if it has a clique on p vertices. A vertex set X of

D is *independent* if no pair of vertices in X are adjacent. A dipath (dicycle) of D with t vertices is denoted by P_t (C_t). We drop the subscript when the order is not specified. A dipath P from v_1 to v_2, denoted by $P_{(v_1,v_2)}$, is often called a (v_1, v_2)-*dipath*. A dipath P is a *Hamilton* dipath if $V(P) = V(D)$. We call C_2 a *digon*. A digraph without digons is called an *oriented graph*. If two digons have and only have one common vertex, then we call this structure a *bidigon*. A *semicomplete* digraph is a digraph D that each pair of vertices has an arc between them. A *tournament* is a semicomplete oriented graph.

In- and out-branchings were defined above. An *out-tree* (*in-tree*) is an out-branching (in-branching) of a subdigraph of D. We use B_s^+ (B_t^-) to denote an out-branching rooted at s (an in-branching rooted at t). The root s (t) is called *out-generator* (*in-generator*) of D. We denote by $\text{Out}(D)$ ($\text{In}(D)$) the set of out-generators (in-denerators) of D. If the root is not specified, then we drop the subscripts of B_s^+ and B_t^-. We also use O_D (I_D) to denote an out-branching (in-branching) of a digraph D. If O_D and I_D are arc-disjoint, then we write (O_D, I_D) to denote a good pair in D.

2 Proofs Outline

In this section, we outline constructions we use to prove our main results. We prove each of them by contradiction. We give the statements of some auxiliary lemmas. Some of their proofs are too complicated and we will not give them in the paper due to the length restriction. For simplicity, when outlining the proof of our main results, we assume that $|D_1| = 7$, $|D_2| = 8$ and $|D_3| = 9$.

2.1 Theorem 2

First we show that the largest clique in D_1 is a tournament by Lemma 6, next we prove that D_1 is an oriented graph in Claim 2.1 by Lemma 7. Lemmas 6 and 7 will be given in Sect. 3. Then we use Proposition 12 to show that D_1 has a Hamilton dipath in Sect. 4. After that, we prove that D_1 has a good pair by Proposition 10, which is shown in Sect. 3.

2.2 Theorem 3

Our proof will follow three steps.

Firstly, we prove that the largest clique R in D_2 has 3 vertices by Lemma 6. Then we show that R is a tournament through Claim 3.1, which is proved by Lemmas 6 and 7.

Our second step is to prove that D_2 is an oriented graph in Claim 3.2 by Lemmas 8, 9 and 10, which are given in Sect. 3.

In the last step, we proceed as follows in Sect. 5. We use Proposition 15 to show that D_2 has a Hamilton dipath. To prove it, we show Proposition 14 first. After that, we prove that D_2 has a good pair by Proposition 10.

2.3 Theorem 4

Our proof will follow four steps.

Firstly, we show that the largest clique R in D_3 has 3 vertices by Claim 4.1, which is proved using Proposition 5 given in Sect. 3, and Lemmas 6 and 7.

Next we show that R has no digons by Claim 4.2, which is proved analogously to Claim 3.1 using Lemmas 7, 8, 9 and 10.

Our third step is to show that D_3 is an oriented graph in Claim 4.3. To do this we need Lemmas 11 and 13 given in Sect. 6. For the first lemma, we give a generalization of Proposition 6 as Proposition 16, and for the second one, we prove Lemma 12 first.

Then we use Lemma 14 to show that D_3 has a Hamilton dipath in Sect. 6. To prove it, we show Proposition 17 first. After that, we prove that D_3 has a good pair by Proposition 10.

3 Preliminaries and Useful Lemmas

Proposition 5. *Let D be a digraph with $\lambda(D) \geq 2$ and with a good pair (B_s^+, B_s^-). If there exists a vertex t in D such that $D[\{s,t\}]$ is a digon, then D has a good pair (B_t^+, B_t^-).*

Proof. Let $B_t^+ = ts + B_s^+ - e_1$ and $B_t^- = B_s^- + st - e_2$, where e_1 (e_2) is the only in-arc (out-arc) of t in B_s^+ (B_s^-). Observe that B_t^+ (B_t^-) is an out-branching (in-branching) rooted at t in D. Since the root of any out-branching has in-degree zero, if $ts \in B_s^+ \cup B_s^-$, then ts must be in B_s^- and moreover ts is the only out-arc e_2 of t in B_s^-. Similarly, if $st \in B_s^+ \cup B_s^-$, then st must be in B_s^+ and moreover st is the only in-arc e_1 of t in B_s^+. Thus, B_t^+ and B_t^- are arc-disjoint and so (B_t^+, B_t^-) is a good pair of D.

Proposition 6. *Let D be a digraph with a subdigraph Q that has a good pair (O_Q, I_Q). Let $X = N_D^-(Q)$ and $Y = N_D^+(Q)$ with $X \cap Y = \emptyset$ and $X \cup Y = V - V(Q)$. Let X_i (Y_j) be the initial (terminal) strong components in $D[X]$ ($D[Y]$), $i \in [a]$ ($j \in [b]$). If one of the following holds, then D has a good pair. Meanwhile, we can always get two arc-disjoint $\mathcal{P}_X, \mathcal{P}_Y$ and respectively an out- and an in-forest T_X and T_Y in D.*

1. $d_Y^-(X_1) \geq 1$, $d_Y^-(X_i) \geq 2$, $i \in \{2, \dots, a\}$ *and* $d_X^+(Y_j) \geq 2$, $j \in [b]$.
2. $d_X^+(Y_1) \geq 1$, $d_X^+(Y_j) \geq 2$, $j \in \{2, \dots, b\}$ *and* $d_Y^-(X_i) \geq 2$, $i \in [a]$.

Proof. Let B^+ be an out-tree containing O_Q and an in-arc of any vertex in Y from Q. Let B^- be an in-tree containing I_Q and an out-arc of any vertex in X to Q. Set $\mathcal{X} = \{X_i,\ i \in [a]\}$ and $\mathcal{Y} = \{Y_j,\ j \in [b]\}$. By the digraph duality, it suffices to prove that condition 1 implies that D has a good pair.

Now assume that $d_Y^-(X_1) \geq 1$, $d_Y^-(X_i) \geq 2$, $i \in \{2, \dots, a\}$, and $d_X^+(Y_j) \geq 2$, $j \in [b]$. Then there are at least two arcs from Y_j (for each $j \in [b]$) to X, at least two arcs from Y to X_i (for each $i \in \{2, \dots, a\}$) and at least one arc from Y to X_1. Set $X_1' = X_1$. If there is an arc y^1x_1 from Y to X_1' with y^1 in some

Y_j, $j \in [b]$, then we choose such an arc and let $Y_1' = Y_j$, otherwise we choose an arbitrary arc y^1x_1 from Y to X_1' and let Y_1' be an arbitrary strong component in $\mathcal{Y}$. Let $\mathcal{P}_X = \{y^1x_1\}$. There now exists an arc, y_1x^1, out of Y_1' ($x^1 \in X$) which is different from y^1x_1 (as Y_1' has at least two arcs out of it). If there is such an arc y_1x^1 with x^1 in some X_i, $i \in \{2, \ldots, a\}$, then we choose one of these arcs and let $X_2' = X_i$, otherwise we choose such an arbitrary arc y_1x^1 out of Y_1' ($x^1 \in X$) and let X_2' be an arbitrary strong component in $\mathcal{X} - X_1'$. Let $\mathcal{P}_Y = \{y_1x^1\}$. Likewise, for $t \geq 2$, we get an arc y^tx_t into X_t' ($y^t \in Y$) which is different from $y_{t-1}x^{t-1}$ in $\mathcal{P}_Y$. If there is such an arc y^tx_t with y^t in some $Y_j \in \mathcal{Y} - \{Y_1', \ldots, Y_{t-1}'\}$, then choose one of these arcs and let $Y_t' = Y_j$, otherwise we choose such an arbitrary arc y^tx_t and let Y_t' be an arbitrary strong component in $\mathcal{Y} - \{Y_1', \ldots, Y_{t-1}'\}$. Add y^tx_t to $\mathcal{P}_X$. For $s \geq 2$, we get an arc y_sx^s out of Y_s' ($x^s \in X$) which is different from y^sx_s in $\mathcal{P}_X$. If there is such an arc y_sx^s with x^s in some $X_i \in \mathcal{X} - \{X_1', \ldots, X_{s-1}'\}$, then we choose one of these arcs and let $X_s' = X_i$, otherwise we choose such an arbitrary arc y_sx^s and let X_s' be an arbitrary strong component in $\mathcal{X} - \{X_1', \ldots, X_{s-1}'\}$. Add y_sx^s to $\mathcal{P}_Y$. Hence we get two arc sets $\mathcal{P}_X$ and $\mathcal{P}_Y$ with $\mathcal{P}_X \cap \mathcal{P}_Y = \emptyset$.

We will now show that D has a good pair. Let D_X be the digraph obtained from $D[X]$ by adding one new vertex y^* and arcs from y^* to x_i for $i \in [a]$. Analogously let D_Y be the digraph obtained from $D[Y]$ by adding one new vertex x^* and arcs from y_j to x^* for $j \in [b]$. Since Out(D_X)= $\{y^*\}$ and In(D_Y)= $\{x^*\}$, there exists an out-branching $B_{y^*}^+$ in D_X and an in-branching $B_{x^*}^-$ in D_Y. Set $T_X = B_{y^*}^+ - y^*$ and $T_Y = B_{x^*}^- - x^*$.

By construction, (O_D, I_D) is a good pair of D with $O_D = B^+ + \mathcal{P}_X + T_X$ and $I_D = B^- + \mathcal{P}_Y + T_Y$.

Corollary 1. *Let D be a digraph with $\lambda(D) \geq 2$ that contains a subdigraph Q with a good pair. Set $X = N_D^-(Q)$ and $Y = N_D^+(Q)$. If $X \cap Y = \emptyset$ and $X \cup Y = V - V(Q)$, then D has a good pair.*

Proof. Let X_i be the initial strong components in $D[X]$ and Y_j be the terminal strong components in $D[Y]$, $i \in [a]$ and $j \in [b]$. Since $\lambda(D) \geq 2$, $d_Y^-(X_i) \geq 2$ and $d_X^+(Y_j) \geq 2$, for any $i \in [a]$ and $j \in [b]$, which implies that D has a good pair by Proposition 6.

Lemma 1 ([2]). *Let D be a digraph and $X \subset V(D)$ be a set such that every vertex of X has both an in-neighbour and an out-neighbour in $V - X$. If $D - X$ has a good pair, then D has a good pair.*

By Lemma 1, in this paper we will often use the fact that if Q is a maximal subdigraph of D with a good pair and $X = N_D^-(Q), Y = N_D^+(Q)$, then $X \cap Y = \emptyset$.

Lemma 2. *Let D be a 2-arc-strong digraph containing a subdigraph Q with a good pair, $X = N_D^-(Q)$ and $Y = N_D^+(Q)$. If $X \cap Y = \emptyset$ and $X \cup Y = V - V(Q) - \{w\}$, where $w \in V - V(Q)$, then D has a good pair.*

Proof. Assume that Q has a good pair (O_Q, I_Q). Let B^+ be an out-tree containing O_Q with an in-arc of any vertex in Y from Q, while B^- be an in-tree containing I_Q with an out-arc of any vertex in X to Q.

First assume that either $(Y,w)_D \neq \emptyset$ or $(w,X)_D \neq \emptyset$. By the digraph duality, we may assume that $(Y,w)_D \neq \emptyset$, i.e., there exists an arc e from Y to w in D. Let $D' = D - e$. Set $X' = N^-_{D'}(Q) = X$ and $Y' = N^+_{D'}(Q) \cup \{w\} = Y \cup \{w\}$. Let X'_i be the initial strong components in $D'[X']$ and Y'_j be the terminal strong components in $D'[Y']$, $i \in [a]$ and $j \in [b]$. If w has an in-neighbour v in Y with v in some Y'_j, $j \in [b]$, then let $e = vw$ and $Y^*_1 = Y'_j$, otherwise we choose an arbitrary in-neighbour v of w in Y and let $e = vw$ and Y^*_1 be an arbitrary terminal strong component of $D'[Y']$. Since $\lambda(D) \geq 2$, $d^+_{X'}(Y^*_1) \geq 1$, $d^+_{X'}(Y'_j) \geq 2$ and $d^-_{Y'}(X'_i) \geq 2$, for any $Y'_j \neq Y^*_1$, $j \in [b]$ and $i \in [a]$, which implies that we get arc sets $\mathcal{P}_{X'}$ and $\mathcal{P}_{Y'}$ with $\mathcal{P}_{X'} \cap \mathcal{P}_{Y'} = \emptyset$, and digraphs $T_{X'}$ and $T_{Y'}$ by Proposition 6. By construction, D has a good pair $(B^+ + \mathcal{P}_{X'} + T_{X'} + e, B^- + \mathcal{P}_{Y'} + T_{Y'})$.

Now assume that $(Y,w)_D = \emptyset$ and $(w,X)_D = \emptyset$, which implies that $d^-_X(w) \geq 2$ and $d^+_Y(w) \geq 2$. Let X_i be the initial strong components in $D[X]$ and Y_j be the terminal strong components in $D[Y]$, $i \in [a]$ and $j \in [b]$. Since $\lambda(D) \geq 2$ and $(w,X)_D = (Y,w)_D = \emptyset$, $d^-_Y(X_i) \geq 2$ and $d^+_X(Y_j) \geq 2$ for any $i \in [a]$ and $j \in [b]$. By Proposition 6, we get $\mathcal{P}_X, T_X$ and $\mathcal{P}_Y, T_Y$ with $\mathcal{P}_X \cap \mathcal{P}_Y = \emptyset$. It follows that $(B^+ + \mathcal{P}_X + T_X + w^-w, B^- + \mathcal{P}_Y + T_Y + ww^+)$ is a good pair of D, where $w^- \in X$ and $w^+ \in Y$.

Proposition 7 ([2]). *Every digraph on 3 vertices has a good pair if and only if it has at least 4 arcs .*

Following [4], we shall use $\delta_0(D)$ to denote the *minimum semi-degree* of D, which is the minimum over all in- and out-degrees of vertices of D.

Proposition 8 ([2]). *Let D be a digraph on 4 vertices with at least 6 arcs except E_4 (see Fig. 1). If $\delta^0(D) \geq 1$ or D is a semicomplete digraph, then D has a good pair.*

Fig. 1. E_4.

Lemma 3 (**[5], p. 354**). *Let $D = (V, A)$ be a digraph. Then D is k-arc-strong if and only if it contains k arc-disjoint (s, t)-paths for every choice of distinct vertices $s, t \in V$.*

Lemma 4 (**Edmonds' branching theorem** [4]). *A directed multigraph $D = (V, A)$ with a special vertex z has k arc-disjoint out-branchings rooted at z if and only if $d^-(X) \geq k$ for all $\emptyset \neq X \subseteq V - z$.*

Lemma 5 (**[2]**). *If D is a 2-arc-strong digraph on n vertices that contains a subdigraph on $n - 3$ vertices with a good pair, then D has a good pair.*

Lemma 6. *If D is a 2-arc-strong digraph on n vertices that contains a subdigraph Q on $n - 4$ vertices with a good pair, then D has a good pair.*

Lemma 7. *Let D be a 2-arc-strong digraph on n vertices that contains a subdigraph Q on $n - 5$ vertices with a good pair, $X = N_D^-(Q)$, $Y = N_D^+(Q)$ and $X \cap Y = \emptyset$. If $|X| \geq 2$ or $|Y| \geq 2$, then D has a good pair.*

Lemma 8. *Let D be a 2-arc-strong digraph on n vertices that contains a subdigraph Q on $n - 6$ vertices with a good pair. Let $X = N_D^-(Q)$ and $Y = N_D^+(Q)$ with $X \cap Y = \emptyset$. If $|X| = |Y| = 2$ and at most one of X and Y is an independent set, then D has a good pair.*

Lemma 9. *Let $D = (V, A)$ be a 2-arc-strong digraph on n vertices that contains a subdigraph Q on $n-6$ vertices with a good pair. Set $X = N_D^-(Q) = \{x_1, x_2\}$ and $Y = N_D^+(Q) = \{y_1, y_2\}$ with $X \cap Y = \emptyset$, and $W = V - X - Y - V(Q) = \{w_1, w_2\}$. If X, Y are both independent sets, then D has a good pair except for the case below:*

(∗) $(Y, X)_D = \{y_j x_i, y_{3-j} x_{3-i}\}$ for some $i, j \in [2]$, $D[W] = C_2$ and $N_W^+(y_j) \cap N_W^+(y_{3-j}) = N_W^-(x_i) \cap N_W^-(x_{3-i}) = \emptyset$ while $N_W^+(y_j) \cap N_W^-(x_i) \neq \emptyset$ and $N_W^+(y_{3-j}) \cap N_W^-(x_{3-i}) \neq \emptyset$.

We use $D \supseteq E_3$ ($D \not\supseteq E_3$) to denote that D contains an arbitrary orientation (no orientation) of E_3 as a subdigraph. (E_3 is a mixed graph and only the two edges are to be oriented.) E_3 is shown in Fig. 2.

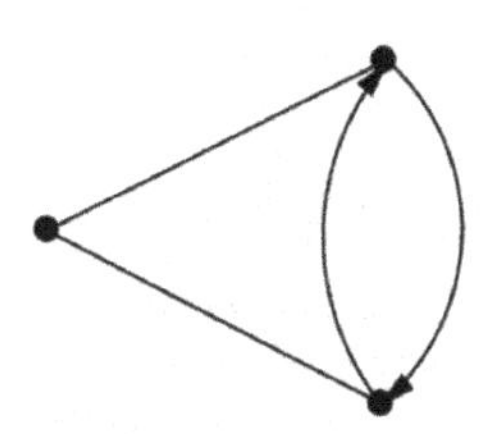

Fig. 2. E_3.

Lemma 10. *Let $D = (V, A)$ be a 2-arc-strong digraph on n vertices that contains a subdigraph Q on $n-6$ vertices with a good pair. Set $X = N_D^-(Q) = \{x_1, x_2\}$ and $Y = N_D^+(Q) = \{y_1, y_2\}$ with $X \cap Y = \emptyset$, and $W = V - X - Y - V(Q) = \{w_1, w_2\}$. If $n = 8$ or 9 and X, Y are both independent, then D has a good pair.*

Proposition 9. ([3]). *A digraph D has an out-branching (resp. in-branching) if and only if it has precisely one initial (resp. terminal) strong component. In that case every vertex of the initial (resp. terminal) strong component can be the root of an out-branching (resp. in-branching) in D.*

We use T_x^+ (resp. T_x^-) to denote an out-tree (resp. in-tree) rooted at x.

Proposition 10. *Let D be an oriented graph on n vertices. Let $P_D = x_1x_2\ldots x_n$ be the Hamilton dipath of D and $D' = D - A(P)$. Assume that there are exactly two non-adjacent strong components I_1 and I_2 in D'. Set $q \in \{2, 3, n-1, n\}$. If for some q, x_{q-1} and x_q are respectively in I_1 and I_2, then D has a good pair.*

Proof. W.l.o.g., assume that $x_{q-1} \in I_1$ and $x_q \in I_2$. Since I_i is strong, $\delta^0(I_i) \geq 1$, for any $i \in [2]$.

First assume $q \in \{n-1, n\}$. Let x be an in-neighbour of x_q in I_2. We get an out-branching of D as $B_{x_1}^+ = P_D - x_{q-1}x_q + xx_q$. Then we will show that there is an in-branching B_x^- in $D - A(B_{x_1}^+)$. Since I_2 is strong, $I_2 - xx_q$ is connected and has only one terminal strong component which contains x. This implies that there is an in-branching T_x^- in $I_2 - xx_q$. Note that there exists an in-branching $T_{x_{q-1}}^-$ in I_1, as I_1 is strong. Then $B_x^- = T_x^- + x_{q-1}x_q + T_{x_{q-1}}^-$, which implies that $(B_{x_1}^+, B_x^-)$ is a good pair of D.

Now we assume $q \in \{2, 3\}$. Let y be an out-neighbour of x_{q-1} in I_1. We get an in-branching of D as $B_{x_n}^- = P_D - x_{q-1}x_q + x_{q-1}y$. Then we will show that there is an out-branching B_y^+ in $D - A(B_{x_n}^-)$. Since I_1 is strong, $I_1 - x_{q-1}y$ is connected and has only one initial strong component which contains y. This implies that there is an out-branching T_y^+ in $I_1 - x_{q-1}y$. Note that there exists an out-branching $T_{x_q}^+$ in I_2, as I_2 is strong. Then $B_y^+ = T_y^+ + x_{q-1}x_q + T_{x_q}^+$. So, $(B_y^+, B_{x_n}^-)$ is a good pair of D.

Proposition 11. *Let D be a 2-arc-strong oriented graph on at least seven vertices. Then D has a dipath P_6.*

Proof. Suppose that there is no P_6 in D. Assume that P_t is the longest dipath in D, then $t \geq 4$, as there is no digon in D and $\lambda(D) \geq 2$. Observe that there is no C_t in D, otherwise D has a longer dipath P_{t+1}.

First assume that $t = 4$ and set $P_4 = x_1x_2x_3x_4$. Since $d_D^+(x_4) \geq 2$ and D has no digon, the out-neighbourhood of x_4 either contains x_1 or contains a vertex in $V - V(P_4)$. This implies that there is a P_5 in D, a contradiction.

Henceforth we may assume that $t = 5$ and set $P_5 = x_1x_2x_3x_4x_5$. Since $\lambda(D) \geq 2$, $d_D^+(x_5) \geq 2$ and $d_D^-(x_1) \geq 2$. Then we get $N_D^+(x_5) = \{x_2, x_3\}$ and

$N_D^-(x_1) = \{x_3, x_4\}$, as P_5 is the longest dipath in D and D has no digon. Observe that there exists a different 4-length dipath, $x_4x_5x_3x_1x_2$, in D. Likewise, $N_D^+(x_2) = \{x_3, x_5\}$, which implies that $D[\{x_2, x_5\}]$ is a digon, a contradiction.

4 Good Pairs in Digraphs of Order 7

Proposition 12. *A 2-arc-strong oriented graph D on n vertices has a P_7, where $7 \leq n \leq 9$.*

Proof. Suppose to the contrary that P is the longest dipath in D, where $|P| = 6$. Obviously D has no C_6 by Proposition 11. Set $P = x_1x_2x_3x_4x_5x_6$. Since $\lambda(D) = 2$, we have $d_D^+(x_6) \geq 2$ and $d_D^-(x_1) \geq 2$. Note that $N_D^+(x_6) \subsetneq \{x_2, x_3, x_4\}$ and $N_D^-(x_1) \subsetneq \{x_3, x_4, x_5\}$.

Assume first that $N_D^+(x_6) \cap N_D^-(x_1) = \emptyset$.

If $N_D^+(x_6) = \{x_2, x_3\}$ and $N_D^-(x_1) = \{x_4, x_5\}$, then we can get a new P_6 in D as $x_6x_2x_3x_4x_5x_1$. Likewise, we have $x_3 \in N_D^+(x_1)$ and $x_4 \in N_D^-(x_6)$. Then there exists a good pair $(B_{x_4}^+, B_{x_4}^-)$ in $D[P]$ with $B_{x_4}^+ = x_4x_1x_2 + x_4x_5 + x_4x_6x_3$ and $B_{x_4}^- = x_5x_6x_2x_3x_4 + x_1x_3$, which implies that D has a good pair by Lemma 5 and Lemma 1.

If $N_D^+(x_6) = \{x_2, x_4\}$ and $N_D^-(x_1) = \{x_3, x_5\}$, then we can get a new P_6 in D as $x_6x_4x_5x_1x_2x_3$. Likewise, we have $x_1 \in N_D^-(x_6)$, which implies that there is a C_6 as $x_6x_2x_3x_4x_5x_1x_6$, a contradiction.

Henceforth, we may assume that there is at least one common vertex in $N_D^+(x_6)$ and $N_D^-(x_1)$. Without loss of generality, assume that x_3 is one of the common vertices. The case when $x_4 \in N_D^+(x_6) \cap N_D^-(x_1)$ can be proved analogously by reversing all arcs of D. Then we can get a new P_6 in D as $x_4x_5x_6x_3x_1x_2$. Likewise, we have

$$N_{D-x_3}^+(x_2) \subsetneq \{x_5, x_6\} \text{ and } N_{D-x_3}^-(x_4) \subsetneq \{x_1, x_6\}. \tag{1}$$

Note that $x_4 \in N_D^-(x_1)$ and $x_5 \in N_D^+(x_2)$ will not hold at the same time, or there will exist a C_6 as $x_1x_2x_5x_6x_3x_4x_1$, a contradiction.

If $x_2 \in N_D^+(x_6)$, then we have $x_5 \in N_D^+(x_2)$ as D is an oriented graph. This implies that x_5 is an in-neighbour of x_1. Observe a new P_6 as $x_4x_5x_6x_2x_3x_1$, we can get that $x_6 \in N_D^+(x_1)$. Thus there exists a C_6 as $x_6x_2x_3x_4x_5x_1x_6$, a contradiction.

Thus we have $N_D^+(x_6) = \{x_3, x_4\}$. If $x_4 \in N_D^-(x_1)$, then x_6 is an out-neighbour of x_2. Observe a new P_6 as $x_5x_6x_4x_1x_2x_3$, we can get that $x_1 \in N_D^-(x_5)$. Thus there exists a good pair $(B_{x_2}^+, B_{x_6}^-)$ in $D[P]$ with $B_{x_2}^+ = x_2x_6x_4 + x_6x_3x_1x_5$ and $B_{x_6}^- = P_D$, which implies that D has a good pair by Lemma 5 and Lemma 1. If $x_5 \in N_D^-(x_1)$, then we can get a new P_6 as $x_6x_3x_4x_5x_1x_2$. Likewise, we have $N_{D-x_3}^+(x_2) \subsetneq \{x_4, x_5\}$ and $N_{D-x_5}^-(x_6) \subsetneq \cup\{x_1, x_4\}$. By (1) and $x_4 \in N_D^+(x_6)$, we can get that $x_5 \in N_D^+(x_2)$ and $x_1 \in N_D^-(x_6)$. Then there exists a good pair $(B_{x_2}^+, B_{x_6}^-)$ in $D[P]$ with $B_{x_2}^+ = x_2x_5x_1x_6x_3 + x_6x_4$ and $B_{x_6}^- = P_D$, which implies that D has a good pair as $\lambda(D) = 2$, a contradiction.

Now we are ready to prove Theorem 2. For convenience, we restate it here.

Theorem 2. *Every 2-arc-strong digraph on 7 vertices has a good pair.*

Proof. Suppose that D has no good pair. Let R be a largest clique in D. By Lemma 5 and Proposition 8, $|R| = 3$. Moreover, R is a tournament by Lemma 6 and Proposition 7.

Claim 2.1. *D is an oriented graph.*

Proof. Suppose that there is a digon Q in D with $V(Q) = \{s, t\}$. Observe that Q has a good pair. Since R is a tournament with three vertices, both in- and out-neibourhoods of Q in D have at least two vertices. This implies that D has a good pair by Lemma 7, a contradiction.

Assume that $P_D = x_1x_2 \ldots x_7$ is a Hamilton dipath of D by Proposition 12. Set $D' = D - A(P_D)$. Let I_i and T_j respectively be the initial and terminal strong component in D', where $i \in [a]$ and $j \in [b]$. Note that $a, b \geq 2$ by Proposition 9. Since D is an oriented graph and $\lambda(D) \geq 2$, $|I_i|, |T_j| \geq 3$, for any $i \in [a], j \in [b]$. Thus there are only two non-adjacent strong components in D', say I_1 and I_2, with $|I_1| = 3$ and $|I_2| = 4$. Note that $|N^-_{D'}(x_1)| \geq 2$ and $|N^+_{D'}(x_7)| \geq 2$ as $\lambda(D) \geq 2$, which implies that $x_1, x_7 \in I_2$. Moreover, $x_2, x_6 \in I_1$ by Claim 2.1. Then D has a good pair by Proposition 10.

5 Good Pairs in Digraphs of Order 8

The digraph E_3 used in the next proposition is shown in Fig. 2.

Proposition 13 ([2]). *Let D be a 2-arc-strong digraph without any subdigraph on order 4 that has a good pair. If D contains an orientation Q of E_3 as a subdigraph, then $N^+_{\bar{D}}(Q) \cap N^-_{\bar{D}}(Q) = \emptyset$, $|N^+_{\bar{D}}(Q)| \geq 2$ and $|N^-_{\bar{D}}(Q)| \geq 2$.*

Proposition 14. *Let D be a 2-arc-strong oriented graph on n vertices without K_4 as a subdigraph, where $8 \leq n \leq 9$. If D has two disjoint cycles C^1 and C^2 which cover exactly 7 vertices, then D contains a P_8.*

Proof. Suppose that P_7 is the longest dipath of D by Proposition 12. In fact there exist arcs between C^1 and C^2 from both directions, otherwise D has a P_8 as $\lambda(D) \geq 2$. W.l.o.g., assume $|C^1| \geq |C^2|$. Then $|C^1| = 4$ and $|C^2| = 3$. Let $C^1 = x_1x_2x_3x_4x_1$, $C^2 = x_5x_6x_7x_5$, $P_7 = x_1x_2 \ldots x_7$ and y_j be the vertex in $V - V(C^1 \cup C^2)$, where $j = 1$ when $n = 8$ and $j \in [2]$ when $n = 9$. From the maximality of P_7 in D, we have the following facts.

Fact 14.1. For any j, at least one of $(C^i, y_j)_D$ and $(y_j, C^{3-i})_D$ is empty for any $i \in [2]$.

Fact 14.2. For any j, at least one of arcs x_iy_j and y_jx_{i+1} is not in A for any $i \in [6]$.

Fact 14.3. For $n = 9$, let $y_j y_{3-j} \in A$. If $x_i y_j \in A$, then $y_{3-j}x_{i+1}, y_{3-j}x_{i+2} \notin A$, where $j \in [2]$ and $i \in [5]$.

Since D is oriented, there are at least three arcs between y_j and C^i, for some i, by Fact 14.1. W.l.o.g., assume $i = 1$. Note that $d^+_{C^1}(y_j) \geq 1$ and $d^-_{C^1}(y_j) \geq 1$. Then $N(y_j) \subset \{y_{3-j}\} \cup C^1$ when $n = 9$ and $N(y_j) \subset C^1$ when $n = 8$.

If y_j is not adjacent to y_{3-j} or $n = 8$, then $N^+(y_j) = \{x_1, x_2\}$ and $N^-(y_{3-j}) = \{x_3, x_4\}$ by Fact 14.2, which implies that D has a P_8 as $y_j x_1 \in A$, a contradiction.

Hence $n = 9$ and y_1 is adjacent to y_2. W.l.o.g., assume that $y_1 y_2 \in A$. If $x_1 y_1 \in A$, then $N^+(y_2) = \{x_1, x_4\}$ by Fact 14.3 and $\lambda(D) \geq 2$, which implies that D has a Hamilton dipath as $y_2 x_1 \in A$, a contradiction. Hence x_1 is not adjacent to y_1. By Fact 14.2, $N^+(y_1) = \{x_2, y_2\}$ and $N^-(y_1) = \{x_3, x_4\}$. By Fact 14.3 and the longestness of P_7, $N^+(y_2) = \{x_2, x_3\}$. It implies that $D[\{x_2, x_3, y_1, y_2\}]$ is a K_4, a contradiction.

Proposition 15. *Let $D = (V, A)$ be a 2-arc-strong digraph on n vertices without good pair, where $8 \leq n \leq 9$. If D is an oriented graph without K_4 as a subdigraph, then D has a P_8.*

Now we are ready to show Theorem 3. For convenience, we restate it here.

Theorem 3. *Every 2-arc-strong digraph on 8 vertices has a good pair.*

Proof. Suppose that D has no good pair. Let R be a largest clique in D. By Lemma 6 and Proposition 8, $|R| = 3$.

Claim 3.1. *No subdigraph of D of order at least 3 has a good pair.*

Proof. By Lemma 6, it suffices to show that there is no $Q \subset D$ on 3 vertices with a good pair. Suppose that Q has a good pair. If Q is an orientation of E_3, then we use Lemma 7 to find a good pair of D by Proposition 13, a contradiction. Now assume that Q is a bidigon. Set $V(Q) = \{x, y, z\}$ with $Q[\{x, y\}] = C_2$ and $Q[\{y, z\}] = C_2$. If there exists a vertex w in $N^+_D(Q) \cap N^-_D(Q)$, then $D[Q \cup \{w\}]$ has a good pair by Lemma 1. Thus $N^+_D(Q) \cap N^-_D(Q) = \emptyset$. If $N^-_D(Q) = \{w\}$, then $D[Q \cup \{w\}]$ has a good pair as $B^+_w = wzyx$ and $B^-_z = wxyz$. By symmetry, this implies that $|N^+_D(Q)| \geq 2$ and $|N^-_D(Q)| \geq 2$. Thus by Lemma 7, D has a good pair, a contradiction. ◊

By the claim above, R is a tournament.

Claim 3.2. *D is an oriented graph.*

Proof. Suppose that there is a digon Q in D with $V(Q) = \{s, t\}$. Observe that Q has a good pair. Since R is a tournament with 3 vertices, both in- and out-neibourhoods of Q in D have at least two vertices with $N^+_D(Q) \cap N^-_D(Q) = \emptyset$. This implies that D has a good pair by Lemmas 2, 8, 9 and 10, and Corollary 1, a contradiction. ◊

By Proposition 15, assume that $P_D = x_1x_2\dots x_8$ is a Hamilton dipath of D. Set $D' = D - A(P_D)$. Let I_i and T_j respectively be the initial and terminal strong component in D', where $i \in [a]$ and $j \in [b]$. Note that $a, b \geq 2$ by Proposition 9. Since D is an oriented graph and $\lambda(D) \geq 2$, $|I_i|, |T_j| \geq 3$ for any $i \in [a], j \in [b]$. Thus there are only two non-adjacent strong components in D', say I_1 and I_2, as $n = 8$. Since $\lambda(D) \geq 2$, x_1 has at least two in-neighbours and one out-neighbour in D', while x_8 has at least two out-neighbours and one in-neighbour in D'. If $|I_1| = 3$ and $|I_2| = 5$, then $x_1, x_8 \in I_2$ and $|A(I_2)| \geq 6$. Note that at least one of x_2 and x_7 is in I_1 as $|R| = 3$. Then we use Proposition 10 to get a good pair of D. Now assume $|I_1| = |I_2| = 4$. If $x_8 \in I_1$ then $x_7 \in I_2$ by Claim 3.2. By Proposition 10, D has a good pair.

6 Good Pairs in Digraphs of Order 9

We have several generalizations of Proposition 6 here, which are easy to check as they satisfy the conditions in Proposition 6.

Proposition 16. *Let $D = (V, A)$ be a digraph and Q be a subdigraph of D with good pair (O_Q, I_Q). Set $X = N_D^-(Q)$ and $Y = N_D^+(Q)$ with $X \cap Y = \emptyset$ and $X \cup Y = V - V(Q) - W$, where $W = \{w_1, w_2\}$. Let e_1 be an arc from w_1 to X and e_2 be an arc from Y to w_2. Set $X' = X \cup w_1$, $Y' = Y \cup w_2$ and $D' = (V, A')$ with $A' = A - \{e_1, e_2\}$. Let $\mathcal{X}$ be the set of initial strong components in $D'[X']$ and $\mathcal{Y}$ be the set of terminal strong components in $D'[Y']$. Assume that there exists X_0 and Y_0 in $\mathcal{X}$ and $\mathcal{Y}$ respectively such that $d_Y^-(X_0) = 1$ and $d_X^+(Y_0) = 1$. Let e_x and e_y be arcs from Y to X_0 and from Y_0 to X respectively. If one of the following holds, then D has a good pair.*

1. *$e_x \neq e_y$, but at least one of $\mathcal{X}$ or $\mathcal{Y}$ has only one element.*
2. *e_x (or e_y) is adjacent to some Y_x (or X_y) in $\mathcal{Y}$ (or $\mathcal{X}$), such that $d_X^+(Y_x) \geq 3$ (or $d_Y^-(X_y) \geq 3$).*
3. *e_x (or e_y) is adjacent to $Y' - V(\mathcal{Y})$ (or $X' - V(\mathcal{X})$).*
4. *e_x (or e_y) is adjacent to some $Y_x \neq Y_0$ (or $X_y \neq X_0$) in $\mathcal{Y}$ (or $\mathcal{X}$), such that there exists an arc from Y_x (or X_y) to $X' - V(\mathcal{X})$ (or $Y' - V(\mathcal{Y})$).*

Lemma 11. *Let D be a 2-arc-strong digraph on 9 vertices that contains a digon Q. Assume that D has no subdigraph with a good pair on 3 or 4 vertices. Set $X = N_D^-(Q)$ and $Y = N_D^+(Q)$ with $X \cap Y = \emptyset$. If $|X| = 3$ and $|Y| = 2$, then D has a good pair.*

Lemma 12. *Let $D = (V, A)$ be a 2-arc-strong digraph on 9 vertices that contains a digon Q. Assume that D has no subdigraph with a good pair on at least 3 vertices. Set $X = N_D^-(Q)$ and $Y = N_D^+(Q)$ with $X \cap Y = \emptyset$ and $W = V - V(Q) - X - Y$. Assume that $|X| = |Y| = 2$ and there is an arc $e = st \in A$ such that $s \in Y$ and $t \in W$ (resp. $s \in W$ and $t \in X$). If there are at least three arcs in $D[Y \cup \{t\}]$ (resp. $D[X \cup \{s\}]$), then D has a good pair.*

Lemma 13. *Let D be a 2-arc-strong digraph on 9 vertices that contains a digon Q. Assume that D has no subdigraph with a good pair on 3 or 4 vertices. Set $X = N_D^-(Q)$ and $Y = N_D^+(Q)$ with $X \cap Y = \emptyset$. If $|X| = 2$ and $|Y| = 2$, then D has a good pair.*

Proposition 17. *Let $D = (V, A)$ be a 2-arc-strong oriented graph on 9 vertices without K_4 as a subdigraph. If D have two cycles C^1 and C^2 with $C^1 \cap C^2 = \emptyset$ which cover 8 vertices, then D contains a Hamilton dipath.*

Lemma 14. *Let $D = (V, A)$ be a 2-arc-strong digraph on 9 vertices without good pair. If D is an oriented graph without K_4 as a subdigraph, then D has a Hamilton dipath.*

Now we are ready to show Theorem 4. For convenience, we restate it here.

Theorem 4. *Every 2-arc-strong digraph on 9 vertices has a good pair.*

Proof. By contradiction, suppose that D has no good pair.

Claim 4.1. *No subdigraph of D of order at least 4 has a good pair.*

Let R be a largest clique in D. Then R has three vertices by Claim 4.1 and Proposition 8.

Claim 4.2. *No subdigraph of D of order at least 3 has a good pair.*

Proof. By Lemma 7, it suffices to show that there is no $Q \subset D$ on 3 vertices with good pair. Suppose to the contrary that Q has a good pair. Analogous to Claim 3.1, $|N_D^+(Q)| \geq 2$ and $|N_D^-(Q)| \geq 2$ with $N_D^+(Q) \cap N_D^-(Q) = \emptyset$. Thus by Lemma 8, D has a good pair, a contradiction. $\Diamond$

By the claim above, R is a tournament.

Claim 4.3. *D is an oriented graph.*

Proof. Suppose that D has a digon Q. Set $X = N_D^-(Q)$ and $Y = N_D^+(Q)$. By Claim 4.2, $X \cap Y = \emptyset$. Since $\lambda(D) \geq 2$, both X and Y have at least two vertices. If $|X|+|Y| = 4$, then D has a good pair by Lemma 13, a contradiction. If $|X|+|Y| = 5$, then D has a good pair by Lemma 11 and the digraph duality, a contradiction. If $|X| + |Y| = 6$, then D has a good pair by Lemma 2, a contradiction. If $|X| + |Y| = 7$, then D has a good pair by Corollary 1, a contradiction. $\Diamond$

Now we are ready to finish the proof of Theorem 3. By Lemma 14, assume that $P_D = x_1x_2 \ldots x_9$ is a Hamilton dipath of D. Set $D' = D - A(P_D)$. Let I_i, $i \in [a]$, be the initial strong components in D' and let T_j, $j \in [b]$, be the terminal strong components in D'. Note that $a, b \geq 2$ by Proposition 9. Since D is an oriented graph and $\lambda(D) \geq 2$, $|I_i|, |T_j| \geq 3$, for any $i \in [a], j \in [b]$. Since $\lambda(D) \geq 2$, x_1 has at least two in-neighbours and one out-neighbour in D' and x_9 has at least two out-neighbours and one in-neighbour in D'. Thus there are only two non-adjacent strong components in D', say I_1 and I_2, as $n = 9$ and D is an oriented graph. We distinguish two cases below.

Case 1: $|I_1| = 4$ and $|I_2| = 5$.
If $x_9 \in I_1$, then $x_8 \in I_2$ as $|R| = 3$. Analogously, if $x_1 \in I_1$, then $x_2 \in I_2$. By Proposition 10, D has a good pair for each cases. Henceforth, both x_1 and x_9 are in I_2. Note that at least one of x_2 and x_8 is in I_1 as $|R| = 3$. By Proposition 10, D has a good pair, a contradiction.

Case 2: $|I_1| = 3$ and $|I_2| = 6$.
In this case, $x_1, x_9 \in I_2$ and $|A(I_2)| \geq 7$. If one of x_2 and x_8 is in I_1, then D has a good pair by Proposition 10. Thus both x_2 and x_8 are in I_2. Then $V(I_1) = \{x_3, x_5, x_7\}$, which implies that D has a good pair by Proposition 10, a contradiction.

This completes the proof of Theorem 4.

References

1. Bang-Jensen, J.: Edge-disjoint in- and out-branchings in tournaments and related path problems. J. Combin. Theory Ser. B **51**(1), 1–23 (1991)
2. Bang-Jensen J., Bessy S., Havet F., Yeo, A.: Arc-disjoint in- and out-branchings in digraphs of independence number at most 2. arXiv:2003.02107 (2020)
3. Bang-Jensen J., Bessy, S., Yeo, A.: Non-separating spanning trees and out-branchings in digraphs of independence number 2. arXiv:2007.02834v1 (2020)
4. Bang-Jensen, J., Gutin, G. (eds.): Classes of Directed Graphs. SMM, Springer, Cham (2018). https://doi.org/10.1007/978-3-319-71840-8
5. Bang-Jensen, J., Gutin, G.: Digraphs: Theory, Algorithms and Applications, 2nd edn. Springer Verlag, London (2009). https://doi.org/10.1007/978-1-84800-998-1
6. Bang-Jensen, J., Gutin, G., Yeo, A.: Arc-disjoint strong spanning subdigraphs of semicomplete compositions. J. Graph Theory **95**(2), 267–289 (2020)
7. Bang-Jensen, J., Huang, J.: Decomposing locally semicomplete digraphs into strong spanning subdigraphs. J. Combin. Theory Ser. B **102**(3), 701–714 (2010)
8. Bang-Jensen, J., Saurabh, S., Simonsen, S.: Parameterized algorithms for non-separating trees and branchings in digraphs. Algorithmica **76**(1), 279–296 (2016). https://doi.org/10.1007/s00453-015-0037-3
9. Gu R., Gutin G., Li S., Shi Y., Taoqiu Z.: The smallest number of vertices in a 2-arc-strong digraph which has no good pair. arXiv:2012.03742 (2020)
10. Gutin, G., Reidl, F., Wahlström, M.: *k*-distinct in- and out-branchings in digraphs. J. Comput. Syst. Sci. **95**, 86–97 (2018)
11. Lovász, L.: On two min-max theorems in graph theory. J. Comb. Theory Ser. B **21**, 96–103 (1976)
12. Thomassen, C.: Configurations in graphs of large minimum degree, connectivity, or chromatic number. Ann. New York Acad. Sci. **555**, 402–412 (1989)

Generalized Self-profit Maximization in Attribute Networks

Liman Du, Wenguo Yang(✉), and Suixiang Gao

School of Mathematical Sciences, University of Chinese Academy of Science, Beijing 100049, China
duliman18@mails.ucas.edu.cn, {yangwg,sxg}@ucas.ac.cn

Abstract. Profit maximization, an extension of classical Influence Maximization, asks for a small set of early adopters to maximize the expected total profit generated by all the adopters. This is a meaningful optimization problem in social network which has attracted many researchers' attention. Nevertheless, most of the related works are based on pure networks and one-entity diffusion model without considering the relationship of different promotional products and the influence of both emotion tendency and interest classification (label) in real marketing process. In this paper, we propose a novel nonsubmodular optimization problem, Generalized Self-profit Maximization in Attribute networks (GSMA), which is based on a community structure in attribute networks and captures the unilateral complementary relationship, a setting with complementary entities. To solve GSMA, the R-GSMA algorithm which is inspired by sampling method and martingale analysis is designed. We evaluate our proposed algorithm by conducting experiments on randomly generated and real datasets and show that R-GSMA is superior in effectiveness and accuracy comparing with other baseline algorithms.

Keywords: Profit maximization · Nonsubmodularity · Attribute networks · Random algorithm · Martingale

1 Introduction

Nowadays, a growing number of people spend more time communicating with their friends through large-scale communication platforms, such as Facebook, Twitter and Wechat. Each of these platforms can be modeled as a social network containing the information of relationships and interaction among users. Motivated by viral marketing, recommend system and so on, there has been a large amount of research into the information propagation in social networks.

A great quantity of studies focus on the classical Influence Maximization (IM) which is modeled as a combinatorial optimization problem in [10] and aims

Supported by the National Natural Science Foundation of China under grant numbers 12071249 and 11991022 and the Fundamental Research Funds for the Central Universities under Grant Number E1E40107.

D.-Z. Du et al. (Eds.): COCOA 2021, LNCS 13135, pp. 333–347, 2021.
https://doi.org/10.1007/978-3-030-92681-6_27

to select at most k influential social network users (named as seeds) to maximize the expected number of users who receive the information initially supplied by seeds and spreading through social interactions. The large body of work in IM research can be classified into the diffusion models, classical IM solutions and extended IM problems. All the efforts that extend the classical IM by incorporating various realistic issues associated with information diffusion are classified as extended IM problems, including Robust Influence Maximization [1,5], Activity Maximization [19], Profit Maximization [16,17], CoFIM [14] and so on. The Profit maximization (PM) extends the influence spread to profit spread which is defined as the difference between benefit and cost generated by all the active nodes. It aims to find all the seed nodes to maximize the return value of profit spread function while the Constrained Profit Maximization (CPM) proposed in [11] confines the number of seed nodes. As is shown in [16] and [11] respectively, both PM and CPM belong to nonsubmodular optimization instead of submodular optimization. Although extensive work on submodular optimization has been carried out, nonsubmodular optimization has attracted more and more researchers in recent years. Many methods to address this problem are well known to us, such as supermodular degree, Difference of Submodular (DS) functions and so on [3,4,7–9,13,20].

Back to the studies about IM problem, Lu et al. firstly highlighted the relationship between entities and extended the classical IC model to Com-IC model which covers the full spectrum of entity interactions from competition to complementarity in [12]. Besides, a new model which makes full use of topology structure similarity and attribute similarity to predict influence strength between nodes in attribute networks is proposed in [6]. And Song et al. put forward the Emotion Independent Cascade (EIC) model and Influence Maximization Problem based on Emotion (IMPE) in [15], taking the impact of emotion on influence spread process into consideration.

In real life, there are various factors that may be present and influence the decision of potential adopters in marketing process and we take into account some of them. At first, it is difficult for modern consumers to exclude the influence of both others' and their own emotions because no one is of an intelligent and rational turn of mind all the time. Different from the assumption under which EIC model is proposed, we underline the influence of emotion tendency on interaction probability between potential adopters and exclude the process of updating emotion. One reason is consumers' impression of a product or its manufacturer is always inherent regardless of whether they purchase the product or not in actual marketing activities. And the other is that we focus on how to reach the maximum of profit instead of emotion. Secondly, we try to reflect the influence of label similarity on interaction between users of social networks. As is known to all, when uploading photos and text to communicate with others on online social platforms, users tend to tag them. Those tags may help others to understand their interest. In addition, they often use labels in self-introduction, such as "sports enthusiast", "music-lover", "amateur photographer" and others. Labels assist users in finding people of the same taste or hobby quickly. And it is beyond doubt that the label similarity affects the interaction frequency. Last but not least, under the assumption that more than one entity is involved, in order to maximize the total profit generated by all the active nodes after information spread, special stress is laid on the interrelation between entities. In this paper, we pay attention to the unilateral instead of mutual complementary relationship between entities.

In this paper, we present Generalized Self-profit Maximization in Attribute Networks for unilateral complementary entities with an unpredicted seed set for one among them. The objective function of GSMA problem is nonsubmodular and nonmonotone, and the R-GSMA algorithm inspired by sampling method and martingale analysis in [18] is proposed. Lastly, we experimentally evaluate the performance of R-GSMA and the results indicate that the R-GSMA algorithm obtains the best results in comparison with some other baseline algorithms.

The remainder of this paper is organized as follows: Sect. 2 presents Attribute networks and the Profit Complementary Independent Cascade Model. The GSMA problem is formulated in this part and related properties are introduced. The detail of the R-GSMA algorithm is given in Sect. 3. Section 4 shows the experimental results and we conclude in Sect. 5.

2 Problem Formulation

2.1 Attribute Networks

Inspired by the study of attribute networks in [6] and Influence Maximization Problem based on Emotion (IMPE) in [15], we propose an attribute network $G = (V, E, W, M, L)$, where V and E represent users and social ties between each pair of them respectively. For each $(u, v) \in E$, let $w_{u,v} \in W$ represent the influence of u on v. M and L reflect the emotion tendency and labels of nodes respectively. For each node $v \in V$, the value of M_v shows v's emotion tendency which can be divided into three cases. $M_v \in [-1, 0)$ ($M_v \in (0, 1]$) means that node v holds negative (positive) emotion while $M_v = 0$ represents that v is neural. Given t different kinds of labels which are used in the attribute network to describe the characters of users, L is a binary-valued label matrix. For $i = 1, \ldots, t$, $l_{v,i} = 1$ when node v associated with label i and $l_{v,i} = 0$ if the association does not exist.

Now, we redefine the influence probability between two users and show how to calculate it based on two kinds of influence strength related to emotion tendency and social interaction respectively.

Emotion Tendency. In social platforms, people tend to increase the frequency of interaction with friends who can provoke emotional resonance with them, and subconsciously avoid communicating with those whose emotion is different from them. If they have the same emotional tendency, the more similar the extent of individuals' liking or aversion is, the stronger the influence between them is. If two people have opposite emotional tendencies, the influence strength will increase with the emotional intensity gap. In other words, if one is very emotional and the other is gentle, the influence strength between them will be larger. So, for a pair of neighbor nodes $u, v \in V$, define the emotion-based influence strength

$$Pe_{u,v} = \begin{cases} 1 - ||M(u)| - |M(v)||, & M(u) \cdot M(v) > 0 \\ ||M(u)| - |M(v)||, & \text{otherwise} \end{cases} \quad (1)$$

where $M(u), M(v) \in [-1, 1]$ represent the emotion tendency of u and v respectively. In addition, define $Pe_{u,u} = 0$.

Social Interaction. Denote $C = \{C_1, C_2, \ldots, C_p\}$ as a non-overlapping community structure of given social network, where $\bigcup_{k=1}^{p} C_k = V$ and $C_i \cap C_j = \emptyset$ for each $i \neq j$. We cite formulas proposed by [6] to calculate influence strength based on social interaction.

- For a pair of neighbor nodes $u, v \in V$, $IS_{u,v} = \frac{Is_{u,v} - Is_{min}}{Is_{max} - Is_{min}}$, where $Is_{u,v} = w_{u,v}$, $Is_{max} = \max_{u,v \in V} w_{u,v}$ and $Is_{min} = \min_{u,v \in V} w_{u,v}$.
- For two adjacent nodes u and v that are in the same community, based on logistic function $\sigma(X) = (1 + e^{-x})^{-1}$ and the definition of transposed vector, topology structure similarity and label similarity between u and v are defined as $TS_{u,v} = \sigma(ts_u^T \cdot ts_v)$ and $Ls_{u,v} = \sigma(ls_u^T \cdot ls_v)$, where $ts_u = \{w_{u,1}, \ldots, w_{u,|V|}\}$, $ts_v = \{w_{v,1}, \ldots, w_{v,|V|}\}$, $ls_u = \{l_{u,1}, \ldots, l_{u,t}\}$ and $ls_v = \{l_{v,1}, \ldots, l_{v,t}\}$.

To calculate influence strength between u and v from the perspective of social interaction, define

$$Pi_{u,v} = \left(1 + e^{6 - 4(IS_{u,v} + TS_{u,v} + LS_{u,v})}\right)^{-1}. \quad (2)$$

Based on the property of logistic function, the influence strength increases with similarity and the value range of $Pi_{u,v}$ is approximately between 0 and 1.

Definition 1 (Influence Probability). *The influence probability between adjacent nodes u and v in attribute networks is defined as*

$$p_{u,v} = a \cdot Pe_{u,v} + (1 - a) \cdot Pi_{u,v}, \quad (3)$$

where $a \in [0, 1]$ is a weight parameter.

Remark 1. The influence probability in attribute networks is similar to the influence probability proposed in pure networks which represents the success probability of information dissemination between nodes.

2.2 Profit Complementary Independent Cascade Model and Diffusion Dynamics

Profit Complementary Independent Cascade Model. Profit Complementary Independent Cascade (PCIC) Model focused in this paper is generated from

Comparative Independent Cascade (Com-IC) Model proposed in [12] and only considers the unilateral complementary relationship between entities.

There are at least two entities involved in the PCIC model, and we focus on a special case which only considers two entities $\mathcal{A}$ and $\mathcal{B}$. The network is abstracted as a directed graph $G = (V, E, P, Q, B, C)$ where V is the set of user and E contains all the social relations between each pair of users. For each edge $(u, v) \in E$, $p_{u,v} \in P$ is the influence probability that the information propagates along the edge from node u to v. For each $q(v) \in Q$, it consists of four parameters related to node v which can influence whether v adopts the target entity or not. B and C respectively represent the benefit and cost of each node with respect to each entity when it is activated.

During the information dissemination, each node $v \in V$ can be in any of the states {idle, suspended, adopted, rejected} with respect to every entity. All the nodes are initially stay in the joint state of ($\mathcal{A}$-idle, $\mathcal{B}$-idle) when only $\mathcal{A}$ and $\mathcal{B}$ are considered. Both the benefit and cost generated by nodes are 0. Edges only control the information diffusion while the state transition is decided by both node's current state with regard to $\mathcal{A}$ or $\mathcal{B}$ and a set of parameters $q(v) = \{q^v_{\mathcal{A}|\emptyset}, q^v_{\mathcal{A}|\mathcal{B}}, q^v_{\mathcal{B}|\emptyset}, q^v_{\mathcal{B}|\mathcal{A}}\}$. For node $v \in V$, $q^v_{\mathcal{A}|\emptyset}$ is the probability that user v adopts $\mathcal{A}$ when $\mathcal{B}$ is not adopted by v. If user v has adopted $\mathcal{B}$ and is informed of $\mathcal{A}$, it adopts $\mathcal{A}$ with $q^v_{\mathcal{A}|\mathcal{B}}$. The meaning of $q^v_{\mathcal{B}|\emptyset}$ and $q^v_{\mathcal{B}|\mathcal{A}}$ are similar to those of $q^v_{\mathcal{A}|\emptyset}$ and $q^v_{\mathcal{A}|\mathcal{B}}$. What should be emphasized is that in order to represent the unilateral complementary relationship between $\mathcal{A}$ and $\mathcal{B}$ which means $\mathcal{B}$-diffusion is independent of $\mathcal{A}$, it is assumed that for every node, $q_{\mathcal{A}|\emptyset} \leq q_{\mathcal{A}|\mathcal{B}}$ and $q_{\mathcal{B}|\emptyset} = q_{\mathcal{B}|\mathcal{A}}$.

Diffusion Dynamics. Let $G = (V, E, P, Q, B, C)$ be a directed social graph. Before the information diffusion, each node stays in the joint state of ($\mathcal{A}$-idle, $\mathcal{B}$-idle). Let $S_\mathcal{A}, S_\mathcal{B} \subset V$ be seed sets for two different entities $\mathcal{A}$ and $\mathcal{B}$ respectively. Then, nodes which belong to $S_\mathcal{A} \backslash S_\mathcal{B}$ become $\mathcal{A}$-adopted and those belonging to $S_\mathcal{B} \backslash S_\mathcal{A}$ are $\mathcal{B}$-adopted at $t = 0$. They only generates profit for corresponding entity. If $u \in S_\mathcal{A} \cap S_\mathcal{B}$, the order of u adopting these entities is randomly decided with a fair coin and u generates profit for two entities. At every time step $t \geq 1$, for a node u that becomes $\mathcal{A}$-adopted or $\mathcal{B}$-adopted at $t-1$, the information spreads to one of its neighbor v with probability $p_{u,v} \in P$. Therefore, whether v makes an adoption depends on both the influence probability for information spread and the state transition parameter set $q(v)$ of v. Figure 1 shows a concise representation of state transition.

2.3 Problem Statements

In this paper, we propose an extension of Constrained Profit Maximization problem studied in [11] and name it as Generalized Self-profit Maximization in Attribute networks (GSMA) problem. It is inspired by Self Influence Maximization which is proposed in [12].

Let $B_\mathcal{A}$ and $C_\mathcal{A}$ represent the benefit and cost generated by $\mathcal{A}$-adopted nodes. For each node $v \in V$ adopting $\mathcal{A}$, $b_\mathcal{A}(v) \in B_\mathcal{A}$ is the benefit while

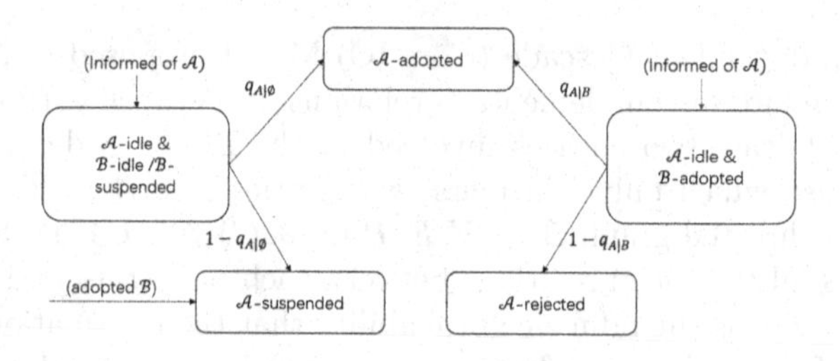

Fig. 1. State transition for $\mathcal{A}$

$c_{\mathcal{A}}(v) \in C_{\mathcal{A}}$ represents the cost. Without lose of generality, the value of both $b'_{\mathcal{A}}(v)$ and $c'_{\mathcal{A}}(v)$ can be confined into $[0,1]$ by setting $b'_{\mathcal{A}}(v) = \frac{b_{\mathcal{A}}(v)}{\max_{v\in V} b'_{\mathcal{A}}(v)}$ and $c'_{\mathcal{A}}(v) = \frac{c_{\mathcal{A}}(v)}{\max_{v\in V} c_{\mathcal{A}}(v)}$. For node v, define $\bar{b_v} = \max\{0, b'_{\mathcal{A}}(v) - c'_{\mathcal{A}}(v)\}$ and $\bar{c_v} = \max\{0, c'_{\mathcal{A}}(v) - b'_{\mathcal{A}}(v)\}$ as its normalized benefit and cost.

Given the seed sets $S_{\mathcal{A}}$ and $S_{\mathcal{B}}$, denote the expected profit of $\mathcal{A}$-adopted nodes under PCIC model by $\phi_{\mathcal{A}}(S_{\mathcal{A}}, S_{\mathcal{B}})$. The profit metric is defined as the difference between benefit and cost, i.e., $\phi_{\mathcal{A}}(S_{\mathcal{A}}, S_{\mathcal{B}}) = \beta_{\mathcal{A}}(S_{\mathcal{A}}, S_{\mathcal{B}}) - \gamma_{\mathcal{A}}(S_{\mathcal{A}}, S_{\mathcal{B}})$. $\beta_{\mathcal{A}}(S_{\mathcal{A}}, S_{\mathcal{B}})$ is the expected normalized benefit of nodes which receive information spread from $(S_{\mathcal{A}}, S_{\mathcal{B}})$ and adopt $\mathcal{A}$. And the interpretation of $\gamma_{\mathcal{A}}(S_{\mathcal{A}}, S_{\mathcal{B}})$ is similar to that of $\beta_{\mathcal{A}}(S_{\mathcal{A}}, S_{\mathcal{B}})$ while it focuses on the expected cost.

Definition 2 (Generalized Self-profit Maximization in Attribute networks). *Given a directed graph $G = (V, E, W)$, emotion tendency vector M, Label matrix L, parameter set Q, benefit and cost of each node with respect to each entities, a probability distribution over $\mathbf{B} = \{S_{\mathcal{B}} | S_{\mathcal{B}} \subset V, |S_{\mathcal{B}}| = r\}$ which collects all possible $\mathcal{B}$ seed set, and constant k, Generalized Self-profit Maximization in Attribute networks problem aims to find $\mathcal{A}$-seed set $S^*_{\mathcal{A}} \subset V$ such that the expected profit of $\mathcal{A}$-adopted nodes is maximized under PCIC model, i.e.*

$$S^*_{\mathcal{A}} \in \arg\max_{S_{\mathcal{A}} \subseteq V, |S_{\mathcal{A}}| \le k} \phi_{\mathcal{A}}(S_{\mathcal{A}}, S_{\mathcal{B}}) \tag{4}$$

Theorem 1. *GSMA is NP-hard.*

Proof. Because of the interpretation of influence probability proposed in Remark 1, it is obviously that GSMA problem can subsume Influence Maximization problem under the classic IC model when $S_{\mathcal{B}} = \emptyset$, $q_{A|\emptyset} = q_{A|B} = 1$ and $b_{\mathcal{A}}(v) = 1, c_{\mathcal{A}}(v) = 0$ for every node $v \in V$. Therefore, it is easy to see that GSMA is NP-hard based on a conclusion about the property of classical Influence Maximization problem proposed in [10]. And it is #P-hard to compute the exact value of $\phi_{\mathcal{A}}(S_{\mathcal{A}}, S_{\mathcal{B}})$ for any given $S_{\mathcal{A}}$ and $S_{\mathcal{B}}$ due to the research shown in [2].

Theorem 2. *For any fixed $\mathcal{B}$-seed set $S_{\mathcal{B}}$, $\phi_{\mathcal{A}}(S_{\mathcal{A}}, S_{\mathcal{B}})$ is nonmonotone with respect to $S_{\mathcal{A}}$ under PCIC model in general.*

Proof. Taking one further step from Theorem 3 in [12], we can come to the conclusion that $\beta_{\mathcal{A}}(S_{\mathcal{A}}, S_{\mathcal{B}})$ and $\gamma_{\mathcal{A}}(S_{\mathcal{A}}, S_{\mathcal{B}})$ are monotonically increasing with

$S_{\mathcal{A}}$ and $S_{\mathcal{B}}$ respectively. However, $\phi_{\mathcal{A}}(S_{\mathcal{A}}, S_{\mathcal{B}})$ may be nonmonotone as it is generally known that a function which can be expressed as the difference of two monotone function no longer remain monotone in general.

Theorem 3. *For any instance of PCIC model, $\phi_{\mathcal{A}}$ is a nonsubmodular function. And $\phi_{\mathcal{A}}$ can be expressed as a difference between two self-submodular functions w.r.t. $S_{\mathcal{A}}$ when $\mathcal{B}$-seed set $S_{\mathcal{B}}$ is fixed.*

Proof. It is widely known that a set function which can be expressed as the difference between two submodular functions may not carry on the special property. So even in the mentioned special cases where $\beta_{\mathcal{A}}$ and $\gamma_{\mathcal{A}}$ are self-submodular, $\phi_{\mathcal{A}}$ may still be a nonsubmodular function.

3 R-GSMA Algorithm

To address the GSMA problem, which is a nonsubmodular optimization problem and difficult to solve, the R-GSMA algorithm including three main steps is proposed in this paper. After randomly sampling a size-r $\mathcal{B}$-adopted seed set $S_{\mathcal{B}}$ and an outcome g by removing each edge $(u, v) \in E$ with probability $1 - P_{uv}$ respectively, let the r-RR set of a selected node v consist of all nodes u such that the singleton set $\{u\}$ would activate v. Denote $f_{\mathcal{R}_\beta}(S) = \sum_{v \in V} b'_{\mathcal{A}}(v) \cdot f_\beta$ as the benefit brought by $\mathcal{A}$-adopted seed set S where f_β is the fraction of r-RR sets covered by S in $\mathcal{R}_\beta$ which is a collection of r-RR sets. And the interpretation of $f_{\mathcal{R}_\beta}(S) = \sum_{v \in V} b'_{\mathcal{A}}(v) \cdot f_\beta$ is similar to that of $f_{\mathcal{R}_\gamma}(S)$.

First of all, based on the result of Louvain Algorithm which aims to get a non-overlapping partition, the influence probability is recalculated by Algorithm 1. Then, Algorithm 2 determines $\mathcal{B}$ adoption and generates r-RR sets by running two backward BFS. Algorithm 3 selects one node which can maximize the marginal value of GSMA's objective function at recent iteration until the number of selected nodes reaches k. Due to the definition of profit, there may exist a node v such that $b_{\mathcal{A}}(v) < c_{\mathcal{A}}(v)$. Therefore, potential seed nodes should be checked as shown in the line 4–7 of Algorithm 3.

Algorithm 1. Model Influence probability

Require: $G = (V, E, W, M, L)$, parameter a
Ensure: a influence probability matrix P
1: C = Louvain Algorithm $(G = (V, E, W))$
2: **for** $(u, v) \in E$ **do**
3: calculate $Pe_{u,v}$ and $IS_{u,v}$
4: **if** u and v are in the same community **then**
5: calculate $TS_{u,v}$ and $LS_{u,v}$ respectively.
6: calculate $Pi_{u,v}$
7: calculate $P_{u,v}$ according to the Eq. 3.

Algorithm 4 begins with a directed graph $G(V, E, B, C)$, influence probability matrix P and four parameters k, r, ε and σ. Based on Algorithm 2 and Algorithm 3, it searches various lower bounds of OPT to get a sufficiently precise one such that the break condition can be satisfied. In addition, it generates some r-RR sets to make sure the size of output $\mathcal{R}$ is big enough. At the third step of R-GSMA,

Algorithm 2. Generate r-RR set

Require: $G(V, E)$, $v \in V$, $S_{\mathcal{B}}$ and influence probability matrix P
Ensure: r-RR set

1: create an FIFO queue Q, empty sets R, T I and A;
2: Q.enqueue(v);
3: **while** Q is not empty **do**
4: u= Q.dequeue(), $T = T \cup \{u\}$;
5: **for** $w \in N^-(u)$ **do**
6: **if** (w, u) is live **then**
7: Q.enqueue(w), $I = I \cup \{w\}$;
8: **if** $T \cap S_{\mathcal{B}} \neq \emptyset$ **then**
9: clear Q;
10: **for** $a \in T \cap S_{\mathcal{B}}$ **do**
11: Q.enqueue(a);
12: **while** Q is not empty **do**
13: u=Q.dequeue(), $A = A \cup \{u\}$;
14: **for** $v \in N^+(u)$ **do**
15: **if** (u, v) is live, $\alpha_{\mathcal{B}}^{v,g} \leq q_{\mathcal{B}|\emptyset}$ and $I \cap \{v\} = \emptyset$ **then**
16: Q.enqueue(v);
17: clear Q;
18: Q.enqueue(v);
19: **while** Q is not empty **do**
20: u=Q.dequeue(), $R = R \cup \{u\}$;
21: **if** ($\{u\} \cap A \neq \emptyset$ and $\alpha_{\mathcal{B}}^{v,g} \leq q_{\mathcal{B}|\emptyset}$) or ($\{u\} \cap A = \emptyset$ and $\alpha_{\mathcal{B}}^{v,g} \leq q_{\mathcal{B}|\emptyset}$) **then**
22: **for** $w \in N^-(u)$ **do**
23: **if** (w, u) is live and $\{w\} \cap I = \emptyset$ **then**
24: Q.enqueue(w), $I = I \cup \{w\}$;
25: **Return** R as the r-RR set

Algorithm 3 is used to generate a $\mathcal{A}$-adopted node set for GSMA problem. The framework of R-GSMA is shown as Algorithm 5.

4 Experiment

We conduct some experiments on two different data sets to test the effectiveness of R-GSMA algorithm proposed above and the analysis of parameters' effect is also based on experiment results. What should be emphasized is that there is a special case in which the profit function is defined as $\phi(X) = \sum_{v \in V} b_v^X - \sum_{u \in X} c(u)$, resulting in the change of profit function's property from nonsubmodular to submodular.

Algorithm 3. Node selection

Require: sampling results $\mathcal{R}_\beta$ and $\mathcal{R}_\gamma$, parameters k
Ensure: a $\mathcal{A}$-adopted seed set S^*

1: Initialize: $S^* = \emptyset$
2: **while** $|S^*| < k$ **do**
3: $\quad u = \arg\max_{u \in V \setminus S^*}(f_{\mathcal{R}_\beta}(S^* \cup \{u\}) - f_{\mathcal{R}_\beta}(S^*)) - (f_{\mathcal{R}_\gamma}(S^* \cup \{u\}) - f_{\mathcal{R}_\gamma}(S^*))$;
4: $\quad$ **if** $(f_{\mathcal{R}_\beta}(S^* \cup \{u\}) - f_{\mathcal{R}_\beta}(S^*)) - (f_{\mathcal{R}_\gamma}(S^* \cup \{u\}) - f_{\mathcal{R}_\gamma}(S^*)) < 0$ **then**
5: $\quad\quad$ return S^*;
6: $\quad$ **else**
7: $\quad\quad$ $S^* = S^* \cup \{u\}$;
8: **Return** S^*

4.1 Experimental Settings

Data Set. We use two data sets in this paper. One is derived from a randomly generated graph of a social network with 2708 nodes and 5278 edges, named as Synthetic. The weight of each edge is randomly selected from $\{0.1, 0.01, 0.001\}$. The other is the NetHEPT academic collaboration network. There are 15233 nodes and 32235 edges. Edges indicate cooperation between authors.

Parameter Settings. In order to compare the experimental results, the following assumptions are proposed.

- Set benefit $b = 0.5$ and cost $c = 1$ for a small part of nodes and set $b = 1, c = 0.5$ for others.
- L is a randomly generated binary-valued matrix and the maximum value of t is 30. Extract the first ten columns and first twenty columns as the new matrix L when $t = 10$ and $t = 20$ respectively.
- Emotion tendency vector M is randomly generated and weight parameter a is chosen from $\{0.25, 0.5, 0.75\}$.
- Denote the number of nodes which adopt $\mathcal{B}$ and $\mathcal{A}$ as r and k respectively. The value of r is set as 10, 20 and 30 while k varies from 1 to 20.
- Unless otherwise specified, we set $\varepsilon = 0.1$ and $\sigma = 1$.

Algorithms. We conduct experiments based on four algorithms to compare their performance.

- R-GSMA. This is the algorithm proposed in this paper.
- Copy. It randomly samples a $\mathcal{B}$-adopted seed set $S_\mathcal{B}$ from **B** and selects all the nodes in $S_\mathcal{B}$ as the seed nodes for $\mathcal{A}$
- Degree. This algorithm sorts nodes according to their degree and selects top k as seed nodes to adopt $\mathcal{A}$.
- Random. This is a classical baseline method where the seed nodes for $\mathcal{A}$ is randomly selected.

Algorithm 4. Sampling of r-RR sets

Require: $G = (V, E, B, C)$, influence probability matrix P, parameters k, r, ε and l
Ensure: collections of r-RR sets

1: Initialize: $\mathcal{R}_\beta = \emptyset$, $\mathcal{R}_\gamma = \emptyset$, NPT $= 1$, $\varepsilon_0 = \sqrt{2}\varepsilon$;
2: $\alpha = \sqrt{l \log n + \log 2}$, $\beta = \sqrt{(1 - \frac{1}{e}) \cdot (\log \binom{n}{k} + l \log n + \log 2)}$;
3: $\mu_0 = (2 + \frac{2}{3}\varepsilon_0) \cdot (\log \binom{n}{k} - l \log n + \log \log_2 n) \cdot n \cdot \varepsilon_0^{-2}$;
4: $\bar{\mu} = 2n \cdot ((1 - \frac{1}{e})\alpha + \beta)^2 \cdot \varepsilon^{-2}$;
5: **for** $i = 1$ to $\log_2 n - 1$ **do**
6: $\quad \xi_i = n \cdot 2^{-i}$, $\rho_i = \frac{\mu_0}{\xi_i}$
7: $\quad$ **while** $|\mathcal{R}_\beta| \le \rho_i$ and $|\mathcal{R}_\gamma| \le \rho_i$ **do**
8: $\quad\quad$ Randomly sample a size-r $\mathcal{B}$-adopted seed set $S_\mathcal{B}$ from $\mathbf{B}$ and an outcome g by removing each edge $(u, v) \in E$ with probability $1 - P_{uv}$;
9: $\quad\quad$ Choose $\alpha_\mathcal{B}^{v,g}$ uniformly at random from $[0, 1]$ for every node $v \in V$ for adoption decision;
10: $\quad\quad$ Select $u \in V$ with probability $\frac{b_v}{\sum_{v \in V} b_v}$ and generate a r-RR set and put it into $\mathcal{R}_\beta$;
11: $\quad\quad$ Select $v \in V$ with probability $\frac{c_v}{\sum_{v \in V} c_v}$ and generate a r-RR set and put it into $\mathcal{R}_\gamma$;
12: $\quad S_i$=Algorithm 3;
13: $\quad$ **if** $n \cdot f_{\mathcal{R}_\beta}(S_i) \ge (1 + \varepsilon_0) \cdot \xi_i$ and $n \cdot f_{\mathcal{R}_\gamma}(S_i) \ge (1 + \varepsilon_0) \cdot \xi_i$ **then**
14: $\quad\quad$ NPT $= n \cdot \frac{\min\{f_{\mathcal{R}_\beta}(S_i), f_{\mathcal{R}_\gamma}(S_i)\}}{1+\varepsilon_0}$
15: $\quad\quad$ **break**
16: $\rho = \frac{\bar{\mu}}{NPT}$;
17: **while** $|\mathcal{R}_\beta| \le \rho$ and $|\mathcal{R}_\gamma| \le \rho$ **do**
18: $\quad$ Randomly sample a size-r $\mathcal{B}$-adopted seed set $S_\mathcal{B}$ from $\mathbf{B}$ and an outcome g by removing each edge $(u, v) \in E$ with probability $1 - P_{uv}$;
19: $\quad$ Choose $\alpha_\mathcal{B}^{v,g}$ uniformly at random from $[0, 1]$ for every node $v \in V$ for adoption decision;
20: $\quad$ Select $u \in V$ with probability $\frac{b_v}{\sum_{v \in V} b_v}$ and generate a r-RR set and put it into $\mathcal{R}_\beta$;
21: $\quad$ Select $v \in V$ with probability $\frac{c_v}{\sum_{v \in V} c_v}$ and generate a r-RR set and put it into $\mathcal{R}_\gamma$;
22: **return** $\mathcal{R}_\beta$ and $\mathcal{R}_\gamma$

Algorithm 5. R-GSMA

Require: $G = (V, E, W, M, L, B, C)$, parameters a, k, r, ε and σ
Ensure: seed node set S^* for R-GSMA problem

1: P = Model influence probability $(G = (V, E, W, M, L), a)$
2: $\sigma = \sigma \cdot \log 2 / \log n$
3: $(\mathcal{R}_\beta, \mathcal{R}_\gamma)$ = Sampling of r-RR sets $(G = (V, E, B, C), P, k, r, \varepsilon, \sigma)$
4: S^* = Node selection $(\mathcal{R}_\beta, \mathcal{R}_\gamma, k)$
5: **Return** S^*

4.2 Experimental Results

First of all, we introduce some experimental results of R-GSMA on Synthetic network and analyze the influence of parameters respectively.

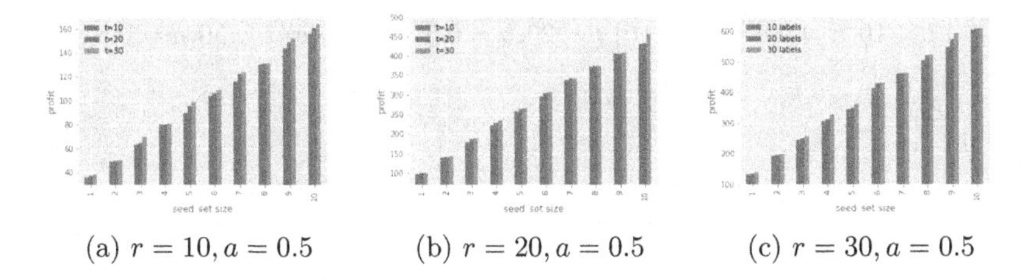

(a) $r = 10, a = 0.5$ (b) $r = 20, a = 0.5$ (c) $r = 30, a = 0.5$

Fig. 2. The influence of label number t on Synthetic

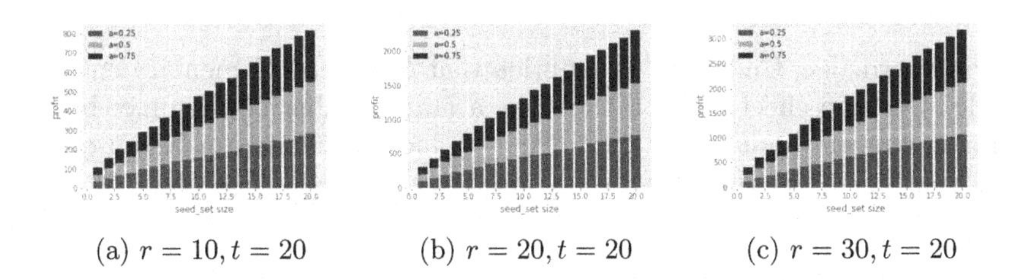

(a) $r = 10, t = 20$ (b) $r = 20, t = 20$ (c) $r = 30, t = 20$

Fig. 3. The influence of weight parameter a on Synthetic

Figure 2 shows how the total profit changes with the number of label under three different conditions in which the weight parameter is set as 0.5 while the number of $\mathcal{B}$-adopted nodes varies. Even though the difference is not so large, it still leads us to a conclusion that the profit always increases with the number of labels, even the size of $\mathcal{B}$-adopted nodes changes. The reason may be that a large number of labels can demonstrate more details and increase the similarity. Based on the property of logistic function, the influence probability between nodes increase with similarity and ultimately more profit are generated.

Similar to Fig. 2, Fig. 3 reflects the influence of weight parameter a on profit generated by all the active nodes when the number of label is set as 20. Due to the definition of influence probability, emotion tendency and social interaction have equal weighting when $a = 0.5$. Given that a can be adjusted as needed, we only take $a = 0.25$, $a = 0.5$ and $a = 0.75$ as examples. Under our settings introduced above, the profit increases with the value of a regardless of the value of r.

After analyzing the influence of two parameters t and a, we focus on another more important parameter r which is the number of $\mathcal{B}$-adopted nodes. In Fig. 4, it is obvious that the profit increases with the value of r. It is consistent with our

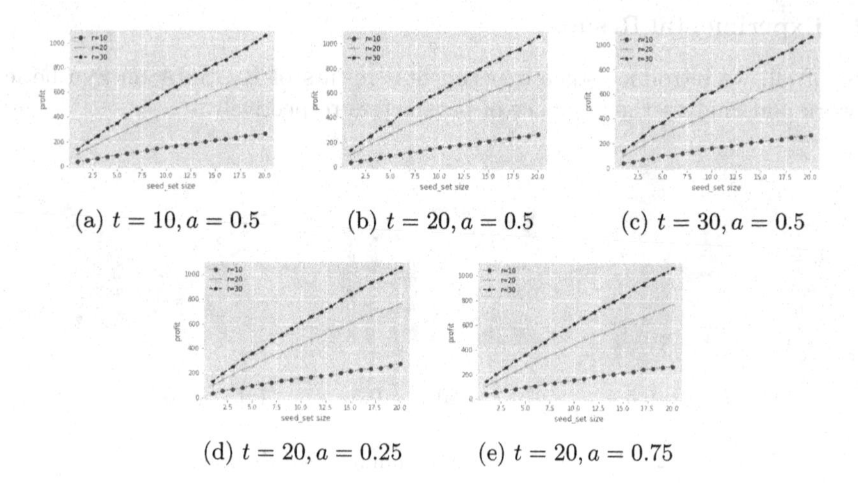

(a) $t = 10, a = 0.5$ (b) $t = 20, a = 0.5$ (c) $t = 30, a = 0.5$

(d) $t = 20, a = 0.25$ (e) $t = 20, a = 0.75$

Fig. 4. The influence of r on Synthetic

common knowledge. Under the assumption that $\mathcal{B}$ is a complementary entity to $\mathcal{A}$ while $\mathcal{A}$ has no effect on $\mathcal{B}$, if there are a large number of consumer buying $\mathcal{B}$, the amount of $\mathcal{A}$-adopter is more likely to increase. This conclusion can be drawn from these experimental results and also suitable for real life.

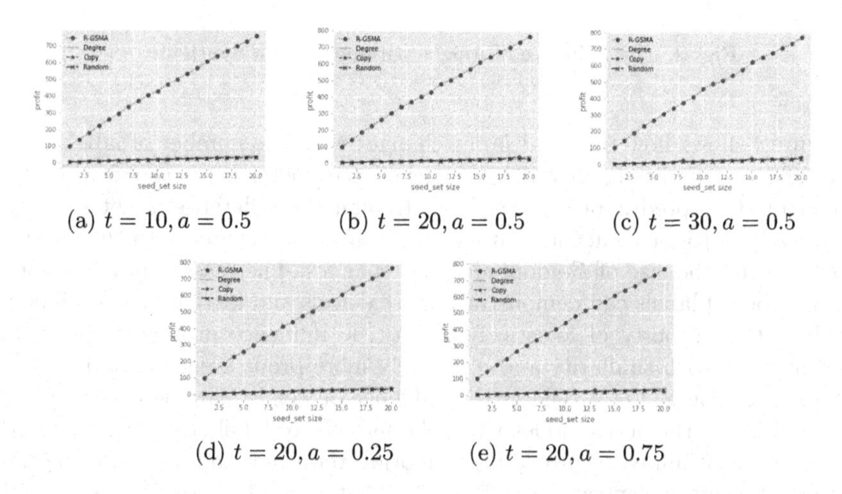

(a) $t = 10, a = 0.5$ (b) $t = 20, a = 0.5$ (c) $t = 30, a = 0.5$

(d) $t = 20, a = 0.25$ (e) $t = 20, a = 0.75$

Fig. 5. Algorithm compare on Synthetic

In the last set of experiments conducted on Synthetic network, we compare R-GSMA with other three algorithms, named as Degree, Copy and Random. Figure 5 illustrates the expected profit generated by all the active nodes on the four algorithms. The performance of algorithms are compared in five different cases with varying the number of seed nodes from 1 to 20. The subtitles show the value of parameters in different cases. Note that r is always set to be 20 due to the design of Copy algorithm. The performance of R-GSMA is much better than other algorithms in all the cases.

Some results of experiments conducted on NetHEPT are shown in Fig. 6. In this part, we set $t = 20, a = 0.5$ and change the data set from a synthetic network to a real network. Figure 6a depicts the influence of r. Although the total profit is smaller than that in the synthetic network mentioned above, we still can form the conclusion that the profit increases with r. And for the comparison between four algorithms, Fig. 6b highlights the superiority of R-GSMA. So, combining conclusions drawn from the experimental results shown in Fig. 5 and Fig. 6b, we can acclaim that R-GSMA proposed in this paper outperforms other algorithms.

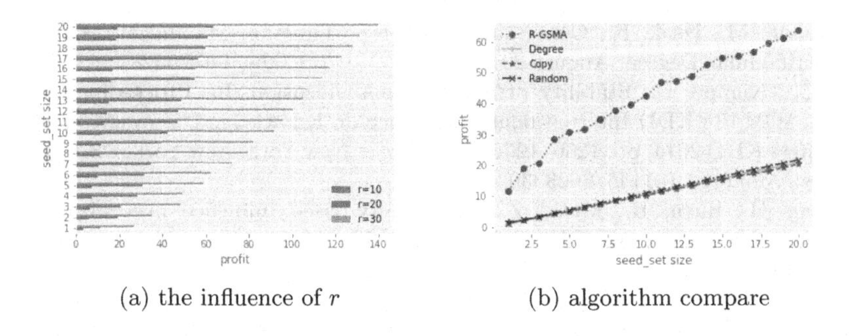

(a) the influence of r (b) algorithm compare

Fig. 6. Experimental results on NetHEPT

5 Conclusion and Future Works

In this paper, we propose Generalized Self-profit Maximization in Attribute networks (GSMA) problem, which involves an unpredicted complementary entity adopter seed set, profit generated by adopters and the influence of label and emotion tendency. The goal of GSMA is to select k seed nodes to adopt an entity such that the total profit generated by all the activated nodes reaches maximum. We redefine the influence probability which considers the attributes and topology structures in attribute network and show that the objective function is neither submodular or monotone in general. To address the GSMA problem, we propose an effective algorithm based on sampling and martingale, i.e. R-GSMA. Finally, we conduct experiments on both artificial and real-world networks to evaluate its effectiveness. The simulation results show the superiority of R-GSMA algorithm.

The number of r-RR sets generated in R-GSMA algorithm is large so the expected running time is quite long. How to improve the performance of R-GSMA in high influence networks deserves further study. What's more, this paper does not consider the mutual complementary or competitive relationships between entities. The related research is also a interesting direction in the future.

References

1. Chen, W., Lin, T., Tan, Z., Zhao, M., Zhou, X.: Robust influence maximization. In: Proceedings of the 22nd ACM SIGKDD International Conference on Knowledge Discovery and Data Mining, pp. 795–804. ACM, San Francisco, California, USA, August 2016. https://doi.org/10.1145/2939672.2939745
2. Chen, W., Wang, C., Wang, Y.: Scalable influence maximization for prevalent viral marketing in large-scale social networks. In: ACM SIGKDD International Conference on Knowledge Discovery and Data Mining, pp. 1029–1038. ACM, New York (2010). https://doi.org/10.1145/1835804.1835934
3. Feige, U., Izsak, R.: Welfare maximization and the supermodular degree. In: Proceedings of the 4th Conference on Innovations in Theoretical Computer Science - ITCS 2013, p. 247. ACM Press, Berkeley, California, USA (2013). https://doi.org/10.1145/2422436.2422466
4. Feldman, M., Izsak, R.: Constrained Monotone Function Maximization and the Supermodular Degree, August 2014. http://arxiv.org/abs/1407.6328
5. He, X., Kempe, D.: Stability of influence maximization. In: Proceedings of the 20th ACM SIGKDD International Conference on Knowledge Discovery and Data Mining - KDD 2014, pp. 1256–1265. ACM Press, New York, New York, USA (2014). https://doi.org/10.1145/2623330.2623746
6. Huang, H., Shen, H., Meng, Z.: Community-based influence maximization in attributed networks. Appl. Intell. **50**(2), 354–364 (2019). https://doi.org/10.1007/s10489-019-01529-x
7. Iyer, R., Jegelka, S., Bilmes, J.: Curvature and Optimal Algorithms for Learning and Minimizing Submodular Functions, November 2013. http://arxiv.org/abs/1311.2110
8. Iyer, R.K., Bilmes, J.A.: Submodular optimization with submodular cover and submodular knapsack constraints. CoRR abs/1311.2106 (2013). http://arxiv.org/abs/1311.2106
9. Iyer, R.K., Bilmes, J.A.: Algorithms for approximate minimization of the difference between submodular functions, with applications. CoRR (2014). http://arxiv.org/abs/1408.2051
10. Kempe, D., Kleinberg, J., Tardos, E.: Maximizing the spread of influence through a social network. Proceedings of the ACM SIGKDD International Conference on Knowledge Discovery and Data Mining, pp. 137–146 (2003). https://doi.org/10.1145/956750.956769
11. Du, L., Chen, S., Gao, S., Yang, W.: Nonsubmodular constrained profit maximization from increment perspective. J. Comb. Optim. **2**, 1–28 (2021). https://doi.org/10.1007/s10878-021-00774-6
12. Lu, W., Chen, W., Lakshmanan, L.V.S.: From competition to complementarity: comparative influence diffusion and maximization. Proc. VLDB Endow. **9**(2), 60–71 (2015). https://doi.org/10.14778/2850578.2850581

13. Narasimhan, M., Bilmes, J.A.: A submodular-supermodular procedure with applications to discriminative structure learning. CoRR (2012). http://arxiv.org/abs/1207.1404
14. Shang, J., Zhou, S., Li, X., Liu, L., Wu, H.: CoFIM: a community-based framework for influence maximization on large-scale networks. Knowl. Based Syst. **117**, 88–100 (2017). https://doi.org/10.1016/j.knosys.2016.09.029
15. Song, J., Liu, Y., Guo, L., Xuan, P.: Research on social network propagation model and influence maximization algorithm based on emotion. J. Comb. Optim. **055**(013), 85–92 (2019)
16. Tang, J., Tang, X., Yuan, J.: Towards Profit Maximization for Online Social Network Providers, December 2017. http://arxiv.org/abs/1712.08963
17. Tang, J., Tang, X., Yuan, J.: Profit maximization for viral marketing in online social networks: algorithms and analysis. IEEE Trans. Knowl. Data Eng. **30**(6), 1095–1108 (2018). https://doi.org/10.1109/TKDE.2017.2787757
18. Tang, Y., Shi, Y., Xiao, X.: Influence maximization in near-linear time: a martingale approach. In: Proceedings of the 2015 ACM SIGMOD International Conference on Management of Data, pp. 1539–1554. SIGMOD 2015. Association for Computing Machinery, New York, NY, USA (2015). https://doi.org/10.1145/2723372.2723734
19. Wang, Z., Yang, Y., Pei, J., Chu, L., Chen, E.: Activity maximization by effective information diffusion in social networks. IEEE Trans. Knowl. Data Eng. **29**(11), 2374–2387 (2017). https://doi.org/10.1109/TKDE.2017.2740284
20. Wu, W.-L., Zhang, Z., Du, D.-Z.: Set function optimization. J. Oper. Res. Soc. China **7**(2), 183–193 (2018). https://doi.org/10.1007/s40305-018-0233-3

Parameterized Complexity Classes Defined by Threshold Circuits: Using Sorting Networks to Show Collapses with W-hierarchy Classes

Raffael M. Paranhos[1], Janio Carlos Nascimento Silva[1,2(✉)], Uéverton S. Souza[1], and Luiz Satoru Ochi[1]

[1] Instituto de Computação, Universidade Federal Fluminense, Niterói, RJ, Brazil
raffaelmp@id.uff.br, {ueverton,satoru}@ic.uff.br
[2] Instituto Federal do Tocantins, Campus Porto Nacional, Porto Nacional, TO, Brazil
janio.carlos@ifto.edu.br

Abstract. The main complexity classes of the Parameterized Intractability Theory are based on weighted Boolean circuit satisfiability problems and organized into a hierarchy so-called W-hierarchy. The W-hierarchy enables fine-grained complexity analyses of parameterized problems that are unlikely to belong to the FPT class. In this paper, we introduce the *Th-hierarchy*, a natural generalization of the W-hierarchy defined by weighted *threshold* circuit satisfiability problems. Investigating the relationship between Th-hierarchy and W-hierarchy, we discuss the complexity of transforming Threshold circuits into Boolean circuits, and observe that sorting networks are powerful tools to handle such transformations. First, we show that these hierarchies collapse at the last level (W[P] = Th[P]). After that, we present a time complexity analysis of an AKS sorting network construction, which supports some of our results. Finally, we prove that $\mathrm{Th}[t] \subseteq \mathrm{W[SAT]}$ for every $t \in \mathbb{N}$.

Keywords: Threshold circuits · AKS sorting network · Parameterized complexity · W-hierarchy

1 Introduction

A decidable problem is a *parameterized problem* when coupled to its instance, some additional information representing particular aspects of the input (that constitute the parameter) are also given. Typically, the parameters measure structural properties of the input or the solution size to be found. In general, the parameters are aspects of the input and/or question to be answered that are isolated for further specialized analysis, where it is assumed that the size of these parameters is much smaller than the size of the input.

Considering an instance I of a problem Π parameterized by k, we say that Π is *fixed-parameter tractable* (it belongs to the class FPT) if Π can be solved

Supported by CAPES, CNPq and FAPERJ.

D.-Z. Du et al. (Eds.): COCOA 2021, LNCS 13135, pp. 348–363, 2021.
https://doi.org/10.1007/978-3-030-92681-6_28

by an algorithm (called FPT algorithm) in $f(k) \cdot poly(|I|)$ time. Alternatively, a parameterized problem Π is *slice-wise polynomial* ($\Pi \in$ XP) if there is an algorithm that solves any instance I of Π in $f(k) \cdot |I|^{g(k)}$ time.

The main goal in a parameterized complexity analysis is to design FPT algorithms for the target problem. However, some problems have a higher level of intractability that brings to us the concepts related to the *W-hierarchy.*

A decision Boolean circuit is a Boolean circuit consisting of small and large gates[1] with a single output line, and no restriction on the fan-out of gates. For such a circuit, the depth is the maximum number of gates on any path from the input variables to the output line, and the weft is the maximum number of large gates on any path from the input variables to the output line.

The W-hierarchy was originally motivated by considering the "circuit representations" of parameterized problems, and terms the union of parameterized complexity classes defined by the weft of decision Boolean circuits.

Before defining the W-hierarchy classes, we consider the following definitions.

Definition 1 (fixed-parameter reduction). *Let $A, B \subseteq \Sigma^* \times \mathbb{N}$ be two parameterized problems. A fixed-parameter (or parameterized) reduction from A to B is an algorithm that, given an instance (x, k) of A, outputs an instance (x', k') of B such that*

- *(x, k) is a yes-instance of A if and only if (x', k') is a yes-instance of B,*
- *$k' \leq g(k)$ for some computable function g, and*
- *the running time is $f(k) \cdot |x|^{O(1)}$ for some computable function f.*

WEIGHTED WEFT t DEPTH h CIRCUIT SATISFIABILITY – WCS(t, h)
Instance: A Boolean decision circuit C with weft t and depth h.
Parameter: A positive integer k.
Question: Does C have a weight k satisfying assignment?

Definition 2. *A parameterized problem Π belongs to the class W[t] if and only if Π is fixed-parameter reducible to WCS(t, h) for some constant h.*

For instance, k-INDEPENDENT SET (parameterized by k) is FPT-reducible to WCS(1, 2), thus it belongs to W[1]. On the other hand, k-DOMINATING SET (parameterized by k) is FPT-reducible to WCS(2, 2), which implies that it is in W[2]. In addition, it is conjectured that k-DOMINATING SET cannot be fixed-parameter reducible to WCS$(1, h)$ for some h, since it is complete for the class W[2]. Therefore, it is assumed that k-DOMINANTING SET has higher parameterized complexity than k-INDEPENDENT SET, since it seems to admit only more complex circuit representations (i.e., circuit representations of bounded depth with greater weft).

Based on this, several parameterized problems are classified according to their parameterized complexity level. Recall that FPT $\subseteq$ W[1] $\subseteq$ W[2] $\subseteq \ldots \subseteq$ XP, and it is conjectured that each of the containment is proper [5].

[1] A gate is called large if its fan-in exceeds some bound, which is typically considered to be two.

The W[t] classes are defined by satisfiability problems of circuits with bounded depth. Additionally, it is also considered parameterized complexity classes defined by circuits having no bound on the depth, so-called W[P] and W[SAT]. These classes are generated by the following problems.

WEIGHTED CIRCUIT SATISFIABILITY – WCS
Instance: A decision Boolean Circuit C.
Parameter: A positive integer k.
Question: Does C has a satisfying assignment of weight k?

WEIGHTED SATISFIABILITY – WSAT
Instance: A decision circuit C corresponding to a Boolean formula (or alternatively, just a Boolean formula C).
Parameter: A positive integer k.
Question: Does C has a satisfying assignment of weight k?

Definition 3. *The class W[P] is the class of parameterized problems that are fixed-parameter reducible to* WEIGHTED CIRCUIT SATISFIABILITY.

Definition 4. *The class W[SAT] is the class of parameterized problems that are fixed-parameter reducible to* WEIGHTED SATISFIABILITY.

Although W[P] and W[SAT] are both defined by unbounded depth circuits, it is worth mentioning that circuits corresponding to Boolean formulas are tree-like circuits, i.e., circuits whose graph induced by its gates is isomorphic to a tree. Besides, general decision Boolean circuits can be transformed into treelike circuits; however, the time complexity for such a transformation typically takes exponential time on their number of gates and depth. Thus, it is conjectured that W[SAT] ⊂ W[P], i.e., the containment is proper.

Thus, the W-hierarchy is organized as follow:

$$W[1] \subseteq W[2] \subseteq \ldots \subseteq W[SAT] \subseteq W[P].$$

Besides, the W-hierarchy classes are defined by circuits restricted to conventional Boolean operators (`AND`, `OR` and `NOT`). By allowing another kind of circuits, potentially, it is possible to represent more decision parameterized intractable problems. In Circuit Complexity, the characterization of complexity classes concerning threshold circuits is well-known. In 1987, Hajnal et al. [7] defined TC^0, the class of all languages which are decided by threshold circuits with constant depth and polynomial size. Analogously, we can use similar reasoning to establish a hierarchy of parameterized complexity classes generated by weighted satisfiability problems on non-conventional circuits. In [6], the authors constructed a hierarchy of classes called W($\mathcal{C}$) as an alternative to W-hierarchy classes. However, in [6], the main discussion is restricted to bounded connectives gates (including threshold gates with bounded threshold). In addition, they also presented some results regarding majority gates.

In this paper, we focus on general threshold gates and define the *Th-hierarchy* as an analogue of the W-hierarchy by replacing decision Boolean circuits by decision *threshold* circuits. The primary tool used in this paper is *sorting networks*, which are used to transform threshold gates in Boolean circuits efficiently.

In Sect. 2, we present some preliminaries about threshold circuits and sorting networks. Also, in Sect. 2, the Th-hierarchy is formally defined. In Sect. 3, we made the first comparisons between W- and Th-hierarchies, especially about W[P] and Th[P]; here, we use sorting networks to support our conclusions. In Sect. 4, we face a challenge in relating the Th-hierarchy classes with the W[SAT] class. As the W[SAT] class deals with treelike circuits, trivial conversions using sorting networks are not helpful because converting a circuit $C(V, E)$ with depth h into a equivalent treelike circuit takes exponential time with respect to h. Finally, by analyzing the time complexity to construct a particular sorting network with $O(\log n)$ depth, called AKS, we show that $\text{Th}[t] \subseteq \text{W[SAT]}$ for every $t \in \mathbb{N}$.

2 Satisfiability of Threshold Circuits

The characterization of decision problems as $\text{WCS}(t, h)$ in standard Boolean circuits has widespread attention, especially considering the enormous advance in Parameterized Complexity Theory. When classifying a problem in the W-hierarchy, in short, we are encapsulating the parameterized intractability of this problem in terms of satisfiability of a corresponding circuit based on Boolean functions (`AND`, `OR` and `NOT`). Thus, some questions emerge. Is W-hierarchy comprehensive enough? Are there problems complete considering a more general basis of functions?

Due to these questions, naturally, our curiosity turns into the *threshold circuits.* Threshold circuits are circuits that admit threshold gates, i.e., gates that emulate threshold functions (See Definition 5). In Sect. 2.1, we provide some notations about threshold circuits.

2.1 Preliminaries

First, we present some conventions and preliminaries that are important for the sequence of this paper.

Definition 5 (Threshold function). *Given a set $A = \{a_1, a_2, \ldots, a_n\}$ of inputs (with $a_i \in \{0, 1\}$, for any $1 \leq i \leq n$), a set $W = \{w_1, w_2, \ldots, w_n\}$ of weights (with $w_i \in \mathbb{Z}$, for any $i \leq n$) and an integer value t called threshold, then a threshold function $T_t^n(A, W)$ holds as follows:*

$$T_t^n(A, W) = \texttt{true} \iff \sum_i^n (a_i \times w_i) \geq t, \text{ \textit{otherwise} } T_t^n(A, W) = \texttt{false}.$$

We can specialize the Definition 5 for functions where every $w_i \in W$ is equal to 1, such functions are called unweighted threshold functions. In practice, an unweighted function evaluates `true` when exactly t inputs $a_i \in A$ are set to be 1. A particular unweighted threshold function is the *majority* function, which has threshold t equals to $n/2$.

Definition 6 (Decision threshold circuits). *A decision threshold circuit is a decision circuit containing* `AND` *gates,* `OR` *gates,* `NOT` *gates, and* **`unweighted threshold`** *gates, where every* **`unweighted threshold`** *gate computes an unweighted threshold function.*

Note that one can consider circuits having **`weighted threshold`** gates; however, in this paper we are dealing only with **`unweighted threshold`** gates.

2.2 The Th-hierarchy

For convenience, we present a generalization of WCS(t, h) by considering decision threshold circuits.

WEIGHTED WEFT t DEPTH h THRESHOLD CIRCUIT SATISFIABILITY – WTCS(t, h) **Instance**: A decision threshold circuit C with weft t and depth h. **Parameter**: A positive integer k. **Question**: Does C has a satisfying assignment of weight k?

Similarly, we present the complexity classes Th[t].

Definition 7. *A parameterized problem Π belongs to the class Th[t] if and only if Π is fixed-parameter reducible to WTCS(t, h) for some constant h.*

Analogously, we define WEIGHTED THRESHOLD CIRCUIT SATISFIABILITY (WTCS) and Th[P] as a generalization of WCS and W[P] by considering decision threshold circuits instead of decision Boolean circuits. In addition, we define the generalization of WSAT as follows.

WEIGHTED TREELIKE THRESHOLD CIRCUIT SATISFIABILITY – WTTSAT **Instance**: A decision threshold circuit C whose graph induced by its gates is isomorphic to a tree (treelike circuit). **Parameter**: A positive integer k. **Question**: Does C has a satisfying assignment of weight k?

Hence, the *Th-hierarchy* is as follow:

$$Th[1] \subseteq Th[2] \subseteq \cdots \subseteq Th[SAT] \subseteq Th[P].$$

By definition, it holds that W[t] $\subseteq$ Th[t] (for every $t \in \mathbb{N}$) as well as W[SAT] $\subseteq$ Th[SAT] and W[P] $\subseteq$ Th[P].

At this point, some questions emerge such as "W[t] = Th[t], for each t?", "W[P] = Th[P]?", "W[SAT] = Th[SAT]?", and "W[1] = Th[1]?". In Sects. 3 and 4, we explore some of these issues.

To understand the relationship between these classes (at the highest levels), we revisit the Sorting Network field and present a time complexity analysis for the construction of AKS sorting networks [11].

3 W-hierarchy Versus Th-hierarchy

Although the W-hierarchy is a set of infinite classes of parameterized problems, it may possible that the W-hierarchy is not complete in the sense that may exist a parameterized problem Π such that $\Pi \in \mathrm{W}[t+1]$; $\Pi \notin \mathrm{W}[t]$; Π is hard for $\mathrm{W}[t]$; but, Π is not complete for $\mathrm{W}[t+1]$, for some t. Then, it seems possible that there are classes of problems between $\mathrm{W}[t]$ and $\mathrm{W}[t+1]$, or between W[SAT] and W[P].

Therefore, one of the motivation of this work is consider classes based on threshold circuits to identify potential gaps in the W-hierarchy.

By definition, the following proposition is clear.

Lemma 1. *$W[t] \subseteq Th[t]$, for every $t \in \mathbb{N}$.*

In contrast, it is not clear if $Th[t] \subseteq W[t]$ for some t. Figure 1 depicts the current snapshot of these classes.

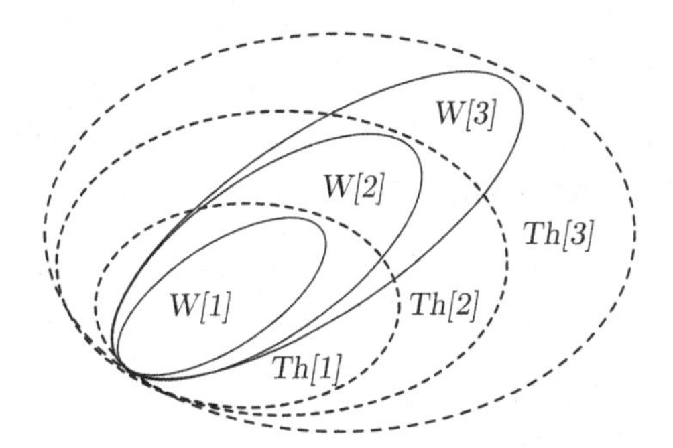

Fig. 1. Relationship between $\mathrm{W}[t]$ and $\mathrm{Th}[t]$ classes.

We begin our discussion by disregarding structural constraints on circuits, which leads us to the *W[P] versus Th[P]* dilemma.

To show that Th[P] $\subseteq$ W[P] it is enough to present a fixed-parameter reduction from WEIGHTED THRESHOLD CIRCUIT SATISFIABILITY to WEIGHTED CIRCUIT SATISFIABILITY. In this case, it suffices to provide a way to locally replace each unweighted threshold gate for an equivalent Boolean circuit. For that, we consider the *Sorting Networks.*

Sorting Networks

A *sorting network* is a comparison network circuit with n inputs and n outputs, where the outputs are monotonically ordered (using AND and OR gates only). Such circuits are represented by directed graphs having n inputs (bits to be ordered) and n outputs (ordered bits).

The first implementation of a sorting network was proposed by Daniel G. O'Connor and Raymond J. Nelson in 1954, patented three years later [10]. In 1968, Kenneth E. Batcher [4] presented some fundamental concepts about Sorting Networks. One of these concepts is the *comparison element* idea (See Fig. 2a), which consists in a node of the network that receives two inputs A and B and it returns the outputs L and H, such that $L = min(A, B)$ and $H = max(A, B)$. For Boolean values, a comparison elements can be constructed with two Boolean gates as shown in Fig. 2b.

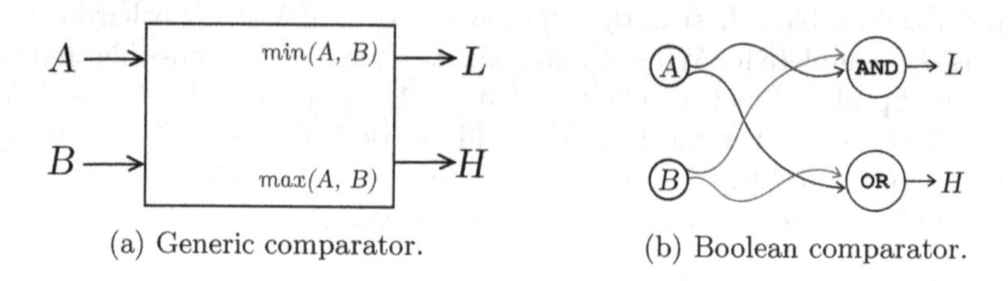

(a) Generic comparator. (b) Boolean comparator.

Fig. 2. Comparison elements.

Therefore, we can define sorting networks as circuits with n inputs that flow by multiple comparison elements, resulting in ordering those n values in a deterministic sequence of steps.

It is possible to construct a sorting network that simulates famous sort algorithms, e.g., Bubble Sort. In [4], for instance, was described the *Bitonic Sorter* inspired by the Merge Sort algorithm.

Depending on the strategy adopted to organize a sorting network, we can have circuits with different depths. While a sorting network based on bubble sort has $O(n)$ depth, a bitonic sorter has $O(\log^2 n)$ depth. Furthermore, the most popular sorting networks can be constructed in polynomial time. As example, Bitonic Mergesort can be constructed in $O(n \times \log^2 n)$ time [4]. Thus, Lemma 2 is supported by a vast literature.

Lemma 2. *Given an unweighted threshold gate T with n inputs and threshold t, in polynomial time with respect to n, one can construct an decision Boolean circuit C_T such that C_T computes the same function of T.*

Proof. Let I be the set of inputs of T. It is enough to construct in polynomial time a sorting network for the set I of inputs (sorting from highest to lowest value) and then connect the network outputs in such a way that the output gate returns `true` if and only if t-th output of the sorting network is `true`. □

Theorem 1. *It holds that $Th[P] = W[P]$.*

Proof. By definition, it is clear that W[P] ⊆ Th[P]. To show that Th[P] ⊆ W[P], it is enough to present a fixed-parameter reduction from WEIGHTED THRESHOLD CIRCUIT SATISFIABILITY to WEIGHTED CIRCUIT SATISFIABILITY by locally replacing each unweighted threshold gate for an equivalent Boolean decision circuit. Therefore, by Lemma 1 the claim holds. □

4 On the Classes Th[t], Th[SAT], and W[SAT]

Due to the depth constraints, Lemma 2 does not implies that $\mathrm{W}[t] = \mathrm{Th}[t]$, for any $t \in \mathbb{N}$. In addition, correlated with the structural issues of each class, we also have to worry about the algorithmic time needed for converting a threshold circuit into a Boolean one, respecting such restrictions.

Also, the discussion between W[SAT] and Th[SAT] seems challenging. The WSAT problem does not have restrictions on weft or depth, but it considers only treelike circuits. This constraint brings us an alert about the depth when using sorting networks to convert threshold gates into a Boolean circuit, because after this conversion, it is still necessary to convert the resulting sub-circuit into a treelike one.

Since treelike decision circuits are exactly the decision circuits where each gate has fan-out equal to one (each gate has a single parent), by duplicating gates, it is easy to see that from a decision circuit G one can obtain an equivalent treelike decision circuit in $|V(G)|^{O(h)}$ time, where h is the depth of G and $V(G)$ is the set containing gates and input variables of G. However, when each gate has fan-out bounded by a constant c, by duplicating the gates in top-down manner according to the gates' depth (the length of the longest path from each gate to the output line), one can duplicate each gate g_i at most c^{h_i} times, where h_i is the length of the longest path from g_i to the output line. Therefore, in this setting one can construct an equivalent treelike decision circuit in $c^h \cdot |V(G)|^{O(1)}$ time.

Note that sorting networks tend to have only bounded fan-out gates, since its construction is typically based on Boolean comparators. This motivates us to revisit an in-depth discussion in Theory of Computation about the existence and construction of sorting networks with $O(\log n)$ depth.

However, even if we have an algorithm that converts a threshold gate in Boolean circuits with $O(\log n)$ depth, we still have trouble with the cascade effect of the local gate-replacement/tree-conversion process, i.e., by replacing a threshold gate by an equivalent treelike decision circuit, one can increase the fan-out of gates that were seen as inputs by this threshold gate, which implies duplications in the level inter such sorting networks. Note that such a process may take $|V(G)|^{O(h)}$ time, since after the replacement of threshold gates the fan-out of some gates may be unbounded. Since instances of WSAT may have unbounded depth, we left open the question "W[SAT] = Th[SAT]?". However, our sorting network framework is able to show that $\mathrm{Th}[t] \subseteq \mathrm{W}[\mathrm{SAT}]$ (for every $t \in \mathbb{N}$), if one can construct in polynomial time sorting networks with depth $O(\log n)$ and bounded maximum fan-out.

Note that in Lemma 2 we do not ensure the existence of sorting networks with logarithmic depth. Besides, the only known sorting network that satisfies such a condition is the *AKS Sorting Networks*. However, to the best of our knowledge, there are no time complexity analysis of an explicit construction of an AKS sorting network. In Sect. 4.1, we detail this particular type of sorting network and we address the possibility of using AKS sorting networks to prove that $\mathrm{Th}[t] \subseteq \mathrm{W}[\mathrm{SAT}]$, for every $t \in \mathbb{N}$.

4.1 AKS Sorting Networks

Proposed by M. Ajtai, J. Komlos, and E. Szemerédi [1], AKS sorting networks are originally based on a probabilistic error-recovery structure called *separator*. A separator is a network composed of structures called *ϵ-halvers*. A separator, in short, partitions its inputs into four parts semi-ordered where each partition has a tolerance for *strangers*. Hence, constructing an AKS sorting network is an arrangement of separators in an efficient manner that the values flow in it with a low depth. The core of AKS sorting networks is a kind of graphs called *expander graphs*, which are graphs endowed with good connectivity properties. We refer to the following definitions to better describe AKS sorting networks.

In this work, for the construction of an AKS network we consider a particular type of expander: a Bipartite balanced k-regular graph based on a Ramanujan graphs. Thus, for convenience, we define only one specific configuration of an expander in Definition 8.

Let G be a bipartite graph with a bipartition of $V(G)$ into V_1 and V_2. We denote by $\Gamma(W)$ the neighborhood of a subset $W \subset V(G)$.

Definition 8 (Bipartite (k, ϵ)-expander graphs). *A bipartite graph $G = ([V_1, V_2], E)$ with n vertices is (k, ϵ)-expander, if and only if*

- *the sets V_1 and V_2 contain exactly $n/2$ each one;*
- *every vertex has degree k, and*
- *for every subset $W \neq \emptyset$ such that either $W \subseteq V_1$, or $W \subseteq V_2$ it holds that*

$$|\Gamma(W)| \times \epsilon \geq \min(\epsilon/2, |W|) \times (1 - \epsilon)$$

where $0 < \epsilon < 0.5$.

Figure 3 illustrates a bipartite $(3, 1/4)$-expander graph.

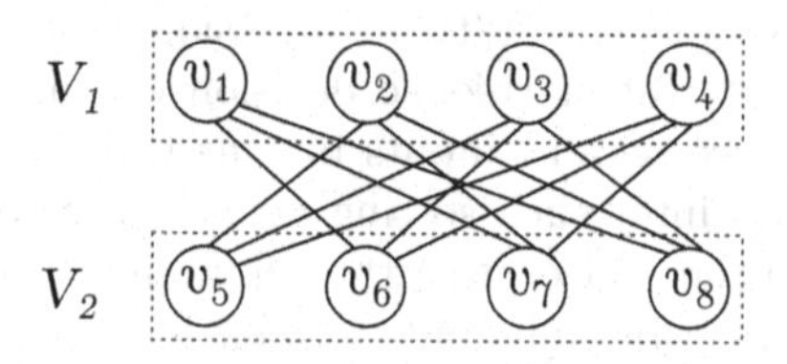

Fig. 3. A bipartite $(3, 1/4)$-expander graph G.

One of the AKS networks "tricks" is the tolerance with elements out of position (*strangers*) after a comparison stage. Therefore, AKS sorting networks uses ϵ-halvers instead of perfect halvers (See Definition 9).

Definition 9 (Halvers). *A circuit C with n inputs (where n is even) is a* perfect halver *if C return the input values in two output sets: One with exactly $n/2$ larger inputs; and other with $n/2$ smaller inputs. We say that a circuit is an* ϵ-halver *when these two output sets has at most $\epsilon \times n$ strangers (inputs which were directed to wrong output set).*

Due to the expansion properties of a (k, ϵ)-expander, it is possible to extract several perfect matchings. Each perfect matching divides the n inputs into two sets with the same size. Thus, it is possible to perform $n/2$ swaps for each perfect matching and the pairs in matchings are natural comparators (see Fig. 4). From (k, ϵ)-expanders we design ϵ-halvers with k swap stages (see Fig. 5).

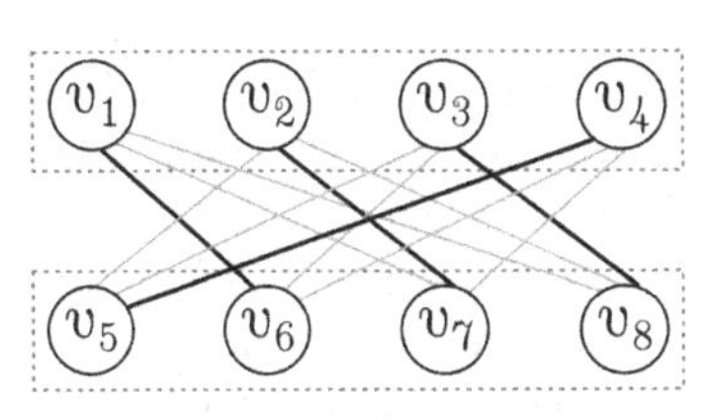

(a) A perfect matching M in G.

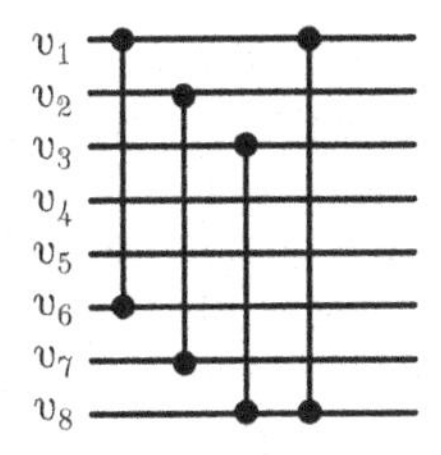

(b) Swap based in M.

Fig. 4. A perfect matching in a (k, ϵ)-expander determining swap stages. Each vertical wire in (b) represents a comparator between its endpoints.

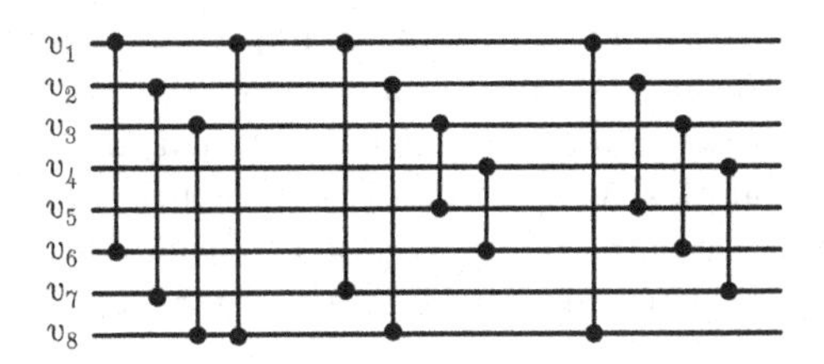

Fig. 5. A (1/4)-halver constructed from G with three swap stages.

Considering that an ϵ-halver H outputs the sets V_1 and V_2 such that V_1 has the smallest values and V_2 has the largest values, with an ϵ-tolerance to *strangers*. Using these semi-ordered values outputted by H to construct another ϵ-halver H', we have a new round of swaps. Here, it is expected that H' outputs the sets V_1' and V_2' with less *strangers*. It is easy to see that a well-structured web of ϵ-halvers, without doubt, can order n inputs in few stages. To this web, we call the notion of *separator*.

Definition 10 (Separator). *A circuit C with n inputs is a $(\lambda, \sigma, \epsilon)$-separator if C returns four output sets G_1, G_2, G_3 and $G4$ (a partition of the inputs) of sizes $\lambda \times (n/2), (1-\lambda) \times (n/2), (1-\lambda) \times (n/2)$ and $\lambda \times (n/2)$, respectively. In addition, G_1 and G_4 have at most $\sigma \times \lambda \times (n/2)$ strangers; and G_2 and G_3 have at most $\epsilon \times (n/2)$ strangers.*

Applying the same reasoning of the ϵ-halvers to separators, it is easy to see that flowing values through a network of separators, in few stages, the partitioning of G_1, G_2, G_3 and G_4 will dissipate the presence of its *strangers*, and consequently, this order the input values. That is a summary of the idea behind the AKS sorting network: a chain of separators, which is, in its turn, a chain of halvers. Note that we are not providing all details about AKS sorting networks. A survey about this topic can be found in [3].

For a while, the existence of a $O(\log n)$-depth sorting network was an open question. The first version of an AKS sorting network [1] solved this question. However, behind the asymptotic expression $O(\log n)$ there is a constant factor of approximately 2^{100}. This huge constant factor inhibits the practical implementation of AKS sorting networks, and there are just a few explicit algorithms describing the construction of an AKS network. In 1990, Paterson [11] presented an improved construction of AKS networks with a depth substantially smaller, but still impracticable. After all, several works address some slight improvements in the AKS construction and Paterson algorithm. However, none of those works achieve a substantial decrease in the constant factor of AKS networks' depth.

In [13], Hang Xie presented an explicit description of the Paterson's strategy. However, although Hang Xie presented a constructive proof to obtain AKS networks, to the best of our knowledge, there is no work presenting a time complexity analysis of the construction of this sorting networks. The main reason for the absence of this analysis is due to the "galactic" size of these networks, which makes their practical implementation impossible. However, in this work our interest in AKS networks is different, since our focus is on complexity classes.

Nevertheless, from the explicit algorithm detailed in [13], we have all tools to address the time complexity of such a construction. As we remark in Sect. 4.2 that the construction presented in [13] can be performed in polynomial time, we can conclude the proper containment of $\text{Th}[t] \subseteq \text{W[SAT]}$, for every $t \in \mathbb{N}$.

4.2 Th[t] $\subseteq$ W[SAT]

Now, we present a description of the construction of an AKS network detailed in [13]. We focused only on algorithmic time to perform each step of the construction. For more details on the construction see [13].

Given a value ϵ and n (number of keys to be ordered), we can construct a ϵ-halver according to Algorithm 1.

Algorithm 1 **(ϵ–Halver construction** [13])

1. Let
$$K = \frac{2(1-\epsilon)(1-\epsilon+\sqrt{1-2\epsilon})}{\epsilon^2}$$
2. Pick the minimum prime p congruent to 1 mod 4($\geq K-1$). Let $k = p+1$.
3. Find the minimum prime q congruent to 1 mod 4, such that $q \geq n/2$.
4. Construct a k-regular Ramanujan graph with q vertices.
5. Construct a balanced bipartite k-regular graph G_B with $2q$ vertices.
6. Find a perfect matching of the graph using Hungarian algorithm and transform each pair in a sequence of comparators; remove this matching from the graph to get a $(k-1)$-regular graph;

7. Repeat 6 until G_B has no perfect matchings.

In this construction, ϵ-halvers are organized as balanced bipartite k-regular graphs generated from k-regular Ramanujan graphs of q vertices, where $q \geq n/2$. Additionally, the method used to construct Ramanujan graphs was based on [8,9] (also known as LPS graphs). Thus, the first three steps in Algorithm 1 were dedicated to finding k and q based on the desirable K.

In order to find an appropriated value of K, Hang Xie was based on [12] to calculate the depth of an associated ϵ-halver providing the Lemma 3.

Lemma 3. *(From [13]) Let*

$$K = \frac{2(1-\epsilon)(1-\epsilon+\sqrt{1-2\epsilon})}{\epsilon^2}$$

For every k-regular LPS graph, if $k \geq K$ then every subset X of size ϵn has at least $(1-\epsilon)n$ neighbors.

The choice of ϵ impacts the depth and size of this ϵ-halver. For example, if we decide to construct a $(1/72)$-halver, we have $K \approx 20162.99$. By steps 2 and 3, we can find $p = 20173$, which indicates the creation of a 20174-regular Ramanujan graph in step 4 [13].

For a smaller (or larger) Ramanujan graph, we need to exploit the equation in Lemma 3 to discover another ϵ. This choice can be guided by the size of the input to be ordered. If we already know the desirable ϵ, then the first step costs $O(1)$. Also, p and q are values used to construct Ramanujan graphs called LPS graphs (more details in [8,13]). Be prime and congruent to 1 mod 4 are specific conditions for this construction. By number theory, we know that for a value $m \geq 13$ then there is a prime number x congruent to 1 mod 4 such that $m \leq x \leq 2m$. Hence, since p need to be larger than K then in steps 2 find p by iterating from K to $2K$ until find the first number that fit the conditions. Hence, steps 2 takes polynomial time on K, since for each candidate value we have to check if it is prime. In addition, K depends only on ϵ. For q we need to find a prime number congruent 1 mod 4 and greater than $n/2$. By iterating q from $n/2$ to n, and verifying for each q if it is prime, we perform in polynomial time on n.

For *Step 4*, it is well-known that a graph with a small second eigenvalue of its adjacency matrix is a good expander [13]. By Alon-Boppana theorem [2], we know that $2\sqrt{k-1}$ is the lower bound for the second eigenvalue of a k-regular graph adjacency matrix. By Definition 11, we can observe that the Ramanujan graph is a family of k-regular graphs with the best possible second eigenvalue of its adjacency matrix, which guarantees Ramanujan graphs as good expanders.

Definition 11. *A Ramanujan graph is a connected k-regular graph whose eigenvalues are at most $2\sqrt{k-1}$ in absolute value.*

The explicit construction of a k-regular Ramanujan graph was presented in [8,9] and takes polynomial time on k.

For *Step 5*, in order to create a balanced bipartite k-regular graph G_b from the Ramanujan graph G created in step 4, we present the Algorithm 2, which takes $O(|E(G)|)$ time.

Algorithm 2

1. Given a $G(V, E)$ with q vertices $V = \{v_1, v_2, \ldots, v_q\}$, create a new graph $G_b([U, W], E_b)$ with $2q$ vertices divided in $U = \{u_1, u_2, \ldots, u_q\}$ and $W = \{w_1, w_2, \ldots, w_q\}$;
2. For each pair of positive integers (i, j), with $i \leq n$ and $j \leq n$, create an edge from u_i to w_j in E_b iff $(v_i, v_j) \in E$.

Finally, for *Step 6 and 7*, we perform k executions of the well-known Hungarian algorithm, which also can be done in polynomial time.

After analyzing each step of Algorithm 1, we conclude that this explicit construction of an ϵ-halver can be done in polynomial-time.

Now, it remains to verify the time complexity in arranging a separator.

Here, we present a simplified construction of a $(\lambda, \sigma, \epsilon)$-separator. In [11,13], there is more sophisticated constructions. But, our purpose is only to show it is possible to perform it with a polynomial construction.

Algorithm 3 *($(\lambda, \sigma, \epsilon)$-separator construction)* [3].

1. Given an input with m keys to be ordered, create an array S with m positions in the bottom level d.
2. Construct a ϵ-halver in the first level (level 0) of the separator. Apply the m keys to this ϵ-halver and send the output sets L_0 and R_0 with $m/2$ (each) to level 1.
3. For each level $0 < i < d$, construct two ϵ-halver h_i^L and h_i^R and then:
   - Apply the $m/2^i$ keys in L_{i-1} to h_i^L and send the output left half to level $i+1$ as L_i.
   - Send the output right half of h_i^L to the positions $S[(m/2^{i+1}) + 1]$ to $S[m/2^i]$ in bottom level.
   - Apply the $m/2^i$ keys in R_{i-1} to h_i^R and send the right half of the output to level $i+1$ as R_i.
   - Send the output left half of h_i^R to the positions $S[(m/2^i)+1]$ to $S[(m/2^i)+(m/2^{i+1})]$ in bottom level.

Note that by this construction $\lambda = 2^{(1-d)}$ and $\sigma = d \times \epsilon$ (the presence of $\epsilon \times m$ *strangers* dissipates across the levels resulting in σn). For a depth $d = 4$ (See Fig. 6), we can construct a $(1/8, 1/18, 1/72)$-separator using $(1/72)$-halvers.

The sorting network has $O(\log n)$ layers with a complete binary tree on each layer with $O(n)$ nodes. Each node has a $(\lambda, \sigma, \epsilon)$-separator, and the layers are interconnected so that the complete circuit has depth $O(\log n)$ (See a detailed construction in [3], Sect. 11.4). Then, Proposition 1 holds.

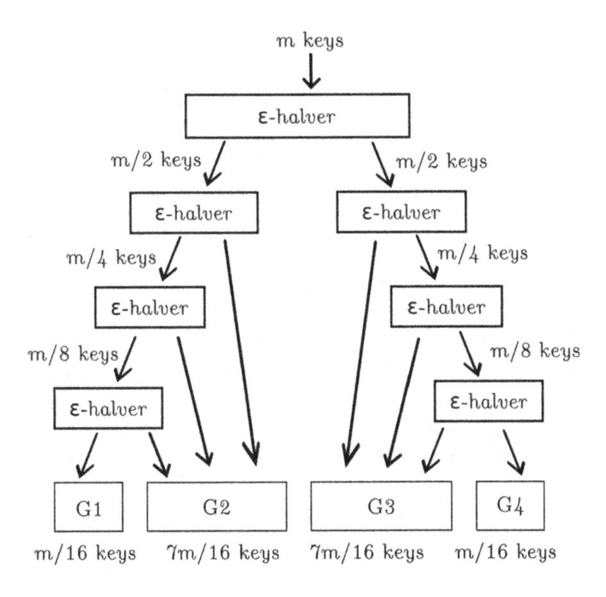

Fig. 6. A $(1/8, 1/18, 1/72)$-separator. The array S is formed by $\{G1, G2, G3, G4\}$.

Proposition 1. *One can construct an AKS sorting network in polynomial time.*

Lemma 4. *For each $t \in \mathbb{N}$, it holds that $Th[t] \subseteq W[SAT]$.*

Proof. To show the W[SAT]-membership of WTCS(t, h), for every $t \in \mathbb{N}$, we first ensure the existence of a polynomial-time algorithm that converts a threshold gate (with fan-in n) in Boolean circuit with $O(\log n)$ depth. We know that the AKS sorting networks have logarithmic depth, and by construction its gates has bounded fan-out. In addition, AKS sorting networks can be constructed in polynomial time on the number of inputs as described in this section.

Therefore, we can replace each unweighted threshold gate for an equivalent decision Boolean circuit by first construct an AKS network and then converting it into a treelike sub-circuit. After that, to obtain a complete treelike circuit, it is enough to handle with the gates inter sorting networks, which can be done since instances of WTCS(t, h) have depth bounded by h, which is a constant.

Hence, in polynomial time one can take an instance of WTCS(t, h) and outputs an equivalent treelike Boolean circuit C, i.e., a decision circuit corresponding to a Boolean formula. Thus, WTCS(t, h) is fixed-parameter reducible to WSAT and Th$[t] \subseteq$ W[SAT], for each $t \in \mathbb{N}$. □

Figure 7 shows the inclusion relationships we know between classes of the W-hierarchy and Th-hierarchy.

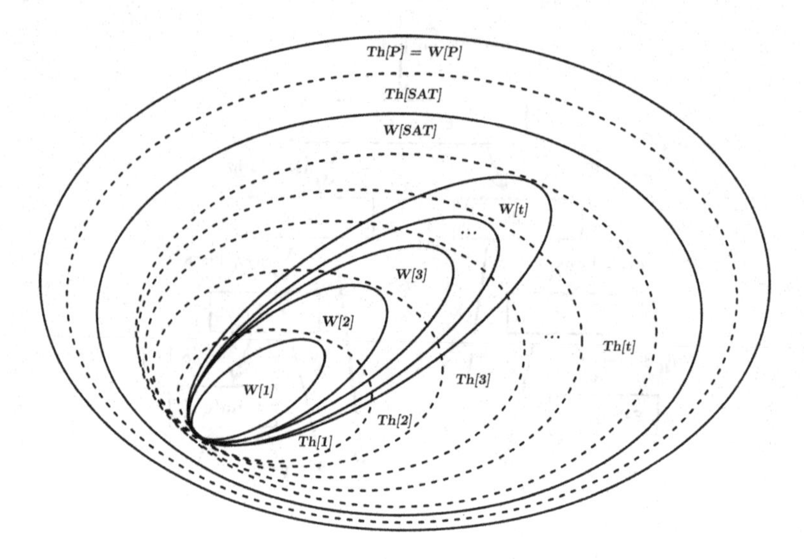

Fig. 7. Relationship between W-hierarchy and Th-hierarchy.

References

1. Ajtai, M., Komlós, J., Szemerédi, E.: An 0 (n log n) sorting network. In: Proceedings of the Fifteenth Annual ACM Symposium on Theory of Computing, pp. 1–9 (1983)
2. Alon, N.: Eigenvalues and expanders. Combinatorica **6**(2), 83–96 (1986). https://doi.org/10.1007/BF02579166
3. Al-Haj Baddar, S.W., Batcher, K.E.: The AKS sorting network. In: Designing Sorting Networks. Springer, New York. https://doi.org/10.1007/978-1-4614-1851-1_11
4. Batcher, K.E.: Sorting networks and their applications. In: Proceedings of the Spring Joint Computer Conference, April 30 – 2 May 1968, pp. 307–314 (1968)
5. Downey, R.G., Fellows, M.R.: Parameterized Complexity. Springer Science & Business Media (2012)
6. Fellows, M., Flum, J., Hermelin, D., Müller, M., Rosamond, F.: W-hierarchies defined by symmetric gates. Theory Comput. Syst. **46**(2), 311–339 (2010). https://doi.org/10.1007/s00224-008-9138-6
7. Hajnal, A., Maass, W., Pudlák, P., Szegedy, M., Turán, G.: Threshold circuits of bounded depth. J. Comput. Syst. Sci. **46**(2), 129–154 (1993)
8. Lubotzky, A., Phillips, R., Sarnak, P.: Explicit expanders and the Ramanujan conjectures. In: Proceedings of the Eighteenth Annual ACM Symposium on Theory of Computing, pp. 240–246 (1986)
9. Lubotzky, A., Phillips, R., Sarnak, P.: Ramanujan graphs. Combinatorica **8**(3), 261–277 (1988)
10. O'connor, D.G., Nelson, R.J.: Sorting system with nu-line sorting switch, 10 April 1962, US Patent 3,029,413
11. Paterson, M.S.: Improved sorting networks with o (log n) depth. Algorithmica **5**(1), 75–92 (1990)

12. Tanner, R.M.: Explicit concentrators from generalized N-GONS. SIAM J. Algebraic Discrete Methods **5**(3), 287–293 (1984)
13. Xie, H.: Studies on sorting networks and expanders. Ph.D. thesis, Ohio University (1998)

Maximization of Monotone Non-submodular Functions with a Knapsack Constraint over the Integer Lattice

Jingjing Tan[1], Fengmin Wang[2], Xiaoqing Zhang[3], and Yang Zhou[4](✉)

[1] School of Mathematics and Information Science, Weifang University, Weifang 261061, People's Republic of China
[2] Beijing Jinghang Research Institute of Computing and Communication, Beijing 100074, People's Republic of China
[3] Department of Operations Research and Information Engineering, Beijing University of Technology, Beijing 100124, People's Republic of China
[4] School of Mathematics and Statistics, Shandong Normal University, Jinan 250014, People's Republic of China
zhouyang@sdnu.edu.cn

Abstract. The problem of submodular maximization on the integer lattice has attracted more and more attention due to its deeply applications in many areas. In this paper, we consider maximizing a non-negative monotone non-submodular function with knapsack constraint on massive data. We combine two proposed algorithms called StreamingKnapsack and BinarySearch for this problem by introducing the DR ratio γ_d and the weak DR ratio γ_w of the non-submodular objective function. Finally, we obtain the performance guarantee of the StreamingKnapsack supplemented by a simple one-pass algorithm, with the approximation ratio of the better output of them as $\min\{\gamma_d^2(1-\varepsilon)/2^{\gamma_d+1}, 1-1/\gamma_w 2^{\gamma_d}-\varepsilon\}$. Meanwhile, both the time complexity and space complexity are dependent on the size of knapsack capacity K and $\varepsilon \in (0,1)$.

Keywords: Integer lattice · Non-submodular · Knapsack constraint · Streaming algorithm

1 Introduction

Submodularity is a concept that describes the diminishing-return property, that is, adding an element to a small set produces more gain than adding it to a larger one. Since the submodularity has this property, and many practical scenes, such as, virus marketing, diversification of search results, active learning, network reasoning and so on [1–6,8–12,14–16,22,29,34]. For traditional maximizing submodular function defined on a set, Nemhauser et al. [25] pioneered designed a greedy algorithm with $(1-1/e)$-approximation subject to a cardinality constraint. Sviridenko [28] proposed a $(1-1/e)$-approximation algorithm under a knapsack constraint. Călinescuet al. [6] used the methods of multilinear extension to design a

D.-Z. Du et al. (Eds.): COCOA 2021, LNCS 13135, pp. 364–373, 2021.
https://doi.org/10.1007/978-3-030-92681-6_29

continuous greedy algorithm with $(1-1/e)$-approximation for the matroid constraint. However, there are still many practical problems that can not be characterized by pure submodular optimization problem, such as some subset selection problems in data science. When selecting subsets of training data in machine learning system, there may be not only redundancy but also complementarity among some subsets of elements. All the collective utilities of these elements can be seen only when they are used together, this requires the use of non-submodular functions for modeling. This idea is also widely used in economics and social sciences, which are aroused a lot of scholars interesting [7,17–21,23,26,27,33,35,36].

However, in the emerging applications, such as exemplar clustering to sensor placement and machine learning problems in feature selection [31,32], we need to maximize a submodular and non-submodular function on the integer lattice. For this kind of problem, Soma et al. [30] proposed a generalization of the problem of submodular optimization in integer lattice. Ene et al. [13] studied the problem of maximizing a DR-submodular function with a budget constraint. For non-submodular maximization problems, Kuhnle et al. [24] proposed Threshold-Greedy algorithm under the cardinality constraint. Zhang et al. [37] presented a threshold Greedy algorithm for maximizing DR-submodular plus supermodular functions with a cardinality constraint.

In this paper, we concentrate on streaming algorithms for maximizing a non-negative monotone non-submodular function with knapsack constraint on massive data.

1.1 Preliminaries

We will first introduce some notations, definitions and lemmas in this part.

Let $K \in \mathbb{N}^+$, and denote $[K]$ be the set of all the positive integer between one and k. $E = \{e_1, e_2, \cdots, e_n\}$ be the ground set. Let $\boldsymbol{u}$ be a n-dimensional vector in $\mathbb{N}^E$, and denote the component of coordinate $e_i \in E$ of $\boldsymbol{u}$ to be $\boldsymbol{u}(e_i)$. We use $\mathbf{0}$ to denote the zero vector, χ_{e_i} denote the standard unit vector, that is, all the components has a value of 0 except for the i-th component has a value of 1. For $U \subseteq E$, we denotes the characteristic vector of U as χ_U, and $\boldsymbol{u}(U) := \sum_{e_i \in U} \boldsymbol{u}(e_i)$. For $\boldsymbol{u} \in \mathbb{N}^E$, $supp^+(\boldsymbol{u}) = \{e \in E | \boldsymbol{u}(e) > 0\}$ is the supporting set of $\boldsymbol{u}$. Let $\{\boldsymbol{u}\}$ be the multi-set where the frequency of occurrence for e is $\boldsymbol{u}(e)$, and $|\{\boldsymbol{u}\}| := \boldsymbol{u}(E)$. Let $\boldsymbol{u} \vee \boldsymbol{w}$ and $\boldsymbol{u} \wedge \boldsymbol{w}$ denote the coordinate wise maximum and minimum of $\boldsymbol{u}$ and $\boldsymbol{w}$, respectively. That is,

$$(\boldsymbol{u} \wedge \boldsymbol{w})(e) = \min\{\boldsymbol{u}(e), \boldsymbol{w}(e)\},$$

and

$$(\boldsymbol{u} \vee \boldsymbol{w})(e) = \max\{\boldsymbol{u}(e), \boldsymbol{w}(e)\},$$

for each element $e \in E$. Let $\{\boldsymbol{u}\} \setminus \{\boldsymbol{w}\} := \{(\boldsymbol{u} \setminus \boldsymbol{w}) \vee \mathbf{0}\}$, where $\{\boldsymbol{u}\}$ and $\{\boldsymbol{w}\}$ are arbitrary two multi-sets.

A function $g : \mathbb{N}^E \to \mathbb{R}_+$ is monotone non-decreasing if $g(\boldsymbol{u}) \leq g(\boldsymbol{w})$ for any $\boldsymbol{u} \leq \boldsymbol{w}$. The nonegative and normalized of the function g means $g(\boldsymbol{u}) \geq 0$ for any $\boldsymbol{u} \in \mathbb{N}^E$ and $g(\mathbf{0}) = 0$. Next, we give the definitions of (lattice) submodularity and diminishing return submodularity.

Definition 1 (DR-submodular). *We call that a function g is DR-submodular if for any $e \in E$, $\boldsymbol{u}, \boldsymbol{w} \in \mathbb{N}^V$ with $\boldsymbol{u} \leq \boldsymbol{w}$, there holds*

$$g(\boldsymbol{w} + \chi_e) - g(\boldsymbol{w}) \leq g(\boldsymbol{u} + \chi_e) - g(\boldsymbol{u}).$$

Definition 2 (Lattice submodular). *We call that a function g is lattice submodular, if for all $\boldsymbol{u}, \boldsymbol{w} \in \mathbb{N}^E$ there holds*

$$g(\boldsymbol{u} \vee \boldsymbol{w}) + g(\boldsymbol{u} \wedge \boldsymbol{w}) \leq g(\boldsymbol{u}) + g(\boldsymbol{w}).$$

Suppose that the function g is defined on $\mathbb{N}^E$. According to Definition 1 and Definition 2, it is obvious that the lattice submodularity of g is not equivalent to the DR-submodularity [32].

Let $\boldsymbol{B} \in \{\mathbb{N} \bigcup \{\infty\}\}^E$ be a positive integer vector and g be defined on $\mathcal{D}_{\boldsymbol{B}} = \{\boldsymbol{u} \in \mathbb{N}^E : \boldsymbol{u} \leq \boldsymbol{B}\}$ satisfying $g(\boldsymbol{0}) = 0$. Denote $\mathcal{F}_{\boldsymbol{B}}$ as the set of non-negative monotone DR-submodular functions defined on $\mathcal{D}_{\boldsymbol{B}}$. For $g \in \mathcal{F}_{\boldsymbol{B}}$, and vectors $\boldsymbol{u}, \boldsymbol{w} \in \mathbb{N}^E$, define $\Delta_{\boldsymbol{w}}(\boldsymbol{u})$ as the marginal increment of a vector $\boldsymbol{u}$ with $\boldsymbol{w}$, that is,

$$\Delta_{\boldsymbol{w}}(\boldsymbol{u}) = g(\boldsymbol{u} + \boldsymbol{w}) - g(\boldsymbol{u}).$$

In the following we recall the definition of DR ratio, weak DR ratio and generalized curvature on the integer lattice.

Definition 3 (DR ratio). *For a function $g \in \mathcal{F}_{\boldsymbol{B}}$, the diminishing return ratio (briefly called DR ratio) $\gamma_d(g)$ of g is the maximal scalar that satisfies*

$$\gamma_d(g)\Delta_{\chi_e}(\boldsymbol{w}) \leq \Delta_{\chi_e}(\boldsymbol{u}),$$

for any $e \in E$, $\boldsymbol{u}, \boldsymbol{w} \in \mathcal{D}_{\boldsymbol{B}}$ with $\boldsymbol{u} \leq \boldsymbol{w}$ and $\boldsymbol{w} + \chi_e \in \mathcal{D}_{\boldsymbol{B}}$.

Definition 4 (Weak DR ratio). *For a function $g \in \mathcal{F}_{\boldsymbol{B}}$, the weak DR ratio $\gamma_w(g)$ of g is the maximum scalar that satisfies*

$$\gamma_w(g)(g(\boldsymbol{w}) - g(\boldsymbol{u})) \leq \sum_{e \in \{\boldsymbol{w}\} \setminus \{\boldsymbol{u}\}} \Delta_{\chi_e}(\boldsymbol{u}),$$

for all $\boldsymbol{u}, \boldsymbol{w} \in \mathcal{D}_{\boldsymbol{B}}$ with $\boldsymbol{u} \leq \boldsymbol{w}$, where $\{\boldsymbol{w}\}$ and $\{\boldsymbol{u}\}$ are the multisets corresponding to vectors $\boldsymbol{w}$ and $\boldsymbol{u}$ respectively.

Based on the above definition, we partition all the real-valued monotone functions defined on $\mathcal{D}_{\boldsymbol{B}}$ into sets $\mathcal{F}_{\boldsymbol{B}}^{\gamma_d, \gamma_w} = \{g \in \mathcal{F}_{\boldsymbol{B}} : \gamma_d(g) = \gamma_d, \gamma_w(g) = \gamma_w\}$ with respect to the above non-submodularity ratios.

1.2 Problem Formulation

In this paper, we study the maximization of a non-submodular function with knapsack constraint on the integer lattice (briefly called MKC). We denote E as the ground set with size n, $\boldsymbol{y} \in \mathbb{N}^E$ as an n-dimensional vector, $g \in \mathcal{F}_{\boldsymbol{B}}$ as a monotone non-submodular function. Therefore, the problem can be expressed as

$$\max g(\boldsymbol{y}) \text{ subject to } \boldsymbol{h}^T\boldsymbol{y} \leq K,$$

where $K \in \mathbb{N}$ is the total budget, and $\boldsymbol{h} \in \mathbb{R}_+^E$ is the vector of weight with $\boldsymbol{h}_{e_i}$ as the weight of element e_i. Consequently we have an assumption that $\boldsymbol{h}_{e_i} \leq K$ for $i = 1, \ldots, n$ since any e_i which does not satisfy this condition must not be in the feasible solution and thus can be discarded. Also note that multiplying both $\boldsymbol{h}$ and K by a positive scalar has no effect on the optimal value or the optimal solution of the problem. Therefore, WLOG, we can always assume that $\min\{\boldsymbol{h}_{e_i}\}_{i=1}^n \geq 1$. Denote $\boldsymbol{y}^*$ as the optimal solution and OPT as the optimal value of MKC.

1.3 Organization

The rest of the paper is organized as follows. In Sect. 2 we propose the StreamingKnapsack algorithm to solve MKC with its theoretical analysis. In Sect. 3, we summarize our work.

2 Algorithms

In this section, we propose the StreamingKnapsack algorithm for MKC, which is not only a generalization of the MarginalRatioThresholding algorithm in [17] to the non-submodular case, but also extends this algorithm to the setting of integer lattice.

Description. The StreamingKnapsack adds all copies of $e \in E$ when the addition does not violate the knapsack constraint and the average marginal gain exceeds the threshold condition

$$\Delta_{t\chi_e}(\boldsymbol{y}^v) \geq \frac{\gamma_d v - g(\boldsymbol{y}^v)}{2^{\gamma_d} K - \boldsymbol{h}^T \boldsymbol{y}^v}.$$

If t also satisfies the following condition

$$\Delta_{(t+1)\chi_e}(\boldsymbol{y}) < \frac{\gamma_d v - g(\boldsymbol{y}^v)}{2^{\gamma_d} K - \boldsymbol{h}^T \boldsymbol{y}^v},$$

then t is the satisfactory level of the remaining element e. Note that although in BinarySearch the marginal increment density, denoted as $\frac{\Delta_{t\chi_e}(\boldsymbol{y})}{t\boldsymbol{h}_e}$, is not monotone with respect to t, the output of this algorithm still satisfies the above two equalities. In this whole subsection we denote $\tilde{\boldsymbol{y}}$ as the output of StreamingKnapsack. For this algorithm we can deduce the following two lemmas to get further conclusions on its approximation guarantee.

Algorithm 1. StreamingKnapsack

Input: stream of data E, $g \in \mathcal{F}_B$, $\boldsymbol{B} \in \mathbb{N}^E$, $\varepsilon > 0$, weight vector $\boldsymbol{h} \in \mathbb{R}_+^E$, total budget K.

Output: a vector $\boldsymbol{y} \in \mathbb{N}^E$.

1: $M \leftarrow \max_{e \in E} g(\chi_e)$;

2: $\mathcal{V}_\varepsilon = \{(1+\varepsilon)^i | \frac{M}{1+\varepsilon} \leq (1+\varepsilon)^i \leq \frac{KM}{(1-\varepsilon)\gamma_d}\}$;

3: **for** each $v \in \mathcal{V}_\varepsilon$ **do**

4: $\quad \boldsymbol{y}^v \leftarrow \boldsymbol{0}$;

5: $\quad$ **for** $e \in E$ **do**

6: $\quad\quad$ **if** $\boldsymbol{h}^T \boldsymbol{y}^v \leq K$ **then**

7: $\quad\quad\quad t \leftarrow$ **BinarySearch**$\left(g, \boldsymbol{y}^v, \boldsymbol{B}, \boldsymbol{h}, e, K, \frac{\gamma_d v - g(\boldsymbol{y}^v)}{2^{\gamma_d} K - \boldsymbol{h}^T \boldsymbol{y}^v}\right)$;

8: $\quad\quad$ **end if**

9: $\quad$ **end for**

10: **end for**

11: **return** $\boldsymbol{y} = \arg\max_v g(\boldsymbol{y}^v)$

Lemma 1. *For each i-th iteration of StreamingKnapsack, we have*

$$g(\boldsymbol{y}_{i-1} + t_i \chi_{e_i}) \geq \frac{\gamma_d v}{2^{\gamma_d} K} \boldsymbol{h}^T \boldsymbol{y}_i. \tag{1}$$

Lemma 2. *If $\boldsymbol{h}^T \tilde{\boldsymbol{y}} < K$, we have*

$$\Delta_{\chi_e}(\tilde{\boldsymbol{y}}) < \frac{v}{2^{\gamma_d} K} \boldsymbol{h}_e,$$

for any $e \in \{\boldsymbol{y}^*\} \setminus \{\tilde{\boldsymbol{y}}\}$.

Indeed, simply running StreamingKnapsack could not get a constant approximation factor solution. Similar with submodular maximization problems with knapsack constraints in set function settings, the reason is that there may exist some elements in the optimal solution which is of large weight with their marginal increment density not large enough. The existence of such elements will lead to the priority of adding some elements with small weights and high density in the running of the algorithm so that they cannot be added due to the knapsack constraint. We call this kind of elements as "bad elements", whose formal definition is as the following.

Definition 5 (Bad element). *We call $e \in \{\boldsymbol{y}^*\}$ a bad element if it satisfies the threshold condition*

$$\frac{\Delta_{t_e \chi_e} \boldsymbol{y}_e}{t_e \boldsymbol{h}_e} \geq \frac{\gamma_d v - g(\boldsymbol{y}_e)}{2^{\gamma_d} K - \boldsymbol{h}^T \boldsymbol{y}_e}, \tag{2}$$

Algorithm 2. BinarySearch $(g, \boldsymbol{y}, \boldsymbol{B}, \boldsymbol{h}, e, K, \tau)$

Input: stream of data E, $e \in E$, $g \in \mathcal{F}_B$, $\boldsymbol{h} \in \mathbb{N}^E$, $\boldsymbol{y} \in \mathbb{N}^E$, and $\tau \in \mathbb{R}_+$.

Output: t.

1: $t_s \leftarrow 1$;

2: $t_r \leftarrow \min\left\{\boldsymbol{B}_e, \left\lfloor \frac{K-\boldsymbol{h}^T\boldsymbol{y}}{\boldsymbol{h}_e} \right\rfloor\right\}$

3: **if** $\frac{\Delta_{t_r\chi_e}(\boldsymbol{y})}{t_r\boldsymbol{h}_e} \geq \tau$ **then**

4: return t_r;

5: **end if**

6: **if** $\frac{\Delta_{\chi_e}(\boldsymbol{y})}{\boldsymbol{h}_e} < \tau$ **then**

7: return 0;

8: **end if**

9: **while** $t_r < t_s + 1$ **do**

10: $m = \lfloor \frac{t_r+t_s}{2} \rfloor$;

11: **if** $\frac{\Delta_{m\chi_e}(\boldsymbol{y})}{m\boldsymbol{h}(e)} \geq \tau$ **then**

12: $t_s = m$;

13: **else**

14: $t_r = m$;

15: **end if**

16: **end while**

17: **return** t_s

in StreamingKnapsack, but the total budget exceeds K when added. That is, $\boldsymbol{h}^T\boldsymbol{y}_e \leq K$ and $\boldsymbol{h}^T(\boldsymbol{y}_e + t_e\chi_e) > K$, where $\boldsymbol{y}_e$ is the vector just before element e arrives.

The following two lemmas indicates that when there is no bad element in the running of StreamingKnapsack, then a preliminary constant approximation guarantee can be obtained.

Lemma 3. *Assume that $v \leq g(\boldsymbol{y}^*)$ and there has been no bad item in StreamingKnapsack, then we have $g(\tilde{\boldsymbol{y}}) \geq (1 - 1/\gamma_w 2^{\gamma_d})v$.*

Lemma 4. *Suppose that $M = \max_{e\in E} \frac{g(\chi_e)}{h_e}$. Then there must exist a $v \in \mathcal{V}_\varepsilon$ satisfying $(1-\varepsilon)\mathrm{OPT} \leq v \leq \mathrm{OPT}$.*

To deal with instances that may exists bad elements, a one-pass algorithm name as OptimalSingleton in which we choose an optimal element with maximal feasible copies is proposed as a supplement of StreamingKnapsack, as is shown in Algorithm 3.

Algorithm 3. OptimalSingleton

Input: stream of data E, $e \in E$, $g \in \mathcal{F}_B$.

1: $\boldsymbol{y} \leftarrow \mathbf{0}$;
2: **for** the coming element e **do**
3: $\quad t_e \leftarrow \min \left\{ \boldsymbol{B}_e, \left\lfloor \frac{K}{h_e} \right\rfloor \right\}$;
4: $\quad$ **if** $g\left(t_e \chi_e\right) > g(\boldsymbol{y})$ **then**
5: $\quad\quad \boldsymbol{y} \leftarrow t_e \chi_e$;
6: $\quad$ **end if**
7: **end for**
8: **return** $\boldsymbol{y}$

In general, the output of separately running OptimalSingleton is not a constant factor approximation solution either, whereas the better one of both StreamingKnapsack and OptimalSingleton can be. Denote $\hat{\boldsymbol{y}}$ as the solution returned by OptimalSingleton, which is the best singleton of the data. Suppose that there are some e's in the OPT that are bad, then together with $\hat{\boldsymbol{y}}$, we analysis the approximation ratio of StreamingKnapsack as follows.

Theorem 1. *Given an instance of MKC with $g \in \mathcal{F}_B^{\gamma_d, \gamma_w}$, let $\tilde{\boldsymbol{y}}$ be the solution returned by StreamingKnapsack, and $\hat{\boldsymbol{y}}$ be the solution returned by OptimalSingleton. Then we have*

$$\max \{g(\tilde{\boldsymbol{y}}), g(\hat{\boldsymbol{y}})\} \geq \min \left\{ \frac{\gamma_d^2(1-\varepsilon)}{2^{\gamma_d+1}}, 1 - \frac{1}{\gamma_w 2^{\gamma_d}} - \varepsilon \right\} \mathrm{OPT}.$$

The total memory complexity of the two algorithms is $O(\frac{K \log K}{\varepsilon})$, and the total query complexity for each element is $O(\frac{\log^2 K}{\varepsilon})$.

3 Conclusions

In this paper we consider the problem of maximizing a non-submodular function with a knapsack constraint. We design the StreamingKnapscak algorithm combing with the BinarySearch algorithm as a subroutine for this problem. By introducing the DR ratio and weak DR ratio, we obtain the approximation ratio of StreamingKnapscak. Moreover, in order to get one pass algorithm, we can further design a online algorithm by dynamically updates the maximal value of $g(\chi_e)$.

Acknowledgements. The first author is supported by Natural Science Foundation of Shandong Province (Nos. ZR2017LA002, ZR2019MA022), and Doctoral research foundation of Weifang University (No. 2017BS02). The second author is supported by National Natural Science Foundation of China (No. 11901544). The third author is supported by National Natural Science Foundation of China (No. 11871081) and Beijing Natural Science Foundation Project No. Z200002. The fourth author is supported by Natural Science Foundation of Shandong Province of China (No. ZR2019PA004) and National Natural Science Foundation of China (No. 12001335).

References

1. Agrawal, R., Gollapudi, S., Halverson, A., Ieong, S. Diversifying search results. In: Proceedings of WSDM, pp. 5–14 (2009)
2. Alon, N., Gamzu, I., Tennenholtz, M.: Optimizing budget allocation among channels and influencers. In: Proceedings of WWW, pp. 381–388 (2012)
3. Badanidiyuru, A., Mirzasoleiman, B., Karbasi, A., Krause, A.: Streaming submodular maximization: massive data summarization on the fly. In: Proceedings of KDD, pp. 671–680 (2014)
4. Buchbinder, N., Feldman, M., Schwartz, R.: Online submodular maximization with preemption. In: Proceedings of SODA, pp. 1202–1216 (2015)
5. Balkanski, E., Rubinstein, A., Singer, Y.: An exponential speedup in parallel running time for submodular maximization without loss in approximation. In: Proceedings of SODA, pp. 283–302 (2019)
6. Călinescu, G., Chekuri, C., Pál, M., Vondrák, J.: Maximizing a submodular set function subject to a matroid constraint. SIAM J. Comput. **40**, 1740–1766 (2011)
7. Chakrabarti, A., Kale, S.: Submodular maximization meets streaming: matchings, matroids, and more. Math. Program. **154**(1), 225–247 (2015). https://doi.org/10.1007/s10107-015-0900-7
8. Chekuri, C., Quanrud, K.: Submodular function maximization in parallel via the multilinear relaxation. In: Proceedings of SODA, pp. 303–322 (2019)
9. Chekuri, C., Quanrud, K.: Randomize MWU for positive LPs. In: Proceedings of SODA, pp. 358–377 (2018)
10. Das, A., Kempe, D.: Algorithms for subset selection in linear regression. In: Proceedings of STOC, pp. 45–54 (2008)
11. Das, A., Kempe, D.: Submodular meets spectral: greedy algorithms for subset selection, sparse approximation and dictionary selection. In: Proceedings of ICML, pp. 1057–1064 (2011)
12. EI-Arini, K., Guestrin, C.: Beyond keyword search: discovering relevant scientific literature. In: Proceedings of ICKDDM, pp. 439–447 (2011)
13. Ene, A., Nguyen, H.L.: Submodular maximization with nearly-optimal approximation and adaptivity in nearly-linear time. In: Proceedings of SODA, pp. 274–282 (2019)
14. Golovin, D., Krause, A.: Adaptive submodularity: theory and applications in active learning and stochastic optimization. J. Artif. Intell. Res. **42**, 427–486 (2011)
15. Gomez, R.M., Leskovec, J., Krause, A.: Inferring networks of diffusion and influence. ACM Trans. Knowl. Discov. Data **8**, 36–39 (2018)
16. Gong, S., Nong, Q., Liu, W., Fang, Q.: Parametric monotone function maximization with matroid constraints. J. Glob. Optim. **75**(3), 833–849 (2019). https://doi.org/10.1007/s10898-019-00800-2

17. Huang, C.-C., Kakimura, N.: Improved streaming algorithms for maximizing monotone submodular functions under a Knapsack constraint. In: Friggstad, Z., Sack, J.-R., Salavatipour, M.R. (eds.) WADS 2019. LNCS, vol. 11646, pp. 438–451. Springer, Cham (2019). https://doi.org/10.1007/978-3-030-24766-9_32
18. Jiang, Y.J., Wang, Y.S., Xu, D.C., Yang, R.Q., Zhang, Y.: Streaming algorithm for maximizing a monotone non-submodular function under d-knapsack constraint. Optim. Lett. **14**, 1235–1248 (2020)
19. Krause, A., Leskovec, J., Guestrin, C., VanBriesen, J.M., Faloutsos, C.: Efficient sensor placement optimization for securing large water distribution networks. J. Water Resour. Plan. Manag. **134**, 516–526 (2008)
20. Krause, A., Singh, A., Guestrin, C.: Near-optimal sensor placements in gaussian processes: theory, efficient algorithms and empirical studies. J. Mach. Learn. Res. **9**, 235–284 (2008)
21. Kapralov, M., Post, I., Vondrák, J.: Online submodular welfare maximization: greedy is optimal. In: Proceedings of SODA, pp. 1216–1225 (2012)
22. Kempe, D., Kleinberg, J., Tardos, E.: Maximizing the spread of influence through a social network. In: Proceedings of KDD, pp. 137–146 (2003)
23. Khanna, R., Elenberg, E.R., Dimakis, A.G., Negahban S., Ghosh, J.: Scalable greedy feature selection via weak submodularity. In: Proceedings of ICAIS, pp. 1560–1568 (2017)
24. Kuhnle, A., Smith, J.D., Crawford, V.G., Thai, M.T.: Fast maximization of non-submodular, monotonic functions on the integer lattice. In: Proceedings of ICML, pp. 2791–2800 (2018)
25. Nemhauser, G.L., Wolsey, L.A., Fisher, M.L.: An analysis of approximations for maximizing submodular set functions. Math. Program. **14**, 265–294 (1978)
26. Norouzi-Fard, A., Tarnawski, J., Mitrovic, S., Zandieh, A., Mousavifar, A., Svensson, O.: Beyond 1/2-approximation for submodular maximization on massive data streams. In: Proceedings of ICML, pp. 3829–3838 (2018)
27. Simon, I., Snavely, N., Seitz, S.M.: Scene summarization for online image collections. In: Proceedings of ICCV, pp. 1–8 (2007)
28. Sviridenko, M.: A note on maximizing a submodular set function subject to a knapsack constraint. Oper. Res. Lett. **32**, 41–43 (2004)
29. Shioura, A.: On the pipage rounding algorithm for submodular function maximization-a view from discrete convex analysis. Discrete Math. Algorithms Appl. **1**, 1–23 (2009)
30. Soma, T., Kakimura, N., Inaba, K., Kawarabayashi, K.: Optimal budget allocation: theoretical guarantee and efficient algorithm. In: Proceedings of ICML, pp. 351–359 (2014)
31. Soma, T., Yoshida, Y.: A generalization of submodular cover via the diminishing return property on the integer lattice. In: Proceedings of NIPS, pp. 847–855 (2014)
32. Soma, T., Yoshida, Y.: Maximization monotone submodular functions over the integer lattice. Math. Program. **172**, 539–563 (2018)
33. Vondrăk, J.: Optimal approximation for the submodular welfare problem in the value oracle model. In: Proceedings of STOC, pp. 67–74 (2008)
34. Wolsey, L.: Maximising real-valued submodular set function: primal and dual heuristics for location problems. Math. Oper. Res. **7**, 410–425 (1982)
35. Wang, Y., Xu, D., Wang, Y., Zhang, D.: Non-submodular maximization on massive data streams. J. Glob. Optim. **76**(4), 729–743 (2019). https://doi.org/10.1007/s10898-019-00840-8

36. Yang, R.Q., Xu, D.C., Jiang, Y.J., Wang, Y.S., Zhang, D.M.: Approximation robust parameterized submodular function maximization in large-scales. Asia Pac. J. Oper. Res. **36**, 195–220 (2019)
37. Zhang, Z.N., Du, D.L., Jiang, Y.J., Wu, C.C.: Maximizing DR-submodular + supermodular function on the integer lattice subject to a cardinality constraint. J. Glob. Optim. **80**, 595–616 (2021)

Sublinear-Time Reductions for Big Data Computing

Xiangyu Gao[1,2], Jianzhong Li[1,2](✉), and Dongjing Miao[1]

[1] Department of Computer Science and Technology, Harbin Institute of Technology, Harbin, China
{gaoxy,lijzh,miaodongjing}@hit.edu.cn

[2] Faculty of Computer Science and Control Engineering, Shenzhen Institute of Advanced Technology Chinese Academy of Sciences, Shenzhen, China

Abstract. With the rapid popularization of big data, the dichotomy between tractable and intractable problems in big data computing has been shifted. Sublinear time, rather than polynomial time, has recently been regarded as the new standard of tractability in big data computing. This change brings the demand for new methodologies in computational complexity theory in the context of big data. Based on the prior work for sublinear-time complexity classes [9], this paper focuses on sublinear-time reductions specialized for problems in big data computing. First, the pseudo-sublinear-time reduction is proposed and the complexity classes P and PsT are proved to be closed under it. To establish PsT-intractability for certain problems in P, we find the first problem in P\PsT. Using the pseudo-sublinear-time reduction, we prove that the nearest edge query is in PsT but the algebraic equation root problem is not. Then, the pseudo-polylog-time reduction is introduced and the complexity class PsPL is proved to be closed under it. The PsT-completeness under it is regarded as an evidence that some problems can not be solved in polylogarithmic time after a polynomial-time preprocessing, unless PsT = PsPL. We prove that all PsT-complete problems are also P-complete, which gives a further direction for identifying PsT-complete problems.

Keywords: Big data computing · Sublinear-time tractability · Reduction techniques · Preprocessing

1 Introduction

Traditionally, a problem is considered to be tractable if there exists a polynomial-time (PTIME) algorithm for solving it. However, PTIME no more serves as a good yardstick for tractability in the context of big data, and sometimes even linear-time algorithms can be too slow in practice. For example, a linear scan of a 1PB dataset with the fastest Solid State Drives on the market will take 34.7 h [1]. Therefore, sublinear time is considered as the new standard of tractability in big data computing [12]. This change has promoted the development of computational complexity theory specialized for problems in big data computing.

D.-Z. Du et al. (Eds.): COCOA 2021, LNCS 13135, pp. 374–388, 2021.
https://doi.org/10.1007/978-3-030-92681-6_30

In the last few years, many complexity classes were proposed to formalize tractable problems in big data computing [8,9,19]. The first attempt was made by Fan et al. in 2013 [8], which focuses on tractable boolean query classes with the help of preprocessing. They defined a concept of $\sqcap$-tractability for boolean query classes. A boolean query class is $\sqcap$-tractable if it can be processed in parallel polylogarithmic time (NC) after a PTIME preprocessing. They defined a query complexity class $\sqcap \mathsf{T}^0_\mathsf{Q}$ to denote the set of $\sqcap$-tractable query classes. To clarify the difference between $\sqcap\mathsf{T}^0_\mathsf{Q}$ and P, they proposed a form of generalized NC reduction, referred as F-reduction $\leq^{\mathtt{NC}}_F$, and proved that $\sqcap\mathsf{T}^0_\mathsf{Q}$ is closed under F-reduction. They showed that $\mathtt{NC} \subseteq \sqcap\mathsf{T}^0_\mathsf{Q} \subseteq \mathtt{P}$, but $\sqcap_{\mathsf{T}^0_\mathsf{Q}} \neq \mathtt{P}$ unless $\mathtt{P} = \mathtt{NC}$.

Then, Yang et al. introduced a $\sqcap'$-tractability for short query classes, i.e. the query length is bounded by a logarithmic function with respect to the data size [19]. On the basis of $\sqcap$-tractability theory, they placed a logarithmic-size restriction on the preprocessing result and relaxed the query execution time to polynomial. The corresponding query complexity class was denoted as $\sqcap'\mathsf{T}^0_\mathsf{Q}$, including the set of $\sqcap'$-tractable short query classes. They proved that F-reduction is also compatible with $\sqcap'\mathsf{T}^0_\mathsf{Q}$ and any $\sqcap'\mathsf{T}^0_\mathsf{Q}$-complete query class under F-reduction is P-complete query class under NC reduction.

A year ago, to completely describe the scope of sublinear-time tractable problems, the authors of this paper proposed two categories of sublinear-time complexity classes [9]. One kind characterizes the problems that are directly feasible in sublinear time, while the other describes the problems that are solvable in sublinear time after a PTIME preprocessing. However, we only showed that the polylogarithmic-time class PPL is closed under DLOGTIME reduction and the sublinear-time class PT is closed under linear-size DLOGTIME reduction, but left reductions for pseudo-sublinear-time complexity classes as a future work.

Open Question 1. *What kind of reductions are appropriate for pseudo-sublinear-time tractable problems in big data computing?*

On the other, it is also important to identify the problems that are unsolvable in sublinear time. Since, the new tractable standard in big data computing essentially dichotomizes problems in P, it is significant to differentiate hardness of problems in P. The modern approach is to prove *conditional lower bounds* via *fine-grained reductions* [3]. Generally, a fine-grained reduction starts from a key problem such as SETH, 3SUM, APSP, etc.., which has a widely believed *conjecture* about its time complexity, and transfers the conjectured intractability to the reduced problem, yielding a conditional lower bounds on how fast the reduced problem can be solved. The resulting area is referred as *fine-grained complexity theory*, and we refer to the surveys [17,18] for further reading. However, to establish a problem is intractable in the context of big data, an unconditional lower bound, even rough, is also preferred. Thus, the other goal of this paper is to overcome the following barrier.

Open Question 2. *Is there a natural problem belonging to* P *but not to* PsT*?*

1.1 Our Results

The focus of this paper is mainly on pseudo-sublinear-time reductions specialized for problems in big data computing. We reformulate the reduction used in [4], which was originally designed for complexity classes beyond NP. The general description of reductions proposed in this paper is illustrated in Fig. 1. We derive appropriate reductions for different complexity classes by limiting the computational power of functions used in it.

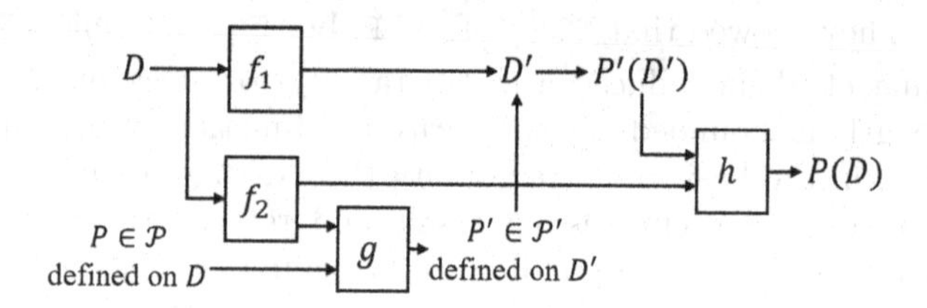

Fig. 1. Illustration of reductions used in this paper.

We first introduce the pseudo-sublinear-time reduction, $\leq_m^{\mathtt{PsT}}$ for problems in $\mathtt{PsT}$. We prove that it is transitive and the complexity classes $\mathtt{P}$ and $\mathtt{PsT}$ are closed under $\leq_m^{\mathtt{PsT}}$. Due to the limitation of the fraction power function, we do not define a new $\mathtt{P}$-completeness under $\leq_m^{\mathtt{PsT}}$ to include the problems in $\mathtt{P}\backslash\mathtt{PsT}$. Instead, we prove a natural problem, the circuit value problem, can not be solved in sublinear time after a $\mathtt{PTIME}$ preprocessing. This also proves that $\mathtt{PsT} \subsetneq \mathtt{P}$. After that, we reduce the algebraic equation root problem to the circuit value problem, which means the former also belongs to $\mathtt{P}\backslash\mathtt{PsT}$. Moreover, we show the nearest neighbor problem is in $\mathtt{PsT}$ by reducing it to the range successor query.

Then, we propose the notion of pseudo-polylog-time reduction, $\leq_m^{\mathtt{PsPL}}$, and show that $\mathtt{PsPL}$ is closed under $\leq_m^{\mathtt{PsPL}}$. We define the $\mathtt{PsT}$-completeness under $\leq_m^{\mathtt{PsPL}}$, which can be treated as an evidence that certain problems are not solvable in polylogarithmic time after a $\mathtt{PTIME}$ preprocessing unless $\mathtt{PsT} = \mathtt{PsPL}$. We prove that all $\mathtt{PsT}$-complete problems are also $\mathtt{P}$-complete. This specifies the range of possible $\mathtt{PsT}$-complete problems.

Moreover, we also extend L-reduction [7] to pseudo-sublinear time and prove that it linearly preserve approximation ratio for pseudo-sublinear-time approximation algorithms. Finally, we give a negative answer to the existence of complete problems in $\mathtt{PPL}$ under $\mathtt{DLOGTIME}$ reduction.

Outline. The remainder of this paper is organized as follows. Necessary preliminaries are stated in Sect. 2. The definitions and properties of pseudo-sublinear-time reduction and pseudo-polylog-time reduction are presented in Sect. 3 and Sect. 4 respectively. The pseudo-sublinear-time L-reduction is introduced in Sect. 5. A negative results for the existence of complete problems in PPL is shown in Sect. 6. The paper is concluded in Sect. 7.

2 Preliminaries

In this section, we briefly review the sublinear-time complexity classes introduced in [9] and the basic concepts of reductions.

We start with some notations.

Notations. To reflect the characteristics in big data computing, the input of a problem is partitioned into data part and problem part. Thus, a decision problem $\mathcal{P}$ can be considered as a binary relation such that for each D and problem P defined on D, $\langle D, P\rangle \in \mathcal{P}$ if and only if $P(D)$ is true. We say that a binary relation is in complexity class $\mathcal{C}$ if it is in $\mathcal{C}$ to decide whether a pair $\langle D, P\rangle \in \mathcal{P}$. Following the convention of complexity theory [14], we assume a finite alphabet Σ of symbols to encode both of them. The length of a string $x \in \Sigma^*$ is denoted by $|x|$. Given an integer n, let $\llcorner n \lrcorner$ denote the binary form of n.

Sublinear-Time Complexity Classes. The computational model is crucial when describing sublinear-time computation procedures. A random-access Turing machine (RATM) M is a k-tape Turing machine including a read-only input tape and $k-1$ work tapes, referred as non-index tape. And M is additionally equipped with k binary index tapes, one for each non-index tape. M has a special *random access* state which, when entered, moves the head of each non-index tape to the cell described by the respective index tape in one step. Based on RATM, a series of pure-sublinear-time complexity classes are proposed in [9] to include problems that are solvable in sublinear time.

Definition 2.1. The class PPL consists of problems that can be solved by a RATM in $O(\text{polylog}(n))$ time, where n is the length of the input. And for each $i \geq 1$, PPL^i consists of problems that can be solved by a RATM in $O(\log^i n)$ time.

Definition 2.2. The class PT consists of problems that can be solved by a RATM in $o(n)$ time, where n is the length of the input.

Moreover, when the data part is fixed and known in advance, it makes sense to perform an off-line preprocessing on it to accelerate the subsequent processing of problem instances defined on it. Hence, some pseudo-sublinear-time complexity classes are also defined to include the problems which are solvable in sublinear time after a PTIME preprocessing on the data part.

Definition 2.3. A problem $\mathcal{P}$ is in PsPL if there exists a PTIME preprocessing function $\Pi(\cdot)$ such that for any pair of strings $\langle D, P\rangle$ it holds that: $P(\Pi(D)) = P(D)$, and $P(\Pi(D))$ can be solved by a RATM in $O(\text{polylog}(|D|))$ time.

Definition 2.4. A problem $\mathcal{P}$ is in PsT if there exists a PTIME preprocessing function $\Pi(\cdot)$ such that for any pair of strings $\langle D, P\rangle$ it holds that: $P(\Pi(D)) = P(D)$, and $P(\Pi(D))$ can be solved by a RATM in $o(|D|)$ time. Moreover, a problem $\mathcal{P}$ is in PsTR (resp. PsTE) if $\mathcal{P} \in$ PsT and the PTIME preprocessing function $\Pi(\cdot)$ satisfies that for all big data D: $|\Pi(D)| < |D|$ (resp. $|\Pi(D)| \geq |D|$).

Reductions. In complexity theory, reductions are always used to both find efficient algorithms for problems, and to provide evidence that finding particularly efficient algorithms for some problems will likely be difficult [11,14]. Two main types of reductions are used in computational complexity theory, the many-one reduction and the Turing reduction. A problem $\mathcal{P}_1$ is *Turing reducible* to a problem $\mathcal{P}_2$, denoted as $\mathcal{P}_1 \leq_T \mathcal{P}_2$ if there is an oracle machine to solve $\mathcal{P}_1$ given an oracle for $\mathcal{P}_2$. That is, there is an algorithm for $\mathcal{P}_1$ if it is available to a subroutine for solving $\mathcal{P}_2$. While, many-one reductions are a special case and stronger form of Turing reductions. A decision problem $\mathcal{P}_1$ is *many-one reducible* to a decision problem $\mathcal{P}_2$, denoted as $\mathcal{P}_1 \leq_m \mathcal{P}_2$, if the oracle that is, the subroutine for $\mathcal{P}_2$ can be only invoked once at the end, and the answer can not be modified.

Reductions define difficulty orders (from different aspects) among problems in a complexity class. Hence, reductions are required to be transitive and easy to compute, relative to the complexity of typical problems in the class. For example, when studying the complexity class NP and harder classes such as the polynomial hierarchy, polynomial-time reductions are used, and when studying classes within P such as NC and NL, log-space reductions are used. We say a complexity class $\mathcal{C}$ is closed under a reduction if problem $\mathcal{P}_1$ is reducible to another problem $\mathcal{P}_2$ and if $\mathcal{P}_2$ is in $\mathcal{C}$, then so must be $\mathcal{P}_1$.

3 Pseudo-sublinear-Time Reduction

In this section, we introduce the notion of pseudo-sublinear-time reduction to tell whether a problem can be solved in sublinear time after a PTIME preprocessing.

Definition 3.1. A decision problem $\mathcal{P}_1$ is *pseudo-sublinear-time reducible* to a decision problem $\mathcal{P}_2$, denoted as $\mathcal{P}_1 \leq_m^{\texttt{PsT}} \mathcal{P}_2$, if there is a triple $\langle f_1(\cdot), f_2(\cdot), g(\cdot,\cdot)\rangle$, where $f_1(\cdot)$ and $f_2(\cdot)$ are linear-size NC computable functions and $g(\cdot,\cdot)$ is a PsT computable function, such that for any pair of strings $\langle D, P\rangle$ it holds that

$$\langle D, P\rangle \in \mathcal{P}_1 \Leftrightarrow \langle f(D), g(D, P)\rangle \in \mathcal{P}_2.$$

Recall the general formalization of reductions specialized for problems in big data computing shown in Fig. 1. In contrast to traditional reductions such as polynomial-time reduction and log-space reduction, the pseudo-sublinear-time reduction is defined for the two parts of problems respectively. Concretely speaking, (1) the data part of $\mathcal{P}_2$ is obtained from the data part of $\mathcal{P}_1$ using $f_1(\cdot)$, and (2) the problem part of $\mathcal{P}_2$ is generated from the problem part of $\mathcal{P}_1$ using $g(\cdot,\cdot)$ with some additional information of the data part of $\mathcal{P}_1$ provided by $f_2(\cdot)$. Intuitively, for different problems defined on the same data D, the computation of $f_2(D)$ can be regarded as an off-line process with a one-time cost. Hence, when talking about the running time of $g(\cdot,\cdot)$, the running time of $f_2(\cdot)$ is excluded. We first prove that $\leq_m^{\texttt{PsT}}$ is transitive.

Theorem 3.1. *If $\mathcal{P}_1 \leq_m^{\texttt{PsT}} \mathcal{P}_2$ and $\mathcal{P}_2 \leq_m^{\texttt{PsT}} \mathcal{P}_3$, then also $\mathcal{P}_1 \leq_m^{\texttt{PsT}} \mathcal{P}_3$.*

Proof. From $\mathcal{P}_1 \leq_m^{\mathtt{PsT}} \mathcal{P}_2$ and $\mathcal{P}_2 \leq_m^{\mathtt{PsT}} \mathcal{P}_3$, it is known that there exist four linear-size NC computable functions $f_1(\cdot)$, $f_2(\cdot)$, $f_1'(\cdot)$, and $f_2'(\cdot)$, and two PsT computable functions $g(\cdot,\cdot), g'(\cdot,\cdot)$ such that for any pair of strings $\langle D_1, P_1\rangle$ and $\langle D_2, P_2\rangle$ it holds that

$$\langle D_1, P_1\rangle \in \mathcal{P}_1 \Leftrightarrow \langle f_1(D_1), g(f_2(D_1), P_1)\rangle \in \mathcal{P}_2,$$

$$\langle D_2, P_2\rangle \in \mathcal{P}_2 \Leftrightarrow \langle f_1'(D_2), g'(f_2'(D_2), P_2)\rangle \in \mathcal{P}_3.$$

To show $\mathcal{P}_1 \leq_m^{\mathtt{PsT}} \mathcal{P}_3$, we define three functions $f_1''(\cdot)$, $f_2''(\cdot)$ and $g''(\cdot)$ as follows. Let $f_1''(x) = f_1'(f_1(x))$, $f_2''(x) = \llcorner|f_2(x)|\lrcorner\#f_2(x)\#f_2'(f_1(x))$ and $g''(x,y) = g'(q, g(p,y))$ if $x = \llcorner|p|\lrcorner\#p\#q$, where $\#$ is a special symbol that is not used anywhere else. Then we have

$$\begin{aligned}
\langle D_1, P_1\rangle \in \mathcal{P}_1 &\Leftrightarrow \langle f_1(D_1), g(f_2(D_1), P_1)\rangle \in \mathcal{P}_2\\
&\Leftrightarrow \langle f_1'(f_1(D_1)), g'(f_2'(f_1(D_1)), g(f_2(D_1), P_1))\rangle \in \mathcal{P}_3\\
&\Leftrightarrow \langle f_1''(D_1), g''(\llcorner|f_2(D_1)|\lrcorner\#f_2(D_1)\#f_2'(f_1(D_1)), P_1)\rangle \in \mathcal{P}_3\\
&\Leftrightarrow \langle f_1''(D_1), g''(f_2''(D_1), P_1)\rangle \in \mathcal{P}_3
\end{aligned}$$

With the fact that the concentration and composition of two linear-size NC computable function are still linear-size NC computable functions, it is easy to verify that $f_1''(\cdot), f_2''(\cdot)$ are linear-size NC computable. As for $g''(\cdot,\cdot)$, the total time needed for computing $g''(f_2''(D), P)$ is bounded by $O(t_g(|f_2(D)|) + t_{g'}(|f_2'(f_1(D))|) + \log|f_2(D)|) = o(|D|)$. This completes the proof. □

The pseudo-sublinear-time reduction is designed as a tool to prove that for some problems in P, there is no algorithm can solve it in sublinear time after a PTIME preprocessing. Hence, in addition to time restriction, we also limit the output size of $f_1(\cdot)$ and $f_2(\cdot)$ to ensure that PsT is closed under $\leq_m^{\mathtt{PsT}}$.

Theorem 3.2. *The complexity classes* P *and* PsT *is closed under* $\leq_m^{\mathtt{PsT}}$.

Proof. To show PsT is closed under $\leq_m^{\mathtt{PsT}}$, we claim that for all $\mathcal{P}_1$ and $\mathcal{P}_2$ if $\mathcal{P}_1 \leq_m^{\mathtt{PsT}} \mathcal{P}_2$ and $\mathcal{P}_2 \in \mathtt{PsT}$, then $\mathcal{P}_1 \in \mathtt{PsT}$. From $\mathcal{P}_1 \leq_m^{\mathtt{PsT}} \mathcal{P}_2$, we know that there exist two linear-size NC computable functions $f_1(\cdot)$ and $f_2(\cdot)$, and a PsT computable function $g(\cdot,\cdot)$ such that for any pair of strings $\langle D_1, P_1\rangle$ it holds that

$$\langle D_1, P_1\rangle \in \mathcal{P}_1 \Leftrightarrow \langle f_1(D_1), g(f_2(D_1), P_1)\rangle \in \mathcal{P}_2.$$

Furthermore, since $\mathcal{P}_2 \in \mathtt{PsT}$, there exists a PTIME preprocessing function $\Pi_2(\cdot)$ such that for any pair of strings $\langle D_2, P_2\rangle$ it holds that: $P_2(\Pi_2(D_2)) = P_2(D_2)$, and $P_2(\Pi_2(D_2))$ can be solved by a RATM M_2 in $o(|D_2|)$ time. Therefore, for any pair of strings $\langle D_1, P_1\rangle$ we have,

$$P_1(D_1) = g(f_2(D_1), P_1)(f_1(D_1)) = g(f_2(D_1), P_1)(\Pi_2(f_1(D_1))).$$

To show $\mathcal{P}_1 \in \mathtt{PsT}$, we define a PTIME preprocessing function $\Pi_1(\cdot)$ for $\mathcal{P}_1$ such that $P_1(D_1) = P_1(\Pi_1(D_1))$ and a RATM for $P_1(\Pi_1(D_1))$ running in sublinear time with respect to $|D_1|$. First, let $\Pi_1(x) = \llcorner|f_2(x)|\lrcorner\#f_2(x)\#\Pi_2(f_1(x))$, where # is a special symbol that is not used anywhere else. It is remarkable to see that $\llcorner|f_2(x)|\lrcorner$ is used to help us to distinguish the two parts of the input in logarithmic time. Then we construct a RATM M_1 by appending a pre-procedure to M_2. More concretely, with input $\Pi_1(D_1)$ and P_1, M_1 first copies $\llcorner|f_2(D_1)|\lrcorner$ to its work tap and computes the index of the second #, which equals to $|f_2(D_1)|+|\llcorner|f_2(D_1)|\lrcorner|+1$. Then M_1 generates $g(f_2(D_1), P_1)$ according to the information between the two #s. Finally, M_1 simulates the computation of M_2 with input $\Pi_2(f_1(D_1))$, the information behind the second #, and $g(f_2(D_1), P_1)$, then outputs the result returned by M_2.

Since $\Pi_2(\cdot)$ is PTIME computable, both $f_1(\cdot)$ and $f_2(\cdot)$ are NC computable, and the length of a string is logarithmic time computable, the running time of $\Pi_1(\cdot)$ can bounded by a polynomial. The time required by computing the index of the second # is $t_I = O(\log|f_2(D_1)|)$ And, $g(\cdot,\cdot)$ is computable in $o(|f_2(D_1)|)$ time. As both $f_1(\cdot)$ and $f_2(\cdot)$ are linear-size functions, the running time of M_1 is bounded by $t_I + t_g + t_{M_2} = O(\log|f_2(D_1)| + o(|f_2(D_1)|) + o(|f_1(D_1)|) = o(|D_1|)$. Thus, $\mathcal{P}_1 \in$ PsT.

As for P, we can consider another characterization for problems in P. That is, there is a PTIME preprocessing function $\Pi(\cdot)$ and a PTIME RATM M such that for any pair of strings $\langle D, P\rangle$ it holds that: $P(\Pi(D)) = P(D)$ and $P(\Pi(D))$ can be solved by M. Then with similar construction as above, it is easy to prove that P is closed under $\leq_m^{\texttt{PsT}}$. □

The reduction defines a partial order of computational difficulty of problems in a complexity class, and the complete problems are regarded as the hardest ones. Analogous to NP-completeness, the P-complete problems under $\leq_m^{\texttt{PsT}}$ can be considered as intractable problems in P\PsT if P ≠ PsT. However, we don't think it is appropriate to define that new P-completeness for the following reason. According to the proofs of the first complete problem of P (under NC reduction) and NP, we notice that the size of the resulted instance is always related to the running time of the Turing machine for the origin problem. Hence, the linear-size restriction of $f_1(\cdot)$ and $f_2(\cdot)$ may be too strict to hold. Nevertheless, we succeeded to find a natural problem in P\PsT. Then, based on it, we can establish the unconditional pseudo-sublinear-time intractability for problems in P\PsT.

Circuit Value Problem (CVP)

- **Given:** A Boolean circuit α, and inputs $x_1, \cdots, x_d$.
- **Problem:** Is the output of α is TRUE on inputs $x_1, \cdots, x_d$?

Theorem 3.3. *There is no algorithm can preprocess a circuit α in polynomial time and subsequently answer whether the output of α on the input $x_1, \cdots, x_d$ is* TRUE *in sublinear time. That is,* CVP ∈ P\PsT.

Proof. As stated in [10], given d variables, there are 2^{2^d} distinct boolean functions can be constructed in total. And each of them can be written as a full disjunctive normal from its truth table, which can easily represented by a circuit. Suppose CVP belongs to PsT, i.e., there is a PTIME preprocessing function $\Pi(\cdot)$ on α

such that for all interpretations of $x_1, \cdots, x_d$, $\alpha(x_1, \cdots, x_d) = \Pi(\alpha)(x_1, \cdots, x_d)$ can be computed in sublinear time with respect to $|\alpha|$. Consider any two distinct circuits α_1 and α_2 with the same variables $x_1, \cdots, x_d$. There exists an interpretation for $x_1, \cdots, x_d$ such that $\alpha_1(x_1, \cdots, x_d) \neq \alpha_2(x_1, \cdots, x_d)$. Consequently, $\Pi(\alpha_1) \neq \Pi(\alpha_2)$. Therefore, all these circuits have different outputs of the function $\Pi(\cdot)$. Since there are totally 2^{2^d} different circuits, then there should be at least 2^{2^d} different outputs of $\Pi(\cdot)$ on all these circuits. To denote these, the length of $\Pi(\alpha)$ should be at least $\log 2^{2^d} = 2^d$. This contradicts to $\Pi(\cdot)$ is PTIME computable by choosing $d = \omega(\log|\alpha|)$. □

Algebraic Equation Root Problem (AERP)

- **Given:** An algebraic equation P with variables $x_1, \cdots, x_d$, and an assignment $A = (a_1, \cdots, a_d)$.
- **Problem:** Is A a root of P?

Theorem 3.4. CVP $\leq_m^{\texttt{PsT}}$ AERP.

Proof. Assume we are given a boolean circuit α, we define a transformation of α into an equation P such that the output of α is TRUE on inputs $x_1, \cdots, x_d$ if and only if $A = (x_1, \cdots, x_n)$ is a root of P. First, let $f_1(\cdot)$, $f_2(\cdot)$ express the following procedure. Traverse α in a topological order: (1) if an AND gate with input u, v is met, represent it by $u \times v$, (2) if an OR gate with input u, v is met, represent it by $u + v - u \times v$, (3) if a NOT gate with input u, is met, represent it by $1 - u$, (4) if the final output gate z is met, represent it by $z = 1$. Then, for each x_i if the input x_i is TRUE, $g(f_2(\alpha), x_i) = 1$, otherwise, $g(f_2(\alpha), x_i) = 0$.

It is easy to see that the output of α is TRUE on inputs $x_1, \cdots, x_d$ if and only if $A = (g(f_2(\alpha), x_1), \cdots, g(f_2(\alpha), x_n))$ is a root of $f_1(\alpha)$. And as stated in [6], the topological traversal of a DAG can be computed in NC. Moreover, both $|f_1(\alpha)|$ and $|f_2(\alpha)|$ are less than $7|\alpha|$. And let $d = o(|\alpha|)$, $g(\cdot, \cdot)$ is PsT computable. □

Corollary 3.1. *There is no algorithm can preprocess an algebraic equation P in polynomial time and subsequently answer whether a given assignment A is a root of P in sublinear time.*

Also, $\leq_m^{\texttt{PsT}}$ can also be used to derive efficient algorithms for problems in PsT. In the breakthrough work of dynamic DFS on undirected graphs [2], Baswana et al. defined a nearest edge query between a subtree and an ancestor-descendant path in the procedure of rerooting a DFS tree, which was used in almost all subsequent work. Chen et al. showed that this query could be solved by running a range successor query [5]. We refine the procedure as a pseudo-sublinear-time reduction. The definitions of these two problems are given as follows.

Nearest Edge Query (NEQ)

- **Given:** A DFS tree T of graph G, the endpoints x, y of an ancestor-descendant path, the root w of a subtree $T(w)$ such that $par(w) \in path(x, y)$.

- **Problem:** Find the edge e that is incident nearest to x among all edges between $T(w)$ and $path(x, y)$.

Range Successor Query (RSQ)

- **Given:** A set of d-dimensional points S, a query rectangle $Q = \Pi_{i=1}^{d}[a_i, b_i]$.
- **Problem:** Find the point p with smallest x-coordinate among all points that are in the rectangle Q.

Theorem 3.5 [5]. NEQ $\leq_m^{\texttt{PsT}}$ RSQ.

Proof. Given a graph $G = (V, E)$ and a DFS tree T of G, define $f_1 : E \rightarrow S$ as follows, where S is a set of 2-dimensional points. Denote the preorder traversal sequence of T by ρ, note that every subtree of T can be represented by a continuous interval of ρ. Let $\rho(v)$ denote the index of vertex v in this sequence that is if v is the i-th element in ρ, then $\rho(v) = i$. For each edge $(u, v) \in E$, $f_1((u, v)) = (\rho(u), \rho(v))$. That is for each edge (u, v), a point $(\rho(u), \rho(v))$ is added into S. Notice that for each point $p \in S$, there exists exactly one edge (u, v) associated with p. Next we state the information provided by $f_2(\cdot)$. For each vertex v, let $\gamma(v) = \max_{w \in T(v)} \rho(w)$, i.e., the maximum index of vertices in $T(v)$. Thus, define $f_2(v)$ as $\rho(v)\#\gamma(v)$ for each $v \in V$.

Then, to answer an arbitrary query instance $T(w), p(x, y)$, let g be the function mapping w, x, y to a rectangles $\Omega = [\rho(x), \rho(w) - 1] \times [\rho(w), \gamma(w)]$. Finally, given a point $p \in S$ as the final result of RSQ, let $h(\cdot, \cdot)$ be reverse function of $f_1(\cdot)$, i.e., it returns the edge of G corresponding to p. It is easily to verify that the edge corresponding to the point with minimum x-coordinate is the edge nearest to x among all edges between $T(w)$ and $path(x, y)$ [5].

The preorder traversal sequence of T can be obtained by performing a DFS on it, which can be done in `NC` as stated in [16]. Therefore, both $f_1(\cdot)$ and $f_2(\cdot)$ are `NC` computable. Moreover, since each $e \in E$, there is a point $p = f_1(e)$ in S and for each point $p \in S$, there is exactly one edge e associated with p, we have $|f_1(G)| \in O(|G|)$. Similarly, for each vertex v, $f_2(v)$ records two values for it. Hence, $|f_2(G)| \in O(|G|)$. As for $g(\cdot, \cdot)$ and $h(\cdot, \cdot)$, with the mapping provided by $f_2(\cdot)$, both of them can be computed in sublinear time. □

Notice that for optimization problems, we need not only the functions converting the data part and problem part of $\mathcal{P}_1$ to corresponding part of $\mathcal{P}_2$, but also a function $h(\cdot, \cdot)$ mapping the solution of $\mathcal{P}_2$ back to the solution of $\mathcal{P}_1$. The resources restriction of $h(\cdot, \cdot)$ is set to be the same as $g(\cdot, \cdot)$. There is numerous work showing that RSQ belongs to `PsT` [13]. Hence, with the fact that the complexity class `PsT` is closed under $\leq_m^{\texttt{PsT}}$, the following corollary is obtained.

Corollary 3.2. NEQ $\in$ `PsT`.

4 Pseudo-polylog-Time Reduction

In this section, we introduce the notion of pseudo-polylog-time reduction, which will be used to clarify the difference between `PsT` and `PsPL`.

Definition 4.1. A decision problem $\mathcal{P}_1$ is *pseudo-polylog-time reducible* to a decision problem $\mathcal{P}_2$, denoted as $\mathcal{P}_1 \leq_m^{\mathtt{PsPL}} \mathcal{P}_2$, if there is a triple $\langle f_1(\cdot), f_2(\cdot), g(\cdot,\cdot)\rangle$, where $f_1(\cdot)$ and $f_2(\cdot)$ are NC computable functions and $g(\cdot,\cdot)$ is a PPL computable function, such that for any pair of strings $\langle D, P\rangle$ it holds that

$$\langle D, P\rangle \in \mathcal{P}_1 \Leftrightarrow \langle f_1(D), g(f_2(D), P)\rangle \in \mathcal{P}_2.$$

With similar proof of Theorem 3.1 and Theorem 3.2, we can show that $\leq_m^{\mathtt{PsPL}}$ is transitive and the complexity class PsPL is closed under $\leq_m^{\mathtt{PsPL}}$.

Theorem 4.1. *If* $\mathcal{P}_1 \leq_m^{\mathtt{PsPL}} \mathcal{P}_2$ *and* $\mathcal{P}_2 \leq_m^{\mathtt{PsPL}} \mathcal{P}_3$*, then also* $\mathcal{P}_1 \leq_m^{\mathtt{PsPL}} \mathcal{P}_3$*.*

Proof. From $\mathcal{P}_1 \leq_m^{\mathtt{PsPL}} \mathcal{P}_2$ and $\mathcal{P}_2 \leq_m^{\mathtt{PsPL}} \mathcal{P}_3$, it is known that there exist four NC computable functions $f_1(\cdot), f_1'(\cdot)$, $f_2(\cdot)$ $f_2'(\cdot)$, and two PPL computable functions $g(\cdot,\cdot), g'(\cdot,\cdot)$ such that for any pair of strings $\langle D_1, P_1\rangle$ and $\langle D_2, P_2\rangle$ it holds that

$$\langle D_1, P_1\rangle \in \mathcal{P}_1 \Leftrightarrow \langle f_1(D_1), g(f_2(D_1), P_1)\rangle \in \mathcal{P}_2,$$

$$\langle D_2, P_2\rangle \in \mathcal{P}_2 \Leftrightarrow \langle f_1'(D_2), g'(f_2'(D_2), P_2)\rangle \in \mathcal{P}_3.$$

To show $\mathcal{P}_1 \leq_m^{\mathtt{PsPL}} \mathcal{P}_3$, we define two NC computable functions $f_1''(\cdot)$, $f_2''(\cdot)$ and a PPL computable function $g''(\cdot)$ as follows. Let $f_1''(x) = f_1'(f_1(x))$, $f_2''(x) = \llcorner|f_2(x)|\lrcorner\#f_2(x)\#f_2'(f_1(x))$ and $g''(x,y) = g'(p, g(q,y))$ if $x = \llcorner|p|\lrcorner\#p\#q$, where $\#$ is a special that is not used anywhere else. Then we have

$$\begin{aligned}\langle D_1, P_1\rangle \in \mathcal{P}_1 &\Leftrightarrow \langle f_1(D_1), g(f_2(D_1), P_1)\rangle \in \mathcal{P}_2\\ &\Leftrightarrow \langle f_1'(f_1(D_1)), g'(f_2'(f_1(D_1)), g(f_2(D_1), P_1))\rangle \in \mathcal{P}_3\\ &\Leftrightarrow \langle f_1''(D_1), g''(\llcorner|f_2(D_1)|\lrcorner\#f_2(D_1)\#f_2'(f_1(D_1)), P_1)\rangle \in \mathcal{P}_3\\ &\Leftrightarrow \langle f_1''(D_1), g''(f_2''(D_1), P_1)\rangle \in \mathcal{P}_3\end{aligned}$$

It is easy to verify that $f_1''(\cdot), f_2''(\cdot)$ are in NC and $g''(\cdot,\cdot)$ is in PPL. □

Theorem 4.2. *The complexity class PsPL is closed under* $\leq_m^{\mathtt{PsPL}}$*.*

Proof. From $\mathcal{P}_1 \leq_m^{\mathtt{PsPL}} \mathcal{P}_2$, we know that there exist two NC computable functions $f_1(\cdot)$, $f_2(\cdot)$, and a PPL computable function $g(\cdot,\cdot)$ such that for any pair of strings $\langle D_1, P_1\rangle$ it holds that

$$\langle D_1, P_1\rangle \in \mathcal{P}_1 \Leftrightarrow \langle f_1(D_1), g(f_2(D_1), P_1)\rangle \in \mathcal{P}_2.$$

Furthermore, since $\mathcal{P}_2 \in$ PsPL, there exists a PTIME preprocessing function $\Pi_2(\cdot)$ such that for any pair of strings $\langle D_2, P_2\rangle$ it holds that: $P_2(\Pi_2(D_2)) = P_2(D_2)$, and $P_2(\Pi_2(D_2))$ can be solved by a RATM M_2 in $O(\log^{c_2}|D_2|)$ for some $c_2 \geq 1$. Therefore, for any pair of strings $\langle D_1, P_1\rangle$ we have,

$$P_1(D_1) = g(f_2(D_1), P_1)(f_1(D_1)) = g(f_2(D_1), P_1)(\Pi_2(f_1(D_1))).$$

To show $\mathcal{P}_1 \in$ PsPL, we claim that there exist a PTIME preprocessing function $\Pi_1(\cdot)$ for $\mathcal{P}_1$ such that $P_1(D_1) = P_1(\Pi_1(D_1))$ and a RATM for $P_1(\Pi_1(D_1))$

running in polylogarithmic time as required in Definition 2.3. First, let $\Pi_1(x) = \llcorner|f_2(x)|\lrcorner\#f_2(x)\#\Pi_2(f_1(x))$, where # is a special symbol that is not used anywhere else. Then we construct a RATM M_1 by appending a pre-procedure to M_2. More concretely, with input $\Pi_1(D_1)$ and P_1, M_1 first copies $\llcorner|f_2(x)|\lrcorner$ to one of its work tapes and computes the index of the second #, which equals to $|f_2(x)| + |\llcorner|f_2(x)|\lrcorner| + 1$. Then M_1 generates $g(f_2(D_1), P_1)$ according to the information between the two #s. Finally, M_1 simulates the computation of M_2 with input $\Pi_2(f_1(D_1))$ behind the second # and $g(f_2(D_1), P_1)$, then outputs the result returned by M_2.

Since $\Pi_2(\cdot)$ is in PTIME, $f_1(\cdot)$ and $f_2(\cdot)$ are in NC, and the length of a string is logarithmic time computable $\Pi_1(\cdot)$ is obviously in PTIME. Notice that computing the index of the second # requires $t_I = O(\log|f_2(D_1)|)$ time and $g(\cdot,\cdot)$ is computable in time $O(\log^{c_3}|f_2(D_1)|)$ for some $c_3 \geq 1$. Therefore, the total running time of M_1 is bounded by $t_I + t_g + t_{M_2} = O(\log^{c_3}|f_2(D_1)| + \log^{c_2}|f_1(D_1)|) = O(\log^{c_1}|D_1|)$ where $c_1 = \max\{c_2, c_3\}$. Thus, $\mathcal{P}_1 \in$ PsPL. □

Due to the limitations of fractional power functions, the complexity class PsT is not closed under $\leq_m^{\text{PsPL}}$ unless we add an addition linear-size restriction of function $f_1(\cdot)$. Fortunately, this does not prevent us from defining PsT-completeness.

Definition 4.2. A problem $\mathcal{P}$ is PsT-hard under $\leq_m^{\text{PsPL}}$ if $\mathcal{P}' \leq_m^{\text{PsPL}} \mathcal{P}$ for all $\mathcal{P}' \in$ PsT. A problem $\mathcal{P}$ is PsT-complete under $\leq_m^{\text{PsPL}}$ if $\mathcal{P}$ is PsT-hard and $\mathcal{P} \in$ PsT.

Identifying the PsT-complete problems may help us to separate PsT and PsPL. That is if there is a PsT-complete problem belonging to PsPL, then PsPL = PsT. In the following, we give a specified range of possible complete problems for PsT, by relating them to a well-known P-complete problem. Given a graph G, a depth-first search(DFS) traverses G in a particular order by picking an unvisited vertex v from the neighbors of the most recently visited vertex u to search, and backtracks to the vertex from where it came when a vertex u has explored all possible ways to search further.

Ordered Depth-First Search (ODFS)

- **Given:** A graph $G = (V, E)$ with fixed adjacent lists, fixed starting vertex s, and vertices u and v.
- **Problem:** Does vertex u get visited before vertex v in the DFS traversal of G starting from s?

Theorem 4.3 [15]. ODFS *is* P*-complete under* NC *reduction.*

Theorem 4.4. *Given a problem $\mathcal{P}$, if $\mathcal{P}$ is* PsT*-complete, then $\mathcal{P}$ is* P*-complete.*

Proof. It is easy to see that ODFS is in PsT. Since $\mathcal{P}$ is PsT-complete, ODFS $\leq_m^{\text{PsPL}} \mathcal{P}$. That is, there exist two NC computable functions $f_1(\cdot)$, $f_2(\cdot)$ and a PPL computable function $g(\cdot,\cdot)$ such that for all $\langle[G, s], [u, v]\rangle$ it holds that

$$\langle[G, s], [u, v]\rangle \in \text{ODFS} \Leftrightarrow \langle f_1([G, s]), g(f_2([G, s]), [u, v])\rangle \in \mathcal{P}.$$

As stated in Theorem 4.3, ODFS is $\mathtt{P}$-complete under $\mathtt{NC}$ reduction. For any problem $L \in \mathtt{P}$, there is a $\mathtt{NC}$ computable function $h(\cdot)$ such that

$$x \in L \Leftrightarrow h(x) \in \text{ODFS}.$$

Recall that the input of ODFS consists of a graph G, a starting point s, and two vertices u, v. It is easy to modify the output format of $h(x)$ to $\llcorner|[G,s]|\lrcorner\#[G,s]\#[u,v]$ in $\mathtt{NC}$, where $\#$ is a new symbol that is not used anywhere else. Now let $f_1'(x) = f_1(y)$ and $g'(x) = g(f_2(y), z)$, if $x = \llcorner|y|\lrcorner\#y\#z$. The two separators $\#$ can be founded in logarithmic time. Consequently, it follows that

$$x \in L \Leftrightarrow \langle h(x).[G,s], h(x).[u,v]\rangle \in \text{ODFS} \Leftrightarrow \langle f_1'(h(x)), g'(h(x))\rangle \in \mathcal{P}.$$

Let $h'(x) = f_1'(h(x)) \circ g'(h(x))$ to denote the concentration of two parts of $\mathcal{P}$ we can see that L is $\mathtt{NC}$ reducible to $\mathcal{P}$. Therefore, $\mathcal{P}$ is $\mathtt{P}$-complete. □

5 Approximation Preserving Pseudo-sublinear-Time Reduction

A natural approach to cope with problems in $\mathtt{P}\backslash\mathtt{PsT}$ or that are $\mathtt{PsT}$-complete is to design pseudo-sublinear-time approximation algorithm. Hence, in this section, we propose the pseudo-sublinear-time L-reduction, and prove that it linearly preserves approximation ratio for pseudo-sublinear-time approximation algorithms.

Let $\mathcal{P}$ be a big data optimization problem, given a dataset D and a problem instance $P \in \mathcal{P}$ defined on D, let $P(D)$ denote the set of feasible solutions of P, and for any feasible solution $y \in P(D)$, let $\tau_{\mathcal{P}}(y)$ denote the positive measure of y, which is called the objective function. The goal of an optimization problem with respect to a problem instance $P \in \mathcal{P}$ is to find an optimum solution, that is, a feasible solution y such that $\tau_{\mathcal{P}}(y) = \{\max, \min\}\{\tau_{\mathcal{P}}(y') : y' \in P(D)\}$. In the following, $\mathsf{opt}_{\mathcal{P}}$ will denote the function mapping an instance $P \in \mathcal{P}$ defined on D to the measure of an optimum solution.

What's more, for each feasible solution y of D, P, the *approximation ratio* of y with respect to D, P is defined as $\rho(D, P, y) = \max\left\{\frac{\tau_{\mathcal{P}}(y)}{\mathsf{opt}_{\mathcal{P}}(D,P)}, \frac{\mathsf{opt}_{\mathcal{P}}(D,P)}{\tau_{\mathcal{P}}(y)}\right\}$. The approximation ratio is always a number greater than or equal to 1 and is as close to 1 as the value of the feasible solution is close to the optimum value. Let $\mathcal{A}$ be an algorithm that for any D and problem instance $P \in \mathcal{P}$ defined on D, returns a feasible solution $\mathcal{A}(\Pi(D), P)$ in sublinear time after a $\mathtt{PTIME}$ preprocessing $\Pi(\cdot)$. Given a rational $r \geq 1$, we say that $\mathcal{A}$ is an r-approximation algorithm for $\mathcal{P}$ if the approximation ratio of the feasible solution $\mathcal{A}(\Pi(D), P)$ with respect to D, P satisfies $\rho_{\mathcal{A}}(D, P, \mathcal{A}(\Pi(D), P)) \leq r$.

Definition 5.1. A problem $\mathcal{P}_1$ is *pseudo-polylog-time L-reducible* to a problem $\mathcal{P}_2$, denoted as $\mathcal{P}_1 \leq_L^{\mathtt{PsPL}} \mathcal{P}_2$, if there is a pseudo-polylog-time reduction $\langle f_1(\cdot), f_2(\cdot), g(\cdot,\cdot), h(\cdot,\cdot)\rangle$ from $\mathcal{P}_1$ to $\mathcal{P}_2$ such that for all D and $P \in \mathcal{P}_1$ defined on D it holds that:

1. $\mathsf{opt}_{\mathcal{P}_2}(f_1(D), g(f_2(D), P)) \le \alpha \cdot \mathsf{opt}_{\mathcal{P}_1}(D, P)$
2. for any $y \in \mathsf{sol}_{\mathcal{P}_2}(f_1(D), g(f_2(D), P))$,

$$|\mathsf{opt}_{\mathcal{P}_1}(D, P) - \tau_{\mathcal{P}_1}(h(f_2(D), y))| \le \beta \cdot |\mathsf{opt}_{\mathcal{P}_2}(f_1(D), g(f_2(D), P)) - \tau_{\mathcal{P}_2}(y)|.$$

Theorem 5.1. *Given two problems $\mathcal{P}_1$ and $\mathcal{P}_2$, if $\mathcal{P}_1 \le_L^{\mathtt{PsPL}} \mathcal{P}_2$ with parameter α and β and there is a pseudo-polylog-time $(1+\delta)$-approximation algorithm for $\mathcal{P}_2$, then there is a pseudo-polylog-time $(1+\gamma)$-approximation algorithm for $\mathcal{P}_1$, where $\gamma = \alpha\beta \cdot \delta$ if $\mathcal{P}_1$ is a minimization problem and $\gamma = \frac{\alpha\beta\delta}{1-\alpha\beta\delta}$ if $\mathcal{P}_1$ is a maximization problem.*

Proof. The algorithm for $\mathcal{P}_1$ is constructed as stated in the proof of Theorem 4.2. Then, if $\mathcal{P}_1$ is a minimization problem, it holds that

$$\begin{aligned}\frac{\tau_{\mathcal{P}_1}(h(D,P,y))}{\mathtt{opt}_{\mathcal{P}_1}(D,P)} &= \frac{\mathtt{opt}_{\mathcal{P}_1}(D,P) + \tau_{\mathcal{P}_1}(h(D,P,y)) - \mathtt{opt}_{\mathcal{P}_1}(D,P)}{\mathtt{opt}_{\mathcal{P}_1}(D,P)} \\ &\le \frac{\mathtt{opt}_{\mathcal{P}_1}(D,P) + \beta \cdot |\tau_{\mathcal{P}_2}(y) - \mathtt{opt}_{\mathcal{P}_2}(f(D), g(D,P))|}{\mathtt{opt}_{\mathcal{P}_1}(D,P)} \\ &\le 1 + \alpha\beta \cdot \left|\frac{\tau_{\mathcal{P}_2}(y) - \mathtt{opt}_{\mathcal{P}_2}(f(D), g(D,P)))}{\mathtt{opt}_{\mathcal{P}_2}(f(D), g(D,P))}\right|\end{aligned}$$

Thus we obtain a $(1+\alpha\beta \cdot \delta)$-approximation algorithm for $\mathcal{P}_1$. And, if $\mathcal{P}_1$ is a maximization problem, it holds that

$$\begin{aligned}\frac{\tau_{\mathcal{P}_1}(h(D,P,y))}{\mathtt{opt}_{\mathcal{P}_1}(D,P)} &= \frac{\mathtt{opt}_{\mathcal{P}_1}(D,P) + \tau_{\mathcal{P}_1}(h(D,P,y)) - \mathtt{opt}_{\mathcal{P}_1}(D,P)}{\mathtt{opt}_{\mathcal{P}_1}(D,P)} \\ &\ge \frac{\mathtt{opt}_{\mathcal{P}_1}(D,P) - \beta \cdot |\mathtt{opt}_{\mathcal{P}_2}(f(D), g(D,P)) - \tau_{\mathcal{P}_2}(y)|}{\mathtt{opt}_{\mathcal{P}_1}(D,P)} \\ &\ge 1 - \alpha\beta \cdot \left|\frac{\mathtt{opt}_{\mathcal{P}_2}(f(D), g(D,P)) - \tau_{\mathcal{P}_2}(y))}{\mathtt{opt}_{\mathcal{P}_2}(f(D), g(D,P))}\right|\end{aligned}$$

Thus the algorithm is a $(1+\frac{\alpha\beta\delta}{1-\alpha\beta\delta})$-approximation algorithm for $\mathcal{P}_1$. □

It is easy to extend the above definition in the context of pseudo-sublinear-time reduction. Hence, the following theorem is derived.

Theorem 5.2. *Given two problems $\mathcal{P}_1$ and $\mathcal{P}_2$, if $\mathcal{P}_1 \le_L^{\mathtt{PsT}} \mathcal{P}_2$ with parameter α and β and there is a pseudo-sublinear-time $(1+\delta)$-approximation algorithm for $\mathcal{P}_2$, then there is a pseudo-sublinear-time $(1+\gamma)$-approximation algorithm for $\mathcal{P}_1$, where $\gamma = \alpha\beta \cdot \delta$ if $\mathcal{P}_1$ is a minimization problem and $\gamma = \frac{\alpha\beta\delta}{1-\alpha\beta\delta}$ if $\mathcal{P}_1$ is a maximization problem.*

6 Complete Problems in PPL

We have shown that PPL is closed under DLOGTIME reduction and defined PPL-completeness in [9]. However, we did not manage to find the first natural PPL-complete problem. In this section, we give a negative answer to the existence of PPL-complete problems.

Lemma 6.1 [9]. *For any two problems $\mathcal{P}_1$ and $\mathcal{P}_2$, if $\mathcal{P}_2 \in \mathtt{PPL}^i$, and there is a* DLOGTIME *reduction from $\mathcal{P}_1$ to $\mathcal{P}_2$, then $\mathcal{P}_1 \in \mathtt{PPL}^{i+1}$.*

Theorem 6.1 [9]. *For any $i \in \mathbb{N}$, $\mathtt{PPL}^i \subsetneq \mathtt{PPL}^{i+1}$.*

Theorem 6.2. *There is no* PPL*-complete problem under* DLOGTIME *reduction.*

Proof. For contradiction, suppose there is a PPL-complete problem $\mathcal{P}$ under DLOGTIME reduction. Hence, there is a constant $c \geq 1$ such that $\mathcal{P} \in \mathtt{PPL}^c$. For Theorem 6.1, for any $i \in \mathbb{N}$, there is a problem $\mathcal{P}_{i+1}$ which belongs to $\mathtt{PPL}^{i+1}$ but not to $\mathtt{PPL}^i$. Let $k = c+1$. Since $\mathcal{P}$ is PPL-complete, there is a DLOGTIME reduction from $\mathcal{P}_{k+1}$ to $\mathcal{P}$. From Lemma 6.1, it is derived that $\mathcal{P}_{k+1} \in \mathtt{PPL}^{c+1} = \mathtt{PPL}^k$. This contradicts to the fact that $\mathcal{P}_{k+1} \in \mathtt{PPL}^{k+1} \backslash \mathtt{PPL}^k$. □

Notice that every un-trivial problems in $\mathtt{PPL}^1$ is $\mathtt{PPL}^1$-complete under DLOGTIME reduction. It is still meaningful to find complete problems of each level in PPL hierarchy.

7 Conclusion

This paper studies the pseudo-sublinear-time reductions specialized for problems in big data computing. Two concrete reductions $\leq_m^{\mathtt{PsT}}$ and $\leq_m^{\mathtt{PsPL}}$ are proposed. It is proved that the complexity classes P and PsT are closed under $\leq_m^{\mathtt{PsT}}$, and the complexity class PsPL is closed under $\leq_m^{\mathtt{PsPL}}$. These provide powerful tools not only for designing pseudo-sublinear-time algorithms for some problems, but also for proving certain problems are infeasible in sublinear time after a PTIME preprocessing. More concretely, based on the fact that circuit value problem belongs to $\mathtt{P} \backslash \mathtt{PsT}$, the algebraic equation root problem is proved not in PsT by establish a $\leq_m^{\mathtt{PsT}}$ reduction from CVP to it. Since CVP is P-complete under NC reduction, it may turn out to be an excellent starting point for many results, yielding pseudo-sublinear-time reductions for fundamental problems and giving unconditional pseudo-sublinear intractable results. Then to separate PsT and PsPL, the PsT-completeness is defined under $\leq_m^{\mathtt{PsPL}}$. We give out a range of possible PsT-complete problems by proving that all of them are also P-complete under NC reduction. We also extend the L-reduction to pseudo-sublinear time and prove it linearly preserves approximation ratio for pseudo-sublinear-time approximation algorithms. Finally, we give an negative answer to the existence of PPL-complete problems under DLOGTIME reduction. This may guide the following efforts focusing on finding complete problems for each level of PPL hierarchy.

Acknowledgment. This work was supported by the National Natural Science Foundation of China under grants 61732003, 61832003, 61972110 and U1811461.

References

1. SSD ranking: the fastest solid state drives. https://www.gamingpcbuilder.com/ssd-ranking-the-fastest-solid-state-drives/. Accessed 4 Aug 2021

2. Baswana, S., Chaudhury, S.R., Choudhary, K., Khan, S.: Dynamic DFS in undirected graphs: breaking the $o(m)$ barrier. In: Krauthgamer, R. (ed.) Proceedings of the Twenty-Seventh Annual ACM-SIAM Symposium on Discrete Algorithms, SODA 2016, Arlington, VA, USA, 10–12 January 2016, pp. 730–739. SIAM (2016)
3. Bringmann, K.: Fine-grained complexity theory (tutorial). In: 36th International Symposium on Theoretical Aspects of Computer Science (STACS 2019). Schloss Dagstuhl-Leibniz-Zentrum fuer Informatik (2019)
4. Cadoli, M., Donini, F.M., Liberatore, P., Schaerf, M.: Preprocessing of intractable problems. Inf. Comput. **176**(2), 89–120 (2002)
5. Chen, L., Duan, R., Wang, R., Zhang, H., Zhang, T.: An improved algorithm for incremental DFS tree in undirected graphs. In: Eppstein, D. (ed.) 16th Scandinavian Symposium and Workshops on Algorithm Theory, SWAT 2018. Volume 101 of LIPIcs, Malmö, Sweden, 18–20 June 2018, pp. 16:1–16:12. Schloss Dagstuhl - Leibniz-Zentrum für Informatik (2018)
6. Cook, S.A.: A taxonomy of problems with fast parallel algorithms. Inf. Control **64**(1–3), 2–22 (1985)
7. Crescenzi, P.: A short guide to approximation preserving reductions. In: Proceedings of the Twelfth Annual IEEE Conference on Computational Complexity, Ulm, Germany, 24–27 June 1997, pp. 262–273. IEEE Computer Society (1997)
8. Fan, W., Geerts, F., Neven, F.: Making queries tractable on big data with preprocessing. Proc. VLDB Endow. **6**(9), 685–696 (2013)
9. Gao, X., Li, J., Miao, D., Liu, X.: Recognizing the tractability in big data computing. Theor. Comput. Sci. **838**, 195–207 (2020)
10. Holdsworth, B., Woods, R.C.: Karnaugh maps and function simplification. In: Holdsworth, B., Woods, R.C. (eds.) Digital Logic Design, 4th edn, pp. 43–80. Newnes, Oxford (2002)
11. Rogers, H., Jr.: Theory of Recursive Functions and Effective Computability. MIT Press, Cambridge (1987). (Reprint from 1967)
12. Li, J.: Complexity, algorithms and quality of big data intensive computing. In: Database Systems for Advanced Applications - 19th International Conference, DASFAA 2014, Bali, Indonesia. Springer (2014)
13. Nekrich, Y., Navarro, G.: Sorted range reporting. In: Fomin, F.V., Kaski, P. (eds.) SWAT 2012. LNCS, vol. 7357, pp. 271–282. Springer, Heidelberg (2012). https://doi.org/10.1007/978-3-642-31155-0_24
14. Papadimitriou, C.H.: Computational Complexity. Addison-Wesley, Reading (1994)
15. Reif, J.H.: Depth-first search is inherently sequential. Inf. Process. Lett. **20**(5), 229–234 (1985)
16. Smith, J.R.: Parallel algorithms for depth-first searches I. Planar graphs. SIAM J. Comput. **15**(3), 814–830 (1986)
17. Williams, V.V.: Hardness of easy problems: basing hardness on popular conjectures such as the strong exponential time hypothesis (invited talk). In: Husfeldt, T., Kanj, I.A. (eds.) 10th International Symposium on Parameterized and Exact Computation, IPEC 2015, Volume 43 of LIPIcs, Patras, Greece, 16–18 September 2015, pp. 17–29. Schloss Dagstuhl - Leibniz-Zentrum für Informatik (2015)
18. Williams, V.V.: On some fine-grained questions in algorithms and complexity. In: Proceedings of the International Congress of Mathematicians: Rio de Janeiro 2018, pp. 3447–3487. World Scientific (2018)
19. Yang, J., Wang, H., Cao, Y.: Tractable queries on big data via preprocessing with logarithmic-size output. Knowl. Inf. Syst. **56**(1), 141–163 (2017). https://doi.org/10.1007/s10115-017-1092-7

Capacitated Partial Inverse Maximum Spanning Tree Under the Weighted l_∞-norm

Xianyue Li[1(✉)], Ruowang Yang[1], Heping Zhang[1], and Zhao Zhang[2]

[1] School of Mathematics and Statistics, Lanzhou University, Lanzhou 730000, Gansu, People's Republic of China
lixianyue@lzu.edu.cn

[2] College of Mathematics Physics and Information Engineering, Zhejiang Normal University, Jinhua 321004, Zhejiang, People's Republic of China

Abstract. Given an edge weighted graph, and an acyclic edge set, the goal of the partial inverse maximum spanning tree problem is to modify the weight function as small as possible such that there exists a maximum spanning tree with respect to the new weight function containing the given edge set. In this paper, we consider this problem with capacitated constraint under the weighted l_∞-norm. By studying the properties of the optimal value and a special kind of optimal solutions, combining the algorithm for the decision version of this problem with the Binary search method, we present a strongly polynomial-time algorithm for calculating the optimal value and an optimal solution.

Keywords: Partial inverse problem · Spanning tree · Polynomial time algorithm

1 Introduction

Many classical 0–1 combinatorial optimization problems can be written as $(E, \mathcal{T}, w)$, where E is a ground set, $\mathcal{T} \subseteq 2^E$ is the set of feasible solutions, w is a weight function on E, and for any $T \in 2^E$, $w(T) = \sum_{e \in T} w(e)$. The objective of $(E, \mathcal{T}, w)$ is to find an optimal solution $T^* \in \mathcal{T}$ such that

$$w(T^*) = opt_{T \in \mathcal{T}} w(T),$$

where "opt" can be "max" or "min". Given a combinatorial optimization problem $P = (E, \mathcal{T}, w)$ and a partial solution T' (contained in some feasible solutions), the *partial inverse problem* on P is to find a new weight function w^* such

Supported by National Numerical Windtunnel Project (No. NNW2019ZT5-B16), National Natural Science Foundation of China (Nos. 11771013, 11871256, 12071194, U20A2068), and the Basic Research Project of Qinghai (No. 2021-ZJ-703), Zhejiang Provincial Natural Science Foundation of China (No. LD19A010001).

D.-Z. Du et al. (Eds.): COCOA 2021, LNCS 13135, pp. 389–399, 2021.
https://doi.org/10.1007/978-3-030-92681-6_31

that T' can be extended to an optimal solution with respect to w^* and the difference between w and w^* is minimized. Researchers usually use the Hamming distance $\|\cdot\|_H$ and the l_p-norm $\|\cdot\|_p$ ($p \geq 1$ is an integer or $p = \infty$) to measure the difference. A partial inverse problem is called as *capacitated*, if for any element $e \in E$, $-l(e) \leq w^*(e) - w(e) \leq u(e)$, where l and u are two nonnegative functions on E, called as decreasing and increasing bound functions, respectively.

By studying the relationship between partial inverse linear programming problem and bi-level linear programming, researchers showed that the partial inverse linear programming problem are strongly NP-Hard [1,5]. Yang [11] showed that the capacitated partial inverse assignment problem and the capacitated partial minimum cut problem under the l_1-norm are APX-Hard. Without the capacitated constraint, Gassner [4] showed that the partial inverse minimum cut problem under the l_1-norm is NP-Hard, and Yang and Zhang [12] presented a strongly polynomial time algorithm to solve the partial inverse assignment problem. Yang and Zhang [13] also studied the partial inverse sorting problem and showed that this problem under the l_p-norm ($p = 1, 2, \infty$) can be solved in polynomial-time. For the special case when the partial solution has only one element, Lai and Orlin [6] proved that under the l_∞-norm, decision versions of the partial inverse shortest path problem on acyclic graphs, the partial inverse assignment problem, the partial inverse minimum cost arc (or vertex) disjoint cycle problem, and the partial inverse minimum cut problem are all NP-Complete.

Spanning tree problem is a very classical and famous combinatorial optimization problem. The *capacitated partial inverse maximum spanning tree problem* (abbreviated as CPIMST) have received extensive attention from researchers. Lai and Orlin [6] firstly showed that the decision version of un-capacitated problem under the weighted l_∞-norm can be solved in strongly polynomial-time. Cai et al. [3] considered a special case of CPIMST when $l \equiv 0$ (in that paper, they considered the capacitated partial inverse *minimum* spanning tree problem with $u \equiv 0$, the two problems are equivalent) and they presented a strongly polynomial-time algorithm to solve it. Zhang et al. [14] generalized above algorithm to solve the capacitated partial inverse maximum cost base of matroid problem with $l \equiv 0$. On the other hand, when $u \equiv 0$, Li et al. [7] showed that CPIMST under the l_∞-norm can be solved in polynomial-time. But under the l_p-norm, Li et al. [9] showed that if the partial solution has at least two edges, CPIMST is APX-Hard, and if the partial solution has only one edge, it can be solved in strongly polynomial-time. In [8], the authors showed that CPIMST under the weighted sum Hamming distance is APX-Hard and it can be solved in strongly polynomial-time under the weighted bottleneck Hamming distance. Recently, Li et al. [10] presented approximation algorithms for CPIMST under the weight l_p-norm and the weighted sum Hamming distance. These are the first approximate algorithms for partial inverse and inverse combinatorial optimization problems.

In this paper, we study CPIMST under the weighted l_∞-norm and present a strongly polynomial-time algorithm to solve it. Therefore, the computational complexity of CPIMST is completely solved.

This paper is organized as follows. The main results are presented in Sect. 2. In detail, we firstly present an algorithm for the decision version of this problem. Then, by characterizing the properties of the optimal value and a special kind of optimal solutions of CPIMST under the un-weighted l_∞-norm, we present a strongly polynomial-time algorithm to solve it. Finally, we generalize the results from the un-weighted l_∞-norm to the weighted l_∞-norm. In Sect. 3, we make a conclusion.

2 Main Results

At the beginning of this section, we give the formal definition of CPIMST under the weighted l_∞-norm and introduce some useful notations.

Definition 1. *Given a graph $G = (V, E)$ with an edge weight function w, an edge subset E' such that $G[E']$ is acyclic, a positive norm-weight function c, and non-negative decreasing and increasing bound functions l and u, the goal of* ***CPIMST*** *is to find a new weight function w^* satisfying:*

(1) there exists a maximum spanning tree T of G with respect to w^* such that $E' \subseteq E(T)$;
(2) $-l(e) \leq w^*(e) - w(e) \leq u(e)$, for any edge $e \in E$;
(3) $\| w - w^* \|_\infty = \max_{e \in E}\{c(e)|w^*(e) - w(e)|\}$ is minimum.

A weight function w' is ***feasible*** *if it satisfies the first two conditions.*

Let $G = (V, E)$ be a graph, and T be a spanning tree of G. For any edge $e \in E(G) - E(T)$, $T + e$ has a unique cycle, which is called the *fundamental cycle* with respect to T and e, and denoted by $C(T, e)$. It is easy to see that for any edge $e' \in C(T, e)$, $T + e - e'$ is also a spanning tree of G. On the other hand, for any edge $e \in E(T)$, $T - e$ has exactly two components. The set of edges connecting these two components is called the *fundamental cut* with respect to T and e, and denoted by $K(T, e)$. Similarly, for any edge $e' \in K(T, e)$, $T - e + e'$ is also a spanning tree of G. The following two equations are the well-known necessary and sufficient conditions for a spanning tree T of G to be maximum (see Exercises 6.2.2 and 6.2.3 in [2]),

$$\text{for any edge } e \in E(T),\ w(e) = \max\{w(y) : y \in K(T, e)\};$$
$$\text{for any edge } e \in E(G) \setminus E(T),\ w(e) = \min\{w(y) : y \in C(T, e)\}.$$

Clearly, these two equations are equivalent to the following two inequalities, respectively.

$$\text{for any edge } e \in E(T),\ w(e) \geq \max\{w(y) : y \in K^0(T, e)\}; \tag{1}$$
$$\text{for any edge } e \in E(G) \setminus E(T),\ w(e) \leq \min\{w(y) : y \in C^0(T, e)\}, \tag{2}$$

where $K^0(T, e) = K(T, e) \setminus \{e\}$ and $C^0(T, e) = C(T, e) \setminus \{e\}$.

2.1 Algorithm for the Decision Version of CPIMST

Let $I = (G, w, E', l, u, c)$ be an instance of CPIMST under the weighted l_∞-norm, and $K > 0$ be a given real number. The decision version of I is to ask whether there exists a feasible weight function w' such that $\|w' - w\|_\infty \leq K$. Lai and Orlin [6] showed that if $l = u \equiv \infty$, which is the PIMST problem, can be solved in strongly polynomial-time. The main idea is as follows. Firstly, they defined a new weight function w_K with

$$w_K(e) = \begin{cases} w(e) + K/c(e), e \in E'; \\ w(e) - K/c(e), e \notin E'. \end{cases} \quad (3)$$

Then, they showed that the answer is "YES" if and only if w_K is a feasible solution of I. We can generalize this method for CPIMST directly and obtain the following result and algorithm.

Theorem 1. *Let I be an instance of CPIMST and $K > 0$ be a given real number. There exists a feasible weight function w' with $\|w' - w\|_\infty \leq K$ if and only if w_K is feasible, where*

$$w_K(e) = \begin{cases} w(e) + \min\{K/c(e), u(e)\}, e \in E'; \\ w(e) - \min\{K/c(e), l(e)\}, \ e \notin E'. \end{cases} \quad (4)$$

Algorithm 1: Algorithm for the decision version of CPIMST under the weighted l_∞-norm

---

Input: An instance $I = (G, w, E', l, r, c)$ of CPIMST and a real number $K > 0$.

Output: "YES" or "NO".

1 Calculate w_K by Eq. (4);
2 **if** *there exists a maximum spanning tree of G with respect to w_K containing E'* **then**
3 | **return** "YES";
4 **else**
5 | **return** "NO";
6 **end**

Naturally, combining Algorithm 1 with the Binary search method, we can design an algorithm to solve CPIMST under the weighted l_∞-norm within any prescribed additive error. However, this algorithm has two drawbacks. First, the number of iterations in the Binary search method is related to the size of the input data. Thus, it is not a strongly polynomial-time algorithm. Second, there may exist error between the result obtained by this algorithm and the exactly optimal value. Hence, it is not trivial to design a strongly polynomial-time algorithm for CPIMST under the weighted l_∞-norm.

2.2 Algorithm for CPIMST Under the Un-Weighted l_∞-norm

Although the Binary search method can not generate a strongly polynomial-time algorithm directly, if we can obtain a candidate set of optimal values, the size of which is a polynomial function related to n and m, the algorithm obtained from combining Algorithm 1 with the Binary search method will be strongly polynomial-time. To achieve this goal, we study properties of the optimal value and the optimal solutions of this problem. For the ease of statement, we first consider CPIMST under the (un-weighted) l_∞-norm ($c(e) = 1$ for all edges). Then, we generalize the results to the weighted l_∞-norm in the next subsection.

Let $I = (G, w, E', l, u, c \equiv 1)$ be an instance of CPIMST under the l_∞-norm, and *opt* be the optimal value of I. Clearly, the optimal solution of I may not be unique. Figure 1 illustrates an instance I which has at least two different optimal solutions. We define an optimal solution w^* to be a *minimal optimal solution* if

(i) the size of w^*, defined to be $|R(w^*)| = |\{e \in E : |w^*(e) - w(e)| = opt\}|$, is minimum among all optimal solutions;
(ii) under condition (i), $\| w^* - w \|_1 = \sum_{e \in E} |w(e^*) - w(e)|$ is minimum.

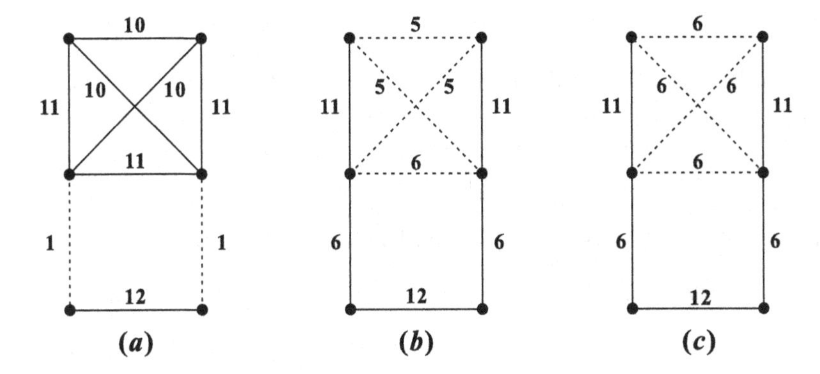

Fig. 1. (a) An instance I of PIMST (CPIMST with $l \equiv u \equiv \infty$) under the l_∞-norm and the E' are the two dashed edges; (b) an optimal solution of I with $opt = 5$ and the solid lines form a maximum spanning tree containing E'; (c) a minimal optimal solution of I.

The following lemma shows that minimal optimal solutions have *separate property*, that is, the weights of all edges in E' are non-decreasing and the weights of all the other edges are non-increasing.

Lemma 1. *Let I be an instance of CPIMST under the l_∞-norm, w^* be a minimal optimal solution of I, and T^* be a maximum spanning tree of G with respect to w^* containing E'. Then,*

$$w^*(e) \begin{cases} \geq w(e), \, e \in E'; \\ = w(e), \, e \in E(T^*) \backslash E'; \\ \leq w(e), \, e \notin E(T^*). \end{cases} \tag{5}$$

Proof. For the first inequality, suppose to the contrary that there is an edge $e' \in E'$ with $w^*(e') < w(e')$. Let

$$w'(e) = \begin{cases} w(e'), \ e = e'; \\ w^*(e), \text{otherwise}. \end{cases} \tag{6}$$

Clearly, $-l(e') \leq w'(e') - w(e') = 0 \leq u(e')$ and $-l(e) \leq w'(e) - w(e) = w^*(e) - w(e) \leq u(e)$ for any other edge e. Thus, the new weight function w' satisfies the capacitated constraint.

Furthermore, by Eq. (1),

$$\begin{aligned} w'(e') = w(e') > w^*(e') & \\ & \geq \max\{w^*(y) : y \in K^0(T^*, e')\} \\ & = \max\{w'(y) : y \in K^0(T^*, e')\}. \end{aligned}$$

And for any edge $e \in E(T^*) \setminus \{e'\}$,

$$w'(e) = w^*(e) \geq \max\{w^*(y) : y \in K^0(T^*, e)\} = \max\{w'(y) : y \in K^0(T^*, e)\},$$

where the last equality holds because $e' \notin K^0(T^*, e)$. So by Eq. (1), T^* is also a maximum spanning tree of G with respect to w' containing E'. Combining this with $|w'(e) - w(e)| \leq |w^*(e) - w(e)|$ for any edge e, w' is also an optimal solution. But, $R(w') \subseteq R(w^*)$ and $\| w' - w \|_1 < \| w^* - w \|_1$, which implies that w^* can not be a minimal optimal solution.

The other inequalities can be proved by similar arguments.

Based on the separate property, we can obtain the following result.

Lemma 2. *Let w^* be a minimal optimal solution of I, and T^* be a maximum spanning tree of G with respect to w^* containing E'. For any edge $e' \in E'$ with $w^*(e') > w(e')$, there exists an edge $\bar{e} \in K^0(T^*, e')$ such that $w^*(\bar{e}) = w^*(e')$. Similarly, for any edge $\bar{e} \notin E(T^*)$ with $w^*(\bar{e}) < w(\bar{e})$, there exists an edge $e' \in C(T^*, e) \cap E'$ such that $w^*(e') = w^*(\bar{e})$.*

Proof. For the first part of this lemma, we suppose that there is an edge $e' \in E'$ with $w^*(e') > w(e')$, but $w^*(\bar{e}) \neq w^*(e')$ for any edge $\bar{e} \in K^0(T^*, e')$. Then, $w^*(\bar{e}) < w^*(e')$ for any $\bar{e} \in K^0(T^*, e')$ since T^* is maximum and because of Eq. (1). Let $q = \max\{w^*(\bar{e}) : \bar{e} \in K^0(T^*, e')\}$ and

$$w'(e) = \begin{cases} \max\{w(e'), q\}, \ e = e'; \\ w^*(e), \qquad\qquad \text{otherwise}. \end{cases} \tag{7}$$

We can obtain that

$$-l(e') \leq 0 \leq w'(e') - w(e') \leq w^*(e') - w(e') \leq u(e).$$

Since $w'(e) \geq q = \max\{w'(y) : y \in K^0(T^*, e')\}$, by a similar argument in the proof of Lemma 1, the spanning tree T^* is also maximum with respect to w'.

If $e' \in R(w^*)$, then either $\| w' - w \|_\infty < \| w^* - w \|_\infty$ or $\| w' - w \|_\infty = \| w^* - w \|_\infty$ and $R(w') \subset R(w^*)$; if $e' \notin R(w^*)$, then $\| w' - w \|_\infty = \| w^* - w \|_\infty$ and $R(w') = R(w^*)$, but $\| w' - w \|_1 < \| w^* - w \|_1$. Both of them contradict with the assumption that w^* is a minimal optimal solution.

To show the second part of this lemma, suppose that $\bar{e} \notin E(T^*)$ has $w(\bar{e}) > w^*(\bar{e})$, and for any edge $e' \in C(T^*, \bar{e}) \cap E'$, $w^*(e') \neq w^*(\bar{e})$. Since T^* is maximum, $w^*(\bar{e}) < w^*(e')$ for any edge $e' \in C(T^*, \bar{e}) \cap E'$ by Eq. (2). Let $t = \min\{w^*(e') : e' \in C(T^*, \bar{e}) \cap E'\}$ and

$$w'(e) = \begin{cases} \min\{w(\bar{e}), t\}, & e = \bar{e}; \\ w^*(e), & \text{otherwise}. \end{cases} \tag{8}$$

Then, $-l(\bar{e}) \leq w^*(\bar{e}) - w(\bar{e}) \leq w'(\bar{e}) - w(\bar{e}) \leq 0 \leq u(\bar{e})$. If w' is a feasible solution, by a similar argument as the first part, w' is better than w^* and w^* cannot be a minimal optimal solution. Hence, we shall show that w' is feasible by proving that w' satisfies Eq. (2). This is obvious that if $w'(\bar{e}) = \min\{w'(y) : y \in C(T^*, \bar{e})\}$.

Next, suppose that $x = \arg\min\{w'(y) : y \in C^0(T^*, \bar{e})\}$ and $w'(\bar{e}) > w'(x)$. Then, $x \notin E'$ and $T' = T^* - x + \bar{e}$ is a spanning tree of G containing E'. Since $C(T', x) = C(T^*, \bar{e})$, we have $w'(x) = \min\{w'(y) : y \in C(T', x)\}$. For any other edge $e \notin E(T')$, if $x \notin C(T^*, e)$, then $C(T', e) = C(T^*, e)$ and

$$w'(e) = \min\{w'(y) : y \in C(T', e)\}. \tag{9}$$

If $x \in C(T^*, e)$, then $C(T', e)$ is contained in the closed walk $C(T^*, e) \cup C(T', x) - x$. Thus,

$$\begin{aligned} w'(e) = w^*(e) &\leq \min\{w^*(y) : y \in C^0(T^*, e)\} \\ &\leq w^*(x) = w'(x) = \min\{w'(y) : y \in C(T', x)\}. \end{aligned} \tag{10}$$

Combining Eq. (9) with (10), we have $w'(e) \leq \min\{w'(y) : y \in C^0(T', e)\}$. Hence, w' is a feasible solution. The proof is completed.

Remark 1. Lemma 1 and Lemma 2 also hold for optimal solutions under the weighted l_p-norm.

Suppose that $E' \cap R(w^*) \neq \emptyset$, that is, there is an edge $e' \in E'$ such that $w^*(e') - w(e') = opt$. Choosing such an edge e', by Lemma 2, there exists an edge $\bar{e} \in K^0(T^*, e')$ such that $w^*(\bar{e}) = w^*(e')$. Notice that $w^*(\bar{e}) \leq w(\bar{e})$ by Lemma 1. If

$$\delta(\bar{e}) = w(\bar{e}) - w^*(\bar{e}) < \min\{l(\bar{e}), opt\}, \tag{11}$$

we can decrease $w^*(e')$ and $w^*(\bar{e})$ a little such that e' and $\bar{e}$ also have the same weight and the new weight also satisfies the capacitated constraint. Figure 2 illustrates this statement.

An intuition is that if all edges in $K^0(T^*, e')$ whose weights satisfy Eq. (11), then we can get a new feasible solution better than w^*. The following lemma gives the detailed statement and strict proof about it.

Lemma 3. *Let opt and w^* be the optimal value and a minimal optimal solution of I, respectively. Let T^* be a maximum spanning tree of G with respect to w^* containing E'. If $e' \in E'$ has $w^*(e') - w(e') = opt$, then there exists an edge $\bar{e} \in K^0(T^*, e')$ such that $w^*(\bar{e}) = w^*(e')$ and $w(\bar{e}) - w^*(\bar{e}) = \min\{l(\bar{e}), opt\}$.*

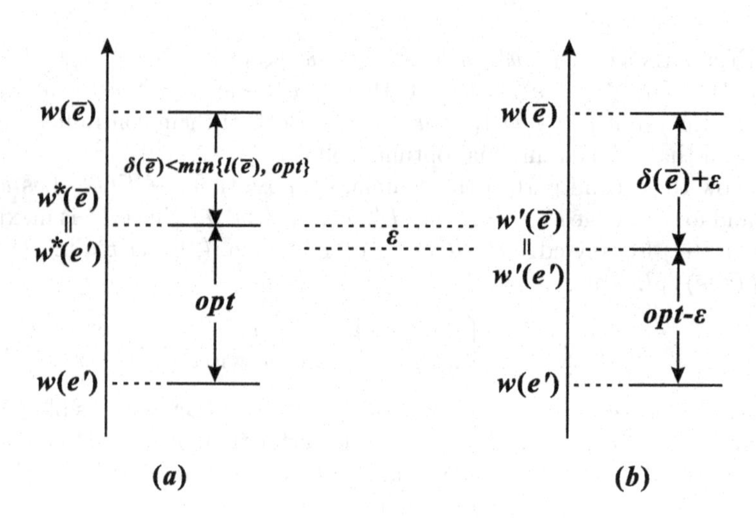

Fig. 2. An illustration of the adjustment.

Proof. Let $M = \{\bar{e} \in K^0(T^*, e') : w^*(\bar{e}) = w^*(e')\}$. By Lemma 2, $M \neq \emptyset$. Suppose to the contrary that for any $\bar{e} \in M$, $w(\bar{e}) - w^*(\bar{e}) \neq \min\{l(\bar{e}), opt\}$. Since w^* is an optimal solution, $w(\bar{e}) - w^*(\bar{e}) < \min\{l(\bar{e}), opt\}$.

Let $\varepsilon_1 = \min_{e \in M}\{\min\{l(e), opt\} - w(e) + w^*(e)\}$ and $\varepsilon_2 = \min\{w^*(e') - w^*(e) : e \in K^0(T^*, e') \setminus M\}$. Then $\varepsilon_1 > 0$ and $\varepsilon_2 > 0$. Now, we define a new weight function w' as follows. Let $\varepsilon = \frac{1}{2}\min\{\varepsilon_1, \varepsilon_2, opt\}$ and

$$w'(e) = \begin{cases} w^*(e) - \varepsilon, & e \in M \cup \{e'\}; \\ w^*(e), & \text{otherwise.} \end{cases} \tag{12}$$

Then, we can obtain that

$$-l(e') \leq 0 < w'(e') - w(e') < opt \leq u(e'), \tag{13}$$

and for any edge $\bar{e} \in M$,

$$-l(\bar{e}) \leq w^*(\bar{e}) - w(\bar{e}) - \varepsilon_1 < w'(\bar{e}) - w(\bar{e}) < w^*(\bar{e}) - w(\bar{e}) \leq u(\bar{e}). \tag{14}$$

Equations (13) and (14) show that w' satisfies the capacitated constraint.

Furthermore, for any edge $\bar{e} \in M$,

$$w'(e') - w'(\bar{e}) = w^*(e') - w^*(\bar{e}) \geq 0; \tag{15}$$

and for any edge $e \in K^0(T^*, e') \setminus M$,

$$w'(e') - w'(e) = w^*(e') - w^*(e) - \varepsilon \geq w^*(e') - w^*(e) - \varepsilon_2 \geq 0. \tag{16}$$

By Eqs. (15) and (16), we have $w'(e') = \max\{w'(y) : y \in K(T^*, e')\}$. For any edge $e \in E(T^*) \setminus \{e'\}$, $e \notin M$ because $M \subset K(T^*, e')$. Thus, we can obtain that

$$w'(e) = w^*(e) \geq \max\{w^*(y) : y \in K^0(T^*, e)\} \geq \max\{w'(y) : y \in K^0(T^*, e)\}.$$

Hence, by Eq. (1), T^* is also a maximum spanning tree of G with respect to w' containing E'. Moreover, w' is a feasible solution of I.

On the other hand, from Eqs. (13) and (14), we can obtain that for any edge $e \in M \cup \{e'\}$,

$$|w'(e) - w(e)| < opt.$$

It implies that either $\| w' - w \|_\infty < \| w^* - w \|_\infty$ or $\| w' - w \|_\infty = \| w^* - w \|_\infty$ and $R(w') \subset R(w^*)$, which contradicts that w^* is a minimal optimal solution.

Using the same technique, we can obtain the following lemma.

Lemma 4. *Let opt and w^* be the optimal value and a minimal optimal solution of I, respectively. Let T^* be a maximum spanning tree of G with respect to w^* containing E'. If $\bar{e} \notin E(T^*)$ has $w(\bar{e}) - w^*(\bar{e}) = opt$, then there exists an edge $e' \in C(T^*, \bar{e}) \cap E'$ such that $w^*(e') = w^*(\bar{e})$ and $w^*(e') - w(e') = \min\{u(e'), opt\}$.*

Combining Lemma 3 with Lemma 4, we can obtain the key result in this subsection directly.

Theorem 2. *Let I be an instance of CPIMST under the l_∞-norm, and opt be the optimal value of I. Then, there exist edges $e' \in E'$ and $\bar{e} \notin E'$ such that*

$$opt = \max\left\{\frac{1}{2}(w(\bar{e}) - w(e')), w(\bar{e}) - w(e') - l(\bar{e}), w(\bar{e}) - w(e') - u(e')\right\}. \quad (17)$$

Theorem 2 indicates that there is a candidate optimal value set whose size is at most $|E'|(m - |E'|) = O(mn)$. Hence, we can present the following algorithm.

Algorithm 2: Algorithm for CPIMST under the un-weighted l_∞-norm

---

Input: An instance $I = (G, w, E', l, u, c \equiv 1)$ of CPIMST.
Output: The optimal value *opt* and an optimal solution w^*.

1 **if** *the output of Alg. 1 on $(I, 0)$ is "YES"* **then**
2 | **return** $opt = 0$ and $w^* = w$;
3 **end**
4 Set $l_{\max} := \max_{e \in E} l(e)$ and $u_{\max} := \max_{e \in E} u(e)$;
5 **if** *the output of Alg. 1 on $(I, \max\{l_{\max}, u_{\max}\})$ is "NO"* **then**
6 | **return** "No Feasible Solutions!";
7 **end**
8 Set $OVS := \emptyset$;
9 **for** $e' \in E'$ *and* $\bar{e} \notin E'$ **do**
10 | **if** $w(e') < w(\bar{e})$ *and* $w(\bar{e}) - w(e') \leq u(e') + l(\bar{e})$ **then**
11 | | Set $OVS := OVS \cup$ $\{\max\{\frac{1}{2}(w(\bar{e}) - w(e')), w(\bar{e}) - w(e') - l(\bar{e}), w(\bar{e}) - w(e') - u(e')\}\}$;
12 | **end**
13 **end**
14 Order the numbers in OVS with increasing ordering;
15 Use the Binary search to find the minimum number *opt* in OVS such that the output of Alg. 1 on (I, opt) is "YES";
16 **return** *opt* and $w^* = w_{opt}$ by Eq. (4).

Remark 2. The running time of Algorithm 2 is $O(mn \log n)$. In fact, since the size of OVS is $O(mn)$, sort the set will cost $O(mn \log mn)$ time and the Binary search will carry out $O(\log mn)$ iterations. Hence, the running time is $O(mn \log mn + \log mn \cdot \min\{(m \log m), n^2\}) = O(mn \log n)$.

2.3 CPIMST Under the Weighted l_∞-norm

To solve this problem, we firstly modify the definition of minimal optimal solution w^* as follows.

(i) The size of w^*, which defined as $|R(w^*)| = |\{e \in E : c(e)|w(e^*) - w(e)| = opt\}|$ is minimum among all optimal solutions;
(ii) Under condition (i), $\| w^* - w \|_1 = \sum_{e \in E} c(e)|w(e^*) - w(e)|$ is minimum.

Using the same proof technique, we can see that Lemma 1 and Lemma 2 also hold for minimal optimal solutions under the weighted l_∞-norm. Hence, Theorem 2 can be generalized to Theorem 3 as follows.

Theorem 3. *Let I be an instance of CPIMST under the weighted l_∞-norm. Then, there exist edges $e' \in E'$ and $\bar{e} \notin E'$ such that the optimal value of I is equal to*

$$\max\left\{\frac{c(e')c(\bar{e})\Delta w(\bar{e}, e')}{c(e') + c(\bar{e})}, c(e')\big(\Delta w(\bar{e}, e') - l(\bar{e})\big), c(\bar{e})\big(\Delta w(\bar{e}, e') - u(e')\big)\right\},$$

where $\Delta w(\bar{e}, e') = w(\bar{e}) - w(e')$.

Therefore, CPIMST under the weight l_∞-norm can also be solved in $O(mn \log n)$ time.

3 Conclusion

In this paper, we study the capacitated partial inverse maximum spanning tree problem under the weighted l_∞-norm. We obtain some properties of the optimal value and minimal optimal solutions of this problem. Based on these properties, combining the algorithm for the decision version with the Binary search method, we present an algorithm to solve this problem with running time $O(mn \log n)$. Thus, combined with the previous results, the computational complexity of CPIMST has been completely solved.

References

1. Ben-Ayed, O., Blair, C.E.: Computational difficulties of bilevel linear programming. Oper. Res. **38**(3), 556–560 (1990)
2. Chvátal, V.: Correction to: a De Bruijn-Erdős theorem in graphs? In: Gera, R., Haynes, T.W., Hedetniemi, S.T. (eds.) Graph Theory. PBM, pp. C1–C2. Springer, Cham (2018). https://doi.org/10.1007/978-3-319-97686-0_15

3. Cai, M.-C., Duin, C.W., Yang, X., Zhang, J.: The partial inverse minimum spanning tree problem when weight increasing is forbidden. Eur. J. Oper. Res. **188**, 348–353 (2008)
4. Gassner, E.: The partial inverse minimum cut problem with L_1-norm is strongly NP-hard. RAIRO Oper. Res. **44**, 241–249 (2010)
5. Hansen, P., Jaumard, B., Savard, G.: New branch-and-bound rules for linear bilevel programming. SIAM J. Sci. Stat. Comput. **13**, 1194–1217 (1992)
6. Lai, T., Orlin, J.: The Complexity of Preprocessing. Research Report of Sloan School of Management. MIT (2003)
7. Li, S., Zhang, Z., Lai, H.-J.: Algorithms for constraint partial inverse matroid problem with weight increase forbidden. Theor. Comput. Sci. **640**, 119–124 (2016)
8. Li, X., Shu, X., Huang, H., Bai, J.: Capacitated partial inverse maximum spanning tree under the weighted Hamming distance. J. Comb. Optim. **38**(4), 1005–1018 (2019). https://doi.org/10.1007/s10878-019-00433-x
9. Li, X., Zhang, Z., Du, D.-Z.: Partial inverse maximum spanning tree in which weight can only be decreased under l_p-norm. J. Glob. Optim. **70**(3), 677–685 (2017). https://doi.org/10.1007/s10898-017-0554-5
10. Li, X., Zhang, Z., Yang, R., Zhang, H., Du, D.-Z.: Approximation algorithms for capacitated partial inverse maximum spanning tree problem. J. Glob. Optim. **77**(2), 319–340 (2019). https://doi.org/10.1007/s10898-019-00852-4
11. Yang, X.: Complexity of partial inverse assignment problem and partial inverse cut problem. RAIRO Oper. Res. **35**, 117–126 (2001)
12. Yang, X., Zhang, J.: Partial inverse assignment problem under l_1 norm. Oper. Res. Lett. **35**, 23–28 (2007)
13. Yang, X., Zhang, J.: Inverse sorting problem by minimizing the total weighted number of changers and partial inverse sorting problem. Comput. Optim. Appl. **36**(1), 55–66 (2007)
14. Zhang, Z., Li, S., Lai, H.-J., Du, D.-Z.: Algorithms for the partial inverse matroid problem in which weights can only be increased. J. Glob. Optim. **65**(4), 801–811 (2016). https://doi.org/10.1007/s10898-016-0412-x

Approximation Algorithms for Some Min-Max and Minimum Stacker Crane Cover Problems

Yuhui Sun, Wei Yu(✉), and Zhaohui Liu

School of Mathematics, East China University of Science and Technology, Shanghai 200237, China
y30190197@mail.ecust.edu.cn, {yuwei,zhliu}@ecust.edu.cn

Abstract. We study two stacker crane cover problems and their variants. Given a mixed graph $G = (V, E, A)$ with vertex set V, edge set E and arc set A. Each edge or arc is associated with a nonnegative weight. The Min-Max Stacker Crane Cover Problem (SCC) aims to find at most k closed walks covering all the arcs in A such that the maximum weight of the closed walks is minimum. The Minimum Stacker Crane Cover Problem (MSCC) is to cover all the arcs in A by a minimum number of closed walks of length at most λ. The Min-Max Stacker Crane Walk Cover Problem (SCWC)/Minimum Stacker Crane Walk Cover Problem (MSCWC) is a variant of the SCC/MSCC problem with closed walks replaced by (open) walks.

For the SCC problem with symmetric arc weights, i.e. for every arc there is a parallel edge of no greater weight, we obtain a 33/5-approximation algorithm. This improves on the previous 37/5-approximation algorithm for a restricted case of the SCC problem with symmetric arc weights. If the arc weights are symmetric, we devise the first constant-factor approximation algorithms for the SCWC problem, the MSCC problem and the MSCWC problem with ratios 5, 5 and 7/2, respectively. Finally, for the (general) MSCWC problem we first propose a 4-approximation algorithm.

Keywords: Approximation algorithm · Stacker Crane Problem · Rural postman problem · Traveling Salesman Problem · Stacker Crane Cover

1 Introduction

Given a mixed graph $G = (V, E, A)$ with vertex set V, edge set E and arc set A, each edge or arc is associated with a nonnegative weight. The Stacker Crane Problem (SCP) [8] is to find a minimum weight closed walk that starts from and ends at a given vertex and traverses all the arcs in A, where the weight of a walk is the sum of the weights of the traversed edges and arcs. The SCP and its variants have a wide range of applications in many related industries, including driving a pick-up and delivery truck [8], the design of the movements of climber

D.-Z. Du et al. (Eds.): COCOA 2021, LNCS 13135, pp. 400–415, 2021.
https://doi.org/10.1007/978-3-030-92681-6_32

robots and cutting plotters [5], ambulatory services [4], street sweeping [7], etc. The SCP satisfying the property that the weight of each arc (u, v) equals the length of the shortest path between u and v in the spanning graph $H = (V, E)$ is also described as the Dial-a-Ride Problem (DARP).

In some practical applications, we are required to design a set of routes that jointly cover the arcs, rather than a single route for one vehicle. Instead of considering the traditional objective of minimizing the total length, many applications also focus on other objective functions, including minimizing the latest finishing time in the snow plow routing problem [13], minimizing the longest traveling time among mobile sensors [18], and minimizing the number of detecting sensors in the wireless sensor networks [10], and so on. This prompts us to study the following extensions of the SCP with multiple vehicles and min-max objective functions. (1) Given a mixed graph $G = (V, E, A)$ with a nonnegative weight function on $E \cup A$ and a positive integer k, the objective is to find at most k closed walks that traverse all the arcs in A such that the maximum weight of the closed walks is minimum. It is called the Min-Max Stacker Crane Cover Problem (SCC). By replacing the closed walks with (open) walks in the SCC problem, we have the Min-Max Stacker Crane Walk Cover Problem (SCWC). (2) Given a mixed graph $G = (V, E, A)$ with a nonnegative weight function on $E \cup A$ and a nonnegative number λ, the goal is to find a set of closed walks which cover all the arcs in A such that the weight of each closed walk is upper bounded by λ and the number of closed walks used is minimum. It is called the Minimum Stacker Crane Cover Problem (MSCC). Analogously, by replacing the closed walks with (open) walks in the MSCC problem, we have the Minimum Stacker Crane Walk Cover Problem (MSCWC).

Since the SCP is a generalization of the well-known Metric Traveling Salesman Problem (Metric TSP), it is NP-hard, as noted by Frederickson et al. [8]. Then the variants of the SCP (the SCC/MSCC problem) are also NP-hard. Similarly, both the SCWC problem and the MSCWC problem are NP-hard since they are extensions of the NP-Complete Hamiltonian Path Problem [11] (pp. 60). Therefore, we mainly consider approximation algorithms for these problems in this paper.

Definition 1. *For a minimization combinatorial optimization problem, an algorithm is called an α-approximation algorithm if for any instance of the problem the algorithm always produces, in polynomial time, a solution of objective value no more than α times the optimal value. α is called the approximation ratio of the algorithm.*

Frederickson et al. [8] first considered the Symmetric SCP, which is a special case of the SCP where the arc weights are symmetric, i.e. for every arc $(u, v) \in A$ there is a parallel edge $[u, v]$ of no greater weight. Note that the symmetric arc weights occur naturally in some real world applications. For example, in electronic printing [16] the weight of arc (u, v) represents the length of a curve from vertex u to vertex v in the plane and the weight of edge $[u, v]$ indicates the length of the straight line between u and v. Since the length of a straight line is no more than that of a curve, the arc weights are symmetric in this case. Frederickson et al. [8] gave an algorithm for the Symmetric SCP which has an approximation ratio of $\frac{9}{5}$ and runs in $O(\max\{|V|^3, |A|^3\})$ time. Frederickson et al. [9] achieved a better approximation ratio of $\frac{5}{4}$ for the DARP on trees, a restricted case of the Symmetric SCP, in which the spanning subgraph $H = (V, E)$ of G is a tree and the weight of the arc (u, v) equals the length of the shortest path between u and v. Moreover, it is proved that the DARP on trees is NP-hard [9] and the DARP on either simple paths or simple cycles is polynomial time solvable [2]. In addition, we mention that Treleaven et al. [17] have proposed asymptotically optimal algorithms for the SCP, which produce, almost surely, a solution approaching the optimal solution as the number of arcs goes to infinity. The SCP/DARP can be extended to a more general model, known as the Pickup and Delivery Problem (PDP). Parragh et al. [14,15] gave a detailed survey on various variants of the PDP, including the online and dynamic versions of the SCP/DARP.

As for the SCP with multiple vehicles, Frederickson et al. [8] considered the Symmetric SCC problem with a single depot, i.e. all the k closed walks must contain a common depot vertex, and derived a $(\rho + 1 - \frac{1}{k})$-approximation algorithm, where ρ is the approximation ratio for the Symmetric SCP. Furthermore, Bao et al. [3] devised a $\frac{37}{5}$-approximation algorithm for a special case of the Symmetric SCC problem where for every arc $(u, v) \in A$ there is a parallel edge $[u, v]$ of the same weight and the edge weights satisfy the triangle inequality. Yu et al. [19] considered another variant of the SCP with k vehicles, called the k-SCP, which aims to find k closed walks including all the arcs such that the total weight of the closed walks is minimized. They gave a 3-approximation algorithm and developed a 2-approximation algorithm for the Symmetric k-SCP. Moreover, they also obtained some approximation algorithms for the k-Depot SCP, where a depot set $D \subseteq V$ with $|D| = k$ is given and each closed walk has to contain a distinct depot vertex.

For the SCC/SCWC problem, there is a closely related problem called the Min-Max Rural Postmen Cover Problem (RPC), where the input is an undirected weighted graph $G = (V, E)$ and the objective is to find at most k closed walks covering a subset $R \subseteq E$ of edges to minimize the length of the longest closed walk. By replacing closed walks with (open) walks in the RPC problem, we have the Min-Max Rural Postmen Walk Cover Problem (RPWC). Arkin et al. [1] first gave a 7-approximation algorithm for the RPWC problem and Yu et al. [20] developed an improved 5-approximation algorithm. Moreover, Yu et al. [20] gave a 6-approximation algorithm for the Metric RPC problem, i.e. a restricted case of the RPC problem with the weights of all the edges satisfying the triangle inequality.

For the MSCC/MSCWC problem, there are also two related problems called the Minimum Rural Postmen Cover Problem (MRPC) and the Minimum Rural Postmen Walk Cover Problem (MRPWC). Arkin et al. [1] first devised a 4-approximation algorithm for the MRPWC problem and gave a 7-approximation algorithm for the MRPC problem. Mao et al. [12] derived two 5-approximation algorithms for the MRPC problem based on graph decomposition.

In this paper, we obtain the following results. For the Symmetric SCC problem, we develop a 33/5-approximation algorithm, improving on the previous 37/5-approximation algorithm designed for a special case of the Symmetric SCC problem. For the Symmetric SCWC problem, the Symmetric MSCC problem and the Symmetric MSCWC problem, we propose the first constant-factor approximation algorithms with ratios 5, 5 and 7/2, respectively. For the (general) MSCWC problem, we first develop a 4-approximation algorithm.

The rest of the paper is organized as follows. We formally describe the problems and give some preliminaries in Sect. 2. In Sect. 3 we give the approximation algorithms for the SCC/SCWC problem, which is followed by the discussion of the MSCC/MSCWC problem in Sect. 4. Finally, we conclude the paper in Sect. 5.

2 Preliminaries

Given a mixed graph $G = (V, E, A)$ with vertex set V, edge set E and arc set A. Each edge $e = [u, v] \in E$ connects two vertices u and v in V, where u, v are called the *end vertices* of e. Each arc $a = (u, v) \in A$ is a directed edge from vertex u to vertex v, where u is called the *tail* of a and v is called the *head* of a. If u is one of the end vertices of edge e (either the head or the tail of arc a), we call u is *incident to* edge e (arc a). The total number of edges and arcs incident to vertex v is called the *degree* of v. The number of arcs directed into (directed out of) vertex v is called the *indegree (outdegree)* of v. If the degree of vertex v is even (odd), v is called *even degree (odd degree)*. Each edge e (arc a) is associated with a nonnegative weight $w(e)$ ($w(a)$). For any $B > 0$, $G[B]$ denotes the subgraph of G obtained by removing all the edges in E (no arcs included) with weight greater than B.

A walk W is a sequence $v_0e_1v_1e_2 \cdots v_{t-1}e_tv_t$ such that $v_i \in V$ for $i = 0, 1, \ldots t$ and either $e_i = [v_{i-1}, v_i] \in E$ or $e_i = (v_{i-1}, v_i) \in A$ for $i = 1, \ldots t$. We call v_0 (v_t) the *initial vertex* (*terminal vertex*) of W. If $v_0, v_1, \ldots, v_l$ are clear in the context, the walk W is denoted simply by $e_1e_2 \cdots e_t$. A *path* is a walk without repeated vertices except for the initial and terminal vertex. Obviously, there are no repeated edges or arcs in a path. A *closed walk* is a walk with identical initial and terminal vertices. Similarly, a *cycle* is defined as a closed path. An undirected path is a path that only uses edges (no arcs included).

For a subgraph G' of G, we say the graph obtained by adding some copies of the edges or arcs of G' a *multi-subgraph* of G. Given a (multi-)subgraph H (e.g. path, cycle, walk) of G, we use $V(H)$, $E(H)$ and $A(H)$ to denote the vertex set, edge (multi-)set and arc (multi-)set of H, respectively. H is called *(weakly) connected* if for any $u, v \in V$, there exists a path between u and v in H after ignoring the directions of the arcs in A. H is called *strongly connected* if for any $u, v \in V$, there exist both a path from u to v and a path from v to u in H. The edge weight of H is defined as $w_E(H) = \sum_{e \in E(H)} w(e)$. The arc weight of H

is defined as $w_A(H) = \sum_{a \in A(H)} w(a)$. The weight of H denotes the sum of the weights of edges and arcs of H, i.e. $w(H) = w_E(H) + w_A(H)$. Note that any edge e appearing t times in $E(H)$ contributes $t \cdot w(e)$ to $w_E(H)$. Similarly, any arc a appearing t times in $A(H)$ contributes $t \cdot w(a)$ to $w_A(H)$. The weight of a path (cycle, walk) is also called its length. A cycle C is also called a tour on $V(C)$. A path (cycle, walk) containing only one vertex and no edges or arcs is a trivial path (cycle, walk) and its length is defined as zero.

For $A' \subseteq A$, we say that a set $\{W_1, \ldots, W_k\}$ of closed walks (some of them may be trivial walks) is a *stacker crane cover* of A' if $A' \subseteq \bigcup_{i=1}^{k} A(W_i)$. The cost of this stacker crane cover of A' is defined as $\max_{1 \le i \le k} w(W_i)$, i.e. the maximum length of the closed walks. By replacing closed walks with (open) walks we can define analogously a *stacker crane walk cover* of A' as well as its cost.

A connected graph is said to be *Eulerian* if there exists a Eulerian tour, i.e. a closed walk which contains each arc and each edge exactly once and each vertex at least once. Eiselt et al. [6] gave necessary and sufficient conditions for a graph to be Eulerian. One of the sufficient conditions is given as follows.

Fact 1. A mixed graph is a Eulerian graph if it is strongly connected, each vertex is even degree and the indegree of each vertex is equal to its outdegree.

Now we formally state the problems considered in this paper:

Min-Max Stacker Crane Cover Problem (SCC)
Input: A mixed graph $G = (V, E, A)$, a nonnegative integer weight function w on $E \cup A$ and a positive integer k.
Output: A stacker crane cover of A with $k' \le k$ closed walks $C_1, \ldots, C_{k'}$.
Goal: Minimize the cost of the stacker crane cover of A, i.e. $\max_{1 \le i \le k'} w(C_i)$.

Min-Max Stacker Crane Walk Cover Problem (SCWC)
Input: A mixed graph $G = (V, E, A)$, a nonnegative integer weight function w on $E \cup A$ and a positive integer k.
Output: A stacker crane walk cover of A with $k' \le k$ walks $W_1, \ldots, W_{k'}$.
Goal: Minimize the cost of the stacker crane walk cover of A, i.e. $\max_{1 \le i \le k'} w(W_i)$.

Minimum Stacker Crane Cover Problem (MSCC)
Input: A mixed graph $G = (V, E, A)$, a nonnegative integer weight function w on $E \cup A$ and a nonnegative integer λ.
Output: A stacker crane cover of A with k closed walks $C_1, \ldots, C_k$ such that $w(C_i) \le \lambda$ for $i = 1, \ldots, k$.
Goal: Minimize k.

Minimum Stacker Crane Walk Cover Problem (MSCWC)
Input: A mixed graph $G = (V, E, A)$, a nonnegative integer weight function w on $E \cup A$ and a nonnegative integer λ.
Output: A stacker crane walk cover of A with k walks $W_1, \ldots, W_k$ such that $w(W_i) \le \lambda$ for $i = 1, \ldots, k$.

Goal: Minimize k.

The Symmetric SCC problem is a special case of the SCC problem where the weights of the arcs are *symmetric*, i.e. for every arc there is a parallel edge of no greater weight. Similarly, we define the Symmetric SCWC problem, the Symmetric MSCC problem and the Symmetric MSCWC problem.

Given an instance I of the SCC/SCWC/MSCC/MSCWC problem defined on a mixed graph $G = (V, E, A)$ with a nonnegative weight function w on $E \cup A$, we may assume w.l.o.g that any two arcs have no common head (tail). Otherwise, let (u, v) and (w, v) be two arcs with a common head v, we can add a new vertex v' connected with v by a zero weight edge $[v, v']$ and replace (w, v) by (w, v') to obtain another instance equivalent to I. The case of common tails can be addressed similarly. From I we can derive a new instance I' with exactly the same input except that G is replaced by a graph $G' = (V', E', A)$ such that:

(i) $V' = V(A)$, i.e., each vertex in V' is either the head or the tail of at least one arc in A;
(ii) For any two vertices $u, v \in V'$, there is an edge $[u, v] \in E'$ whose weight equals the length of the shortest *undirected* path between u and v in G;
(iii) For any arc $(u, v) \in A$ in G, there is an arc $(u, v) \in A$ of equal weight in G'.

We call G' the *reduced graph* of G. Since the weights of edges in E' are generated by the shortest undirected path in G, they clearly satisfy the triangle inequality. In addition, if the arc weights in G are symmetric, the weights of the arcs in G' are also symmetric. Similar to the SCP (see [8]), one can verify that I' is equivalent to I. Therefore, we focus on the instances of the SCC/SCWC/MSCC/MSCWC problem in which G is the reduced graph of some graph, say $\bar{G}$, in the following discussion.

For an instance I of the SCC/SCWC/MSCC/MSCWC problem, we use $OPT(I)$ to represent the optimal value as well as the corresponding optimal solution. Each (closed) walk in $OPT(I)$ is called an optimum (closed) walk. If the instance I is clear we simply write OPT for $OPT(I)$. Note that the optimum (closed) walks may be neither edge-disjoint nor vertex-disjoint.

The following two results on splitting a long walk into several walks of small length are proved implicitly by Arkin et al. [1], which are very useful to design and analyze the algorithms for the SCC/SCWC/MSCC/MSWC problem.

Lemma 1. (Arkin et al. 2006) *Given $B > 0$, $\beta \geq 0$, a walk W and an arc set $A' \subseteq A(W)$ with $\max_{a \in A'} w(a) \leq \beta$, we can decompose W in $O(|E(W) \cup A(W)|)$ time into $t \leq \max\left\{\left\lceil \frac{w(W)}{B} \right\rceil, 1\right\}$ walks $W_1, \ldots, W_t$ of length at most $B + \beta$ such that (i) for each $i = 1, \ldots, t$, either $w(W_i) \leq B$ or W_i consists of a walk W_i' with $w(W_i') \leq B$ and an arc $(v', v) \in A'$ connecting the terminal vertex v' of W_i' with the terminal vertex v of W_i; (ii) these walks contain all the vertices in $V(W)$ and all the arcs in A'.*

Lemma 2. (Arkin et al. 2006) *Given $B > 0$, a walk W and an arc set $A' \subseteq A(W)$ with $\max_{a \in A'} w(a) \leq B$, we can decompose W in $O(|E(W) \cup A(W)|)$ time* *into* $t \leq \max\left\{\left\lceil \frac{w(W)+w(A')}{B} \right\rceil, 1\right\}$ *walks of length at most* B*, which contain all the vertices in* $V(W)$ *and all the arcs in* A'*.*

Based on Lemma 1 we can show the following result.

Lemma 3. *Given* $B > 0$*,* $\beta \geq 0$*, a walk* W *with symmetric arc weights, an arc set* $A' \subseteq A(W)$ *and a path* P_a *with* $w(a) + w(P_a) \leq \beta$ *for any* $a \in A'$*, we can generate in* $O(|E(W) \cup A(W)|)$ *time a stacker crane cover of* A' *that consists of* $t \leq \max\left\{\left\lceil \frac{w(W)}{B} \right\rceil, 1\right\}$ *closed walks and has a cost of at most* $2B + \beta$*.*

The proof idea consists of two steps. First we use Lemma 1 to obtain $t \leq \max\left\{\left\lceil \frac{w(W)}{B} \right\rceil, 1\right\}$ walks $W_1, \ldots, W_t$. For each walk W_i, if $w(W_i) \leq B$, we simply double the edges and add the corresponding parallel edges of the arcs in W_i' to derive a closed walk of length no more than $2w(W_i) \leq 2B$. If W_i consists of a walk W_i' with $w(W_i') \leq B$ and an arc $a \in A'$, we double the edges of W_i', add the corresponding parallel edges of the arcs of W_i', and supplement the path P_a to derive a closed walk of length no more than $2w(W_i') + w(a) + w(P_a) \leq 2B + \beta$.

To design an α-approximation algorithm for the SCC/SCWC problem, it is sufficient to devise an α-subroutine, which is defined as below.

Definition 2. *A polynomial time algorithm is called an* α*-subroutine for the SCC/SCWC problem if for any instance consisting of* $G = (V, E, A)$ *and an integer* $k > 0$*, and any value* $\lambda \in [0, 2w(G)]$*, the algorithm returns either failture or a feasible stacker crane (walk) cover of* A *of cost at most* $\alpha\lambda$*, and it always returns the latter output as long as* $OPT \leq \lambda$*.*

Simiar to the arguments in [20], we can use a binary search to transform an α-subroutine into an α-approximation algorithm for the SCC/SCWC problem.

Lemma 4. *Any* $T(n)$*-time* α*-subroutine for the SCC/SCWC problem yields an* α*-approximation algorithm for the same problem that runs in* $O(T(n) \log w(G))$ *time.*

Based on the sufficient condition for a mixed graph to be a Eulerian graph, the next result obtains a Eulerian graph of bounded weight. This can be used to deal with the Symmetric MSCC problem.

Lemma 5. *Given a mixed connected graph* $G = (V, E, A)$*, let* $G' = (V, E', A')$ *be the graph obtained by doubling all the edges in* E*, adding an edge* $[u, v]$ *for each arc* $(u, v) \in A$ *and associating with* $[u, v]$ *the direction opposite to* (u, v)*, then* G' *is a Eulerian graph. If the weights of the arcs in* A *are symmetric, then* $w(G') \leq 2w(G)$*.*

Proof. Since we double all the edges in E and add in essence a reversed arc for each arc in A, each vertex in G' is even degree and the indegree of each vertex equals its outdegree. By Fact 1, the graph G' is Eulerian. If the weights of the arcs in A are symmetric, the weights of added edges (reversed arcs) are no more than the weights of corresponding arcs in A. Therefore, $w(G') \leq 2w(G)$. □

3 Min-Max Stacker Crane (Walk) Cover

In this section, we consider the Symmetric SCC problem and the Symmetric SCWC problem. Based on a mixed strategy, we first design a $\frac{33}{5}$-approximation algorithm for the Symmetric SCC problem. In the second part, we consider the Symmetric SCWC problem, which is required to find (open) walks instead of closed walks, and present a 5-approximation algorithm.

3.1 Min-Max Stacker Crane Cover Problem

We give the following algorithm for the Symmetric SCC problem, which turns out to be a $\frac{33}{5}$-subroutine.

Algorithm $SymSCC(\lambda)$

Input: An instance of the Symmetric SCC problem consisting of a reduced graph $G = (V, E, A)$, a nonnegative integer weight function w on $E \cup A$ and a positive integer k, and a value $\lambda \in [\max_{a \in A} w(a), 2w(G)]$.

Output: Either failure or a stacker crane cover of A containing at most k closed walks with cost no more than $\frac{33}{5}\lambda$.

Step 1. Delete all the edges with weight greater than $\frac{\lambda}{2}$ in G. Suppose that the resulting graph $G[\frac{\lambda}{2}]$ has p connected components $F_1, \ldots, F_p$.

Step 2. For each $i = 1, \ldots, p$, let $A_i = A \cap A(F_i)$, construct two Eulerian tours C_i^1 and C_i^2 covering all the arcs in A_i, where C_i^1 is derived by Algorithm $SymSCC1$ and C_i^2 is obtained by Algorithm $SymSCC2$.

Step 3. For each $i = 1, \ldots, p$, let C_i^0 be one of C_i^1 and C_i^2 with smaller length. By taking $B = \frac{14}{5}\lambda$, $\beta = \lambda$, $W = C_i^0$, $A' = A_i$ and P_a as the shortest path from the head to the tail of a for any $a \in A_i$ in Lemma 3, we can generate a stacker crane cover of A_i that consists of $k_i \leq \max\left\{\left\lceil \frac{w(C_i^0)}{\frac{14}{5}\lambda} \right\rceil, 1\right\}$ closed walks and has a cost of at most $\frac{33}{5}\lambda$.

Step 4. If $\sum_{i=1}^{p} k_i \leq k$, return the stacker crane cover of A consisting of the $\sum_{i=1}^{p} k_i$ closed walks obtained in Step 3; otherwise, return failure.

Now we give the detailed description of Algorithm $SymSCC1$ and Algorithm $SymSCC2$. First, we describe Algorithm $SymSCC1$ and the steps of this algorithm are shown by an example in Fig. 1.

Algorithm $SymSCC1$

Step 1. Construct a weighted bipartite graph $B_i = (V_i^1 \cup V_i^2, A_0)$ from F_i as follows. The vertex set V_i^1 (V_i^2) consists of all the heads (tails) of the arcs of A_i. For any $u \in V_i^1$ and $v \in V_i^2$, there is an arc $(u, v) \in A_0$. There are no other arcs in A_0. The weight of an arc $a = (u, v) \in A_0$ is defined as the length of the shortest path from u to v in F_i.

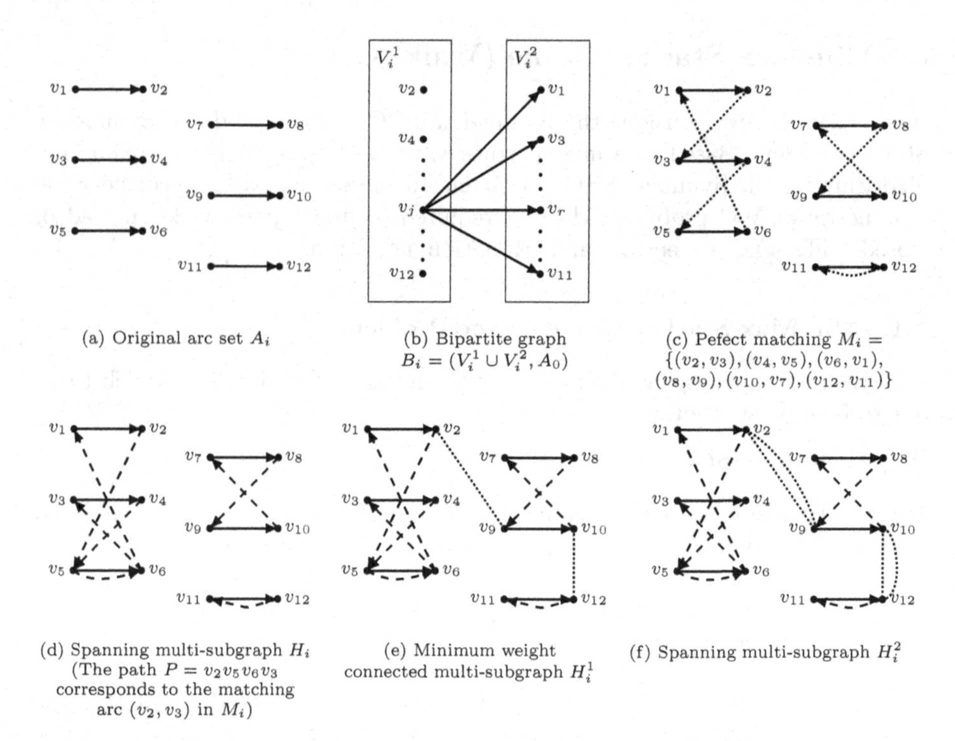

(a) Original arc set A_i

(b) Bipartite graph $B_i = (V_i^1 \cup V_i^2, A_0)$

(c) Pefect matching $M_i = \{(v_2, v_3), (v_4, v_5), (v_6, v_1), (v_8, v_9), (v_{10}, v_7), (v_{12}, v_{11})\}$

(d) Spanning multi-subgraph H_i (The path $P = v_2v_5v_6v_3$ corresponds to the matching arc (v_2, v_3) in M_i)

(e) Minimum weight connected multi-subgraph H_i^1

(f) Spanning multi-subgraph H_i^2

Fig. 1. An example for Algorithm *SymSCC*1. In picture (d), the dashed arrows represent the edges in $E(F_i)$ traversed in the corresponding directions. In picture (e), the dotted lines denote the edges in MST_i.

Step 2. Determine a minimum weight perfect matching M_i of B_i. Each arc $(u, v) \in M_i$ corresponds to a shortest path $P_{u,v}$ from u to v in F_i. Construct a spanning multi-subgraph H_i of F_i composed of the arcs in A_i and the edges and arcs on each path $P_{u,v}$ with $(u, v) \in M_i$. Note that if an arc or edge appears t times in the paths, there are t additional copies of this arc or edge in H_i besides the arcs in A_i. Assume that H_i consists of q strongly connected components $T_i^1, \ldots, T_i^q$, each of which is a Eulerian graph.

Step 3. Find a minimum weight connected multi-subgraph H_i^1 of F_i as follows. Starting with $T_i^1, \ldots, T_i^q$, we keep adding a least weight edge in $E(F_i)$ that connects two distinct connected components, say $\hat{T}_i^j$ and $\hat{T}_i^l$ and stop until a connected graph H_i^1 is obtained. Clearly, H_i^1 contains all the edges and arcs in H_i. Let MST_i be the set of edges added to H_i when deriving H_i^1. By doubling the edges of MST_i in H_i^1, we obtain a spanning multi-subgraph H_i^2 of F_i, which is a Eulerian graph. Then we construct a Eulerian tour C_i^1 of H_i^2, which contains all the arcs in A_i.

Algorithm $SymSCC2$ is described as follows and the steps of this algorithm are demonstrated by an example in Fig. 2.

Algorithm $SymSCC2$

Step 1. Suppose that $A_i = \{(v_1^1, v_1^2), \ldots, (v_{|A_i|}^1, v_{|A_i|}^2)\}$. By contracting all the arcs of A_i in F_i, we derive an undirected complete graph $F_i^0 = (V_A, E_0)$ with $V_A = \{n_1, \ldots, n_{|A_i|}\}$, where vertex n_j $(j = 1, \ldots, |A_i|)$ corresponds to the arc (v_j^1, v_j^2) in A_i. For any pair of vertices $n_j, n_l \in V_A$, the weight of the edge $[n_j, n_l] \in E_0$ is given by

$$w'([n_j, n_l]) = \min\{w([v_j^1, v_l^1]), w([v_j^1, v_l^2]), w([v_j^2, v_l^1]), w([v_j^2, v_l^2])\}.$$

For each edge $[n_j, n_l] \in E_0$, we determine the length of the shortest path between n_j and n_l in F_i^0, denoted by $\tilde{w}(n_j, n_l)$. Obtain a graph $\tilde{G}_i$ from F_i^0 by simply replacing the weight of each edge $[n_j, n_l] \in E_0$ with $\tilde{w}(n_j, n_l)$.

Step 2. Find a minimum weight spanning tree $\tilde{T}_i$ for the graph $\tilde{G}_i$. Let $Odd(\tilde{T}_i)$ be the set of vertices that have odd degree in $\tilde{T}_i$. Compute a minimum weight perfect matching $\tilde{M}_i$ on $Odd(\tilde{T}_i)$. Construct a Eulerian graph $\hat{T}_i$ by adding the matching edges of $\tilde{M}_i$ to $\tilde{T}_i$. We replace each edge of $\hat{T}_i$ by the corresponding shortest path in F_i^0 and then uncontract each vertex $n_j \in V_A$ as the corresponding arc $(v_j^1, v_j^2) \in A_i$ to generate a graph $\hat{F}_i$.

Step 3. In $\hat{F}_i$, consider each arc $(v_j^1, v_j^2) \in A_i$, if it is an *odd arc*, i.e. both v_j^1 and v_j^2 have odd degree, we add a copy of the edge $[v_j^1, v_j^2]$ and direct this edge from v_j^2 to v_j^1. Otherwise, it must be an *even arc*, i.e. both v_j^1 and v_j^2 are of even degree in $\hat{F}_i$, and we add no edges. In this way, we derive an even graph denoted by F_i'. Let A_i^{even} be the set of even arcs in $\hat{F}_i$ (or F_i').

Step 4. In F_i', ignore the directions of the even arcs to obtain a Eulerian graph with Eulerian tour C_i'. If the total weight of the even arcs in F_i' that are traversed backwards in C_i' exceeds $w(A_i^{even})/2$, we reverse the direction of C_i'. For each even arc (v_j^1, v_j^2) traversed in the opposite direction in C_i', we recover the direction of (v_j^1, v_j^2), add two copies of the edge $[v_j^1, v_j^2]$ to C_i' which are oriented from v_j^2 to v_j^1. This results in a Eulerian graph with Eulerian tour C_i^2 covering all the arcs in A_i in the right direction.

We first show the time complexity of Algorithm $SymSCC(\lambda)$.

Lemma 6. *Algorithm* $SymSCC(\lambda)$ *runs in* $O(|V|^3)$ *time.*

We assume w.l.o.g. that the optimal solution consists of exactly k optimum closed walks $C_1^*, \ldots, C_k^*$, since otherwise we may add some trivial closed walks. According to the triangle inequality for all the edges (note that G is a reduced graph) and the arc weights are symmetric, it is easy to obtain the following fact.

Observation 1. If $OPT \leq \lambda$, then $w(e) \leq \frac{OPT}{2} \leq \frac{\lambda}{2}$ for each $e \in \bigcup_{i=1}^k E(C_i^*)$ and $w(a) + w(P_a) \leq OPT \leq \lambda$ for each $a \in \bigcup_{i=1}^k A(C_i^*)$, where P_a denotes the shortest path from the head to the tail of a.

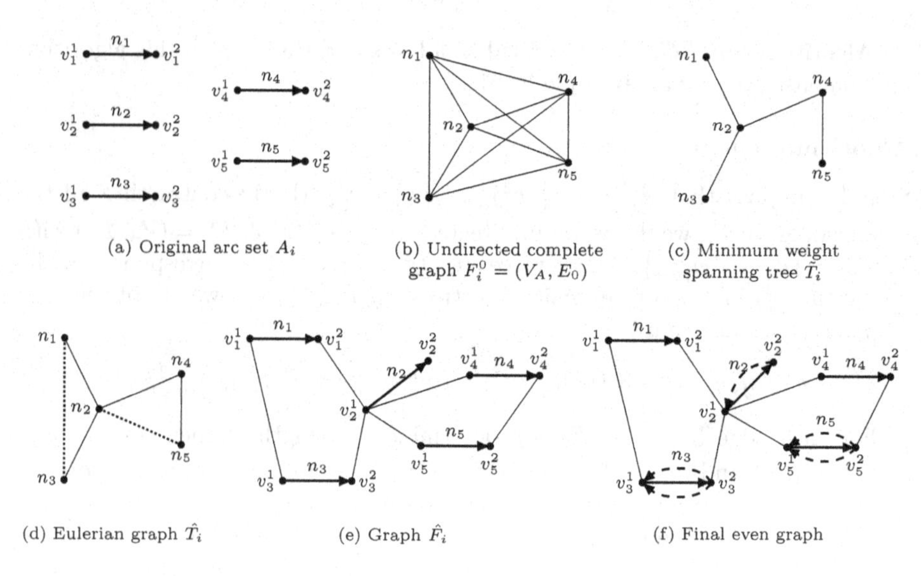

(a) Original arc set A_i (b) Undirected complete graph $F_i^0 = (V_A, E_0)$ (c) Minimum weight spanning tree $\tilde{T}_i$

(d) Eulerian graph $\hat{T}_i$ (e) Graph $\hat{F}_i$ (f) Final even graph

Fig. 2. An example for Algorithm *SymSCC*2. In picture (d), the dotted lines denote the matching edges in $\tilde{M}_i$. In picture (f), the dashed arrows represent the oriented parallel edges of the odd arcs and the even arcs.

Due to Observation 1, if $OPT \leq \lambda$, then in Step 1 Algorithm $SymSCC(\lambda)$ does not delete any edge in the optimal solution and any arc in A. So the vertex set of each optimum closed walk is contained entirely in exactly one of $V(F_1), \ldots, V(F_p)$. Let $k_i^* \geq 1$ $(i = 1, \ldots, p)$ be the number of optimum closed walks whose vertex sets are contained in $V(F_i)$. One can see that these k_i^* optimum closed walks constitute a stacker crane cover of A_i. Furthermore, the cost of this stacker crane cover of A_i is at most OPT, since the length of each optimum closed walk is at most OPT. Therefore, we have the following fact.

Observation 2. If $OPT \leq \lambda$, the $k_i^* \geq 1$ $(i = 1, \ldots, p)$ optimum closed walks whose vertex sets are contained in $V(F_i)$ constitute a stacker crane cover of A_i with cost at most λ.

Based on the above two observations, we can obtain upper bounds on the length of C_i^1 and C_i^2, and hence establish an upper bound on $w(C_i^0)$.

Lemma 7. *If* $OPT \leq \lambda$, *then* $w(C_i^0) \leq \left(\frac{14}{5}k_i^* - 1\right)\lambda$ *for* $i = 1, \ldots, p$.

Now we can show the correctness of Algorithm $SymSCC(\lambda)$.

Lemma 8. *If* $OPT \leq \lambda$, *Algorithm* $SymSCC(\lambda)$ *returns a stacker crane cover of* A *with at most* k *closed walks whose cost is no more than* $\frac{33}{5}\lambda$.

Proof. For each connected component F_i $(i = 1, \ldots, p)$, C_i^0 is a closed walk that contains all the arcs in A_i. By definition the closed walks generated in Step 3 of Algorithm $SymSCC(\lambda)$ can also cover all the arcs in A_i. By Lemmas 7 and 3,

$$\sum_{i=1}^{p} k_i \leq \sum_{i=1}^{p} \max\left\{\left\lceil \frac{w(C_i^0)}{\frac{14}{5}\lambda} \right\rceil, 1\right\} \leq \sum_{i=1}^{p} \max\left\{\left\lceil k_i^* - \frac{5}{14} \right\rceil, 1\right\} \leq \sum_{i=1}^{p} k_i^* = k.$$

This implies that Algorithm $SymSCC(\lambda)$ returns the stacker crane cover of $A = \bigcup_{i=1}^{p} A_i$ in Step 4. Due to Observation 1, it holds that $w(a) + w(P_a) \leq \beta$ for any $a \in A$. So we can indeed take $B = \frac{14}{5}\lambda$ and $\beta = \lambda$ in Lemma 3, the length of each closed walk which generated from C_i^0 is at most $2B + \beta = \frac{33}{5}\lambda$. □

Based on Lemmas 4, 6 and 8, we have the following result.

Theorem 1. *There is an $O(|V|^3 \log w(G))$ time $\frac{33}{5}$-approximation algorithm for the Symmetric SCC problem.*

Remark 1. The algorithm of Bao et al. [3] only applies to a special case of the Symmetric SCC problem. We generalize their algorithm to the (general) Symmetric SCC problem while improve the approximation ratio from $\frac{37}{5}$ to $\frac{33}{5}$. There are mainly two differences between our algorithm and that of Bao et al.

First, the analysis on the algorithm of Bao et al. simply uses the conclusion of the Christofides' algorithm to give an upper bound on $w(C_i^2)$. Instead, we construct two multi-subgraphs of F_i, which contain all the arcs in A_i and spanning $V(F_i)$, to give a more precise analysis. It turns out that this improves the upper bound on $w(C_i^2)$ in Lemma 7 from roughly $\frac{16}{5}k_i^*\lambda$ to $\frac{14}{5}k_i^*\lambda$.

Second, the previous algorithm generates a stacker cover of A_i by first splitting the walk C_i^0, using Lemma 1 (set $B = \frac{16}{5}\lambda$ and $\beta = \frac{\lambda}{2}$), into walks of length at most $B + \beta = \frac{37}{10}\lambda$ and then doubling the edges or adding parallel edges corresponding to the arcs to derive closed walks. Therefore, the resulting stacker crane cover of A_i has a cost of at most $2(B + \beta) = \frac{37}{5}\lambda$. Note that this approach relies on the upper bound $\frac{\lambda}{2}$ on the length of the arcs. For the (general) Symmetric SCC problem, the arc length may be in the interval $(\frac{\lambda}{2}, \lambda]$ and the approach of Bao et al. will result in an approximation ratio of $\frac{42}{5}\lambda$ since the value of β has to be set to λ. In contrast, our algorithm adopts a better approach, i.e. using Lemma 3 (set $B = \frac{14}{5}\lambda$ and $\beta = \lambda$), to derive a stacker cover of A_i of cost at most $2B + \beta = \frac{33}{5}\lambda$. Furthermore, the correctness of this approach is based on the simple but crucial property that the total length of each arc and the shortest path from its head to its tail is at most λ, as stated in Observation 1.

Remark 2. One can see that in Step 4 of Algorithm $SymSCC2$, for each even arc (v_j^1, v_j^2) that traversed in the opposite direction in C_i', we add two copies of the edge $[v_j^1, v_j^2]$ to C_i'. If the weights of the arcs are symmetric, we can upper bound the weight of the added parallel edges in terms of the corresponding arcs. Otherwise, i.e. the arc weights are not symmetric, we can not do this because the weight of the added parallel edges can be arbitrarily large. This makes the Algorithm $SymSCC(\lambda)$ not suitable for the (general) SCC problem. It is also the main difficulty of designing an approximate algorithm for the (general) SCC problem. The same problem also appears in the following discussion on the MSCC problem.

3.2 Min-Max Stacker Crane Walk Cover Problem

We give the following algorithm for the Symmetric SCWC problem, which is a 5-subroutine.

Algorithm $SymSCWC(\lambda)$

Input: An instance of the Symmetric SCWC problem consisting of a reduced graph $G = (V, E, A)$, a nonnegative integer weight function w on $E \cup A$ and a positive integer k, and a value $\lambda \in [\max_{a \in A} w(a), 2w(G)]$.
Output: Either failure or a stacker crane walk cover of A containing at most k walks with cost no more than 5λ.

Step 1. Delete all the edges with weight greater than λ in G. Suppose that the resulting graph $G[\lambda]$ has p connected components $F_1, \ldots, F_p$.
Step 2. For each $i = 1, \ldots, p$, let $A_i = A \cap A(F_i)$, find a minimum weight connected subgraph T_i of F_i such that T_i contains all the arcs in A_i and spans $V(F_i)$. Construct a Eulerian graph T_i' from T_i using Lemma 5. Let C_i be a Eulerian tour of T_i'. By taking $B = 4\lambda$, $\beta = \lambda$, $W = C_i$, and $A' = A_i$ in Lemma 1, we can decompose C_i into $k_i \leq \max\left\{\left\lceil \frac{w(C_i)}{4\lambda} \right\rceil, 1\right\}$ walks, which constitute a stacker crane walk cover of A_i of cost no more than 5λ.
Step 3. If $\sum_{i=1}^{p} k_i \leq k$, return the stacker crane walk cover of A consisting of the $\sum_{i=1}^{p} k_i$ walks obtained in Step 2; otherwise, return failure.

The time complexity of Algorithm $SymSCWC(\lambda)$ can be shown as follows.

Lemma 9. *Algorithm* $SymSCWC(\lambda)$ *runs in* $O(|V|^2 \log |V|)$ *time.*

As before, we assume w.l.o.g. that the optimal solution consists of exactly k optimum walks $W_1^*, \ldots, W_k^*$. It is easy to verify the following fact.

Observation 3. If $OPT \leq \lambda$, then $w(e) \leq OPT \leq \lambda$ for each $e \in \bigcup_{i=1}^{k} E(W_i^*)$ and $w(a) \leq OPT \leq \lambda$ for each $a \in \bigcup_{i=1}^{k} A(W_i^*)$.

By Observation 3, if $OPT \leq \lambda$, in Step 1 Algorithm $SymSCWC(\lambda)$ does not delete any edge in the optimal solution and any arc in A. Therefore, the vertex set of each optimum walk is contained entirely in exactly one of $V(F_1), \ldots, V(F_p)$. Let $k_i^* \geq 1$ $(i = 1, \ldots, p)$ be the number of optimum walks whose vertex sets are contained in $V(F_i)$. Similarly to Observation 2, we have the following fact.

Observation 4. If $OPT \leq \lambda$, the $k_i^* \geq 1$ $(i = 1, \ldots, p)$ optimum walks whose vertex sets are contained in $V(F_i)$ constitute a stacker crane walk cover of A_i with cost at most λ.

Based the above observations, we can obtain an upper bound on $w(C_i)$.

Lemma 10. *If $OPT \leq \lambda$, then $w(C_i) \leq (4k_i^* - 2)\lambda$ for $i = 1, \ldots, p$.*

Using the above lemma, we derive the correctness of Algorithm *SymSCWC*(λ).

Lemma 11. *If $OPT \leq \lambda$, Algorithm SymSCWC(λ) returns a stacker crane walk cover of A with at most k walks whose cost is at most* 5λ.

Based on Lemmas 9, 11 and 4, we have the following result.

Theorem 2. *There is an $O(|V|^2 \log |V| \log w(G))$ time 5-approximation algorithm for the Symmetric SCWC problem.*

4 Minimum Stacker Crane (Walk) Cover Problem

In this section, we deal with the Symmetric MSCC problem and the MSCWC problem. We obtain a 5-approximation algorithm for the Symmetric MSCC problem by using the algorithm for the MRPC problem in [12] as a subroutine. For the MSCWC problem, we devise a 4-approximation algorithm. We also present a better $\frac{7}{2}$-approximation algorithm for the Symmetric MSCWC problem. (Due to page limits, these results are deferred to the full-version of the paper.)

5 Conclusions

In this paper, we deal with two stacker crane cover problems and their variants. We develop a better 33/5-approximation algorithm for the Symmetric SCC problem, which improves on the previous 37/5-approximation algorithm for a special case of the Symmetric SCC problem. We also present the first constant-factor approximation algorithms for the Symmetric SCWC problem, the Symmetric MSCC problem and the Symmetric MSCWC problem with approximation ratios 5, 5 and 7/2, respectively. For the (general) MSCWC problem, we first give a 4-approximation algorithm.

For future research, it is desired to propose approximation algorithms for the above problems with better approximation ratios. Furthermore, one can design good approximation algorithms for the SCC problem, the SCWC problem and the MSCC problem without symmetric arc weights constraint.

Acknowledgement. This research is supported by the National Natural Science Foundation of China under grant numbers 11671135, 11871213 and the Natural Science Foundation of Shanghai under grant number 19ZR1411800.

References

1. Arkin, E.M., Hassin, R., Levin, A.: Approximations for minimum and min-max vehicle routing problems. J. Algorithms **59**(1), 1–18 (2006)
2. Atallah, M.J., Kosaraju, S.R.: Efficient solutions to some transportation problems with applications to minimizing robot arm travel. SIAM J. Comput. **17**(5), 849–869 (1988)
3. Bao, X., Lu, C., Huang, D., Yu, W.: Approximation algorithm for min-max cycle cover problem on a mixed graph. Oper. Res. Trans. **25**(1), 107–113 (2021). [in Chinese]
4. Calvo, R.W., Colorni, A.: An effective and fast heuristic for the Dial-a-Ride problem. 4OR **5**(1), 61–73 (2007). https://doi.org/10.1007/s10288-006-0018-0
5. Corberan, A., Laporte, G.: Arc Routing: problems, methods, and applications. In: SIAM, Philadelphia, pp. 101–127 (2015)
6. Eiselt, H.A., Gendreau, M., Laporte, G.: Arc routing problems, part I: the Chinese postman problem. Oper. Res. **43**, 231–242 (1995)
7. Eiselt, H.A., Gendreau, M., Laporte, G.: Arc routing problems, part II: the rural postman problem. Oper. Res. **43**, 399–414 (1995)
8. Frederickson, G.N., Hecht, M.S., Kim, C.E.: Approximation algorithms for some routing problems. SIAM J. Comput. **7**(2), 178–193 (1978)
9. Frederickson, G.N., Guan, D.J.: Nonpreemptive ensemble motion planning on a tree. J. Algorithms **15**(1), 29–60 (1993)
10. Gao, X., Fan, J., Wu, F., Chen, G.: Approximation algorithms for sweep coverage problem with multiple mobile sensors. IEEE/ACM Trans. Netw. **26**(2), 990–1003 (2018)
11. Garey, M.R., Johnson, D.S.: Computers and Intractability: A Guide to the Theory of NP-Completeness. Freeman, San Francisco (1979)
12. Mao, Y., Yu, W., Liu, Z., Xiong, J.: Approximation algorithms for some minimum postmen cover problems. In: Li, Y., Cardei, M., Huang, Y. (eds.) COCOA 2019. LNCS, vol. 11949, pp. 375–386. Springer, Cham (2019). https://doi.org/10.1007/978-3-030-36412-0_30
13. Olivier, Q., Andre, L., Fabien, L., Olivier, P., Martin, T.: Solving the large-scale min-max k-rural postman problem for snow plowing. Networks **70**(3), 195–215 (2017)
14. Parragh, S.N., Doerner, K.F., Hartl, R.F.: A survey on pickup and delivery problems, Part I: transportation between customers and depot. J. Für Betriebswirtschaft **1**, 21–51 (2008)
15. Parragh, S.N., Doerner, K.F., Hartl, R.F.: A survey on pickup and delivery problems, Part II: transportation between pickup and delivery locations. J. Für Betriebswirtschaft **2**, 81–117 (2008)
16. Safilian, M., Hashemi, S.M., Eghbali, S., Safilian, A.: An approximation algorithm for the Subpath Planning. In: the Proceedings of the 25th International Joint Conference on Artificial Intelligence, pp. 669–675 (2016)
17. Treleaven, K., Pavone, M., Frazzoli, E.: Asymptotically optimal algorithms for one-to-one pickup and delivery problems with applications to transportation systems. IEEE Trans. Autom. Control **58**(9), 2261–2276 (2013)
18. Xu, W., Liang, W., Lin, X.: Approximation algorithms for min-max cycle cover problems. IEEE Trans. Comput. **64**(3), 600–613 (2015)

19. Yu, W., Dai, R., Liu, Z.: Approximation algorithms for multi-vehicle stacker crane problems. J. Oper. Res. Soc. China (2021, to appear)
20. Yu, W., Liu, Z., Bao, X.: Approximation algorithms for some min-max postmen cover problems. Ann. Oper. Res. **300**, 267–287 (2021). https://doi.org/10.1007/s10479-021-03933-4

Succinct Data Structures for Series-Parallel, Block-Cactus and 3-Leaf Power Graphs

Sankardeep Chakraborty[1], Seungbum Jo[2](✉), Kunihiko Sadakane[1], and Srinivasa Rao Satti[3]

[1] The University of Tokyo, Tokyo, Japan
{sankardeep,sada}@mist.i.u-tokyo.ac.jp
[2] Chungnam National University, Daejeon, South Korea
sbjo@cnu.ac.kr
[3] Norwegian University of Science and Technology, Trondheim, Norway
srinivasa.r.satti@ntnu.no

Abstract. We design succinct encodings of *series-parallel, block-cactus* and *3-leaf power* graphs while supporting the basic navigational queries such as degree, adjacency and neighborhood *optimally* in the RAM model with logarithmic word size. One salient feature of our representation is that it can achieve optimal space even though the exact space lower bound for these graph classes is not known. For these graph classes, we provide succinct data structures with optimal query support for the first time in the literature. For series-parallel multigraphs, our work also extends the works of Uno et al. (Disc. Math. Alg. and Appl., 2013) and Blelloch and Farzan (CPM, 2010) to produce optimal bounds.

Keywords: Space efficient data structures · Succinct encoding · Series-parallel graphs · Cactus graphs

1 Introduction

In modern algorithm development, we observe two drastically opposing trends. Even though memory capacities are increasing and their prices are drastically reducing day-by-day, input data sizes that are being stored are growing at a much faster pace, and this is due to the ongoing digital transformation of business and society in general. There are many application areas, e.g., social networks, web mining, and video streaming systems, where already there exists a tremendous amount of data and it is only increasing. In these domains, most often, a natural representation of the underlying data sets is in the form of graphs, and with each passing day, these graphs are becoming massive. To process such huge graphs

S. Chakraborty—Supported by MEXT Quantum Leap Flagship Program (MEXT Q-LEAP) Grant Number JPMXS0120319794.
S. Jo—Supported by the National Research Foundation of Korea (NRF) grant funded by the Korea government (MSIT) (No. NRF-2020R1G1A1101477).

D.-Z. Du et al. (Eds.): COCOA 2021, LNCS 13135, pp. 416–430, 2021.
https://doi.org/10.1007/978-3-030-92681-6_33

and extract useful information from them, we need to answer the following two concrete questions among others: (1) can we store these massive graphs in compressed form using the minimum amount of space? and (2) can we build space-efficient indexes for these huge graphs so that we can extract useful information about them by executing efficient query algorithms on the index itself? The field of *succinct data structures* aims to exactly answer these questions satisfactorily, and it has been one of the key contributions to the algorithm community in the past two decades, both theoretically and practically. More specifically, given a class of certain combinatorial objects, say T, from a universe U, the main objective here is to store any arbitrary member $x \in T$ using the *information-theoretic lower bound* of $\log(|U|)$ bits (in addition to $o(\log(|U|))$ bits)[1] along with efficient support of a relevant set of operations on x.

There exists already a large body of work representing various combinatorial structures succinctly along with fast query support. For example, succinct data structures for rooted ordered trees [16,17,21,22], chordal graphs [18], graphs with treewidth at most k [11], separable graphs [2], interval graphs [1] etc., are some examples of these data structures. Following similar trend, in this work we provide succinct data structures for series-parallel multigraphs [24], block-cactus graphs [14] and 3-leaf power graphs [4]. We defer the definitions of the graph classes to the individual sections where their succinct data structure is proposed. These graphs are important because not only they are theoretically appealing to study but they also show up in important practical application domains, e.g., series-parallel graphs are used to model electrical networks, cacti are useful in computational biology etc. To the best of our knowledge, our work provides succinct data structures with optimal query support for the first time in the literature (although there exists a succinct data structure for simple series-parallel graphs [2], such a structure is not known for series-parallel multigraphs).

1.1 Previous Work

Series-Parallel (SP) Graphs. The information-theoretic lower bound (ITLB) for encoding a simple SP-graph with n vertices is $3.18n + o(n)$ bits [3] whereas the ITLB for encoding an SP multigraph with m edges is $1.84m + o(m)$ bits [27]. Since an SP graph is *separable*, one can obtain a succinct representation of any SP graph by using the result of Blelloch and Farzan [2] while supporting some navigation queries efficiently. However, this only works for simple SP graphs [19] since one cannot store the look-up table for all possible micro-graphs (containing multi SP graphs with any fixed number of vertices) within the limited space (as the number of edges is not bounded)[2]. Also since simple SP graphs are exactly the class of graphs with treewidth 2, one can use the data structure of Farzan and Kamali [11] for representing SP graphs but again, this only works for simple SP

[1] Throughout the paper, we use logarithm to the base 2.

[2] Note that one can encode SP multigraphs by encoding the underlying simple graph using Blelloch and Farzan's encoding, along with a bit string of size m to represent the multiplicities of the edges. However, the space usage is not succinct in this case.

graphs. For multigraph case with m edges, Uno et al. [27] present an encoding for SP multigraphs taking at most 2.53 m bits without supporting any navigational queries efficiently.

Block-Cactus and 3-Leaf Power Graphs. The ITLB for encoding a block-cactus graph and a 3-leaf power graph with n vertices are $2.092n + o(n)$ [29] and $1.35n + o(n)$ [7] bits respectively. Note that the class of Block-cactus graphs contains both *cactus* and *block* graph classes. As any cactus graph is planar, and hence separable, one can again use the result of Blelloch and Farzan [2] to encode them optimally with supporting the navigation queries efficiently. However, this approach doesn't work for block or block-cactus graphs since they are not separable.

1.2 Our Main Contribution

We design succinct data structures for (i) series-parallel multigraphs in Sect. 3 and (ii) block-cactus graphs in Sect. 4, and finally (iii) 3-leaf power graphs in Sect. 5 to support the following queries. Given a graph $G = (V, E)$ and two vertices $u, v \in V$, (i) degree(v) returns the number of edges incident to v in G, (ii) adjacent(u, v) returns true if u and v are adjacent in G, and false otherwise, and finally (iii) neighborhood(v) returns the set of all (distinct) vertices that are adjacent to v in G. The following theorem summarizes our main results on these graphs.

Theorem 1. *There exists a succinct data structure that supports degree(u) and adjacent(u, v) queries in $O(1)$ time, and neighborhood(u) query in O(degree(u)) time, for (1) series-parallel multigraphs, (2) block-cactus graphs, and (3) 3-leaf power graphs.*

The reason for considering these three (seemingly unrelated) graph classes is that any graph in each of these three classes has a corresponding tree-based representation - and hence these graphs can be encoded succinctly by encoding the corresponding tree. In what follows, we briefly discuss a high level idea on how to succinctly represent the graphs of our interest. Roughly speaking, given a graph G (G could be series-parallel, block-cactus or 3-leaf power), we first convert it to a labeled tree T_G which can be used to decode G. We then represent G by encoding T_G using the *tree covering (TC) algorithm* of Farzan and Munro [12], which supports various tree navigation queries in $O(1)$ time. However, we cannot obtain directly the succinct representation of G with efficient navigation queries from the tree covering of T_G. More specifically, the tree covering algorithm first decomposes the input tree and encodes each decomposed tree separately. Thus, a lot of information of G can be lost in each of the decomposed trees. For example, decomposed trees may not even belong to the graph class that we originally started with in the first place (and this is in stark contrast to the situation while designing succinct data structures for trees). Thus, we need to apply non-trivial local changes (catering to each graph class separately) to these decomposed trees and argue that (i) these changes convert them again back to the original graph class, without consuming too much space, and (ii) navigation queries on G can be supported efficiently as tree queries on T_G. As a consequence, one salient feature of our approach is that for the graphs G we consider in this paper, it is not necessary to know the exact information-theoretic lower bound, to design succinct data structures for them if we only know the asymptotic lower bound of the number of non-isomorphic graphs of G with a given number of vertices. Note that the overall idea of 'encoding the graph as a tree-based representation and using the TC algorithm to encode the tree to support the navigation operations on the graph' is subsequently used in [6] to obtain succinct representation for graphs of small clique-width. The other main contribution of this paper is to construct suitable tree-based encodings and showing how to adapt the TC representation to support the operations.

2 Preliminaries and Main Techniques

Throughout our paper, we assume familiarity with succinct/compact data structures (as given in [20]), basic graph theoretic terminology (as given in [9]), and graph algorithms (as given in [8]). All the graphs in our paper are assumed to be connected and unlabeled, i.e., we can number the vertices arbitrarily. Moreover, we assume the usual model of computation, namely a $\Theta(\log n)$-bit word RAM model where n is the size of the input (n is the number of vertices in the case of graphs, and the number of edges in the case of multigraphs). We start by sketching a modification to the tree covering algorithm of Farzan and Munro [13].

2.1 Tree Covering

The high level idea of the tree covering algorithm is to decompose the tree into subtrees called mini-trees (in the rest of the paper, we use *subtree* to denote any connected subgraph of a given tree), and further decompose the mini-trees into yet smaller subtrees called *micro-trees*. The micro-trees are small enough to be stored in a compact table. The root of a mini-tree can be shared by several other mini-trees. To represent the tree, we only have to represent the connections and links between the subtrees. In what follows, we summarize the main result of Farzan and Munro [13] in the following theorem:

Theorem 2. ([13]). *For a rooted ordered tree with n nodes and a positive integer $1 \le L \le n$, we can obtain a tree covering satisfying (1) each subtree contains at most $2L$ nodes, (2) the number of subtrees is $O(n/L)$, (3) each subtree has at most one outgoing edge, apart from those from the root of the subtree.*

For each subtree after the decomposition of Theorem 2, the unique node that has an outgoing edge is called the *boundary node* of the subtree, and the edge is called the *boundary edge* of the subtree. The subtree may have multiple outgoing edges from its root node (in this case, we call it a *shared root node*), and those edges are called *root boundary edges*.

To obtain a succinct representation, we first apply Theorem 2 with $L = \log^2 n$, to obtain $O(n/\log^2 n)$ mini-trees (here and in the rest of the paper, we ignore all floors and ceilings which do not affect the main result). The tree obtained by contracting each mini-tree into a vertex is referred to as the *tree over mini-trees*. If more than one mini-tree shares a common root, we create a dummy node in the tree and make the nodes corresponding to the mini-trees as children of the dummy node. We also set the parent of the dummy node as the node corresponding to the parent mini-tree. (See Fig. 1 for an example.) This tree has $O(n/\log^2 n)$ vertices and therefore can be represented in $O(n/\log n) = o(n)$ bits using a pointer-based representation. Then, for each mini-tree, we again apply Theorem 2 with parameter $\ell = \frac{1}{4}\log n$ to obtain $O(n/\log n)$ micro-trees in total. The tree obtained from each mini-tree by contracting each micro-tree into a node, and adding dummy nodes for micro-trees sharing a common root (as in the case of the tree over mini-trees) is called the *mini-tree over micro-trees*. Each mini-tree over micro-trees has $O(L/\ell) = O(\log n)$ vertices, and can be represented by $O(\log(L/\ell)) = O(\log\log n)$-bit pointers. For each non-root boundary edge of a micro-tree t, we encode from which vertex of t it comes out and the rank among all children of the vertex. One can encode the position where the boundary edge is inserted in $O(\log \ell)$ bits. Note that in our modified tree decomposition, each node in the tree is in exactly one micro-tree.

For each micro-tree, we define its representative as its root node if it is not shared with other micro-trees, or the next node of the root node in preorder if it is shared. Then we mark bits of the balanced parentheses representation [17] of the entire tree corresponding to the representatives. If we extract the marked bits, it forms a balanced parentheses (BP) and it represents the mini-tree over micro-trees. The positions of marked bits are encoded in $O(n \log\log n/\log n)$ bits because there are $O(n/\log n)$ marked bits in the BP representation of $2n$ bits. The BP representation is partitioned into $O(n/\log n)$ many variable-length blocks, each of which is of length $O(\log n)$. We can decode each block in constant time.

To support basic tree navigational operations such as parent, i-th child, child rank, degree, lowest common ancestor (LCA), level ancestor, depth, subtree size, leaf rank, etc. in constant time, we use the data structure of [21]. Note that we slightly change the data structure because now each block is of variable length. We need to store those lengths, but it is done by using the positions of the marked bits.

The total space for all mini-trees over micro-trees is $O(n/\ell \cdot \log \ell) = O(n \log\log n/\log n) = o(n)$ bits. Finally, the micro-trees are stored as two-level pointers (storing the size, and an offset within all possible trees of that size) into a precomputed table that contains the representations of all possible micro-trees. The space for encoding all the micro-trees using this representation can be shown to be $2n + o(n)$ bits.

2.2 Graph Representation Using Tree Covering

This section describes the high-level idea to obtain succinct encodings for the graph classes that we consider. Let $\mathcal{C}$ be one of the graph classes among series-parallel multigraphs, block-cactus graphs, and 3-leaf power graphs. Then the following properties hold.

- the ITLB for representing any graph $G \in \mathcal{C}$ is $kn + o(kn)$ bits for some constant $k > 0$ [7,27,29], where n is the number of vertices (block-cactus, and 3-leaf power graphs) or edges (series-parallel multigraphs) in G.
- For any connected graph $G \in \mathcal{C}$, there exists a labeled tree T_G of $O(n)$ nodes, such that G can be uniquely decoded from T_G.

By the above properties, one can represent any graph $G \in \mathcal{C}$ by encoding the tree covering of T_G (with $L = \log^2 n$ and $\ell = \frac{\log n}{2k}$). Unfortunately, tree covering on T_G does not directly give a succinct encoding of G since the number of all non-isomorphic graphs in $\mathcal{C}$ can be much smaller than the number of all non-isomorphic labeled trees of the same size (for example, multiple labeled trees can correspond to the same graph). To solve this problem, we maintain a precomputed table of all non-isomorphic graphs in $\mathcal{C}$ of size at most ℓ, along with their corresponding trees in *canonical representation*. By representing each micro-tree as an index of the corresponding graph in the precomputed table, we can store all the micro-trees of T_G in succinct space. If a micro-tree t does not have a corresponding graph in $\mathcal{C}$ (i.e., there is no corresponding graph in the precomputed table), we first extend t to T_g where $g \in \mathcal{C}$ of size at most ℓ by adding some *dummy nodes*, and encode t as the index of g, along with the information about dummy nodes. Since we only add a small number of (at most $O(1)$) dummy nodes for each micro-tree, all the additional information can be stored within succinct space. In the following sections, we describe how to add such dummy nodes for series-parallel, block-cactus, and 3-leaf power graphs.

Finally, for the case when $G \in \mathcal{C}$ is not connected, we extend the above idea as follows. We first encode all the connected components of G separately, and encoding the sizes of the connected components using the encoding of [10,26] using at most $O(\sqrt{n})$ additional bits. This implies we can still encode G in succinct space even if G is not connected. In the rest of this paper, we assume that all the graphs are connected.

3 Series-Parallel Graphs

Series-parallel graphs [24] (SP graphs in short) are undirected multi-graphs which are defined recursively as follows.

- A single edge is an SP graph. We call its two end points as terminals.
- Given two SP graphs G_1 with terminals s_1, t_1 and G_2 with terminals s_2, t_2,
  - their *series composition*, the graph made by identifying $t_1 = s_2$, is an SP graph with terminals s_1, t_2; and

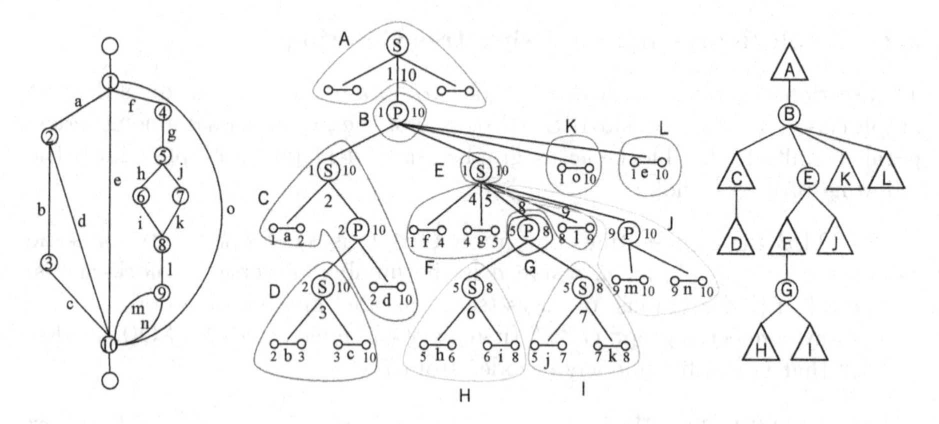

Fig. 1. Example of an SP graph (left), its SP tree representation with tree covering (middle), and the tree over mini-trees (right). The roots of mini-tree G and K are dummy nodes. Numbers below S nodes are inorders. Numbers besides internal nodes of the SP tree are the left and right labels. Leaves of the tree also have the left and right labels, which are vertex labels of the SP graph.

- their *parallel composition*, the graph made by identifying $s_1 = s_2$, and $t_1 = t_2$, is an SP graph with terminals s_1 and t_1.

From this construction, we can obtain the binary tree T representing an SP graph $G = (V, E)$ as follows. Each leaf of the binary tree T corresponds to an edge of G. Each internal node v of T has a label S (or P), which represents an SP graph made by the series (or parallel) composition of the two SP graphs represented by the two child subtrees of v. We convert it into a multi-ary SP tree T_G by merging vertically consecutive nodes with identical labels into a single node. More precisely, while scanning all the nodes in T_G in bottom-up, we contract every edge (v, v') if v and v' have the same labels. Then all the internal nodes at the same depth have the same labels, and the labels alternate between the levels. See Fig. 1 for an example. Note that any two non-isomorphic SP graphs have different SP trees.

Succinct Representation. Let n and m be the number of vertices and edges of G, respectively. Then T_G has m leaves, and $O(m)$ nodes. First, we construct the SP tree T_G from an SP graph $G = (V, E)$. If the root of T_G is a P node, we add a dummy parent r labeled S with three children, and make the original root as the middle child of r. The first and the last children of r correspond to dummy edges. If the root of T_G is an S node, we also add two leaves as the leftmost and rightmost children of the root, corresponding to dummy edges. We refer to this modified tree as T_G. Let $s = O(m)$ be the number of nodes in T_G. Then we apply the tree covering algorithm with parameters $L = \log^2 s$ and $\ell = (\log s)/4$.

It is obvious that each micro-tree without dummy leaf nodes represents an SP graph. For each graph corresponding to a micro-tree, we use a linear time algorithm [28] to obtain a canonical representation of the micro-tree. Note that if the graphs corresponding to two micro-trees are isomorphic, then those two micro-trees have the same canonical representation. We create a table to store all non-isomorphic SP graphs with at most ℓ vertices, and encode each micro-tree as a pointer into this table. To reconstruct the original graph from the graphs corresponding to the micro-trees, we need additional information to combine these graphs. More specifically, assume an SP graph G consists of a series composition of graphs G_1 and G_2, whose terminals are s_1, t_1, and s_2, t_2 respectively. Then one can construct two different graphs G and G' by (i) connecting t_1 and t_2 or (ii) t_1 and s_2. Thus, for each micro tree, we add one extra bit to store this information.

For each S node of T_G, we assign an *inorder number* [25] (we only assign inorder numbers for S nodes). Inorder numbers in a rooted tree are given during a preorder traversal from the root. If a node v is visited from one of its children and another child of v is visited next, we assign one inorder number to v. If a node has k children, we assign $k-1$ inorder numbers to it. (Unary nodes are not assigned any inorder number.) If a node has more than one inorder number, we use the smallest value as its representative inorder number. Now we consider two operations (i) $irank(k, i)$: return the i-th inorder rank of S node k (given as preorder number), and (ii) $iselect(j)$: given an inorder rank j of an S node, return (k, i) where k is the preorder number of the node with inorder rank j and i is the number such that k is the i-th inorder number of the node. The following describes how to support both queries in O(1) time using o(n) bits of additional space.

One can observe that for each micro-tree (or mini-tree) t of T_G, all the inorder numbers corresponding to the S nodes in t form two intervals $I_t^1 = [l_t^1, r_t^1]$ and $I_t^2 = [l_t^2, r_t^2]$. Note that all the intervals corresponding to the mini-trees or micro-trees partition the interval $[1, \mathcal{S}]$ where $\mathcal{S}$ is the largest inorder number in T_G. We construct a dictionary D_M that stores the right end points of all the intervals corresponding to the mini-trees, where with each element of the dictionary, we store a pointer to the mini-tree corresponding to that interval as the satellite information. The number of elements in this dictionary is $O(s/L)$ with universe size at most s, and hence can be represented as an FID [23] using $o(s)$ bits to support membership, rank and select queries in O(1) time. The satellite information can also be stored in $o(s)$ bits, to support O(1)-time access. For each mini-tree M, we also construct a dictionary D_μ^M that stores the right end points of all the intervals corresponding to its micro-trees, where with each element we associate a pointer to the corresponding micro-tree as the satellite information. The space usage of the dictionaries corresponding to all the mini-trees adds up to $o(s)$ bits in total.

In addition, for each mini-tree T of T_G, we store its corresponding intervals I_T^1 and I_T^2 using o(s) bits in total. We call the two values l_T^1 and l_T^2 as the *offsets* corresponding to T. Also, for each micro-tree t contained in the mini-tree T and $i \in \{1, 2\}$, we store $\{[l_t^i - l_T^1, r_t^i - l_T^1]\}$ if $I_t^i \subseteq I_T^1$, and $[l_t^i - l_T^2, r_t^i - l_T^2]$ otherwise (i.e., offsets with respect to the mini-tree intervals they belong to). Since all the endpoints of these intervals are at most L, we can store all such intervals using $o(s)$ bits in total. The total space usage is $o(s)$ bits.

To compute $irank(k, i)$, we first find the micro-tree t which contains the node k. Then, we decode the interval corresponding t using the interval stored at t as well as the offsets corresponding to t, and return the i-th smallest value within the interval. To compute $iselect(k)$, we first find micro-tree t that contains the answer by the rank queries on D_M and D_μ^M. Finally, we compute the answer within the micro-tree t in $O(1)$ time using the intervals stored with t.

Next, we assign labels to the vertices of the graph. Any vertex in the graph corresponds to a common terminal of two SP graphs which are combined by series composition. For each vertex $v \in G$, let S_v be an S node in T_G which represents such series composition. Then we assign one inorder number of S_v as the label of v (note that any two subgraphs which have a common terminal correspond to the subtree at the consecutive child nodes of S_v). For example, vertex 5 in the graph corresponds to the common terminal of the following two subgraphs: (i) the subgraph consisting of the edge g from 4 to 5, and (ii) the subgraph corresponding to the subtree rooted at the mini-tree H (consisting of a single P node), which contains the four edges h, j, i and k. Note that the inorder number 5 is assigned to the S node corresponding to the mini-tree F, when we traverse from subtree corresponding to (i) to the subtree corresponding to (ii) (during the preorder traversal of T).

Also, we define a label for each node v of T_G, which is an ordered pair (l_v, r_v) of the two terminals of the subgraph corresponding to the subtree rooted at that node. We call l_v and r_v the left and the right label of the node v. The label (l_v, r_v) of a P node v can be computed in $O(1)$ time as follows. (1) If v is the leftmost child of its parent p, then r_v is equal to the smallest inorder number of p, given when p is visited from v. To obtain l_v, we traverse the SP tree T_G up from v until we reach an S node q such that v does not belong to the leftmost subtree of q. We can compute the node q in $O(1)$ time as follows. If q is in the same micro-tree as v, then we can find q using a table lookup. Otherwise, if q is in the same mini-tree as v, then we store q with the root of the micro-tree containing v. Finally, if q is not in the same mini-tree, then we explicitly store q with the root of the mini-tree containing v. (2) If v is the rightmost child of its parent p, then l_v is equal to the inorder number of p, given when p is visited the last time before visiting v. To obtain r_v, we traverse the SP tree T_G up from v until we reach an S node p such that v does not belong to the rightmost subtree of p. We use a similar data structure as in (1) to compute the answer. (3) In all other cases, l_v and r_v are the inorder numbers of the parent p of v, defined immediately before visiting v from p, and immediately after visiting the next sibling of v from p, respectively.

The label of an S node is the same as its parent P node (we don't assign a label to the root S node). The label of a leaf can be determined by the same algorithm for P or S nodes depending on whether its parent is an S or P node. Note that, from the above definition, the label of a P node is the same as the label of any of its child S nodes. For an S node v, suppose $v_1, v_2, \ldots, v_k$ be its k

children, and $(\ell_1, r_1), (\ell_2, r_2), \ldots, (\ell_k, r_k)$ be the left and the right labels. Then it holds that $r_1 = \ell_2, r_2 = \ell_3, \ldots, r_{k-1} = \ell_k$, and the label of v is (ℓ_1, r_k).

We also define $b(u)$ and $f(u)$ for each vertex u of the graph, as follows. Suppose that during the preorder traversal of the tree, we visit nodes x, p, y in this order and we assign the inorder number u to p. Then we define $b(u) = x$ and $f(u) = y$. If $iselect(u)$ returns the pair (p, j), then x and y are the j-th and the $(j+1)$-th children of node p, respectively. Thus, $b(u)$ and $f(u)$ can be computed in O(1) time.

This completes the description for encoding of SP graphs.

Supporting Navigation Queries. For SP graphs, we additionally consider *multiplicity* (u, v) queries, which returns the number of edges between u and v.

1. adjacent(u, v): Without loss of generality, assume that $u < v$. We first find the nodes $b(u)$, $f(u)$, $b(v)$ and $f(v)$. (1) If $f(u) = b(v)$, the subgraph corresponding to the node $f(u)$ has terminals with labels u and v. Therefore u and v are adjacent if $f(u)$ is a leaf (this corresponds to the edge (u, v)) or $f(u)$ has a leaf child ($f(u)$ is a P node and it has a leaf child that corresponds to the edge (u, v)).(2) If $depth(b(u)) > depth(b(v))$, find the label of $f(u)$. Let (u, x) be the label of $f(u)$. Then u and v are adjacent iff $x = v$, and $f(u)$ is either a leaf or is a P node with a leaf child. (3) If $depth(b(u)) < depth(b(v))$, find the labels of $b(v)$. Let (y, v) be the label of $b(v)$. Then u and v are adjacent iff $y = u$, and $b(v)$ is either a leaf or is a P node with a leaf child. In all three cases, the query can be supported in $O(1)$ time.
2. multiplicity(u, v): Again, without loss of generality, assume that $u < v$. If adjacent$(u, v) = false$, then we return 0. If not, we consider the three cases above, and describe how to support the multiplicity query. For Case (1), if $f(u)$ is a P node (otherwise, we return 1), we can answer the query by returning the number of leaf children of $f(u)$ (note that this can be supported in $O(1)$ time using the tree covering of T_G). For Case (2), if $f(u)$ is a P node with label (u, v) (if $f(u)$ is a leaf node, we return 1), we answer the query by returning the number of leaf children of $f(u)$. Case (3) is analogous to Case (2).
3. neighborhood(u): First we find $b(u)$ and $f(u)$. Then we apply the following procedure to explore all the neighbors of u by executing the two procedure calls, Explore$(b(u), \mathrm{R})$ and Explore$(f(u), \mathrm{L})$.

   Explore(x, D): if x is a leaf with label (u, w) or (w, u), then output w. If x is an S node, then call Explore(y, D), where y is the leftmost (rightmost) child of x, if $D = \mathrm{L}$ ($D = \mathrm{R}$). If x is a P node, then call Explore(y, D) for all the children y of x.

   The running time of this procedure is proportional to the size of the output. Note that if we do not want to report the same neighbour multiple times, we can define a canonical ordering between the children of P nodes such that all the leaf children appear after the non-leaf children (S nodes), and only report the first leaf child of the node.

4. **degree(u)**: Let μ and M be the micro-tree and mini-tree containing u respectively. Then the degree of u is the summation of (i) the number of adjacent vertices in μ, (ii) the number of adjacent vertices not in μ but in M, and (iii) the number of adjacent vertices not in M, denoted by d_u^1, d_u^2, and d_u^3 respectively. Here an adjacent vertex of u refers to a vertex v such that (u, v) or (v, u) is the label of some leaf node. If u is not one of the labels of the boundary node of μ (of M), then $d_u^2 = 0$ (respectively, $d_u^3 = 0$). Now we consider three cases as follows. First, d_u^1 can be computed in O(1) time using a precomputed table. The value d_u^2 (d_u^3) can be stored with the root of micro-tree (mini-tree) whose parent is the boundary node in μ (M). Note that in the above scheme, we only need to store two values corresponding to the two labels of the root, for each micro-tree/mini-tree root. Thus the space usage for storing these values is $o(n)$ bits.

4 Block/Cactus/Block-Cactus Graphs

A *block graph* (also known as a clique tree or a Husimi tree [15]) is an undirected graph in which every block (i.e., maximal biconnected component) is a clique. A *cactus graph* (same as *almost tree(1)* [14]) is a connected graph in which every two simple cycles have at most one vertex in common (equivalently every block is a cycle). A *block-cactus graph* is a graph in which every block is either a cycle or a complete graph.

Any graph that belongs to one of these three graph classes can be converted into a tree as follows. Replace each block (either a clique or an induced cycle) with k vertices by a star graph $K_{1,k}$ by introducing a dummy node that is connected to the k nodes that correspond to the k vertices of the block. The remaining edges and vertices of the graph are simply copied into the tree. See Fig. 2 for an example. Note that the number of dummy nodes is always less than the number of non-dummy nodes. In the following, we describe a succinct encoding for block-cactus graphs, and note that it is easy to obtain succinct encoding for block graph and cactus graph using these ideas.

Succinct Representation. Let G be the input block-cactus graph, and let T_G be the corresponding tree obtained by replacing each block with a star graph, as described above. We apply the tree covering algorithm of Theorem 2 on T_G with mini-tree and micro-trees of size $L = \log^2 n$ and $\ell = (\log n)/(2\alpha)$ for some constant $\alpha \geq 2.092$ respectively.

It is easy to see that each micro/mini-tree obtained by the tree cover algorithm corresponds to a block-cactus graph, although it may not be a subgraph of the original graph G. And by storing some additional information with each micro/mini-tree along with its representation, we can give a bijective map between the vertices in G and the nodes in T_G, which we use in describing the query algorithms.

We first note that when we convert a block (C_k or K_k) into a star graph ($K_{1,k}$), the neighbors of the dummy node can be ordered in multiple ways when

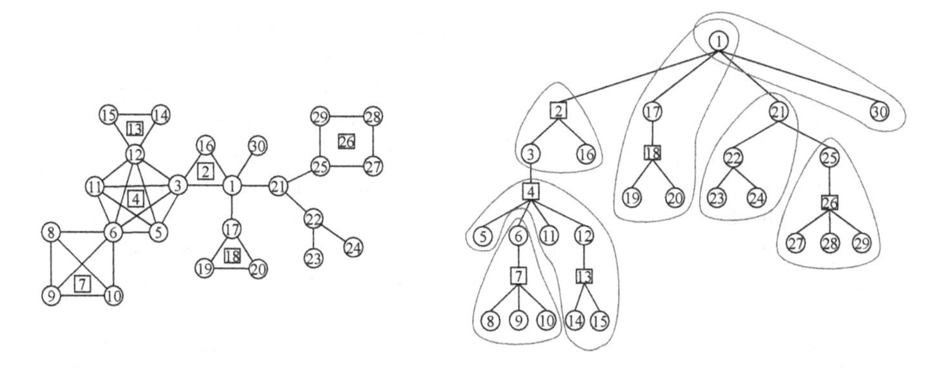

Fig. 2. An example of a block-cactus graph (left) and its tree representation (right). Squares are dummy vertices.

we consider the resulting graph as an ordered tree. In particular, if the ordered tree is rooted at a dummy node corresponding to a cycle, then its children can be ordered in either the clockwise or anti-clockwise order of the cycle, and also the first child can be any vertex on the cycle. When the root of micro-tree t is a dummy node corresponding to a cycle, the cycle corresponding to the dummy node is cut into two or more pieces, and the one inside t represents a shorter cycle. Then the micro-tree t is encoded as a canonical representation of the modified subgraph, and it loses the information of how it was connected to the other part of the graph. To recover this information, for the micro-tree it is enough to store one vertex in the shorter cycle that is connected to the outside and the direction (clockwise or anti-clockwise) of the cycle. The vertex is encoded in $O(\log \ell)$ bits, and the direction in one bit. We need the same information for the non-root boundary node of the micro-tree. This additional information will enable us to reconstruct the cycle in the original graph from the subgraphs corresponding to the micro-trees. Note that if the dummy node corresponds to a clique, then we don't need this information.

Each micro-tree is encoded as a two-level pointer into a precomputed table that stores the set of all possible block-cactus graphs on at most ℓ vertices. Note that the number of dummy nodes is $O(n/\log n)$ since we can delete all the dummy nodes which are not boundary nodes of micro-trees. We also store 1 bit with each of these $O(n/\log n)$ dummy nodes, indicating whether it corresponds to a clique or a cycle. Thus each micro-tree is represented optimally, apart from an $O(\log \ell)$-bit additional information. Hence the overall space usage is succinct. This completes the description for the succinct encoding of block-cactus graphs.

Supporting Navigation Queries

1. $\mathsf{adjacent}(u, v)$: If there is an edge in T_G between the nodes corresponding to u and v, then u and v are adjacent in the graph (since we only delete some edges from the original graph; and all the edges added are incident to some dummy node). Otherwise, u and v are adjacent if they are connected to the

same dummy node x, and either (a) x corresponds to a clique, or (b) u and v are "*adjacent*" in the tree – i.e., if they are adjacent siblings or one of them is the parent of x and the other is either the first or last child of x. Since all these conditions can be checked in O(1) time using the tree representation, we can support the query in O(1) time.

2. neighborhood(u): The algorithm for this follows essentially from the conditions for checking adjacency. More specifically, to report neighborhood(u), we first output all the non-dummy nodes adjacent to u in the tree. And if u is adjacent to any dummy node x, then we also output all the vertices: (a) that are connected to x if x corresponds to a clique, and (b) that are "*adjacent*" to it in the tree if x corresponds to a cycle. This can be done in time proportional to the output size.
3. degree(u): From the algorithm for the neighborhood(u) query, we observe that the degree of a node can be computed by adding the two quantities: (1) the number of non-dummy neighbors of u, and (2) the number of nodes that are adjacent to u through a dummy neighbor. It is easy to compute (1) and (2) within a micro-tree, in constant time using precomputed tables. In addition, we may need to add the contributions from outside the micro-tree, if u is either a boundary node or is adjacent to a boundary node which is dummy. For each such dummy boundary node, we need to add either 1 or 2 (if the dummy node corresponds to a cycle) or k (if the dummy node corresponds to a clique of size k). Since there are at most two such boundary nodes which can be adjacent to u, this can be computed in constant time. Also, for the roots of the mini (micro) trees, which are non-dummy, we store their degrees (within the mini-tree) explicitly. Thus, we can compute the degree(u) query in $O(1)$ time.

5 3-leaf Power Graph

A graph G with n vertices is a *k-leaf power* if there exists a tree T_G with n leaves where each leaf node corresponds to a vertex in the graph G, and any two vertices in G are adjacent if and only if the distance between their corresponding leaves in the tree is at most k. The tree T_G is called a *k-leaf root* of G (see Fig. 3 for an example). Note that for any connected and non-clique 3-leaf power G, T_G has at most O(n) nodes [4]. To obtain a succinct representation of G, we first make T_G as a rooted tree, and apply the tree covering algorithm of Theorem 2 on T_G and show how to support the queries on the graph using the tree covering representation. (We describe all the details in the full version [5]).

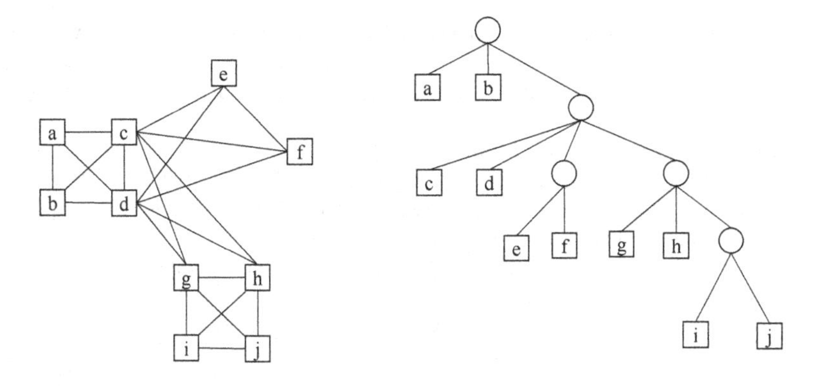

Fig. 3. An example of a 3-leaf power graph (left) and its 3-leaf root (right).

6 Conclusions

We present in this work succinct representations of series-parallel, block-cactus and 3-leaf power graphs along with supporting basic navigational queries optimally. We conclude with some possible future directions for further exploration. Following the works of [1,11], is it possible to support shortest path queries efficiently on these graphs while using same space as in this paper? Is it possible to design space-efficient algorithms for various combinatorial problems for these graphs? Can we generalize the data structure of Sect. 5 to construct a succinct representation of k-leaf power graphs? Finally, can we prove a lower bound between the query time and the extra space i.e., redundancy, for our data structures?

References

1. Acan, H., Chakraborty, S., Jo, S., Satti, S.R.: Succinct data structures for families of interval graphs. In: WADS, pp. 1–13 (2019)
2. Blelloch, G.E., Farzan, A.: Succinct representations of separable graphs. In: CPM, pp. 138–150 (2010)
3. Bodirsky, M., Giménez, O., Kang, M., Noy, M.: Enumeration and limit laws for series-parallel graphs. Eur. J. Comb. **28**(8), 2091–2105 (2007)
4. Brandstädt, A., Le, V.B.: Structure and linear time recognition of 3-leaf powers. Inf. Process. Lett. **98**(4), 133–138 (2006)
5. Chakraborty, S., Jo, S., Sadakane, K., Satti, S.R.: Succinct data structures for series-parallel, block-cactus and 3-leaf power graphs (2021). https://arxiv.org/abs/2108.10776
6. Chakraborty, S., Jo, S., Sadakane, K., Satti, S.R.: Succinct data structures for small clique-width graphs. In: 31st Data Compression Conference, DCC 2021, Snowbird, UT, USA, 23–26 March, 2021, pp. 133–142. IEEE (2021)
7. Chauve, C., Fusy, É., Lumbroso, J.O.: An exact enumeration of distance-hereditary graphs. In: ANALCO 2017, Barcelona, Spain, Hotel Porta Fira, 16–17 January, 2017, pp. 31–45 (2017)

8. Cormen, T.H., Leiserson, C.E., Rivest, R.L., Stein, C.: Introduction to Algorithms, 3rd edn. MIT Press (2009)
9. Diestel, R.: Graph Theory, Graduate texts in mathematics, 4th edn., vol. 173. Springer (2012)
10. El-Zein, H., Lewenstein, M., Munro, J.I., Raman, V., Chan, T.M.: On the succinct representation of equivalence classes. Algorithmica **78**(3), 1020–1040 (2017)
11. Farzan, A., Kamali, S.: Compact navigation and distance Oracles for graphs with small treewidth. Algorithmica **69**(1), 92–116 (2014)
12. Farzan, A., Munro, J.I.: Succinct encoding of arbitrary graphs. Theor. Comput. Sci. **513**, 38–52 (2013)
13. Farzan, A., Munro, J.I.: A uniform paradigm to succinctly encode various families of trees. Algorithmica **68**(1), 16–40 (2014)
14. Gurevich, Y., Stockmeyer, L.J., Vishkin, U.: Solving NP-Hard problems on graphs that are almost trees and an application to facility location problems. J. ACM **31**(3), 459–473 (1984)
15. Husimi, K.: Note on mayers' theory of cluster integrals. J. Chem. Phys. **18**(5), 682–684 (1950)
16. Jacobson, G.J.: Succinct static data structures. Ph.D. thesis, Carnegie Mellon University (1998)
17. Munro, J.I., Raman, V.: Succinct representation of balanced parentheses and static trees. SIAM J. Comput. **31**(3), 762–776 (2001)
18. Munro, J.I., Wu, K.: Succinct data structures for chordal graphs. In: ISAAC, pp. 67:1–67:12 (2018)
19. Munro, J.I., Nicholson, P.K.: Compressed representations of graphs. In: Encyclopedia of Algorithms, pp. 382–386 (2016)
20. Navarro, G.: Compact Data Structures - A Practical Approach. Cambridge University Press, New York (2016)
21. Navarro, G., Sadakane, K.: Fully functional static and dynamic succinct trees. ACM Trans. Algorithms **10**(3), 16:1–16:39 (2014)
22. Raman, R., Rao, S.S.: Succinct representations of ordinal trees. In: Brodnik, A., López-Ortiz, A., Raman, V., Viola, A. (eds.) Space-Efficient Data Structures, Streams, and Algorithms. LNCS, vol. 8066, pp. 319–332. Springer, Heidelberg (2013). https://doi.org/10.1007/978-3-642-40273-9_20
23. Raman, R., Raman, V., Satti, S.R.: Succinct indexable dictionaries with applications to encoding k-ary trees, prefix sums and multisets. ACM Trans. Algorithms **3**(4), 43 (2007)
24. Riordan, J., Shannon, C.E.: The number of two-terminal series-parallel networks. J. Math. Phys. **21**(1–4), 83–93 (1942). https://doi.org/10.1002/sapm194221183
25. Sadakane, K.: Compressed suffix trees with full functionality. Theory Comput. Syst. **41**(4), 589–607 (2007)
26. Sumigawa, K., Sadakane, K.: Storing partitions of integers in sublinear space. Rev. Socionetwork Strateg. **13**(2), 237–252 (2019)
27. Uno, T., Uehara, R., Nakano, S.: Bounding the number of reduced trees, cographs, and series-parallel graphs by compression. Discrete Math. Algorithms Appl. **05**(02), 1360001 (2013)
28. Valdes, J., Tarjan, R.E., Lawler, E.L.: The recognition of series parallel digraphs. SIAM J. Comput. **11**(2), 298–313 (1982)
29. Voblyi, V.A., Meleshko, A.K.: Enumeration of labeled block-cactus graphs. J. Appl. Ind. Math. **8**(3), 422–427 (2014). https://doi.org/10.1134/S1990478914030156

Streaming Submodular Maximization Under Differential Privacy Noise

Di Xiao[1], Longkun Guo[1(✉)], Kewen Liao[2], and Pei Yao[1]

[1] Fuzhou University, Fuzhou, China
lkguo@fzu.edu.cn
[2] Australian Catholic University, Sydney, Australia
kewen.liao@acu.edu.au

Abstract. The era of big data has brought the need of fast data stream analysis. Recently the problem of streaming submodular optimization has attracted much attention due to the importance of both submodular functions and streaming analytics. However, in real practical setting, streaming data often comes with noise which causes difficulties in analysing and optimizing submodular functions. In this paper, we study the problem of submodular maximization with cardinality constraint under a noisy streaming model, where the impact of noise is assumed to be small as inspired by the framework of differential privacy (so we also call it DP noise). For this problem, we eventually give a worst-case approximation ratio of $\frac{1}{\left(2+\left(1+\frac{1}{k}\right)^2\right)\left(1+\frac{1}{k}\right)} - \delta$ in one pass. To complement the theoretical analysis, we also conduct experiments across real datasets to show our algorithm outperforms the baseline streaming methods.

Keywords: Submodular maximization · Streaming algorithms · Differential privacy noise

1 Introduction

Submodular function is a mathematical function over sets naturally derived from many practical observations and experiences. It captures the property of natural diminishing returns, in which the function value gained after repeatly adding a single element into the input set keeps decreasing. This property is deemed to be important in many practical applications and recently the submodular function has continued to show great practicality in a broad topics of machine learning, AI, data mining, and combinatorial optimization [3,9,10,16,19,21,23]. In particular, the submodular maximization problem subject to a cardinality constraint has been widely studied and applied with the simple best greedy algorithms [8,17]. This common variant of the problem asks to select a bounded number of elements from a ground set of elements such that the objective function over the selected set of elements attains the maximum value (details as in

D.-Z. Du et al. (Eds.): COCOA 2021, LNCS 13135, pp. 431–444, 2021.
https://doi.org/10.1007/978-3-030-92681-6_34

the preliminaries section). The greedy solutions are on static sets, but in many cases data arrives in a streaming fashion where the characteristics of data stream determine that the processing of data stream usually requires one access, continuous processing, limited storage, approximate results, and fast response [8]. The aforementioned greedy approaches are not suitable for the streaming model as multiple accesses to the entire dataset are required. A state-of-the-art streaming algorithm at present is SieveStreaming [1].

In addition, the above mentioned submodular optimization algorithms do not consider noise in the data which can naturally arise in many data-driven applications like social media and IoT. In non-streaming setting, noise has been generally considered as additive or multiplicative noise [2,21] and the noisy version of the problem can be transformed into an approximate submodular maximization problem. For instance, an exact submodular function is f but under a noisy environment we can only observe the value of an approximate function F. The concept of approximating the submodular function was introduced by Horel et al. [11]. As an example, for $\varepsilon > 0$, we say $F : 2^V \rightarrow R$ is a ε-multiplicative approximate submodular function, if there exists a non-negative submodular function $f : 2^V \rightarrow R$ such that $(1-\varepsilon) f(S) \leq F(S) \leq (1+\varepsilon) f(S), \forall S \subseteq V$. Similarly the additive noise model dictates that $f(S) - \varepsilon \leq F(S) \leq f(S) + \varepsilon$. Note that these model settings are designed to capture unbounded general noise. However, in client-centric applications such as differential private databases, random noise added to the data creates a small impact only with the purpose of privacy-preserving. Inspired by some pioneer works of differential privacy [7,12,18], we introduce the ε-differential privacy approximate submodular function as follows:

Definition 1. *For $\varepsilon > 0$, we say $F : 2^V \rightarrow R$ is a ε-differential privacy approximate submodular function, if for any non-negative submodular function $f : 2^V \rightarrow R$ there exists a F such that $e^{-\varepsilon} f(S) \leq F(S) \leq e^{\varepsilon} f(S), \forall S \subseteq V$.*

The above function mirrors the outcome of adding differential private noise to the data. A simplest way to simulate the noise is by adding the Laplace noise analyzed by Sarathy et al. [20] which has been shown to preserve differential privacy of numerical data. In addition, the impact of such noise injection is small, so in our analysis ε is kept at a relatively small value.

Our Contribution. In this paper, we study a new streaming submodular maximization problem under cardinality constraint and differential privacy noise with the following results:

- For this problem, we eventually give a streaming approximation algorithm called DP-SS that overcomes the challenge of using approximate submodular function to bound the optimal value of underlying submodular function.
- Our proposed algorithm only needs a single pass over the data stream while the required memory is independent of the size of the dataset. Under the assumption that $\varepsilon = \frac{1}{2} \ln \left(1 + \frac{1}{k}\right)$ where k is the cardinality, the algorithm achieves an approximation ratio of $\frac{1}{\left(2+\left(1+\frac{1}{k}\right)^2\right)\left(1+\frac{1}{k}\right)} - \delta$, and when $\varepsilon \rightarrow 0$ it

can provide an approximation guarantee of $\frac{1}{2} - \delta$, for $\delta > 0$. The algorithm requires $\mathcal{O}\left(k \log\left(e^{2\varepsilon} k\right)/\delta\right)$ memory and $\mathcal{O}\left(\log\left(e^{2\varepsilon} k\right)/\delta\right)$ update time per element.
- We conduct numerical experiments on our proposed method and several baseline methods across three real data sets. Under DP noise, our algorithm DP-SS performs consistently better than SieveStreaming.

2 Preliminaries

The problem studied is to select a subset from a set V data points of size n. We call V the ground set. Now there is a non-negative set function $f : 2^V \rightarrow R_+$ and f is normalized iff $f(\emptyset) = 0$. The function f is non-negative iff $f(S) > 0$ for all $S \subseteq V$. The function f is monotone iff $f(A) \leq f(B)$ for all $A \subseteq B \subseteq V$. For a submodular set function f, its marginal gain can be defined as

$$\nabla_f (a|S) = f(S \cup \{a\}) - f(S)$$

where $S \subseteq V$ and $a \in V$. From this formula, it can be considered that the marginal gain is when a is added to S, the size of the gain brought to function f. Another condition for function f to be monotonic iff $\nabla_f(a|S) \geq 0$ for all a and S. For a set function f is submodular iff $\nabla_f(a|A) \geq \nabla_f(a|B)$ for all $A \subseteq B \subseteq V$ and $a \in V \backslash B$. Another equivalent condition is $f(A)+f(B) \geq f(A \cup B)+f(A \cap B)$, for all $A, B \subseteq V$. The submodularity of the function means that the marginal gain is diminishing, indicates that adding an element to a small set helps more than a large one. Throughout this paper we will use the submodularity and monotonicity of functions.

Here in a noise-free environment, our problem is to maximize the submodular function f subject to the constraint of the base k, that means we need to select a subset of the ground set V whose size cannot exceed k to maximize the function value. The specific expression is

$$\max_{S \subseteq V, |S| \leq k} f(S).$$

The problem we are considering in this paper introduces new constraints. Firstly, the problem's ground set we consider is generated under a streaming model. This means that each element of data stream V arrives at the storage point one by one. The speed of data arrival is very fast, elements will arrive at any time and in a natural order. But our memory is much smaller than the size of the data stream, so we must select and store the most valuable data points. Secondly, the problem we consider is carried out in a noisy environment. Here we still need to maximize f, but we can only estimate the value of f. At the same time, in order to simulate a noisy environment, we pretend that the original data conforms to differential privacy, so we added differential privacy noise to the original data (details explained in the experiments section). Under differential privacy mechanism, it is difficult for us to directly obtain the value of function f, but it's not difficult to calculate the noisy counterpart F. Therefore, we can regard F as an approximate submodular function and F does not necessarily satisfy submodularity. In addition, both submodular function f and noisy function F satisfy monotonicity and normalization rules.

3 Related Work

First we introduce some related work in a noise-free environment, which of course is all about the submodular maximization problem with the cardinality constraint. When it is not a data stream, Nemhauser et al. [17] obtained the approximate ratio of $1-e^{-1}$ through the greedy algorithm of k iterations. Since Feige [5] proved that the submodular function optimization problem is NP-hard, it is difficult to improve the approximation ratio of $1-e^{-1}$ unless $\mathrm{P}=\mathrm{NP}$. Gomes and Krause [8] first gave the StreamGreedy algorithm to solve the problem of submodular maximization on the stream, it provides an approximate value of $0.5-\delta$ and only needs to use $\mathcal{O}(k)$ memory, where ε depends on the submodular function and some user-specified parameters. The best approximation factor can only be obtained after multiple passes through the data, otherwise, the performance of StreamGreedy will decrease arbitrarily with k. Therefore, we believe that StreamGreedy is not a suitable streaming algorithm. For the same problem, Badanidiyuru et al. [1] is the first to give a streaming algorithm that only needs to pass data once, they called SieveStreaming. It also provides $0.5-\delta$ approximate ratio with $\mathcal{O}((k\log k)/\delta)$ memory. Recently, Kazemi et al. [13] extended SieveStreaming and proposed a streaming algorithm called SieveStreaming++, it does not improve the approximation guarantee of SieveStreaming, but only requires $\mathcal{O}(k/\delta)$ memory instead of $\mathcal{O}((k\log k)/\delta)$.

Then we survey the work related to the submodular optimization problem in a noisy environment. Horel et al. [11] proposed the concept of approximate submodular maximization for the first time. They proved that maximizing the monotonic submodular can obtain a better approximation ratio than $\mathcal{O}(1/n^{\beta})$, and this function satisfies the $1/\left(n^{\frac{1}{2}-\beta}\right)$ approximation submodular, where n is the size of the ground set, β is a given positive value. For the problem of maximizing ε approximate submodular function subject to cardinality constraints, they gave a lower bound on query complexity. When $\varepsilon \geq n^{-1/2}$, they show the lower bound of exponential query complexity in the case of general submodular functions. Running the greed algorithm for an ε approximately submodular function results in a constant approximation ratio when $\varepsilon \leq \frac{1}{k}$,and the approximation ratio is $1-e^{-1}-16\delta$, where $\delta = k\varepsilon$ as long as $\delta < 1$. Hassidim and Singer [10] considered a consistent random noise model, for each subset X, except that the first evaluation is randomly drawn from the distribution of $F(X)$, the other evaluations return the same value. They provide a polynomial time algorithm with a constant approximation for certain types of noise distribution. When k is large enough, they designed the SlickGreedy algorithm, which can obtain the guarantee of the approximate ratio of $1-e^{-1}-\delta$ with probability at least $1-n^{-1}$. Qian et al. [19] considered the more general case for the multiplicative noise model and the additive noise model, namely the selection of a noisy subset with a monotonic objective function f. They proved the approximation ratio of the approximation algorithm under the two noise models. Approximate ratios are $\frac{1-\varepsilon}{1+\varepsilon}\left(1-e^{-\gamma}\right)$ and $1-e^{-\gamma}-\mathcal{O}\left(\varepsilon\right)$ respectively, where γ is submodularity ratio. Yang et al. [23] considered the problem of submodular maximization on the stream under the additive noise and multiplicative noise models, when $\varepsilon \rightarrow 0$, the algorithms under both noises have a $2/k$ approximate ratio. For streaming algorithm evaluation, we mainly compared the memory required between algorithms, the update time of each new data element, and the approximate guarantee.

4 The DP-SS Algorithm

In this section, we will first present the submodular optimization streaming algorithm under differential privacy noise (DP-SS) with known OPT_F, and then by guessing the value of OPT_F, we develop another algorithm without knowing OPT_F.

4.1 DP-SS with Known OPT_F

The core of the DP-SS algorithm is to filter the arriving elements by threshold to obtain the final optimal solution set. So the threshold for filtering elements is particularly important. Here we must know $OPT_F = max_{S \subseteq V, |S| \leq k} F\left(S\right)$ and $OPT_f = max_{S \subseteq V, |S| \leq k} f\left(S\right)$. We also assume that we know that OPT_F is used to construct the threshold. Obviously, with the environment of noise, the construction of the threshold must be related to noise. Here we set $Threshold = \frac{OPT_F}{((1+k(e^{2\varepsilon}-1)) \cdot e^{-2\varepsilon}+e^{2\varepsilon})k}$. At the same time, we must limit the size of the noise, here we set $\varepsilon = \frac{1}{2}\ln\left(1+\frac{\theta}{k}\right)$, where $0 < \theta \leq 1$.

Algorithm 1. DP-SS under knowing OPT_F

Input: The value of OPT_F, data streaming $V = \{a_1, a_2, \cdots, a_n\}$, the noise function F, integer k, noise parameter $0 < \varepsilon \leq \frac{1}{2}\ln\left(1+\frac{1}{k}\right)$;

Output: An approximate solution set S.

1: Set $T = \frac{OPT_F}{\left(\left(1+k\left(e^{2\varepsilon}-1\right)\right)e^{-2\varepsilon}+e^{2\varepsilon}\right)k}$;
2: Set $S \leftarrow \emptyset$ and $i \leftarrow 1$;
3: **For** $i = 1$ to n **do**
4: **If** $\nabla_F\left(a_i|S\right) \geq T$ and $|S| < k$ **then**
5: $S \leftarrow S \cup \{a_i\}$;
6: **Endif**
7: **Endfor**
8: **return** S.

In Algorithm 1, we abbreviate the threshold as T, and start selecting elements from $S = \emptyset$. When $|S| < k$, as long as the gain of the element to F is greater than or equal to our threshold, we can select it into the set S, the algorithm can finally output an approximate solution set. After the key parts are set, we can start our algorithm.

Properties of Algorithm 1. a) Obviously it only needs one pass and most store k elements. b) The Algorithm 1 outputs a set S such that $|S| \leq k$, then it have $f(S) \geq \frac{1}{2+\left(1+\frac{1}{k}\right)^2} OPT_f$. See Theorem 2 for details.

From Algorithm 1, we can clearly see that when the output $|S| < k$, if an element wants to enter S, then its marginal gain satisfies at least $\nabla_F(a_i|S) \geq T$. This is a necessary and sufficient condition, so we have Lemma 1.

Lemma 1. *When $|S| < k$, any singleton element $a_i \in V$ is rejected to be elected to the current solution $S_{i-1} \subseteq S$ if and only if $F(S_{i-1} \cup \{a_i\}) - F(S_{i-1}) < T$.*

Theorem 1. *For the submodular optimization under differential privacy noise, for $\varepsilon > 0$, there is an approximate submodular function F through the Algorithm 1 to obtain S such that $|S| \leq k$. The Algorithm 1 achieves an approximation ratio $\frac{1}{(1+k(e^{2\varepsilon}-1))+e^{4\varepsilon}}$.*

InAlgorithm 1, the noise is restricted. Our goal is to make the added noise very small. Therefore, under this restriction, there is Theorem 2.

Theorem 2. *For the submodular optimization under differential privacy noise, assume $\varepsilon = \frac{1}{2}\ln\left(1+\frac{\theta}{k}\right)$, where $0 < \theta \leq 1$. Then the output S of the Algorithm 1 has no more than k elements such that $f(S) \geq \frac{1}{2+\left(1+\frac{1}{k}\right)^2} OPT_f$.*

4.2 The Final DP-SS Under DP Noise

In the previous section, we introduced the streaming algorithm based on the premise of knowing OPT_F, that is to solve the optimal solution set by knowing the optimal solution. In many cases we don't know the value of the optimal solution, but we can estimate the value of the optimal solution, so we must theoretically get the bounds of OPT_F. Here we analyze the construction of the set G capable of estimating the optimal solution, and then we propose the DP-SS algorithm on the basis of expanding the range of G and adding restrictions to the noise. In order to construct the set G we introduced the concept of the largest single element. The function here refers to F with noise. We denote element $a_{max} = arg\ max_{a \in V} F(\{a\})$. We can get the largest single element function value through a data flow calculation. Then we can get the upper and lower bounds of OPT_F under the differential privacy noise.

First of all, it is obvious that can be obtained $F(\{a_{max}\}) \leq OPT_F$ by monotonicity. Next, in order to estimate the upper bound of OPT_F, here we also assume $OPT_F = F(O_F)$, we use the same O_f and O_j as in the proof of Theorem 1. We have

$$\begin{aligned} OPT_F = F(O_F) &\leq e^{\varepsilon} f(O_F) \leq e^{\varepsilon} f(O_f) \\ &= e^{\varepsilon} \sum_{j=1}^{k} \left(f(O_{j-1} \cup \{a_j\}) - f(O_{j-1}) \right) \\ &\leq e^{\varepsilon} \sum_{j=1}^{k} \left(f(\emptyset \cup \{a_j\}) - f(\emptyset) \right) \\ &= e^{\varepsilon} \sum_{j=1}^{k} f(\{a_j\}) \leq e^{2\varepsilon} \sum_{j=1}^{k} F(\{a_j\}) \\ &\leq e^{2\varepsilon} \sum_{j=1}^{k} F(\{a_{max}\}) = e^{2\varepsilon} k F(\{a_{max}\}) \end{aligned}$$

in which the first inequality is right since $F(S) \leq e^{\varepsilon} f(S)$. The following inequality can be easily derived from the submodularity of f and the monotonicity of F. Therefore, we have

$$F(\{a_{max}\}) \leq OPT_F \leq e^{2\varepsilon} \cdot k \cdot F(\{a_{max}\}).$$

But such a rough estimate of the range of OPT_F is not enough to estimate its value, it is not enough to know the specific value to better estimate the OPT_F. So we consider the set

$$G = \left\{ (1+\delta)^i \,|\, i \in Z, m_1 \leq (1+\delta)^i \leq m_2 \right\}$$

where δ is aa tuning parameter and $m_1 = F(\{a_{max}\})$, $m_2 = e^{2\varepsilon} \cdot k \cdot F(\{a_{max}\})$. Obviously from the range of OPT_F given previously, we can know that there is at least one value $v \in G$ that can be a good estimate of the value of OPT_F [1], such that $(1-\delta) OPT_F \leq v \leq OPT_F$. This means that we can run the algorithm one time for each $v \in G$. In order to avoid multiple repeated calculations on the same data stream, we can run multiple copies of Algorithm 1 in parallel. Finally, compare the results of each $v \in G$ to output one optimal solution.

Specifically, we need to calculate the data stream at the beginning of the algorithm to get the maximum singleton element function value, then we can construct the set G. But the algorithm does not yet know which $v \in G$ has a good estimate of OPT_F. Therefore, each $v \in G$ is calculated by Algorithm 1, and then the corresponding candidate solution set S_v is output. Finally, the best one is selected from all the candidate solution sets. But this is not enough. The algorithm needs to calculate the data stream twice. Therefore, here is an improvement to the algorithm, so that the final algorithm only needs to pass the data once.

First, we must construct the current set

$$G_i = (1+\delta)^i, i \in Z, Current\ Lower \leq (1+\delta)^i \leq Current\ Upper$$

Algorithm 2. The DP-SS algorithm

Input: Streaming $V=\{a_1,a_2,\cdots,a_n\}$, the noise function F, integer k, tuning parameter δ, $G=\left\{(1+\delta)^i\,|i\in Z\right\}$, $0<\varepsilon\leq\frac{1}{2}\ln\left(1+\frac{1}{k}\right)$;
Output: S_v.
1: Set $\Omega\leftarrow 0$;
2: **For** each $v\in G$ **do**
3: Set $S_v\leftarrow\emptyset$ and $T_v=\frac{v}{\left(\left(1+k\left(e^{2\varepsilon}-1\right)\right)e^{-2\varepsilon}+e^{2\varepsilon}\right)k}$;
4: **For** $i=1$ to n **do**
5: Set $\Omega\leftarrow max\left\{\Omega,F\left(\{a_i\}\right)\right\}$;
6: Set $Current\ Lower\leftarrow\Omega$, $Current\ Upper\leftarrow 2e^{2\varepsilon}k\Omega$;
7: $G_i=\left\{(1+\delta)^i\,|Current\ Lower\leq(1+\delta)^i\leq Current\ Upper\right\}$;
8: Delete all S_v and T_v such that $v\notin G_i$;
9: **For** $v\in G_i$ **do**
10: **If** $\nabla_F\left(a_i|S_v\right)\geq T_v$ and $S_v|<k$ **then**
11: **Let** $S_v\leftarrow S_v\cup\{a_i\}$;
12: **Endif**
13: **Endfor**
14: **Endfor**
15: **return** $S=argmax_{v\in G}F\left(S_v\right)$.

by keeping the current largest single element. Since we need to see the elements that reach the threshold when or after v is instantiated, here we let

$$Current\ Lower=F\left(\{a_{max}\}\right)$$
$$Current\ Upper=2e^{2\varepsilon}kF\left(\{a_{max}\}\right)$$

We can regard $F\left(\{a_{max}\}\right)$ as a random variable, after observing each element a_i, it will be updated in time. Then a corresponding threshold set G_i will be constructed for the current $F\left(\{a_{max}\}\right)$ and it will only leave all the thresholds that meet $v\in G_i$. This also means to delete all S_v such that $v\notin G_i$. Then run Algorithm 2 on all the remaining v. Finally, we can choose the best one among the many S_v. In this way, the final version of the submodular optimized streaming algorithm under differential privacy noise is obtained. We call Algorithm 1 the DP-SS algorithm.

Properties of Algorithm 2 (DP-SS). a) Note that $|G|=\mathcal{O}\left(\log\left(e^{2\varepsilon}k\right)/\delta\right)$, for each value v, we must keep the size of one set S_v with at most k elements. Thus for differential privacy noise, the total memory size is $\mathcal{O}\left(k\log\left(e^{2\varepsilon}k\right)/\delta\right)$. For each elements, the algorithm has $\mathcal{O}\left(\log\left(e^{2\varepsilon}k\right)/\delta\right)$ update time. b) It need one passes over the data.

The DP-SS algorithm finally outputs a set S such that $|S|\leq k$, and when we set an upper limit on the noise, we can have a good approximate ratio performance.

Theorem 3. *For the submodular optimization under differential privacy noise, for $\varepsilon>0$, there is an approximate submodular function F through the*

Algorithm 2 to obtain S such that $|S| \leq k$. *The DP-SS algorithm achieves an approximation ratio* $\frac{1}{(1+k(e^{2\varepsilon}-1)+e^{4\varepsilon})e^{2\varepsilon}} - \delta$.

Theorem 4. *For the submodular optimization under differential privacy noise, assume* $\varepsilon = \frac{1}{2}\ln\left(1+\frac{\theta}{k}\right)$, *where* $0 < \theta \leq 1$. *Then the output S of the Algorithm 2 has no more than k elements such that* $f(S) \geq \left(\frac{1}{\left(2+\left(1+\frac{1}{k}\right)^2\right)\left(1+\frac{1}{k}\right)} - \delta\right) OPT_f$.

5 Experiments

In order to evaluate the specific effectiveness of Algorithm 2 (DP-SS), we have to perform several sets of numerical experiments on it. The specific experiments are guided by the following ideas.

- First, under the same noise conditions, the experimental results need to show the specific advantages of the approximate solution obtained by Algorithm 2 (DP-SS) over other approximation algorithms.
- Second, numerical experiments need to show the variation of the utility function with the variation of the noise.
- Third, we need to analyze the results of the algorithm under different data sets through numerical experiments.

5.1 Experimental Setup

Here we use three real-world data-sets: ForestCover dataset, Creditfraud dataset and KDDCup99 dataset. The ForestCover dataset [4] includes 286,048 data points with 10 attributes. The Creditfraud dataset [15] consists of 284,807 data points with 29 attributes. And the last one we chose is the KDDCup99 dataset [3], which consists of 60,632 data points with 41 attributes. Baseline methods to compare against are the standard greedy algorithm, the random algorithm [6], and the SieveStreaming [1] algorithm. To simplify our numerical experiments, we randomly select a large enough sample set $W \subseteq V$. Let f_W be the expectation of independent random variables $X_1, X_2, \cdots, X_{|W|}$, and $X_i \in [-1, 1]$ and f be the mean value of random variables, we can use Hoffding's inequality to obtain the upper of the probability of deviation between the mean and the expectation under differential privacy as

$$\begin{aligned} &Pr\left[e^{-\varepsilon} f - e^{\varepsilon} f_W \leq f_W - f \leq e^{\varepsilon} f - e^{-\varepsilon} f_W\right] \\ &\quad \geq 1 - \exp\left(-\frac{|W|}{2}\right) \cdot (\rho_1 + \rho_2) \end{aligned} \tag{1}$$

where

$$\rho_1 = \exp\left(\left(e^{\varepsilon} f - e^{-\varepsilon} f_W\right)^2\right)$$
$$\rho_2 = \exp\left(\left(e^{-\varepsilon} f - e^{\varepsilon} f_W\right)^2\right)$$

Obviously, we can find that the value of f_W under the sample set can be enough to estimate the value of f_v in the real situation by using Hoffding's inequality.

Therefore, it can be seen from Inequality (1) that f_W can be used to estimate f under the differential privacy noise, which means that a sufficient sample W can be used to replace the ground set V. So in our numerical experiment, we select W of size $\frac{1}{10}|V|$ that is similarly [22].

In this paper, our main study is the streaming submodular maximization under differential privacy noise, so we firstly add noise to the selected real data set by using the Laplace mechanism [9,20] to add noise. Firstly of all, we introduce the Laplace probability density function $y\left(x|\mu,b\right)=\frac{1}{2b}exp\left(-\frac{|x-\mu|}{b}\right)$, where μ is the position parameter and b is the scale parameter [14]. Secondly, we can easily obtain its probability density distribution function

$$Y\left(x|\mu,b\right)=\begin{cases}\frac{1}{2}\exp\left(-\frac{\mu-x}{b}\right),x<\mu\\ 1-\frac{1}{2}\exp\left(-\frac{x-\mu}{b}\right),x\geq\mu\end{cases}$$

Obviously, for a given $\mu=0$ and $b>0$, we know that the range of the distribution function is $[0,1]$. We can using the following two steps to obtain Laplace noise: (1) obtaining random values with a uniform distribution in the interval; (2) solving the inverse function of the probability distribution function. Formally, let random variable $\xi\sim Uni\left(0,1\right)$ which means that the random variable ξ satisfies a uniform distribution, we can get the inverse probability distribution function

$$Y^{-1}\left(\xi|\mu,b\right)=\begin{cases}b\ln\left(2\xi\right)+\mu,\xi<\frac{1}{2}\\ \mu-b\ln\left(2\left(1-\xi\right)\right),\xi\geq\frac{1}{2}\end{cases}\tag{2}$$

In our experiments, we will add noise to our data but the availability of data is also needed to be ensured, so we use the privacy protection budget φ in differential privacy. Since the Laplace noise needs satisfy the distribution $Lap\left(0,\triangle f/\varphi\right)$, so we set $\mu=0$, $b=\triangle f/\varphi$. Obviously, the b is proportional to noise that is obtain by Equality (2) that. For the privacy protection budget φ, it can be cleared that the larger the φ, the smaller the noise and the higher the data availability. So to protect the availability of our data, we need to make φ big enough to make noise smaller.

5.2 Experimental Results

The specific results of the experiment are as follows.

- We first analyze the comparison of the results between the algorithm and the benchmark in the case of a non-streaming model. In Fig. 1, we fixed the Laplace noise by setting the scale parameter of the Laplace distribution to no more than 0.1, we selected the Creditfraud dataset to run Algorithm 2 (DP-SS) under the differential privacy model, chose to obtain an approximate solution, and ran the other benchmarks, compare how the values obtained after running them change with the increase of k. We ran the other two datasets under the same operation. In fact, through (a), (b), and (c) in Fig. 1, it can be seen that in the three real data sets, as the base k increases, DP-SS is significantly better than the random algorithm, and it is closer to the result of the greedy algorithm. Here, because the numerical value of the data set is different, the corresponding scale parameters are also different.

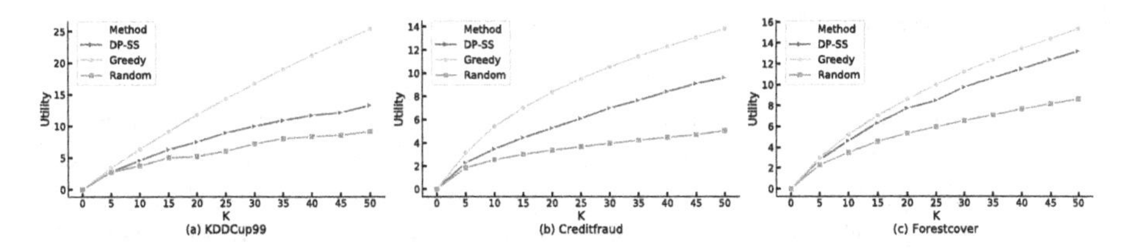

Fig. 1. Performances of the results of Algorithm 2 (DP-SS) is compared with other benchmarks, when the same noise is fixed.

- In Fig. 2, we pretended that all three real datasets are free of noise and brought them into DP-SS and SieveStreaming to ran (Here we just need to not add noise to the dataset). This set of experiments we performed under the streaming model. With the same change of setting k, it can be seen that in the three sets of data, the results of our algorithms are not as good as those of the SieveStreaming algorithm.

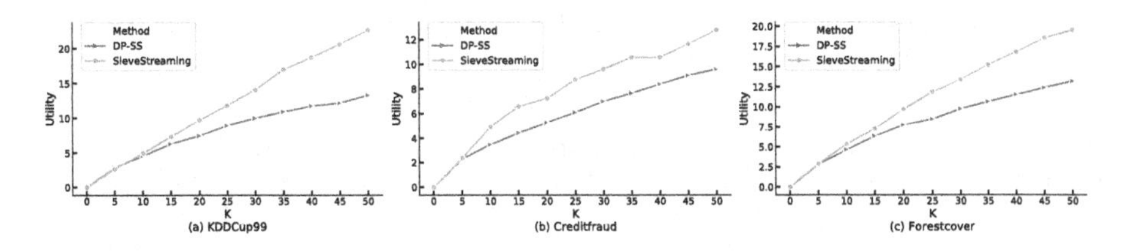

Fig. 2. Comparison of Algorithm 2 (DP-SS) performance and SieveStreaming performance under noise-free environment.

- In the third set of experiments we first set up Algorithm 2 (DP-SS) and SieveStreaming to run in the same data set, the same noise environment, and the same streaming model to compare. We ran Algorithm 2 (DP-SS) and SieveStreaming and compared their results under the change of k. Here we set the scale parameter of the Laplace distribution to be greater than 0.1. In Fig. 3, our algorithm is obviously better than SieveStreaming.

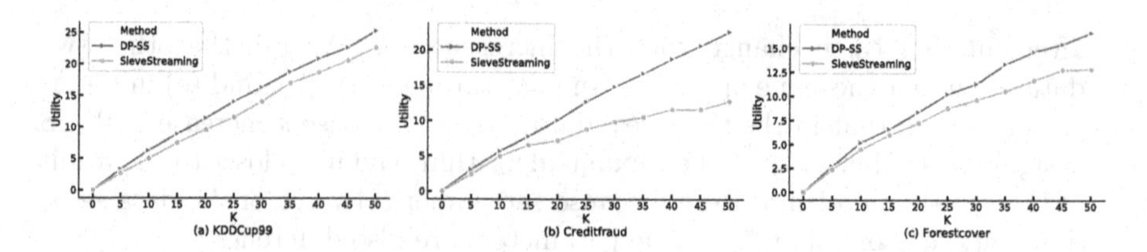

Fig. 3. Comparison of the running results of Algorithm 2 (DP-SS) and SieveStreaming in the same noise environment.

- We need to analyze the situation between the results and the noise under the streaming model. In Fig. 4, under the same three datasets, we set k to 10 and 20 and assume b increases from 0.01 to 0.1. We observed how the results of Algorithm 2 (DP-SS) change as the b increases after k is fixed. Since the scale parameter b is proportional to the noise, combined with the results in Fig. 4 we can see that as b increases, the noise becomes larger, and the value of the truth function utility decreases.

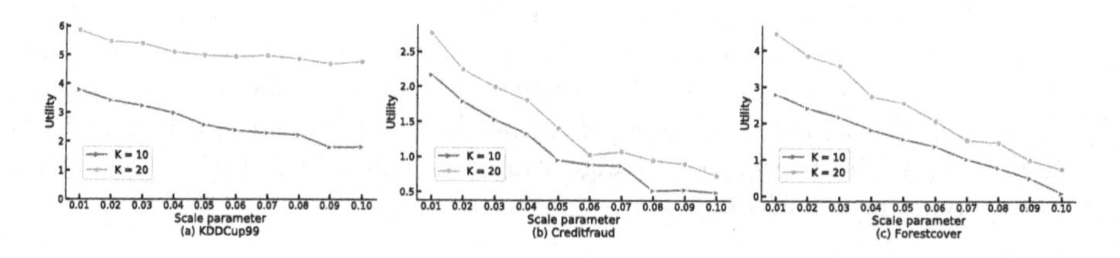

Fig. 4. The performance of Algorithm 2 (DP-SS) under the fluctuation of the scale parameter of Laplace noise, when k is fixed.

Acknowledgements. The authors are supported by Natural Science Foundation of China (No. 61772005), Outstanding Youth Innovation Team Project for Universities of Shandong Province (No. 2020KJN008), Natural Science Foundation of Fujian Province (No. 2020J01845) and Educational Research Project for Young and Middle-aged Teachers of Fujian Provincial Department of Education (No. JAT190613).

References

1. Badanidiyuru, A., Mirzasoleiman, B., Karbasi, A., Krause, A.: Streaming submodular maximization: Massive data summarization on the fly. In: Proceedings of the 20th ACM SIGKDD International Conference on Knowledge Discovery and Data Mining, pp. 671–680 (2014)
2. Belloni, A., Liang, T., Narayanan, H., Rakhlin, A.: Escaping the local minima via simulated annealing: optimization of approximately convex functions. In: Conference on Learning Theory, pp. 240–265. PMLR (2015)
3. Campos, G.O., et al.: On the evaluation of unsupervised outlier detection: measures, datasets, and an empirical study. Data Min. Knowl. Disc. **30**(4), 891–927 (2016)

4. Dal Pozzolo, A., Caelen, O., Johnson, R.A., Bontempi, G.: Calibrating probability with undersampling for unbalanced classification. In: 2015 IEEE Symposium Series on Computational Intelligence, pp. 159–166. IEEE (2015)
5. Feige, U.: A threshold of ln n for approximating set cover. J. ACM (JACM) **45**(4), 634–652 (1998)
6. Feige, U., Mirrokni, V.S., Vondrak, J.: Maximizing non-monotone submodular functions. SIAM J. Comput. **40**(4), 1133–1153 (2011)
7. Feldman, D., Fiat, A., Kaplan, H., Nissim, K.: Private coresets. In: Proceedings of the Forty-first Annual ACM Symposium on Theory of Computing, pp. 361–370 (2009)
8. Gomes, R., Krause, A.: Budgeted nonparametric learning from data streams. In: Fürnkranz, J., Joachims, T. (eds.), Proceedings of the 27th International Conference on Machine Learning (ICML-10), 21–24 June, Haifa, Israel, pp. 391–398. Omnipress (2010)
9. Gupta, A., Ligett, K., McSherry, F., Roth, A., Talwar, K.: Differentially private combinatorial optimization. In: Proceedings of the Twenty-first Annual ACM-SIAM Symposium on Discrete Algorithms, pp. 1106–1125. SIAM (2010)
10. Hassidim, A., Singer, Y.: Submodular optimization under noise. In: Conference on Learning Theory, pp. 1069–1122. PMLR (2017)
11. Horel, T., Singer, Y.: Maximization of approximately submodular functions. In: Lee, D.D., Sugiyama, M., von Luxburg, U., Guyon, I., Garnett, R. (eds.), Advances in Neural Information Processing Systems 29: Annual Conference on Neural Information Processing Systems 2016, pp. 3045–3053, 5–10 December, Barcelona, Spain (2016)
12. Kasiviswanathan, S.P., Lee, H.K., Nissim, K., Raskhodnikova, S., Smith, A.: What can we learn privately? SIAM J. Comput. **40**(3), 793–826 (2011)
13. Kazemi, E., Mitrovic, M., Zadimoghaddam, M., Lattanzi, S. and Karbasi, A.: Submodular streaming in all its glory: tight approximation, minimum memory and low adaptive complexity. In: International Conference on Machine Learning, pp. 3311–3320. PMLR (2019)
14. Kotz, S., Kozubowski, T. and Podgòrski, K.: The Laplace Distribution and Generalizations: A Revisit with Applications to Communications, Economics, Engineering, and Finance. Springer, Cham (2012)
15. Liu, F.T., Ting, K.M., Zhou, Z.H.: Isolation forest. In: Proceedings of the 8th IEEE International Conference on Data Mining (ICDM 2008), 15–19 December, Pisa, Italy, pp. 413–422. IEEE Computer Society (2008)
16. Mirzasoleiman, B., Badanidiyuru, A., Karbasi, A.: Fast constrained submodular maximization: personalized data summarization. In: International Conference on Machine Learning, pp. 1358–1367. PMLR (2016)
17. Nemhauser, G.L., Wolsey, L.A., Fisher, M.L.: An analysis of approximations for maximizing submodular set functions - I. Math. Program. **14**(1), 265–294 (1978)
18. Nissim, K., Raskhodnikova, S., Smith, A.: Smooth sensitivity and sampling in private data analysis. In Proceedings of the thirty-ninth annual ACM symposium on Theory of computing, pp. 75–84 (2007)
19. Qian, C., Shi, J.C., Yu, Y., Tang, K., Zhou, Z.H.: Subset selection under noise. In: NIPS, pp. 3560–3570 (2017)
20. Sarathy, R., Muralidhar, K.: Evaluating Laplace noise addition to satisfy differential privacy for numeric data. Trans. Data Priv. **4**(1), 1–17 (2011)
21. Singer, Y., Vondrák, J.: Information-theoretic lower bounds for convex optimization with erroneous oracles. Adv. Neural Inf. Process. Syst. **28**, 3204–3212 (2015)

22. Vitter, J.S.: Random sampling with a reservoir. ACM Trans. Math. Softw. (TOMS) **11**(1), 37–57 (1985)
23. Yang, R., Xu, D., Cheng, Y., Gao, C., Du, D.Z.: Streaming submodular maximization under noises. In: 2019 IEEE 39th International Conference on Distributed Computing Systems (ICDCS), pp. 348–357. IEEE (2019)

Online Bottleneck Semi-matching

Man Xiao[1], Shu Zhao[1], Weidong Li[1](✉), and Jinhua Yang[2]

[1] School of Mathematics and Statistics, Yunnan University, Kunming, China
[2] Dianchi College, Kunming, China

Abstract. We introduce the online bottleneck semi-matching (OBSM) problem, which is to assign a sequence of requests to a given set of m servers, such that the maximum cost is minimized. We present a lower bound $m + 1$ and an online algorithm with competitive ratio $2m - 1$ for the OBSM problem on a line, where the distance between every pair of adjacent servers is the same. When $m = 2$, we present an optimal online algorithm with competitive ratio 3 for the OBSM problem. When $m = 3$, we present two optimal online algorithms with competitive ratio at most $3 + \sqrt{2}$ for the OBSM problem on a line, which matches the previous best lower bound proposed about thirty years ago.

Keywords: Online bottleneck matching · Online algorithm · Capacity limits · Competitive ratio

1 Introduction

We are given a metric space (X, d), where X is a (possibly infinite) set of points and $d(\cdot, \cdot)$ is a distance function. Let $S = \{s_1, s_2, \ldots, s_m\} \subseteq X$ be a set of servers and $R = \{r_1, r_2, \ldots, r_n\} \subseteq X$ be a set of requests arriving one-by-one in an online fashion. When a request $r_j \in R$ arrives, it must be immediately and irrevocably matched to some previously unmatched server s_i. The cost of matching r_j to s_i is $d(r_j, s_i)$. The classical Online Minimum Matching (OMM) [10] is to find a matching M such that the total cost of matching all requests is minimized. The Online Bottleneck Matching (OBM) [10] is to find a matching M such that the maximum cost of matching all requests is minimized.

We use competitive ratio to evaluate the performance of an online algorithm A. For an input instance I, let $C^A(I)$ (C^A for short) and $C^{OPT}(I)$ (C^{OPT} for short) be the costs of the feasible solution obtained by an online algorithm A and an optimal offline algorithm, respectively. An online algorithm A is *ρ-competitive* (or the *competitive ratio* of A is at most ρ) if $C^A \leq \rho C_{OPT}$ for any input instance I.

Kalyanasundaram and Pruhs [10] introduced the OMM problem and proved that the Permutation algorithm is $(2n-1)$-competitive and optimal. Bansal et al. [5] presented an $O(\log^2 n)$-competitive randomized algorithm for the OMM problem. Kalyanasundaram and Pruhs [11] proposed an interesting question whether one can design an optimal online algorithm for the OMM problem on

D.-Z. Du et al. (Eds.): COCOA 2021, LNCS 13135, pp. 445–455, 2021.
https://doi.org/10.1007/978-3-030-92681-6_35

a line. Gupta and Lewi [7] gave an $O(\log n)$-competitive randomized algorithm for the OMM problem on a line. Fuchs et al. [6] showed that no online algorithm can achieve a competitive ratio strictly less than 9.001 for the OMM problem on a line. Antoniadis et al. [4] designed a deterministic online algorithm with competitive ratio $O(n^{\log(3+\epsilon)-1}/\epsilon)$ for any $\epsilon > 0$ for the OMM problem on a line. Nayyar and Raghvendra [13] proved that the competitive ratio of the deterministic online algorithm proposed in [15] is $O(\log^2 n)$, which is improved to $O(\log n)$ [16], for the OMM problem on a line. Recently, Peserico and Scquizzato [14] proved that the competitive ratio of any randomized online algorithm for the OMM problem on a line exceeds $\sqrt{\log_2(n+1)}/15$.

Kalyanasundaram and Pruhs [10] introduced the OBM problem and proved that the PERMUTATION algorithm is $(2n-1)$-competitive. Idury and Schaffer [8] gave a lower bound approximately $1.44n$ for the OBM problem on a line. Anthony and Cheung [2,3] used resource augmentation analysis to examine the performance of several classic online algorithms for the OBM problem and its variant where a specified number of requests can be reject or skip.

A generalized version of the OMM problem, which is called online b-matching [10], online transportation [10], the fire station problem [12], the school assignment problem [12], or online facility assignment [1], is also considered, where each server can be matched multiple times. Recently, Itoh et al. [9] presented several lower bounds on the competitive ratio for this problem with different number of servers.

In this paper, we consider a variant of the OBM problem, called online bottleneck semi-matching (OBSM), where each server can be matched multiple times and obtain several interesting results. The remainder of the paper is organized as follows. We first introduce the problem definition and theoretical results in Sect. 2. We provide several lower and upper bounds for the OBSM problem with arbitrary number of servers in Sect. 3. We then present several optimal online algorithms for the OBSM problem with 2 or 3 servers in Sect. 4 and Sect. 5, respectively. Finally, we conclude the paper with future research directions in Sect. 6.

2 Preliminaries

We are given a metric space (X, d) and a set $S = \{s_1, s_2, \ldots, s_m\} \subseteq X$ of servers, where X is a (possibly infinite) set of points and $d(\cdot, \cdot)$ is a distance function. Let $R = \{r_1, r_2, \ldots, r_n\} \subseteq X$ be a set of requests arriving one-by-one in an online fashion, where $2 \leq m \leq n$. Each server $s_i \in S$ is characterized by the capacity $c_i \in \mathbb{N}$ that satisfies

$$\sum_{i=1}^{m} c_i = n.$$

It means that the server s_i can be matched with exact c_i requests.

When a request $r_j \in R$ arrives, an online algorithm A must immediately and irrevocably assign a server $s_i = \pi(r_j) \in S$ which is matched less than c_i times to service that request. The cost of matching r_j to $\pi(r_j)$ is $d(r_j, \pi(r_j))$. After all the requests are matched, we obtain a *semi-matching* M where each vertex $r_j \in R$ is matched exactly once. The online bottleneck semi-matching (OBSM) problem is to find an assignment $\pi : R \mapsto S$ such that $|\{r_j|\pi(r_j) = s_i\}| = c_i$ and the maximum cost $\max_j d(r_j, \pi(r_j))$ of matching all requests is minimized. Clearly, the OBSM problem is a generalized version of the OBM problem considered in [2,10]. A most related problem is the online minimum semi-matching (OMSM) problem, which is to find an assignment $\pi : R \mapsto S$ such that $|\{r_j|\pi(r_j) = s_i\}| = c_i$ and the maximum cost $\sum_{j=1}^{n} d(r_j, \pi(r_j))$ of matching all requests is minimized.

It is widely acknowledged that the line is the most interesting metric space for the related online matching problems [4,14]. If X is a line, without loss of generality, assume that the servers are placed in an increasing order of their indices, i.e.,

$$0 = p(s_1) < p(s_2) < \ldots < p(s_m),$$

where $p(s_i)$ is the position of server s_i on the line, for $i = 1, 2, \ldots, m$. Let $p(r)$ be the position of request r. The distance between $a, b \in S \cup R$ can be described as $d(a, b) = |p(a) - p(b)|$. Moreover, for $i = 1, 2, \ldots, m - 1$, let $d_i = p(s_{i+1}) - p(s_i)$ be the distance between two adjacent servers.

3 Online Bottleneck Semi-matching with Arbitrary Number of Servers

In this section, we consider the OBSM problem with arbitrary number of servers. If each server $s_i \in S$ is replaced by c_i servers with capacity 1, the OBSM problem is exactly the OBM problem considered in [2]. Therefore, following from Theorem 3 and Theorem 4 in [2], we have

Theorem 1. The competitive ratio of the PERMUTATION algorithm is at most $(2n - 1)$ for the OBSM problem.

Kalyanasundaram and Pruhs [10] conjectured that the competitive ratio of the PERMUTATION algorithm is $(2m - 1)$ for the OMSM problem. However, this conjecture is still open. Similarly, it is interesting to design an online algorithm for the OBSM problem whose competitive ratio is a function of m. We obtain several related results for the OBSM problem in this section.

Theorem 2. The competitive ratio of any deterministic online algorithm for the OBSM problem on a line is at least

$$2 + \frac{1}{2^{\frac{1}{m-1}} - 1}.$$

Proof. Similarly to the construction in [8]. ∎

Theorem 3. The competitive ratio of any deterministic online algorithm is at least $m+1$ for the OBSM problem on a line, even if $d_i = 1$ for every $i = 1, 2, \ldots, m-1$.

Proof. Let A be any deterministic online algorithm, and π be the corresponding assignment. Our adversary first gives $c_i - 1$ requests at $p(s_i)$ for each $i = 1, 2, \cdots, m$. If A matches some request r with a server not at $p(r)$, then the adversary gives m more requests with one at each position of the server s_i. An optimal offline assignment π^* matches every request r with the server at the same position $p(r)$. Therefore, $C^A > 0$ and $C^{OPT} = 0$, implying that the competitive ratio of A is infinity.

Suppose that A matches each request r with a server at $p(r)$. For convenience, let $r_1, r_2, \ldots, r_m$ be the last m requests and $x = \frac{1}{m}$. The adversary gives the requests r_i at $p(s_{i+1}) - x$ for $i = 1, 2, \ldots, m-1$ one by one (see Fig. 1).

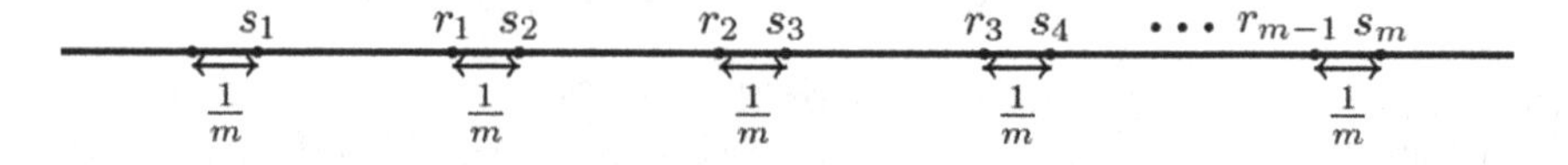

Fig. 1. The $m-1$ requests.

If A matches each request r_i with server s_{i+1} for $1, 2, \ldots, m-1$. The adversary gives the last request r_m at $p(s_m)+1-x$. Clearly, we have $C^{OPT} = 1-x = 1-\frac{1}{m}$, and

$$C^A \geq d(r_m, s_1) = m - 1 + 1 - x = m - \frac{1}{m} \geq (m+1)C^{OPT}.$$

Otherwise, the adversary gives the last request r_m at $p(s_1) - x$. Clearly, we have $C^{OPT} = x = \frac{1}{m}$. We distinguish the follow three cases.

Case 1. A matches request r_1 with server s_1. Clearly, we have

$$C^A \geq d(r_m, \pi(r_m)) \geq 1 + x = 1 + \frac{1}{m} \geq (m+1)C^{OPT}.$$

Case 2. A matches request r_1 with server s_2. Let $k \in \{2, \ldots, m-1\}$ be the minimum index such that A does not match request r_k with server s_{k+1}. If A matches request r_k with server s_l with $l \leq k$, by the minimality of k, r_k must be matched with s_1, implying that

$$C^A \geq d(r_k, s_1) \geq 2 - x \geq 1 + \frac{1}{m} \geq (m+1)C^{OPT}.$$

If A matches r_k with a server s_u with $u \geq k+2$, we have

$$C^A \geq d(r_k, s_u) \geq 1 + x = 1 + \frac{1}{m} \geq (m+1)C^{OPT}.$$

Case 3. A matches request r_1 with a server s_k $(k \geq 3)$. Clearly, we have

$$C^A \geq d(r_1, s_k) \geq 1 + x = 1 + \frac{1}{m} \geq (m+1)C^{OPT}.$$

Therefore, the theorem holds. ∎

As mentioned in [2,10], GREEDY assigns the nearest available server to each request as it arrives. It is proved that the GREEDY algorithm performs exponentially poorly for the OMM problem [10] and the OBM problem [2]. However, GREEDY performs well for the OBSM problem on a line with $d_i = 1$.

Theorem 4. The competitive ratio of GREEDY is at most $2m-1$ for the OBSM problem on a line with $d_i = 1$ for every $i = 1, 2, \ldots, m-1$.

Proof. Without loss of generality, assume that

$$p(s_i) = i - 1, \text{ for } i = 1, 2, \ldots, m.$$

Let $I_1 = (-\infty, \frac{1}{2}]$, $I_m = (m - 1 - \frac{1}{2}, +\infty)$, and $I_i = (i - 1 - \frac{1}{2}, i - 1 + \frac{1}{2}]$ for $i = 2, 3, \ldots, m-1$. Let

$$R_i = \{r_i \in R | p(r_i) \in I_i\}.$$

be the set of requests whose positions lie in $\mathcal{I}_i$ for $i = 1, 2, \ldots, m$.

Clearly, if $|R_i| = c_i$ for every i, GREEDY produces an optimal solution. Else, we have

$$C^{OPT} \geq \frac{1}{2}.$$

Let $r \in R$ be the request attaching the maximum in the feasible solution produced by GREEDY, which implies that $C^A = d(r, \pi(r))$. We distinguish the following three cases.

Case 1. $p(r) \in R_1$.
If $p(r) \in (-\infty, -\frac{1}{2}]$, we have $C^{OPT} \geq d(r, s_1) = |p(r)|$, and

$$C^A \leq |p(r)| + d(s_1, s_m) = |p(r)| + m - 1 \leq |p(r)|(1 + \frac{m-1}{|p(r)|}) \leq (2m-1)C^{OPT}.$$

If $p(r) \in (-\frac{1}{2}, \frac{1}{2}]$, we have

$$C^A \leq d(r, s_m) \leq m - 1 + \frac{1}{2} = (2m-1) \cdot \frac{1}{2} \leq (2m-1)C^{OPT}.$$

Case 2. $p(r) \in R_i$ with $i \in \{2, 3, \ldots, m-1\}$.
Clearly, we have

$$C^A \leq d(r, s_m) \leq m - 1 \leq (2m-1)C^{OPT}.$$

Case 3. $p(r) \in R_m$.
If $p(r) \in (m-1-\frac{1}{2}, m-1+\frac{1}{2}]$, we have

$$C^A \le d(r, s_1) \le m - 1 + \frac{1}{2} = (2m-1) \cdot \frac{1}{2} \le (2m-1)C^{OPT}.$$

If $p(r) \in (m-1+\frac{1}{2}, +\infty)$, we have $C^{OPT} \ge d(r, s_m) = p(r) - (m-1) \ge \frac{1}{2}$ and

$$\begin{aligned} C^A &\le d(r, s_1) \le p(r) = p(r) - (m-1) + (m-1) \\ &\le d(r, s_m) + (2m-2)C^{OPT} \le (2m-1)C^{OPT}. \end{aligned}$$

Therefore, we have $C^A \le (2m-1)C^{OPT}$ in any case. ∎

4 Online Bottleneck Semi-matching with Two Servers

When $m = 2$, it is proved that the PERMUTATION algorithm is an optimal online algorithm with competitive ratio 3 for the OMM problem [10] and the OBM problem [2]. Recently, Itoh et al. [9] proved that GREEDY algorithm is an optimal online algorithm with competitive ratio 3 for the OMSM problem.

By Theorem 2, the lower bound for the OBSM problem is 3 when $m = 2$. Recall that GREEDY assigns the nearest available server to each request as it arrives. We obtain

Theorem 5. The competitive ratio of the GREEDY algorithm is 3.

Proof. Let R be a minimal instance with the least number of requests whose competitive ratio is maximized. For $j = 1, 2, \ldots, n$, if r_j is matched with the server in the offline optimal solution π^* and the feasible solution π produced by the GREEDY algorithm. Without loss of generality, assume that $\pi(r_j) = \pi^*(r_j) = s_1$. If $C^A = d(r_j, s_1)$, we have $C^{OPT} = C^A$, implying that the GREEDY algorithm produces an optimal solution. Else, we construct a new instance $R' = R \setminus \{r_j\}$ with $c_1' = c_1 - 1$. It is easy to verify that $C^A(R') = C^A(R)$ and $C^{OPT}(R') = C^{OPT}(R)$, which contradicts the minimality of R. Therefore, we have

$$\pi(r_j) \neq \pi^*(r_j), \text{ for each } j = 1, 2, \ldots, n. \tag{1}$$

Let r_{j_1} be the request attaching the maximum in the feasible solution produced by the GREEDY algorithm, which means $C^A = d(r_{j_1}, \pi(r_{j_1}))$. Without loss of generality, assume that $\pi(r_{j_1}) = s_1$, which means that

$$C^A = d(r_{j_1}, s_1) \ge d(r_{j_1}, s_2), \text{ and } \pi^*(r_{j_1}) = s_2.$$

Let r_{j_2} be the request attaching the maximum in the optimal solution, which means $C^{OPT} = d(r_{j_2}, \pi^*(r_{j_2}))$. We distinguish the following two cases.

Case 1. $\pi^*(r_{j_2}) = s_1 = \pi(r_{j_1})$.
By (1), we have $j_1 \neq j_2$ and $\pi^*(r_{j_1}) = \pi(r_{j_2}) = s_2$. If $j_1 < j_2$, by the choice of GREEDY, we have $d(r_{j_1}, s_1) \le d(r_{j_1}, s_2)$. Therefore,

$$C^A = d(r_{j_1}, s_1) \leq d(r_{j_1}, s_2) = d(r_{j_1}, \pi^*(r_{j_1})) \leq C^{OPT}.$$

If $j_1 > j_2$, by the choice of GREEDY, we have $d(r_{j_2}, s_2) \leq d(r_{j_2}, s_1)$. Therefore,

$$\begin{aligned} C^A = d(r_{j_1}, s_1) &\leq d(r_{j_1}, s_2) + d(s_1, s_2) \\ &\leq d(r_{j_1}, s_2) + d(s_1, r_{j_2}) + d(r_{j_2}, s_2) \\ &\leq d(r_{j_1}, \pi^*(r_{j_1})) + 2d(r_{j_2}, s_1) \\ &= d(r_{j_1}, \pi^*(r_{j_1})) + 2d(r_{j_2}, \pi^*(r_{j_2})) \\ &\leq 3C^{OPT}. \end{aligned}$$

Case 2. $\pi^*(r_{j_2}) = s_2$.
Clearly, we have $C^{OPT} = d(r_{j_2}, s_2) \leq d(r_{j_2}, s_1)$. Since $c_1 + c_2 = n$, there is a request r_j with $j < j_1$ satisfying that $\pi^*(r_j) = s_1$ and $\pi(r_j) = s_2$. By the choice of GREEDY, we have $d(r_j, s_2) \leq d(r_j, s_1) = d(r_j, \pi^*(r_j)) \leq C^{OPT}$. Therefore,

$$\begin{aligned} C^A = d(r_{j_1}, s_1) &\leq d(r_{j_1}, s_2) + d(s_2, s_1) \\ &\leq d(r_{j_1}, \pi^*(r_{j_1})) + d(s_2, r_j) + d(r_j, s_1) \\ &\leq C^{OPT} + 2d(r_j, s_1) \\ &\leq 3C^{OPT}. \end{aligned}$$

Thus, $C^A \leq 3C^{OPT}$ in any case. ∎

5 Online Bottleneck Semi-matching on a Line with Three Servers

When $m = 3$ and X is a line, Kalyanasundaram and Pruhs [10] claimed that the optimal competitive ratio for the OMM problem on a line is 3.6494359. Recently, Itoh et al. [9] gave a lower bound $1 + \sqrt{6}$ on the competitive ratio for the OMSM problem on a line with $d_1 = d_2 = 1$. However, the optimal competitive ratio is not given. Idury and Schaffer [8] gave a lower bound $3 + \sqrt{2}$ on the competitive ratio for the OBM problem on a line. In this section, we consider the OBSM problem on a line with $m = 3$. Without loss of generality, assume that

$$p(s_1) = 0, p(s_2) = 1, \text{ and } p(s_3) = 1 + \alpha \geq 2.$$

When $\alpha \leq \sqrt{2}$, we design an optimal online algorithm with a competitive ratio of $3 + \alpha$. When $\alpha > \sqrt{2}$, we design an optimal online algorithm with a competitive ratio of $3 + \frac{2}{\alpha}$. Clearly, the upper bound on the competitive ratio for the OBM problem on a line is $3 + \sqrt{2}$, which matches the lower bound given in [8].

Theorem 6. *When $1 \leq \alpha \leq \sqrt{2}$, the competitive ratio of any online algorithm A for the OBSM problem on a line is at least $3 + \alpha$.*

Proof. Let $c_1 = c_2 = c_3 = 1$. The first request r_1 arrives at $p(r_1) = p(s_2) - \frac{1}{2+\alpha}$. We distinguish the following three cases.

Case 1. r_1 is matched with s_1.
The last two requests r_2 and r_3 arrive at $p(r_2) = p(s_1) - \frac{1}{2+\alpha}$ and $p(r_3) = p(s_3)$, respectively. Therefore, $C^{OPT} = \frac{1}{2+\alpha}$ and

$$C^A \geq d(r_2, s_2) = 1 + \frac{1}{2+\alpha} \geq (3+\alpha)C^{OPT}.$$

Case 2. r_1 is matched with s_2.
The second request r_2 arrives at $p(r_2) = p(s_2) + \frac{\alpha}{2}$. If r_2 is matched with s_1, the last request r_3 arrives at $p(r_3) = p(s_1) - \frac{\alpha}{2}$. Since $\frac{1}{2+\alpha} \leq \frac{1}{2} \leq \frac{\alpha}{2}$, we have $C^{OPT} = \frac{\alpha}{2}$ and $C^A = d(r_3, s_3) = 1 + \alpha + \frac{\alpha}{2}$, then

$$C^A \geq d(r_3, s_3) = 1 + \alpha + \frac{\alpha}{2} \geq (3 + \frac{2}{\alpha})C^{OPT} \geq (3+\alpha)C^{OPT},$$

as $\alpha \leq \sqrt{2}$.

If r_2 is matched with s_3, the last request r_3 arrives at $p(r_3) = p(s_3) + \frac{1+\alpha}{2+\alpha}$. Since $\alpha \leq \sqrt{2}$, we have $\frac{1+\alpha}{2+\alpha} \geq \frac{\alpha}{2}$, which implies that $C^{OPT} = d(r_3, s_3) = \frac{1+\alpha}{2+\alpha}$, and

$$C^A = d(r_3, s_1) = 1 + \alpha + \frac{1+\alpha}{2+\alpha} \geq (3+\alpha)C^{OPT}.$$

Case 3. r_1 is matched with s_3.
The last two requests r_2 and r_3 arrive at $p(r_2) = p(s_1)$ and $p(r_3) = p(s_3)$, respectively. Clearly, $C^{OPT} = \frac{1}{2+\alpha}$, and

$$C^A \geq d(r_1, s_3) = \frac{1}{2+\alpha} + \alpha = \frac{\alpha^2 + 2\alpha + 1}{2+\alpha} \geq (3+\alpha)C^{OPT},$$

as $\alpha \geq 1$.

Therefore, the theorem holds. ∎

For convenience, let

$$I_1 = (-\infty, p(s_2) - \frac{1}{2+\alpha}),$$
$$I_2 = [p(s_2) - \frac{1}{2+\alpha}, p(s_2) + \frac{\alpha^2+\alpha-1}{2+\alpha}],$$
$$\text{and } I_3 = (p(s_2) + \frac{\alpha^2+\alpha-1}{2+\alpha}, +\infty).$$

Three intervals are depicted in Fig. 2. For each server s_i, we say that s_i is *available* if it is matched less than c_i times. Otherwise, we say that s_i is *full*.

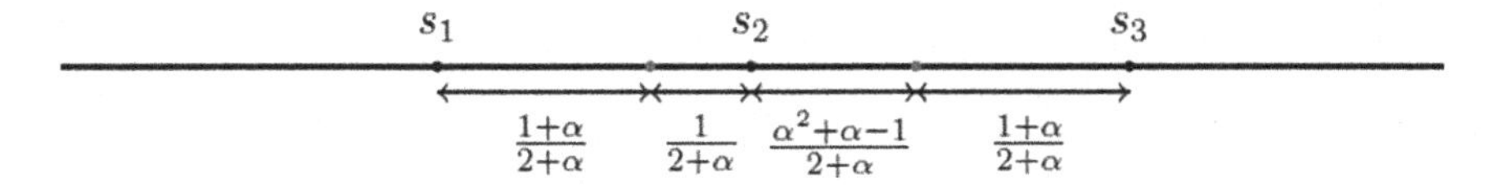

Fig. 2. Three intervals in ALGORITHM A1

ALGORITHM A1:
When a new request r_j arrives, we distinguish the following three cases.

Case 1. $p(r_j) \in I_1$. Match r_j with the first available server in the sequence (s_1, s_2, s_3).

Case 2. $p(r_j) \in I_2$. Match r_j with the first available server in the sequence (s_2, s_1, s_3).

Case 3. $p(r_j) \in I_3$. Match r_j with the first available server in the sequence (s_3, s_2, s_1).

Theorem 7. When $1 \leq \alpha \leq \sqrt{2}$, the competitive ratio of **Algorithm A1** for the OBSM problem on a line is at most $3 + \alpha$.

Proof. We omitted the proof due to space constraints. ∎

Theorem 8. When $\alpha > \sqrt{2}$, the competitive ratio of any online algorithm A for the OBSM problem on a line is at least $3 + \frac{2}{\alpha}$.

Proof. Let $c_1 = c_2 = c_3 = 1$. The first request arrives at $p(r_1) = p(s_2) - \frac{1}{2+\alpha}$. We distinguish the following three cases.

Case 1. r_1 is matched with s_1.
The last two requests arrive at $p(r_2) = p(s_1) - \frac{1}{2+\alpha}$ and $p(r_3) = p(s_3)$, respectively. Therefore, $C^{OPT} = \frac{1}{2+\alpha}$ and

$$C^A \geq d(r_2, s_2) = 1 + \frac{1}{2+\alpha} \geq (3+\alpha)C^{OPT} \geq (3 + \frac{2}{\alpha})C^{OPT},$$

where the last inequality follows from the assumption $\alpha > \sqrt{2}$.

Case 2. r_1 is matched with s_2.
The second request arrives at $p(r_2) = p(s_2) + \frac{\alpha}{2}$. If r_2 is matched with s_1, the last request r_3 arrives at $p(r_3) = p(s_1) - \frac{\alpha}{2}$. Therefore, $C^{OPT} = \frac{\alpha}{2}$ and

$$C^A = d(r_3, s_3) = 1 + \alpha + \frac{\alpha}{2} \geq (3 + \frac{2}{\alpha})C^{OPT}.$$

If r_2 is matched with s_3, the last request r_3 arrives at $p(r_3) = p(s_3) + \frac{\alpha}{2}$. Therefore, $C^{OPT} = \frac{\alpha}{2}$, and

$$C^A = d(r_3, s_1) = 1 + \alpha + \frac{\alpha}{2} \geq (3 + \frac{2}{\alpha})C^{OPT}.$$

Case 3. r_1 is matched with s_3.
The last two requests arrive at $p(r_2) = p(s_1)$ and $p(r_3) = p(s_3)$. Therefore, $C^{OPT} = \frac{1}{2+\alpha}$, and

$$C^A = d(r_1, s_3) \geq \frac{1}{2+\alpha} + \alpha \geq (\alpha^2 + 2\alpha + 1)C^{OPT} \geq (3 + \frac{2}{\alpha})C^{OPT}.$$

as $\alpha > \sqrt{2}$.

Therefore, the theorem holds. ∎

ALGORITHM A2:

When a new request r_j arrives, we distinguish the following three cases.

Case 1. $p(r_j) \in (-\infty, p(s_2) - \frac{\alpha}{2+2\alpha})$. Match r_j with the first available server in the sequence (s_1, s_2, s_3).

Case 2. $p(r_j) \in [p(s_2) - \frac{\alpha}{2+2\alpha}, p(s_2) + \frac{\alpha}{2}]$. Match r_j with the first available server in the sequence (s_2, s_1, s_3).

Case 3. $p(r_j) \in (p(s_2) + \frac{\alpha}{2}, +\infty)$. Match r_j with the first available server in the sequence (s_3, s_2, s_1) (Fig. 3).

Theorem 9. *When $\alpha > \sqrt{2}$, the competitive ratio of* ALGORITHM A2 *for the OBSM problem on a line is at most $3 + \frac{2}{\alpha}$.*

Proof. We omitted the proof due to space constraints. ∎

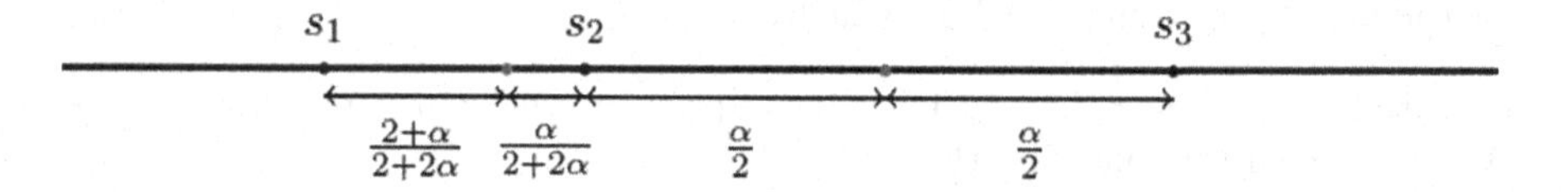

Fig. 3. Three intervals in ALGORITHM A2

6 Conclusion

We propose an online algorithm for the OBSM problem on a line with competitive ratio $2m - 1$, where the distance between every pair of adjacent servers is the same. It is interesting to design an online algorithm with competitive ratio dependent of m for a general metric space. We conjecture that PERMUTATION [2,10] achievers a $(2m - 1)$ competitive ratio for the OBSM problem and the OMSM problem.

When the number of servers is three, we design two optimal online algorithms for the OBSM problem on a line, whose competitive ratio is dependent on the ratio of two distances between two adjacent servers. In addition, we close the gap for the OBM problem with three servers, which has been open about thirty years. Although Kalyanasundaram and Pruhs [10] claimed the optimal competitive ratio of the OMM problem with three servers is 3.6494359, it is interesting to design several optimal online algorithms with competitive ratio dependent on the ratio of two distances between two adjacent servers, as in Sect. 5.

Acknowledgement. The work is supported in part by the National Natural Science Foundation of China [No. 12071417], Program for Excellent Young Talents of Yunnan University, Training Program of National Science Fund for Distinguished Young Scholars, and IRTSTYN.

References

1. Ahmed, A.R., Rahman, M.S., Kobourov, S.: Online facility assignment. Theoret. Comput. Sci. **806**, 455–467 (2020)
2. Anthony, B.M., Chung, C.: Online bottleneck matching. J. Comb. Optim. **27**(1), 100–114 (2012). https://doi.org/10.1007/s10878-012-9581-9
3. Anthony, B.M., Chung, C.: Serve or skip: the power of rejection in online bottleneck matching. J. Comb. Optim. **32**(4), 1232–1253 (2015). https://doi.org/10.1007/s10878-015-9948-9
4. Antoniadis, A., Barcelo, N., Nugent, M., Pruhs, K., Scquizzato, M.: A $o(n)$- competitive deterministic algorithm for online matching on a line. Algorithmica **81**, 2917–2933 (2019)
5. Bansal, N., Buchbinder, N., Gupta, A., Naor, J.S.: A randomized $O(\log^2 k)$-competitive algorithm for metric bipartite matching. Algorithmica **68**, 390–403 (2014)
6. Fuchs, B., Hochstattler, W., Kern, W.: Online matching on a line. Theoret. Comput. Sci. **332**(1), 251–264 (2005)
7. Gupta, A., Lewi, K.: The online metric matching problem for doubling metrics. In: Proceedings of International Colloquium on Automata, Languages, and Programming (ICALP), pp. 424–435 (2012)
8. Idury, R., Schaffer, A.A.: A better lower bound for on-line bottleneck matching, manuscript (1992)
9. Itoh, T., Miyazaki, S., Satake, M.: Competitive analysis for two variants of online metric matching problem, Discrete Mathematics, Algorithms and Applications, ID 2150156 (2021)
10. Kalyanasundaram, B., Pruhs, K.: Online weighted matching. J. Algorithms **14**(3), 478–488 (1993) Preliminary version appeared in Proceedings of the 2nd Annual ACM-SIAM Symposium on Discrete algorithms (SODA), pp. 234–240 (1991)
11. Kalyanasundaram, B., Pruhs, K.: On-line network optimization problems. In: Fiat, A., Woeginger, G.J. (eds.) Online Algorithms. LNCS, vol. 1442, pp. 268–280. Springer, Heidelberg (1998). https://doi.org/10.1007/BFb0029573
12. Kalyanasundaram, B., Pruhs, K.: The online transportation problem. SIAM J. Discret. Math. **13**(3), 370–383 (2000)
13. Nayyar, K., Raghvendra, S.: An input sensitive online algorithm for the metric bipartite matching problem. In: Proceedings of IEEE 58th Annual Symposium on Foundations of Computer Science, pp. 505–515 (2017)
14. Peserico, E., Scquizzato, M.: Matching on the line admits no $o(\sqrt{\log n})$-competitive algorithm. In: Proceedings of the 48th International Colloquium on Automata, Languages, and Programming (ICALP), Article No. 103 (2021)
15. Raghvendra, S.: A robust and optimal online algorithm for minimum metric bipartite matching. In: Proceedings of Approximation, Randomization, and Combinatorial Optimization. Algorithms and Techniques (APPROX/RANDOM 2016), ID 18 (2016)
16. Raghvendra, S.: Optimal analysis of an online algorithm for the bipartite matching problem on a line. In: Proceedings of 34th International Symposium on Computational Geometry, ID 67 (2017)

Optimal Due Date Assignment Without Restriction and Convex Resource Allocation in Group Technology Scheduling

Ying Chen[1] and Yongxi Cheng[1,2(✉)]

[1] School of Management, Xi'an Jiaotong University, Xi'an 710049, China
guolichenying@stu.xjtu.edu.cn, chengyx@mail.xjtu.edu.cn
[2] State Key Lab for Manufacturing Systems Engineering, Xi'an 710049, China

Abstract. We consider a single machine group scheduling problem with convex resource allocation in which the scheduler decides optimal due dates for different jobs under a group technology environment. The jobs are classified into groups in advance due to their production similarities, and jobs in the same group are required to be processed consecutively, to achieve efficiency of high-volume production. The goal is to determine the optimal group sequence and job sequence within each group, together with a due date assignment strategy and resource allocation to minimize an objective function, which includes earliness, tardiness, due date assignment and resource allocation costs. The actual job processing times are resource dependent, and the due date assignment is without restriction, that is, it is allowed to assign different due dates to jobs within one group. We present structural results that characterize the optimal schedule in the case where the number of jobs in each group is identical and the cost ψ_{ij} (the minimum of the due date assignment cost and the tardiness cost) for each job J_{ij} is also identical, and present an $O(n \log n)$ time algorithm to solve this problem optimally.

Keywords: Single machine scheduling · Due date assignment · Group technology · Convex resource allocation

1 Introduction

We consider a due date assignment problem in a group technology (GT) scheduling environment with convex resource allocation. In manufacturing processes, it is well-known that the production efficiency in high-volume production can be increased by grouping various parts and products with similar designs, production processes and/or from the same order. After classified into groups, products (jobs) within a group are consecutively sequenced. Many advantages have been claimed through the wide applications of group technology. For instance, products spend less time waiting, which results in less work-in-process inventory,

D.-Z. Du et al. (Eds.): COCOA 2021, LNCS 13135, pp. 456–467, 2021.
https://doi.org/10.1007/978-3-030-92681-6_36

products tend to move through production in a direct route, and hence the manufacturing lead time is reduced (see, e.g., Liu et al. [1], Keshavarz et al. [2], Qin et al. [3] and Wang et al. [4]).

The class of due date assignment problems is a challenging topic, and has attracted much attention in the past few decades due to the increasing interest in Just-In-Time systems. Products which complete processing prior to their due date often incur earliness costs. These may include storage costs, insurance fees and so on. Equivalently, products completed past their due date often incur tardiness penalties in the form of compensation of customers and overtime work. An increasing number of studies have viewed due date assignment as part of the scheduling process, motivated by the common real-life situation where the due date is often determined during sales negotiations with the customer. Hence, there is also a cost associated with the due date assignment. Due to limited production capacity, it is often unlikely to complete all jobs exactly on their respective due dates. Thus, it is crucial for manufacturing systems to take into account all associated costs and develop policies which focus on the aggregate cost(see, e.g., Seidmann et al. [5], Panwalkar et al. [6] and Shabtay [7]).

Pioneering research in the field of scheduling with due date assignments was conducted by Seidmann et al. [5] and Panwalkar et al. [6]. Seidmann et al. [5] analyzed a single machine non-preemptive scheduling problem where all jobs are available for processing at time zero. They used a due date assignment method where each job can be assigned a due date without any restriction, and presented an $O(n \log n)$ time optimization algorithm to determine the set of due dates. Panwalkar et al. [6] studied a problem where the scheduler has to assign a common due date to all jobs, to minimize an objective function which is a combination of earliness, tardiness, and due date costs. They provided an $O(n \log n)$ optimization algorithm to solve the problem. Li et al. [8] considered a single machine scheduling problem involving both the due date assignment and job scheduling under a group technology environment, with three different due date assignment methods. The objective is to find an optimal combination of the due date assignment strategy and job schedule, to minimize an objective function that includes earliness, tardiness, due date assignment and flow time costs. Bajwa et al. [9] investigated a single machine problem of minimizing the total number of tardy jobs in a GT environment with individual due dates. They proposed a hybrid heuristic approach to solve the problem. Li et al. [10] and Ji et al. [11] considered group scheduling problem on a single machine with multiple due windows assignment.

In classical scheduling problems, it is usually assumed that job processing times are known and fixed. However, for many modern industrial processes the actual processing time of a job is controllable by the allocation of a continuous and nonrenewable resource such as fuel or manpower to compress the job processing times. Shabtay et al. [12] investigated a due date assignment problem under a group technology environment in which jobs within a family (group) are restricted to be assigned the same due date, while the due dates for different families are allowed to be different. The objective is to find the job schedule and the due date for each group that minimizes an objective function which includes earliness, tardiness and due date assignment costs. They also extended the analysis to the case in which the job processing times are resource dependent. Lv et al. [13] introduced a single machine common flow allowance group scheduling with learning effect under resource allocation. The jobs within a group are assigned the same flow allowance and two types of resource allocation function are considered in the paper. They gave a polynomial time algorithm to solve the problem. In addition, Yin et al. [15] investigated scheduling problems on a single machine with learning effect, deteriorating jobs and resource allocation under group technology assumption. They proved that the problems have polynomial solutions under the condition that the number of jobs in each group are the same.

As far as we know, there is no research on scheduling with convex resource allocation and different due dates under group technology in which the unit due date, earliness, tardiness and resource allocation costs could be different for jobs within one group. Based on the research gap found, in this paper we consider the group scheduling problem with due date assignment and convex resource allocation, in which jobs within one group may have different unit due date, earliness, tardiness and resource allocation costs. Besides it is allowed to assign different due dates to jobs within one group.

The rest of this paper is organized as follows. In Sect. 2, we formally describe the group scheduling problem with due date assignment and convex resource allocation studied in this paper and present some preliminary analysis in Sect. 3. In Sect. 4, structural results of the optimal schedule are presented in the case where the number of jobs in each group is identical and the cost ψ_{ij} (the minimum of the due date assignment cost and the tardiness cost) for each job J_{ij} is also identical, and an $O(n \log n)$ time optimization algorithm for the problem is given. The paper is concluded in Sect. 5.

2 Problem Formulation

A set of n independent and non-preemptive jobs is available for processing at time zero and has been classified into m groups $G_1, G_2, \cdots, G_m$. Each group G_i, for $i = 1, 2, \cdots, m$, consists of a set $\{J_{i1}, J_{i2}, \cdots, J_{in_i}\}$ of n_i jobs, where $\sum_{i=1}^{m} n_i = n$. The jobs within the same group are required to be processed contiguously. A sequence-independent machine setup time s_i precedes the processing of the first job in group G_i. The machine can handle one job at a time and jobs within one group are allowed to be assigned different due dates. Let C_{ij} denote the completion time of job J_{ij}. The earliness and tardiness of job J_{ij} is given by $E_{ij} = \max\{d_{ij} - C_{ij}, 0\}$ and $T_{ij} = \max\{C_{ij} - d_{ij}, 0\}$. Clearly, E_{ij} and T_{ij} cannot both be positive.

In scheduling with controllable processing times, the relationship between resource allocation and actual job processing time is usually presented via a convex resource consumption function. The function is a convex decreasing function of the following form $p_{ij}(u_{ij}) = (\frac{w_{ij}}{u_{ij}})^r$, where $p_{ij}(u_{ij})$ is the processing time of

Table 1. Notations.

Simbol	Definition
$G_{[i]}$	The group scheduled in the ith position of a sequence
$J_{[i][j]}$	The job scheduled in the jth position in group $G_{[i]}$
$p_{[i][j]}$	The processing time of job $J_{[i][j]}$
$s_{[i]}$	The setup time for group $G_{[i]}$
$d_{[i][j]}$	The due date assigned to job $J_{[i][j]}$
$\psi_{[i][j]}$	$\min(\alpha_{[i][j]}, \gamma_{[i][j]})$
$P_{[i]} = \sum_{j=1}^{n_{[i]}} p_{[i][j]}$	The total processing time of all jobs within group $G_{[i]}$
$\Psi_{[i]} = \sum_{j=1}^{n_{[i]}} \psi_{[i][j]}$	The sum of $\psi_{[i][j]}$ for all jobs within group $G_{[i]}$

job J_{ij} and affected by the amount of resource allocated to job J_{ij}, i.e., u_{ij}, w_{ij} is a positive parameter which represents the workload of job J_{ij} and r is a positive constant.

For the single machine due date scheduling with the resource allocation in GT environment, the objective is to determine a schedule π for the group sequence and job sequence within each group G_i for $i = 1, 2, \cdots, m$, the due date assignment vector $d(\pi) = (d_{11}(\pi), d_{12}(\pi), \cdots, d_{1n_1}(\pi), \cdots, d_{m1}(\pi), \cdots, d_{mn_m}(\pi))$ specifying the due date for each job J_{ij} and the optimal resource allocation vector $u(\pi) = (u_{11}(\pi), u_{12}(\pi), \cdots, u_{1n_1}(\pi), \cdots, u_{m1}(\pi), \cdots, u_{mn_m}(\pi))$ to minimize a cost function that includes earliness, tardiness, due date assignment and resource allocation costs, given by the following equation:

$$Z(\pi, d(\pi), u(\pi)) = \sum_{i=1}^{m} \sum_{j=1}^{n_i} (\alpha_{ij} d_{ij} + \beta_{ij} E_{ij} + \gamma_{ij} T_{ij} + v_{ij} u_{ij}).$$

where α_{ij}, β_{ij}, γ_{ij} and v_{ij} be unit due date, earliness, tardiness and resource allocation costs of job J_{ij}, respectively. All n_i, s_i, α_{ij}, β_{ij}, γ_{ij} and v_{ij} are given parameters.

Following Graham's et al. [17] three-field notation, we denote the problem with the convex resource consumption function by

$$1|GT, DIF, CONV| \sum_{i=1}^{m} \sum_{j=1}^{n_i} (\alpha_{ij} d_{ij} + \beta_{ij} E_{ij} + \gamma_{ij} T_{ij} + v_{ij} u_{ij}),$$

where GT denotes group technology, DIF denotes that each job can be assigned a different due date without restriction and CONV denotes the convex resource allocation. Please refer to Table 1 for notations that will be used in this paper.

3 Preliminary Analysis

In this section, the following two lemmas will be given for the later analysis.

Lemma 1. *Fix a schedule π for the group sequence and job sequence within each group, the optimal due date assignment vector, $d^*(\pi)$, for problem*

$$1|GT, DIF, CONV|\sum_{i=1}^{m}\sum_{j=1}^{n_i}(\alpha_{ij}d_{ij} + \beta_{ij}E_{ij} + \gamma_{ij}T_{ij} + v_{ij}u_{ij})$$

can be determined as follows:

$$d^*_{[i][j]}(\pi) = \begin{cases} C_{[i][j]}(\pi) & \text{if } \alpha_{[i][j]} < \gamma_{[i][j]} \\ 0 & \text{if } \alpha_{[i][j]} > \gamma_{[i][j]} \\ \text{any value in } [0, C_{[i][j]}(\pi)] & \text{if } \alpha_{[i][j]} = \gamma_{[i][j]} \end{cases}$$

for $i = 1, 2, \cdots, m$, and $j = 1, 2, \cdots, n_i$.

Proof. According to the study in [18], we can know that the optimal due date assignment vector for the problem $1|GT, DIF|\sum_{i=1}^{m}\sum_{j=1}^{n[i]}(\alpha_{ij}d_{ij} + \beta_{ij}E_{ij} + \gamma_{ij}T_{ij})$ is same to the results in Lemma 1. As the optimal due date assignment vector is independent of the resource allocation u_{ij}, then the result can be immediately generalized to the problem $1|GT, DIF, CONV|\sum_{i=1}^{m}\sum_{j=1}^{n_i}(\alpha_{ij}d_{ij} + \beta_{ij}E_{ij} + \gamma_{ij}T_{ij} + v_{ij}u_{ij})$. The proof is completed. □

Lemma 2. *An optimal schedule does not have idle times.*

Proof. It is omitted due to the similarity to the proof in [8].

Given a schedule π, let $d^*(\pi)$ be an optimal due date assignment as given in Lemma 1, and let $Z_{[i][j]}(\pi, d^*(\pi), u(\pi))$ be the contribution to the objective function from job $J_{[i][j]}$. According to Lemma 1, we can analyze $Z_{[i][j]}(\pi, d^*(\pi), u(\pi))$ according to the following three cases.

Case 1. $\alpha_{[i][j]} > \gamma_{[i][j]}$. According to Lemma 1, in this case we have $d^*(\pi)_{[i][j]} = E_{[i][j]} = 0$ and $T_{[i][j]} = C_{[i][j]}$, for $i = 1, 2, \cdots, m$ and $j = 1, 2, \cdots, n_i$. Hence,

$$Z_{[i][j]}(\pi, d^*(\pi), u(\pi)) = \gamma_{[i][j]}C_{[i][j]} + v_{[i][j]}u_{[i][j]}.$$

Case 2. $\alpha_{[i][j]} < \gamma_{[i][j]}$. In this case, we have $d^*(\pi)_{[i][j]} = C_{[i][j]}$ and $E_{[i][j]} = T_{[i][j]} = 0$, for $i = 1, 2, \cdots, m$ and $j = 1, 2, \cdots, n_i$. Hence,

$$Z_{[i][j]}(\pi, d^*(\pi), u(\pi)) = \alpha_{[i][j]}C_{[i][j]} + v_{[i][j]}u_{[i][j]}.$$

Case 3. $\alpha_{[i][j]} = \gamma_{[i][j]}$. Similarly, according to Lemma 1 in this case we have

$$\begin{aligned} Z_{[i][j]}(\pi, d^*(\pi), u(\pi)) &= \alpha_{[i][j]}d_{[i][j]} + \gamma_{[i][j]}(C_{[i][j]} - d_{[i][j]}) + v_{[i][j]}u_{[i][j]} \\ &= \alpha_{[i][j]}C_{[i][j]} + v_{[i][j]}u_{[i][j]}. \end{aligned}$$

Recall that $\psi_{[i][j]} = \min(\alpha_{[i][j]}, \gamma_{[i][j]})$, for $i = 1, 2, \cdots, m$, and $j = 1, 2, \cdots, n_i$. Hence, for all the above three cases $Z_{[i][j]}(\pi, d^*(\pi), u(\pi))$ can be written in a unified way as follows:

$$Z_{[i][j]}(\pi, d^*(\pi), u(\pi)) = \psi_{[i][j]}C_{[i][j]} + v_{[i][j]}u_{[i][j]}.$$

It follows that $Z(\pi, d^*(\pi), u(\pi))$, the objective value of the problem, can be written as

$$Z(\pi, d^*(\pi), u(\pi)) = \sum_{i=1}^{m} \sum_{j=1}^{n_{[i]}} \psi_{[i][j]} C_{[i][j]} + v_{[i][j]} u_{[i][j]}. \tag{1}$$

4 The Optimal Schedule

In this section, we will analyze the structure of an optimal schedule for problem $1|GT, DIF, CONV|\sum_{i=1}^{m}\sum_{j=1}^{n_i}(\alpha_{ij}d_{ij} + \beta_{ij}E_{ij} + \gamma_{ij}T_{ij} + v_{ij}u_{ij})$. By Lemma 2, we can restrict our attention to schedules without idle times. Hence, for any given schedule π, the completion time for job $J_{[i][j]}$ can be calculated by the following equation:

$$C_{[i][j]} = \sum_{k=1}^{i-1} P_{[k]} + \sum_{k=1}^{i} s_{[k]} + \sum_{l=1}^{j} p_{[i][l]}.$$

Thus, by Eq. (1), the objective value can be written as

$$\begin{aligned}
&Z(\pi, d^*(\pi), u(\pi)) \\
&= \sum_{i=1}^{m} \sum_{j=1}^{n_{[i]}} \psi_{[i][j]} C_{[i][j]} + v_{[i][j]} u_{[i][j]} \\
&= \sum_{i=1}^{m} \sum_{j=1}^{n_{[i]}} \psi_{[i][j]} \left(\sum_{k=1}^{i-1} P_{[k]} + \sum_{k=1}^{i} s_{[k]} + \sum_{l=1}^{j} p_{[i][l]} \right) + v_{[i][j]} u_{[i][j]} \\
&= \sum_{i=1}^{m} \sum_{j=1}^{n_{[i]}} \psi_{[i][j]} \left(\sum_{k=1}^{i-1} \sum_{l=1}^{n_{[k]}} p_{[k][l]} + \sum_{k=1}^{i} s_{[k]} + \sum_{l=1}^{j} p_{[i][l]} \right) + v_{[i][j]} u_{[i][j]} \\
&= \sum_{i=1}^{m} \Psi_{[i]} \left(\sum_{k=1}^{i} s_{[k]} \right) + \sum_{i=1}^{m} \sum_{j=1}^{n_{[i]}} v_{[i][j]} u_{[i][j]} + \sum_{i=1}^{m} \sum_{j=1}^{n_{[i]}} \psi_{[i][j]} \left(\sum_{l=1}^{j} p_{[i][l]} \right) + \sum_{i=1}^{m} \sum_{j=1}^{n_{[i]}} \psi_{[i][j]} \left(\sum_{k=1}^{i-1} \sum_{l=1}^{n_{[k]}} p_{[k][l]} \right) \\
&= \sum_{i=1}^{m} \Psi_{[i]} \left(\sum_{k=1}^{i} s_{[k]} \right) + \sum_{i=1}^{m} \sum_{j=1}^{n_{[i]}} v_{[i][j]} u_{[i][j]} + \sum_{i=1}^{m} \sum_{j=1}^{n_{[i]}} \left(\sum_{l=j}^{n_{[i]}} \psi_{[i][l]} \right) p_{[i][j]} + \sum_{i=1}^{m} \Psi_{[i]} \left(\sum_{k=1}^{i-1} \sum_{l=1}^{n_{[k]}} p_{[k][l]} \right) \\
&= \sum_{i=1}^{m} \Psi_{[i]} \left(\sum_{k=1}^{i} s_{[k]} \right) + \sum_{i=1}^{m} \sum_{j=1}^{n_{[i]}} v_{[i][j]} u_{[i][j]} + \sum_{i=1}^{m} \sum_{j=1}^{n_{[i]}} \left(\sum_{l=j}^{n_{[i]}} \psi_{[i][l]} \right) p_{[i][j]} + \sum_{i=1}^{m} \left(\sum_{k=i+1}^{m} \Psi_{[k]} \right) \left(\sum_{l=1}^{n_{[i]}} p_{[i][l]} \right) \\
&= \sum_{i=1}^{m} \Psi_{[i]} \left(\sum_{k=1}^{i} s_{[k]} \right) + \sum_{i=1}^{m} \sum_{j=1}^{n_{[i]}} v_{[i][j]} u_{[i][j]} + \sum_{i=1}^{m} \sum_{j=1}^{n_{[i]}} \left(\sum_{k=i+1}^{m} \Psi_{[k]} + \sum_{l=j}^{n_{[i]}} \psi_{[i][l]} \right) p_{[i][j]}.
\end{aligned} \tag{2}$$

Considering the relationship between the processing time and resource allocation and substituting $p_{ij}(u_{ij}) = (\frac{w_{ij}}{u_{ij}})^r$ into the objective function in Eq. (2), we obtain the following expression

$$\begin{aligned}
Z(\pi, d^*(\pi), u(\pi)) = &\sum_{i=1}^{m} \Psi_{[i]} \left(\sum_{k=1}^{i} s_{[k]} \right) + \sum_{i=1}^{m} \sum_{j=1}^{n_{[i]}} v_{[i][j]} u_{[i][j]} \\
&+ \sum_{i=1}^{m} \sum_{j=1}^{n_{[i]}} \left(\sum_{k=i+1}^{m} \Psi_{[k]} + \sum_{l=j}^{n_{[i]}} \psi_{[i][l]} \right) \left(\frac{w_{[i][j]}}{u_{[i][j]}} \right)^r
\end{aligned} \tag{3}$$

Lemma 3. *The optimal resource allocation for problem*

$$1|GT, DIF, CONV|\sum_{i=1}^{m}\sum_{j=1}^{n_i}(\alpha_{ij}d_{ij} + \beta_{ij}E_{ij} + \gamma_{ij}T_{ij} + v_{ij}u_{ij})$$

can be determined by

$$u^*_{[i][j]}(\pi) = \left(\frac{r(\sum_{k=i+1}^{m}\Psi_{[k]} + \sum_{l=j}^{n_{[i]}}\psi_{[i][l]})}{v_{[i][j]}}\right)^{1/(r+1)} \times w_{[i][j]}^{r/(r+1)}. \quad (4)$$

for $i = 1, 2, \cdots, m$, and $j = 1, 2, \cdots, n_i$.

Proof. By taking the derivative of the objective function given by Eq. (3) with respect to $u_{[i][j]}$ for $i = 1, 2, \cdots, m$, and $j = 1, 2, \cdots, n_i$, equating it to zero and solving it for $u^*_{[i][j]}$, we obtain Eq. (4). Since the objective is a convex function, Eq. (4) provides necessary and sufficient conditions for optimality. □

By substituting Eq. (4) into Eq. (3), we obtain the following new expression for the cost function under an optimal resource allocation and due date assignment.

$$Z(\pi, d^*(\pi), u^*(\pi)) = \sum_{i=1}^{m}\Psi_{[i]}\left(\sum_{k=1}^{i}s_{[k]}\right) + (r^{-r/(r+1)} + r^{1/(r+1)})\sum_{i=1}^{m}\sum_{j=1}^{n_{[i]}}\theta_{[i][j]}\left(\sum_{k=i+1}^{m}\Psi_{[k]} + \sum_{l=j}^{n_{[i]}}\psi_{[i][l]}\right)^{1/(r+1)} \quad (5)$$

where $\theta_{ij} = (w_{ij} \times v_{ij})^{r/(r+1)}$, $i = 1, 2, \cdots, m$, and $j = 1, 2, \cdots, n_i$.

Lemma 4. *Let there be two sequences of numbers x_i and y_i. The sum $\sum_i x_i y_i$ of products of the corresponding elements is the least(largest) if the sequences are monotonic in the opposite (same) sense ([19], 1967).*

Lemma 5. *The optimal job sequence within each group G_i ($i = 1, 2, \cdots, m$) is obtained by sequencing jobs in non-decreasing order of θ_{ij}.*

Proof. As shown in Eq. (5), the values of $\sum_{i=1}^{m}\Psi_{[i]}\left(\sum_{k=1}^{i}s_{[k]}\right)$ and $\sum_{k=i+1}^{m}\Psi_{[k]}$ are independent of the internal job sequence within each group. In addition, the value of the term $r^{-r/(r+1)} + r^{1/(r+1)}$ is a constant. Stemming from $\sum_{l=1}^{n_{[i]}}\psi_{[i][l]} \geq \sum_{l=2}^{n_{[i]}}\psi_{[i][l]} \geq \cdots \geq \sum_{l=n_{[i]}}^{n_{[i]}}\psi_{[i][l]}$ and Lemma 4, the optimal job sequence within a group can be obtained by assigning the largest $\sum_{l=1}^{n_{[i]}}\psi_{[i][l]}$ to the smallest $\theta_{[i][1]}$, the second largest $\sum_{l=2}^{n_{[i]}}\psi_{[i][l]}$ to the second smallest $\theta_{[i][2]}$ and so on. Hence the jobs within a group are arranged in non-decreasing order of θ_{ij} in optimal schedule. □

After obtaining the optimal job sequence within each group G_i ($i = 1, 2, \cdots, m$) according to Lemma 5, the original problem is reduced to finding the optimal group sequence to minimize Eq. (5). The complexity of this problem remains an open question. However, in the following we show that if the value of ψ_{ij} for each job J_{ij} for $i = 1, 2, \cdots, m$ and $j = 1, 2, \cdots, n_i$ is identical and $n_i = \overline{n} = n/m$ for $i = 1, 2, \cdots, m$, then the optimal group sequence can be obtained in $O(m \times max(n, m^2))$ time by optimally assigning the groups into the m possible positions. This can be done by solving a linear assignment problem as described in the following lemma.

Lemma 6. *For $n_1 = n_2 = \cdots = n_m = n/m = \overline{n}$ and $\psi_{ij} = \overline{\psi}$ for each job J_{ij}, where $i = 1, 2, \cdots, m$ and $j = 1, 2, \cdots, n_i$, the optimal group sequence can be determined in $O(m \times max(n, m^2))$ time.*

Proof. For the case where $n_1 = n_2 = \cdots = n_m = n/m = \overline{n}$ and $\psi_{ij} = \overline{\psi}$ for each job J_{ij}, where $i = 1, 2, \cdots, m$ and $j = 1, 2, \cdots, n_i$, recall that $\Psi_{[i]} = \sum_{j=1}^{n_{[i]}} \psi_{[i][j]}$, then the value of Ψ_i for each group is identical, i.e. $\Psi_1 = \Psi_2 = \cdots = \Psi_m = \overline{\Psi}$. Therefore, the objective function in Eq. (5) becomes

$$Z(\pi, u^*(\pi), d^*(\pi)) = \overline{\Psi} \sum_{i=1}^{m} (m - i + 1) s_{[i]} + \left(r^{-r/(r+1)} + r^{1/(r+1)}\right) \sum_{i=1}^{m} \sum_{j=1}^{\overline{n}} \theta_{[i][j]} \left((m - i)\overline{\Psi} + (\overline{n} - j + 1)\overline{\psi}\right)^{1/(r+1)} \tag{6}$$

As the optimal job sequence within each group can be predetermined by Lemma 5, then the total cost with optimal job sequence is just dependent on the position of each group in π.

Let t_{il} be the minimal penalty incurred by assigning group G_i to position l in π for $i = 1, 2, \cdots, m$. According to Eq. (6), we obtain that:

$$t_{il} = \overline{\Psi}(m - l + 1) s_i + \left(r^{-r/(r+1)} + r^{1/(r+1)}\right) \sum_{j=1}^{\overline{n}} \theta_{i[j]} \left((m - l)\overline{\Psi} + (\overline{n} - j + 1)\overline{\psi}\right)^{1/(r+1)} \tag{7}$$

where the job sequence within each group G_i for $i = 1, 2, \cdots, m$ can be determined by Lemma 5.

It is straightforward that computing the value of t_{il} requires $O(\overline{n})$ time. Since there are m groups and m positions for each group, computing all the t_{il} values requires $O(\overline{n} m^2) = O(nm)$ time.

Let us now define $x_{il} = 1$ if group G_i is assigned to position l in π and $x_{il} = 0$ otherwise. The sequencing problem of determining the optimal group can be formulated as the following linear assignment problem:

$$
\begin{aligned}
(P1)\min \quad & \sum_{i=1}^{m}\sum_{l=1}^{m} t_{il} \times x_{il} \\
s.t. \quad & \textstyle\sum_{l=1}^{m} x_{il} = 1, i = 1,2,3,\ldots,m. \\
& \textstyle\sum_{i=1}^{m} x_{il} = 1, l = 1,2,3,\ldots,m. \\
& x_{il} = \ 0 \ or \ 1, i, l = 1,2,3,\ldots,m.
\end{aligned}
$$

The first set of constraints in the formulation ensures that each group will be assigned only to one position, the second set ensures that each position will be assigned only once, and the penalty for each assignment under an optimal resource allocation appears in the objective function.

Since a linear assignment problem can be solved in $O(m^3)$ time, the complexity of determining the optimal group sequence for the case where $n_1 = n_2 = \cdots = n_m = n/m = \overline{n}$ and $\psi_{ij} = \overline{\psi}$ for each job J_{ij} for $i = 1, 2, \cdots, m$ and $j = 1, 2, \cdots, n_i$, is $O(nm) + O(m^3) = O(m \times max(n, m^2))$. Since $m = O(n)$, then the overall complexity is $O(n^3)$. □

Now we are ready to present an $O(n^3)$ time optimal algorithm for problem $1|GT, DIF, CONV, n_i = \overline{n}, \psi_{ij} = \overline{\psi}|\sum_{i=1}^{m}\sum_{j=1}^{n_i}(\alpha_{ij}d_{ij}+\beta_{ij}E_{ij}+\gamma_{ij}T_{ij}+v_{ij}u_{ij})$.

Algorithm 1

1: Calculate the values of θ_{ij} and order the jobs within each group G_i $(i = 1, 2, \cdots, m)$ in non-decreasing order of θ_{ij}.
2: Calculate all t_{il} values according to Eq. (7).
3: Determine the optimal group sequence by solving the linear assignment problem $P1$. Let π be the schedule obtained.
4: Determine the optimal resource allocation $u^*(\pi)$ by using Eq. (4) with optimal schedule π.
5: Assign the optimal due dates for jobs according to Lemma 1.

Theorem 1. *Algorithm 1 optimally solves* $1|GT, DIF, CONV, n_i = \overline{n}, \psi_{ij} = \overline{\psi}|\sum_{i=1}^{m}\sum_{j=1}^{n_i}(\alpha_{ij}d_{ij} + \beta_{ij}E_{ij} + \gamma_{ij}T_{ij} + v_{ij}u_{ij})$ *in* $O(n^3)$ *time.*

Proof. The correctness of Algorithm 1 follows from Lemmas 1–6. Calculating the job sequence in Step 1 requires $O(\sum_{i=1}^{m} n_i \log n_i) = O(n \log n)$ time, and Step 2 requires $O(nm)$ time. Solving a linear assignment problem in Step 3 requires $O(m^3)$ time, and determining the optimal resource allocation in Step 4 requires $O(n)$ time. Step 5 requires $O(n)$ time. Thus the overall time complexity of the algorithm is $O(max(nlogn), m \times max(n, m^2))$. Since $m = O(n)$ and $\sum_{i=1}^{m} n_i = n$, the complexity is bounded by $O(n^3)$. □

4.1 Numerical Example

We consider the following numerical example to illustrate the solution procedure of the problem with $n = 9, m = 3, n_i = \overline{n} = 3, \psi_{ij} = \overline{\psi} = 2, r = 1$. The other parameters, such as the cost parameters and set up times are shown in Table 2 to illustrate our algorithm.

Table 2. Parameters of the numerical examples.

Group	J_{ij}	α_{ij}	β_{ij}	γ_{ij}	v_{ij}	w_{ij}	s_i
G_1	J_{11}	2	3	3	2	3	3
	J_{12}	4	5	2	4	2	
	J_{13}	2	2	2	2	6	
G_2	J_{21}	3	4	2	1	5	5
	J_{22}	5	3	2	2	3	
	J_{23}	2	6	4	2	1	
G_3	J_{31}	4	5	2	4	2	6
	J_{32}	2	3	3	1	4	
	J_{33}	2	3	6	3	5	

Step 1. Calculate the value of θ_{ij} according to $\theta_{ij} = (w_{ij} \times v_{ij})^{r/(r+1)}$, then the results can be shown as follows.

$i, j(\theta_{ij})$	1	2	3
1	$\sqrt{6}$	$\sqrt{8}$	$\sqrt{12}$
2	$\sqrt{5}$	$\sqrt{6}$	$\sqrt{2}$
3	$\sqrt{8}$	$\sqrt{4}$	$\sqrt{15}$

According to Lemma 5, we can get the optimal job sequence within each group is $G_1 : J_{11} \rightarrow J_{12} \rightarrow J_{13}$, $G_2 : J_{23} \rightarrow J_{21} \rightarrow J_{22}$, $G_3 : J_{32} \rightarrow J_{31} \rightarrow J_{33}$.

Step 2. The values of all t_{il} can be calculated as follows:

$i, l(t_{il})$	1	2	3
1	123.335	90.455	51.112
2	138.219	97.797	52.801
3	176.581	125.654	68.066

Step 3. By solving the linear assignment problem $P1$, we can determine the optimal group sequence is $\{G_1, G_2, G_3\}$. Then $\pi = \{J_{11}, J_{12}, J_{13}, J_{23}, J_{21}, J_{22}, J_{32}, J_{31}, J_{33}\}$ is the schedule obtained.

Step 4. Determine the optimal resource allocation $u^*(\pi)$ by using Eq. (4) with optimal schedule π.

Step 5. Assign the optimal due dates for jobs according to Lemma 1.

$[i],[j](u^*_{[i][j]}(\pi))$	1	2	3
1	5.196	2.828	6.481
2	2.449	7.071	3.464
3	4.899	1.414	1.826

$[i],[j](d^*_{[i][j]}(\pi))$	1	2	3
1	3.577	0	5.210
2	10.618	0	0
3	21.007	0	25.160

5 Conclusion

In this paper, we study a single machine group scheduling problem with convex resource allocation, in which the scheduler determines optimal due dates for different jobs under a group technology environment. The due date assignment is without restriction, that is, it is allowed to assign different due dates to jobs within one group. The actual processing time of each job depends on the resource allocated to the job. The objective is to determine the optimal group sequence and job sequence within each group, together with the optimal due date assignment strategy and resource allocation, to minimize an objective function which includes earliness, tardiness, due date assignment and resource allocation costs. We determine the optimal job sequence within each group and show that it is independent of the group sequence. However, the problem of determining the optimal group sequence remains an open question. We present structural results that characterize the optimal schedule in the case where the number of jobs in each group is identical and the cost ψ_{ij} (the minimum of the due date assignment cost and the tardiness cost) for each job J_{ij} is the same, and present an $O(n \log n)$ time algorithm to solve this problem optimally.

Acknowledgements. This work was partially supported by the National Natural Science Foundation of China under Grant No. 11771346.

References

1. Liu, L., Xu, Y., Yin, N., Wang, J.: Single machine group scheduling problem with deteriorating jobs and release dates. Appl. Mech. Mater. **513–517**, 2145–2148 (2014)
2. Keshavarz, T., Savelsbergh, M., Salmasi, N.: A branch-and-bound algorithm for the single machine sequence-dependent group scheduling problem with earliness and tardiness penalties. Appl. Math. Model. **39**(20), 6410–6424 (2015)
3. Qin, H., Zhang, Z., Bai, D.: Permutation flowshop group scheduling with position-based learning effect. Comput. Ind. Eng. **92**, 1–15 (2016)
4. Wang, L., Liu, M., Wang, J., Lu, Y., Liu, W.: Optimization for Due-Date Assignment Single-Machine Scheduling under Group Technology. Complexity, vol. 2021 (2021). https://doi.org/10.1155/2021/6656261
5. Seidmann, A., Panwalkar, S.S., Smith, M.L.: Optimal assignment of due-dates for a single processor scheduling problem. Int. J. Prod. Res. **19**(4), 393–399 (1981)

6. Panwalkar, S.S., Smith, M.L., Seidmann, A.: Common due date assignment to minimize total penalty for the one machine scheduling problem. Oper. Res. **30**(2), 391–399 (1982)
7. Shabtay, D.: Optimal restricted due date assignment in scheduling. Eur. J. Oper. Res. **252**(1), 79–89 (2016)
8. Li, S., Ng, C.T., Yuan, J.: Group scheduling and due date assignment on a single machine. Int. J. Prod. Econ. **130**(2), 230–235 (2011)
9. Bajwa, N., Melouk, S., Bryant, P.: A hybrid heuristic approach to minimize number of tardy jobs in group technology systems. Int. Trans. Oper. Res. **26**(5), 1847–1867 (2019)
10. Li, W.-X., Zhao, C.-L.: Single machine scheduling problem with multiple due windows assignment in a group technology. J. Appl. Math. Comput., 477–494 (2014). https://doi.org/10.1007/s12190-014-0814-1
11. Ji, M., Zhang, X., Tang, X., Cheng, T.C.E., Wei, G., Tan, Y.: Group scheduling with group-dependent multiple due windows assignment. Int. J. Prod. Res. **54**(4), 1244–1256 (2016)
12. Shabtay, D., Itskovich, Y., Yedidsion, L., Oron, D.: Optimal due date assignment and resource allocation in a group technology scheduling environment. Comput. Oper. Res. **37**(12), 2218–2228 (2010)
13. Lv, D., Luo, S., Xue, J., Xu, J., Wang, J.: A note on single machine common flow allowance group scheduling with learning effect and resource allocation. Computers & Industrial Engineering, vol. 151 (2021). https://doi.org/10.1016/j.cie.2020.106941
14. Yin, N., Kang, L., Sun, T., Wang, X.: Unrelated parallel machines scheduling with deteriorating jobs and resource dependent processing times. Appl. Math. Model. **38**(19–20), 4747–4755 (2014)
15. Yin, N., Kang, L., Wang, X.: Single machine group scheduling with processing times dependent on position, starting time and allotted resource. Appl. Math. Model. **38**(19–20), 4602–4613 (2014)
16. Pei, J., Liu, X., Liao, B., Panos, P.M., Kong, M.: Single machine scheduling with learning effect and resource-dependent processing times in the serial-batching production. Appl. Math. Model. **58**, 245–253 (2018)
17. Graham, R.L., Lawler, E.L., Lenstra, J.K., et al.: Optimization and approximation in deterministic sequencing and scheduling: a survey. Ann. Discrete Math. **5**, 287–326 (1979)
18. Chen, Y., Cheng, Y.: Group scheduling and due date assignment without restriction on a single machine. Adv. Prod. Manag. Syst. **632**, 250–257 (2021)
19. Hardy, G.H., Littlewood, J.E., Polya, G.: Inequalities. Cambridge University Press, London (1967)

Constrained Stable Marriage with Free Edges or Few Blocking Pairs

Yinghui Wen(✉) and Jiong Guo(✉)

School of Computer Science and Technology, Shandong University, Qingdao, China
yhwen@mail.sdu.edu.cn, jguo@sdu.edu.cn

Abstract. Given two disjoint sets U and W, where the members (also called agents) of U and W are called men and women, respectively, and each agent is associated with an ordered preference list that ranks a subset of the agents from the opposite gender, a stable matching is a set of pairwise disjoint woman-man pairs admitting no blocking pairs. A blocking pair refers to a woman and a man, who prefer each other to their partners in the matching. Gale and Shapley proved that a stable matching exists for every instance. Since then, a lot of stable matching variants have been introduced. For instance, the π-stable marriage problem asks for a stable matching satisfying a given constraint π. Unlike in the unconstrained case, a given instance may not admit a π-stable matching and one has to accept a semi-stable matching satisfying π, for instance, a matching satisfying π and admitting few blocking pairs. In this paper, we study two such problems, namely, π-Stable Marriage with Free Edges (π-SMFE) and π-Stable Marriage with t-Blocking Pairs (π-SMtBP). π-SMFE seeks for a matching M satisfying π and the condition that all blocking pairs occurring in M are from a given set F of woman-man pairs, while the solution matchings of π-SMtBP need to satisfy π and admit at most t blocking pairs. We examine four constraints, Regret, Egalitarian, Forced, and Forbidden, and prove that both π-SMFE and π-SMtBP are NP-hard for all four constraints even with complete preference lists. Concerning parameterized complexity, we establish a series of fixed-parameter tractable and intractable results for π-SMFE and π-SMtBP with respect to some structural parameters such as the number of agents and the number of free edges/blocking pairs.

Keywords: Social choice · Stable matching · Computational complexity · Parameterized complexity

1 Introduction

Matching problems have received a considerable amount of attention from both economics and computer science communities and have been studied for several

Supported by NSFC 61772314, 61761136017 and 62072275.

D.-Z. Du et al. (Eds.): COCOA 2021, LNCS 13135, pp. 468–483, 2021.
https://doi.org/10.1007/978-3-030-92681-6_37

decades, due to their rich applications in the real world, for instance, assignment of students to colleges [1], kidney patients to donors [19], refugees to host countries [3].

One of the most prominent matching models is the STABLE MARRIAGE problem, which was introduced by Gale and Shapley [12]. Given two disjoint sets U and W, we denote the members of U as *men* and the members of W as *women*. Each member $u \in U$ (resp. $w \in W$) is associated with an ordered preference list that ranks a subset of the members in W (resp. U), called the *preference list* of u (resp. w) and denoted as $\succ_u$ (resp. $\succ_w$). A *matching* M is a one-to-one assignment of $u \in U$ to $w \in W$. We call a woman $w \in W$ is the *partner* of a man $u \in U$ if M matches them together, denoted as $M(u)$, and vice versa. The STABLE MARRIAGE problem seeks for a matching M which is *stable*, that is, there is no *blocking pair* in M. Herein, a blocking pair is a pair of two agents $u \in U$ and $w \in W$ such that u and w are not matched by M but u prefers w to $M(u)$ in $\succ_u$ and w prefers u to $M(w)$ in $\succ_w$. The preference list of an agent $a \in U \cup W$ could be *incomplete*, meaning that some agents in the opposite gender are not acceptable for a, or could contain *ties*, meaning that two agents are considered to be equally good as a partner of a.

A lot of variants of STABLE MARRIAGE have been introduced, which seek for a stable matching satisfying some constraints. Some try to find a stable matching satisfying a score bound, such as Egalitarian [15], Regret [14], Balanced [11], and Sex-equal [16], etc. Some variants focus on finding a matching with restricted edges, such as Forced [7–9], Forbidden [7–9], and Distinguished [20]. Let π denote a constraint. We say a stable matching M is *π-stable* if M satisfies π. When π being Egalitarian/Regret/Forced/Forbidden, it is polynomial-time to decide the existence of a π-stable matching [9,14,15]. Note that, unlike the unconstrained case, π-stable matching may not exist for some instances. In this case, one has to accept a matching close to being stable, for instance, a matching containing a few blocking pairs. This line of research mainly consists of two directions. The first direction asks, given a subset of agent pairs, denoted as F, whether there is a matching M such the set of blocking pairs of M is a subset of F. Each element in F is called a *free edge* of the instance. We name this setting as the *free edges* setting. Cechlárová et al. [6] investigated STABLE ROOMMATES WITH FREE EDGES and provided that finding a stable roommate matching with free edges is NP-hard, where STABLE ROOMMATES is a well-known generalization of STABLE MARRIAGE. Cseh and Heeger [7] studied STRONGLY/SUPER STABLE MARRIAGE WITH FREE EDGES, where strong stable and super stable are two well-known extensive definitions of stable, and derived a similar classical computational complexity result as Cechlárová et al. The second direction decides, given an integer t, whether there is a matching M such that the number of blocking pairs in M is at most t. In this paper, we name this setting as the *t-blocking pairs* setting.[1] Abraham et al. [2] investigated STABLE ROOMMATES WITH t-BLOCKING PAIRS. Biró

[1] In most papers, this setting is named as "almost stable" matchings. Considering a matching in a free edge setting can also be seen as a matching which is "almost stable", we rename this setting as t-blocking pairs.

et al. [4] extended the work of [2] to incomplete preference lists. They found out that even with $t = 3$, STABLE ROOMMATES WITH t-BLOCKING PAIRS becomes NP-hard. Biró et al. [5] introduced t-blocking pairs to STABLE MARRIAGE WITH INCOMPLETE PREFERENCE LISTS and achieved a similar classical computational complexity result. Cseh and Manlove [8] combined t-blocking pairs with STABLE MARRIAGE AND ROOMMATES WITH RESTRICTED EDGES and showed that when forbidden edges are allowed, it is NP-hard to decide the existence of stable matching with t blocking pairs. Manlove et al. [18] studied the combination of t-blocking pairs and HOSPITALS/RESIDENTS PROBLEM WITH COUPLES and proved that find such a matching is NP-hard.

We study the combination of π-STABLE MARRIAGE and free edges/t-blocking pairs. That is, given two sets of agents U and W with each agent having a preference list and a pair set F (for π-STABLE MARRIAGE WITH FREE EDGES (π-SMFE)) or an integer t (for π-STABLE MARRIAGE WITH t-BLOCKING PAIRS (π-SMtBP)), we seek for a matching satisfying π and the condition that the set of blocking pairs is a subset of F (for π-SMFE) or the number of blocking pairs is at most t (for π-SMtBP). There are four constraints studied in this paper, namely, Egalitarian (Egal), Regret (Reg), Forced, and Forbidden. We will formally define the constraints in Sect. 2. We use π-STABLE MARRIAGE WITH TIES (π-SMT) to name the variant of STABLE MARRIAGE, where ties are allowed in the preference lists and the target is to find a π-stable matching. We first show an interesting connection between π-SMT and π-SMFE (resp. π-SMtBP). That is, each instance of π-SMT can be reduced in polynomial time to an equivalent π-SMFE (resp. π-SMtBP) instance without ties. Thus, we can conclude that π-SMFE (resp. π-SMtBP) is NP-hard with π being Egal/Reg/Forced/Forbidden, since their corresponding π-SMT problems are NP-hard[2]. Then we investigate the parameterized complexity of π-SMFE and π-SMtBP and achieve the following results. First, we find problem kernels for Reg-SMFE, Reg-SMtBP and Forced-SMFE, which implies that the three problems are *fixed-parameter tractable (FPT)* with respect to k, where k is the cardinality of a maximum matching of the bipartite graph constructed from the matching instance. Second, we design an FPT algorithm for π-SMFE and an XP algorithm for π-SMtBP with respect to t with t being the number of blocking pairs or the size of the free edge set F. Third, we prove that Egal-SMtBP and Forbidden-SMtBP are W[1]-hard with respect to $n - t$ with n being the number of women (or men). Fourth, we consider the case with short preference lists and prove that even with l being a constant, π-SMFE and π-SMtBP remain NP-hard for π being Egal/Reg/Forced/Forbidden, where l equals the maximum length of all preference lists. Refer to Table 1 for an overview of parameterized complexity results. Due to lack of space, the proofs of the theorems marked with (*) are moved to Appendix.

[2] The NP-hardness of π-SMT with $\pi \in \{$Reg, Egal, Forced$\}$ has been proved by Manlove et al. [17], and the NP-hardness of SMT-Forbidden is proved in Theorem 1.

Table 1. Parameterized complexity results. The FPT results with respect to n are trivial. "Para-NP-hard" stands for NP-hardness even with the corresponding parameter being a constant.

	t	k	n	$n-t$	l
π-SMFE	FPT (Theorem 5)	FPT (Reg/Forced (Theorem 4))	FPT	?	Para-NP-h (Reg/Forced (Theorem 9), Egal/Forbidden (Theorem 10))
π-SMtBP	XP (Theorem 6)	FPT (Reg (Theorem 4))	FPT	W[1]-hard (Egal (Theorem 8), Forbidden (Theorem 7))	Para-NP-h (Reg/Egal/Forced (Theorem 11)))

2 Preliminaries

Let $U = \{u_1, \cdots, u_n\}$ and $W = \{w_1, \cdots, w_n\}$ be two n-elements disjoint sets of agents. We call the members in U *men*, and the members in W *women*. The *preference list* of $u \in U$ is an ordered subset that ranks a subset of the members in W, denoted as $\succ_u$. If the length of $\succ_u$ is less than n, we say $\succ_u$ is *incomplete*. If there are two women w_i and w_j are considered equally good as a partner of u, we say $\succ_u$ contains *tie* and use $w_i \sim w_j$ to denote the relation of w_i and w_j in $\succ_u$. For instance, if there are three women w_1, w_2, w_3, and a man $u \in U$ prefers w_1 to w_2 and w_3, and considers w_2 as good as w_3, then the preference list of u is defined as $\succ_u$: $w_1 \succ w_2 \sim w_3$. The preference list $\succ_w$ of $w \in W$ is defined analogously. A *matching* $M \subseteq \{\{u, w\} | u \in U \wedge w \in W\}$ is a set of pairwise disjoint pairs with u in w's preference list and vice versa. We say M is a *perfect matching* if $|M| = n$. If $\{u, w\} \in M$, we say that w is the *partner* of u matched by M, denoted as $M(u)$, and vice versa. If u has no partner, we say $M(u) = \emptyset$. Given an agent a, $P_a(b)$ stands for the position of b in $\succ_a$ with b being an agent from the opposite gender.[3] Given a matching M, we define "the score of a" as $P_a(M(a))$, and define "the π-score of a set $A \subseteq U \cup W$" as $\sum_{a \in A} P_a(M(a))$ for π = Egal, or $\max_{a \in A} P_a(M(a))$ for π = Reg. A *preference profile* L is the set of all preference lists. We say L contains ties if at least one preference list in L contains ties and L is incomplete if at least one preference list in L is incomplete. A matching M is *stable* if M does not contain *blocking pairs*; a blocking pair is a pair $\{u, w\} \notin M$ such that u prefers w to $M(u)$ and w prefers u to $M(w)$. Given a set B, $\overrightarrow{B}$ denotes an arbitrary but fixed ordering of the elements in B. Given two agents a and c from the same gender with $\succ_a$: $b \succ (\succ_c) \succ d$ and $\succ_c$: $f \succ h$, the notation $(\succ_c)$ in $\succ_a$ means that the agents between b and d in $\succ_a$ are the same agents in $\succ_c$ and have the same order as in $\succ_c$. That is, $\succ_a$: $b \succ f \succ h \succ d$.

Constraints and Problems. There are four constraints studied in this paper, namely, Reg, Egal, Forced, and Forbidden. Each constrained stable marriage

[3] Note that the positions of the agents connected by a tie are set equal to the minimum position of the agents in this tie. For instance, in $\succ_u$: $w_1 \succ w_2 \sim w_3$, the positions of w_2 and w_3 are 2. If an agent has no partner, the position equals to $n + 1$.

Table 2. The four constrains studied in this paper.

π	π_i	π_r
Reg	An integer d (Reg-score bound)	$\max_{a \in U \cup W} P_a(M(a)) \leq d$
Egal	An integer d (Egal-score bound)	$\sum_{a \in U \cup W} P_a(M(a)) \leq d$
Forced	A set $S \subseteq U \times W$ (Forced pairs set)	$S \subseteq M$
Forbidden	A set $S \subseteq U \times W$ (Forbidden pairs set)	$M \cap S = \emptyset$

problem with a constraint π has an additional input π_i, and an additional requirement π_r for the solution matching M. We define them in Table 2.

Given a constraint π, a matching M is π-stable if M is a stable matching and satisfies π, that is, M meets the additional requirement π_r with taking π_i as an additional input. The π-Stable Marriage (π-SM) problem asks for a π-stable matching, given two sets U and W, a profile L, and π_i. The π-Stable Marriage with Ties (π-SMT) problem represents the variant of π-SM, where ties occur in the preference lists. Note that π-stable matchings do not exist for some instances. Thus, we focus on two "relaxed" version of π-SM. Let $\pi \in$ {Egal, Reg, Forced, Forbidden}.

> **π-Stable Marriage with Free Edges** *(* **π-SMFE** *)*
> **Input:** Two sets of agents U and W of n agents each, a preference profile L without ties, a set of pairs $F \subseteq U \times W$, and π_i.
> **Question:** Is there a matching M satisfying π_r and the condition that the set of blocking pairs in M is a subset of F?

> **π-Stable Marriage with t-Blocking Pairs** *(* SMtBP *)*
> **Input:** Two sets of agents U and W of n agents each, a preference profile L without ties, an integer $t \geq 0$, and π_i.
> **Question:** Is there a matching M satisfying π_r and the condition that the number of blocking pairs in M is at most t?

Note that, we also consider incomplete preference lists. Each instance of π-SMFE or π-SMtBP can be seen as a bipartite graph G. That is, each vertex denotes an agent. G has an edge $e = \{v_u, v_w\}$ if v_u and v_w are the vertices corresponding to $u \in U$ and $w \in W$, respectively, and u and w appear in each other's preference list. Each neighbor of v_u (resp. v_w) has an index with respect to v_u (resp. v_w), which is set equal to the position of the corresponding woman (resp. man) in $\succ_u$ (resp. $\succ_w$).

Parameterized Complexity. Parameterized complexity provides a refined exploration of the connection between problem complexity and various problem-specific parameters. A parameterized problem is *fixed-parameter tractable (FPT)* with respect to a parameter k, if there is an $O(f(k) \cdot |I|^{O(1)})$-time algorithm solving the problem, where I denotes the whole input instance and f can be any computable function. Parameterized problems can be classified into many classes with W[1] and W[2] being the basic fixed-parameter intractability classes. For more details on parameterized complexity, we refer to [10,21]. We study the parameterized complexity of π-SMFE and π-SMtBP, and consider the following parameters: t, k, $n = |U| = |W|$, $n - t$, and l, where t is the number of free edges in F or the number of blocking pairs, k is the cardinality of the maximum matching of the bipartite graph constructed from the instance, and l is the maximum length of all preference lists.

3 Classical Complexity

In this section, we prove the classical complexity of π-SMFE and π-SMtBP with $\pi \in$ {Reg, Egal, Forced, Forbidden}. To do this, we reduce π-Stable Marriage with Ties (π-SMT) to π-SMFE and π-SMtBP and show π-SMT with $\pi \in$ {Reg, Egal, Forced, Forbidden} is NP-hard even with all preference lists being complete. We first show the NP-hardness of Forbidden-SMT. The NP-hardness of π-SMT with $\pi \in$ {Reg, Egal, Forced} has been proved by Manlove et al. [17].

Theorem 1. *(*) Forbidden-SMT is NP-hard even with all preference lists being complete.*

3.1 π-SMFE

We introduce some notations which will be used in the following reduction. Given a preference list $\succ_a$ with $a \in U \cup W$, a *tie-set* in $\succ_a$ is a set of agents who occur in a tie in $\succ_a$. A preference list can have several tie-sets. The *tie-length* of $\succ_a$ is defined as the maximum size of all tie-sets in $\succ_a$. For instance, let $\succ_a$ be $b_1 \sim b_2 \sim b_3 \succ b_4 \sim b_5$. There are two tie-sets T_a^1 and T_a^2 in $\succ_a$ with $T_a^1 = \{b_1, b_2, b_3\}$ and $T_a^2 = \{b_4, b_5\}$. The tie-length of $\succ_a$ is 3, the size of T_a^1. Note that a preference list with strict order has no tie-set.

Theorem 2. *π-SMFE with $\pi \in$ {Reg, Egal, Forced, Forbidden} is NP-hard, even with all preference lists being complete.*

Proof. Here, we only prove the NP-hardness of Forced-SMFE; the proofs of other cases are in Appendix. For a given Forced-SMT instance (U, W, L, S), we assume that the preference lists of all women in W contain no tie. For the case that the preference lists of both genders contain ties, we can "eliminate" the ties in the preference lists of U as shown in the following, and then apply the same process to the women side. We construct the Forced-SMFE instance in two steps. First, we create an instance (U', W', L', F, S') with incomplete preference lists, and then, transform the incomplete lists to complete lists, resulting in a new Forced-SMFE instance (U'', W'', L'', F, S''). Given a preference list $\succ_u$ in L with $u \in U$, let $T_u^1, \cdots, T_u^m$ be the tie-sets in $\succ_u$ and d^* be the tie-length of $\succ_u$. Give an arbitrary ordering $\overrightarrow{T_u^j}$ to the elements in T_u^j and let $T_u^j[i]$ be the woman at the i-th position in $\overrightarrow{T_u^j}$. We first create two auxiliary agent sets

$P = \{p_1, \cdots, p_{(m-1)d^*}\}$ and $Q = \{q_1, \cdots, q_{(m-1)d^*}\}$. If the size of a tie-set T_u^j is less than d^*, we add q_x to T_u^j with $[(j-1)d^* + |T_u^j|] < x \leq jd^*$. Thus, all T_u^j's have the same size d^*. Create d^*-1 men $X_u = \{x_u^1, \cdots, x_u^{d^*-1}\}$ and d^*-1 women $Y_u = \{y_u^1, \cdots, y_u^{d^*-1}\}$ for each $u \in U$. Thus, $U' = U \cup (\bigcup_{u \in U} X_u) \cup P$ and $W' = W \cup (\bigcup_{u \in U} Y_u) \cup Q$.

Next, we set the preference lists, where $\succ_{p_i}$ and $\succ_{q_i}$ with $1 \leq i \leq (m-1)d^*$ have the following form:

$$\succ_{p_i}: q_i \succ \overrightarrow{Q \setminus \{q_i\}} \succ \overrightarrow{W' \setminus Q} \qquad \succ_{q_i}: p_i \succ \overrightarrow{P \setminus \{p_i\}} \succ \overrightarrow{U' \setminus P}.$$

We construct the preference lists for $\{u\} \cup X_u$ from $\succ_u$: $(\succ_u^1) \succ T_u^1 \succ (\succ_u^2) \succ T_u^2 \succ \cdots \succ T_u^m \succ (\succ_u^{m+1})$, where we abuse T_u^j to denote the ties in $\succ_u$ and $(\succ_u^j)$'s denote the suborders of $\succ_u$ with no tie. Let $\overrightarrow{T_u^j \setminus B}$ be the suborder of $\overrightarrow{T_u^j}$ resulting by deleting the elements in $B \subseteq T_u^j$ from $\overrightarrow{T_u^j}$. The preference lists of $\{u\} \cup X_u$ in L' are set as follows:

$$\succ'_u: \boxed{y_u^1} \succ \boxed{y_u^2} \succ \boxed{y_u^3} \succ \boxed{\cdots} \succ \boxed{y_u^{d^*-1}} \succ (\succ_u^1) \succ T_u^1[1] \succ \overrightarrow{T_u^1 \setminus \{T_u^1[1]\}} \succ$$
$$(\succ_u^2) \succ T_u^2[1] \succ \overrightarrow{T_u^2 \setminus \{T_u^2[1]\}} \succ \cdots \succ T_u^m[1] \succ \overrightarrow{T_u^m \setminus \{T_u^m[1]\}} \succ (\succ_u^{m+1})$$
$$\succ_{x_u^1}: y_u^1 \succ \boxed{y_u^2} \succ \boxed{y_u^3} \succ \boxed{\cdots} \succ \boxed{y_u^{d^*-1}} \succ (\succ_u^1) \succ T_u^1[2] \succ \overrightarrow{T_u^1 \setminus \{T_u^1[2]\}} \succ$$
$$(\succ_u^2) \succ T_u^2[2] \succ \overrightarrow{T_u^2 \setminus \{T_u^2[2]\}} \succ \cdots \succ T_u^m[2] \succ \overrightarrow{T_u^m \setminus \{T_u^m[2]\}} \succ (\succ_u^{m+1})$$
$$\succ_{x_u^2}: y_u^1 \succ y_u^2 \succ \boxed{y_u^3} \succ \boxed{\cdots} \succ \boxed{y_u^{d^*-1}} \succ (\succ_u^1) \succ T_u^1[3] \succ \overrightarrow{T_u^1 \setminus \{T_u^1[3]\}} \succ$$
$$(\succ_u^2) \succ T_u^2[3] \succ \overrightarrow{T_u^2 \setminus \{T_u^2[3]\}} \succ \cdots \succ T_u^m[3] \succ \overrightarrow{T_u^m \setminus \{T_u^m[3]\}} \succ (\succ_u^{m+1})$$
$$\vdots$$
$$\succ_{x_u^{d^*-1}}: y_u^1 \succ y_u^2 \succ y_u^3 \succ \cdots \succ y_u^{d^*-1} \succ (\succ_u^1) \succ T_u^1[d^*] \succ \overrightarrow{T_u^1 \setminus \{T_u^1[d^*]\}} \succ$$
$$(\succ_u^2) \succ T_u^2[d^*] \succ \overrightarrow{T_u^2 \setminus \{T_u^2[d^*]\}} \succ \cdots \succ T_u^m[d^*] \succ \overrightarrow{T_u^m \setminus \{T_u^m[d^*]\}} \succ (\succ_u^{m+1}).$$

The $\boxed{box}$ in a preference list means that there are free edges formed by the agents inside the $\boxed{box}$ and the agent to whom the preference list is constructed. For instance, we add the free edges $\{u, y_u^i\}$ with $1 \leq i \leq d^*-1$ to F according to $\succ'_u$. Then we set the preference lists of the agents in Y_u as below:

$$\succ_{y_u^1}: \boxed{u} \succ x_u^1 \succ x_u^2 \succ \cdots \succ x_u^{d^*-1}$$
$$\succ_{y_u^2}: \boxed{u} \succ \boxed{x_u^1} \succ x_u^2 \succ \cdots \succ x_u^{d^*-1}$$
$$\vdots$$
$$\succ_{y_u^{d^*-1}}: \boxed{u} \succ \boxed{x_u^1} \succ \boxed{x_u^2} \succ \boxed{\cdots} \succ x_u^{d^*-1}.$$

Next, we construct the preference lists in L' for $w \in W$. Suppose w occurs in the tie-sets $T_{u_1}^{j_1}, \cdots, T_{u_b}^{j_b}$ of $u_1, \cdots, u_b \in U$ with $w = T_{u_r}^{j_r}[i_r]$ and $\succ_w \in L$ has the form: $(\succ_w^1) \succ u_1 \succ (\succ_w^2) \succ u_2 \succ \cdots \succ u_b \succ (\succ_w^{b+1})$. Let $z_{u_r} = u_r$ if $w = T_{u_r}^{j_r}[1]$; otherwise, $z_{u_r} = x_{u_r}^{i_r - 1}$. Define $Z_{u_r} = (\{u_r\} \cup X_{u_r}) \setminus \{z_{u_r}\}$. Then, the preference list of w in L' has the following form: $(\succ_w^1) \succ z_{u_1} \succ (\succ_w^2) \succ z_{u_2} \succ \cdots \succ z_{u_b} \succ (\succ_w^{b+1}) \succ \overrightarrow{Z_{u_1}} \succ \overrightarrow{Z_{u_2}} \succ \cdots \succ \overrightarrow{Z_{u_b}}$. Set the forced pair set S' as follows. For each $\{u, w\} \in S$, if $w = T_u^j[i]$ with $1 \leq j \leq m$, let $\{x_u^i, w\} \in S'$ with $x_u^0 = u$; otherwise, let $\{u, w\} \in S'$.

Note that the preference lists of the agents in $(U' \setminus P) \cup (W' \setminus Q)$ are incomplete. Now, we enter the second step, transforming the lists to complete lists. We create one pair of auxiliary agents p^* and q^*. Set $U'' = U' \cup \{p^*\}$ and $W'' = W' \cup \{q^*\}$. Set the preference list of p^* in L'' as $\succ_{p^*}: \overrightarrow{W'} \succ q^*$ and set the preference list of q^* in L'' analogously. Then set the preference list of $u \in U'$ in L'' as $\succ_u'': (\succ_u') \succ q^* \succ \overrightarrow{\bar{W}_u'}$ with $\bar{W}_u'$ denoting the set of women who are not in $\succ_u'$. Set the preference list of $w \in W'$ in L'' analogously. Define a forced pairs set as $S^+ = \{\{p^*, q^*\}\}$. Obviously, $u \in U'$ cannot be matched to any $w \in \bar{W}_u'$, since, otherwise, $\{u, q^*\}$ forms a blocking pair. Set the forced set as $S'' = S' \cup S^+$. Now, the construction of the new instance is complete. The reduction is clearly doable in polynomial time.

To better illustration, we show a concrete example. Suppose there is an agent $u \in U$ with $\succ_u: w_1 \sim w_2 \sim w_3 \succ w_4 \succ w_5 \sim w_6$. The forced set is $S = \{\{u, w_2\}\}$. There are two tie-sets T_u^1 and T_u^2 in $\succ_u$ and the tie-length of $\succ_u$ is 3. We create two auxiliary set P and Q with $|P| = |Q| = 3$. Thus, we have $T_u^1[1] = w_1$, $T_u^1[2] = w_2$, $T_u^1[3] = w_3$, $T_u^2[1] = w_5$, $T_u^2[2] = w_6$, $T_u^2[3] = q_3$. We create X_u and Y_u with $|X_u| = |Y_u| = 2$. Then we set the preference list of u, X_u, Y_u, and W as below. Herein, the "$\cdots$" denotes that this part of preference is set as the original preference list and $S' = \{\{x_u^1, w_2\}\}$:

$$\succ_u': \boxed{y_u^1} \succ \boxed{y_u^2} \succ \boldsymbol{w_1} \succ w_2 \succ w_3 \succ w_4 \succ \boldsymbol{w_5} \succ w_6 \succ q_3 \succ \overrightarrow{Q \setminus \{q_3\}}$$

$$\succ_{x_u^1}: y_u^1 \succ \boxed{y_u^2} \succ \boldsymbol{w_2} \succ w_1 \succ w_3 \succ w_4 \succ \boldsymbol{w_6} \succ w_5 \succ q_3 \succ \overrightarrow{Q \setminus \{q_3\}}$$

$$\succ_{x_u^2}: y_u^1 \succ y_u^2 \succ \boldsymbol{w_3} \succ w_1 \succ w_2 \succ w_4 \succ \boldsymbol{q_3} \succ w_5 \succ w_6 \succ \overrightarrow{Q \setminus \{q_3\}}$$

$$\succ_{y_u^1}: \boxed{u} \succ x_u^1 \succ x_u^2 \succ \overrightarrow{U' \setminus X_u \cup \{u\}}$$

$$\succ_{y_u^2}: \boxed{u} \succ \boxed{x_u^1} \succ x_u^2 \succ \overrightarrow{U' \setminus X_u \cup \{u\}}$$

$$\succ_{w_1}': \cdots \succ \boldsymbol{u} \succ x_u^1 \succ x_u^2 \succ \cdots$$

$$\succ_{w_2}': \cdots \succ \boldsymbol{x_u^1} \succ u \succ x_u^2 \succ \cdots$$

$$\succ_{w_3}': \cdots \succ \boldsymbol{x_u^2} \succ u \succ x_u^1 \succ \cdots$$

$$\succ_{w_4}': \cdots \succ \boldsymbol{u} \succ x_u^1 \succ x_u^2 \succ \cdots$$

$$\succ_{w_5}': \cdots \succ \boldsymbol{u} \succ x_u^1 \succ x_u^2 \succ \cdots$$

$$\succ_{w_6}': \cdots \succ \boldsymbol{x_u^1} \succ u \succ x_u^2 \succ \cdots .$$

To show the equivalence of the instances, we prove several properties of the constructed instance.

First, $p_i \in P$ can only be matched to $q_i \in Q$, or the solution is not a stable matching. Since each p_i prefers q_i to all other women and each q_i prefers p_i to all other men, if we do not match them together, $\{p_i, q_i\}$ will form a blocking pair.

Second, M' must be a perfect matching. Suppose not, there is a pair of agents having no partner, denoted as $\{a_i, b_j\}$. $\{a_i, b_j\}$ must be connected by a free edge, that is, $a_i \in \{u\} \cup X_u \setminus \{x_u^{d^*-1}\}$ and $b_j \in Y_u$. Since a_j has no partner, then either no woman $w \in W$ prefers a_i to her partner, or $\{a_i, w\}$ has a free edge. b_j meets the same situation. Thus, we can add $\{a_i, b_j\}$ to M' safely. Then, $y \in Y_u$ can only be matched to u or $x \in X_u$, since, otherwise, $\{p^*, y\}$ forms a blocking pair.

Third, for $1 \leq i \leq d^* - 1$, x_u^i can only be matched to $T_u^j[i+1]$ with $1 \leq j \leq m$ if he is not the partner of $y \in Y_u$. Suppose not, there exists a pair $\{x_u^a, T_u^j[b]\}$ with $a \neq b - 1$. We remove all the pairs associated with X_u, u, and Y_u from the solution M', and add the following pairs to M' with $x_u^0 = u$:

- Let $\{x_u^{i-1}, y_u^i\} \in M'$ with $1 \leq i < b$.
- Let $\{x_u^{i-1}, T_u^j[b]\} \in M'$ with $i = b$.
- Let $\{x_u^i, y_u^i\} \in M'$ with $b < i \leq d^* - 1$.

After the modification, M' is still stable. Since each x_u^i prefers y_u^{i+1} to y_u^{i+2} with $0 \leq i < d^* - 1$, and each y_u^i prefers x_u^{i-1} to x_u^i with $1 \leq i \leq d^* - 1$. Although x_u^{b-1} prefers y_u^{i-1} to $T_u^j[b]$, there is a free edge between x_u^{b-1} and y_u^{i-1}, which does not form a blocking pair. Then, M' is still stable. Thus, u can only be matched to a woman who is not in $\{T_u^j[i]\}$ with $2 \leq i \leq d^* - 1$ and $1 \leq j \leq m$. All pairs $\{x, y\}$ with $x \in \{u\} \cup X_u$ and $y \in Y$ cannot be blocking pairs in M', which implies that blocking pairs in M' can only be pairs in $\{\{a, b\} | a \in U \cup X_u, b \in W\}$. Next, we prove the equivalence of the instances.

"$\Rightarrow$": Suppose that M' has a blocking pair, denoted as $\{a, b\}$ with $b \in W$ and $a \in \{u\} \cup X_u$. The only difference between $\succ_u$ and $\succ_a$ is that $\succ_a$ has no ties and $\succ_a$ contains all the agents of Y_u. By the third property, there is no blocking pair containing agent in Y_u. Thus, b must be a woman in T_u^j with $1 \leq j \leq m$. Considering M does not contain this blocking pair, a must be matched to a woman who is no worse than b in $\succ_u$, that is, a is matched to a woman $T_u^j[i]$ who is in the same tie-set as b and has the position being less than b in $\succ_a$, a contradiction of that $\{a, b\}$ forms a blocking pair.

"$\Leftarrow$": Suppose that M contains a blocking pair, denoted as $\{a, b\}$. Then $\{a, b\}$ must be a free edge in M', that is, $b \in Y_u$, which cannot form a blocking pair in M since b is not a member in W, a contradiction. □

3.2 π-SMtBP

First, we show that PERFECT π-SMtBP can be reduced to π-SMtBP with $\pi \in \{\text{Reg, Egal, Forced, Forbidden}\}$. Here, PERFECT π-SMtBP denotes the variant of π-SMtBP with an additional requirement that the solution matching must be perfect. Then, we show that each π-SMT instance can be reduced to an equivalent PERFECT SMtBP instance with $\pi \in$ {Reg, Egal, Forced, Forbidden}. The constructed PERFECT SMtBP instance has complete preference lists with no tie. Thus, π-SMtBP is NP-hard with $\pi \in$ {Reg, Egal, Forced, Forbidden} even if the preference lists are complete.

Theorem 3. *(*) π-SMtBP with $\pi \in$ {Reg, Egal, Forced, Forbidden} is NP-hard, even with all preference lists being complete.*

4 Parameterized Complexity

In this section, we study the parameterized complexity of π-SMFE and π-SMtBP. We first show that Reg-SMFE, Reg-SMtBP, and Forced-SMFE admit problem kernels of size $\mathcal{O}(k^4)$ with k being the cardinality of the maximum matching of the bipartite graph corresponding to the instance of Reg-SMFE, Reg-SMtBP, or Forced-SMFE. The main idea behind the kernelization is similar to the one by Gupta et al. [13]. Then we show an FPT algorithm for π-SMFE and an XP algorithm for π-SMtBP. We also get some intractable results. Egal-SMtBP and Forbidden-SMtBP are W[1]-hard with respect to $n - t$, where t is the number of free edges or blocking pairs. Given incomplete preference lists, both problems are Para-NP-hard with respect to l for all four constraints, where l is the maximum length of the preference lists.

4.1 Tractable Results

Theorem 4. *Reg-SMFE, Reg-SMtBP, and Forced-SMFE admit problem kernels of size $\mathcal{O}(k^4)$.*

Proof. Given a π-SMFE or π-SMtBP instance I, let G be the bipartite graph constructed from I. We first compute a maximum matching of G, denoted as M_a, and thus $k = |M_a|$. We use A_I to denote the set of the agents whose corresponding vertices are in M_a, and A_O the other agents. For Reg-SMFE and Reg-SMtBP, we need two reduction rules:

Reduction Rule 1. *Remove a_o from A_O, if $P_{a_i}(a_o) > 2k + 1$ for all $a_i \in A_I$ who are from the opposite gender of a_o.*

Next we will prove that Reduction Rule 1 is safe. Let I be the original instance, and I' be the resulting instance from which no agent can be removed by Reduction Rule 1. Let U' and W' be the sets of men and women in I'. Let M be a solution of I. Suppose that M is not a solution of I', which means that, (1) M is not stable in I' or (2) M contains a pair $\{u, w\}$ such that at least one of u and w is not in I'. The first one is impossible, since $U' \cup W' \subseteq U \cup W$. If M does not admit blocking pairs in I, then it admits no blocking pairs in I'. For the second one, if $u \notin U'$ and $w \notin W'$, we set $M' = M \setminus \{u, w\}$. M' must be a solution for I', since the only difference between M and M' is $\{u, w\} \in M$ but

$u \notin U'$ and $w \notin W'$. Consider the case that either $u \notin U'$ or $w \notin W'$, saying $u \in U'$ and $w \notin W'$. The number of the agents in $\succ_u$ must be $2k+1$, since, otherwise, w would be one of the first $2k+1$ agents in $\succ_u$, and would not be removed by Reduction Rule 1. Thus, there is at least one woman in $\succ_u$, denoted as w', who is not matched by M, since $|M| = k$. Let $M' = \{u, w'\} \cup M \setminus \{u, w\}$. We claim that M' must be a solution for I', since w' is better than w for u and the blocking pairs associated with $\{u, w\}$ must be blocking pairs associated with $\{u, w'\}$. Thus, M' must be a solution for I'. Let M' be a solution of I'. Then, M' must be a solution for I. Suppose this is not true. M' must be blocked by a pair $\{u, w\}$ such that (1) either $u \notin U'$ or $w \notin W'$ or (2) $u \notin U'$ and $w \notin W'$. For the first case, we suppose that there is a pair $\{u, w'\}$ blocked by $\{u, w\}$. Thus, the position of w in $\succ_u$ must be less than the position of w', a contradiction to $w \notin W'$. The second case implies that in the corresponding bipartite graph, there is an edge between v_u and v_w and both vertices are not in any pair in M_a. This is impossible, since M_a is a maximum matching.

Reduction Rule 2. *Remove a_o from A_O, if $P_{a_i}(a_o) > d$ for all $a_i \in A_I$ with a_i from the opposite gender of a_o.*

This reduction rule is safe. Suppose this is not true. Then, there is an agent $a_o \in A_O$ who (1) is in a pair block the solution matching M or (2) is in the solution matching M. For the first case, we assume that $\{a_o, b\} \notin M$ blocks M. Then, both a_o and b have partners in M worse than each other. That is, b is matched to an agent whose position in $\succ_b$ is greater than d. Then, M cannot satisfy the Reg-score bound, a contradiction. For the second case, M cannot satisfy the Reg-score bound either, since no agent places a_i at the first d positions in her/his preference list. this reduction rule can shrink the length of preference list of $a_i \in A_I$ if $d < 2k+1$.

After applying the two reduction rules, the number of the remaining agents can be bounded by $\mathcal{O}(k^2)$ for Reg-SMFE and Reg-SMtBP, since Reduction Rule 2 is applied after Reduction Rule 1, after which only V_I and $v_o \in V_O$ with v_o being at one of the first $2k+1$ positions in $\succ_{v_i}$ for at least one $v_i \in V_I$. Then, the length of all preference lists is $\mathcal{O}(k^4)$. Thus, Reg-SMFE and Reg-SMtBP admit a size-$\mathcal{O}(k^4)$ problem kernel.

For the case of Forced-SMFE, we need also two reduction rules, first applying the following Reduction Rule 3 and then the above Reduction Rule 1. Reduction Rule 3 shrinks the instance size and makes no pair can block any forced pairs in S.

Reduction Rule 3. *For each $\{u, w\} \in S$,*

(1) remove u' from $\succ_{w'}$, if $P_u(w') < P_u(w)$, $P_{w'}(u') > P_{w'}(u)$, and $\{u', w'\} \notin F$;
(2) remove w' from $\succ_{u'}$, if $P_w(u') < P_w(u)$, $P_{u'}(w') > P_{u'}(w)$, and $\{u', w'\} \notin F$.

This means that, if u prefers w' to w with $\{u, w\} \in S$, we remove agents who are worse than u in the preference list of w', since $\{u, w'\}$ forms a blocking pair if $P_{w'}(u) < P_{w'}(M(w'))$. Similarly, we remove agents who are worse than w in the preference list of u' if w prefers u' to u with $\{u, w\} \in S$. Then, by Reduction

Rule 1, the number of the remaining agents can be bounded by $\mathcal{O}(k^2)$ and the length of all preference lists is $\mathcal{O}(k^4)$. Forced-SMFE admits a size-$\mathcal{O}(k^4)$ problem kernel. □

Next, we present two algorithms for π-SMFE and π-SMtBP. To this end, we show a lemma about $\{\pi_1, \pi_2\}$-SM, the problem that seeks for a stable matching satisfying two constraints π_1 and π_2.

Lemma 1. *(*) $\{\pi$, Forced$\}$-SM with $\pi \in \{$Reg, Egal, Forbidden$\}$ can be solved in polynomial time.*

Theorem 5. *(*) π-SMFE is solvable in $\mathcal{O}(2^t)$ time, where $t = |F|$ and $\pi \in\{$Reg, Egal, Forced, Forbidden$\}$.*

Theorem 6. *(*) π-SMtBP is solvable in $\mathcal{O}(n^{2t})$ time, where t is the number of blocking pairs and $\pi \in\{$Reg, Egal, Forced, Forbidden$\}$.*

4.2 Intractable Results

First, we show that Forbidden-SMtBP and Egal-SMtBP are W[1]-hard with respect to $n - t$.

Theorem 7. *(*) Forbidden-SMtBP is W[1]-hard with respect to $n - t$.*

Theorem 8. *(*) Egal-SMtBP is W[1]-hard with respect to $n - t$.*

In Sect. 3, we proved that π-SMFE and π-SMtBP with $\pi \in \{$Reg, Egal, Forced, Forbidden$\}$ are NP-hard even if all preference lists are complete. Here, we consider the other extreme case, that is, all preference lists being very short. We show that, even with the maximum length l of all preference lists being bounded by a constant, π-SMFE and π-SMtBP with $\pi \in \{$Reg, Egal, Forced, Forbidden$\}$ are NP-hard, which implies that both problems are fixed-parameter intractable with respect to l.

Theorem 9. *Even with $l = 5$, π-SMFE is NP-hard with $\pi \in \{$Reg, Forced$\}$.*

Proof. We first this theorem for Reg-SMFE. We establish this theorem by a reduction from 1-IN-3 POSITIVE 3-SAT. Given a set of variables $V = \{v_1, \cdots, v_{n'}\}$ and a collection $C = \{c_1, \cdots, c_{m'}\}$ of triples of variables, in which all variables occur positively and every variable occurs in exactly three triples, 1-IN-3 POSITIVE 3-SAT asks whether there exists a truth assignment to the variables so that each triple contains exactly one true literal and two false literals. The NP-hardness of this problem has been proved by Porschen et al. [22]. Given a 1-IN-3 POSITIVE 3-SAT instance, we construct a Reg-SMFE instance as follows. For each variable v_i in V, we create twelve agents: $U(v_i) = \{y_{v_i}, n_{v_i}, a^1_{v_i}, a^2_{v_i}, x^1_{v_i}, x^2_{v_i}\}$ and $W(v_i) = \{w_{v_i}, e_{v_i}, b^1_{v_i}, b^2_{v_i}, d^1_{v_i}, d^2_{v_i}\}$, and create twelve pairs of auxiliary agents with one pair for each agent created for v_i, denoted as $p^j_{v_i}$ and $q^j_{v_i}$ with $1 \leq j \leq 12$. Next, for each triple $c_i \in C$, we create six agents per side: $U(c_i) = \{u^1_{c_i}, e^1_{c_i}, e^2_{c_i}, a^1_{c_i}, a^2_{c_i}, a^3_{c_i}\}$ and

$W(c_i) = \{w^1_{c_i}, w^2_{c_i}, w^3_{c_i}, b^1_{c_i}, b^2_{c_i}, b^3_{c_i}\}$. Then create twelve pairs of auxiliary agents with one pair for each agent created for c_i, denoted as $p^j_{c_i}$ and $q^j_{c_i}$ with $1 \le j \le 12$. Finally, we create two pairs of auxiliary agents p^+, q^+, p^-, q^-. There are totally $(6+12)n' + (6+12)m' + 2$ agents per side. Next, we set the preference lists. Set the preference lists of p^+, q^+ as $\succ_{p^+}: q^+$ and $\succ_{q^+}: p^+$ and the preference list of p^-, q^- accordingly. Let $X_\theta[i]$ denote the i-th agent in $U(\theta)$ with $\theta \in V \cup C$ and $1 \le i \le 6$, and $Y_\theta[i]$ is defined analogously. For each p^i_θ and q^i_θ with $\theta \in V \cup C$, we set their preference lists as follows:

$$\succ_{p^i_\theta}: q^i_\theta \qquad \succ_{q^i_\theta}: X_\theta[i] \succ p^i_\theta \qquad 1 \le i \le 6$$
$$\succ_{p^i_\theta}: Y_\theta[i-6] \succ q^i_\theta \qquad \succ_{q^i_\theta}: p^i_\theta \qquad 7 \le i \le 12.$$

Next, for each variable $v_i \in V$, we set the preference lists of its corresponding agents as follows. Herein, assume that variable v_i occurs in the clauses c_j, c_k, c_h:

$$\succ_{y_{v_i}}: b^1_{v_i} \succ w^1_{c_j} \succ w_{v_i} \succ q^+ \succ q^1_{v_i} \qquad \succ_{w_{v_i}}: y_{v_i} \succ n_{v_i} \succ p^+ \succ p^- \succ p^7_{v_i}$$
$$\succ_{a^1_{v_i}}: b^1_{v_i} \succ w^2_{c_k} \succ b^2_{v_i} \succ q^+ \succ q^2_{v_i} \qquad \succ_{b^1_{v_i}}: y_{v_i} \succ a^1_{v_i} \succ p^+ \succ p^- \succ p^8_{v_i}$$
$$\succ_{a^2_{v_i}}: b^2_{v_i} \succ w^3_{c_h} \succ e_{v_i} \succ q^+ \succ q^3_{v_i} \qquad \succ_{b^2_{v_i}}: a^1_{v_i} \succ a^2_{v_i} \succ p^+ \succ p^- \succ p^9_{v_i}$$
$$\succ_{n_{v_i}}: d^1_{v_i} \succ b^1_{c_j} \succ w_{v_i} \succ q^+ \succ q^4_{v_i} \qquad \succ_{d^1_{v_i}}: n_{v_i} \succ x^1_{v_i} \succ p^+ \succ p^- \succ p^{10}_{v_i}$$
$$\succ_{x^1_{v_i}}: d^1_{v_i} \succ b^2_{c_k} \succ d^2_{v_i} \succ q^+ \succ q^5_{v_i} \qquad \succ_{d^2_{v_i}}: x^1_{v_i} \succ x^2_{v_i} \succ p^+ \succ p^- \succ p^{11}_{v_i}$$
$$\succ_{x^2_{v_i}}: d^2_{v_i} \succ b^3_{c_h} \succ e_{v_i} \succ q^+ \succ q^6_{v_i} \qquad \succ_{e_{v_i}}: a^2_{v_i} \succ x^2_{v_i} \succ p^+ \succ p^- \succ p^{12}_{v_i}.$$

Next, for each triple $c_i \in C$ with $c_i = \{v_j, v_k, v_h\}$, we set the preference lists of its corresponding agents as follows:

$$\succ_{u_{c_i}}: w^1_{c_i} \succ w^2_{c_i} \succ w^3_{c_i} \succ q^+ \succ q^1_{c_i} \qquad \succ_{w^1_{c_i}}: a^1_{c_i} \succ y_{v_j} \succ u_{c_i} \succ p^+ \succ p^7_{c_i}$$
$$\succ_{e^1_{c_i}}: b^1_{c_i} \succ b^2_{c_i} \succ b^3_{c_i} \succ q^+ \succ q^2_{c_i} \qquad \succ_{w^2_{c_i}}: a^2_{c_i} \succ a^1_{v_k} \succ u_{c_i} \succ p^+ \succ p^8_{c_i}$$
$$\succ_{e^2_{c_i}}: b^1_{c_i} \succ b^2_{c_i} \succ b^3_{c_i} \succ q^+ \succ q^3_{c_i} \qquad \succ_{w^3_{c_i}}: a^3_{c_i} \succ a^2_{v_h} \succ u_{c_i} \succ p^+ \succ p^9_{c_i}$$
$$\succ_{a^1_{c_i}}: w^1_{c_i} \succ b^1_{c_i} \succ q^+ \succ q^- \succ q^4_{c_i} \qquad \succ_{b^1_{c_i}}: a^1_{c_i} \succ n_{v_j} \succ e^1_{c_i} \succ e^2_{c_i} \succ p^{10}_{c_i}$$
$$\succ_{a^2_{c_i}}: w^2_{c_i} \succ b^2_{c_i} \succ q^+ \succ q^- \succ q^5_{c_i} \qquad \succ_{b^2_{c_i}}: a^2_{c_i} \succ b^1_{v_k} \succ e^1_{c_i} \succ e^2_{c_i} \succ p^{11}_{c_i}$$
$$\succ_{a^3_{c_i}}: w^3_{c_i} \succ b^3_{c_i} \succ q^+ \succ q^- \succ q^6_{c_i} \qquad \succ_{b^3_{c_i}}: a^3_{c_i} \succ b^2_{v_h} \succ e^1_{c_i} \succ e^2_{c_i} \succ p^{12}_{c_i}.$$

Finally, set the free edge set $F = \{\{a^1_{c_i}, w^1_{c_i}\}, \{a^2_{c_i}, w^2_{c_i}\}, \{a^3_{c_i}, w^3_{c_i}\} | c_i \in C\} \cup \{\{y_{v_i}, b^1_{v_i}\}, \{n_{v_i}, d^1_{v_i}\} | v_i \in V\}$. Set the Reg-score bound $d = 4$. The maximum length of the preference lists is clearly bounded by $l = 5$. The proof of Before showing the equivalence of the instances, we prove several properties of the constructed instance.

First, each $p^i_\theta \in P$ can only be matched to $q^i_\theta \in Q$ for each $\theta \in V \cup C$, since, otherwise, the Reg-score of at least one of $M(p^i_\theta)$ and $M(q^i_\theta)$ is equal to 5, which is greater than d. Thus, the solution matching must be a perfect matching. If this is not true, there is a man $u \in U$ who has no partner. Then, $\{u, q\}$ forms a blocking pair with $q \in Q \setminus \{q^+, q^-\}$ being the auxiliary agent in $\succ_u$. $M(w) \neq \emptyset$ with $w \in W$ can be proved in a similar way.

Second, there are only two possible matchings for the agents created for $v_i \in V$.

1. $\{\{y_{v_i}, w_{v_i}\}, \{a^1_{v_i}, b^1_{v_i}\}, \{a^2_{v_i}, b^2_{v_i}\}, \{n_{v_i}, d^1_{v_i}\}, \{x^1_{v_i}, d^2_{v_i}\}, \{x^2_{v_i}, e_{v_i}\}\} \subset M$
2. $\{\{y_{v_i}, b^1_{v_i}\}, \{a^1_{v_i}, b^2_{v_i}\}, \{a^2_{v_i}, e_{v_i}\}, \{n_{v_i}, w_{v_i}\}, \{x^1_{v_i}, d^1_{v_i}\}, \{x^2_{v_i}, d^2_{v_i}\}\} \subset M$

We call the first matching as "TRUE"-matching and the second as "FALSE"-matching.

Third, given a triple c_i, u_{c_i} is matched to one agent in $\{w^1_{c_i}, w^2_{c_i}, w^2_{c_i}\}$. Thus, the agents created for v_k with $1 \leq k \leq n'$ must be matched as "TRUE"-matching, if u_{v_k}, $a^1_{v_k}$, or $a^2_{v_k}$ occurs in the preference list of $M(u_{c_i})$. The other two agents in $\{w^1_{c_i}, w^2_{c_i}, w^2_{c_i}\}$ are matched to the agents in $\{a^1_{c_i}, a^2_{c_i}, a^3_{c_i}\}$. That is, $\{a^j_{c_i}, w^j_{c_i}\} \in M$ if $\{u_{c_i}, w^j_{c_i}\} \notin M$ with $1 \leq j \leq 3$. Thus, $b^j_{c_i}$ can only be matched to $e^1_{c_i}$ or $e^2_{c_i}$ if $\{u_{c_i}, w^j_{c_i}\} \notin M$ with $1 \leq j \leq 3$. Since M cannot admit blocking pairs other than the pairs in Q, the agents created for v_k with $1 \leq k \leq n'$ must be matched as "FALSE"-matching, if n_{v_k}, $c^1_{v_k}$, or $c^2_{v_k}$ occurs in the preference list of $b^j_{c_i}$ and $\{u_{c_i}, w^j_{c_i}\} \notin M$ with $1 \leq j \leq 3$. For instance, if $\{u_{c_i}, w^1_{c_i}\} \in M$, then v_j must be matched as "TRUE", and v_k and v_h must be matched as "FALSE". Now, we show the equivalence between the two instances.

"$\Rightarrow$": Given a truth assignment to the variables, we construct a matching M as follows. For each v_i assigned TRUE, we match the agents created for v_i as "TRUE"-matching, and for v_i assigned FALSE, as "FALSE"-matching. For each $c_i \in C$, if y_{v_i}, $a^1_{v_i}$, or $a^2_{v_i}$ occurs in the preference list of $\{w^j_{c_i}\}$ with $1 \leq j \leq 3$ and v_i being assigned TRUE, we let $\{\{u_{c_i}, w^j_{c_i}\}, \{a^j_{c_i}, b^j_{c_i}\}\} \subset M$, and let $\{\{a^{j^*}_{c_i}, w^{j^*}_{c_i}\} | 1 \leq j^* \leq 3 \wedge j^* \neq j\} \subset M$ as well as $\{\{e^{j^*}_{c_i}, b^{j^*}_{c_i}\} | 1 \leq j^* \leq 3 \wedge j^* \neq j\} \subset M$. Since the original instance has a truth assignment, there is no blocking pair formed by men created for variables and women created for clauses, which implies M is a stable matching.

"$\Leftarrow$": Let M be the solution. By the third property, there is no blocking pair formed by men created for variables and women created for clauses. That is, in each triple, exactly one literal is TRUE and the others are FALSE, which form a truth assignment to the variables of the original instance.

For Forced-SMFE, we use almost the same reduction as above. The only difference is that Forced-SMFE does not have the Reg-score d. We set the forced pair set as $S = \{\{p^i_\theta, q^i_\theta\} | \theta \in V \cup U\}$. Thus, the solution matching must be perfect and each agent in $U(v_i)$ can only be matched to $W(v_i)$ with $v_i \in V$ and each agent in $U(c_i)$ can only be matched to $W(c_i)$ with $c_i \in C$. Then we can prove the equivalence in a similar way. □

Theorem 10. *(*) Even with $l = 6$, π-SMFE is NP-hard with $\pi \in$ {Egal, Forbidden}.*

Theorem 11. *(*) Even with $l = 6$, π-SMtBP is NP-hard with $\pi \in$ {Reg, Egal, Forced}*

5 Concluding Remarks

In this paper, we study π-Stable Marriage with Free Edges (π-SMFE) and π-Stable Marriage with t-Blocking Pairs (π-SMtBP) with $\pi \in$\{Reg, Egal, Forced, Forbidden\}. We first show a connection between π-SMT and π-SMFE and π-SMtBP, which implies that π-SMFE and π-SMtBP are NP-hard if its corresponding π-SMT is NP-hard. This connection also implies solving algorithms for π-SMT if there is an algorithm for π-SMFE or π-SMtBP for some constraints π. We then study the parameterized complexity of π-SMFE or π-SMtBP and provide FPT results as well as intractable results.

There are still some open problems for π-SMFE and π-SMtBP. First, the parameterized complexity of π-SMFE and π-SMtBP with respect to $n - t$ other than Egal-SMtBP and Forbidden-SMtBP remains open. Second, it is interesting to explore other structural parameters such as the one used by Gupta et al. [13].

References

1. Abdulkadiroğlu, A., Pathak, P.A., Roth, A.E.: The New York city high school match. Am. Econ. Rev. **95**(2), 364–367 (2005)
2. Abraham, D.J., Biró, P., Manlove, D.F.: "Almost stable" matchings in the roommates problem. In: Proceedings of the 3rd International Workshop on Approximation and Online Algorithms, pp. 1–14 (2005)
3. Aziz, H., Chen, J., Gaspers, S., Sun, Z.: Stability and pareto optimality in refugee allocation matchings. In: Proceedings of the 17th International Conference on Autonomous Agents and MultiAgent Systems, pp. 964–972 (2018)
4. Biró, P., Manlove, D.F., McDermid, E.: "Almost stable" matchings in the roommates problem with bounded preference lists. Theoret. Comput. Sci. **432**, 10–20 (2012)
5. Biró, P., Manlove, D.F., Mittal, S.: Size versus stability in the marriage problem. Theoret. Comput. Sci. **411**(16–18), 1828–1841 (2010)
6. Cechlárová, K., Fleiner, T.: Stable roommates with free edges. Technical report, Egerváry Research Group on Combinatorial (2009)
7. Cseh, Á., Heeger, K.: The stable marriage problem with ties and restricted edges. Discrete Optim. **36**, 100571 (2020)
8. Cseh, Á., Manlove, D.F.: Stable marriage and roommates problems with restricted edges: complexity and approximability. Discrete Optim. **20**, 62–89 (2016)
9. Dias, V.M.F., da Fonseca, G.D., de Figueiredo, C.M.H., Szwarcfiter, J.L.: The stable marriage problem with restricted pairs. Theoret. Comput. Sci. **306**(1–3), 391–405 (2003)
10. Downey, R., Fellows, M.: Parameterized Complexity. Springer Science & Business Media (2012). https://doi.org/10.1007/978-1-4612-0515-9
11. Feder, T.: Stable Networks and Product Graphs. Stanford University Press, Palo Alto (1995)
12. Gale, D., Shapley, L.S.: College admissions and the stability of marriage. Am. Math. Monthly **69**(1), 9–15 (1962)
13. Gupta, S., Jain, P., Roy, S., Saurabh, S., Zehavi, M.: On the (parameterized) complexity of almost stable marriage. In: Proceedings of the 40th IARCS Annual Conference on Foundations of Software Technology and Theoretical Computer Science, pp. 1–17 (2020)

14. Gusfield, D.: Three fast algorithms for four problems in stable marriage. SIAM J. Comput. **16**(1), 111–128 (1987)
15. Irving, R.W., Leather, P., Gusfield, D.: An efficient algorithm for the "optimal" stable marriage. J. ACM **34**(3), 532–543 (1987)
16. Kato, A.: Complexity of the sex-equal stable marriage problem. Jpn. J. Ind. Appl. Math. **10**(1), 1–19 (1993)
17. Manlove, D.F., Irving, R.W., Iwama, K., Miyazaki, S., Morita, Y.: Hard variants of stable marriage. Theoret. Comput. Sci. **276**(1–2), 261–279 (2002)
18. Manlove, D.F., McBride, I., Trimble, J.: "Almost-stable" matchings in the hospitals/residents problem with couples. Constraints - Int. J. **22**(1), 50–72 (2017)
19. Manlove, D.F., O'Malley, G.: Paired and altruistic kidney donation in the UK: algorithms and experimentation. In: Proceedings of 11th International Symposium on Experimental Algorithms, pp. 271–282 (2012)
20. Mnich, M., Schlotter, I.: Stable marriage with covering constraints-a complete computational trichotomy. In: Proceedings of the 10th International Symposium of Algorithmic Game Theory, pp. 320–332 (2017)
21. Niedermeier, R.: Invitation to Fixed-Parameter Algorithms. Oxford University Press, Oxford (2006)
22. Porschen, S., Schmidt, T., Speckenmeyer, E., Wotzlaw, A.: XSAT and NAE-SAT of linear CNF classes. Discr. Appl. Math. **167**, 1–14 (2014)

Backgammon Is Hard

R. Teal Witter(✉)

NYU Tandon, Brooklyn, NY 11201, USA
rtealwitter@nyu.edu

Abstract. We study the computational complexity of the popular board game backgammon. We show that deciding whether a player can win from a given board configuration is NP-Hard, PSPACE-Hard, and EXPTIME-Hard under different settings of known and unknown opponents' strategies and dice rolls. Our work answers an open question posed by Erik Demaine in 2001. In particular, for the real life setting where the opponent's strategy and dice rolls are unknown, we prove that determining whether a player can win is EXPTIME-Hard. Interestingly, it is not clear what complexity class strictly contains each problem we consider because backgammon games can theoretically continue indefinitely as a result of the capture rule.

Keywords: Computational complexity · Games

1 Introduction

Backgammon is a popular board game played by two players. Each player has 15 pieces that lie on 24 points evenly spaced on a board. The pieces move in opposing directions according to the rolls of two dice. A player wins if they are the first to move all of their pieces to their home and then off the board.

The quantitative study of backgammon began in the early 1970's and algorithms for the game progressed quickly. By 1979, a computer program had beat the World Backgammon Champion 7 to 1 [2]. This event marked the first time a computer program bested a reigning human player in a recognized intellectual activity. Since then, advances in backgammon programs continue especially through the use of neural networks [11,17,18].

On the theoretical side, backgammon has been studied from a probabilistic perspective as a continuous process and random walk [10,19]. However, the computational complexity of backgammon remains an open problem two decades after it was first posed [5]. One possible explanation (given in online resources) is that the generalization of backgammon is unclear.

From a complexity standpoint, backgammon stands in glaring contrast to many other popular games. Researchers have established the complexity of numerous games including those listed in Table 1 but we are not aware of any work on the complexity of backgammon.

In this paper, we study the computational complexity of backgammon. In order to discuss the complexity of the game, we propose a natural generalization

D.-Z. Du et al. (Eds.): COCOA 2021, LNCS 13135, pp. 484–496, 2021.
https://doi.org/10.1007/978-3-030-92681-6_38

Table 1. A selection of popular games and computational complexity results.

Game	Complexity class
Tic-Tac-Toe	PSPACE-complete [12]
Checkers	EXPTIME-complete [14]
Chess	EXPTIME-complete [6]
Bejeweled	NP-hard [7]
Go	EXPTIME-complete [13]
Hanabi	NP-Hard [1]
Mario Kart	PSPACE-complete [3]

of backgammon. Inevitably, though, we have to make arbitrary choices such as the number or size of dice in the generalized game. Nonetheless, we make every effort to structure our reductions so that they apply to as many generalizations as possible.

There are two main technical issues that make backgammon particularly challenging to analyze. The first is the difficulty in forcing a player into a specific move. All backgammon pieces follow the same rules of movement and there are at least 15 unique combinations of dice rolls (possibly more for different generalizations) per turn. For other games, this problem has been solved by more complicated reductions and extensive reasoning about why a player has to follow specified moves [4]. In our work, we frame the backgammon problem from the perspective of a single player and use the opponent and dice rolls to force the player into predetermined moves.

The second challenge is that the backgammon board is one-dimensional. Most other games with computational complexity results have at least two dimensions of play which creates more structure in the reductions [8]. We avoid using multiple dimensions by carefully picking Boolean satisfiability problems to reduce from.

We show that deciding whether a player can win is NP-Hard, PSPACE-Hard, and EXPTIME-Hard for different settings of known or unknown dice rolls and opponent strategies. In particular, in the setting most similar to the way backgammon is actually played where the opponent's strategy and dice rolls are unknown, we show that deciding whether a player can win is EXPTIME-Hard. Our work answers an open problem posed by Demaine in 2001 [5].

In Sect. 2, we introduce the relevant rules of backgammon and generalize it from a finite board to a board of arbitrary dimension. In Sect. 3, we prove that deciding whether a player can win even when all future dice rolls and the opponent's strategy are known is NP-Hard. In Sect. 4, we prove that deciding whether a player can win when dice rolls are known and the opponent's strategy is unknown is PSPACE-Hard. Finally in Sect. 5, we prove that deciding whether a player can win when dice rolls and the opponent's strategy are unknown is EXPTIME-Hard.

2 Backgammon and Its Generalization

We begin by describing the rules of backgammon relevant to our reductions. When played in practice, the backgammon board consists of 24 points where 12 points lie on Player 1's side and 12 points lie on Player 2's side. However, without modifying the structure of the game, we will think of the board as a line of 24 points where Player 1's home consists of the rightmost six points and Player 2's home consists of the leftmost six points. Figure 1 shows the relationship between the regular board and our equivalent model. Player 1 moves pieces right according to dice rolls while Player 2 moves pieces left. The goal is to move all of one's pieces home and then off the board.

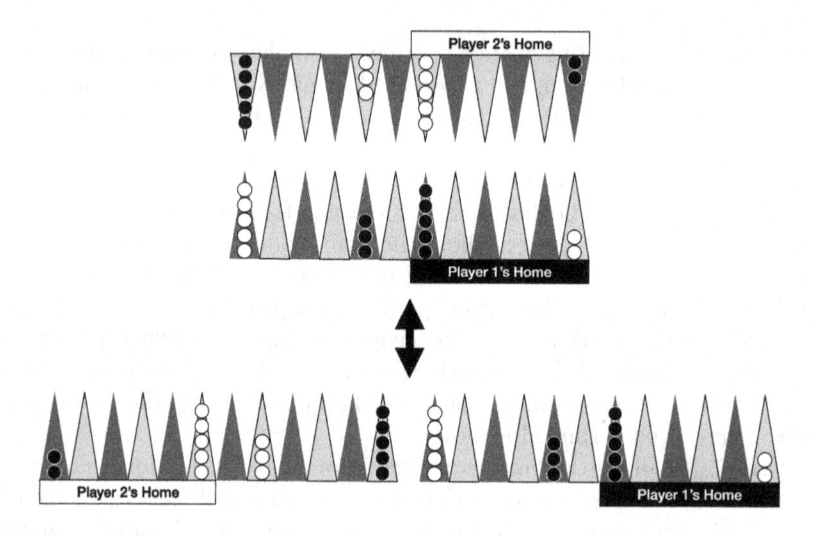

Fig. 1. Backgammon board in normal play (top); equivalent board 'unfolded' (bottom).

Players move their pieces by taking turns rolling dice. On their roll, a player may move one or more pieces 'forward' (right for Player 1 and left for Player 2) by the numbers on the dice provided that the new points are not blocked. A point is blocked if the opponent has at least two pieces on it. The turn ends when either the player has moved their pieces or all moves are blocked. Note that a player must always use as many dice rolls as possible and if the same number appears on two dice then the roll 'doubles' so a player now has four moves (rather than two) of the number.

If only one of a player's pieces is on a point, the opponent may capture it by moving a piece to the point. The captured piece is moved off the board and must be rolled in from the opponent's home *before* any other move may be made. This sets back the piece and can prove particularly disadvantageous if all of the points in the opponent's home are blocked.

The obvious way to generalize the backgammon board used in practice is to concatenate multiple boards together, keeping the top right as Player 1's home and the bottom right as Player 2's home. In the line interpretation, we can equivalently view this procedure as adding more points between the respective homes. The rules we described above naturally extend. We formalize this generalization in Definition 1.

Definition 1 (Generalized Backgammon). *Let m be a positive integer given as input. We define constants $h \geq 6$, and $d \geq 2$ $s \geq 6$ where the lower bounds originate from traditional backgammon. Then a generalized backgammon instance consists of n points on a line with the leftmost h and rightmost h marked as each player's home and d dice each with s sides. We require the number of pieces p to be polynomial in m and specify our choice of p in the reductions.*

In our proofs, we fix the constants without loss of generality by modifying our reductions. We assume $h = 6$ home points by blocking additional points in the opponent's home. We also assume $d = 2$ dice and $s = 6$ sides by rolling blocked pieces for the player and using dummy moves for the opponent.

3 NP-Hardness

In this section, we show that determining whether a player can win against a known opponent's strategy and known dice rolls (KSKR) is NP-Hard. We begin with formal definitions of Backgammon KSKR and the NP-Complete problem we reduce from.

Definition 2 (Backgammon KSKR). *The input is a configuration on a generalized backgammon board, a complete description of the opponent's strategy, and all future dice rolls both for the player and opponent. The problem is to determine whether a player can win the backgammon game from the backgammon board against the opponent's strategy and with the specified dice rolls.*

We do not require that the configuration is easily reachable from the start state. However, one can imagine that given sufficient time and collaboration between two players, any configuration is reachable using the capture rule.

An opponent's strategy is known if the player knows the moves the opponent will make from all possible positions in the resulting game. Notice that such a description can be very large. However, in our reduction, we limit the number of possible positions by forcing the player to make specific moves and predetermining the dice rolls. Therefore the reduction stays polynomial in the size of the 3SAT instance. We formalize this intuition in Lemma 1.

Definition 3 (3SAT). *The input is a Boolean expression in Conjunctive Normal Form (CNF) where each clause has at most three variables. The problem is to determine whether a satisfying assignment to the CNF exists.*

Given any 3SAT instance, we construct a backgammon board configuration, an opponent strategy, and dice rolls so that the solution to Backgammon KSKR yields the solution to 3SAT. Since 3SAT is NP-Complete [9], Backgammon KSKR must be NP-Hard. We state the result formally below.

Theorem 1. *Backgammon KSKR is NP-Hard.*

Proof. Our proof consists of a reduction from 3SAT to Backgammon KSKR. Assume we are given an arbitrary 3SAT instance with n variables and k clauses. First, we force Player 1 (black) to choose either x_i or $\neg x_i$ for every $i \in [n]$. Then, we propagate their choice into the appropriate clauses in the Boolean expression from the 3SAT instance. Finally, we reach a board configuration where Player 1 wins if and only if they have chosen an assignment of bits that satisfies the Boolean expression.

We compartmentalize the process into "gadgets." Each gadget simulates the behavior of a part of the 3SAT problem: There is an assignment gadget for every variable that forces Player 1 to choose either x_i or $\neg x_i$ for every $i \in [n]$. There is a clause gadget for every clause that records whether the assignment Player 1 chose satisfies c_j for every $j \in [k]$.

Player 1 wins if and only if their assignment satisfies all clauses. We ensure this by putting a single black piece in each clause. Player 1 satisfies the clause by protecting their piece. Once the assignment has been propagated to the clauses, Player 2 (white) captures any unprotected piece. If even a single clause is unsatisfied (i.e. a single piece is open), Player 2 traps the piece and moves all the white pieces home before Player 1 can make a single additional move. We block Player 2's home and use the rule that a captured piece must be rolled in the board before any other move can be made.

If, on the other hand, Player 1 satisfies every clause then Player 1 will win since we will feed rolls with larger numbers to Player 1 and smaller numbers to Player 2. Player 1 will then beat Player 2 given their material advantage in the number of pieces on the board.

We now describe the variable and clause gadgets in Fig. 2. In order to simplify the concepts, we explain the gadgets in the context of their function in the reduction rather than providing a complicated, technical definition. There are n variable gadgets followed by k clause gadgets arrayed in increasing order of index from left to right.

For each x_i, we repeat the following process: There are initially two white pieces each on point 4 and point 16 (the top of Fig. 2). We move these pieces to 1 and 13 respectively while feeding Player 1 blocked rolls e.g. one. We then give Player 1 a two and a three. The only moves they can make are from 2 to 4 to 7 or from 14 to 16 to 19. This choice corresponds to setting x_i. Without loss of generality, Player 1 chooses $\overline{x_i}$ and Player 2 blocks 4 from 6 and 7 from 9. We feed Player 1 rolls of two and three until all the $\overline{x_i}$ pieces are on 19; the number of these pieces is exactly the number of times $\overline{x_i}$ appears in clauses. We give Player 1 enough rolls to move all the pieces corresponding to either x_i or $\overline{x_i}$. We then move to the next variable.

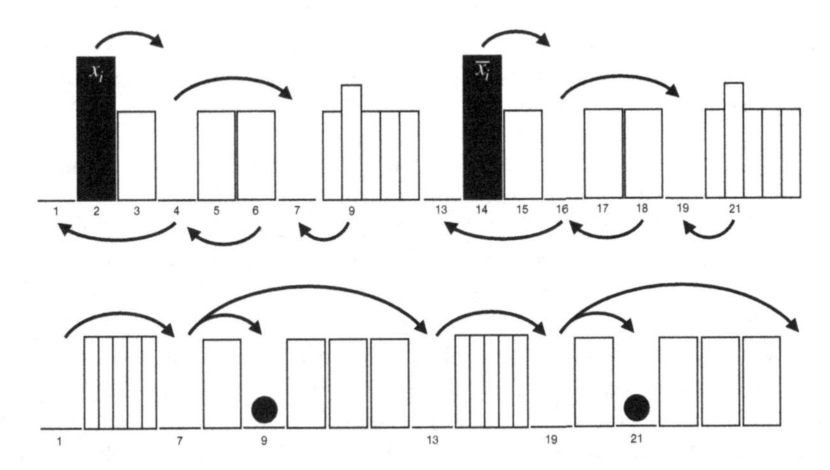

Fig. 2. Reduction from 3SAT: variable gadget (top) for setting x_i and clause gadget (bottom).

Once all the variables have been set, we move down the variable gadgets from x_n to x_1 propagating the choice of x_i to the appropriate clauses. We use Player 2's pieces to block 1 and 13 for all variable gadgets with lower indices so only the pieces in gadget x_i can move. (Variable gadgets with higher indices have already been emptied to clauses.) For each x_i, we move the pieces on 19 through the variable gadgets $x_{i+1}, \ldots, x_n$ with rolls of sixes. Once we reach the clause gadgets, we move one piece at a time with sixes until we reach a clause that contains $\overline{x_i}$. Once it reaches its clause, we give the piece a two to protect the open piece.

We use a similar set of rolls for the x_i pieces on 7. Since the rolls have to be deterministic, we give the rolls for both the x_i and $\overline{x_i}$ before moving on to the x_{i-1} variable gadget. The rolls for whichever of x_i and $\overline{x_i}$ Player 1 did not choose simply cannot be used.

Notice that every roll we give Player 1 can be played by exactly one piece (except when Player 1 sets x_i). While Player 1 receives rolls, we give Player 2 'dummy' rolls of one and two to be used on a stack of pieces near Player 2's home.

Once all variables have been set and the choices propagated to the clauses, we give Player 2 a one to capture any unprotected pieces. If Player 2 captures the unprotected piece, Player 2's home is blocked so Player 1 cannot make any additional moves until all of Player 2's pieces are in their home. At this point, Player 2 easily wins. Otherwise, none of Player 1's pieces are captured and the game becomes a race to the finish. We give Player 2 low rolls and Player 1 high rolls so Player 1 quickly advances and wins.

Since Player 1 wins if and only if they find a satisfying assignment, determining whether Player 1 can win determines whether a satisfying assignment to the 3SAT instance exists. Then, with Lemma 1, Backgammon KSKR reduces from 3SAT in polynomial space and time so Backgammon KSKR is NP-Hard.

Lemma 1. *The reduction from 3SAT to Backgammon KSKR is polynomial in space and time complexity with respect to the number of variables and the number of clauses in the 3SAT instance.*

Proof. The length of the board is linear with respect to the variables and clauses plus some constant buffer on either end. Player 1 has at most twice the number of clauses for each variable while Player 2 has at most a constant number of pieces per variable and clause. Only one piece is captured per reduction so the number of moves is at most the product of the length of the board and the number of pieces.

While it is potentially exponential with respect to the input, the description of the dice rolls and opponent's move may be stored in polynomial space due to their simplicity. In the assignment stage, the rolls and opponent moves are the same for each variable gadget and can be stored in constant space plus a pointer for the current index. In the propagation stage, the rolls and opponent moves are almost the same for each variable gadget and clause gadget except that the number of rolls necessary to move between the variable and clause gadgets varies. However, we can store the number of rolls by the index in addition to constant space for the rules. In the end game, Player 1 and Player 2 both move with doubles if able and the rolls are repeated until one player wins.

4 PSPACE-Hardness

In this section, we show that determining whether a player can win against an unknown opponent's strategy and known dice rolls (USKR) is PSPACE-Hard. We begin with formal definitions of Backgammon USKR and the PSPACE-Complete problem we reduce from.

Definition 4 (Backgammon USKR). *The input is a configuration on a generalized backgammon board, an opponent's strategy which is unknown to the player, and known dice rolls. The problem is to determine whether a player can win the backgammon game from the backgammon board against the opponent's unknown strategy and with the specified dice rolls.*

An opponent's strategy is unknown if the player does not know what the opponent will play given a possible position and dice rolls in the resulting game. The opponent's strategy is not necessarily deterministic; it can be adaptive or stochastic.

Definition 5. ($\mathbf{G_{pos}}$ [15]). *The input is a positive CNF formula (without negations) on which two players will play a game. Player 1 and Player 2 alternate setting exactly one variable of their choosing. Once it has been set, a variable may not be set again. Player 1 wins if and only if the formula evaluates to True after all variables have been set. The problem is to determine whether Player 1 can win.*

Given any G_{pos} instance, we construct a backgammon board configuration, an unknown opponent's strategy, and known dice rolls so that the solution to Backgammon USKR yields the solution to G_{pos}. Since G_{pos} is PSPACE-Complete [15], Backgammon USKR must be PSPACE-Hard. We state the result formally below.

Theorem 2. *Backgammon USKR is PSPACE-Hard.*

Proof. The reduction from G_{pos} to Backgammon USKR closely follows the reduction from $3SAT$ to Backgammon KSKR so we primarily focus on the differences. Assume we are given an arbitrary G_{pos} instance with n variables and k clauses. The key observation is that, since the CNF is positive, Player 1 will always set variables to True while Player 2 will always set variables to False. We can therefore equivalently think about the game as Player 1 moving variables to a True position while Player 2 blocks variables from becoming True. Once all the variables have been set, we propagate the choices to the clause gadgets as we did in the 3SAT reduction.

The winning conditions also remain the same. Player 1 wins if and only if they are able to cover the open piece in each clause. We require the opponent's strategy to be unknown so they can adversarially set variables.

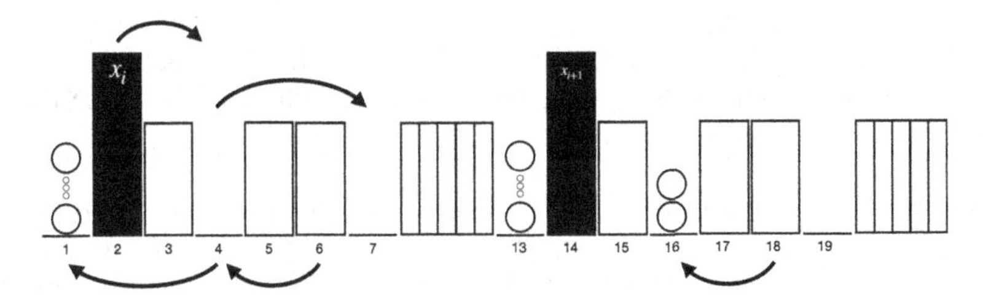

Fig. 3. Reduction from G_{pos}: variable gadgets. Player 1 sets x_i to True while Player 2 has already set x_{i+1} to False.

We now describe the variable gadgets in Fig. 3. In the 3SAT reduction, we needed a stack of x_i pieces and a stack of $\overline{x_i}$ pieces since Player 1 could set x_i to True or False. Here, since the CNF is positive, Player 1 will only set variables to True and so only require an x_i stack.

At the beginning of Player 1's turn, all unset variables are blocked by two pieces on point 4. Then, while Player 1 receives dummy rolls of one, Player 2 moves all the blocking pieces to 1. Next, Player 1 receives a roll of two and three and must choose which unset variable to set to True. Once the variable is chosen, Player 2 blocks all other unset variables by moving two pieces from 6 to 4 while Player 1 again receives dummy rolls. The remaining pieces for the chosen variable are then moved from 2 to 4 to 7.

Player 2's turn is more simple. They choose which variable to set to False and do so by blocking 16 from 18. For the remainder of Player 1's turns, the blocking pieces on 16 will not be moved.

After all the variables have been set to True or False, we move the pieces set to True to the clauses they appear in. We again move from x_n to x_1, removing the blocking pieces on 13 and then 1 as we go. The process and clause gadgets are the same as in the 3SAT reduction.

Player 1 wins in Backgammon USKR if and only if the positive CNF instance in G_{pos} is True after alternating setting variables with Player 2. Therefore the solution to Backgammon USKR yields an answer to G_{pos} and, by Lemma 2, G_{pos} reduces in polynomial space to Backgammon USKR.

Lemma 2. *The reduction from G_{pos} to Backgammon USKR is polynomial in space complexity with respect to the number of clauses and variables in G_{pos}.*

Proof. As in the reduction from 3SAT to Backgammon KSKR, the size of the board is linear in the number of clauses and variables plus some constant buffer. By similar arguments, the size of the description is polynomial because it depends only on the stage of the reduction and the index of the current variable and clause gadgets.

5 EXPTIME-Hardness

In this section, we show that determining whether a player can win against an unknown opponent strategy and unknown dice rolls (USUR) is EXPTIME-Hard. We begin with formal definitions of Backgammon USUR and the EXPTIME-Complete problem we reduce from.

Definition 6 (Backgammon USUR). *The input is a configuration on a generalized backgammon board. The opponent's strategy and dice rolls are unknown to the player. The problem is to determine whether a player can win the backgammon game from the backgammon configuration against the unknown strategy and dice rolls.*

Definition 7. (G_6 [16]). *The input is a CNF formula on sets of variables X and Y and an initial assignment of the variables. Player 1 and Player 2 alternate changing at most one variable. Player 1 may only change variables in X while Player 2 may only change variables in Y. Player 1 wins if the formula ever becomes true. The problem is to determine whether Player 1 can win.*

Given any G_6 instance, we construct a backgammon board configuration and exhibit an opponent's strategy and dice rolls such that the solution to Backgammon USUR yields the solution to G_6. Since G_6 is EXPTIME-Complete [16], Backgammon USUR must be EXPTIME-Hard.

Theorem 3. *Backgammon USUR is EXPTIME-Hard.*

Proof. The proof consists of a reduction from G_6 to Backgammon USUR. Assume we are given a CNF formula with n_x variables X, n_y variables Y, k clauses, and an initial assignment to X and Y. First, Player 1 and Player 2 take turns changing variables in their respective sets X and Y. Then, once Player 1 gives the signal, the game progresses to a board state where Player 1 wins if and only if the CNF formula is True on the current assignment.

Player 1 changes variable x_i by moving a signal piece corresponding to x_i. Then, with Player 2's help, we feed Player 1 dice rolls that update the clause gadgets that contain x_i. We require that the dice rolls adaptively respond to Player 1 and that Player 2 can adversarially set variables in Y.

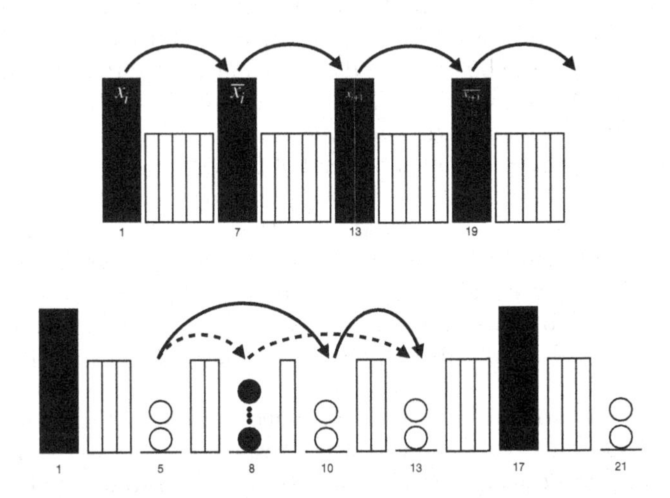

Fig. 4. Reduction from G_6: variable gadget (top) and clause gadget (bottom).

We next describe the gadgets in Fig. 4. The variable gadgets consist of stacks of pieces corresponding to x_i and $\overline{x_i}$ for $i \in [n_x]$. On their turn, Player 1 changes a variable by using their six to move the appropriate piece. For example, Player 1 can change x_i to False by moving a piece on point 7 to 13 as shown at the top of Fig. 4. If x_i is already False, Player 1 has effectively skipped their turn (which is an acceptable move in G_6).

Once Player 1 changes a variable, Player 2 and the dice rolls work together to update the appropriate clauses. The key insight of a clause is that it is True if at least one variable in the clause is True in either X or Y. We represent this on the backgammon board as shown at the bottom of Fig. 4. Point 8 is empty or contains Player 1's pieces if at least one of the X variables in the clause is True and 10 is empty if at least one of the Y variables in the clause is True. Therefore Player 1 can progress a piece from 5 to 13 on rolls three and five if and only if either a variable in X or Y in the clause is True.

We update the clause when Player 1 sets a variable in one of two ways: If the variable is True in the clause, we move two pieces from 1 to 5 to 8. If the variable is False in the clause, we move two pieces from 8 to 13 to 17. If 8 becomes empty, we move two pieces from a white stack to 8 in order to block Player 1 from unfairly using it to bypass the clause. By using blocking pieces on 5 and 13 we ensure the correct clause is modified.

We update the clause when Player 2 sets a variable in one of two ways: If the variable is True in the clause and all other Y variables in the clause are True, we move two pieces from 10 to another white stack. If the variable is False in the clause and all other Y variables in the clause are True, we move two pieces from another white stack to 10. In all other cases, 10 should remain in its current 'open' or 'closed' position.

Notice that the process of changing variables could continue indefinitely. We make sure that we do not run out of pieces by using the capture rule. If Player 1 needs more pieces in the variable or clause gadgets, we feed them rolls to move excess pieces through the board towards their home where Player 2 will capture them one by one. We perform an analogous process if Player 2 needs more pieces.

The variable changing process ends when, instead of moving a variable piece, Player 1 moves a specified signal at the end of the variable gadgets. Then Player 1 will receive enough six and 4-3-5-4 rolls to move all of their pieces home while Player 2 makes slow progress. If all of the clauses are True, Player 1 can successfully get all their pieces home and win the game. Otherwise, they will be blocked at a False clause and Player 2 will continue their slow progress until all white pieces except for those in the False clause remain. We will then feed small rolls to Player 1 and large rolls to Player 2 so Player 2 can capitalize on their advantage and win.

We have therefore simulated the G_6 instance and Player 1 can win Backgammon USUR if and only if they can win the corresponding G_6 game. The reduction is polynomial in the G_6 input size since there are a constant number of pieces and points for every clause and variable.

Lemma 3. *The reduction from G_6 to Backgammon USUR is exponential in time and space complexity with respect to the number of clauses and variables in G_6.*

Proof. As before, the board is still polynomial in the number of clauses. The game may now continue

Notice that Backgammon USUR is not obviously EXPTIME-Complete because the game can progress indefinitely as a result of the capture rule.

6 Conclusion

We show that deciding whether a player can win a backgammon game under different settings of known or unknown opponent strategies and dice rolls is NP-Hard, PSPACE-Hard, and EXPTIME-Hard. It would seem that our results show backgammon is hard even when it is a one-player game. However, in the settings for our PSPACE-Hardness and EXPTIME-Hardness results, the second player is hidden in the unknown nature of the opponent's strategy and dice rolls.

Despite the popularity of backgammon and academic interest in the computational complexity of games, to the best of our knowledge our work is the first to address the complexity of backgammon. One possible explanation is the apparent ambiguity in generalizing backgammon. We contend, however, that the backgammon generalization we use is as natural as those for other games like checkers or chess. Another explanation is the difficulty in forcing backgammon moves as needed for a reduction.

Interestingly, it is not clear that the problems we consider are in EXPTIME because backgammon games can theoretically continue indefinitely. One natural follow up question to our work is what complexity class contains these backgammon problems.

References

1. Baffier, J.F., et al.: Hanabi is NP-hard, even for cheaters who look at their cards. Theoret. Comput. Sci. **675**, 43–55 (2017)
2. Berliner, H.J.: Backgammon computer program beats world champion. Artif. Intell. **14**(2), 205–220 (1980)
3. Bosboom, J., Demaine, E.D., Hesterberg, A., Lynch, J., Waingarten, E.: Mario kart is hard. In: Japanese Conference on Discrete and Computational Geometry and Graphs, pp. 49–59. Springer (2015). https://doi.org/10.1007/978-3-319-48532-4_5
4. Buchin, K., Hagedoorn, M., Kostitsyna, I., van Mulken, M.: Dots & boxes is pspace-complete. arXiv preprint arXiv:2105.02837 (2021)
5. Demaine, E.D.: Playing games with algorithms: algorithmic combinatorial game theory. In: Sgall, J., Pultr, A., Kolman, P. (eds.) MFCS 2001. LNCS, vol. 2136, pp. 18–33. Springer, Heidelberg (2001). https://doi.org/10.1007/3-540-44683-4_3
6. Fraenkel, A.S., Lichtenstein, D.: Computing a perfect strategy for n × n chess requires time exponential in n. In: Even, S., Kariv, O. (eds.) ICALP 1981. LNCS, vol. 115, pp. 278–293. Springer, Heidelberg (1981). https://doi.org/10.1007/3-540-10843-2_23
7. Guala, L., Leucci, S., Natale, E.: Bejeweled, candy crush and other match-three games are (NP-) hard. In: 2014 IEEE Conference on Computational Intelligence and Games, pp. 1–8. IEEE (2014)
8. Hearn, R.A., Demaine, E.D.: Games, Puzzles, and Computation. CRC Press, Boca Raton (2009)
9. Karp, R.M.: Reducibility among combinatorial problems. In: Complexity of Computer Computations, pp. 85–103. Springer (1972). https://doi.org/10.1007/978-1-4684-2001-2_9
10. Keeler, E.B., Spencer, J.: Optimal doubling in backgammon. Oper. Res. **23**(6), 1063–1071 (1975)
11. Pollack, J.B., Blair, A.D.: Co-evolution in the successful learning of backgammon strategy. Mach. Learn. **32**(3), 225–240 (1998)
12. Reisch, S.: Hex ist pspace-vollständig. Acta Informatica **15**(2), 167–191 (1981)
13. Robson, J.M.: The complexity of go. In: Proceedings of the 9th World Computer Congress on Information Processing 1983, pp. 413–417 (1983)
14. Robson, J.M.: N by N checkers is Exptime complete. SIAM J. Comput. **13**(2), 252–267 (1984)
15. Schaefer, T.J.: On the complexity of some two-person perfect-information games. J. Comput. Syst. Sci. **16**(2), 185–225 (1978)
16. Stockmeyer, L.J., Chandra, A.K.: Provably difficult combinatorial games. SIAM J. Comput. **8**(2), 151–174 (1979)

17. Tesauro, G.: TD-Gammon, a self-teaching backgammon program, achieves master-level play. Neural Comput. **6**(2), 215–219 (1994)
18. Tesauro, G.: Programming backgammon using self-teaching neural nets. Artif. Intell. **134**(1–2), 181–199 (2002)
19. Thorp, E.O.: Backgammon: the optimal strategy for the pure running game. Optimal Play: Mathematical Studies of Games and Gambling. Institute for the Study of Gambling and Commercial Gaming, University of Nevada, Reno, pp. 237–265 (2007)

Two-Facility Location Games with a Minimum Distance Requirement on a Circle

Xiaoyu Wu[1], Lili Mei[2](✉), and Guochuan Zhang[3]

[1] School of Mathematical Sciences, Zhejiang University, Hanghzou, China
xiaoyu_wu@zju.edu.cn
[2] College of Sciences, China Jiliang University, Hangzhou, China
meilili@zju.edu.cn
[3] College of Computer Science, Zhejiang University, Hanghzou, China
zgc@zju.edu.cn

Abstract. We consider the games of locating two facilities with a minimum distance requirement, which was first introduced by Duan et al. 2019. In the setting, a mechanism maps the locations reported by strategic agents to two facilities, and the distance between them is at least d. The cost or utility of an agent is the sum of his distances to two facilities given that both facilities are favorite or obnoxious. One aims at designing strategyproof mechanisms, meanwhile achieving good approximation bounds on minimizing the total cost/maximizing the total utility or minimizing the maximum cost/maximizing the minimum utility.

This paper is mainly focused on a circle network. We devise optimal strategyproof mechanisms for minimizing the maximum cost and maximizing the minimum utility, respectively. A group strategyproof mechanism with approximation ratio of $1/(2d)$ is designed for minimizing the total cost. And for maximizing the total utility, we establish a group strategyproof $(2-2d)$-approximation mechanism.

We also revisit the line interval, while we propose a strategyproof mechanism towards maximizing the total utility, improving upon the previous bounds.

Keywords: Facility location games · Strategyproofness · Approximation mechanism design

1 Introduction

We study a model consisting of a network where a set of strategic agents stay. Two facilities with a minimum distance requirement are located, in a way based on the locations reported by all the agents. Here, the minimum distance requirement means that the distance between the two facilities must be at least a certain

This work was supported by Shanghai Key Laboratory of Pure Mathematics and Mathematical Practice [Project NO. 18dz2271000], and National Natural Science Foundation of China [Project NO. 11771365].

D.-Z. Du et al. (Eds.): COCOA 2021, LNCS 13135, pp. 497–511, 2021.
https://doi.org/10.1007/978-3-030-92681-6_39

value. In this paper, we deal with two distinct scenarios in which either the two facilities are both favorite or both obnoxious to all agents. In the sense that an agent is willing to be close to a facility, it is favorite, while in the obnoxious case, an agent wants to keep away from the facility. For both scenarios, each agent takes care of the total distance to both facilities, which is defined as the cost (utility) with respect to the favourite (obnoxious) facilities.

A *mechanism* in the above setting is a map that takes the locations reported by all agents as input, and outputs a pair of facility locations which satisfy the minimum distance requirement. Our game-theoretic goal is to design mechanisms that are *strategyproof*, which says that an agent cannot benefit by lying, regardless of the reports of other agents. Moreover, we wish our mechanism has a good approximation ratio, where the approximation ratio is defined by the worst-case bound between the mechanism's solution and the optimal solution.

The setting was first proposed by Duan et al. [7], where the network is a closed interval. As they said, many real life facility locating scenarios either favorite or obnoxious can be modeled to this setting. For example, when building a school and a theater, it is inappropriate to build them close to each other; while building a dumping ground and a chemical plant, to comply with the environmental regulations, we should not build them close to each other. Moreover, they proposed that it is interesting to consider other metric spaces. In this paper, we mainly focus on a circle network.

1.1 Related Work

Procaccia and Tennenholtz [17] initialized the seminal study of approximation mechanism design without money. In the literature, researchers discussed the characterization of deterministic strategyproof mechanisms such as [16,18]. Procaccia and Tennenholtz mainly studied the model of locating one or two favorite facilities in the line network. For the 1-facility location games, they designed deterministic and randomized strategyproof mechanisms. Alon et al. [1] extended mechanisms to the tree and circle networks.

For the 2-facility location games, researchers first considered the cost of each agent is his distance to the nearest facility. For the line network, Procaccia and Tennenholtz [17] showed a mechanism outputting two extreme points is group strategyproof with the approximation ratio of $(n-2)$ for minimizing the total cost, where n is the number of agents. Furthermore, Fotakis and Tzamos [9] showed this mechanism is best possible. The settings for other metric networks were studied by Lu et al. [13]. They mainly discussed deterministic and randomized strategyproof mechanisms in a circle and the general metric space.

As mentioned above, Duan et al. [7] first proposed the model where the cost of each agent is the total distance to two facilities. It is easy to see that an optimal mechanism locates two facilities at the same point if there is no constraint on the distance of the facilities. They thus proposed a minimum distance requirement for two facilities, which is available for scores of real life applications. They provided optimal strategyproof mechanisms for the objectives of minimizing the total cost and the maximum cost. Many other models dealing with different agent costs and different objective functions have been addressed as well. See also [2,3,8,11,12,19].

Cheng et al. [5] first proposed a model of locating one obnoxious facility in a closed interval, where each agent wants to maximize his distance to the facility, coined by the utility of each agent. They presented a 3-approximation deterministic group strategyproof mechanism for maximizing the total utility. In [6], they further studied mechanisms in tree and circle networks. Ibara and Nagamochi [10] completely characterized strategyproof deterministic mechanisms in the general metric space, which implies that the mechanism proposed in [5] is best possible. Similarly to favorite facility location games, a lot of work exists for other utility functions. See also [4,14,15,21].

Table 1. Our results and the state of the art[2]

		Circle	Line interval
2-FLG	SC	UB : $\mathbf{\frac{1}{2d}}$	UB : 1 SP [20]
		LB : NA	
	MC	UB : **1 SP**	UB : 1 SP [20]
2-OFLG	SU	UB : $\mathbf{2-2d}$	UB : $\leq \sqrt{3}$[1]
		LB : NA	LB: $\frac{7-d}{6}$ [20]
	MU	UB : **1 SP**	UB : 1 SP [20]

[1]The upper bound is related to the distance constraint, which will be specified in Sect. 4.
[2]Our results are in bold. UB refers to upper bound; LB refers to lower bound.

1.2 Our Results

In this paper, we mainly study 2-facility location games with a minimum distance requirement d on a circle, where the cost (resp. utility) of each agent is the sum of distances to both facilities. For two-facility location games (2-FLG), we investigate the objectives of minimizing the total cost (SC) and the maximum cost (MC). Meanwhile, the setting of locating two obnoxious facilities (2-OFLG) is also considered in this paper. For this model, we discuss the objectives of maximizing the total utility (SU) and the minimum utility (MU).

On the circle, we present optimal strategyproof mechanisms for minimizing the MC and maximizing the MU, respectively. If the objectives are minimizing the SC and maximizing the SU, we first illustrate an instance to demonstrate that the optimal solutions are not strategyproof, whereas on the line interval, one optimal solution for minimizing the SC is strategyproof [7]. Finally, we establish strategyproof mechanisms with the approximation ratios of $\frac{1}{2d}$ and $2-2d$ for minimizing the SC and maximizing the SU, respectively. Besides, on the line interval, for maximizing the SU, we present a group strategyproof mechanism which improves the approximation ratio in [7].

We summarize our results and the state of the art in Table 1.

1.3 Organization of the Paper

In Sect. 2, we present some definitions and notations. At the end of this section, a relationship between 2-facility location games and obnoxious 2-facility games is established. In Sect. 3, we design strategyproof mechanisms for obnoxious 2-facility location games with respect to two objectives of maximizing the total utility and the minimum utility. Then according to the relationship between the obnoxious version and the favorite one, it is easy to obtain strategyproof mechanisms for minimizing the total cost and the maximum cost. In Sect. 4, we study the obnoxious 2-facility location games on the line interval, and present a strategyproof mechanism which improves the approximation ratio in [7]. The conclusion of this paper is presented in the last section.

2 Preliminaries

Let $N = \{1, 2, \ldots, n\}$ be a set of agents. Each agent has a location x_i on a network G. Let $\mathbf{x} = (x_1, x_2, \ldots, x_n)$ be a *location profile.* We use $d(x, y)$ to denote the distance between two points $x, y \in G$.

Given a location profile $\mathbf{x}$, a deterministic mechanism outputs two facility locations $(y_1, y_2) \in G^2$ satisfying the minimum distance requirement d, i.e., $d(y_1, y_2) \geq d$.

In the 2-facility location games, the cost of agent i is the sum of his distances to both facilities, i.e.,

$$c_i((y_1, y_2), x_i) = d(y_1, x_i) + d(y_2, x_i).$$

Analogously, in the obnoxious 2-facility games, the utility of agent i is,

$$u_i((y_1, y_2), x_i) = d(y_1, x_i) + d(y_2, x_i).$$

In the 2-facility location games (resp. the obnoxious 2-facility games), a mechanism f is *strategyproof* if no agent can benefit by reporting a false location, regardless of the other agents. Formally, for any agent i, any location profile $\mathbf{x} = (x_1, \ldots, x_i, \ldots, x_n) \in G^n, i \in N$ and any location $x'_i \in G$, it holds that

$$c_i(f(\mathbf{x}), x_i) \leq c_i(f(x'_i, \mathbf{x}_{-i}), x_i) (\text{resp. } u_i(f(\mathbf{x}), x_i) \geq u_i(f(x'_i, \mathbf{x}_{-i}), x_i)),$$

where $\mathbf{x}_{-i} = (x_1, \ldots, x_{i-1}, x_{i+1}, \ldots, x_n)$ is the location profile of all agents without agent i. Furthermore, a mechanism is *group strategyproof* if for any group of agents, at least one of them cannot benefit if they misreport simultaneously.

For the 2-facility location games, we study the objectives of *minimizing the total cost* and *the maximum cost.* Given two facility locations (y_1, y_2), the *the total cost* is

$$sc((y_1, y_2), \mathbf{x}) = \sum_{i \in N} c_i((y_1, y_2), x_i),$$

and *maximum cost* is

$$mc((y_1, y_2), \mathbf{x}) = \max_{i \in N} c_i((y_1, y_2), x_i).$$

Meanwhile, for the obnoxious 2-facility games, we discuss the objectives of *maximizing the total utility* denoted by $su((y_1, y_2), \mathbf{x}) = \sum_{i \in N} u_i((y_1, y_2), x_i)$, and *maximizing the minimum utility* that is $mu((y_1, y_2), \mathbf{x}) = \min_{i \in N} u_i((y_1, y_2), x_i)$.

A mechanism f is *ρ-approximate* ($\rho \geq 1$) if the ratio between the optimal solution and the mechanism solution for a maximizing objective is at most ρ. For a minimizing objective, the ratio is reciprocal. Let (y_1^*, y_2^*) denote the pair of the optimal facility locations. Formally, $sc(f(\mathbf{x}), \mathbf{x}) \leq \rho \cdot sc((y_1^*, y_2^*), \mathbf{x})$ (resp. $su((y_1^*, y_2^*), \mathbf{x}) \leq \rho \cdot sc(f(\mathbf{x}), \mathbf{x})$).

A single circle C is mainly interested in this paper. Without loss of generality, the circumference of C can be normalized to 1, and the minimum distance constraint d is between 0 and $1/2$. For each point $x \in C$, the antipodal point of x is denoted by $\hat{x}$. Given $x, y \in C$, let $d(x, y)$ be the length of the shorter arc between x and y.

In the context of an arc of length less than 1, we introduce $\succeq$ as "clockwise operator ", i.e., given $x, y \in C$, $x \succeq y$ means that y lies beyond x in a clockwise direction. We replace location profile $(x_1, x_2, \ldots, x_n)$ with $< x_1, x_2, \ldots, x_n >$, which contains direction relationship among points. Formally, $x_i \succeq x_{i+1} (i = 0, \ldots, n)$ and $x_{n+1} = x_1; x_0 = x_n$. Particularly, $< x, y >$ not only represents the location profile of x and y, but also represents the arc spanned from x to y in a clockwise direction. Let $d(< x, y >)$ be the length of the arc $< x, y >$. We denote δ as the longest length of the clockwise arc between two adjacent agents, i.e., $\delta = \max_{i \in N} d(< x_i, x_{i+1} >)$. And let $\lambda = 1 - \delta$.

For simplicity, let $[x, y]$, (x, y), $[x, y)$ and $(x, y]$ denote the closed, open, left open right closed and left closed right open clockwise arc $< x, y >$, respectively.

Observation 1. *For any points $x, y, \hat{y} \in C$, $d(x, y) + d(x, \hat{y}) = \frac{1}{2}$. Furthermore, for two pairs of facility locations (y_1, y_2) and $(\hat{y}_1, \hat{y}_2)$, it holds that $d(x, y_1) + d(x, y_2) + d(x, \hat{y}_1) + d(x, \hat{y}_2) = 1$.*

Remark. The above observation implies if f is a (group) strategyproof mechanism for the 2-facility location games, then given arbitrary location profile $\mathbf{x}$, a mechanism $\hat{f}$ outputting $f(\hat{\mathbf{x}})$ is (group) strategyproof for the obnoxious 2-facility games, vice versa. Here $\hat{\mathbf{x}}$ is the location profile constructed by all antipodals of points in $\mathbf{x}$. Especially, if f is an optimal (group) strategyproof mechanism for minimizing the maximum cost (resp. the total cost), then $\hat{f}$ is an optimal (group) strategyproof mechanism for maximizing the minimum cost (resp. the total utility).

3 The Circle

In this section, we study mechanisms for obnoxious 2-facility location games. According to the relationship between the obnoxious version and the favorite one described in the last section, it is easy to establish mechanisms for 2-facility location games.

3.1 Maximizing the Total Utility

First, we take an instance to illustrate that a mechanism which outputs the optimal facility locations is not strategyproof even if $d = 0$.

Given a location profile $\mathbf{x} =< x_1, x_2, x_3 >$ such that $d(x_1, x_2) = \frac{1}{3} - 2\epsilon - \theta$ and $d(x_2, x_3) = \frac{1}{3} + \epsilon + \theta$, where ϵ, θ are extremely small positive numbers. The optimal solution locates the two facilities both at $\hat{x}_1$. Note that currently, the cost of agent 2 at x_2 is $1/3 + 4\epsilon + 2\theta$. Then consider another location profile $\mathbf{x}'$ that agent 2 moves $\epsilon/2 + \theta$ distance clockwise. Let x'_2 denote the new location of agent 2. For location profile $\mathbf{x}'$, the optimal location of facilities are both $\hat{x}'_2$. Now, the cost of the agent at x_2 is $1 - \epsilon - 2\theta > 1/3 + 4\epsilon + 2\theta$, which implies that agent 2 at x_2 can benefit by misreporting to x'_2.

Here, we discuss a simple mechanism locating two antipodal facilities, which naturally satisfies the distance constraint.

Mechanism 1. *Given a location profile* $\mathbf{x} =< x_1, x_2, \ldots, x_n >$, *outputs* $(x_1, \hat{x}_1)$.

Theorem 1. *Mechanism 1 is group strategyproof with the approximation ratio of* $2 - 2d$ *for maximizing the total utility.*

Proof. From Observation 1, one can know wherever a coalition of agents deviate, the utility/cost of each agent is still $\frac{1}{2}$. Thus, Mechanism 1 is group strategyproof.

Now we turn to prove the approximation ratio. Let f denote Mechanism 1. Since the utility of each agent is $\frac{1}{2}$, the total utility of Mechanism 1 is

$$su(f(\mathbf{x}), \mathbf{x}) = \frac{1}{2}n.$$

By the minimum distance constraint d, the utility of each agent is at most $1 - d$. Therefore,

$$su((y_1^*, y_2^*), \mathbf{x}) \leq (1 - d)n.$$

Hence, the approximation ratio is

$$\frac{su((y_1^*, y_2^*), \mathbf{x})}{su(f(\mathbf{x}), \mathbf{x})} \leq \frac{(1 - d)n}{n/2} = 2 - 2d,$$

which is tight. Consider a location profile that all the agents are at the same location. The optimal solution is $(1 - d)n$, while the mechanism solution is $\frac{1}{2}n$.

□

By the relationship between the obnoxious setting and the favorite one, we can know that Mechanism 1 is also group strategyproof for minimizing the total cost. And the approximation ratio is revealed in the following corollary.

Corollary 1. *Mechanism 1 is group strategyproof with the approximation ratio of* $\frac{1}{2d}$ *for minimizing the total cost.*

3.2 Maximizing the Minimum Utility

Unlike the previous objective, we first give a complete characterization of the optimal facility locations if the addresses of all the agents are public. Before discussing the full characterization of the optimal facility locations, we illustrate the following observation.

Observation 2. *Given two facility locations* (y_1, y_2)*, w.l.o.g., suppose that the clockwise arc* $< y_1, y_2 >$ *satisfies* $d(< y_1, y_2 >) \leq \frac{1}{2}$*. Denote* y_m *as the midpoint of arc* $< y_1, y_2 >$*. The utility of agent* i *on the circle is*

$$u_i((y_1, y_2), x_i) = \begin{cases} d(y_1, y_2) & \textit{if agent } i \textit{ is on arc}[y_1, y_2], \\ 2d(x_i, y_m) & \textit{if agent } i \textit{ is on arc}[y_2, \hat{y}_1] \textit{or arc}[\hat{y}_2, y_1], \\ 1 - d(y_1, y_2) & \textit{if agent } i \textit{ is on arc}[\hat{y}_1, \hat{y}_2]. \end{cases}$$

Remark. Due to $d(< y_1, y_2 >) \leq \frac{1}{2}$, we can find that agents on arc $[y_1, y_2]$ get the smallest utility and agents on arc $[\hat{y}_1, \hat{y}_2]$ get the largest utility.

Given a location profile $\mathbf{x}$, recall that δ is the longest length of the clockwise arc between two adjacent agents and $\lambda = 1 - \delta$. Let x_j, x_{j+1} be two adjacent agents who admit δ, that is, $d(< x_j, x_{j+1} >) = \delta$. Let z_m be the midpoint of arc $< x_j, x_{j+1} >$. Note that there may be more than one pair of points admit δ, and this scenario could only happen for $\lambda \geq \frac{1}{2}$.

The following theorem gives a full characterization of the optimal facility locations for maximizing the minimum utility.

Theorem 2. *Given a location profile* $\mathbf{x}$*, let* (y_1^*, y_2^*) *be any pair of optimal facility locations for maximizing the minimum utility. Without loss of generality, assume that the clockwise arc* $< y_1^*, y_2^* >$ *satisfies* $d(< y_1^*, y_2^* >) \leq \frac{1}{2}$*.*

1. *If* $d = \frac{1}{2}$ *or* $\lambda > \frac{1}{2}$*, then* (y_1^*, y_2^*) *satisfy that* y_2^* *is the antipodal point of* y_1^**. (Condition 1)*
2. *If* $\lambda < d < \frac{1}{2}$*, then* (y_1^*, y_2^*) *satisfy that* $d(y_1^*, y_2^*) = d$ *and* $[y_1^*, y_2^*] \cap [\hat{x}_{j+1}, \hat{x}_j] = [\hat{x}_{j+1}, \hat{x}_j]$*. (Condition 2)*
3. *If* $d \leq \lambda < \frac{1}{2}$*, then* (y_1^*, y_2^*) *satisfy that* y_1^* *and* y_2^* *are symmetric on* z_m *with the distance in the range of* $[\frac{d}{2}, \frac{\lambda}{2}]$*, i.e.,* $d(< y_1^*, z_m >) = d(< y_2^*, z_m >) \in [\frac{d}{2}, \frac{\lambda}{2}]$*. (Condition 3)*
4. *If* $\lambda = \frac{1}{2}$ *and* $d < \frac{1}{2}$*, then* (y_1^*, y_2^*) *satisfy Condition 1 or Condition 3.*

Proof. We discuss four cases described in the theorem. We first calculate the minimum utility of (y_1^*, y_2^*) if they satisfy the conditions. Then it is sufficient to show that for any pair of facility locations (y_1, y_2) which do not satisfy the conditions, w. l. o. g., assume that $< y_1, y_2 >$ satisfies $d(< y_1, y_2 >) \leq \frac{1}{2}$, then the minimum utility is strictly less than the minimum utility of (y_1^*, y_2^*).

Case 1: $d = \frac{1}{2}$ **or** $\lambda > \frac{1}{2}$**.**

If $d = \frac{1}{2}$, then by Observation 1, it is easy to verify that any two antipodal points are a pair of optimal facility locations.

Then we discuss the scenario that $\lambda > \frac{1}{2}$. If (y_1^*, y_2^*) satisfy Condition 1, then for any agent $i \in N$, $u_i((y_1^*, y_2^*), x_i) = \frac{1}{2}$. It is sufficient to show that there always exists some agent $i \in N$ such that $u_i((y_1, y_2), x_i) < \frac{1}{2}$.

If there exists an agent i on the arc $< y_1, y_2 >$, then by Observation 2, $u_i((y_1, y_2), x_i) = d(y_1, y_2) < \frac{1}{2}$. Now consider the scenario that there are no agents on the arc $< y_1, y_2 >$. Let c be the midpoint of $< y_1, y_2 >$. We claim that there exists some agent i on the semicircle with the midpoint of c excluding the boundary points, otherwise, all the agents are on a semicircle, which is a contradiction of $\lambda > \frac{1}{2}$. Then by Observation 2, $u_i((y_1, y_2), x_i) = 2d(x_i, c)) < \frac{1}{2}$. The last inequality holds since x_i and c is on a quarter of circle arc (Fig. 1).

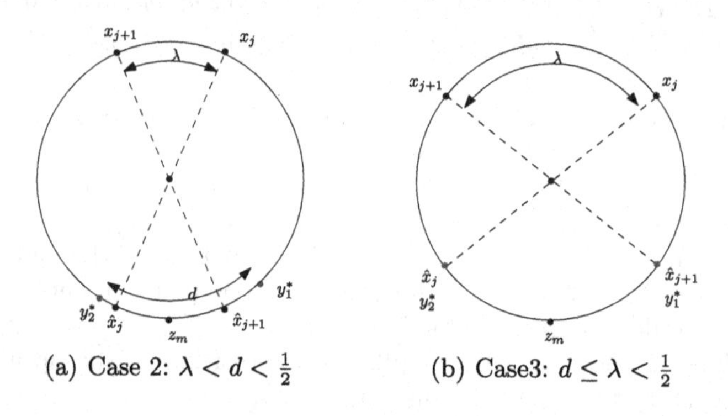

Fig. 1. Illustrations for Case 2 and Case 3.

Case 2: $\lambda < d < \frac{1}{2}$. Note that if (y_1^*, y_2^*) satisfy Condition 2, then the utility of each agent is $1 - d$.

By Observation 2, if $d(y_1, y_2) > d$, the utility of each agent is strictly less than $1 - d$. Hence, we consider the scenario that $d(y_1, y_2) = d$.

If $[y_1, y_2] \cap [x_{j+1}, x_j] \neq \emptyset$, then at least one of agent j and $j + 1$ is on the arc $[y_1, y_2]$. Due to Observation 2, it holds that the utility of this agent is $d(y_1, y_2) = d < 1 - d$. Now, we only need to consider the case that $[y_1, y_2] \cap [x_{j+1}, x_j] = \emptyset$. In this case, by the symmetry, suppose that y_2 is on the arc $(\hat{x}_j, x_{j+1})$ and y_1 is on the arc $[\hat{x}_j, x_{j+1})$ or $(\hat{x}_{j+1}, \hat{x}_j)$. Then we obtain that x_{j+1} is always on the arc $(y_2, \hat{y}_1)$. The utility of agent $j+1$ is $d(< y_1, x_{j+1} >) + d(< \hat{y}_2, \hat{x}_{j+1} >) < 1 - d(< x_{j+1}, \hat{y}_2 >) < 1 - d(< \hat{y}_1, \hat{y}_2 >) = 1 - d$. The inequalities hold since the location y_1 cannot coincide with $\hat{x}_{j+1}$, otherwise, (y_1, y_2) satisfy Condition 2.

Case 3: $d \leq \lambda < \frac{1}{2}$. By Observation 2, we can know that if (y_1^*, y_2^*) satisfy Condition 3, then agents j and $j + 1$ both obtain the minimum utility. And $u_j((y_1^*, y_2^*), x_j) = u_{j+1}((y_1^*, y_2^*), x_{j+1}) = 2d(x_j, z_m) = 1 - \lambda$.

Similarly as the analysis for case 2, if $d(y_1, y_2) > \lambda$ or $[y_1, y_2] \cap [x_{j+1}, x_j] \neq \emptyset$ there always exists an agent whose utility is equal to $d(y_1, y_2) < 1 - \lambda$. Hence, we only need to consider the scenario that $d(y_1, y_2) \leq \lambda$ and $[y_1, y_2] \cap [x_{j+1}, x_j] = \emptyset$.

Assume that $[y_1, y_2] \cap [\hat{x}_{j+1}, \hat{x}_j] \neq [y_1, y_2]$. Without loss of generality, suppose that y_2 is on the arc $(\hat{x}_j, x_{j+1})$ and y_1 is on the arc $[\hat{x}_j, x_{j+1})$ or $(\hat{x}_{j+1}, \hat{x}_j)$. The utility of agent $j+1$ is $d(< y_1, x_{j+1} >) + d(< \hat{y}_2, \hat{x}_{j+1} >) < 1 - d(< x_{j+1}, \hat{y}_2 >) < 1 - \lambda$. The last inequality holds since arc $[x_{j+1}, \hat{y}_2] \cap [x_{j+1}, x_j] = [x_{j+1}, x_j]$.

Finally, we study the case that $[y_1, y_2] \cap [\hat{x}_{j+1}, \hat{x}_j] = [y_1, y_2]$, i.e., the arc $< y_1, y_2 >$ is totally on the arc $< \hat{x}_{j+1}, \hat{x}_j >$. By Observation 2, we know that one of agents j and $j+1$ gets the minimum utility. By the fact that $u_j((y_1, y_2), x_j) = 2d(y_2, x_j) - d(y_1, y_2)$, $u_{j+1}((y_1, y_2), x_{j+1}) = 2d(y_1, x_{j+1}) - d(y_1, y_2)$ and $u_j((y_1, y_2), x_j) + u_{j+1}((y_1, y_2), x_{j+1}) = 2(1-\lambda)$, if y_1 and y_2 are not symmetric on z_m, the minimum utility is less than $1 - \lambda$.

Case 4: $\lambda = \frac{1}{2}$ **and** $d < \frac{1}{2}$**.**

Combining the analysis of case 1 and case 3, we can know that any pair of optimal facilities (y_1^*, y_2^*) satisfies Condition 1 or Condition 3. It worth to mention that, for location profile $\mathbf{x} = < x, \ldots, x, \hat{x}, \ldots, \hat{x} >$, there are two midpoints. □

Then, we present the following optimal mechanism.

Mechanism 2. *Given a location profile* $\mathbf{x} = < x_1, x_2, \ldots, x_n >$, *if* $\lambda \geq \frac{1}{2}$, *output* $(x_1, \hat{x}_1)$, *otherwise, output two facilities that are symmetric on* z_m *with the distance of* $\max\{\frac{\lambda}{2}, \frac{d}{2}\}$. *Recall that* x_j, x_{j+1} *are two adjacent points with the largest gap and* z_m *is the midpoint of arc*$< x_j, x_{j+1} >$.

Theorem 3. *Mechanism 2 is strategyproof and outputs a pair of optimal facilities for maximizing the minimum utility.*

Proof. By Theorem 2, it is easy to check that Mechanism 2 outputs a pair of optimal facilities.

For simple statement, denote Mechanism 2 by f. For any location profile $\mathbf{x}$, let $(y_1, y_2) = f(\mathbf{x})$. Without loss of generality, suppose $d(< y_1, y_2 >) \leq \frac{1}{2}$. Let $\mathbf{x}'$ denote the location profile that some agent i deviates from x_i to x_i'. Let $(y_1', y_2') = f(\mathbf{x}')$ and $d(< y_1', y_2' >) \leq \frac{1}{2}$. Denote the longest length of the clockwise arc between two adjacent agents for location profile $\mathbf{x}'$ by δ', and let $\lambda' = 1 - \delta'$.

Consider the following two cases.

Case 1: $\lambda \geq \frac{1}{2}$**.**

Note that in this case, the utility of each agent is $\frac{1}{2}$. If $\lambda' \geq \frac{1}{2}$, the utility of each agent is still $\frac{1}{2}$. Hence, the agent will not misreport. Then the case $\lambda' < \frac{1}{2}$ will be discussed. This can only happen in the case that there exists an agent i such that $d(< x_{i-1}, x_{i+1}) >) > \frac{1}{2}$, x_i is on arc $< \hat{x}_{i+1}, \hat{x}_{i-1} >$ and x_i' is on arc $< \hat{x}_{i-1}, \hat{x}_{i+1} >$. By Mechanism 2, we have $[\hat{x}_{i+1}, \hat{x}_{i-1}] \subseteq [y_1', y_2']$, then x_i is on the arc $< y_1', y_2' >$. Therefore, by Observation 2, $u_i(f(\mathbf{x}'), x_i) = \max\{\lambda', d\} \leq \frac{1}{2}$, which implies that agent i will not deviate.

Case 2: $\lambda < \frac{1}{2}$**.**

By Observation 2, the utility of each agent i is

$$u_i((y_1, y_2), x_i) = 1 - d(y_1, y_2) = \min\{1 - \lambda, 1 - d\} \geq \frac{1}{2}.$$

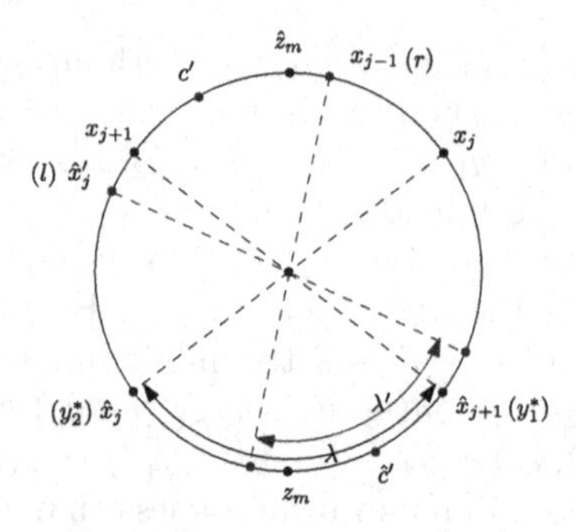

Fig. 2. An illustration for case of $\lambda > d, \lambda' < \lambda$ and $[x_{j+1}, x_j] \nsubseteq [l, r]$

For location profile $\mathbf{x}'$ with $\lambda' \geq \frac{1}{2}$, the utility of each agent in location profile $\mathbf{x}$ is $\frac{1}{2}$, which implies that agent i has no incentive to lie. Hence, we only need to consider the case that $\lambda' < \frac{1}{2}$.

If $d(y_1, y_2) \leq d(y_1', y_2')$, then agent i will have no incentive to lie by the fact that $u_i((y_1, y_2), x_i)$ already gets the maximum utility he can obtain. Meanwhile, if $\lambda' \geq \lambda$ or $d \geq \lambda$, we have that $1 - d(y_1, y_2) \geq 1 - d(y_1', y_2')$. Thus, we will discuss the scenario that $\lambda' < \lambda$ and $\lambda > d$.

Denote two adjacent points who admit largest gap δ' of $\mathbf{x}'$ by points l and r. Suppose that $d(< l, r >) = \lambda'$. If $[x_{j+1}, x_j] \subseteq [l, r]$, then we have $\lambda \leq \lambda'$, by the above statement, then $u_i((y_1', y_2'), x_i) \leq u_i((y_1, y_2), x_i)$.

Then, we only need to consider the case that $[x_{j+1}, x_j] \nsubseteq [l, r]$, which can only happen when agents j or $j+1$ deviate. By the symmetry, assume that agent j deviates to a location on the arc $< \hat{x}_{j-1}, x_j >$. (See Fig. 2 for an illustration.) Recall that we now discuss the case that $\lambda' < \lambda$ and $\lambda > d$. Hence, we have $\hat{y}_2 = x_j$ and $d(\hat{z}_m, x_j) = \frac{\lambda}{2}$. It is worth to notice that $\hat{z}_m$ is the midpoint of the arc $< x_{j+1}, x_j >$. Let c' denote the midpoint of $< l, r >$. Then $d(< c', x_j >) > d(\hat{z}_m, x_j) = \frac{\lambda}{2}$. Actually, $\hat{y}_1'$ and $\hat{y}_2'$ are symmetric on c' with distance of $\max\{\frac{\lambda'}{2}, \frac{d}{2}\}$. Hence, we know that x_j cannot on the arc $[\hat{y}_1', \hat{y}_2']$.

If x_j is on the arc $[y_1', y_2']$, it is obvious that $u_j((y_1', y_2'), x_j) = d(y_1', y_2') < u_j((y_1, y_2), x_j)$. Finally, consider the case that x_j is on the arc $[\hat{y}_2, y_1']$. The utility of x_j is

$$u_j((y_1', y_2'), x_j) = 2d(x_j, \hat{c}').$$

Note that $u_j((y_1, y_2), x_j) = 2d(x_j, z_m)$ and $d(< c', x_j >) > d(< \hat{z}_m, x_j >)$. Therefore, we have

$$\begin{aligned} u_j((y_1', y_2'), x_j) &= 2d(x_j, \hat{c}') = 1 - 2d(< c', x_j >) \\ &< 1 - 2d(< \hat{z}_m, x_j >) = 2d(x_j, z_m) = u_j((y_1, y_2), x_j), \end{aligned}$$

which completes the proof. □

Note that not all mechanisms that output a pair of optimal facility locations are strategyproof. Only carefully choosing a pair of optimal facility locations can be strategyproof. For example, let an optimal mechanism be outputting $(x_1, \hat{x}_1)$

if $\lambda \geq \frac{1}{2}$, otherwise, outputting two facilities that are symmetric on z_m with the distance of $\frac{d}{2}$. Consider an instance if $\frac{1}{2} > \lambda > d$. The utility of agent j is $2d(x_j, \hat{z}_m)$. Then if he misreports his location to $(x_j, \hat{x}_{j+1}]$, his utility will increase.

For minimizing the maximum cost, we have the following corollary.

Corollary 2. *A mechanism which outputs the antipodal points of Mechanism 2 is strategyproof and optimal for minimizing the maximum cost.*

4 A Line Interval

In this section, we consider the obnoxious 2-facility location games where all the agents are on a closed line interval $I = [0, 1]$, which was studied in [7]. We present a group strategyproof mechanism for maximizing the total utility, which improves the approximation ratio in [7].

Mechanism 3. *Let* $l_1 = \frac{1}{2}(1-d)$ *and* $l_2 = \frac{1}{2}(1+d)$. *Given a location profile* $\mathbf{x}$, *let* n_l, n_r *and* n_m *denote the number of agents in* $[0, l_1)$, $(l_2, 1]$ *and* $[l_1, l_2]$, *respectively. If* $n_l > \alpha n$, *then output* $(1-d, 1)$, *if* $n_r > \alpha n$, *output* $(0, d)$, *and otherwise, output* $(0, 1)$, *where* $\alpha \geq \frac{1}{2}$ *will be given later.*

Before discussing the strategyproofness and the approximation ratio of Mechanism 3, we import the following useful results, which is showed in [7].

Theorem 4 ([7])**.** *For any* $x_i \in [0, l_1)$, *we have* $u_i((1-d, 1), x_i) \geq u_i((0, 1), x_i) \geq u_i((0, d), x_i)$; *for any* $x_i \in (l_2, 1]$, *we have* $u_i((0, d), x_i) \geq u_i((0, 1), x_i) \geq u_i((1-d, 1), x_i)$; *for any* $x_i \in [l_1, l_2]$, *we have* $u_i((0, 1), x_i) \geq u_i((0, d), x_i), u_i((1-d, 1), x_i)$.

Theorem 5 ([7])**.** *One of* $(0, d), (1-d, 1)$ *and* $(0, 1)$ *is optimal for maximizing the total utility.*

Theorem 6. *Mechanism 3 is group strategyproof with an approximation ratio of*

$$\rho = \begin{cases} \max\{\frac{(2-d)-(1+d)\alpha}{\alpha(1-d)+d}, \frac{1}{\alpha(1-d)+d}, (1-d)\alpha + 1\} & \text{if } d \leq \frac{1}{3}, \\ \max\{\frac{1}{\alpha(1-d)+d}, (1-d)\alpha + 1\} & \text{otherwise.} \end{cases}$$

Moreover, if $\frac{1-d}{1+d} \geq \frac{\sqrt{d^2-2d+5}-(1+d)}{2(1-d)}$, *let* $\alpha = \frac{\sqrt{(1+d)^2(2-d)^2+8(1-d)^3}-(2-d)(1+d)}{2(1-d)^2}$, *then the ratio is* $\frac{\sqrt{(1+d)^2(2-d)^2+8(1-d)^3}+d^2-3d}{2(1-d)}$, *which does not exceed* $\sqrt{3}$; *otherwise, let* $\alpha = \frac{\sqrt{d^2-2d+5}-(1+d)}{2(1-d)}$, *then the approximation ratio is* $\frac{\sqrt{d^2-2d+5}+1-d}{2}$, *which is at most* $\frac{\sqrt{5}+1}{2}$.

Before proving this theorem, we give a brief illustration of the approximation ratio. If d is approximately in $[0, 0.258]$ (reserve three decimals), then $\frac{1-d}{1+d} \geq \frac{\sqrt{d^2-2d+5}-(1+d)}{2(1-d)}$. Successively, set the corresponding α and the approximation ratio is at most $\sqrt{3}(\approx 1.732)$. While d is somewhere more than 0.259, the approximation ratio is at most $\frac{\sqrt{5}+1}{2}(\approx 1.618)$.

Proof. Denote Mechanism 3 as f. We first show the strategyproofness. Consider another location profile $\mathbf{x}'$ that each agent $i \in S$ in location profile $\mathbf{x}$ deviates from x_i to x_i'. Similarly, let n_l', n_r', n_m' denote the number of agents in $[0, l_1)$, $(l_2, 1]$, $[l_1, l_2]$ for location profile $\mathbf{x}'$, respectively. For simplicity, let $(y_1', y_2') = f(\mathbf{x}')$. We then show the strategyproofness according to the outputs of $f(\mathbf{x})$. Due to the symmetry, we only need to consider that $f(\mathbf{x}) = (1-d, 1)$ or $(0, 1)$.

We first consider the scenario that $f(\mathbf{x}) = (1-d, 1)$.

If $f(\mathbf{x}')$ is still $(1-d, 1)$, then $f(\mathbf{x}) = f(\mathbf{x}')$. Thus, agent i has no incentive to misreport. If $f(\mathbf{x}') = (0, 1)$, then $n_l' \le \alpha n$. Hence, there exists at least one agent in $[0, l_1)$ misreports his location to a point in $[l_1, 1]$. By Theorem 4, we conclude that $u_i(f(\mathbf{x}), x_i) \ge u_i(f(\mathbf{x}'), x_i)$. Finally, if $f(\mathbf{x}') = (0, d)$, then in this case, we can easily find that there exists at least one agent in $[0, l_1)$ misreports his location to $x_i' \in (l_2, 1]$, since the total number of agents in $[l_1, 1]$ is at most $(1-\alpha)n \le \alpha n$. Analogously, by Theorem 4, that agent cannot benefit by lying.

Finally, for $f(\mathbf{x}) = (0, 1)$, using similar analysis as above, we can draw the same conclusion that f is group strategyproof.

Now we turn to discuss the approximation ratio. We show the approximation ratio by three cases according to the outputs of Mechanism 3. Given any location profile $\mathbf{x}$, let (y_1^*, y_2^*) be a pair of optimal facility locations. Denote $f(\mathbf{x})$ by (y_1, y_2).

Case 1: $(y_1, y_2) = (1-d, 1)$.

By Theorem 5, if the optimal facility locations are $(1-d, 1)$, then Mechanism 3 obtains the optimal solution.

First, consider the scenario that $(y_1^*, y_2^*) = (0, d)$. If $d \le \frac{1}{3}$, then $l_1 \ge d$. For location profile $\mathbf{x}$, we first move all the agents in $[l_1, 1]$ to 1, then move all the agents in $[0, l_1)$ to l_1. In the moving process, $su((y_1^*, y_2^*), \mathbf{x})$ will be better and $su((y_1, y_2), \mathbf{x})$ will not be better. Then the approximation ratio of $\mathbf{x}$ is

$$\frac{su((y_1^*, y_2^*), \mathbf{x})}{su((y_1, y_2), \mathbf{x})} = \frac{su((0, d), \mathbf{x})}{su((1-d, 1), \mathbf{x})} \le \frac{(2-d)n - (1+d)n_l}{n_l(1-d) + nd} \le \frac{(2-d) - (1+d)\alpha}{\alpha(1-d) + d},$$

where the last inequality holds since the left-hand side is decreasing with n_l and $n_l > \alpha n$.

Similarly, for $d > \frac{1}{3}$, we can first move all the agents in $[l_1, 1]$ to 1, then move all the agents in $[0, l_1)$ to l_1. The approximation ratio of $\mathbf{x}$ is

$$\begin{aligned}\frac{su((y_1^*, y_2^*), \mathbf{x})}{su((y_1, y_2), \mathbf{x})} &= \frac{su((0, d), \mathbf{x})}{su((1-d, 1), \mathbf{x})} \le \frac{(2-d)n - 2(1-d)n_l}{n_l(1-d) + nd} \\ &\le \frac{(2-d) - 2(1-d)\alpha}{\alpha(1-d) + d} \le \frac{1}{\alpha(1-d) + d}.\end{aligned}$$

The last inequality holds since $\alpha \ge \frac{1}{2}$.

Finally, if $(y_1^*, y_2^*) = (0, 1)$, the approximation ratio of $\mathbf{x}$ is

$$\frac{su((y_1^*, y_2^*), \mathbf{x})}{su((y_1, y_2), \mathbf{x})} = \frac{su((0, 1), \mathbf{x})}{su((1-d, 1), \mathbf{x})} \le \frac{n}{n_l + (n - n_l)d} \le \frac{1}{\alpha(1-d) + d}.$$

The first inequality hold since the utility of each agent in $[l_1, 1]$ is at least d. The last inequality holds since $n_l > \alpha n$.

Case 2: $(y_1, y_2) = (0, 1)$.

Due to the symmetry, we only need to consider the case that $(y_1^*, y_2^*) = (0, d)$.

If $d \leq \frac{1}{3}$, we can first move all the agents in $(l_2, 1]$ to 1, then move all the agents in $[l_1, l_2]$ to l_2, finally move all the agents in $[0, l_1)$ to l_1. The approximation ratio of $\mathbf{x}$ is

$$\frac{su((y_1^*, y_2^*), \mathbf{x})}{su((y_1, y_2), \mathbf{x})} = \frac{su((0, d), \mathbf{x})}{su((0, 1), \mathbf{x})} \leq \frac{n_l(1 - 2d) + n_m + n_r(2 - d)}{n} \leq (1 - d)\alpha + 1,$$

where the last inequality holds since that the total utility admits its maximum value when exact αn agents are at 1, and the remaining $(1 - \alpha)n$ agents are at l_2.

Similarly, when $d > \frac{1}{3}$, it still holds that $\frac{su((y_1^*, y_2^*), \mathbf{x})}{su((y_1, y_2), \mathbf{x})} \leq (1 - d)\alpha + 1$.

Case 3: $(y_1, y_2) = (0, d)$. Due to the symmetry, the analysis is similar to Case 1.

Therefore, we can see if $d \leq \frac{1}{3}$, the approximation ratio is

$$\rho = \max\{\frac{(2 - d) - (1 + d)\alpha}{\alpha(1 - d) + d}, \frac{1}{\alpha(1 - d) + d}, (1 - d)\alpha + 1\}. \tag{1}$$

View the above three formulas as functions on α. It is easy to see that the second and the third function are decreasing and increasing on α, respectively. Meanwhile, the first and the second functions also have unique intersection $\frac{1-d}{1+d}$. And if $\alpha < \frac{1-d}{1+d}$, then $\frac{(2-d)-(1+d)\alpha}{\alpha(1-d)+d} > \frac{1}{\alpha(1-d)+d}$. Additionally, the intersection of the second and third functions is $\frac{\sqrt{d^2-2d+5}-(1+d)}{2(1-d)}$. Moreover, the minimal value of Eq. (1) is according to the relative positions of the above two intersections. Finally, the intersection of the second and third functions is $\frac{\sqrt{(1+d)^2(2-d)^2+8(1-d)^3}-(2-d)(1+d)}{2(1-d)^2}$.

Therefore, if $\frac{1-d}{1+d} \geq \frac{\sqrt{d^2-2d+5}-(1+d)}{2(1-d)}$, Eq. (1) reaches its minimum value at $\alpha = \frac{\sqrt{(1+d)^2(2-d)^2+8(1-d)^3}-(2-d)(1+d)}{2(1-d)^2}$, and the value is at most $\sqrt{3}$. Otherwise, Eq. (1) gets its minimum value at $\alpha = \frac{\sqrt{d^2-2d+5}-(1+d)}{2(1-d)}$. And the value is $\frac{\sqrt{d^2-2d+5}+1-d}{2} \leq \frac{\sqrt{5}+1}{2}$.

If $d > \frac{1}{3}$, the approximation ratio is

$$\rho = \max\left\{\frac{1}{\alpha(1 - d) + d}, (1 - d)\alpha + 1\right\}. \tag{2}$$

Using the analogous analysis, we can obtain the minimum approximation ratio when $\alpha = \frac{\sqrt{d^2-2d+5}-(1+d)}{2(1-d)}$. It is worth to mention that $\frac{1-d}{1+d} < \frac{\sqrt{d^2-2d+5}-(1+d)}{2(1-d)}$ always holds for $d > \frac{1}{3}$, which completes our proof. □

5 Conclusions

In this paper, we study the 2-facility location games with the minimum distance requirement, and investigate strategyproof mechanisms with respect to two objectives: minimizing the total cost/maximizing the total utility and minimizing the maximum cost/maximizing the minimum utility. We are mainly focused on a circle network, and obtain optimal strategyproof mechanisms for minimizing the maximum cost/maximizing the minimum utility. Meanwhile, a group strategyproof mechanism with the approximation ratio of $\frac{1}{2d}$(resp. $2-2d$) is designed for minimizing the total cost(resp. maximizing the total utility). On the line interval, for maximizing the total utility, we obtain a group strategyproof mechanism which improves the approximation ratio in [7].

It is meaningful to improve the lower bound for 2-facility location games with minimum distance requirement on the circle. Besides, our work could be extended in several ways, such as considering randomized mechanisms for two-facility location games with minimum distance requirement; extending facility location games to more general metric spaces.

References

1. Alon, N., Feldman, M., Procaccia, A.D., Tennenholtz, M.: Strategyproof approximation mechanisms for location on networks. CoRR, abs/0907.2049 (2009)
2. Chen, X., Fang, Q., Liu, W., Ding, Y., Nong, Q.: Strategyproof mechanisms for 2-facility location games with minimax envy. J. Comb. Optim., 1–17 (2021)
3. Chen, Z., Fong, K.C.K., Li, M., Wang, K., Yuan, H., Zhang, Y.: Facility location games with optional preference. Theoret. Comput. Sci. **847**, 185–197 (2020)
4. Cheng, Y., Han, Q., Wei, Yu., Zhang, G.: Strategy-proof mechanisms for obnoxious facility game with bounded service range. J. Comb. Optim. **37**(2), 737–755 (2019)
5. Cheng, Y., Yu, W., Zhang, G.: Mechanisms for obnoxious facility game on a path. In: Proceedings of the 5th International Conference on Combinatorial Optimization and Applications, pp. 262–271 (2011)
6. Cheng, Y., Wei, Yu., Zhang, G.: Strategy-proof approximation mechanisms for an obnoxious facility game on networks. Theoret. Comput. Sci. **497**, 154–163 (2013)
7. Duan, L., Li, B., Li, M., Xu, X.: Heterogeneous two-facility location games with minimum distance requirement. In: Proceedings of the 18th International Conference on Autonomous Agents and MultiAgent Systems, pp. 1461–1469 (2019)
8. Fong, C.K.K., Li, M., Lu, P., Todo, T., Yokoo, M.: Facility location games with fractional preferences. In: Proceedings of the AAAI Conference on Artificial Intelligence, vol. **32** (2018)
9. Fotakis, D., Tzamos, C.: On the power of deterministic mechanisms for facility location games. ACM Trans. Econ. Comput. **2**(4), 1–37 (2014)
10. Ibara, K., Nagamochi, H.: Characterizing mechanisms in obnoxious facility game. In: Proceedings of the 6th International Conference on Combinatorial Optimization and Applications, pp. 301–311 (2012)
11. Li, M., Lu, P., Yao, Y., Zhang, J.: Strategyproof mechanism for two heterogeneous facilities with constant approximation ratio. In: Proceedings of the 29th International Joint Conference on Artificial Intelligence, pp. 238–245 (2020)
12. Liu, W., Ding, Y., Chen, X., Fang, Q., Nong, Q.: Multiple facility location games with envy ratio. Theoret. Comput. Sci. **864**, 1–9 (2021)
13. Lu, P., Sun, X., Wang, Y., Zhu, Z.A.: Asymptotically optimal strategy-proof mechanisms for two-facility games. In: Proceedings of the 11th ACM Conference on Electronic Commerce, pp. 315–324 (2010)

14. Mei, L., Li, M., Ye, D., Zhang, G.: Facility location games with distinct desires. Discret. Appl. Math. **264**, 148–160 (2019)
15. Mei, L., Ye, D., Zhang, G.: Mechanism design for one-facility location game with obnoxious effects on a line. Theoret. Comput. Sci. **734**, 46–57 (2018)
16. Moulin, H.: On strategy-proofness and single peakedness. Public Choice **35**(4), 437–455 (1980)
17. Procaccia, A.D., Tennenholtz, M.: Approximate mechanism design without money. ACM Trans. Econ. Comput. **1**(4), 1–26 (2013)
18. Schummer, J., Vohra, R.V.: Strategy-proof location on a network. J. Econ. Theory **104**(2), 405–428 (2002)
19. Serafino, P., Ventre, C.: Truthful mechanisms without money for non-utilitarian heterogeneous facility location. In: Proceedings of the 29th AAAI Conference on Artificial Intelligence, pp. 25–30 (2015)
20. Xu, X., Li, B., Li, M., Duan, L.: Two-facility location games with minimum distance requirement. J. Artif. Intell. Res. **70**, 719–756 (2021)
21. Zou, S., Li, M.: Facility location games with dual preference. In: Proceedings of the 2015 International Conference on Autonomous Agents and Multiagent Systems, pp. 615–623 (2015)

Open Shop Scheduling Problem with a Non-resumable Flexible Maintenance Period

Yuan Yuan[1], Xin Han[1](✉), Xinbo Liu[2], and Yan Lan[3]

[1] School of Software, Dalian University of Technology, Dalian 116620, Liaoning, China
hanxin@dlut.edu.cn
[2] SolBridge International School of Business, Woosong University, Uam-ro 128, Daejeon 34613, South Korea
xliu215@student.solbridge.ac.kr
[3] School of Information and Communication Engineering, Dalian Minzu University, Dalian 116600, Liaoning, China
lanyan@dlnu.edu.cn

Abstract. This paper investigates two-machine open shop scheduling problem in which one flexible maintenance period is imposed on the second machine, where the maintenance period has to start within a given time window and its duration is constant. The objective is to minimize the makespan. Mosheiov et al. present a 3/2-approximation algorithm for this problem. We propose a 4/3-approximation algorithm with $O(n)$ time complexity.

Keywords: Scheduling · Machine maintenance · Approximation algorithm

1 Introduction

Machine scheduling problems with machine maintenance constraints have been attracting many research interests in the field of operational research. This paper investigates the two-machine open shop scheduling problem in which the second machine is subject to maintenance period, this problem was first introduced in [1].

There are two common kinds of maintenance period (MP): *fixed* MP and *flexible* MP. The former is the case where the starting time and completion time of the MP are fixed, while the latter is the case where the MP starts within a given window and the duration is constant.

In the scheduling literature, there are three scenarios handling scheduling problems with the fixed MP. During the MP no job can be processed on that machine. Suppose that some job fails to finish on a certain machine prior to the MP. If the job can be continued after the MP without any penalty, Lee [2] calls this scheduling model *resumable*; if the uncompleted job must restart from scratch after the completion time of the MP, the model is named *non-resumable* [2]; and if the job will have to partially restart after the machine has become available again, the model is called *semi-resumable* [3].

The two-machine open shop scheduling problem with one fixed MP is NP-hard [4]. We prefer to look for an approximation solution rather than an exact optimal solution for these problems. Therefore, this paper concentrates on designing approximation

D.-Z. Du et al. (Eds.): COCOA 2021, LNCS 13135, pp. 512–526, 2021.
https://doi.org/10.1007/978-3-030-92681-6_40

algorithms and analyzing their performance. A polynomial-time algorithm is called a *ρ-approximation algorithm*, if it creates a schedule with makespan at most ρ times the optimal value (where $\rho \geq 1$); where the value of ρ is called a *worst-case ratio bound*. A family of ρ-approximation algorithms is called a *polynomial-time approximation scheme*, or a PTAS for short, if $\rho = 1 + \varepsilon$ for any fixed $\varepsilon > 0$ and the running time is polynomial with respect to the length of problem input.

While interesting results have been proposed for two-machine open shop scheduling problems with fixed MPs, there has been few results on the shop scheduling problem with flexible MPs. For the problem with one fixed MP, Breit et al. [4] gave a $\frac{4}{3}$-approximation algorithm considering the resumable MP, while Breit et al. [6] presented a $\frac{4}{3}$-approximation algorithm with the non-resumable MP. Yuan et al. [7] gave a PTAS for open shop scheduling problem with one fixed MP under the non-resumable scenario. Kubzin et al. [8] presented two PTAS, where one deals with the problem with one fixed MP on each machine and the other tackles the problem with several fixed MPs on one of machines, considering the resumable scenario. Mosheiov et al. [1] established two $\frac{3}{2}$-approximation algorithms, one for flow shop, and the other for open shop, provided that one flexible MP is imposed on the second machine.

For the two-machine open shop scheduling problem with one flexible MP on the second machine, we present a $\frac{4}{3}$-approximation algorithm which outperforms the previous approximation algorithm proposed in [1], under the same time complexity. The main idea of our approximation algorithm is as follows: we first define two big jobs, then try several possible schedules for big jobs, finally run the greedy strategy to schedule the remaining jobs.

The remainder of this paper is organized as follows. Section 2 gives a formal description of the problem. We present a $\frac{4}{3}$-approximation algorithm for two-machine open shop scheduling problem with a flexible MP in Sect. 3. In Sect. 4, we conclude with a short summary.

2 Preliminaries

In the two-machine open shop scheduling problem, we are given two machines M_1, M_2 and a set of jobs $N = \{1, \cdots, n\}$. Each job j consists of two operations: $O_{j,1}$ and $O_{j,2}$, where operation $O_{j,1}$ is processed on M_1 with a_j time units, and operation $O_{j,2}$ is performed on M_2 with b_j time units. Each job is processed on at most one machine at a time, while each machine processes no more than one job at a time. No preemption is allowed in processing of any operation. If the processing routes are not given in advance, but have to be chosen, the processing system is called the *open shop*. The objective mentioned in this paper is to minimize the *makespan*, i.e., the completion time of all jobs on two machines.

Following the notation of [9], the classical two-machine open shop scheduling problem is denoted by $O2||C_{\max}$, which can be solved in $O(n)$ time due to Gonzalez and Sahni [10]. We define $a(N) = \Sigma_{j=1}^{n} a_j$, $b(N) = \Sigma_{j=1}^{n} b_j$ and denote the optimal makespan of problem $O2||C_{\max}$ by $C_{\max}(\widehat{S}^*)$, according to [10], it is

$$C_{\max}(\widehat{S}^*) = \max\{a(N), b(N), \max_{j \in N}\{a_j + b_j\}\}. \quad (1)$$

In this paper, we study an extension of the classical two-machine open shop scheduling problem, in which there is a flexible MP on machine M_2. The MP starts within a given time interval $[T_1, T_2]$ and the duration of the MP is Δ which is irrelevant to T_1 and T_2, and during the MP no job can be processed on the corresponding machine. We denote the two-machine open shop scheduling problem with a flexible MP on M_2 by $O2|nr - f(M_2)|C_{\max}$, where nr denotes the non-resumable MP and $f(M_2)$ denotes the MP is imposed on machine M_2.

For a feasible schedule S for problem $O2|nr - f(M_2)|C_{\max}$, let $C_{j,i}(S)$ denote the completion time of job j on machine M_i in schedule S, and $C_{\max}(S)$ denote the makespan of schedule S. Let S^* denote the optimal schedule and $C_{\max}(S^*)$ the optimal makespan. Let LB denote the lower bound of the optimal makespan, and we define $LB = \max\{a(N), b(N) + \Delta, T_1 + \Delta, \max_{j \in N}\{a_j + b_j\}\}$, we can obtain

$$C_{\max}(S^*) \geq LB. \tag{2}$$

Observation 1. $C_{\max}(\widehat{S^*}) \leq C_{\max}(S^*)$.

We prove that Observation 1 is true. Firstly, we can obtain that any one feasible schedule for problem $O2|nr - f(M_2)|C_{\max}$ is also feasible for problem $O2||C_{\max}$; secondly, for the same feasible schedule, the makespan of problem $O2||C_{\max}$ is no worse than that of problem $O2|nr - f(M_2)|C_{\max}$; finally, for problem $O2||C_{\max}$, the optimal makespan is no more than the makespan of any feasible schedule. Therefore, the conclusion is valid.

3 A $\frac{4}{3}$-Approximation Algorithm

In this section, we present a $\frac{4}{3}$-approximation algorithm for $O2|nr - f(M_2)|C_{\max}$ whose time complexity is $O(n)$.

3.1 The Greedy Algorithm

Before introducing the formal algorithm, we first state the standard greedy algorithm in [11], which is used in the subsequent algorithm. Sevastianov and Woeginger [12] apply the greedy algorithm as a part of their PTAS for $Om||C_{\max}$. The greedy algorithm is involved in approximation algorithms proposed by [1,6,8]. When $m = 2$, the standard greedy algorithm is stated below:

Algorithm 1: The Greedy Algorithm *GA*

Input: The set of jobs, two machines M_1 and M_2.
Output: A feasible schedule S.
1 At any time t, when machine M_i, for $i = 1, 2$, becomes available, arbitrarily chooses operation $O_{k,i}$ which is unscheduled, such that operation $O_{k,3-i}$ is not being scheduled on M_{3-i}.

For problem $O2|nr-f(M_2)|C_{\max}$, we introduce a modified greedy algorithm $GA(c)$.

Algorithm 2: The Greedy Algorithm $GA(c)$

Input: The set of jobs, the starting time of MP $c \in [T_1, T_2]$.
Output: A feasible schedule S.

1 At any time t, when machine M_i, for $i = 1, 2$, becomes available, arbitrarily chooses unscheduled operation $O_{k,i}$, such that operation $O_{k,3-i}$ is not being scheduled on M_{3-i}. In particular, if machine M_2 is available and $t < c$, process operation $O_{k,2}$ only if it can be completed before time c.

The running time of the $GA(c)$ is $O(n)$.

Observation 2. *When we apply $GA(c)$ for problem $O2|nr-f(M_2)|C_{\max}$, there is at most one idle interval W on M_1 or after the MP on M_2. Only one job k is left to be processed after W, and during W the other operation of job k is being processed and whose length is more than the length of W. See Fig. 1(a) and (b).*

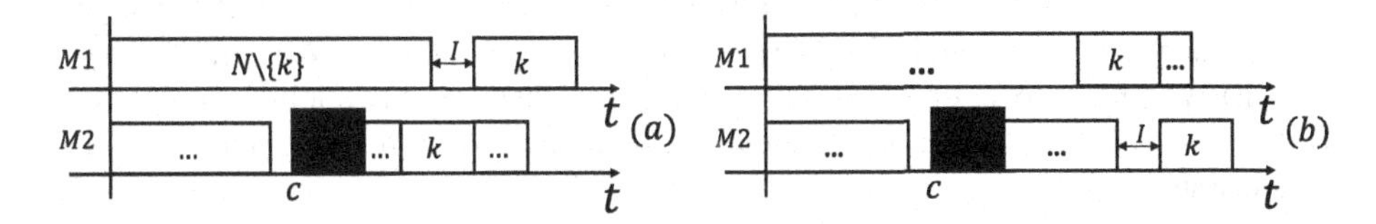

Fig. 1. Possible schedules found by $GA(c)$ for $O2|nr-f(M_2)|C_{\max}$.

Definition 1. *We denote by N_1 the operations completed before the MP on M_2, and by N_2 the operations processed after the MP on M_2.*

Lemma 1. *When we apply $GA(c)$ for problem $O2|nr-f(M_2)|C_{\max}$, we have the following properties:*

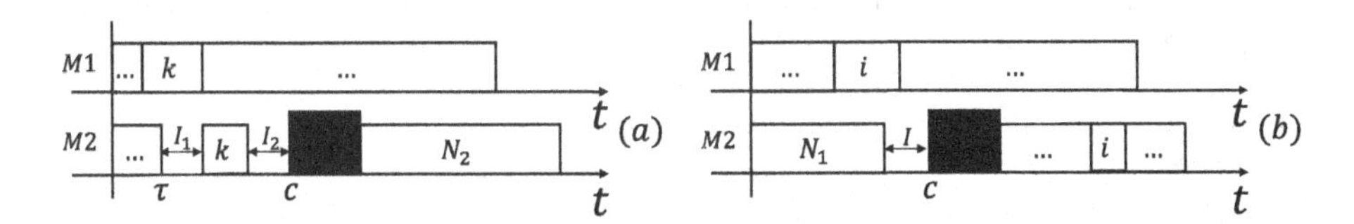

Fig. 2. Possible schedules found by $GA(c)$ for $O2|nr-f(M_2)|C_{\max}$.

(i) if no idle interval occurs after the MP on M_2, then there are at most two idle intervals before the MP.

(ii) if there are two idle intervals whose length are I_1 and I_2, respectively, before the MP on M_2, then the total length of two intervals is less than the processing time of any one job $j \in N_2$ on M_2, i.e., $I_1 + I_2 < b_j$, see Fig. 2(a).

(iii) if there is only one idle interval whose length is I before the MP on M_2, then there exists at most one job $i \in N_2$ such that $b_i \leq I$. See Fig. 2(b).

Proof. (i) We prove it by contradiction. Assume there are three idle intervals before the MP on M_2, let I_1, I_2 and I_3 denote the length of three intervals, respectively, and let τ denote the starting time of the first idle interval, See Fig. 3. During the first and second interval, jobs k and i are being processed on M_1, respectively. According to $GA(c)$, job i can be processed at time τ rather than after job k on M_2. Therefore, the second idle interval will not appear.

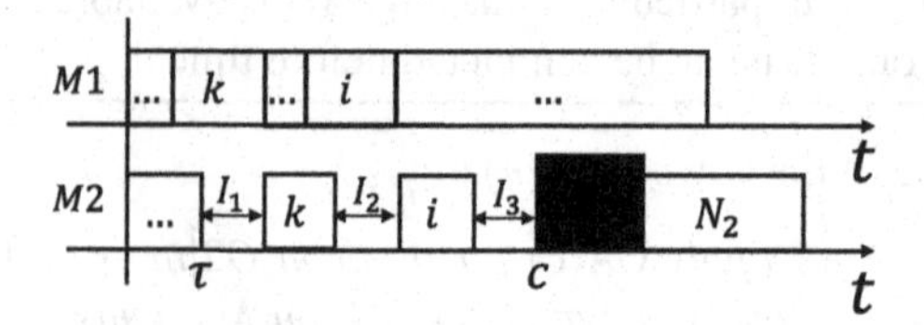

Fig. 3. A counterexample of Lemma 1 **(i)**.

(ii) Let τ denote the starting time of the first idle interval, set $\delta = c - \tau$. Any one job $j \in N_2$ cannot be processed before the MP since $\delta < b_j$, so $I_1 + I_2 < \delta < b_j$, $\forall j \in N_2$.

(iii) We again prove it by contradiction. Assume there exist two jobs $i, j \in N_2$ such that $b_i \leq I$ and $b_j \leq I$, then one of them can be processed in the idle interval. So the conclusion is valid. □

3.2 Approximation Algorithm and Analysis

In this section, we present and analyze a $\frac{4}{3}$-approximation algorithm for problem $O2|nr-f(M_2)|C_{\max}$. The running time of this algorithm is linear in the size of the instance. The main idea of the algorithm is as follows: we first define two big jobs, then try several possible schedules for big jobs, finally run $GA(c)$ to schedule the remaining jobs.

We define jobs p and q such that $b_p \geq b_q \geq \max\{b_j | j \in N \backslash \{p,q\}\}$, and refer to p as the first big job and q as the second big job.

Lemma 2. *Due to the definition of jobs p and q and Eq. (2), we can get* $\forall j \in N \backslash \{p,q\}$, $b_j \leq \frac{1}{3}LB$.

Proof. We prove the result by contradiction. Assume there exists a job $j \in N \backslash \{p,q\}$ such that $b_j > \frac{1}{3}LB$, which implies $b_p \geq b_q \geq b_j > \frac{1}{3}LB$. So, we can obtain $b(N) \geq b_p + b_q + b_j > LB$, which contradicts (2). □

Our approximation algorithm is given below.

Algorithm 3: $\frac{4}{3}$-approximation algorithm

Input: A set of jobs, the starting time window of MP is $[T_1, T_2]$, the duration of MP is Δ.

Output: A feasible schedule S

1 Find jobs p and q, set $Q = \emptyset$;
2 **if** $b_p \leq T_2$ **then**
3 Generate schedule S_1: first schedule job q on M_1 and p on M_2 at time zero, then run $GA(\max\{b_p, T_1\})$ for the remaining jobs; $Q \leftarrow Q \cup \{S_1\}$;
4 If $b_p + b_q \leq T_2$, generate schedule S_2: first schedule job q on M_2 at time zero, then process job q on M_2 at time b_q, finally run $GA(\max\{b_p + b_q, T_1\})$ for the remaining jobs; $Q \leftarrow Q \cup \{S_2\}$;
5 If $\max\{b_p + b_q, a_q + b_q\} \leq T_2$, generate schedule S_3: first schedule job q on M_1 and p on M_2 at time zero, then process job q on M_2 at time $\max\{b_p, a_q\}$, finally run $GA(\max\{b_p + b_q, a_q + b_q, T_1\})$ for the remaining jobs; $Q \leftarrow Q \cup \{S_3\}$;
6 If $\max\{b_p + b_q, a_p + b_p\} \leq T_2$, generate schedule S_4: reverse the roles of jobs p and q in schedule S_3; $Q \leftarrow Q \cup \{S_4\}$;
7 Choose the schedule with the minimal makespan from Q;
8 **else**
9 **if** $b_q \leq T_2$ **then**
10 Generate schedule S_5: first schedule job p on M_1 and q on M_2 at time zero, then run $GA(\max\{b_q, T_1\})$ for the remaining jobs;
11 **else**
12 Generate schedule S_6: run Gonzalez-Sahni algorithm for all jobs, provided that two machines are available all the time from time $T_1 + \Delta$;
13 Generate schedule S_7: first schedule job p on M_1 at time zero, then process job q on M_1 at time a_p, finally run $GA(T_1)$ for the remaining jobs;
14 Choose the schedule with the minimal makespan from S_6 and S_7;

The running time of Algorithm 3 is $O(n)$.

Theorem 1. *Algorithm 3 outputs schedule S which satisfies*

$$C_{\max}(S) \leq \frac{4}{3} C_{\max}(S^*), \tag{3}$$

and this bound is tight.

Proof. We are going to branch in two cases depending on $b_p \leq T_2$.

Case 1: $b_p \leq T_2$. It is not difficult to verify the following lower bound of the optimal makespan hold

$$C_{\max}(S^*) \geq \max\{b_p + b_q, T_1\} + \Delta \tag{4}$$

Case 1-1: $b_q \le \frac{1}{3}LB$. We take **Schedule** S_1 into consideration. In this case, we can get a better lower bound of the optimal makespan than (4)

$$C_{\max}(S^*) \ge \max\{T_1, b_p\} + \Delta. \tag{5}$$

If the makespan in schedule S_1 occurs on M_1, then either there is no idle interval on M_1 and S_1 is optimal or there is an idle interval W on M_1. In the latter case, according to Observation 2, there is only one job to be processed after W. The last job is either job p or $j \in N \backslash \{p,q\}$, see Fig. 4(a) and (b), respectively. So we have

$$C_{\max}(S_1) \le \max\{a_p + b_p, a(N) + b_j\} \le \frac{4}{3}C_{\max}(S^*),$$

where the last inequality is due to $b_j \le b_q \le \frac{1}{3}LB$ and (2).

If the makespan in schedule S_1 occurs on M_2, we need to consider whether there is an idle interval after the MP on M_2 or not. If there is an idle time W after the MP, due to Observation 2, there is only one job either $j \in N \backslash \{p,q\}$ see Fig. 4(c), or q see Fig. 4(d), to be processed after W. So we have

$$C_{\max}(S_1) = \max\{C_{j,1}(S_1) + b_j, a_q + b_q\} \le \frac{4}{3}C_{\max}(S^*),$$

where the last inequality is due to $b_j \le b_q \le \frac{1}{3}LB$ and (2).

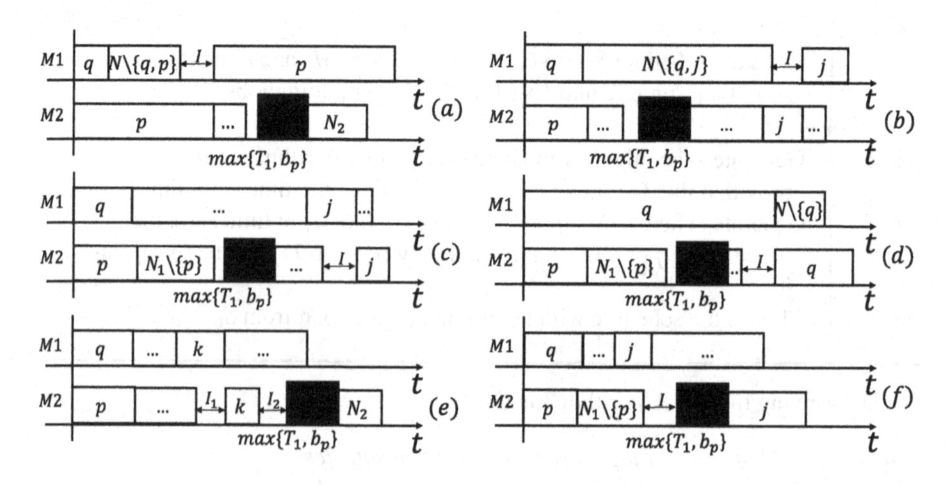

Fig. 4. Schedule S_1.

If there is no idle time after the MP, but there are idle intervals before the MP on M_2. According to Lemma 1, there are at most two gaps before the MP on M_2. If there are two idle intervals before the MP on M_2, let the lengths of two intervals be I_1 and I_2, respectively, see Fig. 4(e), then we have $I_1 + I_2 \le b_j$ $(\forall j \in N_2)$. Therefore, we can obtain

$$C_{\max}(S_1) = b(N_1) + I_1 + I_2 + \Delta + b(N_2) < b(N) + \Delta + b_j \le \frac{4}{3}C_{\max}(S^*),$$

where the last inequality is due to (2) and $b_j \leq b_q \leq \frac{1}{3}LB$. If there is only one idle interval before the MP on M_2, let I denote the length of the idle interval, we may assume that $I > \frac{1}{3}LB$; otherwise we have

$$C_{\max}(S_1) = b(N_1) + I + \Delta + b(N_2) = b(N) + \Delta + I \leq \frac{4}{3}C_{\max}(S^*).$$

We claim that there is at most one job $j \in N_2$ to be processed after the MP, see Fig. 4(f). So we obtain

$$C_{\max}(S_1) = \max\{T_1, b_p\} + \Delta + b_j \leq \frac{4}{3}C_{\max}(S^*),$$

where the last inequality is due to $b_j \leq \frac{1}{3}LB$, (5) and (2).

Case 1-2: $b_q > \frac{1}{3}LB$. We first note that $b(N \backslash \{p,q\}) \leq \frac{1}{3}LB$ due to (2). Then we consider the following cases.

Case 1-2-1: $b_p + b_q > T_2$. We still consider **Schedule** S_1. The conditions $b_p \leq T_2$ and $b_p + b_q > T_2$ hold, which means that in any feasible schedule there is at most one job in p and q to be processed before the MP, so we deduce a new lower bound of the optimal makespan

$$C_{\max}(S^*) \geq \max\{T_1, b_p\} + \Delta + b_q. \tag{6}$$

If the makespan in schedule S_1 occurs on M_1, same as **Case 1-1**, we can obtain $C_{\max}(S_1) \leq \max\{a(N), a_p + b_p, a(N) + b_j\} \leq \frac{4}{3}C_{\max}(S^*)$.

If the makespan in schedule S_1 occurs on M_2 and there is an idle interval W after the MP, similarly as **Case 1-1**, we have $C_{\max}(S_1) = \max\{a_q + b_q, C_{j,1}(S_1) + b_j\} \leq \frac{4}{3}C_{\max}(S^*)$.

If the makespan in schedule S_1 occurs on M_2 but no idle interval occurs after the MP, we notice that $p \notin N_2$ and $b(N_2 \backslash \{q\}) \leq b(N \backslash \{p,q\}) \leq \frac{1}{3}LB$, so it is sufficient to consider Fig. 4(e). Thus, we get

$$C_{\max}(S_1) = \max\{T_1, b_p\} + \Delta + b_q + b(N_2 \backslash \{q\}) \leq \frac{4}{3}C_{\max}(S^*),$$

where the last inequality is due to (2) and (6).

Case 1-2-2: $b_p + b_q \leq T_2$.

Case 1-2-2-1: $a_p + b_p > T_2$ and $a_q + b_q > T_2$. If there is at most one job in p and q to be processed before the MP in the optimal schedule, **Schedule** S_1 is considered. Similarly as **Case 1-2-1**, we can obtain the derived result.

If both p and q are processed before the MP in the optimal schedule, we consider **Schedule** S_2. According to conditions $a_p + b_p > T_2$ and $a_q + b_q > T_2$, it is not hard to get $a_p > b_q > \frac{1}{3}LB$ and $a_q > b_p > \frac{1}{3}LB$. We further have $a(N \backslash \{p,q\}) \leq \frac{1}{3}LB$ due to (2).

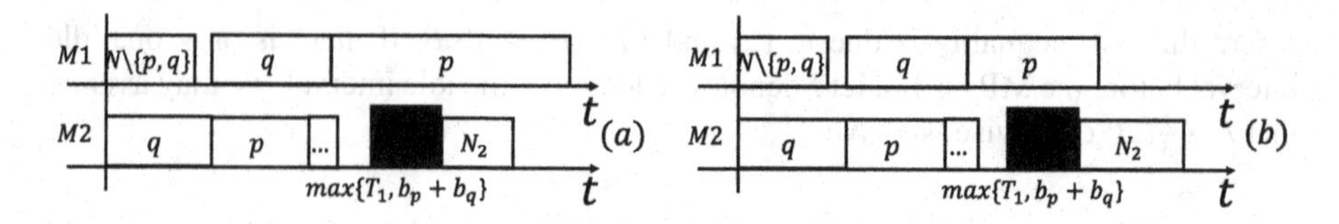

Fig. 5. Schedules S_2.

If the makespan in schedule S_2 occurs on M_1, since $a(N\backslash\{p,q\}) \leq \frac{1}{3}LB < b_q$, there must exist an idle interval on M_1 in schedule S_2, see Fig. 5(a), so we have $C_{\max}(S_2) = b_q + b_p + a_p$. According to conditions $a_p + b_p > T_2$, $a_q + b_q > T_2$ and $b_p + b_q \leq T_2$, we know that the order of jobs p and q on two machines in the optimal schedule is the same as that in schedule S_2, so we have $C_{\max}(S_2) \leq C_{\max}(S^*)$.

If the makespan in schedule S_2 occurs on M_2, we know that scheduling set $N\backslash\{p,q\}$ on two machines will produce no clashes, see Fig. 5(b). Consequently, it is impossible to induce idle time after the MP on M_2, so we have

$$C_{\max}(S_2) = \max\{b_p + b_q, T_1\} + \Delta + b(N_2) \leq \frac{4}{3}C_{\max}(S^*),$$

due to $p, q \notin N_2$, $b(N_2) \leq b(N\backslash\{p,q\}) \leq \frac{1}{3}LB$, (2) and (4).

Case 1-2-2-2: $a_p + b_p \leq T_2$ or $a_q + b_q \leq T_2$.

When $a_q + b_q \leq T_2$, **Schedule** S_2 or S_3 is considered. Without loss of generality, we assume that $a_q + b_q \leq b_p + b_q$ and consider schedule S_3. In the same way, we can prove the case $a_q + b_q > b_p + b_q$.

If the makespan in schedule S_3 occurs on M_1, then either there is no idle interval on M_1 and S_3 is optimal, or there is an idle interval on M_1. In the latter case, see Fig. 6(a) and (b), we have $C_{\max}(S_3) \leq \max\{a_p + b_p, a(N) + b_j\} \leq \frac{4}{3}C_{\max}(S^*)$, due to (2) and Lemma 2.

If the makespan in schedule S_3 occurs on M_2, we need to consider whether there is an idle interval after the MP on M_2 or not. If there is an idle interval after the MP on M_2, then only one job $j \in N\backslash\{p,q\}$ is processed after W on M_2, see Fig. 6(c). So we have $C_{\max}(S_3) = C_{j,1}(S_3) + b_j \leq \frac{4}{3}C_{\max}(S^*)$ due to (2) and Lemma 2.

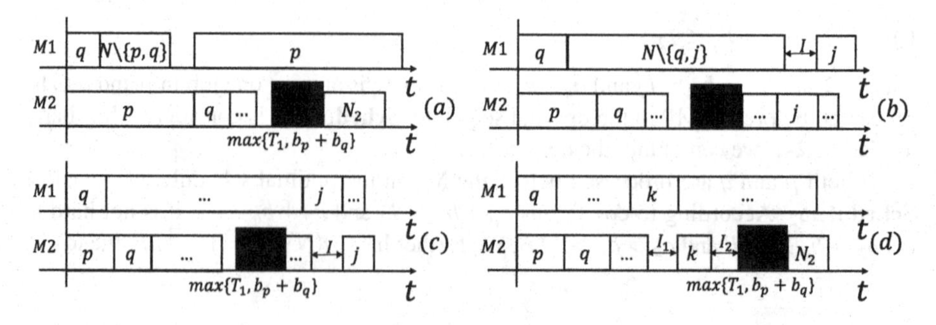

Fig. 6. Schedule S_3.

When there is no idle interval after the MP, see Fig. 6(d), then we have

$$C_{\max}(S_3) = \max\{b_p + b_q, T_1\} + \Delta + b(N_2) \le \frac{4}{3}C_{\max}(S^*),$$

where the last inequality is due to $b(N_2) \le b(N\backslash\{p,q\}) \le \frac{1}{3}LB$, (2) and (4).

When $a_p + b_p \le T_2$, **Schedule** S_2 or S_4 is considered. Without loss of generality, we assume that $a_p + b_p > b_p + b_q$. In the same way, we can prove the case $a_p + b_p \le b_p + b_q$. If in an optimal schedule jobs p and q have the same order on two machines, **Schedule** S_2 is considered. We can verify that (3) is valid similar to **Case 1-2-2-1**. If in an optimal schedule jobs p and q are assigned the opposite route, we can get a new lower bound of the optimal makespan

$$C_{\max}(S^*) \ge \max\{a_p + b_p, T_1\} + \Delta. \tag{7}$$

If the makespan in schedule S_4 occurs on M_1, then either there is no idle interval on M_1 and S_4 is optimal, or there is an idle interval on M_1. In the latter case, see Fig. 7(a), we have $C_{\max}(S_4) \le a(N) + b_j \le \frac{4}{3}C_{\max}(S^*)$.

If the makespan in schedule S_4 occurs on M_2 and there is an idle interval after the MP on M_2, then only one job $j \in N\backslash\{p,q\}$ is processed after W on M_2, see Fig. 7(b). So we have $C_{\max}(S_4) = C_{j,1}(S_4) + b_j \le \frac{4}{3}C_{\max}(S^*)$ due to (2) and Lemma 2.

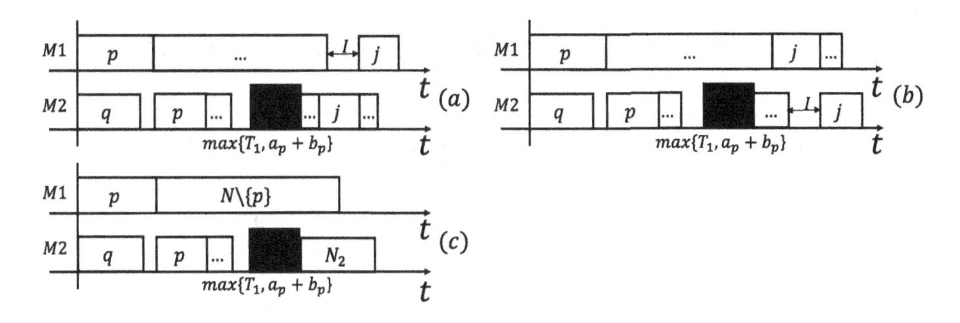

Fig. 7. Schedule S_4.

When there is no idle interval after the MP, see Fig. 7(c), then we have

$$C_{\max}(S_4) = \max\{a_p + b_p, T_1\} + \Delta + b(N_2) \le \frac{4}{3}C_{\max}(S^*),$$

where the last inequality is due to $b(N_2) \le b(N\backslash\{p,q\}) \le \frac{1}{3}LB$, (2) and (7).

Case 2: $b_p > T_2$. This condition means that in any schedule job p must be processed after the MP. We still need to judge whether $b_q \le T_2$ or not.

Case 2-1: $b_q \leq T_2$. We have a new lower bound of the optimal makespan

$$C_{\max}(S^*) \geq \max\{T_1, b_q\} + \Delta + b_p. \tag{8}$$

In this situation, we consider **Schedule** S_5.

If the makespan in schedule S_5 occurs on M_1 and no idle interval occurs on M_1, then S_5 is optimal. If there is one idle interval occurs on M_1, see Fig. 8(a) and (b), we have

$$C_{\max}(S_5) \leq \max\{a_q + b_q, a(N) + b_j\} \leq \frac{4}{3} C_{\max}(S^*),$$

where the last inequality is due to (2) and Lemma 2.

If the makespan in schedule S_5 occurs on M_2 and there is an idle time W after the MP on M_2, according to Observation 2, only one job either p, see Fig. 8(c), or $j \in N \backslash \{p, q\}$, see Fig. 8(d), is processed after W on M_2. We have

$$C_{\max}(S_5) = \max\{a_p + b_p, C_{j,1}(S_5) + b_j\} \leq \frac{4}{3} C_{\max}(S^*),$$

where the last inequality is due to (2) and Lemma 2.

If there is no idle time after the MP on M_2, we need to consider idle intervals before the MP. According to Lemma 1, if there are two idle intervals before the MP on M_2, see Fig. 8(e), let the length of two intervals be I_1 and I_2, respectively, then the total length of two idle intervals is no more than b_j, $j \in N_2 \backslash \{p\}$, i.e., $I_1 + I_2 < b_j$. So we can get

$$C_{\max}(S_5) = b(N_1) + I_1 + I_2 + \Delta + b(N_2) < b(N) + \Delta + b_j \leq \frac{4}{3} C_{\max}(S^*),$$

where the last inequality is due to (2) and Lemma 2. If there is only one idle interval before the MP on M_2, let I denote the length of the idle interval. We assume $I > \frac{1}{3} LB$; otherwise we have

$$C_{\max}(S_5) = b(N_1) + I + \Delta + b(N_2) = b(N) + \Delta + I \leq \frac{4}{3} C_{\max}(S^*).$$

We claim that there are at most two jobs p and $j \in N_2 \backslash \{p\}$ to be processed after the MP on M_2, see Fig. 8(f). Thus, we deduce

$$C_{\max}(S_5) = \max\{T_1, b_q\} + \Delta + b_p + b_j \leq \frac{4}{3} C_{\max}(S^*),$$

where the last inequality is due to $q \notin N_2$, (8) and Lemma 2.

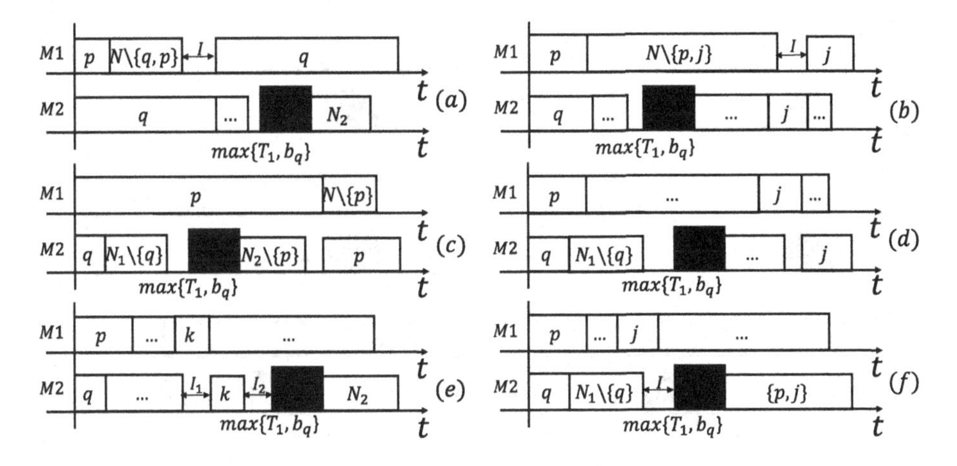

Fig. 8. Schedule S_5.

Case 2-2: $b_q > T_2$. It is not difficult to imply in any feasible schedule jobs p and q are processed after the MP, so that a new lower bound of the optimal makespan

$$C_{\max}(S^*) \geq T_1 + \Delta + b_p + b_q \tag{9}$$

follows. We will choose the best one from **Schedule** S_6 and S_7.

Case 2-2-1: If $b_q > T_1 + \Delta$, we consider **Schedule** S_6. We first claim $T_1 + \Delta \leq \frac{1}{3}C_{\max}(S^*)$, otherwise we would imply $T_1 + \Delta + b_p + b_q > C_{\max}(S^*)$, which contradicts (9). So, we have

$$C_{\max}(S_6) = T_1 + \Delta + C_{\max}(\widehat{S^*}) \leq \frac{4}{3}C_{\max}(S^*),$$

due to the fact that $\widehat{S^*}$ denotes the optimal makespan for $O2||C_{\max}$ and Observation 1.

Case 2-2-2: If $b_q \leq T_1 + \Delta$, we consider **Schedule** S_7. We claim $b_q \leq \frac{1}{3}C_{\max}(S^*)$, otherwise, we would have $T_1 + \Delta + b_p + b_q > C_{\max}(S^*)$, which also contradicts (9).

If the makespan in schedule S_7 occurs on M_1 and no idle interval occurs on M_1, then S_7 is optimal. If there is one idle interval occurs on M_1, according to Observation 2, only one job $j \in N\backslash\{p,q\}$ is processed after the idle interval, see Fig. 9(a). So we have

$$C_{\max}(S_7) = a(N) + I \leq a(N) + b_j \leq \frac{4}{3}C_{\max}(S^*),$$

due to $b_j \leq \frac{1}{3}LB$ and (2).

If the makespan in schedule S_5 occurs on M_2, we need to consider whether there is an idle interval after the MP on M_2 or not. If there is an idle time W after the MP on M_2, according to Observation 2, only one job either p, or $j \in N\backslash\{p,q\}$, or q, is processed after W on M_2. If job p or $j \in N\backslash\{p,q\}$ is the last job on M_2, see Fig. 9(b) and (c), it is easy to deduce

$$C_{\max}(S_7) = \max\{a_p + b_p, C_{j,1}(S_7) + b_j\} \leq \frac{4}{3}C_{\max}(S^*)$$

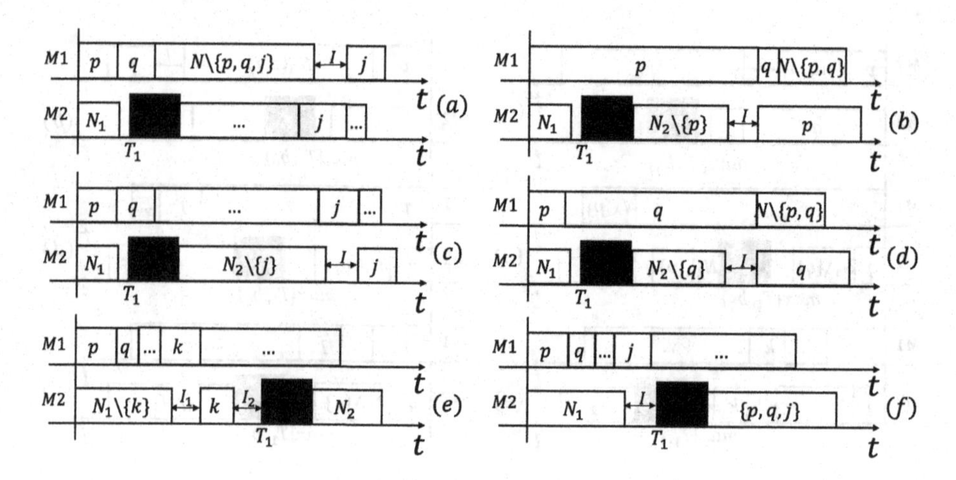

Fig. 9. Schedule S_7.

due to (2) and Lemma 2. If job q is processed after the idle interval on M_2, see Fig. 9(d), we have

$$C_{\max}(S_7) = a_p + a_q + b_q \leq \frac{4}{3}C_{\max}(S^*),$$

where the last inequality is due to (2) and $b_q \leq \frac{1}{3}C_{\max}(S^*)$.

When there is no idle time after the MP on M_2, we consider the idle time before the MP on M_2. According to Lemma 1, if there are two idle intervals before the MP on M_2, see Fig. 9(e), then the total length of two idle intervals is no more than b_j, $j \in N_2\backslash\{p,q\}$. Let the length of two intervals be I_1 and I_2, respectively. So we have

$$C_{\max}(S_7) = b(N_1) + I_1 + I_2 + \Delta + b(N_2) < b(N) + \Delta + b_j \leq \frac{4}{3}C_{\max}(S^*),$$

where the last inequality is due to (2) and Lemma 2. If there is only one idle interval before the MP on M_2, let I denote the length of the idle interval, we assume $I > \frac{1}{3}LB$; otherwise we have

$$C_{\max}(S_7) = b(N_1) + I + \Delta + b(N_2) = b(N) + \Delta + I \leq \frac{4}{3}C_{\max}(S^*).$$

We claim that there are at most three jobs p, q and $j \in N_2\backslash\{p,q\}$ to be processed after the MP, see Fig. 9(f). Thus, we can obtain

$$C_{\max}(S_7) = T_1 + \Delta + b_p + b_q + b_j \leq \frac{4}{3}C_{\max}(S^*),$$

where the last inequality is due to (9) and Lemma 2.

Table 1. A tight example for Algorithm 3

	1	2	3	4
a_j	1	1	1	1
b_j	$U+1$	$U+1$	1	$U-1$

Thus, we have proved that Algorithm 3 derives a $\frac{4}{3}$-approximation ratio. To see that the bound is tight, consider an instance that consists of four jobs, the processing times are given in Table 1, where $U > 4$ is a large number. The MP start window on machine M_2 is defined by $T_1 = U-2$ and $T_2 = U-1$, while the duration of MP is 1, i.e., $\Delta = 1$. It follows that $LB = \max\{a(N), b(N)+\Delta, T_1+\Delta, \max_{j\in N}\{a_j+b_j\}\} = b(N)+\Delta = 3U+3$, and jobs p and q are 1 and 2, respectively.

Since $b_1 = b_2 = U+1 > U-1 = T_2$, so both jobs 1 and 2 are processed after the MP on M_2, Algorithm 3 creates schedules S_6 and S_7. We first consider schedule S_6, see Fig. 10(a), it is not difficult to obtain that the optimal makespan of problem $O2||C_{\max}$ is $C_{\max}(\widehat{S}^*) = \max\{a(N), b(N), \max_{j\in N}\{a_j+b_j\}\} = 3U+2$, so we have $C_{\max}(S_6) = T_1+\Delta+C_{\max}(\widehat{S}^*) = 4U+1$.

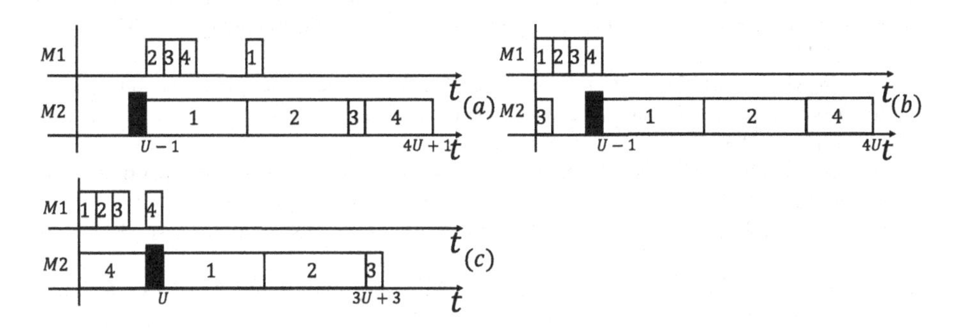

Fig. 10. (a) Schedule S_6, (b) Schedule S_7, (c) Schedule S^*.

Then we consider schedule S_7. Since the jobs besides jobs 1 and 2 to be assigned on M_2 can be taken arbitrarily, it is possible that the order of jobs on M_2 is $(3,1,2,4)$, while the order on machine M_1 is $(1,2,3,4)$, see Fig. 10(b). Consequently, $C_{\max}(S_7) = 4U$. So that $C_{\max}(S) = \min\{C_{\max}(S_6), C_{\max}(S_7)\} = C_{\max}(S_7) = 4U$.

On the other hand, there exists an optimal schedule S^* with $C_{\max}(S^*) = 3U+3$, in which machine M_2 processed the jobs in the sequence $(4,1,2,3)$ since job 4 can be completed before the MP, see Fig. 10(c). As U increases, the ratio $\frac{C_{\max}(S)}{C_{\max}(S^*)}$ approaches $\frac{4}{3}$. □

4 Conclusion

This paper investigates two-machine open shop scheduling problem with one flexible maintenance window under the non-resumable scenario. For this problem, a fast

$\frac{4}{3}$-approximation algorithm is proposed, which solves the open question in [1]. No paper studies the two-machine flow shop scheduling problem with one flexible MP on machine M_1, which is interesting goal in the future research.

References

1. Mosheiov, G., Sarig, A., Strusevich, V.A., Mosheiff, J.: Two-machine flow shop and open shop scheduling problems with a single maintenance window. Eur. J. Oper. Res. **271**(2), 388–400 (2018)
2. Lee, C.: Machine scheduling with an availability constraint. J. Global Optim. **9**(3–4), 395–416 (1996)
3. Lee, C.: Two-machine flowshop scheduling with availability constraints. Eur. J. Oper. Res. **114**(2), 420–429 (1999)
4. Breit, J., Schmidt, G., Strusevich, V.A.: Two-machine open shop scheduling with an availability constraint. Oper. Res. Lett. **29**(2), 65–77 (2001)
5. Kubzin, M., Potts, C., Strusevich, V.A.: Approximation results for flow shop scheduling problems with machine availability constraints. Comput. Oper. Res. **36**(2), 379–390 (2009)
6. Breit, J., Schmidt, G., Strusevich, V.A.: Non-preemptive two-machine open shop scheduling with non-availability constraints. Math. Methods Oper. Res. **57**(2), 217–234 (2003)
7. Yuan, Y., Lan, Y., Ding, N., Han, X.: A PTAS for non-resumable open shop scheduling with an availability constraint. J. Comb. Optim., 1–13 (2021). https://doi.org/10.1007/s10878-021-00773-7
8. Kubzin, M., Strusevich, V., Breit, J., Schmidt, G.: Polynomial-time approximation schemes for two-machine open shop scheduling with nonavailability constraints. Nav. Res. Logist. **53**(1), 16–23 (2006)
9. Lawler, E., Lenstra, J., Kan, A., Shmoys, D.: Sequencing and scheduling: algorithms and complexity. Handbooks Oper. Res. Management Sci. **4**, 445–522 (1993)
10. Gonzalez, T., Sahni, S.: Open shop scheduling to minimize finish time. J. ACM **23**(4), 665–679 (1976)
11. Strusevich, V.A.: A greedy open shop heuristic with job priorities. Ann. Oper. Res. **83**, 253–270 (1998)
12. Sevastianov, S., Woeginger, G.: Makespan minimization in open shops: a polynomial time approximation scheme. Math. Program. **82**(1–2), 191–198 (1998)

Parallel Algorithm for Minimum Partial Dominating Set in Unit Disk Graph

Weizhi Hong, Zhao Zhang(✉), and Yingli Ran(✉)

College of Mathematics and Computer Sciences, Zhejiang Normal University, Jinhua 321004, Zhejiang, China
hxhzz@sina.com, ranyingli@zjnu.edu.cn

Abstract. In this paper, we consider the minimum partial dominating set problem in unit disk graphs (MinPDS-UD). Given a set of points V on the plane with $|V| = n$, two points in V are adjacent in the unit disk graph if their Euclidean distance is no larger than one unit length. A point dominates itself and all its neighboring points. For an integer $k \leq n$, the goal of MinPDS-UD is to find a minimum subset of points $D \subseteq V$ such that at least k points are dominated by D. We present the first parallel algorithm for MinPDS-UD. It runs in $O(\log n)$ rounds on $O(n)$ machines, and achieves a constant approximation ratio.

Keywords: Partial dominating set · Unit disk graph · Parallel algorithm · Approximation ratio

1 Introduction

For a graph $G = (V, E)$ with vertex set V and edge set E, a vertex v is *dominated* by a vertex set $D \subseteq V$ if either $v \in D$ or v has a neighbor in D. A *dominating set* (DS) of G is a subset $D \subseteq V$ which dominates every vertex of G. The *minimum dominating set problem* (MinDS) is to compute a dominating set of the smallest size. It is widely used in many fields such as wireless networks [23]. MinDS is a well-known NP-hard problem [11] and there are extensive studies on its approximation algorithms [7]. This paper considers the partial version which requires D to dominate at least k vertices instead of all vertices. This is called the *minimum partial dominating set problem* (MinPDS).

A *unit disk graph* is a graph in which every vertex corresponds to a point on the plane and there is an edge between two vertices of the graph if the Euclidean distance between the two corresponding points is no greater than one unit. It is a widely adopted topology in a homogeneous wireless sensor network [23]. This paper studies the MinPDS problem on a unit disk graph (MinPDS-UD).

MinPDS is a special case of the *minimum partial set cover problem* (MinPSC). Given a ground set U with n elements, a collection $\mathcal{S}$ of subsets of U and an

This research work is supported in part by NSFC (11901533, U20A2068, 11771013), and ZJNSFC (LD19A010001).

D.-Z. Du et al. (Eds.): COCOA 2021, LNCS 13135, pp. 527–537, 2021.
https://doi.org/10.1007/978-3-030-92681-6_41

integer $k \leq n$, the goal of MinPSC is to find a minimum size sub-collection of $\mathcal{S}$ that covers at least k elements of U. The *minimum set cover problem* (MinSC) is a special case of MinPSC with $k = n$. A MinPDS instance can be viewed as a MinPSC instance by setting $U = V$ and $\mathcal{S} = \{S_v : v \in V\}$, where S_v is the set of neighbors of v including v itself. There are a lot of studies on parallel algorithms for MinSC [2,18,21]. For the partial version MinPSC, Ran *et al.* [21] presented a parallel algorithm with approximation ratio at most $\frac{f}{1-2\varepsilon}$ in $O(\frac{1}{\varepsilon}\log\frac{mn}{\varepsilon})$ rounds, where f is the maximum number of sets containing a common element, $0 < \varepsilon < \frac{1}{2}$ is a constant, and m is the number of the sets. This is the only paper we know on parallel algorithm for the partial cover problem with a theoretical guarantee of performance. Note that f may be as big as $\Theta(n)$, which is very large. The question is, using the speciality of the dominating set problem and the geometric structure of a unit disk graph can we design a parallel algorithm for MinPDS-UD with a constant approximation ratio?

1.1 Related Works

For MinDS, Bar-Yehuda and Moran proved that MinDS on general graphs is polynomially equivalent to MinSC [1]. Thus no polynomial time algorithm can achieve an approximation ratio within $(1 - \varepsilon)\log n$ for any real number $\varepsilon > 0$ unless $P = NP$ [6,8], where n is the number of vertices. Clark *et al.* [4] proved that MinDS is NP-hard even on unit disk graphs. Extensive studies on approximation algorithms for MinDS can be found in the monograph [7]. In particular, MinDS-UD has much better approximation due to the geometric structure of unit disk graphs. Hunt *et al.* [15] presented a PTAS for MinDS-UD using partition and shifting technique, the running time is $n^{O(\frac{1}{\varepsilon^2})}$. Nieberg and Hurink [20] presented a PTAS for MinDS-UD without requiring a geometric representation of the unit disk graph, the running time is $n^{O(\frac{1}{\varepsilon}\log\frac{1}{\varepsilon})}$. For the partial verstion, Joachim *et al.* [17] presented an exact algorithm for MinPDS-UD whose running time is $O(n(16+\varepsilon)^k)$.

As will be seen from Sect. 3, the MinDS-UD problem is related with the *minimum unit disk cover problem* (MinUDC), the goal of which is to find the minimum number unit disks (that is, disks of the same size) to cover all points. The *continuous version* of MinUDC (in which unit disks can be placed anywhere on the plane) is NP-hard [9] and has been known to admit PTAS [13,14]. The *discrete version* of MinUDC is much more tricky. Das *et al.* [5] presented an 18-approximation algorithm for the discrete MinUDC problem. The ratio was improved to $(9 + \varepsilon)$ by Raslimisnata *et al.* [22]. For the weighted version of MinUDC, constant approximation ratio is known [3] based on LP rounding, and Li and Jin [19] found a PTAS using a complicated guessing and dynamic programming technique.

Most of the above algorithms are sequential, which have very high running time, especially for the LP-based methods and dynamic programming methods. Although divide and conquer technique has some parallel mechanism, the above algorithms using this technique run in time at least $n^{O(\frac{1}{\varepsilon}\log\frac{1}{\varepsilon})}$, and thus are not parallel algorithms in the real sense.

There are a lot of studies on parallel algorithms for MinSC. Khuller *et al.* [18] presented a parallel $(f+\varepsilon)$-approximation algorithm in $O(f \log n \log \frac{1}{\varepsilon})$ rounds, where f is the maximum number of sets containing a common element. For the MinPSC problem, Gandhi *et al.* [10] designed a parallel algorithm with approximation ratio $\frac{f}{1-\varepsilon}$ in $(1+f\log\frac{1}{\varepsilon})(1+\log n)$ rounds. Since f might be as large as $\Theta(n)$, the number of rounds is not log-polynomial in the input size. Recently, Ran *et al.* [21] improved the result to approximation ratio at most $\frac{f}{1-2\varepsilon}$ in $O(\frac{1}{\varepsilon}\log\frac{mn}{\varepsilon})$ rounds. As we have noted before, these ratios for the general problem might be too large for a geometric setting. This motivates us to find a better parallel algorithm for MinPDS-UD.

1.2 Our Contributions

In this paper, we design a parallel algorithm for MinPDS-UD. Although MinPDS is a special case of MinPSC, compared with [21], which has approximation ratio $\frac{f}{1-2\varepsilon}$ for MinPSC in $O(\frac{1}{\varepsilon}\log\frac{mn}{\varepsilon})$ rounds, special structure indeed brings new benefit: a constant approximation ratio can be achieved in $O(\log n)$ rounds for MinDS-UD, where n is the number of vertices of the unit-disk graph. This is the first parallel constant approximation algorithm for MinPDS-UD.

In Sect. 2, we present an algorithm for MinPDS-UD by utilizing a relation between maximal independent set and dominating set, obtaining approximation ratio at most 80 in $O(\log n)$ rounds on $O(n)$ machines. Then in Sect. 3, we propose another algorithm by exploring a relation between unit disk cover and dominating set, improving the ratio to 14 in $O(\log n)$ rounds on $O(n)$ machines. A big challenge brought by the "partial" consideration is to determine which points are to be dominated. We employ a greedy idea in a parallelized manner.

2 A Constant Approximation Algorithm for MinPDS-UD

In this section, we make use of maximal independent set to design a parallel algorithm for MinPDS-UD. An *independent set* (IS) in a graph is a set of mutually nonadjacent vertices. A maximal independent set (MIS) is an IS such that adding any vertex is no longer independent. Note that an MIS is also a DS.

A unit disk graph G can also be viewed as an intersection graph of unit disks, that is, every vertex corresponds to a point on the plane and a disk of diameter 1 centered at this point. Two vertices are adjacent in G if and only if their corresponding unit disks have nonempty intersection. From such a point of view, a DS of a unit disk graph G is a set of unit disks D such that every other unit disk of $V(G)\setminus D$ has a nonempty intersection with some unit disk in D, and an IS of G is a set of disjoint unit disks.

For a MinPDS-UD instance on unit disk graph G with a geometric representation on the plane, suppose the n points are contained in a square Q. Partition Q into blocks of side-length 2×2, yielding a partition P. If block b in P contains no point, then b is called an empty block. Otherwise, b is called a nonempty block. Let $block(P)$ be the set of all nonempty blocks in partition P, i.e. $block(P) = \{b$: there exists at least one point in b, $b \in P\}$. For a $b \in block(P)$, denote by $V_P(b)$ the set of points contained in b with respect to partition P, sort all nonempty blocks as $b_1, b_2, \cdots, b_q$ such that $|V_P(b_1)| \geq |V_P(b_2)| \geq \cdots \geq |V_P(b_q)|$. Since $|V| = n$ and $q \leq n$, there exists an index n_P such that

$$\sum_{i=1}^{n_P} |V_P(b_i)| \geq k \text{ and } \sum_{i=1}^{n_P-1} |V_P(b_i)| < k. \tag{1}$$

The algorithm is described in Algorithm 1. It returns the better one between two solutions A_{P_1} and A_{P_2} with respect to two partition P_1 and P_2. Assume, without loss of generality, that both P_1 and P_2 contain all points. For each partition P, sort all nonempty blocks b in decreasing order of $|V_P(b)|$ and find the index n_P satisfying inequality (1). Our next step is to find, for each $b \in block(P)$, a dominating set dominating all the points in b. Note that a point p in b might be dominated by a vertex v whose center is outside of b. Since two vertices are adjacent if and only if their corresponding points have distance no larger than 1, such a v must have its center in the *extended block* b' of b, which is obtained from b through extending its four boundaries by 1 (see Fig. 1). Compute a maximal independent set $\mathcal{I}(b)$ in b' to serve as a DS of b, using a parallel algorithm for MIS such as the one described in [12]. By the choice of n_P, $\bigcup_{b \in block(P)} \mathcal{I}(b)$ covers at least k points.

Algorithm 1. Algorithm for MinPDS-UD by MIS

Input: A geometric representation of a unit disk graph G.
Output: A set A of vertices dominating at least k vertices of G.

1: $P_1 \leftarrow$ a partition of the area containing all points into blocks of side-length 2×2
2: $P_2 \leftarrow$ a shifting of P_1 to north-east by 1 unit up and 1 unit right
3: $j \leftarrow 0$
4: **for** $j = 1$ to 2 **do**
5:     Sort blocks in $block(P_j)$ as $b_1, \ldots, b_q$ such that $|V_{P_j}(b_1)| \geq \cdots \geq |V_{P_j}(b_q)|$.
6:     $n_{P_j} \leftarrow \arg\min_{j'} \{j' : |\bigcup_{i=1}^{j'} V_{P_j}(b_i)| \geq k\}$
7:     **for** any b_i with $i \leq n_{P_j}$ in parallel **do**
8:         $\mathcal{I}(b_i) \leftarrow$ a maximal independent set in b'_i
9:     **end for**
10:     $A_{P_j} \leftarrow \bigcup_{i=1}^{n_{P_j}} \mathcal{I}(b_i)$
11: **end for**
12: **if** $|A_{P_1}| \leq |A_{P_2}|$ **then**
13:     $A \leftarrow A_{P_1}$
14: **else**
15:     $A \leftarrow A_{P_2}$
16: **end if**
17: **return** A

The next lemma evaluates the size of an MIS in an extended block (Fig. 1).

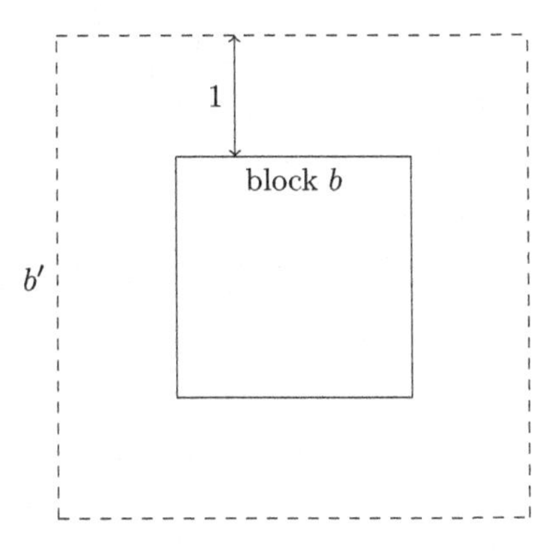

Fig. 1. Block b and its extended block b'.

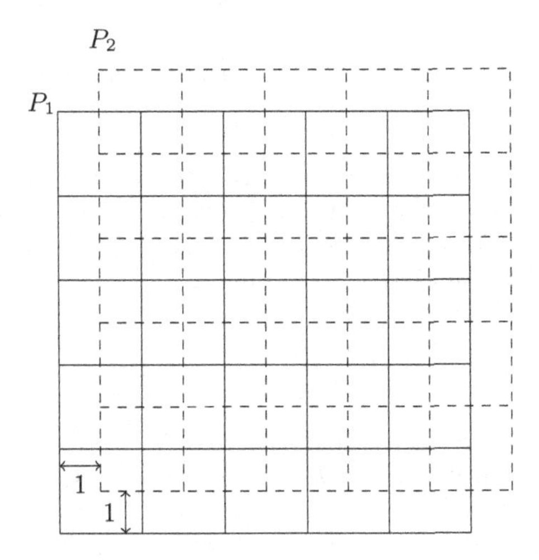

Fig. 2. Partition P_1 and partition P_2.

Lemma 1. *The size of a maximal independent set $\mathcal{I}(b)$ computed in line 8 of Algorithm 1 is at most 32.*

Proof. Let b' be the extended block of b and b'' be the block extending the boundaries of b by $\frac{3}{2}$ (see Fig. 3). Since every unit disk in $\mathcal{I}(b)$ has its center located in b', it must be completely contained in b''. Combining this observation with the fact that an independent set corresponds to a set of mutually disjoint unit disks, so $|\mathcal{I}(b)|$ is upper bound by $\lceil \frac{(2+3)^2}{\pi/4} \rceil \leq 32$.

Next we estimate the approximation ratio of Algorithm 1.

Theorem 1. *Algorithm 1 achieves approximation ratio at most 80 and runs in $O(\log n)$ rounds on $O(n)$ machines.*

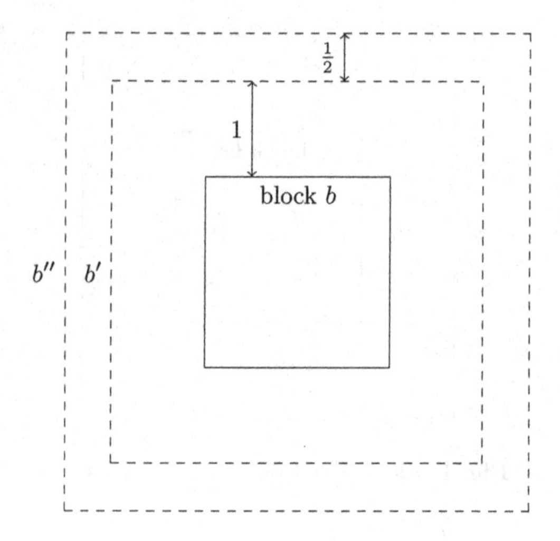

Fig. 3. Extending the boundaries of the block by 1 and $\frac{3}{2}$

Proof. Let OPT be an optimal solution and *opt* be the value of the OPT. For a partition P, denote by n_{op} the number of blocks in $block(P)$ intersecting the unit disks in OPT. For a block $b \in block(P)$, denote $OPT_P(b) = \{d \in OPT : d \cap b \neq \emptyset\}$ the set of unit disks of OPT intersecting b. Then

$$n_{op} \leq \sum_{b \in block(P)} |OPT_P(b)|. \tag{2}$$

Denote by H_P and Y_P the set of unit disks in OPT that intersect two horizontal strips and two vertical strips of P, respectively. Note that if a disk intersects more than two blocks in P, it must belong to both H_P and Y_P. Furthermore, a unit disk can intersect at most four blocks of P. Therefore,

$$\sum_{b \in block(P)} |OPT_P(b)| \leq opt + |H_P| + 2|Y_P|. \tag{3}$$

Note that a unit disk cannot belong to both H_{P_1} and H_{P_2}. Therefore, $H_{P_1} \cap H_{P_2} = \emptyset$ and thus

$$|H_{P_1}| + |H_{P_2}| \leq opt. \tag{4}$$

Similarly

$$|Y_{P_1}| + |Y_{P_2}| \leq opt. \tag{5}$$

By the greedy method in line 6 of Algorithm 1, we have

$$n_P \leq n_{op}. \tag{6}$$

Combining Lemma 1 with inequalities (2), (3) and (6), for any partition P_j,

$$|A_P| \leq 32 n_P \leq 32 n_{op} \leq 32(opt + |H_P| + 2|Y_P|). \tag{7}$$

Combining inequalities (4), (5) and (7),

$$|A_{P_1}| + |A_{P_2}| \leq 32(2opt + |H_{P_1}| + 2|Y_{P_1}| + |H_{P_2}| + 2|Y_{P_2}|) \leq 160opt.$$

Since Algorithm 1 chooses the minimum of $|A_{P_1}|$ and $|A_{P_2}|$, we have $|A| \leq 80opt$.

Next we estimate the number of rounds and the number of machines needed by Algorithm 1. Line 5 and 6 can be done in $O(\log n)$ rounds on $O(n)$ machine by a parallel sorting method a parallel selecting method in [16], Using the algorithm in [12] to compute an MIS in parallel needs $O(\log n)$ rounds on $O(n)$ machines. The other operations can be done in $O(1)$ rounds on $O(n)$ machines. So the adaptive complexity follows.

3 An Improved Approximation Algorithm for MinPDS-UD

In this section, we propose another parallel algorithm for MinPDS-UD, which improves the approximation ratio as well as the adaptive complexity.

The algorithm makes use of a relation between MinPDS-UD and a restricted version of the partial unit disk cover problem. Given a set V of n points and a set $\mathcal{D}$ of disks of the same size on the plane, the goal of the *Minimum Partial Unit Disk Cover problem* (MinPUDC) is to find a minimum number of disks to cover at least k points. The meaning of "restricted" is that the centers of those disks in $\mathcal{D}$ coincide with the points in V. In fact, for a MinPDS-UD instance on unit disk graph G, let the points corresponding to the vertices of G to be the points to be covered, as well as the centers of disks of diameter 2. Note that a disk of diameter 2 centered at point u covers point v if and only if the Euclidean distance between u and v is at most 1, that is, v is dominated by u in G. Hence we may focus on such a restricted MinPUDC problem. It should be emphasized that in the following, a unit disk refers to a disk of diameter 2, not 1.

The algorithm is described in Algorithm 2. It differs from Algorithm 1 in three aspects. First, instead of taking the better one between two solutions, it only computes a solution for one partition. Second, the size of each block is $\frac{\sqrt{2}}{2} \times \frac{\sqrt{2}}{2}$ (not 2×2). Third, a DS in block $b \in block(P)$ is no longer approximated by an MIS, but is computed by selecting an arbitrary point in b; denote the unit disk (of diameter 2) centered at this point as $D(b)$. The other steps are the same as Algorithm 1. Note that $D(b)$ covers all the points in block b (see Fig. 4 for an illustration). By the choice of n_P in line 3 of Algorithm 2, $\bigcup_{i=1}^{n_P} D(b)$ covers at least k points.

Lemma 2. *Every $D(b)$ can intersect at most 14 blocks.*

Proof. Note that a disk of diameter 2 is completely contained in a square of side-length $2\sqrt{2} \times 2\sqrt{2}$, and thus it can intersect at most $4 \times 4 = 16$ blocks. To further reduce the number, we divide block b into four cells of side-length $\frac{\sqrt{2}}{4} \times \frac{\sqrt{2}}{4}$, and denote them as $e_1, \ldots, e_4$ (see Fig. 5 for an illustration). Without loss of generality, we assume that the selected point v is located in cell e_1 (see

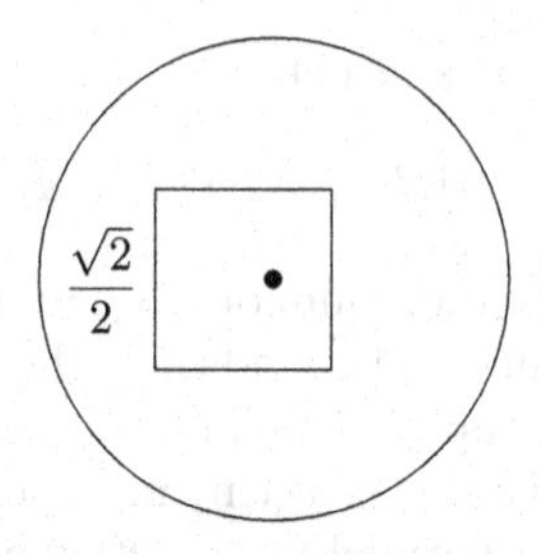

Fig. 4. A disk centered in a $\frac{\sqrt{2}}{2} \times \frac{\sqrt{2}}{2}$ block can cover this block.

Algorithm 2. Algorithm for restricted MinPUDC

Input: Area Q containing all points in V.
Output: A disks set A, which covers at least k points of V.

1: $P \leftarrow$ a partition of Q into blocks of side-length $\frac{\sqrt{2}}{2} \times \frac{\sqrt{2}}{2}$
2: Sort nonempty blocks as $b_1, \cdots, b_q$ such that $|V_P(b_1)| \geq |V_P(b_2)| \geq \cdots \geq |V_P(b_q)|$.
3: $n_P \leftarrow \arg\min_{j'} \{j' : \bigcup_{i=1}^{j'} |V_P(b_i)| \geq k\}$
4: **for** any block b_i with $i \leq n_P$ in parallel **do**
5: $\quad D(b_i) \leftarrow$ the disk centered at an arbitrarily selected point in b_i
6: **end for**
7: **return** $A \leftarrow \bigcup_{i=1}^{n_P} D(b_i)$

Fig. 6). The following two facts can be observed. First, $D(b)$ does not interest b_{11}. Otherwise, the radius of the disk is larger than 1 since the distance between any point in b_{11} and v is larger than 1, contradicting the fact that the diameter of a disk is 2. Second, $D(b)$ does not simultaneously interest both b_{14} and b_{41}, since the distance between any point in b_{41} and any point in b_{14} is larger than 2. It follows that $D(b)$ can intersect at most 14 blocks.

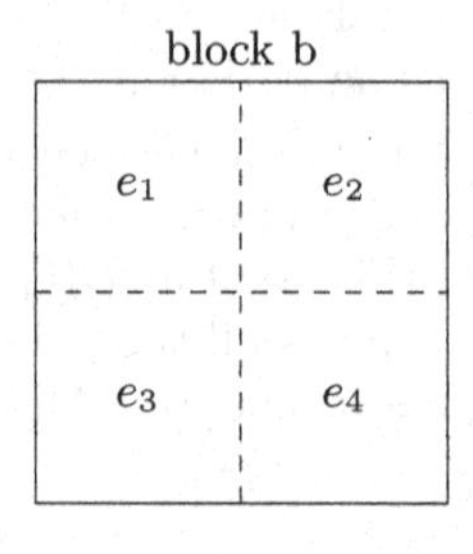

Fig. 5. Divide a block into four cells

Next we evaluate $|A|$ in Algorithm 2.

Theorem 2. *The approximation ratio of Algorithm 2 is at most 14 and Algorithm 2 runs $O(\log n)$ rounds on $O(n)$ machines.*

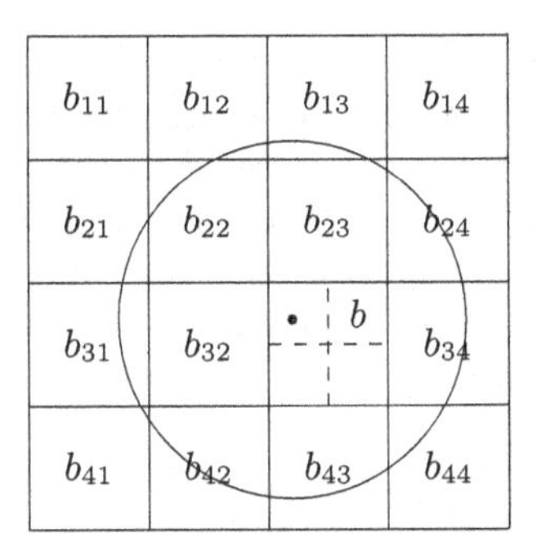

Fig. 6. $D(b)$ can intersect at most 14 blocks

Proof. Since we only selects *one* point in each of the n_P nonempty blocks,

$$|A| = n_P. \tag{8}$$

Using the same notation n_{O_P} as in the proof of Lemma 1, we also have inequality (6). Combining this with Lemma 2, we have

$$n_P \leq n_{O_P} \leq 14opt. \tag{9}$$

Combining (8) and (9), approximation ratio 14 is proved.

Since line 2 and line 3 of Algorithm 2 can be done in $O(\log n)$ rounds on $O(n)$ machines (see [16]), and line 5 can be done in constant time on each machine, the adaptive complexity follows.

4 Conclusion

This paper presented two parallel approximation algorithms for MinPDS-UD. The first one makes use of a relation between a maximal independent set and a dominating set, achieving approximation ratio 80 in $O(\log n)$ rounds on $O(n)$ machines. The second one transforms the MinPDS-UD problem into a restricted partial unit disk cover problem, achieving approximation ratio 14 in $O(\log n)$ rounds on $O(n)$ machines. These are the first parallel algorithms for MinPDS-UD achieving constant approximation ratio in log-polynomial rounds.

The first method is much more complicated, while its approximation ratio is worse. The reason might be: using inequality (2) to bridge the computed solution and an optimal solution might be too loose, it is tight only when every block intersects very few disks from the optimal solution, but the estimation for the number of vertices in an MIS only depends on the size of the block. In fact, the larger the block size is, the looser inequality (2) will be, and the larger the number of disks in a MIS will be. We believe that the first method might yield a better solution if a more delicate relation between MIS and PDS can be found. One difficulty lies in the fact that in a partial cover problem, one does not know which points should be covered.

Note that our method in Sect. 3 can be applied to the *minimum partial continuous unit disk cover problem*, the goal of which is to find the minimum number of disks with diameter 2, that *can be located anywhere* on the plane, to cover at least k points. Similar method yields a parallel algorithm with approximation ratio at most 9 in $O(\log n)$ rounds on $O(n)$ machines.

Note that our method can only be used for the cardinality case. New techniques have to be further explored for the weighted version.

References

1. Bar-Yehuda, R., Moran, S.: On approximation problems related to the independent set and vertex cover problem. Disc. Appl. Math. **9**(1), 1–10 (1984)
2. Berger, B., Rompel, J., Shor, P.W.: Efficient NC algorithms for set cover with applications to learning and geometry. J. Comput. Syst. Sci. **49**(3), 454–477 (1994)
3. Chan, T.M., Grant, E.: Weighted capacitated, priority, and geometric set cover via improved quasi-uniform sampling. In: ACM-SIAM Symposium on Discrete Algorithms, pp. 1576–1585. SIAM (2012)
4. Clark, B.N., Colbourn, C.J., Johnson, D.S.: Unit disk graphs. Disc. Math. **86**(1–3), 165–177 (1990)
5. Das, G.K., Fraser, R., Lopez-ortiz, A., Nickerson, B.G.: On the discrete unit disk cover problem. Int. J. Comput. Geometry Appl. **22**(5), 407–419 (2012)
6. Dinur, I., Steurer, D.: Analytical approach to parallel repetition. In: 46th Annual ACM Symposium on Theory of Computing, pp. 624–633, New York (2014)
7. Du, D.Z., Wang, P.J.: Connected Dominating Set: Theory and Applications. Springer, New York (2013). https://doi.org/10.1007/978-1-4614-5242-3
8. Feige, U.: A threshold of ln n for approximating set cover. In: 28th International Proceedings on ACM Symposium on Theory of Computing, pp. 314–318. ACM, New York (1996)
9. Fowler, R.J., Paterson, M.S., Tanimoto, S.L.: Optimal packing and covering in the plane are NP-complete. Inf. Process. Lett. **12**(3), 133–137 (1981)
10. Gandhi, R., Khuller, S., Srinivasan, A.: Approximation algorithms for partial covering problems. J. Algorithms **53**(1), 55–84 (2004)
11. Garey, M.R., Johnson, D.S.: Computers and Intractability: A Guide to the Theory of NP-Completeness. W. H. Freeman & Co.Subs. of Scientific American, Inc. 41 Madison Avenue, 37th Fl. New York, NY, United States (1979)
12. Ghaffari, M., Haeupler, B.: A Time-optimal randomized parallel algorithm for MIS. In: Proceedings of the 2021 ACM-SIAM Symposium on Discrete Algorithms, pp. 2892–2903. SIAM (2021)
13. Gonzalez, T.F.: Covering a set of points in multidimensional space. Inf. Process. Lett. **40**(4), 181–188 (1991)
14. Hochbaum, D.S., Maass, W.: Approximation schemes for covering and packing problems in image processing and VLSI. J. ACM **32**(1), 130–136 (1985)
15. Hunt, H.B., III., Marathe, M.V., Radhakrishnan, V., Ravi, S.S., Rosenkrantz, D.J., Stearns, R.E.: NC-approximation schemes for NP- and PSPACE-hard problems for geometric graphs. J. Algorithms **26**(2), 238–274 (1998)
16. JáJá J.: An Introduction to Parallel Algorithms. Addison Wesley Longman Publishing Co., Inc. 350 Bridge Pkwy suite 208 Redwood City, CA, United State (1992)

17. Joachim, K., Daniel, M., Peter, R.: Partial vs. complete domination: t-dominating set. In: International Conference on Current Trends in Theory and Practice of Computer Science, vol. 4362(1), pp. 367–376 (2007)
18. Khuller, S., Vishkin, U., Young, N.: A primal-dual parallel approximation technique applied to weighted set and vertex covers. J. Algorithms **17**(2), 280–289 (1994)
19. Li, J., Jin, Y.: A PTAS for the weighted unit disk cover problem. In: Halldórsson, M.M., Iwama, K., Kobayashi, N., Speckmann, B. (eds.) ICALP 2015. LNCS, vol. 9134, pp. 898–909. Springer, Heidelberg (2015). https://doi.org/10.1007/978-3-662-47672-7_73
20. Nieberg, T., Hurink, J.: A PTAS for the minimum dominating set problem in unit disk graphs. In: Erlebach, T., Persinao, G. (eds.) WAOA 2005. LNCS, vol. 3879, pp. 296–306. Springer, Heidelberg (2006). https://doi.org/10.1007/11671411_23
21. Ran, Y.L., Zhang, Y., Zhang, Z.: Parallel approximation for partial set cover. Appl. Math. Comput. **408**, 126358 (2021)
22. Rashmisnata, A., Manjanna, B., Gautam, K.D.: Unit disk cover problem in 2D. J. Discrete Algorithms **33**, 193–201 (2015)
23. Wu, J., Dai, F., Gao, M., Stojmenovic, I.: On calculating power-aware connected dominating sets for efficient routing in Ad Hoc wireless networks. J. Commun. **4**(1), 59–70 (2002)

An Improved Approximation Algorithm for Squared Metric k-Facility Location

Zhen Zhang[1,2] and Qilong Feng[1(✉)]

[1] School of Computer Science and Engineering, Central South University, Changsha 410000, People's Republic of China
csufeng@mail.csu.edu.cn

[2] School of Frontier Crossover Studies, Hunan University of Technology and Business, Changsha 410000, People's Republic of China

Abstract. In this paper, we study the squared metric k-facility location problem, which generalizes the k-means problem in that each facility has a specific cost of opening it in the solution. The current best approximation guarantee for the squared metric k-facility location problem is a ratio of $44.473 + \epsilon$ based on a local search algorithm. We give a $(36.343 + \epsilon)$-approximation for the problem using the techniques of primal-dual and Lagrangian relaxation. We propose a new rounding approach that exploits the properties of the squared metric, which is the crucial step in getting the improved approximation ratio.

Keywords: Approximation algorithm · Squared metric k-facility location · k-means

1 Introduction

k-means is a commonly studied clustering problem. This problem considers a set $\mathcal{D}$ of clients and a set $\mathcal{F}$ of facilities located in a metric space. The goal is to open no more than k facilities and assign each client to an opened facility, such that the sum of the squared distance from each client to the corresponding facility is minimized. The k-means problem was known to be NP-hard [2,15], which leads to considerable attentions paid on the design of its approximation algorithms [1,3,9,13,17,21,22]. The current best approximation guarantee for the k-means problem is a ratio of $9 + \epsilon$ [1], which was obtained based on a primal-dual approach.

The k-means problem inherently assumes that each facility can be opened without paying opening cost. However, the facilities are associated with non-uniform opening costs in many clustering applications, such as data placement [5, 14], network design [19,25], and warehouse location [10,24]. In this paper, we consider the squared metric k-facility location problem (SM-k-FL), which is

This work was supported by National Natural Science Foundation of China (61872450 and 62172446).

D.-Z. Du et al. (Eds.): COCOA 2021, LNCS 13135, pp. 538–552, 2021.
https://doi.org/10.1007/978-3-030-92681-6_42

similar to the k-means problem, except that it involves non-uniform facility opening costs. The problem is formally defined as follows.

Definition 1 (squared metric k-facility location). *The squared metric k-facility location problem considers a set $\mathcal{D}$ of clients and a set $\mathcal{F}$ of facilities in a metric space, and an integer $k > 0$, where each $i \in \mathcal{F}$ has an opening cost $f(i) > 0$. The goal is to open a set $\mathcal{S} \subseteq \mathcal{F}$ of no more than k facilities, such that the objective function $\sum_{i \in \mathcal{S}} f(i) + \sum_{j \in \mathcal{D}} \Delta(j, \mathcal{S})$ is minimized, where $\Delta(j, \mathcal{S})$ denotes the squared distance for j to the nearest facility from $\mathcal{S}$.*

Jain and Vazirani [21] showed that the techniques of primal-dual and Lagrangian relaxation yield a $(54 + \epsilon)$-approximation for SM-k-FL. Zhang *et al.* [27] later gave a $(44.473 + \epsilon)$-approximation based on local search, which is currently the best approximation guarantee for SM-k-FL. A closely related problem to SM-k-FL is the k-facility location problem, where the assignment cost of each client is measured by its distance to the corresponding facility. Jain and Vazirani [21] gave a $(6 + \epsilon)$-approximation for k-facility location using a primal-dual algorithm. The approximation guarantee for the problem was later improved by a series of work [20,28] to the current best ratio of 3.25 [8].

We briefly remark on the commonly used techniques for the related problems to SM-k-FL, such as k-means and k-facility location, to show the challenges in obtaining approximation ratios better than $44.473 + \epsilon$ for SM-k-FL.

- Local search has been successfully applied to solve the problems of k-facility location [28] and k-means [17,22,29]. Starting with an arbitrary feasible solution, local search-based algorithms iteratively improve the solution by closing a set of opened facilities and opening a set of closed facilities. The desired approximation guarantee is obtained by the fact that none of such swaps can significantly improve a local optimum. Gupta and Tangwongsan [17] showed that local search yields a $(25 + \epsilon)$-approximation for the k-means problem. For SM-k-FL, the non-uniform facility opening costs make the approximation ratio of a local optimum much more difficult to bound. It is hard to improve the $44.473 + \epsilon$ ratio using more refined analysis of local search.
- Charikar and Li [8] gave a 3.25-approximation for the k-facility location problem using an LP-rounding algorithm. Unfortunately, it was known that under the squared metric, the current LP-rounding techniques yield quite large approximation ratios, and some commonly used linear programming formulations in clustering problems have unbounded integrality gaps [12]. Thus, it seems quite difficult to improve the performance guarantee of SM-k-FL based on LP-rounding.
- Jain and Vazirani [21] gave a $(6+\epsilon)$-approximation algorithm for the k-facility location problem based on the techniques of primal-dual and Lagrangian relaxation. The algorithm uses a rounding method that relies on triangle inequality. This significantly deteriorates the performance of the method in the squared metric. Indeed, Jain and Vazirani [21] showed that a similar algorithm yields a $(54 + \epsilon)$-approximation for SM-k-FL. Most recently, improved

rounding methods that behave well in the squared metric have been introduced. For example, Ahmadian *et al.* [1] gave a $(9+\epsilon)$-approximation algorithm for the k-means problem, and Feng *et al.* [11] presented a $(19.849+\epsilon)$-approximation algorithm for the k-means with penalties problem, both of which are based on the framework outlined in [21]. However, these algorithms rely heavily on the assumption that the facilities can be arbitrarily opened without paying opening cost, and can induce unbounded cost when applied to solve SM-k-FL.

1.1 Our Results

In this paper, we obtain a $(36.343+\epsilon)$-approximation for SM-k-FL.

Theorem 1. *For any constant $\epsilon > 0$, there is a $(36.343+\epsilon)$-approximation algorithm for* SM-k-FL *that runs in polynomial time.*

Given an instance of SM-k-FL, our algorithm first relaxes the constraint on the number of opened facilities, and then obtains two solutions $\mathcal{H}_1$ and $\mathcal{H}_2$ to the relaxed problem. Here, $\mathcal{H}_1$ opens $k_1 < k$ facilities, $\mathcal{H}_2$ opens $k_2 > k$ facilities, and we have $ak_1 + (1-a)k_2 = k$ for a real number $a \in (0,1)$. Let $\mathcal{S}_1$ and $\mathcal{S}_2$ denote the sets of the facilities opened in $\mathcal{H}_1$ and $\mathcal{H}_2$, respectively. A convex combination of $\mathcal{H}_1$ and $\mathcal{H}_2$ is a feasible fractional solution to SM-k-FL, which we want to round to an integral solution.

We apply different strategies depending on the values of a and $k - k_1$.

- Case (1): $a \in [\frac{1}{4}, 1)$. $\mathcal{H}_1$ opens no more than k facilities and thus is a feasible solution to SM-k-FL. For this case, we show that the cost of $\mathcal{H}_1$ is quite small and directly return $\mathcal{H}_1$ as the final solution.
- Case (2): $a \in [\frac{1}{36}, \frac{1}{4})$ and $k - k_1 \leq \frac{1}{\epsilon a}$. For this case, we show that a set of k facilities from $\mathcal{S}_2$ yields the desired approximation for the problem. Such k facilities can be found in polynomial time by enumerating all the subsets of $\mathcal{S}_2$ of size k since k_2 is close to k.
- Case (3): $a \in [\frac{1}{36}, \frac{1}{4})$ and $k - k_1 > \frac{1}{\epsilon a}$. We first open each facility from $\mathcal{S}_1$, and then improve the solution by swapping in some facilities from $\mathcal{S}_2$. A linear program that maximizes the reduced cost is considered.
- Case (4): $a \in (0, \frac{1}{36})$. For this case, $\mathcal{H}_1$ has unbounded cost, and we prefer to select the facilities from $\mathcal{S}_2$ to open. We show that k facilities selected from $\mathcal{S}_2$ using a greedy method achieve the desired approximation for SM-k-FL.

In the past two decades, the technique of Lagrangian relaxation has been extensively studied and used in clustering problems. Specifically, the best known approximation ratios for many problems, such as k-median [6], k-means [1], and k-median with uniform penalties [26], are based on Lagrangian relaxation, which significantly improved the ratios of the local search-based approaches [4, 17,18,22]. However, this technique has not yielded similar improvements for SM-k-FL, and the current best approximation guarantee for SM-k-FL is based on local search [27]. This paper overcomes this barrier and gives an improved approximation ratio for SM-k-FL. The main technical contribution of the paper is a deterministic rounding approach that is quite different from the previously used randomized rounding methods [7,21]. The approach exploits the properties of the squared metric, which is the crucial step in getting the improved ratio.

2 Preliminaries

Let $\mathcal{I} = (\mathcal{D}, \mathcal{F}, k, f)$ be an instance of SM-k-FL. Given two points $i, j \in \mathcal{D} \cup \mathcal{F}$, let $\delta(i,j)$ and $\Delta(i,j)$ be the distance and squared distance from i to j, respectively. For any $\mathcal{A} \subseteq \mathcal{D} \cup \mathcal{F}$, define $\delta(i, \mathcal{A}) = \min_{l \in \mathcal{A}} \delta(i,l)$ and $\Delta(i, \mathcal{A}) = \min_{l \in \mathcal{A}} \Delta(i,l)$. We formalize $\mathcal{I}$ as an integer program and relax the integrality constraints to get the following linear program.

$$\min \quad \sum_{i \in \mathcal{F}} y_i f(i) + \sum_{i \in \mathcal{F}, j \in \mathcal{D}} x_{ij} \Delta(j,i) \qquad \text{LP1}$$

$$\text{s.t.} \quad \sum_{i \in \mathcal{F}} x_{ij} = 1 \qquad \forall j \in \mathcal{D} \qquad (1)$$

$$x_{ij} \leq y_i \qquad \forall j \in \mathcal{D}, i \in \mathcal{F} \qquad (2)$$

$$\sum_{i \in \mathcal{F}} y_i \leq k \qquad (3)$$

$$x_{ij}, y_i \geq 0 \qquad \forall j \in \mathcal{D}, i \in \mathcal{F} \qquad (4)$$

LP1 associates a variable x_{ij} with each $i \in \mathcal{F}$ and $j \in \mathcal{D}$ indicating whether j is assigned to i, and a variable y_i with each $i \in \mathcal{F}$ indicating whether i is opened. Constraints (1) and (2) enforce that each client should be assigned to an opened facility, and constraint (3) says that at most k facilities can be opened. An integral solution to LP1 exactly corresponds to a solution to $\mathcal{I}$. Let opt^* be the cost of an optimal integral solution to LP1.

Let $\Delta_{\min}$ and $\Delta_{\max}$ be the minimum and maximum squared distances between any $i, j \in \mathcal{D} \cup \mathcal{F}$, repectively. Similarly, let $f_{\min} = \min_{i \in \mathcal{F}} f(i)$ and $f_{\max} = \max_{i \in \mathcal{F}} f(i)$. The following result shows that both $\Delta_{\max}/\Delta_{\min}$ and $f_{\max}/f_{\min}$ can be polynomially bounded, which induces an arbitrarily small loss in the approximation guarantee.

Lemma 1. *Given a real number $\epsilon > 0$ and an instance $\mathcal{I} = (\mathcal{D}, \mathcal{F}, k, f)$ of* SM-k-FL, *we can assume that $\Delta_{\max}/\Delta_{\min} \leq |\mathcal{D}|^2 \epsilon^{-1}$ and $f_{\max}/f_{\min} \leq |\mathcal{D}| k \epsilon^{-1}$ with losing a factor $1 + O(\epsilon)$ in the approximation guarantee.*

The following algebraic fact is useful in analyzing the assignment costs.

Lemma 2. *For any three positive numbers μ, ν, and ρ, we have $(\mu + \nu)^2 \leq (1+\rho)\mu^2 + (1 + \frac{1}{\rho})\nu^2$.*

Proof. Observe that

$$(\mu + \nu)^2 = \mu^2 + \nu^2 + 2\sqrt{\rho}\mu \frac{1}{\sqrt{\rho}} \nu \leq \mu^2 + \nu^2 + \rho\mu^2 + \frac{1}{\rho}\nu^2,$$

as desired. □

3 A Fractional Solution

The constraint on the number of opened facilities is one of the main obstacles in finding integral solutions to LP1. The framework of Lagrangian relaxation outlined in [21] is frequently used to overcome such kind of obstacles [1,6,16,23, 26,30]. The idea is to remove the constraint but pay the penalty for its violation. This motivates the following relaxation of LP1, where $\tau \geq 0$.

$$\begin{aligned} \min \quad & \sum_{i\in\mathcal{F}} y_i f(i) + \sum_{i\in\mathcal{F},j\in\mathcal{D}} x_{ij}\Delta(j,i) + \tau\Big(\sum_{i\in\mathcal{F}} y_i - k\Big) \qquad \text{LP2}(\tau) \\ \text{s.t.} \quad & (1), (2), \text{and } (4) \end{aligned}$$

The dual program of LP2(τ) is

$$\begin{aligned} \max \quad & \sum_{j\in\mathcal{D}} \alpha_j - \tau k & & \text{DUAL}(\tau) \\ \text{s.t.} \quad & \alpha_j \leq \Delta(j,i) + \beta_{ij} & \forall j \in \mathcal{D}, i \in \mathcal{F} \quad & (5) \\ & \sum_{j\in\mathcal{D}} \beta_{ij} \leq f(i) + \tau & \forall i \in \mathcal{F} \quad & (6) \\ & \alpha_j, \beta_{ij} \geq 0 & \forall j \in \mathcal{D}, i \in \mathcal{F} \quad & (7) \end{aligned}$$

For any $\tau \geq 0$, let $opt(\tau)$ denote the cost of an optimal solution to LP2(τ). Given a solution $\mathcal{H}' = (x', y')$ to LP2(τ), define $F(\mathcal{H}') = \sum_{i\in\mathcal{F}} y'_i f(i)$ and $S(\mathcal{H}') = \sum_{i\in\mathcal{F},j\in\mathcal{D}} x'_{i,j}\Delta(i,j)$ for brevity, and let $C(\mathcal{H}') = F(\mathcal{H}') + S(\mathcal{H}')$. If $\mathcal{H}'$ is a feasible solution to LP1, then $C(\mathcal{H}')$ is its cost for LP1. It can be seen that LP2(τ) is a linear program of the squared metric facility location problem [12], where each facility $i \in \mathcal{F}$ is associated with an opening cost $f(i) + \tau$, and the objective function has a constant term of $-\tau k$. Based on the primal-dual algorithm given by Jain and Vazirani [21], integral solutions to LP2(τ) with the following guarantee can be obtained.

Lemma 3 ([21]). *Given an instance $\mathcal{I} = (\mathcal{D}, \mathcal{F}, k, f)$ of* SM-*k*-FL *and a real number $\tau \geq 0$, there is an algorithm that yields an integral solution $\mathcal{H} = (x, y)$ to* LP2(τ) *and the corresponding dual solution (α, β) in polynomial time, such that $9\big(F(\mathcal{H}) + \tau\sum_{i\in\mathcal{F}} y_i\big) + S(\mathcal{H}) \leq 9\sum_{j\in\mathcal{D}} \alpha_j$.*

As a corollary of Lemma 3, we have the following result. This corollary says that we can obtain two integral solutions to LP2(τ) for some $\tau > 0$, such that a convex combination of the two solutions is a well-behaved fractional solution to LP1, which we round to an integral one in Sect. 4.

Corollary 1. *Given an instance $\mathcal{I} = (\mathcal{D}, \mathcal{F}, k, f)$ of* SM-*k*-FL *and a real number $\epsilon > 0$, there exists a polynomial-time algorithm that returns either a 9-approximation solution to $\mathcal{I}$, or two integral solutions $\mathcal{H}_1 = (x^1, y^1)$ and $\mathcal{H}_2 = (x^2, y^2)$ to* LP2(τ) *for some $\tau > 0$, such that $\sum_{i\in\mathcal{F}} y^1_i = k_1 < k$, $\sum_{i\in\mathcal{F}} y^2_i = k_2 > k$, and $a\big(9F(\mathcal{H}_1) + S(\mathcal{H}_1)\big) + b\big(9F(\mathcal{H}_2) + S(\mathcal{H}_2)\big) \leq (9 + O(\epsilon))opt^*$, where $a = \frac{k_2-k}{k_2-k_1}$ and $b = 1 - a$.*

The essential idea of Corollary 1 follows the Lagrangian relaxation approach given by Jain and Vazirani [21], but it implies a slightly stronger performance guarantee for the solutions to LP2(τ). Jain and Vazirani [21] gave that we can find two solutions $\mathcal{H}_1$ and $\mathcal{H}_2$ to LP2(τ) for some $\tau > 0$, such that the value of $aC(\mathcal{H}_1) + (1-a)C(\mathcal{H}_2)$ is within a constant times opt^* for some $a \in (0,1)$. Corollary 1 further indicates that the facility opening costs induced by such two solutions can be relatively small compared with opt^*, which is quite valuable in selecting opened facilities in the rounding phase.

4 Rounding

In this section, we give a rounding approach that yields the desired integral solution to SM-k-FL. We first introduce some notations to help with analysis. Given a constant $\epsilon > 0$ and an instance $\mathcal{I} = \{\mathcal{D}, \mathcal{F}, k, f\}$ of SM-k-FL, if Corollary 1 does not yield a 9-approximation solution to $\mathcal{I}$, then let $\mathcal{H}_1 = \{x^1, y^1\}$ and $\mathcal{H}_2 = \{x^2, y^2\}$ be the two integral solutions given by Corollary 1, where $\sum_{i\in\mathcal{F}} y_i^1 = k_1 < k$ and $\sum_{i\in\mathcal{F}} y_i^2 = k_2 > k$. Denote by $\mathcal{S}_1 = \{i \in \mathcal{F} : y_i^1 = 1\}$ and $\mathcal{S}_2 = \{i \in \mathcal{F} : y_i^2 = 1\}$ the sets of the facilities opened in $\mathcal{H}_1$ and $\mathcal{H}_2$, respectively. For each $j \in \mathcal{D}$, let $i_1(j)$ be the nearest facility to j from $\mathcal{S}_1$, and let $\Delta_1(j) = \Delta(j, i_1(j))$. Similarly, define $i_2(j)$ as the nearest facility to j from $\mathcal{S}_2$, and let $\Delta_2(j) = \Delta(j, i_2(j))$. We can assume that $S(\mathcal{H}_1) = \sum_{j\in\mathcal{D}} \Delta_1(j)$ and $S(\mathcal{H}_2) = \sum_{j\in\mathcal{D}} \Delta_2(j)$, which is without loss of generalization since reassigning each client to the nearest opened facility can only improve the solutions. Define $\gamma_1(i) = \{j \in \mathcal{D} : i_1(j) = i\}$ for each $i \in \mathcal{S}_1$ and $\gamma_2(i) = \{j \in \mathcal{D} : i_2(j) = i\}$ for each $i \in \mathcal{S}_2$. For any $\mathcal{A}_1 \subseteq \mathcal{S}_1$ and $\mathcal{A}_2 \subseteq \mathcal{S}_2$, define $\gamma_1(\mathcal{A}_1) = \bigcup_{i\in\mathcal{A}_1} \gamma_1(i)$ and $\gamma_2(\mathcal{A}_2) = \bigcup_{i\in\mathcal{A}_2} \gamma_2(i)$. For each $i \in \mathcal{S}_1$, denote by $\eta_2(i)$ the nearest facility to i from $\mathcal{S}_2$. Similarly, for each $i \in \mathcal{S}_2$, let $\eta_1(i)$ be the nearest facility to i from $\mathcal{S}_1$. The following result implies that both $\Delta(j, \eta_1(i_2(j)))$ and $\Delta(j, \eta_2(i_1(j)))$ can be bounded by a combination of $\Delta_1(j)$ and $\Delta_2(j)$ for each $j \in \mathcal{D}$.

Lemma 4. *For any real number $\rho > 0$ and any $j \in \mathcal{D}$, it is the case that $\Delta(j, \eta_1(i_2(j))) \le 4(1+\rho)\Delta_2(j) + (1+\frac{1}{\rho})\Delta_1(j)$ and $\Delta(j, \eta_2(i_1(j))) \le 4(1+\rho)\Delta_1(j) + (1+\frac{1}{\rho})\Delta_2(j)$.*

We now show how the desired solution to $\mathcal{I}$ can be obtained based on $\mathcal{H}_1$ and $\mathcal{H}_2$. Define $c_1 = 9F(\mathcal{H}_1) + S(\mathcal{H}_1)$ and $c_2 = 9F(\mathcal{H}_2) + S(\mathcal{H}_2)$ for brevity. Define $c_f = ac_1 + bc_2$, where $a = \frac{k_2 - k}{k_2 - k_1}$ and $b = 1 - a$. Corollary 1 implies that $c_f \le \big(9 + O(\epsilon)\big)opt^*$. We consider the following four cases: (1) $a \in [\frac{1}{4}, 1)$, (2) $a \in [\frac{1}{36}, \frac{1}{4})$ and $k - k_1 \le \frac{1}{\epsilon a}$, (3) $a \in [\frac{1}{36}, \frac{1}{4})$ and $k - k_1 > \frac{1}{\epsilon a}$, and (4) $a \in (0, \frac{1}{36})$. The simplest case is that of $a \in [\frac{1}{4}, 1)$, where we have $c_1 \le \frac{1}{a}(ac_1 + bc_2) \le \big(36 + O(\epsilon)\big)opt^*$ due to Corollary 1. Observe that $\mathcal{H}_1$ opens less than k facilities and is feasible for SM-k-FL. Thus, directly returning $\mathcal{H}_1$ as the final solution yields a $\big(36 + O(\epsilon)\big)$-approximation for the problem for case (1).

We prefer to select the facilities from $\mathcal{S}_2$ to open for the rest three cases since the cost of $\mathcal{H}_1$ might be quite large. For cases (2) and (4), we show that a set of k facilities from $\mathcal{S}_2$ yields the desired approximation ratio for the problem, and can be found in polynomial time. For case (3), we start with opening each facility from $\mathcal{S}_1$, and then improve the solution by swapping in some facilities from $\mathcal{S}_2$. A linear program is considered to maximize the reduced cost.

Case (2): $\boldsymbol{a \in [\frac{1}{36}, \frac{1}{4})}$ **and** $\boldsymbol{k - k_1 \leq \frac{1}{\epsilon a}}$.

For this case, we show that k facilities from $\mathcal{S}_2$ yield the desired approximation. Such k facilities are found by enumerating all the subsets of $\mathcal{S}_2$ of size k and selecting the one with the minimum cost for SM-k-FL. By the fact that $a = \frac{k_2 - k}{k_2 - k_1}$ and the assumption that $a \in [\frac{1}{36}, \frac{1}{4})$ and $k - k_1 \leq \frac{1}{\epsilon a}$, we have $k_2 - k = (k - k_1)\frac{a}{1-a} = O(\frac{1}{\epsilon})$, which implies that the enumeration of the facilities can be completed in polynomial time. We now consider the cost induced by the k facilities selected from $\mathcal{S}_2$.

Lemma 5. *If $c_1 > c_f$, then there exists a set $\mathcal{C} \subset \mathcal{S}_2$ of k facilities, such that $\sum_{i \in \mathcal{C}} f(i) + \sum_{j \in \mathcal{D}} \Delta(j, \mathcal{C}) < (26 + O(\epsilon))opt^*$.*

Proof. Denote by $\mathcal{S}^*$ the set of the facilities opened in an optimal solution to $\mathcal{I}$. For each $i \in \mathcal{S}^*$, let $\pi(i)$ be the nearest facility to i from $\mathcal{S}_2$. Define $\mathcal{S}' = \{\pi(i) : i \in \mathcal{S}^*\}$. For each $j \in \mathcal{D}$, let $i^*(j)$ denote its nearest facility from $\mathcal{S}^*$. We have

$$\begin{aligned}\delta(j, \mathcal{S}') &\leq \delta(j, \pi(i^*(j))) \leq \delta(j, i^*(j)) + \delta(i^*(j), \pi(i^*(j))) \\ &\leq \delta(j, i^*(j)) + \delta(i^*(j), i_2(j)) \leq 2\delta(j, i^*(j)) + \delta(j, i_2(j)),\end{aligned} \tag{8}$$

where the first step follows from the fact that $\pi(i^*(j)) \in \mathcal{S}'$, the second and last steps are derived from triangle inequality, and the third step follows from the fact that $\pi(i^*(j))$ is the nearest facility to $i^*(j)$ in $\mathcal{S}_2$. Using inequality (8) and Lemma 2, we get

$$\begin{aligned}\Delta(j, \mathcal{S}') &\leq \left(2\delta(j, i^*(j)) + \delta(j, i_2(j))\right)^2 \leq 8\Delta(j, i^*(j)) + 2\Delta_2(j) \\ &= 8\Delta(j, \mathcal{S}^*) + 2\Delta_2(j).\end{aligned}$$

Summing both sides of the inequality over $j \in \mathcal{D}$, we have

$$\sum_{j \in \mathcal{D}} \Delta(j, \mathcal{S}') \leq 8 \sum_{j \in \mathcal{D}} \Delta(j, \mathcal{S}^*) + 2\, S(\mathcal{H}_2) < 8opt^* + 2\, S(\mathcal{H}_2). \tag{9}$$

Consequently, for any $\mathcal{C} \subset \mathcal{S}_2$ that satisfies $|\mathcal{C}| = k$ and $\mathcal{S}' \subseteq \mathcal{C}$, we have

$$\begin{aligned}\sum_{i \in \mathcal{C}} f(i) + \sum_{j \in \mathcal{D}} \Delta(j, \mathcal{C}) &\leq \sum_{i \in \mathcal{C}} f(i) + \sum_{j \in \mathcal{D}} \Delta(j, \mathcal{S}') \\ &< 8opt^* + \sum_{i \in \mathcal{C}} f(i) + 2\, S(\mathcal{H}_2) \\ &< 8opt^* + F(\mathcal{H}_2) + 2\, S(\mathcal{H}_2) \\ &< 8opt^* + 2c_2 < 8opt^* + 2c_f \\ &\leq (26 + O(\epsilon))opt^*,\end{aligned}$$

where the first step is due to $\mathcal{S}' \subseteq \mathcal{C}$, the second step is due to inequality (9), the third step follows from the fact that $\mathcal{C} \subset \mathcal{S}_2$, the fourth step is derived from the definition of c_2, the fifth step follows from the fact that $c_f = ac_1 + (1-a)c_2$ and the assumption that $c_1 > c_f$, and the last step is due to Corollary 1. Thus, Lemma 5 is true. □

Lemma 5 implies that either the k facilities selected from $\mathcal{S}_2$ yield a $\big(26 + O(\epsilon)\big)$-approximation for SM-$k$-FL, or $c_1 \le c_f$, in which case $\mathcal{H}_1$ is a $\big(9+O(\epsilon)\big)$-approximation solution due to Corollary 1.

Case (3): $\boldsymbol{a \in [\frac{1}{36}, \frac{1}{4})}$ and $\boldsymbol{k - k_1 > \frac{1}{\epsilon a}}$.

For this case, we show how $\mathcal{H}_1$ and $\mathcal{H}_2$ can be combined into the desired solution to $\mathcal{I}$. We first consider a straightforward solution, which opens all the facilities from $\mathcal{S}_1$ and assigns each $j \in \mathcal{D}$ to $\eta_1(i_2(j))$. Lemma 4 implies that the total assignment cost of the clients induced by this solution is no more than $\sum_{j\in\mathcal{D}} \big(6\Delta_2(j) + 3\Delta_1(j)\big) = 6S(\mathcal{H}_2) + 3S(\mathcal{H}_1)$. We reduce the assignment cost by opening some facilities from $\mathcal{S}_2$. Given a facility $i \in \mathcal{S}_1$, define $\mathcal{Q}_i = \{i' \in \mathcal{S}_2 : \eta_1(i') = i\}$. If we open a facility $i' \in \mathcal{Q}_i$ and reassign each $j \in \gamma_2(i')$ to i', then the provable assignment cost of the solution can be reduced by $\sum_{j\in\gamma_2(i')} \big(6\Delta_2(j)+3\Delta_1(j)\big) - \sum_{j\in\gamma_2(i')} \Delta_2(j) = \sum_{j\in\gamma_2(i')} \big(5\Delta_2(j)+3\Delta_1(j)\big)$. Moreover, if we close i, open all the facilities from $\mathcal{Q}_i$, and reassign each $j \in \gamma_2(\mathcal{Q}_i)$ to $i_2(j)$, then we can reduce the provable assignment cost by $\sum_{j\in\gamma_2(\mathcal{Q}_i)} \big(5\Delta_2(j) + 3\Delta_1(j)\big)$. Define $\Phi_2(i') = \sum_{j\in\gamma_2(i')} \big(5\Delta_2(j) + 3\Delta_1(j)\big)$ for each $i' \in \mathcal{S}_2$ and $\Phi_1(i) = \sum_{j\in\gamma_2(\mathcal{Q}_i)} \big(5\Delta_2(j) + 3\Delta_1(j)\big) = \sum_{i'\in\mathcal{Q}_i} \Phi_2(i')$ for each $i \in \mathcal{S}_1$. We want to maximize the reduced assignment cost, which motivates the following linear program.

$$\begin{aligned} \max \quad & \sum_{i\in\mathcal{S}_1} \Phi_1(i) z_i & & \text{LP3} \\ \text{s.t.} \quad & \sum_{i\in\mathcal{S}_1} z_i(|\mathcal{Q}_i| - 1) = k - k_1 & & (10) \\ & 0 \le z_i \le 1 \qquad \forall i \in \mathcal{S}_1 & & (11) \end{aligned}$$

LP3 has a variable z_i for each $i \in \mathcal{S}_1$. $z_i = 1$ implies that we close i, open each facility from $\mathcal{Q}_i$, and reassign each $j \in \gamma_2(\mathcal{Q}_i)$ from i to $i_2(j)$. By the argument above, this reduces the provable assignment cost by $\Phi_1(i)$ compared with the aforementioned straightforward solution, and increases the number of opened facilities by $|\mathcal{Q}_i| - 1$. Constraint (10) says that the number of opened facilities can be increased to k.

LP3 is similar to the Knapsack-type linear programs considered in [6,23], which are solved to select opened facilities for the k-median problem. Unfortunately, the algorithms given in [6,23] can only yield pseudo-solutions that open $k + 2$ facilities in the worst case. Moreover, the algorithms need to open some specific facilities that minimize the total assignment cost of a given set of clients, and can induce unbounded facility opening cost in SM-k-FL. In this section, we give a more refined analysis and obtain the desired approximation solution for SM-k-FL that opens at most k facilities.

Lemma 6. *There is a polynomial-time algorithm that yields an optimal solution to* LP3*. The solution has at most one fractional variable, which is associated with a facility* $l \in \mathcal{S}_1$ *satisfying* $|\mathcal{Q}_l| > 1$.

We find an optimal solution z^* to LP3 using Lemma 6. Let $\mathcal{L}_0 = \{i \in \mathcal{S}_1 : z_i^* = 0\}$ and $\mathcal{L}_1 = \{i \in \mathcal{S}_1 : z_i^* = 1\}$. If z^* has a fractional variable, then let $l \in \mathcal{S}_1$ denote the facility associated with this variable. Our solution to $\mathcal{I}$ is constructed as follows.

- If facility $l \in \mathcal{S}_1$ is associated with a fractional variable z_l^*, then define $\mathcal{Q}^\dagger = \arg\max_{\mathcal{Q} \subset \mathcal{Q}_l \wedge |\mathcal{Q}| = \lfloor z_l^* |\mathcal{Q}_l| \rfloor} \sum_{i' \in \mathcal{Q}} \Phi_2(i')$. We open l and each facility from $\mathcal{Q}^\dagger$. We assign each $j \in \gamma_2(\mathcal{Q}^\dagger)$ to $i_2(j)$ and each $j \in \gamma_2(\mathcal{Q}_l \backslash \mathcal{Q}^\dagger)$ to l. Lemma 4 implies that the total assignment cost of the clients from $\gamma_2(\mathcal{Q}_l)$ is no more than $\sum_{j \in \gamma_2(\mathcal{Q}^\dagger)} \Delta_2(j) + \sum_{j \in \gamma_2(\mathcal{Q}_l \backslash \mathcal{Q}^\dagger)} (6\Delta_2(j) + 3\Delta_1(j))$. If z^* does not involve a fractional variable, then let $\mathcal{Q}_l = \mathcal{Q}^\dagger = \emptyset$.
- For each $i \in \mathcal{L}_1$ and $i' \in \mathcal{Q}_i$, we open i' and assign each $j \in \gamma_2(i')$ to i'. The total assignment cost of the clients from $\gamma_2(\bigcup_{i \in \mathcal{L}_1} \mathcal{Q}_i)$ is $\sum_{j \in \gamma_2(\bigcup_{i \in \mathcal{L}_1} \mathcal{Q}_i)} \Delta_2(j)$.
- For each $i \in \mathcal{L}_0$, we open i and assign each $j \in \gamma_2(\mathcal{Q}_i)$ to i. Lemma 4 implies that the total assignment cost of the clients from $\gamma_2(\bigcup_{i \in \mathcal{L}_0} \mathcal{Q}_i)$ is at most $\sum_{j \in \gamma_2(\bigcup_{i \in \mathcal{L}_0} \mathcal{Q}_i)} (6\Delta_2(j) + 3\Delta_1(j))$.

We first show that our solution opens no more than k facilities and is feasible for SM-k-FL. If z^* does not have a fractional variable, then the number of the facilities opened in the solution is

$$|\mathcal{L}_0| + \sum_{i \in \mathcal{L}_1} |\mathcal{Q}_i| = \sum_{i \in \mathcal{L}_1} (|\mathcal{Q}_i| - 1) + k_1 = \sum_{i \in \mathcal{S}_1} z_i^* (|\mathcal{Q}_i| - 1) + k_1,$$

which is exactly k due to constraint (10). For the case where z^* involves a fractional variable z_l^*, the facilities opened in our solution is no more than

$$\begin{aligned} 1 + z_l^* |\mathcal{Q}_l| + |\mathcal{L}_0| + \sum_{i \in \mathcal{L}_1} |\mathcal{Q}_i| &= z_l^* |\mathcal{Q}_l| + \sum_{i \in \mathcal{L}_1} (|\mathcal{Q}_i| - 1) + k_1 \\ &= \sum_{i \in \mathcal{S}_1} z_i^* (|\mathcal{Q}_i| - 1) + k_1 + z_l^* \\ &= k + z_l^* < k + 1, \end{aligned}$$

where the third step is due to constraint (10), and the last step follows from the fact that z_l^* is a fractional variable. Thus, our solution opens at most k facilities and is a feasible solution to $\mathcal{I}$ in both cases.

We now consider the cost induced by our solution. Denote by R_s and R_f the total assignment cost of the clients and the total opening cost of the facilities induced by the solution, respectively. Define $R = R_s + R_f$. By the argument above, we know that

$$R_s \leq \sum_{j \in \gamma_2(\bigcup_{i \in \mathcal{L}_0} \mathcal{Q}_i) \cup \gamma_2(\mathcal{Q}_l \setminus \mathcal{Q}^\dagger)} (6\Delta_2(j) + 3\Delta_1(j)) + \sum_{j \in \gamma_2(\bigcup_{i \in \mathcal{L}_1} \mathcal{Q}_i) \cup \gamma_2(\mathcal{Q}^\dagger)} \Delta_2(j)$$
$$= 6\, S(\mathcal{H}_2) + 3S(\mathcal{H}_1) - \sum_{i \in \mathcal{L}_1} \Phi_1(i) - \sum_{i \in \mathcal{Q}^\dagger} \Phi_2(i), \tag{12}$$

where the second step is due to the definitions of $\Phi_1(i)$ and $\Phi_2(i)$. Compared with the straightforward solution that opens all the facilities from $\mathcal{S}_1$ and assigns each $j \in \mathcal{D}$ to $\eta_1(i_2(j))$ (whose total assignment cost is at most $6S(\mathcal{H}_2) + 3S(\mathcal{H}_1)$), our solution reduces the provable assignment cost by $\sum_{i \in \mathcal{L}_1} \Phi_1(i) + \sum_{i \in \mathcal{Q}^\dagger} \Phi_2(i)$. The following result shows a lower bound on this difference value.

Lemma 7. *If* $k - k_1 > \frac{1}{\epsilon a}$, *then* $\sum_{i \in \mathcal{L}_1} \Phi_1(i) + \sum_{i \in \mathcal{Q}^\dagger} \Phi_2(i) > (1 - \epsilon a) b \big(5S(\mathcal{H}_2) + 3S(\mathcal{H}_1)\big)$.

Proof. Let *opt'* be the value of z^* for LP3. We first show that $\sum_{i \in \mathcal{L}_1} \Phi_1(i) + \sum_{i \in \mathcal{Q}^\dagger} \Phi_2(i)$ is close to *opt'*. If z^* does not have a fractional variable, then we have $\mathcal{Q}^\dagger = \emptyset$, and $\sum_{i \in \mathcal{L}_1} \Phi_1(i) + \sum_{i \in \mathcal{Q}^\dagger} \Phi_2(i) = \sum_{i \in \mathcal{L}_1} \Phi_1(i) = opt'$ due to the definition of $\mathcal{L}_1$. We now consider the case where z^* has a fractional variable z_l^*. For this case, we have

$$opt' = \sum_{i \in \mathcal{L}_1} \Phi_1(i) + z_l^* \Phi_1(l). \tag{13}$$

This implies that

$$opt' - \sum_{i \in \mathcal{L}_1} \Phi_1(i) - \sum_{i \in \mathcal{Q}^\dagger} \Phi_2(i) = z_l^* \Phi_1(l) - \sum_{i \in \mathcal{Q}^\dagger} \Phi_2(i)$$
$$\leq z_l^* \Phi_1(l) - \frac{z_l^* |\mathcal{Q}_l| - 1}{|\mathcal{Q}_l|} \Phi_1(l)$$
$$= \frac{1}{|\mathcal{Q}_l|} \Phi_1(l) \leq \frac{1}{|\mathcal{Q}_l| - 1} \Phi_1(l), \tag{14}$$

where the second step is derived from the fact that $\Phi_1(l) = \sum_{i \in \mathcal{Q}_l} \Phi_2(i)$ and $\mathcal{Q}^\dagger = \underset{\mathcal{Q} \subset \mathcal{Q}_l \wedge |\mathcal{Q}| = \lfloor z_l^* |\mathcal{Q}_l| \rfloor}{\arg\max} \sum_{i' \in \mathcal{Q}} \Phi_2(i')$, and the last step follows from the fact that $|\mathcal{Q}_l| > 1$, which is due to Lemma 6.

Define $\mathcal{A}_1 = \{i \in \mathcal{L}_1 : |\mathcal{Q}_i| \leq 1\}$ and $\mathcal{A}_2 = \{i \in \mathcal{L}_1 : |\mathcal{Q}_i| > 1\}$. It can be seen that inequality $\Phi_1(i)/(|\mathcal{Q}_i| - 1) \geq \Phi_1(l)/(|\mathcal{Q}_l| - 1)$ holds for any $i \in \mathcal{A}_2$. Otherwise, we can simultaneously increase z_l^* and decrease z_i^* to obtain an improved solution to LP3, which contradicts the fact that z^* is an optimal solution to LP3. Consequently, we have

$$
\begin{aligned}
\epsilon a \cdot opt' &= \epsilon a\big(\sum_{i \in \mathcal{L}_1} \Phi_1(i) + z_l^* \Phi_1(l)\big) > \frac{1}{k-k_1}\big(\sum_{i \in \mathcal{L}_1} \Phi_1(i) + z_l^* \Phi_1(l)\big) \\
&= \frac{\sum_{i \in \mathcal{L}_1} \Phi_1(i) + z_l^* \Phi_1(l)}{\sum_{i \in \mathcal{L}_1}(|\mathcal{Q}_i|-1) + z_l^*(|\mathcal{Q}_l|-1)} \\
&= \frac{\sum_{i \in \mathcal{A}_1} \Phi_1(i) + \sum_{i \in \mathcal{A}_2} \Phi_1(i) + z_l^* \Phi_1(l)}{\sum_{i \in \mathcal{A}_1}(|\mathcal{Q}_i|-1) + \sum_{i \in \mathcal{A}_2}(|\mathcal{Q}_i|-1) + z_l^*(|\mathcal{Q}_l|-1)} \\
&\geq \frac{\sum_{i \in \mathcal{A}_2} \Phi_1(i) + z_l^* \Phi_1(l)}{\sum_{i \in \mathcal{A}_2}(|\mathcal{Q}_i|-1) + z_l^*(|\mathcal{Q}_l|-1)} \geq \frac{1}{|\mathcal{Q}_l|-1}\Phi_1(l),
\end{aligned} \tag{15}
$$

where the first step is due to equality (13), the second step follows from the assumption that $k - k_1 > \frac{1}{\epsilon a}$, the third step is due to constraint (10), the fifth step is derived from the fact that $|\mathcal{Q}_i| - 1 \leq 0$ for each $i \in \mathcal{A}_1$, and the last step follows from the fact that $\Phi_1(i)/(|\mathcal{Q}_i| - 1) \geq \Phi_1(l)/(|\mathcal{Q}_l| - 1)$ for each $i \in \mathcal{A}_2$. Inequalities (14) and (15) imply that

$$
\sum_{i \in \mathcal{L}_1} \Phi_1(i) + \sum_{i \in \mathcal{Q}^\dagger} \Phi_2(i) \geq opt' - \frac{1}{|\mathcal{Q}_l|-1}\Phi_1(l) > (1-\epsilon a)opt'. \tag{16}
$$

If we can show that

$$
opt' \geq b\big(5S(\mathcal{H}_2) + 3S(\mathcal{H}_1)\big), \tag{17}
$$

then Lemma 7 can be proven using inequality (16). It remains to show inequality (17). Observe that $\sum_{i \in \mathcal{S}_1} b(|\mathcal{Q}_i| - 1) = b(k_2 - k_1) = k - k_1$, which implies that the solution taking $z_i = b$ for each $i \in \mathcal{S}_1$ is a feasible solution to LP3, whose value is $b\sum_{i \in \mathcal{S}_1} \Phi_1(i) = b\big(5S(\mathcal{H}_2) + 3S(\mathcal{H}_1)\big)$ due to the definition of $\Phi_1(i)$. By the fact that *opt'* is the value of an optimal solution to LP3, we complete the proof of inequality (17). Thus, Lemma 7 is true. □

We now show that our solution achieves a $\big(36 + O(\epsilon)\big)$-approximation for SM-$k$-FL in case (3).

Lemma 8. *If $a \in [\frac{1}{36}, \frac{1}{4})$ and $k - k_1 > \frac{1}{\epsilon a}$, then $R < \big(36 + O(\epsilon)\big)opt^*$.*

Proof. Inequality (12) and Lemma 7 imply that

$$
\begin{aligned}
R_s &\leq 6S(\mathcal{H}_2) + 3S(\mathcal{H}_1) - \sum_{i \in \mathcal{L}_1} \Phi_1(i) - \sum_{i \in \mathcal{Q}^\dagger} \Phi_2(i) \\
&< 6S(\mathcal{H}_2) + 3S(\mathcal{H}_1) - (1-\epsilon a)b\big(5S(\mathcal{H}_2) + 3S(\mathcal{H}_1)\big) \\
&= (1+5a)S(\mathcal{H}_2) + 3aS(\mathcal{H}_1) + \epsilon ab\big(5S(\mathcal{H}_2) + 3S(\mathcal{H}_1)\big).
\end{aligned}
$$

Thus, we have

$$
\begin{aligned}
\frac{R_s}{aS(\mathcal{H}_1) + bS(\mathcal{H}_2)} &< \frac{(1+5a)S(\mathcal{H}_2) + 3aS(\mathcal{H}_1)}{(1-a)S(\mathcal{H}_2) + aS(\mathcal{H}_1)} + 5\epsilon \\
&< \frac{(3-3a)S(\mathcal{H}_2) + 3aS(\mathcal{H}_1)}{(1-a)S(\mathcal{H}_2) + aS(\mathcal{H}_1)} + 5\epsilon \\
&= 3 + O(\epsilon),
\end{aligned} \tag{18}
$$

where the second step is due to the assumption that $a \in [\frac{1}{36}, \frac{1}{4})$. By the same assumption and the fact that our solution only selects the facilities from $\mathcal{S}_1 \cup \mathcal{S}_2$ to open, we have

$$\frac{R_f}{9aF(\mathcal{H}_1) + 9bF(\mathcal{H}_2)} < \frac{F(\mathcal{H}_1) + F(\mathcal{H}_2)}{9aF(\mathcal{H}_1) + 9bF(\mathcal{H}_2)} < 4. \tag{19}$$

Using inequalities (18) and (19), we get

$$\begin{aligned} R_f + R_s &< \big(4 + O(\epsilon)\big)\big(aS(\mathcal{H}_1) + 9aF(\mathcal{H}_1) + bS(\mathcal{H}_2) + 9bF(\mathcal{H}_2)\big) \\ &\leq \big(36 + O(\epsilon)\big)opt^*, \end{aligned}$$

where the last step is due to Corollary 1. Thus, Lemma 8 is true. □

Case (4): $\boldsymbol{a \in (0, \frac{1}{36})}$.

For this case, we select a set of facilities from $\mathcal{S}_2$ to open. Let ρ be an arbitrary positive number. Lemma 4 implies that $\Delta(j, \eta_2(i_1(j))) \leq 4(1+\rho)\Delta_1(j) + (1+\frac{1}{\rho})\Delta_2(j)$ for each $j \in \mathcal{D}$. Our idea is to ensure that $\eta_2(i)$ is opened in the solution for each $i \in \mathcal{S}_1$, such that each $j \in \mathcal{D}$ can always be assigned to either $i_2(j)$ or $\eta_2(i_1(j))$. This implies that the cost induced by assigning j is $\Delta_2(j)$ if $i_2(j)$ is opened, and can be upper-bounded by $4(1+\rho)\Delta_1(j) + (1+\frac{1}{\rho})\Delta_2(j)$ otherwise.

Let $\mathcal{S}_2^\dagger = \{\eta_2(i) : i \in \mathcal{S}_1\}$. It is the case that $|\mathcal{S}_2^\dagger| \leq k_1$. For each $i \in \mathcal{S}_2 \backslash \mathcal{S}_2^\dagger$, define $\Gamma(i) = \sum_{j \in \gamma_2(i)} \big(4(1+\rho)\Delta_1(j) + \frac{1}{\rho}\Delta_2(j)\big)$ for brevity, and let $\mathcal{S}_2^\ddagger = \underset{\mathcal{S} \subset \mathcal{S}_2 \backslash \mathcal{S}_2^\dagger \wedge |\mathcal{S}| = k - |\mathcal{S}_2^\dagger|}{\arg\max} \sum_{i \in \mathcal{S}} \Gamma(i)$. Our solution for case (4) opens each facility from $\mathcal{S}_2^\dagger \cup \mathcal{S}_2^\ddagger$ and assigns each client to the nearest opened facility. We now show that this solution achieves the desired approximation for SM-k-FL.

Lemma 9. *If $a \in (0, \frac{1}{36})$, then we have $\sum_{i \in \mathcal{S}_2^\dagger \cup \mathcal{S}_2^\ddagger} f(i) + \sum_{j \in \mathcal{D}} \Delta(j, \mathcal{S}_2^\dagger \cup \mathcal{S}_2^\ddagger) < \max\{4 + 4\rho, \frac{36}{35} + \frac{1}{35\rho}\}c_f$ for any $\rho > 0$.*

Proof. We consider the following strategy for assigning clients to obtain an upper bound on $\sum_{j \in \mathcal{D}} \Delta(j, \mathcal{S}_2^\dagger \cup \mathcal{S}_2^\ddagger)$. We open each facility from $\mathcal{S}_2^\dagger \cup \mathcal{S}_2^\ddagger$. For each $j \in \gamma_2(\mathcal{S}_2^\dagger \cup \mathcal{S}_2^\ddagger)$, we assign j to $i_2(j)$. The total assignment cost of the clients from $\gamma_2(\mathcal{S}_2^\dagger \cup \mathcal{S}_2^\ddagger)$ is $\sum_{j \in \gamma_2(\mathcal{S}_2^\dagger \cup \mathcal{S}_2^\ddagger)} \Delta_2(j)$. For each $j \in \gamma_2(\mathcal{S}_2 \backslash (\mathcal{S}_2^\dagger \cup \mathcal{S}_2^\ddagger))$, the definition of $\mathcal{S}_2^\dagger$ implies that $\eta_2(i_1(j))$ is guaranteed to be opened, and we can assign j to $\eta_2(i_1(j))$. Lemma 4 implies that the total assignment cost of the clients from $\gamma_2(\mathcal{S}_2 \backslash (\mathcal{S}_2^\dagger \cup \mathcal{S}_2^\ddagger))$ is no more than $\sum_{j \in \gamma_2(\mathcal{S}_2 \backslash (\mathcal{S}_2^\dagger \cup \mathcal{S}_2^\ddagger))} 4(1+\rho)\Delta_1(j) + (1+\frac{1}{\rho})\Delta_2(j)$. Thus, we have

$$\begin{aligned}
&\sum_{j\in\mathcal{D}}\Delta(j,\mathcal{S}_2^{\dagger}\cup\mathcal{S}_2^{\ddagger})\\
&\qquad\le\sum_{j\in\gamma_2(\mathcal{S}_2^{\dagger}\cup\mathcal{S}_2^{\ddagger})}\Delta_2(j)+\sum_{j\in\gamma_2(\mathcal{S}_2\setminus(\mathcal{S}_2^{\dagger}\cup\mathcal{S}_2^{\ddagger}))}4(1+\rho)\Delta_1(j)+(1+\frac{1}{\rho})\Delta_2(j)\\
&\qquad=\sum_{j\in\mathcal{D}}\Delta_2(j)+\sum_{j\in\gamma_2(\mathcal{S}_2\setminus(\mathcal{S}_2^{\dagger}\cup\mathcal{S}_2^{\ddagger}))}4(1+\rho)\Delta_1(j)+\frac{1}{\rho}\Delta_2(j)\\
&\qquad=\sum_{j\in\mathcal{D}}\Delta_2(j)+\sum_{i\in\mathcal{S}_2\setminus(\mathcal{S}_2^{\dagger}\cup\mathcal{S}_2^{\ddagger})}\Gamma(i),
\end{aligned}\tag{20}$$

where the last step follows from the definition of $\Gamma(i)$.

Observe that

$$\begin{aligned}
\sum_{i\in\mathcal{S}_2\setminus(\mathcal{S}_2^{\dagger}\cup\mathcal{S}_2^{\ddagger})}\Gamma(i)&\le\frac{|\mathcal{S}_2\setminus(\mathcal{S}_2^{\dagger}\cup\mathcal{S}_2^{\ddagger})|}{|\mathcal{S}_2\setminus\mathcal{S}_2^{\dagger}|}\sum_{i\in\mathcal{S}_2\setminus\mathcal{S}_2^{\dagger}}\Gamma(i)=\frac{k_2-k}{k_2-|\mathcal{S}_2^{\dagger}|}\sum_{i\in\mathcal{S}_2\setminus\mathcal{S}_2^{\dagger}}\Gamma(i)\\
&\le\frac{k_2-k}{k_2-k_1}\sum_{i\in\mathcal{S}_2\setminus\mathcal{S}_2^{\dagger}}\Gamma(i)=a\sum_{i\in\mathcal{S}_2\setminus\mathcal{S}_2^{\dagger}}\Gamma(i)\le a\sum_{i\in\mathcal{S}_2}\Gamma(i)\\
&=4a(1+\rho)\sum_{j\in\mathcal{D}}\Delta_1(j)+\frac{a}{\rho}\sum_{j\in\mathcal{D}}\Delta_2(j),
\end{aligned}\tag{21}$$

where the first step is due to the fact that $\mathcal{S}_2^{\ddagger}=\arg\max_{\mathcal{S}\subset\mathcal{S}_2\setminus\mathcal{S}_2^{\dagger}\wedge|\mathcal{S}|=k-|\mathcal{S}_2^{\dagger}|}\sum_{i\in\mathcal{S}}\Gamma(i)$, the third step is due to the fact that $|\mathcal{S}_2^{\dagger}|\le k_1$, and the last step is derived from the definition of $\Gamma(i)$. Using inequalities (20) and (21), we have

$$\begin{aligned}
\sum_{j\in\mathcal{D}}\Delta(j,\mathcal{S}_2^{\dagger}\cup\mathcal{S}_2^{\ddagger})&\le\sum_{j\in\mathcal{D}}\Delta_2(j)+4a(1+\rho)\sum_{j\in\mathcal{D}}\Delta_1(j)+\frac{a}{\rho}\sum_{j\in\mathcal{D}}\Delta_2(j)\\
&=4a(1+\rho)S(\mathcal{H}_1)+(1+\frac{a}{\rho})S(\mathcal{H}_2).
\end{aligned}\tag{22}$$

Consequently, we get

$$\begin{aligned}
&\frac{1}{c_f}\Big(\sum_{i\in\mathcal{S}_2^{\dagger}\cup\mathcal{S}_2^{\ddagger}}f(i)+\sum_{j\in\mathcal{D}}\Delta(j,\mathcal{S}_2^{\dagger}\cup\mathcal{S}_2^{\ddagger})\Big)\\
&\qquad=\frac{\sum_{i\in\mathcal{S}_2^{\dagger}\cup\mathcal{S}_2^{\ddagger}}f(i)+\sum_{j\in\mathcal{D}}\Delta(j,\mathcal{S}_2^{\dagger}\cup\mathcal{S}_2^{\ddagger})}{aS(\mathcal{H}_1)+9aF(\mathcal{H}_1)+bS(\mathcal{H}_2)+9bF(\mathcal{H}_2)}\\
&\qquad<\frac{F(\mathcal{H}_2)+4a(1+\rho)S(\mathcal{H}_1)+(1+\frac{a}{\rho})S(\mathcal{H}_2)}{aS(\mathcal{H}_1)+9aF(\mathcal{H}_1)+bS(\mathcal{H}_2)+9bF(\mathcal{H}_2)}\\
&\qquad<\max\{\frac{1}{9b},4(1+\rho),(1+\frac{a}{\rho})\frac{1}{b}\}<\max\{4+4\rho,\frac{36}{35}+\frac{1}{35\rho}\},
\end{aligned}$$

where the first step follows from the definition of c_f, the second step is due to inequality (22) and the fact that $\mathcal{S}_2^{\dagger}\cup\mathcal{S}_2^{\ddagger}\subset\mathcal{S}_2$, and the last step is due to the assumption that $a\in(0,\frac{1}{36})$. This completes the proof of Lemma 9. □

Let $\rho = \frac{1}{105}$. Using Lemma 9 and Corollary 1, we have $\sum_{i \in \mathcal{S}_2^\dagger \cup \mathcal{S}_2^\ddagger} f(i) + \sum_{j \in \mathcal{D}} \Delta(j, \mathcal{S}_2^\dagger \cup \mathcal{S}_2^\ddagger) < \frac{424}{105} c_f < \big(36.343 + O(\epsilon)\big) opt^*$. This implies that we get a $\big(36.343 + O(\epsilon)\big)$-approximation solution to $\mathcal{I}$ in case (4). Recall that $\mathcal{H}_1$ is a $\big(36 + O(\epsilon)\big)$-approximation solution in case (1), the k facilities selected from $\mathcal{S}_2$ form a $\big(26+O(\epsilon)\big)$-approximation solution in case (2), and the solution constructed based on LP3 achieves a $\big(36+O(\epsilon)\big)$-approximation in case (3). Putting everything together, we obtain a $\big(36.343 + O(\epsilon)\big)$-approximation algorithm for the problem.

5 Conclusions

In this paper, we give a new approximation algorithm for the squared metric k-facility location problem based on the techniques of primal-dual and Lagrangian relaxation, which has the guarantee of yielding a $(36.343 + \epsilon)$-approximation solution. This improves the current best approximation ratio of $44.473 + \epsilon$ given in [27]. Our main technical contribution is a new rounding approach that exploits the properties of the squared metric.

References

1. Ahmadian, S., Norouzi-Fard, A., Svensson, O., Ward, J.: Better guarantees for k-means and Euclidean k-median by primal-dual algorithms. SIAM J. Comput. **49**(4), FOCS17:97–FOCS17:156 (2020)
2. Aloise, D., Deshpande, A., Hansen, P., Popat, P.: NP-hardness of Euclidean sum-of-squares clustering. Mach. Learn. **75**(2), 245–248 (2009)
3. Arthur, D., Vassilvitskii, S.: k-means++: the advantages of careful seeding. In: Proceedings of the 18th Annual ACM-SIAM Symposium on Discrete Algorithms (SODA), pp. 1027–1035 (2007)
4. Arya, V., Garg, N., Khandekar, R., Meyerson, A., Munagala, K., Pandit, V.: Local search heuristics for k-median and facility location problems. SIAM J. Comput. **33**(3), 544–562 (2004)
5. Baev, I.D., Rajaraman, R., Swamy, C.: Approximation algorithms for data placement problems. SIAM J. Comput. **38**(4), 1411–1429 (2008)
6. Byrka, J., Pensyl, T., Rybicki, B., Srinivasan, A., Trinh, K.: An improved approximation for k-median and positive correlation in budgeted optimization. ACM Trans. Algor. **13**(2), 23:1–23:31 (2017)
7. Charikar, M., Khuller, S., Mount, D.M., Narasimhan, G.: Algorithms for facility location problems with outliers. In: Proceedings of the 12th Annual ACM-SIAM Symposium on Discrete Algorithms (SODA), pp. 642–651 (2001)
8. Charikar, M., Li, S.: A dependent LP-rounding approach for the k-median problem. In: Proceedings of the 39th International Colloquium on Automata, Languages, and Programming (ICALP), pp. 194–205 (2012)
9. Cohen-Addad, V., Klein, P.N., Mathieu, C.: Local search yields approximation schemes for k-means and k-median in Euclidean and minor-free metrics. SIAM J. Comput. **48**(2), 644–667 (2019)
10. Cura, T.: A parallel local search approach to solving the uncapacitated warehouse location problem. Comput. Ind. Eng. **59**(4), 1000–1009 (2010)

11. Feng, Q., Zhang, Z., Shi, F., Wang, J.: An improved approximation algorithm for the k-means problem with penalties. In: Proceedings of the 13th International Workshop on Frontiers in Algorithmics (FAW), pp. 170–181 (2019)
12. Fernandes, C.G., Meira, L.A.A., Miyazawa, F.K., Pedrosa, L.L.C.: A systematic approach to bound factor-revealing LPs and its application to the metric and squared metric facility location problems. Math. Program. **153**(2), 655–685 (2015)
13. Friggstad, Z., Rezapour, M., Salavatipour, M.R.: Local search yields a PTAS for k-means in doubling metrics. SIAM J. Comput. **48**(2), 452–480 (2019)
14. Golubchik, L., Khanna, S., Khuller, S., Thurimella, R., Zhu, A.: Approximation algorithms for data placement on parallel disks. ACM Trans. Algor. **5**(4), 34:1–34:26 (2009)
15. Guha, S., Khuller, S.: Greedy strikes back: Improved facility location algorithms. J. Algor. **31**(1), 228–248 (1999)
16. Gupta, A., Guruganesh, G., Schmidt, M.: Approximation algorithms for aversion k-clustering via local k-median. In: Proceedings of the 43rd International Colloquium on Automata, Languages and Programming (ICALP), pp. 1–13 (2016)
17. Gupta, A., Tangwongsan, K.: Simpler analyses of local search algorithms for facility location. CoRR abs/0809.2554 (2008)
18. Hajiaghayi, M., Khandekar, R., Kortsarz, G.: Local search algorithms for the red-blue median problem. Algorithmica **63**(4), 795–814 (2012)
19. Hayrapetyan, A., Swamy, C., Tardos, É.: Network design for information networks. In: Proceedings of the 16th Annual ACM-SIAM Symposium on Discrete Algorithms (SODA), pp. 933–942 (2005)
20. Jain, K., Mahdian, M., Markakis, E., Saberi, A., Vazirani, V.V.: Greedy facility location algorithms analyzed using dual fitting with factor-revealing LP. J. ACM **50**(6), 795–824 (2003)
21. Jain, K., Vazirani, V.V.: Approximation algorithms for metric facility location and k-median problems using the primal-dual schema and Lagrangian relaxation. J. ACM **48**(2), 274–296 (2001)
22. Kanungo, T., Mount, D.M., Netanyahu, N.S., Piatko, C.D., Silverman, R., Wu, A.Y.: A local search approximation algorithm for k-means clustering. Comput. Geom. **28**, 89–112 (2004)
23. Li, S., Svensson, O.: Approximating k-median via pseudo-approximation. SIAM J. Comput. **45**(2), 530–547 (2016)
24. Michel, L., Hentenryck, P.V.: A simple tabu search for warehouse location. Eur. J. Oper. Res. **157**(3), 576–591 (2004)
25. Wang, S., Bi, J., Wu, J., Vasilakos, A.V.: CPHR: In-network caching for information-centric networking with partitioning and hash-routing. IEEE/ACM Trans. Netw. **24**(5), 2742–2755 (2016)
26. Wu, C., Du, D., Xu, D.: An approximation algorithm for the k-median problem with uniform penalties via pseudo-solution. Theor. Comput. Sci. **749**, 80–92 (2018)
27. Zhang, D., Xu, D., Wang, Y., Zhang, P., Zhang, Z.: A local search approximation algorithm for a squared metric k-facility location problem. J. Comb. Optim. **35**(4), 1168–1184 (2018)
28. Zhang, P.: A new approximation algorithm for the k-facility location problem. Theor. Comput. Sci. **384**(1), 126–135 (2007)
29. Zhang, Z., Feng, Q., Huang, J., Guo, Y., Xu, J., Wang, J.: A local search algorithm for k-means with outliers. Neurocomputing **450**, 230–241 (2021)
30. Zhang, Z., Feng, Q., Xu, J., Wang, J.: An approximation algorithm for k-median with priorities. Sci. China Inf. Sci. **64**(5), 150104 (2021)

Parameterized Algorithms for Linear Layouts of Graphs with Respect to the Vertex Cover Number

Yunlong Liu(✉), Yixuan Li, and Jingui Huang(✉)

College of Information Science and Engineering, Hunan Provincial Key Laboratory of Intelligent Computing and Language Information Processing, Hunan Normal University, Changsha 410081, People's Republic of China
{ylliu,yxlee,hjg}@hunnu.edu.cn

Abstract. The LINEAR LAYOUT OF GRAPHS problem asks, given a graph $G = (V, E)$ and an integer k, whether G admits a linear layout consisting of a linear order of V and a partition of E into k sets such that the edges in each set satisfy some restrictions. In this paper, we study parameterized algorithms for a series of specific linear layout problems with respect to the vertex cover number τ of the input graph. We first focus on the MIXED s-STACK q-QUEUE LAYOUT problem and show that it admits a kernel of size $2^{\mathcal{O}(\tau)}$ and an algorithm running in time $\mathcal{O}(2^{2^{\mathcal{O}(\tau)}} + \tau \cdot |V|)$, where $|V|$ denotes the size of the input graph. Our work does not only confirm the existence of a fixed-parameter tractable algorithm for this problem which was mentioned by Bhore et al. (GD 2020), but also derives new results improving that for the k-STACK LAYOUT problem (J. Graph Alg. Appl. 2020), that for the UPWARD k-STACK LAYOUT problem (GD 2021), and that for the k-QUEUE LAYOUT problem (GD 2020) respectively. We also generalize our techniques to the k-ARCH LAYOUT problem and obtain a similar result.

Keywords: Linear layouts · Vertex cover number · Cloneable vertex

1 Introduction

Linear layouts of graphs form a main topic for drawing graphs. In a linear order of the vertices of a graph, there are three possible layouts of two independent edges, that is, they can be crossing, nested, or disjoint. Correspondingly, three basic linear layouts can be defined as follows. A *k-stack (respectively, k-queue, k-arch) layout* of a graph $G = (V, E)$ is a linear layout $\langle \prec, \sigma \rangle$ of G consisting of a linear order $\prec$ of its vertices along a spine and an assignment σ of its edges to k subsets of pairwise non-crossing (respectively, non-nested, non-disjoint) edges [11]. Note that edges with a common endpoint do not cross, do not nest, and are not

This research was supported in part by the National Natural Science Foundation of China under Grant No. 61572190 and Hunan Provincial Science and Technology Program under Grant No. 2018TP1018.

D.-Z. Du et al. (Eds.): COCOA 2021, LNCS 13135, pp. 553–567, 2021.
https://doi.org/10.1007/978-3-030-92681-6_43

disjoint. A family of edges assigned to the same subset accordingly forms a *stack (respectively, queue, arch) page*. See Fig. 1 for an illustration. Linear layouts of graphs have a wide rang of applications including sorting permutations, parallel process scheduling, fault tolerant VLSI design, matrix computations, and so on (see, e.g., [3,9,22] and refer also to [11] for a survey).

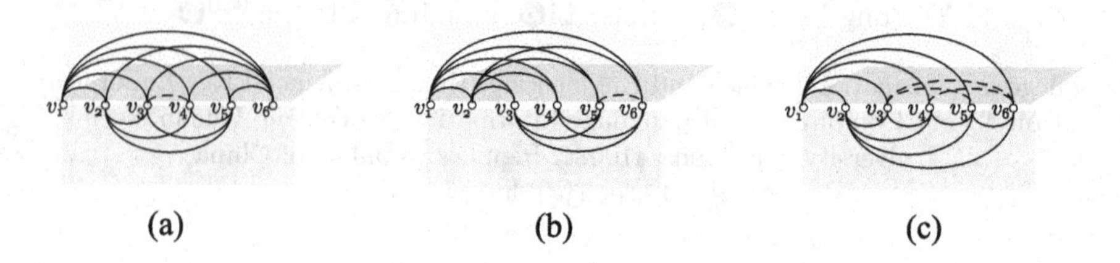

Fig. 1. Layouts of K_6: (a) 3-queue, (b) 3-arch, and (c) mixed 2-stack 1-queue.

There is a growing interest in studying various computation problems about linear layouts of graphs (see, e.g., [18,21,24] and refer also to [11] for an overview). In particular, a series of parameterized algorithms with respect to the vertex cover number of the input graph have been presented in recent years (see, e.g., [1,4,5,14,17,19,20]).

Stack layouts are also commonly called book embeddings [2,4]. The k-STACK LAYOUT problem deciding, given a graph $G = (V, E)$ and a positive integer k, whether G admits a k-stack layout, has been extensively studied (see, e.g., [4, 9,24]). The 2-STACK LAYOUT problem is known to be **NP**-complete [9]. Just because of this, Bhore et al. [4] studied parameterized algorithms for k-STACK LAYOUT. Specially, they presented an algorithm running in time $\mathcal{O}(2^{\tau^{\mathcal{O}(\tau)}} + \tau \cdot |V|)$ with respect to the vertex cover number τ [4].

Upward stack layouts are a natural extension of stack layouts to directed acyclic graphs with the additional requirement that the linear order $\prec$ of vertices respects the directions of all edges [6]. The UPWARD k-STACK LAYOUT problem is generally **NP**-complete, even for fixed values of $k \geq 3$ [9]. Recently, Bhore et al. [6] also presented an algorithm running in time $\mathcal{O}(\tau^{\tau^{\mathcal{O}(\tau)}} + \tau \cdot |V|)$ with respect to the vertex cover number τ.

Queue layouts were introduced by Heath et al. [15,16]. The k-QUEUE LAYOUT problem asks, given a graph $G = (V, E)$ and a positive integer k, whether G admits a k-queue layout. The 1-QUEUE LAYOUT problem is known to be **NP**-complete [7]. Bhore et al. [5] also studied parameterized algorithms for it with respect to the vertex cover number τ. They presented an algorithm running in time $\mathcal{O}(2^{\tau^{\mathcal{O}(\tau)}} + \tau \log \tau \cdot |V|)$ for the optimization version of the k-QUEUE LAYOUT problem [5].

Arch layouts were formally introduced by Dujmović and Wood [11]. The k-ARCH LAYOUT problem determines, given a graph G and a positive integer k, whether G admits a k-arch layout. When $k \geq 2$, the k-ARCH LAYOUT problem is known to be **NP**-complete [11].

Besides these basic linear layouts, the mixed linear layouts, which combine $s \geq 1$ stack pages and $q \geq 1$ queue pages, have also attracted much attention (see, e.g., [10,12,13,16,23]). It is **NP**-complete to decide if a given graph G admits a 2-stack 1-queue layout [10]. The parameterized complexity for mixed s-stack q-queue layouts was mentioned as an interesting question by Bhore et al. in [5].

In this paper, we further study these linear layout problems parameterized by the vertex cover number τ of the input graph. The target is to develop some uniform problem-solving techniques for them. For ease of presentation, we assume that $s \geq 0, q \geq 0$ but $s + q \neq 0$ in the definition of MIXED s-STACK q-QUEUE LAYOUT. By this assumption, the k-STACK LAYOUT problem and the k-QUEUE LAYOUT problem can be seen as two special cases of the MIXED s-STACK q-QUEUE LAYOUT problem. Thus, we first focus on the MIXED s-STACK q-QUEUE LAYOUT problem and show that it admits a kernel of size $2^{\mathcal{O}(\tau)}$ and an algorithm running in time $\mathcal{O}(2^{2^{\mathcal{O}(\tau)}} + \tau \cdot |V|)$. Our work does not only confirm the existence of a fixed-parameter tractable algorithm for this problem mentioned by Bhore et al. [5], but also derives new results improving that for the k-STACK LAYOUT problem in [4], that for the UPWARD k-STACK LAYOUT problem in [6], and that for the k-QUEUE LAYOUT problem in [5] respectively. For the k-ARCH LAYOUT problem, by using our techniques, we also obtain a kernel of size $2^{\mathcal{O}(\tau)}$ and an algorithm running in time $\mathcal{O}(2^{2^{\mathcal{O}(\tau)}} + \tau \cdot |V|)$.

Our kernelization is inspired by the "suitable vertex" used in [4,5], by which the reduced vertices together with their incident edges can follow up in the same way. However, we introduce the concept of *uncloneable/cloneable vertex* into our algorithm. The cloneable vertex corresponds to the "suitable vertex", but the concept of uncloneable vertex plays a key role in deriving new reduction rules. Given a yes-instance of the considered problem, we argue that the number of uncloneable vertices can be bounded by a function of τ. Based on this fact, we give a new approach to locate the cloneable vertex. In our kernelization procedure, the primary step is to identify the uncloneable/cloneable vertex according to specific problem. It also turns out that with this concept the proving correctness of the algorithms becomes a rather simple task.

2 Preliminaries

We consider only simple graphs. Some notations only for the linear layouts of undirected graphs are given as follows. On those for the upward stack layouts of directed acyclic graphs, refer to [6] for details.

For a graph $G = (V, E)$, let $n = |V|$, and we denote by $V(G)$ the vertex set of G and by $E(G)$ the edge set of G respectively. For two vertices u and v in $V(G)$, we denote by uv the edge between u and v. Two edges are called *independent edges* if they do not share an endpoint. For $r \in \mathbb{N}$, we use $[1, r]$ to denote the set $\{1, \ldots, r\}$.

Given a graph $G = (V, E)$, we use $\langle \prec, \sigma \rangle$ to denote a k-stack (respectively, k-queue, k-arch) layout, where $\prec$ is a linear order of V, and $\sigma : E \rightarrow [1, k]$ is a function that maps each edge of E to one of k stack pages (respectively, queue pages, arch pages).

In a vertex ordering $\prec$ of $V(G)$, let $L(e)$ and $R(e)$ denote the endpoints of each edge $e \in E(G)$ such that $L(e) \prec R(e)$. Let $e_1, e_2 \in E(G)$ be two independent edges. Without loss of generality, $L(e_1) \prec L(e_2)$. The three basic layouts for e_1 and e_2 are defined as follows [11]. (1) e_1 and e_2 cross if $L(e_1) \prec L(e_2) \prec R(e_1) \prec R(e_2)$; (2) e_1 and e_2 nest and e_2 is nested inside e_1 if $L(e_1) \prec L(e_2) \prec R(e_2) \prec R(e_1)$; and (3) e_1 and e_2 are disjoint if $L(e_1) \prec R(e_1) \prec L(e_2) \prec R(e_2)$.

A *vertex cover* C of a graph $G = (V, E)$ is a subset $C \subseteq V$ such that each edge in E has at least one endpoint in C. The *vertex cover number* of G, denoted by τ, is the size of a minimum vertex cover of G. Given a graph G, a vertex cover with size τ can be computed in time $\mathcal{O}(2^{\tau} + \tau \cdot n)$ [8]. In the rest of this paper, we will use C to denote a minimum vertex cover of the input graph.

Let $W \subseteq C$. A vertex in $V(G) \setminus C$ is of type W if its set of neighbors is equal to W [4]. By this definition, the vertices in $V(G) \setminus C$ are partitioned into at most $2^{\tau} - 1$ distinct types. We denote by V_W the set of vertices of type W.

Because of the space limit, several proofs are deferred to a full version of this paper.

3 Mixed s-Stack q-Queue Layout

The MIXED s-STACK q-QUEUE LAYOUT problem parameterized by the vertex cover number τ can be formally described as follows.

> **Parameterized mixed s-stack q-queue layout (abbreviated as SQ-Lay)**
> **Input:** an undirected graph $G = (V, E)$, two non-negative integers s, q ;
> **Parameter:** τ ;
> **Question:** does there exist a mixed s-stack q-queue layout $\langle \prec, \sigma \rangle$ of G?

In this section, we will show that the SQ-LAY problem admits a kernel of size $2^{\mathcal{O}(\tau)}$ and an algorithm running in time $\mathcal{O}(2^{2^{\mathcal{O}(\tau)}} + \tau \cdot |V|)$. Our result confirms the existence of a fixed-parameter tractable algorithm for this problem mentioned by Bhore et al. [5].

3.1 Two Kinds of Vertices in V_W

Assume that (G, τ, s, q) is a yes-instance of the SQ-LAY problem. Let $\langle \prec, \sigma \rangle$ be a mixed s-stack q-queue layout for (G, τ, s, q) and let $W \subseteq C$. Without loss of generality, we denote by $[1, s]$ the s stack pages (if $s \neq 0$) and by $[s+1, s+q]$ the q queue pages (if $q \neq 0$) in $\langle \prec, \sigma \rangle$. Observe that if $|W| = 1$, then the edges incident to the unique vertex in W can be simultaneously assigned to an arbitrary page. In the following, we assume that $|W| \geq 2$ and identify two kinds of vertices in V_W (i.e., the uncloneable vertices and the cloneable vertices) as follows.

When $s \neq 0$, let $p \in [1, s]$. A vertex v in V_W is called an *uncloneable vertex* with respect to stack page p if v has two edges, say w_iv and w_jv, assigned to page p simultaneously. Vertices w_i and w_j are accordingly called *a pair of barrier vertices* for v with respect to stack page p. Similarly, when $q \neq 0$, let $h \in [s+1, s+q]$. A vertex v in V_W is called an *uncloneable vertex* with respect to queue page h if v has two edges, say w_av and w_bv, assigned to page h simultaneously such that the ordering of vertices w_a, w_b, and v in $\prec$ is either $w_a \prec w_b \prec v$ or $v \prec w_a \prec w_b$. Vertices w_a and w_b are accordingly called *a pair of barrier vertices* for v with respect to queue page h. Furthermore, a vertex v in V_W is called an *uncloneable vertex* with respect to $\langle \prec, \sigma \rangle$ if it is an uncloneable vertex with respect to either at least one stack page or at least one queue page in $\langle \prec, \sigma \rangle$. Otherwise, vertex v is called a *cloneable vertex* with respect to $\langle \prec, \sigma \rangle$ (see Fig. 2 for examples). For simplicity, we also call v *uncloneable (resp. cloneable)* if it is an uncloneable (resp. a cloneable) vertex.

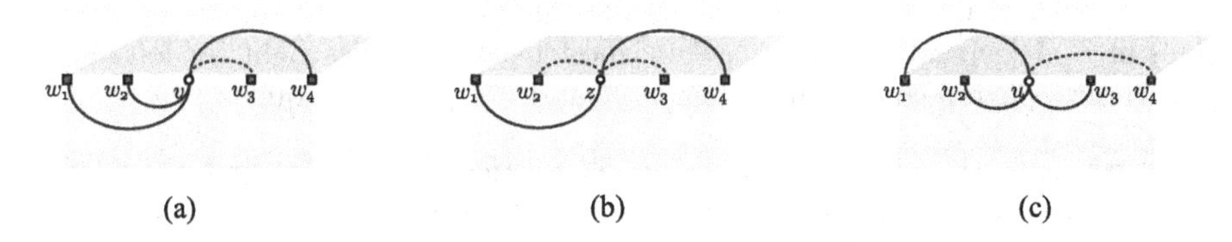

Fig. 2. An uncloneable vertex v in (a), an uncloneable vertex z in (b), and a cloneable vertex u in (c) are shown in three mixed 2-stack 1-queue layouts respectively. In each layout, the upper vertical page and the horizontal page are all stack pages, but the lower vertical page is a queue page.

From the definition above, the following property for the cloneable vertex holds.

Lemma 1. *Let $\langle \prec, \sigma \rangle$ be a mixed s-stack q-queue layout of a given graph. If v is a cloneable vertex with respect to $\langle \prec, \sigma \rangle$, then inserting a vertex v^* of type W right (or left) next to v and assigning each edge v^*w (for $w \in W$) to the same page as vw will result in a valid mixed s-stack q-queue layout of the resulting graph.*

Proof. Let p be an arbitrary page to which v has some edge assigned and let E_p be the set of edges assigned to page p. We distinguish two cases based on whether p is a stack page or not.

Case 1: p is a stack page. Since v is a cloneable vertex with respect to $\langle \prec, \sigma \rangle$, v has at most one edge assigned to any stack page in $[1, s]$. Without loss of generality, assume that an edge vw_i (for some $w_i \in W$) is assigned to page p. After assigning the corresponding edge v^*w_i to page p, edges v^*w_i and vw_i would not cross on page p because they have a common endpoint. Moreover, no two independent edges in $E_p \setminus \{vw_i\} \cup \{v^*w_i\}$ cross because v^* lies right (or left) next to v. Thus, the resulting page is still a stack page.

Case 2: p is a queue page. First, we show that the cloneable vertex v has at most two edges assigned to page p simultaneously. Suppose by contradiction that v has at least three edges assigned to page p simultaneously. By the pigeonhole principle, vertex v must have at least two edges, say vw_i and vw_j ($w_i \prec w_j$), assigned to the same side of v on page p. Namely, the ordering of vertices w_i, w_j and v is either $v \prec w_i \prec w_j$ or $w_i \prec w_j \prec v$. Hence, vertex v is an uncloneable vertex, leading to a contradiction. Next, we show that the resulting page is still a queue page. If v has exactly one edge vw assigned to page p, by employing the same argument used in the proof of case 1, then page p containing v^*w is still a queue page. Assume now that v has exactly two edges, say w_iv and vw_j ($w_i \prec w_j$), assigned to page p simultaneously. Then, it holds that $w_i \prec v \prec w_j$. After inserting v^* right (or left) next to v and assigning edges w_iv^* and v^*w_j to page p, no two independent edges in $\{w_iv, w_iv^*, vw_j, v^*w_j\}$ nest on page p because both v and v^* lie between w_i and w_j. Moreover, no two independent edges in $E_p \setminus \{w_iv, vw_j\} \cup \{w_iv^*, v^*w_j\}$ nest because v^* lies right (or left) next to v. Hence, page p containing edges w_iv^* and v^*w_j is still a queue page. □

To estimate the number of uncloneable vertices in V_W, we first consider a special case, i.e., the number of uncloneable vertices for which two fixed vertices in W form a pair of barrier vertices.

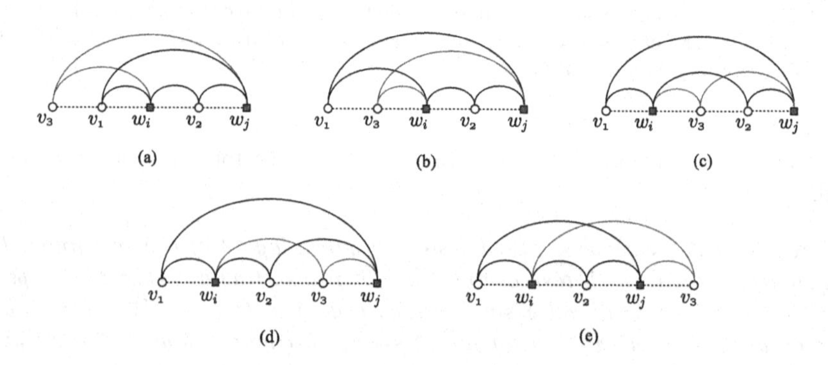

Fig. 3. Illustration for having at most 2 uncloneable vertices for which w_i and w_j form a pair of barrier vertices with respect to a stack page.

Lemma 2. *Assume that $\langle \prec, \sigma \rangle$ is a mixed s-stack q-queue layout for (G, τ, s, q) and $W \subseteq C$. Let w_i and w_j be two arbitrary vertices in W. Then there exist at most $2(s+q)$ uncloneable vertices of type W for which w_i and w_j form a pair of barrier vertices with respect to at least one page in $\langle \prec, \sigma \rangle$.*

Proof. When $s \neq 0$, we will show that V_W contains at most $2s$ uncloneable vertices for which w_i and w_j form a pair of barrier vertices for which w_i and w_j form a pair of barrier vertices with respect to at least one stack page in $\langle \prec, \sigma \rangle$. Suppose by contradiction that V_W contains at least $2s + 1$ uncloneable vertices. By the pigeonhole principle, there must be at least one page $p \in [1, s]$ such that V_W contains at least 3 uncloneable vertices, say v_1, v_2 and v_3, for which w_i and w_j form a pair of barrier vertices with respect to page p. Without loss of generality, assume that $w_i \prec w_j$, $v_1 \prec v_2$, and that no two independent edges in $\{v_1w_i, v_1w_j, v_2w_i, v_2w_j\}$ cross. Then, for the linear ordering of vertices in $\{v_1, v_2, w_i, w_j\}$, there are only two possible layouts: $v_1 \prec w_i \prec v_2 \prec w_j$ and $w_i \prec v_1 \prec w_j \prec v_2$.

Considering the ordering $v_1 \prec w_i \prec v_2 \prec w_j$, we distinguish five cases based on the position of v_3 (refer to Fig. 3 for an illustration).

Case (a): $v_3 \prec v_1$, then v_3w_i and v_1w_j are crossing.

Case (b): $v_1 \prec v_3 \prec w_i \prec v_2 \prec w_j$, then v_3w_j and v_1w_i are crossing.

Case (c): $v_1 \prec w_i \prec v_3 \prec v_2 \prec w_j$, then v_3w_j and w_iv_2 are crossing.

Case (d): $v_1 \prec w_i \prec v_2 \prec v_3 \prec w_j$, then w_iv_3 and v_2w_j are crossing.

Case (e): $w_j \prec v_3$, then v_1w_j and w_iv_3 are crossing.

Hence, no matter where vertex v_3 lies, there must be an edge-crossing caused by two independent edges on page p, thus contradicting the hypothesis that p is a stack page. For the ordering $w_i \prec v_1 \prec w_j \prec v_2$, the proof is similar.

When $q \neq 0$, we use a similar proof to show that V_W contains at most $2q$ uncloneable vertices for which w_i and w_j form a pair of barrier vertices with respect to at least one queue page in $\langle \prec, \sigma \rangle$. Suppose by contradiction that V_W contains at least $2q + 1$ uncloneable vertices. By the pigeonhole principle, there must be at least one queue page $h \in [s + 1, s + q]$ such that V_W contains at least 3 uncloneable vertices, say v_1, v_2 and v_3, for which w_i and w_j form a pair of barrier vertices with respect to page h. Assume w.l.o.g. that $w_i \prec w_j$, $v_1 \prec v_2$, and that no two independent edges in $\{v_1w_i, v_1w_j, v_2w_i, v_2w_j\}$ nest. Then, vertices v_1, v_2, w_i and w_j must be ordered by ordering $v_1 \prec w_i \prec w_j \prec v_2$ in $\prec$. We distinguish four cases based on the position of v_3 (refer to Fig. 4 for an illustration).

Case (a): $v_3 \prec v_1$, then v_1w_i is nested inside v_3w_j.

Case (b): $v_1 \prec v_3 \prec w_i \prec w_j \prec v_2$, then v_3w_i is nested inside v_1w_j.

Case (c): $v_1 \prec w_i \prec w_j \prec v_3 \prec v_2$, then w_jv_3 is nested inside w_iv_2.

Case (d): $v_2 \prec v_3$, then w_jv_2 is nested inside w_iv_3.

Thus, in each case, page h is not a queue page, leading to a contradiction. Note that we need not consider the order $v_1 \prec w_i \prec v_3 \prec w_j \prec v_2$ because this case means v_3 is not uncloneable. □

Using Lemma 2, we can give an upper bound on the number of uncloneable vertices in V_W.

Lemma 3. *Assume that $\langle \prec, \sigma \rangle$ is a mixed s-stack q-queue layout for (G, τ, s, q) and $W \subseteq C$. Then there are at most $|W| \cdot (|W| - 1) \cdot (s + q)$ uncloneable vertices of type W with respect to $\langle \prec, \sigma \rangle$.*

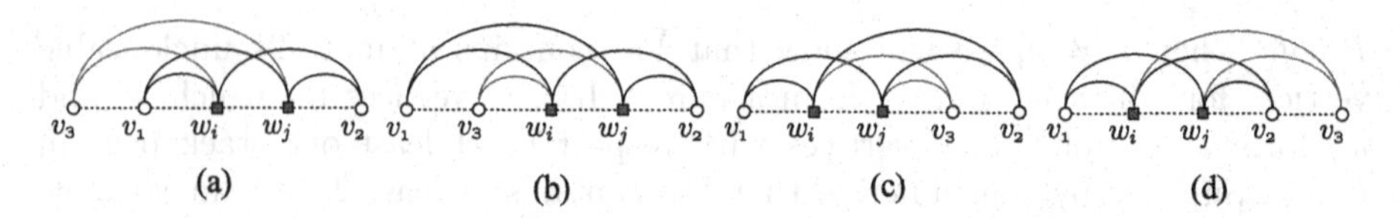

Fig. 4. Illustration for having at most 2 uncloneable vertices for which w_i and w_j form a pair of barrier vertices with respect to a queue page.

Proof. Let V_W^* be the set of all uncloneable vertices of type W and let $v \in V_W^*$. Then, there must exist at least one page $p \in [1, s+q]$ and two vertices in W, say w_i and w_j, such that w_i and w_j form a pair of barrier vertices to v with respect page p. From Lemma 2, there are at most $2(s+q)$ uncloneable vertices for which w_i and w_j form a pair of barrier vertices in $\langle \prec, \sigma \rangle$. Furthermore, for the vertices in W, there are in total $\binom{|W|}{2}$ combinations of two vertices. Thus, $|V_W^*| \leq 2(s+q) \cdot \binom{|W|}{2} = |W| \cdot (|W|-1) \cdot (s+q)$. □

3.2 An Algorithm Based on Kernelization

Now, we present an algorithm based on kernelization for the SQ-LAY problem. Given an instance (G, τ, s, q), we first locate a minimum vertex cover C and then, for each subset $W \subseteq C$, deal with the vertices in the corresponding set V_W by some reduction rule (this strategy has been used in [4,5]). By Lemma 3, our specific reduction rule is described as follows. If $|V_W| > |W| \cdot (|W|-1) \cdot (s+q)+1$ then delete all but $|W| \cdot (|W|-1) \cdot (s+q)+1$ vertices in V_W. Note that when $|W| = 1$ and $|V_W| > 1$, this reduction rule means keeping only one vertex in V_W, which is sufficient for our purposes. After executing this preprocessing procedure, we obtain a reduced instance denoted by (G^*, τ, s, q).

Theorem 1. *(G, τ, s, q) is a yes-instance of the* SQ-LAY *problem if and only if (G^*, τ, s, q) is a yes-instance of the* SQ-LAY *problem. Moreover, the size of G^* can be bounded by $2^{\mathcal{O}(\tau)}$.*

Proof. ($\Rightarrow$) Assume that (G, τ, s, q) is a yes-instance of the SQ-LAY problem. Then (G^*, τ, s, q) must be a yes-instance of the SQ-LAY problem because deleting some vertex from a mixed s-stack q-queue layout keeps the property of being a mixed s-stack q-queue layout.

($\Leftarrow$) Assume that (G^*, τ, s, q) is a yes-instance of the SQ-LAY problem. Let $\langle \prec, \sigma \rangle$ be a mixed s-stack q-queue layout for the instance (G^*, τ, s, q). Assume that C is a minimum vertex cover of G^* and $W \subseteq C$. Let $v \in V_W$. By our definition, the vertex v is either cloneable or uncloneable. From Lemma 3, the number of uncloneable vertices in V_W is at most $|W| \cdot (|W|-1) \cdot (s+q)$. Thus, if there are at least $|W| \cdot (|W|-1) \cdot (s+q)+1$ vertices in V_W, then there must exist at least one cloneable vertex, say u. By Lemma 1, we can extend $\prec$ by inserting the reduced vertices right next to u one by one, and by assigning the edges of each reduced vertex to the same pages as the corresponding edges of

u. Obviously, the extended assignment is a mixed s-stack q-queue layout for the graph G.

Finally, we can easily estimate the size of G^*. Given a vertex cover C with size τ, there are at most $2^\tau - 1$ nonempty subsets. Since $W \subseteq C$, $|W| \leq \tau$. Moreover, we can assume that $s < \tau$ and $q < \tau$ because if $s \geq \tau$ (resp. $q \geq \tau$), then we can immediately construct an s-stack (resp. q-queue) layout of G [4,5]. Hence, the size of G^* can be bounded by $2^\tau \cdot (2\tau^3 + 1) + \tau = 2^{\mathcal{O}(\tau)}$. □

By Theorem 1, the reduced instance (G^*, τ, s, q) is exactly a kernel of the original instance (G, τ, s, q). Thus, we can obtain the following conclusion.

Theorem 2. *The* SQ-LAY *problem admits an algorithm running in time* $\mathcal{O}(2^{2^{\mathcal{O}(\tau)}} + \tau \cdot |V|)$*, where* τ *and* $|V|$ *denote the vertex cover number and the size of the input graph, respectively. If* (G, τ, s, q) *is a yes-instance, this algorithm can also return a mixed* s*-stack* q*-queue layout of* G*.*

Proof. From Theorem 1, we have an equivalent instance (G^*, τ, s, q) with size $2^{\mathcal{O}(\tau)}$. The next step is to solve the equivalent instance by guessing all possible linear orders of $V(G^*)$ and by enumerating all possible edge assignments for each fixed linear order.

The running time can be analyzed as follows. The time for computing a vertex cover C of size τ is $\mathcal{O}(2^\tau + \tau \cdot |V|)$ [8]. All subsets of C can be enumerated in time $\mathcal{O}(2^\tau)$. Partitioning the vertices in $V \setminus C$ into 2^τ types and deleting all redundant vertices can be done in time $\mathcal{O}(\tau \cdot |V|)$ respectively. Hence, the running time for our kernelization procedure is $\mathcal{O}(2^\tau + \tau \cdot |V|)$. From Theorem 1, the size of $V(G^*)$ is $2^{\mathcal{O}(\tau)}$. Since the position of two vertices of the same type can be exchanged in a linear order [4], the number of fixed linear orders on $V(G^*)$ can be bounded by $(2^\tau)^{2^{\mathcal{O}(\tau)}} = 2^{2^{\mathcal{O}(\tau)}}$. For each fixed linear order, the mixed s-stack q-queue layout can be checked by enumerating at most $(2\tau)^{2^{\mathcal{O}(\tau)}} = 2^{2^{\mathcal{O}(\tau)}}$ assignments, where 2τ denotes the sum of at most τ stack pages and at most τ queue pages. Therefore, whether G^* admits a mixed s-stack q-queue layout can be determined in time $\mathcal{O}(2^{2^{\mathcal{O}(\tau)}} \cdot 2^{2^{\mathcal{O}(\tau)}}) = \mathcal{O}(2^{2^{\mathcal{O}(\tau)}})$. If (G^*, τ, s, q) is a yes-instance, then we can obtain a mixed s-stack q-queue layout of G by extending the mixed s-stack q-queue layout of G^* in time $O(\tau \cdot |V|)$. □

3.3 The Derived Results for Some Related Problems

We further consider the k-STACK LAYOUT problem, the UPWARD k-STACK LAYOUT problem, and the k-QUEUE LAYOUT problem respectively.

The k-STACK LAYOUT problem parameterized by the vertex cover number τ can be formally described as follows.

Parameterized k-stack layout (abbreviated as S-Lay)
Input: an undirected graph $G = (V, E)$ and a positive integer k ;
Parameter: τ ;
Question: does there exist a k-stack layout $\langle \prec, \sigma \rangle$ of G?

For the S-Lay problem, Bhore et al. [4] presented an algorithm running in time $\mathcal{O}(2^{\tau^{\mathcal{O}(\tau)}} + \tau \cdot |V|)$ based on a kernel of size $k^{\mathcal{O}(\tau)}$. Since a k-stack layout of G can be seen as a mixed k-stack 0-queue layout of G, we can immediately obtain improved results for it. From Theorem 1, the S-Lay problem admits a kernel of size $2^{\mathcal{O}(\tau)}$. By Theorem 2, the following conclusions hold.

Theorem 3. *The* S-Lay *problem admits an algorithm running in time* $\mathcal{O}(2^{2^{\mathcal{O}(\tau)}} + \tau \cdot |V|)$, *where* τ *and* $|V|$ *denote the vertex cover number and the size of the input graph, respectively. If* (G, τ, k) *is a yes-instance, this algorithm can also return a* k*-stack layout of* G.

Corollary 1. *Let* $G = (V, E)$ *be a graph with vertex cover number* τ. *A stack layout of* G *with minimum number of stack pages can be computed in* $\mathcal{O}(2^{2^{\mathcal{O}(\tau)}} + \tau \log \tau \cdot |V|)$ *time.*

The upward k-stack layout problem parameterized by the vertex cover number τ can be formally described as follows.

Parameterized upward k-stack layout (abbreviated as U-S-Lay)
Input: a directed acyclic graph $D = (V, E)$ and a positive integer k;
Parameter: τ;
Question: does there exist an upward k-stack layout $\langle \prec, \sigma \rangle$ of D?

For the U-S-Lay problem, Bhore et al. [6] presented an algorithm running in time $\mathcal{O}(\tau^{\tau^{\mathcal{O}(\tau)}} + \tau \cdot |V|)$ based on a kernel of size $k^{\mathcal{O}(\tau)}$. Since the U-S-Lay problem can be seen as a variant of the S-Lay problem, our technique can also be applied to it with minor adjustments. Assume that (D, τ, k) is a yes-instance of the U-S-Lay problem. Let C be a minimum vertex cover set of D. The vertices in $V(D) \setminus C$ are firstly classified into at most $2^{2\tau}$ distinct types [6]. Let $W \subseteq C$. Then we estimate the number of uncloneable vertices in V_W. Let w_i, w_j be two vertices in W, let v_1, v_2 be two vertices in V_W, and let p be an arbitrary page in $\langle \prec, \sigma \rangle$. Since $v_1 w$ and $v_2 w$ for $w \in \{w_i, w_j\}$ have the same orientation, at most *one* vertex in $\{v_1, v_2\}$ is an uncloneable vertex for which w_i and w_j form a pair of barrier vertices with respect to page p. It follows that there are at most $\frac{1}{2}|W| \cdot (|W| - 1) \cdot k$ uncloneable vertices in V_W with respect to $\langle \prec, \sigma \rangle$. Thus, the size of kernel for the U-S-Lay problem can also be bounded by $2^{\mathcal{O}(\tau)}$. The following conclusions hold.

Theorem 4. *The* U-S-Lay *problem admits an algorithm running in time* $\mathcal{O}(2^{2^{\mathcal{O}(\tau)}} + \tau \cdot |V|)$, *where* τ *and* $|V|$ *denote the vertex cover number and the size of the input graph, respectively. If* (D, τ, k) *is a yes-instance, this algorithm can also return an upward* k*-stack layout of* D.

Corollary 2. *Let* $D = (V, E)$ *be a directed acyclic graph with vertex cover number* τ. *An upward stack layout of* D *with minimum number of stack pages can be computed in* $\mathcal{O}(2^{2^{\mathcal{O}(\tau)}} + \tau \log \tau \cdot |V|)$ *time.*

The k-QUEUE LAYOUT problem parameterized by the vertex cover number τ can be formally described as follows.

Parameterized k-queue layout (abbreviated as Q-Lay)
Input: an undirected graph $G = (V, E)$ and a positive integer k;
Parameter: τ;
Question: does there exist a k-queue layout $\langle \prec, \sigma \rangle$ of G?

For the optimization version of the k-QUEUE LAYOUT problem, Bhore et al. [5] presented a parameterized algorithm running in time $\mathcal{O}(2^{\tau^{\mathcal{O}(\tau)}} + \tau \log \tau \cdot |V|)$ based on a kernel of size $k^{\mathcal{O}(\tau)}$. Since a k-queue layout of G can be seen as a mixed 0-stack k-queue layout of G, we can immediately obtain improved results for it. From Theorem 1, the Q-LAY problem admits a kernel of size $2^{\mathcal{O}(\tau)}$. By Theorem 2, the following conclusions are sound.

Theorem 5. *The* Q-LAY *problem admits an algorithm running in time* $\mathcal{O}(2^{2^{\mathcal{O}(\tau)}} + \tau \cdot |V|)$*, where* τ *and* $|V|$ *denote the vertex cover number and the size of the input graph, respectively. If* (G, τ, k) *is a yes-instance, this algorithm can also return a* k*-queue layout of* G.

Corollary 3. *Let* $G = (V, E)$ *be a graph with vertex cover number* τ*. A queue layout of* G *with minimum number of queue pages can be computed in* $\mathcal{O}(2^{2^{\mathcal{O}(\tau)}} + \tau \log \tau \cdot |V|)$ *time.*

4 k-Arch Layout

The k-ARCH LAYOUT problem parameterized by the vertex cover number τ can be formally defined as follows.

Parameterized k-arch layout (abbreviated as A-Lay)
Input: an undirected graph $G = (V, E)$, a positive integer k ;
Parameter: τ ;
Question: does there exist a k-arch layout $\langle \prec, \sigma \rangle$ of G?

In this section, we will generalize the techniques used in the SQ-LAY problem to the A-LAY problem. Our result is that the A-LAY problem admits a kernel of size $2^{\mathcal{O}(\tau)}$ and an algorithm running in time $\mathcal{O}(2^{2^{\mathcal{O}(\tau)}} + \tau \cdot |V|)$. The key step in our work is to redefine the uncloneable/cloneable vertex.

We start by giving an observation, which matches the analogous observations in [4,5].

Lemma 4. *Every graph* $G = (V, E)$ *with a vertex cover* C *of size* τ *admits a* τ*-arch layout. Moreover, if* G *and* C *are given as input, such a* τ*-arch layout can be computed in* $\mathcal{O}(|V| + \tau \cdot |V|)$ *time.*

By Lemma 4, if $k \geq \tau$, we can immediately construct a k-arch layout of G. Thus, we assume that $k < \tau$ in the rest of this paper.

4.1 Two Kinds of Vertices in V_W

Recall that an arch page in $\langle \prec, \sigma \rangle$ is a set of edges $F \subseteq E(G)$ such that no two independent edges in F are disjoint. By this definition, we can identify the uncloneable/cloneable vertices in V_W (for $|W| \geq 2$) as follows.

Assume that (G, τ, k) is a yes-instance of the A-LAY problem. Let $\langle \prec, \sigma \rangle$ be a k-arch layout of G and let $W \subseteq C$. A vertex v of type W is called an *uncloneable vertex* with respect to $\langle \prec, \sigma \rangle$ if there exist at least one page $p \in [1, k]$ such that v has two edges $w_i v$ and $v w_j$ with order $w_i \prec v \prec w_j$ assigned to page p. The edges $w_i v$ and $v w_j$ are accordingly called *a pair of barrier edges* for v respect to page p. Otherwise, the vertex v of type W is called a *cloneable vertex* with respect to $\langle \prec, \sigma \rangle$ (refer to Fig. 5 for examples).

Fig. 5. Two uncloneable vertices v_1, v_2 (left) and two cloneable vertices u_1, u_2 (right) are shown in two 2-arch layouts respectively, where $W = \{w_1, w_2, w_3\}$.

For the cloneable vertex, the following property holds.

Lemma 5. *Let $\langle \prec, \sigma \rangle$ be a k-arch layout of a given graph. If v is a cloneable vertex with respect to $\langle \prec, \sigma \rangle$, then inserting a vertex v^* of type W right (or left) next to v and assigning each edge $v^* w$ (for $w \in W$) to the same arch page as vw will result in a valid k-arch layout of the resulting graph.*

Proof. Assume that v is a *cloneable* vertex with respect to $\langle \prec, \sigma \rangle$. Then with respect to each arch page $p \in [1, k]$, the vertex v has no pair of barrier edges. Suppose that v has t $(1 \leq t \leq |W|)$ edges, say $w_p^1 v, w_p^2 v, \ldots, w_p^t v$, assigned to page p such that the vertices $w_p^1, w_p^2, \ldots, w_p^t$ uniformly lie on the left of v in $\prec$. After inserting v^* right next (or left next) to v and assigning all edges $w_p^i v^*$ for $i = 1, 2, \ldots, t$ to page p, no pair of disjoint edges occur on page p because v^* is right next (or left next) to v. Thus, the resulting page is still an arch page. □

The next lemma shows an upper bound on the number of uncloneable vertices in V_W.

Lemma 6. *Assume that $\langle \prec, \sigma \rangle$ is a k-arch layout for (G, τ, k) and $W \subseteq C$. Then there exist at most k uncloneable vertices of type W with respect to $\langle \prec, \sigma \rangle$.*

Proof. Suppose by contradiction that V_W contains at least $k+1$ uncloneable vertices with respect to $\langle \prec, \sigma \rangle$. By the pigeonhole principle, there must be at least one page $p \in [1, k]$ such that V_W contains an uncloneable vertex, say v_1, having a pair of barrier edges $w_{i_1}v$ and vw_{j_1} ($w_{i_1} \prec v_1 \prec w_{j_1}$) assigned to page p, and another uncloneable vertex, say v_2, having a pair of barrier edges $w_{i_2}v_2, v_2w_{j_2}$ ($w_{i_2} \prec v_2 \prec w_{j_2}$) assigned to page p. Note that vertices w_{i_2} and w_{i_1} (resp. w_{j_2} and w_{j_1}) may be the same one. If $v_1 \prec v_2$ then the edges $w_{i_1}v_1$ and $v_2w_{j_2}$ are disjoint; otherwise the edges $w_{i_2}v_2$ and $v_1w_{j_1}$ are disjoint. Consequently, the resulting page is no longer an arch page. Therefore, there are at most k uncloneable vertices of type W with respect to $\langle \prec, \sigma \rangle$. □

4.2 An Algorithm Based on Kernelization

Now, we propose an algorithm based on kernelization for the A-LAY problem. The kernelization framework is the same as that for the SQ-LAY problem described in Sect. 3. By Lemma 6, our specific reduction rule for the A-LAY problem is described as follows. If $|V_W| > k+1$ then delete all but $k+1$ vertices in V_W. We denote by (G^*, τ, k) the reduced instance.

Theorem 6. *(G, τ, k) is a yes-instance of the* A-LAY *problem if and only if (G^*, τ, k) is a yes-instance of the* A-LAY *problem. Moreover, the size of G^* can be bounded by $2^{\mathcal{O}(\tau)}$.*

Proof. ($\Rightarrow$) Assume that (G, τ, k) is a yes-instance of the A-LAY problem. Then (G^*, τ, k) must be a yes-instance of the A-LAY problem because deleting some vertex from a k-arch layout keeps the property of being a k-arch layout.

($\Leftarrow$) Assume that (G^*, τ, k) is a yes-instance of the A-LAY problem. Let $\langle \prec, \sigma \rangle$ be a k-arch layout for (G^*, τ, k). Assume that C is a minimum vertex cover of G^* and $W \subseteq C$. From Lemma 6, the number of uncloneable vertices is at most k in V_W. Thus, if there are at least $k+1$ vertices in V_W, then there must exist at least one vertex, say u, that is cloneable. By Lemma 5, we can extend $\prec$ by inserting the reduced vertices right next to u one by one, and by assigning the edges of each reduced vertex to the same arch pages as the corresponding edges of u. Obviously, the extended assignment is exactly a k-arch layout of G.

Finally, we can easily estimate the size of G^*. For a vertex cover C with size τ, there are at most $2^\tau - 1$ nonempty subsets. Since $W \subseteq C$, $|W| \le \tau$. Moreover, from Lemma 4, we can assume that $k < \tau$. Thus, the size of G^* can be bounded by $2^\tau \cdot (k+1) + \tau \le 2^\tau \cdot \tau + \tau = 2^{\mathcal{O}(\tau)}$. □

Lemma 7 *([11]). Given a vertex ordering $\prec$ of an n-vertex m-edge graph G, there is an algorithm $\mathcal{A}$ that assigns the edges of G to the minimum number of arches with respect to $\prec$ in $\mathcal{O}(n+m)$ time.*

Using Theorem 6 and Lemma 7, we give the following.

Theorem 7. *The* A-LAY *problem admits an algorithm running in time $\mathcal{O}(2^{2^{\mathcal{O}(\tau)}} + \tau \cdot |V|)$, where τ and $|V|$ denote the vertex cover number and the size of the input graph, respectively. If (G, τ, k) is a yes-instance, this algorithm can also return a k-arch layout of G.*

An arch layout with the minimum number of arches can be obtained by trying all possible choices for $k \in [1, \tau]$. Herein, by applying a binary search on the number of pages k (this technique has been used in [4,5]), we obtain the following result.

Corollary 4. *Let $G = (V, E)$ be a graph with vertex cover number τ. An arch layout of G with minimum number of arch pages can be computed in $\mathcal{O}(2^{2^{\mathcal{O}(\tau)}} + \tau \log \tau \cdot |V|)$ time.*

5 Conclusion

In this work, we study a series of linear layouts of graphs parameterized by the vertex cover number τ of the input graph $G = (V, E)$. By introducing the concept of uncloneable/cloneable vertex, we propose some novel reduction rules in kernelization. By these rules, we show that each of parameterized problems we considered admits a kernel of size $2^{\mathcal{O}(\tau)}$ and an algorithm running in time $\mathcal{O}(2^{2^{\mathcal{O}(\tau)}} + \tau \cdot |V|)$.

Some questions are interesting and deserve further research. (1) Do the problems we considered admit kernels of polynomial size with respect to the vertex cover number of the input graph? (2) The key ingredient in our kernelization lies in the concept of uncloneable/cloneable vertex. We believe that this tool will find further utility in other graph drawing problems.

Acknowledgements. The authors thank the anonymous referees for their valuable comments and suggestions.

References

1. Bannister, M.J., Cabello, S., Eppstein, D.: Parameterized complexity of 1-planarity. J. Graph Alg. Appl. **22**(1), 23–49 (2018)
2. Bekos, M.A., Gronemann, M., Raftopoulou, C.N.: Two-page book embeddings of 4-planar graphs. Algorithmica **75**(1), 158–185 (2016)
3. Bhatt, S.N., Chung, F.R.K., Leighton, F.T., Rosenberg, A.L.: Scheduling tree-dags using FIFO queues: a control-memory trade-off. J. Parallel Distrib. Comput. **33**, 56–68 (1996)
4. Bhore, S., Ganian, R., Montecchiani, F., Nöllenburg, M.: Parameterized algorithms for book embedding problems. J. Graph Alg. Appl. **24**(4), 603–620 (2020)
5. Bhore, S., Ganian, R., Montecchiani, F., Nöllenburg, M.: Parameterized algorithms for queue layouts. In: Auber, D., Valtr, P., et al. (eds.) GD 2020. LNCS, vol. 12590, pp. 40–54. Springer, Cham (2020). https://doi.org/10.1007/978-3-030-68766-3_4
6. Bhore, S., Da Lozzo, G., Montecchiani, F., Nöllenburg, M.: On the upward book thickness problem: combinatorial and complexity results. arXiv: 2108.12327v1 [cs.DM], 27 August 2021. GD 2021 (in press)
7. Binucci, C., Da Lozzo, G., Di Giacomo, E., Didimo, W., Mchedlidze, T., Patrignani, M.: Upward book embeddings of st-graphs. In: Barequet, G., Wang, Y. (eds.) SoCG 2019. LIPIcs, vol. 129, pp. 13:1–13:22 (2019). https://doi.org/10.4230/LIPIcs.SoCG.2019.13

8. Chen, J., Kanj, I.A., Xia, G.: Improved upper bounds for vertex cover. Theor. Comput. Sci. **411**(40–42), 3736–3756 (2010)
9. Chung, F., Leighton, F., Rosenberg, A.: Embedding graphs in books: a layout problem with applications to VLSI design. SIAM J. Alg. Discr. Meth. **8**(1), 33–58 (1987)
10. de Col, P., Klute, F., Nöllenburg, M.: Mixed linear layouts: complexity, heuristics, and experiments. In: Archambault, D., Tóth, C.D. (eds.) GD 2019. LNCS, vol. 11904, pp. 460–467. Springer, Cham (2019). https://doi.org/10.1007/978-3-030-35802-0_35
11. Dujmović, V., Wood, D.R.: On linear layouts of graphs. Discrete Math. Theor. Comput. Sci. **6**, 339–358 (2004)
12. Dujmović, V., Wood, D.R.: Stacks, queues and tracks: layouts of graph subdivisions. Discrete Math. Theor. Comput. Sci. **7**(1), 155–202 (2005)
13. Enomoto, H., Miyauchi, M.: Stack-queue mixed layouts of graph subdivisions. In: Forum on Information Technology, pp. 47–56 (2014)
14. Fellows, M.R., Lokshtanov, D., Misra, N., Rosamond, F.A., Saurabh, S.: Graph layout problems parameterized by vertex cover. In: Hong, S.-H., Nagamochi, H., Fukunaga, T. (eds.) ISAAC 2008. LNCS, vol. 5369, pp. 294–305. Springer, Heidelberg (2008). https://doi.org/10.1007/978-3-540-92182-0_28
15. Heath, L.S., Leighton, F.T., Rosenberg, A.L.: Comparing queues and stacks as mechanisms for laying out graphs. SIAM J. Discrete Math. **5**(3), 398–412 (1992)
16. Heath, L.S., Rosenberg, A.L.: Laying out graphs using queues. SIAM J. Comput. **21**(5), 927–958 (1992)
17. Hliněný, P., Sankaran, A.: Exact crossing number parameterized by vertex cover. In: Archambault, D., Tóth, C.D. (eds.) GD 2019. LNCS, vol. 11904, pp. 307–319. Springer, Cham (2019). https://doi.org/10.1007/978-3-030-35802-0_24
18. Klawitter, J., Mchedlidze, T., Nöllenburg, M.: Experimental evaluation of book drawing algorithms. In: Frati, F., Ma, K.-L. (eds.) GD 2017. LNCS, vol. 10692, pp. 224–238. Springer, Cham (2018). https://doi.org/10.1007/978-3-319-73915-1_19
19. Liu, Y., Chen, J., Huang, J.: Fixed-order book thickness with respect to the vertex-cover number: new observations and further analysis. In: Chen, J., Feng, Q., Xu, J. (eds.) TAMC 2020. LNCS, vol. 12337, pp. 414–425. Springer, Cham (2020). https://doi.org/10.1007/978-3-030-59267-7_35
20. Liu, Y., Chen, J., Huang, J.: Parameterized algorithms for fixed-order book drawing with bounded number of crossings per edge. In: Wu, W., Zhang, Z. (eds.) COCOA 2020. LNCS, vol. 12577, pp. 562–576. Springer, Cham (2020). https://doi.org/10.1007/978-3-030-64843-5_38
21. Liu, Y., Chen, J., Huang, J., Wang, J.: On parameterized algorithms for fixed-order book thickness with respect to the pathwidth of the vertex ordering. Theor. Comput. Sci. **873**, 16–24 (2021)
22. Pemmaraju, S.V.: Exploring the powers of stacks and queues via graph layouts. Ph.D. thesis, Virginia Tech (1992)
23. Pupyrev, S.: Mixed linear layouts of planar graphs. In: Frati, F., Ma, K.-L. (eds.) GD 2017. LNCS, vol. 10692, pp. 197–209. Springer, Cham (2018). https://doi.org/10.1007/978-3-319-73915-1_17
24. Yannakakis, M.: Linear and book embeddings of graphs. In: Makedon, F., Mehlhorn, K., Papatheodorou, T., Spirakis, P. (eds.) AWOC 1986. LNCS, vol. 227, pp. 226–235. Springer, Heidelberg (1986). https://doi.org/10.1007/3-540-16766-8_20

The Fractional k-truncated Metric Dimension of Graphs

Eunjeong Yi(✉)

Texas A&M University at Galveston, Galveston, TX 77553, USA
yie@tamug.edu

Abstract. Let G be a graph with vertex set $V(G)$, and let $d(x, y)$ denote the length of a shortest $x - y$ path in G. Let k be a positive integer. For any $x, y \in V(G)$, let $d_k(x, y) = \min\{d(x, y), k+1\}$ and let $R_k\{x, y\} = \{z \in V(G) : d_k(x, z) \neq d_k(y, z)\}$. A set $S \subseteq V(G)$ is a *k-truncated resolving set* of G if $|S \cap R_k\{x, y\}| \geq 1$ for any distinct $x, y \in V(G)$, and the *k-truncated metric dimension* $\dim_k(G)$ of G is the minimum cardinality over all k-truncated resolving sets of G. For a function g defined on $V(G)$ and for $U \subseteq V(G)$, let $g(U) = \sum_{s \in U} g(s)$. A real-valued function $g : V(G) \to [0, 1]$ is a *k-truncated resolving function* of G if $g(R_k\{x, y\}) \geq 1$ for any distinct $x, y \in V(G)$, and the *fractional k-truncated metric dimension* $\dim_{k,f}(G)$ of G is $\min\{g(V(G)) : g$ is a k-truncated resolving function of $G\}$. Note that $\dim_{k,f}(G)$ reduces to $\dim_k(G)$ if the codomain of k-truncated resolving functions is restricted to $\{0, 1\}$. In this paper, we initiate the study of the fractional k-truncated metric dimension of graphs. For any connected graph G of order $n \geq 2$, we show that $1 \leq \dim_{k,f}(G) \leq \frac{n}{2}$; we characterize G satisfying $\dim_{k,f}(G)$ equals 1 and $\frac{n}{2}$, respectively. We also examine $\dim_{k,f}(G)$ of some graph classes and conclude with some open problems.

Keywords: (Fractional) metric dimension · k-truncated metric dimension · Distance-k dimension · Fractional k-truncated metric dimension · Fractional distance-k dimension

1 Introduction

Let G be a finite, simple, undirected, and connected graph with vertex set $V(G)$ and edge set $E(G)$. The *distance* between two vertices $x, y \in V(G)$, denoted by $d(x, y)$, is the minimum number of edges on a path connecting x and y in G. The *diameter*, $\text{diam}(G)$, of G is $\max\{d(x, y) : x, y \in V(G)\}$. Let $\mathbb{Z}^+$ denote the set of positive integers. For $k \in \mathbb{Z}^+$ and for two vertices $x, y \in V(G)$, let $d_k(x, y) = \min\{d(x, y), k+1\}$.

Metric dimension, introduced in [15] and [23], is a graph parameter that has been studied extensively. But, what if the landmarks have a limited range – capable of sending signals to nodes no further than k-units away from themselves? For distinct $x, y \in V(G)$, let $R\{x, y\} = \{z \in V(G) : d(x, z) \neq d(y, z)\}$.

D.-Z. Du et al. (Eds.): COCOA 2021, LNCS 13135, pp. 568–578, 2021.
https://doi.org/10.1007/978-3-030-92681-6_44

A vertex subset $S \subseteq V(G)$ is a *resolving set* of G if $|S \cap R\{x, y\}| \geq 1$ for any pair of distinct $x, y \in V(G)$, and the *metric dimension* $\dim(G)$ of G is the minimum cardinality over all resolving sets of G. For $k \in \mathbb{Z}^+$ and for distinct $x, y \in V(G)$, let $R_k\{x, y\} = \{z \in V(G) : d_k(x, z) \neq d_k(y, z)\}$. A vertex subset $S \subseteq V(G)$ is a *k-truncated resolving set* (also called a distance-k resolving set) of G if $|S \cap R_k\{x, y\}| \geq 1$ for any pair of distinct $x, y \in V(G)$, and the *k-truncated metric dimension* (also called the distance-k dimension) $\dim_k(G)$ of G is the minimum cardinality over all k-truncated resolving sets of G. Notice that $\dim_k(G) = \dim(G)$ if $k \geq \text{diam}(G) - 1$. The metric dimension of a metric space (V, d_k) is studied in [3]. The k-truncated metric dimension corresponds to the $(1, k+1)$-metric dimension in [7] and [8]. We note that $\dim_1(G)$ is also called the adjacency dimension of G in [17]. For detailed results on $\dim_k(G)$, we refer to [11], which is formed from merging the two papers [14] and [25]. It is known that determining the metric dimension and the k-truncated metric dimension of a general graph are NP-hard problems; see [8,10,12] and [20].

The fractionalization of various graph parameters has been extensively studied (see [22]). For definition and a formulation of fractional metric dimension as the optimal solution to a linear programming problem by relaxing a condition of the integer programming problem for metric dimension, see [5] and [9]. The fractional metric dimension of graphs was officially studied in [1]. For a function g defined on $V(G)$ and for $U \subseteq V(G)$, let $g(U) = \sum_{s \in U} g(s)$. A real-valued function $g : V(G) \to [0, 1]$ is a *resolving function* of G if $g(R\{x, y\}) \geq 1$ for any distinct vertices $x, y \in V(G)$. The *fractional metric dimension*, $\dim_f(G)$, of G is $\min\{g(V(G)) : g \text{ is a resolving function of } G\}$.

For $k \in \mathbb{Z}^+$, a real-valued function $h : V(G) \to [0, 1]$ is a *k-truncated resolving function* (also called a distance-k resolving function) of G if $h(R_k\{x, y\}) \geq 1$ for any pair of distinct $x, y \in V(G)$. The *fractional k-truncated metric dimension* (also called the fractional distance-k dimension) of G, denoted by $\dim_{k,f}(G)$, is $\min\{h(V(G)) : h \text{ is a } k\text{-truncated resolving function of } G\}$. Note that $\dim_{k,f}(G) = \dim_k(G)$ if the codomain of k-truncated resolving functions is restricted to $\{0, 1\}$, $\dim_{k,f}(G) = \dim_f(G)$ if $k \geq \text{diam}(G) - 1$, and $\dim_{k,f}(G) = \dim(G)$ if $k \geq \text{diam}(G) - 1$ and the codomain of k-truncated resolving functions is restricted to $\{0, 1\}$.

In this paper, we initiate the study of the fractional k-truncated metric dimension of graphs. In Sect. 2, for any connected graph G of order $n \geq 2$ and for any $k \in \mathbb{Z}^+$, we show that $1 \leq \dim_{k,f}(G) \leq \frac{n}{2}$ and we characterize G satisfying $\dim_{k,f}(G)$ equals 1 and $\frac{n}{2}$, respectively. In Sect. 3, we examine $\dim_{k,f}(G)$ of some graph classes. In Sect. 4, we examine the relation among $\dim(G)$, $\dim_f(G)$, $\dim_k(G)$ and $\dim_{k,f}(G)$. We show the existence of non-isomorphic graphs G and H such that $\dim_k(G) = \dim_k(H)$ and $\dim_{k,f}(G) \neq \dim_{k,f}(H)$. Based on the construction in [13], we also show the existence of two connected graphs H and G with $H \subset G$ such that $\frac{\dim_{k,f}(H)}{\dim_{k,f}(G)}$ can be arbitrarily large. We conclude the paper with some open problems. Throughout the paper, let P_n, C_n and K_n, respectively, denote the path, the cycle and the complete graph on n vertices.

2 Some Observations and Bounds on $\dim_{k,f}(G)$

In this section, for any connected graph G of order $n \ge 2$ and for any $k \in \mathbb{Z}^+$, we show that $1 \le \dim_{k,f}(G) \le \frac{n}{2}$; we characterize G satisfying $\dim_{k,f}(G)$ equals 1 and $\frac{n}{2}$, respectively.

We begin with some observations. For $v \in V(G)$, let $N(v) = \{w \in V(G) : vw \in E(G)\}$. Two vertices $x, y \in V(G)$ are called *twins* if $N(x) - \{y\} = N(y) - \{x\}$. It was observed in [26] that, if x and y are distinct twin vertices, then $g(x) + g(y) \ge 1$ for any resolving function g of G. Hernando et al. [16] observed that the twin relation is an equivalence relation and that an equivalence class under it, called a twin equivalence class, induces a clique or an independent set.

Observation 1. *Let G be a non-trivial connected graph, and let $k, k' \in \mathbb{Z}^+$. Then*

(a) [1] $\dim_f(G) \le \dim(G)$;
(b) [3, 7] if $k > k'$, then $\dim(G) \le \dim_k(G) \le \dim_{k'}(G) \le \dim_1(G)$;
(c) $\dim_f(G) \le \dim_{k,f}(G) \le \dim_k(G)$;
(d) if $k > k'$, then $\dim_f(G) \le \dim_{k,f}(G) \le \dim_{k',f}(G) \le \dim_{1,f}(G) \le \dim_1(G)$.

Observation 2. *Let G be a connected graph with $\text{diam}(G) = d$, and let $k \in \mathbb{Z}^+$.*

(a) [7] If $k \ge d - 1$, then $\dim_k(G) = \dim(G)$.
(b) If $k \ge d - 1$, then $\dim_{k,f}(G) = \dim_f(G)$.

Next, we recall some results involving the bounds of the k-truncated metric dimension and the fractional metric dimension of graphs.

Theorem 3. *Let G be a connected graph of order $n \ge 2$, and let $k \in \mathbb{Z}^+$. Then*

(a) [7] $1 \le \dim_k(G) \le n - 1$, and $\dim_k(G) = 1$ if and only if $G \in \cup_{i=2}^{k+2}\{P_i\}$;
(b) [1, 18] $1 \le \dim_f(G) \le \frac{n}{2}$, and $\dim_f(G) = \frac{n}{2}$ if and only if there exists a bijection $\phi : V(G) \to V(G)$ such that $\phi(v) \ne v$ and $|R\{v, \phi(v)\}| = 2$ for all $v \in V(G)$;
(c) [19] $\dim_f(G) = 1$ if and only if $G = P_n$.

For the characterization of connected graphs G of order n satisfying $\dim_k(G) = n - 2$ and $\dim_k(G) = n - 1$ respectively, see [11] and [24]. For an explicit characterization of graphs G satisfying $\dim_f(G) = \frac{|V(G)|}{2}$, we recall the following construction from [2]. Let $\mathcal{K} = \{K_a : a \ge 2\}$ and $\overline{\mathcal{K}} = \{\overline{K}_b : b \ge 2\}$, where $\overline{G}$ denotes the complement of a graph G. Let $H[\mathcal{K} \cup \overline{\mathcal{K}}]$ be the family of graphs obtained from a connected graph H by replacing each vertex $u_i \in V(H)$ by a graph $H_i \in \mathcal{K} \cup \overline{\mathcal{K}}$, and each vertex in H_i is adjacent to each vertex in H_j if and only if $u_iu_j \in E(H)$.

Theorem 4. *[2] Let G be a connected graph of order at least two. Then $\dim_f(G) = \frac{|V(G)|}{2}$ if and only if $G \in H[\mathcal{K} \cup \overline{\mathcal{K}}]$ for some connected graph H.*

Next, we obtain the bounds on $\dim_{k,f}(G)$.

Proposition 1. *For any connected graph G of order $n \geq 2$ and for any $k \in \mathbb{Z}^+$, $1 \leq \dim_{k,f}(G) \leq \frac{n}{2}$.*

Proof. Let $k \in \mathbb{Z}^+$, and let G be a connected graph of order $n \geq 2$. By definition, $\dim_{k,f}(G) \geq 1$. If $g : V(G) \to [0,1]$ is a function defined by $g(v) = \frac{1}{2}$ for each $v \in V(G)$, then $R_k\{x,y\} \supseteq \{x,y\}$ and $g(R_k\{x,y\}) \geq g(x) + g(y) = 1$ for any distinct $x, y \in V(G)$; thus, g is a k-truncated resolving function of G with $g(V(G)) = \frac{n}{2}$. So, $\dim_{k,f}(G) \leq \frac{n}{2}$. □

Next, we characterize connected graphs G satisfying $\dim_{k,f}(G) = 1$ for any $k \in \mathbb{Z}^+$.

Theorem 5. *Let G be a non-trivial connected graph, and let $k \in \mathbb{Z}^+$. Then $\dim_{k,f}(G) = 1$ if and only if $G \in \cup_{i=2}^{k+2}\{P_i\}$.*

Proof. Let G be a connected graph of order $n \geq 2$, and let $k \in \mathbb{Z}^+$.

($\Leftarrow$) Let $G \in \cup_{i=2}^{k+2}\{P_i\}$. Then $1 = \dim_f(G) \leq \dim_{k,f}(G) \leq \dim_k(G) = 1$ by Observation 1(c) and Theorem 3(a)(c). So, $\dim_{k,f}(G) = 1$.

($\Rightarrow$) Let $\dim_{k,f}(G) = 1$. By Observation 1(c) and Theorem 3(c), $\dim_{k,f}(G) \geq \dim_f(G) \geq 1$ and $\dim_f(G) = 1$ if and only if $G = P_n$. So, if $G \neq P_n$, then $\dim_{k,f}(G) > 1$. Now, suppose $G = P_n$, and let P_n be a path given by $u_1, u_2, \ldots, u_n$. Let $g : V(P_n) \to [0,1]$ be any minimum k-truncated resolving function of P_n. If $n \leq k+2$, then $\dim_{k,f}(P_n) = 1$ as shown above. So, suppose $n \geq k+3$; we show that $\dim_{k,f}(P_n) > 1$.

First, let $k+3 \leq n \leq 2k+3$. Then $R_k\{u_1,u_2\} = \cup_{i=1}^{k+2}\{u_i\}$, $R_k\{u_{n-1},u_n\} = \cup_{i=n-(k+1)}^{n}\{u_i\}$, and $R_k\{u_{i-1},u_{i+1}\} = V(P_n) - \{u_i\}$ for $u_i \in R_k\{u_1,u_2\} \cap R_k\{u_{n-1},u_n\}$ (i.e., $n-k-1 \leq i \leq k+2$). So, $g(R_k\{u_1,u_2\}) = \sum_{i=1}^{k+2} g(u_i) \geq 1$, $g(R_k\{u_{n-1},u_n\}) = \sum_{i=n-(k+1)}^{n} g(u_i) \geq 1$, and $g(R_k\{u_{i-1},u_{i+1}\}) = g(V(P_n)) - g(u_i) \geq 1$ for each $i \in \{n-(k+1), \ldots, k+2\}$. By summing over the $(6+2k-n)$ inequalities, we have $(5+2k-n)g(V(P_n)) \geq 6+2k-n$, i.e., $g(V(P_n)) \geq \frac{6+2k-n}{5+2k-n}$; thus, $\dim_{k,f}(P_n) \geq \frac{6+2k-n}{5+2k-n}$. On the other hand, let $h : V(P_n) \to [0,1]$ be a function defined by

$$h(u_i) = \begin{cases} \frac{1}{5+2k-n} & \text{if } u_i \in \{u_1,u_n\} \cup (R_k\{u_1,u_2\} \cap R_k\{u_{n-1},u_n\}), \\ 0 & \text{otherwise.} \end{cases}$$

Then h is a k-truncated resolving function of P_n with $h(V(P_n)) = \frac{6+2k-n}{5+2k-n}$; thus, $\dim_{k,f}(P_n) \leq \frac{6+2k-n}{5+2k-n}$. Therefore, $\dim_{k,f}(P_n) = \frac{6+2k-n}{5+2k-n}$ for $k+3 \leq n \leq 2k+3$.

Second, let $n \geq 2k+4$. Then $R_k\{u_1,u_2\} = \cup_{i=1}^{k+2}\{u_i\}$ and $R_k\{u_{n-1},u_n\} = \cup_{i=n-k-1}^{n}\{u_i\}$. So, $g(R_k\{u_1,u_2\}) = \sum_{i=1}^{k+2} g(u_i) \geq 1$ and $g(R_k\{u_{n-1},u_n\}) = \sum_{i=n-k-1}^{n} g(u_i) \geq 1$. Since $n-k-1 \geq 2k+4-k-1 = k+3$, $g(V(P_n)) \geq \sum_{i=1}^{k+2} g(u_i) + \sum_{i=n-k-1}^{n} g(u_i) \geq 2$, which implies $\dim_{k,f}(P_n) \geq 2$. □

Next, via a proof technique used in [2], we characterize connected graphs G satisfying $\dim_{k,f}(G) = \frac{|V(G)|}{2}$ for any $k \in \mathbb{Z}^+$.

Theorem 6. *Let G be a connected graph of order $n \geq 2$. Then $\dim_{1,f}(G) = \frac{n}{2}$ if and only if $G \in H[\mathcal{K} \cup \overline{\mathcal{K}}]$ for some connected graph H.*

Proof. Let G be a connected graph of order $n \geq 2$.

($\Leftarrow$) Let $G \in H[\mathcal{K} \cup \overline{\mathcal{K}}]$ for some connected graph H. Then $\dim_{1,f}(G) \geq \dim_f(G) = \frac{n}{2}$ by Observation 1(d) and Theorem 4. By Proposition 1, $\dim_{1,f}(G) = \frac{n}{2}$.

($\Rightarrow$) Let $\dim_{1,f}(G) = \frac{n}{2}$. It suffices to show that each twin equivalence class of $V(G)$ has cardinality at least two. Assume, to the contrary, that there exists a twin equivalence class $Q \subset V(G)$ consisting of exactly one element; let $z \in Q$. Let $h : V(G) \to [0, 1]$ be a function defined by $h(z) = 0$ and $h(v) = \frac{1}{2}$ for each $v \in V(G) - \{z\}$. Since $|R_1\{z, u\}| \geq 3$ for any $u \in V(G) - \{z\}$, h is a 1-truncated resolving function of G with $h(V(G)) = \frac{n-1}{2}$, and hence $\dim_{1,f}(G) \leq \frac{n-1}{2}$, a contradiction. So, each twin equivalence class of $V(G)$ must have cardinality at least two. By the connectedness of G, we conclude that $G \in H[\mathcal{K} \cup \overline{\mathcal{K}}]$ for some connected graph H. □

Observation 1(d), Theorems 4 and 6 imply the following

Corollary 1. *Let G be a connected graph of order $n \geq 2$, and let $k \in \mathbb{Z}^+$. Then $\dim_{k,f}(G) = \frac{n}{2}$ if and only if $G \in H[\mathcal{K} \cup \overline{\mathcal{K}}]$ for some connected graph H.*

3 $\dim_{k,f}(G)$ of Some Graph Classes

In this section, we examine $\dim_{k,f}(G)$ for some classes of graphs.

3.1 Cycles and Graphs G with $\text{diam}(G) \leq 2$

First, we determine $\dim_{k,f}(C_n)$ for any $k \in \mathbb{Z}^+$ and for $n \geq 3$.

Theorem 7. *[1] For $n \geq 3$, $\dim_f(C_n) = \begin{cases} \frac{n}{n-1} & \text{if } n \text{ is odd,} \\ \frac{n}{n-2} & \text{if } n \text{ is even.} \end{cases}$*

Theorem 8. *For any $k \in \mathbb{Z}^+$ and for $n \geq 3$,*

$$\dim_{k,f}(C_n) = \begin{cases} \frac{n}{n-1} & \text{if } n \leq 2k+3 \text{ and } n \text{ is odd,} \\ \frac{n}{n-2} & \text{if } n \leq 2k+3 \text{ and } n \text{ is even,} \\ \frac{n}{2(k+1)} & \text{if } n \geq 2k+4. \end{cases}$$

Proof. Let $k \in \mathbb{Z}^+$. For $n \geq 3$, let C_n be given by $u_0, u_1, \ldots, u_{n-1}, u_0$. Let $g : V(C_n) \to [0, 1]$ be any minimum k-truncated resolving function of C_n.

First, let $n \leq 2k+3$; then $\text{diam}(C_n) \leq \lfloor \frac{2k+3}{2} \rfloor = k+1$. By Observation 2(b), $\dim_{k,f}(C_n) = \dim_f(C_n)$. So, by Theorem 7, $\dim_{k,f}(C_n) = \frac{n}{n-1}$ for an odd n and $\dim_{k,f}(C_n) = \frac{n}{n-2}$ for an even n.

Second, let $n \geq 2k+4$. Note that, for each $i \in \{0, 1, \ldots, n-1\}$, $R_k\{u_i, u_{i+2}\} = \cup_{j=0}^{k}\{u_{i-j}, u_{i+2+j}\}$, where the subscript is taken modulo n; thus

$\sum_{j=0}^{k}(g(u_{i-j}) + g(u_{i+2+j})) \geq 1$. By summing over n such inequalities, we have $2(k+1)g(V(C_n)) \geq n$ since each vertex appears $2(k+1)$ times in the n inequalities. So, $g(V(C_n)) \geq \frac{n}{2(k+1)}$, and hence $\dim_{k,f}(C_n) \geq \frac{n}{2(k+1)}$. On the other hand, if we let $h : V(C_n) \to [0,1]$ be a function defined by $h(u_i) = \frac{1}{2(k+1)}$ for each $i \in \{0, 1, \ldots, n-1\}$, then h is a k-truncated resolving function of C_n with $h(V(C_n)) = \frac{n}{2(k+1)}$. To see this, for any distinct $x, y \in \{0, 1, \ldots, n-1\}$, note that $|R_k\{u_x, u_y\}| \geq 2(k+1)$ and $h(R_k\{u_x, u_y\}) \geq \frac{1}{2(k+1)} \cdot 2(k+1) = 1$. So, $\dim_{k,f}(C_n) \leq h(V(C_n)) = \frac{n}{2(k+1)}$. Therefore, $\dim_{k,f}(C_n) = \frac{n}{2(k+1)}$. □

Next, we consider graphs G with $\text{diam}(G) \leq 2$. The *join* of two graphs G and H, denoted by $G+H$, is the graph obtained from the disjoint union of G and H by joining an edge between each vertex of G and each vertex of H. If $\text{diam}(G) \leq 2$, then $\dim_{k,f}(G) = \dim_f(G)$ for any $k \in \mathbb{Z}^+$. So, $\dim_{k,f}(G) = \dim_f(G)$ when G is the Petersen graph, the join graph (a wheel graph or a fan graph, for examples) or a complete multipartite graph. See [1] for $\dim_f(G)$ when G is the Petersen graph or a wheel graph; see [26] for $\dim_f(G)$ when G is a complete multipartite graph. Next, we state the following result on $\dim_{k,f}(P_n + K_1)$; we refer to [27] for its proof.

Theorem 9. *Let $k \in \mathbb{Z}^+$.*

(a) For $n \geq 1$,

$$\dim_{k,f}(P_n + K_1) = \dim_f(P_n + K_1) = \begin{cases} \frac{n+1}{2} & \text{if } n \in \{1,2,3\}, \\ \frac{5}{3} & \text{if } n \in \{4,5\}, \\ \frac{n+1}{4} & \text{if } n \geq 6 \text{ and } n \equiv 1,3 \pmod 4, \\ \frac{n+2}{4} & \text{if } n \geq 6 \text{ and } n \equiv 2 \pmod 4. \end{cases}$$

(b) If $n \geq 8$ and $n \equiv 0 (\text{mod } 4)$, then $\frac{n}{4} \leq \dim_{k,f}(P_n + K_1) = \dim_f(P_n + K_1) \leq \frac{n+2}{4}$.

3.2 Grid Graphs

For $s, t \geq 2$, we examine $\dim_{k,f}(P_s \times P_t)$. We recall some notations. For two functions $f(x)$ and $g(x)$ defined on $\mathbb{R}$, $f(x) = O(g(x))$ if there exist positive constants N and C such that $|f(x)| \leq C|g(x)|$ for all $x > N$, $f(x) = \Omega(g(x))$ if $g(x) = O(f(x))$, and $f(x) = \Theta(g(x))$ if $f(x) = O(g(x))$ and $f(x) = \Omega(g(x))$. We note that $\frac{\dim_1(G)}{\dim(G)}$ can be arbitrarily large (see [13]) and that both $\frac{\dim_1(G)}{\dim_k(G)}$ and $\frac{\dim_k(G)}{\dim(G)}$ can be arbitrarily large for $k > 1$ (see [11]).

First, we show that $\frac{\dim_{1,f}(G)}{\dim_{k,f}(G)}$ and $\frac{\dim_{k,f}(G)}{\dim_f(G)}$ can be arbitrarily large for $k > 1$.

Proposition 2. *[1] If $G = P_s \times P_t$ $(s, t \geq 2)$, then* $\dim_f(G) = 2$.

Proposition 3. *[13] If $G = P_m \times P_m$ for $m \geq 2$, then* $\dim_1(G) = \Theta(m^2)$.

Proposition 4. *[11] If $G = P_{k^2} \times P_{k^2}$ for $k > 1$, then* $\dim_k(G) = \Theta(k^2)$.

Proposition 5. *If $G = P_{4m} \times P_{3m}$ for $m \geq 1$, then $\dim_{1,f}(G) = \Theta(m^2)$.*

Proof. By Proposition 3 and Observation 1(d), $\dim_{1,f}(G) = O(m^2)$. To see that $\dim_{1,f}(G) = \Omega(m^2)$, suppose that the grid graph $G = P_{4m} \times P_{3m}$ is drawn in the xy-plane with the four corners at $(1,1)$, $(4m,1)$, $(1,3m)$ and $(4m,3m)$ with horizontal and vertical edges of equal lengths, and let $g : V(G) \rightarrow [0,1]$ be any 1-truncated resolving function of G. Then, for every $P_4 \times P_3$ subgraph, say $B_{i,j}$, of G with the four corners $(1+4i, 1+3j)$, $(4+4i, 1+3j)$, $(1+4i, 3+3j)$ and $(4+4i, 3+3j)$, where $i, j \in \{0, 1, \ldots, m-1\}$, we have $R_1\{(2+4i, 2+3j), (3+4i, 2+3j)\} \subset V(B_{i,j})$, and thus $g(V(B_{i,j})) \geq 1$. So, $\dim_{1,f}(G) \geq \sum_{j=0}^{m-1} \sum_{i=0}^{m-1} g(V(B_{i,j})) \geq m^2$, and hence $\dim_{1,f}(G) = \Omega(m^2)$. Therefore, $\dim_{1,f}(G) = \Theta(m^2)$. □

Theorem 10. *For $k > 1$, let $G = P_{(2k+2)^2} \times P_{(2k+1)^2}$. Then $\dim_{k,f}(G) = \Theta(k^2)$, and thus both $\frac{\dim_{1,f}(G)}{\dim_{k,f}(G)}$ and $\frac{\dim_{k,f}(G)}{\dim_f(G)}$ can be arbitrarily large.*

Proof. For $k > 1$, let $G = P_{(2k+2)^2} \times P_{(2k+1)^2}$. Then $\dim_f(G) = 2$ by Proposition 2, and $\dim_{1,f}(G) = \Theta(k^4)$ by Proposition 5. Next, we show that $\dim_{k,f}(G) = \Theta(k^2)$. By Proposition 4 and Observation 1(c), $\dim_{k,f}(G) = O(k^2)$. To see that $\dim_{k,f}(G) = \Omega(k^2)$, notice that G contains disjoint union of $(2k+2)(2k+1)$ copies of $P_{2k+2} \times P_{2k+1}$. Let $g : V(G) \rightarrow [0,1]$ be any k-truncated resolving function of G. For each subgraph $P_{2k+2} \times P_{2k+1}$ of G, if x and y are the two adjacent central vertices of $P_{2k+2} \times P_{2k+1}$, then $R_k\{x,y\} \subseteq V(P_{2k+2} \times P_{2k+1})$, and hence $g(V(P_{2k+2} \times P_{2k+1})) \geq g(R_k\{x,y\}) \geq 1$; thus $\dim_{k,f}(G) \geq (2k+2)(2k+1)$. So, $\dim_{k,f}(G) = \Omega(k^2)$. Therefore, $\dim_{k,f}(G) = \Theta(k^2)$ for $k > 1$, and both $\frac{\dim_{1,f}(G)}{\dim_{k,f}(G)}$ and $\frac{\dim_{k,f}(G)}{\dim_f(G)}$ can be arbitrarily large. □

Next, we state the following result on grid graphs G satisfying $\dim_{1,f}(G) = \dim_f(G)$; we refer to [27] for its proof.

Proposition 6. *For the grid graph $G = P_s \times P_t$ with $s \geq t \geq 2$, $\dim_{1,f}(G) = \dim_f(G)$ if and only if $G \in \{P_2 \times P_2, P_3 \times P_2, P_4 \times P_2, P_3 \times P_3\}$.*

3.3 Trees

We examine $\dim_{k,f}(T)$ for non-trivial trees T. For $n \geq 2$, $\dim_{k,f}(P_n) = \dim_f(P_n)$ if and only if $n \in \{2, 3, \ldots, k+2\}$ by Theorems 3(c) and 5. We first state the following result on $\dim_{k,f}(P_n)$; we refer to [27] for its proof.

Theorem 11. *Let $k \in \mathbb{Z}^+$ and $n \geq 2$.*

(a) If $n \leq k+2$, then $\dim_{k,f}(P_n) = 1$.
(b) If $k+3 \leq n \leq 2k+3$, then $\dim_{k,f}(P_n) = \frac{6+2k-n}{5+2k-n}$.
(c) Let $n \geq 2k+4$.

(i) If $n \equiv 1 (\text{mod } (2k+2))$, then $\dim_{k,f}(P_n) = \frac{n+k}{2k+2}$.
(ii) If $n \equiv 2, 3, \ldots, k+2 (\text{mod } (2k+2))$, then $\dim_{k,f}(P_n) = \lceil \frac{n}{2k+2} \rceil$.

(iii) If $n \equiv 0 (\mathrm{mod}\ (2k+2))$ *or* $n \equiv k+3, k+4, \ldots, 2k+1 (\mathrm{mod}\ (2k+2))$*, then* $\lceil \frac{n}{2k+2} \rceil \le \dim_{k,f}(P_n) \le \lceil \frac{n}{2k+2} \rceil + \frac{1}{2}$.

Next, we examine non-trivial trees T satisfying $\dim_{1,f}(T) = \dim_f(T)$. We recall some terminology and notation. The *degree* of a vertex $v \in V(G)$ is $|N(v)|$; a *leaf* is a vertex of degree one and a *major vertex* is a vertex of degree at least three. Fix a tree T. A leaf ℓ is called a *terminal vertex* of a major vertex v if $d(\ell, v) < d(\ell, w)$ for every other major vertex w in T. The *terminal degree*, $ter(v)$, of a major vertex v is the number of terminal vertices of v in T, and an *exterior major vertex* is a major vertex that has positive terminal degree. An *exterior degree-two vertex* is a vertex of degree 2 that lies on a path from a terminal vertex to its major vertex, and an *interior degree-two vertex* is a vertex of degree 2 such that the shortest path to any terminal vertex includes a major vertex. Let $M(T)$ be the set of exterior major vertices of T and let $L(T)$ be the set of leaves of T. Let $M_1(T) = \{w \in M(T) : ter(w) = 1\}$, $M_2(T) = \{w \in M(T) : ter(w) \ge 2\}$; then $M(T) = M_1(T) \cup M_2(T)$. Let $\sigma(T) = |L(T)|$, $ex(T) = |M(T)|$ and $ex_1(T) = |M_1(T)|$.

Theorem 12. *[26] For any non-trivial tree* T, $\dim_f(T) = \frac{1}{2}(\sigma(T) - ex_1(T))$.

The following lemma is used in proving Proposition 7; see [27] for proofs of the next two statements.

Lemma 1. *Let* T *be a tree with* $ex(T) \ge 1$ *satisfying* $\dim_{1,f}(T) = \dim_f(T)$.

(a) If $v \in M_2(T)$*, then every terminal vertex of* v *is adjacent to* v *in* T.
(b) T *contains no major vertex of terminal degree one.*
(c) T *contains neither a major vertex of terminal degree zero nor an interior degree-two vertex.*

Proposition 7. *Let* T *be a non-trivial tree. Then* $\dim_{1,f}(T) = \dim_f(T)$ *if and only if* $T \in \{P_2, P_3\}$*, or* $ex(T) \ge 1$ *and* $V(T) = M_2(T) \cup L(T)$.

For any $k \in \mathbb{Z}^+$, it is an interesting yet challenging task to characterize all connected graphs G satisfying $\dim_{k,f}(G) = \dim_f(G)$ even when G is restricted to trees. For trees T with $ex(T) = 1$, we state the following result on T satisfying $\dim_{k,f}(T) = \dim_f(T)$; we refer to [27] for its proof.

Proposition 8. *Let* $k \in \mathbb{Z}^+$*, and let* T *be a tree with* $ex(T) = 1$ *such that* $\ell_1, \ell_2, \ldots, \ell_\alpha$ *are the terminal vertices of the exterior major vertex* v *in* T*. Then* $\dim_{k,f}(T) = \dim_f(T)$ *if and only if* $d(v, \ell_i) \le k$ *for each* $i \in \{1, 2, \ldots, \alpha\}$.

4 Some Remarks and Open Problems

In this section, we examine the relation among $\dim_f(G)$, $\dim_{k,f}(G)$, $\dim(G)$ and $\dim_k(G)$ for $k \in \mathbb{Z}^+$ in conjunction with Observation 1. We show that, for two connected graphs H and G with $H \subset G$, $\frac{\dim_{k,f}(H)}{\dim_{k,f}(G)}$ can be arbitrarily large. We also show the existence of non-isomorphic graphs G and H with $\dim_k(G) = \dim_k(H)$ and $\dim_{k,f}(G) \neq \dim_{k,f}(H)$. We conclude the paper with some open problems.

It is known that metric dimension is not a monotone parameter on subgraph inclusion (see [6]), and the following results were obtained in [13] and [11].

Theorem 13. *Let H and G be connected graphs with $H \subset G$. Then*
(a) [13] $\frac{\dim(H)}{\dim(G)}$ and $\frac{\dim_1(H)}{\dim_1(G)}$ can be arbitrarily large;
(b) [11] for any $k \in \mathbb{Z}^+$, $\frac{\dim_k(H)}{\dim_k(G)}$ can be arbitrarily large.

We recall the following construction from [13]. For $m \geq 3$, let $H = K_{\frac{m(m+1)}{2}}$; let $V(H)$ be partitioned into $V_1, V_2, \ldots, V_m$ such that $V_i = \{w_{i,1}, w_{i,2}, \cdots, w_{i,i}\}$ with $|V_i| = i$, where $i \in \{1, 2, \ldots, m\}$. Let G be the graph obtained from H and m isolated vertices $u_1, u_2, \ldots, u_m$ such that, for each $i \in \{1, 2, \ldots, m\}$, u_i is joined by an edge to each vertex of $V_i \cup (\cup_{j=i+1}^{m}\{w_{j,i}\})$; notice $H \subset G$. Since $\text{diam}(H) = 1$ and $\text{diam}(G) = 2$, by Observation 2(b), $\dim_{k,f}(H) = \dim_f(H)$ and $\dim_{k,f}(G) = \dim_f(G)$ for any $k \in \mathbb{Z}^+$. Note that $\dim_f(H) = \frac{m(m+1)}{4}$ by Theorem 4, and $\dim_f(G) \leq m$ by Observation 1(a) since $\{u_1, u_2, \ldots, u_m\}$ forms a resolving set of G. So, $\frac{\dim_{k,f}(H)}{\dim_{k,f}(G)} = \frac{\dim_f(H)}{\dim_f(G)} \geq \frac{m+1}{4}$ for any $k \in \mathbb{Z}^+$, which implies the following

Corollary 2. *For any $k \in \mathbb{Z}^+$, there exist connected graphs H and G such that $H \subset G$ and both $\frac{\dim_f(H)}{\dim_f(G)}$ and $\frac{\dim_{k,f}(H)}{\dim_{k,f}(G)}$ can be arbitrarily large.*

It was shown that $\frac{\dim_k(G)}{\dim(G)}$ and $\frac{\dim_{k,f}(G)}{\dim_f(G)}$ can be arbitrarily large (see [11] and Theorem 10, respectively). In view of Observation 1, we obtain the following

Remark 1. Let $k \in \mathbb{Z}^+$, and let G be a non-trivial connected graph. Then

(a) $\dim(G) - \dim_f(G)$ can be arbitrarily large;
(b) $\dim_k(G) - \dim_{k,f}(G)$ can be arbitrarily large;
(c) $\dim(G) - \dim_{k,f}(G)$ can be arbitrarily large;
(d) $\frac{\dim_{k,f}(G)}{\dim(G)}$ can be arbitrarily large.

Proof. Let $k \in \mathbb{Z}^+$. It is known that, for any tree T that is not a path, $\dim(T) = \sigma(T) - ex(T)$ (see [4,20,21]). For (a), (b) and (c), let G be a tree with $V(G) = M_2(G) \cup L(G)$ such that $M_2(G) = \{v_1, v_2, \ldots, v_x\}$ with $ter(v_i) = \alpha \geq 3$ for each $i \in \{1, 2, \ldots, x\}$, where $x \geq 1$. Then $\dim_k(G) \geq \dim(G) = x(\alpha - 1)$ by Observation 1(b). Also, note that $\dim_{k,f}(G) \geq \dim_f(G) = \frac{x\alpha}{2}$ by Observation 1(c) and Theorem 12. Since a function $g : V(G) \rightarrow [0, 1]$ defined by $g(u) = \frac{1}{2}$ for each $u \in L(G)$ and $g(w) = 0$ for each $w \in M_2(G) = V(G) - L(G)$ is a k-truncated resolving function for G with $g(V(G)) = \frac{x\alpha}{2}$, $\dim_{k,f}(G) \leq \frac{x\alpha}{2}$. So, $\dim_{k,f}(G) = \dim_f(G) = \frac{x\alpha}{2}$. Thus, $\dim_k(G) - \dim_{k,f}(G) \geq \dim(G) - \dim_{k,f}(G) = \dim(G) - \dim_f(G) = x(\alpha - 1) - \frac{x\alpha}{2} = \frac{x(\alpha-2)}{2} \rightarrow \infty$ as $x \rightarrow \infty$ or $\alpha \rightarrow \infty$.

For (d), let $G = C_{x(k+1)(k+5)}$ for some $x \in \mathbb{Z}^+$. Then $\dim_{k,f}(G) = \frac{x(k+5)}{2}$ by Theorem 8 and $\dim(G) = 2$. So, $\frac{\dim_{k,f}(G)}{\dim(G)} = \frac{x(k+5)}{4} \rightarrow \infty$ as $x \rightarrow \infty$. □

Next, we show the existence of non-isomorphic graphs G and H with $\dim_k(G) = \dim_k(H)$ and $\dim_{k,f}(G) \neq \dim_{k,f}(H)$ for each $k \in \mathbb{Z}^+$. We recall the following result.

Theorem 14. *[11] Let $k \in \mathbb{Z}^+$. Then*

(a) $\dim_k(P_n) = 1$ *for* $2 \leq n \leq k+2$;
(b) $\dim_k(C_n) = 2$ *for* $3 \leq n \leq 3k+3$, *and* $\dim_k(P_n) = 2$ *for* $k+3 \leq n \leq 3k+3$;
(c) for $n \geq 3k+4$,

$$\dim_k(C_n) = \dim_k(P_n) = \begin{cases} \lfloor \frac{2n+3k-1}{3k+2} \rfloor & \text{if } n \equiv 0, 1, \ldots, k+2 \pmod{(3k+2)}, \\ \lfloor \frac{2n+4k-1}{3k+2} \rfloor & \text{if } n \equiv k+3, \ldots, \lceil \frac{3k+5}{2} \rceil - 1 \pmod{(3k+2)}, \\ \lfloor \frac{2n+3k-1}{3k+2} \rfloor & \text{if } n \equiv \lceil \frac{3k+5}{2} \rceil, \ldots, 3k+1 \pmod{(3k+2)}. \end{cases}$$

Remark 2. Let $k \in \mathbb{Z}^+$. There exist non-isomorphic graphs G and H such that $\dim_k(G) = \dim_k(H)$ and $\dim_{k,f}(G) \neq \dim_{k,f}(H)$. For $n \geq 3k+4$, $\dim_k(C_n) = \dim_k(P_n)$ by Theorem 14 and $\dim_{k,f}(C_n) \neq \dim_{k,f}(P_n)$ for $n \equiv 1 (\text{mod } (2k+2))$ by Theorems 8 and 11(c).

We conclude the paper with some open problems.

Question 1. For any tree T and for any $k \in Z^+$, can we determine $\dim_{k,f}(T)$?

Question 2. For any $k \in \mathbb{Z}^+$, can we characterize all connected graphs G satisfying $\dim_{k,f}(G) = \dim_f(G)$?

Question 3. For any $k \in \mathbb{Z}^+$, can we characterize all connected graphs G satisfying $\dim_k(G) = \dim_{k,f}(G)$?

Acknowledgements. The author thanks the anonymous referees for some helpful comments.

References

1. Arumugam, S., Mathew, V.: The fractional metric dimension of graphs. Discrete Math. **312**, 1584–1590 (2012)
2. Arumugam, S., Mathew, V., Shen, J.: On fractional metric dimension of graphs. Discrete Math. Algorithms Appl. **5**, 1350037 (2013)
3. Beardon, A.F., Rodríguez-Velázquez, J.A.: On the k-metric dimension of metric spaces. Ars Math. Contemp. **16**, 25–38 (2019)
4. Chartrand, G., Eroh, L., Johnson, M.A., Oellermann, O.R.: Resolvability in graphs and the metric dimension of a graph. Discrete Appl. Math. **105**, 99–113 (2000)
5. Currie, J., Oellermann, O.R.: The metric dimension and metric independence of a graph. J. Combin. Math. Combin. Comput. **39**, 157–167 (2001)
6. Eroh, L., Kang, C.X., Yi, E.: Metric dimension and zero forcing number of two families of line graphs. Math. Bohem. **139**(3), 467–483 (2014)
7. Estrada-Moreno, A.: On the (k,t)-metric dimension of a graph. Dissertation, Universitat Rovira i Virgili (2016)

8. Estrada-Moreno, A., Yero, I.G., Rodríguez-Velázquez, J.A.: On the (k, t)-metric dimension of graphs. Comput. J. **64**(5), 707–720 (2021)
9. Fehr, M., Gosselin, S., Oellermann, O.R.: The metric dimension of Cayley digraphs. Discrete Math. **306**, 31–41 (2006)
10. Fernau, H., Rodríguez-Velázquez, J.A.: On the (adjacency) metric dimension of corona and strong product graphs and their local variants: combinatorial and computational results. Discrete Appl. Math. **236**, 183–202 (2018)
11. Frongillo, R.M., Geneson, J., Lladser, M.E., Tillquist, R.C., Yi, E.: Truncated metric dimension for finite graphs (2021, Submitted)
12. Garey, M.R., Johnson, D.S.: Computers and Intractability: A Guide to the Theory of NP-Completeness. Freeman, New York (1979)
13. Geneson, J., Yi, E.: Broadcast dimension of graphs. arXiv:2005.07311v1 (2020). https://arxiv.org/abs/2005.07311
14. Geneson, J., Yi, E.: The distance-k dimension of graphs. arXiv:2106.08303v2 (2021). https://arxiv.org/abs/2106.08303
15. Harary, F., Melter, R.A.: On the metric dimension of a graph. Ars Combin. **2**, 191–195 (1976)
16. Hernando, C., Mora, M., Pelayo, I.M., Seara, C., Wood, D.R.: Extremal graph theory for metric dimension and diameter. Electron. J. Combin. **17**, 1–28 (2010). Article no. R30
17. Jannesari, M., Omoomi, B.: The metric dimension of the lexicographic product of graphs. Discrete Math. **312**(22), 3349–3356 (2012)
18. Kang, C.X.: On the fractional strong metric dimension of graphs. Discrete Appl. Math. **213**, 153–161 (2016)
19. Kang, C.X., Yi, E.: The fractional strong metric dimension of graphs. In: Widmayer, P., Xu, Y., Zhu, B. (eds.) COCOA 2013. LNCS, vol. 8287, pp. 84–95. Springer, Cham (2013). https://doi.org/10.1007/978-3-319-03780-6_8
20. Khuller, S., Raghavachari, B., Rosenfeld, A.: Landmarks in graphs. Discrete Appl. Math. **70**, 217–229 (1996)
21. Poisson, C., Zhang, P.: The metric dimension of unicyclic graphs. J. Combin. Math. Combin. Comput. **40**, 17–32 (2002)
22. Scheinerman, E.R., Ullman, D.H.: Fractional Graph Theory: A Rational Approach to the Theory of Graphs. Wiley, New York (1997)
23. Slater, P.J.: Leaves of trees. Congr. Numer. **14**, 549–559 (1975)
24. Tillquist, R.C.: Low-dimensional embeddings for symbolic data science. Dissertation, University of Colorado, Boulder (2020)
25. Tillquist, R.C., Frongillo, R.M., Lladser, M.E.: Truncated metric dimension for finite graphs. arXiv:2106.14314v1 (2021). https://arxiv.org/abs/2106.14314
26. Yi, E.: The fractional metric dimension of permutation graphs. Acta Math. Sin. Engl. Ser. **31**, 367–382 (2015)
27. Yi, E.: The fractional k-truncated metric dimension of graphs. arXiv:2108.02745v1 (2021). https://arxiv.org/abs/2108.02745

On Structural Parameterizations of the Offensive Alliance Problem

Ajinkya Gaikwad and Soumen Maity(✉)

Indian Institute of Science Education and Research, Pune, India
ajinkya.gaikwad@students.iiserpune.ac.in, soumen@iiserpune.ac.in

Abstract. The OFFENSIVE ALLIANCE problem has been studied extensively during the last twenty years. A set $S \subseteq V$ of vertices is an offensive alliance in an undirected graph $G = (V, E)$ if each $v \in N(S)$ has at least as many neighbours in S as it has neighbours (including itself) outside S. We study the parameterzied complexity of the OFFENSIVE ALLIANCE problem, where the aim is to find a minimum size offensive alliance. Our focus here lies on parameters that measure the structural properties of the input instance. We enhance our understanding of the problem from the viewpoint of parameterized complexity by showing that the OFFENSIVE ALLIANCE problem is W[1]-hard parameterized by a wide range of fairly restrictive structural parameters such as the feedback vertex set number, treewidth, pathwidth, and treedepth of the input graph. We also prove that the STRONG OFFENSIVE ALLIANCE problem parameterized by the vertex cover number of the input graph does not admit a polynomial compression unless coNP $\subseteq$ NP/poly.

Keywords: Defensive and offensive alliance · Parameterized complexity · FPT · W[1]-hard · Treewidth

1 Introduction

An alliance is a collection of people, groups, or states such that the union is stronger than individual. The alliance can be either to achieve some common purpose, to protect against attack, or to assert collective will against others. This motivates the definitions of defensive and offensive alliances in graphs. The properties of alliances in graphs were first studied by Kristiansen, Hedetniemi, and Hedetniemi [11]. They introduced defensive, offensive and powerful alliances. The alliance problems have been studied extensively during last twenty years [2,6,13,15,16], and generalizations called r-alliances are also studied [14]. Throughout this article, $G = (V, E)$ denotes a finite, simple and undirected

A. Gaikwad—Gratefully acknowledges support from the Ministry of Human Resource Development, Government of India, under Prime Minister's Research Fellowship Scheme (No. MRF-192002-211).

S. Maity—Research was supported in part by the Science and Engineering Research Board (SERB), Govt. of India, under Sanction Order No. MTR/2018/001025.

D.-Z. Du et al. (Eds.): COCOA 2021, LNCS 13135, pp. 579–586, 2021.
https://doi.org/10.1007/978-3-030-92681-6_45

graph of order $|V| = n$. The subgraph induced by $S \subseteq V(G)$ is denoted by $G[S]$. We use $d_S(v)$ to denote the degree of vertex v in $G[S]$. The complement of the vertex set S in V is denoted by S^c.

Definition 1. A non-empty set $S \subseteq V$ is a strong offensive alliance in G if $d_S(v) \geq d_{S^c}(v) + 2$ for all $v \in N(S)$.

In this paper, we consider OFFENSIVE ALLIANCE, EXACT OFFENSIVE ALLIANCE and STRONG OFFENSIVE ALLIANCE problems under structural parameters. We define these problems as follows:

OFFENSIVE ALLIANCE
Input: An undirected graph $G = (V, E)$ and an integer $r \geq 1$.
Question: Is there an offensive alliance $S \subseteq V(G)$ such that $1 \leq |S| \leq r$?

STRONG OFFENSIVE ALLIANCE
Input: An undirected graph $G = (V, E)$ and an integer $r \geq 1$.
Question: Is there a strong offensive alliance $S \subseteq V(G)$ such that $1 \leq |S| \leq r$?

For the standard concepts in parameterized complexity, see the recent textbook by Cygan et al. [3]. The graph parameters we explicitly use in this paper are vertex cover number, feedback vertex set number, pathwidth, treewidth and treedepth [3].

1.1 Known Results

Fernau and Raible showed in [4] that the defensive and offensive alliance problems and their global variants are fixed parameter tractable when parameterized by the solution size k. Kiyomi and Otachi showed in [9], the problems of finding smallest alliances of all kinds are fixed-parameter tractable when parameteried by the vertex cover number. The problems of finding smallest defensive and offensive alliances are also fixed-parameter tractable when parameteried by the neighbourhood diversity [7]. Bliem and Woltran [1] proved that deciding if a graph contains a defensive alliance of size at most k is W[1]-hard when parameterized by treewidth of the input graph. This puts it among the few problems that are FPT when parameterized by solution size but not when parameterized by treewidth (unless FPT = W[1]).

2 Hardness Results

In this section we show that OFFENSIVE ALLIANCE is W[1]-hard parameterized by a vertex deletion set to trees of height at most seven, that is, a subset D of the vertices of the graph such that every component in the graph, after removing D, is a tree of height at most seven. On the way towards this result, we provide hardness results for several interesting versions of the OFFENSIVE ALLIANCE problem which we require in our proofs. The problem OFFENSIVE ALLIANCE^{F} generalizes OFFENSIVE ALLIANCE where some vertices are forced to be outside the solution; these vertices are called forbidden vertices. This variant can be formalized as follows:

> OFFENSIVE ALLIANCE^{F}
> **Input:** An undirected graph $G = (V, E)$, an integer r and a set $V_{\square} \subseteq V(G)$ of forbidden vertices such that each degree one forbidden vertex is adjacent to another forbidden vertex and each forbidden vertex of degree greater than one is adjacent to a degree one forbidden vertex.
> **Question:** Is there an offensive alliance $S \subseteq V$ such that (i) $1 \leq |S| \leq r$, and (ii) $S \cap V_{\square} = \emptyset$?

STRONG OFFENSIVE ALLIANCE^{FN} is a generalization of STRONG OFFENSIVE ALLIANCE^{F} that, in addition, requires some "necessary" vertices to be in S. This variant can be formalized as follows:

> STRONG OFFENSIVE ALLIANCE^{FN}
> **Input:** An undirected graph $G = (V, E)$, an integer r, a set $V_{\triangle} \subseteq V$, and a set $V_{\square} \subseteq V(G)$ of forbidden vertices such that each degree one forbidden vertex is adjacent to another forbidden vertex and each forbidden vertex of degree greater than one is adjacent to a degree one forbidden vertex.
> **Question:** Is there a strong offensive alliance $S \subseteq V$ such that (i) $1 \leq |S| \leq r$, (ii) $S \cap V_{\square} = \emptyset$, and (iii) $V_{\triangle} \subseteq S$?

While the OFFENSIVE ALLIANCE problem asks for offensive alliance of size at most r, we also consider the EXACT OFFENSIVE ALLIANCE problem that concerns offensive alliance of size exactly r. Analogously, we also define exact versions of STRONG OFFENSIVE ALLIANCE presented above. To prove Lemma 2, we consider the following problem:

> MULTIDIMENSIONAL RELAXED SUBSET SUM (MRSS)
> **Input:** Two integers k and k', a set $S = \{s_1, \ldots, s_n\}$ of vectors with $s_i \in \mathbb{N}^k$ for every i with $1 \leq i \leq n$ and a target vector $t \in \mathbb{N}^k$.
> **Parameter**: $k + k'$
> **Question**: Is there a subset $S' \subseteq S$ with $|S'| \leq k'$ such that $\sum_{s \in S'} s \geq t$?

Lemma 1. [8] MRSS is W[1]-hard when parameterized by the combined parameter $k + k'$, even if all integers in the input are given in unary.

We now show that the STRONG OFFENSIVE ALLIANCE^{FN} problem is W[1]-hard parameterized by the size of a vertex deletion set into trees of height at most 5, via a reduction from MRSS.

Lemma 2 ($\star$[1]). The Strong Offensive Alliance^{FN} problem is W[1]-hard when parameterized by the size of a vertex deletion set into trees of height at most 5.

We have the following corollary from Lemma 2.

Corollary 1. The Strong Offensive Alliance^{FN} problem is W[1]-hard when parameterized by the size of a vertex deletion set into trees of height at most 5, even when $|V_{\triangle}| = 1$.

Next, we give an FPT reduction without proof that eliminates necessary vertices.

Lemma 3 ($\star$). The Offensive Alliance^{F} problem is W[1]-hard when parameterized by the size of a vertex deletion set into trees of height at most 5.

We are now ready to show our main hardness result for Offensive Alliance using a reduction from Offensive Alliance^{F}.

Theorem 1. The Offensive Alliance problem is W[1]-hard when parameterized by the size of a vertex deletion set into trees of height at most 7.

Proof. We give a parameterized reduction from Offensive Alliance^{F} which is W[1]-hard when parameterized by the size of a vertex deletion set into trees of height at most 5. Let $I = (G, r, V_{\square})$ be an instance of Offensive Alliance^{F}. Let $n = |V(G)|$. We construct an instance $I' = (G', r')$ of Offensive Alliance the following way. We set $r' = r$. Recall that each degree one forbidden vertex is adjacent to another forbidden vertex and each forbidden vertex of degree greater than one is adjacent to a degree one forbidden vertex. Let u be a degree one forbidden vertex in G and u is adjacent to another forbidden vertex v. For each degree one forbidden vertex $u \in V_{\square}$, we introduce a tree T_u rooted at u of height 2 as shown in Fig. 1. The forbidden vertex v has additional neighbours from the original graph G which are not shown here. We define G' as follows: $V(G') = V(G) \bigcup\limits_{u \in V_{\square}} \Big\{V(T_u) \mid \text{where } u \text{ is a degree one forbidden vertex in } G\Big\}$ and $E(G') = E(G) \bigcup\limits_{u' \in V_{\square}} E(T_u)$. We claim I is a yes instance if and only if

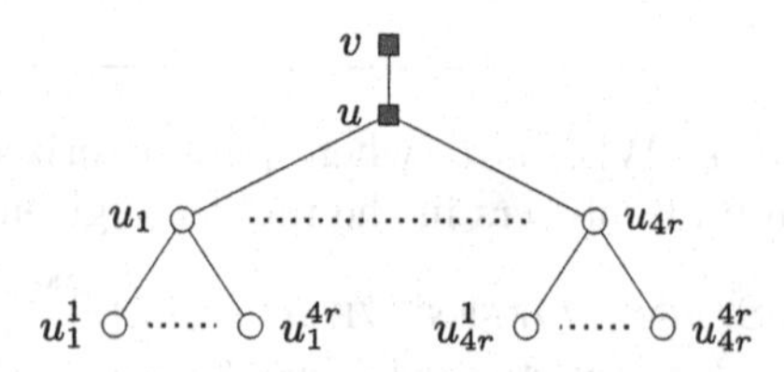

Fig. 1. Our tree gadget T_u for each degree one forbidden vertex $u \in V_{\square}$

[1] Due to paucity of space, the proofs of statements marked with a $\star$ have been omitted.

I' is a yes instance. It is easy to see that if R is an offensive alliance of size at most r in G then it is also an offensive alliance of size at most $r' = r$ in G'.

To prove the reverse direction of the equivalence, suppose that G' has an offensive alliance R' of size at most $r' = r$. We claim that no vertex from the set $V_\square \bigcup\limits_{u \in V_\square} V(T_u)$ is part of R'. It is easy to see that if any vertex from the set $V_\square \bigcup\limits_{u \in V_\square} V(T_u)$ is in R' then the size of R' exceeds $2r$. This implies that $R = R' \cap G$ is an offensive alliance such that $R \cap V_\square = \emptyset$ and $|R| \leq r$. This shows that I is a yes instance. □

We have the following consequences.

Corollary 2. The Exact Offensive Alliance problem is W[1]-hard when parameterized by the size of a vertex deletion set into trees of height at most 7.

Clearly trees of height at most seven are trivially acyclic. Moreover, it is easy to verify that such trees have pathwidth [10] and treedepth [12] at most seven, which implies:

Theorem 2. The Offensive Alliance and Exact Offensive Alliance problems are W[1]-hard when parameterized by any of the following parameters:

- the feedback vertex set number,
- the treewidth and pathwidth of the input graph,
- the treedepth of the input graph.

3 No Polynomial Kernel When Parameterized by Vertex Cover Number

A set $C \subseteq V$ is a vertex cover of $G = (V, E)$ if each edge $e \in E$ has at least one endpoint in X. The minimum size of a vertex cover in G is the *vertex cover number* of G, denoted by $vc(G)$. Parameterized by vertex cover number vc, the Offensive Alliance problem is FPT [9] and in this section we prove the following kernelization hardness of the Strong Offensive Alliance problem.

Theorem 3. The Strong Offensive Alliance problem parameterized by the vertex cover number of the input graph does not admit a polynomial compression unless coNP $\subseteq$ NP/poly.

To prove Theorem 3, we give a polynomial parameter transformation (PPT) from the well-known Red Blue Dominating Set problem (RBDS) to Strong Offensive Alliance parameterized by vertex cover number. Recall that in RBDS we are given a bipartite graph $G = (R \cup B, E)$ and an integer k, and we are asked whether there exists a vertex set $S \subseteq B$ of size at most k such that $N(S) = R$. The following theorem is known:

Theorem 4. [5] RBDS parameterized by $|R|$ does not admit a polynomial compression unless coNP $\subseteq$ NP/poly.

3.1 Proof of Theorem 3

By Theorem 4, RBDS parameterized by $|R|$ does not admit a polynomial compression unless coNP $\subseteq$ NP/poly. To prove Theorem 3, we give a PPT from RBDS parameterized by $|R|$ to Strong Offensive Alliance parameterized by the vertex cover number. Given an instance $(G = (R \cup B, E), k)$ of RBDS, we construct an instance (G', k') of Strong Offensive Alliance as follows. First we duplicate the vertices of B. That is, G' contains both B and B' where $B' = \{b' \mid b \in B\}$. For $r \in R$ and $b \in B$, we have $(r, b), (r, b') \in E(G')$ if and only if $(r, b) \in E(G)$. Next we add a vertex a and make it adjacent to all the vertices of B. We also add a set $V_a = \{a_1, \ldots, a_{|B|}\}$ of vertices and make them adjacent to a. We introduce two vertices x_1, x_2 and make them adjacent to all the vertices of B. Make x_2 adjacent to all the vertices of R. We also make a adjacent to x_1 and x_2. Finally we add a set V_x of $6|B|$ vertices and make x_1 and x_2 adjacent to all the vertices of V_x. We observe that $R \cup \{a, x_1, x_2\}$ is a vertex cover of G'. Therefore the vertex cover size of G' is bounded by $|R| + 3$. We set $k' = |B| + k + 2$. See Fig. 2 for an illustration.

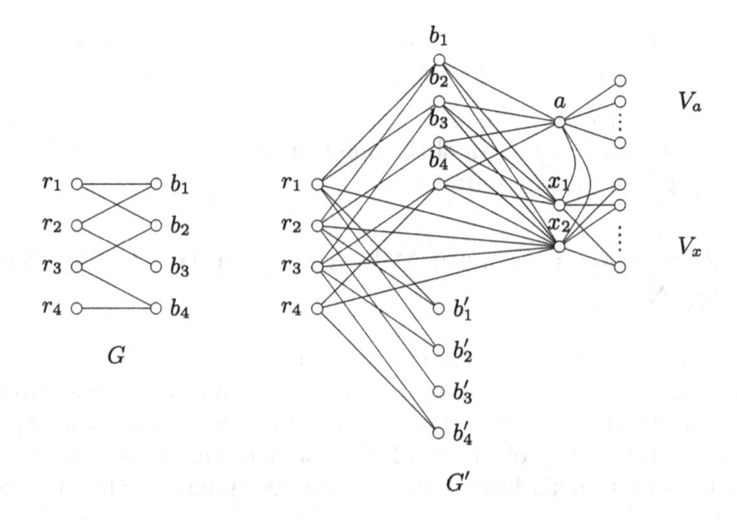

Fig. 2. The graph G' produced by the reduction algorithm from G in Theorem 3.

We now claim that (G, k) is a yes instance of RBDS if and only if (G', k') is a yes instance of Strong Offensive Alliance. Suppose there exists a vertex set $S \subseteq B$ of size at most k in G such that $N(S) = R$. We claim that the set $D = B \cup \{b' \in B' \mid b \in S\} \cup \{x_1, x_2\}$ is a strong offensive alliance in G'. Observe that $N(D) = R \cup V_x \cup \{a\}$. Next, we show that each $v \in N(D)$ satisfies the inequality $d_D(v) \geq d_{D^c}(v) + 2$.

Case 1: Suppose $v \in R$ has d neighbours in B. As v has at least one neighbour in $\{b' \in B' \mid b \in S\}$ and v is adjacent to x_2, we get $|N_D(v)| \geq d + 2$. Note that $|N_{D^c}(v)| \leq d - 1$. This implies $d_D(v) \geq d_{D^c}(v) + 2$.

Case 2: Observe that for a, we have $N_{D^c}(a) = V_a$. This implies that $|N_{D^c}(a)| = |B|$. It is easy to see that $N_D(a) = B \cup \{x_1, x_2\}$. Therefore, we have $|N_D(a)| = |B| + 2$. This implies that $d_D(a) \geq d_{D^c}(a) + 2$. For each $v \in V_x$, it easy to see that $d_{D^c}(v) = 0$ and $d_D(v) = 2$.

Conversely, suppose there exists a strong offensive alliance D of size at most k' in graph G'. First, we show that $\{x_1, x_2\} \subseteq D$. It is easy to see that $\{x_1, x_2\} \nsubseteq N(D)$ due to the size of V_x as otherwise $|D| \geq 3|B| > k'$. Since a strong offensive alliance is non-empty, it must contain a vertex from one of the sets $\{a\}, V_a, B, B'$, R. Clearly, if D contains a vertex from $B \cup R \cup \{a\}$ then $\{x_1, x_2\} \subseteq D$. If D contains a vertex from V_a then $a \in N(D)$ and a has to satisfy the condition $d_D(a) \geq d_{D^c}(a) + 2$ which requires at least two vertex from $B \cup \{x_1, x_2\}$ inside D. This implies that $\{x_1, x_2\} \subseteq D$. If D contains a vertex from B' then some vertex $r \in R$ is in $N(D)$ and r has to satisfy the condition $d_D(r) \geq d_{D^c}(r) + 2$ which requires at least one vertex from B to be inside D. This implies $\{x_1, x_2\} \subseteq D$. This shows that for any strong offensive alliance D, we have $\{x_1, x_2\} \subseteq D$.

Next, we show that starting from D, we can always construct another strong offensive alliance D' such that $B \subseteq D'$ and $|D'| \leq |D|$. Since $x_1 \in D$, this implies that $a \in D$ or $a \in N(D)$.

Case 1: Observe that if $a \in D$ then $V_a \subseteq D$ as otherwise $V_a \cap N(D) \neq \emptyset$ and the vertices in the set $V_a \cap N(D)$ do not satisfy the inequality $d_D(v) \geq d_{D^c}(v) + 2$. Therefore, we have $V_a \subseteq D$. In this case, we replace $\{a\} \cup V_a$ by B in D'. Note that $N(D') = N(D) \cup \{a\}$ and a satisfies the condition $d_{D'}(v) \geq d_{D'^c}(v) + 2$. The vertices v in $N(D)$ also satisfy the condition $d_{D'}(v) \geq d_{D'^c}(v) + 2$ as we increase their neighbours in D'. In this case, it is easy to see that $|D'| \leq |D|$.

Case 2: Observe that if $a \in N(D)$ then a has to satisfy the condition $d_D(a) \geq d_{D^c}(a) + 2$ which requires at least $|B|$ vertices from $V_a \cup B$ in D. We replace that set of $|B|$ vertices by B in D'. It is easy to see that a satisfies the condition $d_{D'}(a) \geq d_{D'^c}(a) + 2$, in fact, each $v \in N(D')$ satisfies the condition $d_{D'}(a) \geq d_{D'^c}(a) + 2$. Therefore D' is a strong offensive alliance and $|D'| \leq |D|$.

Thus we assume that $\{x_1, x_2\} \cup B \subseteq D'$. Since $|D'| \leq |B| + k + 2$, it implies that D' contains at most k vertices apart from the vertices in $\{x_1, x_2\} \cup B$. Next, we observe that every red vertex $r \in R$ is either in D' or $N(D')$. If $r \in N(D')$, then r has to satisfy the condition $d_{D'}(r) \geq d_{D'^c}(r) + 2$ which requires r having at least one neighbour in $D' \cap B'$. We construct another set $D^\star$ from D' by replacing each $r \in D'$ by an arbitrary neighbour of r in B'. Thus we get $R \cap D^\star = \emptyset$. Note that each $r \in R$ now satisfies the condition $d_{D^\star}(r) \geq d_{D^{\star c}}(r) + 2$. Therefore $D^\star$ is a strong offensive alliance and $|D^\star| \leq |D'|$. This implies that each vertex in R has at least one neighbour in $S' = D^\star \cap B'$ and $|S'| \leq k$. Define $S = \{b \in B \mid b' \in S'\}$. This implies that $S \subseteq B$ is a vertex set of size at most k such that $N(S) = R$. $\square$

4 Conclusions

It would be interesting to consider the parameterized complexity with respect to twin cover. The parameterized complexity of offensive and defensive alliance problems remain unsettled when parameterized by other important structural graph parameters like clique-width and modular-width.

References

1. Bliem, B., Woltran, S.: Defensive alliances in graphs of bounded treewidth. Discret. Appl. Math. **251**, 334–339 (2018)
2. Chellali, M., Haynes, T.W.: Global alliances and independence in trees. Discuss. Math. Graph Theor. **27**(1), 19–27 (2007)
3. Cygan, M., et al.: Parameterized Algorithms. Springer, Cham (2015). https://doi.org/10.1007/978-3-319-21275-3
4. Fernau, H., Raible, D.: Alliances in graphs: a complexity-theoretic study. In: Proceeding Volume II of the 33rd International Conference on Current Trends in Theory and Practice of Computer Science (2007)
5. Fomin, F.V., Lokshtanov, D., Saurabh, S., Zehavi, M.: Kernelization: Theory of Parameterized Preprocessing. Cambridge University Press (2019)
6. Fricke, G., Lawson, L., Haynes, T., Hedetniemi, M., Hedetniemi, S.: A note on defensive alliances in graphs. Bull. Inst. Combin. Appl. **38**, 37–41 (2003)
7. Gaikwad, A., Maity, S., Tripathi, S.K.: Parameterized complexity of defensive and offensive alliances in graphs. In: Goswami, D., Hoang, T.A. (eds.) ICDCIT 2021. LNCS, vol. 12582, pp. 175–187. Springer, Cham (2021). https://doi.org/10.1007/978-3-030-65621-8_11
8. Ganian, R., Klute, F., Ordyniak, S.: On structural parameterizations of the bounded-degree vertex deletion problem. Algorithmica **83**(1), 297–336 (2021)
9. Kiyomi, M., Otachi, Y.: Alliances in graphs of bounded clique-width. Discret. Appl. Math. **223**, 91–97 (2017)
10. Kloks, T. (ed.): Treewidth. Computations and Approximations. LNCS, vol. 842. Springer, Heidelberg (1994). https://doi.org/10.1007/BFb0045375
11. Kristiansen, P., Hedetniemi, M., Hedetniemi, S.: Alliances in graphs. J. Comb. Math. Comb. Comput. **48**, 157–177 (2004)
12. Nešetřil, J., Ossona de Mendez, P.: Sparsity. Graphs, Structures, and Algorithms. AC, vol. 28. Springer, Heidelberg (2012). https://doi.org/10.1007/978-3-642-27875-4
13. Rodríguez-Velázquez, J., Sigarreta, J.: Global offensive alliances in graphs. Electron. Notes Discrete Math. **25**, 157–164 (2006)
14. Sigarreta, J., Bermudo, S., Fernau, H.: On the complement graph and defensive k-alliances. Discret. Appl. Math. **157**(8), 1687–1695 (2009)
15. Sigarreta, J., Rodríguez, J.: On defensive alliances and line graphs. Appl. Math. Lett. **19**(12), 1345–1350 (2006)
16. Sigarreta, J., Rodríguez, J.: On the global offensive alliance number of a graph. Discret. Appl. Math. **157**(2), 219–226 (2009)

On the k-colored Rainbow Sets in Fixed Dimensions

Vahideh Keikha[1(✉)], Hamidreza Keikha[2], and Ali Mohades[3]

[1] Institute of Computer Science, The Czech Academy of Sciences, Pod Vodárenskou věží 2, 182 07 Prague, Czech Republic
keikha@cs.cas.cz

[2] Department of Computer Engineering, Amirkabir University of Technology, Tehran, Iran
keikha@aut.ac.ir

[3] Department of Mathematics and Computer Science, Amirkabir University of Technology, Tehran, Iran
mohades@aut.ac.ir

Abstract. In this paper, we introduce a variant of the minimum diameter color spanning set (MDCSS) problem. Let P be a set of n points of m colors in $\mathbb{R}^d$. For a given k, our objective is to find a set with k points of different colors that admits the minimum possible diameter. Such a set is called a k-rainbow set. This problem has applications in database queries, mostly composed by weighted points (i.e., a positive value is assigned to each point as its weight), and seeking a maximum weight k-rainbow set. We first assume the points have equal weight and design an FPT algorithm, which we generalize to the weighted version. We also solve the decision and the enumeration version of the problem by introducing a reduction to all maximal independent sets of a bipartite graph. We also introduce a 1.154-approximation algorithm for this problem and a 2.236-approximation for the enumeration version, and we perform some experimental studies on a real data-set, as well as providing several analyses of the data-set based on the outputs of our algorithm. Our exact algorithms and the approximation algorithm for the enumeration problem have a complexity being near-linear to n in $\mathbb{R}^2$.

Keywords: Minimum diameter color spanning set · FPT algorithms · Colored points

1 Introduction

Let $P = \{p_1, \ldots, p_n\}$ be a set of n points in $\mathbb{R}^d$, the *diameter* of P is defined as $diameter(P) = \max_{p_i, p_j \in P} d(p_i, p_j)$, and can be computed in $O(dn^2)$ time [26]. In $\mathbb{R}^2$, computing the diameter takes $O(n \log n)$ time [28]. Now, suppose each $p_i \in P$ is assigned a color. The objective of the minimum diameter color spanning set (MDCSS) problem is to find a subset $P^* \subseteq P$ that contains one point from each color, and P^* has the smallest possible diameter among all choices of P^*, where

D.-Z. Du et al. (Eds.): COCOA 2021, LNCS 13135, pp. 587–601, 2021.
https://doi.org/10.1007/978-3-030-92681-6_46

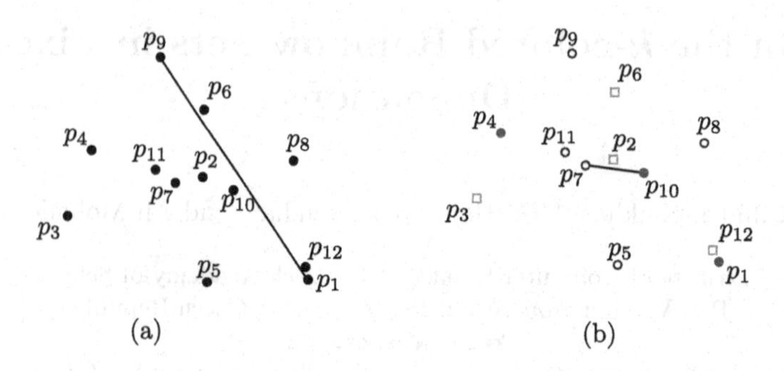

Fig. 1. (a) The diameter of a set P of points in $\mathbb{R}^2$. (b) For a set P with $m = 3$ colors, the rainbow set $P^* = \{p_2, p_7, p_{10}\}$.

the diameter is the maximum distance between any two points in P^*. P^* is called the *color spanning* set or the *rainbow* set; see Fig. 1.

The MDCSS problem can be considered as a database query; consider a spatial database where each tuple is associated with a *keyword* or, equivalently, a *color code* in our setting. The *m-closest* keywords query is the problem of finding the m tuples that match all the keywords chosen by the customer [29]. In our problem, the closeness is measured by the diameter. Now suppose a customer aims at finding some closest keywords of the desired number and his/her maximum willingness. The motivation behind this study is efficiently answering such queries. We note that such queries are introduced in the database literature as *reverse top-k queries* [13], without theoretic analysis, but have recently received considerable attention from the database community.

Related Work. Fleischer and Xu [12] showed that for a large number of colors, the MDCSS problem is NP-hard even in two dimensions but is solvable in polynomial-time for a small number of colors. The fixed-parameter tractability of MDCSS is posed as an open problem in [12], in which they assume that the dimension d is fixed. Recently, Pruente [27] answered this question by proving that MDCSS is W[1]-hard by using a complicated reduction from multi-colored clique graph problems [11], where the dimension d is not fixed. Also, the author shows that the problem does not admit an FPTAS in arbitrarily high dimensional spaces. In the same paper, some algorithms with quadratic dependencies to n are also supporting the result.

Kazemi et al. presented a PTAS in high dimensional space for the MDCSS problem and proved that assuming the Exponential Time Hypothesis (ETH), there is no $(1+\epsilon)$-approximation algorithm with running time $2^{o(\epsilon^{(1-d)/2})}\text{poly}(n)$ to solve the MDCSS problem [17].

Instead of considering a discrete set for the possible locations of a color code, a continuous region of possible locations may determine a color code. Finding a point in each region such that the chosen set admits the smallest diameter is also extensively studied in this model. This formulation is introduced and extensively studied by Löffler and van Kreveld [22] for disks and squares,

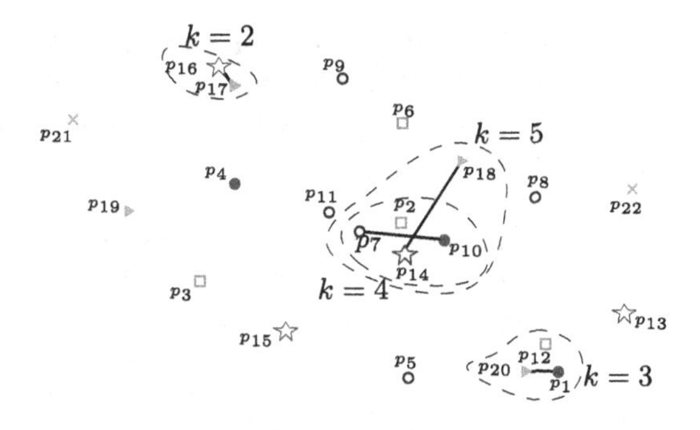

Fig. 2. Problem definition and optimal solutions, with k-rainbow sets for $k = 2, 3, 4, 5$.

and several improvements have been made recently to the complexity of their algorithms by Keikha et al. [19].

Regardless of whether the associated set of each color code is a continuous or a discrete set, the maximum diameter color spanning set problem is to locate a set of points, where the diameter has the largest possible size. This problem usually takes polynomial time as it is involved with the points in convex position. We refer the interested reader to [2,7,15,29] for other related studies on MDCSS.

Our problem is closely related to outlier detection problems, except that their input is a set of monochromatic points: for a given $k < n$, exclude $n - k$ points (referred to as *outliers*) from P, such that the remaining points have the smallest possible diameter. In $\mathbb{R}^2$, the best-known algorithm for this problem, developed by Eppstein et al., runs in $O(n \log n + k^2 n \log^2 k)$ time [9]. There also exists a lower bound $\Omega(n \log n)$ for this problem even for one outlier since the diameter picks the outlier as a vertex [3]. This implies that any fixed-parameter algorithm for computing a k-rainbow set is no better than $\Omega(n \log n)$ in $\mathbb{R}^2$.

We finally note that to the best of our knowledge, no study has been conducted on our problem or the weighted version of the MDCSS problem.

Contribution. In the following, we formally define our problems: Let $P = \{p_1, \ldots, p_n\}$ be a set of n points of m colors in $\mathbb{R}^d$, let t be the maximum frequency of any color in P, and let $1 < k < m$ be a positive integer.

Definition 1. ***Minimum Diameter k-Colored Spanning Set** (MDkCSS). The objective of the MDkCSS problem is to find a subset $P^* \subseteq P$ of size k of distinct colors, such that P^* has the smallest possible diameter among all possible choices. Formally diameter(P^*) $= \min_{\mathcal{P} \in \mathcal{D}(P)}$ diameter($\mathcal{P}$), where $\mathcal{D}(P)$ denote the collection of all k-subsets[1] of P of distinct colors.*

We call P^* a k-rainbow set of P; see Fig. 2 for an illustration. The main application of this problem is in the case where the points have a predefined weight assigned, and the optimal k-rainbow set has the maximum total weight.

[1] i.e., subsets of size k.

Definition 2. ***Maximum Weight Minimum Diameter k-Colored Spanning Set (MWMDkCSS).*** *We define a* maximum weight k-rainbow set P^* *as a k-subset of distinct colors that minimizes* $\frac{diameter(P^*)}{weight(P^*)}$, *where* $weight(P^*)$ *is the total sum of the weights of the points in* P^*.

Results. In this paper, we first focus on the case where all the points have the same weight and then we discuss to what extent our results can be generalized to the weighted version under some restrictions. In particular:

- For the first time, we introduce a relation between the MDCSS problem and higher-order Voronoi diagrams. We first provide a fixed-parameter tractable (FPT) algorithm that has near-linear dependency on n in $\mathbb{R}^2$ (Theorem 1), which is helpful to improve the existing quadratic FPT algorithm (for small k and t) for the MDCSS problem [27].
- We show that MDkCSS is fixed-parameter tractable in $\mathbb{R}^d$ for any fixed d (Sect. 3.1). We then show our FPT algorithm gives an approximation for the MWMDkCSS problem (Sect. 3.2).
- We have implemented our exact algorithm on a real data-set to consider the efficiency of our technique in practice, and we give several analyses on the studied data-set (Sect. 4).
- We then discuss the decision and the enumeration version of the MDkCSS problem for a given value q, and introduce an $O(n(tk)^2((tk)^{2.5} + \alpha))$ time algorithm, where α is the maximum number of the k-rainbow sets of size at most q. We hope these problems are of independent interest in data mining and database inquiries. To solve these problems, we introduce a reduction to all maximal independent sets of a bipartite graph (Sect. 5).
- We introduce a 2.236-approximation with running time $O(mn \log mn)$ for the enumeration version of MDkCSS, and a 1.154-approximation for the MDkCSS problem in $\mathbb{R}^2$ with running time $O(m^3 n)$ (Sect. 6).

Our FPT algorithm is efficient when the parameters t and k are small, which is the common assumption of any FPT algorithm. Note that parametrizing a problem by the number of colors is common in computational geometry. We also remark that in the MDCSS problem if the number of the existing colors in P is a small k (possibly constant), we still do not have any exact algorithm with a running time better than $\binom{n}{k}$. In $\mathbb{R}^2$, our FPT algorithm is near-linear to n.

2 Preliminaries

Maximum Independent Set (MIS). A maximum independent set of a graph $G = (V, E)$ is a subset $X \subseteq V$ with maximum size, in which there is no edge $e \in E$ between any $a, b \in X$. This problem is NP-hard, fixed-parameter intractable, and also hard to approximate [8]. The best algorithm for computing all maximum independent sets of a bipartite graph takes $O(s^{2.5} + \alpha)$ time [16], where s and α are the number of vertices and the total size of the output, respectively.

k-Order Voronoi Diagram. The *Voronoi diagram* of *order* k of P is the partitioning of the plane into a set of Voronoi cells, such that each Voronoi cell c is associated with a set $X \subseteq P$ of k points, and for each point p in the cell c, the k nearest neighbors of p are exactly the elements of X. We denote this diagram by V_k. Such diagrams can be computed in $O(k^2 n + n \log n)$ time and have at most $O(nk)$ cells [20].

Fixed-Parameter Tractable (FPT). In fixed-parameter tractability, we provide some algorithms which no longer are exponential on the input size but on some other parameter related to the problem. These parameters are called the fixed parameter of the problem. Formally, for a given problem Υ, we characterize the input size, n, and some parameter k, and say Υ is fixed-parameter tractable if Υ can be solved by an algorithm that runs in $O(\mathcal{F}(k) \cdot n^c)$ time, where $\mathcal{F}$ is a computable function depending on k, and c is any constant independent of k. Also, it is already known that parameterized complexity can be extended to achieve approximation algorithms for hard problems [24]. We use the same idea to achieve an FPT approximation algorithm for MWMDkCSS in Sect. 3.2.

Minimum Color Spanning Circle. For a set of n colored points of m colors, the smallest color spanning circle is a circle of the smallest radius that is covering m distinct colors [1]. In $\mathbb{R}^2$, the smallest color spanning circle of m colors can be computed in $O(nm \log n)$ time by computing the upper envelope of some Voronoi surfaces [1,14]. This problem becomes NP-hard in $\mathbb{R}^d$, where d is in the input, but admits a $(1+\epsilon)$-approximation in $O(dn^{\lceil 1/\epsilon \rceil + 1})$ time [18].

We first note that MDCSS problem is para-NP-hard[2] for the parameter t since the proof in [12] shows NP-hardness for t bounded by three. It can easily be extended to also show NP-hardness if at most 5 colored points are co-located (if we do a reduction by MAX-E3SAT(5) [10]). Hence, the problem may get easier if the number of colors is large, i.e., more than $\frac{n}{3}$.

3 MDkCSS is in FPT in Any Fixed Dimension

In the following, we assume that the points of P are in general position, that means no four points are co-circular. Recall that a k-rainbow set P^* is a set of points of k distinct colors, where P^* has the smallest possible diameter among all choices. In [12] it is posed as an open question which value of k is the threshold between easy and hard. We partially answer that question, as we do not need to cover all, but only k colors that their instances realize the smallest possible diameter. Our algorithm has a near-linear dependency on the number of points, where its hardness depends on k (and t, but we discussed above that t is not a parameter to determine the hardness). Consequently, we answer the posed

[2] A problem is para-NP-hard if it is NP-hard already for a constant value of the parameter.

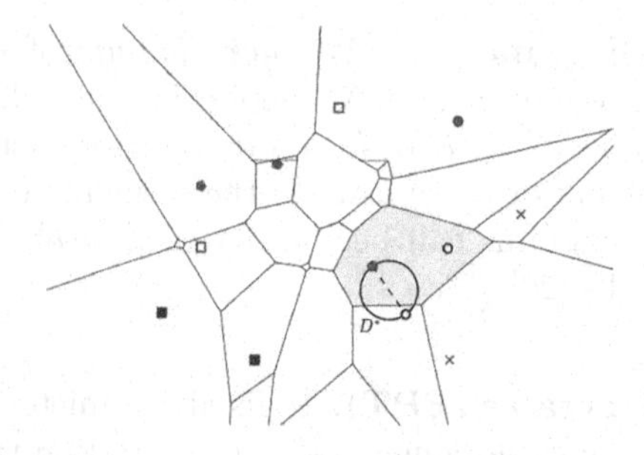

Fig. 3. Illustration of Lemma 1, the case where D^* has two points on its boundary. On a set P of colored points with $t = 2$, the optimal solution with $k = 2$ is associated with a cell c (shown in gray) of $V_{t(k-1)+1}$, and uses a pair of red and blue points (connected by a dashed line segment). Observe that D^* cannot contain more than 3 points of P, otherwise, there must be two points of different colors strictly within D^*, such that they realize a smaller diameter than the diameter of D^*. (Color figure online)

question in [12] partially as follows: for any constant number of colors which we need to span, the MDCSS problem in $\mathbb{R}^2$ can be answered in near-linear time.

In the following, we will show that any set of k colored points of smallest diameter is a subset of the points which are associated to a Voronoi cell of a Voronoi diagram (of P) of order $t(k-1)+1$, or $3t(k-1)+1$.

Lemma 1. *Let P be a set of n colored points, and let P^* be a k-rainbow set of P. Then P^* is a subset of the points which are corresponding to a Voronoi cell of a Voronoi diagram either of order $t(k-1)+1$, or $3t(k-1)+1$.*

Proof. Let $c(P)$ denote a subset of points of P that are associated with only one cell c of a Voronoi diagram $V_{t(k-1)+1}$, or $V_{3t(k-1)+1}$. Recall that for each Voronoi cell c, there exists a disk D having its center within c, where D contains no other point of $P - c(P)$. The set P^* also realizes a disk D^* such that either two or three points of P^* are located on its boundary.

Suppose by contradiction that the lemma is false, and P^* of k colors is not associated with one cell of a Voronoi diagram of order $t(k-1)+1$, or $3t(k-1)+1$.

By definition, in $V_{t(k-1)+1}$, the points of each cell of the diagram have the same $t(k-1)+1$ nearest neighbors. Observe that in the case where there are two points on the boundary of D^*, D^* cannot contain more than $t(k-1)+1$ points. If not, there always exist at least another set P' of k points from k distinct colors, which they all are entirely located within D^*, and the diameter of P' is strictly smaller than the diameter of D^* (i.e., P^*). This gives a contradiction. It follows that D^* cannot contain more than $t(k-1)+1$ points and P^* is contained in some Voronoi cell of a Voronoi diagram of order $t(k-1)+1$. See Fig. 3 for an illustration.

In the case where D^* has three points on its boundary, we partition D^* into three sectors by connecting the center of D^* to the points of P^* on its boundary.

Algorithm 1. Exact Algorithm

Input: $P = \{P_1, \ldots, P_n\}$, and $k > 0$
Output: k-rainbow set of P
1: $d^* = \infty$
2: Compute $V_{3t(k-1)+1}$ (or $V_{t(k-1)+1}$ if $V_{3t(k-1)+1}$ does not exist)
3: **for** each cell c in $V_{3t(k-1)+1}$ **do**
4: P_c= the associated points of P to c
5: **for** any k-subset P_c of c **do**
6: **if** there are k distinct colors in P_c **then**
7: d_c=the diameter of P_c
8: **end if**
9: **if** $d^* > d_c$ **then**
10: $d^* = d_c, P_c^* = P_c$
11: **end if**
12: **end for**
13: **end for**
14: **return** d^* and P_c^*

Then each sector cannot contain more than $t(k-1)$ points since otherwise there would be k points from distinct colors in that sector so that the determined diameter by those points is strictly smaller than the diameter of P^*. Hence, D^* is contained in the associated points of a Voronoi cell of $V_{3t(k-1)+1}$. □

3.1 Algorithm

Observe that we only need to consider $V_{3t(k-1)+1}$ as the associated points in its cells strictly cover all possibilities in $V_{t(k-1)+1}$. From Lemma 1, the smallest diameter among each subset of k points of distinct colors that is associated to a Voronoi cell of $V_{3t(k-1)+1}$ determines an optimal solution. We design our algorithm based on this fact.

In the algorithm, we first make the Voronoi diagram of order $3t(k-1)+1$, of all the n points of P, without considering their colors in the construction. In each step of the algorithm, we consider the associated points of each cell of $V_{3t(k-1)+1}$ independently. Let d^* denote the diameter of the k-rainbow set P^*, and let P_c denote the associated points of a cell c. We use a brute force idea on P_c to find a subset $P_c^* \subseteq P_c$ of k distinct colors with the smallest possible diameter, and remember the P_c^* with the smallest d_c^* among all the cells of $V_{3t(k-1)+1}$.

In Lemma 1 we observed that each set P_c has a reasonable size with only a linear dependency to k and t, which means our algorithm has exponential dependence only in k and t. Since the complexity of the number of the cells of a Voronoi diagram of order tk is $O(ntk)$, our method gives an FPT algorithm with k and t as the parameters; see Algorithm 1.

Running Time Analysis. We will now elaborate on the complexity of the algorithm for a cell of $V_{3t(k-1)+1}$. To analyse the running time of considering all k-subsets of $3tk-3t+1$ points, we use the Stirling's formula: $\binom{3tk-3t+1}{k} = 2^{\log(3tk-3t+1)!-\log k!-\log(3tk-3t-k+1)!}$. Then we have $\log(3tk-3t+1)! - \log k! - \log(3tk-3t-k+1)! = 3tk\ln(3tk-3t+1) - (3tk-3t+1) + O(\ln(3tk-3t+1))$ $-k\ln k + k - O(\ln k) - (3tk-3t+1)\ln(3tk-3t-k+1) + (3tk-3t-k+1) - O(\ln(3tk-3t-k+1)) \in O(tk)$.

Hence, considering all possible k-rainbow sets of one cell of $V_{3t(k-1)+1}$ takes $O(2^{O(tk)})$ time.

For each cell c of $V_{3t(k-1)+1}$, we can find a solution to the MDkCSS problem by finding a k-rainbow set with the smallest possible diameter among the corresponding points of c in $O(k\log k 2^{O(tk)})$ time, in which, in $O(k)$ time we determine whether the selected set contains k distinct colors, and $O(k\log k)$ time is required to find the diameter of this k-subset.

To generalize our FPT to a higher dimension d, we first need to construct the k-order Voronoi diagram in that dimension. Recall that a k-order Voronoi diagram in $\mathbb{R}^d$ can be constructed in $O(n^{\lceil d/2\rceil}k^{\lfloor d/2\rfloor+1})$ time [6].

Theorem 1. *Let P be a set of n colored points in $\mathbb{R}^d$.* MDkCSS *can be solved in $O(n(2^{O(tk)}+\log n)+n^{\lceil d/2\rceil}k^{\lfloor d/2\rfloor+1})$ time.*

Proof. A Voronoi diagram of order $O(tk)$ can be computed in $O((tk)^2n+n\log n)$ time and has at most $O(ntk)$ cells. A k-order Voronoi diagram in the dimension d can be constructed in $O(n^{\lceil d/2\rceil}k^{\lfloor d/2\rfloor+1})$ time, so by repeating the algorithm of Sect. 3.1 for all the cells of the Voronoi diagram $V_{3t(k-1)+1}$ in $\mathbb{R}^d$, the algorithm takes $O(n2^{O(tk)}+n\log n)$ time. Hence, the problem can be solved in $O(n(2^{O(tk)}+\log n)+n^{\lceil d/2\rceil}k^{\lfloor d/2\rfloor+1})$ time. □

Corollary 1. MDkCSS *is in FPT in $\mathbb{R}^d$ for any fixed d, with k and t as the parameters.*

3.2 Maximum Weight k-rainbow Set

For any point $p_i \in P$, let w_i denote the weight of p_i. W.l.o.g, we assume $w_i > 0$, $i = 1,\ldots,n$. It is easy to observe that the problem at which a k-rainbow set P^* (for general values of k) minimizes $\frac{diameter(P^*)}{weight(P^*)}$, where $weight(P^*) = \sum_{p_i\in P^*} w_i$ is NP-hard with the same reduction in [12] for the MDCSS, by using an extra assumption of assigning the same weight to all the points in P. We discuss that Algorithm 1 is applicable on particular cases of this problem, at which the ratio of the weights of any two points in P is at most ω. This is a reasonable assumption since in any environment, the input data are usually relevant and are not that much different in the sense of measurement precisions. Also, we can measure the ratio of the weights in polynomial time. Then Algorithm 1 gives a ω-approximation for the MWMDkCSS problem, as in the worst case the two points of large weight that are far apart from each other, and have to be in an optimal solution, will not land in the same cell. So we may not consider solutions containing both these points. But the ratio of the weight of a point in the reported and the optimal solution is at most within a factor ω. Assuming all the elements

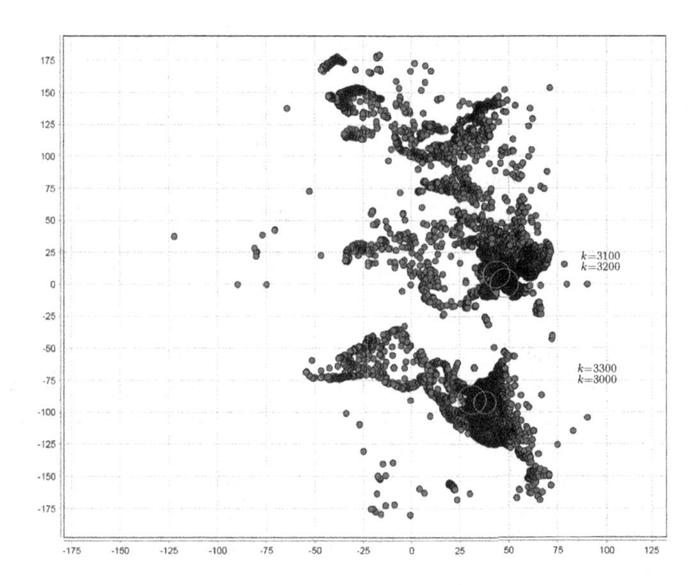

Fig. 4. Illustration of the output of Algorithm 1 on the data-set [5,21], and the ranges at which the optimal solutions for $k = 3000, 3100, 3200, 3300$ appear.

of P^* lands at different cells and summing up the weights of such points gives the approximation factor at most ω. Note that an arbitrary k-rainbow set has the same approximation ratio only for the total sum of the weights of the points. But such a set cannot guarantee to have the minimum possible diameter among all choices for approximating $\frac{diameter(P^*)}{weight(P^*)}$ within a factor ω.

Theorem 2. *If the ratio of the weights of any two points in P is at most ω, Algorithm 1 gives an FPT ω-approximation for* MWMDkCSS *in $\mathbb{R}^d$ for any fixed d, with k and t as the parameters.*

4 Experimental Studies

We discussed an application for our problem in the Introduction. In this section, we discuss another application along with our experimental tests to evaluate the performance of Algorithm 1 in practice. We do our computational tests on a real data-set in $\mathbb{R}^2$.

Our data-set characterizes the locations and times of check-ins of the Brightkite social network, which has a reasonable size and several users with different check-ins each, so that we can model each user as a color code. This network has 58,228 users and 4,491,143 check-ins of these users ranging in the period of April 2008 to October 2010. This data-set is contained in the SNAP network [5,21].

We assign a color to each user, and of course we need the users with at least one check-in (to denote the frequency at least one for each color). This number equals 51,685 in this data-set. The total number of colors (m) is also

Table 1. Experimental results of Algorithm 1 on the Brightkite data-set [5,21].

Input size	k	# cells of $V_{t(k-1)+1}$	Alg 1 Time (s)	BF Time (s)
4,491,143	3000	10272	2.3867	0.1191e+2
	3100	13070	2.8967	0.1259e+2
	3200	14221	3.9683	0.1436e+2
	3300	19744	4.1662	0.1696e+2

51,685, and n equals 4,491,143, that is the total number of the check-ins. Each user had at most 325,821 check-ins which means t =325,821. For a given k, our objective is to find k customers whose target check-ins are as close as possible. One may use this information to locate a facility for at least k customers in the neighbourhood of their check-in places. Our experiments have shown that for $k \leq 2876$, the solution to the MDkCSS was zero in this data-set, which means this number of customers have at least one common check-in station. In our experiments, we set $k = 3000, 3100, 3200, 3300$; see Fig. 4.

We have implemented our algorithm in C++ with Visual Studio 2013. The algorithm is performed on a Core (TM) i9CPU and 8 GB RAM computer with Windows 10 operating system. In some of the computations of the Voronoi diagrams, we have used CGAL-5.1. The reported running time in Table 1 is the elapsed time of searching for a solution on $V_{t(k-1)+1}$, since the condition $3t(k-1)+1 \leq n$ did not hold, and there is no solution on $V_{3t(k-1)+1}$.

In each test, we have verified the output of our algorithm with the brute force algorithm which is trying all k-subsets, as this problem is not considered so far, and the brute force is the only existing current algorithm. We have reported the running times in Table 1. The last column contains the running time of the brute force algorithm which is comparable to the running time of our algorithm in the previous column.

We observe that our algorithm has a reasonable performance in the reported experimental studies in this paper. In our experiments, computing a Voronoi diagram of a high order was the time-consuming part, and was taking at least 67.46% of the reported elapsed times. Based on this, we conclude the other computations were relatively quick; that is because the dependency of the algorithm to the number of points is near-linear in $\mathbb{R}^2$. The results of the implementation are summarized in Table 1.

5 Enumerating All MDkCSS of Diameter at Most q

In this section, we study the following problems: given a set P of colored points and a positive value q, determine whether there is any k-rainbow set in P of diameter at most q, and report all k-rainbow sets of P of diameter at most q.

Let c be any cell of $V_{3t(k-1)+1}$ that has at most $O(tk)$ points, and let $X \subseteq P$ denote the associated points of P to c. For any pair $p_i, p_j \in X$ of distinct colors, let $z = d(p_i, p_j)$ denote their Euclidean distance. Our first objective is to

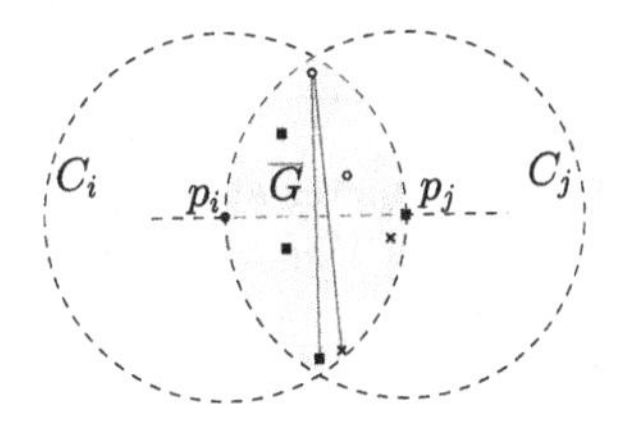

Fig. 5. Illustration of the graph $\overline{G}$ in Lemma 2.

determine whether there is any set of k points of distinct colors in X, where the pairwise distances between the points are at most $z \leq q$. Consider two circles C_i and C_j of radius z, one is centered at p_i and the other at p_j. Let X' denote the set of points in $C_i \cap C_j \cap P$. Construct a graph G on X' by connecting any pair of points with a distance at most z, and let $\overline{G}$ denote the complement of G. Observe that the vertices of X' which are lying on exactly one side of the line through p_ip_j are at a distance less than z. Consequently, in $\overline{G}$, the connected pair of vertices lie on opposite sides of the line through p_ip_j.

Lemma 2. *$\overline{G}$ is a bipartite graph.*

Proof. The vertices of the graph which lie on only one side of the line through p_ip_j have a smaller distance of z. Consequently, the vertices which are already connected in $\overline{G}$, have a further distance than z, and lie at different sides of the line through p_ip_j; see Fig. 5. Hence, the vertices at each side of the line through p_ip_j in $\overline{G}$ determine a part in a bipartite graph. □

Observe that the points forming a clique in G are points such that each pair of points has a distance smaller than z, and thus, such points form an independent set in $\overline{G}$.

Lemma 3. *Any k-rainbow set of diameter at most z is a subset of at least one maximum independent set in $\overline{G}$.*

Proof. Suppose, by contradiction, there is a k-rainbow set S of diameter at most z that is not a subset of any maximum independent set in $\overline{G}$. Every independent set (including the ones in $\overline{G}$) has this property that there is no edge between any pair of the vertices. If S is not a subset of any of the independent sets (including the maximum ones) in $\overline{G}$, there must be an edge between at least one pair of vertices in S. But this means the distance between those vertices is strictly larger than z; contradiction. □

Hence, maximum independent set enumeration algorithm can be used for our problem but only reports the ones having our cardinality and color constraint. We check whether there is any maximum independent set X^* of size at least k in $\overline{G}$, where at least k vertices in X^* has distinct colors. To treat this, for each possible maximum independent set we check whether there is any set of k distinct colors among the reported vertices or not. Using the presented algorithm

Algorithm 2. Enumeration Algorithm

Input: $P = \{P_1, \ldots, P_n\}$, and $k, q > 0$
Output: all k-rainbow set of P of diameter at most q

```
 1: Compute V_{3t(k-1)+1} (or V_{t(k-1)+1} if V_{3t(k-1)+1} does not exist)
 2: for each cell c in V_{3t(k-1)+1} do
 3:     X=the associated points of P to c
 4:     for each pair p_i, p_j ∈ X do
 5:         if they have distinct colors and z = d(p_i, p_j) ≤ q then
 6:             G = (X', E) = ∅
 7:             Construct C_i, C_j
 8:             X' = C_i ∩ C_j ∩ P
 9:         end if
10:         for any pair x, y ∈ X' do
11:             if d(x, y) ≤ z then
12:                 add xy to E
13:             end if
14:         end for
15:         \overline{G} = the complement of G
16:         report any MIS of \overline{G} which has k distinct colors
17:     end for
18: end for
```

in [16] and considering the freedom of p_i and p_j in $O(ntk)$ cells of $V_{3t(k-1)+1}$, the enumeration algorithm takes $O(ntk) \times O(tk)^2 \times O((tk)^{2.5} + \alpha)$ time, where α is the maximum number of the k-subsets of diameter at most q. This procedure is outlined in Algorithm 2. We note that one may use this algorithm combined with a binary search to compute the optimal q among $O(n^2)$ possible candidates. But the asymptotic running time would be worse than what we discussed in Sect. 3.1.

Theorem 3. *The decision or the enumeration version of* MDkCSS *can be solved in $O(n(tk)^2((tk)^{2.5} + \alpha))$ time, where α is the maximum number of k-subsets of diameter at most q.*

6 Approximation Algorithms

In this section, we discuss several approximation algorithms, mostly by geometric reductions to other problems. We first reduce MDkCSS to a well-known problem in *trajectory analysis*, the *discrete popular places* problem. Given is a set Π of polygonal paths with a total of n vertices, that is modelling a set of moving points (so-called *entity*) belonging to m distinct entities in the plane, an integer threshold $k > 0$ and a real value $r > 0$. A *popular place* is a square of side length r, that is visited by at least k distinct entities [4]. This problem can be solved in $O(mn \log mn)$ time and $O(mn)$ space [4]. In our setting, we assign the points of the same color to a single entity. The path between them is arbitrary. Hence, a popular place with a maximum number of entities gives a $\sqrt{2}$-approximation for the MDCSS problem. Also, any algorithm for squares assuming a threshold k as a popular place gives also an approximation for the MDkCSS problem. Reporting all the popular places for rectangles of threshold k takes $O(mn \log mn)$ time and $O(mn)$ space [4]. Reporting all popular places with at least k entities, where the popular places modelled by a rectangle of size 1×2, reports all k-rainbow sets of diameter at most γq for a given $q > 0$, where $\gamma = \sqrt{5} \approx 2.236$.

Theorem 4. *For a given $q > 0$, all the k-rainbow sets of diameter of size at most $\sqrt{5}q$ can be listed in $O(mn \log mn)$ time and $O(mn)$ space.*

A 1.154-Approximation for MDkCSS. We discuss another simple efficient approximation algorithm. We start stating our result with the following lemma.

Lemma 4. *For any set X of points, the diameter of X is longer than $\sqrt{3}$ times the radius of the smallest enclosing circle (SEC) of X.*

Proof. Consider the configuration at which three points on the boundary of the SEC form an equilateral triangle, and the side of the triangle determines the diameter. If one translates any pair of these points on the boundary of the SEC, to get closer, the size of the diameter would be increased between at least one pair. The lemma follows. □

Let r_X and d_X denote the radius of the SEC and the diameter of X, respectively. For a set P of points, let X be the set realizing the smallest color spanning circle with k colors, and let P^* denote the set of points realizing the k-rainbow set of smallest diameter. Using the fact that the radius r_X is smaller than the radius of the color spanning circle of P^*, we have $d_{P^*} \leq d_X \leq 2r_{P^*} \leq 2/3\sqrt{3}(\sqrt{3}r_{P^*}) \leq 2/3\sqrt{3}d_{P^*}$. So, the diameter of X approximates the optimal k-rainbow set within a factor $2/3\sqrt{3} \approx 1.154$. An obvious $O(m^3n)$ time algorithm for computing the smallest color spanning circle of at least k colors considers any pair or triple of points of distinct colors that define a circle. We then have the following result.

Theorem 5. *Let P be a set of n colored points of m colors. In $\mathbb{R}^2$, a 1.154-approximation for the MDkCSS problem can be computed in $O(m^3n)$ time.*

7 Discussions and Open Problems

In this paper, we introduced an easy proof that MDCSS problem is in FPT in $\mathbb{R}^d$ for any fixed d, and we discussed several new variants of this problem, FPT, exact and approximation algorithms along a practical application.

One open question concerns designing efficient algorithms for the general case of the weighted points, and also for the enumeration problem on particular sets of points, in which the bipartite graph G has a bounded tree-width and admits a polynomial time algorithm for computing all MIS's. The tree-width $O(kt)$ for G is obvious. Another direction is to find the attributes on the point sets in which the *maximum colorful independent set* (i.e., an independent set of maximum number of colors) on the bipartite graph G admits a polynomial time algorithm. This problem was recently shown to be NP-hard, but admits polynomial time algorithms on trees and cluster graphs [23]. Another open question is the existence of the FPT algorithms for other parameters of a point set in the MDCSS problem, such as assuming a specific extent measure for the points of any color code.

The possibility of improving the running time of our algorithms also remained open. One possible improvement to our results concerns approximating the MDkCSS in fixed dimensions using LP-formulation. According to Theorem 1.2 in [25], computing a circle of smallest radius that intersects n points can be reformulated to satisfy only k of n constraints, in $O(nk^d)$ time, where d equals the geometric dimension of the original problem, and this would be performed by finding the optimal solution to $O(k^d)$ independent LP-type problems. When we are generating an independent LP-type problem from the original problem, we can rewrite the constraint that counts the number of points to count the number of points of distinct colors; let $x_i = 1$ if the color c_i appears in the solution space, and $x_i = 0$, otherwise. Then we need to satisfy the constraint $\sum_{i=1}^{m} x_i = k$ in any of the independent solution sub-spaces. Thus, we can apply the existing algorithms for computing the smallest color spanning balls in $\mathbb{R}^2$ [1,14] and in $\mathbb{R}^d$ [18], that intersect k colors in each of the solution spaces of the independent LP-type problems. This may slightly improve the approximation ratio and the running time we discussed in Theorem 5.

Acknowledgment. V. Keikha was supported by the Czech Science Foundation, grant number GJ19-06792Y, and with institutional support RVO: 67985807.

References

1. Abellanas, M., et al.: Smallest color-spanning objects. In: auf der Heide, F.M. (ed.) ESA 2001. LNCS, vol. 2161, pp. 278–289. Springer, Heidelberg (2001). https://doi.org/10.1007/3-540-44676-1_23
2. Agarwal, P.K., Har-Peled, S., Varadarajan, K.R.: Approximating extent measures of points. J. ACM **51**(4), 606–635 (2004)
3. Atanassov, R., et al.: Algorithms for optimal outlier removal. J. Discrete Algorithms **7**(2), 239–248 (2009)
4. Benkert, M., Djordjevic, B., Gudmundsson, J., Wolle, T.: Finding popular places. Int. J. Comput. Geom. Appl. **20**(01), 19–42 (2010)
5. Cho, E., Myers, S.A., Leskovec, J.: Friendship and mobility: user movement in location-based social networks. In: Proceedings of the 17th ACM SIGKDD International Conference on Knowledge Discovery and Data Mining, pp. 1082–1090 (2011)
6. Clarkson, K.L., Shor, P.W.: Applications of random sampling in computational geometry, II. Discrete Comput. Geom. **4**(5), 387–421 (1989). https://doi.org/10.1007/BF02187740
7. Consuegra, M.E., Narasimhan, G.: Geometric avatar problems. In: Leibniz International Proceedings in Informatics. LIPIcs, vol. 24. Schloss Dagstuhl-Leibniz-Zentrum fuer Informatik (2013)

8. Cook, S.A.: The complexity of theorem-proving procedures. In: Proceedings of the 3th Annual ACM Symposium on Theory of Computing, pp. 151–158 (1971)
9. Eppstein, D., Erickson, J.: Iterated nearest neighbors and finding minimal polytopes. Discrete Comput. Geom. **11**(3), 321–350 (1994). https://doi.org/10.1007/BF02574012
10. Feige, U.: A threshold of ln n for approximating set cover. J. ACM **45**(4), 634–652 (1998)
11. Fellows, M.R., Hermelin, D., Rosamond, F.A., Vialette, S.: On the parameterized complexity of multiple-interval graph problems. Theoret. Comput. Sci. **410**(1), 53–61 (2009)
12. Fleischer, R., Xu, X.: Computing minimum diameter color-spanning sets is hard. Inf. Process. Lett. **111**(21), 1054–1056 (2011)
13. He, Z., Lo, E.: Answering why-not questions on top-k queries. IEEE Trans. Knowl. Data Eng. **26**(6), 1300–1315 (2012)
14. Huttenlocher, D.P., Kedem, K., Sharir, M.: The upper envelope of Voronoi surfaces and its applications. Discrete Comput. Geom. **9**(3), 267–291 (1993)
15. Ju, W., Fan, C., Luo, J., Zhu, B., Daescu, O.: On some geometric problems of color-spanning sets. J. Combinat. Optim. **26**(2), 266–283 (2013)
16. Kashiwabara, T., Masuda, S., Nakajima, K., Fujisawa, T.: Generation of maximum independent sets of a bipartite graph and maximum cliques of a circular-arc graph. J. Algorithms **13**(1), 161–174 (1992)
17. Kazemi, M.R., Mohades, A., Khanteimouri, P.: Approximation algorithms for color spanning diameter. Inf. Process. Lett. **135**, 53–56 (2018)
18. Kazemi, M.R., Mohades, A., Khanteimouri, P.: On approximability of minimum color-spanning ball in high dimensions. Discret. Appl. Math. **279**, 188–191 (2020)
19. Keikha, V., Löffler, M., Mohades, A.: A fully polynomial time approximation scheme for the smallest diameter of imprecise points. Theoret. Comput. Sci. **814**, 259–270 (2020)
20. Lee, D.-T.: On k-nearest neighbor Voronoi diagrams in the plane. IEEE Trans. Comput. **100**(6), 478–487 (1982)
21. Leskovec, J., Krevl, A.: Snap datasets: Stanford large network dataset collection (2014)
22. Löffler, M., van Kreveld, M.: Largest bounding box, smallest diameter, and related problems on imprecise points. Comput. Geom. **43**(4), 419–433 (2010)
23. Manoussakis, Y., Pham, H.P.: Maximum colorful independent sets in vertex-colored graphs. Electron. Notes Discrete Math. **68**, 251–256 (2018)
24. Marx, D.: Parameterized complexity and approximation algorithms. Comput. J. **51**(1), 60–78 (2008)
25. Matoušek, J.: On geometric optimization with few violated constraints. Discrete Comput. Geom. **14**(4), 365–384 (1995). https://doi.org/10.1007/BF02570713
26. Preparata, F.P., Shamos, M.I.: Convex hulls: basic algorithms. In: Computational Geometry. Texts and Monographs in Computer Science, pp. 95–149. Springer, New York (1985). https://doi.org/10.1007/978-1-4612-1098-6_3
27. Pruente, J.: Minimum diameter color-spanning sets revisited. Discrete Optim. **34**, 100550 (2019)
28. Toussaint, G.T.: Solving geometric problems with the rotating calipers. In: Proceedings of the IEEE MELECON, vol. 83, p. A10 (1983)
29. Zhang, D., Chee, Y.M., Mondal, A., Tung, A.K., Kitsuregawa, M.: Keyword search in spatial databases: towards searching by document. In: Proceedings of the 25th IEEE International Conference on Data Engineering, pp. 688–699 (2009)

Cycle-Connected Mixed Graphs and Related Problems

Junran Lichen(✉)

Institute of Applied Mathematics, Academy of Mathematics and Systems Science, No. 55, Zhongguancun East Road, Beijing 100190, People's Republic of China

Abstract. In this paper, motivated by vertex connectivity of digraphs or graphs, we address the cycle-connected mixed graph (CCMG) problem. Specifically, given a mixed graph $G = (V, A \cup E)$, for every pair x, y of distinct vertices in G, we are asked to find a mixed cycle C in G to contain such two vertices x and y, where C traverses its arc (u, v) along the direction from u to v and its edge uv along one direction either from u to v or from v to u. Similarly, we consider the circuit-connected (or weakly cycle-connected) mixed graph (WCCMG) problem, substituting a mixed circuit for a mixed cycle in the CCMG problem. Whenever the CCMG problem is specialized to either digraphs or graphs, we refer to this version as either the cycle-connected digraph (CCD) problem or the cycle-connected graph (CCG) problem, and we refer to this mixed graph as either a cycle-connected digraph or a cycle-connected graph in related versions. Furthermore, given a graph $G = (V, E)$, the cycle-connectivity, denoted by $\kappa_c(G)$, of G is the smallest number of vertices in G whose deletion causes the reduced subgraph either not to be a cycle-connected graph or to become an isolated vertex; In addition, for every pair x, y of distinct vertices in G, we denote by $\kappa_{sc}(x, y)$ the maximum number of internally vertex-disjoint cycles in G containing x and y, then the strong cycle-connectivity, denoted by $\kappa_{sc}(G)$, of G is the smallest of these numbers $\kappa_{sc}(x, y)$ among all pairs $\{x, y\}$ of distinct vertices in G, *i.e.*, $\kappa_{sc}(G) = \min\{\kappa_{sc}(x, y) \mid x, y \in V\}$.

We obtain the three main results. (1) Using the directed 2-linkage problem which is *NP*-complete, we prove that the CCD problem is *NP*-complete, implying that the CCMG problem still remains *NP*-complete, and however, we can design an exact algorithm in polynomial time to solve the CCG problem; (2) We present a simply exact algorithm in polynomial time to solve the WCCMG problem; (3) Given a graph $G = (V, E)$, we provide twin exact algorithms in polynomial time to compute cycle-connectivity $\kappa_c(G)$ and strong cycle-connectivity $\kappa_{sc}(G)$, respectively.

Keywords: Combinatorial optimization · Cycle-connected mixed graphs · Circuit-connected mixed graphs · Cycle-connectivity · Exact algorithms

This paper is supported by the National Natural Science Foundation of China [Nos. 11861075, 12101593].

D.-Z. Du et al. (Eds.): COCOA 2021, LNCS 13135, pp. 602–614, 2021.
https://doi.org/10.1007/978-3-030-92681-6_47

1 Introduction

Given a graph $G = (V, E)$, the vertex connectivity of G is the smallest number of vertices whose deletion separates this graph G or makes it trivial. Vertex connectivity of graphs is a fundamental parameter that, by the Menger Theorem [1,15], can be characterized by the existence of internally vertex-disjoint paths between every vertex pairs in graphs. So far, much work concerning vertex connectivity has been devoted to the internally vertex-disjoint paths problem, *i.e.*, given a graph $G = (V, E)$ and a positive integer k, for every pair x, y of distinct vertices in G, it is asked to compute k internally vertex-disjoint paths between x and y if such paths exist. Despite all further efforts, the traditional flow-based approach by Even and Tarjan [3] in 1975 provided still an efficient deterministic algorithm in time $O(n^{1/2}m^2)$ for solving the vertex connectivity problem, where $n = |V|$ and $m = |E|$. Using the maximal forest decomposition of Nagamochi and Ibaraki [11] and a generalization of the preflow-push algorithm of Hao and Orlin [5], Henzinger et al. [6] in 2000 presented an algorithm in time $O(\min\{\kappa^3 + n, \kappa n\}\kappa n)$ to compute the vertex connectivity of a graph G, where κ is the vertex connectivity of G.

We are interested in designing an algorithm to determine whether a graph is 2-connected. The Menger Theorem [1,15] tells us that a graph $G = (V, E)$ is 2-connected if and only if, for every pair x, y of distinct vertices in G, there exist two internally vertex-disjoint paths P^1_{xy} and P^2_{xy} connecting x and y. In addition, we can construct a cycle $P^1_{xy} \circ Q^2_{yx}$ to contain x and y using a combination of these two paths P^1_{xy} and Q^2_{yx}, where Q^2_{yx} is a path connecting y and x using the path P^2_{xy} along the inverse direction from y to x. Furthermore, when we increase vertex connectivity of G, we can construct more internally vertex-disjoint cycles in G containing x and y.

Given a digraph $D = (V, A)$, D is called to be strongly connected if, for every pair x, y of distinct vertices in D, there exists an (x, y)-walk from x to y and a (y, x)-walk from y to x. In this sense, D is called a strongly connected digraph. Given a digraph $D = (V, A)$, the vertex connectivity of D is the smallest number of vertices whose deletion causes the reduced directed subgraph either not to be a strongly connected digraph or to become an isolated vertex. Vertex connectivity of digraphs is an important and fundamental parameter that, also by the (directed) Menger Theorem [1,15], can be characterized by the existence of internally vertex-disjoint directed paths from every vertex to any other vertex in a digraph. Using the Menger Theorem [1,15] and the traditional flow-based methods, Even and Tarjan [3] in 1975 also presented a simple deterministic algorithm in time $O(n^{1/2}m^2)$ for computing vertex connectivity of a digraph D, where $n = |V|$ and $m = |E|$. Combining two previous vertex connectivity algorithms and a generalization of the preflow-push algorithm of Hao and Orlin [5], Henzinger et al.[6] in 2000 also presented an algorithm in time $O(\min\{\kappa^3 + n, \kappa n\}m)$ to compute the vertex connectivity of a digraph D, where κ is the vertex connectivity of D.

When we are interested in determining whether a digraph $D = (V, A)$ is a strongly 2-connected digraph, the Menger Theorem [1,15] tells us that a digraph

D is strongly 2-connected if and only if, for every pair x, y of distinct vertices in D, there exist two internally vertex-disjoint directed paths P_{xy}^1 and P_{xy}^2 from x to y, and meanwhile two internally vertex-disjoint directed paths Q_{yx}^1 and Q_{yx}^2 from y to x. However, being different from a 2-connected graph, where we may construct a cycle to contain u and v using two internally vertex-disjoint paths connecting any two distinct vertices u, v in G, for every pair x, y of distinct vertices in a digraph D, we can not find an algorithm to construct a directed cycle to contain x and y using two internally vertex-disjoint directed paths P_{xy}^1 and P_{xy}^2 from x to y plus two internally vertex-disjoint paths Q_{yx}^1 and Q_{yx}^2 from y to x in D, further we can neither find an algorithm in other ways to determine a directed cycle to contain x and y in D. Whenever we increase the vertex connectivity of D, can we design an algorithm to determine a directed cycle to contain every pair x, y of distinct vertices in D? Unfortunately, Thomassen [17] in 1991 constructed a planar digraph D_k to show that, for every positive integer k, there exists a strongly k-connected digraph which contains two vertices through which there is no directed cycle.

Motivated by vertex connectivity of digraphs or graphs, where we can not construct a directed cycle to contain every pair of distinct vertices in a digraph and meanwhile we are able to do the same affair in a graph, we think that the vertex connectivity of a digraph has no relationship with the existence of a directed cycle passing through every pair of distinct vertices in this digraph. In this paper, we should address the cycle-connected mixed graph (CCMG) problem, which is modelled as follows. Given a mixed graph $G = (V, A \cup E)$, for every pair x, y of distinct vertices in G, we are asked to determine a mixed cycle C in G to contain such two vertices x and y, where C traverses its arc (u, v) along the direction only from u to v and its edge uv along one direction either from u to v or from v to u. Similarly, we consider the circuit-connected (or weakly cycle-connected) mixed graph (WCCMG) problem, substituting a mixed circuit for a mixed cycle in the CCMG problem. Whenever the CCMG problem is specialized to either digraphs or graphs, we refer to this version of the CCMG problem as either the cycle-connected digraph (CCD) problem or the cycle-connected graph (CCG) problem. For convenience, we refer to this mixed graph as either a cycle-connected digraph or a cycle-connected graph in related versions.

It is different from computing the vertex connectivity of a digraph or a graph by executing some exact algorithms in polynomial time [3,6], using the directed 2-linkage problem which is *NP*-complete [4], we prove that the CCD problem is *NP*-complete, implying that the CCMG problem still remains *NP*-complete, and however, we design an exact algorithm in polynomial time to solve the CCG problem, then we can present a simply exact algorithm in polynomial time to solve the WCCMG problem.

Since there exists a planar digraph D_k, which is constructed in 1991 by Thomassen [17], with highly strong vertex connectivity k such that there is no directed cycle to pass through some two vertices in D_k, we may consider a new parameter of a graph, which is defined in the twin versions as follows. Given a graph $G = (V, E)$, the cycle-connectivity, denoted by $\kappa_c(G)$, of G is the smallest number of vertices in G whose deletion causes the reduced subgraph either not to be a cycle-connected graph or to become an isolated vertex; In addition, for every pair x, y of distinct vertices in G, we denote by $\kappa_{sc}(x, y)$ the maximum number of internally vertex-disjoint cycles in G containing such two vertices x and y, then the strong cycle-connectivity, denoted by $\kappa_{sc}(G)$, of G is the smallest of these numbers $\kappa_{sc}(x, y)$ among all pairs $\{x, y\}$ of distinct vertices in G, *i.e.*, $\kappa_{sc}(G) = \min\{\kappa_{sc}(x, y) \mid x, y \in V\}$.

It is easy to see that this new parameter in the twin versions, saying cycle-connectivity or strong cycle-connectivity of graphs, is different from either vertex connectivity or edge connectivity of graphs. Given a graph $G = (V, E)$, we shall provide twin exact algorithms in polynomial time to compute cycle-connectivity $\kappa_c(G)$ and strong cycle-connectivity $\kappa_{sc}(G)$, respectively.

We have noticed other parameter called as cyclic connectivity of graphs, which is not only different from vertex connectivity of digraphs or graphs, but also different from cycle-connectivity or strong cycle-connectivity of graphs. Peroche [13] in 1983 considered the cyclic connectivity of a graph $G = (V, E)$ so as to plan to present the relations among several sorts of connectivity. A set S of vertices in G is a cyclic vertex cut-set if the reduced subgraph $G[V \backslash S]$ is not connected and at least two components of $G[V \backslash S]$ contain a cycle, respectively. The cyclic connectivity, denoted by $c\kappa(G)$, of G is the minimum cardinality taken over all vertex cyclic cut-sets of G. There is little previous work on polynomial-time algorithms for cyclic vertex connectivity of general graphs, but only a few algorithms in polynomial time for cyclic vertex connectivity of some regular graphs [8–10].

In the sequel sections, we successively consider cycle-connected mixed graphs and new related parameters. We hope that these new parameters and related problems will have many further applications in our reality life. We shall design some algorithms to solve them and related problems, respectively.

This paper is organized as follows. In Sect. 2, we present some terminologies and fundamental lemmas to ensure the correctness of our algorithms. In Sect. 3, using the directed 2-linkage problem which is *NP*-complete [4], we prove that the CCD problem is *NP*-complete, implying that the CCMG problem still remains *NP*-complete, and however, we design an exact algorithm in polynomial time to solve the CCG problem. In Sect. 4, we present a simply exact algorithm in polynomial time to solve the WCCMG problem and propose a conjecture concerning k-circuit-connected mixed graphs. In Sect. 5, given a graph $G = (V, E)$, we provide twin exact algorithms in polynomial time to compute cycle-connectivity $\kappa_c(G)$ and strong cycle-connectivity $\kappa_{sc}(G)$, respectively. In Sect. 6, we provide our conclusion and further research.

2 Terminologies and Fundamental Lemmas

In this section, we present some terminologies and fundamental lemmas in order to prove our results in the sequel, and other terminologies and notations not defined can be found in those references [1,7,15].

Given a mixed graph $G = (V, A \cup E)$, a walk P from a vertex v_{i_1} to a vertex $v_{i_{k+1}}$ is an alternating sequence $\pi = v_{i_1} e_{i_1} v_{i_2} e_{i_2} v_{i_3} \cdots v_{i_k} e_{i_k} v_{i_{k+1}}$ such that, for each integer $1 \leq j \leq k$, e_{i_j} is either an arc from v_{i_j} to $v_{i_{j+1}}$ or an edge connecting v_{i_j} and $v_{i_{j+1}}$, where P traverses its arc $(v_{i_j}, v_{i_{j+1}})$ along the direction only from v_{i_j} to $v_{i_{j+1}}$ and its edge e_{i_j} along one direction either from v_{i_j} to $v_{i_{j+1}}$ or from $v_{i_{j+1}}$ to v_{i_j}, and the integer k is called as the length of this walk P. For convenience, we may define this walk as $P_{v_{i_1} v_{i_{k+1}}} = v_{i_1} v_{i_2} v_{i_3} \cdots v_{i_k} v_{i_{k+1}}$ or $(v_{i_1}, v_{i_{k+1}})$-walk. A walk P is called as a circuit or tour with k edges if $v_{i_1} = v_{i_{k+1}}$. In addition, a walk P is called as a path if the vertices in P are all distinct, and we may simply refer to this path P as a v_{i_1}-$v_{i_{k+1}}$ path or a $(v_{i_1}, v_{i_{k+1}})$-path in G. Similarly, a circuit C is called as a cycle if the vertices in C are all distinct. Particularly, whenever a mixed graph $G = (V, A \cup E)$ becomes a directed graph (simply, digraph), *i.e.*, $E = \emptyset$, we refer to path, cycle, circuit in such a digraph as directed path, directed cycle, directed circuit.

Given a mixed graph $G = (V, A \cup E)$, for every pair of distinct vertices x, y in G, if there exists a mixed path P (in G) from x to y, then G is called as a connected mixed graph. In addition, whenever $E = \emptyset$ in $G = (V, A \cup E)$, we call this mixed graph as a strongly connected digraph, denoted by $D = (V, A)$; whenever $A = \emptyset$ in $G = (V, A \cup E)$, we call this mixed graph as a connected graph, denoted by $G = (V, E)$. According to the definitions mentioned-above in Sect. 1, a cycle-connected mixed graph must be a connected mixed graph, however, the reverse result is not true (seeing contents in Sect. 3).

Given a mixed graph $G = (V, A \cup E)$ and $V' \subseteq V$, we construct a mixed graph $H = (V', A' \cup E')$ equipped with the vertex-set V', the arc-set $A' = \{(x, y) \in A \mid x, y \in V'\}$ and the edge-set $E' = \{xy \in E \mid x, y \in V'\}$, then we denote this graph $H = (V', A' \cup E')$ as the subgraph (of G) reduced by the set V', simply a reduced subgraph of G, and denoted by $G[V']$.

For a mixed graph $G = (V, A \cup E)$, there are many exact algorithms [7,12,14,15] to find a mixed path P_{xy} from x to y for every pair x, y of distinct vertices in G, then we may determine whether G is a connected mixed graph or not. We restate such an algorithm, denoted by the Prim algorithm, in the following lemma.

Lemma 1 [14]. *Given a mixed graph $G = (V, A \cup E)$, for every pair x, y of distinct vertices in G, the Prim algorithm determines a shortest mixed path from x to y, and this algorithm runs in time $O(|A \cup E|)$.*

In addition, for a fixed vertex x in a mixed graph $G = (V, A \cup E)$, we can modify the Prim algorithm [14] to determine whether there exists a shortest mixed path from x to every other vertex y in G or not, and this modified algorithm still runs in time $O(|A \cup E|)$. For convenience, we denote this modified algorithm as the Prim algorithm-modified-1. At the same time, we can revise the Prim algorithm-modified-1 to determine whether there exists a shortest mixed path from every other vertex y in G to x or not, and that new algorithm still runs in time $O(|A \cup E|)$. Again for convenience, we still denote that modified algorithm as the Prim algorithm-modified-2.

In the point of algorithmic view, we need the following definitions to determine whether a mixed graph $G = (V, A \cup E)$ to be a cycle-connected mixed graph or not.

Definition 1. *Given a digraph $D = (V, A)$ equipped with $2k$ distinct vertices $x_1, x_2, \ldots, x_k, y_1, y_2, \ldots, y_k$, if there exist k directed paths $P_1, P_2, \ldots, P_k$ in D such that P_i is a directed path from x_i to y_i for each $i = 1, 2, \ldots, k$ and that $V(P_i) \cap V(P_j) = \emptyset$ for $i \neq j$, then these k directed paths are called as k vertex-disjoint directed paths in D from $(x_1, x_2, \ldots, x_k)$ to $(y_1, y_2, \ldots, y_k)$.*

For convenience, these k vertex-disjoint directed paths in Definition 1 is also a k-linkage from $(x_1, x_2, \ldots, x_k)$ to $(y_1, y_2, \ldots, y_k)$ in Thomassen [16,17].

Definition 2. *Given a digraph $D = (V, A)$ equipped with two distinct vertices x, y, if there exist k directed paths $P_1, P_2, \ldots, P_k$ in D from x to y such that $V(P_i) \cap V(P_j) = \{x, y\}$ for $i \neq j$, then these k directed paths are called as k internally vertex-disjoint directed paths in D from x to y.*

Definition 3. *Given a graph $G = (V, E)$ equipped with two distinct vertices x, y, if there exist k paths $P_1, P_2, \ldots, P_k$ in G connecting x and y such that $V(P_i) \cap V(P_j) = \{x, y\}$ for $i \neq j$, these k paths are called as k internally vertex-disjoint paths in G connecting x and y.*

In order to compute connectivity of digraphs or graphs, we need the Menger Theorem [1,15] restated as follows.

Lemma 2 [1,15] (Menger Theorem)

(1) *Given a digraph $D = (V, A)$, D is a strongly k-connected digraph if and only if, for every pair x, y of distinct vertices in D, there exist at least k internally vertex-disjoint directed paths (in D) from x to y.*
(2) *Given a graph $G = (V, E)$, G is a k-connected graph if and only if, for every pair x, y of distinct vertices in G, there exist at least k internally vertex-disjoint paths (in G) connecting x and y.*

Using the Menger Theorem [1,15] as a wonderful bridge between computing connectivity of digraphs or graphs and traditional flow-based methods [7], Even and Tarjan [3] in 1975 presented an efficient deterministic algorithm for finding k internally vertex-disjoint paths, then we obtain the following lemma.

Lemma 3 [3]. *Given a network $N = (V, A; c, \widetilde{c}; s, t)$, where $c : A \to Z^+$ and $\widetilde{c} : V \to Z^+$, we obtain the following*

(1) *If the vertex capacities are all equal to one, the Dinic algorithm [2] requires at most $O(n^{1/2}m)$ time;*

(2) *If the edge capacities are all equal to one, the Dinic algorithm [2] requires at most $O(n^{2/3}m)$ time.*

These results are used to test the vertex connectivity of a graph in $O(n^{1/2}m^2)$ time and the edge connectivity in $O(n^{5/3}m)$ time, where $n = |V|$ and $m = |E|$.

For convenience, we denote the Connectivity algorithm in Lemma 3 to present an algorithm to compute the vertex connectivity of D or the vertex connectivity of G, depending on the related version of digraphs or graphs.

In order to determine the feasible solutions for our problems, we need the following definitions, which are similarly defined as Definitions 2 and 3.

Definition 4. *Given a digraph $D = (V, A)$ equipped with two distinct vertices x, y, if there exist k directed cycles C_1, C_2, ..., C_k in D to contain x and y such that $V(C_i) \cap V(C_j) = \{x, y\}$ for $i \neq j$, then these k directed cycles are called as k internally vertex-disjoint directed cycles in D.*

Definition 5. *Given a graph $G = (V, E)$ equipped with two distinct vertices x, y, if there exist k cycles C_1, C_2, ..., C_k in G to contain x and y such that $V(C_i) \cap V(C_j) = \{x, y\}$ for $i \neq j$, then these k cycle are called as k internally vertex-disjoint cycles in G.*

We describe the directed 2-linkage problem [4]. Given a digraph $D = (V, A)$ with four distinct vertices x_1, x_2, y_1, y_2, it is asked to determine whether D contains two vertex-disjoint directed paths P_1, P_2 such that P_i is an x_i-y_i directed path in D for each $i \in \{1, 2\}$.

Lemma 4 [4]. *The 2-linkage problem is NP-complete.*

3 The Cycle-Connected Mixed Graph Problem

In this section, we consider the cycle-connected mixed graph (CCMG) problem, and when the CCMG problem is specialized to either digraphs or graphs, we refer to this version as either the cycle-connected digraph (CCD) problem or the cycle-connected graph (CCG) problem, respectively.

By Lemma 1, the connected mixed graph (CMG) problem is solvable in polynomial time. However, we shall prove in the following way that the CCD problem is *NP*-complete, implying that the CCMG problem still remains *NP*-complete.

Now, we may consider a special version of the CCD problem, denoted by the SV-CCD problem, which is modelled as follows. Given a digraph $D = (V, A)$ with the two fixed distinct vertices x^* and y^*, we are asked to determine a directed cycle C in D to contain such two vertices x^* and y^*.

We obtain the following result concerning the SV-CCD problem.

Theorem 1. *Unless $\mathcal{P} = \mathcal{NP}$, the SV-CCD problem is NP-complete.*

Proof. It is easy to see that the SV-CCD problem is in $\mathcal{NP}$. Using a transformation from the directed 2-linkage problem [4], we shall prove that the SV-CCD problem is *NP*-complete.

Given an instance $\mathcal{I}$ of the directed 2-linkage problem [4], *i.e.*, given a digraph $H = (V, A)$ with four distinct vertices x_1, x_2, y_1, y_2, it is asked to find two vertex-disjoint directed paths P_1, P_2 in H such that P_i is a directed path in H from x_i to y_i for each $i \in \{1, 2\}$.

We may construct an instance $\tau(\mathcal{I})$ of the SV-CCD problem as follows, *i.e.*, a digraph $D = (V \cup \{x^*, y^*\}, A \cup A^*)$, where x^*, y^* are two new vertices which are not in V and $A^* = \{(x^*, x_1), (y_1, y^*), (y^*, x_2), (y_2, x^*)\}$, it is asked to determine a directed cycle C in D to contain such two vertices x^* and y^*.

For the instance $\mathcal{I}$ and the instance $\tau(\mathcal{I})$, we can obtain the following claim.
Claim 1. The instance $\mathcal{I}$ of the directed 2-linkage problem has a feasible solution, *i.e.*, a 2-linkage from (x_1, x_2) to (y_1, y_2) in H, if and only if the instance $\tau(\mathcal{I})$ of the V-CCD problem has a feasible solution, *i.e.*, a directed cycle C in D to contain such two fixed vertices x^* and y^*.

Since the directed 2-linkage problem [4] is *NP*-complete, using a transformation from the instance $\mathcal{I}$ to the instance $\tau(\mathcal{I})$ and Claim 1, we prove that the SV-CCD problem is *NP*-complete, too.

This completes a proof of this theorem. ∎

Since the CCD problem is a generalization of the SV-CCD problem, using Theorem 1, we can obtain the following result.

Corollary 1. *Unless $\mathcal{P} = \mathcal{NP}$, the CCD problem is NP-complete. In addition, the CCMG problem is NP-complete.*

At the end of this section, we consider the cycle-connected graph (CCG) problem. Using Menger Theorem [1,15] (seeing Lemma 2), we can obtain the following lemma.

Lemma 5 [1,15]**.** *Given a graph $G = (V, E)$, the following statements are equivalent from a computational point of view.*

(1) This graph G is a 2-connected graph;
(2) For every pair x, y of distinct vertices in G, there exist two internally vertex-disjoint paths connecting x and y;
(3) For every pair x, y of distinct vertices in G, there exists a cycle containing x and y.

Using Lemma 5, we design an exact algorithm to solve the CCG problem as follows.

Algorithm 1: CCG
Input: A graph $G = (V, E)$;
Output: G is whether a cycle-connected graph or not.
Begin

Step 1. Given a graph $G = (V, E)$, we construct a weighted digraph $D = (V, A; c, \widetilde{c}; b)$ equipped with the arc-set $A = \{(u, v), (v, u) \mid uv \in E\}$, define an arc-capacity function $c : A \rightarrow Z^+$ by $c(x, y) = 1$ for each arc $(x, y) \in A$ and a capacity function $\widetilde{c} : V \rightarrow Z^+$ by $\widetilde{c}(u) = 1$ for each vertex $u \in V$, respectively, and a cost function $b : A \rightarrow Z^+$ by $b(x, y) = 1$ for each arc $(x, y) \in A$.
Step 2. For every pair s, t of distinct vertices in G, do
2.1 We construct a network $N_{st} = (V, A; c, \widetilde{c}; b; s, t)$, where s is a source, t is a sink and the digraph $D = (V, A; c, \widetilde{c}; b)$ is constructed in Step 1.
2.2 Using the Buildup algorithm [12] to find a minimum-cost integral flow f_{st} from s to t, having its value $v(f_{st}) = 2$.
2.3 If (there is no such flow f_{st} in Step 2.2) then
Output "G is not a cycle-connected graph", stop.
Step 3. Output "G is a cycle-connected graph", stop.
End

Using Lemmas 1 and 5, we can obtain the following result.

Theorem 2. *The CCG algorithm is an exact algorithm to solve the CCG problem, and this algorithm runs in time $O(n^2m)$, where n is the number of vertices and m is the number of edges of a graph $G = (V, E)$.*

Proof. Given a graph $G = (V, E)$, for every pair s, t of distinct vertices in G, by executing Step 2 in the CCG algorithm, if we obtain a minimum-cost integral flow f_{st} from s to t, having its flow value $v(f_{st}) = 2$, we can construct two internally vertex-disjoint paths P_{st}, Q_{st} in G connecting x and y (the minimum cost of this integral flow f_{st} from s to t can maintain this property that these two paths are internally vertex-disjoint in G), then the combination of two paths P_{st} and Q_{st} in G becomes a cycle to contain such two vertices s and t. This shows that G is a cycle-connected graph.

On the other hand, if there exists a pair s^*, t^* of distinct vertices in G such that there is no flow $f_{s^*t^*}$ from s^* to t^* to have its flow value $v(f_{s^*t^*}) = 2$, by using Lemma 5, this shows that G is not a cycle-connected graph.

The complexity of the CCG algorithm (*i.e.*, Algorithm 1) can be determined as follows. (1) Step 1 needs time $O(m)$ to construct a weighted digraph $D = (V, A; c, \widetilde{c}; b)$. (2) For two fixed vertices s and t in G, since each arc in the network $N_{st} = (V, A; c, \widetilde{c}; s, t)$ has its capacity 1, by executing the Buildup algorithm [12] at Step 2.2 per time, it needs time $O(m)$ to increase an unit flow from s to t along a shortest directed path in the auxiliary network N_f [12] (by using Lemma 1), where f is a current flow at present. This process lasts two times to obtain a current flow f_{st} from s to t having its flow value $v(f_{st}) = 2$, implying that it needs time $O(m)$ to construct the minimum-cost integral flow f_{st} having its flow value $v(f_{st}) = 2$, thus it needs time $O(n^2m)$ to execute Step 2. Hence, the total time of the CCG algorithm is at most $O(n^2m)$.

This establishes the theorem. ∎

4 The Circuit-Connected Mixed Graph Problem

In this section, we consider the circuit-connected (or weakly cycle-connected) mixed graph (WCCMG) problem. We need the following lemma to present an exact algorithm to solve the circuit-connected mixed graph (WCCMG) problem.

Lemma 6. *For a mixed graph $G = (V, E \cup A)$, the following statements are equivalent from a computational point of view.*

(1) *This mixed graph G is a connected graph;*
(2) *For every pair x, y of distinct vertices in G, there exists a mixed path from x to y and other path from y and x;*
(3) *For every pair x, y of distinct vertices in G, there exists a mixed circuit to contain x and y.*

Using Lemma 6 and the Prim algorithm (seeing Lemma 1) to find a shortest mixed path from x to y, it is easy for us to present an exact algorithm in time $O(n^2m)$ to solve the WCCMG problem.

To more efficiently decease the complexity of the algorithm mentioned-above to solve the WCCMG problem, we may first choose a fixed vertex x^* in G and use the Prim algorithm-modified-1 (seeing Lemma 1) to find a shortest mixed path from x^* to every other vertex y in G, then we use the Prim algorithm-modified-2 (seeing Lemma 1) to find a shortest mixed path from every other vertex y in G to x^*.

We describe our new algorithm to solve the WCCMG problem as follows.

Algorithm 2: WCCMG
Input: A mixed graph $G = (V, E \cup A)$;
Output: G is whether a circuit-connected mixed graph or not.
Begin
Step 1. Choose a fixed vertex x^* in G;
Step 2. Use the Prim algorithm-modified-1 (seeing Lemma 1) to find a shortest mixed path P_{x^*y} from x^* to y, where y is every other vertex in G; If there is no such path $P_{x^*y^*}$ for some other vertex y^* in G, then output "G is not a circuit-connected graph", and stop.
Step 3. Use the Prim algorithm-modified-2 (seeing Lemma 1) to find a shortest mixed path Q_{yx^*} from y to x^*, where y is every other vertex in G; If there is no such path $Q_{y^*x^*}$ for some other vertex y^* in G, then output "G is not a circuit-connected graph", and stop.
Step 4. Output "G is a circuit-connected graph".
End

Executing the WCCMG algorithm and using Lemma 6, we can obtain the following result, where we may omit the correct proof.

Theorem 3. *The WCCMG algorithm is an exact algorithm to solve the WCCMG problem, and this algorithm runs in time $O(m)$, where n is the order and m is the size of a graph $G = (V, E)$.*

5 Cycle-Connectivity and Strong Cycle-Connectivity

In this section, given a graph $G = (V, E)$, we consider the twin problems of computing cycle-connectivity $\kappa_c(G)$ and strong cycle-connectivity $\kappa_{sc}(G)$ of G, respectively.

We need the following two results, whose proofs are omitted in the sequel.

Lemma 7. *For a graph $G = (V, E)$ and a positive integer k, the following statements are equivalent from a computational point of view.*

(1) *This graph G is a k-connected graph, where the vertex connectivity $\kappa(G)$ of G is k;*
(2) *This graph G is a $(k-1)$-cycle-connected graph, where the cycle-connectivity $\kappa_c(G)$ of G is $k-1$.*

Lemma 8. *Given a graph $G = (V, E)$ and a positive integer k, the following statements are equivalent from a computational point of view.*

(1) *This graph G is a $2k$-connected graph, where the vertex connectivity $\kappa(G)$ of G is either $2k$ or $2k + 1$;*
(2) *This graph G is a k-cycle-connected graph, where the strong cycle-connectivity $\kappa_{sc}(G)$ of G is k.*

Given of a graph G, we design an algorithm, denoted by Cycle-Connectivity, to compute the cycle-connectivity $\kappa_c(G)$ and the strong cycle-connectivity $\kappa_{sc}(G)$.

Algorithm 3: Cycle-Connectivity
Input: A graph $G = (V, E)$;
Output: Cycle-connectivity $\kappa_c(G)$ and strong cycle-connectivity $\kappa_{sc}(G)$.
Begin
Step 1. Executing the Connectivity algorithm [1,3,7,15] on the graph $G = (V, E)$, compute the vertex connectivity $\kappa(G)$ of G.
Step 2. Denote $\kappa_c(G) = \kappa(G) - 1$ and $\kappa_{sc}(G) = \lfloor \kappa(G)/2 \rfloor$.
Step 3. Output "Cycle-connectivity $\kappa_c(G)$ and strong cycle-connectivity $\kappa_{sc}(G)$"
End

Using Lemmas 7–8 and executing the Cycle-Connectivity algorithm, we can easily obtain the following result.

Theorem 4. *Given a graph $G = (V, E)$, the Cycle-Connectivity algorithm is an exact algorithm to compute the cycle-connectivity $\kappa_c(G)$ and the strong cycle-connectivity $\kappa_{sc}(G)$ of G, and its complexity is as same as that of the Connectivity algorithm [1, 3, 7, 15].*

6 Conclusion and Further Research

In this paper, we consider the cycle-connected mixed graph (CCMG) problem and related problems, where the CCMG problem is indeed different from the connected mixed graph (CMG) problem. We obtain the following three main results.

(1) We prove that the CCD problem is *NP*-complete, implying that the CCMG problem still remains *NP*-complete, and we design an exact algorithm in polynomial time to solve the CCG problem.
(2) We present an exact algorithm in polynomial time to solve the WCCMG problem, and we propose a conjecture concerning the k-circuit-connected mixed graph problem.
(3) Given a graph $G = (V, E)$, we provide twin exact algorithms in polynomial time to compute cycle-connectivity $\kappa_c(G)$ and strong cycle-connectivity $\kappa_{sc}(G)$, respectively.

A challenging task for further research is to determine whether the cycle-connected mixed graph problem specialized to planar mixed graphs is *NP*-complete or not.

References

1. Bondy, J.A., Murty, U.S.R.: Graph Theory. Springer, New York (2008)
2. Dinic, E.A.: Algorithm for solution of a problem of maximum flow in a network with power estimation. Soviet Math. Dokl. **11**, 1277–1280 (1970)
3. Even, S., Tarjan, R.E.: Network flow and testing graph connectivity. SIAM J. Comput. **4**(4), 507–518 (1975)
4. Fortune, S., Hopcroft, J.E., Wyllie, J.: The directed subgraph homeomorphism problem. Theoret. Comput. Sci. **10**, 111–121 (1980)
5. Hao, J., Orlin, J.B.: A faster algorithm for finding the minimum cut in a directed graph. J. Algorithms **17**, 424–446 (1994)
6. Henzinger, M., Rao, S., Gabow, H.N.: Computing vertex connectivity: new bounds from old techniques. J. Algorithms **34**(2), 222–250 (2000)
7. Korte, B., Vygen, J.: Combinatorial Optimization: Theory and Algorithms. AC, vol. 21, 5th edn. Springer, Heidelberg (2012). https://doi.org/10.1007/978-3-642-24488-9
8. Liang, J., Lou, D.J.: A polynomial algorithm determining cyclic vertex connectivity of k-regular graphs with fixed k. J. Comb. Optim. **37**, 1000–1010 (2019)
9. Liang, J., Lou, D.J., Qin, Z.R., Yu, Q.L.: A polynomial algorithm determining cyclic vertex connectivity of 4-regular graphs. J. Comb. Optim. **38**(2), 589–607 (2019)
10. Liang, J., Lou, D.J., Zhang, Z.B.: The cubic graphs with finite cyclic vertex connectivity larger than girth. Discrete Math. **344**(2) (2021). Paper No. 112197, 18pp
11. Nagamochi, H., Ibaraki, T.: A linear-time algorithm for finding a sparse k-connected spanning subgraph of a k-connected graph. Algorithmica **7**, 583–596 (1992)

12. Papadimitriou, C., Steiglitz, D.K.: Combinatorial Optimization: Algorithms and Complexity. Dover Publications Inc., New York (1998)
13. Peroche, B.: On several sorts of connectivity. Discret. Math. **46**, 267–277 (1983)
14. Prim, R.C.: Shortest connecting networks and some generalizations. Bell Syst. Tech. J. **36**, 1389–1401 (1957)
15. Schrijver, A.: Combinatorial Optimization: Polyhedra and Efficiency. Springer, Heidelberg (2003)
16. Thomassen, C.: 2-linked graphs. Eur. J. Comb. **1**, 371–378 (1980)
17. Thomassen, C.: Highly connected non-2-linked digraphs. Combinatorica **11**, 393–395 (1991)

Directed Width Parameters on Semicomplete Digraphs

Frank Gurski[1], Dominique Komander[1](✉), Carolin Rehs[1], and Sebastian Wiederrecht[2]

[1] Institute of Computer Science, Algorithmics for Hard Problems Group, Heinrich-Heine-University Düsseldorf, 40225 Düsseldorf, Germany
{frank.gurski,dominique.komander,carolin.rehs}@hhu.de
[2] Département Informatique, Université de Montpellier, 34095 Montpellier, France

Abstract. The map of relations between the different directed width measures in general still has some blank spots. In this work we fill in many of these open relations for the restricted class of semicomplete digraphs. To do this we show the parametrical equivalence between directed path-width, directed tree-width, DAG-width, and Kelly-width. Moreover, we show that directed (linear) clique-width is upper bounded in a function of directed tree-width on semicomplete digraphs. In general the algorithmic use of many of these parameters is fairly restricted. On semicomplete digraphs our results allow to combine the nice computability of directed tree-width with the algorithmic power of directed clique-width. Moreover, our results have the effect that on semicomplete digraphs every digraph problem, which is describable in monadic second-order logic on quantification over vertices and vertex sets, is fixed parameter tractable for all of the shown width measures if a decomposition of bounded width is given.

Keywords: Algorithmic meta theorem · Semicomplete digraphs · Directed clique-width · Directed tree-width

1 Introduction

The rediscovery of path-width and tree-width in the graph minors project by Robertson and Seymour [23] has led to a wide range of algorithmic results. In the wake of this success several possible generalizations to directed graphs have since emerged, among which are *directed path-width* (d-pw), *directed tree-width* (d-tw) [20], *DAG-width* (dagw) [5] and *Kelly-width* (kw) [18].

While all of these parameters are related, directed path-width and directed tree-width are not parametrically equivalent to either of the other parameters and the equivalence of DAG-width and Kelly-width is an open conjecture.

All these width parameters correspond to different variants of so-called cops and robber games. Width parameters corresponding to variants of the cops and robber game

This work is partly funded by the Deutsche Forschungsgemeinschaft (DFG, German Research Foundation) - 388221852.

D.-Z. Du et al. (Eds.): COCOA 2021, LNCS 13135, pp. 615–628, 2021.
https://doi.org/10.1007/978-3-030-92681-6_48

have the inherent advantage of coming with an XP-time (approximation) algorithm for finding a decomposition of (almost) optimal width. They also tend to correlate with structural properties and thus, as exemplified by tree-width, make for great tools for structure theory. However, there exists strong evidence that for digraphs no such parameter can, in addition to these advantages, replicate the algorithmic power of tree-width in undirected graphs [13].

An algorithmically stronger parameter is directed clique-width (d-cw) which is related to the parameter clique-width by Courcelle et al. [8]. This width measure is, in essence, defined for relational structures and its algorithmic properties do not distinguish between graphs and digraphs. Hence directed clique-width does not suffer from the sudden increase in complexity when transitioning from graphs to digraphs and the existence of a powerful algorithmic meta theorem is preserved: Every problem expressible in monadic second-order logic on quantification over vertices and vertex sets, MSO_1 for short, is fixed parameter tractable with respect to the parameter directed clique-width [8]. Still, directed clique-width has its drawbacks, as there is no known direct way to compute a bounded width expression. The current method to obtain such an expression is by approximating birank-width which leads to an exponential approximation of directed clique-width [22]. Unfortunately, directed clique-width is in general incomparable to the previously mentioned tree-width inspired parameters. So in general the nice computability properties of the decompositions relating to variants of the cops and robber game cannot be used to obtain bounded width expression for directed clique-width.

Table 1. Relations between digraph parameters on **semicomplete digraphs**. The parameter of the left column is bounded by the respective parameter of the top row by the specified function where k is the corresponding width. We use '∞' if the relation is unbounded, that is if $h_{f,g}$ does not exist.

g	f					
	d-pw	d-tw	dagw	kw	d-lcw	d-cw
d-pw	k	$4k^2+15k+10$	$k-1$	$4k^2+7k$	∞	∞
d-tw	k	k	$k-1$	$6k-2$	∞	∞
dagw	$k+1$	$4k^2+15k+11$	k	k^2	∞	∞
kw	$k+1$	$4k^2+15k+11$	k	k	∞	∞
d-lcw	$k+2$	$4k^2+15k+12$	$k+1$	k^2+2	k	∞
d-cw	$k+2$	$4k^2+15k+12$	$k+1$	k^2+2	k	k

A digraph $G=(V,E)$ is called *semicomplete*, if for every two vertices $u,v \in V$ it holds that at least one of the two arcs (u,v) and (v,u) is in E. This leads to a superclass of tournaments which received significant attention in the past [7,21]. In this paper we show that on semicomplete digraphs, all of the path-width and tree-width inspired parameters are equivalent. Indeed, all of these equivalences are realized by relatively tame functions obtained without complicated proofs.

As by [10] for a semicomplete digraph G it holds that d-cw(G) is at most d-pw$(G)+2$, we finally conclude that all above mentioned parameters are upper bounds to directed clique-width. This result is even extendable to directed linear clique-width (d-lcw). More precisely we show that, for any choice of functions $f,g \in \{\text{d-pw}, \text{d-tw}, \text{dagw}, \text{kw}, \text{d-lcw}, \text{d-cw}\}$, there exists a function $h_{f,g}$ such that, if G is a semicomplete digraph with $f(G) \leq k$ then $g(G) \leq h_{f,g}(k)$ where the functions $h_{f,g}$ are presented in Table 1.

Theorem 1. *Let G be a semicomplete digraph and $f,g \in \{d\text{-}pw, d\text{-}tw, dagw, kw, d\text{-}lcw, d\text{-}cw\}$. If $f(G) \leq k$, then $g(G) \leq h_{f,g}(k)$ where $h_{f,g}\colon \mathbb{N} \to \mathbb{N}$ is given by Table 1 if a function exists.*

Combining these results with the above mentioned theorem of Courcelle et al. on bounded clique-width [8] and the FPT-algorithm for approximating directed tree-width within a linear factor by Campos et al. [6], we have the following result:

Theorem 2. *Every problem expressible in MSO_1 logic is fixed parameter tractable on semicomplete digraphs with respect to the parameter directed tree-width.*

2 Preliminaries

For basic definitions of digraphs we refer to [3]. For digraph $G=(V,E)$ let $und(G)$ be the underlying undirected graph of G without multiple edges. A subdigraph induced by $V' \subseteq V$ of G is denoted by $G[V']$. The sets $N^+(v) = \{u \in V \mid (v,u) \in E\}$ and $N^-(v) = \{u \in V \mid (u,v) \in E\}$ are called the *out-neighbors* and the *in-neighbors* of a vertex $v \in V$. The *out-degree* of vertex v is $|N^+(v)|$ while the *in-degree* is $|N^-(v)|$. An *acyclic* digraph (*DAG* for short) is a digraph without any directed cycles. We say that G is *bioriented* if for every $(u,v) \in E$ there is also $(v,u) \in E$, we call G the *complete biorientation* of $und(G)$. A bioriented clique is the complete biorientation of K_n, where K_n is the undirected graph on n vertices and all possible edges. A *directed walk* in G is an alternating sequence $W = (u_1, e_1, u_2, e_2, u_3, \ldots, e_{k-1}, u_k)$ of vertices $v_i \in V$, $1 \leq i \leq k$, and edges $e_i \in E$, $1 \leq i \leq k-1$, such that $e_i = (u_i, u_{i+1})$, with $1 \leq i \leq k-1$. If the vertices of the directed walk W are mutually distinct, then W is a *directed path*.

Let $D=(V,E)$ be a DAG. A vertex $r \in V$ is called a *source* if $N^-(r) = \emptyset$ and a *sink* if $N^+(r) = \emptyset$. We say that a vertex $u \in V$ *reaches* a vertex $v \in V$ if there exists a directed path from u to v in D. In this case we write $u \preccurlyeq_D v$. Note that $\preccurlyeq_D$ defines a partial order on V. A DAG which is an orientation of a tree and has a unique source is called an *arborescence*.

A *(directed) graph parameter* of a (directed) graph G is a function α that maps from (di)graph G to an integer. We call two graph parameters α and β *equivalent*, if there exist functions f,g such that for every digraph G it holds that $\alpha(G) \leq f(\beta(G))$ and $\beta(G) \leq g(\alpha(G))$.

3 Directed Width Parameters

In this section we give a more formal introduction to the above mentioned directed width parameters. All presented width parameters except for directed (linear) clique-width correspond to different variants of cops and robber games. So we start with introducing the notion of cops and robber games, especially two variants will come up in our proofs. Then we formally define different parameters by so-called decompositions. Please note that many definitions also occur in our papers [15] and [14].

3.1 Directed Cops and Robber Games

A cops and robbers game on a (directed) graph is a pursuit-evasion game with two teams of players, the cops, which can move unrestrictedly to every vertex with their helicopters and the robbers moving from vertex to vertex along the arcs/edges of the graph. The cops try to "catch" the robbers by moving onto the vertices where the robbers are positioned, while the robbers try to evade this capture.

Let $G = (V,E)$ be a directed graph with one robber and a set of cops. A position in the game is a pair (C,r) where $C \subseteq V$ is the current position of the cops and $r \in V$ is the current position of the robber. Initially, there is no cop on the graph, i.e., $C_0 = \emptyset$ and in the first round the robber can choose a start position r_0. In every round $i+1$, (C_i, r_i) is the current position of the cops and robber. The game is then played as follows: The cops announce their new position C_{i+1}. Then the robber can chose any vertex r_{i+1} as a new position, that is reachable from r_i in the graph $G-(C_i \cap C_{i+1})$. There are two variations of reachability: In strong component searching, the robber can move to every vertex in the same strong component of $G-(C_i \cap C_{i+1})$. In reachability searching, the robber can move to any vertex r_{i+1} such that there is a directed walk from r_i to r_{i+1}.

If $r_i \in C_i$ after any round i, then the cops capture the robber and win the game. Otherwise, the game never ends and the robber wins the game. Clearly, the game can always be won by the cops, by positioning a cop on every vertex of G. However, an interesting question is, how many cops are needed for a graph G, such that there is always a winning strategy for the cops.

By varying the rules, many different cops and robber games can be defined. The best known modification is, if the cops know the current robber position (visible CnR-Game) or do not know the current robber position (invisible CnR-Game). Another variant is a so-called inert robber: This robber is only allowed to move, if $r_i \in C_{i+1}$, i.e., if the robber would be captured in the next round.

3.2 Directed Path-Width

In the following we define some width parameters, starting with a very common one which got many different equivalent definitions over time.

Definition 1 (directed path-width). *Let $G = (V,E)$ be a digraph. A* directed path-decomposition *of G is a sequence $(X_1,\ldots,X_r)$ of subsets of V, called* bags, *such that the following three conditions hold.*

(dpw-1) $X_1 \cup \cdots \cup X_r = V$,
(dpw-2) for each $(u,v) \in E$ there is a pair $i \leq j$ such that $u \in X_i$ and $v \in X_j$, and
(dpw-3) for all i,j,ℓ with $1 \leq i < j < \ell \leq r$ it holds $X_i \cap X_\ell \subseteq X_j$.

The width *of a directed path-decomposition $\mathcal{X} = (X_1,\ldots,X_r)$ is $\max_{1 \leq i \leq r} |X_i| - 1$. The* directed path-width *of G, d-pw(G) for short, is the smallest integer w such that there is a directed path-decomposition for G of width w.*

3.3 Directed Tree-Width

There exist many definitions for directed tree-width which differ in technical details but are otherwise parametrically equivalent. In this paper we use the definition by Johnson et al. from [20].

Let $G = (V,E)$ be a digraph and $Z \subseteq V$. The digraph $G[V \setminus Z]$ which is obtained from G by deleting Z is denoted by $G - Z$. A vertex set $S \subseteq V \setminus Z$ is *Z-normal* if every directed walk which leaves and again enters S must contain a vertex from Z. For two vertices u, v of an out-tree T we write $u \leq v$ if there is a directed path from u to v or $u = v$.

Definition 2 (directed tree-width, [20]**).** *A (-n arboreal) tree-decomposition of a digraph $G = (V_G, E_G)$ is a triple $(T, \mathcal{X}, \mathcal{W})$. Here $T = (V_T, E_T)$ is an arborescence, $\mathcal{X} = \{X_e \mid e \in E_T\}$ and $\mathcal{W} = \{W_r \mid r \in V_T\}$ are sets of subsets of V_G, such that the following two conditions hold.*

(dtw-1) $\mathcal{W} = \{W_r \mid r \in V_T\}$ is a partition of V_G into nonempty subsets.[1]
(dtw-2) For every $(u,v) \in E_T$ the set $\bigcup\{W_r \mid r \in V_T, v \leq r\}$ is $X_{(u,v)}$-normal.

The width *of a (-n arboreal) tree-decomposition $(T, \mathcal{X}, \mathcal{W})$ is $\max_{r \in V_T} |W_r \cup \bigcup_{e \sim r} X_e| - 1$. Here $e \sim r$ means that r is one of the two vertices of arc e. The* directed tree-width *of G, d-tw(G) for short, is the smallest integer k such that there is a (-n arboreal) tree-decomposition $(T, \mathcal{X}, \mathcal{W})$ of G of width k.*

As mentioned in the introduction, there is a strong link between a variant of the cops and robber game and directed tree-width. Indeed, this link played an important role in finding the definition of directed tree-width in the first place.

Proposition 1 ([20]). *If G has directed tree-width of at most k, then $k+1$ cops have a robber monotone winning strategy in the visible strong component cops and robber game on G. If k cops have a winning strategy in this game, then the directed tree-width of G is at most $3k+2$.*

3.4 DAG-Width

The main difference between directed tree-width and DAG-width is that the separations in an arboreal decomposition only destroy strong connectivity, while those in a DAG-decomposition block all directed paths leaving the bags of a sub-DAG. Since directed separations are more restricted than strong separations, the model graph which is used for the decomposition needs to be relaxed from an arborescence to a DAG. DAG-width has been defined in [4].

For a digraph $G = (V_G, E_G)$ let $V' \subseteq V_G$, then a set $W \subseteq V_G$ *guards* V' if for all $(u,v) \in E_G$ it holds that if $u \in V'$ then $v \in V' \cup W$.

Definition 3 (DAG-width). *A* DAG-decomposition *of a digraph $G = (V_G, E_G)$ is a pair $(D, \mathcal{X})$ where $D = (V_D, E_D)$ is a directed acyclic graph (DAG) and $\mathcal{X} = \{X_u \mid X_u \subseteq V_G, u \in V_D\}$ is a family of subsets of V_G such that:*

[1] A remarkable difference to the undirected tree-width [23] is that the sets W_r have to be disjoint and non-empty.

(dagw-1) $\bigcup_{u \in V_D} X_u = V_G$.
(dagw-2) For all vertices $u, v, w \in V_D$ *with* $u \succcurlyeq_D v \succcurlyeq_D w$, *it holds that* $X_u \cap X_w \subseteq X_v$.
(dagw-3) For all edges $(u,v) \in E_D$ *it holds that* $X_u \cap X_v$ *guards* $X_{\succcurlyeq_v} \setminus X_u$, *where* $X_{\succcurlyeq_v} = \bigcup_{v \succcurlyeq_D w} X_w$. *For any source u,* $X_{\succcurlyeq_u}$ *is guarded by* $\emptyset$.

The width *of a DAG-decomposition* $(D, \mathcal{X})$ *is the number* $\max_{u \in V_D} |X_u|$. *The* DAG-width *of a digraph G, dagw(G) for short, is the smallest width of all possible DAG-decompositions for G.*

It is straightforward that a DAG-decomposition where D is a path can also be seen as a directed path decomposition, as it meets the same conditions.

3.5 Kelly-Width

We now come to Kelly-width, which has originally been introduced in [18]. The original definition of Kelly-width bears some resemblance to the definition of DAG-width, but it is more technical. In [18] it was conjectured that Kelly-width and DAG-width are indeed parametrically equivalent, but so far only one of the two relations has been proven [2]. The actual definition of Kelly-width is based on a decomposition [18].

Definition 4 (Kelly-width). *A* Kelly decomposition *of a digraph* $G = (V_G, E_G)$ *is a triple* $(\mathcal{W}, \mathcal{X}, D)$ *where D is a directed acyclic graph,* $\mathcal{X} = \{X_u \mid X_u \subseteq V_G, u \in V_D\}$ *and* $\mathcal{W} = \{W_u \mid W_u \subseteq V_G, u \in V_D\}$ *are families of subsets of* V_G *such that:*

(kw-1) $\mathcal{W}$ *is a partition for* V_G.
(kw-2) For all vertices $v \in V_G$, X_v *guards* $W_{\succcurlyeq_v}$.
(kw-3) For all vertices $v \in V_G$, *there is a linear order* $u_1, \ldots, u_s$ *on the successors of v such that for every* u_i *it holds that* $X_{u_i} \subseteq W_i \cup X_i \cup \bigcup_{j<i} W_{\succcurlyeq_{u_j}}$. *Similarly, there is a linear order* $r_1, r_2, \ldots$ *on the roots of D such that for each root* r_i *it holds that* $W_{r_i} \subseteq \bigcup_{j<i} W_{\succcurlyeq_{r_j}}$.

The width *of a Kelly decomposition* $(\mathcal{W}, \mathcal{X}, D)$ *is the number* $\max_{u \in V_D} |X_u| + |W_u|$. *The* Kelly-width *of a digraph G, denoted with kw(G), is the smallest width of all possible Kelly decompositions for G.*

For our purpose we need a relation of Kelly-width to a variant of the cops and robber game given by the following characterization.

Proposition 2 ([18]). *A digraph G has Kelly-width of at most* $k+1$ *if and only if* $k+1$ *cops have a winning strategy to capture an invisible and inert robber in the reachability searching game.*

3.6 Directed (Linear) Clique-Width

Directed clique-width has been introduced together with clique-width on undirected graphs, see by Courcelle and Olariu in [9]. The linear clique-width for undirected graphs was introduced in [16] as a parameter by restricting the clique-width, to an underlying path-structure.

Definition 5 (directed clique-width). *The* directed clique-width *of a vertex-labeled digraph G, d-cw(G) for short, is the minimum number of labels needed to define G using the following four operations:*

1. *Creation of a new vertex with label a (denoted by $\bullet_a$).*
2. *Disjoint union of two labeled digraphs G and H (denoted by $G \oplus H$).*
3. *Inserting an arc from every vertex with label a to every vertex with label b ($a \neq b$, denoted by $\alpha_{a,b}$).*
4. *Change label a into label b (denoted by $\rho_{a \to b}$).*

The directed clique-width *of an unlabeled digraph $G = (V,E)$, d-cw(G) for short, is the smallest integer k, such that there is a mapping $lab : V \to \{1,\ldots,k\}$ such that the labeled digraph (V,E,lab) has directed clique-width at most k.*

Directed linear clique-width can be obtained, when the disjoint union operation is only allowed for one digraph and one labeled vertex, i.e., in the above definition, the graph H contains exactly one vertex.

4 Comparison of Directed Width Parameters on Semicomplete Digraphs

Before we start with the comparisons between different parameters which eventually lead to Theorem 1, let us introduce the current landscape of bounding functions between these parameters on general digraphs.

Proposition 3. *Let G be a digraph and $f,g \in \{$d-pw, d-tw, dagw, kw, d-lcw, d-cw$\}$. If $f(G) \leq k$, then $g(G) \leq h'_{f,g}(k)$ where $h'_{f,g} : \mathbb{N} \to \mathbb{N}$ is given by Table 2 if a function exist.*

Proof. 1. d-pw is unbounded in terms of kw: In [3] the example of a complete biorientation of an undirected binary tree of height h is considered. This digraph has directed path-width h while it has the fixed Kelly-width of 2.
2. d-pw is unbounded in terms of d-tw: Holds with the example from 1 which is inspired by the undirected comparisons of path-width and tree-width. Increasing h, the directed tree-width is 1, while the directed path-width increases.
3. d-tw is bounded by d-pw: This follows immediately from the definition.
4. d-pw, d-tw, dagw and kw are unbounded in terms of d-lcw and thus in d-cw: The set of all bioriented cliques is a counterexample.
5. kw is unbounded in terms of d-tw: As an example consider a binary tree, where all edges are oriented from the root to the leaves. Additionally, every vertex has a backward edge to each of its predecessors on the unique path from the root to itself.
6. d-cw and thus also d-lcw is unbounded in terms of d-pw, d-tw, dagw and kw. An acyclic orientation of a grid graph is a counterexample.
7. d-lcw is unbounded in terms of d-cw: In Lemma 11 of [16] it has been shown that for $G_1 = \bullet$ and $G_{i+1} = (G_i \cup G_i) \times (G_i \cup G_i)$ for $i \geq 1$, graph G_i has linear NLC-width at least i. The proof idea can be used to show that for $G_1 = \bullet$ and $G_{i+1} = (G_i \oslash G_i) \otimes (G_i \oslash G_i)$ for $i \geq 1$, digraph G_i has directed linear NLC-width at least i.

All graphs G_i are directed co-graphs which implies that they have directed clique-width at most 2, see [17]. Since directed linear clique-width is greater or equal to directed linear NLC-width [15], the result follows.

8. d-cw is bounded by d-lcw: This follows immediately from the definition. □

Table 2. Relations between digraph parameters on **digraphs**. The parameter of the left column is bounded by the respective parameter of the top row by the specified function where k is the corresponding width. We use '∞' if the relation is unbounded, that is if $h'_{f,g}$ does not exist. The cell with '???' represents the remaining relation of the conjecture on DAG-width and Kelly-width.

g	f					
	d-pw	d-tw	dagw	kw	d-lcw	d-cw
d-pw	k	∞	∞ [5]	∞	∞	∞
d-tw	k	k	$3k+1$ [5]	$6k-2$ [18]	∞	∞
dagw	$k+1$ [5]	∞ [5]	k	$72k^2$ [2]	∞	∞
kw	$k+1$ [12]	∞	??? [18]	k	∞	∞
d-lcw	∞	∞	∞	∞	k	∞
d-cw	∞	∞	∞	∞	k	k

Please note that Proposition 3 contains in particular the known fact that directed path-width poses as an upper bound for all tree-width inspired width parameters. Moreover, on semicomplete digraphs, by Proposition 7 it also is an upper bound on directed clique-width. It therefore suffices, towards a proof of Theorem 1, to establish upper bounds on directed path-width in terms of directed tree-width, DAG-width, and Kelly-width, as well as proving that Proposition 7 can be extended to also include directed linear clique-width.

4.1 DAG-Width and Directed Path-Width on Semicomplete Digraphs

As a first step towards Theorem 1 we show that DAG-width plus 1 and directed path-width are equal on the class of semicomplete digraphs, which leads also to the fact that computing DAG-width of a semicomplete digraph is in NP.

This later fact might be of independent interest since DAG-width is PSPACE-complete in general [2], but, it is one of only few known parameters from the tree-width inspired family which allows for an efficient solving of parity games [4].

As a tool we need a normalized version of DAG-decompositions.

Definition 6 (Nice DAG-decomposition). *A DAG-decomposition $(D, \mathcal{X})$ of a digraph G is* nice, *if the following properties are fulfilled.*

1. *D has exactly one source r.*
2. *Every vertex in D has at most two successors.*
3. *If vertex d has two successors d' and d'', then $X_d = X_{d'} = X_{d''}$.*

4. *If vertex d has one successors d', then* $|(X_d \setminus X_{d'}) \cup (X_{d'} \setminus X_d)| = 1$.

Berwanger et al. [4] showed that if digraph G has a DAG-decomposition of width k, it also has a nice DAG-decomposition of width k. Moreover, since deleting transitive edges from D does neither destroy any of the properties of a DAG-decomposition, nor increase the width of the DAG-decomposition, we get the following property.

Lemma 1. *If digraph G has a DAG-decomposition of width k, it also has a nice DAG-decomposition $(D, \mathcal{X})$ of width k such that D has no transitive edges.*

Proposition 4. *For every semicomplete digraph G it holds that*

$$\text{d-pw}(G) \leq \text{dagw}(G) - 1.$$

Proof. Let G be a semicomplete digraph and let $(D, \mathcal{X})$ be a nice DAG-decomposition for G of width k with digraph D, vertex set V_D and $\mathcal{X} = \{X_u \mid u \in V_D\}$. By Lemma 1 we can assume that D has exactly one source, every vertex in D has at most two successors and no transitive edges. We show that in case D is not a path, we can convert it into a path without increasing the width. Assume D is not a path. For any vertex r let V_{D_r} is the set of vertices of D which are reachable from r. Let D_t be the maximal subdigraph of D with unique source t. Consider vertex $q \in V_D$ with two successors s and t. We differentiate three cases: All vertices from G which are in bags of D_s are also in the bags of D_t (Case 1.a), the opposite inclusion (Case 1.b) or, at last none of these inclusions (Case 2) occur.

Case 1.a: $(\bigcup_{u \in V_{D_s}} X_u) \cup X_q \subseteq (\bigcup_{u \in V_{D_t}} X_u) \cup X_q$.

In order to define a new DAG-decomposition $(D', \mathcal{X}')$ for G, we simply remove all vertices $V_{D_s} \backslash V_{D_t}$ from D and forget all bags associated with removed vertices. We now show that $(D', \mathcal{X}')$ is a DAG-decomposition for G by checking the conditions of Definition 3.

– (dagw-1) Is satisfied since

$$\bigcup_{u \in V_{D'}} X_u = \bigcup_{u \in V_D \backslash V_{D_s}} X_u \cup \bigcup_{u \in V_{D_t}} X_u \overset{(*)}{\supseteq} \bigcup_{u \in V_D \backslash V_{D_s}} X_u \cup \bigcup_{u \in V_{D_s}} X_u = \bigcup_{u \in V_D} X_u = V_G$$

The inclusion in $(*)$ holds by assumption of case $1a)$ since $q \in V_D \backslash V_{D_s}$.

– (dagw-2) is still satisfied since for every $a, b, c \in V_{D'}$ it holds that if $a \preccurlyeq_{D'} b \preccurlyeq_{D'} c$ then

$$X'_a \cap X'_b = X_a \cap X_c \subseteq X_b = X'_b$$

– (dagw-3) Let $(a, b) \in E_{D'}$, then it follows that $(a, b) \in E_D$. Therefore, it must hold that $X_a \cap X_b$ guards $X_{\succcurlyeq b} \backslash X_a$. It holds that $X'_a = X_a$ and $X'_b = X_b$. Further, $X'_{\succcurlyeq b}$ is the union of all bags of vertices that we can reach from vertex b in D', such that $X'_{\succcurlyeq b} = \bigcup_{b \preccurlyeq_{D'} u} X_u$.

(i) If $b \preccurlyeq_{D'} t$, then:

$$X'_{\succcurlyeq b} = \bigcup_{b \preccurlyeq_{D'} u \preccurlyeq_{D'} t} X_u \cup \bigcup_{t \preccurlyeq_{D'} u} X'_u = \bigcup_{b \preccurlyeq_D u \preccurlyeq_D t} X_u \cup \bigcup_{t \preccurlyeq_D u} X_u$$

$$(\text{since } X_q \subseteq \bigcup_{b \preccurlyeq_D u \preccurlyeq_D t} X_u)$$

$$= \bigcup_{b \preccurlyeq_D u \preccurlyeq_D t} X_u \cup \bigcup_{t \preccurlyeq_D u} X_u \cup \bigcup_{s \preccurlyeq_D u} X_u = \bigcup_{b \preccurlyeq_D u} X_u = X_{\succcurlyeq b}$$

(ii) Else $t \prec_{D'} b$, then: Since every successor of b in D is also in D' it holds that

$$X'_{\succcurlyeq b} = \bigcup_{b \preccurlyeq_{D'} u} X_u = \bigcup_{b \preccurlyeq_D u} X_u = X_{\succcurlyeq b}$$

This leads to $X'_a \cap X'_b = X_a \cap X_b$ guards $X'_{\succcurlyeq b} \backslash X'_a = X_{\succcurlyeq b} \backslash X_a$.
Thus, all requirements of a DAG-decomposition are met by $(D', \mathcal{X}')$.

Case 1.b: $(\bigcup_{u \in V_{D_t}} X_u) \cup X_q \subseteq (\bigcup_{u \in V_{D_s}} X_u) \cup X_q$ can be handled analogously to case 1.a.
Case 2: $(\bigcup_{u \in V_{D_s}} X_u) \cup X_q \not\subseteq (\bigcup_{u \in V_{D_t}} X_u) \cup X_q$ and $(\bigcup_{u \in V_{D_t}} X_u) \cup X_q \not\subseteq (\bigcup_{u \in V_{D_s}} X_u) \cup X_q$. More informally, this means that there exist vertices from G that are only represented in bags of D_s but not in bags of D_t. We show now, that this case cannot occur. There are x, y such that

$$x \in X_q \cup \bigcup_{u \in V_{D \geq s}} X_u, x \notin X_q \cup \bigcup_{u \in V_{D \geq t}} X_u \tag{1}$$

$$y \notin X_q \cup \bigcup_{u \in V_{D \geq s}} X_u, y \in X_q \cup \bigcup_{u \in V_{D \geq t}} X_u \tag{2}$$

Since G is semicomplete, there is an arc between x and y in G. W.l.o.g. let $(x, y) \in E_G$. By the connectivity property given by (dagw-2) it holds that $x, y \notin \bigcup_{u \preccurlyeq_D q} X_u$, since $x, y \notin X_q$. Let $w \in V_D, x \in X_w, x \notin X_u$ and $u \preccurlyeq_D w$. As Eq. (1) holds, this leads to $s \preccurlyeq_D w$. By (dagw-3) it further holds that $X_{w'} \cap X_w$ guards $X_{\succcurlyeq w} \backslash X_{w'}$ for a predecessor w' of w in D with $w' \neq s$. This means that for all $(z, z') \in E_G$ with $z \in X_{\succcurlyeq w} \backslash X_{w'}$ it holds that $z' \in (X_{\succcurlyeq w} \backslash X_{w'}) \cup (X_{w'} \cap X_w)$.
As assumed before, it holds that $(x, y) \in E_G$ with $x \in X_{\succcurlyeq w} \backslash X_{w'}$. By Eq. (2) it holds that $y \notin X_{w'} \cap X_w \Rightarrow y \in X_{\succcurlyeq w} \backslash X_{w'}$. By Eq. (2) it holds that $y \notin X_{w'} \Rightarrow y \in X_{\succcurlyeq w} = \bigcup_{w \preccurlyeq_D u} X_u$. But since $s \preccurlyeq_D w$ it holds that $\bigcup_{w \preccurlyeq_D u} X_u \subseteq \bigcup_{s \preccurlyeq_D u} X_u$. This contradicts that by Eq. (2) it holds that $y \notin \bigcup_{s \preccurlyeq_D u} X_u$. This leads to the conclusion that case 2 cannot occur.

Consequently, starting at the root, we can transform every DAG D of a DAG-decomposition of the semicomplete digraph G into a directed path. Since directed path-width is exactly the path variant of DAG-width, $\text{d-pw}(G) \leq \text{dagw}(G) - 1$ holds. □

By Proposition 4 we can conclude that on semicomplete digraphs, DAG-width plus 1 and path-width are equal.

Corollary 1. *For every semicomplete digraph G it holds that*

$$\text{d-pw}(G) + 1 = \text{dagw}(G).$$

4.2 Escaping Pursuit in the Jungle: Directed Path-Width, Directed Tree-Width and Kelly-Width

Fradkin and Seymour [11] gave a description of semicomplete digraphs of bounded directed path-width. Indeed, they proved that every semicomplete digraph of huge directed path-width must contain a subdivision of a large bioriented clique [11]. While this result immediately implies that directed path-width acts, parametrically, as a lower bound for all tree-width inspired directed width measures discussed in this paper, the proof uses a Ramsey argument and thus, for G to contain a subdivision of the complete biorientation of K_t, the directed path-width must be exponential in t. However, Fradkin and Seymour introduced another obstruction to small directed path-width on semicomplete digraphs which is similar to the idea of well linked sets. With a bit of more careful analysis we are able to obtain the quadratic bounds of Theorem 1.

Note that [11] could also be used for comparisons between directed path-width and DAG-width, but this would only lead to equivalence between those parameters, whereas we could prove equality (plus 1).

Two vertices u, v are *k-connected*, if there are at least k internally-disjoint paths from u to v and from v to u. For digraph $G = (V, E)$ a set $U \subseteq V$ is a *k-jungle* in G if $|U| = k$ and for all $u, v \in U$ it holds that u and v are k-connected.

For both directed tree-width and Kelly-width, we show that the existence of a $k+1$-jungle is enough to ensure a winning strategy for the robber against k cops in the respective variants of cops and robber game. Let us start with directed tree-width.

Proposition 5. *Let G be a semicomplete digraph. If d-pw$(G) \geq 4(k+1)^2 + 7(k+1)$ then d-tw$(G) \geq k$.*

Proof. Let us assume d-pw$(G) \geq 4(k+1)^2 + 7(k+1)$. Then, by the results from [11], we know that $G = (V, E)$ contains a $k+1$-jungle $J \subseteq V$. If we can show that the existence of J is enough to ensure that k-cops cannot catch the robber in the visible strong component cops and robber game on G, it follows from Proposition 1 that the directed tree-width of G must be at least k and thus the assertion follows. Hence what is left to do is describe a winning strategy for the robber against k cops on a $k+1$-jungle J. For the first position (C_0, r_0) we have $C_0 = \emptyset$ and the robber may select r_0 to be any vertex of J. Now suppose the game has been going on for i rounds and in each round the robber was able to select a vertex of J as her position. Let (C_{i-1}, r_{i-1}) be the current state of the game and let $C_i \subseteq V$ be the next position of the cops. In case $r_{i-1} \notin C_i$ there is nothing to do for the robber and she can stay where she is i.e. $r_i := r_{i-1}$. So we may assume $r_{i-1} \in C_i$. In this case we know $|C_i \setminus \{r_{i-1}\}| \leq k-1$ and thus $|C_{i-1} \cap C_i| \leq k-1$. Hence there must exist a vertex $v \in J \setminus C_i$. As $r_{i-1} \neq v$ we know from J being a $k+1$-jungle that there exist $k+1$ pairwise internally disjoint paths from r_{i-1} to v and vice versa. As $|C_i| \leq k$ in $G - (C_{i-1} \cap C_i)$ at least one path from r_{i-1} to u and one from u to r_{i-1} must be left and thus both vertices belong to the same strong component of $G - (C_{i-1} \cap C_i)$. Thus v is reachable from r_{i-1} and we may set $r_i := v$. As the robber was able to flee to another vertex of J our claim now follows by induction. □

From [2] and Corollary 1 we already know an upper bound on directed path-width in terms of Kelly-width, which is d-pw$(G) \leq 72$kw$(G)^2 + 1$. We can improve this bound following the same general idea as given above. Indeed, since in the strategy as described in the proof of Proposition 5 the robber only changed her position if she was threatened to be caught if she did not, the strategy above is already a strategy for a visible robber in the strong component game. Since the reachability searching game is a relaxation of the strong component game and the (in)visibility of the robber does not play a role in this strategy it is straightforward to see that using the same technique, an invisible and inert robber can also avoid being caught by k cops in the reachability searching game. From these arguments we obtain the following result.

Proposition 6. *Let G be a semicomplete digraph. If $d\text{-}pw(G) \geq 4(k+1)^2+7(k+1)$ then $kw(G) \geq k$.*

4.3 Directed (Linear) Clique-Width and Directed Path-Width on Semicomplete Digraphs

In [10], the authors prove that on semicomplete digraphs, directed path-width can be used to give an upper bound for directed clique-width. The main idea of the proof of [10, Lemma 2.14] is to define a directed clique-width expression along a nice path-decomposition. Since this proof only uses linear clique-width operations, we can restrict their result to the following result:

Proposition 7 ([10]). *For every semicomplete digraph G it holds that*

$$d\text{-}cw(G) \leq d\text{-}lcw(G) \leq d\text{-}pw(G)+2.$$

Note that the other direction, i.e., using directed (linear) clique-width as an upper bound of directed path-width, is not possible for semicomplete digraphs in general. That follows directly from the proof of Proposition 3, as the counterexample, a bioriented clique, is a semicomplete digraph.

Using the results from this and previous subsections, it is possible to improve the general results for the comparison of directed width parameters on semicomplete digraphs.

By using Propositions 4, 5, 6 and 7 we improve also other bounds between directed width parameters on semicomplete digraphs.

5 Conclusion

The landscape of directed width measures is a wild one. Started by the introduction of directed tree-width many different generalizations of undirected tree-width have been invented and received different amounts of attention. Some of these parameters were considered very little; possibly because of the results of [13], which essentially rule out any algorithmic application of these parameters beyond some specialized routing problems. So while the search for 'good' digraph width parameters inspired by tree-width does not seem very promising, one could turn to the logic based parameters instead. Here directed clique-width reigns supreme, but recently other attempts at finding interesting parameters such as a directed version of *maximum induced matching width* [19] have been made.

In this paper we have shown the equivalence of directed path-width, directed tree-width, Kelly-width and DAG-width on semicomplete digraphs. In particular this implies that each of these measures acts as an upper bound on directed clique-width and thus, the algorithmic power of directed clique-width can now be accessed by any of the other parameters.

Hence as a consequence of our results on semicomplete digraphs every digraph problem, which is describable in MSO_1 logic is fixed parameter tractable for these width measures if a decomposition of bounded width is given.

Our result, that computing DAG-width is in NP on semicomplete digraphs while it is PSPACE-hard in general [1], recalls the question if computing directed path-width and thus, DAG-width is NP-hard on semicomplete digraphs, though there are FPT algorithms to solve this problem [10].

References

1. Amiri, S.A., Kreutzer, S., Rabinovich, R.: DAG-width is PSPACE-complete. Theoret. Comput. Sci. **655**, 78–89 (2016)
2. Amiri, S.A., Kaiser, L., Kreutzer, S., Rabinovich, R., Siebertz, S.: Graph searching games and width measures for directed graphs. In: 32nd International Symposium on Theoretical Aspects of Computer Science (STACS), volume 30 of Leibniz International Proceedings in Informatics (LIPIcs), pp. 34–47. Schloss Dagstuhl-Leibniz-Zentrum fuer Informatik (2015)
3. Bang-Jensen, J., Gutin, G. (eds.): Classes of Directed Graphs. SMM, Springer, Cham (2018). https://doi.org/10.1007/978-3-319-71840-8
4. Berwanger, D., Dawar, A., Hunter, P., Kreutzer, S.: DAG-width and parity games. In: Durand, B., Thomas, W. (eds.) STACS 2006. LNCS, vol. 3884, pp. 524–536. Springer, Heidelberg (2006). https://doi.org/10.1007/11672142_43
5. Berwanger, D., Dawar, A., Hunter, P., Kreutzer, S., Obdržálek, J.: The DAG-width of directed graphs. J. Comb. Theory Ser. B **102**(4), 900–923 (2012)
6. Campos, V., Lopes, R., Maia, A.K., Sau, I.: Adapting the directed grid theorem into an FPT algorithm. Electron. Notes Theoret. Comput. Sci. **346**, 229–240 (2019)
7. Chudnovsky, M., Seymour, P.D.: A well-quasi-order for tournaments. J. Comb. Theory Ser. B **101**(1), 47–53 (2011)
8. Courcelle, B., Makowsky, J.A., Rotics, U.: Linear time solvable optimization problems on graphs of bounded clique-width. Theory Comput. Syst. **33**(2), 125–150 (2000)
9. Courcelle, B., Olariu, S.: Upper bounds to the clique width of graphs. Discret. Appl. Math. **101**, 77–114 (2000)
10. Fomin, F.V., Pilipczuk, M.: On width measures and topological problems on semi-complete digraphs. J. Comb. Theory Ser. B **138**, 78–165 (2019)
11. Fradkin, A., Seymour, P.D.: Tournament pathwidth and topological containment. J. Comb. Theory Ser. B **103**, 374–384 (2013)
12. Ganian, R., Hlinený, P., Kneis, J., Langer, A., Obdržálek, J., Rossmanith, P.: Digraph width measures in parameterized algorithmics. Discret. Appl. Math. **168**, 88–107 (2014)
13. Ganian, R., et al.: Are there any good digraph width measures? J. Comb. Theory Ser. B **116**, 250–286 (2016)
14. Gurski, F., Komander, D., Rehs, C.: How to compute digraph width measures on directed co-graphs. Theoret. Comput. Sci. **855**, 161–185 (2021)
15. Gurski, F., Rehs, C.: Comparing linear width parameters for directed graphs. Theory Comput. Syst. **63**(6), 1358–1387 (2019)

16. Gurski, F., Wanke, E.: On the relationship between NLC-width and linear NLC-width. Theoret. Comput. Sci. **347**(1–2), 76–89 (2005)
17. Gurski, F., Wanke, E., Yilmaz, E.: Directed NLC-width. Theoret. Comput. Sci. **616**, 1–17 (2016)
18. Hunter, P., Kreutzer, S.: Digraph measures: Kelly decompositions, games, and orderings. Theoret. Comput. Sci. **399**(3), 206–219 (2008)
19. Jaffke, L., Kwon, O., Telle, J.A.: Classes of intersection digraphs with good algorithmic properties. ACM Computing Research Repository (CoRR), abs/2105.01413 (2021)
20. Johnson, T., Robertson, N., Seymour, P.D., Thomas, R.: Directed tree-width. J. Comb. Theory Ser. B **82**, 138–155 (2001)
21. Kim, I., Seymour, P.D.: Tournament minors. J. Comb. Theory Ser. B **112**(C), 138–153 (2015)
22. Oum, S., Seymour, P.D.: Approximating clique-width and branch-width. J. Comb. Theory Ser. B **96**(4), 514–528 (2006)
23. Robertson, N., Seymour, P.D.: Graph minors II. Algorithmic aspects of tree width. J. Algorithms **7**, 309–322 (1986)

Improved Parameterized Approximation for Balanced k-Median

Zhen Zhang[1,2] and Qilong Feng[2(✉)]

[1] School of Frontier Crossover Studies, Hunan University of Technology and Business, Changsha 410000, People's Republic of China

[2] School of Computer Science and Engineering, Central South University, Changsha 410000, People's Republic of China

csufeng@mail.csu.edu.cn

Abstract. Balanced k-median is a frequently encountered problem in applications requiring balanced clustering results, which generalizes the standard k-median problem in that the number of clients connected to each facility is constrained by the given lower and upper bounds. This problem is known to be W[2]-hard if parameterized by k, implying that the existence of an FPT(k)-time exact algorithm is unlikely. In this paper, we give a $(3 + \epsilon)$-approximation algorithm for balanced k-median that runs in FPT(k) time, improving upon the previous best approximation ratio of $7.2 + \epsilon$ obtained in the same time. The crucial step in getting the improved ratio and our main technical contribution is a different random sampling method for selecting opened facilities.

Keywords: Approximation algorithm · Parameterized algorithm · k-median

1 Introduction

Clustering is a common task in computer science, which aims to partition a given set of data points into several clusters, such that the points assigned to the same cluster are relatively similar. Many objective functions have been proposed to estimate the clustering quality, among which the k-median cost function is one of the most ordinary versions. Given a set of facilities and a set of clients in a metric space, the goal of the k-median problem is to open at most k facilities and connect each client to an opened facility, such that the sum of the distance from each client to the corresponding facility is minimized. Designing approximation algorithms for this problem remains an active area of research due to its applications in many fields [2,6,7,16,17,20]. The current best approximation ratio for the k-median problem is $2.675 + \epsilon$ [6], which was obtained based on the pseudo-approximation technique outlined in [20].

This work was supported by National Natural Science Foundation of China (61872450 and 62172446).

D.-Z. Du et al. (Eds.): COCOA 2021, LNCS 13135, pp. 629–640, 2021.
https://doi.org/10.1007/978-3-030-92681-6_49

In many clustering applications, the data points are required to be regularly assigned to each cluster, such that the sizes of the clusters are similar to each other. Such examples can be found in the design of wireless sensor networks, where balanced clustering is utilized to avoid unbalanced energy consumption [23], and distributed computation, where balancedness needs to be taken into account when dispatching data to multiple machines [3,4,11]. The balanced k-median problem has been extensively studied due to its important role in these applications [5,11,12,21]. This problem generalizes the k-median problem in that the size of each cluster is constrained within a given interval, which can be formally defined as follows.

Definition 1 (balanced k-median [12]). *The balanced k-median problem considers a set $\mathcal{C}$ of clients and a set $\mathcal{F}$ of facilities located in a metric space, an integer $k > 0$, a positive number $B_l \leq |\mathcal{C}|$ called lower bound, and a number $B_u \geq \max\{|\mathcal{C}|k^{-1}, B_l\}$ called upper bound. The goal is to open a set $\mathcal{H} \subseteq \mathcal{F}$ of no more than k facilities and connect each $j \in \mathcal{C}$ to an opened facility $\tau(j) \in \mathcal{H}$, such that $B_l \leq |\tau^{-1}(i)| \leq B_u$ for each $i \in \mathcal{H}$, and the objective function $\sum_{j\in\mathcal{C}} \Delta(j, \tau(j))$ is minimized, where $\Delta(j, \tau(j))$ denotes the distance from j to $\tau(j)$.*

The balanced k-median problem is much more poorly understood than the standard k-median problem. While practical algorithms for the balanced k-median problem are known [5,21], these approaches are at best heuristics and have no performance guarantee. Dick *et al.* [11] gave a constant factor approximation algorithm for balanced k-median based on the techniques of min cost flow and LP-rounding, but it violates the upper bound by a constant factor. Ding [12] gave a $(7.2 + \epsilon)$-approximation algorithm for the problem under the assumption that k is a fixed parameter. Whether a constant factor approximation for balanced k-median can be obtained in polynomial time without such assumption is still an opened problem.

1.1 Our Contributions

In this paper we consider the balanced k-median problem for the case where k is small. Given an instance $(\mathcal{C}, \mathcal{F}, k, B_l, B_u)$ of the problem, brute force searching yields an optimal solution in $(|\mathcal{C}||\mathcal{F}|)^{O(k)}$ time, but the question we consider is: what can be done in FPT(k) time (*i.e.*, $f(k)(|\mathcal{C}||\mathcal{F}|)^{O(1)}$)?

We cannot hope to obtain an optimal solution to the balanced k-median problem in FPT(k) time: Guha and Khuller [13] showed that the problem is W[2]-hard if parameterized by k, and Cohen-Addad *et al.* [8] proved that if the Gap-Exponential Time Hypothesis [22] is true, then the approximation ratios of FPT(k)-time algorithms for the problem cannot be better than $1 + 2/e - \epsilon$. However, these negative results do not rule out the possibility of approximating balanced k-median by a constant factor in FPT(k) time. Indeed, Ding [12] showed that an FPT(k)-time algorithm yields a $(7.2 + \epsilon)$-approximation for the problem. We go a step further in this direction and give a $(3 + \epsilon)$-approximation algorithm that runs in FPT(k) time.

Fig. 1. (a) The clusters in a solution to the k-median problem; (b) the clusters in a solution to the balanced k-median problem, where $B_l = 6$ and $B_u = 8$.

Theorem 1. *Given a real number $\epsilon \in (0, 1]$ and an instance $(\mathcal{C}, \mathcal{F}, k, B_l, B_u)$ of balanced k-median, there exists a $(3 + O(\epsilon))$-approximation algorithm that runs in $(|\mathcal{C}||\mathcal{F}|)^{O(1)}(k\epsilon^{-1})^{O(k)}$ time.*

A crucial property utilized in the algorithms developed for the standard k-median problem is that each client from a cluster is located in the Voronoi cell defined by the corresponding facility. The main obstacle in solving the balanced k-median problem lies in the deficiency of this property (the clients are correlated with each other to satisfy the constraint on the size of each cluster, which is illustrated in Fig. 1). We propose a random sampling-based approach to deal with this obstacle, which involves the following two steps.

- Instead of directly selecting the opened facilities, our algorithm first identifies a set $\mathcal{R}$ that involves k clients close to the facilities opened in an optimal solution. Such k clients can be found with a factor of $|\mathcal{R}|^k$ multiplied on runtime. To ensure that our algorithm runs in FPT(k) time, the size of client set $\mathcal{R}$ should be independent of the total number of the clients and facilities. We prove that randomly sampling k clients yields the desired client set with certain probability.
- The facilities opened in our solution are selected based on the k clients identified in the first step. When solving the standard k-median problem, we can directly connect each client to the nearest opened facility to minimize the clustering cost. However, for the case of balanced k-median, connecting clients is not as trivial as exhibited in the k-median problem due to the constraint on the sizes of clusters. We give a connection algorithm that ensures the legitimacy of the clustering result.

1.2 Other Related Work

The variants of the k-median problem that take into account upper or lower bounds on cluster sizes have been paid lots of attentions. For the case where the cluster sizes have upper bounds, the current best approximation guarantee is a ratio of $O(\log k)$ obtained by tree embedding of the underlying metric [1], and a constant factor approximation can be obtained with violating the constraint on the cluster sizes [10,19] or cluster number [18]. For the case with lower bounds, Han *et al.* [15] and Guo *et al.* [14] gave an $O(1)$-approximation that violates the lower bounds by some constant factor. Based on such a bi-criteria approximation solution, Guo *et al.* [14] gave a 516-approximation algorithm that firmly satisfies the constraints.

Most recently, the technique of coreset construction has been refined and used for designing parameterized approximation algorithms for many clustering problems. These problems include k-median [8], k-means [8], facility location [8], capacitated k-median [1,9], and capacitated k-means [9]. A coreset is a small subset of the input such that minimizing the objective function on it is sufficient to obtain the desired solution. In [1,8,9], the coresets are constructed by partitioning the space into a set of cells and selecting a weighted representative point from each cell. The main challenge in constructing such coresets for the balanced k-median problem lies in that the loss in the approximation guarantee induced by each cell is quite difficult to analyze due to the additional constraint on the cluster sizes. How to approximate the input data by a small coreset in the balanced k-median problem is still not clear.

1.3 Preliminaries

Denote by $\mathcal{I} = \{\mathcal{C}, \mathcal{F}, k, B_l, B_u\}$ an instance of the balanced k-median problem. For each $x, y \in \mathcal{C} \cup \mathcal{F}$ and $\mathcal{S}_1, \mathcal{S}_2 \subseteq \mathcal{C} \cup \mathcal{F}$, let $\Delta(x, y)$ denote the distance from x to y, and define $\Delta(x, \mathcal{S}_2) = \min_{z \in \mathcal{S}_2} \Delta(x, z)$ and $\Delta(\mathcal{S}_1, y) = \sum_{z \in \mathcal{S}_1} \Delta(z, y)$. The following algebraic fact is useful in analyzing the running time of our algorithm.

Lemma 1. *Given two numbers i and j larger than 1, we have $\log^j i \leq i j^{O(j)}$.*

Proof. We consider the following two cases: (1) $j < \frac{\log i}{\log \log i}$, and (2) $j \geq \frac{\log i}{\log \log i}$. For case (1), we have $\log^j i < \log^{\frac{\log i}{\log \log i}} i = i$, as desired. For case (2), we have $\log i \leq O(j \log j)$, which implies that $\log^j i \leq j^{O(j)}$. This completes the proof of Lemma 1. □

2 The Sampling Algorithm

In this section, we give a sampling algorithm that identifies a set of clients close to the facilities opened in an optimal solution, which is described in Algorithm 1 and Algorithm 2. Algorithm 1 invokes Algorithm 2 to obtain a set of clients in each iteration. The input parameters of Algorithm 2 involve an integer k and three sets $\mathcal{R}$, $\mathcal{C}^\dagger$, and $\mathbb{R}$, where k denotes the maximum allowable quantity of opened facilities, $\mathcal{R}$ is used to record the clients that has been selected, $\mathcal{C}^\dagger$ is a subset of $\mathcal{C}$ that is sampled from, and $\mathbb{R}$ contains all the client sets sampled by Algorithm 1. The idea of Algorithm 2 is to select the clients close to the opened facilities in optimal solutions by random sampling. For the case where such clients are hard to be selected, the algorithm narrows the sample range to increase the probability of finding the desired clients.

We now introduce some notations to help with analysis. Let k^* denote the number of facilities opened in an optimal solution to $\mathcal{I}$, let $\mathbb{C} = \{\mathcal{C}_1, \ldots, \mathcal{C}_{k^*}\}$ denote the set of clusters in the solution, where $|\mathcal{C}_1| \geq |\mathcal{C}_2| \geq \ldots \geq |\mathcal{C}_{k^*}|$, and

Algorithm 1: The Sampling Algorithm

Input: An instance $\mathcal{I} = (\mathcal{C}, \mathcal{F}, k, B_l, B_u)$ of the balanced k-median problem and a real number $\epsilon \in (0, 1]$;
Output: A set $\mathbb{R}$ of client sets;
1 $\mathbb{R} \Leftarrow \emptyset$;
2 **for** $i \Leftarrow 1$ *to* $(k\epsilon^{-1})^{O(k)}$ **do**
3 **Median**$(k, \emptyset, \mathcal{C}, \mathbb{R})$;
4 **return** $\mathbb{R}$.

Algorithm 2: Median$(k, \mathcal{R}, \mathcal{C}^{\dagger}, \mathbb{R})$

if $|\mathcal{R}| = k$ **then**
  $\mathbb{R} \Leftarrow \mathbb{R} \cup \{\mathcal{R}\}$;
else
  Randomly and uniformly select a client $c \in \mathcal{C}^{\dagger}$;
  Median$(k, \mathcal{R} \cup \{c\}, \mathcal{C}^{\dagger}, \mathbb{R})$;
  $\mathcal{C}^{\ddagger} \Leftarrow \underset{\mathcal{C}' \subseteq \mathcal{C}^{\dagger}, |\mathcal{C}'| = \lceil |\mathcal{C}^{\dagger}|/2 \rceil}{\arg\max} \sum_{x \in \mathcal{C}'} \Delta(x, \mathcal{R})$;
  Median$(k, \mathcal{R}, \mathcal{C}^{\ddagger}, \mathbb{R})$.

let $f_1, \ldots, f_{k^*}$ be the corresponding opened facilities, where the clients from $\mathcal{C}_t$ are connected to f_t for each $t \in [1, k^*]$. For each $\mathcal{C}_t \in \mathbb{C}$, define $opt_t = \Delta(\mathcal{C}_t, f_t)$, $r_t = opt_t/|\mathcal{C}_t|$, and $\mathbf{b}_\alpha(\mathcal{C}_t) = \{c \in \mathcal{C}_t : \Delta(c, f_t) \leq \alpha r_t\}$ for any $\alpha > 0$. $\mathbf{b}_\alpha(\mathcal{C}_t)$ is the set of clients from $\mathcal{C}_t$ that lie on a closed ball with radius αr_t centered at f_t. Define $opt = \sum_{t=1}^{k^*} opt_t$ as the cost of an optimal solution to $\mathcal{I}$. The following result shows a lower bound on the proportion of $\mathbf{b}_\alpha(\mathcal{C}_t)$ in $\mathcal{C}_t$ for each $\alpha \geq 1$ and $\mathcal{C}_t \in \mathbb{C}$.

Lemma 2. *Given a cluster $\mathcal{C}_t \in \mathbb{C}$ and a real number $\alpha \geq 1$, we have $|\mathbf{b}_\alpha(\mathcal{C}_t)| \geq (1 - \frac{1}{\alpha})|\mathcal{C}_t|$.*

Proof. The definition of $\mathbf{b}_\alpha(\mathcal{C}_t)$ implies that

$$\Delta(\mathcal{C}_t \backslash \mathbf{b}_\alpha(\mathcal{C}_t), f_t) > |\mathcal{C}_t \backslash \mathbf{b}_\alpha(\mathcal{C}_t)| \alpha r_t = \alpha \frac{|\mathcal{C}_t \backslash \mathbf{b}_\alpha(\mathcal{C}_t)|}{|\mathcal{C}_t|} opt_t. \quad (1)$$

By the fact that $\Delta(\mathcal{C}_t \backslash \mathbf{b}_\alpha(\mathcal{C}_t), f_t) \leq opt_t$ and inequality (1), we know that $|\mathcal{C}_t \backslash \mathbf{b}_\alpha(\mathcal{C}_t)| < \frac{1}{\alpha}|\mathcal{C}_t|$, and thus $|\mathbf{b}_\alpha(\mathcal{C}_t)| \geq (1 - \frac{1}{\alpha})|\mathcal{C}_t|$. This completes the proof of Lemma 2. □

Let $\mathbb{R}$ denote the set given by Algorithm 1. It can be seen that $|\mathbb{R}| = (k\epsilon^{-1})^{O(k)} \log^k |\mathcal{C}|$, which is upper-bounded by $(k\epsilon^{-1})^{O(k)}|\mathcal{C}|$ due to Lemma 1. We will show the following result in this section.

Lemma 3. *With a constant probability, there exists a client set $\mathcal{R} \in \mathbb{R}$ satisfying $\sum_{t=1}^{k^*} |\mathcal{C}_t|(f_t, \mathcal{R}) \leq (1 + O(\epsilon))opt$.*

Lemma 3 says that $\mathbb{R}$ involves a client set containing k^* clients close to $f_1, \ldots, f_{k^*}$ with high probability. We first give a high-level idea for proving Lemma 3. Given a cluster $\mathcal{C}_t \in \mathbb{C}$ and a real number $\alpha > 0$, it can be seen that each client from $\mathbf{b}_\alpha(\mathcal{C}_t)$ is close to f_t if α is small. We want to find a client from $\mathbf{b}_\alpha(\mathcal{C}_t)$ for each $t \in [1, k^*]$. Lemma 2 implies that a certain number of clients from $\mathcal{C}_t$ lie on $\mathbf{b}_\alpha(\mathcal{C}_t)$, and thus $\mathbf{b}_\alpha(\mathcal{C}_t)$ has a good chance to be sampled from if $\mathcal{C}_t$ accounts for a relatively large part of the sample range. For the case where $\mathcal{C}_t$ contains only a tiny portion of the sample range, we show that Algorithm 2 can successfully narrow the sample range and increase the possibility of finding a client from $\mathbf{b}_\alpha(\mathcal{C}_t)$.

Denote by $\mathbb{R}^\dagger$ the set of client sets obtained in the first iteration of Algorithm 1. For each $\mathcal{R} \in \mathbb{R}^\dagger$ and $t \in [1, k^*]$, let $\mathcal{R}_t$ denote the set of the first t clients added to $\mathcal{R}$ by Algorithm 2. Rather than immediately proving Lemma 3, we first consider the following variant, which will be shown to be maintained for $t \in [1, k^*]$.

$\kappa(t)$: With probability higher than $(\epsilon k^{-1})^{O(t)}$, there exists a client set $\mathcal{R} \in \mathbb{R}^\dagger$, such that $\sum_{s=1}^{t} |\mathcal{C}_s| \Delta(f_s, \mathcal{R}_s) \leq (1 + O(\epsilon)) \sum_{s=1}^{t} opt_s + \frac{\epsilon t}{k^*} opt$.

If $\kappa(k^*)$ is true, then the probability that Algorithm 1 finds the desired client set of Lemma 3 in each iteration can be lower-bounded by $(\epsilon k^{-1})^{O(k^*)} \geq (\epsilon k^{-1})^{O(k)}$. Since the algorithm iterates $(k\epsilon^{-1})^{O(k)}$ times, the probability can be boosted to a constant, which implies that Lemma 3 is true.

It remains to show the correctness of invariant $\kappa(t)$. The invariant is proved by induction on t. We first consider the case of $t = 1$. Observe that the client from $\mathcal{R}_1$ is randomly and uniformly selected from $\mathcal{C}$. We have

$$\frac{|\mathbf{b}_{1+\epsilon}(\mathcal{C}_1)|}{|\mathcal{C}|} \geq \frac{|\mathbf{b}_{1+\epsilon}(\mathcal{C}_1)|}{k^* |\mathcal{C}_1|} \geq \frac{|\mathbf{b}_{1+\epsilon}(\mathcal{C}_1)|}{k |\mathcal{C}_1|} \geq (\frac{\epsilon}{k})^{O(1)}, \tag{2}$$

where the first step follows from the fact that $|\mathcal{C}_1| \geq |\mathcal{C}_2| \geq \ldots \geq |\mathcal{C}_{k^*}|$, and the last step is due to Lemma 2. The definition of $\mathbf{b}_{1+\epsilon}(\mathcal{C}_1)$ and inequality (2) imply that with probability higher than $(\frac{\epsilon}{k})^{O(1)}$, the client $c \in \mathcal{R}_1$ satisfies $|\mathcal{C}_1| \Delta(f_1, c) \leq (1 + \epsilon) opt_1$, which implies that $\kappa(1)$ is true.

Given an integer $t \in (1, k^*]$, we prove the correctness of $\kappa(t)$ under the assumption that $\kappa(t')$ holds for each $t' \in [1, t-1]$. Define $\mathcal{B}_t = \{c \in \mathcal{C} : \Delta(c, \mathcal{R}_{t-1}) \leq \epsilon \frac{opt}{k^* |\mathcal{C}_t|}\}$. We break the analysis into the following two cases: (1) $\mathbf{b}_{1+\epsilon}(\mathcal{C}_t) \cap \mathcal{B}_t \neq \emptyset$, and (2) $\mathbf{b}_{1+\epsilon}(\mathcal{C}_t) \cap \mathcal{B}_t = \emptyset$.

We first consider case (1). Let c denote a client from $\mathbf{b}_{1+\epsilon}(\mathcal{C}_t) \cap \mathcal{B}_t$, and let $r(c)$ be its nearest client from $\mathcal{R}_{t-1}$. It is the case that

$$\Delta(f_t, r(c)) \leq \Delta(f_t, c) + \Delta(c, r(c)) \leq (1 + \epsilon) r_t + \epsilon \frac{opt}{k^* |\mathcal{C}_t|}, \tag{3}$$

where the first step is due to triangle inequality, and the second step follows from the definitions of $\mathbf{b}_{1+\epsilon}(\mathcal{C}_t)$ and $\mathcal{B}_t$. Consequently, we know that inequality

$$\begin{aligned}\sum_{s=1}^{t}|\mathcal{C}_s|\Delta(f_s,\mathcal{R}_s) &\leq \sum_{s=1}^{t-1}|\mathcal{C}_s|\Delta(f_s,\mathcal{R}_s)+|\mathcal{C}_t|\Delta(f_t,r(c))\\ &\leq (1+O(\epsilon))\sum_{s=1}^{t-1}opt_s+\frac{\epsilon(t-1)}{k^*}opt+|\mathcal{C}_t|\Delta(f_t,r(c))\\ &\leq (1+O(\epsilon))\sum_{s=1}^{t}opt_s+\frac{\epsilon t}{k^*}opt\end{aligned}$$

holds with probability higher than $(\epsilon k^{-1})^{O(t)}$, where the second step follows from the assumption that $\kappa(t-1)$ is true, and the last step is derived from inequality (3). This implies that $\kappa(t)$ holds for case (1).

We now consider case (2). For this case, we prove $\kappa(t)$ by showing that a client near to f_t can be selected with high probability. Consider a client $c\in\mathbf{b}_{1+\epsilon}(\mathcal{C}_t)$, the definition of $\mathbf{b}_{1+\epsilon}(\mathcal{C}_t)$ implies that $|\mathcal{C}_t|\Delta(f_t,c)\leq(1+\epsilon)|\mathcal{C}_t|r_t=(1+\epsilon)opt_t$. By this inequality and the assumption that $\kappa(t-1)$ is true, we know that inequality

$$\begin{aligned}\sum_{s=1}^{t-1}|\mathcal{C}_s|\Delta(f_s,\mathcal{R}_s)+|\mathcal{C}_t|\Delta(f_t,c) &\leq (1+O(\epsilon))\sum_{s=1}^{t-1}opt_s+\frac{\epsilon(t-1)}{k^*}opt+(1+\epsilon)opt_t\\ &\leq (1+O(\epsilon))\sum_{s=1}^{t}opt_s+\frac{\epsilon t}{k^*}opt \qquad (4)\end{aligned}$$

holds with probability at least $(\epsilon k^{-1})^{O(t)}$. Thus, proving that a client $c\in\mathbf{b}_{1+\epsilon}(\mathcal{C}_t)$ can be selected with high probability is sufficient to ensure the correctness of $\kappa(t)$. We now show that $\mathbf{b}_{1+\epsilon}(\mathcal{C}_t)$ indeed contains a substantial portion of the sample range.

Lemma 4. *If $\kappa(t-1)$ is true, then we have $|\mathcal{C}\backslash\mathcal{B}_t|\leq O(k\epsilon^{-2})|\mathbf{b}_{1+\epsilon}(\mathcal{C}_t)|$.*

Proof. For each $s\in[1,t-1]$, let $h_s\in\mathcal{R}_{t-1}$ denote the client nearest to f_s. The definition of $\mathcal{B}_t$ implies that

$$\Delta(\mathcal{C}_s,h_s)\geq\frac{\epsilon|\mathcal{C}_s\backslash\mathcal{B}_t|opt}{k^*|\mathcal{C}_t|}. \qquad (5)$$

Using inequality (5), we get

$$\begin{aligned}\sum_{s=1}^{t-1}|\mathcal{C}_s\backslash\mathcal{B}_t| &\le \frac{k^*|\mathcal{C}_t|}{\epsilon\cdot opt}\sum_{s=1}^{t-1}\Delta(\mathcal{C}_s,h_s)\\ &\le \frac{k^*|\mathcal{C}_t|}{\epsilon\cdot opt}\sum_{s=1}^{t-1}(opt_s+|\mathcal{C}_s|\Delta(f_s,h_s))\\ &\le \frac{k^*|\mathcal{C}_t|}{\epsilon\cdot opt}((2+O(\epsilon))\sum_{s=1}^{t-1}opt_s+\epsilon\cdot opt)\\ &\le O(k^*\epsilon^{-1})|\mathcal{C}_t|\le O(k\epsilon^{-1})|\mathcal{C}_t|, \end{aligned}\tag{6}$$

where the second step is due to triangle inequality, and the third step follows from the assumption that $\kappa(t-1)$ is true.

Observe that

$$\begin{aligned}|\mathcal{C}\backslash\mathcal{B}_t| &= \sum_{s=1}^{t-1}|\mathcal{C}_s\backslash\mathcal{B}_t|+|\mathcal{C}_t\backslash\mathcal{B}_t|+\sum_{s=t+1}^{k^*}|\mathcal{C}_s\backslash\mathcal{B}_t|\\ &\le O(k\epsilon^{-1})|\mathcal{C}_t|+|\mathcal{C}_t\backslash\mathcal{B}_t|+\sum_{s=t+1}^{k^*}|\mathcal{C}_s|\\ &= O(k\epsilon^{-1})|\mathcal{C}_t|, \end{aligned}\tag{7}$$

where the second step is derived from inequality (6), and the last step is due to the fact that $|\mathcal{C}_1|\ge|\mathcal{C}_2|\ge\ldots\ge|\mathcal{C}_{k^*}|$. Consequently, we get

$$\frac{|\mathbf{b}_{1+\epsilon}(\mathcal{C}_t)|}{|\mathcal{C}\backslash\mathcal{B}_t|}=\frac{|\mathcal{C}_t|}{|\mathcal{C}\backslash\mathcal{B}_t|}\cdot\frac{|\mathbf{b}_{1+\epsilon}(\mathcal{C}_t)|}{|\mathcal{C}_t|}\ge O(k\epsilon^{-2}),$$

where the last step follows from Lemma 2 and inequality (7). Thus, Lemma 4 is true. □

We sort each $c\in\mathcal{C}$ by increasing value of $\Delta(c,\mathcal{R}_{t-1})$, and let c_l denote the l-th client in this order for each $l\in[1,|\mathcal{C}|]$. Given an integer $x\in[0,\lfloor\log|\mathcal{C}|\rfloor]$, define $\mathcal{Q}_x=\{c_{|\mathcal{C}|+1-\lfloor 2^{-x}|\mathcal{C}|\rfloor},\ldots,c_{|\mathcal{C}|}\}$. In order to find a client close to f_t, Algorithm 2 invokes **Median**$(k,\mathcal{R}_{t-1},\mathcal{Q}_x,\mathbb{R})$ and selects clients from $\mathcal{Q}_x$ for each $x\in[0,\lfloor\log|\mathcal{C}|\rfloor]$. The definition of $\mathcal{B}_t$ implies that there exists an integer $x^*\in[0,\lfloor\log|\mathcal{C}|\rfloor]$, such that $\mathcal{C}\backslash\mathcal{B}_t\subseteq\mathcal{Q}_{x^*}$ and $|\mathcal{Q}_{x^*}|\le 2|\mathcal{C}\backslash\mathcal{B}_t|$. Using Lemma 4, we have

$$\frac{|\mathbf{b}_{1+\epsilon}(\mathcal{C}_t)|}{|\mathcal{Q}_{x^*}|}=\frac{|\mathbf{b}_{1+\epsilon}(\mathcal{C}_t)|}{|\mathcal{C}\backslash\mathcal{B}_t|}\cdot\frac{|\mathcal{C}\backslash\mathcal{B}_t|}{|\mathcal{Q}_{x^*}|}\ge O(k\epsilon^{-2}).$$

Moreover, the assumption that $\mathbf{b}_{1+\epsilon}(\mathcal{C}_t)\cap\mathcal{B}_t=\emptyset$ implies that $\mathbf{b}_{1+\epsilon}(\mathcal{C}_t)\subseteq\mathcal{C}\backslash\mathcal{B}_t\subseteq\mathcal{Q}_{x^*}$. Consequently, we know that a client from $\mathbf{b}_{1+\epsilon}(\mathcal{C}_t)$ can be found and added to $\mathcal{R}$ with probability at least $O(k\epsilon^{-2})$. Using inequality (4), we complete the proof of $\kappa(t)$. Thus, $\kappa(k^*)$ and Lemma 3 are true.

3 The Connection Algorithm

Given a real number $\epsilon \in (0,1]$ and an instance $\mathcal{I} = \{\mathcal{C}, \mathcal{F}, k, B_l, B_u\}$ of balanced k-median, denote by k^* the number of facilities opened in an optimal solution to $\mathcal{I}$, let $\mathbb{C} = \{\mathcal{C}_1, \ldots, \mathcal{C}_{k^*}\}$ be the set of clusters in the solution, and denote by $f_1, \ldots, f_{k^*}$ the corresponding opened facilities. Let $\mathbb{R}$ denote the set returned by Algorithm 1, and let $\mathcal{R} \in \mathbb{R}$ denote the desired client set in Lemma 3. We define a multi-set $\mathcal{R}^\dagger = \{h_t : t \in [1, k^*]\}$, where $h_t \in \mathcal{R}$ denote the client nearest to f_t for each $t \in [1, k^*]$. By multiplying the running time of our algorithm by a factor of $|\mathbb{R}|k^{O(k)} = (k\epsilon^{-1})^{O(k)}|\mathcal{C}|$, we can assume that both $\mathcal{R}^\dagger$ and k^* are known.

In this section, we show how to construct the solution to $\mathcal{I}$ based on $\mathcal{R}^\dagger$. Given a set $\mathcal{H} \subseteq \mathcal{F}$ of opened facilities, we connect the clients from $\mathcal{C}$ based on the following Integer program.

$$\min \sum_{f \in \mathcal{H}, c \in \mathcal{C}} \Delta(c, f) x_{cf} \tag{IP1}$$

$$\text{s.t.} \quad \sum_{f \in \mathcal{H}} x_{cf} = 1 \qquad \forall c \in \mathcal{C} \tag{8}$$

$$\sum_{c \in \mathcal{C}} x_{cf} \geq B_l \qquad \forall f \in \mathcal{H} \tag{9}$$

$$\sum_{c \in \mathcal{C}} x_{cf} \leq B_u \qquad \forall f \in \mathcal{H} \tag{10}$$

$$x_{cf} \in \{0, 1\} \qquad \forall f \in \mathcal{H}, c \in \mathcal{C} \tag{11}$$

IP1 associates each $f \in \mathcal{H}$ and $c \in \mathcal{C}$ with a variable x_{cf}, which indicates whether c is connected to f. Constraint (8) says that each client should be connected to an opened facility, and constraints (9) and (10) ensure that the upper and lower bounds on the cluster sizes are satisfied. It can be seen that IP1 is an integer program of the minimum cost circulation problem, where each client corresponds to a vertex with the supply 1, and each facility corresponds to a vertex with the demand B_l and capacity B_u. Using the rounding approach given in [12], an optimal solution to IP1 can be found in $(|\mathcal{C}|k)^{O(1)}$ time.

It remains to show how to select the opened facilities. A straightforward approach is to open the nearest facility to each $h \in \mathcal{R}^\dagger$. For the case of balanced k-median, the issue with this approach lies in that a facility may be selected for more than once, which can make the constraint on cluster sizes violated. We open the facilities using a color-coding method to deal with this issue. We assign a random label from $[1, k^*]$ to each $f \in \mathcal{F}$. With probability higher than k^{-k}, we can associate label t with f_t for each $t \in [1, k^*]$, and the probability can be boosted to a constant by repeating the algorithm for k^k times. We now open the nearest facility $m_t \in \mathcal{F}$ to h_t that is labeled with t for each $t \in [1, k^*]$. Such a selection method is described in Algorithm 3. Since we need to guess k^* and $\mathcal{R}^\dagger$, the algorithm runs in $(|\mathcal{C}||\mathcal{F}|)^{O(1)}(k\epsilon^{-1})^{O(k)}$ time.

Algorithm 3: The Selection of Opened Facilities

Input: An instance $\mathcal{I} = (\mathcal{C}, \mathcal{F}, k, B_l, B_u)$ of the balanced k-median problem, client set $\mathcal{R}^\dagger = \{h_1, \ldots, h_{t^*}\}$, and the number of opened facilities k^*;
Output: A set $\mathcal{H} \subseteq \mathcal{F}$ of opened facilities and a connection function τ;

1 $\mathbb{U} = \emptyset$;
2 **for** $i \Leftarrow 1$ *to* k^k **do**
3 $\quad \mathcal{H} = \emptyset$;
4 $\quad$ **for** *each* $f \in \mathcal{F}$ **do**
5 $\quad\quad$ Randomly and uniformly select an integer $x \in [1, k^*]$, $c(f) \Leftarrow x$;
6 $\quad$ **for** $t \Leftarrow 1$ *to* k^* **do**
7 $\quad\quad \mathcal{H} \Leftarrow \mathcal{H} \cup \{\underset{m \in \mathcal{F}, c(m)=t}{\arg\min} \ \Delta(h_t, m)\}$;
8 $\quad$ Compute the connection function τ using $\mathcal{H}$ and IP1;
9 $\quad \mathbb{U} \Leftarrow \mathbb{U} \cup \{(\mathcal{H}, \tau)\}$;
10 **return** $(\mathcal{H}, \tau) \Leftarrow \underset{(\mathcal{H}', \tau') \in \mathbb{U}}{\arg\min} \sum_{c \in \mathcal{C}} \Delta(c, \tau'(c))$.

Let $(\mathcal{H}, \tau)$ be the solution given by Algorithm 3, where $\mathcal{H} = \{m_1, \ldots, m_{k^*}\}$, and let $(\mathcal{H}^*, \tau^*)$ be the optimal solution to $\mathcal{I}$ that opens facilities $f_1, \ldots, f_{k^*}$. We have

$$\begin{aligned}
\sum_{c \in C} \Delta(c, \tau(c)) &\le \sum_{t=1}^{k^*} \Delta(C_t, m_t) \le \sum_{t=1}^{k^*} (\Delta(C_t, h_t) + |C_t| \Delta(h_t, m_t)) \\
&\le \sum_{t=1}^{k^*} (\Delta(C_t, f_t) + |C_t| \Delta(h_t, f_t) + |C_t| \Delta(h_t, m_t)) \\
&\le \sum_{t=1}^{k^*} (\Delta(C_t, f_t) + 2|C_t| \Delta(h_t, f_t)) \\
&\le (3 + O(\epsilon)) \sum_{t=1}^{k^*} \Delta(C_t, f_t),
\end{aligned}$$

where the second and third steps are due to triangle inequality, the fourth step is due to the definition of m_t, and the last step follows from Lemma 3. This inequality implies that a $(3 + O(\epsilon))$-approximation solution to $\mathcal{I}$ is obtained.

4 Conclusions

In this paper we give an FPT(k)-time algorithm for the balanced k-median problem, which has the guarantee of yielding a $(3 + \epsilon)$-approximation solution. This improves the current best approximation ratio of $7.2 + \epsilon$ [12] obtained in the same time. Our main technical contribution is a different sampling approach for selecting opened facilities, which we think is of independent interests.

References

1. Adamczyk, M., Byrka, J., Marcinkowski, J., Meesum, S.M., Wlodarczyk, M.: Constant-factor FPT approximation for capacitated k-median. In: Proceedings of the 27th Annual European Symposium on Algorithms (ESA), pp. 1:1–1:14 (2019)
2. Arya, V., Garg, N., Khandekar, R., Meyerson, A., Munagala, K., Pandit, V.: Local search heuristics for k-median and facility location problems. SIAM J. Comput. **33**(3), 544–562 (2004)
3. Aydin, K., Bateni, M., Mirrokni, V.S.: Distributed balanced partitioning via linear embedding. In: Proceedings of the 9th ACM International Conference on Web Search and Data Mining (WSDM), pp. 387–396 (2016)
4. Bateni, M., Bhaskara, A., Lattanzi, S., Mirrokni, V.S.: Distributed balanced clustering via mapping coresets. In: Proceedings of the 27th Annual Conference on Neural Information Processing Systems (NeurIPS), pp. 2591–2599 (2014)
5. Borgwardt, S., Brieden, A., Gritzmann, P.: An LP-based k-means algorithm for balancing weighted point sets. Eur. J. Oper. Res. **263**(2), 349–355 (2017)
6. Byrka, J., Pensyl, T., Rybicki, B., Srinivasan, A., Trinh, K.: An improved approximation for k-median and positive correlation in budgeted optimization. ACM Trans. Algorithms **13**(2), 23:1–23:31 (2017)
7. Charikar, M., Li, S.: A dependent LP-rounding approach for the k-median problem. In: Proceedings of the 39th International Colloquium on Automata, Languages, and Programming (ICALP), pp. 194–205 (2012)
8. Cohen-Addad, V., Gupta, A., Kumar, A., Lee, E., Li, J.: Tight FPT approximations for k-median and k-means. In: Proceedings of the 46th International Colloquium on Automata, Languages, and Programming (ICALP), pp. 42:1–42:14 (2019)
9. Cohen-Addad, V., Li, J.: On the fixed-parameter tractability of capacitated clustering. In: Proceedings of the 46th International Colloquium on Automata, Languages, and Programming (ICALP), pp. 41:1–41:14 (2019)
10. Demirci, H.G., Li, S.: Constant approximation for capacitated k-median with $(1+\epsilon)$-capacity violation. In: Proceedings of the 43rd International Colloquium on Automata, Languages, and Programming (ICALP), pp. 73:1–73:14 (2016)
11. Dick, T., Li, M., Pillutla, V.K., White, C., Balcan, N., Smola, A.J.: Data driven resource allocation for distributed learning. In: Proceedings of the 20th International Conference on Artificial Intelligence and Statistics (AISTATS), pp. 662–671 (2017)
12. Ding, H.: Faster balanced clusterings in high dimension. Theor. Comput. Sci. **842**, 28–40 (2020)
13. Guha, S., Khuller, S.: Greedy strikes back: improved facility location algorithms. J. Algorithms **31**(1), 228–248 (1999)
14. Guo, Y., Huang, J., Zhang, Z.: A constant factor approximation for lower-bounded k-median. In: Proceedings of the 16th International Conference on Theory and Applications of Models of Computation (TAMC), pp. 119–131 (2020)
15. Han, L., Hao, C., Wu, C., Zhang, Z.: Approximation algorithms for the lower-bounded k-median and its generalizations. In: Proceedings of the 26th International Conference on Computing and Combinatorics (COCOON), pp. 627–639 (2020)
16. Jain, K., Mahdian, M., Markakis, E., Saberi, A., Vazirani, V.V.: Greedy facility location algorithms analyzed using dual fitting with factor-revealing LP. J. ACM **50**(6), 795–824 (2003)

17. Jain, K., Vazirani, V.V.: Approximation algorithms for metric facility location and k-median problems using the primal-dual schema and Lagrangian relaxation. J. ACM **48**(2), 274–296 (2001)
18. Li, S.: Approximating capacitated k-median with $(1 + \epsilon)k$ open facilities. In: Proceedings of the 27th Annual ACM-SIAM Symposium on Discrete Algorithms (SODA), pp. 786–796 (2016)
19. Li, S.: On uniform capacitated k-median beyond the natural LP relaxation. ACM Trans. Algorithms **13**(2), 22:1–22:18 (2017)
20. Li, S., Svensson, O.: Approximating k-median via pseudo-approximation. SIAM J. Comput. **45**(2), 530–547 (2016)
21. Lin, W., He, Z., Xiao, M.: Balanced clustering: a uniform model and fast algorithm. In: Proceedings of the 28th International Joint Conference on Artificial Intelligence (IJCAI), pp. 2987–2993 (2019)
22. Manurangsi, P., Raghavendra, P.: A birthday repetition theorem and complexity of approximating dense CSPs. In: Proceedings of the 44th International Colloquium on Automata, Languages, and Programming (ICALP), pp. 78:1–78:15 (2017)
23. Siavoshi, S., Kavian, Y.S., Sharif, H.: Load-balanced energy efficient clustering protocol for wireless sensor networks. IET Wirel. Sens. Syst. **6**(3), 67–73 (2016)

A LP-based Approximation Algorithm for generalized Traveling Salesperson Path Problem

Jian Sun[1], Gregory Gutin[2], and Xiaoyan Zhang[3](✉)

[1] Department of Operations Research and Information Engineering, Beijing University of Technology, Beijing 100124, People's Republic of China
B201806011@emails.bjut.edu.cn

[2] Department of Computer Science Royal Holloway, University of London, Egham, Surrey TW200EX, UK
g.gutin@rhul.ac.uk

[3] School of Mathematical Science and Institute of Mathematics, Nanjing Normal University, Nanjing 210023, Jiangsu, China
zhangxiaoyan@njnu.edu.cn

Abstract. Hamiltonian path problem is one of the fundamental problems in graph theory, the aim is to find a path in the graph that visits each vertex exactly once. In this paper, we consider a generalizedized problem: given a complete weighted undirected graph $G = (V, E, c)$, two specified vertices s and t, let V' and E' be subsets of V and E, respectively. We aim to find an s-t path which visits each vertex of V' and each edge of E' exactly once with minimum cost. Based on LP rounding technique, we propose a $\frac{9-\sqrt{33}}{2}$-approximation algorithm.

Keywords: Hamiltonian path · LP rounding · Generalized TSP path problem · Approximation algorithm

1 Introduction

The traveling salesperson problem (TSP) is one of the most important problems in graph theory and computer science [7,10,14]: given a series of cities and the distance between each pair of cities, find the shortest circuit to visit each city once and return to the starting city. In graph theoretical terms, the problem can be modeled as follows: there is an undirected complete graph $G = (V, E, c)$ in which $c : E \rightarrow \mathbb{R}_+$ is a metric edge-cost function, the task is to find a minimum-cost Hamilton cycle.

Even if the edge-cost function is metric, TSP is still NP-hard [14]. Therefore, approximation algorithms are appropriate tools for solving this problem (and other NP-hard problems); their approximation ratios are often used to measure the performance of approximation algorithms. Christofides [2] designed a

X. Zhang–This research is supported or partially supported by the National Natural Science Foundation of China (Grant Nos. 11871280 and 11871081) and Qinglan Project.

D.-Z. Du et al. (Eds.): COCOA 2021, LNCS 13135, pp. 641–652, 2021.
https://doi.org/10.1007/978-3-030-92681-6_50

well known approximation algorithm for metric TSP with ratio 1.5. Recently, Karlin et al. made a breakthrough on this issue [13], they designed a $(1.5 - \epsilon)$-approximation algorithm for metric TSP where ϵ is a constant greater than 10^{-36}. If the edge-cost function is not metric, there is no polynomial time constant factor approximation algorithm, under the assumption that $P \neq NP$ [18].

The earliest mathematical programming formulation of TSP was proposed by Dantzig et al. [4]. The traveling salesperson problem was studied in depth in the field of discrete optimization and many optimization methods use as a benchmark for testing. Although the problem is computationally difficult, there are a number of heuristic algorithms and exect methods available to solve instances with tens of thousands of vertices with error within 1%.

TSP on graphic metrics (graph-TSP), where cost function over the vertex pair is the minimum number of edges on the path between vertices in the underlying graph, has recently received a great deal of attention and is the subject of current work. Using a sophisticated probabilistic analysis, Gharan, Saberi and Singh [17] presented a $(1.5 - \epsilon)$-approximation for $\epsilon > 0$. Mömke and Svensson [15] obtained a 1.461-approximation by a simple and clever polyhedral idea, which easily yields the ratio $\frac{4}{3}$ for cubic (actually subcubic) graphs. The technique of [15] was used in subsequent studies. Mucha [16] proposed an approximation ratio of $\frac{13}{9}$ via redefining their analysis. Sebö and Vygen [21] proved new results for approximating the graph-TSP and some related problems, in particular they improved the approximation ratio to 1.4 for graph-TSP.

Compared to TSP, the *traveling salesperson path problem (TSPP)* is a more general model and has also received much attention [8,11,12,19,20,22]. The only difference between TSP and TSPP is that while the TSP's goal is to find the Hamilton cycle with minimum cost, the TSPP's goal is to search for the minimum-cost Hamilton path.

Depending on whether the endpoints are given, Hoogeveen [12] considered three subproblems of TSPP as follows and presented a modification of Christofides' algorithm.

(1) neither endpoints is pre-specified;
(2) one of the endpoints is pre-specified;
(3) both the endpoints are pre-specified.

Property 1. It was proved in [12] that for cases (1) and (2), the approximation ratio of the modified algorithm is $\frac{3}{2}$.

Case (3) is more difficult than cases (1) and (2), and the modified algorithm in [12] is a $\frac{5}{3}$-approximation algorithm. This result has not been improved in nearly two decades. In 2012, An et al. [1] presented an approximation algorithm with ratio $\frac{1+\sqrt{5}}{2}$ for the metric s-t path TSP. Traub and Vygen [22] obtained an improvement and proposed a $1.5 + \epsilon$-approximation algorithm for any fixed $\epsilon > 0$. Zenklusen [23] pointed out that a variation of the dynamic programming idea recently introduced by Traub and Vygen [22] was sufficient to handle larger cuts by utilizing Karger's groundbreaking results on near minimum cut numbers.

Based on this observation, he designed an approximation algorithm with ratio 1.5 which matches the ratio of cases (1) and (2).

In this paper, we introduce and consider the *generalized traveling salesperson path problem* (GTSPP), a generalization of traveling salesperson path problem. In GTSPP, there is an edge-weighted undirected graph $G = (V, E, c)$ where the weight function c is metric, and two distinct specified endpoints s and t. Given a vertex subset $V' \subseteq V$ and a disjoint edge subset $E' \subseteq E$ (i.e., any two edges in E' have no common vertices), we aim to find a minimum cost s-t path which visits each vertex in V' and each edge in E' exactly once. Using the LP-rounding technique, we further utilize some structural features of the graph, and then design an approximation algorithm for the problem. We use the concept of a narrow cut to analyze the approximation ratio of this approximation algorithm. First, we prove that the approximation ratio is 1.6577, and then using further analysis, we show that the ratio is no more than $\frac{9-\sqrt{33}}{2} \simeq 1.6277$. Both bounds improve our previous work [24].

In a transportation network, edges correspond to streets or parts of streets and vertices correspond to street intersections and locations on a street. Vertices can generally be divided into two types: one represents actual street intersections on the transportation network, called junctions; the other represents the required vehicle visits (pickup or delivery points), called stops. In real life, due to restrictions on vehicle types or license plates, vehicles have to detour through certain routes in the distribution process. The edges corresponding to these routes are required edges in E', and the vertices corresponding to the stops belong to V'. s and t represent the starting and ending points of the distribution route respectively. Then the vehicle transport problem can be modeled as GTSPP.

The remainder of this paper is organized as follows. We provide some preliminaries and the LP model of GTSPP in Sect. 2. We study the algorithm for the GTSPP and propose its analysis in Sect. 3. We conclude in Sect. 4.

2 Preliminaries

In this section we introduce some basic notation and terminology which will be used throughout this paper. For $U \subset V$, $(U, \bar{U})$ is the cut defined by U; $\delta(U) := \{\{u, v\} | u \in U, v \in \bar{U}\}$ denotes the set of cut edges. If one of endpoints belongs to U, i.e., $|U \cap \{s, t\}| = 1$, then $(U, \bar{U})$ is called an s-t cut; otherwise it is called a non-separating cut.

The problem has a feasible solution if and only if E' is the union of disjoint paths (and s and t have degree at most 1 in E'). Thus edges in $E \setminus E'$ that are incident to an interior vertex of one of these paths can never be used; so we can delete them and we can essentially treat each paths in E' as if it was a single edge from its startpoint to its endpoint. In the absence of confusion, we still call the new edge set E'. In the following, we only study the case where feasible solutions exist.

We can delete vertices that are neither contained in V', nor in $\{s,t\}$, and that are not an endpoint of any edge in E'. This is because the edge weights are metric (and thus we could shortcut any tour that visits one of these vertices). Thus instead of considering the problem on G, we construct a new graph $G_1 = (V_1, E_1)$, where $V_1 := V' \cup V(E') \cup \{s,t\}$ and $G_1 = G[V_1]$ is the induced subgraph of G on V_1. Without loss of generality, we may assume that $\{s,t\} \cap V(E') = \emptyset$. Otherwise if $s \in V(E')$, i.e., there exists an edge $e_1 = (s, s_1) \in E'$, we can construct an s_1-t path P_1 that visits each vertex of V' and each edge of $E' \setminus \{e_1\}$ exactly once; then $e_1 \cup P_1$ is a feasible s-t path for original problem.

Based on the Held-Karp relaxation of TSP and Path TSP, we present the LP-relaxation of our problem as follows:

$$
\begin{aligned}
\min \quad & \sum_{e \in E_1} c(e)x(e) & & \\
\text{s.t.} \quad & x(\delta(S)) \geq 1 & & \forall S \subsetneq V_1, |S \cap \{s,t\}| = 1 \\
& x(\delta(S)) \geq 2 & & \forall S \subsetneq V_1, |S \cap \{s,t\}| \neq 1, S \neq \emptyset \\
& x(\delta(\{v\})) = 2 & & \forall v \in V_1 \setminus \{s,t\} \\
& x(\delta(\{s\})) = x(\delta(\{t\})) = 1 & & \\
& x(e) = 1 & & \forall e \in E' \\
& x(e) \geq 0 & & \forall e \in E_1 \setminus E',
\end{aligned}
\tag{1}
$$

where $x(\delta(S)) = \sum_{e \in \delta(S)} x(e)$. The first and second constraints are referred to as sub-path and sub-tour removal, respectively. Though this linear program contains exponential multiple constraints, it can be solved in polynomial time via the ellipsoid method using a min-cut algorithm to solve the separation problem [9].

In the algorithm and analysis, we will use the following terms.

Definition 1 *(Q-join). Let Q be a subset of V with even cardinality. Then an edge set J with odd degrees precisely for the vertices in Q is called an Q-join, i.e., odd$(J) = Q$ where odd(J) denotes the set of odd-degree vertices in J.*

In order to facilitate the understanding of the concept of Q-join, a simple example is given in Fig. 1:

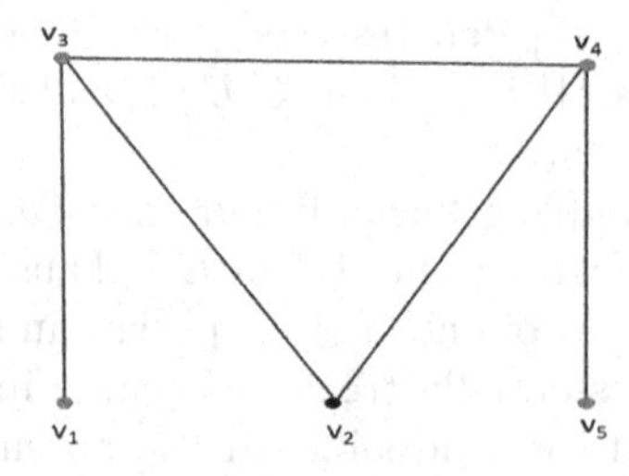

Fig. 1. An example of Q-join, where $Q = \{v_1, v_3, v_4, v_5\}$.

Definition 2 *(wrong-degree set). Let T be a spanning tree of G and $s \neq t \in V$. Then the vertex subset $odd(T) \triangle \{s,t\} := (odd(T) \setminus \{s,t\}) \cup (\{s,t\} \setminus odd(T))$ is called the wrong-degree set.*

Definition 3 *[6]. (Q-join polytope)*

$$P^{\uparrow}_{Q-join} := \{y \in \mathbb{R}^E_{\geq 0} | y(\delta(C)) \geq 1 \ \forall |C \cap Q| \equiv 1 (\mod 2)\}.$$

Observe that the cost of a feasible solution of Q-join polytope is an upper bound on the cost of a minimum Q-join.

Definition 4 *(spanning tree polytope [5]). The spanning tree polytope of $G = (V,E)$ can be described as follows:*

$$P_{ST} = \begin{cases} x(E) = |V| - 1, \\ x(E[S]) \leq |S| - 1, \\ x \in \mathbb{R}^E_{\geq 0}. \end{cases} \tag{2}$$

It is not difficult to show that any feasible solution of (1) is in the spanning tree polytope of G_1. Thus for any given feasible solution x^* of (1), it can be expressed as $x^* := \sum_{i=1}^k \gamma_i \chi_{T_i}$, where T_i is a spanning tree of G_1 and $\gamma_i \geq 0$ $\forall i \in [k]$ and k is bounded by a polynomial. Besides, $\sum_{i=1}^k \gamma_i = 1$, i.e., x^* equals to a convex combination of k spanning trees of G_1. In particular, such a convex combination can be found in polynomial time [3,9].

In the above convex combination of x^*, each spanning tree satisfies the following:

Lemma 1. *For each $e \in E'$ and $i \in [k]$, $\chi_{T_i}(e) = 1$.*

Proof. Since $\forall e \in E'$ $(x^*(e) = 1)$ and $x^* := \sum_{i=1}^k \gamma_i \chi_{T_i}$, for each $e \in E'$ we have $\sum_{i=1}^k \gamma_i \chi_{T_i}(e) = 1$. Besides, $\forall e \in E'$ $(0 \leq \chi_{T_i}(e) \leq 1)$, and hence

$$1 = x^*(e) = \sum_{i=1}^k \gamma_i \chi_{T_i}(e) \leq \sum_i^k \gamma_i = 1.$$

Note that the inequality is in fact an equality. Thus for each $e \in E'$ and $i \in [k]$, $\chi_{T_i}(e) = 1$. □

3 Algorithm and Analysis

The immediate idea is to use Zenklusen's algorithm to deal with this generalized problem and expect to get the same approximate ratio. However, due to the graph structure characteristics of the problem itself, we cannot simply call Zenklusen's algorithm. To better explain the question, let's briefly review the algorithmic framework for solving TSP and TSPP.

The framework of TSP or TSPP is generally divided into three steps: the first step is to construct the spanning tree; the second step is to modify the spanning tree constructed in the first step to obtain an Euler tour or Euler trail. The third step uses the shortcutting algorithm to transform the Euler tour or Euler trail into Hamiltonian cycle or path. Compared to the classical TSPP, the main difference is that in GTSPP, the edges in E' must be passed exactly once. Therefore, in order to ensure the feasibility of the final solution, the algorithm for solving GTSPP needs to ensure that the edge in E' has been included in the obtained Euler trail after completing the first and second steps. In addition, it is also necessary to ensure that the edge in E' will not be deleted after the third step of shortcutting operation (this can be achieved by the method we show in Fig. 2).

In Zenklusen's algorithm, the spanning tree T is constructed by the support of a special solution y of the programming (1), then it computes a minimum Q-join where $Q := odd(T)\Delta\{s,t\}$ such that the cost of T and Q does not exceed one and half of the cost of the optimal solution, respectively. Thus the ratio of this algorithm is 1.5. (Theorem 2.1 of [23]). If we call this algorithm directly, there is no guarantee that the edge in E' will be included in the Euler trail since some edges in E' may be not in T.

Thus in this section, we propose the following approximation algorithm for GTSPP and its detailed analysis.

3.1 Approximation Algorithm

The details of our algorithm is presented in following:

Algorithm 1. Algorithm of GTSPP

Input:
1: An edge-weighted undirected graph $G = (V, E, c)$.
2: Endpoints s and t.
3: $V' \subseteq V$, $E' \subseteq E$ are required vertex subset and edge subset, respectively.
Output: A generalized travelling salesperson path.
begin:
4: Construct a new graph $G_1 := G[V_1]$ where $V_1 = \{v|v \in e, e \in E'\} \cup \{s,t\} \cup V'$.
5: Solve the LP-relaxation (1) and obtain a optimal solution x^*.
6: Construct a convex combination of $x^* = \sum_{i=1}^{k} \gamma_i \chi_{T_i}$ with polynomial number of spanning trees $T_1, T_2, \cdots, T_k$.
7: Select the spanning tree T_i with probability γ_i and denote the tree by T.
8: Find all the wrong-degree vertices in T, denoted by Q.
9: Compute a minimum Q-join J.
10: Construct a solution P via shortcutting the Eulerian s-t trail in $T \cup J$.
11: **output** P.
end

Lemma 2. *The solution P output by Algorithm 1 is a feasible solution of GTSPP.*

Proof. In GTSPP, it is required that all the vertices in V' and edges in E' must be visited exactly once. Note that P is an s-t path in G_1, thus it must traverse each vertices exactly once. So the point of the proof is that P goes through each edge in E' exactly once.

If there exists no cycle in $T \cup J$, then P happens to be the Eulerian s-t trail and thus it visits all the edges in E' exactly once. Otherwise, consider the two cases as illustrated in Fig. 2:

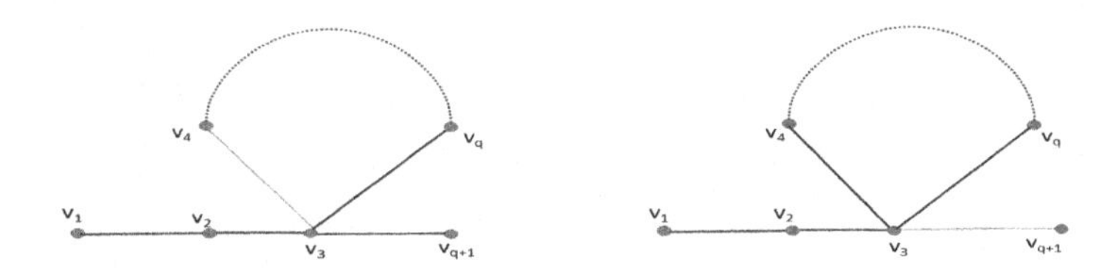

Fig. 2. An example illustrating the existence of cycle as subgraph of $T \cup J$, where every purple edge belongs to E'.

For the left case, we preform the shortcutting to obtain a subpath v_1-v_2-v_3-v_4-$\cdots$-v_q-v_{q+1}; for the right case, we preform the shortcutting to obtain a subpath v_1-v_2-v_4-$\cdots$-v_q-v_3-v_{q+1}. For both two cases, we construct a feasible subpath via shortcutting. □

3.2 An Improvement on $\frac{5}{3}$

We designed a $\frac{5}{3}$-approximation algorithm for GTSPP via the discrete technique [24]. In this subsection we will prove that the ratio of Algorithm 1 is less that $\frac{5}{3}$. Inspired by the work of [1], we use the combination of x^* and χ_T to obtain an upper bound on the minimum Q-join cost.

In order to construct a nice combination, we need to take advantage of some structural features especially the narrow cut.

Lemma 3 *[1]. Suppose that $(U, \bar{U})$ is an odd s-t cut with respect to Q, i.e., $|U \cap Q| \equiv 1 (\mod 2)$, then we have $|\delta(U) \cap T| \geq 2$.*

The probability that $(U, \bar{U})$ is an odd s-t cut with respect to Q can be formulated by the following inequality.

Lemma 4.
$$Pr[|U \cap Q| \equiv 1(\mod 2)] \leq x^*(\delta(U)) - 1.$$

Proof. Let $|\delta(U) \cap T| = X$, $p(x)$ be the density function of the random variable X. Then we have that

$$E[X] - 1 = \int_0^\infty xp(x)dx - \int_0^\infty p(x)dx$$
$$= \int_0^\infty (x-1)p(x)dx.$$

Due to the connectivity of T, $|\delta(U) \cap T| \geq 1$ always holds. Thus

$$\begin{aligned} E[X] - 1 &= \int_1^\infty (x-1)p(x)dx \\ &= \int_1^2 (x-1)p(x)dx + \int_2^\infty (x-1)p(x)dx \\ &\geq \int_2^\infty p(x)dx \\ &= Pr[X \geq 2]. \end{aligned}$$

From Lemma 3, it can be derived that $Pr[|U \cap Q| \equiv 1(\mod 2)] \leq Pr[X \geq 2]$, thus

$$Pr[|U \cap Q| \equiv 1(\mod 2)] \leq E[X] - 1 = x^*(\delta(U)) - 1,$$

the last equality holds from the fact $Pr[e \in T] = x^*(e)$. □

Initially set $y := \alpha x^* + \beta\chi_T$, then $y(\delta(U)) \geq 2\alpha + \beta$ for non-separating cut $(U, \bar{U})$. If $(U, \bar{U})$ is an odd s-t cut with respect to Q, then $y(\delta(U)) \geq \alpha x^*(\delta(U)) + 2\beta$.

The key to improving the approximation ratio is to adjust α and β while keeping the resulting y still in Q-join polytope. If we increase α by ϵ and decrease β by 2ϵ, then $2\alpha + \beta$ is not changed while $E[c(y)]$ is decreased by $\epsilon c(x^*)$.

If $(U, \bar{U})$ is an odd s-t cut with large capacities, then $\alpha x^*(\delta(U)) + 2\beta$ will still be larger than 1 after a small adjustment on α. If $x^*(\delta(U)) = 1$, then we have that $E[|\delta(U) \cap T|] = x^*(\delta(U)) = 1$. According to Lemma 3, $|U \cap Q|$ is even, thus this is not a concern since the Q-join polytope only has constraint on odd cut with respect to Q.

To ensure that y is in Q-join polytope, we will add small fractions of the deficient odd s-t cuts. Note that an edge may belong to several different s-t cuts, we need to be careful to address this issue. Based on the concept of a narrow cut, we show that s-t cuts of small capacities are "almost" disjoint.

Definition 5 *[1]. For $0 < \tau \leq 1$, an s-t cut $(U, \bar{U})$ is called a τ-narrow cut if $x^*(\delta(U)) < 1 + \tau$.*

An, Kleinberg, and Shmoys [1] proved that the τ-narrow cuts form a chain, and there exists a partition $\{L_i\}_{i=1}^k$ of V_1 such that:

1. $L_1 = \{s\}$, $L_k = \{t\}$;
2. $\{U|(U, \bar{U})$ is τ-narrow, $s \in U\}$=$\{U_i|1 \leq i < k\}$ in which $U_i := \cup_{j=1}^i L_j$.

Let $L_{\leq i} := \cup_{j=1}^{i} L_j$, $L_{\geq i} = \cup_{j=i}^{k} L_j$ and $F_i := E(L_i, L_{\geq i+1})$, it is easily to see that F_i's are disjoint and $F_i \subseteq \delta(U_i)$ for all i. For each τ-narrow cut $(U_i, \bar{U_i})$, they obtain an lower bound on $x^*(F_i)$ as follows:

$$x^*(F_i) > \frac{1 - \tau + x^*(\delta(U_i))}{2} \geq 1 - \frac{\tau}{2}.$$

For each τ-narrow cut, we define an incident vector $g^*_{U_i}$:

$$(g^*_{U_i})_e = \begin{cases} x^*(e), & \text{if } e \in F_i; \\ 0, & \text{otherwise.} \end{cases}$$

Now we are ready to obtain the first result in this paper:

Theorem 1. $E[c(P)] \leq 1.6577c(x^*)$.

Proof. Set

$$y := \alpha x^* + \beta \chi_T + \sum_{i:|U_i \cap Q| \text{ is odd, } 1 \leq i < k} \frac{1 - (\alpha + 2\beta)}{1 - 0.5\tau} g^*_{U_i},$$

where $\alpha = 0.35$, $\beta = 0.3$ and $\tau = \frac{1-2\beta}{\alpha} - 1$. Then we show that y is in the Q-join polytope. It is easy to see that $y \geq 0$, besides we have claimed that $y(\delta(U)) \geq 1$ for non-separating cut $(U, \bar{U})$ as $2\alpha + \beta$ still equals to 1.

Suppose $(U, \bar{U})$ is an odd s-t cut with respect to Q, if $x^*(\delta(U)) \geq 1 + \tau$, then

$$\begin{aligned} y(\delta(U)) &\geq \alpha x^*(\delta(U)) + \beta|\delta(U) \cap Q| \\ &\geq \alpha(1 + \tau) + 2\beta \\ &= 1. \end{aligned}$$

If $x^*(\delta(U)) < 1 + \tau$, i.e., there exists $1 \leq i < k$ such that $U = U_i$, then

$$\begin{aligned} y(\delta(U)) &\geq \alpha x^*(\delta(U)) + \beta|\delta(U) \cap Q| + \frac{1 - (\alpha + 2\beta)}{1 - 0.5\tau} g^*_U \\ &\geq \alpha + 2\beta + \frac{1 - (\alpha + 2\beta)}{1 - 0.5\tau}(1 - 0.5\tau) \\ &= 1. \end{aligned}$$

Thus y is in the Q-join polytope. The next task is to analyze the upper bound on the cost of P.

$$\begin{aligned}
E[c(P)] &\leq E(c(T)) + E[c(J)] \\
&\leq E(c(T)) + E[c(y)] \\
&= E(c(T)) + \alpha E[c(x^*)] + \beta E[c(\chi_T)] + E\left[c\left(\sum_{i:|U_i \cap Q| \text{ is odd}, 1 \leq i < k} A \cdot g^*_{U_i}\right)\right] \\
&= (1+\alpha+\beta)c(x^*) + c\left(\sum_{i=1}^{k-1} Pr[|U_i \cap Q| \text{ is odd}] \cdot A \cdot g^*_{U_i}\right) \\
&\leq (1+\alpha+\beta)c(x^*) + \tau \cdot A \cdot c\left(\sum_{i=1}^{k-1} g^*_{U_i}\right) \\
&\leq (1+\alpha+\beta+\tau \cdot A)c(x^*),
\end{aligned}$$

where $A := \frac{1-(\alpha+2\beta)}{1-0.5\tau}$.

The third inequality holds according to the conclusion of Lemma 4 as for each τ-narrow cut $(U, \bar{U})$ $Pr[|U \cap Q| \text{ is odd}] \leq x^*(\delta(U)) - 1 < \tau$. Based on the disjointness of F_i's, the last inequality is obvious.

Thus we have that $E[c(P)] \leq 1.6577c(x^*)$. □

3.3 Tighter Analysis

In this subsection, we aim to propose a tighter analysis for the algorithm so as to obtain a better approximation ratio. Now, unlike the previous analysis, we use $\frac{1-\tau+x^*(\delta(U_i))}{2}$ instead of $1-\frac{\tau}{2}$ to represent the lower bound of $x^*(F_i)$. Then we can get the following:

Theorem 2. $E[c(P)] \leq \frac{9-\sqrt{33}}{2}c(x^*)$.

Proof. Set

$$y := \alpha x^* + \beta \chi_T + \sum_{i:|U_i \cap Q| \text{ is odd}, 1 \leq i < k} \frac{1-(\alpha x^*(\delta(U_i))+2\beta)}{b_i} g^*_{U_i},$$

where $b_i := \frac{1-\tau+x^*(\delta(U_i))}{2}$, $\alpha = \frac{1}{2} - \frac{1}{2\sqrt{33}}$ and $\beta = \frac{1}{\sqrt{33}}$.

Similarly to Theorem 1, we can prove that y is in the Q-join polytope. We have that

$$\begin{aligned}
E[c(P)] &\leq (1+\alpha+\beta)c(x^*) + c\left(\sum_{i=1}^{k-1} Pr[|U_i \cap Q| \text{ is odd}] \frac{1-(\alpha x^*(\delta(U_i))+2\beta)}{b_i} g^*_{U_i}\right) \\
&\leq (1+\alpha+\beta)c(x^*) + c\left(\sum_{i=1}^{k-1} (x^*(\delta(U_i))-1) \frac{1-(\alpha x^*(\delta(U_i))+2\beta)}{b_i} g^*_{U_i}\right) \\
&\leq (1+\alpha+\beta)c(x^*) + \left[\max_{0 \leq \eta \leq \tau} \left(\eta \frac{1-(2\beta+\alpha(1+\eta))}{1-\frac{\tau}{2}+\frac{\eta}{2}}\right)\right] c\left(\sum_{i=1}^{k-1} g^*_{U_i}\right) \\
&\leq \left(1+\alpha+\beta+\max_{0 \leq \eta \leq \tau}\left(\eta \frac{1-(2\beta+\alpha(1+\eta))}{1-\frac{\tau}{2}+\frac{\eta}{2}}\right)\right) c(x^*).
\end{aligned}$$

Let $R(\eta) := \eta \frac{1-(2\beta+\alpha(1+\eta))}{1-\frac{\tau}{2}+\frac{\eta}{2}}$; then using differentiation we obtain that $R(\eta)$ attains its maximum value at

$$\eta_0 = \frac{1}{\alpha}(1 - 3\alpha - 2\beta + \sqrt{(-2\alpha)(1 - 3\alpha - 2\beta)}),$$

implying that $E[c(P)] \leq (11\alpha + 5\beta - 1 - 4\sqrt{(-2\alpha)(1 - 3\alpha - 2\beta)})c(x^*)$. Then we have that $E[c(P)] \leq \frac{9-\sqrt{33}}{2} c(x^*)$. □

4 Conclusion

In this paper, we consider a variant of traveling salesperson path problem and design a constant approximation algorithm for this problem. First we prove that the approximation ratio of this algorithm is 1.6577, then based on some observations we further analyze and prove that the ratio is $\frac{9-\sqrt{33}}{2}$.

References

1. An, H.-C., Kleinberg, R., Shmoys, D.-B.: Improving Christofides' algorithm for the *s*-*t* path TSP. J. ACM **62**(5), 34 (2015)
2. Christofides, N.: Worst-case analysis of a new heuristic for the traveling salesman problem. Carnegie-Mellon University of Pittsburgh Pa Management Sciences Research Group (1976)
3. Cunningham, W.-H.: Testing membership in matroid polyhedra. J. Comb. Theory Ser. B **36**(2), 161–188 (1984)
4. Dantzig, G., Fulkerson, R., Johnson, S.: Solution of a large-scale traveling-salesman problem. J. Oper. Res. Soc. Am. **2**(4), 393–410 (1954)
5. Edmonds, J.: Matroids and the greedy algorithm. Math. Program. **1**(1), 127–136 (1971)
6. Edmonds, J., Johnson, E.-L.: Matching, Euler tours and the Chinese postman. Math. Program. **5**(1), 88–124 (1973)
7. Frederickson, G.-N.: Approximation algorithms for some postman problems. J. ACM **26**, 538–554 (1979)
8. Fumei, L., Alantha, N.: Traveling salesman path problems. Math. Progrom. **13**, 39–59 (2008)
9. Grötschel, M., Lovász, L., Schrijver, A.: The ellipsoid method and its consequences in combinatorial optimization. Combinatorica **1**(2), 169–197 (1981)
10. Gutin, G., Punnen, A.: The Traveling Salesman Problem and its Variations. Kluwer, Dordrecht (2002)
11. Guttmann-Beck, N., Hassin, R., Khuller, S., Raghavachari, B.: Approximation algorithms with bounded performance guarantees for the clustered traveling salesman problem. Algorithmica **28**, 422–437 (2000)
12. Hoogeveen, J.-A.: Analysis of Christofides' heuristic: some paths are more difficult than cycles. Oper. Res. Lett. **10**, 291–295 (1991)
13. Karlin, A.-R., Klein, N., Gharan, S.-O.: A (slightly) improved approximation algorithm for metric TSP. In: Proceedings of the 53rd Annual ACM SIGACT Symposium on Theory of Computing, pp. 32–45 (2021)

14. Karp, R.-M.: Reducibility among combinatorial problems. Complex. Comput. Comput. **2**, 85–103 (1972)
15. Mömke, T., Svensson, O.: Removing and adding edges for the traveling salesman problem. J. ACM **63**(1), 2 (2016)
16. Mucha, M.: 13/9-approximation for graphic TSP. Theory Comput. Syst. **55**, 640–657 (2014)
17. Gharan, S.-O., Saberi, A., Singh, M.: A randomized rounding approach to the traveling salesman problem (2011)
18. Sahni, S., Gonzales, T.: *P*-complete approximation problems. J. ACM **23**(3), 555–565 (1976)
19. Sebő, A.: Eight-Fifth approximation for the path TSP. In: Goemans, M., Correa, J. (eds.) IPCO 2013. LNCS, vol. 7801, pp. 362–374. Springer, Heidelberg (2013). https://doi.org/10.1007/978-3-642-36694-9_31
20. Sebö, A., Van Zuylen, A.: The salesman's improved paths through forests. J. ACM **66**(4), 28 (2019)
21. Sebő, A., Vygen, J.: Shorter tours by nicer ears: 7/5-Approximation for the graph-TSP, 3/2 for the path version, and 4/3 for two-edge-connected subgraphs. Combinatorica **34**(5), 597–629 (2014). https://doi.org/10.1007/s00493-014-2960-3
22. Traub, V., Vygen, J.: Approaching $\frac{3}{2}$ for the s-t path TSP. J. ACM **66**(2), 14 (2019)
23. Zenklusen, R.-A.: 1.5-Approximation for path TSP. In: Proceedings of the 30th Annual ACM-SIAM Symposium on Discrete Algorithms, pp. 1539–1549 (2019)
24. Zhang, X., Du, D., Gutin, G., Ming, Q., Sun, J.: Approximation algorithms for general cluster routing problem. In: Kim, D., Uma, R.N., Cai, Z., Lee, D.H. (eds.) COCOON 2020. LNCS, vol. 12273, pp. 472–483. Springer, Cham (2020). https://doi.org/10.1007/978-3-030-58150-3_38

Hardness Results of Connected Power Domination for Bipartite Graphs and Chordal Graphs

Pooja Goyal and B. S. Panda(✉)

Computer Science and Application Group, Department of Mathematics, Indian Institute of Technology Delhi, Hauz Khas, New Delhi 110016, India
{Pooja.Goyal,bspanda}@maths.iitd.ac.in

Abstract. A set $D \subseteq V$ of a graph $G = (V, E)$ is called a connected power dominating set of G if $G[D]$, the subgraph induced by D, is connected and every vertex in the graph can be observed from D, following the two observation rules for power system monitoring: Rule 1: if $v \in D$, then v can observe itself and all its neighbors, and Rule 2: for an already observed vertex whose all neighbors except one are observed, then the only unobserved neighbor becomes observed as well. Minimum Connected Power Domination Problem is to find a connected power dominating set of minimum cardinality of a given graph G and Decide Connected Power Domination Problem is the decision version of Minimum Connected Power Domination Problem. Decide Connected Power Domination Problem is known to be NP-complete for general graphs. In this paper, we strengthen this result by proving that Decide Connected Power Domination Problem remains NP-complete for perfect elimination bipartite graph, a proper subclass of bipartite graphs, and split graphs, a proper subclass of chordal graphs. On the positive side, we show that Minimum Connected Power Domination Problem is polynomial-time solvable for chain graphs, a proper subclass of perfect elimination bipartite graph, and for threshold graphs, a proper subclass of split graphs. Further, we show that Minimum Connected Power Domination Problem cannot be approximated within $(1 - \epsilon) \ln |V|$ for any $\epsilon > 0$ unless $\mathsf{P} = \mathsf{NP}$, for bipartite graphs as well as for chordal graphs.

Keywords: Connected power domination · NP-complete · Graph algorithm

1 Introduction

A set $D \subseteq V$ is called a dominating set of G, if every vertex $v \in V \setminus D$ is adjacent to at least one vertex in D. The domination and its variations have been widely studied in the literature (see [7,8]).

D.-Z. Du et al. (Eds.): COCOA 2021, LNCS 13135, pp. 653–667, 2021.
https://doi.org/10.1007/978-3-030-92681-6_51

A power dominating set (PD-set) $D \subseteq V$ of a graph $G = (V, E)$ is obtained by considering the following two observation rules:

OR1: if $v \in D$, then v can observe itself and all its neighbors.
OR2: for an already observed vertex whose all neighbors except one are observed, then the only unobserved neighbor becomes observed as well.

The goal is to get all vertices observed by a minimum number of observers. If only $OR1$ is considered, the power dominating set problem is equivalent to the dominating set problem. The minimum cardinality of a PD-set in a graph G is known as *power domination number* of G and is denoted by $\gamma_p(G)$. Haynes et al. [9] introduced the concept of power domination and also studied the problem from algorithmic point of view. Further, this problem has been studied in [6,12, 13]. The definition of power domination presented here, was defined by Kneis et al. [12]. MINIMUM POWER DOMINATION PROBLEM is to find a power dominating set of minimum cardinality and DECIDE POWER DOMINATION PROBLEM is the decision version of MINIMUM POWER DOMINATION PROBLEM.

A connected power dominating set (CPD-set) of a graph $G = (V, E)$ is a set $D \subseteq V$ such that D is a power dominating set of G and subgraph induced by D is connected in G. The minimum cardinality of a CPD-set in a graph G is known as *connected power domination number* of G and is denoted by $\gamma_{P,c}(G)$. We formalize MINIMUM CONNECTED POWER DOMINATION PROBLEM and its decision version as follows:

MINIMUM CONNECTED POWER DOMINATION PROBLEM
Instance: A graph $G = (V, E)$.
Solution: A connected power dominating set D of G.
Measure: Cardinality of the set D.

DECIDE CONNECTED POWER DOMINATION PROBLEM
Instance: A graph $G = (V, E)$ and a positive integer r.
Question: Deciding $\gamma_{P,c}(G) \leq r$?

The Fig. 1 illustrates the difference between the definitions of dominating set, power dominating set and connected power dominating set. In graph G, we clearly see that $D_1 = \{v_4, v_9\}$ forms a minimum power dominating set of G whereas $G[D_1]$ is not connected. Add $\{v_5, v_6, v_7, v_8\}$ to D_1 to make it connected. Thus, $D_1 \cup \{v_5, v_6, v_7, v_8\}$ is a minimum connected power dominating set of G. However, neither D_1 nor $D_1 \cup \{v_5, v_6, v_7, v_8\}$ dominates all the vertices of G. It can be easily observed that $\{v_2, v_5, v_8, v_{11}\}$ forms a minimum dominating set of G.

The concept of connected power domination was introduced by Fan and Watson [4]. Further, Brimkov et al. [1] studied the connected power domination from algorithmic point of view and showed that MINIMUM CONNECTED POWER DOMINATION PROBLEM is polynomial-time solvable for trees, cactus graphs and block graphs. They also obtained various structural results about connected power domination. Brimkov et al. [1] proved that DECIDE CONNECTED POWER DOMINATION PROBLEM is NP-complete for general graphs and posed the following problem.

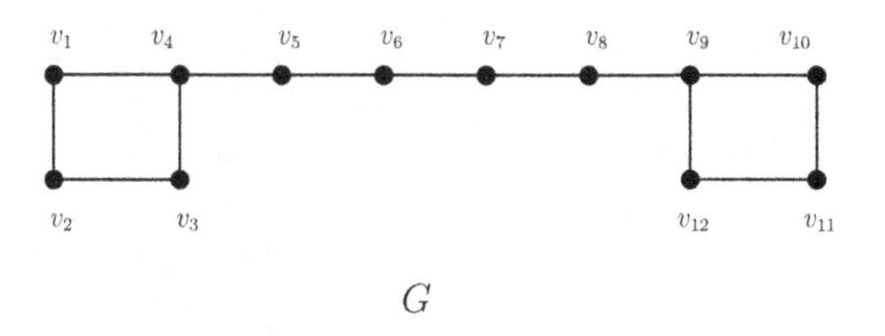

Fig. 1. A graph G.

Problem: Is DECIDE CONNECTED POWER DOMINATION PROBLEM NP-complete even for bipartite graphs, chordal graphs and split graphs?

In this paper, we answer some of the open problem proposed by Brimkov et al. [1] and extend the algorithmic study of MINIMUM CONNECTED POWER DOMINATION PROBLEM. The rest of the paper is organised as follows:

In Sect. 2, we present some pertinent definitions and some preliminary results. In Sect. 3, we strengthen the NP-completeness result of DECIDE CONNECTED POWER DOMINATION PROBLEM by showing that this problem remains NP-complete for perfect elimination bipartite graphs and split graphs. In Sect. 4, we show that minimum CPD-set of a given chain graph and a threshold graph can be computed in polynomial time. In Sect. 5, we show that MINIMUM CONNECTED POWER DOMINATION PROBLEM cannot be approximated within $(1 - \epsilon) \ln |V|$ for any $\epsilon > 0$ unless $\mathsf{P} = \mathsf{NP}$ for bipartite graphs as well as chordal graphs.

2 Definitions and Preliminary Results

Let $G = (V, E)$ be a finite, simple, and undirected graph with vertex set V and edge set E. For a vertex $v \in V$, the *open neighborhood* and the *closed neighborhood* of v in G are defined as $N_G(v) = \{u \in V | uv \in E\}$ and $N_G[v] = N_G(v) \cup \{v\}$, respectively. The degree of a vertex v is $|N_G(v)|$ and is denoted by $d_G(v)$. If $d_G(v) = 1$, then v is called *pendant* vertex or *leaf* of G and the unique neighbor of v in G is called *support* vertex. The number of connected components of G will be denoted by $c(G)$. If the graph G is clear from the context, then we often omit it in our notations. For example, we write V and E instead of writing $V(G)$ and $E(G)$, respectively. For $D \subseteq V$, let $G[D]$ denote the subgraph of G induced by D. For any $C \subseteq V$, if $G[C]$ is a complete subgraph of G, then C is called a *clique* of G. For any $I \subseteq V$, if $G[I]$ has no edge, then I is called an *independent set* of G. We use the standard notation $[k] = \{1, 2, \ldots, k\}$.

A *chord* of a cycle is an edge joining two non-consecutive vertices of the cycle. A graph is called a *chordal graph* if every cycle of length at least 4 has a chord. A vertex $v \in V(G)$ is a simplicial vertex of G if $N_G[v]$ is a clique of G. An ordering $\sigma = (v_1, v_2, \ldots, v_n)$ is a perfect elimination ordering (PEO) of G if v_i is a simplicial vertex of $G_i = G[\{v_i, v_{i+1}, \ldots, v_n\}]$ for each $i \in [n]$. It is characterized that a graph G is chordal if and only if it has a PEO [5]. A chordal graph $G = (V, E)$ is a *split* graph if V can be partitioned into two sets I and C such that C is a clique and I is an independent set.

A *bipartite graph* is a graph $G = (V, E)$ whose vertices can be partitioned into two disjoint sets X and Y such that every edge has one end vertex in X and other in Y. We denote a bipartite graph with bi-partition X and Y of V as $G = (X \cup Y, E)$. Let $G = (X \cup Y, E)$ be a bipartite graph. An edge $e = xy$ is called bisimplicial edge if $N(x) \cup N(y)$ induces a complete bipartite subgraph. Let $\sigma = (x_1y_1, x_2y_2, \ldots, x_ky_k)$ be a sequence of pairwise non-adjacent edges of G. Denote $S_j = \{x_1, x_2, \ldots, x_j\} \cup \{y_1, y_2, \ldots, y_j\}$ and let $S_0 = \emptyset$. Then σ is said to be a perfect edge elimination scheme for G if each edge $x_{j+1}y_{j+1}$ is bisimplicial in $G_j = [(X \cup Y) \setminus S_j]$ for $j \in \{0, 1, \ldots, k-1\}$ and $G_k = [(X \cup Y) \setminus S_k]$ has no edge. A graph for which there exists a perfect edge elimination scheme is called a *perfect elimination* bipartite graph.

We now recall some terminology and notation from [1]. Let $G = (V, E)$ be a connected graph different from path and v be a vertex of degree at least 3. A *pendant path* attached to v is a maximal set $P \subset V$ such that $G[P]$ is a connected component of $G - v$ which is a path, one of whose ends is adjacent to v in G. The neighbor of v in P will be called the base of the path, and $p(v)$ will denote the number of pendant paths attached to $v \in V$. Finally, for a connected graph $G = (V, E)$ different from a path, define:

$$\begin{aligned}
R_1(G) &= \{v \in V : c(G - v) = 2, p(v) = 1\} \\
R_2(G) &= \{v \in V : c(G - v) = 2, p(v) = 0\} \\
R_3(G) &= \{v \in V : c(G - v) \geq 3\} \\
M(G) &= R_2(G) \cup R_3(G).
\end{aligned}$$

The following observations will be used in the rest of the paper.

Observation 1 [1]. *Let $G = (V, E)$ be a connected graph different from a path and D be an arbitrary connected power dominating set of G. Then $M(G) \subset D$.*

Observation 2 [1]. *Let G be a graph different from a path. Then, no minimum connected power dominating set of G contains a leaf of G.*

3 NP-completeness Results

It is known that for any graph G, Decide Connected Power Domination Problem is NP-complete [1]. In this section, we strengthen the NP-completeness result by showing that problem remains NP-complete for perfect elimination bipartite graphs and split graphs.

3.1 Result for Perfect Elimination Bipartite Graphs

Theorem 3. Decide Connected Power Domination Problem *is NP-complete for perfect elimination bipartite graphs.*

Proof. Clearly, DECIDE CONNECTED POWER DOMINATION PROBLEM is in NP for perfect elimination bipartite graphs. To show the hardness of DECIDE CONNECTED POWER DOMINATION PROBLEM on perfect elimination bipartite graphs, we give a polynomial reduction from DECIDE X3C, which is already known to be NP-complete (see [10]). Given an arbitrary instance $(X, \mathcal{C})$ of $X3C$, $X = \{x_1, x_2, \dots, x_{3q}\}$ and $\mathcal{C} = \{C_1, C_2, \dots, C_t\}$. We construct a perfect elimination bipartite graph $G = (V, E)$ from the system $(X, \mathcal{C})$ as follows:

- For each vertex $x_j \in X$, add a path of length 4, $P_4(j) : x_j u_j v_j w_j$.
- For each $C_i \in \mathcal{C}$, add two vertices b_i and c_i. Also join vertex b_i with c_i.
- Also add 4 more vertices $\{p, q, r, s\}$ and add edges pq, pr, ps and pc_i, for every $C_i \in \mathcal{C}$.
- Finally add edges $x_j c_i$ if and only if $x_j \in C_i$.

Clearly, G is a bipartite graph and the ordering $\sigma(G) = (v_1w_1, v_2w_2, \dots, v_{3q}w_{3q}, x_1u_1, x_2u_2, \dots, x_{3q}u_{3q}, b_1c_1, b_2c_2, \dots, b_tc_t, pq)$ is a perfect edge elimination ordering of G. We show an example in Fig. 2, perfect elimination bipartite graph G is obtained from the system $(X, \mathcal{C})$, where $X = \{x_1, x_2, x_3, x_4, x_5, x_6\}$ and $\mathcal{C} = \{\{x_1, x_2, x_3\}, \{x_2, x_4, x_5\}, \{x_3, x_5, x_6\}, \{x_4, x_5, x_6\}\}$. Now to complete the proof, it suffices to prove the following claim:

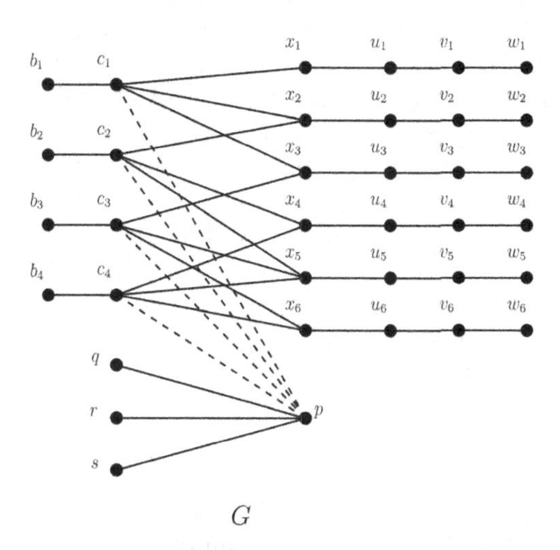

Fig. 2. An illustration of the construction of G from system $(X, \mathcal{C})$ in the proof of Theorem 3.

Claim. The system $(X, \mathcal{C})$ has an exact cover if and only if G has a connected power dominating set of cardinality at most $q + 1$.

Proof. Suppose that the instance $(X, \mathcal{C})$ of DECIDE X3C has a solution $\mathcal{C}'$. Define $D = \{p\} \cup \{c_i \mid C_i \in \mathcal{C}'\}$. Every element of the set $\{c_i, x_j \mid i \in [t], j \in$

$[3q]\} \cup \{p, q, r, s\} \cup \{b_k \mid c_k \in \mathcal{C}'\}$ can be observed by applying $OR1$ to the set D. Further every u_j can be observed by applying $OR2$ to the corresponding x_j, every v_j can be observed by applying $OR2$ to u_j and every w_j can be observed by applying $OR2$ to v_j. Also, remaining b_i's can be observed by applying $OR2$ to the corresponding c_i's. Hence, $D = \{p\} \cup \{c_i \mid C_i \in \mathcal{C}'\}$ is a CPD-set for G of cardinality $q + 1$.

Conversely, suppose that G has a CPD-set of cardinality at most $q + 1$. Now, we will show that if D is a CPD-set of G, then there is a CPD-set D' of G with $|D'| \leq |D|$, $p \in D'$ and $D' \setminus \{p\} \subseteq \{c_i \mid i \in [t]\}$. Let us assume that D is a CPD-set of G. Since vertex p is a cut vertex of G and belongs to $R_3(G)$, by Observation 1, $p \in D$. By Observation 2, we assume that no leaf vertex of G belongs to D. Hence, $D \cap \{q, r, s, b_1, b_2, \ldots, b_t, w_1, w_2, \ldots, w_{3q}\} = \emptyset$. Now, if $u_j \in D$, then $x_j \in D$ because $G[D]$ is connected. If $u_j, x_j \in D$, then $D' = D \setminus \{u_j\}$ is a CPD-set of G with $|D'| < |D|$. So continuing this way, we get a CPD-set D' of G with $|D'| \leq |D|$ such that $D' \cap \{u_1, u_2, \ldots, u_{3q}\} = \emptyset$. Similarly. $D' \cap \{v_1, v_2, \ldots, v_{3q}\} = \emptyset$. Further if $x_k \in D$, then there exists a $c_i \in D$ where $x_k \in C_i$. Such a c_i belongs to D because $G[D]$ is connected. If such $x_k, c_i \in D$, then $D' = D \setminus \{x_k\}$ is again a CPD-set of G with $|D'| < |D|$. So continuing this way, we get a CPD-set D' of G with $|D'| \leq |D|$ such that $D' \subseteq \{c_i \mid i \in [t]\}$.

Further we will show that each vertex of the set $\{x_j \mid j \in [3q]\}$ will be observed by applying $OR1$ to some vertex of D'. On contrary we assume that there exists at least one vertex x_k of G such that x_k cannot be observed by applying $OR1$ to any vertex of D'. That means $T \cap D' = \emptyset$, where $T = \{c_i \mid x_k c_i \in E\}$. The set T can be observed by some element of D' and every element of the set T has at least two unobserved vertices $\{b_i, x_k\}$. Hence we get a contradiction to the fact that D' is a CPD-set of G and it is due to our assumption that vertex x_k cannot be observed by applying $OR1$ to the set D'. Hence, we can conclude that every vertex of the set $\{x_j \mid j \in [3q]\}$ will be observed by applying $OR1$ to some vertex of D'. That means, every vertex x_j has a neighbor in D'. Now let $\mathcal{C}' = \{C_i \in \mathcal{C} \mid c_i \in D'\}$ and $|\mathcal{C}'| \leq q$. Then $\mathcal{C}'$ is an exact cover of $(X, \mathcal{C})$. This completes the proof of the claim. □

Therefore, DECIDE CONNECTED POWER DOMINATION PROBLEM is NP-complete for perfect elimination bipartite graphs. □

3.2 Result for Split Graphs

We next strengthen the NP-completeness result of DECIDE CONNECTED POWER DOMINATION PROBLEM for chordal graphs by showing that this problem remains NP-complete for split graphs, a subclass of chordal graphs.

Theorem 4. DECIDE CONNECTED POWER DOMINATION PROBLEM *is NP-complete for split graphs.*

Proof. Clearly, DECIDE CONNECTED POWER DOMINATION PROBLEM is in NP for split graphs. To show the hardness of DECIDE CONNECTED POWER DOMINATION PROBLEM on split graphs, we give a polynomial reduction from DECIDE

X3C, which is already known to be NP-complete (see [10]). Given an arbitrary instance $(X, \mathcal{C})$ of $X3C$, $X = \{x_1, x_2, \ldots, x_{3q}\}$ and $\mathcal{C} = \{C_1, C_2, \ldots, C_t\}$. We construct a split graph $G = (V, E)$ with split partition (K, I), where K is a clique and I is an independent set, from the system $(X, \mathcal{C})$ as follows:

- For each vertex $x_j \in X$, add a vertex x_j in I.
- For each $C_i \in \mathcal{C}$, we add a vertex c_i in K and a vertex a_i in I. Add edges $c_i a_i$ for every $i \in [t]$ and $c_i c_k$ for every $i, k \in [t]$ and $i \neq k$.
- Finally add edges $x_j c_i$ if and only if $x_j \in C_i$.

Clearly, G is a split graph with split partition (K, I), where $K = \{c_i \mid C_i \in \mathcal{C}\}$ and $I = \{a_i \mid C_i \in \mathcal{C}\} \cup \{x_j \mid j \in [3q]\}$. We show an example in Fig. 3, split graph G is obtained from the system $(X, \mathcal{C})$, where $X = \{x_1, x_2, x_3, x_4, x_5, x_6\}$ and $\mathcal{C} = \{\{x_1, x_2, x_3\}, \{x_2, x_4, x_5\}, \{x_3, x_5, x_6\}, \{x_4, x_5, x_6\}\}$. Now to complete the proof, it suffices to prove the following claim:

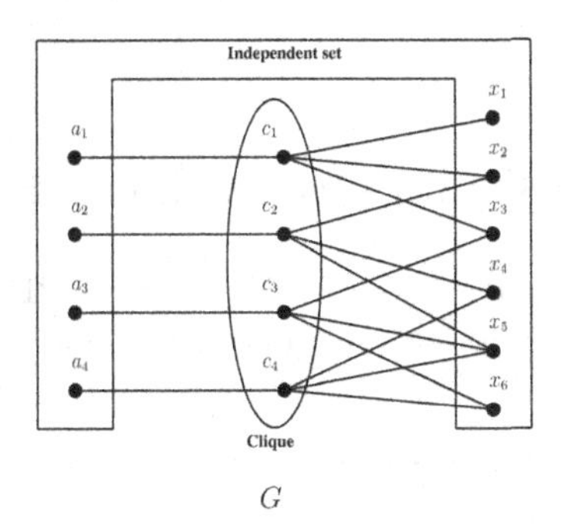

Fig. 3. An illustration of the construction of G from system $(X, \mathcal{C})$ in the proof of Theorem 4.

Claim. The system $(X, \mathcal{C})$ has an exact cover if and only if G has a connected power dominating set of cardinality at most q.

Proof. Suppose that the instance $(X, \mathcal{C})$ of DECIDE X3C has a solution $\mathcal{C}'$. Define $D = \{c_i \mid C_i \in \mathcal{C}'\}$. Every element of the set $\{c_i, x_j \mid i \in [t], j \in [3q]\} \cup \{a_i \mid c_i \in D\}$ can be observed by applying $OR1$ to the set D and remaining vertices of G can be observed by applying $OR2$. Hence, $D = \{c_i \mid C_i \in \mathcal{C}'\}$ is a CPD-set for G of cardinality q.

Conversely, suppose that G has a CPD-set of cardinality at most q. Now, we will show that if D is a CPD-set of G, then there is a CPD-set D' of G with $|D'| \leq |D|$ and $D' \subseteq \{c_i \mid i \in [t]\}$. Suppose that D is a CPD-set of G. By Observation 2, we assume that no leaf vertex of G belongs to D. Hence, $D \cap \{a_1, a_2, \ldots, a_t\} = \emptyset$. Further if $x_k \in D$, then there exists a $c_i \in D$ where $x_k \in C_i$. Such a c_i belongs to D because $G[D]$ is connected. If such $x_k, c_i \in D$, then $D' = D \setminus \{x_k\}$ is again a CPD-set of G with $|D'| < |D|$. So continuing this way, we get a CPD-set D' of G with $|D'| \leq |D|$ such that $D' \cap \{a_1, a_2, \ldots, a_t\} = \emptyset$ and $D' \cap \{x_1, x_2, \ldots, x_{3q}\} = \emptyset$. Hence we get a CPD-set D' of G with $|D'| \leq |D|$ and $D' \subseteq \{c_i \mid i \in [t]\}$.

Further we will show that each vertex of the set $\{x_j \mid j \in [3q]\}$ will be observed by applying $OR1$ to some vertex of D'. On contrary we assume that there exists at least one vertex x_k of G such that x_k cannot be observed by applying $OR1$ to any vertex of D'. That means $T \cap D' = \emptyset$, where $T = \{c_i \mid x_k c_i \in E\}$. The set T can be observed by some element of D' and every element of the set T has at least two unobserved vertices $\{a_i, x_k\}$. Hence we get a contradiction to the fact that D' is a CPD-set of G and it is due to our assumption that vertex x_k cannot be observed by applying $OR1$ to the set D'. Hence, we can conclude that every vertex of the set $\{x_j \mid j \in [3q]\}$ will be observed by applying $OR1$ to some vertex of D'. That means, every vertex x_j has a neighbor in D'. Now let $C' = \{C_i \in \mathcal{C} \mid c_i \in D'\}$ and $|C'| \leq q$. Then C' is an exact cover of $(X, \mathcal{C})$. This completes the proof of the claim. □

Therefore, Decide Connected Power Domination Problem is NP-complete for split graphs. □

4 Algorithms for Chain Graphs and Threshold Graphs

In this section, we show that minimum cardinality CPD-set for chain graph and threshold graph can be computed in polynomial time.

4.1 Connected Power Domination for Chain Graphs

In this paper, we have shown that Decide Connected Power Domination Problem remains NP-complete for perfect elimination bipartite graphs. In this section, we present a positive result by proposing a polynomial-time algorithm to solve Minimum Connected Power Domination Problem in chain graphs, a subclass of perfect elimination bipartite graphs.

A bipartite graph $G = (X \cup Y, E)$ is called a chain graph if the neighborhoods of the vertices of X form a chain, that is, the vertices of X can be linearly ordered, say $\{x_1, x_2, \ldots, x_p\}$, such that $N_G(x_1) \subseteq N_G(x_2) \subseteq \ldots \subseteq N_G(x_p)$. If $G = (X \cup Y, E)$ is a chain graph, then the neighborhoods of the vertices of Y also form a chain [16]. An ordering $\alpha = (x_1, x_2, \ldots, x_p, y_1, y_2, \ldots, y_q)$ of $X \cup Y$ is called a chain ordering if $N_G(x_1) \subseteq N_G(x_2) \subseteq \ldots \subseteq N_G(x_p)$ and $N_G(y_1) \supseteq N_G(y_2) \supseteq \ldots \supseteq N_G(y_q)$. It is well known that every chain graph admits a chain ordering [11,16].

First we prove the following lemma, which will be helpful in proving the main result of this section.

Lemma 1. *Let $G = (X \cup Y, E)$ is a chain graph. Then $\gamma_{P,c}(G) \leq 2$.*

Proof. Given $G = (X \cup Y, E)$ be a chain graph, where $V = X \cup Y$. It can be easily observed that vertex $x_p \in X$ is adjacent to every vertex of the set Y. Thus, every vertex of the set Y can be observed by x_p by applying $OR1$. Similarly, vertex $y_1 \in Y$ is adjacent to every vertex of the set X. Thus, every vertex of the set X can be observed by y_1 by applying $OR1$. Thus, $D = \{x_p, y_1\}$ is a CPD-set of G. Hence, $\gamma_{P,c}(G) \leq |D| = |\{x_p, y_1\}| = 2$. □

Lemma 2. *Let $G = (X \cup Y, E)$ be a chain graph such that $|N_G(x_1)| \leq 2$. For every $i \in [p-1]$, $|N_G(x_{i+1}) - N_G(x_i)| \leq 1$ if and only if $D^*_{P,c} = \{y_1\}$, where $D^*_{P,c}$ is a minimum* CPD*-set of G.*

Proof. It can be easily observed that vertex $y_1 \in Y$ is adjacent to every vertex of the set X. Thus, every vertex of the set X can be observed by y_1 by applying $OR1$. Now we have to observe every vertex of set $Y \setminus \{y_1\}$. Let $y_2 \in N_G(x_1)$ and $y_2 \neq y_1$. Such a vertex y_2 may exist because $|N_G(x_1)| \leq 2$. Then vertex y_2 can be observed after applying $OR2$ to vertex x_1. That means every neighbor of x_1 has been observed. Since $|N_G(x_{i+1}) - N_G(x_i)| \leq 1$, there can exist at most one unobserved neighbor of x_2 and that can be observed by x_2 after applying $OR2$. Hence, every neighbor of x_2 has been observed. Similarly, x_3 can have at most one unobserved neighbor in Y and that can be observed by x_3 after applying $OR2$. Hence, every neighbor of x_3 has been observed. Continuing this way, every vertex of $Y \setminus \{y_1\}$ can be observed. Thus, $D^*_{P,c} = \{y_1\}$ will be a minimum CPD-set of G.

Conversely, suppose that $D^*_{P,c} = \{y_1\}$ is a minimum CPD-set of G. Then y_1 observed every vertex of set X by applying $OR1$. Let us assume that there exists a $k \in [p-1]$ such that $|N_G(x_{k+1}) - N_G(x_k)| \geq 2$. Let y_j, y_{j+1} be two consecutive vertices belonging to $N_G(x_{k+1}) \setminus N_G(x_k)$. Then y_j, y_{j+1} are adjacent to every vertex in the set $\{x_{k+1}, x_{k+2}, \ldots, x_p\}$. Also both y_j and y_{j+1} are not adjacent to any vertex in the set $\{x_1, x_2, \ldots, x_k\}$. That means vertex x_{k+1} has two unobserved neighbors y_j and y_{j+1}. Hence, we get a contradiction to the fact that $D^*_{P,c} = \{y_1\}$ is a minimum CPD-set of G. So there does not exist $k \in [p-1]$ such that $|N_G(x_{k+1}) - N_G(y_k)| \geq 2$. Hence, $|N_G(x_{i+1}) - N_G(x_i)| \leq 1$ for every $i \in [p-1]$. □

Lemma 3. *Let $G = (X \cup Y, E)$ be a chain graph such that $|N_G(y_q)| \leq 2$. For every $j \in [q-1]$, $|N_G(y_j) - N_G(y_{j+1})| \leq 1$ if and only if $D^*_{P,c} = \{x_p\}$, where $D^*_{P,c}$ is a minimum* CPD*-set of G.*

Proof. The proof is similar as in above lemma.

A chain ordering of a chain graph $G = (X \cup Y, E)$ can be computed in linear time [15]. Then we checked $d_G(x_1) \leq 2$ and $d_G(x_{i+1}) - d_G(x_i) \leq 1$ for every $i \in [p-1]$. Also we have to check $d_G(y_q) \leq 2$ and $d_G(y_j) - d_G(y_{j+1}) \leq 1$ for every $j \in [q-1]$. All these can be tested in linear time. Based on above discussion and lemmas, we present following linear time algorithm to compute a minimum CPD-set of a chain graph.

Hence, we have the following theorem.

Theorem 5. *A minimum* CPD*-set of a chain graph can be computed in linear time.*

Algorithm 1. MIN-CPD-CHAIN(G)

```
Input: A chain graph G = (X ∪ Y, E) and chain ordering α =
(x_1, x_2, ..., x_p, y_1, y_2, ..., y_q) of X ∪ Y.
Output: A minimum CPD-set of graph G.
begin
    Set D*_{P,c} = ∅;
    if (|N_G(x_1)| ≤ 2) then
        set i = 1;
        while (|N_G(x_{i+1}) − N_G(x_i)| ≤ 1 and i ≤ p − 1) do
            i++;
        if (i == p) then
            D*_{P,c} = {y_1};
    if (D*_{P,c} = ∅ and |N_G(y_q)| ≤ 2) then
        set j = 1;
        while (|N_G(y_j) − N_G(y_{j+1})| ≤ 1 and j ≤ q − 1) do
            j++;
        if (j == q) then
            D*_{P,c} = {x_p};
    if (D*_{P,c} = ∅) then
        D*_{P,c} = {y_1, x_p};
    return D*_{P,c}
```

4.2 Connected Power Domination for Threshold Graphs

In this paper, we have shown that DECIDE CONNECTED POWER DOMINATION PROBLEM remains NP-complete for split graphs. In this section, we present a positive result by proposing a polynomial-time algorithm to solve MINIMUM CONNECTED POWER DOMINATION PROBLEM in threshold graphs, a subclass of split graphs. Firstly, we will define threshold graphs.

A graph $G = (V, E)$ is called a *threshold* graph if there is a real number T and a real number $w(v)$ for every $v \in V$ such that a set $S \subseteq V$ is independent if and only if $\Sigma_{v \in S} w(v) \leq T$ [2]. Many characterizations of threshold graphs are available in the literature. An important characterization of threshold graph, which is used in designing polynomial-time algorithms is following: A graph G is *threshold* graph if and only if it is a split graph and, for any split partition (C, I) of G, there is an ordering $(x_1, x_2, \ldots, x_p)$ of the vertices of C such that $N_G[x_1] \subseteq N_G[x_2] \subseteq \ldots \subseteq N_G[x_p]$, and there is an ordering $(y_1, y_2, \cdots, y_q)$ of the vertices of I such that $N_G(y_1) \supseteq N_G(y_2) \supseteq \ldots \supseteq N_G(y_q)$ [14].

Theorem 6. *Let $G = (V, E)$ be a threshold graph with split partition (C, I) as defined above, then G has a minimum connected power dominating set $D^*_{P,c} = \{x_p\}$.*

Proof. It can be easily observed that vertex $x_p \in C$ is adjacent to every vertex of the independent set I. Thus, every vertex of independent set I can be observed by x_p by applying $OR1$. Since C is clique, vertex x_p observed every vertex of C by applying $OR1$. Hence, $D^*_{P,c} = \{x_p\}$ is a minimum connected power dominating set of G.

5 Lower Bound on Approximation Ratio

In this subsection, we obtain lower bounds on the approximation ratio of Minimum Connected Power Domination Problem for bipartite graphs and chordal graphs. To obtain lower bound for bipartite graphs we give an approximation preserving reduction from Minimum Set Cover. For this, we need the following theorem proved in [3].

Theorem 7 [3]. Minimum Set cover *for set system* $(U, \mathcal{C})$ *cannot be approximated within* $(1-\epsilon)\ln|U|$ *for any* $\epsilon > 0$ *unless* $\mathsf{P} = \mathsf{NP}$.

We are ready to prove the inapproximability of Minimum Connected Power Domination Problem for bipartite graphs.

Theorem 8. Minimum Connected Power Domination Problem *for bipartite graphs cannot be approximated within* $(1-\varepsilon)\ln|V|$ *for any* $\varepsilon > 0$ *unless* $\mathsf{P} = \mathsf{NP}$.

Proof. Given an instance $(U, \mathcal{C})$ of Minimum Set Cover, where $U = \{u_1, u_2, \ldots, u_q\}$. $\mathcal{C} = \{C_1, C_2, \ldots, C_t\}$. Now we construct a bipartite graph $G = (X \cup Y, E)$ in polynomial time as follows.

- For each element u_j in the set U, add two vertices x_j and y_j in partite set X of G.
- For each set C_i in the collection $\mathcal{C}$, add a vertex c_i in partite set Y of G.
- Add a vertex r in X, a set of vertices $\{u, v, w\}$ in Y, and set of edges $\{ru, rv, rw, rc_i \mid i \in [t]\}$ in E.
- If an element u_j belongs to set C_i, then add edges x_jc_i and y_jc_i in G.

Formally, $X = \{x_1, x_2, \ldots, x_q, y_1, y_2, \ldots, y_q, r\}$, $Y = \{c_1, c_2, \ldots, c_t, u, v, w\}$ and $E = \{x_jc_i, y_jc_i \mid u_j \in C_i\} \cup \{rc_i, ru, rv, rw \mid i \in [t]\}$. We show an example in Fig. 4, where G is a bipartite graph obtained from set system $(U, \mathcal{C})$ with $U = \{u_1, u_2, u_3\}$ and $\mathcal{C} = \{\{u_1, u_2\}, \{u_2, u_3\}, \{u_3, u_1\}, \{u_3\}\}$.

Claim. $\gamma_{P,c}(G) = |S^*| + 1$, where S^* is the minimum cardinality set cover of system $(U, \mathcal{C})$.

Proof. Due to space constraint, we omit the proof of claim.

Now, for the resulting bipartite graph G one can now confine to CPD-set D consisting of r and a subset S of $\mathcal{C}$ corresponding to a set cover, hence we have $|D| = |S| + 1$.

Suppose that Minimum Connected Power Domination Problem can be approximated within a ratio of α, where $\alpha = (1-\epsilon)\ln(|V|)$ for some fixed

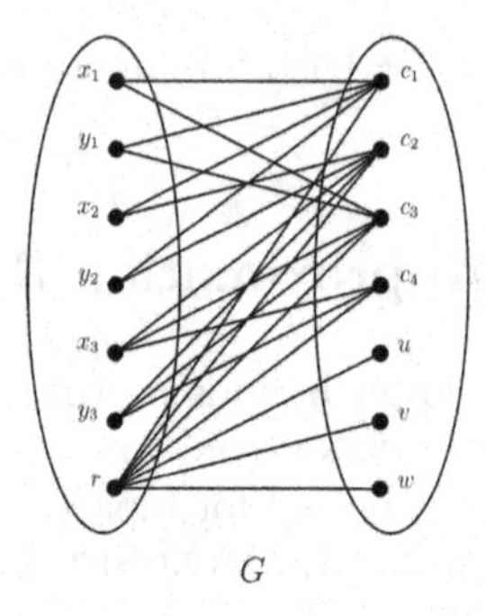

Fig. 4. An illustration of the construction of G from system $(U, \mathcal{C})$ in the proof of Theorem 9.

Algorithm 2. A': Approximation Algorithm for MINIMUM SET COVER

Input: A set system $(U, \mathcal{C})$
Output: A minimum set cover S of $(U, \mathcal{C})$.
begin
  if *(there exists a minimum set cover S of $(U, \mathcal{C})$ of cardinality $< l$)* **then**
    return S;
  else
    Construct a bipartite graph G as described above;
    Compute a CPD-set D of G using algorithm A;
    $S = \{C_i \in \mathcal{C} \mid c_i \in D\}$;
    return S;

$\epsilon > 0$, by some polynomial-time approximation algorithm, say Algorithm A. Next, we propose an algorithm A' to compute a set cover of a given set system $(U, \mathcal{C})$ in polynomial time. Clearly, A' is a polynomial-time algorithm as A is a polynomial-time algorithm. Since l is a constant, step 1 of the algorithm can be executed in polynomial time. Note that if S is computed in Step 1, then S is optimal. So we analyze the case where $|S| \geq l$.

Let D^* be an optimal CPD-set in G and S^* be an optimal set cover in $(U, \mathcal{C})$. It is clear that $|S^*| \geq l$. Let S be the set cover computed by algorithm A'. Then $|S| = |D| - 1 \leq |D| \leq \alpha|D^*| \leq \alpha(|S^*| + 1) \leq \alpha\Big(1 + \frac{1}{|S^*|}\Big)|S^*| \leq \alpha\Big(1 + \frac{1}{l}\Big)|S^*|$ Hence, algorithm A' approximates MINIMUM SET COVER for given set system $(U, \mathcal{C})$ within the ratio $\alpha(1 + \frac{1}{l})$.

Let l be a positive integer such that $\frac{1}{l} < \frac{\epsilon}{2}$. Then algorithm A' approximates MINIMUM SET COVER for given set system $(U, \mathcal{C})$ within the ratio $\alpha(1+\frac{1}{l}) \leq (1-\epsilon)(1+\frac{\epsilon}{2}) \ln |V| = (1-\epsilon') \ln |U|$ for $\epsilon' = \frac{\epsilon^2}{2} + \frac{\epsilon}{2}$ as $\ln |V| = \ln(2|U|+|\mathcal{C}|+4) \approx \ln |U|$ for sufficiently large value of $|U|$.

Therefore, the Algorithm A' approximates MINIMUM SET COVER within ratio $(1-\epsilon)\ln(|U|)$ for some $\epsilon > 0$. By Theorem 7, if MINIMUM SET COVER can be approximated within ratio $(1-\epsilon)\ln(|U|)$ for some $\epsilon > 0$, then $\mathsf{P} = \mathsf{NP}$. Hence, if MINIMUM CONNECTED POWER DOMINATION PROBLEM can be approximated within ratio $(1-\epsilon)\ln(|V|)$ for some $\epsilon > 0$, then $\mathsf{P} = \mathsf{NP}$. This proves that MINIMUM CONNECTED POWER DOMINATION PROBLEM for bipartite graphs cannot be approximated within $(1-\epsilon)\ln(|V|)$ for any $\epsilon > 0$ unless $\mathsf{P} = \mathsf{NP}$. □

Next, we prove inapproximability of MINIMUM CONNECTED POWER DOMINATION PROBLEM for chordal graphs by giving an approximation preserving reduction from MINIMUM SET COVER. We are ready to prove the inapproximability of MINIMUM CONNECTED POWER DOMINATION PROBLEM for chordal graphs.

Theorem 9. MINIMUM CONNECTED POWER DOMINATION PROBLEM *for chordal graph* $G = (V, E)$ *cannot be approximated within* $(1-\epsilon)\ln|V|$ *for any* $\epsilon > 0$ *unless* $P = NP$.

Proof. Given an instance $(U, \mathcal{C})$ of MINIMUM SET COVER, where $U = \{u_1, u_2, \ldots, u_q\}$ and $\mathcal{C} = \{C_1, C_2, \ldots, C_t\}$. Now we construct a chordal graph $G = (V, E)$ in polynomial time as follows.

- For each element u_j in the set U, we add two vertices x_j and y_j, and add edges $x_j y_j$ in G.
- For each set C_i in the collection $\mathcal{C}$, we add two vertices c_i and d_i, and add edges $c_i d_i$ in G. Also, add edges $c_i c_k$ for every $i, k \in [t]$ and $i \neq k$.
- If an element u_j belongs to set C_i, then add an edge between vertices x_j and c_i in G.

$V = \{x_j, y_j \mid u_j \in U\} \cup \{c_i, d_i \mid C_i \in \mathcal{C}\}$, and $E = \{x_j y_j \mid u_j \in U\} \cup \{c_i d_i \mid C_i \in \mathcal{C}\} \cup \{x_j c_i \mid u_j \in C_i\} \cup \{c_i c_k \mid i, k \in [t], i \neq k\}$. Since $|V| = 2(|U| + |\mathcal{C}|)$, graph G can be constructed in polynomial time. It can be easily verified that G is a chordal graph with PEO $\sigma(G) = (d_1, d_2, \ldots, d_t, y_1, y_2, \ldots, y_q, x_1, x_2, \ldots, x_q, c_1, c_2, \ldots, c_t)$. We show an example in Fig. 5, where G is obtained from the system $(U, \mathcal{C})$, where $U = \{u_1, u_2, u_3, u_4, u_5, u_6\}$ and $\mathcal{C} = \{\{u_1, u_2, u_3\}, \{u_3, u_5, u_6\}, \{u_4, u_5\}, \{u_5, u_6\}\}$.

Claim. $\gamma_{P,c}(G) = |S^*|$, where S^* is the minimum cardinality set cover of system $(U, \mathcal{C})$.

Proof. Due to space constraint, we omit the proof of claim.

Now, for the resulting chordal graph G one can now confine to CPD-set D and a subset S of $\mathcal{C}$ corresponding to a set cover, hence we have $|D| = |S|$.

Suppose that MINIMUM CONNECTED POWER DOMINATION PROBLEM can be approximated within a ratio of α, where $\alpha = (1-\epsilon)\ln(|V|)$ for some fixed $\epsilon > 0$, by some polynomial-time approximation algorithm, say Algorithm B. Next, we propose an algorithm B' to compute a set cover of the given set system $(U, \mathcal{C})$ in polynomial time.

Clearly, B' is a polynomial-time algorithm as B is a polynomial-time algorithm.

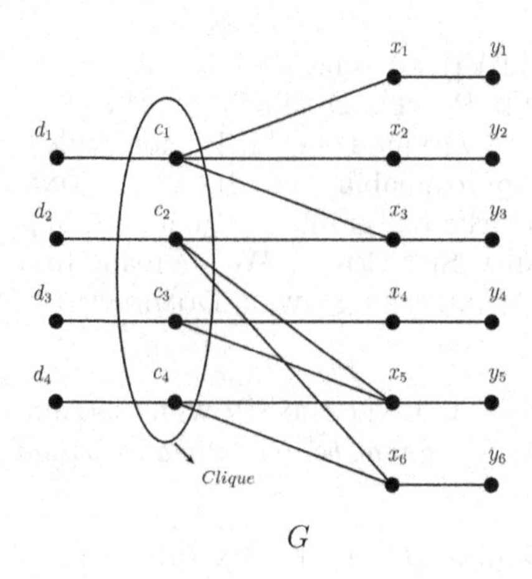

Fig. 5. An illustration of the construction of G from system $(U, \mathcal{C})$ in the proof of Theorem 9.

Algorithm 3. B': Approximation Algorithm for MINIMUM SET COVER

Input: A set system $(U, \mathcal{C})$
Output: A minimum set cover S of $(U, \mathcal{C})$.
begin
  Construct a chordal graph G as described above;
  Compute a CPD-set D of G using algorithm B;
  $S = \{C_i \in \mathcal{C} \mid c_i \in D\}$;
  return S;

Let D^* be an optimal CPD-set in G and S^* be an optimal set cover in $(U, \mathcal{C})$. Let S be the set cover computed by algorithm B'. Then $|S| = |D| \leq \alpha|D^*| \leq \alpha|S^*|$ Hence, algorithm B' approximates MINIMUM SET COVER for given set system $(U, \mathcal{C})$ within the ratio $\alpha = (1-\epsilon)\ln|V| = (1-\epsilon)\ln(2|U| + 2|\mathcal{C}|) \approx (1-\epsilon)\ln|U|$ for sufficiently large value of $|U|$.

Therefore, the Algorithm B' approximates MINIMUM SET COVER within ratio $(1-\epsilon)\ln(|U|)$ for some $\epsilon > 0$. By Theorem 7, if MINIMUM SET COVER can be approximated within ratio $(1-\epsilon)\ln(|U|)$ for some $\epsilon > 0$, then $\mathsf{P} = \mathsf{NP}$. Hence, if MINIMUM CONNECTED POWER DOMINATION PROBLEM can be approximated within ratio $(1-\epsilon)\ln(|V|)$ for some $\epsilon > 0$, then $\mathsf{P} = \mathsf{NP}$. This proves that MINIMUM CONNECTED POWER DOMINATION PROBLEM for chordal graphs cannot be approximated within $(1-\epsilon)\ln(|V|)$ for any $\epsilon > 0$ unless $\mathsf{P} = \mathsf{NP}$. □

6 Conclusion

In this paper, we have shown that DECIDE CONNECTED POWER DOMINATION PROBLEM is NP-complete for perfect elimination bipartite graphs and split graphs which answers some of the open question left by Brimkov et al. [1]. On the positive side, we have shown that MINIMUM CONNECTED POWER DOMINATION PROBLEM can be solved in polynomial time for chain graphs and threshold graphs. Apart from these, we have then presented inapproximability results of MINIMUM CONNECTED POWER DOMINATION PROBLEM for bipartite graphs and chordal graphs. It would be interesting to design an approximation algorithm for MINIMUM CONNECTED POWER DOMINATION PROBLEM with good approximation ratio.

References

1. Brimkov, B., Mikesell, D., Smith, L.: Connected power domination in graphs. J. Comb. Optim. **38**(1), 292–315 (2019). https://doi.org/10.1007/s10878-019-00380-7
2. Chvátal, V., Hammer, P.L.: Aggregations of inequalities. Stud. Integer Program. Ann. Discret. Math. **1**, 145–162 (1977)
3. Dinur, I., Steurer, D.: Analytical approach to parallel repetition. In: Proceedings of 46th ACM STOC, pp. 624–633 (2014)
4. Fan, N., Watson, J.-P.: Solving the connected dominating set problem and power dominating set problem by integer programming. In: Lin, G. (ed.) COCOA 2012. LNCS, vol. 7402, pp. 371–383. Springer, Heidelberg (2012). https://doi.org/10.1007/978-3-642-31770-5_33
5. Fulkerson, D., Gross, O.: Incidence matrices and interval graphs. Pac. J. Math. **15**, 835–855 (1965)
6. Guo, J., Niedermeier, R., Raible, D.: Improved algorithms and complexity results for power domination in graphs. Algorithmica **52**(2), 177–202 (2008)
7. Haynes, T., Hedetniemi, S., Slater, P.: Domination in Graphs: Advanced Topics. Marcel Dekker, New York (1998)
8. Haynes, T., Hedetniemi, S., Slater, P.: Fundamentals of Domination in Graphs. Marcel Dekker, New York (1998)
9. Haynes, T., Hedetniemi, S.M., Hedetniemi, S.T., Henning, M.A.: Domination in graphs applied to electric power networks. SIAM J. Discrete Math. **15**(4), 519–529 (2002)
10. Johnson, D.S., Garey, M.R.: Computers and Intractability: A Guide to the Theory of NP-Completeness. WH Freeman, New York (1979)
11. Kloks, T., Kratsch, D., Müller, H.: Bandwidth of chain graphs. Inf. Process. Lett. **68**(6), 313–315 (1998)
12. Kneis, J., Mölle, D., Richter, S., Rossmanith, P.: Parameterized power domination complexity. Inf. Process. Lett. **98**(4), 145–149 (2006)
13. Liao, C.-S., Lee, D.-T.: Power domination problem in graphs. In: Wang, L. (ed.) COCOON 2005. LNCS, vol. 3595, pp. 818–828. Springer, Heidelberg (2005). https://doi.org/10.1007/11533719_83
14. Mahadev, N.V.R., Peled, U.N.: Threshold Graphs and Related Topics. Elsevier, Amsterdam (1995)
15. Uehara, R., Uno, Y.: Efficient algorithms for the longest path problem. In: Fleischer, R., Trippen, G. (eds.) ISAAC 2004. LNCS, vol. 3341, pp. 871–883. Springer, Heidelberg (2004). https://doi.org/10.1007/978-3-540-30551-4_74
16. Yannakakis, M.: Node-and edge-deletion NP-complete problems. In: Proceedings of 10th ACM STOC, pp. 253–264 (1978)

Approximation Algorithm for Min-Max Correlation Clustering Problem with Outliers

Sai Ji[1], Min Li[2], Mei Liang[3], and Zhenning Zhang[4(✉)]

[1] Academy of Mathematics and Systems Science, Chinese Academy of Sciences, Beijing 100190, People's Republic of China
[2] School of Mathematics and Statistics, Shandong Normal University, Jinan 250358, People's Republic of China
[3] College of Statistics and Data Science, Beijing University of Technology, Beijing 100124, People's Republic of China
[4] Department of Operations Research and Information Engineering, Beijing University of Technology, Beijing 100124, People's Republic of China
zhangzhenning@bjut.edu.cn

Abstract. In this paper, we investigate the min-max correlation clustering problem with outliers, which is a combination of the min-max correlation clustering problem with the robust clustering. We first prove that the problem is NP-hard to obtain any finite approximation algorithm. Then we design an approximation algorithm based on LP-rounding technique and receive a bi-criteria guarantee.

Keywords: Min-max clustering · Correlation clustering · Outliers · Approximation algorithm · LP-rounding

1 Introduction

Arising from cut problems by Bansal et al. [3] such as min s-t cut and multiway cut, correlation clustering problem has received much attention recently [12–14,17,24,27], and has been widely applied in machine learning, computer vision, data mining and so on.

For given a complete graph $G = (V, E)$, each edge (u, v) is labeled by positive or negative based on the similarity of the two nodes u and v. The goal of the correlation clustering problem is to partition the vertex set into several clusters so that the number of disagreements is minimized. Notice that disagreements are the positive edges between different clusters and negative edges within clusters based on the partition. Since the correlation clustering problem is NP-hard, people usually use combinatorial techniques and LP-rounding techniques to design approximation algorithms [1,2,4,7,16,23]. The first constant-factor approximation algorithm was provided by Bansal et al. [3] based on combinatorial technique. Until now, the algorithm with the best 2.06-approximation ratio was designed by Chawla et al. [8] based on LP-rounding technique algorithm.

There are many interesting variants of correlation clustering problem, such as min-max correlation clustering problem [6,21], higher-order correlation clustering problem [10,14], robust correlation clustering problem [15], hierarchical correlation clustering

D.-Z. Du et al. (Eds.): COCOA 2021, LNCS 13135, pp. 668–675, 2021.
https://doi.org/10.1007/978-3-030-92681-6_52

[9,25], correlation clustering problem with noisy input [19,20], correlation clustering with a fixed number of clusters [11], and so on. Here, we mainly concern with the min-max correlation clustering problem and robust correlation clustering problem.

The min-max correlation clustering problem was first introduced by Puleo and Milenkovic [21], whose research perspective is essentially different from the traditional correlation clustering problem. It focuses on individual vertex, and its goal is to minimize the number of disagreements at the worst vertex. In the following, Puleo and Milenkovic [21] gave a 48-approximation algorithm based on LP-rounding technique. Charikar et al. [6] proposed a 7-approximation algorithm, which is still the best approximation ratio until now.

Robust correlation clustering problem is a generalization of the correlation clustering problem, which was introduced by Krishnaswamy and Rajaraman [15]. Given a complete graph $G = (V, E)$ with an integer r, the goal of this problem is to find a deleted set R and partition the vertex set $V \backslash R$ into several clusters such as to minimize the disagreements generated by the partition. This problem can also be vividly called as correlation clustering problem with outliers. In [15], Krishnaswamy and Rajaraman also proved that the problem is NP-hard to obtain any finite approximation factor, unless the number of deleted vertices is violated. Thus, they gave a bi-criteria $(6, 6)$-approximation algorithm. Finally, they provided a bi-criteria $(O(\log n), O(\log^2 n))$-approximation algorithm for the correlation clustering problem with outliers on general graphs, where n is the number of vertices in graph G.

Recently, since a single variant of clustering problem can not accurately describe some practical problems, the combination of two different variants has attracted much attention [5,18,22,26]. In present paper, we explore the min-max clustering problem combining with outliers. The problem is stated as follows. Given a complete graph $G = (V, E)$ as well as an integer r, the goal is to find a deleted set R and partition the vertex set $V \backslash R$ into several clusters so that the number of disagreements at the worst vertex is minimized. There are two contributions of this paper: (1) We prove that the min-max correlation clustering problem with outliers is NP-hard to obtain any finite approximation algorithm; (2) We propose a bi-criteria approximation algorithm based on LP-rounding technique in [6].

The rest of this paper is organized as follows. In Sect. 2, we introduce a detailed description and the integer programming of the min-max correlation clustering problem with outliers. In Sect. 3, we present our approximation algorithm and the corresponding theoretical analysis.

2 Preliminaries

In this section, we introduce some definitions and terminology used throughout this paper. Meanwhile, we describe the problem we consider here , and transform it into an integer programming. Moreover, we provide its relaxation LP. For each integer m, denote $[m] = \{1, 2, \ldots, m\}$.

Definition 1 (Min-max correlation clustering problem). *Given a labeled complete graph $G = (V, E)$, the goal of the min-max correlation clustering problem is to find a partition $V_1, V_2, \ldots, V_k (k \in [|V|])$ of V such that*

$$\max_{v \in V_i, i \in [k]} (|\{(u,v) \in E^+, u \in V \backslash V_i\}| + |\{(u,v) \in E^-, u \in V_i\}|)$$

is minimized, where E^+ is the set of positive edges and E^- is the set of negative edges.

Definition 2 (Correlation clustering problem with outliers). *Given a labeled complete graph $G = (V, E)$ and an integer r. The goal of the correlation clustering problem with outliers is to find a set $R \subseteq V$ with size r as well as a partition $V_1, V_2, \ldots, V_k, (k \in [|V| - r])$ of $V \backslash R$ such that*

$$\frac{1}{2} \sum_{v \in V_i, i \in [k]} (|\{(u,v) \in E^+, u \in V \backslash (V_i \cup R)\}| + |\{(u,v) \in E^-, u \in V_i\}|)$$

is minimized, where E^+ is the set of positive edges and E^- is the set of negative edges.

Definition 3 (Min-max correlation clustering problem with outliers). *Given a labeled complete graph $G = (V, E)$ and an integer r. The goal of the min-max correlation clustering problem with outliers is to find a set $R \subseteq V$ with size r as well as a partition $V_1, V_2, \ldots, V_k, (k \in [|V| - r])$ of $V \backslash R$ such that*

$$\max_{v \in V_i, i \in [k]} (|\{(u,v) \in E^+, u \in V \backslash (V_i \cup R)\}| + |\{(u,v) \in E^-, u \in V_i\}|)$$

is minimized, where E^+ is the set of positive edges and E^- is the set of negative edges.

From Definition 2 and Definition 3, we can obtain the following property.

Property 1. If algorithm A is an α-approximation algorithm for the min-max correlation clustering problem with outliers, then algorithm A is also an $\alpha/2$-approximation algorithm for the correlation clustering problem with outliers.

Combining Theorem 6 of [15] and Property 1, we can obtain Theorem 1.

Theorem 1. *It is NP-hard to obtain any finite approximation factor for the min-max correlation clustering problem with outliers, unless the constraint of the number of deleted vertices is violated.*

Before giving the integer programming formulation for the min-max correlation clustering problem with outliers, let us first introduce the following three kinds of 0-1 variables:

- For each edge $(u, v) \in E$, variable x_{uv} indicates whether two vertices u and v are in a same cluster. To be specific, if u and v lie in a same cluster, then $x_{uv} = 0$; otherwise $x_{uv} = 1$.
- For each vertex $v \in V$, variable y_v indicates whether the vertex v is deleted, that is, if vertex v is deleted, then $y_v = 1$; otherwise $y_v = 0$.
- For each edge $(u, v) \in E$, variable z_{uv} indicates whether the edge (u, v) is a disagreement, i.e., if edge (u, v) is a disagreement, then $z_{uv} = 1$; otherwise $z_{uv} = 0$.

Thus, the min-max correlation clustering problem with outliers can be formulated by the following IP:

$$\begin{aligned} \min \ & \max_{v\in V} \sum_{u\in V} z_{uv} \\ \text{s. t. } & x_{uv} + x_{vw} \geq x_{uw}, && \forall u,v,w \in V, \\ & y_u + y_v + z_{uv} \geq 1 - x_{uv}, && \forall (u,v) \in E^-, \\ & y_u + y_v + z_{uv} \geq x_{uv}, && \forall (u,v) \in E^+, \\ & \sum_{u\in V} y_u = r, \\ & x_{uv}, z_{uv}, y_u \in \{0,1\}, && \forall u,v \in V. \end{aligned} \tag{1}$$

The value of the objective function is the number of disagreements. There are four types of constraints in Programming (1). The first one is the triangle inequality, which guarantees the solution of Programming (1) to be a feasible solution of the correlation clustering problem. The second and third constraints ensure the edge to be a disagreement. The fourth one describes that the number of deleted vertices is exactly r. By relaxing the variables, we can obtain the following LP relaxation of Programming (1):

$$\begin{aligned} \min \ & \max_{v\in V} \sum_{u\in V} z_{uv} \\ \text{s. t. } & x_{uv} + x_{vw} \geq x_{uw}, && \forall u,v,w \in V, \\ & y_u + y_v + z_{uv} \geq 1 - x_{uv}, && \forall (u,v) \in E^-, \\ & y_u + y_v + z_{uv} \geq x_{uv}, && \forall (u,v) \in E^+, \\ & \sum_{u\in V} y_u = r, \\ & x_{uv}, z_{uv}, y_u \in [0,1], && \forall u,v \in V. \end{aligned} \tag{2}$$

3 Algorithm and Analysis

In this section, we present our algorithm in Subsect. 3.1 and the theoretical analysis has been discussed in Subsect. 3.2.

3.1 Algorithm

To obtain an approximation algorithm for the min-max correlation clustering problem with outliers, we first solve Programming (2) to receive the optimal fractional solution (x^*, y^*, z^*), where the value x^*_{uv} is viewed as the distance between the two vertices u and v. Then we need to consider the following two problems: (i) What criteria are used to select the deleted vertices? (ii) How do we partition the rest of the vertices? For the first problem, whether the vertex is deleted is decided by its y^* value and a parameter γ. For each vertex v, v is deleted if $y^*_v \geq \gamma$. Otherwise, we remain vertex v. The number of deleted vertices may be greater than r. However, we can prove that the

Algorithm 1

Input: A labeled complete graph $G = (V, E)$, positive integer r, parameter $\gamma \in (0, 1/14)$
Output: A partition of vertices
1: Let $S := V, C := \emptyset, R := \emptyset$
2: Solve (2) to obtain the optimal fractional solution (x^*, y^*, z^*)
3: Update $R := \{v \in V : y_v^* \geq \gamma\}$, $S := S \backslash R$
4: **while** $S \neq \emptyset$ **do**
5: **for** each vertex $v \in S$ **do**
6: Set $T_v^* := \{u \in S : x_{uv}^* \leq 1/7\}$ and $T_v := \{u \in S : x_{uv}^* \leq 3/7\}$
7: **end for**
8: Choose vertex

$$v^* := \arg\max_{v \in S} |T_v^*|$$

9: Let T_{v^*} be a cluster
10: Update $C := C \cup \{v^*\}, S := S - T_{v^*}$.
11: **end while**
12: **return** $\{T_{v^*} : v^* \in C\}$ and set R

number is not more than a constant multiple of r. Notice that r is the desired number of the deleted vertices and γ is just a parameter appeared in Algorithm 1. For the second problem, we adopt an iterative clustering method. In each iteration, for every vertex v, we construct two neighbor sets T_v^* and T_v. They contain all the un-clustered vertices with the distance to the vertex v no more than $1/7$ and $3/7$, respectively. We select the vertex v^* with the largest $|T_{v^*}^*|$ as a center vertex, and make T_{v^*} to be a cluster. Then, we update the un-clustered set and repeat above iterative processes until all the vertices in the graph are clustered.

3.2 Theoretical Analysis

In this subsection, we give a theoretical analysis of Algorithm 1. Without loss of generality, we assume that Algorithm 1 contains exactly k iterations. The center set is $C = \{v_1^*, v_2^*, \ldots, v_k^*\}$, the set of outliers is R, and the corresponding partition of $V \backslash R$ is $\{T_{v_1^*}, T_{v_2^*}, \ldots, T_{v_k^*}\}$. From the construction of R, we have the following properties, which play an important role in the proof of Lemmas 1–5.

(1) For each $v \in R$, $y_v^* \geq \gamma$ holds.
(2) For each $v \in V \backslash R$, $y_v^* < \gamma$ holds.

Lemma 1. *There are at most $\frac{r}{\gamma}$ vertices in set R.*

This lemma can be concluded by the fourth constraint of (2) and the construction of set R, we omit the proof.

As shown in Fig. 1, for each $i \in [k]$ and $j \in [i]$, let $A_j = T_{v_j^*}^*, B_j = \{t \in V \backslash (\cup_{s \in [j-1]} T_{v_s^*} \cup R) : x_{ut}^* \leq 1/7\} \cap T_{v_j^*}, D_j = T_{v_j^*} \backslash (A_j \cup B_j)$, $F_i = \{t \in V \backslash (\cup_{s \in [i]} T_{v_s^*} \cup R) : x_{ut}^* \leq 1/7\}$ and $M_i = \{t \in V \backslash (\cup_{s \in [i]} T_{v_s^*} \cup R) : x_{ut}^* > 1/7\}$. The number of disagreements caused by positive edges is analyzed by Lemma 2–Lemma 4.

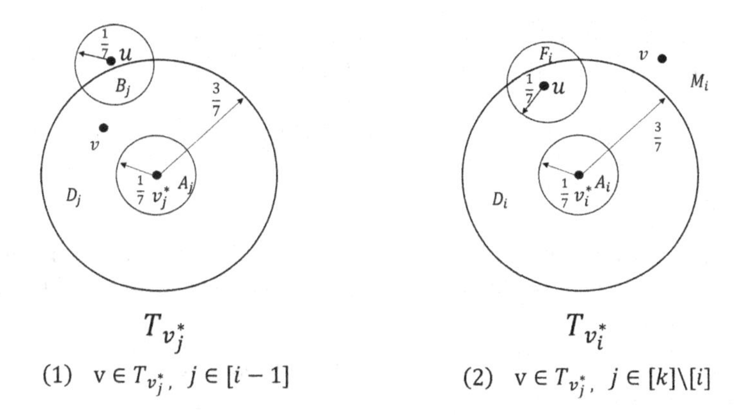

Fig. 1. Partition of clustered vertices.

The number of disagreements caused by negative edges is analyzed by Lemma 5. Finally, the total number of disagreements is shown by Lemma 6. All these proofs will be presented in journal edition.

Disagreements Caused by Positive Edges. Similar to the proof of [6], for each $u \in T_{v_i^*}, i \in [k]$, we analyze the upper bound of the number of disagreements in the following two cases:

(1) $(u, v) \in E^+, v \in T_{v_j^*}, j \in [i-1]$.
(2) $(u, v) \in E^+, v \in T_{v_j^*}, j \in [k]\backslash[i]$.

Lemma 2. *For each vertex $u \in T_{v_i^*}, i \in [k]$, the number of disagreement caused by each edge $(u, v) \in E^+$, $v \in M_i \cup D_j$, $j \in [i-1]$ can be bounded by $7z_{uv}^*/(1-14\gamma)$.*

Lemma 3. *For each vertex $u \in T_{v_i^*}, i \in [k]$, the number of disagreements generated by edges $(u, v) \in E^+$, $v \in A_j \cup B_j$, $j \in [i-1]$ can be bounded by*

$$\sum_{(u,v)\in E^+, v\in A_j} \frac{7}{1-7\gamma} z_{uv}^* + \sum_{(u,v)\in E^-, v\in A_j} \frac{7}{2-14\gamma} z_{uv}^*.$$

Lemma 4. *For each vertex $u \in T_{v_i^*}, i \in [k]$, the number of disagreements generated by edges $(u, v) \in E^+$, $v \in F_i$ can be bounded by*

$$\sum_{(u,v)\in E^+, v\in A_i} \frac{7}{1-14\gamma} z_{uv}^* + \sum_{(u,v)\in E^-, v\in A_i} \frac{7}{3-14\gamma} z_{uv}^*.$$

Disagreements Caused by Negative Edges. For each vertex $u \in T_{v_i^*}, i \in [k]$, the number of disagreements caused by negative edges equals $|\{(u, v) \in E^- : v \in T_{v_i^*}\}|$, and the upper bound can be bounded by Lemma 5.

Lemma 5. *For each vertex* $u \in T_{v_i^*}, i \in [k]$, *the number of disagreements generated by negative edges* (u, v) *with* $v \in T_{v_i^*}$ *can be bounded by*

$$\sum_{(u,v)\in E^-, v\in D_i} \frac{7}{1-14\gamma} z_{uv}^* + \sum_{(u,v)\in E^-, v\in A_i} \frac{7}{3-14\gamma} z_{uv}^*.$$

Total Disagreements. Combining Lemma 2–Lemma 5, we can obtain Lemma 6 to analyze the upper bound on the total number of disagreements generated by partition $\{T_{v_1^*}, T_{v_2^*}, \ldots, T_{v_k^*}\}$.

Lemma 6. *For each vertex* $u \in T_{v_i^*}, i \in [k]$, *the number of disagreements caused by vertex* u *can be bounded by*

$$\frac{7}{1-14\gamma} \sum_{(u,v)\in E} z_{uv}^*.$$

From Lemma 1 and Lemma 6, we can obtain our main result of this paper.

Theorem 2. *Algorithm 1 is a* $(\frac{1}{\gamma}, \frac{7}{1-14\gamma})$*-bi-criteria approximation algorithm for the min-max correlation clustering problem with outliers.*

4 Conclusions

In this paper, we study the min-max correlation clustering problem with outliers and give a bi-criteria approximation algorithm. There are two interesting future works for the min-max correlation clustering problem with outliers. One is to design an algorithm which can improve the current ratio. The other is to study the generalization of the problem, such as capacitated min-max correlation clustering problem with outliers and the lower bounded min-max correlation clustering problem with outliers.

Acknowledgements. The first author is supported by National Natural Science Foundation of China (No. 12101594) and the Project funded by China Postdoctoral Science Foundation (No. 2021M693337). The second author is supported by Natural Science Foundation of Shandong Province (No. ZR2020MA029) of China. The third author is supported by Beijing Natural Science Foundation (No. Z200002). The fourth author is supported by Nationa Natural Science Foundation of China (Nos. 12131003, 12001025).

References

1. Ailon, N., Avigdor-Elgrabli, N., Liberty, E., Zuylen, A.V.: Improved approximation algorithms for bipartite correlation clustering. SIAM J. Comput. **41**(5), 1110–1121 (2012)
2. Ailon, N., Charikar, M., Newman, A.: Aggregating inconsistent information: ranking and clustering. J. ACM **55**(5), Article No. 23 (2008)
3. Bansal, N., Blum, A., Chawla, S.: Correlation clustering. Mach. Learn. **56**(1–3), 89–113 (2004)

4. Bressan, M., Cesa-Bianchi, N., Paudice, A., Vitale, F.: Correlation clustering with adaptive similarity queries. In: Proceedings of NeurIPS, pp. 12510–12519 (2019)
5. Byrka, J., Fleszar, K., Rybicki, B., Spoerhase, J.: Bi-factor approximation algorithms for hard capacitated k-median problems. In: Proceedings of SODA, pp. 722–736 (2014)
6. Charikar, M., Gupta, N., Schwartz, R.: Local guarantees in graph cuts and clustering. In: Proceedings of IPCO, pp. 136–147 (2017)
7. Charikar, M., Guruswami, V., Wirth, A.: Clustering with qualitative information. J. Comput. Syst. Sci. **71**(3), 360–383 (2005)
8. Chawla, S., Makarychev, K., Schramm, T., Yaroslavtsev, G.: Near optimal LP rounding algorithm for correlation clustering on complete and complete k-partite graphs. In: Proceedings of the 47th ACM Symposium on Theory of Computing, pp. 219–228 (2015)
9. Chehreghani, M.H.: Hierarchical correlation clustering and tree preserving embedding (2020). ArXiv preprint arXiv: 2002.07756
10. Fukunaga, T.: LP-based pivoting algorithm for higher-order correlation clustering. J. Comb. Optim. **37**(4), 1312–1326 (2018). https://doi.org/10.1007/s10878-018-0354-y
11. Giotis, I., Guruswami, V.: Correlation clustering with a fixed number of clusters. In: Proceedings of SODA, pp. 1167–1176 (2006)
12. Hou, J.P., Emad, A., Puleo, G.J., Ma, J., Milenkovic, O.: A new correlation clustering method for cancer mutation analysis. Bioinformatics **32**(24), 3717–3728 (2016)
13. Jafarov, J., Kalhan, S., Makarychev, K., Makarychev, Y.: Correlation clustering with asymmetric classification errors. In: Proceedings of ICML, pp. 4641–4650 (2020)
14. Kim, S., Yoo, C.D., Nowozin, S., Kohli, P.: Image segmentation using higher-order correlation clustering. IEEE Trans. Pattern Anal. Mach. Intell. **36**(9), 1761–1774 (2014)
15. Krishnaswamy, R., Rajaraman, N.: Robust correlation clustering. In: Proceedings of APPROX/RANDOM, pp. 33:1–33:18 (2019)
16. Lange, J.H., Karrenbauer, A., Andres, B.: Partial optimality and fast lower bounds for weighted correlation clustering. In: Proceedings of ICML, pp. 2892–2901 (2018)
17. Li, P., Puleo, G.J., Milenkovic, O.: Motif and hypergraph correlation clustering. IEEE Trans. Inf. Theory **66**(5), 3065–3078 (2019)
18. Lv, W., Wu, C.: An LP-rounding based algorithm for a capacitated uniform facility location problem with penalties. J. Comb. Optim. **41**(4), 888–904 (2021). https://doi.org/10.1007/s10878-021-00726-0
19. Makarychev, K., Makarychev, Y., Vijayaraghavan, A.: Correlation clustering with noisy partial information. In: Proceedings of COLT, pp. 1321–1342 (2015)
20. Mathieu, C., Schudy, W.: Correlation clustering with noisy input. In: Proceedings of SODA, pp. 712–728 (2010)
21. Puleo, G.J., Milenkovic, O.: Correlation clustering and biclustering with locally bounded errors. IEEE Trans. Inf. Theory **64**(6), 4105–4119 (2018)
22. Saif, A., Delage, E.: Data-driven distributionally robust capacitated facility location problem. Eur. J. Oper. Res. **291**(3), 995–1007 (2021)
23. Thiel, E., Chehreghani, M.H., Dubhashi, D.: A non-convex optimization approach to correlation clustering. In: Proceedings of AAAI, pp. 5159–5166 (2019)
24. Ukkonen, A.: Crowdsourced correlation clustering with relative distance comparisons. In: Proceedings of ICDM, pp. 1117–1122 (2017)
25. Vainstein, D., Chatziafratis, V., Citovsky, G., Rajagopalan, A., Mahdian, M., Azar, Y.: Hierarchical clustering via sketches and hierarchical correlation clustering (2021). ArXiv preprint arXiv: 2101.10639
26. Vasilyev, I., Ushakov, A.V., Maltugueva, N., Sforza, A.: An effective heuristic for large-scale fault-tolerant k-median problem. Soft Comput. **23**(9), 2959–2967 (2019)
27. Veldt, N., Gleich, D.F., Wirth, A.: A correlation clustering framework for community detection. In: Proceedings of WWW, pp. 439–448 (2018)

Delay-Constrained Minimum Shortest Path Trees and Related Problems

Junran Lichen[1], Lijian Cai[2], Jianping Li[2(✉)], Suding Liu[2], Pengxiang Pan[2], and Wencheng Wang[2]

[1] Institute of Applied Mathematics, Academy of Mathematics and Systems Science, No. 55, Zhongguancun East Road, Beijing 100190, People's Republic of China
[2] Department of Mathematics, Yunnan University, East Outer Ring South Road, University Town, Kunming 650504, People's Republic of China
jianping@ynu.edu.cn

Abstract. Motivated by applications in communication networks of the diameter-constrained minimum spanning tree problem, we consider the delay-constrained minimum shortest path tree (DcMSPT) problem. Specifically, given a weighted graph $G = (V, E; w, c)$ and a constant d_0, where length function $w : E \rightarrow R^+$ and cost function $c : E \rightarrow R^+$, we are asked to find a minimum cost shortest path tree among all shortest path trees (in G) whose delays are no more than d_0, where the delay of a shortest path tree is the maximum distance (depending on $w(\cdot)$) from its source to every other leaves in that tree, and the cost of a shortest path tree is the sum of costs of all edges (depending on $c(\cdot)$) in that tree. Particularly, when a constant d_0 is exactly the radius of G, we refer to this version of the DcMSPT problem as the minimum radius minimum shortest path tree (MRMSPT) problem. Similarly, the maximum delay minimum shortest path tree (MDMSPT) problem is asked to find a minimum cost shortest path tree among all shortest path trees (in G) whose delays are exactly the diameter of G.

We obtain the following two main results. (1) We design an exact algorithm in time $O(n^3)$ to solve the DcMSPT problem, and we provide the similar algorithm to solve the MRMSPT problem; (2) We present an exact algorithm in time $O(n^3)$ to solve the MDMSPT problem.

Keywords: Combinatorial optimization · Distances · Delay-constrained shortest path trees · Exact algorithms · Complexity

This paper is supported by the National Natural Science Foundation of China [Nos. 11861075, 12101593], Project for Innovation Team (Cultivation) of Yunnan Province [No. 202005AE160006], Key Project of Yunnan Provincial Science and Technology Department and Yunnan University [No. 2018FY001014] and Program for Innovative Research Team (in Science and Technology) in Universities of Yunnan Province [C176240111009]. Jianping Li is also supported by Project of Yunling Scholars Training of Yunnan Province.

D.-Z. Du et al. (Eds.): COCOA 2021, LNCS 13135, pp. 676–686, 2021.
https://doi.org/10.1007/978-3-030-92681-6_53

1 Introduction and Problem Description

Many graph optimization problems are motivated from applications in our reality life, for example, the minimum spanning tree problem, the shortest path problem and the minimum Steiner tree problem [17]. In addition, there exist several optimization problems which can be regarded as combinations of some classic graph optimization problems, for example, the single-source shortest path tree problem, briefly as the shortest path tree (SPT) problem, which was first raised in 1957 by Dantzig [6], can be regarded as a combination of the shortest path problem [7] and the minimum arborescence problem [5,9]. In the past five decades, these problems mentioned-above have been deeply studied in the literature and have many wide applications in our reality life, and there exist many polynomial-time exact or approximation algorithms to solve these problems [17].

Spanning trees related problems in weighted graphs have been well studied in theory and widely applied in our reality life [4,17]. So has the diameter problem in which the diameter is originally measured in terms of the number of edges, instead of the total weight of a spanning tree [1,13,17]. In recent decades, the diameter of a weighted graph $G = (V, E; w)$, however, is defined as the longest of the shortest paths among all pairs of distinct vertices in G, *i.e.*, $diam(G) = \max\{\sum_{e \in P_{st}} w(e) \mid P_{st}$ is a shortest path connecting every pair s, t of distinct vertices in $G\}$. In particular, since the path to connect every pair of distinct vertices in a weighted tree $T = (V, E; w)$ is unique, the diameter of T is the maximum weight of a path connecting any two leaves of T. What motivates this investigation is that we want to find a communication network among n vertices, where the communication delay is measured in terms of the total weight of a shortest path between them. A desirable communication network is naturally one tree that has a minimum diameter. Different from studying the minimum spanning tree problem, Ho et al. [13] in 1991 considered the minimum diameter spanning tree (MDST) problem which is formally defined as follows.

Problem 1 (the MDST problem [13]). Given a weighted graph $G = (V, E; w)$ with length function $w : E \rightarrow R^+$, it is asked to find a spanning tree T of G such that the objective is to minimize $\max\{\sum_{e \in P} w(e) \mid P$ is a path connecting any two leaves in $T\}$, *i.e.*, T has a minimum diameter among all spanning trees of G.

Meanwhile, Ho et al. [13] indeed considered the minimum diameter minimum spanning tree (MDMST) problem which is formally defined as follows.

Problem 2 (the MDMST problem [13]). Given a weighted graph $G = (V, E; w)$ with length function $w : E \rightarrow R^+$, it is asked to find a spanning tree $T = (V, E_T)$ of G, the objective is to minimize the total weight $\sum_{e \in E_T} w(e)$ among the all spanning trees that have their diameter values as $diam(G)$ of that graph G, *i.e.*, the all spanning trees considered in this problem have the diameter $diam(G)$ of that graph G.

Ho et al. [13] in 1991 design an exact algorithm in time $O(n^3)$ to find a minimum diameter spanning tree (MDST) of a special graph, called an Euclidean graph, induced by a set of n points in the Euclidean plane, also referred to as a geometric MDST problem. On the other hand, they proved that the MDMST problem is *NP*-complete, using a reduction from the 3SAT problem [10]. Hassin and Tamir [12] in 1995 observed an important fact that the MDST problem is identical to the well studied absolute 1-center problem introduced in 1964 by Hakimi [11], imply that the existing algorithms [8,11], which solves the absolute 1-center problem, also solves the MDST problem on a general graph in time $O(mn + n^2 \log n)$.

At present, we may firmly believe that it would be better to describe the minimum diameter minimum spanning tree (MDMST) problem using the following definition, which involves two different functions.

Problem 3 (the MDMST problem). Given a weighted graph $G = (V, E; w, c)$ with length function $w : E \rightarrow R^+$ and cost function $c : E \rightarrow R^+$, it is asked to find a spanning tree $T = (V, E_T)$ of G, the objective is to minimize the total cost $\sum_{e \in E_T} c(e)$ among the all spanning trees that have their diameter values as $diam(G)$ of that graph G, where the distance between any two vertices depends on computing of length function $w(\cdot)$ and the diameter of G is defined as mentioned-above.

A similar problem, which is called as the diameter-constrained minimum spanning tree (DcMST) problem [10], is formally defined as follows.

Problem 4 (the DcMST problem). Given a weighted connected graph $G = (V, E; w)$ and a positive integer d, where $w : E \rightarrow R^+$, it is asked to seek a spanning tree T on G of minimum weight among all the spanning trees in which no path in T between any two vertices (actually, two leaves) contains more than d edges.

The DcMST problem is sometimes called as the bounded diameter minimum spanning tree problem [14,19,20], and it was shown to be *NP*-complete by Garey and Johnson [10]. In the DcMST problem, the measure of the diameter is in terms of the maximum number of edges in any path of the spanning tree. It is easy to see that the DcMST problem may roughly be treated as a generalization of the MDMST problem (*i.e.*, Problem 2), where no path in spanning tree between any two leaves contains more than d edges for the DcMST problem and the diameter is exactly the longest of a shortest weighted path between any two leaves in spanning tree for the MDMST problem. The DcMST problem arises in various contexts in communication network design, and it has also been given some applications in the area of information retrieval in [2,3]. For the DcMST problem, Kortsarz and Peleg [16] in 1999 showed that, unless $\mathcal{P} = \mathcal{NP}$, no polynomial-time approximation algorithm can be guaranteed to find a solution whose weight is within $\log(n)$ of the optimum. Although Ho et al. [13] in 1991 proved that the MDMST problem is *NP*-complete, using a reduction from the 3SAT problem [10], Seo et al. [18] in 2009 showed that the MDMST problem specialized to Euclidean graphs remains *NP*-complete, using a reduction from the PARTITION problem [10].

In a centralized communication network in which there is a vertex as source, Ho et al. [13] in 1991 defined the minimum radius spanning tree (MRST) problem in a similar manner using the radius instead of the diameter of that weighted graph, the objective is to minimize the maximum communication delay from a source vertex to other vertices in that weighted graph. The minimum radius minimum spanning tree (MRMST) problem is similarly defined as the MDMST problem. The same authors [13] in 1991 proved that the MRMST problem is NP-complete.

We have known the fact that, given a weighted connected graph $G = (V, E; w)$ with length function $w : E \rightarrow R^{+}$, messages are transmitted along a shortest path from vertex to other vertex in G, and shortest path trees will play an important role in such a transmitting system in G. Then we have the following two facts. (1) When messages are transmitted from a source vertex to any other vertices by using shortest path trees, we may consider the delay of messages along a shortest path tree from its source vertex to any other vertices not beyond the expected time d_0, particularly not beyond the radius of that graph G; (2) When we implement this mechanism in (1), we should consider minimum cost to construct such a communication network from the original weighted graph G.

Motivated by the problems and related mechanisms mentioned-above to transmit messages in communication networks, we should consider the following problem and its related variations.

Problem 5 (the DcMSPT problem). Given a weighted graph $G = (V, E; w, c)$ and a constant d_0, where length function $w : E \rightarrow R^{+}$ and cost function $c : E \rightarrow R^{+}$, it is asked to find a minimum cost shortest path tree among all shortest path trees (in G) whose delays are no more than d_0, where the delay of a shortest path tree is the maximum distance (depending on $w(\cdot)$) from its source to every other leaves in that tree, and the cost of a shortest path tree is the sum of costs of all edges (depending on $c(\cdot)$) in that tree.

For convenience, we refer to Problem 5 as the delay-constrained minimum shortest path tree (DcMSPT) problem. Particularly, when a constant d_0 is exactly the radius of a weighted graph G, we refer to this version of the DcM-SPT problem as the minimum radius minimum shortest path tree (MRMSPT) problem. Similarly, the maximum delay minimum shortest path tree (MDMSPT) problem is asked to find a minimum cost shortest path tree among all shortest path trees (in G) whose delays are exactly the diameter of G.

So far as what we have known, although the DcMSPT problem and its related variations have many important applications implied in our reality life, where messages are transmitted along shortest paths from a source vertex to every other vertices, these optimization problems have not been studied deeply both in theory and in practice, and there are no exact or approximation algorithms in polynomial time to solve them. For the DcMST problem (*i.e.*, Problem 4), Kortsarz and Peleg [16] in 1999 showed that, unless $\mathcal{P} = \mathcal{NP}$, no polynomial-time approximation algorithm can be guaranteed to find a solution whose weight is within $\log(n)$ of the optimum. However, we hope to design some exact algorithms in polynomial time to optimally solve the DcMSPT problem and its related variations, respectively.

This paper is organized as follows. In Sect. 2, we present some terminologies for easily describing our algorithms and provide key lemmas to ensure the correctness of algorithms. In Sect. 3, we design an exact algorithm to solve the DcMSPT problem, and the similar algorithm solves the MRMSPT problem, where this algorithm runs in time $O(n^3)$; In Sect. 4, we present an exact algorithm to solve the MDMSPT problem, where that algorithm runs in time $O(n^3)$; In Sect. 5, we provide our conclusion and further work.

2 Terminologies and Key Lemmas

In this section, we present some notations and terminologies to solve the delay-constrained minimum shortest path tree (DcMSPT) problem and its related variations, respectively, and other terminologies and notations not defined can be found in those references [1,15,17].

Given a weighted graph $G = (V, E; w, c)$ with length function $w : E \to R^+$ and cost function $c : E \to R^+$, for any two vertices $s, t \in V$, the distance between s and t, denoted by $dist_G(s,t)$, is the minimum length of a path connecting s and t if such a path exists, and otherwise $dist_G(s,t) = +\infty$. Concretely, $dist_G(s,t) = \min\{\sum_{e\in P_{st}} w(e) \mid P_{st}$ is a path connecting s and t in $G\}$. And if this path P_{st} satisfies $\sum_{e\in P_{st}} w(e) = dist_G(s,t)$, P_{st} is called as a shortest s-t path or a shortest path connecting s and t. For each vertex $s \in V$, we define the eccentricity of s, denoted by $e_G(s)$, is the maximum of all distances from s to other vertices in G, *i.e.*, $e_G(s, V) = \max\{dist_G(s,t) \mid t \in V\}$. In addition, the diameter of a weighted graph G, denoted by $diam(G)$, is the maximum of all distances between pairs of vertices in G, *i.e.*, $diam(G) = \max\{dist_G(s,t) \mid s, t \in V\}$, and the radius of a weighted graph G, denoted by $rad(G)$, is the minimum of eccentricities of vertices in G, *i.e.*, $rad(G) = \min\{e_G(s) \mid s \in V\}$, meanwhile we refer to this vertex s as a center of G if this vertex s satisfies $e_G(s) = rad(G)$. Furthermore, if $T = (V, E_T; w, c)$ is a spanning tree of a weighted graph $G = (V, E; w, c)$, then the diameter of T is the maximum length of a shortest path connecting any two leaves in T, *i.e.*, $diam(T) = \max\{dist_T(s,t) \mid s$ and t are two leaves of $T\}$, and meanwhile the cost of T is defined as $c(T) = \sum_{e\in E_T} c(e)$.

Dantzig [6,17] in 1957 observed the following. Let $D = (V, A; w)$ be a weighted digraph with a fixed source vertex $s \in V$, where length function $w : E \to R^+$. An arborescence $T = (V', A')$ rooted at s is called a single source shortest path tree (rooted at s) if V' is the set of vertices in D reachable from s and $A' \subseteq A$, such that for each vertex $t \in V'$, the s-t path in T is a shortest s-t path in D, depending on computing of length function $w(\cdot)$. In particular, we call $T = (V, A')$ as a spanning shortest path tree rooted at s, briefly $T = (V, A')$ as a shortest path tree of D if no ambiguity.

With the similar arguments, we define a shortest path tree in a weighted graph as follows. Let $G = (V, E; w)$ be a weighted graph with a fixed source vertex $s \in V$, where length function $w : E \to R^+$. A spanning tree $T = (V, E')$

is called a single source shortest path tree with the source vertex s of G, if $dist_T(s,t) = dist_G(s,t)$ holds for every other vertex $t \in V$, *i.e.*, the s-t path in T is a shortest s-t path in G, depending on computing of length function $w(\cdot)$. Briefly, we refer to such a tree $T = (V, E')$ as a shortest path tree with the source vertex s of G, simply as a shortest path tree of G if no ambiguity.

Now, we address the minimum shortest path tree (MSPT) problem as follows.

Problem 6 (the MSPT problem). Given a weighted graph $G = (V, E; w, c)$ with a fixed source s, where length function $w : E \to R^+$ and cost function $c : E \to R^+$, it is asked to find a shortest path tree $T = (V, E_T; w, c)$ rooted at s, the objective is to minimize the cost $\sum_{e \in E_T} c(e)$ among all shortest path trees $T = (V, E_T; w)$ rooted at s, where the distance from s to every other vertex t in G depends on computing of length function $w(\cdot)$.

We call a shortest path tree T as to be a minimum cost shortest path tree T in G if the cost of T attains the minimum value among all shortest path trees of G, where distance depends on computing of length function $w(\cdot)$. We present some remarks to the MSPT problem. For an instance of the MSPT problem, a weighted graph $G = (V, E; w, c)$ generally involves two different functions, saying $w(\cdot)$ and $c(\cdot)$, which are with no relationships. We have known the fact that the MSPT problem originally appeared in the literature [1,6,17], where the two functions $w(\cdot)$ and $c(\cdot)$ are essentially the same function.

The strategy to solve the MSPT problem (*i.e.*, Problem 6) is executed as follows. (1) Given a weighted graph $G = (V, E; w, c)$ equipped with a source vertex s for the MSPT problem, depending on computing of length function $w(\cdot)$ in G, we can modify the Dijkstra algorithm [7] to construct an auxiliary acyclic digraph $D_s = (V, A_s; w, c)$ that consists of the union of all shortest s-t paths in G for every other vertex t in $V \setminus \{s\}$. (2) Depending on computing of cost function $c(\cdot)$ in D_s, construct a minimum-cost arborescence at a root s in $D_s = (V, A_s; w, c)$. In fact, we can construct a minimum-cost arborescence at a root s choosing a minimum-cost arc to enter every other vertex in the acyclic digraph $D_s = (V, A_s; w, c)$, without executing an algorithm to solve the minimum arborescence problem specialized to weighted graphs. For convenience, we still denoted such an algorithm in (1) as the Dijkstra algorithm-modified.

Using the Dijkstra algorithm-modified and the strategy mentioned-above, we can obtain the following lemma.

Lemma 1. *There exists a polynomial-time exact algorithm, denoted by the MSPT algorithm, to optimally solve the MSPT problem, and it runs in time* $O(n^2)$*, where* n *is the order of weighted graph* G.

3 An Exact Algorithm to Solve the DcMSPT Problem

In this section, we consider the delay-constrained minimum shortest path tree (RcMSPT) problem. Modifying the strategy to solve the MSPT problem (see the Problem 6), we execute the strategy to solve the DcMSPT problem as follows.

(1) For each vertex v in a weighted graph $G = (V, E; w, c)$, depending on computing of length function $w(\cdot)$, use the Dijkstra algorithm-modified [7] to construct an auxiliary acyclic digraph $D_v = (V, A_v; w, c)$ that consists of the union of all shortest paths in G to connect this vertex v to all other vertices v_i in $V \setminus \{v\}$, and in addition, if $dist_{D_v}(v, v_i) \leq d_0$ holds for each vertex v_i in V, depending on cost function $c(\cdot)$, we construct a minimum-cost shortest path tree T_v at the source v in the digraph D_v, otherwise we ignore this vertex v.
(2) Choose a minimum-cost shortest path tree from all shortest path trees constructed in (1).

We describe our algorithm to solve the DcMSPT problem as follows.

Algorithm 1: DcMSPT
INPUT: a weighted graph $G = (V, E; w, c)$ and a constant d_0;
OUTPUT: a delay-constrained minimum cost shortest path tree, rooted at a source $v^* \in V$.
Begin
Step 1. For each vertex $s \in V$, do
(1.1) Depending on computing of length function $w(\cdot)$, execute the Dijkstra algorithm-modified [7] to construct an auxiliary acyclic digraph $D_s = (V, A_s; w, c)$, where A_s consists of arcs $(x, y) \in A$ that lies on a shortest (s, v_i)-path from s to every other vertex $v_i \in V$. For convenience, we may assume that all vertices in $D_s = (V, A_s)$ are topologically sorted in the order $v_{j_1}, v_{j_2}, \ldots, v_{j_n}$, where $v_{j_1} = s$.
(1.2) If $(e_{D_s}(s) \leq r_0)$ then
(1.2.1) For each vertex $v_{j_t} \in V$, $t = 2, 3, \ldots, n$, depending on computing of cost function $c(\cdot)$, choose a minimum cost arc $e_{i_t} = (v_{i_t}, v_{j_t})$ in D_s to enter the vertex v_{j_t}, where $v_{i_t} \in \{v_{j_1}, v_{j_2}, \cdots, v_{j_n}\}$ (for two distinct integers $t, t' \in \{2, 3, \ldots, n\}$, we may have the same vertex $v_{i_t} = v_{i_{t'}}$).
(1.2.2) Construct a shortest path tree $T_s = (V, A_{T_s})$ with the edge set $A_{T_s} = \{e_{i_2}, e_{i_3}, \cdots, e_{i_n}\}$.
Step 2. Choose a minimum-cost shortest path tree $T_{v^*} = (V, A_{T_{v^*}})$ from all shortest path trees constructed at Step (1.2.2), *i.e.*, satisfying $c(A_{T_{v^*}}) = \min\{c(A_{T_s}) \mid T_s = (V, A_{T_s})$ is a shortest path tree in G, having $e_{T_s}(s) \leq r_0\}$.
Step 3. Output the minimum-cost shortest path tree $T_{v^*} = (V, A_{T_{v^*}})$ obtained at Step 2.
End

We can use the DcMSPT algorithm to obtain the following result to optimally solve the DcMSPT problem.

Theorem 1. *The DcMSPT algorithm (i.e., Algorithm 1) is an exact algorithm to solve the DcMSPT problem, and it runs in time $O(n^3)$, where n is the order of weighted graph G.*

Proof. We may assume, without loss of generality, that this weighted graph $G = (V, E; w, c)$ is connected. For each vertex $s \in V$, we shall prove that either $T_s = (V, A_{T_s})$ produced at Step 1 of Algorithm 1 is a minimum-cost shortest path tree at the source vertex s to satisfy $e_{T_s}(v) = e_G(s) \leq r_0$ or this graph G contains no such shortest path trees. For the former, $T_s = (V, A_{T_s})$ is a feasible solution to an instance $G = (V, E; w, c)$ of the DcMSPT problem, and for the latter, there is no feasible solution at the source vertex s to the instance $G = (V, E; w, c)$ of the DcMSPT problem.

Given a fixed source vertex s, since this vertex s is the fixed vertex in G such that each vertex v_r is reachable from s, Step 1.1 at Algorithm 1 executes the Dijkstra algorithm-modified [7] to construct the auxiliary acyclic digraph $D_s = (V, A_s)$ to keep $d_{D_s}(s, v_r) = d_G(s, v_r)$ for every other vertex $v_r \in V$, then this subgraph $D_s = (V, A_s)$ of G contains all shortest path trees at the source vertex s in D_s plus some other edges in G, satisfying $d_{T_s}(s, v_r) = d_{D_s}(s, v_r) = d_G(s, v_r)$. In the view of the choices of arcs at Step 1.2, when $e_{D_s}(s) \leq r_0$, we can indeed prove by induction that the subgraph $T_s = (V, A_{T_s})$ produced at Step 1.2 is a shortest path tree at the source vertex s in D_s, indeed also in G, satisfying $d_{T_s}(s, v_r) = d_{D_s}(s, v_r) = d_G(s, v_r)$ and $dist_{T_s}(s, v_r) \leq r_0$ for every other vertex $v_r \in V$. Thus, this shows that $T_s = (V, A_{T_s})$ produced at Step 1 of Algorithm 1 is a feasible solution to the instance $G = (V, E; w, c)$.

Using greedy technique at Step 1.2 of Algorithm 1 to choose some suitable edges from D_s to be added into T_s, we can indeed prove by induction that $T_s = (V, A_{T_s})$ is a minimum-cost shortest path tree among all shortest path trees in G at the source vertex s, where $A_{T_s} = \{e_{i_2}, e_{i_3}, \ldots, e_{i_n}\}$ and $c(T_s) = \sum_{k=2}^{n} w(e_{i_k})$.

Now, we may assume that $T^*_{s^*} = (V, A_{T^*_{s^*}})$ is a minimum-cost shortest path tree for an instance $G = (V, E; w, c)$ of the the DcMSPT problem, where some vertex $s^* \in V$ satisfies $rad(T^*_{s^*}) \leq r_0$. Since the minimum value outputted at Step 2 is attained by executing the DcMSPT algorithm, enumerating the all minimum-cost shortest path trees at all distinct source vertices in V, we can obtain a minimum-cost shortest path tree $T_{s^*} = (V, A_{T_{s^*}})$ at the source vertex $s^* \in V$ to satisfy $w(T^*_{s^*}) = w(T_{s^*})$ where $rad(T^*_{s^*}) \leq r_0$, implying that $T_{s^*} = (V, A_{T_{s^*}})$ is a minimum-cost shortest path tree and $rad(T^*_{s^*}) \leq r_0$ for an instance $G = (V, E; w, c)$ of the the RMSPT problem.

The complexity of the DcMSPT algorithm (*i.e.*, Algorithm 1) can be determined as follows. (1) For each vertex $s \in V$, the Dijkstra algorithm-modified [7] implies that Step 1.1 needs time $O(n^2)$ to compute the auxiliary acyclic digraph $D_s = (V, A_s)$, and for $e_{D_s}(s) \leq r_0$, Step 1.2 needs time $O(|E|)$ to find such a minimum-cost arborescence T_s at the source vertex s in D_s, showing that the running time at Step 1 is in total $O(n^3)$; (2) Step 2 needs at most time $O(n)$. Hence, the running time of the RMSPT algorithm is in total $O(n^3)$.

This establishes the theorem. ∎

Given a weighted graph $G = (V, E; w, c)$, when we add a step that compute the radius $rad(G)$ of G as the first step in the DcMSPT algorithm, and denoting $d_0 = rad(G)$, we can provide an algorithm to solve the MRMSPT problem. We only present the following conclusion, no description of the MRMSPT algorithm in details to to save a room.

Corollary 1. *The MRMSPT algorithm is an exact algorithm to solve the MRMSPT problem, and it runs in time $O(n^3)$, where n is the order of weighted graph G.*

4 An Exact Algorithm to Solve the MDMSPT Problem

In this section, we consider the maximum delay minimum shortest path tree (MDMSPT) problem, where maximum delay is exactly the diameter of a weighted graph $G = (V, E; w, c)$.

Using the Dijkstra algorithm [7] for many times and modifying the strategy to solve the DcMSPT problem (seeing the DcMSPT algorithm), we can design our algorithm to solve the MDMSPT problem as follows.

Algorithm 2: MDMSPT
INPUT: a weighted graph $G = (V, E; w, c)$;
OUTPUT: a maximum delay minimum shortest path tree, rooted at a source $v^* \in V$.
Begin
Step 1. For each vertex $x \in V$, do
(1.1) Depending on computing of length function $w(\cdot)$, use the Dijkstra algorithm [7] to compute the eccentricity of x, *i.e.*, $e_G(x, V) = \max\{dist_G(x, y) \mid y \in V\}$.
Step 2. Compute the diameter of G as maximum of the eccentricities of all vertices in G, *i.e.*, $diam(G) = \max\{e_G(x, V) \mid x \in V\}$, and denote the set $V_{diam} = \{x \in V \mid e_G(x, V) = diam(G)\}$.
Step 3. For each vertex $s \in V_{diam}$, do
(3.1) Depending on computing of length function $w(\cdot)$, execute the Dijkstra algorithm-modified [7] to construct an auxiliary acyclic digraph $D_s = (V, A_s; w, c)$, where A_s consists of arcs $(x, y) \in A$ that lies on a shortest (s, v_i)-path from s to all other vertices $v_i \in V$. For convenience, we may assume that all vertices in $D_s = (V, A_s)$ are topologically sorted in the order $v_{j_1}, v_{j_2}, \ldots, v_{j_n}$, where $v_{j_1} = s$.
(3.2) For each vertex $v_{j_t} \in V$, $t = 2, 3, \ldots, n$, depending on computing of cost function $c(\cdot)$, choose a minimum cost arc $e_{i_t} = (v_{i_t}, v_{j_t})$ in D_s to enter the vertex v_{j_t}, where $v_{i_t} \in \{v_{j_1}, v_{j_2}, \cdots, v_{j_n}\}$ (for two distinct integers $t, t' \in \{2, 3, \ldots, n\}$, we may have the same vertex $v_{i_t} = v_{i_{t'}}$).
(3.3) Construct a shortest path tree $T_s = (V, A_{T_s})$ with the edge set $A_{T_s} = \{e_{i_2}, e_{i_3}, \ldots, e_{i_n}\}$.
Step 4. Choose a minimum-cost shortest path tree $T_{v^*} = (V, A_{T_{v^*}})$ from all shortest path trees constructed at Step (1.2.2), *i.e.*, satisfying $c(A_{T_{v^*}}) = \min\{c(A_{T_s}) \mid T_s = (V, A_{T_s})$ is a shortest path tree in G, having $e_{T_s}(s) \leq r_0\}$.
Step 5. Output the minimum-cost shortest path tree $T_{v^*} = (V, A_{T_{v^*}})$ obtained at Step 4.
End

Using the MDMSPT algorithm, we can obtain the following result, whose correct proof is similar to the arguments in the proof of Theorem 1, and we omit the details.

Theorem 2. *The MDMSPT algorithm (i.e., Algorithm 2) is a polynomial-time exact algorithm to solve the MDMSPT problem, and its complexity is $O(n^3)$, where n is the order of weighted graph G.*

5 Conclusion and Further Research

In this paper, we consider the delay-constrained minimum shortest path tree (DcMSPT) problem and its related variations, respectively, then we obtain two main results

(1) We design an exact algorithm to solve the DcMSPT problem, we provide the similar algorithm to solve the MRMSPT problem, and both algorithms run in time $O(n^3)$.
(2) We present an exact algorithm to solve the MDMSPT problem, and this algorithm runs in time $O(n^3)$.

A challenging task for further research is to design other exact algorithms in lower running times to solve the DcMSPT problem and its related variations.

References

1. Bondy, J.A., Murty, U.S.R.: Graph theory. In: Graduate Texts in Mathematics, vol. 244. Springer, Heidelberg (2008)
2. Bookstein, A., Klein, S.T.: Construction of optimal graphs for bit-vector compression. In: Proceedings of the 13th Annual International ACM SIGIR Conference on Research and Development in Information Retrieval, pp. 327–342 (1990)
3. Bookstein, A., Klein, S.T.: Compression of correlated bit-vectors. Inf. Syst. **16**, 387–400 (1991)
4. Cheriton, D., Tarjan, R.: Finding minimum spanning trees. SIAM J. Comput. **5**, 724–742 (1976)
5. Chu, Y.J., Liu, Z.H.: On the shortest arborescence of a directed graph. Scientia Sinica **14**, 1396–1400 (1965)
6. Dantzig, G.B.: Discrete-variable extremum problems. Oper. Res. **5**, 266–277 (1957)
7. Dijkstra, E.W.: A note on two problems in connexion with graphs. Numerische Mathematik **1**, 269–271 (1959)
8. Dvir, D., Handler, G.Y.: The absolute center of a network. Networks **43**(2), 109–118 (2004)
9. Edmonds, J.: Optimum branchings. J. Res. Natl. Bureau Stand. Sect. B **71**, 233–240 (1967)
10. Garey, M.R., Johnson, D.S.: Computers and Intractability: A Guide to the Theory of NP-Completeness. Freeman, San Francisco (1979)
11. Hakimi, S.L.: Optimal locations of switching centers and medians of a graph. Oper. Res. **12**, 450–459 (1964)

12. Hassin, R., Tamir, A.: On the minimum diameter spanning tree problem. Inf. Process. Lett. **53**, 109–111 (1995)
13. Ho, J.M., Lee, D.T., Chang, C.H., Wong, C.K.: Minimum diameter spanning trees and related problems. SIAM J. Comput. **20**(5), 987–997 (1991)
14. Julstrom, B.A.: Greedy heuristics for the bounded diameter minimum spanning tree problem, ACM J. Exp. Algorithmics **14**, Article No. 1.1 (2009)
15. Korte, B., Vygen, J.: Combinatorial Optimization: Theory and Algorithms, 5th edn. Springer, Berlin (2012)
16. Kortsarz, G., Peleg, D.: Approximating the weight of shallow Steiner trees. Discrete Appl. Math. **93**, 265–285 (1999)
17. Schrijver, A.: Combinatorial Optimization: Polyhedra and Efficiency. Springer, The Netherlands (2003)
18. Seo, D.Y., Lee, D.T., Lin, T.-C.: Geometric minimum diameter minimum cost spanning tree problem. In: Dong, Y., Du, D.-Z., Ibarra, O. (eds.) ISAAC 2009. LNCS, vol. 5878, pp. 283–292. Springer, Heidelberg (2009). https://doi.org/10.1007/978-3-642-10631-6_30
19. Singh, A., Gupta, A.K.: Improved heuristics for the bounded-diameter minimum spanning tree problem. Soft Comput. **11**, 911–921 (2007)
20. Torkestani, J.A.: An adaptive heuristic to the bounded-diameter minimum spanning tree problem. Soft Comput. **16**, 1977–1988 (2012)

On the Feedback Number of 3-Uniform Linear Extremal Hypergraphs

Zhongzheng Tang[1], Yucong Tang[2,3], and Zhuo Diao[4(✉)]

[1] School of Science, Beijing University of Posts and Telecommunications, Beijing 100876, China
tangzhongzheng@amss.ac.cn

[2] Department of Mathematics, Nanjing University of Aeronautics and Astronautics, Nanjing 211106, China
tangyucong@nuaa.edu.cn

[3] Key Laboratory of Mathematical Modelling and High Performance Computing of Air Vehicles (NUAA), MIIT, Nanjing 211106, China

[4] School of Statistics and Mathematics, Central University of Finance and Economics, Beijing 100081, China
diaozhuo@amss.ac.cn

Abstract. Let $H = (V, E)$ be a hypergraph with vertex set V and edge set E. $S \subseteq V$ is a feedback vertex set (FVS) of H if $H \backslash S$ has no cycle and $\tau_c(H)$ denote the minimum cardinality of a FVS of H. Chen et al. [IWOCA, 2016] has proven if H is a 3-uniform linear hypergraph with m edges, then $\tau_c(H) \leq m/3$. In this paper, we furthermore characterize all the extremal hypergraphs with $\tau_c(H) = m/3$ holds.

Keywords: Feedback Vertex Set (FVS) · 3-uniform linear hypergraph · Extremal hypergraph

1 Introduction

A feedback vertex set (FVS) in a graph G is a vertex subset S such that $G \backslash S$ is acyclic. In the case of directed graphs, it means $G \backslash S$ is a directed acyclic graph (DAG). In the (Directed) Feedback Vertex Set ((D)FVS) problem we are given as input a (directed) graph G, and the objective is to find a minimum cardinality of FVS S. Both the directed and undirected version of the problem are NP-complete [19] and have been extensively studied from the perspective of approximation algorithms [3,15], parameterized algorithms [8,11,23], exact exponential time algorithms [25,28] as well as graph theory [14,26].

Supported by National Natural Science Foundation of China under Grant No. 11901605, No. 11901292, No. 71801232, No. 12101069, the disciplinary funding of Central University of Finance and Economics, the Emerging Interdisciplinary Project of CUFE, the Fundamental Research Funds for the Central Universities and Innovation Foundation of BUPT for Youth (500421358).

D.-Z. Du et al. (Eds.): COCOA 2021, LNCS 13135, pp. 687–700, 2021.
https://doi.org/10.1007/978-3-030-92681-6_54

There are several reasons why minimizing FVS is the one of the most central problem in algorithm design and parameterized complexity: First and foremost, the main point of parameterized complexity, being that in many instance the parameter k of the FVS's size is small, is very applicable for FVS: In the instances arising from e.g. resolving deadlocks in systems of processors [4], or from Bayesian inference or constraint satisfaction, one is only interested in whether small FVS's exist [5,12,27]. Second, minimizing FVS is a very natural graph modification problem (remove/add few vertices/edges to make the graph satisfy a certain property) that serves as excellent starting point for many other graph modification problems such a planarization or treewidth-deletion (see e.g. [20] for a recent overview). Third, FVS and many of its variants (see e.g. [22]) admit elegant duality theorems such as the Erdös-Pośa property; understanding their use in designing algorithms can be instrumental to solve many problems different from FVS faster. The popularity of FVS also led to work on a broad spectrum of its variations such as Subset, Group, Connected, Simultaneous, or Independent FVS (see for example [1] and the references therein).

The combinatorial bound on the number of FVS's is of independent interest. One of the very natural questions in graph theory is: how many minimal (maximal) vertex subsets satisfying a given property can be contained in a graph on n vertices? The trivial bound is $O(2^n/\sqrt{n})$ (which is roughly the maximum number of subsets of an n-set such that none of them is contained in the other). For general directed graphs, no non-trivial upper bounds on the number of minimal FVSs are known. For undirected graphs, Fomin et al. [17] showed that any undirected graph on n vertices contains at most $1.8638n$ minimal FVSs, and that infinitely many graphs have $105^{n/10} > 1.5926^n$ minimal FVSs. Lower bounds of roughly $\log n$ on the size of a maximum-size acyclic subtournament have been obtained by Reid and Parker [21] and Neumann-Lara [24].

The main objective of this paper is a study of FVS on hypergraphs. A hypergraph is a set family H with a universe $V(H)$ and a family of hyperedges $E(H)$, where each hyperedge is a subset of $V(H)$. If every hyperedge in $E(H)$ is of size at most d, it is known as a d-hypergraph. Observe that if each hyperedge is of size exactly two, we get an undirected graph. There are many classical books in hypergraph theory [2,3,6,7,13,14,16,17]. The natural question is, how does FVS generalize to hypergraphs. Formally, Let $H = (V, E)$ be a hypergraph with vertex set V and edge set E. $S \subseteq V$ is a feedback vertex set (FVS) of H if $H \backslash S$ has no cycle and $\tau_c(H)$ denote the minimum cardinality of a FVS in H. However, there is very little study of FVS on hypergraphs. The only known algorithmic result is a factor d approximation for FVS on d-hypergraphs [18]. Upper bounds on minimum FVS in 3-uniform linear hypergraphs are studied in [9,10].

In this paper, we consider the feedback vertex set (FVS) in hypergraphs. Chen et al. [9,10] has proven if H is a 3-uniform linear hypergraph with m edges, then $\tau_c(H) \leq m/3$. In this paper, we furthermore characterize all the extremal hypergraphs with $\tau_c(H) = m/3$ holds (as shown in Fig. 1).

The main content of the article is organized as follows:

- In Sect. 2, the basic concepts and the symbols in hypergraphs are introduced.

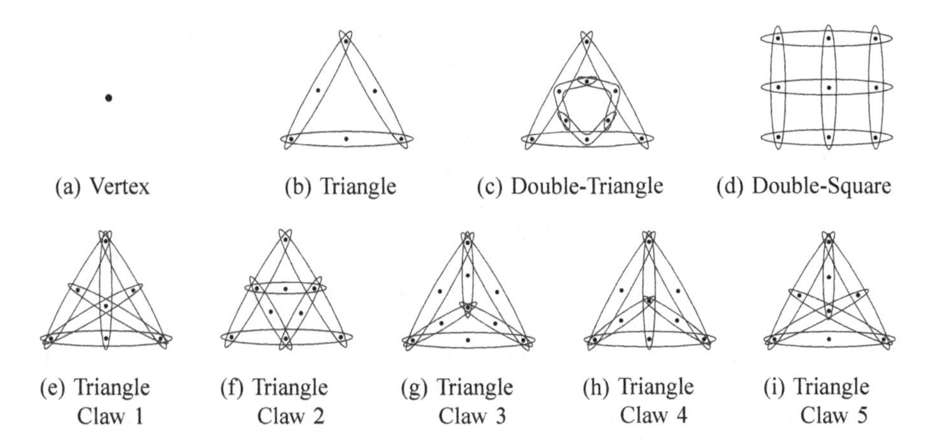

Fig. 1. Nine possible hypergraphs of every component

- In Sect. 3, we characterize all the extremal hypergraphs with $\tau_c(H) = m/3$ holds. Actually, it is easy to prove that for any extremal hypergraph, the maximum degree is no more than 3.
  - In Subsect. 3.1, we characterize all the extremal hypergraphs with maximum degree smaller than 3 (as shown in Fig. 1(a)–(d)). The basic idea is as follows: For any extremal hypergraph with maximum degree smaller than 3, let us do a series of edge deletion operations and keep the properties of extremal hypergraphs in the process. The edge deletion operations run recursively and the number of edges is decreasing, thus finally we get some isolated vertices. Because the extremal hypergraphs are always kept during the execution of the process, we trace the process back and get all the extremal hypergraphs with maximum degree smaller than 3.
  - In Subsect. 3.2, we characterize all the extremal hypergraphs with maximum degree 3 (as shown in Fig. 1). The basic idea is as follows: For any extremal hypergraph with maximum degree 3, let us do a series of 3-degree vertex deletion operations and keep the properties of extremal hypergraphs in the process. The 3-degree vertex deletion operations run recursively and finally we get an extremal hypergraph with maximum degree smaller than 3 (as shown in Fig. 1(a)–(d)). Then we trace the process back and get all the extremal hypergraphs with maximum degree 3. It is worth noting that in the tracing back process, graph theoretical analysis and numerical calculation are combined.
- In Sect. 4, the conclusions are summarized and some future works are proposed.

2 Hypergraphs

Let $H = (V, E)$ be a hypergraph with vertex set V and edge set E. For each vertex $v \in V$, the degree $d(v)$ is the number of edges in E that contains v.

We say v is an isolated vertex of H if $d(v) = 0$. Hypergraph H is k-regular if each vertex's degree is k ($d(v) = k, \forall v \in V$). Hypergraph H is k-uniform if each edge contains exactly k vertices ($|e| = k, \forall e \in E$). Hypergraph H is called *linear* if any two distinct edges have at most one common vertex ($|e \cap f| \leq 1, \forall e, f \in E$).

Let $k \geq 2$ be an integer. A cycle of length k, denoted as k-cycle, is a vertex-edge sequence $C = v_1e_1v_2e_2 \cdots v_ke_kv_1$ with: (1) $\{e_1, e_2, \ldots, e_k\}$ are distinct edges of H. (2) $\{v_1, v_2, \ldots, v_k\}$ are distinct vertices of H. (3) $\{v_i, v_{i+1}\} \subseteq e_i$ for each $i \in [k]$, here $v_{k+1} = v_1$. We consider the cycle C as a sub-hypergraph of H with vertex set $\{v_i, i \in [k]\}$ and edge set $\{e_j, j \in [k]\}$. For any vertex set $S \subseteq V$, we write $H \backslash S$ for the sub-hypergraph of H obtained from H by deleting all vertices in S and all edges incident with some vertices in S. For any edge set $A \subseteq V$, we write $H \backslash A$ for the sub-hypergraph of H obtained from H by deleting all edges in A and keeping vertices. If S is a singleton set s, we write $H \backslash s$ instead of $H \backslash \{s\}$.

We say $S \subseteq V$ is a feedback vertex set (FVS) of H if it intersects every cycle's vertex set in H. This is equivalent to say that $H \backslash S$ has no cycle and let us denote $\tau_c(H)$ as the minimum cardinality of a FVS in H. In this paper, we consider the feedback vertex set (FVS) in 3-uniform linear hypergraphs.

3 The 3-Uniform Linear Extremal Hypergraphs

Chen et al. [9,10] has proven if H is a 3-uniform linear hypergraph with m edges, then $\tau_c(H) \leq m/3$. In this paper, we furthermore characterize all the extremal hypergraphs with $\tau_c(H) = m/3$ holds. Actually, it is easy to prove for any extremal hypergraph, the maximum degree is no more than 3. In Subsect. 3.1, we characterize all the extremal hypergraphs with maximum degree smaller than 3. In Subsect. 3.2, we characterize all the extremal hypergraphs with maximum degree 3.

Theorem 1 [9,10]. *Let H be a 3-uniform linear hypergraph with m edges. Then $\tau_c(H) \leq m/3$.*

Corollary 1. *Let H be a 3-uniform linear hypergraph with m edges and $\tau_c(H) = m/3$, then the maximum degree is no more than* 3.

Proof. Suppose the corollary fails. Let us take out a counterexample $H = (V, E)$ with $\tau_c(H) = m/3$ and there is a vertex $v \in V, d(v) \geq 4$. Then $\tau_c(H \backslash v) \leq (m - d(v))/3 \leq (m - 4)/3$. Considering a minimum FVS S of $H' = H \backslash v$, we have $S \subseteq V \backslash v$ and $|S| \leq (m - 4)/3$. Thus $S \cup \{v\}$ is a FVS for H and $|S \cup \{v\}| = |S| + 1 \leq (m - 1)/3$, this is a contradiction with $\tau_c(H) = m/3$.

3.1 The Extremal Hypergraphs with Maximum Degree Smaller Than 3

In Subsect. 3.1, we characterize all the extremal hypergraphs with maximum degree smaller than 3 (as shown in Fig. 1(a)–(d)). The basic idea is as follows: For any extremal hypergraph $H(V, E)$, we do a series of edge deletion operations and keep the properties of extremal hypergraphs in the process. The edge deletion operations run recursively and the number of edges is decreasing, thus finally we derive some isolated vertices. Since the extremal hypergraph is always kept during the execution of the process, then we trace the process back and get all the extremal hypergraphs with maximum degree smaller than 3.

Our discussion will frequently use the trivial observation that given a hypergraph $H(V, E)$ and an edge subset A in E, if no cycle of H contains any edge in A, then H and $H \backslash A$ have the same FVS set and $\tau_c(H) = \tau_c(H \backslash A)$. Next, we will state a useful lemma and begin the proof of our main theorem.

Lemma 1. *Two triangles are added together to form a connected 3-uniform linear hypergraph $H(V, E)$ with maximum degree smaller than 3, then $H(V, E)$ is an extremal hypergraph if and only if $H(V, E)$ is a 2-regular double-triangle (as shown in Fig. 1(c)).*

Proof. $H(V, E)$ is formed by adding two triangles. Let us denote two triangles in order of addition as T_1, T_2. Because $H(V, E)$ is a connected 3-uniform linear hypergraph with maximum degree smaller than 3, there are only 3 possibilities as shown in Fig. 2. The only extremal hypergraph is a 2-regular double-triangle.

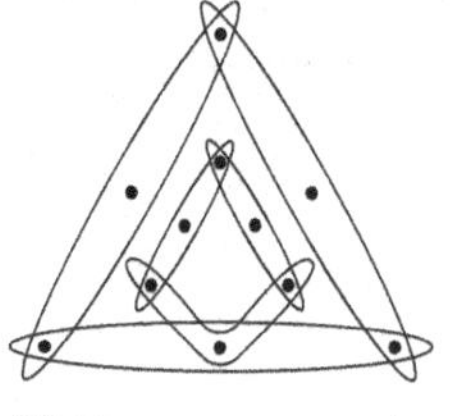

(1) One common vertex

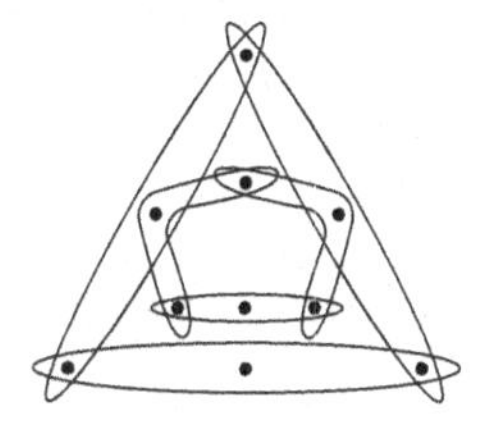

(2) Two common vertices

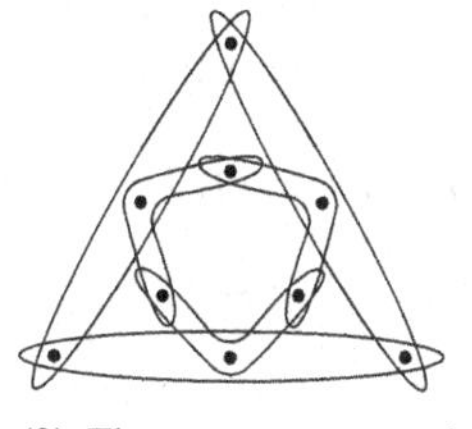

(3) Three common vertices

Fig. 2. Three kinds of double-triangles, classified according to the number of vertices shared by the two triangles

Theorem 2. *Let H be a 3-uniform linear hypergraph with m edges and $\tau_c(H) = m/3$. If the maximum degree is smaller than 3, then every component of H is an isolated vertex, a triangle, a 2-regular double-triangle or a 2-regular square as shown in Fig. 3.*

Proof. Let H be a 3-uniform linear hypergraph with m edges and $\tau_c(H) = m/3$. The maximum degree is smaller than 3. We will break the proof into a series of operations.

Observation 1. *Every edge in E is contained in some cycle in H.*

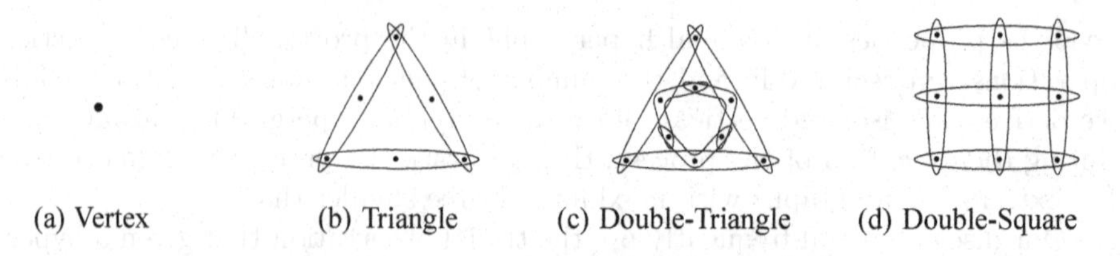

Fig. 3. Four possible hypergraphs of every component

If there exists $e \in E$ which doesn't belong to any cycle of H. Then $\tau_c(H) = \tau_c(H \backslash e)$. According to Theorem 1, we have $\tau_c(H \backslash e) \leq (m-1)/3$. Thus $\tau_c(H) \leq (m-1)/3$, this is a contradiction with $\tau_c(H) = m/3$.

Let us do a series of edge deletion operations recursively and construct a sub-hypergraph H' of H, which is also an extremal hypergraph with $\tau_c(H') = m'/3$.

Operation 1. *Deleting triangles.*

Let $C = v_1e_1v_2e_2v_3e_3v_1$ be a triangle. Denote $H' = H \backslash \{e_1, e_2, e_3\}$ and according to Theorem 1, $\tau_c(H') \leq (m-3)/3$. Considering a minimum FVS S of H', we have $S \subseteq V$ and $|S| \leq (m-3)/3$. Because every vertex's degree is no more than 2, $S \cup \{v_1\}$ is a FVS for H (as shown in Fig. 4). Thus we have

$$m/3 = \tau_c(H) \leq |S \cup \{v_1\}| \leq |S| + 1 = \tau_c(H') + 1 \leq (m-3)/3 + 1 = m/3$$

This means $\tau_c(H') = (m-3)/3$ and H' is also an extremal hypergraph.

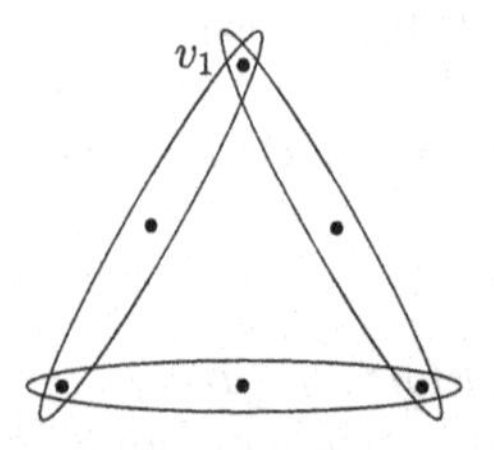

Fig. 4. The schematic diagram in Operation 1

We do Operation 1 on H until the resulting hypergraph contains no triangles. For the convenience of description, we still denote the new triangle-free hypergraph as H.

Operation 2. *Deleting* 1*-degree vertices.*

If there exists $v \in V$ with $d(v) = 1$ in H, we could assume $v \in e_1$. Due to Observation 1 and Operation 1, we can assume there exists $C = v_1e_1v_2e_2v_3e_3v_4 \cdots e_kv_1$ with $k \geq 4$ in H and $e_1 = \{v_1, v, v_2\}$. Denote $H' =$

$H\backslash\{e_1, e_2, e_3\}$ and according to Theorem 1, $\tau_c(H') \leq (m-3)/3$. Considering a minimum FVS S of $H' = H\backslash\{e_1, e_2, e_3\}$, we have $S \subseteq V$ and $|S| \leq (m-3)/3$. Because every vertex's degree is no more than 2 and $d(v) = 1$, we have $S \cup \{v_3\}$ is a FVS for H (as shown in Fig. 5). Thus we have

$$m/3 = \tau_c(H) \leq |S \cup \{v_3\}| \leq |S| + 1 = \tau_c(H') + 1 \leq (m-3)/3 + 1 = m/3$$

This means $\tau_c(H') = (m-3)/3$ and H' is also an extremal hypergraph.

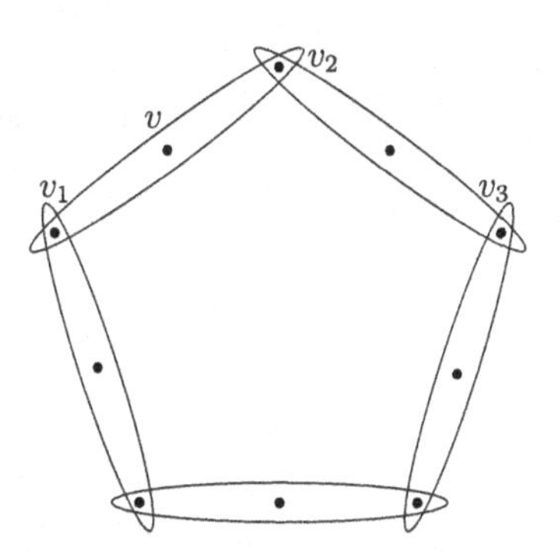

Fig. 5. The schematic diagram in Operation 2

Operation 2 may be executed repeatedly. When other operations are executed, we need to check whether Operation 2 can be executed.

Operation 3. *Deleting 4-cycles.*

Let $C = v_1e_1v_2e_2v_3e_3v_4e_4v_1$ be a 4-cycle in H, we have $e_1 \cap e_3 = e_2 \cap e_4 = \emptyset$ due to Operation 1. We can assume $e_1 = \{v_1, u_1, v_2\}, e_2 = \{v_2, u_2, v_3\}, e_3 = \{v_3, u_3, v_4\}, e_4 = \{v_4, u_4, v_1\}$ and these vertices are distinct. Due to Operation 2, we can assume $u_1 \in e_5 \neq e_1$, $u_2 \in e_6 \neq e_2$, $u_3 \in e_7 \neq e_3$ and $u_4 \in e_8 \neq e_4$. Due to Operation 1, we have $e_5 \neq e_6, e_6 \neq e_7, e_7 \neq e_8, e_8 \neq e_5$.

a If $e_5 = e_7$ and $e_6 = e_8$, there are two edges e_5, e_6 connecting u_1, u_3 and u_2, u_4 (as shown in Fig. 6). Let us denote $H' = H\backslash\{e_1, e_2, e_3, e_4, e_5, e_6\}$ and according to Theorem 1, $\tau_c(H') \leq (m-6)/3$. Considering a minimum FVS S of H', we have $S \subseteq V$ and $|S| \leq (m-6)/3$. Because every vertex's degree is no more than 2, $S \cup \{u_1, u_2\}$ is a FVS for H. Thus we have

$$m/3 = \tau_c(H) \leq |S \cup \{u_1, u_2\}| \leq |S| + 2 = \tau_c(H') + 2 \leq (m-6)/3 + 2 = m/3$$

This means $\tau_c(H') = (m-6)/3$ and H' is also an extremal hypergraph.

b If $e_5 \neq e_7$, there is no edge connecting u_1, u_3 (as shown in Fig. 7**b**). Let us denote $H' = H\backslash\{e_1, e_2, e_3, e_4, e_5, e_7\}$ and according to Theorem 1, $\tau_c(H') \leq (m-6)/3$. Considering a minimum FVS S of H', we have $S \subseteq V$ and $|S| \leq$

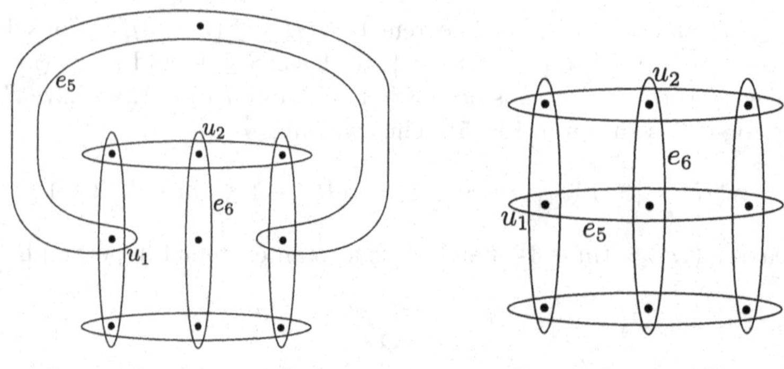

a1. e_5 and e_6 are not adjacent a2. e_5 and e_6 are adjacent

Fig. 6. The schematic diagrams of **a** in Operation 3

$(m-6)/3$. Because every vertex's degree is no more than 2, $S \cup \{u_1, u_3\}$ is a FVS for H. Thus we have

$$m/3 = \tau_c(H) \leq |S \cup \{u_1, u_3\}| \leq |S| + 2 = \tau_c(H') + 2 \leq (m-6)/3 + 2 = m/3$$

This means $\tau_c(H') = (m-6)/3$ and H' is also an extremal hypergraph.

c If $e_6 \neq e_8$, there is no edge connecting u_2, u_4 (as shown in Fig. 7**c**). Let us denote $H' = H \backslash \{e_1, e_2, e_3, e_4, e_6, e_8\}$ and according to Theorem 1, $\tau_c(H') \leq (m-6)/3$. Considering a minimum FVS S of H', we have $S \subseteq V$ and $|S| \leq (m-6)/3$. Because every vertex's degree is no more than 2, $S\{\cup u_2, u_4\}$ is a FVS for H. Thus we have

$$m/3 = \tau_c(H) \leq |S \cup \{u_2, u_4\}| \leq |S| + 2 = \tau_c(H') + 2 \leq (m-6)/3 + 2 = m/3$$

This means $\tau_c(H') = (m-6)/3$ and H' is also an extremal hypergraph.

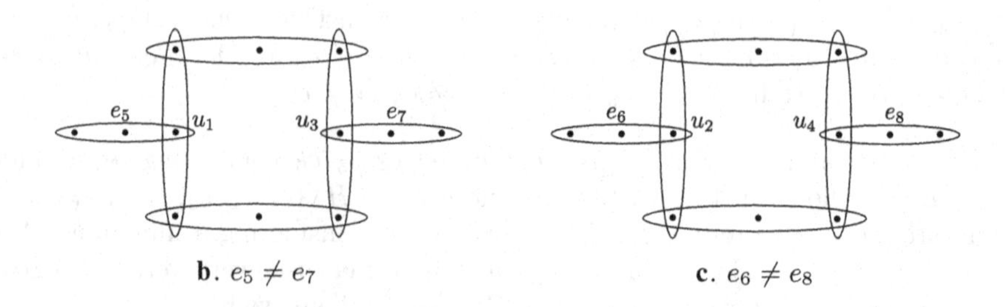

b. $e_5 \neq e_7$ c. $e_6 \neq e_8$

Fig. 7. The schematic diagrams of **b** and **c** in Operation 3

Let $C = v_1 e_1 v_2 e_2 \cdots v_k e_k v_1$ be a shortest cycle in H. For each $i \in [k]$, suppose that $e_i = \{v_i, u_i, v_{i+1}\}$, where $v_{k+1} = v_1$. Due to Operation 1 and 3, we have $k \geq 5$. Because C is the shortest cycle, for each index pair $\{i \neq j\} \subseteq [k]$, if e_i and e_j are not adjacent in C, we have $e_i \cap e_j = \emptyset$.

Operation 4. *Deleting k-cycles with $k \equiv 0 \pmod 3$.*

This means $k = 3t$, $t \geq 2$. Let us denote $H' = H \backslash \{e_1, e_2, \ldots, e_k\}$ and according to Theorem 1, $\tau_c(H') \leq (m-k)/3$. Considering a minimum FVS S of H', we have $S \subseteq V$ and $|S| \leq (m-k)/3$. Because every vertex's degree is no more than 2, $S \cup \{v_3, v_6, \ldots, v_{3t}\}$ is a FVS for H as shown in Fig. 8(1). Thus we have

$$\begin{aligned} m/3 &= \tau_c(H) \leq |S \cup \{v_3, v_6, \ldots, v_{3t}\}| \leq |S| + t = \tau_c(H') + t \leq (m-k)/3 + t \\ &= (m-3t)/3 + t = m/3 \end{aligned}$$

This means $\tau_c(H') = (m-k)/3$ and H' is also an extremal hypergraph.

Operation 5. *Deleting k-cycles with $k \equiv 1 \pmod 3$.*

This means $k = 3t+1$, $t \geq 2$. Due to Operation 1, 2 and 3, we have $u_1 \in e_{3t+2} \neq e_1$, $u_3 \in e_{3t+3} \neq e_3$ and $e_1, e_2, e_3, \ldots, e_k, e_{3t+2}, e_{3t+3}$ are distinct. Let us denote $H' = H \backslash \{e_1, e_2, \ldots, e_k, e_{3t+2}, e_{3t+3}\}$ and according to Theorem 1, $\tau_c(H') \leq (m-k-2)/3$. Considering a minimum FVS S of H', we have $S \subseteq V$ and $|S| \leq (m-k-2)/3$. Because every vertex's degree is no more than 2, $S \cup \{u_1, u_3, v_6, \ldots, v_{3t}\}$ is a FVS for H as shown in Fig. 8(2). Thus we have

$$\begin{aligned} m/3 &= \tau_c(H) \leq |S \cup \{u_1, u_3, v_6, \ldots, v_{3t}\}| \leq |S| + t + 1 = \tau_c(H') + t + 1 \\ &\leq (m-k-2)/3 + t + 1 = (m-3t-3)/3 + t + 1 = m/3 \end{aligned}$$

This means $\tau_c(H') = (m-k-2)/3$ and H' is also an extremal hypergraph.

Operation 6. *Deleting k-cycles with $k \equiv 2 \pmod 3$.*

This means $k = 3t+2$, $t \geq 2$. Due to Operation 1, 2 and 3, we have $u_1 \in e_{3t+3} \neq e_1$ and $e_1, e_2, e_3, \ldots, e_k, e_{3t+3}$ are distinct. Let us denote $H' = H \backslash \{e_1, e_2, \ldots, e_k, e_{3t+3}\}$ and according to Theorem 1, $\tau_c(H') \leq (m-k-1)/3$. Considering a minimum FVS S of H', we have $S \subseteq V$ and $|S| \leq (m-k-1)/3$. Because every vertex's degree is no more than 2, $S \cup \{u_1, v_4, \ldots, v_{3t+1}\}$ is a FVS for H as shown in Fig. 8(3). Thus we have

$$\begin{aligned} m/3 &= \tau_c(H) \leq |S \cup \{u_1, v_4, \ldots, v_{3t+1}\}| \leq |S| + t + 1 = \tau_c(H') + t + 1 \\ &\leq (m-k-1)/3 + t + 1 = (m-3t-3)/3 + t + 1 = m/3 \end{aligned}$$

This means $\tau_c(H') = (m-k-1)/3$ and H' is also an extremal hypergraph.

The algorithm runs recursively and the number of edges is decreasing, thus finally we get some isolated vertices.

Because the extremal hypergraph is always kept during the execution of the algorithm, now we trace the algorithm process back. The first backtracking step tells us the extremal hypergraph is a triangle or a 2-regular square(as shown in Fig. 3(b)(d)). In other cases, The deleting edges do not form an extremal hypergraph.

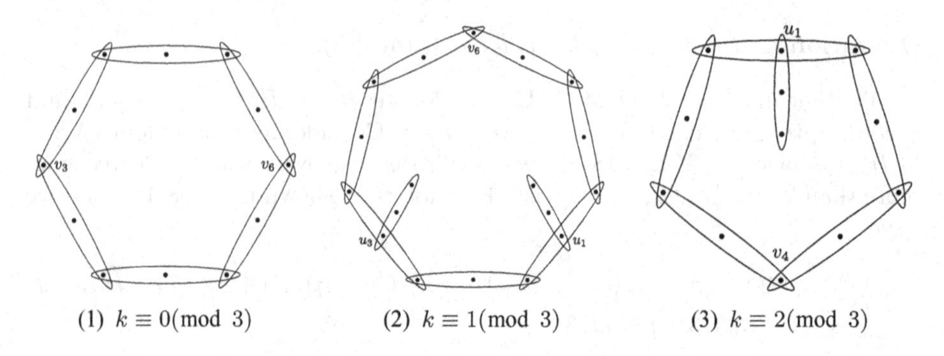

Fig. 8. The schematic diagrams in Operation 4, 5 and 6

a The first backtracking step is a 2-regular square. Because every vertex's degree is no more than 2, the 2-regular square is a component of H. The second backtracking step is the same as the first backtracking step, which tells us the extremal hypergraph is also a triangle or a 2-regular square(as shown in Fig. 3(b)(d)).

b The first backtracking step is a triangle. **Because the algorithm process initially deletes all triangles, more backtracking steps are done by deleting triangles.** According to Lemma 1, each component of H is a triangle or a 2-regular double-triangle (as shown in Fig. 3(b)(c)).

After all steps of the backtracking process, the hypergraph H is restored and every component of H is an isolated vertex, a triangle, a 2-regular double-triangle or a 2-regular square (as shown in Fig. 3).

Procedure 1 demonstrates the complete process of the above proof. The input is an extremal hypergraph $H(V, E)$ with maximum degree of no more than 2. Then, we can repeatedly do six edge deletion operations in the proof of Theorem 2, and finally output some isolated vertices.

3.2 The Extremal Hypergraphs with Maximum Degree of 3

This subsection aims to characterize all possible components of extremal hypergraphs with maximum degree of 3.

Due to space limitations, we briefly introduce the proof ideas of the main theorem and omit the proof. First, we can repeatedly delete the 3-degree vertices and their associated edges until we derive a hypergraph whose maximum degree does not exceed 2. It is easy to show that the extremal properties of the resulting hypergraphs are always maintained during the deletion process. Next, we get an extremal hypergraph with maximum degree smaller than 3 (as shown in Fig. 3). Furthermore, we trace the process back and get all the extremal hypergraphs with maximum degree 3 by using the graph theoretical analysis and the numerical calculation.

Procedure 1. Convert to a hypergraph consisting of isolated vertices

Input: A 3-uniform linear hypergraph H with m edges that satisfies $\tau_c(H) = m/3$ and has the maximum degree of no more than 2.

Output: A hypergraph consisting of isolated vertices.

1: **if** H is a hypergraph consisting of isolated vertices **then**
2:     **return** H
3: **while** H contains triangles **do**
4:     Do Operation 1 on H.
5: **while** H contains 1-degree vertices **do**
6:     Do Operation 2 on H.
7: **if** H contains 4-cycles **then**
8:     Do Operation 3 on H.
9:     **return** Procedure 1(H)
10: **if** H contains k-cycles ($k \geq 5$) **then**
11:     **if** $k \equiv 0 (\bmod 3)$ **then**
12:         Do Operation 4 on H.
13:     **else if** $k \equiv 1 (\bmod 3)$ **then**
14:         Do Operation 5 on H.
15:     **else**
16:         Do Operation 6 on H.
17:     **return** Procedure 1(H)

Theorem 3. *Let $H(V, E)$ be a 3-uniform linear hypergraph, H is an extremal hypergraph if and only if each component is an isolated vertex, a triangle, a 2-regular double-triangle, a 2-regular square or one of 3-degree extremal hypergraphs(as shown in Fig. 9).*

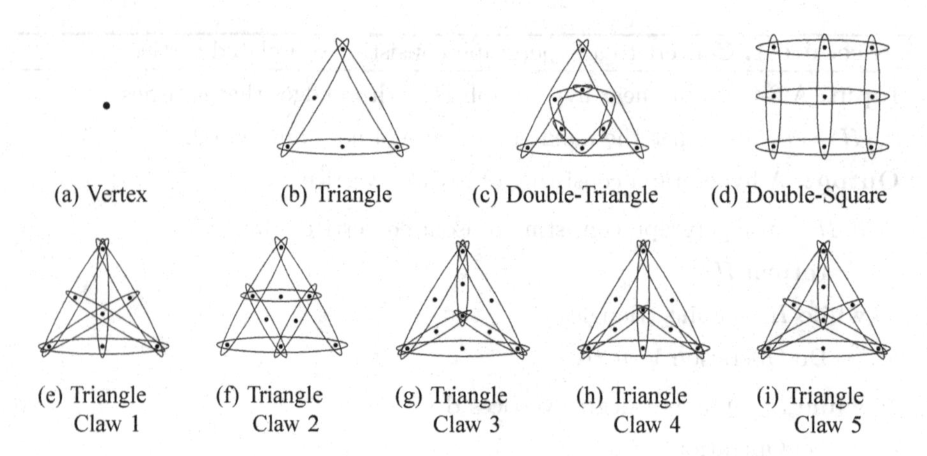

Fig. 9. Nine possible hypergraphs of every component

4 Conclusion and Future Work

In this paper, we consider the feedback vertex set (FVS) in 3-uniform linear hypergraphs. We characterize all the extremal hypergraphs with $\tau_c(H) = m/3$ holds (as shown in Fig. 9). This is a supplement for the result in [9,10] which states for any 3-uniform linear hypergraph H,$\tau_c(H) \leq m/3$ holds. It is interesting and worthwhile to consider the similar bounds for k-uniform linear hypergraphs with $k \geq 4$ and also characterize all the extremal hypergraphs.

Acknowledges. The authors are very indebted to Professor Xujin Chen and Professor Xiaodong Hu for their invaluable suggestions and comments.

References

1. Agrawal, A., Gupta, S., Saurabh, S., Sharma, R.: Improved algorithms and combinatorial bounds for independent feedback vertex set. In: Guo, J., Hermelin, D. (eds.) 11th International Symposium on Parameterized and Exact Computation (IPEC 2016). LIPIcs, Aarhus, Denmark, vol. 63, pp. 2:1–2:14. Schloss Dagstuhl - Leibniz-Zentrum für Informatik (2016)
2. Alon, N., Spencer, J.H.: The Probabilistic Method. Wiley-Interscience Series in Discrete Mathematics and Optimization, 3rd edn. Wiley, New York (2008)
3. Bafna, V., Berman, P., Fujito, T.: A 2-approximation algorithm for the undirected feedback vertex set problem. SIAM J. Discrete Math. **12**(3), 289–297 (1999)
4. Bar-Yehuda, R., Geiger, D., Naor, J., Roth, R.M.: Approximation algorithms for the feedback vertex set problem with applications to constraint satisfaction and Bayesian inference. SIAM J. Comput. **27**(4), 942–959 (1998)
5. Becker, A., Bar-Yehuda, R., Geiger, D.: Randomized algorithms for the loop cutset problem. J. Artif. Intell. Res. **12**, 219–234 (2000)
6. Berge, C.: Hypergraphs - Combinatorics of Finite Sets. North-Holland Mathematical Library, vol. 45. North-Holland, Amsterdam (1989)

7. Brualdi, R.A.: Introductory Combinatorics, 5th edn. Pearson Education, London (2009)
8. Chen, J., Liu, Y., Lu, S., O'Sullivan, B., Razgon, I.: A fixed-parameter algorithm for the directed feedback vertex set problem. J. ACM **55**(5), 21:1–21:19 (2008)
9. Chen, X., Diao, Z., Hu, X., Tang, Z.: Sufficient conditions for Tuza's conjecture on packing and covering triangles. In: Mäkinen, V., Puglisi, S.J., Salmela, L. (eds.) IWOCA 2016. LNCS, vol. 9843, pp. 266–277. Springer, Cham (2016). https://doi.org/10.1007/978-3-319-44543-4_21
10. Chen, X., Diao, Z., Hu, X., Tang, Z.: Covering triangles in edge-weighted graphs. Theory Comput. Syst. **62**(6), 1525–1552 (2018). https://doi.org/10.1007/s00224-018-9860-7
11. Cygan, M., Nederlof, J., Pilipczuk, M., Pilipczuk, M., van Rooij, J.M.M., Wojtaszczyk, J.O.: Solving connectivity problems parameterized by treewidth in single exponential time. In: Ostrovsky, R. (ed.) IEEE 52nd Annual Symposium on Foundations of Computer Science (FOCS 2011), Palm Springs, CA, USA, pp. 150–159. IEEE Computer Society (2011)
12. Dechter, R.: Enhancement schemes for constraint processing: backjumping, learning, and cutset decomposition. Artif. Intell. **41**(3), 273–312 (1990)
13. Diestel, R.: Graph Theory. Graduate Texts in Mathematics, vol. 173, 4th edn. Springer, Heidelberg (2012)
14. Erdös, P., Pósa, L.: On independent circuits contained in a graph. Can. J. Math. **17**, 347–352 (1965)
15. Even, G., Naor, J.S., Schieber, B., Sudan, M.: Approximating minimum feedback sets and multi-cuts in directed graphs. In: Balas, E., Clausen, J. (eds.) IPCO 1995. LNCS, vol. 920, pp. 14–28. Springer, Heidelberg (1995). https://doi.org/10.1007/3-540-59408-6_38
16. Festa, P., Pardalos, P.M., Resende, M.G.C.: Feedback set problems. In: Du, D.Z., Pardalos, P.M. (eds.) Handbook of Combinatorial Optimization, pp. 209–258. Springer, Boston (1999). https://doi.org/10.1007/978-1-4757-3023-4_4
17. Fomin, F.V., Gaspers, S., Pyatkin, A.V., Razgon, I.: On the minimum feedback vertex set problem: exact and enumeration algorithms. Algorithmica **52**(2), 293–307 (2008). https://doi.org/10.1007/s00453-007-9152-0
18. Fujito, T.: Approximating minimum feedback vertex sets in hypergraphs. Theor. Comput. Sci. **246**(1–2), 107–116 (2000)
19. Garey, M.R., Johnson, D.S.: Computers and Intractability: A Guide to the Theory of NP-Completeness. W. H. Freeman, New York (1979)
20. Gupta, A., Lee, E., Li, J., Manurangsi, P., Wlodarczyk, M.: Losing treewidth by separating subsets. In: Chan, T.M. (ed.) Proceedings of the Thirtieth Annual ACM-SIAM Symposium on Discrete Algorithms (SODA 2019), San Diego, California, USA, pp. 1731–1749. SIAM (2019)
21. Kenneth Brooks, R., Ernest Tilden, P.: Disproof of a conjecture of Erdös and Moser on tournaments. J. Comb. Theory **9**(3), 225–238 (1970)
22. Kim, E.J., Kwon, O.: Erdős-Pósa property of chordless cycles and its applications. J. Comb. Theory Ser. B **145**, 65–112 (2020)
23. Kociumaka, T., Pilipczuk, M.: Faster deterministic feedback vertex set. Inf. Process. Lett. **114**(10), 556–560 (2014)
24. Neumann-Lara, V.: A short proof of a theorem of Reid and Parker on tournaments. Graphs Comb. **10**(2–4), 363–366 (1994). https://doi.org/10.1007/BF02986686
25. Razgon, I.: Computing minimum directed feedback vertex set in $o(1.9977^{n})$. In: Italiano, G.F., Moggi, E., Laura, L. (eds.) 10th Italian Conference on Theoretical Computer Science (ICTCS 2007), Rome, Italy, pp. 70–81. World Scientific (2007)

26. Reed, B.A., Robertson, N., Seymour, P.D., Thomas, R.: Packing directed circuits. Combinatorica **16**(4), 535–554 (1996). https://doi.org/10.1007/BF01271272
27. Wang, C., Lloyd, E.L., Soffa, M.L.: Feedback vertex sets and cyclically reducible graphs. J. ACM **32**(2), 296–313 (1985)
28. Xiao, M., Nagamochi, H.: An improved exact algorithm for undirected feedback vertex set. J. Comb. Optim. **30**(2), 214–241 (2014). https://doi.org/10.1007/s10878-014-9737-x

A Multi-pass Streaming Algorithm for Regularized Submodular Maximization

Qinqin Gong[1], Suixiang Gao[2], Fengmin Wang[3], and Ruiqi Yang[2](✉)

[1] Department of Operations Research and Information Engineering, Beijing University of Technology, Beijing 100124, People's Republic of China

[2] School of Mathematical Sciences, University of Chinese Academy Sciences, Beijing 100049, People's Republic of China

{sxgao,yangruiqi}@ucas.ac.cn

[3] Beijing Jinghang Research Institute of Computing and Communication, Beijing 100074, People's Republic of China

Abstract. In this paper, we consider a problem of maximizing regularized submodular functions with a k-cardinality constraint under streaming fashion. In the model, the utility function $f(\cdot) = g(\cdot) - \ell(\cdot)$ is expressed as the difference between a non-negative monotone non-decreasing submodular function g and a non-negative modular function ℓ. In addition, the elements are revealed in a streaming setting, that is to say, an element is visited in one time slot. The problem asks to find a subset of size at most k such that the regularized utility value is maximized. Most of the existing algorithms for the submodular maximization heavily rely on the non-negativity assumption of the utility function, which may not be applicable for our regularized scenario. Indeed, determining if the maximum is positive or not is NP-hard, which implies that no multiplicative factor approximation is existed for this problem. To circumvent this challenge, several works paid attention to more meaningful guarantees by introducing a slightly weaker notion of approximation, and any developed algorithm is aim to construct a solution S satisfying $f(S) \geq \rho \cdot g(OPT) - \ell(OPT)$ for some $\rho > 0$. In this work, assume there is a weak ρ-approximation for the k-cardinality constrained regularized submodular maximization, we develop Distorted-Threshold-Streaming, a multi-pass bicriteria algorithm for the streaming regularized submodular maximization with the k-cardinality constraint (SRSMCC), which produces a $(\rho/\lambda, 1/\lambda)$-bicriteria approximation by making over $O(\log(\lambda/\rho)/\varepsilon)$ passes, consuming $O(k)$ memory and using $O(\log(\lambda/\rho)/\varepsilon)$ queries per element, where $\lambda = \frac{2-(2\rho+2)/(3+\sqrt{5-4\rho})}{(3+\sqrt{5-4\rho})/(2\rho+2)-1}$ and any accuracy parameter $\varepsilon > 0$.

Keywords: Submodular maximization · Stream model · Streaming algorithms · Threshold-based · Multi-pass

1 Introduction

Submodularity, an intuitive notion of diminishing returns, plays an important role in set function optimization and has been extensively studied in previous

D.-Z. Du et al. (Eds.): COCOA 2021, LNCS 13135, pp. 701–711, 2021.
https://doi.org/10.1007/978-3-030-92681-6_55

literature. Data summarization, a central task considered in machine learning, involves maximizing a utility function, which selects representative subsets of manageable size out of large data sets. From the optimization perspective, the data summarization can be turned into the problem of selecting a subset of data elements that equipped with a submodular utility function that quantifies representativeness of the selected set.

We investigate the previous works for the constrained submodular maximization in this part. Since the unconstrained submodular maximization in monotonic case can be readily implemented, we ignore it in the following sections. In rest of this paper, we mainly pay attention to the monotonic setting throughout our context and will not always mention it for clarity. The k-cardinality constrained submodular maximization has been well studied and has deep theoretical and practical results. The submodular maximization subject to a cardinality constraint (SMCC), which can be stated as $\arg\max_{S\subseteq\Omega,|S|\leq k} g(S)$, where $g: 2^{\Omega} \rightarrow \mathbb{R}_{+}$ is a non-negative monotone non-decreasing submodular function and k denotes an input parameter of size. An elegant greedy algorithm was proposed by Nemhauser et al. [16], which starts with an empty set and iteratively locates the element with maximal marginal benefit. Finally, it returns a $(1-e^{-1})$-approximation in $O(nk)$ time. The hardness was proved by Feige [4], that is to say, for any $\varepsilon > 0$, there exists no $(1-e^{-1}+\varepsilon)$-approximation algorithm for SMCC unless P = NP. Knapsack extends the cardinality constraint by encoding each element with a size and a profit values and the submodular maximization with a knapsack constraint (SMKC), is formally denoted by $\arg\max_{S\subseteq\Omega,c(S)\leq K} g(S)$, where $c(S)$ denotes the size of subset S and K denotes the input parameter of knapsack budget. Sviridenko [19] combined the enumeration with greedy technique and novelly presented a tight $(1-e^{-1})$-approximation in $O(n^5 k)$ time for SMKC. A more general matroid system is denoted by a two-tuple $(\Omega, \mathcal{I})$, which satisfies the following three properties:

1. I is a collection of subsets chosen from Ω, further $\emptyset \in \mathcal{I}$.
2. Consider $A \in \mathcal{I}$, then any $A' \subseteq A$ holds $A' \in \mathcal{I}$.
3. For any two $A, B \in \mathcal{I}$ and $|A| < |B|$, there must exist an element $u \in B\backslash A$ satisfying $A \cup \{u\} \in \mathcal{I}$.

Formally, the two-tuple $(\Omega, \mathcal{I})$ is defined as an independence system if it justly holds the first two properties and each subset belong to $\mathcal{I}$ is called as an independent set. The base of an independence system is shortly defined as the maximal size of independent sets. The k-cardinality is usually described as the union matroid, since its basis are exactly equal to k. For the submodular maximization with a matroid constraint (SMMC), it means to say the family of the feasible solution sets constructing a matroid system. Fisher et al. [8] provided a greed-based determined 0.5-approximation for the SMMC. Further, Calinescu et al. [3] developed a continuous greedy algorithm, which produces a randomized $(1-e^{-1})$-approximation. A similar result presented by Lee et al. [12], which built up a local search by introducing a novelly potential function. Recently, a breakout work was given by Buchbinder et al. [2], which developed a new split and derandomization techniques and obtained a determined 0.5008-approximation for the SMMC. With the development of the study and the extensive application, researchers are interesting more general or mixed constraints, such as k-exchange system [7], k-system [6], knapsack mixing together matroid system [18], k-system merging with knapsack [14], and just name a few.

Notice that the aforementioned studies are optimized under the non-negativity assumption. Indeed, with the help of the monotonicity assumption in prior, it may give rise to a danger of over-fitting to the solution since adding more elements can never hurt the utility. Kazemi et al. [10] formally defined a version of regularized submodular maximization by adding a modular penalty or regularizer term, which is denoted by $\arg\max_{S \subseteq \Omega, S \in \mathcal{I}} g(S) - \ell(S)$, where $\ell : 2^{\Omega} \to \mathbb{R}_+$ denotes a non-negative modular function and $\mathcal{I}$ denotes a some constraint, such as k-cardinality, knapsack, matroid constraints and so on. Consider the regularized submodular maximization with a k-cardinality constraint (RSMCC), as the objective loses non-negativity and monotonicity, the previous algorithms do not readily apply to this regularized setting. In fact, the maximum can not be determined in polynomial time under the assumption P $\neq$ NP, which implies no multiplicative factor approximation is existed. Then a meaningful weaker notion of approximation was considered and we say an algorithm gets a weak ρ-approximation for the RSMCC, if it can construct a solution set S such that

$$g(S) - \ell(S) \geq \rho \cdot g(OPT) - \ell(OPT),$$

where $OPT \in \arg\max_{S \subseteq \Omega, |S| \leq k}[g(S) - \ell(S)]$ denotes an optimum solution. To incorporate our setting, we consider a bicriteria approximation introduced in [22], to evaluate the quality of a solution. We restate the definition as follows.

Definition 1 [22]. *For any $\lambda_1, \lambda_2 > 0$, an algorithm is a (λ_1, λ_2)-bicriteria approximation for the constrained regularized submodular maximization, if it can produce a solution S satisfying*

$$g(S) - \ell(S) \geq \lambda_1 \cdot g(OPT) - \lambda_2 \cdot \ell(OPT),$$

where $OPT \in \arg\max_{S \subseteq \Omega, S \in \mathcal{I}} g(S) - \ell(S)$ is an optimum solution.

Obviously, the above weak ρ-approximation is $(\rho, 1)$-bicriteria approximation in terms of the bicriteria approximation view. In addition, we focus on the submodular maximization with an another streaming twist. Assuming elements arrive in a streaming fashion is a popular style in dealing with the submodular maximization at scale and it also has been extensively studied in literatures. It is very different from the previous investigated offline centralized scenarios. At any point of time, one has access only to a small fraction of elements stored in primary memory. The performance guarantees of any streaming algorithm are formally introduced in [1], we restate the four basic parameters as following: pass made by algorithm to access the entire stream, the approximation ratio, memory and query complexities produced by the algorithm.

Our Contribution. In this work, we consider the streaming regularized submodular maximization with a k-cardinality constraint (SRSMCC), which is asked to find a subset of size at most k from the stream that maximizes $g(\cdot)-\ell(\cdot)$, where g is a non-negative monotone submodular function and ℓ denotes a non-negative modular function.

Notice that the threshold-based technique has been successfully applied to the streaming submodular optimization setting. Indeed, most of the existing threshold strategies are determined by the optimum, but which can not be accessed a prior. To implement the procedures, the previous works found in [1,11,15], preferred to lazily guess the optimal threshold values, which gave rise to the increment of memory complexity. In our method, we assume one can access a weak ρ-approximation for the SRSMCC for some $\rho > 0$. Namely, we have an approximate value Γ such that $\Gamma \geq \rho \cdot g(OPT) - \ell(OPT)$, where OPT represents any optimum solution. We initially set threshold value as $\Gamma/(\rho k)$ and utilize a threshold decreasing strategy with a $(1-\varepsilon)$ fraction for any $\varepsilon > 0$ during any iteration. In order to ensure the algorithm can be effectively terminated, we also install a lower bound of the thresholds, denoted by $\Gamma/(\lambda k)$, where λ is a parameter to be determined in following sections. Based on the above ideas, we develop a multi-pass streaming algorithm named as Threshold-Decrease-Streaming, which is summarized as Algorithm 1. Theorem 1 guarantees the performance of Threshold-Decrease-Streaming.

Theorem 1. *Let* $\lambda = \frac{2-(2\rho+2)/(3+\sqrt{5-4\rho})}{(3+\sqrt{5-4\rho})/(2\rho+2)-1}$. *Assume there exists a ρ-approximation $\mathcal{A}$ for RSMCC, Threshold-Decrease-Streaming gets a $(\rho/\lambda, 1/\lambda)$-bicriteria approximation for the SRSMCC. That is to say, there exists a streaming algorithm, which produces a solution S of size at most k such that*

$$g(S) - \ell(S) \geq \frac{\rho}{\lambda} \cdot g(OPT) - \frac{1}{\lambda} \cdot \ell(OPT).$$

We restate the previous studies for SRSMCC and readily derive the following corollary by the result of Theorem 1.

Corollary 1. *Consider $\rho = 0.384$, presented in [10], then Threshold-Decrease-Streaming gives a solution $S \subseteq \Omega$ of size at most k satisfying*

$$g(S) - \ell(S) \geq 0.203 \cdot g(OPT) - 0.529 \cdot \ell(OPT).$$

In addition, consider $\rho = 0.632$, presented in [9], then Threshold-Decrease-Streaming gives a solution $S \subseteq \Omega$ of size at most k satisfying

$$g(S) - \ell(S) \geq 0.199 \cdot g(OPT) - 0.317 \cdot \ell(OPT).$$

1.1 Related Work

Streaming model is a popular topic for large scale optimization and has been extensively studied in submodular maximization. The works developed in prior crucially depended on the threshold techniques, which usually initiated a proper (adaptive) threshold for any arriving element and filtered the elements that are lower than the beforehand threshold values.

Streaming Submodular Maximization.. For maximizing the streaming submodular with a k-cardinality constraint, Badanidiyuru et al. [1] provided a one pass $(1/2-\varepsilon)$-approximation with $O(k\log(k)/\varepsilon)$ memory and $O(k\log k)$ queries per element. In practical, they used a threshold value of $1/(2k)$ fraction to the optimum in their threshold-based algorithm. Although the optimum can not be accessed in advance, they developed a lazily guessing step by increasing a $O(\log k)$ factor to memory complexity. By guessing a more tighter threshold value, Kazemi et al. [11] presented a streaming algorithm with an improved memory complexity of $O(k/\varepsilon)$ and the other parameters are maintained. Norouzi-Fard et al. [17] developed a piecewise threshold strategy and yielded a multi-pass streaming algorithm for this streaming SMCC model.

Regularized Submodular Maximization (RSM). We further consider a another regularized twist. Most of the existing algorithms for submodular maximization emphasized that the utility function to take only non-negative values, which may not be applicable for the regularized scenario. Now we give a briefly investigation for the developing of algorithms for the regularized submodular maximization. Following from the fact of the regularized objective function is potentially negative, there must exist no multiplicative factor approximation algorithm for the RSM. Researchers mainly pay attention to develop weak approximation algorithms stated in previous section.

A prior work presented in [20], first studied an equivalent regularized submodular maximization with the matroid constraint. It can be formally restated as $\arg\max_{S\subseteq\Omega,S\in\mathcal{I}}[g(S)+\ell(S)]$, where $(\Omega,\mathcal{I})$ constructs a matroid system, $g:2^{\Omega}\rightarrow\mathbb{R}_{+}$ denotes a non-negative monotone submodular function, but $\ell:2^{\Omega}\rightarrow\mathbb{R}_{+}$ denotes a modular function that may be negative. They individually presented two modified continuous greedy and non-oblivious local search, both of which can attain a same weak $(1-e^{-1})$-approximation with a tail term of $O(\varepsilon)$ for any $\varepsilon>0$. Feldman [5] provided a distorted continuous greedy that avoids the guessing step and keeps the same weak $(1-e^{-1})$-approximation. For the RSMCC, Harshaw et al. [9] introduced a novelty distorted greedy, which greedily selects elements during iterations with the maximum distorted marginal gains. The distorted algorithm produced a weak approximation with $\rho=(1-e^{-1})$ in time of $O(nk)$. In addition, they also gave a faster randomized distorted algorithm which obtains a weak $(1-e^{-1}-\varepsilon)$-approximation in expectation with time of $O(n/\varepsilon\log^2(1/\varepsilon))$. Further, a hardness result of approximation was proved, namely, for any $\varepsilon>0$, there exists no ρ-approximation for some $\rho\geq 1-e^{-1}+\varepsilon$ for RSMCC unless P = NP. Lu et al. [13] studied a regularized non-monotone submodular maximization with a matroid constraint denoted as $\arg\max_{S\subseteq\Omega,S\in\mathcal{I}}[g(S)-\ell(S)]$, where $g:2^{\Omega}\rightarrow\mathbb{R}_{+}$ denotes a non-negative non-monotone submodular function. They presented a continuous greedy which can construct a weak e^{-1}-approximation in expectation with a tail of $O(\varepsilon)$. When the matroid is reduced to a cardinality constraint, they gave a much faster $(e^{-1}-\varepsilon)$-approximation in expectation in time of $O((n/\varepsilon^2)\ln(1/\varepsilon))$. In addition, they also derived a randomized e^{-1}-approximation in expectation with $O(n)$ time. The aforementioned algorithms were addressed to the non-adaptive setting where one must select a group of elements all at once, Tang and Yuan [21] introduced an adaptive regularized submodular under the k-cardinality constraint. They individually provided $(1-e^{-1})$-approximation policy and e^{-1}-approximation policy in expectation according to the cases of $g(\cdot)$ is adaptive monotone submodular and $g(\cdot)$ is general adaptive submodular. Kazemi et al. [10] first studied the streaming submodular maximization with the k-cardinality constraint and provided a distorted threshold-based streaming algorithm, which derives a weak approximation with $\rho = 0.384$. Recently, Wang et al. [22] considered an extended regularized γ-weakly submodular maximization without any constraint and presented a series of bicriteria algorithms.

Organization. The rest of the paper is organized as follows. Section 2 gives necessary preliminaries. Section 3 provides Threshold-Decrease-Streaming and the theoretical analysis are summarized in Sect. 4. In last, Sect. 5 offers a conclusion for our work.

2 Preliminaries

We consider a ground set Ω, which is a collection of elements. We study a streaming fashion in this model. Here the input is revealed element-by-element to an algorithm that has a limited memory capacity. We restate a non-negative monotone submodular function aforementioned in lots of previous work. For any two sets $A, B \subseteq \Omega$, the function $g : 2^{\Omega} \to \mathbb{R}_+$ is defined as

$$g(A) + g(B) \geq g(A \cup B) + g(A \cap B).$$

In addition, the marginal gain of A with respect to B is denoted by $g(B|A) = g(A \cup B) - g(A)$ and $g(u|A) = g(A \cup \{u\}) - g(A)$ for clarity. We say g is monotonic if for any element $u \in \Omega$ and any set $A \subseteq \Omega$, it holds that $g(u|A) \geq 0$.

Formally, we consider a problem of regularized submodular maximization, which is casted as

$$\max_{S \subseteq \Omega, |S| \leq k} g(S) - \ell(S), \tag{1}$$

where $\ell(A) = \sum_{u \in A} \ell(\{u\})$ is a non-negative monotone modular function.

Throughout our paper, we assume that g and ℓ are given in terms of a value oracle which compute $g(A)$ and $\ell(A)$ for any set A.

3 Algorithm

In this section, we present our streaming algorithm for SRSMCC. The proposed algorithm is summarized as Algorithm 1.

Let $\mathcal{A}$ be a weak ρ-approximation algorithm for maximizing regularized submodular maximization under a k-cardinality constraint. Threshold-Decrease-Streaming runs $\mathcal{A}$ to construct a solution set $\mathcal{A}(S)$ such that $\Gamma =$

Algorithm 1. Threshold-Decrease-Streaming

1: **Initialization** Evaluation oracles $g : 2^{\Omega} \rightarrow \mathbb{R}_+$ and $\ell : 2^{\Omega} \rightarrow \mathbb{R}_+$, integer k, ρ-approximation $\mathcal{A}$, and an approximation value Γ satisfying $\Gamma \geq \rho \cdot g(OPT) - \ell(OPT)$. Set parameters of $\beta = \frac{3+\sqrt{5-4\rho}}{2(\rho+1)} (> 1), \lambda = \frac{2-(2\rho+2)/(3+\sqrt{5-4\rho})}{(3+\sqrt{5-4\rho})/(2\rho+2)-1}$.
2: Let $S \leftarrow \emptyset, \tau \leftarrow \Gamma/(\rho k)$
3: **while** $\tau \geq (1-\varepsilon)\Gamma/(\lambda k)$ **do**
4: $\quad \tau \leftarrow (1-\varepsilon)\tau$
5: $\quad$ **for** each element $u \in \Omega$ **do**
6: $\quad\quad$ **if** $g(u|S) - \beta \cdot \ell(u) \geq \tau$ **then**
7: $\quad\quad\quad S \leftarrow S + u$
8: $\quad\quad$ **end if**
9: $\quad\quad$ **if** $|S| = k$ **then**
10: $\quad\quad\quad$ **return** S
11: $\quad\quad$ **end if**
12: $\quad$ **end for**
13: **end while**
14: **return** S

$g(\mathcal{A}(S)) - \ell(\mathcal{A}(S))$. As input, the algorithm Threshold-Decrease-Streaming takes an instance (g, ℓ, k) of Problem (1), a weak approximate value Γ obeying

$$g(OPT) - \ell(OPT) \geq \Gamma \geq \rho \cdot g(OPT) - \ell(OPT),$$

where $OPT \in \arg\max_{S \subseteq \Omega, |S| \leq k} g(S) - \ell(S)$ denotes for the regularized submodular maximization under a k-cardinality constraint.

The algorithm works by making one pass through the ground set Ω for each threshold value τ and any element u with distorted marginal gain

$$g(u|S) - \beta \cdot \ell(u) \geq \tau$$

will be added to solution set S, where $\beta = \frac{3+\sqrt{5-4\rho}}{2(\rho+1)}$. The maximum and minimum threshold values are determined by Γ, ρ, and k. The procedure initializes $\tau = \Gamma/(\rho k)$ and terminates if $\tau < (1-\varepsilon)\Gamma/(\lambda k)$, where λ is setting by $\lambda = \frac{2-(2\rho+2)/(3+\sqrt{5-4\rho})}{(3+\sqrt{5-4\rho})/(2\rho+2)-1}$. The algorithm starts with an empty set $S = \emptyset$ and beaks if $|S| = k$. Otherwise, the algorithm will terminate and return the solution S at most $O(\log(\lambda/\rho)/\varepsilon)$ passes when the lower bound of the thresholds is reached.

4 Theoretical Analysis

Our analysis mainly depends on the following two cases. Consider the case that at termination $|S| = k$, in which the solution S reaches the maximum possible size k. In this case, following by submodularity of g, we know that each of these elements has a large marginal contribution. Now we formally describe this case as the following lemma.

Lemma 1. *If* $|S| = k$, *then* $g(S) - \ell(S) \geq \frac{1}{\lambda} \cdot [(\rho - \varepsilon) \cdot g(OPT) - \ell(OPT)]$.

Proof. W.l.o.g., $S = \{u_1, ..., u_k\}$, which is ordered by the addition of elements to S, and let $S_0 = \emptyset$. Observe that

$$\begin{aligned} g(S) - \beta \cdot \ell(S) &= \sum_{i=1}^{k} [g(u_i|S_{i-1}) - \beta \cdot \ell(u_i)] \\ &\geq \sum_{i=1}^{k} \tau_i \\ &\geq \frac{1-\varepsilon}{\lambda} \cdot \Gamma \\ &\geq \frac{1-\varepsilon}{\lambda} \cdot [\rho \cdot g(OPT) - \ell(OPT)] \\ &\geq \frac{1}{\lambda} \cdot [(\rho - \varepsilon) \cdot g(OPT) - \ell(OPT)], \end{aligned}$$

where τ_i denotes the threshold encountered by adding element u_i. Note that there may exist many added elements met a same threshold. Since $\beta = \frac{3+\sqrt{5-4\rho}}{2(\rho+1)} \geq 1$, we now get

$$g(S) - \ell(S) \geq g(S) - \beta \cdot \ell(S) \geq \frac{1}{\lambda} \cdot [(\rho - \varepsilon) \cdot g(OPT) - \ell(OPT)].$$

The following lemma gives the approximation of Algorithm 1 for the case $|S| < k$, in which the solution S dose not reach its maximum size k when the stream finishes.

Lemma 2. *If* $|S| < k$, *then* $g(S) - \ell(S) \geq \frac{\rho}{\lambda} \cdot g(OPT) - \frac{1}{\lambda} \cdot \ell(OPT)$.

Proof. Consider any arbitrary element $u \in OPT \backslash S$. From the fact that the element u was not added to S and $|S| < k$, for any pass, one can conclude that

$$g(u|S_u) - \beta \cdot \ell(u) \leq \tau_0,$$

where S_u represents the set S at the time of encountering u and $\tau_0 = (1 - \varepsilon)[\rho \cdot g(OPT) - \ell(OPT)]/(\lambda k)$ as the lower bound of thresholds. Following submodularity, we further get

$$g(u|S) - \beta \cdot \ell(u) \leq \tau_0.$$

Adding the above inequality over all elements $u \in OPT \backslash S$ implies

$$\begin{aligned} g(OPT) - g(S) - \beta \cdot \ell(OPT) &\leq g(OPT|S) - \beta \cdot \ell(OPT) \\ &\leq \sum_{u \in OPT \backslash S} g(u|S) - \beta \cdot \ell(u) \\ &\leq |OPT \backslash S| \cdot \tau_0 \\ &\leq \frac{1}{\lambda} \cdot [g(OPT) - \ell(OPT)]. \end{aligned} \tag{2}$$

The first inequality follows by the monotonicity of g, the second derives by the submodularity of g and the non-negativity of ℓ. Rearranging the inequality (2), we obtain

$$g(S) \geq \left(1-\frac{1}{\lambda}\right) \cdot g(OPT) - \left(\beta - \frac{1}{\lambda}\right) \cdot \ell(OPT). \tag{3}$$

In addition, for any $\tau > 0$, we easily yield

$$g(S) - \beta \cdot \ell(S) \geq \sum_{i=1}^{|S|} \tau_i \geq 0. \tag{4}$$

Adding a fraction $1/\beta$ of the inequality (4) to a $1-1/\beta$ fraction of the inequality (3) yields

$$\begin{aligned} g(S) - \ell(S) &\geq \left(1-\frac{1}{\beta}\right)\left(1-\frac{1}{\lambda}\right) \cdot g(OPT) - \left(1-\frac{1}{\beta}\right)\left(\beta-\frac{1}{\lambda}\right) \cdot \ell(OPT) \\ &= \frac{\rho}{\lambda} \cdot g(OPT) - \frac{1}{\lambda} \cdot \ell(OPT), \end{aligned}$$

where the equality holds by the setting of β.

Based on the above two cases presented by Lemmas 1 and 2, we consequently yield the main result, Theorem 1. In addition, Threshold-Decrease-Streaming will stop at most $O(\log(\lambda/\rho)/\varepsilon)$ passes, consume the $O(k)$ memory, and have at most $O(\log(\lambda/\rho)/\varepsilon)$ queries per element.

5 Conclusion

In this paper, we provided a bicriteria approximation for maximizing a streaming regularized submodular function with a k-cardinality constraint, in which the utility function was expressed as the difference between a non-negative monotone non-decreasing submodular function g and a modular function ℓ. The discussed regularized model has been formally casted as Problem (1). Utilizing a threshold-based decreasing technique, we developed a multi-pass streaming algorithm, which produced a bicriteria approximation with a theoretical performance guarantee. In our method, we assume there exists a weak approximate value with ρ-approximation and instigate the threshold values by the approximate value instead of the optimum discussed in the previous algorithms. And thus we efficiently avoid the guessing steps of the optimum, which may give rise to the increment of memory complexity. Consider the following two weak 0.384-approximation and 0.632-approximation, individually presented in [9] and [10], then there accordingly exist two $(0.203, 0.529)$-bicriteria and $(0.199, 0.317)$-bicriteria approximations for the streaming regularized submodular maximization with the k-cardinality constraint.

Acknowledgements. The third author is supported by National Natural Science Foundation of China (No. 11901544). The fourth author is supported by National Natural Science Foundation of China (No. 12101587), China Postdoctoral Science Foundation (No. 2021M703167) and Fundamental Research Funds for the Central Universities (No. EIE40108X2).

References

1. Badanidiyuru, A., Mirzasoleiman, B., Karbasi, A., Krause, A.: Streaming submodular maximization: massive data summarization on the fly. In: Proceedings of SIGKDD, pp. 671–680 (2014)
2. Buchbinder, N., Feldman, M., Garg, M.: Deterministic $(1/2+\varepsilon)$-approximation for submodular maximization over a matroid. In: Proceedings of SODA, pp. 241–254 (2019)
3. Calinescu, G., Chekuri, C., Pál, M., Vondrák, J.: Maximizing a monotone submodular function subject to a matroid constraint. SIAM J. Comput. **40**, 1740–1766 (2011)
4. Feige, U.: A threshold of $\ln(n)$ for approximating set cover. J. ACM **45**(4), 634–652 (1998)
5. Feldman, M.: Guess free maximization of submodular and linear sums. In: Proceedings of WADS, pp. 380–394 (2019)
6. Feldman, M., Harshaw, C., Karbasi, A.: Greed is good: near-optimal submodular maximization via greedy optimization. In: Proceedings of COLT, pp. 758–784 (2017)
7. Feldman, M., Naor, J.S., Schwartz, R., Ward, J.: Improved approximations for k-exchange systems. In: Demetrescu, C., Halldórsson, M.M. (eds.) ESA 2011. LNCS, vol. 6942, pp. 784–798. Springer, Heidelberg (2011). https://doi.org/10.1007/978-3-642-23719-5_66
8. Fisher, M.L., Nemhauser, G.L., Wolsey, L.A.: An analysis of approximations for maximizing submodular set functions-II. In: Balinski, M.L., Hoffman, A.J. (eds.) Polyhedral Combinatorics. MATHPROGRAMM, vol. 8, pp. 73–87. Springer, Heidelberg (1978). https://doi.org/10.1007/BFb012119
9. Harshaw, C., Feldman, M., Ward, J., Karbasi, A.: Submodular maximization beyond non-negativity: guarantees, fast algorithms, and applications. In: Proceedings of ICML, pp. 2634–2643 (2019)
10. Kazemi, E., Minaee, S., Feldman, M., Karbasi, A.: Regularized submodular maximization at scale. In: Proceedings of ICML, pp. 5356–5366 (2021)
11. Kazemi, E., Mitrovic, M., Zadimoghaddam, M., Lattanzi, S., Karbasi, A.: Submodular streaming in all its glory: tight approximation, minimum memory and low adaptive complexity. In: Proceedings of ICML, pp. 3311–3320 (2019)
12. Lee, J., Sviridenko, M., Vondrák, J.: Submodular maximization over multiple matroids via generalized exchange properties. Math. Oper. Res. **35**, 795–806 (2010)
13. Lu, C., Yang, W., Gao, S.: Regularized non-monotone submodular maximization. arXiv:2103.10008
14. Mirzasoleiman, B., Jegelka, S., Krause, A.: Streaming non-monotone submodular maximization: personalized video summarization on the fly. In: Proceedings of AAAI, pp. 1379–1386 (2018)
15. Mitrovic, M., Kazemi, E., Zadimoghaddam, M., Karbasi, A.: Data summarization at scale: a two-stage submodular approach. In: Proceedings of ICML, pp. 3593–3602 (2018)

16. Nemhauser, G.L., Wolsey, L.A., Fisher, M.L.: An analysis of approximations for maximizing submodular set functions-I. Math. Program. **14**, 265–294 (1978). https://doi.org/10.1007/BF01588971
17. Norouzi-Fard, A., Tarnawski, J., Mitrovic, S., Zandieh, A., Mousavifar, A., Svensson, O.: Beyond 1/2-approximation for submodular maximization on massive data streams. In: Proceedings of ICML, pp. 3826–3835 (2018)
18. Sarpatwar, K.K., Schieber, B., Shachnai, H.: Constrained submodular maximization via greedy local search. Oper. Res. Lett. **41**(1), 1–6 (2019)
19. Sviridenko, M.: A note on maximizing a submodular set function subject to a knapsack constraint. Oper. Res. Lett. **32**, 41–43 (2004)
20. Sviridenko, M., Vondrák, J., Ward, J.: Optimal approximation for submodular and supermodular optimization with bounded curvature. Math. Oper. Res. **42**(4), 1197–1218 (2017)
21. Tang, S., Yuan, J.: Adaptive regularized submodular maximization. arXiv:2103.00384
22. Wang, Y., Xu, D., Du, D., Ma, R.: Bicriteria algorithms to balance coverage and cost in team formation under online model. Theor. Comput. Sci. **854**, 68–76 (2021)

Author Index

Printed in the United States
by Baker & Taylor Publisher Services